Handbook of Environmental Engineering

Handbook of Environmental Engineering

Edited by

Myer Kutz
Myer Kutz Associates, Delmar, NY, USA

WILEY

This edition first published 2018.

Registered Office
John Wiley & Sons, Inc., 111 River Street, Hoboken, NJ 07030, USA

Editorial Office
111 River Street, Hoboken, NJ 07030, USA

For details of our global editorial offices, customer services, and more information about Wiley products visit us at www.wiley.com.

Wiley also publishes its books in a variety of electronic formats and by print-on-demand. Some content that appears in standard print versions of this book may not be available in other formats.

Library of Congress Cataloging-in-Publication Data

Names: Kutz, Myer, editor.
Title: Handbook of environmental engineering / edited by Myer Kutz.
Description: First edition. | Hoboken, NJ : John Wiley & Sons, 2018. | Includes bibliographical references and index. |
Identifiers: LCCN 2018015512 (print) | LCCN 2018028239 (ebook) | ISBN 9781119304401 (pdf) |
ISBN 9781119304432 (epub) | ISBN 9781118712948 (cloth)
Subjects: LCSH: Environmental engineering–Handbooks, manuals, etc.
Classification: LCC TA170 (ebook) | LCC TA170 .H359 2018 (print) | DDC 628–dc23
LC record available at https://lccn.loc.gov/2018015512

Cover design by Wiley
Cover image: © Jonutis/Shutterstock

Set in 10/12pt Warnock by SPi Global, Pondicherry, India

Printed in the United States of America

V10004332_091118

To Rick Giardino and to all the other contributors to this handbook

Contents

List of Contributors

Georgios N. Anastassakis
School of Mining and Metallurgical Engineering, National Technical University of Athens, Athens, Greece

Mason Andrews
Department of Architecture, Hampton University, Hampton, VA, USA

Emmanuel Arthur
Department of Agroecology, Aarhus University, Tjele, Denmark

Adisa Azapagic
School of Chemical Engineering and Analytical Science, The University of Manchester, Manchester, UK

Ebrahim Babaeian
Department of Soil, Water and Environmental Science, The University of Arizona, Tucson, AZ, USA

Christian M. Carrico
Department of Civil and Environmental Engineering, New Mexico Institute of Mining and Technology, Socorro, NM, USA

Bora Cetin
Department of Civil, Construction, and Environmental Engineering, Iowa State University, Ames, IA, USA

Steven C. Chapra
Department of Civil & Environmental Engineering, Tufts University, Medford, MA, USA

Clayton J. Clark II
Department of Civil & Environmental Engineering, FAMU-FSU College of Engineering, Florida A&M University, Tallahassee, FL, USA

Ardeshir M. Ebtehaj
Department of Civil, Environmental and Geo-Engineering, Saint Anthony Falls Laboratory, University of Minnesota, Minneapolis, MN, USA

Ali Elkamel
Department of Chemical Engineering, University of Waterloo, Waterloo, Ontario, Canada
Department of Chemical Engineering, Khalifa University of Science and Technology, Abu Dhabi, UAE

Mujde Erten-Unal
Department of Civil and Environmental Engineering, Old Dominion University, Norfolk, VA, USA

Kevin R. Gamache
Water Management and Hydrological Science Program and High Alpine and Arctic Research Program (HAARP), The Bush School of Government and Public Service, Texas A&M University, College Station, TX, USA

Kumar Ganesan
Department of Environmental Engineering, Montana Tech, Butte, MT, USA

John R. Giardino
Water Management and Hydrological Science Program and High Alpine and Arctic Research Program (HAARP), Department of Geology and Geophysics, Texas A&M University, College Station, TX, USA

Sharad Gokhale
Civil Engineering Department, Indian Institute of Technology Guwahati, Guwahati, India

Huei-Ping Huang
School for Engineering of Matter, Transport, and Energy, Arizona State University, Tempe, AZ, USA

Scott B. Jones
Department of Plants, Soils and Climate, Utah State University, Logan, UT, USA

Nigel B. Kaye
Glenn Department of Civil Engineering, Clemson University, Clemson, SC, USA

Abdul A. Khan
Glenn Department of Civil Engineering, Clemson University, Clemson, SC, USA

Cheng Seong Khor
Chemical Engineering Department, Universiti Teknologi PETRONAS, Seri Iskandar, Perak Darul Ridzuan, Malaysia

Lin Li
Department of Civil and Environmental Engineering, Jackson State University, Jackson, MS, USA

Stephanie Luster-Teasley
Department of Civil, Architectural, & Environmental Engineering, College of Engineering, North Carolina A&T State University, Greensboro, NC, USA

Chandra Mouli R. Madhuranthakam
Chemical Engineering Department, Abu Dhabi University, Abu Dhabi, UAE

Will Martin
General Engineering Department, Clemson University, Clemson, SC, USA

Jana B. Milford
Department of Mechanical Engineering and Environmental Engineering Program, University of Colorado, Boulder, CO, USA

Shelly L. Miller
Department of Mechanical Engineering, University of Colorado, Boulder, CO, USA

Carl C. Nesbitt
Department of Chemical Engineering, Michigan Technological University, Houghton, MI, USA

Say Kee Ong
Department of Civil, Construction, and Environmental Engineering, Iowa State University, Ames, IA, USA

Sibel Pamukcu
Department of Civil and Environmental Engineering, Lehigh University, Bethlehem, PA, USA

Brad Putman
Glenn Department of Civil Engineering, Clemson University, Clemson, SC, USA

Morteza Sadeghi
Department of Plants, Soils and Climate, Utah State University, Logan, UT, USA

Judith L. Sims
Utah Water Research Laboratory, Utah State University, Logan, UT, USA

Kirsten M. Sims
WesTech Engineering, Inc., Salt Lake City, UT, USA

Milani Sumanasooriya
Department of Civil & Environmental Engineering, Clarkson University, Potsdam, NY, USA

Firat Y. Testik
Civil and Environmental Engineering Department, University of Texas at San Antonio, San Antonio, TX, USA

Louis Theodore
Professor Emeritus, Manhattan College, New York, NY, USA

Markus Tuller
Department of Soil, Water and Environmental Science, The University of Arizona, Tucson, AZ, USA

Daniel A. Vallero
Department of Civil and Environmental Engineering, Duke University, Durham, NC, USA

C. Visvanathan
Environmental Engineering and Management Program, Asian Institute of Technology, Khlong Luang, Thailand

Panshu Zhao
Water Management and Hydrological Science Graduate Program and High Alpine and Arctic Research Program (HAARP), Texas A&M University, College Station, TX, USA

Preface

The discipline of environmental engineering deals with solutions to problems whose neglect would be harmful to society's well-being. The discipline plays a vital role in a world where human activity has affected the Earth's climate, the levels of the seas, the air we breathe, and the cleanliness of water and soil. It is hardly a stretch, in my view, to assert that the work of environmental engineers can contribute to mitigating problems caused by extreme weather events; protecting populations in coastal areas; reducing illnesses caused by polluted air, soil, and water from improperly regulated industrial and transportation activities; and promoting the safety of the food supply. Environmental engineers do not need to rely on political stands on climate change or pollution sources for motivation. As perceptive theoreticians and practitioners, they need to merely observe where problems exist. Then they can use their knowledge and experience to analyze elements of problems, recommend solutions, and enable effective action.

This environmental engineering handbook provides sources of information for students and practitioners interested in both fundamentals and real-world applications of environmental engineering. The handbook is organized around the assertions highlighted above. The first major section is composed of six wide-ranging chapters that cover methods for analyzing environmental systems and making measurements within those systems, legal issues that environmental engineers have to know about, methods for modeling the Earth's climate and analyzing impacts of climate change, and lastly ways designed to respond to rise in sea levels.

The next three major sections address, in order, pollution in soils, with three chapters focusing on the physics of soils, remediation methods for polluted soils and sediments, and remote sensing techniques; water quality issues, with five chapters dealing with fundamentals of environmental fluid mechanics, water quality assessment, wastewater treatment, and design of porous pavement systems (which can mitigate flooding); air pollution issues, with three chapters covering air pollution control methods, measuring disbursement of aerosols into the atmosphere, and mitigating indoor air pollution; and finally, there is a chapter on noise pollution, another serious environmental problem.

The handbook's final section is devoted to confronting issues of contaminants and waste. The six chapters in this section provide information crucial for disposing of, and where possible, recycling solid and hazardous wastes and for assessing pollution created by metals manufacturing and chemical processes and plants. Crucial to success of these solutions is not only the active involvement of industry but also the participation of academia and government. The handbook is written at a level that allows upper-level students and practitioners and researchers, including environmental scientists and engineers, urban planners, government administrators, and environmental lawyers, to understand major environmental issues.

My heartfelt thanks to the contributors to this handbook, all of them recognized experts in their fields. It's a miracle that contributors, with their taxing professional lives, are able to produce well-written, cogently presented, and useful chapters. Contributors write, as one of them told me recently,

> because it is a good way to organize one's thoughts and because it is part of my duty as a scientist to publish my work so that others can learn from it. I spend valuable time writing because it allows me the opportunity to access a wide audience. It is an investment. The time I spend writing today is the time I don't have to spend educating someone 1 : 1 in the future.

Or as another contributor noted,

> for a handbook of this kind, the deciding factor [of whether to contribute a chapter] is the desire of the author to share his/her expertise with others who have a more general or superficial interest in the chapter topic. I use handbooks of this kind if I have (or are part of a team that has) to solve a complex multi-facetted problem and need to quickly come up to speed on parts of the solution that I am not familiar with.

In keeping with this idea about handbook usage, this volume is replete with illustrations throughout the text and extensive lists of references at the end of chapters. Guides to sources of information on the Internet and in library stacks are provided by experts, thereby improving research results.

A final word of thanks, to my wife, Arlene, whose very presence in my life makes my work all that much easier.

April 2018 *Delmar, NY*

1

Environmental Systems Analysis

Adisa Azapagic

School of Chemical Engineering and Analytical Science, The University of Manchester, Manchester, UK

1.1 Introduction

Throughout history, engineers were always expected to provide innovative solutions to various societal challenges, and these expectations continue to the present day. However, nowadays, we are facing some unprecedented challenges, such as climate change, growing energy demand, resource scarcity, and inadequate access to food and water, to name but a few. With a fast-growing population, it is increasingly clear that the lifestyles of modern society cannot be sustained indefinitely. Growing scientific evidence shows that we are exceeding the Earth's capacity to provide many of the resources we use and to accommodate our emissions to the environment (IPCC, 2013; UNEP, 2012).

Engineers have a significant role to play in addressing these sustainability challenges by helping meet human needs through provision of technologies, products, and services that are economically viable, environmentally benign, and socially beneficial (Azapagic and Perdan, 2014). However, one of the challenges is determining what technologies, products, and services are sustainable and which metrics to use to ascertain that.

Environmental systems analysis (ESA) can be used for these purposes. ESA takes a systems approach to describe and evaluate the impacts of various human activities on the environment. A systems approach is essential for this as it enables consideration of the complex interrelationships among different elements of the system, recognizing that the behavior of the whole system is quite different from its individual elements when considered in isolation from each other. The "system" in this context can be a product, process, project, organization, or a whole country.

Many methods are used in ESA, including:

- Energy and exergy analysis
- Material and substance flow analysis (SFA)
- Environmental risk assessment (ERA)
- Environmental management systems (EMS)
- Environmental input–output analysis (EIOA)
- Life cycle assessment (LCA)
- Life cycle costing (LCC)
- Social life cycle assessment (S-LCA)
- Cost–benefit analysis (CBA).

These methods are discussed in the rest of this chapter.

1.2 Environmental Systems Analysis Methods

In addition to the methodologies that underpin them, ESA methods differ in many other respects, including the focus, scope, application, and sustainability aspects considered. This is summarized in Table 1.1 and discussed in the sections that follow.

1.2.1 Energy and Exergy Analysis

Energy analysis is used to quantify the total amount of energy used by a system and to determine its efficiency. It can also be used to identify energy "hot spots" and opportunities for improvements. Exergy analysis goes a step further, and, instead of focusing on the quantity, it measures the quality of energy or the maximum amount of work that can be theoretically obtained from a system as it comes into equilibrium with its environment. Exergy analysis can be used to determine the efficiency of resource utilization and how it can be improved.

Although energy analysis has traditionally focused on production processes, it is also used in other applications, including energy analysis at the sectorial and national levels. However, the usefulness of exergy analysis is questionable for non-energy systems. Furthermore,

Handbook of Environmental Engineering, First Edition. Edited by Myer Kutz.

Table 1.1 An overview of methods used in environmental systems analysis.

Method	Focus	Scope/system boundary	Sustainability aspects	Application
Energy/exergy analysis	Production processes, supply chains, regions, countries	Production process, sectorial, regional, national	Energy	Process or project analysis, energy efficiency, identification of energy "hot spots"
Material flow analysis	Materials	Regional, national, global	Natural resources	Environmental accounting, preservation of resources, policy
Substance flow analysis	Chemical substances	Regional, national, global	Environmental pollution	Environmental accounting and protection, strategic management of chemicals, policy
Environmental risk assessment	Products, installations	Product or installation, local, regional, national	Environmental, health and safety	Risk analysis, evaluation of risk mitigation measures, financial planning, regulation
Environmental management systems	Organizations	Organization	Environmental	Environmental management
Environmental input–output analysis	Product groups, sectors, national economy	Sectors, supply chains, national economy	Environmental and economic	Environmental accounting, policy
Life cycle assessment	Products, processes, services, activities	Life cycle/supply chain	Environmental	Benchmarking, identification of opportunities for improvements, eco-design, policy
Life cycle costing	Products, processes, services, activities	Life cycle/supply chain	Economic	Benchmarking, identification of opportunities for improvements
Social life cycle assessment	Products, processes, services, activities	Life cycle/supply chain	Social	Benchmarking, identification of opportunities for improvements, policy
Cost–benefit analysis	Projects, activities	Project, activity	Socioeconomic and environmental	Appraisal of costs and benefits of different projects or activities

many users find it difficult to estimate and interpret the meaning of exergy (Jeswani et al., 2010).

1.2.2 Material Flow Analysis

MFA enables systematic accounting of the flows and stocks of different materials over a certain time period in a certain region (Brunner and Rechberger, 2004). The term "materials" is defined quite broadly, spanning single chemical elements, compounds, and produced goods. Examples of materials often studied through MFA include aluminum, steel, copper, and uranium. MFA is based on the mass balance principle, derived from the law of mass conservation. This means that inputs and outputs of materials must be balanced, including any losses or stocks (i.e. accumulation).

As indicated in Figure 1.1, MFA can include the entire life cycle of a material, including its mining, production use, and waste management. In addition to the material flows, MFA also considers material stocks, making it suitable for analysis of resource scarcity. Material flows are typically tracked over a number of years enabling evaluation of long-term trends in the use of materials. MFA can also serve as a basis for quantifying the resource productivity of an economy, but it is not suitable for consideration of single production systems (Jeswani et al., 2010).

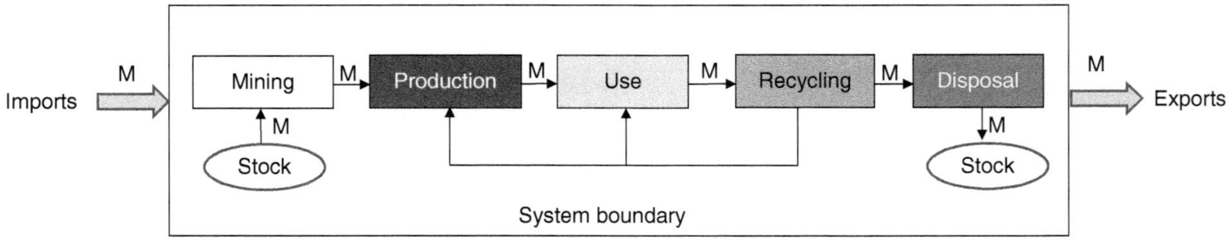

Figure 1.1 Material flow analysis tracks flows of materials through an economy from "cradle to grave." (M – flows of material under consideration).

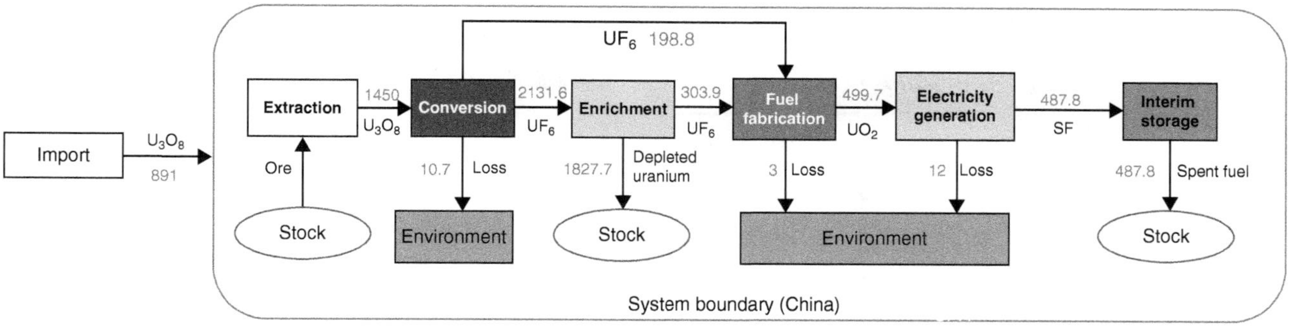

Figure 1.2 Material flow analysis of uranium flows and stocks in China in tonnes per year. *Source*: Adapted from Yue et al. (2016).

An example of MFA applied to uranium in China is given in Figure 1.2. As can be seen, the annual flows and stocks of uranium, which is used as a fuel in nuclear power plants, are tracked within the country along the whole fuel life cycle. This includes extraction of the ore, conversion and enrichment of uranium, fuel fabrication, and electricity generation. Thus, MFA helps to quantify the total consumption of uranium over time and stocks of depleted uranium that could be used for fuel reprocessing. It can also help with the projections of future demand and estimates of how much uranium can be supplied from indigenous reserves and how much needs to be imported.

1.2.3 Substance Flow Analysis

SFA is a specific type of MFA, focusing on chemical substances or compounds. The main aim of most SFA studies is to provide information for strategic management of chemical substances at a regional or national level (van der Voet, 2002). SFA can be also applied to track environmental pollution over time in a certain region. The latter is illustrated in Figure 1.3, which shows emissions of the pollutant of interest from different sources to air, water, and land in a defined region. However, the distinction between MFA and SFA is often blurred, and sometimes the two terms are used interchangeably.

1.2.4 Environmental Risk Assessment

ERA is used to assess environmental risks posed to ecosystems, animals, and humans by chemicals, industrial installations, or human activities. The risks can be physical, biological, or chemical (Fairman et al., 1998).

Many types of ERA are used, including pollution, natural disaster, and chemical risk assessment. The assessment covers emissions and related environmental impacts in the whole life cycle of a chemical or an installation. For chemicals, this includes their production, formulation, use, and end-of-life management. For industrial installations, construction, operation, and decommissioning must be considered. ERA aims to protect the atmosphere, aquatic, and soil organisms as well as mammals and birds further up in the food chain. It is used by industry not only to comply with regulations but also to improve product safety, financial planning, and evaluation of risk mitigation measures.

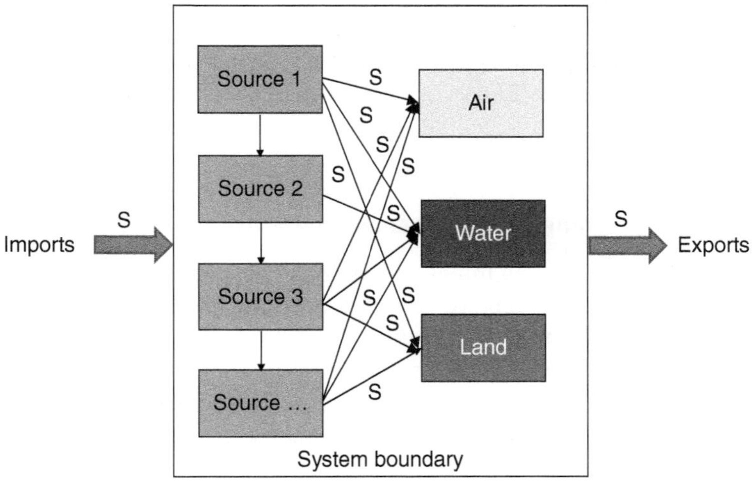

Figure 1.3 Substance flow analysis tracks the flows of pollutants into, within and out of a region (S – flows of substance under consideration). *Source*: Adapted from Azapagic et al. (2007).

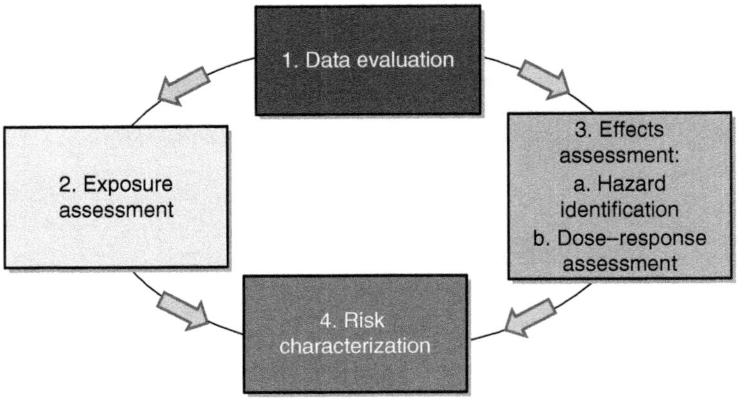

Figure 1.4 Environmental risk assessment steps according to the EUSES. *Source*: Based on Lijzen and Rikken (2004).

There are many methods and tools for carrying out an ERA. One such tool used in Europe is the European Union System for the Evaluation of Substances (EUSES) that enables rapid assessments of risks posed by chemical substances (EC, 2016). As indicated in Figure 1.4, EUSES comprises the following steps (Lijzen and Rikken, 2004):

1) Data collection and evaluation
2) Exposure assessment: estimation of the concentrations/doses to which the humans and the environment are exposed
3) Effects assessment comprising:
   a) Hazard identification: identification of the adverse effects that a substance has an inherent capacity to cause
   b) Dose–response assessment: estimation of the relationship between the level of exposure to a substance (dose, concentration) and the incidence and severity of an effect
4) Risk characterization: estimation of the incidence and severity of the adverse effects likely to occur in a human population or the environment due to actual or predicted exposure to a substance.

EUSES is intended mainly for initial rather than comprehensive risk assessments. The EUSES software is available freely and can be downloaded from the European Commission's website (EC, 2016). In the United States, ERA is regulated by the US Environmental Protection Agency (EPA); for various methods, consult the EPA guidelines (EPA, 2017). For a review of other ERA methods, see Manuilova (2003).

1.2.5 Environmental Management Systems

An EMS represents an integrated program for managing environmental impacts of an organization, with the ultimate aim of helping it improve the environmental performance. The most widely used EMS standard is ISO 14001 (ISO, 2015). This EMS follows the concept of plan–do–check–act, an iterative process aimed at achieving continual improvement.

The main steps of the ISO 14001 EMS outlined in Figure 1.5 are:

1) Planning
2) Support and operation
3) Performance evaluation
4) Implementation.

The EMS is set up and driven by the organization's leadership who are responsible for its implementation. The EMS must be congruent with and follow the organization's environmental policy.

1) *Planning*: In the planning step, the organization must determine the environmental aspects that are relevant to its activities, products, and services. The aspects include both those the organization can control and those that it can influence, and their associated environmental impacts, considering a life cycle perspective (ISO, 2015). Significant environmental impacts must be addressed through appropriate action, also ensuring compliance with legislation.
2) *Support and operation*: This step involves providing adequate resources for the implementation of the EMS and appropriate internal and external communication. The organization must also establish and control the processes needed to meet EMS requirements. Consistent with a life cycle perspective, this must cover all relevant life cycle stages, including procurement of materials and energy, production of product(s) or provision of services, transportation, use, end-of-life treatment, and final disposal of its product(s) or services.
3) *Performance evaluation*: This step involves monitoring, measurement, analysis, and evaluation of the environmental performance. This is typically carried out over the period of one year.
4) *Implementation*: The information obtained in the previous step is then used to identify and implement improvement opportunities across the organization's

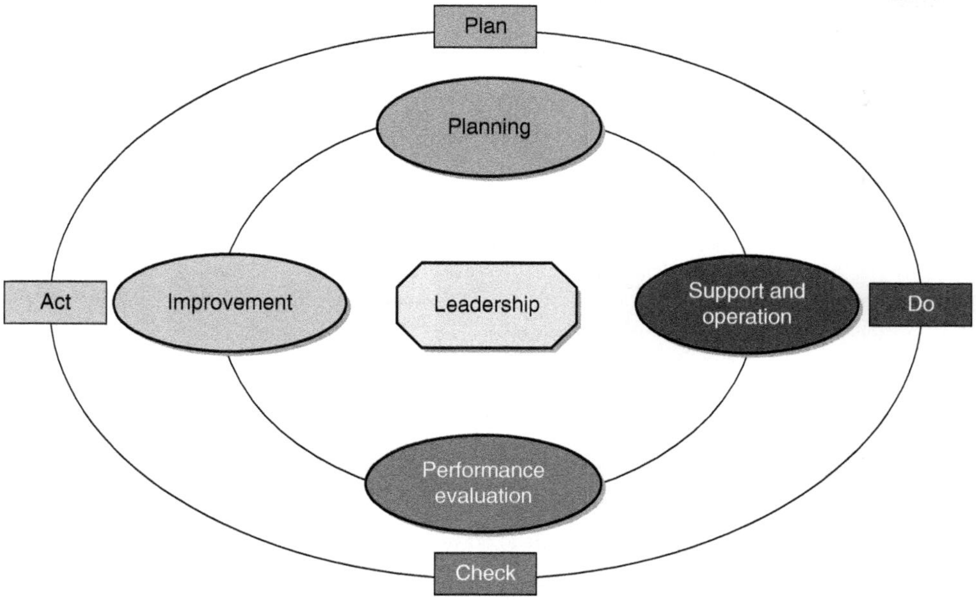

Figure 1.5 Main steps in the ISO 14001 environmental management system. *Source*: Based on ISO (2015).

activities (Figure 1.5). This whole process is repeated iteratively, typically on an annual basis, helping toward continuous improvement of environmental performance.

An alternative to the ISO 14001 is the Eco-Management and Audit Scheme (EMAS) developed by the European Commission. For details, see EC (2013).

1.2.6 Environmental Input–Output Analysis

EIOA represents an expansion of conventional input–output analysis (IOA). While the latter considers monetary flows within an economic system, EIOA combines environmental impacts with the conventional economic analysis carried out in IOA. Environmental impacts are considered either by adding environmental indicators to IOA or by replacing the monetary input–output matrices with those based on physical flows (Jeswani et al., 2010). Different environmental indicators can be considered in EIOA, including material and energy inputs as well as emissions to air and water, and waste. Social aspects, such as employment, can also be integrated into EIOA (Finnveden et al., 2003).

EIOA is suitable for determining the environmental impacts of product groups, sectors, or national economies. While this can be useful for environmental accounting and at a policy level, EIOA has many limitations. First, the data are too aggregated to be useful at the level of specific supply chains, products, or activities. It also often assumes an identical production technology for imported and domestic products, that each sector produces a single product, and that a single technology is used in the production process (Jeswani et al., 2010). Furthermore, allocation of environmental impacts between different sectors, products, and services is proportional to the economic flows.

1.2.7 Life Cycle Assessment

LCA applies life cycle thinking to quantify environmental sustainability of products, processes, or human activities on a life cycle basis. As shown in Figure 1.6, the following stages in the life cycle of a product or an activity can be considered in LCA:

- Extraction and processing of raw materials
- Manufacture
- Use, including any maintenance
- Re-use and recycling
- Final disposal
- Transportation and distribution.

LCA is a well-established tool used by industry, researchers, and policy makers. Some of the applications of LCA include (Azapagic, 2011):

- Measuring environmental sustainability
- Comparison of alternatives to identify environmentally sustainable options
- Identification of hot spots and improvement opportunities
- Product design and process optimization
- Product labeling.

The LCA methodology is standardized by the ISO 14040/44 standards (ISO, 2006a, b) that define LCA as

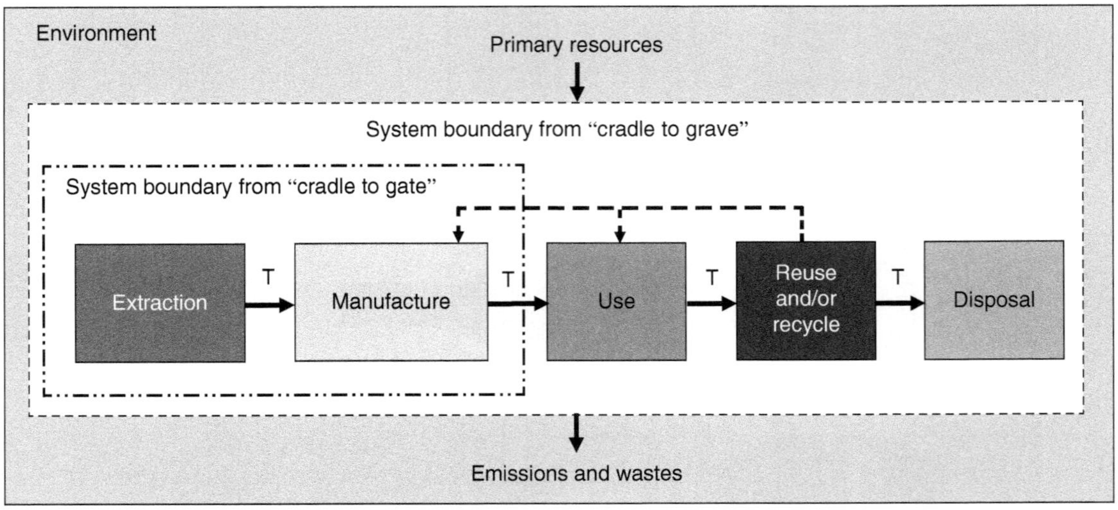

Figure 1.6 The life cycle of a product or an activity from "cradle to gate" and "cradle to grave." *Source*: Based on Azapagic (2011).

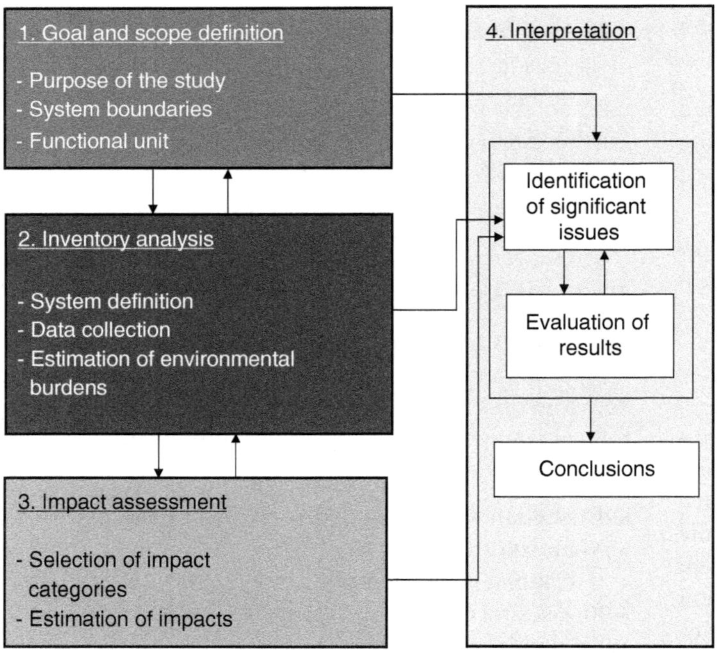

Figure 1.7 LCA methodology according to ISO 14040 (ISO, 2006a).

"…a compilation and evaluation of the inputs, outputs and the potential environmental impacts of a product throughout its life cycle." According to these standards, LCA comprises four phases (Figure 1.7):

1) Goal and scope definition
2) Inventory analysis
3) Impact assessment
4) Interpretation.

1) *Goal and scope definition*: An LCA starts with a goal and scope definition that includes definition of the purpose of the study, system boundaries, and the functional unit (unit of analysis). As indicated in Figure 1.6, the system boundary can be from "cradle to grave" or "cradle to gate." The former considers all stages in the life cycle from extraction of primary resources to end-of-life waste management. The "cradle-to-gate" study stops at the factory "gate" where the product of interest is manufactured, excluding its use and end-of-life waste management. Definition of the system boundary depends on the goal and scope of the study. For example, the goal of the study may be to identify the hot spots in the life cycle of a product or to select environmentally the most sustainable option among alternative products delivering the same function.

Defining the function of the system is one of the most important elements of an LCA study as that determines the functional unit, or unit of analysis, to be used in the study. The functional unit represents a quantitative measure of the outputs that the system delivers (Azapagic, 2011). In comparative LCA studies it is essential that systems are compared on the basis of an equivalent function, i.e. the functional unit. For example, comparison of different types of drinks packaging should be based on their equivalent function that is to contain a certain amount of drink. The functional unit is then defined as "the quantity of packaging material necessary to contain a specified volume of a drink."

2) *Inventory analysis*: This phase involves detailed specification of the system under study and collection of data. The latter includes quantities of materials and energy used in the system and emissions to air, water, and land throughout the life cycle. These are known as environmental burdens. If the system has several functional outputs, e.g. produces several products, the environmental burdens must be allocated among them. Different methods are used for this purpose, including allocation on a mass and economic basis (ISO, 2006b).
3) *Impact assessment*: In this phase, the environmental impacts are translated into different environmental impacts. Example impacts considered in LCA include global warming, acidification, eutrophication, ozone layer depletion, human toxicity, and ecotoxicity. A number of life cycle impact assessment methods are available but the most widely used are CML 2 (Guinee et al., 2001) and Eco-indicator 99 (Goedkoop and Spriensma, 2001). The former is based on a "midpoint" approach, linking the environmental burdens somewhere in between the point of their occurrence (e.g. emissions of CO_2) and the ultimate damage caused (e.g. global warming). Ecoinvent 99 follows a damage-oriented approach that considers the "endpoint" damage caused by environmental burdens to human health, ecosystem, and natural resources. An overview of the CML 2 and Eco-indicator 99 methods can be found in Boxes 1.1 and 1.2. The ReCiPe method (Goedkoop et al., 2009) is gradually superseding CML 2 as its updated and broadened version. In addition to the midpoint approach, ReCiPe also enables calculation of endpoint impacts, thus combining the approaches in CML 2 and Eco-indicator 99.
4) *Interpretation*: The final LCA phase involves evaluation of LCA findings, including identification of significant environmental impacts and hot spots that can then be targeted for system improvements or innovation. Sensitivity analysis is also carried out in this phase to help identify the effects that data gaps and uncertainties have on the results of the study. Further details on the LCA methodology can be found in the ISO 14040 and 14044 standards (ISO, 2006a, b).

Numerous LCA databases and software packages are available, including CCaLC (2016) and Gemis (Öko Institute, 2016), which are freely available, and Ecoinvent (Ecoinvent Centre, 2016), Gabi (Thinkstep, 2016), and SimaPro (PRé Consultants, 2016), which are available at a cost.

1.2.8 Life Cycle Costing

Like LCA, LCC also applies life cycle thinking, but, instead of environmental impacts, it estimates total costs of a product, process, or an activity over its life cycle. Thus, as indicated in Figure 1.8, LCC follows the usual life cycle stages considered in LCA. LCC can be used for benchmarking, ranking of different investment alternatives, or identification of opportunities for cost improvements. However, unlike LCA, LCC is yet to become a mainstream tool – while microeconomic costing is used routinely as a basis for investment decisions, estimations of costs on a life cycle basis, including costs to consumers and society, are still rare.

Although there is no standardized LCC methodology, the *code of practice* developed by Swarr et al. (2011) and largely followed by practitioners is congruent with the ISO 14040 LCA methodology, involving definition of the goal and scope of the study, inventory analysis, impact assessment, and interpretation of results. Inventory data are similar to those used in LCA, but in addition they include costs and revenues associated with the inputs into and outputs from different activities in the life cycle (Figure 1.8).

The comparable structure, data, system boundaries, and life cycle models provide the possibility of integrating LCA and LCC to assess simultaneously the economic and environmental sustainability of the system of interest and to identify any trade-offs. This also enables estimations of the eco-efficiency of products or processes by expressing environmental impacts per unit of life cycle cost or vice versa (Udo de Haes et al., 2004).

1.2.9 Social Life Cycle Assessment

S-LCA can be used to assess social and sociological aspects of products and supply chains, considering both their positive and negative impacts (UNEP and SETAC, 2009). There is no standardized methodology for S-LCA. In an attempt to ease implementation of S-LCA and make it congruent with LCA, UNEP and SETAC

Box 1.1 CML 2 method: Definition of environmental impact categories (Azapagic, 2011)

Abiotic resource depletion potential represents depletion of fossil fuels, metals, and minerals. The total impact is calculated as:

$$\mathrm{ADP} = \sum_{j=1}^{J} \mathrm{ADP}_j B_j \quad (\text{kg Sb eq.})$$

where B_j is the quantity of abiotic resource j used and ADP_j represents the abiotic depletion potential of that resource. This impact category is expressed in kg of antimony used, which is taken as the reference substance. Alternatively, kg oil eq. can be used instead for fossil resources.

Impacts of land use are calculated by multiplying the area of land used (A) by its occupation time (t):

$$\mathrm{ILU} = A \times t \ \left(\mathrm{m}^2.\mathrm{yr}\right)$$

Climate change represents the total global warming potential (GWP) of different greenhouse gases (GHG), such as carbon dioxide (CO_2), methane (CH_4), nitrous oxide (N_2O), etc. GWP is calculated as the sum of GHG emissions multiplied by their respective GWP factors, GWP_j:

$$\mathrm{GWP} = \sum_{j=1}^{J} \mathrm{GWP}_j B_j \quad (\text{kg } CO_2 \text{ eq.})$$

where B_j represents the emission of GHG j. GWP factors for different GHGs are expressed relative to the GWP of CO_2, which is defined as unity. The values of GWP depend on the time horizon over which the global warming effect is assessed. GWP factors for shorter times (20 and 50 years) provide an indication of the short-term effects of GHG on the climate, while GWP for longer periods (100 and 500 years) are used to predict the cumulative effects of these gases on the global climate.

Stratospheric ozone depletion potential (ODP) indicates the potential of emissions of chlorofluorohydrocarbons (CFCs) and other halogenated hydrocarbons to deplete the ozone layer and is expressed as:

$$\mathrm{ODP} = \sum_{j=1}^{J} \mathrm{ODP}_j B_j \quad (\text{kg CFC-11 eq.})$$

where B_j is the emission of ozone depleting gas j. The ODP factors are expressed relative to the ozone depletion potential of CFC-11.

Human toxicity potential (HTP) is calculated by taking into account releases toxic to humans to three different media, i.e. air, water, and soil:

$$\mathrm{HTP} = \sum_{j=1}^{J} \mathrm{HTP}_{jA} B_{jA} + \sum_{j=1}^{J} \mathrm{HTP}_{jW} B_{jW} + \sum_{j=1}^{J} \mathrm{HTP}_{jS} B_{jS} \quad (\text{kg 1,4-DB eq.})$$

where HTP_{jA}, HTP_{jW}, and HTP_{jS} are toxicological classification factors for substances emitted to air, water, and soil, respectively, and B_{jA}, B_{jW}, and B_{jS} represent the respective emissions of different toxic substances into the three environmental media. The reference substance for this impact category is 1,4-dichlorobenzene.

Ecotoxicity potential (ETP) is also calculated for all three environmental media and comprises five indicators ETP_n:

$$\mathrm{ETP}_n = \sum_{j}^{J} \sum_{i=1}^{I} \mathrm{ETP}_{i,j} B_{i,j} \quad (\text{kg 1,4-DB eq.})$$

where n (n = 1–5) represents freshwater and marine aquatic toxicity, freshwater and marine sediment toxicity, and terrestrial ecotoxicity, respectively. $\mathrm{ETP}_{i,j}$ represents the ecotoxicity classification factor for toxic substance j in the compartment i (air, water, soil), and $B_{i,j}$ is the emission of substance j to compartment i. ETP is based on the maximum tolerable concentrations of different toxic substances in the environment by different organisms. The reference substance for this impact category is also 1,4-dichlorobenzene.

Photochemical oxidants creation potential (POCP) is related to the potential of volatile organic compounds (VOCs) and nitrogen oxides (NO_x) to generate photochemical or summer smog. It is usually expressed relative to the POCP of ethylene and can be calculated as:

$$\mathrm{POCP} = \sum_{j=1}^{J} \mathrm{POCP}_j B_j \quad (\text{kg ethylene eq.})$$

where B_j is the emission of species j participating in the formation of summer smog and POCP_j is its classification factor for photochemical oxidation formation.

Acidification potential (AP) is based on the contribution of sulfur dioxide (SO_2), NO_x and ammonia (NH_3) to the potential acid deposition. AP is calculated according to the equation:

$$\mathrm{AP} = \sum_{j=1}^{J} \mathrm{AP}_j B_j \quad (\text{kg } SO_2 \text{ eq.})$$

where AP_j represents the AP of gas j expressed relative to the AP of SO_2 and B_j is its emission in kg.

Eutrophication potential (EP) is defined as the potential of nutrients to cause over-fertilization of water and soil, which can result in increased growth of biomass (algae). It is calculated as:

$$\mathrm{EP} = \sum_{j=1}^{J} \mathrm{EP}_j B_j \quad \left(\text{kg } PO_4^{3-} \text{ eq.}\right)$$

where B_j is an emission of species such as N, NO_x, NH_4^+, PO_4^{3-}, P, and chemical oxygen demand (COD); EP_j represent their respective EPs. EP is expressed relative to PO_4^{3-}.

See Guinée et al. (2001) for a full description of the methodology.

Source: Reproduced with permission of John Wiley & Sons.

Box 1.2 Eco-indicator 99: Definition of the damage (endpoint) categories (Azapagic, 2011)

1. Damage to Human Health

This damage category comprises the following indicators:

- Carcinogenesis
- Respiratory effects
- Ionizing radiation
- Ozone layer depletion
- Climate change.

They are all expressed in disability-adjusted life years (DALYs) and calculated by carrying out:

1) Fate analysis, to link an emission (expressed in kg) to a temporary change in concentration
2) Exposure analysis, to link the temporary concentration change to a dose
3) Effect analysis, to link the dose to a number of health effects, such as occurrence and type of cancers
4) Damage analysis, to link health effects to DALYs, using the estimates of the number of years lived disabled (YLD) and years of life lost (YLL).

For example, if a cancer causes a 10-year premature death, this is counted as 10 YLL and expressed as 10 DALYs. Similarly, hospital treatment due to air pollution has a value of 0.392 DALYs/year; if the treatment lasted 3 days (or 0.008 years), then the health damage is equal to 0.003 DALYs.

2. Damage to Ecosystem Quality

The indicators within this damage category are expressed in terms of potentially disappeared fraction (PDF) of plant species due to the environmental load in a certain area over certain time. Therefore, damage to ecosystem quality is expressed as PDF·m^2·year. The following indicators are considered:

- Ecotoxicity is expressed as the percentage of all species present in the environment living under toxic stress (potentially affected fraction [PAF]). As this is not an observable damage, a rather crude conversion factor is used to translate toxic stress into real observable damage, i.e. convert PAF into PDF.
- Acidification and eutrophication are treated as one single impact category. Damage to target species (vascular plants) in natural areas is modeled. The model used is for the Netherlands only, and it is not suitable to model phosphates.
- Land use and land transformation are based on empirical data of occurrence of vascular plants as a function of land use types and area size. Both local damages in the area occupied or transformed and regional damage to ecosystems are taken into account.

For ecosystem quality, two different approaches are used:

1) Toxic, acid, and the emissions of nutrients go through the following three steps:
   a) Fate analysis, linking the emissions to concentrations.
   b) Effect analysis, linking concentrations to toxic stress or increased nutrient or acidity levels.
   c) Damage analysis, linking these effects with the PDF of plant species.
2) Land use and transformation are modeled on the basis of empirical data on the quality of ecosystems, as a function of the type of land use and area size.

3. Damage to Resources

Two indicators are included here: depletion of minerals and fossil fuels. They are expressed as additional energy in MJ that will be needed for extraction in the future due to a decreasing amount of minerals and fuels. Geostatical models are used to relate availability of a mineral resource to its remaining amount or concentration. For fossil fuels, the additional energy is based on the future use of oil shale and tar sands.

Resource extraction is modeled in two steps:

1) Resource analysis, which is similar to fate analysis, as it links an extraction of a resource to a decrease in its concentrations (through geostatical models)
2) Damage analysis, linking decreased concentrations of resources to the increased effort for their extraction in the future.

More detail on Eco-indicator 99 can be found in Goedkoop and Spriensma (2001).

Source: Reproduced with permission of John Wiley & Sons.

(2009) have developed an S-LCA method that follows the ISO 14040 structure. Therefore, according to this method, S-LCA involves the same methodological steps as LCA: goal and scope definition, inventory, impact assessment, and interpretation. However, while the impacts in LCA represent quantitative indicators, S-LCA also includes qualitative indicators. In total, there are 194 social indicators, grouped around five groups of stakeholder: workers, consumers, local community, society, and value chain actors. The main impact categories applicable to different stakeholders are listed in Table 1.2, with each impact category comprising a number of social indicators; for the details of the latter, see UNEP and SETAC (2009).

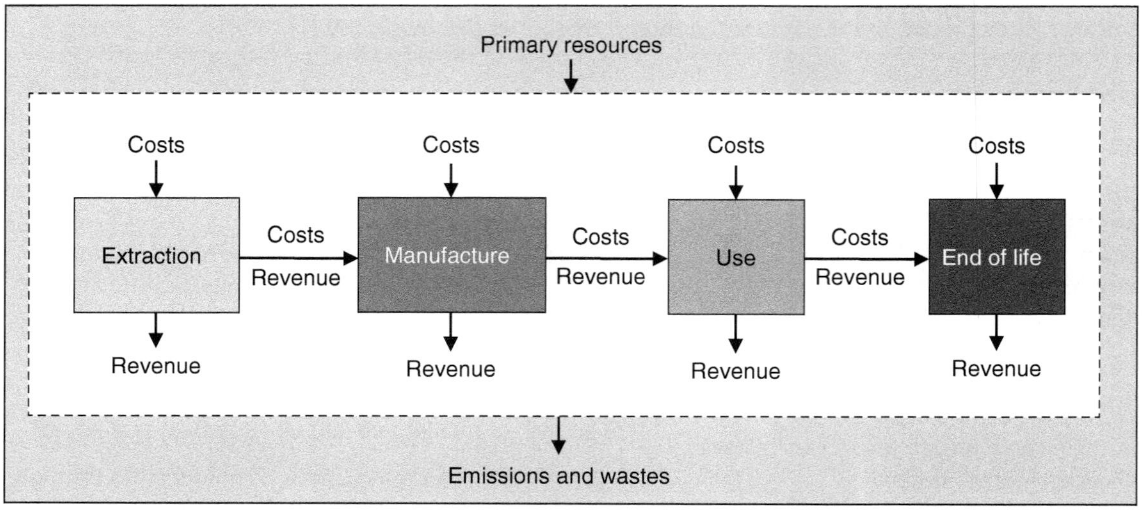

Figure 1.8 Life cycle costing estimates total costs in the life cycle of a product or an activity.

Table 1.2 The UNEP–SETAC framework for social impact categories (UNEP and SETAC, 2009).

Stakeholder group	Impact category
Workers	Freedom of association and collective bargaining
	Child labor
	Fair salary
	Working hours
	Forced labor
	Equal opportunities/discrimination
	Health and safety
	Social benefits/social security
	Feedback mechanism
Consumers	Consumer privacy
	Transparency
	End-of-life responsibility
	Access to material resources
	Access to immaterial resources
	Delocalization and migration
	Cultural heritage
Local community	Safe and healthy living conditions
	Respect of indigenous rights
	Community engagement
	Local employment
	Secure living conditions
Society	Public commitments to sustainability issues
	Contribution to economic development
	Prevention and mitigations of armed conflicts
	Technology development
	Corruption
Value chain actors	Fair competition
	Promoting social responsibility
	Supplier relationships
	Respect of intellectual property rights

As can be inferred from Table 1.2, a significant proportion of the indicators are qualitative and could be highly subjective; hence, their assessment poses a challenge. Another challenge associated with S-LCA is data availability and reliability, particularly for complex supply chains. Furthermore, geographic location of different parts of the supply chain of interest is fundamental for the assessment of social impacts, requiring specific data as generic data may be a poor substitute (Jeswani et al., 2010). However, collecting site-specific data is resource demanding and may hinder a wider adoption of the method.

1.2.10 Cost–Benefit Analysis

CBA is used widely for assessing costs and benefits of a project or an activity and to guide investment decisions. In ESA it is used for weighing environmental and socioeconomic costs and benefits of different alternatives (Jeswani et al., 2010).

CBA is based on the idea of maximum net gain – it reduces aggregate social welfare to the monetary unit of net economic benefit. So, for example, given several alternatives, the CBA approach would favor the one in which the difference between benefits and costs is the greatest. CBA has some similarities with LCC when applied to products (Finnveden and Moberg, 2005).

The most widely applied CBA technique in ESA is contingent valuation (CV). In CV, participants are asked to say how much they would be prepared to pay to protect an environmental asset. This is known as the "willingness to pay" approach. Alternatively, participants can be asked how much they would be willing to accept for loss of that asset, which is known as the "willingness to accept" method.

One of the advantages of CBA is that it presents the results as a single criterion – money – that can be easily communicated (Jeswani et al., 2010). However, measuring the expected benefits, or placing monetary value on the benefits in a simplistic way is often problematic (Ness et al., 2007). In particular, the results of the analysis largely depend on the way the questions are asked and whether the participants are familiar with the environmental asset in question. It is more likely that people who know nothing about the asset will place a nil value on it, although the life of others may depend on it. Furthermore, the value that people place on the environment strongly depends on their individual preferences and self-interest that does not serve as a firm foundation for environmental decision-making.

1.3 Summary

This chapter has presented and discussed various methods used in ESA. Broadly, they can be divided into those that take a life cycle approach and those that are more narrow in their perspective. They can also be distinguished by their focus and application, with some tools being applicable to individual products, technologies or organizations, and others to regional or national-level analyses. A further distinguishing feature is the sustainability aspect they consider: environmental, economic, and social, or their combination. Which method is used in the end will depend on the specific decision-making context and on the question(s) being asked by those carrying out the analysis. Nevertheless, the general trend in legislation and engineering practice is toward application of life cycle methods that integrate all three aspects of sustainability – the environment, economy, and society – in an attempt to balance them and drive sustainable development. Different approaches can be used to help integrate environmental, economic, and social indicators used in different ESA methods. One of the probably most useful approaches is multicriteria decision analysis (MCDA). In MCDA, relevant stakeholders are asked to state their preferences for different sustainability aspects that are then used to aggregate the considered sustainability indicators into an overall sustainability score, allowing easy comparisons of alternative products, technologies, etc. For further details on MCDA used in ESA, see Azapagic and Perdan (2005a, b).

References

Azapagic, A. (2011). Assessing environmental sustainability: Life cycle thinking and life cycle assessment. In: *Sustainable Development in Practice: Case Studies for Engineers and Scientists* (eds. A. Azapagic and S. Perdan), Chapter 3. Chichester: Wiley.

Azapagic, A. and Perdan, S. (2005a). An integrated sustainability decision-support framework: problem structuring, part I. *International Journal of Sustainable Development & World Ecology* 12 (2): 98–111.

Azapagic, A. and Perdan, S. (2005b). An integrated sustainability decision-support framework: methods and tools for problem analysis, part II. *International Journal of Sustainable Development & World Ecology* 12 (2): 112–131.

Azapagic, A. and Perdan, S. (2014). Sustainable chemical engineering: dealing with wicked sustainability problems. *AIChE Journal* 60 (12): 3998–4007.

Azapagic, A., Pettit, C., and Sinclair, P. (2007). A life cycle approach to mapping the flows of pollutants in the urban environment. *Clean Technologies and Environmental Policy* 9 (3): 199–214.

Brunner, P. and Rechberger, H. (2004). *Practical Handbook of Material Flow Analysis.* Lewis Publishers.

CCaLC (2016). CCaLC Tool and Database. The University of Manchester. www.ccalc.org.uk (accessed 5 January 2018).

EC (2013). *2013/131/EU: Commission Decision of 4 March 2013 on Eco-management and Audit Scheme (EMAS).* Brussels: European Commission. http://eur-lex.europa.eu/legal-content/EN/TXT/?qid=1405520310854&uri=CELEX:32013D0131 (accessed 5 January 2018).

EC (2016). *The European Union System for the Evaluation of Substances.* Brussels: European Commission. https://ec.europa.eu/jrc/en/scientific-tool/european-union-system-evaluation-substances (accessed 5 January 2018).

Ecoinvent Centre (2016). Ecoinvent Database. Ecoinvent Centre. http://www.ecoinvent.ch/ (accessed 5 January 2018).

EPA (2017). *Risk Assessment Guidelines.* US Environmental Protection Agency. https://www.epa.gov/risk/risk-assessment-guidelines (accessed 5 January 2018).

Fairman, R., Williams, W., and Mead, C. (1998). *Environmental Risk Assessment: Approaches, Experiences and Information Sources.* European Environment Agency: Copenhagen.

Finnveden, G. and Moberg, A. (2005). Environmental systems analysis tools – an overview. *Journal of Cleaner Production* 13: 1165–1173. doi: 10.1016/j.jclepro.2004.06.004.

Finnveden, G., Nilsson, M., Johansson, J. et al. (2003). Strategic environmental assessment methodologies - applications within the energy sector. *Environmental Impact Assessment Review* 23 (1): 91–123.

Goedkoop, M., Heijungs, R., Huijbregts, M. et al. (2009). A life cycle impact assessment method which comprises harmonised category indicators at the midpoint and the endpoint level. https://www.leidenuniv.nl/cml/ssp/publications/recipe_characterisation.pdf (accessed 2 February 2018).

Goedkoop, M. and Spriensma, R. (2001). *The Eco-Indicator 99: A Damage Oriented Method for Life Cycle Assessment, Methodology Report,* 3e. Amersfoort: Pré Consultants.

Guinée, J.B., Gorrée, M., Heijungs, R. et al. (2001). *Life Cycle Assessment: An Operational Guide to the ISO Standards. Parts 1, 2a & 2b.* Dordrecht: Kluwer Academic Publishers.

IPCC (2013). *Climate Change 2013 – the Physical Science Basis.* Working Group I Contribution to the Fifth Assessment Report of the Intergovernmental Panel on Climate Change. Cambridge University Press.

ISO (2006a). *ISO/DIS 14040: Environmental Management – Life Cycle Assessment – Principles and Framework.* Geneva: ISO.

ISO (2006b). *ISO/DIS 14044: Environmental Management – Life Cycle Assessment – Requirements and Guidelines.* Geneva: ISO.

ISO (2015). *ISO 14001:2015 – Environmental Management Systems. Requirements with Guidance for Use.* Geneva: ISO. doi: 10.3389/fphys.2015.00416.

Jeswani, H., Azapagic, A., Schepelmann, P., and Ritthoff, M. (2010). Options for broadening and deepening the LCA approaches. *Journal of Cleaner Production* 18 (2): 120–127. doi: 10.1016/j.jclepro.2009.09.023.

Lijzen, J.P.A. and Rikken, M.G.J. (2004). *EUSES – European Union System for Evaluation of Substances.* Background report. Bilthoven (January).

Manuilova, A. (2003). Methods and tools for assessment of environmental risk. DANTES Life EC project. www.dantes.info/Publications/Publication-doc/An%20overview%20of%20ERA%20-methods%20and%20tools.pdf (accessed 5 January 2018).

Ness, B., Urbel-Piirsalu, E., Anderberg, S., and Olsson, L. (2007). Categorising tools for sustainability assessment. *Ecological Economics* 60 (3): 498–508.

Öko Institute (2016). Global Emission Model for Integrated Systems (GEMIS). Germany. http://iinas.org/gemis.html (accessed 2 February 2018).

PRé Consultants (2016). *SimaPro Database and Software.* The Netherlands: PRé Consultants.

Swarr, T., Hunkeler, D., Klopffer, W. et al. (2011). *Environmental Life Cycle Costing: A Code of Practice.* Brussels: Society of Environmental Toxicology and Chemistry.

Thinkstep (2016). *Gabi LCA Software and Database.* Stuttgart: Thinkstep.

Udo de Haes, H., Heijungs, R., Suh, S., and Huppes, G. (2004). Three strategies to overcome the limitations of life-cycle assessment. *Journal of Industrial Ecology* 8 (3): 19–32.

UNEP (2012). GEO-5, Global Environmental Outlook – Environment for the Future We Want. United Nations Environmental Programme, Nairobi.

UNEP and SETAC (2009). *Guidelines for Social Life Cycle Assessment of Products.* UNEP/SETAC. www.unep.fr/shared/publications/pdf/DTIx1164xPA-guidelines_sLCA.pdf (accessed 5 January 2018).

van der Voet, E. (2002). Substance flow analysis methodology. In: *A Handbook of Industrial Ecology* (ed. R. Ayres and L. Ayres). Northampton, MA: Edward Elgar Publishing.

Yue, Q., He, J., Stamford, L., and Azapagic, A. (2016). Nuclear power in China: an analysis of the current and near-future uranium flows. *Energy Technology* 5 (5): 681–691.

2

Measurements in Environmental Engineering

Daniel A. Vallero

Department of Civil and Environmental Engineering, Duke University, Durham, NC, USA

Summary

The environment consists of very complex systems ranging in scale from the cell to the planet. These systems are comprised of matrices of nonliving (i.e. abiotic) and living (biotic) components. To determine the condition of such systems calls for various means of measurement. Many of these measurements direct physical and chemical measurements, e.g. temperature, density, and pH of soil and water. Others are indirect, such as light scattering as indication of the number of aerosols in the atmosphere. This chapter provides an overview of some of the most important measurement methods in use today. In addition, the chapter introduces some of the techniques available for sampling, analysis, and extrapolation and interpolation of measured results, using various types of models.

2.1 Introduction

According to their codes of ethics and practice, engineers must hold paramount the public's safety, health, and welfare (National Society of Professional Engineers, 2016). Engineers apply the sciences to address societal needs. Environmental engineers are particularly interested in protecting public health and ecosystem conditions.

Myriad human activities, such as energy generation and transmission, transportation, food production, and housing, generate wastes and pollute environmental media, i.e. air, water, soil, sediment, and biota. Environmental engineers must find ways to reduce or eliminate risks posed by these wastes. The first step in assessing and managing risks is to determine the condition of the environment, which includes estimates of the amount of contaminants in each environmental medium (Vallero, 2015). Such estimates must be based on with reliable data and information, beginning by characterizing the release of a substance, e.g. an emission from a stack, the substance's movement and transformation in the environment, and its concentrations near or within an organism, i.e. the receptor (see Figure 2.1).

Environmental measurement is an encompassing term, which includes developing methods, applying those methods, deploying monitoring technologies, and interpreting the results from these technologies. An environmental assessment can address chemical, physical, and biological factors (U.S. Environmental Protection Agency, 2015c). This article addresses measurements of concentrations of substances in the environment. Such measurements are one part of health, exposure, and risk assessments, but not everything that needs to be measured for such assessments. For example, exposure assessments require information about the pollutant concentrations in the locations where specific human activities occur. A measurement of a pollutant at a central monitoring site, therefore, is not an exposure measurement, since it does not reflect the concentration where the activity takes place. A personal monitor worn during a day would be a more precise and accurate measurement of exposure for that particular day if it were matched with the person's activities, e.g. using a diary.

Measurements may be direct or indirect. Direct measurements are those in which the substance of concern is what is actually collected and analyzed. For example, a measurement of particulate matter (PM) would be directly measured by pumping air through a PM monitor and collecting particles on a filter. The particles would then be measured, e.g. weighed, sized, and chemically analyzed. Direct measurements can be in situ, i.e. taken in the environment, or ex situ, collected and taken elsewhere for measurement.[1]

An indirect PM measurement is one where the substance itself is not collected, but would be characterized by an indicator, e.g. light scattering in a nephelometer

Handbook of Environmental Engineering, First Edition. Edited by Myer Kutz.

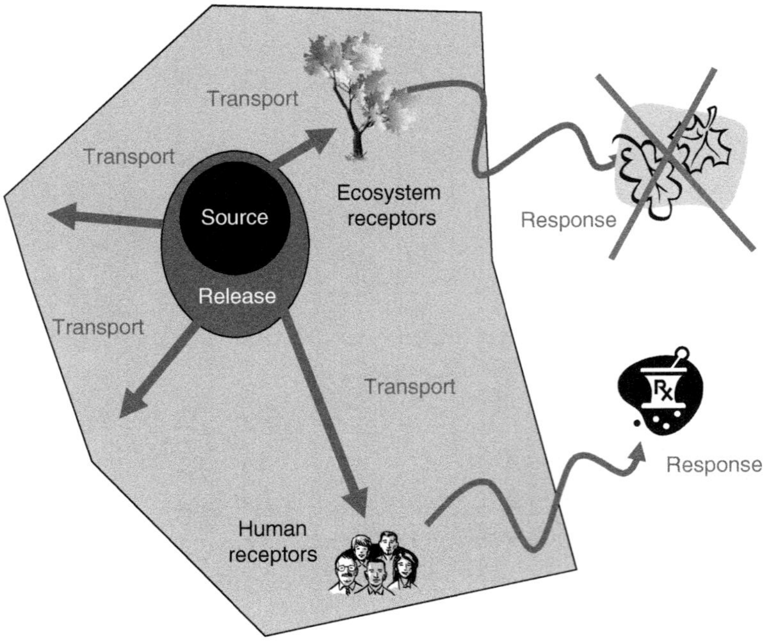

Figure 2.1 Sites of environmental measurements. *Source*: Letcher and Vallero (2011). Reproduced with permission of Elsevier.

(Vallero, 2014; Whitby and Willeke, 1979). The amount and type of scattering would indicate the quantity and size of particles. Remote sensing of pollutants relies on indirect measurements, e.g. using a laser to backscatter specific electromagnetic wavelengths is used to characterize aerosols in the atmosphere, including particles in the stratosphere. The principal method for aerosol profiling is light detection and ranging, i.e. LIDAR, which uses a pulsed laser with a system to detect the backscattered radiation (De Tomasi and Perrone, 2014).

The monitoring underpinning the assessment is dependent upon the quality of sample collection, preparation, and analysis. Sampling is a statistical term, and usually a geostatistical term. An environmental sample is a fraction of air, water, soil, biota, or other environmental media (e.g. paint chips, food, etc. for indoor monitoring) that represents a larger population or body. For example, a sample of air may consist of a canister or bag that holds a defined quantity or air that will be subsequently analyzed. The sample is representative of a portion of an air mass. The number of samples must be collected and results aggregated to ascertain with defined certainty the quality of an air mass. More samples will be needed for a large urban air shed than for that of a small town. Intensive sampling is often needed for highly toxic contaminants and for sites that may be particularly critical, e.g. near a hazardous waste site or in an "at risk" neighborhood (such as one near a manufacturing facility that uses large quantities of potentially toxic materials). Similar to other statistical measures, environmental samples allow for statistical inference. In case, inferences are made regarding the condition of an ecosystem and the extent and severity of exposure of a human population.

For example, to estimate the amount of a chemical compound in a lake near a chemical plant, an engineer gathers a 500 ml sample in the middle of the lake that contains 1 million liters of water. Thus, the sample represents only 5×10^{-7} of the lake's water. This is known as a "grab" sample, i.e. a single sample taken to represent an entire system. Such a sample is limited in location vertically and horizontally, so there is much uncertainty. However, if 10 samples are taken at 10 spatially distributed sites, the inferences are improved. Furthermore, if the samples were taken in each season, then there would be some improvement to understanding of intra-annual variability. If the sampling is continued for several years, the inter-annual variability is better characterized. Indeed, this approach can be used in media other than water, e.g. soil, sediment, and air.

2.1.1 Data Quality Objectives

A monitoring plan must be in place before samples are collected and arrive at the laboratory. The plan includes quality assurance (QA) provisions and describes the procedures to be employed. These procedures must be strictly followed to investigate environmental conditions. The plan describes in detail the sampling apparatus (e.g. real-time probes, sample bags, bottles, and soil cores), the number of samples needed, and the sample handling and transportation. The quality and quantity of samples are determined by data quality objectives (DQOs), which are defined by the objectives of the overall contaminant assessment plan. DQOs are qualitative and quantitative statements that translate nontechnical project goals into scientific and engineering outputs need to answer technical questions (U.S. Environmental Protection Agency, 2006).

Quantitative DQOs specify a required level of scientific and data certainty, while qualitative DQOs express decisions goals without specifying those goals in a quantitative manner. Even when expressed in technical terms, DQOs must specify the decision that the data will ultimately support, but not the manner that the data will be collected. DQOs guide the determination of the data quality that is needed in both the sampling and analytical efforts. The U.S. Environmental Protection Agency has listed three examples of the range of detail of quantitative and qualitative DQOs (Crumbling, 2001; U.S. Environmental Protection Agency, 1994):

1) *Example of a less detailed, quantitative DQO*: Determine with greater than 95% confidence that contaminated surface soil will not pose a human exposure hazard.
2) *Example of a more detailed, quantitative DQO*: Determine to a 90% degree of statistical certainty whether or not the concentration of mercury in each bin of soil is less than 96 ppm
3) *Example of a detailed, qualitative DQO:* Determine the proper disposition of each bin of soil in real-time using a dynamic work plan and a field method able to turnaround lead (Pb) results on the soil samples within 2 h of sample collection.

Thus, if the condition in question is tightly defined, e.g. the seasonal change in pH near a fish hatchery, a small number of samples using simple pH probes would be defined as the DQO. Conversely, if the environmental assessment is more complex and larger in scale, e.g. the characterization of year-round water quality for trout in the stream, the sampling plan's DQO may dictate that numerous samples at various points be continuously sampled for inorganic and organic contaminants, turbidity, nutrients, and ionic strength. This is even more complicated for biotic systems, which may also require microbiological monitoring.

The sampling plan must include all environmental media, e.g. soil, air, water, and biota, which are needed to characterize the exposure and risk of any biotechnological operation. The sampling and analysis plan should explicitly point out which methods will be used. For example, if toxic chemicals are being monitored, the US EPA specifies specific sampling and analysis methods (U.S. Environmental Protection Agency, 1999, 2007).

The geographic area where data are to be collected is defined by distinctive physical features such as volume or area, e.g. metropolitan city limits, the soil within the property boundaries down to a depth of 6 cm, a specific water body, length along a shoreline, or the natural habitat range of a particular animal species. Care should be taken to define boundaries. For example, Figure 2.2 shows a sampling grid, with a sample taken from each cell in the grid (U.S. Environmental Protection Agency,

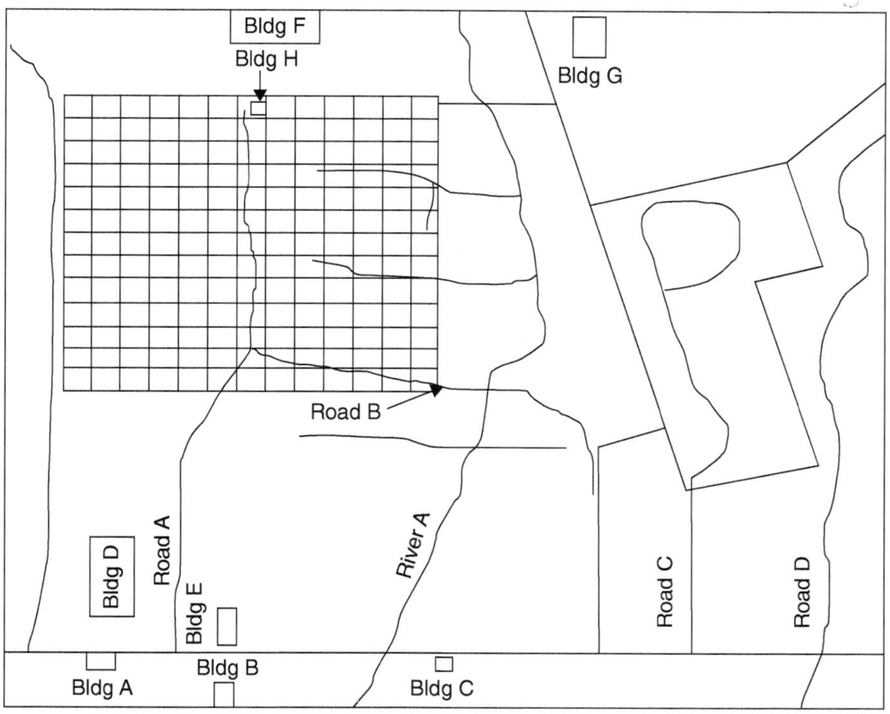

Figure 2.2 Environmental assessment area delineated by map boundaries. *Source*: U.S. Environmental Protection Agency (2002).

2002). The target population may be divided into relatively homogeneous subpopulations within each area or subunit. This can reduce the number of samples needed to meet the tolerable limits on decision errors and to improve efficiency.

Time is another essential parameter that determines the type and extent of monitoring needed. Conditions vary over the course of a study due to changes in weather conditions, seasons, operation of equipment, and human activities. These include seasonal changes in groundwater levels, seasonal differences in farming practices, daily or hourly changes in airborne contaminant levels, and intermittent pollutant discharges from industrial sources. Such variations must be considered during data collection and in the interpretation of results. Some examples of environmental time sensitivity are:

- Concentrations of lead in dust on windowsills may show higher concentrations during the summer when windows are raised and paint/dust accumulates on the windowsill.
- Terrestrial background radiation levels may change due to shielding effects related to soil dampness.
- Amount of pesticides on surfaces may show greater variations in the summer because of higher temperatures and volatilization.
- Instruments that may not give accurate measurements when temperatures are colder.
- Airborne PM measurements that may not be accurate if the sampling is conducted in the wetter winter months rather than the drier summer months.

Feasibility should also be considered. This includes gaining legal and physical access to the properties, equipment acquisition and operation, environmental conditions, and times and conditions when sampling is prohibited (e.g. freezing temperatures, high humidity, and noise).

2.1.2 Monitoring Plan Example

Consider a plan to measure mobile source air toxic (MSAT) concentrations and variations in concentrations as a function of distance from the highway and to establish relationships between MSAT concentrations as related to highway traffic flows including traffic count, vehicle types and speeds, and meteorological conditions such as wind speed and wind direction. Specifically, the monitoring plan has the following goals (Kimbrough et al., 2008):

1) Identify the existence and extent of elevated air pollutants near roads.
2) Determine how vehicle operations and local meteorology influence near road air quality for criteria and toxic air pollutants.
3) Collect data that will be useful in ground truthing, evaluating, and refining models to determine the emissions and dispersion of motor vehicle-related pollutants near roadways.

A complex monitoring effort requires management and technical staff with a diversity of skills that can be brought to bear on the implementation of this project. This diverse skill set includes program management, contracts administration, field monitoring experience, laboratory expertise, and QA oversight.

The purpose of any site selection process is to gather and analyze sufficient data that would lead one to draw informed conclusions regarding the selection of the most appropriate site for the monitoring at a specific location. Moreover, the site selection process needs to include programmatic issues to ensure an informed decision is reached.

2.1.3 Selection of a Monitoring Site

Selecting a monitoring site must be based on scientific and feasibility factors, as shown in Table 2.1 and Figure 2.3. Each step has varying degrees of complexity due to "real-world" issues. The first step was to determine site selection criteria (see Table 2.2). The follow-on steps include (ii) develop list of candidate sites and supporting information, (iii) apply site selection filter ("coarse" and "fine"), (iv) site visit, (v) select candidate site(s) via team discussion, (vi) obtain site access permission(s), and (vii) implement site logistics.

A list of candidate sites based on these criteria can then be developed. Geographic information system (GIS) data, tools and techniques, and on-site visits would be used to compare various sites that meet these criteria.

Quite commonly, even a well-designed environmental monitoring plan will need to be adjusted during the implementation phase. For example, investigators may discover barriers or differing conditions from what was observed in the planning phase (e.g. different daily traffic counts or new road construction).

After applying site selection criteria as a set of "filters," candidate sites are incrementally eliminated. For example, the first filter would be sites with low traffic counts; the next filter, the presence of extensive sound barriers, eliminates additional sites; and other filters, e.g. complex geometric design or lack of available traffic volume data, eliminates additional sites. Next, feasibility considerations would eliminate additional candidate sites.

An important component of "ground truthing" or site visit is to obtain information from local sources. Local businesses and residents can provide important information needed in a decision process, such as types of

Table 2.1 Example of steps in selecting an air quality monitoring site.

Step	Site selection steps	Method	Comment
1	Determine site selection criteria	Monitoring protocol	
2	Develop list of candidate sites	Geographic information system (GIS) data; on-site visit(s)	Additional sites added as information is developed
3	Apply coarse site selection filter	Team discussions, management input	Eliminate sites below acceptable minimums
4	Site visit	Field trip	Application of fine site selection filter
5	Select candidate site(s)	Team discussions, management input	
6	Obtain site access permissions	Contact property owners	If property owners do not grant permission, then the site is dropped from further consideration
7	Site logistics (i.e. physical access, utilities – electrical and communications)	Site visit(s), contact utility companies	

Source: Kimbrough et al. (2008). Reproduced with permission of Elsevier.

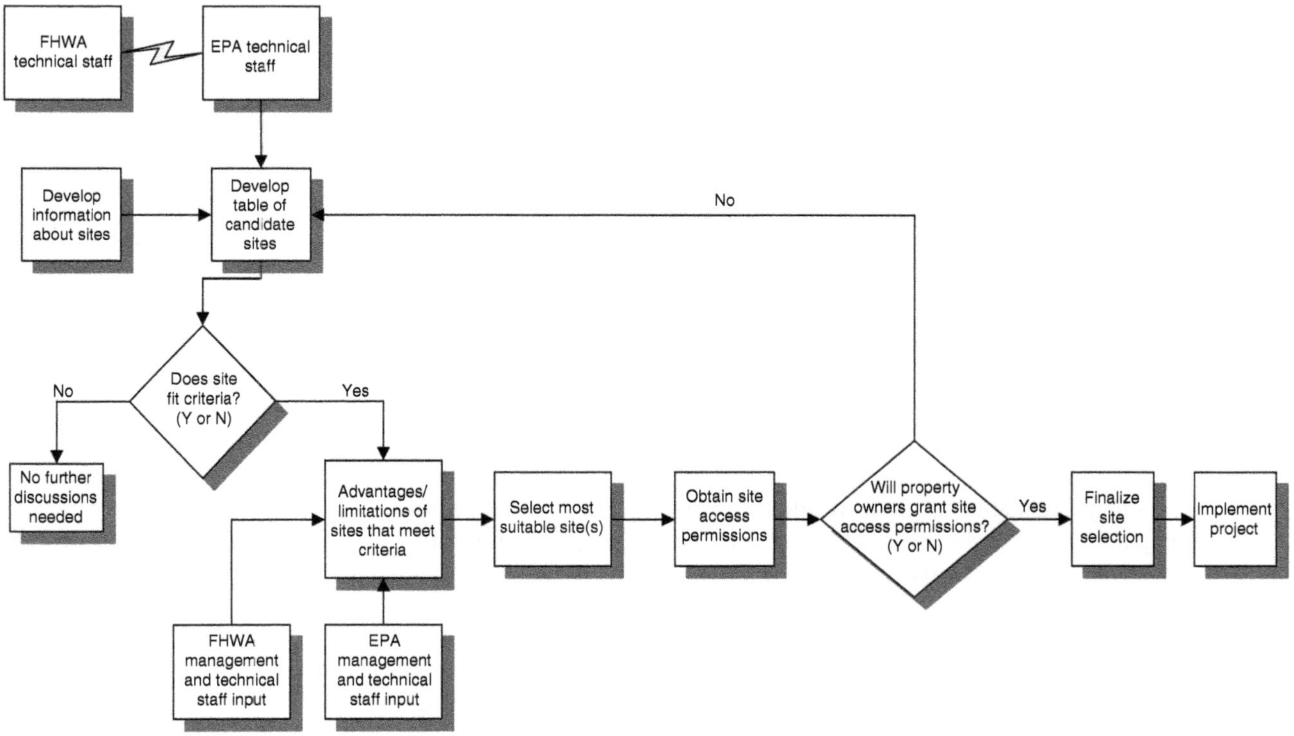

Figure 2.3 Monitoring location selection decision flowchart. *Source*: Kimbrough et al. (2008). Reproduced with permission of Elsevier.

chemicals stored previously at a site, changes in vegetation, or even ownership histories.

Spatial tools are an important part of the environmental engineer's daily work. They are very useful in making and explaining environmental decisions (Malczewski, 1999; Sumathi et al., 2008). Until recently, the use of GIS and other spatial tools in decision processes have required the acquisition of large amounts of the data. In addition, the software has not been user-friendly. GIS data have now become more readily available in both quantity and quality, and GIS exists in common operating system environments. Indeed, environmental regulatory agencies increasingly use data layers to assess and describe environmental conditions. For example, the US EPA has developed the EnviroAtlas, a system of interactive tools to support and to document "ecosystem goods and

Table 2.2 Example selection considerations and criteria.

Selection considerations	Monitoring protocol criteria
Essential criteria for this monitoring study	
AADT (>150 000)	Only sites with more than 150 000 annual average daily traffic (AADT) are considered as candidates
Geometric design	The geometric design of the facility, including the layout of ramps, interchanges, and similar facilities, will be taken into account. Where geometric design impedes effective data collection on MSATs and $PM_{2.5}$, those sites will be excluded from further consideration. All sites have a "clean" geometric design
Topology (i.e. sound barriers, road elevation)	Sites located in terrain making measurement of MSAT concentrations difficult or that raise questions of interpretation of any results will not be considered. For example, sharply sloping terrain away from a roadway could result in under representation of MSAT and $PM_{2.5}$ concentration levels on monitors in close proximity to the roadway simply because the plume misses the monitor as it disperses
Geographic location	Criteria applicable to representing geographic diversity within the United States as opposed to within any given city
Availability of data (traffic volume data)	Any location where data, including automated traffic monitoring data, meteorological, or MSAT concentration data, is not readily available or instrumentation cannot be brought in to collect such data will not be considered for inclusion in the study
Meteorology	Sites will be selected based on their local climates to assess the impact of climate on dispersion of emissions and atmospheric processes that affect chemical reactions and phase changes in the ambient air
Desirable, but not essential criteria	
Downwind sampling	Any location where proper siting of downwind sampling sites is restricted due to topology, existing structures, meteorology, etc. may exclude otherwise suitable sites for consideration and inclusion in this study
Potentially confounding air pollutant sources	The presence of confounding emission sources may exclude otherwise suitable sites for consideration and inclusion in this study
Site access (admin/ physical)	Any location where site access is restricted or prohibited either due to administrative or physical issues, will not be considered for inclusion in the study

Source: Kimbrough et al. (2008). Reproduced with permission of Elsevier.

services," i.e. ecological benefits to humans from nature, including food supply, water supply, flood control, security, public health, and economy (U.S. Environmental Protection Agency, 2015a). Table 2.3 shows some of the map layers that underpin the EnviroAtlas (U.S. Environmental Protection Agency, 2015b).

The GIS data layers that are commonly needed in environmental engineering include the location of suitable soils, wells, surface water sources, residential areas, schools, airports, roads, etc. From these data, layers queries are formulated to provide the most suitable sites (e.g. depth to water table may help identify sources of pollution). Typically, quantitative weighting criteria are associated with the siting criteria as well as elements of the data layers, e.g. certain types of soils would be more suitable than others and thus would have applicable quantitative values (Environmental Systems Research Institute, 1995).

2.2 Environmental Sampling Approaches

Engineers use various methods of collecting environmental samples. As mentioned, the grab sample is simply a measurement at a site at a single point in time. Composite sampling physically combines and mixes multiple grab samples (from different locations or times) to allow for physical, instead of mathematical, averaging. The acceptable composite provides a single value of contaminant concentration measurement that can be used in statistical calculations.

Multiple composite samples can provide improved sampling precision and reduce the total number of analyses required compared to non-composite sampling (U.S. Environmental Protection Agency, 2015d), e.g. "grab" or integrated soil sample of x mass or y volume, the number of samples needed (e.g. for statistical significance) the minimum acceptable quality as defined by the QA plan and sampling standard operating procedures (SOPs), and sample handling after collection.

A weakness of composite sampling is the false negative effect. Consider, for example, samples collected from an evenly distributed grid of homes to represent a neighbor exposure to a contaminant, as shown in Figure 2.4. The assessment found the values of 3, 1, 2, 12, and 2 $mg\,l^{-1}$, and the mean contamination concentration is only 4 $mg\,l^{-1}$. If cleanup is needed above the threshold of 5 $mg\,l^{-1}$, the mean concentration would indicate the area does not need remediation and would be reported below the threshold level. However, the fourth home is well

Table 2.3 National scale map layers used in the EnviroAtlas (page 1 of 20 pages).

Map layer title	Description	Metadata link	Biodiversity conservation	Clean air	Clean and plentiful water	Climate stabilization	Food, fuel, and materials	Natural hazard mitigation	Recreation, culture, and aesthetics
Acres of crops that have no nearby pollinator habitat	This map layer depicts the total acres of agricultural crops within a subwatershed (12-digit HUC) that require or would benefit from the presence of pollinators, but are without any nearby supporting habitat	Meta data					◆		
Agricultural water use (million gal d^{-1})	This map estimates the millions of gallons of water used daily for agricultural irrigation for each subwatershed (HUC-12) in the contiguous United States. Estimates include self-supplied surface and groundwater, as well as water supplied by irrigation water providers, which may include governments, companies, or other organizations	Meta data			◆				
Area of solar energy (km^2)	This map estimates the square kilometers of area within each subwatershed (12-digit HUC) that offers the potential for harvesting solar energy. This map does not take into account land use or ownership	Meta data					◆		
Average annual daily potential ($kWh\,m^{-2}\,d^{-1}$)	This map estimates the average daily potential kilowatt hours of solar energy that could be harvested per square meter within each subwatershed (12-digit HUC). This calculation is based on environmental factors and does not take into account land ownership or viability of installing solar harvesting systems	Meta data					◆		
Average annual precipitation (in. yr^{-1})	This map estimates the average number of inches of precipitation that fall within a subwatershed (12-digit HUC) each year. Precipitation includes snow and rain accumulation	Meta data		◆	◆	◆		◆	◆
Carbon storage by tree biomass ($kg\,m^{-2}$)	This map estimates the kilograms of dry carbon stored per square meter of above ground biomass of trees and forests in each subwatershed (12-digit HUC)	Meta data				◆			
Carbon storage by tree root biomass ($kg\,m^{-2}$)	This map estimates the kilograms of dry carbon stored per square meter in below ground biomass in each subwatershed (12-digit HUC). Biomass below ground includes tree root biomass and soils	Meta data				◆			
Cotton yields (thousand tons yr^{-1})	This map depicts the thousands of tons of cotton that are grown annually within each subwatershed (12-digit HUC)	Meta data					◆		
Cultivated biological nitrogen fixation ($kg\,N\,ha^{-1}\,yr^{-1}$)	This map depicts the mean rate of biological nitrogen fixation from the cultivation of crops within each subwatershed (12-digit HUC) in $kg\,N\,ha^{-1}\,yr^{-1}$	Meta data		◆		◆			
Domestic water use (million gal d^{-1})	This map estimates the millions of gallons of water used daily for domestic purposes in each subwatershed (HUC-12). For the purposes of this map, domestic or residential water use includes all indoor and outdoor uses, such as for drinking, bathing, cleaning, landscaping, and pools for primary residences	Meta data			◆				

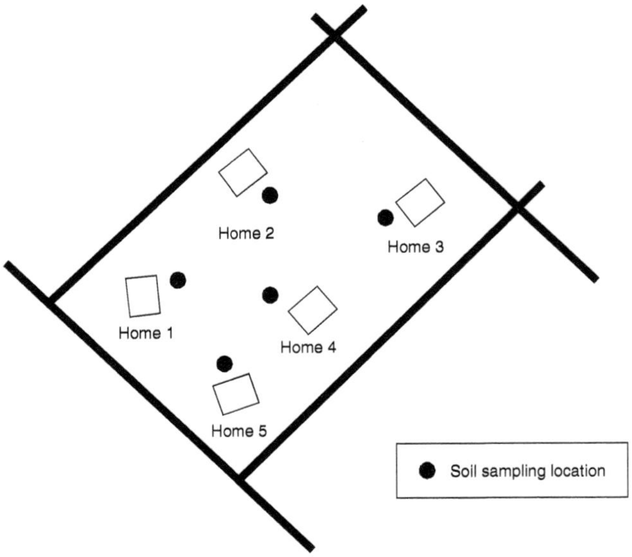

Figure 2.4 Hypothetical composite sampling grid for a neighborhood. *Source*: Vallero (2015). Reproduced with permission of Elsevier.

above the safety level. This could also have a false positive effect. For example if the mean concentration were 6 mg l^{-1} in the example, the whole neighborhood may not need cleanup if the source is isolated to a confined area in the yard of home 5.

Another example of where geographic composites may not be representative is in cleaning up and monitoring the success of cleanup actions. For example, if a grid is laid out over a contaminated groundwater plume (Figure 2.5), it may not take into account horizontal and vertical impervious layers, unknown sources (e.g. tanks), and flow differences among strata, so that some of the plume is eliminated but pockets are left (as shown in Figure 2.5b).

Soil is highly heterogeneous in its texture, chemical composition, moisture content, organic matter content, sorption coefficients, and other physical and biological characteristics (Vallero, 2000; Vallero and Peirce, 2002; Wang et al., 2015). Thus, it is often good practice to assume that a contaminated site will have a heterogeneous

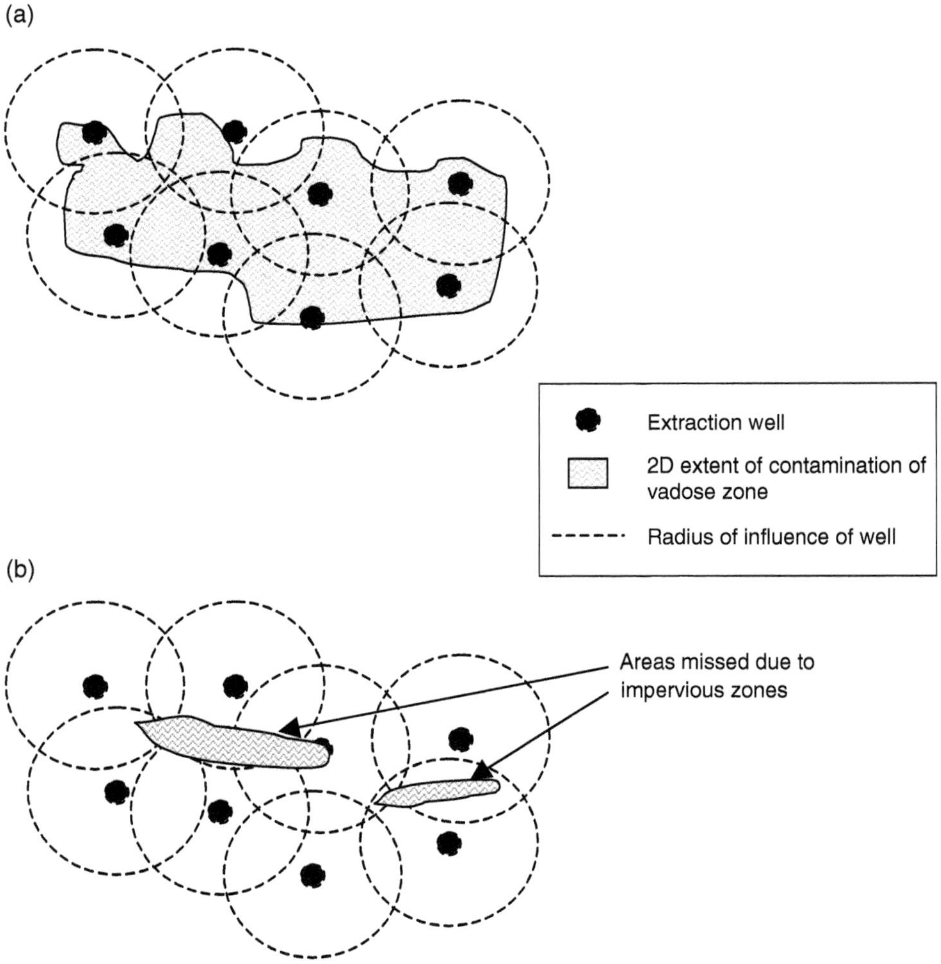

Figure 2.5 Extraction well locations on a geometric grid, showing hypothetical cleanup after 6 months (a) Before treatment. (b) After treatment. *Source*: Vallero (2015). Reproduced with permission of Elsevier.

distribution of contamination. Selecting appropriate sampling methods and considerations on their use are key parts of any study design and environmental assessment, since the results will be the basis for exposure models, risk assessments, feasibility studies, land use and zoning maps, and other information used by fellow engineers, clients, and regulators (Environmental Health Australia, 2012b). As indicated in Table 2.4, measurement errors and uncertainties will accompany the results and even be compounded as the data are translated into information (U.S. Environmental Protection Agency, 2004), so it is important to include all necessary metadata to ensure others may deconstruct, quality assure, and ensure appropriate applications.

Table 2.4 Types of uncertainty and contributing errors in environmental engineering.

Type of uncertainty	Type of error causing uncertainty	Description or example
Emissions	Misclassification and miscalculation	Reliance on third party and other sources of information with little or no metadata regarding quality. Confusing actual emission measurements with reported estimates
Transport and transformation	Incorrect model application	Applying a model for the wrong chemistry, atmospheric, terrain, and other conditions, e.g. using a simple dispersion model in a complex terrain. Applying a quantitative structure–activity relationship (QSAR) to inappropriate compounds, e.g. for metals when the QSAR is only for organic compounds, or for semi-volatile compounds when the QSAR is only for volatile organic species
Exposure scenario	Misclassification	Failure to adequately identify exposure routes, exposure media, and exposed population
Sampling or measurement (parameter uncertainty)	Measurement: random	Random errors in analytical devices (e.g. imprecision of continuous monitors that measure stack emissions)
	Measurement: systemic	Systemic bias (e.g. estimating inhalation from indoor ambient air without considering the effect of volatilization of contaminants from hot water during showers)
	Surrogate data	Use of alternate data for a parameter instead of direct analysis of exposure (e.g. use of population figures as a surrogate for population exposure)
	Misclassification	Incorrect assignment of exposures of subjects in historical epidemiologic studies resulting from faulty or ambiguous information
	Random sampling error	Use of a small sample of individuals to estimate risk to exposed workers
	Nonrepresentativeness	Developing exposure estimates for a population in a rural area based on exposure estimates for a population in a city
Observational or modeling	Relationship errors	Incorrectly inferring the basis of correlations between environmental concentrations and urinary output
	Oversimplification	Misrepresentations of reality (e.g. representing a three-dimensional aquifer with a two-dimensional mathematical model)
	Incompleteness	Exclusion of one or more relevant variables (e.g. relating a biomarker of exposure measured in a biological matrix without considering the presence of the metabolite in the environment)
	Surrogate variables	Use of alternate variables for ones that cannot be measured (e.g. using wind speed at the nearest airport as a proxy for wind speed at the facility site)
	Failure to account for correlations	Not accounting for correlations that cause seemingly unrelated events to occur more frequently than expected by chance (e.g. two separate components of a nuclear plant are missing a particular washer because the same newly hired assembler put them together)
	Model disaggregation	Extent of (dis)aggregation used in the model (e.g. separately considering subcutaneous and abdominal fat in the fat compartment of a physiologically based pharmacokinetic [PBPK] model)
	Biological plausibility	Applying a PBPK model to chemicals for which it has not been coded, e.g. a one-compartment model for a compound with known metabolites

Source: U.S. Environmental Protection Agency (2004) and Vallero (2014).

2.2.1 Sampling Approaches

Random sampling: While it has the value of statistical representativeness, with a sufficient number of samples for the defined confidence levels (e.g. x samples needed for 95% confidence, random sampling may lead to large areas of the site being missed for sampling because due to chance distribution of results. It also neglects prior knowledge of the site. For example, if maps show an old tank that may have stored contaminants, a purely random sample will not give any preference to samples near the tank.

Stratified random sampling: By dividing the site into areas and randomly sampling within each area, avoiding the omission problems of random sampling alone.

Stratified sampling: Contaminants or other parameters are targeted. The site is subdivided and sampling patterns and densities varied in different areas. Stratified sampling can be used for complex and large sites, such as mining.

Grid or systematic sampling: The whole site is covered. Sampling locations are readily identifiable, which is valuable for follow-on sampling, if necessary. The grid does not have to be rectilinear. In fact, rectangles are not the best polygon to use in the value to be a representative of a cell. Circles provide equidistant representation, but overlap. Hexagons are sometimes used as a close approximation to the circle. The US Environmental Monitoring and Assessment Program (EMAP) has used a hexagonal grid pattern, for example.

Judgmental sampling: Samples are collected base upon knowledge of the site. This overcomes the problem of ignoring sources or sensitive areas but is vulnerable to bias of both inclusion and exclusion. Obviously, this would not be used for spatial representation, but for pollutant transport, plume characterization, or monitoring near a sensitive site (e.g. a day care center).

At every stage of monitoring from sample collection through analysis and archiving, only qualified and authorized persons should be in possession of the samples. This is usually assured by requiring chain-of-custody manifests. Sample handling includes specifications on the temperature range needed to preserve the sample, the maximum amount of time the sample can be held before analysis, special storage provisions (e.g. some samples need to be stored in certain solvents), and chain-of-custody provisions (only certain, authorized persons should be in possession of samples after collection).

Each person in possession of the samples must require that recipient sign and date the chain-of-custody form before transferring the samples. This is because samples have evidentiary and forensic content, so any compromising of the sample integrity must be avoided.

2.3 Laboratory Analysis

Although real-time analysis of air and other media is becoming more commonplace, most environmental samples must be brought to a laboratory analyzed after collection. The steps that must be taken to interpret the concentration of a chemical in the sample are known as "wet chemistry."

2.3.1 Extraction

When an environmental sample arrives at the laboratory, the next step may be "extraction." Extraction is needed for two reasons. First, the environmental sample may be in sediment or soil, where the chemicals of concern are sorbed to particles and must be freed for analysis to take place. Second, the actual collection may have been by trapping the chemicals onto sorbents, meaning that the chemicals must first be freed from the sorbent matrix. Numerous toxic chemicals have low vapor pressures and may not be readily dissolved in water. Thus, they may be found in various media, e.g. sorbed to particles, in the gas phase, or in the water column suspended to colloids (and very small amounts dissolved in the water itself). To collect such chemicals in the gas phase, a common method calls for trapping it on polyurethane foam (PUF). Thus, to analyze dioxins in the air, the PUF and particle matter must first be extracted, and to analyze dioxins in soil and sediment, those particles must also be extracted.

Extraction makes use of physics and chemistry. For example, many compounds can be simply extracted with solvents, usually at elevated temperatures. A common solvent extraction is the Soxhlet extractor, named after the German food chemist, Franz Soxhlet (1848–1913). The Soxhlet extractor (the US EPA Method 3540) removes sorbed chemicals by passing a boiling solvent through the media. Cooling water condenses the heated solvent and the extract is collected over an extended period, usually several hours. Other automated techniques apply some of the same principals as solvent extraction but allow for more precise and consistent extraction, especially when large volumes of samples are involved. For example, supercritical fluid extraction (SFE) brings a solvent, usually carbon dioxide to the pressure and temperature near its critical point of the solvent, where the solvent's properties are rapidly altered with very slight variations of pressure (Ekhtera et al., 1997). Solid phase extraction (SPE), which uses a solid and a liquid phase to isolate a chemical from a solution, is often used to clean up a sample before analysis. Combinations of various extraction methods can enhance the extraction efficiencies, depending upon the chemical and the media in which it is found. Ultrasonic and microwave extractions may be used alone or in combination with solvent extraction. For example, the US EPA Method 3546 provides a procedure for extracting hydrophobic (that is, not soluble in water) or slightly water-soluble organic compounds from particles such as soils, sediments, sludges, and solid wastes. In this method, microwave energy elevates the temperature and pressure conditions (i.e. 100–115 °C and 50–175 psi) in a closed extraction vessel containing the sample and solvent(s). This combination can improve recoveries of chemical analytes and can reduce the time needed compared than the Soxhlet procedure alone.

2.3.2 Separation Science

Not every sample needs to be extracted. For example, air monitoring using canisters and bags allows the air to flow directly into the analyzer. Water samples may also be directly injected. Surface methods, such as fluorescence, sputtering, and atomic absorption (AA), require only that the sample be mounted on specific media (e.g. filters). Also, continuous monitors like the chemiluminescent system mentioned in the next section provide ongoing measurements.

An environmental or tissue sample is a complex mixture, so before a compound can be detected, it must first be separated from the mixture. Thus, separation science embodies the techniques for separating complex mixtures of analytes, which is the first stage of chromatography. The second is detection. Separation makes use of the chemicals' different affinities for certain surfaces under various temperature and pressure conditions. The first step, injection, introduces the extract to a "column." The term column is derived from the time when columns were packed with sorbents of varying characteristics, sometimes meters in length, and the extract was poured down the packed column to separate the various analytes. Today, columns are of two major types, gas and liquid. Gas chromatography (GC) makes use of hollow tubes ("columns") coated inside with compounds that hold organic chemicals. The columns are in an oven, so that after the extract is injected into the column, the temperature is increased, as well as the pressure, and the various organic compounds in the extract are released from the column surface differentially, whereupon they are collected by a carrier gas (e.g. helium) and transported to the detector. Generally, the more volatile compounds are released first (they have the shortest retention times), followed by the semi-volatile organic compounds. So, boiling point is often a very useful indicator as to when a compound will come off a column. This is not always the case, since other characteristics such as polarity can greatly influence a compound's resistance to be freed from the column surface. For this reason, numerous GC columns are available to the chromatographer (different coatings, interior diameters, and lengths). Rather than coated columns, liquid chromatography (LC) makes use of columns packed with different sorbing materials with differing affinities for compounds. Also, instead of a carrier gas, LC uses a solvent or blend of solvents to carry the compounds to the detector. In the high performance LC (HPLC), pressures are also varied.

Detection is the final step for quantifying the chemicals in a sample. The type of detector needed depends upon the kinds of pollutants of interest. Detection gives the "peaks" that are used to identify compounds (Figure 2.6). For example, if hydrocarbons are of concern, GC with flame ionization detection (FID) may be used. GC-FID gives a count of the number of carbon atoms, so, for example, long chains can be distinguished from short chains. The short chains come off the column first and have peaks that appear before the long-chain peaks. However, if pesticides or other halogenated compounds are of concern, electron capture detection (ECD) is a better choice.

A number of detection approaches are also available for LC. Probably the most common is absorption. Chemical compounds absorb energy at various levels, depending upon their size, shape, bonds, and other structural characteristics. Chemicals also vary in whether they will absorb light or how much light they can absorb depending upon wavelength. Some absorb very well in the ultraviolet (UV) range, while others do not. Diode arrays help to identify compounds by giving a number of absorption ranges in the same scan. Some molecules can be excited and will fluoresce. The *Beer–Lambert law* tells us that energy absorption is proportional to chemical concentration:

$$A = eb[C] \tag{2.1}$$

where A is the absorbency of the molecule, e is the molar absorptivity (proportionality constant for the molecule), b is the light's path length, and $[C]$ is the chemical concentration of the molecule. Thus, the concentration of the chemical can be ascertained by measuring the light absorbed.

One of the most popular detection methods for environmental pollutants is mass spectrometry (MS), which can be used with either GC or LC separation. The MS detection is highly sensitive for organic compounds and works by using a stream of electrons to consistently break apart compounds into fragments. The positive ions resulting from the fragmentation are separated according to their masses. This is referred to as the "mass to charge ratio" or m/z. No matter which detection device is used, software is used to decipher the peaks and to perform the quantitation of the amount of each contaminant in the sample.

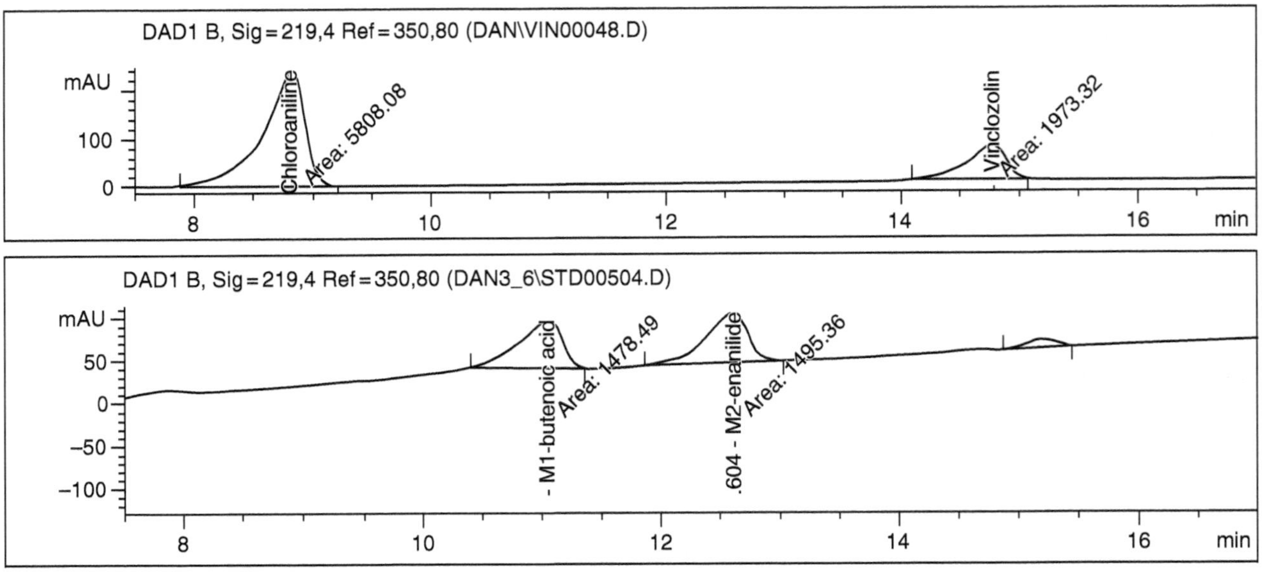

Figure 2.6 Chromatogram. *Source:* Vallero and Peirce (2003). Reproduced with permission of Elsevier.

For inorganic substances and metals, the additional extraction step may not be necessary. The actual measured media (e.g. collected airborne particles) may be measured by surface techniques like AA, X-ray fluorescence (XRF), inductively coupled plasma (ICP), or sputtering. As for organic compounds, the detection approaches can vary. For example ICP may be used with absorption or MS. If all one needs to know is elemental information, for example, to determine total lead or nickel in a sample, AA or XRF may be sufficient. However, if speciation (i.e. knowing the various compounds of a metal), then significant sample preparation is needed, including a process known as "derivatization." Derivatizing a sample is performed by adding a chemical agent that transforms the compound in question into one that can be recognized by the detector. This is done for both organic and inorganic compounds, for example, when the compound in question is too polar to be recognized by MS.

The physical and chemical characteristics of the compounds being analyzed must be considered before visiting the field and throughout all the steps in the laboratory. Although it is beyond the scope of this book to go into detail, it is worth mentioning that the quality of results generated about contamination depends upon the sensitivity and selectivity of the analytical equipment. Table 2.5 defines some of the most important analytical chemistry threshold values.

Table 2.5 Expressions of chemical analytical limits.

Type of limit	Description
Limit of detection (LOD)	Lowest concentration or mass that can be differentiated from a blank with statistical confidence. This is a function of sample handling and preparation, sample extraction efficiencies, chemical separation efficiencies, and capacity and specifications of all analytical equipment being used (see IDL below)
Instrument detection limit (IDL)	The minimum signal greater than noise detectable by an instrument. The IDL is an expression of the piece of equipment, not the chemical of concern. It is expressed as a signal to noise (S : N) ratio. This is mainly important to the analytical chemists, but the engineer should be aware of the different IDLs for various instruments measuring the same compounds, so as to provide professional judgment in contracting or selecting laboratories and deciding on procuring for appropriate instrumentation for all phases of remediation
Limit of quantitation (LOQ)	The concentration or mass above which the amount can be quantified with statistical confidence. This is an important limit because it goes beyond the "presence–absence" of the LOD and allows for calculating chemical concentration or mass gradients in the environmental media (air, water, soil, sediment, and biota)
Practical quantitation limit (PQL)	The combination of LOQ and the precision and accuracy limits of a specific laboratory, as expressed in the laboratory's quality assurance/quality control (QA/QC) plans and standard operating procedures (SOPs) for routine runs. The PQL is the concentration or mass that the engineer can consistently expect to have reported reliably

Source: Vallero and Peirce (2003). Reproduced with permission of Elsevier.

2.4 Sources of Uncertainty

Contaminant assessments have numerous sources of uncertainty. There are two basic types of uncertainty: type A and type B. Type A uncertainties result from the inherent unpredictability of complex processes that occur in nature. These uncertainties cannot be eliminated by increasing data collection or enhancing analysis. The scientist and engineer must simply recognize that type A uncertainty exists, but must not confuse it with type B uncertainties, which can be reduced by collecting and analyzing additional scientific data.

The first step in an uncertainty analysis is to identify and describe as many uncertainties that may be encountered. Sources of type B uncertainty take many forms (Finkel, 1990). There can be substantial uncertainty concerning the numerical values of the attributes being studied (e.g. contaminant concentrations, wind speed, discharge rates, groundwater flow, and other variables). Modeling generates its own uncertainties, including errors in selecting the variables to be included in the model, such as surrogate contaminants that represent whole classes of compounds (e.g. does benzene represent the behavior or toxicity of other aromatic compounds?). The application of the findings, even if the results themselves have tolerable uncertainty, may lead to the propagation of uncertainties when ambiguity arises regarding their meaning. For example, a decision rule is a statement about which alternative will be selected, e.g. for cleanup, based on the characteristics of the decision situation. A "decision-rule uncertainty" occurs when there are disagreements or poor specification of objectives (i.e. is our study really addressing the client's needs?).

Variability and uncertainty must not be confused. Variability consists of measurable factors that differ across populations such as soil type, vegetative cover, or body mass of individuals in a population. Uncertainty consists of unknown or not fully known factors that are difficult to measure, such as the inability to access an ideal site that would be representative because it is on private property.

Modeling uncertainties, for example, may consist of extrapolations from a single value to represent a whole population, i.e. a point estimate (e.g. 70 kg as the weight of an adult male). Such estimates can be typical values for a population or an estimate of an upper end of the population's value, e.g. 70 years as the duration of exposure used as a "worse-case" scenario. Another approach is known as the Monte Carlo technique (Figure 2.7). The Monte Carlo-type exposure assessments use probability distribution functions, which are statistical distributions of the possible values of each population characteristic according to the probability of the occurrence of each

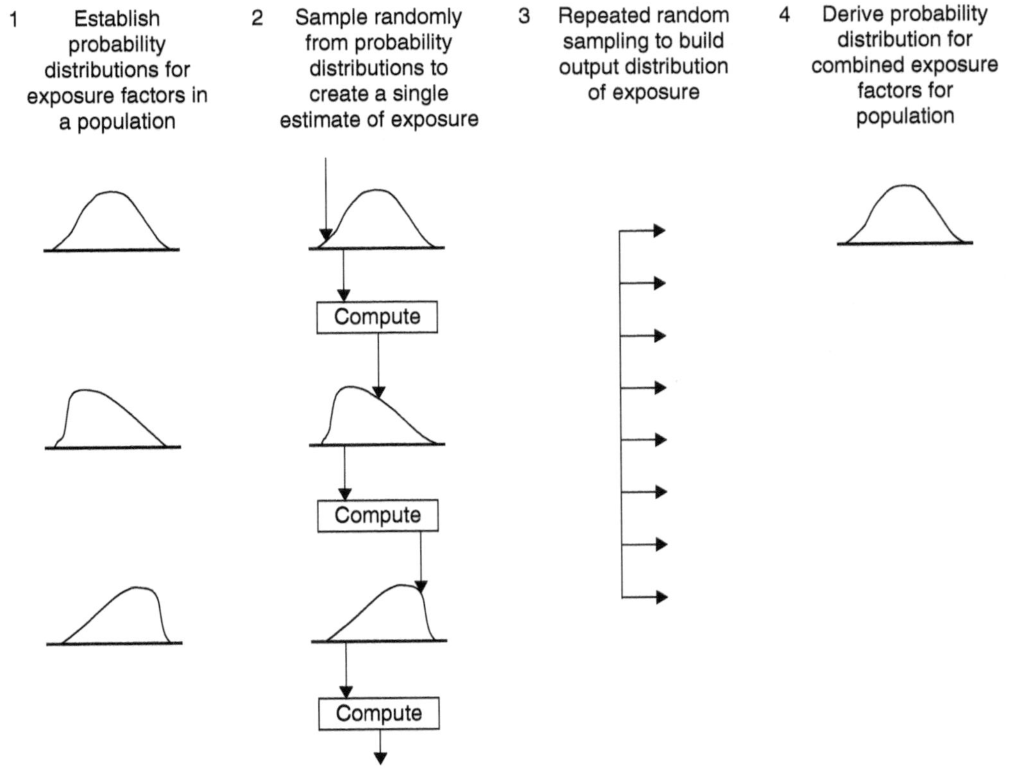

Figure 2.7 Principles of the Monte Carlo method for aggregating data. *Source*: From enHealth Council (2018).

value (Environmental Health Australia, 2012a). These are derived using iterations of values for each population characteristic. While the Monte Carlo technique may help to deal with the point estimate limitation, it can suffer from confusing variability with uncertainty.

Other data interpretation uncertainties can result from the oversimplification of complex entities. For example, assessments consist of an aggregation of measurement data, modeling, and combinations of sampling and modeling results. However, these complicated models are providing only a snapshot of highly dynamic human and environmental systems. The use of more complex models does not necessarily increase precision, and extreme values can be improperly characterized. For example, a 50th percentile value can always be estimated with more certainty than a 99th percentile value.

The bottom line is that uncertainty is always present in sampling, analysis, and data interpretation, so the monitoring and data reduction plan should be systematic and rigorous. The uncertainty analysis must be addressed for each step of the contaminant assessment process, including any propagation and enlargement of cumulative error (e.g. an incorrect pH value that goes into an index where pH is weighted heavily, and then used in another algorithm for sustainability). The characterization of the uncertainty of the assessment includes selecting and rejecting data and information ultimately used to make environmental decisions and includes both qualitative and quantitative methods (see Table 2.6).

Uncertainty factors (UFs) are applied to the address, both the inherent and study uncertainties upon which to establish safe levels of exposure to contaminants. The UFs consider the uncertainties resulting from the variation in sensitivity among the members of the populations, including interhuman and intraspecies variability; the extrapolation of animal data to humans (i.e. interspecies variability); the extrapolation from data gathered in a study with less-than-lifetime exposure to lifetime exposure, i.e. extrapolating from acute or subchronic to chronic exposure; the extrapolation from different thresholds, such as the LOAEL rather than from a NOAEL; and the extrapolation from an incomplete data base is incomplete. Note that most of these sources of uncertainty have a component associated with measurement and analysis.

The numerical value uncertainties are directly related to the quality and representativeness of the sampling design and the analytical expressions described in Table 2.6. When these values are used in environmental models, they are known as "parameter uncertainties." They account for imprecision and inaccuracy associated with the measurement and analytical equipment, systemic weaknesses in data gathering (i.e. bias).

Table 2.6 Example of an uncertainty table for exposure assessment.

	Effect on exposure[a]		
Assumption	**Potential magnitude for over-estimation of exposure**	**Potential magnitude for under-estimation of exposure**	**Potential magnitude for over- or under-estimation of exposure**
Environmental sampling and analysis			
Sufficient samples may not have been taken to characterize the media being evaluated, especially with respect to currently available soil data			Moderate
Systematic or random errors in the chemical analyses may yield erroneous data			Low to high
Exposure parameter estimation			
The standard assumptions regarding body weight, period exposed, life expectancy, population characteristics, and lifestyle may not be representative of any actual exposure situation			Moderate
The amount of media in take is assumed to be constant and representative of the exposed population	Moderate		
Assumption of daily lifetime exposure for residents	Moderate to high		

Source: From enHealth Council (2018).

[a] As a general guideline, assumptions marked as "low," may affect estimates of exposure by less than one order of magnitude; assumptions marked "moderate" may affect estimates of exposure by between one and two orders of magnitude; and assumptions marked "high" may affect estimates of exposure by more than two orders of magnitude.

2.5 Measurements and Models

Generally, environmental science may be seen as a two-by-two table of focus and approach (Table 2.7). For example, cell A may include measuring the concentrations of lead (Pb) in drinking water at the taps in 1000 homes. Developing a model to extrapolate concentrations in a million homes would be an example of cell B. Environmental scientists would operate in cell C if they attempt to understand the relationship between the release of iron into a wetland by designing a study that collects samples at the source and in the wetland. Using a model to interpolate iron concentrations in that same wetland between measurement locations is an example of cell D.

2.6 Contaminants of Concern

Environmental contaminants may be physical, chemical, or biological. Heat is physical contaminant since every species has a unique range of temperature tolerance. Chemical contaminants are probably the first type that comes to mind, since they are often measured in surface and well water, soil, air, and bodily fluids. Biological contaminants may be pathogenic microbes (e.g. fecal coliform bacteria in water), but they may also include organisms that upset environmental conditions, e.g. animals like the zebra mussels the destroy biodiversity in the Great Lakes or plants like kudzu that cover large swaths of ecosystems in the Southeastern United States.

The value of a water body can be directly related to water temperature since it is directly proportional to dissolved oxygen (DO) content, which is a limiting factor of the type of fish communities that can be supported by a water body (see Tables 2.8 and 2.9). A trout stream is a highly valued resource that is adversely impacted if mean temperatures increase. Rougher, less valued fish (e.g. carp and catfish) can live at much lower ambient water body temperatures than can salmon, trout, and other cold-water fish populations. Thus, net increase in heat may directly stress the game fish population. That is, fish species vary in their ability to tolerate higher temperatures, meaning that the less tolerant, higher value fish will be inordinately threatened.

The threat may not be completely explained as heat stress due directly to the increase in temperature (Dohner et al., 1997). Much can be explained by the concomitant decrease in the stream's DO concentrations (see Figures 2.8 and 2.9), which deems the water body hostile to the fish. Even if the adult fish can survive at the reduced DO levels, their reproductive capacities decreases. Or, the reproduction is not adversely affected, but the survival of juvenile fish can be reduced.

The increased temperature can also increase the solubility of substances toxic to organisms, which increases the exposure. For example, greater concentrations of mercury and other toxic metals will occur with elevated temperatures. The lower DO concentrations will lead to a reduced environment where the metals and compounds will for sulfides and other compounds that can be toxic to the fish. Thus, the change in temperature, the resulting decrease in DO and increasing metal concentrations, and the synergistic impact of the combining the hypoxic water and reduced metal compounds is a cascade of harm to the stream's ecosystems (Figure 2.10).

Table 2.7 Focus and approaches for providing environmental information.

		Approach	
		Measurement	Modeling
Focus	Human health	A	B
	Ecosystems	C	D

Table 2.8 Relationship between water temperature and maximum dissolved oxygen (DO) concentration in water (at 1 atm).

Temperature (°C)	Dissolved oxygen (mg l^{-1})	Temperature (°C)	Dissolved oxygen (mg l^{-1})
0	14.60	23	8.56
1	14.19	24	8.40
2	13.81	25	8.24
3	13.44	26	8.09
4	13.09	27	7.95
5	12.75	28	7.81
6	12.43	29	7.67
7	12.12	30	7.54
8	11.83	31	7.41
9	11.55	32	7.28
10	11.27	33	7.16
11	11.01	34	7.16
12	10.76	35	6.93
13	10.52	36	6.82
14	10.29	37	6.71
15	10.07	38	6.61
16	9.85	39	6.51
17	9.65	40	6.41
18	9.45	41	6.41
19	9.26	42	6.22
20	9.07	43	6.13
21	8.90	44	6.04
22	8.72	45	5.95

Source: Data from Vallero (2015) and Dohner et al. (1997).

Table 2.9 Normal temperature tolerances of aquatic organisms.

Organism	Taxonomy	Range in temperature tolerance (°C)	Minimum dissolved oxygen (mg l^{-1})
Trout	*Salma, Oncorhynchus and Salvelinus* spp.	5–20	6.5
Smallmouth bass	*Micopterus dolomieu*	5–28	6.5
Caddisfly larvae	*Brachycentrus* spp.	10–25	4.0
Mayfly larvae	*Ephemerella invaria*	10–25	4.0
Stonefly larvae	*Pteronarcys* spp.	10–25	4.0
Catfish	Order Siluriformes	20–25	2.5
Carp	Cyprinus spp.	10–25	2.0
Water boatmen	*Notonecta* spp.	10–25	2.0
Mosquito larvae	Family Culicidae	10–25	1.0

Source: Data from Vallero (2010) and Vernier Corporation (2009).

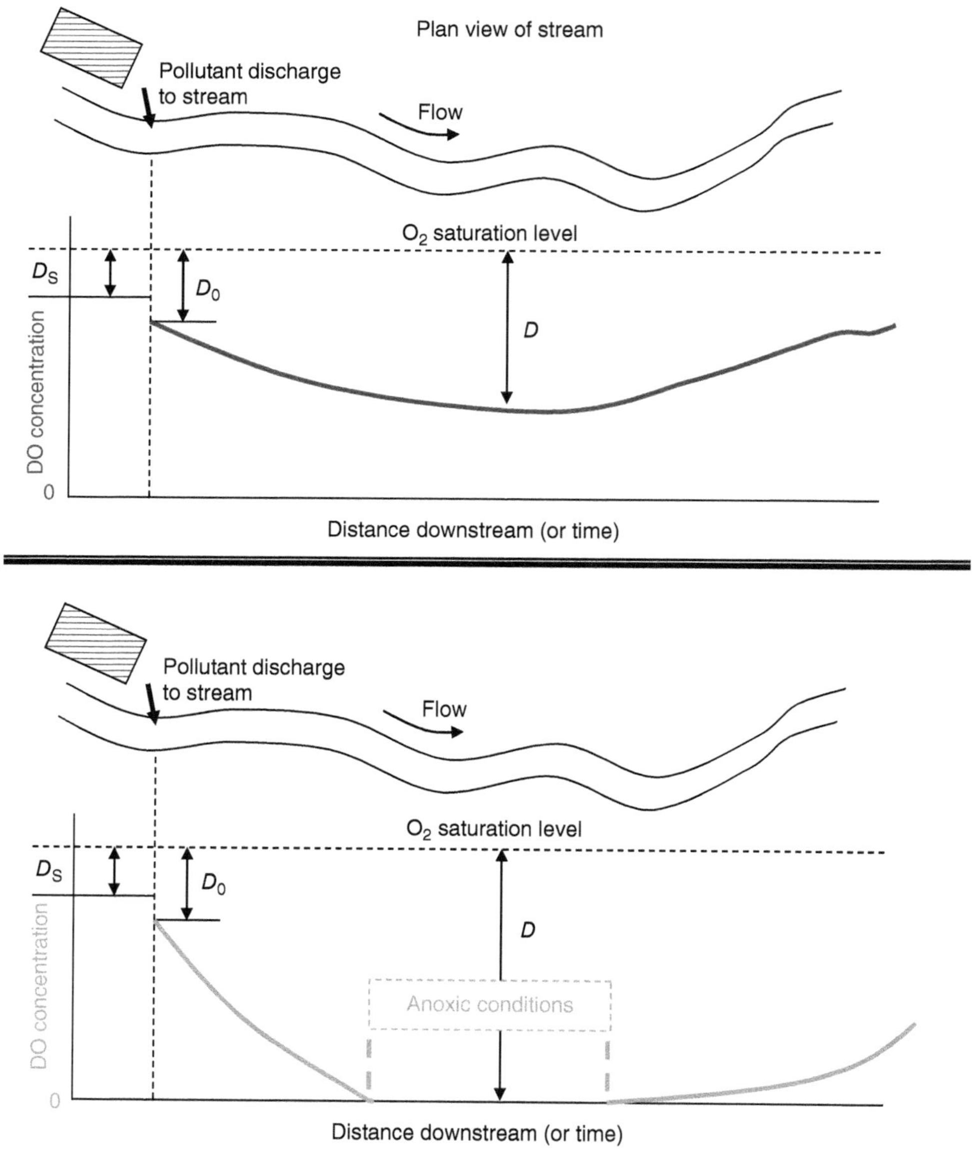

Figure 2.8 Dissolved oxygen (DO) deficit downstream from a heated effluent. The increased temperature can result in an increase in microbial kinetics, as well as more rapid abiotic chemical reactions, both consuming DO. The concentration of dissolved oxygen in the top curve remains above 0, so although the DO decreases, the overall system DO recovers. The bottom sags where dissolved oxygen falls to 0, and anoxic conditions result and continue until the DO concentrations begin to increase. D_S is the background oxygen deficit before the pollutants enter the stream. D_O is the oxygen deficit after the pollutant is mixed. D is the deficit for contaminant A that may be measured at any point downstream. The deficit is overcome more slowly in the lower curve (smaller slope) because the reoxygenation is dampened by the higher temperatures and changes to microbial system, which means the system has become more vulnerable to another insult, e.g. another downstream source could cause the system to return to anoxic conditions. *Source*: Vallero (2015). Reproduced with permission of Elsevier.

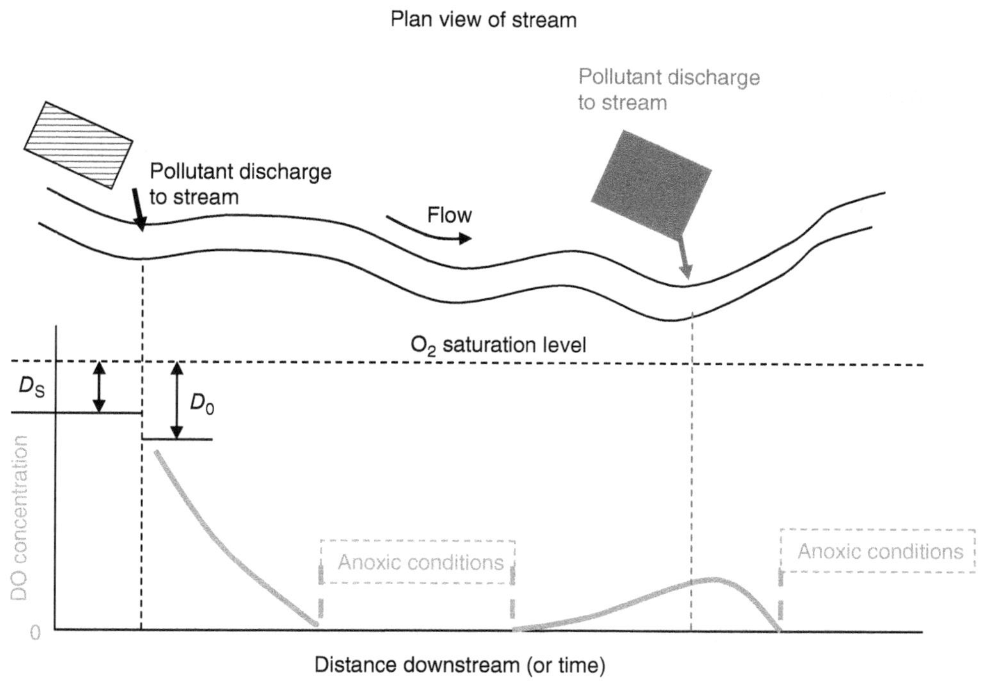

Figure 2.9 Cumulative effect of a second heat source, causing an overall system to become more vulnerable. The rate of reoxygenation is suppressed, with a return to anoxic conditions. *Source*: Vallero (2015). Reproduced with permission of Elsevier.

Figure 2.10 Adverse effects in the real world usually result from a combination of conditions. In this example, the added heat results in an abiotic response (i.e. decreased dissolved oxygen [DO] concentrations in the water). *Source*: Vallero (2015). Reproduced with permission of Elsevier.

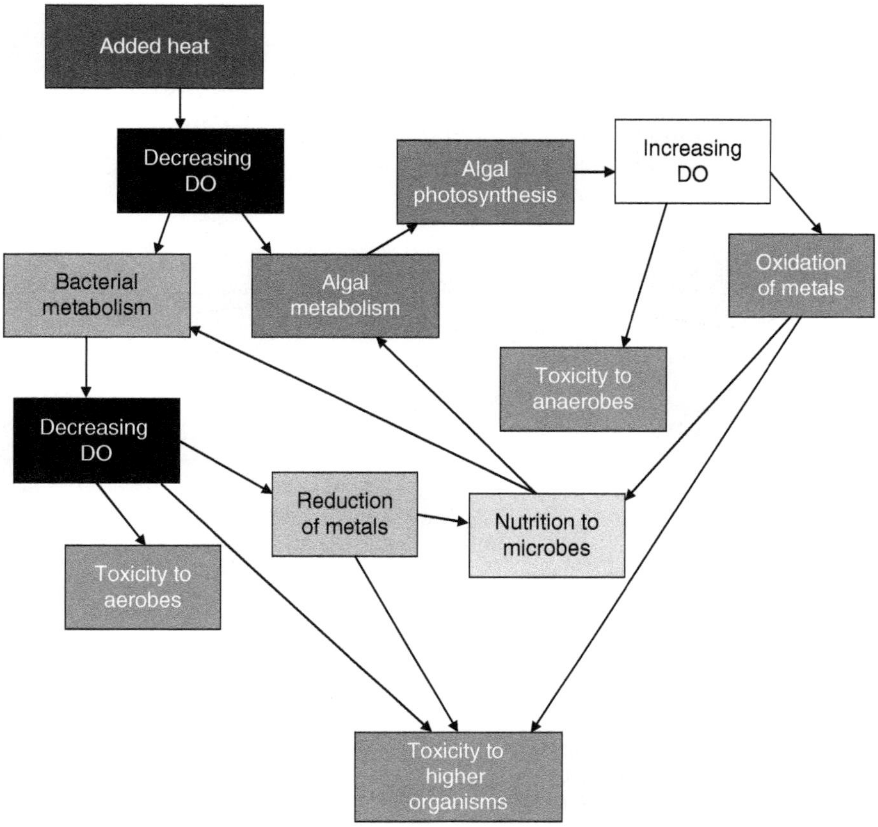

Biota also play a role in the heat-initiated effect. Combined abiotic and biotic responses occur. Notably, the growth and metabolism of the bacteria results in even more rapidly decreasing DO levels. Algae both consume DO for metabolism and produce DO by photosynthesis. The increase in temperature increases their aqueous solubility and the decrease in DO is accompanied by redox changes, e.g. formation of reduced metal species, such as metal sulfides. This is also being mediated by the bacteria, some of which will begin reducing the metals as the oxygen levels drop (reduced conditions in the water and sediment). However, the opposite is true in the more oxidized regions, i.e. the metals are forming oxides. The increase in the metal compounds combined with the reduced DO and combined with the increased temperatures can act synergistically to make the conditions toxic for higher animals, e.g. a fish kill.

The first-order abiotic effect (i.e. increased temperature) results in an increased microbial population. However, microbial populations may affect oxygen differently. For example, the growth and metabolism of both algae and bacteria decrease DO levels, but algae also undergo photosynthesis, which adds oxygen to the water. So, increasing the bacterial biomass should also depress DO, while the same biomass growth and metabolism in algae would depress DO levels much less, and may even result in a net increase if photosynthetic O_2 additions exceed metabolic O_2 demand. Meanwhile a combined abiotic and biotic response occurs with the metals. The increase in temperature increases their aqueous solubility, and the decrease in DO is accompanied by redox changes, e.g. formation of reduced metal species, such as metal sulfides. This is also being mediated by the bacteria, some of which will begin reducing the metals as the oxygen levels drop (reduced conditions in the water and sediment). However, the opposite is true in the more oxidized regions, i.e. the metals are forming oxides. The increase in the metal compounds combined with the reduced DO, combined with the increased temperatures can act synergistically to make the conditions toxic for higher animals, e.g. a fish kill (Vallero et al., 2007). Predicting the likelihood of a fish kill can be quite complicated, with many factors that either mitigate or exacerbate the outcome (see Figures 2.11 and 2.12).

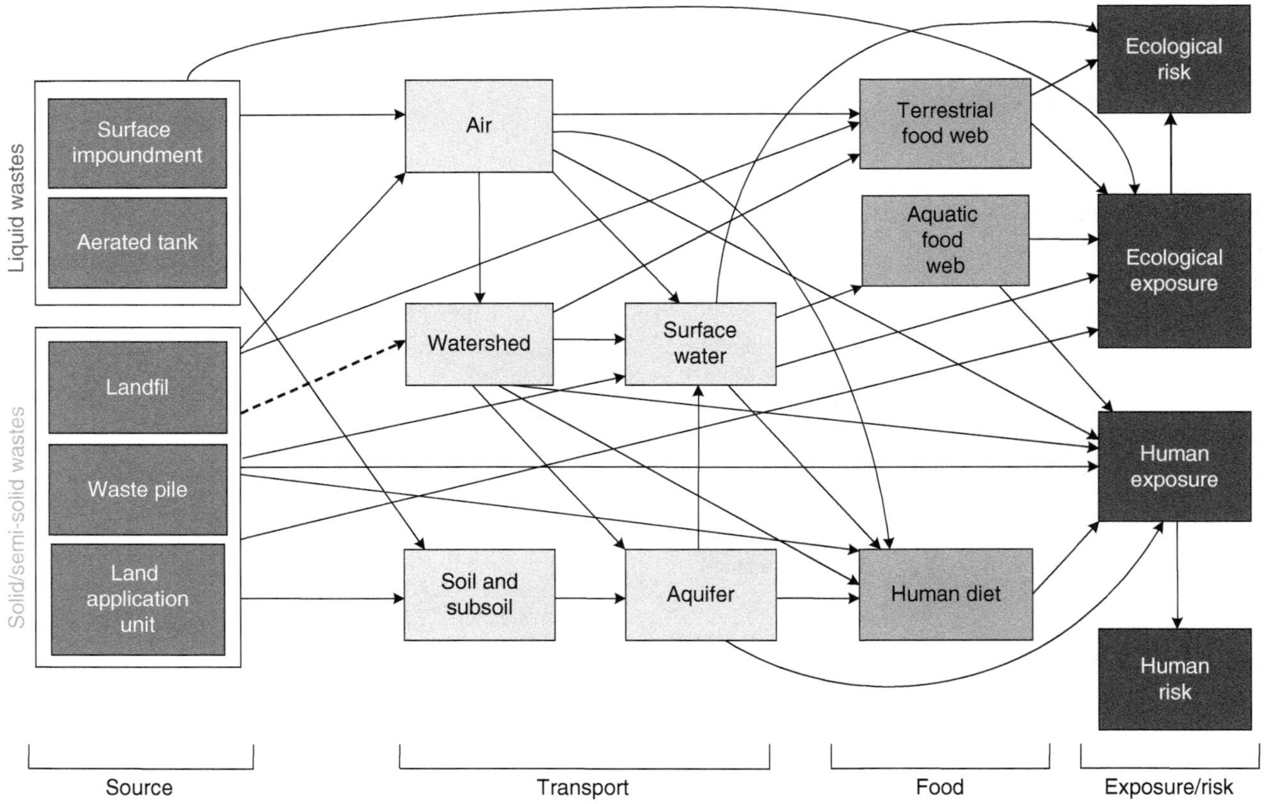

Figure 2.11 Environmental transport pathways can be affected by net heat gain. Compounds (nutrients, contaminants), microbes, and energy (e.g. heat) follow the path through the environment indicated by arrows. The residence time within in any of the boxes is affected by conditions, including temperature. *Source*: Adapted from Vallero et al. (2007) and U.S. Environmental Protection Agency.

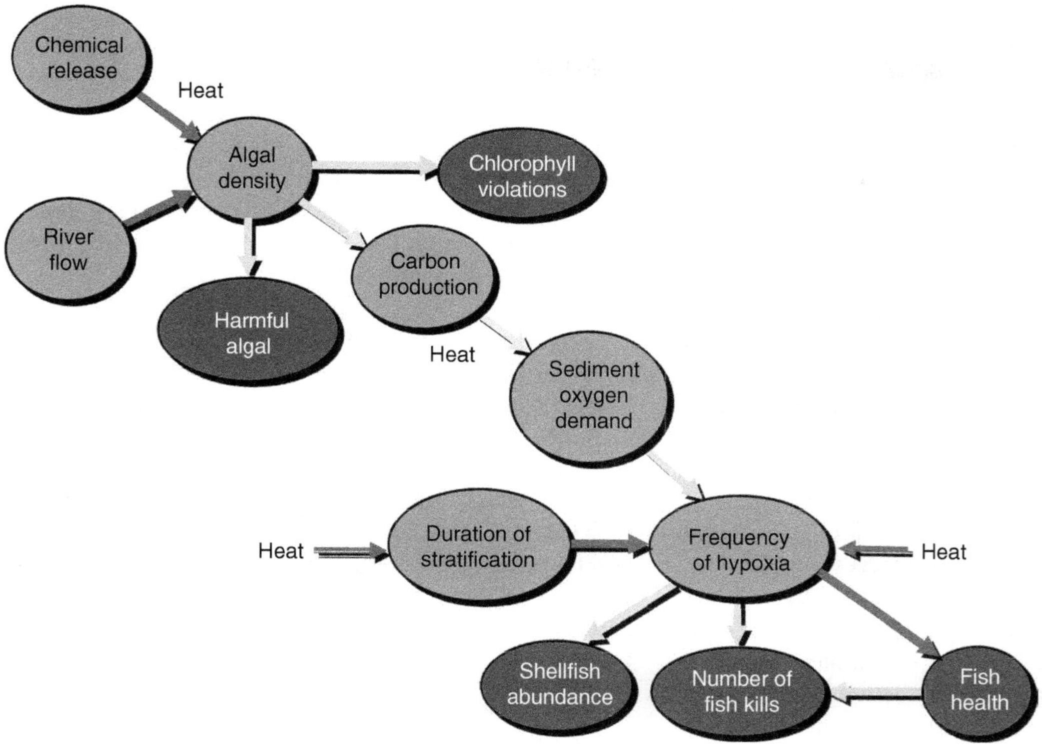

Figure 2.12 Flow of events and conditions leading to fish kills, indicating some of the points where added heat can exacerbate the likelihood of a fish kill or other adverse environmental event. *Source*: From Vallero et al. (2007).

2.7 Environmental Indicators

2.7.1 Oxygen

The biochemical oxygen demand (BOD) is the amount of oxygen that bacteria will consume in the process of decomposing organic matter under aerobic conditions. The BOD is measured by incubating a sealed sample of water for 5 days and measuring the loss of oxygen by comparing the O_2 concentration of the sample at time = 0 (just before the sample is sealed) to the concentration at time = 5 days (known specifically as BOD_5). Samples are commonly diluted before incubation to prevent the bacteria from depleting all of the oxygen in the sample before the test is complete (Vallero, 2015). BOD_5 is merely the measured DO at the beginning time (i.e. the initial DO [D_0], measured immediately after it is taken from the source) minus the DO of the same water measured exactly 5 days after D_1, i.e. D_5:

$$\mathrm{BOD} = \frac{D_1 - D_5}{P} \tag{2.2}$$

where P = decimal volumetric fraction of water utilized. D units are in mg l^{-1}. If the dilution water is seeded, the calculation becomes

$$\mathrm{BOD} = \frac{(D_1 - D_5) - (B_1 - B_5)f}{P} \tag{2.3}$$

where B_1 = initial DO of seed control, B_5 = final DO of seed control, and f = the ratio of seed in sample to seed in control = (%seed in D_1)/(%seed in B_1). B units are in mg l^{-1}.

For example, to find the BOD_5 value for a 10 ml water sample added to 300 ml of dilution water with a measured DO of 7 mg l^{-1} and a measured DO of 4 mg l^{-1} 5 days later:

$$P = \frac{10}{300} = 0.03$$

$$\mathrm{BOD}_5 = \frac{7-4}{0.03} = 100\,\mathrm{mg\,l^{-1}}$$

Thus, the microbial population in this water is demanding 100 mg l^{-1} DO over the 5-day period. So, if a conventional municipal wastewater treatment system is achieving 95% treatment efficiency, the effluent discharged from this plant would be 5 mg l^{-1}.

Chemical oxygen demand (COD) does not differentiate between biologically available and inert organic matter, and it is a measure of the total quantity of oxygen required to oxidize all organic material completely to carbon dioxide and water. COD values always exceed BOD values for the same sample. COD ($mg\,l^{-1}$) is measured by oxidation using potassium dichromate ($K_2Cr_2O_7$) in the presence of sulfuric acid (H_2SO_4) and silver. By convention, 1 g of carbohydrate or 1 g of protein accounts for about 1 g of COD. On average, the ratio BOD : COD is 0.5. If the ratio is less than 0.3, the water sample likely contains elevated concentrations of *recalcitrant* organic compounds, i.e. compounds that resist biodegradation (Gerba and Pepper, 2009). That is, there are numerous carbon-based compounds in the sample, but the microbial populations are not efficiently using them for carbon and energy sources. This is the advantage of having both BOD and COD measurements. Sometimes, however, COD measurements are conducted simply because they require only a few hours compared with the 5 days for BOD.

Since available carbon is a limiting factor, the carbonaceous BOD reaches a plateau, i.e. ultimate carbonaceous BOD (see Figure 2.13). However, carbonaceous compounds are the only substances demanding oxygen. Microbial populations will continue to demand O_2 from the water to degrade other compounds, especially nitrogenous compounds, which account for the bump in the BOD curve. Thus, in addition to serving as an indication of the amount of molecular oxygen (O_2) needed for biological treatment of the organic matter, BOD also provides a guide to sizing a treatment process, assign its efficiency and giving operators and regulators information about whether the facility is meeting its design criteria and is complying with pollution control permits.

If effluent with high BOD concentrations reaches surface waters, it may diminish DO to levels lethal for some fish and many aquatic insects. As the water body re-aerates as a result of mixing with the atmosphere and by algal photosynthesis, O_2 is added to the water, the oxygen levels will slowly increase downstream. The drop and rise in DO concentrations downstream from a source of BOD is known as the DO sag curve, because the concentration of DO "sags" as the microbes deplete it. So, the falling O_2 concentrations fall with both time and distance from the point where the high BOD substances enter the water.

The stress from decreasing DO is usually indicated by the BOD. Like most environmental systems, the water bodies that receive sediment loads are complex in their response to increased input of materials. The DO will respond both positively and negatively to increased nutrient levels, since the biota have unique optimal ranges of growth and metabolism that varies among species (e.g. green plants, algae, bacteria, and fungi have different O_2 demands, and green plants and algae will add some O_2 by photosynthesis).

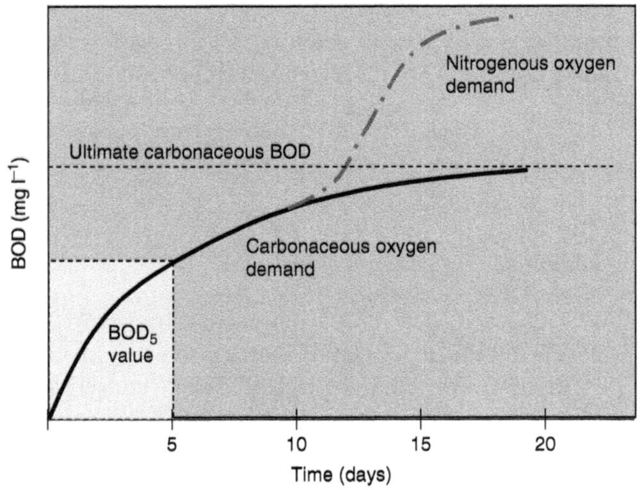

Figure 2.13 Biochemical oxygen demand (BOD) curve, showing ultimate carbonaceous BOD and nitrogenous BOD.
Source: Adapted from Gerba and Pepper (2009).

2.7.2 Indices

The most widely applied environmental indices are those that follow the framework of an index of biological integrity. In biological systems, integrity is the capacity of a system to sustain a balanced and healthy community. This means the community of organisms in that system meets certain criteria for species composition, diversity, and adaptability, often compared with a reference site that is a benchmark for integrity. As such, biological integrity indices are designed to integrate the relationships of chemical and physical parameters with each other and across various levels of biological organization. They are now used to evaluate the integrity of environmental systems using a range of metrics to describe the system.

Thus, environmental indices combine attributes to determine a system's condition (e.g. diversity and productivity) and to estimate stresses. The original index of biotic integrity (Karr, 1981) was based on fish fauna attributes and has provided predictions of how well a system will respond to a combination of stresses. In fact, the index is completely biological, with no direct chemical measurements. However, the metrics (see Table 2.10) are indirect indicators of physicochemical factors (e.g. the abundance of game fish is directly related to DO concentrations). The metrics provide descriptions of a system's structure and function (Karr et al., 1986).

An example of the data that is gathered to characterize a system is provided in Table 2.11. The information that is gleaned from these data is tailored to the physical, chemical, and biological conditions of an area, e.g. for large spatial regions. The information from a biologically

Table 2.10 Biological metrics used in the original index of biological integrity (IBI).

Integrity aspect	Biological metric
Species richness and composition	Total number of fish species (total taxa)
	Number of *Catostomidae* species (suckers)
	Number of darter species
	Number of sunfish species
	Number of intolerant or sensitive species
Indicator species metrics	Percent of individuals that are *Lepomis cyanellus* (Centrarchidae)
	Percent of individuals that are omnivores
Trophic function metrics	Percent of individuals that are insectivorous Cyprinidae
	Percent of individuals that are top carnivores or piscivores
	Percent of individuals that are hybrids
Reproductive function metrics	Abundance or catch per effort of fish
Abundance and condition metrics	Percent of individuals that are diseased, deformed, or that have eroded fins, lesions, or tumors (DELTs)

Source: Karr (1981). Reproduced with permission of Taylor & Francis.

based index can be used to evaluate a system, as shown in Figure 2.14.

Systems involve scale and complexities in both biology and chemistry. For example, a fish's direct aqueous exposure (AE in $\mu g\,day^{-1}$) is the product of the organism's ventilation volume, i.e. the flow Q (in $ml\,day^{-1}$), and the compound's aqueous concentration, C_W ($\mu g\,ml^{-1}$). The fish's exposure by its diet (DE, in $\mu g\,day^{-1}$) is the product of its feeding rate, F_w (g wet weight day^{-1}) and the compound's concentration in the fish's prey, C_p ($\mu g\,g^{-1}$ wet weight). If the fish's food consist of single type of prey that is at equilibrium with the water, fish's aqueous and dietary exposures and the bioconcentration factor (BCF) can be calculated when they are equal:

$$\mathrm{AE} = \mathrm{DE};\quad QC_w = F_w C_p;\quad \mathrm{BCF} = \frac{Q}{F_w} \tag{2.4}$$

The ventilation-to-feeding ratio for a 1 kg trout has been found (Erickson and McKim, 1990) to be on the order of $10^{4.3}$ $ml\,g^{-1}$. Assuming the quantitative structure–activity relationship (QSAR) for the trout's prey is BCF = 0.048 times the octanol–water coefficient (K_{ow}); it appears that the trout's predominant route of exposure for any chemical with a $K_{ow} > 10^{5.6}$. Exposure must also account for the organism's assimilation of compounds in food, which for very lipophilic compounds, will probably account for the majority of exposure compared with that from the water. Even though chemical exchange occurs from both food and water via passive diffusion (Fick's law relationships), the uptake from food, unlike direct uptake from water, does not necessarily relax the diffusion gradient into the fish. The difference between digestion and assimilation of food can result in higher contaminant concentrations in the fish's gut.

Predicting expected uptake where the principal route of exchange is dietary can be further complicated by the fact that most fish species exhibit well-defined size-dependent, taxonomic, and temporal trends regarding their prey. Thus, a single bioaccumulation factor (BAF) may not universally be useful for risk assessments for all fish species. Indeed, the BAF may not even apply to different sizes of the same species.

The systematic biological exchange of materials between the organism, in this case various species of fishes, is known as uptake, which can be expressed by the following three differential equations for each age class or cohort of fish (Barber, 2001):

$$\frac{dB_f}{dt} = J_g + J_i + J_M \tag{2.5}$$

where B is body burden; J_g represents the net chemical exchange ($\mu g\,day^{-1}$) across the fish's gills from the water; J_i represents the net chemical exchange ($\mu g\,day^{-1}$) across the fish's intestine from food; and J_M represents the chemical compound's biotransformation rate ($\mu g\,day^{-1}$).

Physiologically based models for fish growth are often formulated in terms of energy content and flow (e.g. $kcal\,fish^{-1}$ and $kcal\,day^{-1}$), Eq. (2.4) is basically the same as such bioenergetics models because energy densities of fish depend on their dry weight (Hartman and Brandt, 1995; Kushlan et al., 1986; Schreckenbach et al., 2001). Obviously, feeding depends on the availability of suitable prey, so the mortality of the fish is a function of the individual feeding levels and population densities of its predators. Thus, the fish's dietary exposure is directly related to the organism's feeding rate and the concentrations chemicals in its prey.

2.8 Emerging Trends in Measurement

2.8.1 Sensors

In addition to the traditional monitoring equipment discussed here, sensors have become important assets in environmental measurements (see Table 2.12). As mentioned, sensors are often used to provide remote sensing, e.g. soil moisture, which is needed to cover large areas (e.g. farms) and/or distant regions (e.g. upper atmosphere, water and waste distribution systems, and isolated locations). Such technologies are important at all scales,

Table 2.11 Biological metrics that apply to various regions of North America.

Alternative IBI metrics	Midwestern United States	Central Appalachians	Sacramento-San Joaquin	Colorado Front Range	Western Oregon Ohio	Ohio Headwater sites	Ohio	Northeastern United States	Ontario	Central Corn Belt Plain	Wisconsin-Warmwater	Wisconsin-Coldwater	Maryland Coastal Plain	Maryland Non-Tidal
1. Total number of species	X	X	X	X				X		X			X	X
No. native fish species					X	X	X		X		X			
No. salmonid age classes[a]				X	X									
2. Number of darter species	X	X		X		X				X	X			
No. sculpin species					X									
No. benthic insectivore species								X						
No. darter and sculpin, species							X							
No. darter, sculpin, and madtom species											X			
No. salmonid juveniles (individuals)[a]			X		X			X						
% Round-bodied suckers						X[b]								
No. sculpins (individuals)			X											
No. benthic species													X	X
3. Number of sunfish species	X			X		X				X	X			
No. cyprinid species					X									
No. water column species								X						
No. sunfish and trout species									X					

No. salmonid species			X							X				
No. headwater species							X							
% Headwater species							X			X				
4. Number of sucker species	X				X	X		X		X	X			
No. adult trout species[a]			X		X									
No. minnow species				X			X			X				
No. sucker and catfish species									X					
5. Number of intolerant species	X			X	X	X		X			X	X	X	X
No. sensitive species							X			X				
No. amphibian species			X											
Presence of brook trout									X					
% Stenothermal cool and cold water species												X		
% of salmonid ind. as brook trout												X		
6. % Green sunfish	X													
% Common carp					X									
% White sucker				X				X						
% Tolerant species						X	X			X	X	X	X	X
% Creek chub		X												
% Dace species									X					
% Eastern mudminnow													X	

(Continued)

Table 2.11 (Continued)

Alternative IBI metrics	Midwestern United States	Central Appalachians	Sacramento-San Joaquin	Colorado Front Range	Western Oregon Ohio	Ohio Headwater sites	Ohio	Northeastern United States	Ontario	Central Corn Belt Plain	Wisconsin-Warmwater	Wisconsin-Coldwater	Maryland Coastal Plain	Maryland Non-Tidal
7. % Omnivores	X			X		X	X	X	X	X	X			
% Generalist feeders		X												
% Generalists, omnivores, and invertivores														X
8. % Insectivorous Cyprinids	X												X	
% Insectivores					X			X		X	X		X	X[c]
% Specialized insectivores		X		X										
No. juvenile trout			X											
% Insectivorous species						X	X							
9. % Top carnivores	X					X		X	X	X	X	X		
% Catchable salmonids					X									
% Catchable trout			X											
% Pioneering species							X			X			X	
Density catchable wild trout			X											
10. Number of individuals (or catch per effort)	X	X	X	X	X	X[d]	X[d]		X	X	X[d]		X	
Density of individuals								X						X
% Abundance of dominant species													X	X
Biomass (per m^2)														X[e]

11. % Hybrids	X							X					
% Introduced species				X	X								
% Simple lithophills						X				X	X		X
No. simple lithophills species							X						
% Native species			X										
% Native wild individuals			X										
% Silt-intolerant spawners												X	
12. % Diseased individuals (deformities, eroded fins, lesions, and tumors)	X	X		X	X	X	X	X	X	X	X	X	X

Source: From Barbour et al. (1999). Taken from Karr et al. (1986), Leonard and Orth (1986), Moyle et al. (1986), Fausch and Schrader (1987), Hughes and Gammon (1987), Ohio EPA (1987), Miller et al. (1988), Steedman (1988), Simon (1991), Lyons (1992), Barbour et al. (1995), Simon and Lyons (1995), Hall et al. (1996), Lyons et al. (1996), Roth et al. (1997).

Note: X = metric used in region. Many of these variations are applicable elsewhere.

[a] Metric suggested by Moyle et al. (1986) or Hughes and Gammon (1987) as a provisional replacement metric in small western salmonid streams.

[b] Boat sampling methods only (i.e. larger streams/rivers).

[c] Excluding individuals of tolerant species.

[d] Non-coastal Plain streams only.

[e] Coastal Plain streams only.

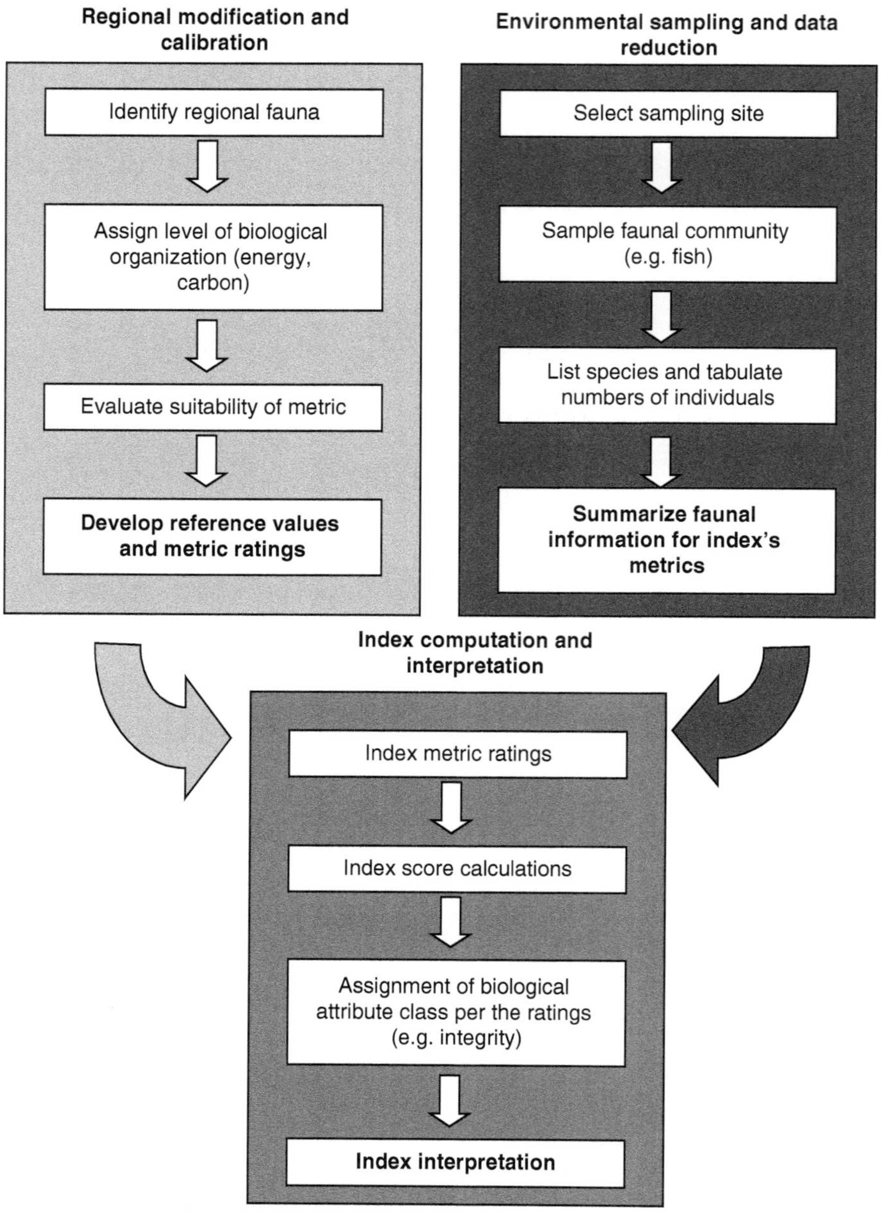

Figure 2.14 Sequence of activities involved in calculating and interpreting an index of biotic integrity (IBI). *Source*: Adapted from Barbour et al. (1999) and Karr (1987).

including continental and global, e.g. to assess conditions (e.g. soil carbon measurements) that can lead to changes in climate (Gehl and Rice, 2007).

So-called next-generation, or "next-gen," technologies and software are improving the timeliness and spatial and temporal representation of environmental measurements. Local communities can use a variety of new tools tailored to neighborhoods, which can support their management of facilities, decisions on potential new businesses, and support of land use and other planning decisions (Watkins, 2013).

Sensors that produce near real-time, portable measurements with resolutions near those of "wet laboratory" levels are becoming increasingly economical (Snyder et al., 2013). Notable technologies in current use include infrared cameras that characterized emissions of toxic air pollutants rising from storage tanks and other equipment, as well as remote and fence line monitoring near pollution sources. Since these are often imperceptible by normal human senses, such fugitive emissions can continue indefinitely (Watkins, 2013). The applications of sensors are increasing. For example, farmers are increasingly

Table 2.12 Selected environmental engineering applications of sensors.

Application	Description	Example
Research	Scientific studies aimed at discovering new information about air pollution	A network of air sensors is used to measure particulate matter variation across a city
Personal exposure monitoring	Monitoring the air quality that a single individual is exposed to while doing normal activities	An individual having a clinical condition increasing sensitivity to air pollution wears a sensor to identify when and where he or she is exposed to pollutants potentially impacting their health
Supplementing existing monitoring data	Placing sensors within an existing state/local regulatory monitoring area to fill in coverage	A sensor is placed in an area between regulatory monitors to better characterize the concentration gradient between the different locations
Source identification and characterization	Establishing possible emission sources by monitoring near the suspected source	A sensor is placed downwind of an industrial facility to monitor variations in air pollutant concentrations over time
Education	Using sensors in educational settings for science, technology, engineering, and math lessons	Sensors are provided to students to monitor and understand air quality issues
Information/awareness	Using sensors for informal air quality awareness	A sensor is used to compare air quality at people's home or work, in their car, or at their child's school

Source: From Williams et al. (2014).

using sensors to stay apprised of soil conditions, e.g. moisture and temperature, and to manage crops accordingly, e.g. optimize irrigation. The technologies are becoming more widely available and economical as sensors and actuators require less power and maintenance (Khriji et al., 2014). Therefore, the utility of environmental sensors includes not only government inspectors but also anyone wishing to identify environmental conditions and pollution sources. This phenomenon has been coined "citizen science."

2.8.2 Big Data and the New Decision-Making Paradigm

The next-gen technologies are key parts of the new environmental engineering measurement paradigm. Engineers must be prepared to incorporate data from numerous sources. Engineers, like numerous other professionals, are increasingly applying so-called "big data" to their projects. Big data can be conceived as "any collection of data sets which volume and complexity make data management and processing difficult to perform using traditional tools," e.g. using structured query language (SQL) in relational databases (Vitolo et al., 2015). Individuals are increasingly gaining computational power and access to large data files, so engineers should be expected to be challenged regarding precision, accuracy, and representativeness of their measurements. Indeed, engineers are encountering a new measurement paradigm.

Figure 2.15 provides examples of both the current approach, which relies on sophisticated equipment sited and used by governments, industry, and researchers for compliance monitoring, enforcement, status and trend reports, and specific, targeted research. The new measurement paradigm is based on a more expansive used by individuals and communities, using new and/or adapted technologies. This increases data availability and provides much wider access to these data via the Internet and social media, key aspects of citizen science (Snyder et al., 2013).

2.8.3 Biological Measurements

As mentioned, risk assessments require much information beyond environmental measurements, including those indicating dose and endogenous transformations within the body. To explain risks, the totality of a person's biological makeup, activities, and locations must be understood (Rappaport, 2013; Wild, 2012), including the complex pathways involving both genetic and exposures to hazards in the environment (Patel et al., 2010; Rappaport, 2012). Most environmental engineers do not conduct such studies, but should be aware that major changes are occurring that will change environmental risk assessments.

One of the key features of risk assessments is the biomarker. Upon contact with a chemical compound, for example, the organism must first absorb and distribute it before the chemical is metabolized. After this, the parent compound and any new metabolites are further metabolized, stored, and/or eliminated (Vallero, 2015). Recently, the pathways leading to a biomarker have become quantified, e.g. in adverse outcome pathway (AOP), with each

Figure 2.15 Monitoring station used to measurement air pollutants near a roadway in Las Vegas, Nevada. Rooftop monitors include both traditional and "next-gen" systems, e.g. aerosol monitors and cameras, respectively. Shelter includes real-time measurement equipment for oxides of nitrogen, carbon monoxide and benzene, and meteorological conditions (tower shown in collapsed position). In addition to the real-time devices, the shelter contains sample collection systems from which samples will be transferred to the laboratory for later analysis, including particulate filters and manifolds connected to stainless steel canisters to collect volatile organic compounds. *Source*: Courtesy of U.S. Environmental Protection Agency.

event characterized by various informatics and "omics" tools (e.g. genomics, proteomics, and metabolomics). These can serve as early warning systems, such as when an AOP shows that a particular genetic change has a chance of leading to an adverse outcome (e.g. an allergy in a human subpopulation or a loss of a sensitive species in an ecosystem), regulators may prohibit this use (Tan et al., 2014).

2.9 Measurement Ethics

Engineers who design and implement measurement studies must ensure not only that the measurements are scientifically credible but also that the investigations are conducted in an ethical manner. This is particularly important when studies involve individuals, such as when measurements are taken in and near residences. Environmental engineers usually do not conduct direct biological sampling, such as blood and urine collection. However, engineers may be team members of studies in which biomedical researchers and practitioners take such samples. In addition, engineers may be privy to private or controlled information and must follow all protocols designed to protect such information.

Any study that involves humans must ensure that the person is respected, that studies provide a tangible benefit, and that justice is ensured. This not only includes studies where a person knows that he or she is involved, but those where people may be indirectly affected. For example, the engineer must inform and obtain permission from homeowners before collecting samples for a study that will be conducted to determine the quality of water in an aquifer that serves as a town's water supply. In addition to the homes whose tap water and well water will be collected, others will also need to be informed, e.g. at a town meeting. The town engineer is a good point of contact during the planning stages. This is also a way to prevent problems, such as finding that a similar study had already been recently conducted or that the town has more than one source of drinking water. It is also a good way to determine special monitoring needs, such as ways to sample in homes with vulnerable persons (e.g. the elderly) and historically underrepresented neighborhoods, i.e. environmental justice communities (VanDerslice, 2011).

Indeed, any research involving humans must meet rigorous ethical standards, including approval of a committee whose sole purpose is to conduct ethical review of proposed research, known as an institutional review board (Department of Health, 2014). Therefore, the engineer must be certain at the outset that a measurement study protects privacy, properly obtains permissions, completely informs participants (including necessary consent forms), and does not allow personal information to be identified in any unauthorized way. A particularly useful resource for identifying potential ethical problems and improving protections of the confidentiality of individuals' information is the document: Scientific and Ethical Approaches for Observational Exposure Studies (U.S. Environmental Protection Agency, 2008).

Note

1 Incidentally, these are the same terms used for treatment and remediation, i.e. contaminated soil or groundwater may be treated in situ, e.g. by air stripping, or ex situ, e.g. removed and thermally treated in an incinerator.

References

Barber, M.C. (2001). *Bioaccumulation and Aquatic System Simulator (BASS) User's Manual Beta Test Version 2.1*. Athens, GA: U.S. Environmental Protection Agency.

Barbour, M. T., Gerritsen, J., Snyder, B. D., and Stribling, J. B. (1999). *Rapid Bioassessment Protocols for Use in Streams and Wadeable Rivers: Periphyton, Benthic Macroinvertebrates and Fish*, Vol. 339 (EPA 841-B-99-002). Washington, DC: U.S. Environmental Protection Agency, Office of Water.

Barbour, M.T., Stribling, J.B., and Karr, J.R. (1995). Multimetric approach for establishing biocriteria and measuring biological condition. In: *Biological Assessment and Criteria. Tools for Water Resource Planning and Decision Making* (ed. W.S. Davis and T.P. Simon), 63–77. Boca Raton, FL: Lewis Publishers.

Crumbling, D. (2001). *Clarifying DQO Terminology Usage to Support Modernization of Site Cleanup Practice* (EPA 542-R-01-014). Washington, DC: US EPA.

De Tomasi, F. and Perrone, M.R. (2014). Multiwavelengths lidar to detect atmospheric aerosol properties. *IET Science, Measurement & Technology* 8 (3): 143–149.

Department of Health (2014). The Belmont report. Ethical principles and guidelines for the protection of human subjects of research. *The Journal of the American College of Dentists* 81 (3): 4.

Dohner, E., Markowitz, A., Barbour, M. et al. (1997). Water quality conditions. In: *Volunteer Stream Monitoring: A Methods Manual* (Report No. EPA 841-B-97-003), chapter 5. Washington, DC: Office of Water, U.S. Environmental Protection Agency

Ekhtera, M.R., Mansoori, G.A., Mensinger, M.C. et al. (1997). Supercritical fluid extraction for remediation of contaminated soil. In: *Supercritical Fluids: Extraction and Pollution Prevention*, vol. 670 (ed. M.A. Abraham and A.K. Sunol), 280–298. Washington, DC: American Chemical Society.

enHealth Council (2018). *Environmental Health Risk Assessment: Guidelines for Assessing Human Health Risks from Environmental Hazards*. Department of Health and Ageing and EnHealth. http://www.eh.org.au/documents/item/916 (accessed 17 April 2018).

Environmental Health Australia (2012a). *Australian Exposure Factor Guidance*. Canberra, ACT: Commonwealth of Australia.

Environmental Health Australia (2012b). *Environmental Health Risk Assessment: Guidelines for Assessing Human Health Risks from Environmental Hazards*. Canberra, ACT: Commonwealth of Australia.

Environmental Systems Research Institute (1995). *Understanding GIS: The ARC/INFO Method: Self Study Workbook: Version 7 for UNIX and OpenVMS*. Esri Press.

Erickson, R.J. and McKim, J.M. (1990). A model for exchange of organic chemicals at fish gills: flow and diffusion limitations. *Aquatic Toxicology* 18 (4): 175–197.

Fausch, K.D. and Schrader, L.H. (1987). *Use of the Index of Biotic Integrity to Evaluate the Effects of Habitat, Flow, and Water Quality on Fish Communities in three Colorado Front Range Streams*. Final Report to the Kodak-Colorado Division and the Cities of Fort Collins, Loveland, Greeley, Longmont, and Windsor. Department of Fishery and Wildlife Biology, Colorado State University: Fort Collins.

Finkel, A.M. (1990). *Confronting Uncertainty in Risk Management*. Washington, DC: Center for Risk Management, Resources for the Future.

Gehl, R.J. and Rice, C.W. (2007). Emerging technologies for in situ measurement of soil carbon. *Climatic Change* 80 (1–2): 43–54. doi: 10.1007/s10584-006-9150-2.

Gerba, C.P. and Pepper, I.L. (2009). Wastewater treatment and biosolids reuse. In: *Environmental Microbiology*, 2e (ed. R.M. Maier, I.L. Pepper and C.P. Gerba), 503–530. Burlington, MA: Elsevier Academic Press.

Hall, L.W., M.C. Scott, and Killen, W.D. (1996). *Development of Biological Indicators Based on Fish Assemblages in Maryland Coastal Plain Streams* (CBWP-MANTA-EA-96-1). Maryland Department of Natural Resources, Chesapeake Bay and Watershed Programs, Annapolis, Maryland

Hartman, K.J. and Brandt, S.B. (1995). Estimating energy density of fish. *Transactions of the American Fisheries Society* 124 (3): 347–355.

Hughes, R.M. and Gammon, J.R. (1987). Longitudinal changes in fish assemblages and water quality in the Willamette River, Oregon. *Transactions of the American Fisheries Society* 116 (2): 196–209.

Karr, J.R. (1981). Assessment of biotic integrity using fish communities. *Fisheries* 6 (6): 21–27.

Karr, J.R. (1987). Biological monitoring and environmental assessment: a conceptual framework. *Environmental Management* 11: 249–256.

Karr, J.R., Fausch, K.D., Angermeier, P.L. et al. (1986). *Assessment of Biological Integrity in Running Waters: A Method and Its Rationale*. Special Publication No. 5. Champaign, IL: Illinois Natural History Survey.

Khriji, S., El Houssaini, D., Jmal, M.W. et al. (2014). Precision irrigation based on wireless sensor network. *IET Science, Measurement & Technology* 8 (3): 98–106.

Kimbrough, S., Vallero, D., Shores, R. et al. (2008). Multi-criteria decision analysis for the selection of a near road ambient air monitoring site for the measurement of mobile source air toxics. *Transportation Research Part D: Transport and Environment* 13 (8): 505–515.

Kushlan, J.A., Voorhees, S.A., Loftus, W.F., and Frohring, P. (1986). Length, mass, and calorific relationships of Everglades animals. *Florida Scientist (USA)* 49: 65–79.

Leonard, P.M. and Orth, D.J. (1986). Application and testing of an index of biotic integrity in small, coolwater streams. *Transactions of the American Fisheries Society* 115: 401–414.

Letcher, T. and Vallero, D. (2011). *Waste: A Managers Handbook*. Amsterdam, NV: Elsevier.

Lyons, J. 1992. Using the index of biotic integrity (IBI) to measure environmental quality in warmwater streams of Wisconsin. General Technical Report, NC-149. St. Paul, Minnesota: U.S. Department of Agriculture, Forest Service.

Lyons, J., Wang, L., and Simonson, T.D. (1996). Development and validation of an index of biotic integrity for coldwater streams in Wisconsin. *North American Journal of Fisheries Management* 16: 241–256.

Malczewski, J. (1999). *GIS and Multicriteria Decision Analysis*. New York: Wiley.

Miller, D.L., Leonard, P.M., Hughes, R.M. et al. (1988). Regional applications of an index of biotic integrity for use in water resource management. *Fisheries* 13 (5): 12–20.

Moyle, P.B., L.R. Brown, and Herbold, B. (1986). *Final Report on Development and Preliminary Tests of Indices of Biotic Integrity for California*. Final report to the U.S. Environmental Protection Agency. Environmental Research Laboratory: Corvallis, Oregon.

National Society of Professional Engineers (2016). NSPE Code of Ethics for Engineers. Retrieved from http://www.nspe.org/resources/ethics/code-ethics (accessed 16 January 2018).

Ohio EPA (1987). *Biological Criteria for Protection of Aquatic Life. Vol. 2. Users Manual for Biological Field Assessment of Ohio Surface Waters*. Columbus, OH: Division of Water Quality Planning and Assessment, Ohio EPA.

Patel, C.J., Bhattacharya, J., and Butte, A.J. (2010). An environment-wide association study (EWAS) on type 2 diabetes mellitus. *PloS One* 5 (5): e10746, e10746.

Rappaport, S.M. (2012). Biomarkers intersect with the exposome. *Biomarkers* 17 (6): 483–489.

Rappaport, S. (2013). *The Exposome*. Berkeley, CA: Center for Exposure Biology, University of California – Berkeley.

Roth, N.E., M.T. Southerland, J.C. Chaillou, et al. (1997). *Maryland Biological Stream Survey: Ecological Status of Non-tidal Streams in six Basins Sampled in 1995* (CBWP-MANTA-EA-97-2). Maryland Department of Natural Resources, Chesapeake Bay and Watershed Programs, Monitoring and Non-tidal Assessment, Annapolis, Maryland.

Schreckenbach, K., Knosche, R., and Ebert, K. (2001). Nutrient and energy content of freshwater fishes. *Journal of Applied Ichthyology* 17 (3): 142–144.

Simon, T.P. (1991). *Development of Index of Biotic Integrity Expectations for the Ecoregions of Indiana Central Corn Belt Plain. I.* (EPA 905/9-91/025). Chicago, IL: U.S. Environmental Protection Agency, Region V.

Simon, T.P. and Lyons, J. (1995). Application of the index of biotic integrity to evaluate water resource integrity in freshwater ecosystems. In: *Biological Assessment and Criteria: Tools for Water Resource Planning and Decision Making* (ed. W.S. Davis and T.P. Simon), 245–262. Boca Raton, FL: Lewis Publishers.

Snyder, E.G., Watkins, T.H., Solomon, P.A. et al. (2013). The changing paradigm of air pollution monitoring. *Environmental Science & Technology* 47 (20): 11369–11377.

Steedman, R.J. (1988). Modification and assessment of an index of biotic integrity to quantify stream quality in southern Ontario. *Canadian Journal of Fisheries and Aquatic Science* 45: 492–501.

Sumathi, V., Natesan, U., and Sarkar, C. (2008). GIS-based approach for optimized siting of municipal solid waste landfill. *Waste Management* 28 (11): 2146–2160.

Tan, Y., Chang, D., Phillips, M. et al. (2014). Biomarkers in computational toxicology. In: *Biomarkers in Toxicology* (ed. R. Gupta). Waltham, MA: Elsevier.

U.S. Environmental Protection Agency (1994). Method 1613: Tetra-through Octa-Chlorinated Dioxins and Furans by Isotope Dilution HRGC/HRMS (Rev. B). Washington, DC: U.S. Environmental Protection Agency.

U.S. Environmental Protection Agency (1999). Method TO-9A in compendium of methods for the determination of toxic organic compounds in ambient air, 2 (EPA/625/R-96/010b). Washington, DC: U.S. Environmental Protection Agency.

U.S. Environmental Protection Agency (2002). *Guidance for the Data Quality Objectives Process* (EPA QA/G-4 [EPA/600/R-96/055]). Washington, DC: U.S. Environmental Protection Agency.

U.S. Environmental Protection Agency (2004). *An Examination of EPA Risk Assessment Principles and Practices* (EPA 100/B-04/001). Washington, DC: U.S. Environmental Protection Agency.

U.S. Environmental Protection Agency (2006). *Data Quality Objectives Guidance* (EPA/240/B-06/001). Washington, DC: U.S. Environmental Protection Agency.

U.S. Environmental Protection Agency (2007). SW-846 Test Method 8290A: Polychlorinated Dibenzodioxins (PCDDs) and Polychlorinated Dibenzofurans (PCDFs) by High Resolution Gas Chromatography/High Resolution Mass Spectrometry (HRGC/HRMS). https://www.epa.gov/sites/production/files/2016-01/documents/sw846method8290a.pdf (accessed 14 February 2018).

U.S. Environmental Protection Agency (2008). *Scientific and Ethical Approaches for Observational Exposure Studies* (EPA/600/R-08/062 [NTIS PB2008-112239]). Research Triangle Park, NC: National Exposure Research Laboratory, Office of Research and Development, U.S. Environmental Protection Agency. https://cfpub.epa.gov/si/si_public_record_Report.cfm?dirEntryId=191443 (accessed 17 April 2018).

U.S. Environmental Protection Agency (2015a). Ecosystem services in EnviroAtlas. *EnviroAtlas*. Retrieved from http://www.epa.gov/enviroatlas/ecosystem-services-enviroatlas (accessed 16 January 2018).

U.S. Environmental Protection Agency (2015b). EnviroAtlas data layer matrix. *EnviroAtlas*. Retrieved from http://www.epa.gov/enviroatlas/enviroatlas-data-layer-matrix (accessed 16 January 2018).

U.S. Environmental Protection Agency (2015c). Environmental measurement. Retrieved from http://www.epa.gov/measurements (accessed 16 January 2018).

U.S. Environmental Protection Agency (2015d). Test methods: frequent questions. *Hazardous Waste*. https://waste.zendesk.com/hc/en-us?faq=true (accessed 14 February 2018).

Vallero, D. A. (2000). *Dicarboximide Fungicide Flux from an Aquic Hapludult Soil to the Lower Troposphere*. Durham, NC: Duke University. Retrieved from https://books.google.com/books?id=OexGHAAACAAJ (accessed 16 January 2018).

Vallero, D.A. (2010). *Environmental Biotechnology: A Systems Approach*. Amsterdam, NV: Elsevier Academic Press.

Vallero, D.A. (2014). *Fundamentals of Air Pollution*, 5e. Waltham, MA: Elsevier Academic Press.

Vallero, D.A. (2015). *Environmental Biotechnology: A Biosystems Approach*, 2e. Amsterdam, NV: Elsevier Academic Press.

Vallero, D.A. and Peirce, J.J. (2002). Transformation and transport of vinclozolin from soil to air. *Journal of Environmental Engineering* 128 (3): 261–268.

Vallero, D.A. and Peirce, J.J. (2003). *Engineering the Risks of Hazardous Wastes*. Burlington, MA: Butterworth-Heinemann.

Vallero, D. A., Reckhow, K. H., and Gronewold, A. D. (2007). Application of multimedia models for human and ecological exposure analysis. *International Conference on Environmental Epidemiology and Exposure*, Durham, NC (17 October).

VanDerslice, J. (2011). Drinking water infrastructure and environmental disparities: evidence and methodological considerations. *American Journal of Public Health* 101 (S1): S109–S114.

Vernier Corporation (2009). Computer 19: Dissolved Oxygen in Water. http://www2.vernier.com/sample_labs/BWV-19-COMP-dissolved_oxygen.pdf (accessed 19 October 2009).

Vitolo, C., Elkhatib, Y., Reusser, D. et al. (2015). Web technologies for environmental big data. *Environmental Modelling & Software* 63: 185–198.

Wang, L., Liu, C., Alves, D.G. et al. (2015). Plant diversity is associated with the amount and spatial structure of soil heterogeneity in meadow steppe of China. *Landscape Ecology* 30 (9): 1713–1721.

Watkins, T. (2013). The US EPA roadmap for next generation air monitoring. Paper Presented at the EuNetAir Second Scientific Meeting, Queens' College, Cambridge.

Whitby, K. and Willeke, T. (1979). Single particle optical counters: principles and field use. In: *Aerosol Measurement*, 145–182. Gainesville, FL: University Press of Florida.

Wild, C.P. (2012). The exposome: from concept to utility. *International Journal of Epidemiology* 41 (1): 24–32.

Williams, R., Kilaru, V., Snyder, E. et al. (2014). *Air Sensor Guidebook* (EPA 600/R-14/159). Research Triangle Park, NC, USA: National Exposure Research Laboratory, Office of Research and Development, US Environmental Protection Agency.

3

Environmental Law for Engineers

Jana B. Milford

Department of Mechanical Engineering and Environmental Engineering Program, University of Colorado, Boulder, CO, USA

3.1 Introduction and General Principles

In the United States since the 1960s, public concern for ensuring a healthy environment has given rise to a complex multilevel system of environmental laws that includes federal, state, tribal, and municipal statutes, regulations, and court decisions. The system of environmental law addresses a wide and evolving array of concerns, from air and water pollution to biodiversity and ecosystem protection. Environmental engineers and scientists and environmental managers in private and public sector organizations play critical roles in ensuring that their organizations comply with these legal requirements and in shaping the requirements through public rulemaking processes and through legally binding agreements that are tailored to individual parties.

3.1.1 Sources of Law

For purposes of this chapter, we define environmental law as the system of laws and legal procedures that aim to prevent, minimize, remedy, or compensate for actions that could harm the environment or harm public health and welfare through environmental pathways. This system stems from the following sources of law:

- Federal, state, and tribal statutes and local ordinances that are enacted by elected legislative bodies.
- Regulations promulgated by administrative agencies at the federal, state, tribal, and local level.
- Legally binding provisions in permits, leases, or licenses issued by government authorities.
- International treaties and US treaties with Native American tribes.
- Court decisions interpreting laws, regulations, and legal agreements.
- The common law, which is judge-made law protecting customarily recognized rights, including torts, property, and contracts.

These laws are embedded in the broader legal system that is framed by the US Constitution and state and tribal constitutions and includes statutes and judicial decisions governing procedure for administrative rulemaking and for adjudication of rights and responsibilities.

International environmental law, including international treaties and conventions, and environmental laws in countries other than the United States are beyond the scope of this chapter. The following websites are recommended for interested readers:

- Information on the United Nations Framework Convention on Climate Change is available at http://unfccc.int.
- Information on scientific and policy aspects of climate change is available from the Intergovernmental Panel on Climate Change at www.ipcc.ch.
- The United Nations Environment Program (UNEP) website at www.unep.org has information on a wide array of global environmental issues.
- The American Society of International Law maintains a website with primary source documents for international treaties and conventions at www.eisil.org.
- The ECOLEX website at http://www.ecolex.org is operated by the UNEP, Food and Agriculture Organization (FAO), and the International Union for the Conservation of Nature (IUCN) and provides a database of international agreements, legislation, court cases, and literature on international environmental law.

Handbook of Environmental Engineering, First Edition. Edited by Myer Kutz.

3.1.2 Environmental Statutes

Starting in the 1960s, the US Congress responded to public concern for environmental protection by enacting a series of far-reaching laws. Early pieces of landmark legislation included the 1964 Wilderness Act (Pub. L. 88-577) and the Wild and Scenic Rivers Act of 1968 (Pub. L. 90-542). The National Environmental Policy Act (NEPA) (Pub. L. 91-190)[17] was signed in 1970, launching a flurry of environmental legislation in the following decade. To keep a reasonable scope, we focus here on the broadest and most far-reaching US federal statutes and implementing regulations that address pollution concerns, namely, the Clean Air Act (CAA), the Clean Water Act (CWA), the Resource Conservation and Recovery Act (RCRA), the Comprehensive Environmental Response, Compensation, and Liability Act (CERCLA), and the NEPA.

Other significant US laws that address environmental issues but are beyond the scope of this chapter include the following statutes:

- The Marine Protection, Research, and Sanctuaries Act of 1972 (Pub. L. 92-532) regulates the intentional disposal of materials such as dredged sediments into the ocean.
- The Federal Insecticide, Fungicide, and Rodenticide Act (FIFRA) as amended by the 1972 Federal Environmental Pesticide Control Act (Pub. L. 92-516) prohibits the distribution, sale, and use of pesticides unless registered by EPA based on a showing they "will not generally cause unreasonable adverse effects on the environment." The 1996 Food Quality Protection Act (Pub. L. 104-170) amended FIFRA in order to better address concerns about pesticide residues on food.
- The Safe Drinking Water Act passed in 1974 (Pub. L. 93-523) with major amendments in 1986 and 1996, requires EPA to establish minimum standards to protect tap water, to set minimum standards for states to regulate underground injection of fluids, and to administer a groundwater protection program.
- The Toxic Substances Control Act passed in 1976 (Pub. L. 94-469) and substantially amended in 2016 (Pub. L. 114-182) authorizes EPA to require premanufacture reporting and (as needed) testing of new chemicals and to regulate or ban those that pose unreasonable risks.
- The Endangered Species Act passed in 1973 (Pub. L. 93-205) prohibits federal action that would jeopardize an endangered species or destroy critical habitat and prohibits anyone from taking such species.
- The Emergency Planning and Community Right-to-Know Act of 1986 (EPCRA) (Pub. L. 99-499) required the establishment of local and state emergency planning and response committees and preparation of emergency response plans. EPCRA also set up the Toxics Release Inventory (TRI) and requires facilities that meet specified thresholds to annually report discharges and emissions of listed chemicals.

The *Environmental Law Handbook*[1] published by Government Institutes covers most of these statutes and key features of their implementing regulations.

The US federal environmental laws employ a wide range of approaches for environmental management, including command and control regulatory regimes that require compliance with specified performance, equipment, or work practice standards and market-based trading systems that issue allowances for emissions or resource use and then allow the entities that are required to hold allowances to buy and sell them through an allowance market. Many of the statutes include requirements for testing or monitoring, record keeping, and disclosure of environmental attributes or performance; some rely primarily on this approach to environmental protection. Most of the major environmental laws enacted in the United States since 1970 are distinguished from the prior common law regime by being preventive in their focus – requiring action or providing incentives to avoid environmental harm. However some have also established new liability regimes, defining legal responsibility for corrective action and compensation after environmental damage has occurred.

Technically oriented environmental professionals will benefit from gaining an overview of this part of the federal legal system and the state and tribal requirements that flow from it. However, environmental professionals should realize that a given action in a particular location may also be legally restricted or governed by state, tribal, and local laws and ordinances or by restrictions or obligations in permits, leases, and other legally binding agreements. In addition, state common law may also apply to impose liability for or redress environmental harm. For environmental professionals who have responsibility for complying with environmental law, there is a lot to know, especially because the field is continually evolving. Where questions arise about specific requirements, or about individual or organizational liability, it is important to get guidance from staff of the relevant government agencies and may also be necessary to retain legal counsel.

3.1.3 US Federal System

The United States has a tripartite system of government with three coequal branches: the executive, the legislative, and the judicial branch. Each of these branches plays a major role in environmental law, as do the counterparts of these branches in tribal and state governments.

The mechanics of enacting legislation are similar for the US Congress and for state legislatures. In general at the federal level, members of either the House of Representatives or the Senate can introduce bills for consideration. Once introduced, a bill is referred to one of the standing committees of the legislative body for study, including through committee or subcommittee hearings with invited testimony. A bill must be released out of committee, either in its original form or as revised, before the full House or Senate can consider it. In both houses, bills are passed by simple majority vote. Once passed in both chambers, the House and Senate versions of the bill are sent to a conference committee to reconcile any differences. The reconciled version of the bill is then sent back to the respective chambers for final votes. If passed, the bill is sent to the president for his or her consideration; it becomes a law if signed by the president. A bill can be enacted without the president's signature only if the veto is overridden by a two-thirds vote of the House of Representatives and a two-thirds vote of the Senate.

The US Congress derives its main authority to enact legislation from Article I of the US Constitution. Article I, Section 8, enumerates Congress' legislative powers. In particular, Section 8 grants Congress the power "to regulate commerce...among the several states." The interstate commerce clause provides the primary constitutional authorization for the major federal laws regulating activities or products that could harm human health and the environment. Legislation aimed at environmental protection on federal lands, including lands administered by the US Department of Interior and the US Department of Agriculture, is authorized by Article IV of the Constitution, which gives Congress "power to dispose of and make all needful Rules and Regulations respecting the Territory or other Property belonging to the United States."

The Constitution has also been held to impose limits on Congress' power that come into play with regard to environmental issues. The Supreme Court has held that the 10th Amendment bars Congress from "commandeering the states," i.e. from directly requiring that states administer or enforce a federal program.[2] Congress can, however, condition federal grants to states in order to provide incentives for them to administer federal programs. And as done in the CAA, CWA, and RCRA, Congress can allow states to accept authority to administer and/or enforce regulations, with the incentive that this provides the states with some flexibility to fill in implementation details in ways that are best suited to local conditions. In addition, some environmental statutes, including the CAA, CWA, CERCLA, and the Safe Drinking Water Act, expressly authorize EPA to treat "tribes as states" for administering at least some regulations in areas over which they have jurisdiction. EPA has interpreted other environmental laws that are silent on the issue of tribal authority as allowing their participation.

Federal statutes lay out relatively broad frameworks for environmental law. The executive branch departments or independent agencies that the statutes charge with implementing these laws fill in the details. The EPA is the implementing agency for many provisions of the major environmental laws. President Richard Nixon created the EPA by executive order in 1970. Congress subsequently gave EPA statutory authority to develop and enforce a myriad of detailed regulations under the CAA, CWA, and other environmental laws. The US Army Corps of Engineers, the US Department of Agriculture, the Department of Interior, the Department of Transportation (DOT), the Nuclear Regulatory Commission, the Occupational Safety and Health Administration under the Department of Labor, and the Council on Environmental Quality (CEQ) also have significant environmental responsibilities under US environmental statutes.

3.1.4 Administrative Law and Rulemaking Procedure

In promulgating regulations under the CAA, CWA, and other federal environmental statutes, EPA and the other agencies are obligated to follow procedural requirements specified under the Administrative Procedure Act of 1946 (APA) (Pub. L. 79-404)[3] as well as any particular procedures specified in the authorizing legislation. The APA recognizes two distinct approaches to rulemaking: formal rulemaking "on the record" through a trial-like agency hearing and informal or "notice and comment" rulemaking. Most environmental regulations are promulgated using informal rulemaking. Section 553 of the APA establishes minimum procedural requirements for informal rulemaking, including issuance of a notice of proposed rulemaking that references the legal authority for the rule; an opportunity for public comment, which must be open for at least 30 days; consideration of all comments received; and publication of the final rule accompanied by a statement of basis and purpose that responds to comments and justifies any policy choices.

Beyond the APA, Congress has passed a number of laws specifying additional procedural requirements for agency rulemaking. The Regulatory Flexibility Act (RFA) (Pub. L. 96-354),[4] which was passed in 1980, requires agencies to assess the impact of new rules on small entities. The Unfunded Mandates Reform Act (Pub. L. 104-4),[5] passed in 1995, requires agencies to consider less burdensome alternatives if a rule would impose costs in excess of $100 million on state, local, or tribal governments or the private sector. The Small Business Regulatory Enforcement Fairness Act of 1996 (Pub. L. 104-121) amended the RFA to allow for judicial review of agency findings regarding impacts on small business. A component of that law, the Congressional Review Act of 1996,[6] provides a streamlined mechanism for Congress to pass a joint resolution of disapproval of an agency rule. However, such resolutions are still subject to presidential veto.

A number of executive orders also establish procedural requirements for agency rulemakings. In 1981, President Reagan issued EO 12291, which required all rules to be reviewed by the Office of Management and Budget (OMB) before they could be published in the Federal Register and required that regulatory impact assessments be conducted for all rules with economic impact greater than $100 million.[7] In 1993, President Clinton replaced Reagan's order with EO 12866.[8] The new order limited OMB review to significant rules, generally those with greater than $100 million economic impact. EO 12866 requires agencies to disclose changes made in response to OMB's review. It also requires cost–benefit analysis. President Clinton issued EO 12898[9] in 1994, requiring federal agencies to identify and address "disproportionately high and adverse human health or environmental effects of its programs, policies, and activities on minority populations and low-income populations."

As do other federal agencies, EPA publishes proposed regulations in the Federal Register (www.federalregister.gov). These notices include information on how to submit written comments and plans for public hearings, if applicable. The federal government maintains an electronic docket of comments on pending regulations at www.regulations.gov. Under the APA, agencies must respond to the comments received on proposed rules before finalizing the regulations. Final regulations are announced in the Federal Register with a preamble that discusses the agency's rationale for adopting them. The final regulations themselves are incorporated on an annual basis into the Code of Federal Regulations (CFR), which provides a comprehensive listing of the regulations in effect at a given time. An electronic version is available at www.ecfr.gov. Regulations issued by EPA are compiled in Title 40 of the CFR. In addition to legally enforceable regulations, EPA also issues nonbinding guidance to assist its own staff, state, tribal, and local regulators and regulated entities in interpreting and complying with regulatory requirements.

3.1.5 Judicial Review

The federal courts have multiple roles in the system of federal environmental law. First, the courts may be asked to consider the legality of federal statutes under the US Constitution. Second, they are often asked to determine the legality of regulations promulgated under environmental statutes in the face of challenges to the regulations that may be based on either statutory or constitutional grounds. Courts may also be asked to force federal agencies such as EPA to promulgate new regulations or revise existing ones based on mandatory duties prescribed by statute (see, for example, CAA §304(a)(2) and CWA§505(a)(2)). Finally, the federal courts are often engaged to help enforce environmental laws, either when the government takes action against private parties or when citizens file suit alleging that federal environmental statutes are or have been violated.

Judicial review of agency regulations may be sought either under the review provisions of the authorizing statute or under the APA. Under §706 of the APA, petitioners with standing can challenge final agency actions in court and may request an injunction or declaratory relief. To show standing to contest a final agency rule, petitioners must allege a legally cognizable injury that is traceable to the defendant's conduct and that a favorable decision by the court could redress that injury.[10] Petitioners' claims must also be within the zone of interest the statute is meant to address.

Petitioners with standing can also ask the court to compel agency action that has been unlawfully withheld or unreasonably delayed.

To prevail in challenging a final rule, the plaintiff must prove one or more of the following (APA §706):

- A violation of substantive law – that the rule is not consistent with the authorizing statute or is contrary to the US Constitution.
- That the rule was adopted without observing required rulemaking procedure.
- That the agency's decision was arbitrary and capricious or unsupported by substantial evidence or unwarranted by the facts.

3.2 Common Law

Up until the 1970s, the common law system of torts was the main legal mechanism available for addressing environmental harm. The common law is the body of law that has been developed over centuries of custom and judicial precedent. This is in contrast with statutory law, which is derived from legislative enactments. The law of torts is the body of common law that governs legal liability and remedies for wrongful acts, whether intentional or unintentional, which cause injury to another person. In environmental cases, the tort claim of nuisance is most common, although claims of trespass, negligence, and strict liability for abnormally dangerous activities are also used. Principles developed through the common law of torts inform many of the statutes that are now seen as the main body of US environmental law. Furthermore, suits based on tort claims are still used to seek remedies in cases where injury is alleged to have occurred notwithstanding the modern system of preventative environmental statutes.

The American Law Institute's (ALI) Restatement of Torts[11] is an influential summary of the principles of tort law in the United States. The ALI Restatements provide common definitions of tort claims and explain the elements a plaintiff must generally prove to prevail in court. Restatements are periodically updated to reflect new case law and legal scholarship. Tort claims are generally heard in state courts, however, and rules can vary from state to state.

The Restatement of Torts (Second) Chapter 40 defines private nuisance as "a nontrespassory invasion of another's interest in the private use and enjoyment of land." Dust from a cement plant, vibrations from excavating activities, and feedlot odors are examples of "nontrespassory" invasions that have been addressed through private nuisance claims. To be held liable, the invasion must be either "(a) intentional and unreasonable, or (b) unintentional and otherwise actionable under the rules controlling liability for negligent or reckless conduct, or for abnormally dangerous conditions or activities."[12] As explained in the Restatement, the conduct giving rise to a nuisance claim can be either an affirmative act or a failure to take action to prevent or abate the invasion of the private property interest.[13] The invasion may be deemed intentional if the actor "(a) acts for the purpose of causing it, or (b) knows that it is resulting or substantially certain to result from his conduct."[14] And the invasion may be deemed unreasonable if "(a) the gravity of the harm outweighs the utility of the actor's conduct or (b) the harm caused by the conduct is serious and the financial burden of compensating for this and similar harm to others would not make the continuation of the conduct not feasible."[15]

To prevail with a private nuisance claim, the plaintiff must prove several elements of her case by a preponderance of the evidence. She must establish that:

1) She has a right to enjoyment of the land (generally property ownership).
2) She has suffered harm caused by the defendant.
3) The harm was substantial.
4) The defendant's action was intentional and unreasonable (or negligent or reckless or abnormally dangerous).

To avoid liability, defendants may try to argue that their activity was appropriate for the location (e.g. by virtue of zoning), in compliance with relevant regulations, and/or that the plaintiff "came to the nuisance" because the activity was going on before the plaintiff's arrival on the scene. Courts differ on the weight given to these arguments.

Remedies for successful private nuisance claims include the award of monetary damages to compensate for harm suffered by the plaintiff or a court-ordered injunction for the defendant to modify or cease the conduct that caused the invasion. In the case of nuisances that are substantial and continuing, an injunction may be the preferred remedy. However, courts have also declined to issue injunctions based on finding that the utility of the defendant's activities outweighs the harm they cause. Classic cases demonstrating this effort by the courts to balance benefits and harm include *Madison v. Ducktown Sulphur, Copper & Iron Co.*, 83 S.W. 658 (Tenn. 1904) and *Boomer v. Atlantic Cement Co.*, 257 N.E.2d 870 (N.Y. 1970). Even if they do not determine liability, arguments about zoning, compliance with regulations, and whether the plaintiff came to the nuisance are factors courts consider in assigning penalties.

In contrast to private nuisance, a public nuisance is "an unreasonable interference with a right common to the general public."[16] This tort claim may apply when a nuisance interferes with public property or with the health or welfare of many people. Government authorities are often the ones who pursue public nuisance claims. Although private parties can also bring these claims, they can only recover damages if they can show they have suffered a particular harm.

Some of the most challenging environmental cases that arise under common law are "toxic torts" cases, in which the plaintiff alleges he or she suffered harmful health effects due to exposure to a toxic chemical. In addition to environmental exposures through air or water contamination, these cases also arise with pharmaceuticals and consumer products. To prevail in a toxic torts case, the plaintiff must generally show (i) that the substance was dangerous, (ii) that he or she was exposed to the substance, and (iii) that harm resulted. Toxic tort claims may be filed under a theory of negligence, in which case the plaintiff must also show that the defendant's actions (or failure to act, including failure to warn) fell below a specified standard of care. Toxic tort claims may also be filed on a theory of strict liability. In that case, the plaintiff does not have to address the level of care exercised by the defendant, but can rely on a showing that the defendant's actions would have carried an abnormally high degree of risk, even if reasonable care were exercised in carrying them out.

A fundamental limitation of tort law is that it is remedial in nature. The law can provide a remedy after harm has occurred, but not before. The law of private nuisance is further limited in its application by the requirement that the plaintiff has a private right in the land where the "interference" occurs. In all of these common law causes of action, the plaintiff bears a heavy burden of demonstrating that the defendant's action caused his or her injury. Demonstrating causation is hard enough when an isolated source of pollution releases air or water pollution but may be impossible when multiple sources release pollution in close proximity to each other. The burst of environmental legislation in the United States in the 1970s responded to these and other limitations in the common law with a proactive framework for preventing environmental harm before it occurs.

3.3 The National Environmental Policy Act

NEPA (Pub. L. 91-190)[17] was signed into law on 1 January 1970, kicking off the decade when most of the major environmental statutes in the United States were enacted. In NEPA, the US Congress expressed the need for a broad, integrated view of the environmental consequences of actions undertaken by the federal government. NEPA seeks to ensure that federal decision makers have access to and consider high quality information on potential environmental consequences before decisions are made and that the public is given an opportunity to participate in the environmental assessment process. NEPA requirements differ from those of the CAA, CWA, and most other federal environmental statutes in the United States in that they do not specify outcomes but rather require decision makers to give serious consideration to environmental impacts and justify their decisions in light of this information.[18]

3.3.1 Federal Agency Planning under NEPA

NEPA's key environmental planning provisions are found in §102(2)(c), which requires all agencies of the federal government to "include in every recommendation or report on proposals for legislation and other major Federal actions significantly affecting the quality of the human environment, a detailed statement by the responsible official on–

(i) the environmental impact of the proposed action,
(ii) any adverse environmental effects which cannot be avoided should the proposal be implemented,
(iii) alternatives to the proposed action,
(iv) the relationship between local short-term uses of man's environment and the maintenance and enhancement of long-term productivity, and
(v) any irreversible and irretrievable commitments of resources which would be involved in the proposed action should it be implemented."

Section 202 of NEPA established the CEQ within the Executive Office of the President. CEQ is comprised of three members who are appointed by the president and are supported by staff. CEQ regulations for implementing NEPA's environmental planning requirements can be found at 40 CFR §§1500–1508. In addition to the legally binding regulations, CEQ issues and periodically updates nonbinding guidance for federal agencies on NEPA practice.[19]

To illustrate the scale of the NEPA enterprise in the United States, CEQ reports that in 2010 a total of 487 environmental impact statements (EIS) were filed, including 123 from the US Department of Interior, 106 from the US Forest Service, 81 from the DOT, and 76 from the Department of Defense. In addition, hundreds of other federal actions are covered by less exhaustive environmental assessments (EA), which provide a threshold analysis to determine whether a full EIS is needed. Most countries around the world, including China, India, and Russia, have adopted environmental impact assessment requirements similar to those imposed by NEPA. Multilateral lending agencies also utilize a similar environmental impact assessment process to inform their decision making. A number of US states, including California, New York, Virginia, and Washington, have environmental impact assessment requirements that apply to state-level decisions.[20]

Section 102 of NEPA requires that EIS be prepared for federal actions "significantly impacting the environment," so an important preliminary step is to consider whether the effects of a proposed action rise to this level. The threshold test is streamlined if the action at issue falls under a categorical exclusion, meaning that the agency has previously determined that the type of action does not have a significant effect. Lists of categorical exclusions are agency specific but commonly include administrative or personnel actions and minor renovation or reconstruction projects. If the action does not fall under a categorical exclusion, the agency must conduct an EA to consider whether the effects might be significant. EAs commonly include consideration of alternatives that could minimize negative environmental effects and thus avoid the need for an EIS. Agencies have discretion over the extent of public involvement in the conduct of an EA, but the EA and corresponding Finding of No Significant Impact (FONSI) (if applicable) must be published.

When an agency determines that an EIS is required, it follows several prescribed steps to complete the EIS. The process begins when the agency publishes a notice of intent (NOI) in the Federal Register, providing basic information on the proposed action and inviting public participation in the scoping process. In the scoping process, the agency identifies significant issues, interested parties, cooperating agencies, data needs, and information gaps. The agency must engage public participation at the scoping stage, ensuring that public input is invited early in the EIS process. After scoping is complete, the agency produces a draft EIS. Agencies must publish a notice of availability of the draft in the Federal Register and give a minimum of 45 days for public comment. The agency may hold public meetings to receive comments on the draft. The agency then produces the final EIS, including a response to substantive comments on the draft. Publication in the Federal Register of the notice of availability of the final EIS starts a 30-day waiting period before a final decision can be issued. The final decision is announced in a record of decision (ROD), which reviews the alternatives considered, identifies mitigation requirements, and discusses remaining environmental impacts. The "final" EIS may be followed by a supplemental EIS in cases where circumstances change or an action needs to be modified.[21]

CEQ regulations state that in their EIS agencies should "rigorously explore and objectively evaluate all reasonable alternatives...," including the alternative of no action. The regulations further state that agencies should identify their preferred alternatives in the draft EIS and that they should include appropriate mitigation measures.[22] The regulations specify that EIS include discussion of the following categories of environmental consequences[23]:

"(a) Direct effects and their significance.
(b) Indirect effects and their significance.
(c) Possible conflicts between the proposed action and the objectives of federal, regional, state, and local (and in the case of a reservation, Indian tribe) land use plans, policies, and controls for the area concerned.
(d) The environmental effects of alternatives including the proposed action.
(e) Energy requirements and conservation potential of various alternatives and mitigation measures.
(f) Natural or depletable resource requirements and conservation potential of various alternatives and mitigation measures.
(g) Urban quality, historic and cultural resources, and the design of the built environment, including the reuse and conservation potential of various alternatives and mitigation measures.
(h) Means to mitigate adverse environmental impacts."

Actions of many agencies that manage natural resources are made at different levels, with broad programmatic decisions made at a national scale, followed by decisions covering multistate regions, and on down to decisions on projects at specific locales. In this case, CEQ regulations encourage agencies to "tier" their EIS, referring back to the higher-level assessments without the need to revisit issues covered previously.[24] Additional CEQ guidance on the implementation of NEPA is available at https://ceq.doe.gov/guidance/guidance.html.

Detailed NEPA procedures vary somewhat from agency to agency. Although all agencies follow CEQ's regulations, the details can be tailored to the individual agency's mission. Some agencies issue their own procedures as regulations, published in the CFR, while others issue them as policy manuals or guidance documents. The US Department of Energy maintains a list of links to NEPA procedures for individual federal agencies at https://ceq.doe.gov/laws-regulations/agency_implementing_procedures.html.

As required by Section 309 of the CAA, the US EPA reviews all federal EIS for compliance with NEPA. If an EIS is deemed unsatisfactory, EPA refers the matter to the CEQ for resolution. EPA maintains a database of all EIS it has received, available at www2.epa.gov/nepa. The database includes full EIS documents dating back to 2012 and a record of all the EIS received since 1987.

Citizens with standing can seek review of agency decisions based on procedural or substantive concerns about NEPA compliance. Some agencies, such as the Bureau of Land Management (BLM), have internal administrative appeals processes that may have to be exhausted before citizens can take a complaint to court.[25]

3.3.2 NEPA and Climate Change

Treatment of climate change in EA and EIS is an evolving area of NEPA practice.[26] In their compilation of climate change litigation in the United States, Gerrard et al.[27] list more than a dozen cases that were filed or active in 2014 challenging agencies' inadequate consideration of climate change under NEPA (available at http://www.arnoldporter.com/resources/documents/ClimateChangeLitigationChart.pdf).

On 18 December 2014, CEQ published "Revised Draft Guidance on Consideration of Greenhouse Gas Emissions and the Effects of Climate Change" (available at https://obamawhitehouse.archives.gov/sites/default/files/docs/nepa_revised_draft_ghg_guidance_searchable.pdf). CEQ accepted public comment through 25 March 2015, and issued final guidance on 1 August 2016.[28] The guidance calls for agencies to consider both the impacts of their proposed actions on climate change and the implications of climate change for the environmental consequences of the proposed action. The guidance applies to all agencies, including land and resource management agencies, and calls for quantifying changes to carbon sequestration and storage, as appropriate, in addition to considering other greenhouse gas emissions. The guidance identifies a threshold level of 25 000 metric tons per year of CO_2-equivalent emissions, below which quantitative analysis is generally not recommended. Agencies have some flexibility in determining how to define direct, indirect, and cumulative environmental effects for greenhouse gas emissions under 40 CFR §1502.16, but the guidance provides the example that reasonably foreseeable impacts of a new open-pit coal mine could include emissions from land clearing, access roads and transportation, processing, and use of the mined coal.[29] The guidance was withdrawn by the Trump administration on 5 April 2017,[30] but it still provides insight into how climate change concerns may be treated under NEPA.

3.4 Clean Air Act

The CAA Amendments of 1970 (Pub. L. 91-604)[31] established the main features of the current federal framework for regulation of air pollutants. Congress has since made major modifications and added new requirements in the CAA Amendments of 1977 (Pub. L. 95-95) and 1990 (Pub. L. 101-549). Title I of the CAA charges the EPA with setting National Ambient Air Quality Standards (NAAQS) and the states with developing and carrying out plans to ensure they are met. Title I also authorizes EPA to set nationally uniform emissions standards for stationary sources, including for sources of hazardous air pollutants (HAPs). It also institutes construction permit requirements for new or modified sources and requirements for emissions controls necessary to improve visual air quality, especially in national parks and wilderness areas. Title II requires EPA to set nationally uniform emissions standards for motor vehicles, which are implemented through manufacturer certification requirements. Title II also authorizes EPA to regulate vehicle fuels and fuel additives if they contribute to harmful air pollution or would impair the performance of emissions control devices. Title III contains general provisions, including authorization of citizen suits and providing for judicial review in the DC Circuit. Titles IV, V, and VI were added by the CAA Amendments of 1990. Title IV established a cap and trade program to reduce emissions of sulfur and nitrogen oxides to address the problem of acid deposition. Title V established an operating permit program for large stationary sources as an effort to consolidate requirements for these sources within a single permit framework. Title VI phases out production and consumption of chlorofluorocarbons and other gases that destroy stratospheric ozone.

3.4.1 National Ambient Air Quality Standards and State Implementation Plans

Sections 108 and 109 of the CAA require EPA to set primary and secondary NAAQS for air pollutants that are widespread in outdoor air, come from numerous and diverse sources, and "may reasonably be anticipated to endanger public health or welfare."[32] Primary NAAQS are to be set to protect health, "allowing an adequate margin of safety," while secondary NAAQS are meant to protect welfare, including visual air quality and vegetation.[33] The Supreme Court has held that the CAA requires EPA to set primary NAAQS purely on the basis of health protection, without considering costs.[34] The statute requires EPA to review the standards periodically and to revise them as necessary.[35] EPA has established and periodically revised NAAQS for carbon monoxide, sulfur dioxide, nitrogen dioxide, photochemical oxidants, particulate matter, and lead. The current standards are published at 40 CFR Part 50.

In the process of establishing and revising NAAQS, the CAA requires EPA to produce "criteria" documents that comprehensively review and assess scientific evidence regarding health and welfare effects.[36] Correspondingly, the set of pollutants for which NAAQS are established are often referred to as criteria pollutants. EPA's National Center for Environmental Assessment (NCEA) prepares the review documents, which are now known as integrated science assessments. Under current practice, the science assessment documents are accompanied by risk/exposure assessment and policy assessment documents, which provide additional information for the EPA administrator to consider in reviewing the NAAQS. The CAA also requires the administrator to consult with a body of independent experts, known as the Clean Air Scientific Advisory Committee (CASAC), in setting or revising NAAQS.[37] Information on recent or ongoing NAAQS reviews is available at https://www.epa.gov/isa.

Section 110 of the CAA assigns the states primary responsibility for ensuring that air quality within their borders meets the NAAQS. Within 3 years after promulgation or revision of the NAAQS for a particular criteria pollutant, states are required to submit a State Implementation Plan (SIP) that "provides for implementation, maintenance, and enforcement" of the standard.[38] Attainment deadlines for the NAAQS are set in the statute. They differ by pollutant and for some pollutants by the severity of the nonattainment problem. The SIPs must include enforceable control measures as needed to meet the standards and must provide for ambient air quality monitoring and enforcement of the control measures. The plans must also ensure that emissions within the state do not "contribute significantly to nonattainment in or interfere with maintenance by any other state."[39] In the past decade, EPA has required states in the Eastern United States to amend their SIPs under this provision to reduce power plant emissions of SO_2 and NO_x that contribute to regional problems with PM2.5 (particulate matter less than 2.5 μm in aerodynamic diameter) and ozone.[40]

EPA promulgates detailed requirements for SIPs in conjunction with issuing new or revised NAAQS; these regulations are published at 40 CFR Part 51. The SIP submittal requirements include preparation of emissions inventories and use of air quality models to demonstrate attainment and may include new control or emissions offset requirements for existing or new sources. EPA can require revisions if it finds a SIP inadequate. If a state fails to submit a required plan or cannot correct a deficiency, EPA must promulgate a Federal Implementation Plan for the state.[41] Upon finding that a state has failed to attain a NAAQS by the specified deadline, EPA can require a new SIP submission and require additional control measures. EPA is also authorized to issue sanctions, including withholding federal grants for transportation and increasing emissions offset requirements for new sources.[42]

CAA §301(d) authorizes EPA to "treat Indian tribes as states" for most CAA purposes. To be eligible, tribes must have a governing body carrying out "substantial governmental duties and powers," must be found by EPA to be capable of carrying out the required functions, and must exercise the functions that pertain to management and protection of air resources in areas within the tribe's jurisdiction.[43] As of 2015, 56 tribes had been approved for "tribes as states" status for some CAA programs.[44]

3.4.2 New Source Review

Sections 160–169 (Title I Part C) of the CAA contain provisions to protect and enhance air quality in areas that are already meeting the NAAQS, with heightened protections for national parks, wilderness areas, and other areas of natural or scenic value. These "prevention of significant deterioration" (PSD) requirements include preconstruction review and permitting requirements for large new or modified sources that would be located in or near these areas.[45] The "major emitting facilities" subject to the preconstruction review requirements are defined in CAA §169(1) as any stationary source "with the potential to emit two hundred and fifty tons per year or more of any air pollutant." The permit threshold is 100 tons per year for listed source types including fossil fuel-fired steam electric plants, Portland cement plants, and metal smelters. Note that if a source has potential to emit any one pollutant at levels above the PSD threshold, the PSD permitting requirements must be followed for all pollutants that source emits, unless they are below levels deemed insignificant. Among other requirements imposed, these sources must perform air quality analysis to ensure that air quality is not excessively degraded and must determine and install best available control technology (BACT), as determined by the permitting authority (usually the state) on a case-by-case basis.

Complementing the PSD provisions, §§171–192 (Title I Part D) contain special "nonattainment area" (NAA) requirements for areas that are not meeting the NAAQS, including permit requirements for construction and operation of large new or modified sources that would be located in these areas. The permit threshold for "major" stationary sources in NAAs is generally the potential to emit 100 tons per year, but the threshold can be lower depending on the severity of the air quality problem. Among other requirements, the new or modified sources must obtain emissions offsets (which vary in stringency depending on pollutant and nonattainment severity) and meet the lowest achievable emission rate (LAER) for the pollutants or precursors of pollutants for which the area has been designated nonattainment.[46] Unlike the PSD requirements, the NAA new source review (NSR) provisions only apply to the pollutant(s) for which the area is designated nonattainment. However, PSD requirements may still apply for the other pollutants.

The overall NSR program under the CAA encompasses the PSD and NAA programs for large stationary sources, along with state-tailored programs for smaller "minor" sources such as the storage tanks and compressors used in oil and gas production.[47] The minor source construction permit requirements derive from §110(a)(2)(C), which requires that every SIP include a program to regulate the construction and modification of stationary sources, including a permit program, to ensure attainment and maintenance of the NAAQS. EPA's webpage on NSR guidance displays hundreds of documents testifying to the complexity of the NSR enterprise. A particular point of contention has been defining what constitutes a modification of an existing source that would trigger NSR.[48] EPA regulations for PSD construction permits are published at 40 CFR 51.166 and 52.21. Regulations for NAA construction permits are published at 40 CFR 51.165a. EPA also maintains a clearinghouse of information on BACT and LAER determinations at https://cfpub.epa.gov/rblc/.

3.4.3 National Emissions Standards for Stationary Sources

CAA §111 requires EPA to identify and list categories of stationary sources that "cause or contribute to air pollution which may reasonably be anticipated to endanger health or welfare" and then to issue performance standards for emissions from new or modified sources in the listed categories.[49] The new source performance standards (NSPS) are published at 40 CFR Part 60. Standards have been issued for approximately 90 source categories, ranging from residential wood heaters to hot mix asphalt facilities, beverage can coating operations, and zinc smelters. The CAA requires that the standards "reflect[] the degree of emission limitation achievable through the application of the best system of emission reduction which (taking into account the cost of achieving such reduction and any non air quality health and environmental impact and energy requirements) the Administrator determines has been adequately demonstrated."[50] The standards are to be reviewed at least every 8 years and revised as necessary.[51] Standards issued under Section 111(b) are federal standards that apply across the country.

Section 111(d) also provides for standards of performance to be set for existing sources in the special case of air pollutants that are not regulated as criteria pollutants under CAA §108 or as HAPs under §112. Under this provision, EPA issues regulations, known as "emissions guidelines," for states to follow in setting standards of performance for existing sources under their jurisdiction. Standards of performance under Section 111(d) are also to be set based on the best system of emission reduction (BSER), but EPA's regulations must allow states to take into consideration the remaining useful life of the source in applying the standards of performance.

In the CAA Amendments of 1990, Congress sought to accelerate EPA's progress in setting emissions standards for toxic air pollutants by substantially revising §112. The amended §112(b) listed 189 chemicals or compound classes as HAPs. (EPA now regulates 187 chemicals or compound classes as HAPs.) Section 112(c) directs EPA to list categories of sources of these air pollutants and to establish emissions standards for all of them. These standards are known as National Emission Standards for Hazardous Air Pollutants (NESHAP). Emissions standards for new and existing sources of HAPs are to "require the maximum degree of reduction." For purposes of §112, major sources are those with the potential to emit, considering controls, 10 tons per year or more of any HAP or 25 tons per year or more of any combination of HAPs.[52] For new major sources of HAPs, control requirements cannot be "less stringent than the emission control that is achieved in practice by the best controlled similar source."[53] For existing major sources, they generally cannot be less stringent than "the average emission limitation achieved by the best performing 12% of the existing sources."[54] Section 112 further requires EPA to follow up on the emissions standards by assessing the residual risk remaining from the source category and promulgating further standards if necessary.[55] As of 2015, the agency had issued NESHAPs for more than 100 source categories and had completed risk and technology reviews for more than 40 of them. NESHAPs are published at 40 CFR Part 63. EPA maintains a website on its air toxics rules and risk studies at www3.epa.gov/ttn/atw/.

3.4.4 Motor Vehicles and Fuels

Title II of the CAA contains provisions for mobile sources and fuels. Broadly stated, the statute establishes a framework whereby EPA sets emissions standards for mobile sources including on-road cars and trucks and nonroad vehicles and equipment, including marine engines and aircraft. These standards are implemented as manufacturer certification standards, with preproduction, assembly line, and in-use testing. States other than California are preempted from setting their own standards, although other states can also adopt California's standards. States are responsible for implementing inspection and maintenance requirements and can regulate existing vehicles and equipment. States also influence vehicle emissions through development, operation, and funding of transportation infrastructure and through transportation demand management.

Starting with the 1970 amendments to the CAA, Congress has set explicit standards and compliance deadlines for motor vehicle emissions and fuel characteristics. The 1990 CAA Amendments included extensive provisions for the national emissions standards program, including specific emissions standards for on-road vehicles and also required EPA to study whether further reductions should be required. EPA's "Tier 2" standards for light-duty vehicles and complementary limits on sulfur content of gasoline applied through model year 2016.[56] The agency has also issued "Tier 3" standards that took effect for the 2017 model year.[57] Additional information on EPA's extensive regulatory program for on-road vehicles (including heavy-duty vehicles) and for nonroad vehicles, engines, and equipment is available at https://www3.epa.gov/otaq/.

3.4.5 Regulation of Greenhouse Gases under the Clean Air Act

CAA §202(a)(1) states that "the Administrator shall by regulation prescribe (and from time to time revise)... standards applicable to the emission of any air pollutant from any class or classes of new motor vehicles or new motor vehicle engines, which in his judgment cause, or contribute to, air pollution which may reasonably be anticipated to endanger public health or welfare...." In 1999, a number of states and environmental organizations petitioned EPA to regulate emissions of greenhouse gases from new motor vehicles, asserting that the agency was required to do so under §202(a)(1). EPA denied the petition in 2003. After the DC Circuit Court of Appeals upheld EPA's decision, the petitioners appealed to the Supreme Court. In a landmark case, the Supreme Court reversed, holding that greenhouse gases were encompassed by the CAA's broad definition of "air pollutant" and that EPA was required to regulate greenhouse gases if it found they "may reasonably be anticipated to endanger public health or welfare".[58] On remand, in December 2009, EPA issued its finding that "six greenhouse gases taken in combination endanger both the public health and the public welfare of current and future generations" and that "the combined emissions of these greenhouse gases from new motor vehicles and new motor vehicle engines contribute to the greenhouse gas air pollution that endangers public health and welfare under CAA Section 202(a)".[59] The DC Circuit Court of Appeals upheld EPA's endangerment finding in 2012.[60] In May 2010, EPA and the DOT finalized joint regulations addressing fuel economy and greenhouse gas emissions from light-duty vehicles for model years 2012–2016.[61] A second phase of standards covering model years 2017–2025 was issued in October 2012.[62] Under the Obama administration, EPA also issued greenhouse gas emissions standards for heavy-duty vehicles and proposed an endangerment finding for aircraft. More information and updates are available on EPA's website at https://www3.epa.gov/otaq/climate/index.htm.

Based on its conclusion that regulation of greenhouse gases under CAA §202 triggered NSR requirements for stationary sources under §§165 and 169 and stationary source permit requirements under Title V, EPA issued the "Tailoring Rule" in June 2010.[63] Because the thresholds for PSD NSR and Title V permits start at 100 tons per year, a relatively small quantity in terms of carbon dioxide emissions, EPA proposed in the Tailoring Rule to stage implementation of the requirements. Permit requirements would immediately be extended to greenhouse gases for sources that already had to obtain permits based on emissions of other previously regulated pollutants (referred to as "anyway sources"). For sources that were only subject to the program because of greenhouse gases, the Tailoring Rule limited application starting in 2011 to new sources emitting more than 100 000 tons per year of CO_2-equivalent or modified sources emitting more than 75 000 tons per year. In 2014, the Supreme Court vacated the portion of EPA's regulations that applied PSD and Title V permit requirements to stationary sources based solely on their emissions of greenhouse gases.[64] However, the Court held that EPA could still require "anyway sources" to address greenhouse gases in their PSD and Title V permits. EPA has revised its regulations to comport with this ruling.

In the fall of 2015, EPA issued NSPS under CAA §111(b) limiting CO_2 emissions for new, modified, and reconstructed fossil fuel-fired electric utility generating units.[65] The standards for newly constructed steam generating units limit emissions to 1400 lb. CO_2/MWh electricity output (gross), based on supercritical pulverized coal with partial carbon capture and storage as the BSER. At the same time, EPA used its authority under §111(d) to establish carbon pollution emissions guidelines for existing electric utility generating units.[66] In the Clean Power Plan, EPA estimated BSER CO_2 emission rates for fossil fuel-fired electric steam generating units and stationary combustion turbines and used these rates to set state-by-state goals for emissions. States are required to develop and implement plans to meet these goals and are encouraged to allow emission trading in their programs. EPA expected the Clean Power Plan to reduce CO_2 emissions from the power sector by 32% below 2005 levels, when fully implemented in 2030. The Supreme Court stayed implementation of the Clean Power Plan on 9 February 2016, pending judicial review. Further uncertainty about the status of the Clean Power Plan was created on 16 October 2017, when the Trump administration proposed its repeal (82 Fed. Reg. 48035).

3.5 Clean Water Act

The CWA[67] was adopted into law in 1972 as the Federal Water Pollution Control Act Amendments (FWPCA) (Pub. L. 92-500). The FWPCA was renamed the CWA when it was amended in 1977 (Pub. L. 95-217). The CWA emphasizes technology-based effluent limitations for point source discharges, including industrial and municipal wastewater and stormwater runoff. The CWA also has provisions to address nonpoint source pollution, although implementation of those provisions has generally lagged behind the point source programs.

The objective of the CWA is "to restore and maintain the chemical, physical, and biological integrity of the Nation's waters."[68] To achieve this objective, the goals set by the CWA included the goal that "the discharge of pollutants into the navigable waters be eliminated by 1985" and that "wherever attainable, an interim goal of water quality which provides for the protection and propagation of fish, shellfish, and wildlife and provides for recreation in and on the water be achieved by July 1, 1983."[69] Though not met by the dates specified, these goals remain central to the Act and its implementing regulations.

Section 301 of the CWA prohibits unpermitted discharges of "any pollutant" to "navigable waters." "Navigable waters" is defined in CWA §502(7) as "waters of the United States, including the territorial seas." The scope of this definition, and of Congress' authority to regulate water pollution, has been the subject of extensive litigation.[70] In June 2015, EPA and the US Army Corps of Engineers finalized a new rule defining the scope of their jurisdiction under the CWA.[71] The rule sought to clarify which tributaries, wetlands, and water features that are not literally "navigable" are covered by the CWA. The future scope of CWA applicability is uncertain, however, because on 18 February 2017, President Trump issued Executive Order 13778 calling on EPA to revise or rescind the 2015 rule.

CWA §502(6) and 40 CFR §122.2 define "pollutant" broadly, as "dredged spoil, solid waste, incinerator residue, sewage, garbage, sewage sludge, munitions, chemical wastes, biological materials, radioactive materials, heat, wrecked of discarded equipment, rock, sand, cellar dirt and industrial, municipal, and agricultural waste discharged into water." The statutory definition excludes sewage from vessels and under certain conditions water, gas, or other materials injected into wells to facilitate oil and gas production or "water derived in association with oil or gas production and disposed of in a well."[72]

3.5.1 National Pollutant Discharge Elimination System Permits

Section 301 of the CWA prohibits discharge of pollutants to navigable waters from point sources, except in compliance with the Act. To be in compliance, discharges must be authorized by a permit issued under the National Pollutant Discharge Elimination System (NPDES) program, created by CWA §401 and administered by EPA or by states with EPA-approved programs; or §404 permits for discharge of dredge and fill material, which are administered by the Army Corps of Engineers. Discharges must also comply with effluent limits and other standards developed under §301 for existing sources, §306 for new sources, and §307 for toxic pollutants and pretreatment. Per CWA §301(b)(1)(C), permitted discharges must also comply with water quality-based limitations (see www.epa.gov/waterscience/standards). NPDES requirements apply only to direct dischargers, not to indirect dischargers that send pollutants only to a publicly owned treatment works (POTWs). However, industrial or commercial facilities that are indirect dischargers may be regulated under the national pretreatment program (NPP).[73] The NPP itself is administered as an NPDES permit requirement for POTWs, which must develop and implement local pretreatment programs as conditions of their permits.

States, tribes, or territories can be authorized to administer all or part of the NPDES program in the area under their jurisdiction. EPA lists states and tribes' authorization status at https://www.epa.gov/npdes/npdes-state-program-information. If EPA approves a state program, that jurisdiction becomes the permitting authority and new permit applications are submitted to them. EPA retains authority to review certain permits and may formally object to certain elements, issuing the permit directly if the objection is not resolved. Both the state and EPA have authority to enforce requirements of state-issued permits. Private citizens can also bring a civil action in federal court against an alleged violator or against EPA for failure to enforce permit requirements.[74] NPDES permits can be either individual or general permits. The former are specifically tailored to an individual facility. The latter cover multiple facilities in a specific category and are used for administrative streamlining. As part of the process of issuing a permit, notice and an opportunity to comment must be provided to the public, with consideration given to all comments received.

Regulations related to the NPDES program are published in 40 CFR Parts 121–125 (federal and state NPDES program requirements), 129–131 (toxic pollutant effluent standards), 133 (secondary treatment regulations), and 135–136, 401, 403, and 405–471 (effluent guidelines). EPA's Office of Water maintains a website with technical and regulatory information on NPDES permit requirements at www.epa.gov/npdes.

3.5.2 Technology-Based Effluent Limitations

The core of the CWA's scheme for reducing point source discharges is a series of technology-based effluent limitations. For existing industrial point sources that discharge directly to surface waters, CWA §301 set up a tiered system of increasingly stringent discharge limits and corresponding compliance deadlines. The 1972 FWPCA required EPA to initially establish discharge limits corresponding to best practicable technology and then to set more stringent best available technology (BAT) limits. The limits are required to be uniform across the country for point sources in a given industrial sector. After it became clear that the original deadlines in the 1972 Act would not be met, the 1977 Amendments modified the categories and timelines for discharge limits. Additional amendments in the Water Quality Act of 1987 (Pub. L. 100-4) changed the deadlines again. To establish the point source discharge limits, §304(b) requires EPA to publish and periodically revise effluent guidelines that identify the BAT for a particular industry and set regulatory requirements based on the performance of that technology. Information on EPA's effluent guidelines program is available at http://www.epa.gov/eg.

Different levels of stringency apply to different categories of pollutants and sources being regulated. For conventional pollutants, namely, 5-day biochemical oxygen demand (BOD_5), total suspended solids (TSS), pH, fecal coliform, and oil and grease, CWA §301 requires that industrial source effluents be controlled by the best conventional technology or best practicable technology, defined based on the average performance by the best performers in the source category. For nonconventional pollutants (those not designated as conventional or toxic, such as ammonia, chlorine, nitrogen, and phosphorus) and for listed toxic pollutants, the effluent limits for existing industrial sources correspond to BAT, determined by the single best performer. Section 301(b)(B) requires EPA to set minimum standards for POTWs based on secondary treatment.

The EPA publishes and may from time to time revise the CWA list of toxic pollutants, which is published at 40 CFR 401.15. Along with the list of toxic pollutants, EPA has also established a "Priority Pollutant" list, which is meant to be a more useable form of the toxic pollutant list in that it specifies pollutants by their individual chemical names (instead of listing groups of chemicals) and only includes pollutants for which the agency has published analytical test methods.

Effluent limits for new sources that discharge directly to surface waters are set under CWA §306. These limits are to be based on best demonstrated technology and must be met immediately upon construction. The CWA lists 27 categories of point sources for which EPA was required to establish NSPS and directs the administrator to revise the list as needed.

CWA §§307(b) and (c) require EPA to develop pretreatment standards for existing and new sources that discharge pollutants to POTWs, known as "indirect dischargers." These standards are designed to prevent discharge of pollutants at levels that would "interfere with, pass through, or otherwise be incompatible with" operation of the POTWs. These industry-specific categorical standards apply in addition to nationally uniform "general prohibitions" that forbid discharge of pollutants that cause pass-through or interference[75] and "specific prohibitions" that forbid specific types of discharges such as those that would create a fire or explosion hazard.[76] In addition, POTWs with mandatory pretreatment programs and those where nondomestic dischargers could cause violations must also supplement the national and categorical restrictions with local limits on indirect discharges.

The various effluent limits set by EPA under the CWA are technology based and nationally uniform. Based only on the level of reductions achievable through previously demonstrated technologies, they are set without regard to the impact of a particular discharge on the quality of the local receiving water. This aspect of the CWA has been criticized as economically inefficient. However, impacts on receiving water are addressed under other provisions of the CWA through water quality standards and water quality-based effluent limitations.

3.5.3 Water Quality Standards

The Water Quality Act of 1987 supplemented technology-based effluent limitations with the requirement that all states identify waters not expected to meet water quality standards after implementation of technology-based controls. Industrial source permits for discharges to these waters were subsequently required to include water quality-based limitations, in addition to BAT and BCT limits. The 1987 Amendments also added requirements for municipal separate storm sewer systems (MS4s) to reduce pollutant discharges to the maximum extent practicable (MEP). Phase I of the MS4 program applied to systems serving a population of 100 000 or more and required MEP-based limits. Phase II extended the program to smaller MS4 systems in urbanized areas and required implementation of best management practices. The regulations implementing these requirements are published at 40 CFR Parts 122 and 125.

CWA §303(c) establishes the framework for water quality standards, requiring states to develop and from time to time revise standards for water bodies within their jurisdiction. The standards are comprised of (i) designated uses, (ii) numeric and/or narrative water quality criteria, and (iii) an anti-degradation policy. In setting the standards, states must also consider use of the water body for public water supply, fish and wildlife habitat, recreation, agriculture, industrial purposes, and navigation. Where these goals are attainable, the standards must be set to provide water quality needed to protect fish, shellfish, wildlife, and recreation in and on the water. Per CWA §510, state standards can be more stringent but cannot be less protective than required by the CWA. State standards must be submitted to EPA for review and approval; EPA can promulgate replacement standards if state efforts are deemed inadequate. As of 2015, EPA had also found 51 tribes eligible to administer water quality standards programs and had approved at least initial water quality standards for 42 of them.[77]

Under CWA §304(a), EPA publishes water quality criteria to assist states in developing their standards. These standards take the form of numeric levels that will assure fishable and swimmable water quality. National recommended water quality criteria are specified for aquatic life, human health, and taste and odor and are listed at www.epa.gov/wqc. Updated human health criteria values for 94 chemical pollutants were published in the Federal Register in June 2015.[78]

To meet the location-specific water quality standards, the CWA requires states to employ a system of total maximum daily loads (TMDLs). A TMDL represents the maximum amount of a pollutant that can be discharged into a water body if it is still to meet water quality standards. The TMDL program goes beyond consideration of point source discharges to also address discharges from nonpoint sources such as agricultural or urban runoff or sediments from construction or timber removal sites. CWA §303(d)(1)(A) requires states to identify "impaired" waters for which effluent limits under §301 are insufficient to meet water quality standards and to revise the list of impaired waters every 2 years. Section 303(d)(1)(C) requires them to establish TMDLs for each of these water bodies. Tens of thousands of TMDLs have been issued by states and territories, mostly in the last 10–15 years. TMDLs may spur adjustment of point source discharge limits or development of nonpoint source management or restoration programs. The latter may be regulatory or nonregulatory, involving incentive programs or voluntary efforts and often with engagement of local citizens and environmental groups. While EPA has authority to set TMDLs for impaired waters if states fail to do so, the agency does not have authority to implement them. Regulations for the §303(d) program are published at 40 CFR Part 130.7. EPA publishes the National Summary of Impaired Waters and TMDL Information at https://iaspub.epa.gov/waters10/attains_nation_cy.control?p_report_type=T.

3.6 Resource Conservation and Recovery Act

RCRA[79] was passed in 1976 (Pub. L. 94-580) amending the 1965 Solid Waste Disposal Act. RCRA was subsequently amended through the Hazardous and Solid Waste Amendments of 1984 (Pub. L. 98-616), the 1992 Federal Facilities Compliance Act (Pub. L. 102-386), and the Land Disposal Program Flexibility Act of 1996 (Pub. L. 104-119). The objectives of RCRA are "to promote the protection of health and the environment and to conserve valuable material and energy resources...."[80] The statute declares it the national policy of the United States that "wherever feasible, the generation of hazardous waste is to be reduced or eliminated as expeditiously as possible. Waste that is nevertheless generated should be treated, stored, or disposed of so as to minimize the present and future threat to human health and the environment."[81] RCRA's provisions cover both nonhazardous and hazardous solid waste. The hazardous waste provisions are in Subtitle C (§§3001–3020) and are the major focus of EPA's regulatory authority. Subtitle D (§§4001–4010) contains the nonhazardous solid waste provisions. RCRA Subtitle I (§§9002–9003) addresses regulation of underground storage tanks. RCRA is implemented through regulations developed by the EPA Office of Solid Waste and Emergency Response (OSWER), as applied to specific sites by states, tribes, and EPA's regional offices. The regulations that implement RCRA are published in 40 CFR Parts 239–282.

3.6.1 Identifying and Classifying Wastes

Solid wastes encompass a wide array of materials from a range of sources, including household garbage, industrial refuse, sludge from wastewater treatment plants, ash from power plants and other combustion facilities, crop and forest residues, and mining spoil. Municipal solid waste is a subset of solid waste that includes durable and nondurable goods, containers and packaging, food wastes, and yard waste. Wastes covered by RCRA are not exclusively in the solid phase, but may include liquids and gases as well. The determination of what is covered under RCRA hinges on whether the material is a "waste" in the sense that it has been abandoned, is inherently waste-like due to posing a risk to human health or the environment, is discarded or unusable military munitions, or is being recycled. However, in order to encourage reuse, RCRA excludes waste materials that are directly used as an ingredient or feedstock in a further production process or are used as direct substitutes for commercial products. The legal status of recycled materials that have to be reclaimed prior to reuse depends on the type of material.

Some materials that might be considered solid wastes are explicitly excluded from this classification by the statute. These include domestic sewage, point source discharges that are subject to regulation under the CWA, irrigation return flows, and radioactive wastes regulated under the Atomic Energy Act. Numerous other wastes are also excluded from classification as solid wastes under RCRA, as listed at 40 CFR 261.4(a). EPA provides a compendium of relevant regulations, memoranda, and notices regarding the definition of solid waste at https://archive.epa.gov/epawaste/hazard/web/html/compendium.html.

Once it has been identified as a solid waste, the next question under RCRA is whether the solid waste should be classified as a hazardous waste. Again, some materials that might otherwise be considered hazardous wastes are specifically excluded from this definition. EPA currently excludes more than a dozen types of wastes from categorization as hazardous wastes. These include household hazardous wastes, agricultural wastes from crops and livestock that are returned to the ground as fertilizers or soil conditioners, mining overburden that is returned to the mine site, ash and other fossil fuel combustion wastes, and wastes from exploration and production of oil and gas and geothermal energy. The complete list of solid wastes that are excluded from being classified as hazardous wastes is given at 40 CFR 261.4(b). U.S. Environmental Protection Agency[82] provides a compendium of additional materials detailing these exclusions.

While not subject to the hazardous waste management provisions of RCRA Subtitle C, some of the exempted wastes are covered by other federal regulations. For example, in April 2015, EPA published final requirements for the disposal of coal combustion residuals in landfills and surface impoundments based on RCRA §§1008(a)(3) and 4004(a) (Subtitle D).[83] EPA was prompted to develop these regulations by a catastrophic coal ash spill at a power plant in Kingston, TN, that occurred in 2008. The regulations include location restrictions, structural integrity and liner design criteria, operating criteria, groundwater monitoring and corrective action requirements, closure and post-closure requirements, financial assurance, and record keeping and reporting requirements. States are encouraged but not required to adopt the federal criteria into their own programs; facilities must comply with the regulations either way.

If not exempted from the definition of solid waste or hazardous waste, the next step in determining if a waste is subject to RCRA Subtitle C is to ascertain whether the waste is deemed hazardous either as a listed waste or as a characteristic waste. EPA has developed four lists of wastes, designated F, K, P, and U, which are deemed to be hazardous wastes. The F list[84] was developed based on wastes from common industrial and manufacturing processes such as spent solvents or electroplating wastes; the K list[85] includes wastes from specific industries such as pesticide manufacturing. The P and U lists[86] cover specific commercial chemical products, with P-list chemicals demonstrating acute toxicity and U-list chemicals including those with other hazardous characteristics such as ignitability or reactivity.

Characteristic wastes are those that exhibit measurable properties that warrant regulation as hazardous wastes, corresponding to ignitability, corrosivity, reactivity, or toxicity. EPA specifies testing protocols and criteria corresponding to each of these characteristics in 40 CFR 261.21–24. The toxicity characteristic specifically pertains to the potential for toxic chemicals in the waste to leach into groundwater at dangerous levels when disposed of in an MSW landfill. This is determined via a laboratory test known as the toxicity characteristic leaching procedure (TCLP), which requires the waste generator to produce leachate through a specified protocol and compare the concentrations of 40 toxic chemicals in the leachate with TCLP regulatory levels.[87]

Finally, RCRA allows waste generators to petition for delisting of their site-specific wastes. To succeed, they must demonstrate that the waste does not meet the criteria for which it was listed, does not exhibit any hazardous waste characteristics, and does not pose a threat to human health or the environment. The requirements for a delisting petition are given in 40 CFR §260.22. An EPA evaluation conducted in 2002 found that in the first 20 years of the delisting program, petitions were granted to delist waste streams at 115 facilities. Electroplating wastes were the most commonly delisted type of waste.[88]

3.6.2 Nonhazardous Waste Management

RCRA's nonhazardous solid waste provisions in Subtitle D recognize state and local governments as the primary regulators. Subtitle D charges EPA with providing information and guidance to the states on nonhazardous waste management, including the development of federal criteria for the design and operation of municipal solid waste landfills. RCRA requires EPA to ensure that state programs for regulating MSW landfills meet the federal criteria as a minimum standard.

EPA criteria for MSW landfills are published at 40 CFR Part 258. The criteria address landfill location, operations, design, groundwater monitoring and corrective action, closure and post-closure management, and financial assurance for landfill closure and post-closure care. Landfill design criteria require demonstration that the design will ensure that specified maximum contaminant levels will not be exceeded in the uppermost aquifer at a designated point of compliance or use of a composite liner and leachate collection system.[89] EPA criteria for nonhazardous industrial waste landfills developed under RCRA Subtitle D are published at 40 CFR Part 257.

3.6.3 Hazardous Waste Management

Subtitle C of RCRA establishes a federal program to regulate hazardous wastes from "cradle to grave." The provisions include a system for tracking the wastes from their point of generation to their point of ultimate disposal; standards for waste generators and transporters; standards for operators of treatment, storage, and disposal facilities (TSDFs); and requirements for a permit system to govern the activities of waste generators and handlers. The statute authorizes EPA to delegate implementation of the permit program to the states upon finding that the state has adopted requirements that are at least as stringent as the federal regulations. As of 2009, 48 states were authorized to implement RCRA Subtitle C. According to EPA, in 2009 there were approximately 460 TSDFs, 18 000 transporters, and 15 000 large quantity generators in the United States that were subject to regulation under RCRA Subtitle C.[90]

Hazardous waste generators are those who first produce a hazardous waste or first bring it into the RCRA system by importing it to the United States. Waste generators are responsible for identifying and classifying their waste. EPA's regulations address both large quantity generators and small quantity generators.[91] The former are defined as facilities that produce more than 1000 kg of hazardous waste or more than 1 kg of acutely hazardous waste per month. The latter are defined as facilities that generate between 100 and 1000 kg of hazardous waste per month and accumulate less than 6000 kg of waste on-site at any time.[92] Generators in both categories are responsible for determining if their waste is hazardous, quantifying the amount of hazardous waste being generated, registering with responsible officials and obtaining an EPA ID number, and complying with waste management and storage requirements. Large quantity generators must have a written emergency response plan and an established program for training employees in proper handling of hazardous wastes. Small quantity generators have fewer formal requirements but must ensure employees are familiar with proper handling and emergency response procedures. Generators must also appropriately prepare waste for transport, track shipments and comply with record keeping and reporting requirements. Waste tracking is accomplished through the uniform hazardous waste manifest, which is initiated by the waste generator.

Hazardous waste transporters are jointly regulated by EPA under RCRA and by the DOT under the Hazardous Materials Transportation Act. DOT regulations are published at 49 CFR Parts 171–179. Under RCRA, transporters are required to obtain an EPA ID number, maintain the manifest for each waste shipment, comply with requirements for waste handling and spill response, and comply with RCRA record keeping and reporting requirements.

The most extensive regulations under RCRA Subtitle C apply to TSDFs. TSDFs that handle limited quantities or limited types of wastes, undertake only specific types of treatment operations or are engaged only in emergency response may be exempt from TSDF requirements or subject to reduced requirements. Fully regulated TSDFs are subject to general regulations covering analysis of waste shipments, prevention of unauthorized access, training on waste handling and emergency response, emergency response planning, maintaining manifests, record keeping and reporting, and groundwater monitoring. TSDF facilities are also subject to both design and operating standards for waste storage in containers, tanks, landfills, impoundments, and piles and unit operating standards for thermal, biological, and chemical treatment processes. Underground injection control wells are regulated jointly under RCRA and the Safe Drinking Water Act. Facilities that manage organic wastes are subject to air pollution control requirements for process vents, equipment leaks, containers, and tanks. To prepare for closure, TSDFs must develop and obtain approval for a closure plan, addressing waste removal or post-closure management and monitoring. Post-closure maintenance and monitoring are required for TSDFs designed to permanently dispose of hazardous waste. TSDFs are also required to demonstrate that they have financial resources to cover all closure and post-closure activities, as well as to cover liabilities in the event that property damage or bodily harm results from the facility.

3.6.4 Land Disposal Restrictions

The 1984 Hazardous and Solid Waste Amendments established a new regulatory program under RCRA that was intended to reduce the threat to groundwater, surface water, and air resources from land-based disposal of hazardous wastes.[93] The program, known as land disposal restrictions (LDR), calls for regulations that prescribe treatment methods to reduce the toxicity of hazardous wastes or reduce the likelihood of contaminant migration before the wastes are disposed of in landfills, surface impoundments, injection wells, or other land-based disposal units. Under the LDR provisions, RCRA prohibits the land disposal of hazardous waste that has not been adequately treated. Examples of treatment methods include biodegradation, chemical reduction, combustion, and solidification. Treatment standards are based on the best demonstrated available technology for a particular waste but are often set as concentration limits, allowing the use of alternative treatment technologies. The LDR program prohibits use of dilution to meet the concentration limits. Instead, wastes must be properly treated to reduce the mass or migration potential of the hazardous contaminants. LDR treatment standards are listed in 40 CFR §268.40.

3.6.5 Waste Incineration

Waste combustion or incineration can be a highly effective method of destroying toxic organic compounds and reducing the volume of waste before land disposal. In some cases, combustion can also be used for materials or energy recovery. However, historically hazardous waste combustion has sometimes posed the potential for public health risk and caused significant public concern.

Hazardous waste combustors are consequently subject not only to general TSDF requirements under RCRA but also to additional requirements under both RCRA and the CAA. Under RCRA, EPA has promulgated combustion standards for four categories of pollutants: organic compounds, hydrogen chloride and chlorine gas, particulate matter, and metals. RCRA standards for organics are set in terms of a unit's destruction and removal efficiency (DRE), requiring 99.99% destruction and removal for most organic compounds and 99.9999% removal for dioxin-containing waste streams.[94] Standards for hydrogen chloride, chlorine gas, and metals are specified through a three-tiered system, with the first tier focused on limiting the feed rates, the second tier on limiting stack emissions, and the third tier on limiting exposure in the surrounding environment.[95] The tiers provide an opportunity to trade-off increased monitoring requirements for less stringent limits on waste composition. Permits issued to waste combustors under RCRA specify operating requirements that constrain feed rates, gas flow rates, and combustor temperature ranges, among other parameters.

Under the CAA, EPA has set MACT standards for several types of hazardous waste combustors, including incinerators, boilers, and cement kilns.[96] MACT standards for organic pollutants are similar to the RCRA standards in specifying DRE minimums. The MACT standards for incinerators and cement kilns also specify toxicity-equivalent concentration limits for dioxins and furans at the inlet to the particulate control device used at those facilities. MACT standards for hydrogen chloride and chlorine gas are specified as numerical emissions limits. Metals are addressed through emissions limits for particulate matter and separately for mercury, low-volatility metals, and semi-volatile metals.

3.7 CERCLA

CERCLA (Pub. L. 96-510)[97] was passed on 11 December 1980. It was enacted in response to concern about environmental and health risks from inactive and abandoned hazardous waste sites. Congress amended CERCLA in 1986 with the Superfund Amendments and Reauthorization Act (SARA) (Pub. L. 99-499) and again in 2002 with the Small Business Liability Relief and Brownfields Revitalization Act (Pub. L. 107-118).

CERCLA provides broad authority for EPA to respond to releases or threatened releases of hazardous substances and establishes the procedures and basis for standards for EPA to follow in doing so. CERCLA authorizes EPA to undertake short-term removal actions, when prompt action is needed to address releases or threatened releases, and long-term remedial response actions intended to permanently clean up contaminated sites. Removal actions typically take less than a year, whereas remedial actions may take several years to complete. Waste sites must be listed on EPA's National Priorities List (NPL) before long-term remedial response actions can be undertaken. CERCLA's liability provisions authorize EPA to identify parties responsible for the releases and to hold them liable for cleanup costs. EPA can either compel the responsible parties to undertake clean up or it can undertake remedial actions itself and then recover costs from the identified parties. CERCLA §104(1) authorizes EPA to undertake removal or remedial actions whenever there is a release or substantial threat of release of "any pollutant or contaminant which may present an imminent and substantial danger to the public health or welfare" or of a "hazardous substance." However, EPA can only issue cleanup orders or recover cleanup costs for hazardous substances.

CERCLA applies common law principles of strict liability for abnormally dangerous activities to hazardous waste disposal activities. This means that EPA does not need to prove that the actions were negligent, reckless, or intentional. CERCLA also imposes joint and several liability so each of the defendants in a case can be held responsible for the entire amount of the judgment, unless they can demonstrate that the harm caused is divisible. Unlike other US environmental laws that focus on regulations to prevent future environmental harm, CERCLA's core elements address remediation and liability for past actions. Despite the emphasis on remediation, however, CERCLA's liability provisions are also recognized as having the deterrent effect of giving disposal site owners, waste generators, and handlers an incentive to manage hazardous wastes more carefully by giving them notice of the liability they might incur if something goes wrong.

CERCLA also created the Hazardous Substance Superfund to cover costs of remediation undertaken by EPA. The Superfund was initially funded through taxes on chemical feedstocks and later on petroleum. The tax provisions expired in 1995 and were not reauthorized, so Superfund cleanup efforts now rely on general revenues.

3.7.1 CERCLA Liability

CERCLA §101(14) defines hazardous substance to include hazardous wastes as defined under RCRA Subpart C, hazardous substances defined under CWA §311, toxics as defined by CWA §307, HAPs defined under CAA §112, and imminently hazardous substances identified under TSCA §7. CERCLA §101(14) explicitly excludes petroleum, natural gas, natural gas liquids, liquefied natural gas, and synthetic gas useable as fuel. Consequently, releases of these substances, such as releases from underground storage tanks at gas stations, are cleaned up under RCRA. EPA maintains a list of the CERCLA hazardous substances at 40 CFR Part 302.

CERCLA §101(22) defines "release" as "any spilling, leaking, pumping, pouring, emitting, emptying, discharging, injecting, escaping, leaching, dumping, or disposing into the environment (including the abandonment or discarding of barrels, containers, and other closed receptacles containing any hazardous substance or pollutant or contaminant)." The definition also contains several specific exclusions, including vehicle and engine emissions and the normal application of fertilizer.

CERCLA §107(a) identifies potentially responsible parties as:

- Current owners and operators of the facility where the release occurred or is threatened.
- Owners and operators at the time the hazardous substance was disposed of at the facility.
- Parties who arranged for the hazardous substance to be disposed of at the facility.
- Parties who transported the hazardous substance to the facility.

These parties are potentially liable for:

- All costs of removal or remedial action incurred by the federal government, a state, or an Indian tribe.
- Other necessary costs of response incurred by other parties.
- Damages for injury to natural resources.
- Costs of health assessments or health effects studies.

Section 107(b) lists defenses to liability, specifically that the release was caused by an act of God, an act of war, or under some circumstances an act or omission of a third party other than an employee or agent of the defendant. The 1986 and 2002 amendments to CERCLA expanded the third-party defenses to cover "innocent purchasers" who meet the criteria laid out in §101(35) and "bona fide prospective purchasers" who meet the terms of §101(40) and §107(r).

3.7.2 National Priorities List and Cleanup Process

In order for a site to qualify for long-term remedial action under the Superfund, CERCLA requires that it must first have been placed on the NPL, which is part of the National Contingency Plan for the removal of oil and hazardous substances.[98] The National Contingency Plan, published at 40 CFR Part 300, provides guidance for EPA and private parties in conducting response actions for oil spills as well as hazardous substance releases. Sites are generally placed on the NPL based on scoring by the Hazard Ranking System (HRS), which uses information from initial investigations to assess the potential for the site to pose a human health or environmental threat. EPA provides basic information on NPL sites at www.epa.gov/superfund/npl-site-status-information. Through the Superfund Enterprise Management System (www.epa.gov/superfund/superfund-data-and-reports), EPA provides more comprehensive information on active NPL sites, sites in the screening and assessment phase for possible inclusion on the NPL, and sites that have been remediated and removed from the NPL.

Once a site is placed on the NPL, EPA undertakes a remedial investigation/feasibility study (RI/FS) to determine the nature and extent of contamination, investigate the feasibility of alternative treatment technologies, and assess prospective cleanup costs. During this phase, EPA develops a proposed cleanup plan that is published in its ROD for the site. The next step is remedial design/remedial action (RD/RA), during which detailed cleanup designs are developed and implemented. Sites are deleted from the NPL after cleanup activities have been completed and cleanup goals have been achieved. CERCLA §117 requires that EPA provide notice and take public comment on the proposed plan for site remediation and that the ROD be published before remedial action begins.

3.7.3 Cleanup Standards

Section 121 of CERCLA governs cleanup standards for Superfund sites. The statute establishes a preference for remedial actions that "permanently and significantly reduce[ ] the volume, toxicity or mobility of the hazardous substances."[99] Overall, EPA is required to "select a remedial action that is protective of human health and the environment, that is cost effective, and that utilizes permanent solutions and alternative treatment technologies or resource recovery technologies to the maximum extent practicable."[100] Section 121(d) requires that remedial actions must achieve all "applicable or relevant and appropriate requirements" (known as ARARs), i.e. the remedies must meet the most stringent cleanup levels established by other federal or state standards. EPA has published extensive guidance on remediation and cleanup standards, including guidance organized by contaminant and based on the contaminated medium (see http://www.epa.gov/superfund/superfund-policy-guidance-and-laws for more information).

3.8 Enforcement and Liability

Environmental laws cannot provide meaningful environmental protection unless the entities they seek to regulate comply with them. Environmental professionals need to be aware of enforcement and compliance regimes and remedies or penalties for violations of environmental laws. Legal requirements can originate in self-implementing provisions of state and federal statutes and regulations and also from conditions in permits issued to regulated entities. Past judicial decrees, legal settlements, or administrative enforcement orders may also impose distinct obligations.

Given the potential for harm to human health and the environment from violations of environmental regulations, a main purpose of enforcement is to prevent or minimize such harm, stopping any violations as quickly as possible. Another key aim is to deter further violations, either by past violators or by others. To support aims of deterrence, environmental statutes generally authorize monetary penalties that exceed the cost of compliance and/or capture any economic benefit the violator may have expected from their failure to comply in the first place. Regulatory agencies may also publicize enforcement actions in order to help deter violations by others. On the other hand, EPA and state regulators also undertake extensive compliance assistance efforts to help entities that are operating in good faith meet regulatory requirements.

The division of EPA with responsibility for enforcement is the Office of Enforcement and Compliance Assurance (OECA). Civil judicial actions are filed in court by the US Department of Justice on behalf of EPA. The Department of Justice also prosecutes criminal enforcement actions. OECA maintains a website at http://www.epa.gov/enforcement with updates on EPA's enforcement activities and policies. Many of the federal environmental laws provide for states to implement and enforce their own regulatory programs, as long as they meet the minimum requirements established by the federal law. If EPA approves the state's program, the state normally has lead enforcement authority. However, even in this situation, EPA generally retains authority to enforce against violations based on the federal requirements if it concludes the state has not taken adequate enforcement action on its own.[101]

3.8.1 Citizen Suits

Major environmental statutes allow citizens who are adversely affected by alleged violations to bring enforcement actions in federal court, if government authorities have failed to take sufficient action. Citizens can discover violations through direct observation, tips, or from public records reporting noncompliance. Citizen suit enforcement actions can seek injunctions, civil penalties, and award of the costs of the lawsuit. As an example citizen suit provision, CWA §505 authorizes suits against "any person...who is alleged to be in violation" of an effluent standard or other CWA limitation or a related EPA or state order. In 1987, the Supreme Court held that the phrase "in violation" limits citizen suits under CWA §505 to cases in which the plaintiffs allege that the violations are ongoing, not to wholly past violations where there is no expectation of recurrence.[102] As amended in 1990, the citizen suit provision in the CAA, CAA§304(a), authorizes citizen suits against "any person...who is alleged to have violated (if there is evidence that the alleged violation has been repeated) or to be in violation of an emission standard or limitation" under the CAA or against "any person who proposes to construct or constructs any new or modified major emitting facility without a [required] permit...or who is alleged to be in violation of any condition of such permit." Citizen suits are generally not allowed if the administrator or state is already "diligently prosecuting" a civil action for the same violation (e.g. see CAA§304(b) and CWA§505(b)). Statutory provisions authorizing citizen suits generally require advance notice to EPA and/or the state as well as to the alleged violator.

3.8.2 Penalties for Violating Environmental Laws

Environmental statutes authorize both civil and criminal penalties, as well as injunctive relief (e.g. CWA §309, CAA §113). In civil administrative actions, EPA or state regulatory agencies enforce regulations directly, without necessarily going to court. The action may be issuance of a notice of violation or an enforcement (compliance) order, with or without monetary penalties. The vast majority of enforcement actions are handled through administrative actions. The entity charged with the violation has rights to request review by an adjudicatory board, administrative law judge, or in court, but often the right to such review is waived. As an example, CWA §309(g) authorizes administrative penalties of up to $10 000 per day the violation continues with a cumulative cap of $25 000 for administrative enforcement without an adjudicatory hearing and up to $10 000 per day with a cap of $125 000 for administrative enforcement with an adjudicatory hearing.

Civil judicial actions are filed in court by the US Department of Justice in cases referred by EPA or by state attorneys general in cases referred by state regulatory agencies. Cases seeking an injunction, recovery of response costs, or enforcement of administrative orders must be brought in court. These cases are often settled without going to trial, with the resolution embodied in a consent decree filed with the court. Again using the CWA as an example, §309(d) authorizes civil penalties up to $25 000 per day for each violation. Environmental statutes usually specify the maximum fine that can be levied in a particular situation and list factors that should be considered in determining the amount of a civil penalty. Factors listed in CWA §309(d) include the seriousness of the violation, the history of previous violations, any economic benefit the entity may have received from the violation, and any good faith efforts to comply. Administrative agencies and courts retain discretion to adjust penalties up to the statutory limit. EPA's civil enforcement penalty policies are posted on its website at http://www.epa.gov/enforcement.

In civil cases, the government has the burden of proving the violation based on the preponderance of the evidence. The US Supreme Court has held that the 7th Amendment right to trial by jury applies to the liability phase of suits in federal court for civil penalties under environmental laws but that the trial court, not the jury, should determine the amount of the penalty.[103] Under the US environmental laws, civil liability is generally defined as strict liability, and thus does not require consideration of the defendant's intent or knowledge with respect to the action or inaction that caused the violation.[104] Statutes differ in the affirmative defenses they allow, including whether unintentional equipment malfunctions or upset conditions qualify. Statutes also differ in the weight given to compliance with a permit as a shield to liability for violations of the statute under which the permit was issued.[105]

Criminal convictions under the federal environmental statutes are comparatively rare but can lead to substantial fines and imprisonment. CWA §309(c) authorizes criminal penalties, rising to fines of up to $250 000 and 15 years in prison for knowing endangerment. Maximum penalties under CAA §113(c)(3) are up to 15 years in prison and up to a $1 million fine for each violation. Criminal convictions require proof "beyond a reasonable doubt." And in contrast with civil liability, criminal liability requires a showing of a culpable state of mind. Under most federal environmental statutes, malicious intent is not required. CAA §113(c) and CWA §309(c)(1) authorize criminal penalties for some negligent actions. Additionally, criminal liability can attach to a knowing violation, meaning that a conscious or deliberate action or failure to act brought about the violation. The defendant may be criminally liable even if he or she did not know they were violating a specific regulation or permit condition. The Supreme Court has held that under "public welfare statutes," a conscious or deliberate action suffices for criminal liability whether or not the defendant knew the action was illegal, because "anyone who is aware that he is...dealing with [a dangerous or deleterious material] must be presumed to be aware of the regulation".[106]

Individual employees can be held criminally liable for their companies' violations if they "have a responsible share in the furtherance of the transaction which the statute outlaws".[107] Furthermore, individuals may be held responsible for the conduct of those they supervise, not only for their own direct actions. Corporate officers can be held liable for failure to prevent or promptly correct violations, if they would reasonably have been in a position do so.[108] This concept has been expressly written into the CAA and CWA. For example, CWA §309(c)(6) states "for purposes of enforcement,...'person' means... any responsible corporate officer."

3.8.3 Monitoring Compliance and Discovering Violations

Monitoring compliance with environmental laws is an enormous challenge. Hundreds of thousands of sources hold permits for air or water discharges. Millions of others (e.g. vehicles and engines) are governed by manufacturer certification standards with limited opportunities for in-use testing. To try to address this challenge, the environmental statutes employ a number of different mechanisms to help with compliance monitoring and discovering violations. First, as illustrated by CAA §114, the statutes impose extensive monitoring, record keeping, and reporting requirements on regulated entities, with responsible individuals required to certify their accuracy and completeness. Regulators may prosecute reporting violations as well as violations of discharge or emissions standards in order to maintain the integrity of this system. EPA and state agencies also have statutory authority to enter and inspect regulated facilities and to request relevant information.

Noncompliance may be discovered through routine inspections, anonymous tips, required reporting by the company itself, or voluntary disclosure of a specific instance of noncompliance. Most environmental statutes include whistle-blower protections for employees who report violations. CWA §507, CAA §322, and RCRA §7001 provide examples. Federal whistle-blower protections are administered by the US Department of Labor. Its implementing regulations are published at 29 C.F.R. 24.1 et seq. Additional information is available at www.whistleblowers.gov.

To conduct an inspection, the government must either obtain consent or obtain an administrative warrant or a criminal warrant.[109] To obtain an administrative warrant, the government must provide specific evidence of an existing violation or show that the inspection is being conducted pursuant to a general neutral administrative plan. To obtain a criminal search warrant, the government must show that the search is likely to reveal evidence of a crime.[110]

EPA and many states have sought to incentivize voluntary disclosure of compliance issues by offering reduced civil penalties when regulated entities voluntarily disclose and correct violations identified through a systematic compliance management process. Some states have adopted laws that are qualifying self-audits from being used as evidence in civil or criminal proceedings. Other states go further and provide immunity from penalties that might adhere to violations if they are revealed through qualifying self-audit programs. EPA has opposed these privileges and immunities but has adopted the policy that self-audit programs will be considered as mitigating factors in the exercise of prosecutorial discretion and in setting penalties.[111] For sources to gain these advantages, EPA requires systematic self-audit and compliance management programs to be in place, prompt disclosure, and prompt correction of the violation (see http://www.epa.gov/compliance/epas-audit-policy).

Notes

1 Sullivan, T.F.P. ed. (2011). *Environmental Law Handbook*, 21e. Lanham, MD: Government Institutes.
2 New York v. United States, 505 U.S. 144 (1992).
3 5 U.S.C. §§551–559, 701–706.
4 5 U.S.C. §601 et seq.
5 2 U.S.C. §1501 et seq.
6 5 U.S.C. §§801–808.
7 46 Fed. Reg. 13193 (17 February 1981).
8 58 Fed. Reg. 51735 (4 October 1993).
9 59 Fed. Reg. 7629 (16 February 1994).
10 Friends of the Earth v. Laidlaw Environmental Services, 528 U.S. 167 (2000).
11 American Law Institute, Restatement of the Law, Second: Torts (1965).
12 American Law Institute, Restatement of the Law, Second: Torts, Chapter 40, §822 (1965).
13 American Law Institute, Restatement of the Law, Second: Torts, Chapter 40, §824 (1965).
14 American Law Institute, Restatement of the Law, Second: Torts, Chapter 40, §825 (1965).
15 American Law Institute, Restatement of the Law, Second: Torts, Chapter 40, §826 (1965).
16 American Law Institute, Restatement of the Law, Second: Torts, Chapter 40, §821B (1978).
17 42 U.S.C. §§4321–4347.
18 Calvert Cliffs Coordinating Committee v. United States Atomic Energy Commission, 449 F.2d 1109 (D.C. Cir. 1971); Strycker's Bay Neighborhood Council, Inc. v. Karlen, 444 U.S. 223 (1980).
19 Executive Office of the President, Council on Environmental Quality (n.d.). *NEPA.GOV National Environmental Policy Act.* Available at http://ceq.doe.gov/index.html.
20 Petts, J. ed. (1999). *Handbook of Environmental Impact Assessment*, vol. 1. New York: Wiley.
21 40 CFR §1502.9.
22 40 CFR §1502.14.
23 40 CFR §1502.16.
24 40 CFR §1502.20.
25 Council on Environmental Quality (2007). *A Citizen's Guide to the NEPA: Having Your Voice Heard* (December). https://energy.gov/nepa/downloads/citizens-guide-nepa-having-your-voice-heard-ceq-2007 (28 January 2018).
26 Lee, K. (2015). CEQ's draft guidance on NEPA climate analyses: potential impacts on climate litigation, 45 ELR 10925.
27 Gerrard, M., Cullen Howe, J., and Margaret Barry, L. (n.d.). *Climate Change Litigation in the U.S.*. Available at http://www.arnoldporter.com/resources/documents/ClimateChangeLitigationChart.pdf (accessed 17 January 2018).
28 81 Fed. Reg. 51866 (5 August 2016)
29 Council on Environmental Quality (2014). *Revised Draft Guidance on Consideration of Greenhouse Gas Emissions and the Effects of Climate Change*, p. 12. https://obamawhitehouse.archives.gov/sites/default/files/docs/nepa_revised_draft_ghg_guidance_searchable.pdf (28 January 2018).
30 82 Fed. Reg. 16576 (5 April 2017)
31 42 U.S.C. §§7401 to 7671q.
32 42 U.S.C. §§7408(a)(1)(A).
33 42 U.S.C. §7409(b)(1) and (2).
34 Whitman v. American Trucking Ass'ns, 531 U.S. 457 (2001).
35 42 U.S.C. §7409(d)(1).
36 42 U.S.C. §7408(a)(2).
37 42 U.S.C. §7409(d)(2).
38 42 U.S.C. §7410(a)(1).
39 42 U.S.C. §7410(a)(2)(D).
40 76 Fed. Reg. 48208 (8 August 2011).
41 42 U.S.C. §7410(c)(1).
42 42 U.S.C. §7479(b).
43 42 U.S.C. §7601(d)(2).
44 National Tribal Air Association (2015). *Status of Tribal Air Report.* https://www7.nau.edu/itep/main/ntaa/Resources/StatusTribalAir/ (28 January 2018).
45 42 U.S.C. §7465(a).
46 42 U.S.C. §§7472(c) and 7473.
47 Milford, J.B. (2014). Out in front? State and Federal Regulation of air pollution emissions from oil and gas production activities in the Western United States. *Natural Resources Journal* 55 (1): 1–45.
48 Environmental Defense v. Duke Energy Corp., 549 U.S. 561 (2007).

49 42 U.S.C. §7411(b).
50 42 U.S.C. §7411(a)(1).
51 42 U.S.C. §7411(b)(1)(B).
52 42 U.S.C. §7412(a)(1).
53 42 U.S.C. §7412(d)(3).
54 42 U.S.C. §7412(d)(3)(A).
55 42 U.S.C. §7412(f)(2)(A).
56 65 Fed. Reg. 6698 (10 February 2000).
57 79 Fed. Reg. 23414 (28 April 2014).
58 Massachusetts v. EPA, 549 U.S. 497 (2007).
59 74 Fed. Reg. 66496 (15 December 2009).
60 Coalition for Responsible Regulation v. EPA, 684 F.3d 102 (D.C. Cir. 2012).
61 75 Fed. Reg. 25324 (7 May 2010).
62 77 Fed. Reg. 62624 (15 October 2012).
63 75 Fed. Reg. 31514 (3 June 2010).
64 Utility Air Regulatory Group v. EPA, 134 S.Ct. 2427 (2014).
65 80 Fed. Reg. 64510 (23 October 2015).
66 80 Fed. Reg. 64662 (23 October 2015).
67 33 U.S.C. §§1251–1387.
68 33 U.S.C. §1251(a).
69 33 U.S.C. §1251(a)(1) and (2).
70 United States v. Riverside Bayview Homes, 474 U.S. 121 (1985); Solid Waste Agency of Northern Cook County v. U.S. Army Corps of Engineers, 531 U.S. 159 (2001); Rapanos v. United States, 547 U.S. 715 (2006).
71 80 Fed. Reg. 37054 (29 June 2015). Clean Water Rule: Definition of Waters of the United States.
72 33 U.S.C. §1362(6).
73 40 CFR Part 403.
74 33 U.S.C. §1365.
75 40 CFR Part 403.5(a).
76 40 CFR Part 403.5(b).
77 U.S. Environmental Protection Agency (n.d.). *EPA Approvals of Tribal Water Quality Standards*. Available at https://www.epa.gov/wqs-tech/epa-approvals-tribal-water-quality-standards.
78 80 Fed. Reg. 36986 (29 June 2015). Final Updated Ambient Water Quality Criteria for the Protection of Human Health.
79 42 U.S.C. §6901 et seq.
80 42 U.S.C. §6902(a).
81 42 U.S.C. §6902(b).
82 U.S. Environmental Protection Agency (2009). *Identification and Listing of Hazardous Waste 40 CFR §261.4(b): Exclusions: Solid Wastes Which Are Not Hazardous Wastes*. A User-Friendly Reference Document, Version 1 (October). https://www.epa.gov/sites/production/files/2016-01/documents/rcra2614b-ref.pdf (28 January 2018).
83 80 Fed. Reg. 21302–21501 (17 April 2015); 40 CFR Parts 257 and 261.
84 40 CFR §261.31.
85 40 CFR §261.32.
86 40 CFR §261.33.
87 40 CFR §261.24.
88 U.S. Environmental Protection Agency (2002). *RCRA Hazardous Waste Delisting: The First 20 Years*. Office of Solid Waste (June). https://www.epa.gov/sites/production/files/2016-01/documents/delistingreport.pdf (28 January 2018).
89 40 CFR Part 258.40.
90 U.S. Environmental Protection Agency (2014). *RCRA Orientation Manual 2014, EPA530-F11-003*. Office of Solid Waste and Emergency Response (October). Available at https://www.epa.gov/sites/production/files/2015-07/documents/rom.pdf (accessed 17 January 2018).
91 40 CFR Part 262.
92 U.S. Environmental Protection Agency (2014). *RCRA Orientation Manual 2014, EPA530-F11-003*. Office of Solid Waste and Emergency Response (October). Available at https://www.epa.gov/sites/production/files/2015-07/documents/rom.pdf (accessed 17 January 2018).
93 42 U.S.C. §6924(d)–(m).
94 U.S. Environmental Protection Agency (2014). *RCRA Orientation Manual 2014, EPA530-F11-003*. Office of Solid Waste and Emergency Response (October). Available at https://www.epa.gov/sites/production/files/2015-07/documents/rom.pdf (accessed 17 January 2018).
95 U.S. Environmental Protection Agency (2014). *RCRA Orientation Manual 2014, EPA530-F11-003*. Office of Solid Waste and Emergency Response (October). Available at https://www.epa.gov/sites/production/files/2015-07/documents/rom.pdf (accessed 17 January 2018).
96 40 CFR Part 64 Subpart EEE.
97 42 U.S.C. §§9601–9675.
98 42 U.S.C. §9605(a).
99 42 U.S.C. §9621(b)(1).
100 42 U.S.C. §9621(b)(1).
101 U.S. v. Smithfield Foods Inc., 191 F.3d 516 (4th Cir. 1999).
102 Gwaltney of Smithfield, Ltd. v. Chesapeake Bay Foundation, 484 U.S. 49 (1987).
103 Tull v. United States, 481 U.S. 412 (1987).
104 Sullivan, T.F.P. ed. (2011). *Environmental Law Handbook*, 21e, p. 84. Lanham, MD: Government Institutes.
105 Sullivan, T.F.P. ed. (2011). *Environmental Law Handbook*, 21e, p. 85. Lanham, MD: Government Institutes.
106 United States v. International Minerals and Chem. Corp., 402 U.S. 558 (1971).
107 United States v. Dotterweich, 320 U.S. 277 (1943).
108 United States v. Park, 421 U.S. 658 (1975).
109 Marshall v. Barlow's, Inc., 436 U.S. 307 (1978).
110 Sullivan, T.F.P. ed. (2011). *Environmental Law Handbook*, 21e, p. 115. Lanham, MD: Government Institutes.
111 65 Fed. Reg. 19618 (11 April 2000).

4

Climate Modeling

Huei-Ping Huang

School for Engineering of Matter, Transport, and Energy, Arizona State University, Tempe, AZ, USA

4.1 Introduction

Predicting the Earth's climate has become one of the key challenges in the emerging global trend of sustainable development. Giving the urgency of the problem, scientists and engineers have mobilized in multinational efforts to quantify, using physical theories and computers, the variation of climate from interannual to centennial time scales. An outstanding example of such efforts is the publication of the Fifth Intergovernmental Panel on Climate Change (IPCC) Report (IPCC, 2013) that summarizes state-of-the-art projections of global climate through the end of the twenty-first century. Contrasting the fifth IPCC report with its predecessors, one immediately recognizes the increasing societal demands for climate information and the matching responses by scientists and engineers. For example, until the third IPCC Report (IPCC, 2001), the projections were limited to a few meteorological variables – temperature, wind, and precipitation – and only on global and continental scales. The fifth IPCC Report expanded the projections to a much longer list of variables critical to economic development, resource management, and disaster mitigation. The projections were refined to regional scales. To meet societal demands, climate modeling has become increasingly interdisciplinary. This survey is written for the broader multidisciplinary readership from engineering and physical science perspectives.

A climate model is essentially a suite of computer codes that calculate the transient or equilibrium state of the atmosphere–ocean–land system. The codes are supplemented by input data that defines external forcing and boundary conditions. A good practice in climate model development consists of four stages:

1) Extensive research on the climatic processes to understand their mutual interaction and relative importance to specific applications. Through which, a set of key processes are retained and their governing equations defined.
2) Development of numerical methods to convert the governing equations to discretized forms. This also includes the development of physical parameterization to account for the effects of unresolved processes.
3) Generation of computer codes and improvement of the efficiency of the codes, for example, by advanced parallelization.
4) Validation of the model using available observations. The outcome of stage (4) will, in turn, help refine the work in stages (1)–(3).

Through this developmental cycle, the bias in a climate model is gradually reduced to an acceptable level for specific applications. All four stages of the cycle are important. Our review focuses particularly on stage (2) and (4) of model development and on the practices of climate prediction using numerical models.

4.2 Historical Development

4.2.1 From Weather Forecasting to Climate Modeling

Climate modeling first emerged 50 years ago as a subdiscipline of meteorology, oceanography, and computational fluid dynamics. The Earth's atmosphere and oceans form a complicated fluid dynamical system that involves multiple spatial and time scales. Predictions for such a large system would seem impossible before advanced computers became available. Indeed, the improvement of climate models closely tracks the long-term increase in computer power as famously described by Moore's law.

With limited computer power, in the prehistory of climate modeling, two lines of research helped set the stage

Handbook of Environmental Engineering, First Edition. Edited by Myer Kutz.

of its later development. The first considered the energy balance of a single atmospheric column and studied how a perturbation to the system, for example, by an increase in atmospheric CO_2 concentration, leads to the change of the "climate" in the model (Manabe and Wetherald, 1967). Studying the 1D system allowed scientists to refine the methods of physical parameterization for radiative and chemical processes and understand these processes in the absence of a detailed representation of atmospheric motion other than vertical convection. The second was the development of 3D models for short-term weather forecasting. To track the evolution of the atmosphere for only a few days, many complicated processes in the climate system can be ignored. For example, the change in sea ice concentration or surface vegetation coverage has little impact on the weather over such a short period of time. The decoupling from those "slow" components allows a great simplification of a global climate model into an atmosphere-only model with fixed surface boundary conditions. In this framework, the dynamical core (which computes fluid motion) and basic physical parameterization schemes for radiation and cumulus convection were developed and tested against daily observations.

During the early development of global atmospheric models, the concept of subgrid-scale parameterization was introduced (Smagorinsky, 1963) to allow the use of relatively coarse spatial resolution, a necessary compromise for modeling fluid flows over a very large domain and for a long period of time. Numerical schemes that guarantee conservation of energy and other global conservative quantities were introduced (Arakawa, 1966 and numerous follow-ups) to allow stable long-term integrations of the model. These advances set the stage for a natural transition from weather to climate modeling. The close relation between the developments of weather and climate models is surveyed by Senior et al. (2011).

4.2.2 Beyond an Atmospheric Model

The branching of climate modeling from the classical discipline of weather prediction began in the 1970s by the introduction of atmosphere–ocean coupling (Manabe and Bryan, 1969). Recognizing that the atmosphere alone has very short memory or predictability limit (typically not exceeding 1–2 weeks, Lorenz, 1982), coupling to the slow components is essential for the atmosphere to sustain long-term memory relevant to climate predictions. At the same time, noisy fluctuations in the atmosphere cannot be ignored because they collectively feedback to the ocean to modify the oceanic state or any other slow components of the climate system (Hasselmann, 1976). The successful coupling of the atmosphere and ocean put the fast and slow components on equal footing and helped establish the paradigm for treating the interaction among multiple time scales in climate models.

Coupled atmosphere–ocean models of various levels of sophistication have been used for predicting interannual climate variability related to El Niño (Goddard et al., 2001), decadal climate variability (Keenlyside et al., 2008), and centennial climate changes (IPCC, 2013). At time scales longer than a few years, the influences of the slow components other than the ocean also become important. Developments in climate modeling in the last two decades have increasingly focused on those components. Among them, the most important are (i) the dynamics of sea ice (Hunke et al., 2015) and land ice sheets (Lipscomb et al., 2009) in the *cryosphere*, (ii) land surface processes that determine vegetation coverage and surface hydrology (Chen and Dudhia, 2001; Oleson et al., 2010), and (iii) biogeochemical processes that determine the concentration of various chemical and biological species in the atmosphere–ocean–land system (Bonan, 1995; Moore et al., 2004; Dutkiewicz et al., 2009). These components are surveyed in Section 4.4.

The most advanced climate models today broadly fall under the class of *Earth system model* (ESM) (Taylor et al., 2012), which incorporates the couplings among all of the aforementioned major components. In particular, the ESMs permit a dynamic carbon cycle in the biogeochemical processes that facilitate a more accurate calculation of the impact of anthropogenic greenhouse gas (GHG) emission on future climate. One-third of the models used in the fifth IPCC Report are ESMs (IPCC, 2013; Jones et al., 2013).

4.3 Numerical Architecture of the Dynamical Core

4.3.1 Discretization and Numerical Methods

Fluid flows in the atmosphere and ocean are the facilitators of climate interaction. Without atmospheric and oceanic motion, interactions among different components of the climate system across space and time would be much less efficient. Many important quantities in climate prediction such as temperature, wind, and precipitation are directly tied to the governing equations of fluid motion, namely, the Navier–Stokes equations and the associated thermodynamic equations. Those equations describe the temporal evolution of continuous field variables of velocity, temperature, pressure, and concentration of water vapor. Since computer models are based on discrete mathematics, those equations need to be discretized in space and time. The discretized version of the Navier–Stokes equations for the atmosphere and ocean forms the *dynamical core* of a climate model.

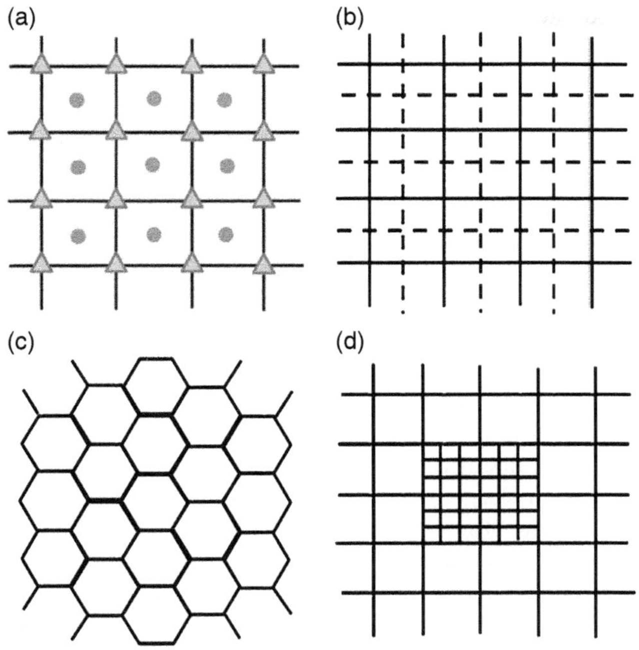

Figure 4.1 Examples of grid systems used in climate models. (a) Regular grid or mesh. (b) Staggered grids. (c) Icosahedral grid. (d) Nested grids.

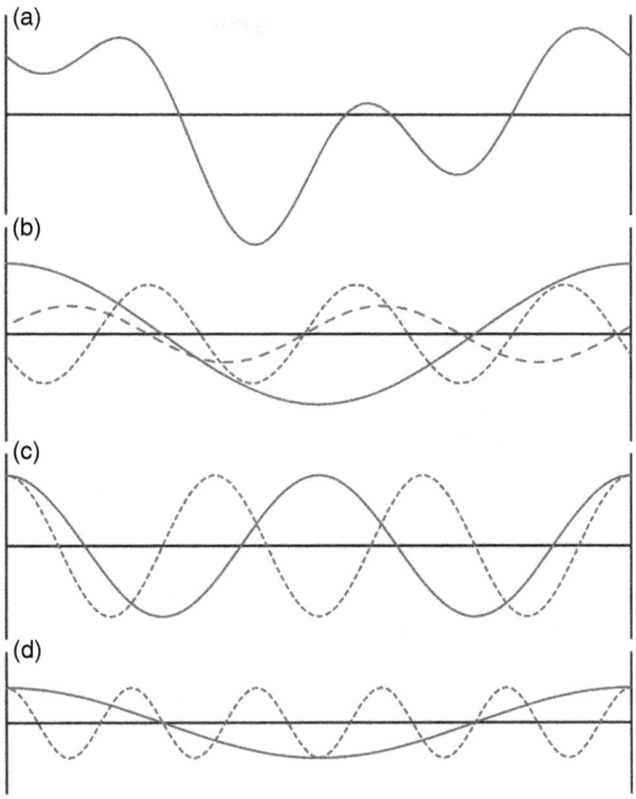

Figure 4.2 An illustration of the concept of spectral methods. A total field in (a) is decomposed into spectral components in (b). The nonlinear product of two modes in (c) gives rise to a pair of longer and shorter waves in (d). In the spectral transform method, the short wave is resolved by a grid with an enhanced resolution to eliminate aliasing.

By spatial discretization, the domain of a climate model is represented by a grid or mesh. Different types of grids have been developed to suit specific numerical algorithms as detailed in standard literature on computational fluid dynamics (Ferziger and Peric, 2002; Durran, 2010). Figure 4.1 illustrates a few selected examples of grid systems:

1) A regular grid in Cartesian or curvilinear coordinate. In *finite difference methods*, the flow variables are computed at the grid points marked by triangles. In *finite volume methods*, they are computed at the nodes marked by filled circles by taking into account the conservation of momentum, mass, and energy for the grid volume.
2) A system of staggered grids. Each of them carries one variable and is shifted from another by a half grid size. For example, one could place the east–west velocity on the solid grid and temperature on the dashed grid. This type of arrangement is widely adopted in atmospheric models (Mesinger and Arakawa, 1976; Skamarock et al., 2008) and ocean models (Griffies et al., 2000) to enhance numerical accuracy and/or stability.
3) An icosahedral grid, which is particularly ideal for numerical discretization on the sphere. It can be used with finite difference or finite element methods (Cullen, 1974; Heikes and Randall, 1995) to circumvent the problem of overclustering of grid points in the polar region.
4) A nested grid system with a high-resolution grid embedded within a coarse-resolution one. It is adopted for regional climate modeling (Skamarock et al., 2008; Prein et al., 2015) to help enhance the resolution of a target region without dramatically increasing the computational cost.

An alternative to discretization in physical space is to represent the flow field by a set of base functions, as commonly used in *spectral methods*. A classic example of the idea is the Fourier series expansion of a function on a periodic domain. An arbitrary function (Figure 4.2a) is decomposed into the sum of Fourier modes with increasing wave number or decreasing spatial scale (Figure 4.2b). The governing equations for the individual Fourier modes are integrated forward in time separately before those modes are reassembled into the full solution. On the sphere, spherical harmonics replace sinusoidal functions as the appropriate base functions. If a partial differential equation is linear, this *spectral representation* will help transform the equation into a set of decoupled linear ordinary differential equations for the expansion coefficients that can be readily solved by standard methods. Since the Navier–Stokes equations are nonlinear, additional treatments are needed to make the spectral method work. By nonlinear interaction, the product of two spectral components, for example, with wave number $n=2$ and 3 as shown in Figure 4.2c, can give rise to a pair of longer ($n=3-2=1$) and shorter ($n=3+2=5$) waves as shown in Figure 4.2d. A grid system that is sufficient for resolving the $n=3$ mode might not be fine enough for resolving $n=5$. This will cause numerical *aliasing*. The solution is to use a refined grid that resolves all the higher modes that arise from nonlinear interaction and effectively perform the nonlinear product in physical space before transforming it back to the spectral domain (Orszag, 1970; Machenhauer, 1979). This procedure is now standard in global climate models that adopt the spectral method (Neale et al., 2012). Although spectral methods work particularly well for constructing the dynamical core of a global model, they can also be used in limited-area models by carefully choosing the base functions and boundary conditions (Juang et al., 1997).

The standard grids discussed heretofore can potentially be deformed or rotated to improve the numerical accuracy or stability for specific applications. For example, by the "cubed sphere" method (Ronchi et al., 1996), the global domain is divided into six subdomains that are mapped to six squares that form the faces of a cube. In this manner, the governing equations can be solved over those squares when suitable matching conditions are imposed at the edges of the cube. For a global ocean model, it is useful to rotate the spherical curvilinear mesh such that the North Pole is located over land and outside the oceanic model domain (Madec and Imbard, 1996). Both strategies help circumvent the problem of a singular distribution of grid points at the poles. These are but two of many examples of recent innovations toward the improvement of the dynamical core.

Along with the technical development, intercomparisons for the dynamical cores of atmospheric models have been actively pursued using several proposed standards (Held and Suarez, 1994; Neale and Hoskins, 2001). Idealized "aquaplanet" models that consist of mainly the dynamical core and optional simplified physical processes are used for testing numerical convergence (Williamson, 2008) and for understanding the dynamical regimes of global circulation in the absence of more complicated physical processes and feedbacks (Medeiros et al., 2008).

4.3.2 Boundary Conditions and Vertical Coordinate

A global atmospheric model does not require lateral boundary conditions in the horizontal direction. A radiation (nonreflecting outflow) boundary condition is commonly used at top of atmosphere. At the surface, exchanges of momentum, heat, and water with ocean and land need to be parameterized as discussed in Section 4.4. A regional atmospheric model is generally constrained by lateral boundary conditions taken from either observation or the output of a global model (Skamarock et al., 2008).

While the numerical architecture of the dynamical core of ocean models is not too different from that of atmospheric models (Madec et al., 2008; Smith et al., 2010), the treatment of lateral boundary conditions is more complicated for the ocean. On the one hand, it is costly to resolve the fine details of the coastlines and nearshore bathymetry. On the other hand, some of those details can critically influence ocean circulation even at much larger scales. For example, the narrow Indonesian throughflow contributes substantially to interbasin exchanges between the Indian and Pacific Ocean (Godfrey, 1996). Even the seemingly small Galapagos Islands have nontrivial influences on the ocean circulation in the equatorial Pacific (Karnauskas et al., 2008). With the complicated lateral boundaries, spatially nonuniform grids are more widely used in ocean models. Spectral methods are rarely used in ocean models due to their requirement of simple or highly symmetric boundary conditions.

A common issue for atmospheric and ocean modeling is the treatment of surface or bottom topography. To circumvent the problem of a coordinate surface intersecting with topography, terrain-following vertical coordinates are widely used. For example, defining $\eta=(P-P_T)/(P-P_S)$ where P is atmospheric pressure and P_T and P_S are pressures at the surface and top of the atmosphere in the model, the η-coordinate will have values of 1 at the surface and 0 at the top of the model (Skamarock et al., 2008). Variations of this strategy, usually by merging η-coordinate in the lower troposphere with pressure coordinate in the upper atmosphere (Neale et al., 2012), are adopted by the majority of atmospheric models. A straightforward application of the equivalent of η-coordinate to the ocean is problematic due to the dramatic variation of depth from the middle of an ocean basin to the shallow continental shelf. The η-levels will be extremely tightly packed over the shallow part of the ocean, potentially causing numerical instability. This problem is circumvented by artificially modifying the bathymetry, for example, turning the continental shelf into a series of steps that match the model levels (Madec et al., 2008). Lastly, alternatives also exist that use thermodynamically conserved quantities as the vertical coordinate. Notable examples are isopycnal (constant density) coordinate used in ocean models (Hallberg, 1997) and isentropic (constant potential temperature) coordinate used in atmospheric models (Hsu and Arakawa, 1990).

4.4 Physical and Subgrid-Scale Parameterization

4.4.1 Radiative Processes

Radiative heating by the sun provides the fundamental source of energy for the entire climate system. The incoming solar radiation, dominated by shortwave in the visible band, is absorbed mainly by the Earth's surface and partially reflected by the surface and clouds. Absorption of sunlight by atmospheric molecules occurs mainly in the ozone layer in the upper atmosphere. The heat loss by infrared radiation emitted by the Earth-atmosphere system counters solar heating and helps maintain an overall radiative energy balance (Trenberth et al., 2009; Mlynczak et al., 2011). The net effects of solar and infrared radiation are parameterized as a diabatic forcing term in the thermodynamic equations in climate models. It is through thermodynamic effects, especially differential heating, that radiative forcing affects atmospheric motion. For example, excessive solar heating of the Earth's surface, as typically occurring in a hot summer day, can trigger vertical convection. On a much larger scale, the contrast between excessive solar heating in the tropics and excessive infrared cooling in the polar region is one of the main driving forces of the global circulation (Randall, 2015).

The detailed solution of the radiative transfer equation, incorporating the effects of molecular absorption and scattering by clouds, requires complicated numerical techniques (Thomas and Stamnes, 2002) that can be too costly to adopt in climate models. A strong dependence of radiative properties for absorption and scattering on the wavelength of radiation is another reason that contributes to the high computational cost. In practice, those details are parameterized in climate models by using approximate but fast algorithms for the radiative transfer calculation and by effectively averaging over the solar and infrared spectra (Stephens, 1984; Edwards and Slingo, 2006). The net outcome of the parameterization scheme is the radiative heating or cooling integrated over the entire solar and infrared spectra, which enter the governing equations of the climate model through the diabatic term in the thermodynamic equation.

The simplified calculation of radiative transfer in climate models is largely based on a 1D framework with no detailed variation of clouds within a grid box. The radiative effect of 3D clouds is significantly more complicated (Wiscomb, 2005) and remains to be properly parameterized. The validation of radiative transfer calculations is also a challenge because of the limited availability of observations. The detailed vertical profiles of radiative variables are not routinely observed but are obtained only through specially designed field experiments (Ackerman and Stokes, 2003). Global validation of the radiation parameterization scheme relies mainly on comparisons with satellite observations that are limited to cloud cover and the energy budget at top of atmosphere (Dolinar et al. 2015).

In the important problem of projecting centennial climate changes driven by anthropogenic emission of GHG, the influence of an increasing GHG concentration enters a climate model through radiation parameterization in the form of excessive absorption and emission of infrared radiation. This leads to an overall warming at the surface and in the troposphere. The direct effect of doubling the CO_2 concentration from present-day level is equivalent to a radiative forcing of approximately 4 W m^{-2} (IPCC, 2013). That this is a small fraction of the solar constant ($\approx$1360 W m^{-2}) underlies the importance of an accurate radiation parameterization scheme in climate models.

4.4.2 Moist Processes and Cloud Physics

Water vapor in the Earth's atmosphere plays an important role in regulating weather and climate. Latent heat release due to condensation is a major contributor to the diabatic forcing that drives atmospheric motion. Since the updraft associated with cumulus convection is typically only a few kilometer wide, it cannot be resolved by global or even regional climate models. *Convective parameterization* is needed to account for the net contribution of unresolved convective clouds to precipitation, diabatic heating, and momentum transport (Arakawa, 2004). In general, the parameterization scheme treats not only deep cumulus convection but also shallow convection (Stensrud, 2007). Global climate models used in the IPCC reports adopt a wide variety of convective parameterization schemes. This potentially contributes to the spread in the hydrological variables simulated by those models (Baker and Huang, 2014).

Related to cumulus convection, the formation of cloud droplets occurs at even smaller scales and involves physical processes not covered by the Navier–Stokes equations for fluid motion. These processes, which determine the concentration and size distribution of cloud droplets and ice crystals, are represented by the parameterization of *cloud microphysics* (Stensrud, 2007). Since the detail of cloud microphysics affects the radiative properties of clouds, a strong interconnection also exists between the parameterization schemes for cloud microphysics and radiation (Baran et al., 2014).

4.4.3 Boundary Layer Turbulence

The turbulent boundary layer of the atmosphere has a typical thickness of about 1 km. Above the boundary layer, large-scale motion in the free atmosphere is close to quasi-two-dimensional under the influence of the Earth's rotation and stable stratification. A large fraction of kinetic and available potential energy in the atmosphere resides in the synoptic-scale and large-scale motions that are approximately in geostrophic balance (i.e. the balance between pressure gradient and Coriolis force). This justifies the use of a relatively coarse horizontal resolution in climate models. This setup breaks down in two places. First, small-scale cumulus convection occurs over areas where vertical stratification is unstable. As already discussed, this is treated by convective parameterization in climate models. Second, within atmospheric boundary layer the turbulent motion is fully 3D and contains eddies with multiple scales due to turbulent energy cascade. Turbulent eddies play an important role in transporting heat and momentum between the surface and free atmosphere. Similar to the atmosphere, the upper ocean also has a turbulent boundary layer within which eddies and internal waves help facilitate the exchange of heat, momentum, and biotas between the surface and the interior of the ocean. Since the individual eddies or waves cannot be resolved by the ocean model, their effects need to be parameterized.

The simplest strategy for *boundary layer parameterization* is to represent the eddy momentum and heat fluxes in terms of the local vertical wind shear and potential temperature gradient, essentially formulating turbulent eddy transport as a diffusive process. The nonlocal version of the diffusive approach has been developed for atmospheric models (Hong and Pan, 1996) and ocean models (Large et al., 1994). Higher-order turbulence closure schemes (Mellor and Yamada, 1982) are also used in some climate models. For an unstable boundary layer, one can understand the effect of turbulent eddy transport by its role in restoring the vertical wind shear or potential temperature gradient to neutrality. Nevertheless, even when the boundary layer is stable, transport of heat and momentum can still occur by the interaction of the broad spectrum of gravity waves. This effect has also been systematically quantified (Sukoriansky et al., 2005) and incorporated into standard boundary layer parameterization.

4.4.4 Land Surface Processes

The exchanges of heat, momentum, and water between land surface and the atmosphere form an essential part of the boundary conditions for an atmospheric model. They are not directly resolved but need to be parameterized. Some of those processes (e.g. evapotranspiration) occur within the atmosphere. The others (e.g. diffusion of moisture within the upper layers of soil) occur underground and require separate calculations by a dynamical model for soil moisture and ground water. Together, they are treated in a unified land surface model (Chen and Dudhia, 2001; Oleson et al., 2010) and coupled to the atmospheric model.

An increasingly important aspect of climatic influences of land surface processes involves anthropogenic modifications of land cover. For example, deforestation and urban development can cause semipermanent changes in the radiative and hydrological properties of the surface, which in turn affect regional climate. To treat land surface processes related to urbanization, a class of *urban canopy model* has been developed (Kusaka and Kimura, 2004; Best and Grimmond, 2015) as an enhanced component of the land surface model. It has been used to model the effect of urban expansion on the local climate of major metropolitan areas (Georgescu et al., 2009; Kusaka et al., 2012; Kamal et al., 2015).

4.4.5 Gravity Wave Drag

The interaction between atmospheric flow and unresolved small-scale topography can produce gravity waves that act to transfer momentum from the atmosphere to the surface. The net effect of this *orographic gravity wave drag* (GWD) is parameterized in climate models by representing the net drag in terms of the strength of the resolved mean flow, vertical stability, and variance of the height of unresolved topography. The net drag is distributed vertically according to the criterion of gravity wave breaking (Palmer et al., 1986; McFarlane, 1987; Kim et al., 2003). The inclusion of GWD has been shown to reduce common model biases in midlatitude zonal wind (Palmer et al., 1986). Nevertheless, diagnostics of the budget of angular momentum reveal that the GWD term degrades the balance of global angular momentum in some models (Huang et al., 1999; Brown, 2004). Since GWD represents the unresolved form drag, ideally the sum of GWD and the resolved form drag should be independent of model resolution. Brown (2004) found a deviation from this behavior for some models. This problem of resolution dependence might not be limited to GWD and is an important aspect to investigate for the parameterization of all subgrid-scale processes.

4.4.6 Biogeochemical Processes

One of the latest developments in climate modeling is the incorporation of biogeochemical processes into atmospheric and ocean models. Among many important examples, those processes are critical for determining the concentration of GHG in the atmosphere and the concentration of biota such as phytoplankton in the ocean. For the atmosphere, since important biogeochemical processes usually occur near the surface, they are parameterized as part of the land surface model (Bonan, 1995; Oleson et al., 2010). Parameterization of biogeochemical and ecological processes has also been developed for ocean models (Moore et al., 2004; Dutkiewicz et al., 2009) and used in climate simulations.

The most advanced class of climate models, broadly named "Earth system models" (ESMs) (Taylor et al., 2012), are characterized by their incorporation of a dynamic carbon cycle in the biogeochemical processes. Before the implementation of a dynamic carbon cycle, long-term climate projections were based on a one-way response of the climate model to an imposed GHG concentration (IPCC, 2000; Meehl et al., 2007). With the ESMs, the GHG concentration is allowed to evolve through the interaction with natural biogeochemical processes (Moss et al., 2010; Taylor et al., 2012). This is one of the notable advances in climate modeling that will likely see a wider adoption in the next IPCC report.

Figure 4.3 summarizes the essential couplings among different components of a climate model. Specifically, Figure 4.3b highlights the connections among the dynamical core of the atmospheric model and its various physical parameterization schemes.

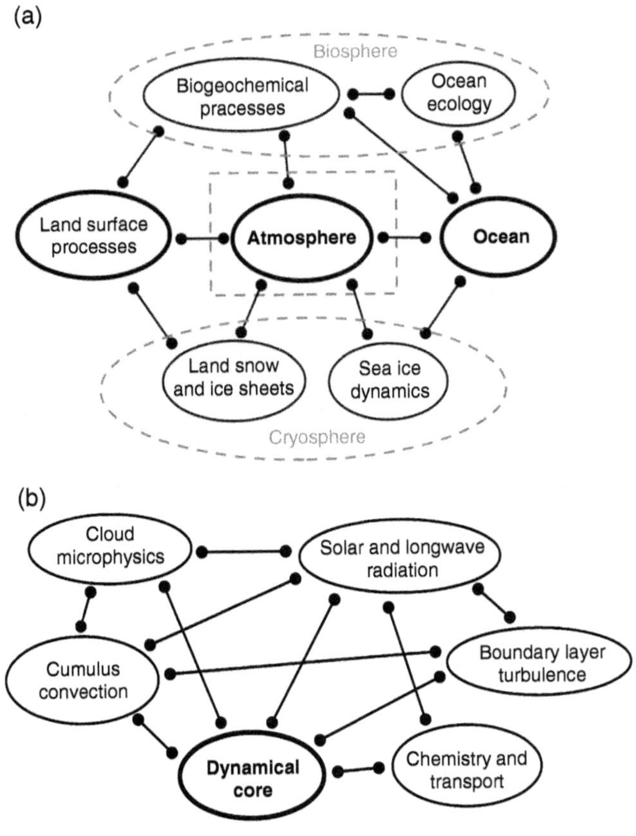

Figure 4.3 (a) A schematic diagram of the couplings among the major components in a climate model. (b) The detail of the atmospheric model (within the dashed box) from (a) which further shows the interactions among the dynamical core and physical parameterization schemes.

4.5 Coupling among the Major Components of the Climate System

The climate system is driven by complicated interactions among its five major components: *atmosphere, ocean, land, cryosphere,* and *biosphere,* as illustrated in Figure 4.3a. Although the atmosphere plays a crucial role in bringing the connections together, in a modern climate model, the five components are treated on equal footing given the increasing importance of the other four components with an increasing time scale of interest. For example, on interdecadal time scales the changes in sea ice concentration, land vegetation coverage, and deep ocean circulation all become essential factors in determining the slow variation of climate.

The atmospheric, ocean, and land models can be regarded as standalone models that can be developed and tested in isolation before their coupling to other components. The two subcomponents in the cryosphere, sea ice model (Hunke et al., 2015) and land ice sheet model (Lipscomb et al., 2009), depend on the dynamics in the atmospheric and ocean models. The subcomponents in the biosphere, namely, the biogeochemical and ecological processes (Bonan, 1995; Moore et al., 2004; Dutkiewicz et al., 2009), are usually parameterized as part of the atmospheric or ocean model. The interaction between two major components is modeled as either the exchange of fluxes across the mutual boundary or diabatic forcing to one of the components. For example, the atmosphere and ocean are coupled through the heat, momentum, and water fluxes at the sea surface. A biogeochemical process that affects the GHG concentration will influence atmospheric dynamics in the form of a diabatic heating through the change in long-wave radiation by the absorption of GHG. Improving the model architecture to allow seamless interconnections among the major components is a new challenge for climate modeling. To efficiently process the complicated interactions, some models have incorporated a separate "coupler" to facilitate all the couplings in the system (CESM Software Engineering Group, 2013).

4.6 The Practice of Climate Prediction and Projection

4.6.1 Validation of Climate Models

Validations of climate models are routinely performed by comparing the simulation of present-day climate with observation. This practice relies critically on the data collected from an existing global network of meteorological observations. About 50 years, upper-air measurements of temperature, humidity, and wind are available from an irregularly distributed (with the typical interstation distance on the order of 100 km) network of stations. Measurements by surface stations are more abundant and stretch further back to the turn of the twentieth century. These observations are quality checked and interpolated onto regular grids to form the so-called *reanalysis* data sets (Kalnay et al., 1996; Saha et al., 2010; Compo et al., 2011; Dee et al., 2011; Rienecker et al., 2011). The oceanic counterparts of reanalysis have also been created (Köhl and Stammer, 2008; Chang et al., 2012; Zuo et al., 2015), although they rely heavily on model-based dynamical adjustment and interpolation constrained by sparse observations for the deep ocean. More reliable observations, from in situ and satellite measurements, are available for the sea surface temperature (Reynolds et al., 2002; Rayner et al., 2003). Together, climate model simulations for the last few decades are compared with these reanalysis data sets to determine the model biases.

Figure 4.4 illustrates a typical comparison between a climate model simulation (a) and observation (b). Both panels are the 30-year (1961–1990) averaged climatology of surface air temperature over the global domain. Contours at the levels from −25 to 25 °C with a 5 °C interval are shown with negative contours dashed. The warmest areas with temperature exceeding 20 and 25 °C are lightly and heavily shaded. Over the ocean, surface air temperature is close to sea surface temperature. The observation is from a reanalysis data set produced by the National Center for Environmental Prediction (NCEP). The simulation is from a twentieth-century run performed with the Canadian Center for Climate Modeling and Analysis (CCCma) model and archived at the IPCC data distribution portal (www.ipcc-data.org). It is chosen because the resolution of the model is comparable with that of the reanalysis. Many obvious features such as cold spots over Tibetan Plateau and the Andes Mountains are readily identified in both maps. Many nontrivial details in the observation are also reproduced by the model. For example, while the tropics as a whole are warmer than the higher latitudes, along the equator a strong east–west contrast exists as characterized by a large "warm pool" over the Western Pacific. In the Eastern Pacific, maximum temperature occurs not on the equator but on the north of it, along the so-called intertropical convergence zone (ITCZ). In the North Pacific, the north–south temperature gradient is greater over the western part of the basin, and it decreases eastward. The climate model allows free evolution of the coupled atmosphere–ocean system driven only by the imposed solar radiation at top of atmosphere and GHG and aerosol concentrations. That the model is capable of reproducing those aforementioned subtle details from observation is a major achievement of climate modeling.

At the same time, one can also identify nontrivial differences between the model simulation and observation. For example, the temperature over Indian subcontinent is about 1 °C cooler in the model. This is likely related to the model bias in the precipitation associated with Indian monsoon, a known issue for many models (Sperber et al., 2013). Over the Eastern Pacific, the model produces a more pronounced temperature minimum on the equator that extends to Central Pacific and a more symmetric "double ITCZ" structure with two maximum of temperature north and south of the equator. This "double ITCZ" bias is also common to climate models and is a subject of ongoing research (Lin, 2007; Oueslati and Bellon, 2015). Model validations as illustrated by Figure 4.4 help users of climate models assess the reliability of the predictions for specific variables and target regions. It also guides model developers to focus on improving the representation of specific processes related to the biases.

The intercomparisons among different climate models and the observation have been systematically carried out under the organization of the coupled model intercomparison project (CMIP) (Taylor et al., 2012) parallel to the activities of climate assessment and projection for IPCC. Based on a set of indicators that measure the model biases, recent studies have shown an encouraging trend of a continued reduction of model biases in successive generations of climate models (Reichler and Kim, 2008; Paek and Huang, 2013).

4.6.2 Climate Prediction on Interannual-to-Decadal Time Scales

On time scales shorter than a decade, internal dynamics of the atmosphere–ocean system plays a key role in driving climate variation. In principle, predictions over those time scales can be formulated as an initial value problem. Namely, given the present (observed) state of atmosphere–ocean as the initial condition, the model is allowed to freely evolve until it reaches the target time for prediction. On seasonal-to-interannual time scales and especially for El Niño, predictions using coupled atmosphere–ocean models have been routinely carried out (Goddard et al., 2001; Wang et al., 2010) and verified to demonstrate useful skills for predicting tropical sea surface temperature and the temperature and precipitation in midlatitudes (Yang et al., 2009).

In numerical predictions on seasonal-to-interannual time scales, the useful signals in the model are embedded mostly in the slow component of the oceans. By ocean–atmosphere interaction, slow variations in sea surface temperature influence the climate of remote regions over land through specific pathways (Ropelewski and Halpert, 1989; Chang et al., 2000; Lau et al., 2006; Huang et al., 2009). For that reason, climate model predictions are

(a)

(b)

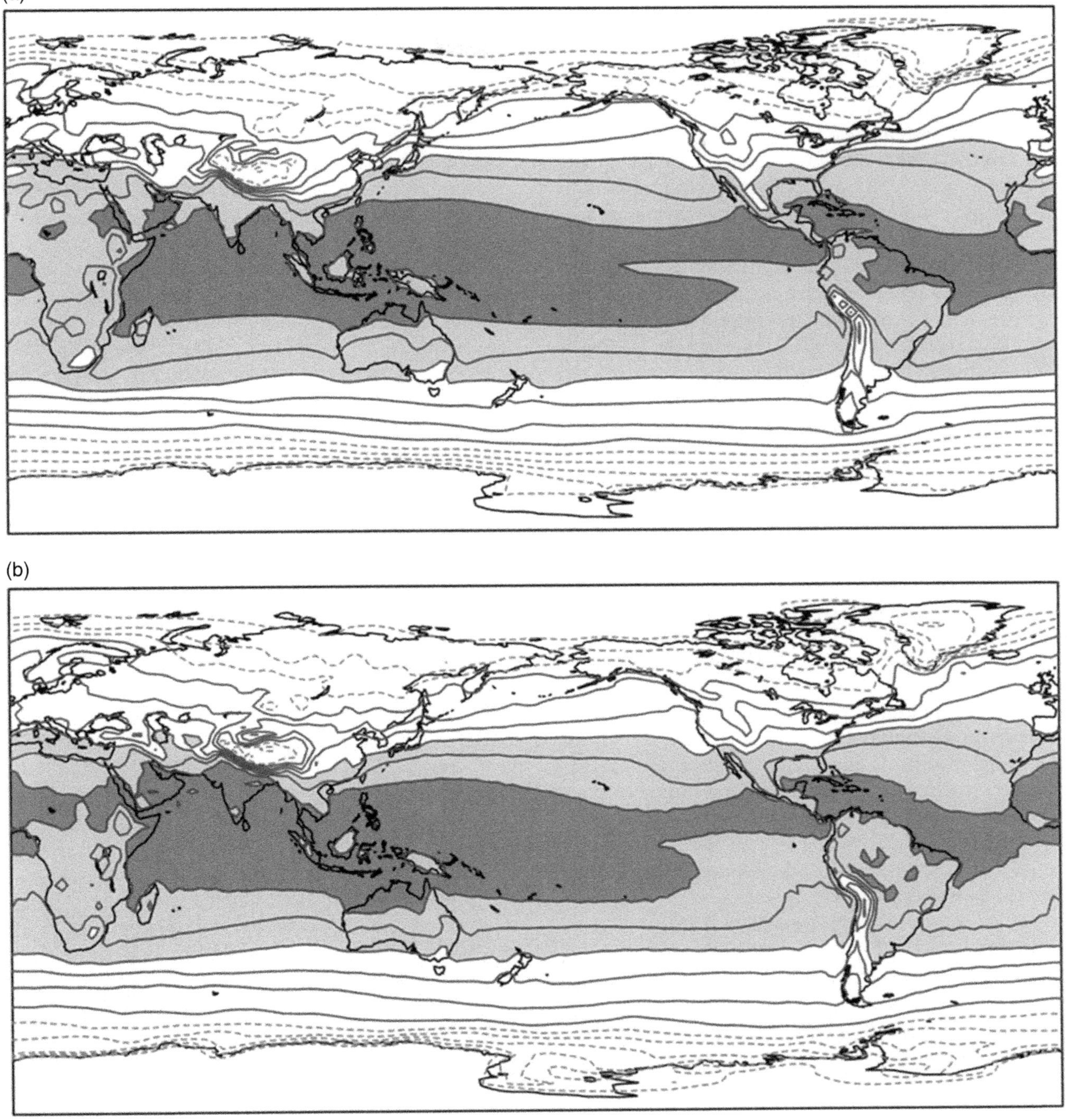

Figure 4.4 The 30-year averaged climatology of surface air temperature. (a) Climate model simulation. (b) Observation from a reanalysis data set. Contour interval is 5 °C with negative values dashed. Light and heavy shadings indicate areas with temperature exceeding 20 and 25 °C, respectively.

skillful only over specific regions that have strong dynamical connections to particular ocean basins.

The initial-value-problem approach has also been adopted for decadal predictions using coupled atmosphere–ocean models (Keenlyside et al., 2008). Those simulations are still experimental. Assessments of the predictive skills, especially for internal variability, of those models on decadal and interdecadal time scales are ongoing (Kim et al., 2012; Van Oldenborgh et al., 2012). The dependence of the decadal prediction on the process of initialization is also an important current research topic (Keenlyside et al., 2008; Magnusson et al., 2013).

4.6.3 Climate Projection on Multidecadal to Centennial Time Scales

On multidecadal to centennial time scales, trends in the external forcing and surface boundary conditions become important in determining the long-term changes of the climate system. The most important components in the external forcing are due to changes in solar activity and increases in GHG and aerosol concentrations of anthropogenic origins. For long-term climate projections, those forcing terms have to be constructed outside the climate model (IPCC, 2000; Moss et al., 2010) and imposed to the model to drive the simulations. In that sense, centennial-scale climate projections as described by the IPCC reports (IPCC, 2013) represent the responses of climate models to the imposed external forcing instead of the solution of a pure initial value problem.

The most comprehensive collection of the model-based projections for the climate of the twenty-first century can be found in the IPCC reports (IPCC, 2013, and its predecessors). While the projected increase in global-mean surface or tropospheric temperature due to increased GHG concentration is well known (as commonly termed "global warming"), other projected changes such as modifications of hydrological cycle over arid regions (Seager et al., 2007; Baker and Huang, 2014), shifts in midlatitude storm tracks (Wu et al., 2011), and thinning of sea ice (Stroeve et al., 2012) also have potentially important implications for human life.

The verification of multidecadal to centennial climate projections is difficult due to the shortness of observational records, especially for 3D variables for the upper atmosphere and deep ocean. Some attempts to compare the projections made for the recent decades with the available observations have produced mixed results (Fyfe et al., 2013). This is likely because the effect of GHG forcing, which produces the "signal" of climate change against the "noise" of internal variability, does not yet stand out over the relatively short period. It is hoped that the predictive skills of climate models at multidecadal to centennial scales will be clarified when more observations become available in the future.

There are currently many climate models based on different numerical methods as reviewed in Sections 4.3 and 4.4. In both simulations of present-day climate and projections for the future, the multimodel ensemble exhibits a substantial spread in major climatic variables (IPCC, 2013). How the diverse results from different models should be combined to produce an optimal prediction remains a challenge in research (Knutti, 2010). Notably, along with a decrease of the bias in the simulated climatology (Reichler and Kim, 2008; Paek and Huang, 2013), there is an indication that the spread of some dynamic variables for the multimodel ensemble has narrowed from previous to current generation of climate models (Paek and Huang, 2013).

4.6.4 Climate Downscaling

We have so far discussed the prediction and projection of global-scale or large-scale climate, using global climate models with relatively coarse resolutions. The median of the horizontal resolution for the atmosphere among the models participating in CMIP5 (Taylor et al., 2012) and used for IPCC AR5 (IPCC, 2013) is around 100 km. At this resolution, details of small-scale topography and inhomogeneity of land cover are not properly represented by the models. This leaves a scale gap between the direct output of climate model simulation and the region-dependent information that is critical for stakeholders. To bridge the gap, various strategies of *climate downscaling* have been developed. Broadly speaking, the output of a global climate model is used as a constraint to impose upon a regional model that includes the needed detail of local topography and land cover and/or additional physical processes such as those associated with an urban environment.

The details of the strategies for downscaling differ in how the large-scale climate information is passed on to the smaller scales. We will discuss *dynamical downscaling* in which a high-resolution regional climate model is nested within a coarse-resolution global climate model. The latter supplies the time-dependent lateral boundary condition for the former. Simulations with this setup have been performed to resolve mesoscale and submesoscale features of local climate using horizontal resolutions from 3 to 15 km (Caldwell et al., 2008; Heikkila et al., 2011; Kusaka et al., 2012; Sharma and Huang, 2012). How this framework is useful for extracting additional information on local climate is illustrated by an example in Figure 4.5. Consider the square in Figure 4.5 as a grid box of a coarse-resolution global model. A mountain range, shown as the elongated oval, falls within the grid box and is not resolved by the "parent" model. If a high-resolution model is embedded within this grid box, even by blowing a uniform flow (from the output of the global model) over the fine topography, one obtains new details as the local response to the large-scale flow. For example, suppose that the global climate model predicts a future shift in the direction of surface wind as indicated in the figure, in the high-resolution regional model, this shift will produce more precipitation along the mountain due to enhanced topographic lifting. This information would otherwise be absent in the global model since it regards the entire grid box as flat.

In dynamical downscaling, the interaction between atmospheric flow and surface boundary conditions in the high-resolution regional model could feedback to the

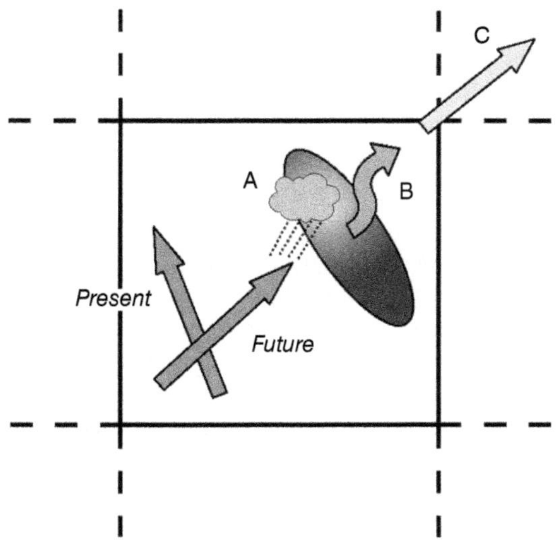

Figure 4.5 An illustration of the effect and consequence of dynamical downscaling.

large-scale flow in the "parent" model. To generalize the example in Figure 4.5, after the large-scale flow interacts with small-scale topography to produce local precipitation (marked by "A"), drier and lighter air might flow over the mountain (marked by "B") and eventually reenter the domain of the "parent" model (marked by "C"). Without dynamical downscaling to produce the happenings at A and B, the air flow at C would be wetter and colder. Thus, for consistency, downscaling should be accompanied by *upscaling* to complete the two-way interaction. This is an important aspect to explore in future development of climate downscaling and multiscale climate simulation.

4.7 Statistical Model

The approaches surveyed so far follow the *Newtonian* thinking of predicting, from the first principles, a future state of a physical system by integrating the governing equations from a known initial state given at $t = 0$ where t is time. A textbook example of such an approach is the tracking of a point-like object in free fall under gravity. Given the initial position and velocity of the object, the future position and velocity at any given time can be calculated from Newton's second law of motion. The history of the object from $t < 0$ is not used for the prediction. This approach relies on the absolute accuracy of the first principles and their numerical realization. For the prediction of complicated environmental systems, an infusion of observational data from $t < 0$ might help offset uncertainties in the governing equations and errors due to numerical discretization. In short-term weather prediction, the use of prior observations for bias correction, for example, by the technique of reforecast (Hamill et al., 2013), is well established. Giving historical observations even more weight, an alternative framework has been developed by climate modelers in which conventional governing equations – Navier–Stokes equations for fluid motion in the atmosphere and ocean – are not used at all. Instead, prediction for the future is generated empirically and based solely on historical observations (Van den Dool, 2007).

The statistical framework of climate modeling can produce meaningful results if the climate system does not behave randomly but has memory in time and correlation in space. Statistical analysis of past observations helps extract the spatiotemporal correlations that can be used for extrapolation into the future (Hasselmann, 1988; Van den Dool, 2007). So far, statistical climate models are most useful for predicting climate variability on interannual time scales, especially that related to El Niño (a coupled atmosphere–ocean oscillation with a typical period of a few years). The phenomenon is approximately cyclic with distinctive growing, mature, and decaying phases. Past observations of over two dozen cycles since the turn of the twentieth century allow a statistical reconstruction of the cyclic behavior. When a new El Niño event emerges, the statistical model can help identify the observed current state with a specific phase, using it as the starting point to fill in the remaining sequence in the expected life cycle. The statistical model is more useful for capturing the evolution of the oceanic state, especially tropical sea surface temperature. The assistance of a dynamical atmospheric model is sometimes useful for connecting the predicted oceanic state to climate variability over land. In some cases, the predictive skills of this type of two-tier systems are comparable with those of fully coupled ocean–atmosphere models (Goddard et al., 2001).

The reliability of the statistical approach depends critically on the availability of long-term observations that can be used to "train" the statistical models. It is thus understandable that purely statistical models have not been widely used for climate prediction on multidecadal and centennial time scales. For that purpose, century-long observational records are likely needed to make statistical models useful. This is the reason that centennial climate projections in the IPCC reports (IPCC, 2013) rely almost exclusively on dynamical, instead of statistical, climate models.

4.8 Outlook

The last two decades have seen a rapid expansion of the enterprise of climate modeling. Its research and development community has become increasingly multidisciplinary. While the addition of more complicated processes helps enhance the realism of climate models, it also increases computational cost and makes the maintenance and revision of the models more challenging. To put it in perspective, one of the most advanced climate models, the NCAR CESM, now contains about one and a half million lines of codes (Baker et al., 2015). This is in contrast to the classic era when a small group of researchers could work alone to create a model from scratch. Newcomers today will find it essential to establish close collaborations with major climate modeling centers and reach beyond their own discipline to contribute to the mainstream efforts of developing the next generation of models.

At the same time, as the archive of climate model simulations become easier to access due to concerted efforts by CMIP (Meehl et al., 2007; Taylor et al., 2012; CMIP5 data portal is http://cmip-pcmdi.llnl.gov/cmip5/) and major climate modeling centers, individual researchers can now work from their laptops to perform detailed diagnostics for the climate models or develop specialized applications using the model output. These are the areas where individual or small groups of researchers can quickly develop new ideas and make impacts, by providing critical feedback to climate model developers and helping them connect to stakeholders. For example, by connecting the projections made with global climate models to a local air quality or water resource model, one can produce customized information for long-term city planning and resource management. Ultimately, this kind of applications is where the economic value of climate modeling lies.

The climate system is one of the most complicated nonlinear systems ever studied. Although climate model development is driven mainly by the more immediate concerns on economy and sustainability, the models so developed also serve as important platforms to advance basic science. For example, studying the interaction among the subcomponents of the model could reveal new insights on the predictability of a multiscale system. The efforts of improving the accuracy and efficiency of climate models have generated many new ideas in computational science, especially computational fluid dynamics. It is hoped that the discipline of climate modeling will continue to attract not only practically minded scientists and engineers but also those who are interested in fundamental sciences.

References

Ackerman, T.P. and Stokes, G.M. (2003). The atmospheric radiation measurement program. *Physics Today* 56 (1): 38–44.

Arakawa, A. (1966). Computational design for long-term numerical integration of the equations of fluid motion: two-dimensional incompressible flow, part I. *Journal of Computational Physics* 1: 119–143.

Arakawa, A. (2004). The cumulus parameterization problem: past, present, and future. *Journal of Climate* 17: 2493–2525.

Baker, A.H., Hammerling, D.M., Levy, M.N. et al. (2015). A new ensemble-based consistency test for the Community Earth System Model (pyCECT v1.0). *Geoscientific Model Development* 8: 2829–2840.

Baker, N.C. and Huang, H.-P. (2014). A comparative study of precipitation and evaporation between CMIP3 and CMIP5 climate model ensembles in semiarid regions. *Journal of Climate* 27: 3731–3749.

Baran, A.J., Hill, P., Furtado, K. et al. (2014). A coupled cloud physics–radiation parameterization of the bulk optical properties of cirrus and its impact on the Met Office Unified Model Global Atmosphere 5.0 configuration. *Journal of Climate* 27: 7725–7752.

Best, M.J. and Grimmond, C.S.B. (2015). Key conclusions of the first international urban land surface model comparison project. *Bulletin of the American Meteorological Society* 96: 805–819.

Bonan, G.B. (1995). Land-atmosphere interactions for climate system models: coupling biophysical, biogeochemical, and ecosystem dynamical processes. *Remote Sensing of Environment* 51: 57–73.

Brown, A.R. (2004). Resolution dependence of orographic torques. *Quartely Journal of the Royal Meteorological Society* 130: 3029–3046.

Caldwell, P., Chin, H.-N.S., Bader, D.C., and Bala, G. (2008). Evaluation of WRF dynamical downscaling simulation. *Climatic Change* 95: 499–521. doi: 10.1007/s10584-009-9583-5.

CESM Software Engineering Group (2013). *CESM User's Guide (CESM 1.2 Series User's Guide)*, 104. National Center for Atmospheric Research.

Chang, P., Saravanan, R., Ji, L., and Hegerl, G.C. (2000). The effect of local sea surface temperatures on atmospheric circulation over the tropical Atlantic sector. *Journal of Climate* 13: 2195–2216.

Chang, Y.-S., Zhang, S., Rosati, A. et al. (2012). An assessment of oceanic variability for 1960–2010 from the GFDL ensemble coupled data assimilation. *Climate Dynamics* 40 (3–4): 775–803. doi: 10.1007/s00382-012-1412-2.

Chen, F. and Dudhia, J. (2001). Coupling an advanced land surface-hydrology model with the Penn State-NCAR MM5 modeling system, part I: model implementation and sensitivity. *Monthly Weather Review* 129: 569–585.

Compo, G.P., Whitaker, J.S., Sardeshmukh, P.D. et al. (2011). The twentieth century reanalysis project. *Quartely Journal of the Royal Meteorological Society* 137: 1–28.

Cullen, M.J.P. (1974). Integrations of the primitive equations on a sphere using the finite element method. *Quartely Journal of the Royal Meteorological Society* 100: 555–562.

Dee, D.P., Uppala, S.M., Simmons, A.J. et al. (2011). The ERA-interim reanalysis: configuration and performance of the data assimilation system. *Quartely Journal of the Royal Meteorological Society* 137: 553–597.

Dolinar, E.K., Dong, X., Xi, B. et al. (2015). Evaluation of CMIP5 simulated clouds and TOA radiation budgets using NASA satellite observations. *Climate Dynamics* 44: 2229–2247.

Durran, D.R. (2010). *Numerical Methods for Fluid Dynamics with Applications to Geophysics*, 2e, 516. New York, NY: Springer.

Dutkiewicz, S., Follows, M.J., and Bragg, J.G. (2009). Modeling the coupling of ocean ecology and biogeochemistry. *Global Biogeochemical Cycles* 23: GB4017. doi: 10.1029/2008GB003405.

Edwards, J.M. and Slingo, A. (2006). Studies with a flexible new radiation code. I: choosing a configuration for a large-scale model. *Quartely Journal of the Royal Meteorological Society* 122: 689–719.

Ferziger, J.H. and Peric, M. (2002). *Computational Methods for Fluid Dynamics*, 3e, 423. New York, NY: Springer.

Fyfe, J.C., Gillett, N.P., and Zwiers, F.W. (2013). Overestimated global warming the past two decades. *Nature Climate Change* 3: 767–769. doi: 10.1038/nclimate1972.

Georgescu, M., Miguez-Macho, G., Steyaert, L.T., and Weaver, C.P. (2009). Climatic effects of 30 years of landscape change over the Greater Phoenix, Arizona, region: 1. Surface energy budget changes. *Journal of Geophysical Research* 114: D05110. doi: 10.1029/2008JD010745.

Goddard, L., Mason, S.J., Zebiak, S.E. et al. (2001). Current approaches to seasonal to interannual climate predictions. *International Journal of Climatology* 21: 1111–1152.

Godfrey, J.S. (1996). The effect of the Indonesian throughflow on ocean circulation and heat exchange with the atmosphere: a review. *Journal of Geophysical Research* 101: 12217–12237.

Griffies, S.M., Boening, C., Bryan, F.O. et al. (2000). Developments in ocean climate modelling. *Ocean Modeling* 2: 123–192.

Hallberg, R. (1997). *HIM: The Hallberg Isopycnal Coordinate Primitive Equation Model*, 39. Princeton, NJ: Geophysical Fluid Dynamics Laboratory.

Hamill, T.M., Bates, G.T., Whitaker, J.S. et al. (2013). NOAA'S second-generation global medium-range ensemble reforecast dataset. *Bulletin of the American Meteorological Society* 94: 1553–1565.

Hasselmann, K. (1976). Stochastic climate models, part I: theory. *Tellus* 28: 473–484.

Hasselmann, K. (1988). PIPs and POPs: the reduction of complex dynamical systems using principle interaction and oscillating patterns. *Journal of Geophysical Reserarch* 93: 11015–11021.

Heikes, R. and Randall, D.A. (1995). Numerical integration of the shallow-water equations on a twisted icosahedral grid. Part I: basic design and results of tests. *Monthly Weather Review* 123: 1862–1880.

Heikkila, U., Sandvik, A., and Sorteberg, A. (2011). Dynamical downscaling of ERA-40 in complex terrain using the WRF regional climate model. *Climate Dynamics* 37: 1551–1564. doi: 10.1007/s00382-010-0928-6.

Held, I.M. and Suarez, M.J. (1994). A proposal for the intercomparison of the dynamical cores of atmospheric general circulation models. *Bulletin of the American Meteorological Society* 75: 1825–1830.

Hong, S.-Y. and Pan, H.-L. (1996). Non-local vertical boundary layer diffusion in a medium-range forecast model. *Monthly Weather Review* 124: 2322–2339.

Hsu, Y.-J.G. and Arakawa, A. (1990). Numerical modeling of the atmosphere with an isentropic vertical coordinate. *Monthly Weather Review* 118: 1933–1959.

Huang, H.-P., Robertson, A.W., Kushnir, Y., and Peng, S. (2009). Hindcast of tropical Atlantic SST gradient and South American precipitation: the influences of the ENSO forcing and the Atlantic preconditioning. *Journal of Climate* 22: 2405–2421.

Huang, H.-P., Sardeshmukh, P.D., and Weickmann, K.M. (1999). The balance of global angular momentum in a long-term atmospheric data set. *Journal of Geophysical Research* 104: 2031–2040.

Hunke, E. C., W. H. Lipscomb, A. K. Turner, et al. 2015: CICE: The Los Alamos Sea Ice Model Documentation and Software User's Manual Version 5.1, Los Alamos National Laboratory Report LA-CC-06-012, 116 pp.

IPCC (2000). *Special Reports on Emissions Scenarios* (ed. N. Nakicenovic and R. Swart), 570. Cambridge, UK: Cambridge University Press.

IPCC (2001). *Climate Change 2001: The Scientific Basis. Contribution of Working Group I to the Third Assessment Report of the Intergovernmental Panel on Climate Change* (ed. J.T. Houghton, Y. Ding, D.J. Griggs, et al.), 881. Cambridge, UK and New York, NY: Cambridge University Press.

IPCC (2013). *Climate Change 2013: The Physical Science Basis. Contribution of Working Group I to the Fifth Assessment Report of the Intergovernmental Panel on Climate Change* (ed. T.F. Stocker, D. Qin, G.-K. Plattner, et al.), 1535. Cambridge, UK and New York, NY: Cambridge University Press.

Jones, C., Robertson, E., Arora, V. et al. (2013). Twenty-first century compatible CO_2 emissions and airborne fraction simulated by CMIP5 Earth system models under four representative concentration pathways. *Journal of Climate* 26: 4398–4413.

Juang, H.-M.H., Hong, S.-Y., and Kanamitsu, M. (1997). The NCEP regional spectral model: an update. *Bulletin of the American Meteorological Society* 78: 2125–2143.

Kalnay, E., Kanamitsu, M., Kistler, R. et al. (1996). The NCEP/NCAR 40-year reanalysis project. *Bulletin of the American Meteorological Society* 77: 437–470.

Kamal, S., Huang, H.-P., and Myint, S.W. (2015). The influence of urbanization on the climate of Las Vegas metropolitan area: a numerical study. *Journal of Applied Meteorology and Climatology* 54: 2157–2177.

Karnauskas, K.B., Murtugudde, R., and Busalacchi, A.J. (2008). The effect of the Galápagos Islands on ENSO in forced ocean and hybrid coupled models. *Journal of Physical Oceanography* 38: 2519–2534.

Keenlyside, N.S., Latif, M., Jungclaus, J. et al. (2008). Advancing decadal-scale climate prediction in the North Atlantic sector. *Nature* 453: 84–88. doi: 10.1038/nature06921.

Kim, Y.-J., Eckermann, S.D., and Chun, H.-Y. (2003). An overview of the past, present and future of gravity-wave drag parametrization for numerical climate and weather prediction models. *Atmosphere-Ocean* 41: 65–98.

Kim, H.-M., Webster, P.J., and Curry, J.A. (2012). Evaluation of short-term climate change prediction in multi-model CMIP5 decadal hindcasts. *Geophysical Research Letters* 39. doi: 10.1029/2012GL051644.

Knutti, R. (2010). The end of model democracy? *Climatic Change* 102: 395–404.

Köhl, A. and Stammer, D. (2008). Variability of the meridional overturning in the North Atlantic from the 50 years GECCO state estimation. *Journal of Physical Oceanography* 38: 1913–1930.

Kusaka, H., Hara, M., and Takane, Y. (2012). Urban climate projection by the WRF model at 3 km horizontal grid increment: dynamical downscaling and predicting heat stress in the 2070's August for Tokyo, Osaka, and Nagoya metropolises. *Journal of the Meteorological Society of Japan* 90B: 47–63. doi: 10.2151/jmsj.2012-B04.

Kusaka, H. and Kimura, F. (2004). Coupling a single-layer urban canopy model with a simple atmospheric model: impact on urban heat island simulation for an idealized case. *Journal of the Meteorological Society of Japan* 82: 67–80.

Large, W., McWilliams, J.C., and Doney, S. (1994). Oceanic vertical mixing: a review and a model with nonlocal boundary layer parameterization. *Reviews of Geophysics* 32: 363–403.

Lau, N.-C., Leetma, A., and Nath, M.J. (2006). Attribution of atmospheric variations in the 1997–2003 period to SST anomalies in the Pacific and Indian Ocean basins. *Journal of Climate* 19: 3507–3628.

Lin, J.-L. (2007). The double-ITCZ problem in IPCC AR4 coupled GCMs: ocean–atmosphere feedback analysis. *Journal of Climate* 20: 4497–4525.

Lipscomb, W., Bindschadler, R., Bueler, E. et al. (2009). A community ice sheet model for sea-level prediction. *Eos, Transactions American Geophysical Union* 90: 23.

Lorenz, E.N. (1982). Atmospheric predictability experiments with a large numerical model. *Tellus* 34: 505–513.

Machenhauer, B. (1979). The spectral method. In: *Numerical Methods Used in Atmospheric Models*, GARP Publication Series No. 17, Global Atmospheric Research Program, vol. 2, 121–275. World Meteorological Organization.

Madec, G. and Imbard, M. (1996). A global ocean mesh to overcome the North Pole singularity. *Climate Dynamics* 12: 381–388.

Madec, G., and the NEMO team, 2008: NEMO Ocean Engine. Note du Pôle de modélisation, Institut Pierre-Simon Laplace (IPSL), France, No 27, ISSN No 1288-1619.

Magnusson, L., Alonso-Balmaseda, M., Corti, S. et al. (2013). Evaluation of forecast strategies for seasonal and decadal forecasts in presence of systematic model errors. *Climate Dynamics* 41: 2393–2409.

Manabe, S. and Bryan, K. (1969). Climate calculations with a combined ocean-atmosphere model. *Journal of the Atmospheric Sciences* 26: 786–789.

Manabe, S. and Wetherald, R.T. (1967). Thermal equilibrium of the atmosphere with a given distribution of relative humidity. *Journal of the Atmospheric Sciences* 24: 241–259.

McFarlane, N.A. (1987). The effect of orographically excited gravity-wave drag on the general circulation of the lower stratosphere and troposphere. *Journal of the Atmospheric Sciences* 44: 1775–1800.

Medeiros, B., Stevens, B., Held, I.M. et al. (2008). Aquaplanets, climate sensitivity, and low clouds. *Journal of Climate* 21: 4974–4991.

Meehl, G.A., Covey, C., Taylor, K.E. et al. (2007). The WCRP CMIP3 multimodel dataset: a new era in climate change research. *Bulletin of the American Meteorological Society* 88: 1383–1394.

Mellor, G.L. and Yamada, T. (1982). Development of a turbulence closure model for geophysical fluid problems. *Reviews of Geophysics* 20: 851–875.

Mesinger, F. and Arakawa, A. (1976). *Numerical Methods Used in Atmospheric Models*, GARP Publication Series No. 17, Global Atmospheric Research Program, vol. 1, 64. World Meteorological Organization.

Mlynczak, P.E., Smith, G.L., and Doelling, D.R. (2011). The annual cycle of Earth radiation budget from clouds and the Earth's radiant energy system (CERES) data. *Journal of Applied Meteorology and Climatology* 50: 2490–2503.

Moore, J.K., Doney, S.C., and Lindsay, K. (2004). Upper ocean ecosystem dynamics and iron cycling in a global three-dimensional model. *Global Biogeochemical Cycles* 18: GB4028. doi: 10.1029/2004GB002220.

Moss, R.H., Edmonds, J.A., Hibbard, K.A. et al. (2010). The next generation of scenarios for climate change research and assessment. *Nature* 463: 747–756. doi: 10.1038/nature08823.

Neale, R. B., Chen, C-C, Gettelman, A., et al., 2012: *Description of the NCAR Community Atmosphere Model (CAM 5.0)*, NCAR Technical Note TN-486+STR. National Center for Atmospheric Research, Boulder, CO, 274 pp.

Neale, R.B. and Hoskins, B.J. (2001). A standard test for AGCMs including their physical parametrizations: I: the proposal. *Atmospheric Science Letters* 1. doi: 10.1006/asle.2000.0019.

Oleson, K. W., Lawrence, D.M, Bonan, G.B., et al., 2010: *Technical Description of Version 4.0 of the Community Land Model (CLM)*, NCAR Technical Note, NCAR/TN-478+STR. National Center for Atmospheric Research, Boulder, CO, 266 pp.

Orszag, S.A. (1970). Transform method for the calculation of vector-coupled sums: application to the spectral form of the vorticity equation. *Journal of Atmospheric Science* 27: 890–895.

Oueslati, B. and Bellon, G. (2015). The double ITCZ bias in CMIP5 models: interaction between SST, large-scale circulation and precipitation. *Climate Dynamics* 44: 585–607.

Paek, H. and Huang, H.-P. (2013). Centennial trend and decadal-to-interdecadal variability of atmospheric angular momentum in CMIP3 and CMIP5 simulations. *Journal of Climate* 26: 3846–3864. doi: 10.1175/JCLI-D-12-00515.1.

Palmer, T.N., Shutts, G.J., and Swinbank, R. (1986). Alleviation of a systematic westerly bias in circulation and numerical weather prediction models through an orographic gravity-wave drag parameterization. *Quartely Journal of the Royal Meteorological Society* 112: 1001–1039.

Prein, A.F., Langhans, W., Fosser, G. et al. (2015). A review on regional convection-permitting climate modeling: demonstrations, prospects, and challenges. *Reviews of Geophysics* 53. doi: 10.1002/2014RG000475.

Randall, D. (2015). *An Introduction to the Global Circulation of the Atmosphere*, 442. Princeton and Oxford: Princeton University Press.

Rayner, N.A., Parker, D.E., Horton, E.B. et al. (2003). Global analyses of sea surface temperature, sea ice, and night marine air temperature since the late nineteenth century. *Journal of Geophysical Research* 108: 4407. doi: 10.1029/2002JD002670.

Reichler, T. and Kim, J. (2008). How well do coupled models simulate today's climate? *Bulletin of the American Meteorological Society* 89: 303–311. doi: 10.1175/BAMS-89-3-303.

Reynolds, R.W., Rayner, T.M., Smith, D.C. et al. (2002). An improved in situ and satellite SST analysis for climate. *Journal of Climate* 15: 1609–1625.

Rienecker, M.M., Suarez, M.J., Gelaro, R. et al. (2011). MERRA: NASA's modern-era retrospective analysis for research and applications. *Journal of Climate* 24: 3624–3648.

Ronchi, C., Iacono, R., and Paolucci, P.S. (1996). The "cubed sphere": a new method for the solution of partial differential equations in spherical geometry. *Journal of Computational Physics* 124: 93–114.

Ropelewski, C.F. and Halpert, M.S. (1989). Precipitation patterns associated with the high index phase of the Southern Oscillation. *Journal of Climate* 2: 268–284.

Saha, S., Moorthi, S., Pan, H.-L. et al. (2010). The NCEP climate forecast system reanalysis. *Bulletin of the American Meteorological Society* 91: 1015–1057.

Seager, R., Ting, M., Held, I.M. et al. (2007). Model projections of an imminent transition to a more arid climate in southwestern North America. *Science* 316: 1181–1184.

Senior, C.A., Arribas, A., Brown, A.R. et al. (2011). Synergies between numerical weather prediction and general circulation climate models. In: *The Development of Atmospheric General Circulation Models* (ed. L. Donner, W. Schubert and R. Somerville), 76–116. Cambridge University Press.

Sharma, A. and Huang, H.-P. (2012). Regional climate simulation for Arizona: impact of resolution on precipitation. *Advances in Meteorology* 505726. doi: 10.1155/2012/505726.

Skamarock, W.C., Klemp, J.B., Dudhia, J. et al. (2008). *A Description of the Advanced Research WRF Version 3, NCAR Technical Note TN-475+STR*, 125. Boulder, CO: National Center for Atmospheric Research.

Smagorinsky, J. (1963). General circulation experiments with the primitive equations, I: the basic experiment. *Monthly Weather Review* 91: 99–164.

Smith, R., Jones, P., Briegleb, B., et al., 2010: *The Parallel Ocean Program (POP) Reference Manual: Ocean Component of the Community Climate System Model (CCSM) and Community Earth System Model (CESM)*, Los Alamos National Laboratory, LAUR-10-01853, 141 pp.

Sperber, K.R., Annamalai, H., Kang, I.-S. et al. (2013). The Asian summer monsoon: an intercomparison of CMIP5 vs. CMIP3 simulations of the late 20th century. *Climate Dynamics* 41: 2711–2744.

Stensrud, D.J. (2007). *Parameterization Schemes: Keys to Understanding Numerical Weather Prediction Models*, 459. Cambridge, UK: Cambridge University Press.

Stephens, G.L. (1984). The parameterization of radiation for numerical weather prediction and climate models. *Monthly Weather Review* 112: 826–867.

Stroeve, J.C., Kattsov, V., Barrett, A. et al. (2012). Trends in Arctic sea ice extent from CMIP5, CMIP3 and observations. *Geophysical Research Letters* 39: L16502. doi: 10.1029/2012GL052676.

Sukoriansky, S., Galperin, B., and Perov, V. (2005). Application of a new spectral theory of stably stratified turbulence to the atmospheric boundary layer over sea ice. *Boundary-Layer Meteorology* 117: 231–257.

Taylor, K.E., Stouffer, R.J., and Meehl, G.A. (2012). An overview of CMIP5 and the experimental design. *Bulletin of the American Meteorological Society* 93: 485–498. doi: 10.1175/BAMS-D-11-00094.1.

Thomas, G.E. and Stamnes, K. (2002). *Radiative Transfer in the Atmosphere and Ocean*, 548. Cambridge, UK: Cambridge University Press.

Trenberth, K.E., Fasullo, J.T., and Kiehl, J. (2009). Earth's global radiation budget. *Bulletin of the American Meteorological Society* 90: 311–323.

Van den Dool, H. (2007). *Empirical Methods in Short-Term Climate Prediction*, 240. Oxford: Oxford University Press.

Van Oldenborgh, G.J., Doblas-Reyes, F.J., Wouters, B., and Hazeleger, W. (2012). Decadal prediction skill in a multi-model ensemble. *Climate Dynamics* 38: 1263–1280.

Wang, W., Chen, M., and Kumar, A. (2010). An assessment of the CFS real-time seasonal forecasts. *Weather and Forecasting* 25: 950–969.

Williamson, D.L. (2008). Convergence of aqua-planet simulations with increasing resolution in the Community Atmospheric Model, version 3. *Tellus A* 60: 848–864.

Wiscomb, W.J. (2005). Scales, tools, and reminiscences. In: *3D Radiative Transfer in Cloudy Atmosphere* (ed. A. Marshak and A.B. Davis), 3–92. Berlin: Springer.

Wu, Y., Ting, M., Seager, R. et al. (2011). Changes in storm tracks in the warmer climate simulated by the GFDL CM2.1 model. *Climate Dynamics* 37: 53–72.

Yang, S., Jiang, Y., Zheng, D. et al. (2009). Variations of US regional precipitation and simulations by the NCEP CFS: focus on the southwest. *Journal of Climate* 22: 3211–3231.

Zuo, H., Balmaseda, M.A., and Mogensen, K. (2015). The new eddy-permitting ORAP5 ocean reanalysis: description, evaluation and uncertainties in climate signals. *Climate Dynamics* 45. doi: 10.1007/s00382-015-2675-1.

5

Climate Change Impact Analysis for the Environmental Engineer

Panshu Zhao[1], John R. Giardino[2], and Kevin R. Gamache[3]

[1] *Water Management and Hydrological Science Graduate Program and High Alpine and Arctic Research Program (HAARP), Texas A&M University, College Station, TX, USA*
[2] *Water Management and Hydrological Science Program and High Alpine and Arctic Research Program (HAARP), Department of Geology and Geophysics, Texas A&M University, College Station, TX, USA*
[3] *Water Management and Hydrological Science Program and High Alpine and Arctic Research Program (HAARP), The Bush School of Government and Public Service, Texas A&M University, College Station, TX, USA*

5.1 Introduction

More than half of the land surface of the Earth has been "plowed, pastured, fertilized, irrigated, drained, fumigated, bulldozed, compacted, eroded, reconstructed, manured, mined, logged, or converted to new uses" (Richter and Mobley, 2009). In less than three centuries, 46 million acres of the virgin landscape in America has been converted to urban uses, and in the next 25 years that number will more than double to 112 million acres (Carbonell and Yaro, 2005). Activities like these have far reaching impacts on life-sustaining processes of the near-surface environment, recently termed the "critical zone" (Richter and Mobley, 2009), and if the current rate of land transformation continues, it is unsustainable.

The biological and human systems of the Earth are already beginning to undergo transformation because of climate change. According to Crutzen (2002), rapid changes occurring on Earth since the beginning of the industrial revolution have steered us into a new geological epoch referred to as the Anthropocene (Amundson et al., 2007). The Anthropocene (~250 years BP to present) encompasses some of the most noticeable changes in the history of the Earth by any measurement: extinction rates, the extent of climate change, and so forth (Amundson et al., 2007).

Driven by these myriad global changes caused by human interaction with the natural environment, the critical zone concept was conceived in 1998 to represent the importance of system science in integrating the research of the four scientific spheres (lithosphere, hydrosphere, biosphere, and atmosphere) at the surface of the Earth by studying the linkages, feedbacks, and processes that occurred in the past, are occurring today, and will operate in the future.

The critical zone is the area of the surface and near-surface systems that extends from bedrock to the atmosphere boundary layer (Anderson et al., 2010; NRC, 2001; Figure 5.1). It lies at the interface of the lithosphere, atmosphere, and hydrosphere (Amundson et al., 2007) and encompasses soils and terrestrial ecosystems. It is a complex mixture of air, water, biota, organic matter, and earth materials (Brantley et al., 2007).

Future global change has some implications for the critical zone because of changes in such phenomena as rates of evapotranspiration, precipitation characteristics, plant distributions, and human responses (Goudie, 2006). Global climate models predict a warmer planet. This prediction could mean changes to our climate – specifically temperature, evaporation, rainfall, and drought (Mace and Wade, 2008). At the same time, rapidly growing demands in urban areas are already straining local and regional water supplies, however, and concerns over urban water scarcity in the United States are becoming more prominent (Levin et al., 2002; Padowski and Jawitz, 2012). Water shortages in Atlanta, GA, in 2008, and San Francisco, CA, in 2006–2007 (Dorfman et al., 2011; Padowski and Jawitz, 2012) are illustrative of the impacts climate change can have on population growth, environmental regulation, and water supplies, among others.

In the United States, more than 80% of the population now lives in urban areas compared with 64% in 1950

Handbook of Environmental Engineering, First Edition. Edited by Myer Kutz.

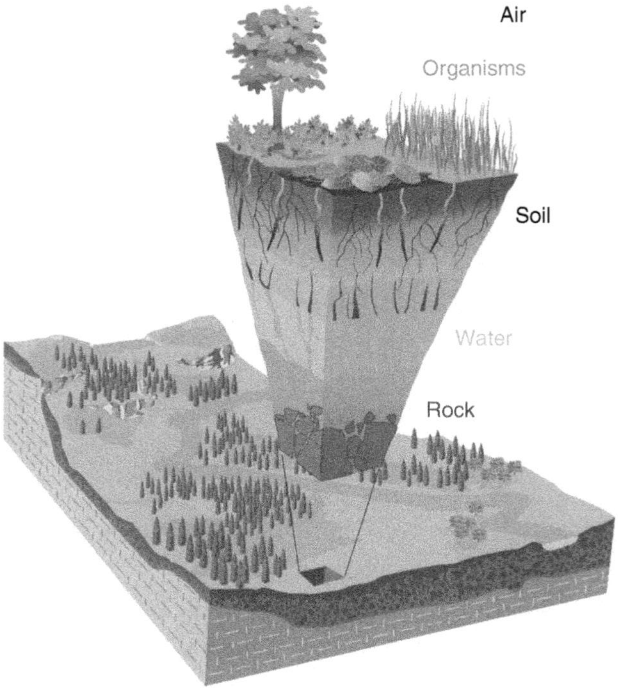

Figure 5.1 The different components of the critical zone (NRC, 2001).

(Padowski and Jawitz, 2012). The US population will likely increase 40% by 2050 with the growth concentrated in 8–10 megaregions (Dewar and Epstein, 2007). A megaregion consists of two or more metropolitan areas linked with interdependent environmental systems, a multimodal transportation infrastructure, and complementary economies (Butler et al., 2009; Zhang et al., 2007). Ensuring that cities have an adequate supply of water and other resources will become increasingly important as human populations continue to concentrate in these highly urbanized megaregions.

Threats to this essential component, the life-support system, have reached an acute level, yet the science of understanding and managing these threats mostly remains embedded within individual disciplines, as well as largely remaining qualitative. Never has a more important time occurred for an international, interdisciplinary approach to accelerate our understanding of processes in the critical zone and how to intervene positively to mitigate threats and to ensure the healthful function of the critical zone. Because without understanding and action, the impact of climate change can catastrophically harm the environment of the Earth. Therefore, this chapter focuses on the interaction between humans and the environment of the Earth through the focus on the concepts, methods, and technologies presently available to analyze the impact of climate change.

5.2 Earth System's Critical Zone

5.2.1 What Is the Critical Zone?

Although the critical zone is a useful concept, it most certainly is not a new or novel idea. The idea of the critical zone is an innovative means of demonstrating the connection and flow of mass and energy from the top of the canopy to the bottom of the aquifer using a general system theory approach.

The critical zone has been defined as "...the heterogeneous, near-surface environment in which complex interactions involving rock, soil, water, air, and living organisms regulate the natural habitat and determine the availability of life-sustaining resources." It has evolved as a dynamic and self-sustaining system (Amundson et al., 2007). This thin, fragile envelope that includes the land surface and its canopy of vegetation, rivers, lakes, and shallow seas (Wilding and Lin, 2006) is critical from a human perspective because it is the environment in which most people live and work (Graf, 2008).

The critical zone sustains most of Earth's surface life including humanity. It is the thin outer veneer at the surface of the Earth extending from the top of the vegetation canopy through the soil to the subsurface depths at which fresh groundwater freely circulates. This is the zone where most terrestrial life resides. The National Research Council (NRC) defined the critical zone in 2001 as "...the heterogeneous, near surface environment in which complex interactions involving rock, soil, water, air and living organisms regulate the natural habitat and determine the availability of life-sustaining resources." It is a constantly evolving boundary layer where rock, soil, water, air, and living organisms interact. These complex interactions regulate the natural habitat and regulate the availability of life-sustaining resources, including the production of food and water quality. Complex biogeochemical–physical processes combine in the critical zone to transform rock and biomass into soil that in turn supports much of the terrestrial biosphere.

The NRC did not create the critical zone term. It has been around since at least 1909 when D.E. Tsakalotos used the term to describe the binary mixture of two fluids. The precursor application of the term critical zone to this relatively thin zone of the Earth was first suggested by Gail Ashley in 1998 when she noted that the term applied where the soil connects the vegetation canopy to the soil, the soil connects to the weathered materials, the weathered materials connect to bedrock, and the bedrock provides the connection to the aquifer.

It is quite possible that the definition of the critical zone will be modified in the future. The depth will not be confined by the bottom of the aquifer, but will be extended to include the depth of human impact. As we search for resources, the depths that humans mine or drill will continue to increase. Thus, the impacts on human society will reach further into the Earth.

5.2.2 Tying the Critical Zone Together

As the critical zone was being defined, three recommendations associated with the critical zone were tenured by the NRC: first, make soil science stronger; second, continue to build hydrological sciences; and finally, strengthen the study of coastal zone processes. With these recommendations, soil, and not water, became the thread that ties the various systems of the critical zone together.

A focus on world hunger and increasing rates of soil erosion provided a strong foundation for soil being viewed as the link between soil erosion, food production, and worldwide human welfare. While soil certainly plays an important role, it is important to understand that the soil profile connects the vegetation canopy to the soil, the soil connects to the weathered materials, the weathered materials connect to bedrock, and the bedrock provides the connection to the aquifer. These pathways facilitate flows (i.e. energy, mass, biogeochemical) from the top to the bottom and vice versa. However, none of these energy and material flows would be possible without water. Thus, we firmly argue that the link between all these components or systems is water rather than soil.

Water is the true thread that stitches these systems together. It is water that travels through the atmosphere, to and through the biosphere, and into the lithosphere where it is stored as surface water, soil water, and groundwater. Water is a key component in chemical weathering processes, a sustainer of life, and is responsible for floods or absent to cause drought. While the National Science Foundation (NSF) never mentioned the above concern, we believe that they also saw the "thread problem" when they recommended that focus needs to be brought to water as the unifying theme in the study of complex environmental systems.

5.2.3 Addressing the Fundamental Questions about Critical Zone Processes

Critical zone is a term that has focused the scientific community and the layperson on the critical role this relatively thin zone plays in the existence of life on Earth, but despite the critical zone's importance to terrestrial life, critical zone processes remain poorly understood. The physical, chemical, and biological processes are highly nonlinear and can range across scales from atomic to global and from seconds to eons. Thus, very little is known about how these processes are coupled and at what spatial and temporal scales they interact. Understanding the complex web of physical, chemical, and biological processes of the critical zone calls for a systems approach across a broad array of sciences: hydrology, geology, soil science, biology, ecology, geochemistry, geomorphology, and more.

The immediate challenge is to develop a robust predictive ability for how the critical zone attributes, processes, and outputs will respond to projected climate and land-use changes to guide societal adaptations toward a more sustainable future. Some of the fundamental questions that must be answered include:

Short-term processes and impacts:

- What controls the resilience, response, and recovery of the critical zone and its integrated geophysical–geochemical–ecological functions to perturbations such as climate and land-use changes, and how can this be quantified by observations and predicted by mathematical modeling of the interconnected physical, chemical, and biological processes and their interactions?
- How can sensing technology, e-infrastructure, and modeling be integrated for simulation and forecasting of essential terrestrial variables for water supplies, food production, biodiversity, and other major benefits?
- How can theory, data, and mathematical models from the natural and social sciences, engineering, and technology be integrated to simulate, value, and manage critical zone goods and services and their benefits to people?

Long-term processes and impacts:

- How has geological evolution and paleobiology of the critical zone established ecosystem functions and the foundations of critical zone sustainability?
- How do molecular-scale interactions between critical zone processes dictate the linkages in flows and transformations of energy, material, and genetic information across the vertical extent of aboveground vegetation, soils, aquatic systems, and regolith and influence the development of watersheds and aquifers as integrated in ecological–geophysical units?
- How can theory and data be combined from molecular to global scales to interpret past transformations of the Earth's surface and forecast critical zone evolution and its planetary impact?

Solving these six big science questions requires integrative research spanning traditionally siloed disciplines and the long-term studies that are hierarchically structured in both space and time.

This predictive ability must be based on broad knowledge of the critical zone system and processes to describe the interactions of the varied climatic, ecologic, and geologic factors distinguishing different geographic regions. This is a primary focus of the scientific and educational efforts at Critical Zone Observatories (Critical Zone Observatory). By identifying shared or "common" research questions across these Critical Zone Observatories, an opportunity arises to advance new understanding on key issues.

5.2.4 Critical Zone Observatories

The NRC's committee on "Basic Research Opportunities in Earth Sciences" and the NSF both took a bold step in addressing this research challenge by establishing the Critical Zone Observatory program in 2005 to provide new information on the short-term, near-surface processes. Today some 68 critical zone observatories have been established around the world by various governments. For a complete listing of these observatories (see Giardino and Houser, 2015).

This newfound focus on the critical zone is resulting in massive amounts of data being collected by a wide range of scientific disciplines. It is resulting in the development of new methods to assemble and model these data. It is also providing new methods for integrating data across Critical Zone Observatories to model large-scale changes in the Earth. Finally, it is engaging researchers from disparate fields to revisit the idea of interdisciplinary research and to view Earth processes from a system's perspective.

A revolution in the way our planet is viewed and data are collected occurs through a truly interdisciplinary approach that expands the spatial extent and connections of critical zone research formalized by geographical zone linkages between and among all the Critical Zone Observation sites around the globe. Critical Zone Observatories provide the ability to transform our understanding of the Earth surface couple processes, the impacts of climate and land-use change, and the value of critical zone functions and services. Research at the Critical Zone Observatory scale seeks to understand these little-known coupled processes through monitoring of streams, climate/weather, and groundwater, along with other processes. COs are instrumented for hydrogeochemical measurements and are sampled for soil, canopy, and bedrock materials, for example. Critical Zone Observatories involve teams of cross-disciplinary scientists whose research is motivated and implemented by both field and theoretical approaches and include substantial education and outreach.

The Critical Zone Observatory research objectives are to identify and measure fluxes that are occurring, determine their cumulative impact, and then use these data to develop models that will be accurate in both short- and long-term forecasts of change.

Critical Zone Observatory research may be summarized by the following three general shared questions:

1) What controls critical zone properties and processes?
2) What will be the response of the critical zone structure, and its stores and fluxes, to climate and land-use change?
3) How can improved understanding of the critical zone be used to enhance ecosystem resilience and sustainability and restore ecosystem function?

Meeting the goals of critical zone science requires researchers with deep disciplinary knowledge as well as collaboration and integration across multiple disciplines. An interdisciplinary approach across a broad array of sciences, especially in the geological and biological sciences, is also applied to better understand the complex interconnection of critical zone processes. Examples include pedology, hydrology, geology, geophysics, ecology, biogeochemistry, and geomorphology.

Close collaboration has required the development of infrastructure and protocols that allow measurements to be directly compared across Critical Zone Observatories. Critical Zone Observatories provide essential datasets and a coordinated community of researchers that integrate hydrological, ecological, geochemical, and geomorphic process sciences from grain to watershed scales and, perhaps as importantly, from deep time to human time scales. The 10 US Critical Zone Observatories represent a wide range of geological, climatic, and land-use settings that provide an opportunity to develop a broad and general understanding of the evolution and function of the critical zone. Within the US Critical Zone Observatory program, the objectives are largely focused on developing methods to quantitatively project the dynamics of the Earth surface processes from the past into the future. Critical Zone Observatories provide the opportunity not only to focus on the processes and fluxes operating today but also to compare these to the record of the processes in the rock and soil and sediment record and then to use quantitative models developed from these observations across scales of space and time to project the future.

Of all the environmental observatories and networks, the Critical Zone Observatories are the only ones to tightly integrate ecological and geological sciences to combine with computational simulation and to project from the deep geologic past to that of human life spans. The network of Critical Zone Observatories will empower an array of new approaches to scientific investigations that were not considered by the traditional NSF funding mechanism. They suggest that this new approach will lead to creative and stimulating efforts in understanding the fundamental operations of the important systems of the Earth. Because of the diversity of the sites, as well as the diversity of the researchers who collectively bring the essential observational and theoretical skills and knowledge to common problems, this network of Critical Zone Observatories can accomplish what no one observatory can provide.

A review of the literature surrounding the critical cone indicating research is being produced by the various Critical Zone Observatories demonstrates that the scientific community has begun to engage in truly interdisciplinary studies of single locations, but the larger effort is still very discipline focused and published in discipline-specific journals.

Human-driven changes cause multiple effects that cascade through the Earth system in complex ways. These effects interact with each other and with local- and regional-scale changes in multidimensional patterns that are difficult to understand and even more difficult to predict. It is in this arena that the Critical Zone Observatories will play a significant role. The Critical Zone Observatories are intended to identify flow pathways and energy transformations at various scales. Through this research, they will be creating a continuum of change that will provide keys to understanding the linkages between the components in Earth systems. Many of the Critical Zone Observatories are employing a hindcasting approach to accomplish a forecasting approach to their research. This type of research will provide the data to predict and understand future change and global impacts.

5.2.5 Summary

The central theme of critical zone science has captured the imagination of scientists worldwide: to learn to measure the myriad processes occurring today in the critical zone and to relate these to the history of these processes that is recorded in the soil and rock record. In the future, we will experience huge pressures to grow food and provide clean water to a growing population. Increased pressure will be put on the importance of the critical zone for the production of food as well as providing water resources. The term "critical zone" will become a common word in both the scientific and the layperson communities, and everyone will come to understand and appreciate the critical role this relatively thin zone performs in the existence of life on Earth, as we know it today.

While much of the science being studied at Critical Zone Observatories around the world is already proven science, the term critical zone is driving a new awareness of interdisciplinary study of these locations. Critical Zone Observatories foster conventional studies, governance, and coordination of data on a global scale. Ultimately the contributions from the Critical Zone Observatory network will be fostering understanding of the relationship between the strength of a response, the time of recovery, and the overall resilience to environmental perturbations of our planet.

Understanding the flows of energy and mass from one subsystem to another will be fundamental in refining the accuracy of the predictive models developed from critical zone research.

While soil plays an important role in critical zone processes, the major connection between all the components or systems is water. For this reason, we believe research concentration on fluxes will begin to show that water is the true thread that stitches these systems together.

Perhaps one of the most significant aspects of the global Critical Zone Observatory network will be the development of reliable data to support informed policy and management of constrained resources of the Earth.

The critical zone approach could prove to play a major role as geoscientists continue their quest for unifying principle and the means to move toward integrative studies of the biophysical environment and the role of humans within that environment.

Human-driven changes cause multiple effects that cascade through the Earth system in complex ways. These effects interact with each other and with local- and regional-scale changes in multidimensional patterns that are difficult to understand and even harder to predict. It is in this arena that the Critical Zone Observatories will play a significant role. The Critical Zone Observatories are intended to identify flow pathways and energy transformations at various scales. Through this research, they will be creating a continuum of change that will provide keys to understanding the linkages between the components in Earth systems. Many of the Critical Zone Observatories are employing a hindcasting approach to accomplish a forecasting approach to their research. This type of the investigation will provide the data to predict and understand future change and global impacts.

5.3 Perception, Risk, and Hazard

5.3.1 Introduction

Understanding the usefulness of impact analysis of climate change requires familiarity with how humans perceive risks and hazards. Thus, the focus of this section is a discussion of risk and various categories of hazards.

A phenomenon that can threaten life, property, or environmental quality can be defined as a hazard, which can be confined by time, location, and frequency of occurrence. The potential that a location can have a specific hazard is referred to as the hazard probability. The risk then is a combination of hazard and its probability (Varnes, 1984).

The lifespan impact of a hazard has four distinct phases: dormant, potential, active, and mitigation. The dormant phase describes an instance in time where the factors that could lead to hazard potential are limited. For example, a hill slope becomes saturated from a heavy rainfall and is prime for failure. However, the stable local geology and abundant vegetation cover mitigate the hazard. The second stage is the potential phase when indicative of time, hazard-inducing factors increase. From the previous example, the possible phase would occur if the strength component of the geology becomes compromised and vegetation cover is eliminated through natural processes or land-use change. The active phase, which is the third stage, is a time when the hazard occurs. The fourth and final stage is the mitigation phase of a hazard, which consists of stakeholders taking actions to reduce hazard potential and damage (MacCollum, 2007).

Hazards can also be classified by their causes, encompassing geophysical, anthropogenic, and climatic. Geophysical hazards include mass movement, land subsidence, earthquakes, floods, and volcanic eruptions.

Some examples of anthropogenic hazards include deforestation, water pollution, haze, increased surface runoff associated with impervious coverings in the built environment, soil erosion related to agricultural practices, and groundwater withdrawal and human-induced desertification.

Climatic-related hazards range from forest fires to floods, avalanches, permafrost degradation, sea level rise, coastal erosion, lightning-caused fires, hurricanes, tornadoes, and blizzards. In this section, this classification approach is used to articulate different hazards (Regmi et al., 2013).

Hazards can also be classified by their seriousness (i.e. the scale of impact) or frequency. The combination of seriousness and frequency can be used in conjunction to help evaluate risk. Geographic information systems (GIS) are used to assess risk from a broad perspective (Carrara et al., 1991).

This section briefly addresses the most common hazards with emphasis on those related to climate. In this section, both fundamental and advanced risk control approaches are introduced, along with risk mitigation strategies.

5.3.2 Hazards

A hazard can cause human demise, lead to poor health conditions, damage infrastructure and investments, endanger various natural species, and disrupt habitats. For each hazard, the cause could be linked to one factor. However, the cause most often comprises various factors. In general, the three distinct categories of causes, including geophysical, anthropogenic, and climatic, cannot satisfy engineering needs when considering the existing close and rapid interactions of these three causes in the critical zone (Giardino and Houser, 2015).

5.3.2.1 Geophysical Hazards

5.3.2.1.1 Earthquakes

Earthquakes are one of the most catastrophic hazards in the world. About 90% of earthquakes occur in the Circum-Pacific belt (i.e. Ring of Fire), whereas 5% of earthquakes take place in the Alpine belt (i.e. Alpine-Himalayan orogenic belt) (USGS, 2012). These two regions are along the tectonic-plate boundaries and thus more disposed to the release of strain energy stored in the crust of Earth. Temporally, the active phase of an earthquake occurs over a very short period without any apparent warning. Also, flooding, fire, tsunamis, landslides, surface depressions, and land subsidence often co-occur with an earthquake (Ohnaka, 2013).

Although researchers have been attempting for many years to predict the occurrence of earthquakes, despite their best efforts, success in these endeavors have been limited at best. Long-term (i.e. 10–100 years) earthquake prediction relies primarily on in situ fault studies and compilation of seismic history. Intermediate-term (i.e. 1–10 years) prediction commonly is based on the use of seismology and geodesy instrument data. Unfortunately, current technology to predict a short-term earthquake is not doable (Uyeda et al., 2009).

5.3.2.1.2 Volcanic Eruptions

Volcanoes are surface expressions of fissures in the crust of the Earth through which gaseous, liquid, or solid materials can flow and erupt. Volcanoes are associated with plate tectonics, specifically through plate divergence and convergence. Volatile materials, such as groundwater, near the surface interact with the magma through plate tectonic activity and create violent eruptions. There are two major eruption styles on the forms of eruption material, i.e. effusive (lava) and explosive (pyroclastic) eruptions. At the vent, effusive eruptions produce lava that slowly flows outward until it cools and forms deposits nearby.

Explosive eruptions, on the other hand, create solid fragments (i.e. pyroclastic flows), which can be transported by atmospheric circulation and settle as airborne deposits.

Debris avalanches are a special case of a pyroclastic flow. On 27 March 1980, Mount St. Helens, in the State of Washington, erupted. This eruption was triggered by series of continuous, pre-eruption earthquakes and resulted in violent eruption and pyroclastic flow. The pyroclastic flow turned into the largest known debris avalanche in history and destroyed vegetation and structures within an area of approximately 600 km^2. Other hazards associated with volcanic eruptions include lava flows, pyroclastic dams and flooding, landslide, lahars, tsunami, and toxic gas emanations (Francis and Oppenheimer, 2004).

5.3.2.1.3 *Landslide*

Landslides are among the most dramatic mass-movement processes associated with the influence of gravity. These hazards often occur when a slope becomes unstable. Destabilizing conditions include climatic change, land-use changes resulting in deforestation or built environment, construction/industrial activity by humans, tectonic uplift, and landform change caused by denudation processes.

A landslide is usually classified based on five key criteria: movement mechanism (flow, slide, fall), the material of underlying slope (debris, earth, rock, soil), shape of surface rupture (planar, curved), velocity, and disruption of mass displacement and presence/absence of water. Thus, the size of hazards from landslide depends on the type of landslide. The threat from rockfalls is much localized as the result of the limited distance a rockfall event can extend. On the other hand, slides, avalanches, and flows can cause significant damage on a large scale. For instance, the death rate associated with 1970 Huascaran rock avalanche was approximately 18 000 (Dikau et al., 1996).

5.3.2.2 Anthropogenic Hazards

5.3.2.2.1 *Deforestation and Desertification*

Deforestation is the process of forest degradation. Changing land use as the result of expanding cultivation and collection of wood fuel are the two major factors that motivate deforestation, especially in developing countries. Strip mining plays an important role is changing land use from forest to rangeland or barren land. Other nonanthropogenic factors could also lead to deforestation, such as climate variations, geological activities, and lightning-caused forest fires.

Desertification is the process of land degradation by adverse human influence in sensitive dry-land areas. The major outcome of the process is that relative dry land becomes increasingly dry with a loss of species diversity. The exploitation of soil is considered the primary cause of desertification, including inappropriate irrigation, overgrazing, and deforestation (Thomas, 2004). In most cases, both deforestation and desertification are irreversible.

5.3.2.2.2 *Air Pollution*

Air pollution has become a worldwide concern in recent decades. Epidemiologists suggest a strong association between particulate air pollution and exacerbation of disease and death in urban areas (Seaton et al., 1995). The rapid rate of urbanization and industrialization both contribute to the acceleration of air pollution. Also, vehicle transmission inefficiency and adoption of fossil fuels as major energy sources also result in accelerated air pollution. Toxic particulates produced in these activities concentrate in the atmosphere and enter the human respiratory system and result in respiratory problems or eventual death.

5.3.2.3 Climate Hazards

5.3.2.3.1 *Wildfire*

The major causes of wildfires are lightning and the wind to transfer sparks and embers, hot and dry conditions, volcano lava, and earthquake tremors. It has been recognized that climate has a dominant influence on wildfire. Climatologists have found positive correlations between fire frequency and warm and dry conditions for the past 8000 years based on carbon traces in alluvial deposits (Pierce et al., 2004). Droughts, heat waves, global climate circulation patterns, such as El Niño, and regional weather patterns such as high-pressure ridges all exacerbate the risk of wildfire.

5.3.2.3.2 *Flooding*

A flood occurs when the carrying capacity of a channel is exceeded by the rapid increase of flow volume (i.e. discharge), commonly resulting from intense rainfall or long-duration rainfall or melting snow. There is sufficient evidence that floods are becoming more severe regarding damage magnitude and frequency of occurrence (Doocy et al., 2013). Besides climatic factors, soil infiltration capacity on flood plains, local topographic conditions (such as slope), and vegetation cover type and density can also enhance the severity of river flooding.

5.3.2.3.3 *Hazards in the Cryosphere*

The cryosphere is the region of the surface of the Earth dominated by permafrost and glaciers. It plays a vital role in controlling the energy budget of the Earth and also furnishes freshwater to various habitats. Hazards in the cryosphere such as permafrost melting, glacial lake outburst flood (GLOF), snow avalanche, and freeze–thaw can cause damages to the built environment.

Permafrost contains a large quantity of methane and melting releases methane – a typical greenhouse gas in the atmosphere, which is known to accelerate global warming. GLOFs occur when lakes dammed by glacial moraines fail. They pose great threats to populated areas situated down the valley. For example, villages in Alpine valleys in Nepal and Bhutan are in harm's way.

Snow avalanches are typically triggered by the snowpack in the upslope when the force exceeds strength.

Figure 5.2 Photograph of wave erosion taken in Shishmaref, Alaska, by John R. Giardino. Source: Photo courtesy of John R. Giardino.

The mass of snow avalanching down a slope can include rocks and debris and can cause damage to built structures and human fatalities. In 2017, an avalanche in the Italian Alps buried a hotel with a loss of life of almost 30 people (Smith-Spark and Messia, 2017).

Freezing is another cryosphere process that normally occurs during fall–winter and winter–spring in middle latitudes and sometimes in subtropical areas. It can damage electrical power cables, interrupt industrial processes, impact building safety, and damage water and sewage distribution and drainage system. In some high latitude areas, water and allocation systems are carried in utilidors (Ref).

5.3.2.3.4 Sea Level Rise

Sea level has risen approximately 120 m since the end of the last ice age (Gornitz, 2009). There are two generic drivers of sea level rise. One is thermal expansion, and the other is water circulation. Seawater volume increases as sea temperature increases. If current freshwater ice stored in the cryosphere melted and all was transferred into the ocean, sea level would rise an additional 65 m compared with the current level (Lemke et al., 2007). Sea level rise can lead to flooding in coastal regions and low-lying islands. Another hazardous impact from sea level rise is coastal erosion, which could damage infrastructure along the coastline. Recent sea level rise combined with climate change has resulted in accelerated erosion along Sarichef Island in Alaska. The village of Shishmaref has been forced to relocate as the degradation of permafrost and wave erosion as the result of lack of freezing of the Chukchi Sea during part of the winter has led to increased wave erosion (Figure 5.2).

5.3.2.4 An Interdisciplinary Thinking Approach about Hazards in the Critical Zone

The critical zone is characterized by four features:

1) Temporal evolution – processes such as erosion and deposition are normally a response to time that is irreversible.
2) System coupling – the integration of different spheres and disciplines makes the critical zone a holistic system in the Earth science.
3) Vertical stratification – system interactions within the critical zone are vertically stratified from the top of the vegetation layer to the bottom of the aquifer.
4) Horizontal heterogeneity – scale dependence well illustrates spatial complexity of the critical zone (Lin, 2010).

Hazards are one part of the critical zone; thus in nature it has inherited the four characteristics of the critical zone. The study of the critical zone, as suggested by US NRC (2001), is one of the most promising interdisciplinary approaches in the geosciences this century. However, unfortunately, the scope and vision of current hazard studies have been confined by human bias and incomplete understanding of critical zone, especially with regard to incomplete and partial recognition of the four characteristics of the critical zone.

Isolating one hazardous cause to another is evidence of ignoring the coupling nature of the critical zone.

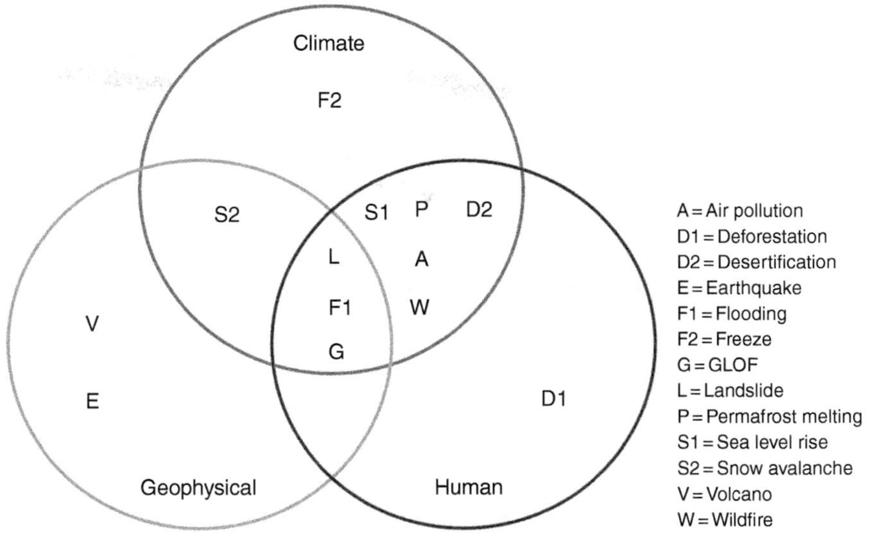

Figure 5.3 Interdisciplinary thinking approach to the causes of hazards.

Table 5.1 NOAA-operated radars adopted for precipitation and cloud observations.

NOAA precipitation and cloud radars		
Radar	**Wavelength**	**Application**
NOAA/D (hydro radar)	3.2 cm	Precipitation, snow, storms, ocean surface
NOAA/K (cloud radar)	8.7 mm	Clouds, boundary layer, ocean surface
NPCO (cloud radar)	8.7 mm	Long-term cloud profiling
MMCR-ARM (cloud radar)	8.7 mm	Long-term cloud monitoring
Ron Brown (precipitation radar)	5 cm	Ocean precipitation measurement
S-PROF (precipitation profiler)	10 cm	Precipitation

We strongly suggest that environmental engineers adopt a more holistic view when dealing with hazard studies. To fulfill a knowledge gap in hazard studies, researchers need the vision regarding causes of hazards as an interaction of different spheres and processes in the critical zone (Giardino and Houser, 2015). Interdisciplinary thinking is inevitable for conducting hazard-related researches (Figure 5.3).

5.3.3 Risk Control

The observational approach is the most direct method for predicting a hazard. It is considered both an empirical approach and an analytical approach. Using precursors for predicting a short-term earthquake is a classic example of the empirical approach, even though the accuracy and efficiency are still pessimistic. Predictions made by the analytical approach based on raw observational data are more reliable and promising for environmental engineering projects.

Numerous instruments have been designed and deployed for collecting data related to atmosphere, the surface of the Earth, ocean, and cryosphere (Vaughan et al., 2013). These observational data once retrieved can be rigorously analyzed employing spatial statistics to predict a hazard. The National Oceanic and Atmospheric Administration (NOAA) has been the main driver of the state-of-art remote sensing technology for monitoring precipitation and clouds for long time periods (Table 5.1). Radar technology can be used to help measure cloud water content and microphysical features such as water droplet size, ice-crystal shape and type, and so forth (Germann et al., 2006). By using radar observations, hazards resulting from severe storms and tornados can be predicted, identified, and further tracked. More information about climate-related data and analytical methods can be found in Section 5.4.

5.3.3.1 Artificial Intelligence

Artificial intelligence (AI) is a knowledge-based technique that can serve as an alternative to traditional approaches when modeling environmental complexity is needed. It facilitates insight toward physical process and simulation like a bright brain, making it most suitable for solving hazard-related problems. To elaborate, there are 12 AI techniques, including artificial neural networks (ANN), cellular automata (CA), fuzzy cognitive map (FCM), case-based reasoning, rule-based systems, decision tree, genetic algorithms, multiagent systems, swarm intelligence, reinforcement learning, hybrid systems, and Bayesian networks (Chen et al., 2008). In this section, we focus on the most popular three, which are ANN, CA, and FCM.

5.3.3.1.1 Artificial Neural Network

ANNs are a machine learning technique, which can learn relationships between specified input and output variables. Neural networks constitute an information processing model that stores empirical knowledge through a learning process and subsequently makes the stored knowledge available for future use. ANN can mimic a human brain to acquire knowledge from the environment through a learning process.

A neuron is the fundamental processing unit in ANNs. A neuron consists of connection links characterized with certain weights (Figure 5.4). Input is passed from one end of the links, multiplied by the connection weight, and transmitted to the summing junction of the neuron (Haykins, 1999). In environmental studies, ANN can facilitate the modeling of the cause–effect relationship such as water quality forecasting (Palani et al., 2008) and rainfall–runoff modeling (Hsu et al., 1995).

5.3.3.1.2 Cellular Automata

CA is designed for modeling the physical processes during which only a local neighborhood (Figure 5.5) interacts. These processes are usually discrete in both space and time; however, the state of each neighborhood is synchronously updated according to the preset rules. The spread of wildfire is a good example of such physical process. When a spot is ignited, the fire propagates in a specific path. Land cover type, wind speed and direction, elevation and slope, and meteorological conditions all are combined to decide the path of fire propagation, which is called the rule in CA. The state of each neighborhood (i.e. burned, burning, and unburned) is the result of the rule and the time (Karafyllidis and Thanilakis, 1997).

5.3.3.1.3 Fuzzy Cognitive Mapping

Fuzzy set theory is designed to exhibit a degree of indeterminate boundaries among target objects that contain members (Sui and Giardino, 1995). FCM are fuzzy graph structures useful for representing causal reasoning. They

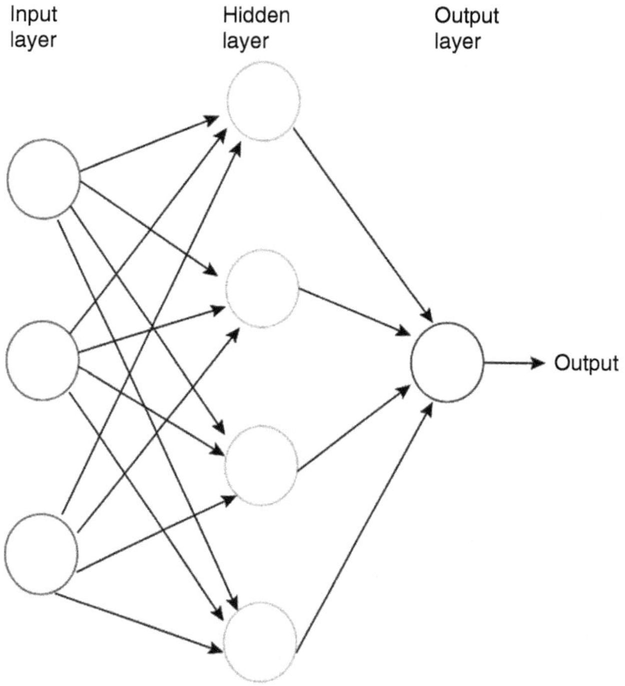

Figure 5.4 An example of an artificial neural network.

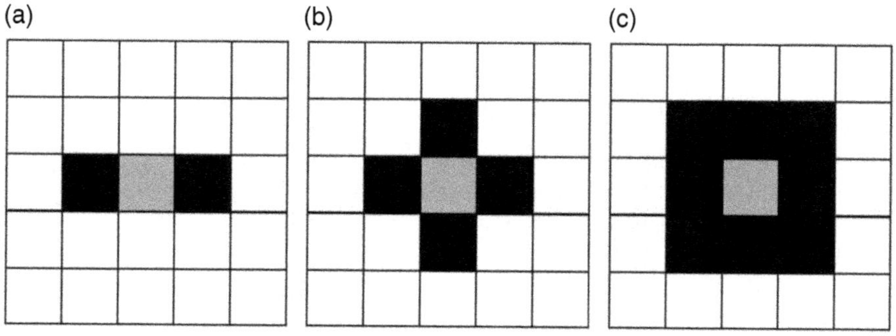

Figure 5.5 Typical neighborhood: (a) 3-cell neighborhood, (b) 5-cell neighborhood, i.e. "von Neumann neighborhood," and (c) 9-cell neighborhood, i.e. "Moore neighborhood."

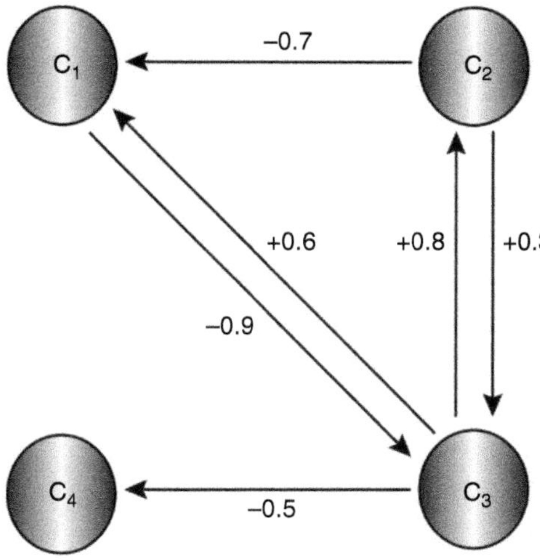

Figure 5.6 A typical FCM architecture. In this example, there are four nodes, which are connected according to the cause–effect relationship. The initial node of each arrow is the factor that can lead to the result at the arrow's end. The degree of influence (−1 to 1) can be quantified by human input. The positive sign indicates a direct causal relationship, whereas a negative sign indicates no direct causal relationship.

represent conceptual nodes that are connected based upon the perceived degree of causality between concepts (Kosko, 1986). Their graph structure allows systematic causal propagation, specifically forward and backward chaining, and it allows knowledge representation and analytical reasoning based upon the strength and direction of causal connections (Figure 5.6). FCM is an integration of ANN and fuzzy set theory. Therefore, it is substantially efficient to adopt FCM to characterize complicated system such as natural hazards. To date, FCM has been widely used in biomedical, industrial, and engineering fields (Papageorgiou, 2012; Papageorgiou and Salmeron, 2013). In environmental studies, FCM has been applied for predicting cryovolcanism in Titan and evaluating eolian stability in Texas (Furfaro et al., 2010; Houser et al., 2015).

5.3.3.2 Mitigation

As we have mentioned earlier, natural hazard processes are the cause of the substantial number of disasters. Hurricanes, earthquakes, tornadoes, floods, landslides, and other natural disasters, unfortunately, cannot be foiled. Although hazards cannot be prevented, many opportunities to reduce the potential impacts on loss of life, serious injury, damage to the built environment, and curtailment of business operations and negative impact on the environment exist. Numerous mitigation strategies exist and can be used to reduce damage from hazards. In this section, we propose a triangular hazard mitigation strategy (THMS) (Figure 5.7), which is a refined version of risk management theory developed by Greiving et al. (2006).

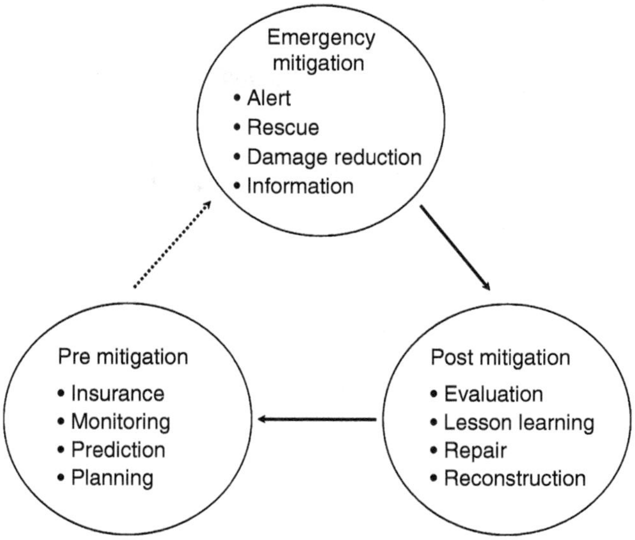

Figure 5.7 The THMS. Since premitigation does not necessarily prevent hazard, a dashed line is used to link premitigation and emergence.

In THMS, an occurring hazard is named as an emergency. Emergency mitigation is the instantaneous action that is needed following an emergency. An efficient emergency mitigation requires prompt alerts for evacuation, timely and effective rescue, maximized damage reduction, and transparent information sharing with the public. Following emergency mitigation, a short-term postmitigation action is applied. Thorough postmitigation response includes a full evaluation report of the emergency, explicit summary for lessons learned, applicable repair, and reconstruction, if necessary. The premitigation stage is a long-term preparedness approach to minimize risk at the lowest level. Premitigation activity includes purchase of insurance to cover financial loss, continuous monitoring, prediction modeling, and sustainable planning for infrastructure location and natural resources usage. A premitigation action does not necessarily prevent the hazard.

5.3.4 Summary

In this section, we addressed the major hazards associated with geophysical, human, and climate. We take this opportunity to strongly support the view that environmental engineers undertake an interdisciplinary approach in their problem solving. Hazard prediction can be achieved by integrating observations with advanced technology such as AI. Rigorous risk mitigation is composed of long-term preparedness, short-term strategy, and emergency responses.

5.4 Climatology Methods

5.4.1 Introduction

Climatologists have long established various approaches for employing current technology for addressing climate-related issues. These approaches, however, are constrained by time and space. All climatology projects begin with data, so in this section, typical climate data are introduced, including observational data and simulation data. Also, statistics such as linear regression, geostatistics, principal component analysis (PCA), empirical orthogonal function (EOF), and wavelet are also addressed in this section.

5.4.2 Analysis in Time and Space

5.4.2.1 Time

Climate changes at all time scales as a response to its forcing factors (Mitchell, 1976). The time interval is a first-order factor that must be identified when conducting climate-related research. Time intervals vary as microscale (seconds to hours), mesoscale (days to years), and macroscale (decades to millennium). Rigorous assessment and adoption of the appropriate time interval can not only be cost-effective for computing purposes but also aid in revealing underlying patterns and trends that other time intervals would not or could not (Gaffen and Ross, 1999; Gallo et al., 1996).

Climate data, such as temperature, fluctuates over time. In most cases, such fluctuations follow a prevailing trend. However, when the "normal" threshold or boundary condition is reached and crossed, an anomaly occurs. In general, natural causes of climatic change and variability include continental drift, Milankovitch cycles, volcanic activity, variations in solar output, and El Niño/La Niña occurrences. Human-introduced climate change, such as rapid urbanization and greenhouse gas emission, can also cause climate change (Rohli and Vega, 2013).

Continental drift: The crustal plates of the surface of the Earth are constantly moving over millions of years. The redistribution of continents and water can alter global climate patterns by (i) strengthening the effect caused by the differences of the specific heat capacities between water and consolidated materials of the surface of the Earth (i.e. rock, soil, etc.), (ii) acting to intervene the heat transfer in both the ocean and the atmosphere, and (iii) creating landform barriers such as Himalayan and Alpine mountains to regulate regional precipitation and, thus, influence global climate (Pichon, 1968).

Milankovitch cycle: It accounts for the climatic cycles in geological scales such as glacial and interglacial cycles. The Milankovitch cycle includes three specific cycles: (i) eccentricity, (ii) tilt, and (iii) precession (Short et al., 1991).

Volcanic activity: Volcanic eruptions contribute the largest volume of CO_2 to the atmosphere, and CO_2 is a greenhouse gas that can alter the energy budget of the Earth. Throughout geological time, eruptions have not been evenly distributed nor linear. Lack of eruptions may cause a relatively cold climate. For instance, a Gondwanan ice age began when continental plates were all part of Pangaea and then readjusting as plates drifted apart. During the readjustment period, there was a lack of volcanic activities, which potentially led to a cooling climate (Crowell, 1978).

Variations in solar output: Solar radiation is the major energy source for the Earth, especially in the biosphere in the critical zone. Solar radiation experiences short-term variations associated with sunspot cycles. Sunspots are a huge magnetic storm on the surface of the sun. These sunspots always appear as darker spots. Existing observations suggest an 11-year sunspot cycle, which can account for solar irradiance variations (Frohlich, 1998).

El Niño and La Niña: El Niño and La Niña are associated with large-scale ocean–atmosphere interaction that leads to sea surface temperature (SST) fluctuations. El Niño means Christian Child or Little Boy in Spanish. It refers to the occurrence of the abnormally warmer SST in South America across the equatorial Pacific Ocean, which was first observed by fishermen in 1600s AD. La Niña means Little Girl in Spanish and is the opposite of El Niño. During a La Niña cycle, below-average SST across the Pacific Ocean occurs. Research suggests that El Niño and La Niña both exhibit periodic characteristics of 3–7 year cycles (Trenberth, 1997).

5.4.2.2 Space

Climate changes across space with specific spatial structures associated with each climate phenomenon; some are evident, whereas others are not apparent. For a climate research topic, a researcher can ask questions related to spatial structure, such as "how is precipitation spatially distributed in Southeastern United States," "are there any recurrent spatial patterns of drought events in Central Texas," or "what are the geographic impacts of climate events such as El Niño and La Niña"?

The degree of complexity associated with the spatial structure of climate phenomena varies. It could be as easy as interpolating a dozen of in situ temperature records into a smooth temperature map covering an entire region, or it could be as difficult as finding the causal connection between a set of different meteorological phenomena that occur distances apart. A better understanding of the essence of climate data and proficient knowledge of climatic skills can facilitate extracting spatial patterns and reveal a cause–effect relationship.

5.4.2.3 Integration of Time and Space

In most cases, the concepts of time and space need to be integrated so that a climate phenomenon can be thoroughly investigated. The study of teleconnection is a good example of such integration (Feldstein, 2000). Teleconnection is a typical type of climate variable that characterizes spatially and temporally large-scale anomalies that could substantially impact the variability of atmospheric circulation, such as El Niño–Southern Oscillation (ENSO).

Other teleconnections include Arctic Oscillation (AO), East Atlantic (EA), East Atlantic/Western Russia (EAWR) pattern, East Pacific/North Pacific (EPNP), North Atlantic Oscillation (NAO), Pacific/North American (PNA) pattern, Polar/Eurasia (POL) pattern, Pacific Transition (PT) pattern, Scandinavia (SCA) pattern, Tropical/Northern Hemisphere (TNH) pattern, West Pacific (WP) pattern, and Pacific Decadal Oscillation (PDO) (Bjerknes, 1969; Wang et al., 2000).

5.4.3 Climate Data

Climate data are essential for climatologists. Acquiring data from appropriate sources requires a cautious approach. Data types include observational data and modeling data. Observed data include in situ data, satellite, and reanalysis data. Climate modeling data are an excellent addition to observed data, such as general circulation model (GCM).

5.4.3.1 Observational Data

In situ data are the most direct way for climate data acquisition. Many climate variables have constantly been measured at meteorological stations, such as precipitation, temperature, wind speed, wind direction, air pressure, air humidity, solar radiance, snow depth, etc. Careful selection of instruments is fundamental for collecting useful data. In addition to the instrument itself, data logger plays vital role for in situ data storage at remote locations. New trends for data loggers include an interconnection with digital sensors, real-time data uploading using telemetry with statistics, and collecting data via the internet.

Satellite data facilitate in situ data collection, because (i) satellite can collect data for remote locations, such as very rugged, isolated terrain, and (ii) satellites offer a large footprint for spatial coverage. Satellites have been used for climate researchers for decades. Examples of mainstream satellites are listed in Table 5.2.

Table 5.2 Six examples of satellites, ASTER, AVHRR, Landsat, MERIS, MODIS, and SPOT.

Name	Resolution	Orbiting
ASTER	15 m	16 d
AVHRR	1.1 km	Daily
Landsat	30 m	16 d
MERIS	300 m	3 d
MODIS	1 km	Daily
SPOT	6 m	26 d

Table 5.3 Biophysical indices can be calculated from remote sensing image.

Acronym	Full name	Field
NDVI	Normalized difference vegetation index	Vegetation
LAI	Leaf area index	
SVI	Simple vegetation index	
SAVI	Soil adjusted vegetation index	
PVI	Perpendicular vegetation index	
NDWI	Normalized difference water index	Water
NDSI	Normalized difference snow index	Snow
SI	Salinity index	Soil
SINDRI	Shortwave infrared normalized difference residue index	
NDTI	Normalized difference tillage index	
LCA	Lignin-cellulose absorption index	

In essence, the data retrieved from the satellite are digital number (DN). Numerous algorithms have been developed to convert the DN into a more complex biophysical parameter for climatological purposes. Table 5.3 summarizes some satellite-derived indices that are widely used in climatology.

Reanalysis data are a data-assimilation product used in climatological research. It produces climate datasets on a global scale with a time step of 6–12 h. During each time step, approximately 8 million observations are assimilated, including ship reports, weather station records, satellite data retrieval, buoy information, and radiosonde data. Reanalysis data can be used to produce continuous data back to 1948. However, this technique has been criticized of its lack of reliability, data inaccuracy, and observation discrepancy. Two major climate reanalysis data sources are NCEP/NCAR and ERA-40.

5.4.3.2 Modeling Data

A simple climate model can be as basic as a Lapse rate model, which can be used to build the linear relationship

Table 5.4 Online climate data sources.

Name	Manager	Data
PRISM	Oregon University	High-resolution climate data
DAYMET	Oak Ridge National Laboratory	Daily climate data
NCEI	NOAA-National Centers for Environmental Information	Historical weather data
ESRL	NOAA-Earth System Research Laboratory	Climate reanalysis data
CPC	NOAA-Climate Prediction Center	Climate predictions and teleconnections
GES DISC	NASA-Goddard Earth Sciences Data and Information Services Center	Observed climate data from space
TRMM	NASA-Goddard Space Flight Center	Observed precipitation in low latitude
NSIDC	National Snow and Ice Data Center	Snow and ice data
USGS	U.S. Geological Survey	Runoff, discharge data
NRCS	National Water and Climate Center	Runoff, discharge data
IPCC	Intergovernmental Panel on Climate Change Data Distribution Centre	GCM data

between air temperature and elevation. More complicated models, however, are numerical simulations of the Earth to investigate the consequence of the climate system as a response to different forcings, which is also known as a GCM. GCMs can be used to predict future climate at both regional and global scales. It can be applied to most climate-related dynamics including the atmosphere, the ocean, sea ice, land surfaces, carbon cycle, hydrology, aerosol, insolation, etc. For any GCM, three factors need to be considered: spatial resolution, temporal resolution, and level of complexity. These three factors altogether influence the model accuracy and computing cost. Most GCMs facilitate temporal and spatial downscaling capacities for specific research needs.

GCMs are numerical modeling of nature, and one must evaluate an existing model's accuracy to improve them, which is crucial in climatological research. Scientists have been collectively working to ensure the models' compatibility. The undergoing Coupled Model Intercomparison Project Phase (CMIP) is a comprehensive collaboration framework for advancing the scientific understanding of climate system. The CMIP has undergone five different phases since 1995, and the current phase is 6 (i.e. CMIP6). It has focused the research agenda as (i) investigating climate responses to forcings, (ii) evaluating model uncertainties, and (iii) assessing future climate change. The progress of CMIP5 has been explicitly documented in the Fifth Assessment Report (AR5) of Intergovernmental Panel on Climate Change (IPCC).

5.4.3.3 Data Sources and Collaborations

Environmental engineers collecting data for climate change impact analysis need to consider the following factors: grant writing for financial support, instrument purchase and calibration, fieldtrip plan making, transportation and lodging, sampling strategy, in-field instrument placement, raw-data assessment, data transfer and storage, fieldwork safety, etc. High quality fieldwork can be time consuming, laborious, and expensive. Thus, use of a reliable climate data sharing platform will substantially reduce individual output. Major online climate data sources are listed in Table 5.4. In addition, these sources are free and can be downloaded via the Internet.

Climate change impact analysis requires collaborations at all societal, academic, and governmental levels. Nonprofitable organizations exist to help scientists and environmental engineers. As addressed previously, IPCC is a good example of global collaboration for solving climate change related issues. Additional organizations include American Geophysical Union (AGU), American Meteorological Society (AMS), American Association of Geographers (AAG), Geological Society of America (GSA), World Meteorological Organization (WMO), etc.

Governmental agencies also play vital roles in building bridges between climate data and user interfaces, such as the National Aeronautics and Space Administration (NASA), United States Environmental Protection Agency (US EPA), United States Geological Survey (USGS), United Nations Development Programme (UNDP), United Nations Educational, Scientific and Cultural Organization (UNESCO).

5.4.3.4 Missing Data

5.4.3.4.1 Filling Temporal Data Gaps

Time series data retrieved from field locations are episodic in nature. Thus, missing data are common encounters in climatic research. For instance, river discharge dataset retrieved from the USGS may include data gaps as the result of instrument malfunction or human misoperation. Understanding the missing mechanism is important to handling missing data so that the right replacement method is used. If the missing data occur randomly, common methods such as multiple regression, expectation maximization, and regression trees can be used in estimating missing values (Kim and Ahn, 2009; Schneider, 2000). However, if the missing data occurrence is conditioned on a deterministic event that affects the data values, such as a flood event, the method of estimation of missing values is more complicated (Little and Rubin, 2002).

5.4.3.4.2 *Filling Spatial Data Gaps*

Biophysical variables, such as surface temperature, exhibit a continuous surface gradient in reality. Field sampling, however, cannot adequately record every location because of limited time and funding. A subset of samples in the form of points can be measured. To fill the spatial gaps or unsampled area, different interpolation methods have been widely adopted in climate research. These methods differ from three perspectives: (i) the mathematic function adopted, (ii) the distance and weight being considered, and (iii) the number of samples taking into account. Five typical interpolation approaches are nearest-neighbor interpolation, fixed radius (i.e. local averaging), inverse-distance-weighted (IDW) interpolation, splines interpolation, and kriging (Jeffrey, 2001). Kriging is also a good geostatistical method for characterizing scale-related issues, which will be detailed in Section 5.4.4.2.

5.4.4 Statistics in Climatology

Climate variables can exhibit collinearity (Carleton, 1999). To employ this type of approach requires specific statistical skills to extract the spatial structure underlying each climate phenomenon. In this section, we introduce the basic concepts of linear regression, geostatistics, and data reduction methods (PCA, EOF) and spectral analysis (wavelet).

5.4.4.1 Linear Regression Analysis and Significance Test

In climatology, linear regression is used for modeling the relationship between the dependent variable Y and various independent variable(s) X(s). If there is only one independent variable, it is a simple linear regression. Otherwise, it is multiple linear regression. One must understand that it is different from multivariate linear regression in which a set of correlated dependent variables (Ys) are predicted. In this section, the focus is placed on simple and multiple linear regressions.

Linear regression has two major applications in climatology: (i) modeling for prediction purpose and (ii) examining correlations among X(s) and Y. For the first application, a fitting function can serve well as a prediction model for linking different climate variables. For instance, Lapse Rate model is a simple linear regression model for predicting air temperature based upon elevation and altitude. For the second application, Pearson-correlation coefficient is the statistical criteria for the degree of dependence between X and Y. The coefficient ranges from −1 to +1, where −1 is a totally negative linear relationship, +1 is a totally positive linear relationship, and 0 shows no linear relationship. In climatology, a significance test is typically followed by correlation analysis, because the correlation relationship could be real in reality or could be just a statistical artifact. However, overemphasis on p-values and significant levels could mislead the result of a climate study (Ambaum, 2010; Ling and Mahadevan, 2013; Marden, 2000; Ziliak and McCloskey, 2008).

5.4.4.2 Geostatistics

Geostatistics is suitable for characterizing spatial variations of climate phenomenon. Kriging is one of the geostatistical methods that can quantify the spatial dependence of natural resources. Matheron (1971) first proposed the theory of geostatistics, and Curran (1988) further expanded the theory into remote sensing studies. As mentioned previously, kriging is an interpolation method. However, it is distinct from other interpolation methods because the core of kriging is a statistical model that adopts spatial autocorrelations, by which spatial structure can be first computed based upon available samples, and new predictions can be made after. Kriging also provides cross-validation capacity to evaluate interpolation accuracy (Krige, 1951).

Spatial autocorrelation is a criterion for examining the statistical relationship among observations. The distance between a pair of sampling points and the direction linking these samples are two major factors that must be considered when computing spatial autocorrelation. A semivariogram summarizes spatial autocorrelation with the x-axis as distance and y-axis as semivariance. Semivariance can be calculated with the following function:

$$\gamma(h)=\frac{1}{2n(h)}\sum_{i=1}^{n(h)}\left[Z(i)-Z(i+h)\right] \quad (5.1)$$

where $n(h)$ is the number of paired sampling points at distance h and $Z(i)$ is the surface value at location i. The complete set of sampling pairs is used to generate a semivariogram (Longley et al., 2004).

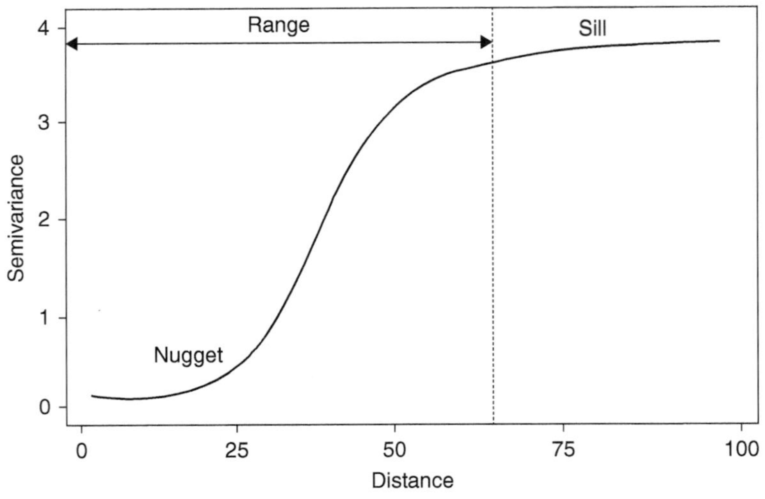

Figure 5.8 A traditional semivariogram.

Figure 5.8 is a classic semivariogram in which closer distance exhibits smaller semivariance, whereas longer distance exhibits larger semivariance. Such a pattern suggests that features closer are more likely to be similar and that is Tobler's first law of geography. Thus, a semivariogram can serve as an excellent tool for investigating scale dependence and spatial variability in climatology (Burrough, 1983).

Another benefit of the semivariogram is that it can be used to help determine the best sampling interval when developing strategies for data collection in the field. As shown in Figure 5.8, a range distance is a distance where semivariance approaches flat. It means as distance increases, semivariance increases; however, spatial autocorrelation decreases until there is no spatial autocorrelation beyond that range (Meentemeyer, 1989). Therefore, a range is the maximum distance that a sampling interval can be used in the field to capture every spatial variability since within the range, one point can be representative of every location if they are spatially autocorrelated (Chiles and Delfiner, 1999; Cressie, 1993).

5.4.4.3 PCA (EOF) and Wavelet Analysis

PCA is a data reduction method used in climatology. It is a technique that can transform an original dataset with potentially correlated variables into a new dataset with linearly uncorrelated variables (Bretherton et al., 1992). This type of orthogonal transformation can significantly reduce data dimensionality because (i) the number of new variables is less than or equal to the previous number of variables, (ii) the first principal component carries the maximum variance of the original dataset, and (iii) all the following principal components carry most variance for the remaining principal components. As a result of the mathematical nature of PCA, it has been widely used for exacting spatial patterns of climate variables at different scales. EOF is an extreme case of PCA because EOF is a PCA when applied to time series data. Thus, EOF is suitable for processing climate data (Lorenz, 1956; Thuiller, 2004).

In recent decades, spectral analysis such as wavelet has been used in numerous applications, such as harmonic analysis, numerical analysis, signal and image processing, nonlinear dynamics, fractal analysis, etc. The wavelet transform is very popular in dealing with time frequency, orthogonality, and scale-space analysis (Foufoula and Kumar, 1994). The essence of wavelet analysis is to transform an array of numbers at a certain dimension from their original digits into an array of wavelet coefficients. A wavelet coefficient informs the correlation between a wavelet function of a certain size and data (Daubechies, 1992). In recent years, wavelet analysis has been employed as a tool for analyzing the power spectra of time series data in climatology (Torrence and Compo, 1997). This analysis can help researchers capture both regional and overall views of the match between a wavelet function and target data by changing the size of wavelet function and shifting the wavelet, which is called localization (Hubbard, 1998).

The variability of a one-dimensional data array can be represented in the form of a two-dimensional (2D) plot showing the variability of amplitude at different scales and how the amplitude is changing at different time frequencies (Lau and Weng, 1995; Torrence and Compo, 1997). For instance, a wavelet analysis can be applied to a precipitation dataset to reveal the periodicity of the high frequency event such as rains during monsoon seasons.

5.4.5 Summary

In this section, various climatic approaches are discussed for environmental engineers to use. Because climate varies in time and space, a better understanding of temporal and spatial scale can facilitate the targeting of impact analysis more clearly and tangibly. In addition, familiarity with climate data access can enhance research efficiency and broaden the research horizon. Statistics is essential for any type of climatic impact analysis, because it is the most suitable tool to examine the collinearity behind different phenomenon. Collaboration at various levels is required as impact analysis of climate change occurs at a global scale.

5.5 Geomorphometry: The Best Approach for Impact Analysis

5.5.1 Introduction

Geomorphometry is a numerical representation of topography, which integrates various disciplines: i.e. mathematics, Earth science, engineering, and computer science (Pike and Park, 1995). Human beings interact with the surface of the Earth every day, so to depict a landform in a qualitative manner is not a big challenge per se. However, it becomes a major obstacle for individuals who lack the domain knowledge to delineate land surface quantitatively or even complex when attempting to characterize some complicated Earth surface process, such as erosion and deposition. Geomorphometry is the best approach to overcome such obstacles.

To be detailed, geomorphometry can help in the understanding of natural processes associated with the surface of the Earth, and also it can help support technological needs of society (Pike and Park, 1995). For instance, geomorphometry can help in extracting some underlying information from various fields of study, such as geology, hydrology, climatology, etc. The extracted information can further be applied in evaluating natural hazards and conserving natural resources. On a societal level, geomorphometry can be used in engineering, transportation, public works, and military operations.

Naturally, the next question is how a surface is categorized in a "geomorphometrical" manner. Traditionally, there are six factors that contribute to the topographic existence: elevation, terrain surface shape, topographic position, topographic context, spatial scale, and landform (object type). The first four factors can be parameterized, whereas the last two can be analyzed (Deng, 2007).

As technology advances, especially the advent of GIS and remote sensing, it significantly accelerates the capacities of calculating the aforementioned factors (Goodchild, 1992). Numerous commercial software that can be used in the field of geomorphometry have unique analytical functionality as well as modularity. It is especially common that different software adapt different algorithm even when calculating one single parameter, e.g. slope and aspect. Thus, it is important to first assess the software before using a specific one but finally find the devotion is worthless.

5.5.2 Geomorphometry Research and Applications

Many applications of geomorphometry exist and include, for example, optimizing crop yields, measuring runway roughness, mapping seafloor terrain types, guiding missiles, assessing soil erosion, analyzing wildfire propagation, mapping ecological regions, modeling climatic changes, etc. (Pike and Park, 1995). Geomorphometry is especially effective when combed with remote sensing and GIS. Remote sensing can provide raw data to GIS, and various GIS functions have been developed to aid in the interpretation of remote sensing data. Thus, it is difficult to separate remote sensing from GIS when employing.

In geomorphometry, remote sensing normally means collecting information from aerial platforms, such as airplanes, unmanned aerial vehicle (UAV), and satellites. Even though remote sensing cannot replace traditional geomorphic field observations, it has become fundamental to use remote sensing because it offers a synoptic overview of a study area.

In the early days of geomorphic research, besides in situ study, geomorphologists primarily interpreted phenomena qualitatively from remotely sensed images. However, quantitative methods are readily used in geomorphometry today. One significant advantage of using remote sensing in geomorphometry is that it facilitates access to data that are inaccessible via in situ observations. A significant advantage of remote sensing is in the recording of information chronologically. Aerial photography records date back to the 1920s. The first Earth resources satellite, i.e. Landsat-1, was launched in 1972 (Jensen, 2005).

To conduct successful geomorphometry research using remote sensing imagery, the spatial, temporal, and spectral resolution need to be carefully scrutinized. The spatial resolution is the smallest cell size of instantaneous field of view (IFOV), which is closely related to the concept of scale dependency. Temporal resolution demines the frequency of retrieving images from the same site. Spectral resolution tells the spectral characteristics the channels used by the sensor.

A 3D view of the landscape can be rendered with the help of remote sensing images, which significantly enhances the interpretability of land surfaces. Stereoscopic measurements can be used to generate topographic maps or digital representations of topography (i.e. DEM).

Currently synthetic-aperture radar (SAR) and light detection and ranging (LiDAR) are two popular technologies used in geomorphometry. SAR is valuable because it can penetrate into dry materials, such as sand or dry snow. LiDAR has many merits in delineating land surface details.

5.5.3 Software Package Evaluation

At present there are numerous software that can be used to conduct geomorphometric research. In this chapter, 13 are examined. These software can be further classified into four categories, i.e. digital image processing software, GIS systems, hydrology software, and geomorphometry software.

Eighteen criteria are assessed, and the evaluations are summarized in Tables 5.5–5.9. Each criterion is further ranked as strong (S), weak (W), or null (N). The current URL for each software package is listed in Table 5.10.

In general, ArcGIS®, ENVI®, Erdas Imagine®, GRASS®, and IDRISI® are the top five software packages that are standard for geomorphometric research. Based on evaluation, the ranks for geomorphometric capacity are as follows (from high to low): GRASS, Erdas Imagine, ENVI, ArcGIS, and IDRISI. Other software have many disadvantages at some point. Some uniquenesses of some software are listed in Table 5.11.

5.5.4 Case Study of Analysis of Climate Change Impact

We use two study areas where we conducted three separate case studies to show how one can analyze the impact of climate change on hydrology of a drainage basin and its impact on a region that currently is being glaciated. The first study area is focused on the San Juan Mountains in Southwestern Colorado, USA. The second study is centered on the Karakoram region in Pakistan.

We selected the first study site to show how remote sensing can be used to study impact on drainage and erosion rates associated with changing climate. We selected the second study site to demonstrate that although climate is changing, its impact can lead to very different results than expected. Although glaciers in the Eastern Himalayas are rapidly retreating, glaciers in the Western Himalayas are advancing, which is in stark contrast to the global trend.

5.5.4.1 Characterization of Rivers Basins in the San Juan Mountains, Colorado

5.5.4.1.1 Background

Alpine streams are not only a vital source of water for human settlements in mountain terrain but also potential hazards associated with spring melt and intense runoff from summer convective storms. Thus, the identification and characterization of alpine river basins is crucial for the optimization of potable water resources, as well as minimizing floods. Second- and third-order drainage basins are an integral part of the geomorphology of the San Juan Mountains. However, with an increase in year-round residents and greater influx of tourists to the region, there is a need to improve the management of water resources and to minimize flood hazards in the San Juan Mountains.

Global change can dramatically impact alpine water resources by changing precipitation patterns and amount of precipitation as wells as warmer winter and summer temperatures. Temperature changes can promote higher

Table 5.5 Evaluation on LiDAR input, GPS tracks, spatial analysis and spatial interpolation.

Software name	Category	LiDAR input	GPS tracks	Spatial analysis	Spatial interpolation
ENVI	Digital image processing	S	S	S	W
Erdas Imagine		S	S	S	S
ER mapper		N	S	W	N
Arc GIS	Geographic information systems	S	S	S	S
GRASS		S	S	S	S
SAGA		S	S	S	S
IDRISI		W	S	S	S
ILWIS		S	S	S	S
PCRaster		N	W	S	S
TAS	Hydrology	S	S	S	S
Surfer	Geomorphometry systems	N	S	S	N
Landserf		N	S	S	N
MicroDEM		S	N	N	N

Table 5.6 Evaluation on geostatistics, image filters, and image transformation (e.g. PCA).

Software name	Category	Geostatistics	Image filters	Image transformation (e.g. PCA)	Spatial modeling
ENVI	Digital processing	W	S	S	N
Erdas Imagine		S	S	S	S
ER mapper		N	S	S	N
Arc GIS	Geographic information systems	S	W	W	S
GRASS		S	S	S	S
SAGA		S	S	N	N
IDRISI		S	S	S	S
ILWIS		S	W	S	N
PCRaster		S	N	N	S
TAS	Hydrology	S	S	W	W
Surfer	Geomorphometry systems	S	N	N	N
Landserf		S	N	N	N
MicroDEM		N	N	N	N

Table 5.7 Evaluation on image enhancement, geodatabase, classification, and segmentation.

Software name	Category	Image enhancement	Geodatabase	Classification	Segmentation
ENVI	Digital processing	S	W	S	S
Erdas Imagine		S	S	S	S
ER mapper		S	N	S	N
Arc GIS	Geographic information systems	W	S	W	N
GRASS		S	S	S	S
SAGA		N	W	W	N
IDRISI		S	S	S	S
ILWIS		S	W	S	N
PCRaster		W	W	W	N
TAS	Hydrology	S	W	W	N
Surfer	Geomorphometry systems	N	N	N	N
Landserf		N	N	W	W
MicroDEM		N	N	N	N

Table 5.8 Evaluation on pattern recognition, terrain analysis, and raster analysis.

Software name	Category	Pattern recognition	Terrain analysis	Raster analysis
ENVI	Digital processing	W	S	S
Erdas Imagine		S	W	S
ER mapper		N	W	W
Arc GIS	Geographic information systems	N	S	W
GRASS		S	S	S
SAGA		N	W	S
IDRISI		W	S	S
ILWIS		N	S	S
PCRaster		N	W	S
TAS	Hydrology	N	W	S
Surfer	Geomorphometry systems	N	S	N
Landserf		N	S	S
MicroDEM		N	S	N

Table 5.9 Evaluation on vector analysis, 3D visualization, scripting functionality, and batch processing.

Software name	Category	Vector analysis	3D visualization	Scripting functionality	Batch processing
ENVI	Digital processing	N	S	S	S
Erdas Imagine		S	S	S	S
ER mapper		W	S	W	W
Arc GIS	Geographic information systems	S	S	S	S
GRASS		S	S	S	S
SAGA		S	S	S	W
IDRISI		W	S	N	N
ILWIS		S	S	W	N
PCRaster		N	N	S	N
TAS	Hydrology	W	W	N	N
Surfer	Geomorphometry systems	N	S	S	S
Landserf		W	S	S	S
MicroDEM		N	S	N	N

Table 5.10 URL for each software package.

Software name	Link
ENVI	www.harrisgeospatial.com/Home.aspx
Erdas Imagine	www.hexagongeospatial.com
ER mapper	www.hexagongeospatial.com
Arc GIS	www.esri.com/
GRASS	https://grass.osgeo.org/
SAGA	http://www.saga-gis.org/en/index.html
IDRISI	https://clarklabs.org/
ILWIS	www.ilwis.org/
PCRaster	http://pcraster.geo.uu.nl/
TAS	https://www.tradeareasystems.com/products/tas-analyst
Surfer	http://www.goldensoftware.com/products/surfer
Landserf	www.landserf.org/
MicroDEM	https://www.usna.edu/Users/oceano/pguth/website/microdem/microdem.htm

Table 5.11 Uniqueness of some software.

Software name	Uniqueness
SAGA	This software has fractal dimension analysis module and pattern analysis module
IDRISI	It offers a triangular wavelet analysis function and trend analysis of time series images
TAS	It is highly dependent upon ArcGIS input
Surfer	It is very powerful of 3D visualization
Landserf	This is very powerful of generating geomorphometric parameters. Fractal dimensions and multiscale topographic parameters can be calculated from it
MicroDEM	It can offer a way to visual 3D images in time series, and It also provide a dynamic recording of 3D images

evaporation rates in summer months and changes in precipitation type from snow to rain. Seasonal change can also occur. The beginning of the winder precipitation season can be pushed to late fall or even mid-winter. And the beginning of spring melt season can be delayed by several weeks to a month.

All of the previously mentioned changes impact surface in groundwater resources. To analyze and understand the impact of climate change on water resource requires the study of the linkage between atmospheric processes and geomorphology of a drainage basin. Thus, this section demonstrates the use of a new time and cost of time and cost-effective technique to determine surface roughness. Surface roughness is a major basin-wide threshold that serves as an indicator of precipitation as well as a controller of drainage pattern development and rate of runoff.

In the San Juan Mountains, streams represent regional topography and local surface roughness. Our study area encompasses the Ironton, Ophir, Ouray, Silverton, and Telluride USGS quadrangles in Southwestern Colorado, covering an areas of 805 km^2 (Figure 5.9).

We have developed an innovative geospatial approach implementing novel geomorphometric indices to efficiently delineate river basins in mountain terrains. To accomplish this fast Fourier transformation (FFT), analysis is employed to confirm that no scale dependence for the regional topography exists. And to supplement the FFT, new topographical-classification approach that addresses surface roughness to better understand hydrological controls in the study area was created. Lithology was coupled with the topographical-classification map. Additionally, divergence and convergence indices were developed based on water flow to improve the existing river channel extraction algorithm. This new method is cost-effective and expedites the process of delineating river basins in mountain terrains to better manage water resources and prepare hazard mitigation plans.

5.5.4.1.2 Scale Characterization

The 2D discrete Fourier transformation (DFT) can benefit the field of digital image processing with respect to image enhancement, image restoration, data compression, data fusion, and other applications (Anuta, 1970; Cooley et al., 1969). DFT transforms the information from an original image with spatial domain to frequency domain. Thus it can aid in revealing information hidden in frequency domains, such as periodicity, orientation, and scale (Gonzalez et al., 2009). FFT algorithm is the implementation of DFT in reality and can be applied by a couple of software packages (e.g. ENVI) and several programming languages (e.g. MATLAB, IDL, and Python).

To examine the scale-dependence issues in the San Juan Mountains, we developed a protocol to adopt FFT techniques and image filter technique. The steps in the protocol are as follows: a DEM of 10-m spatial resolution of the study area (Figure 5.10a) is first converted into Fourier spectrum then centered and enhanced by a log transformation (Figure 5.10b). To check the scale

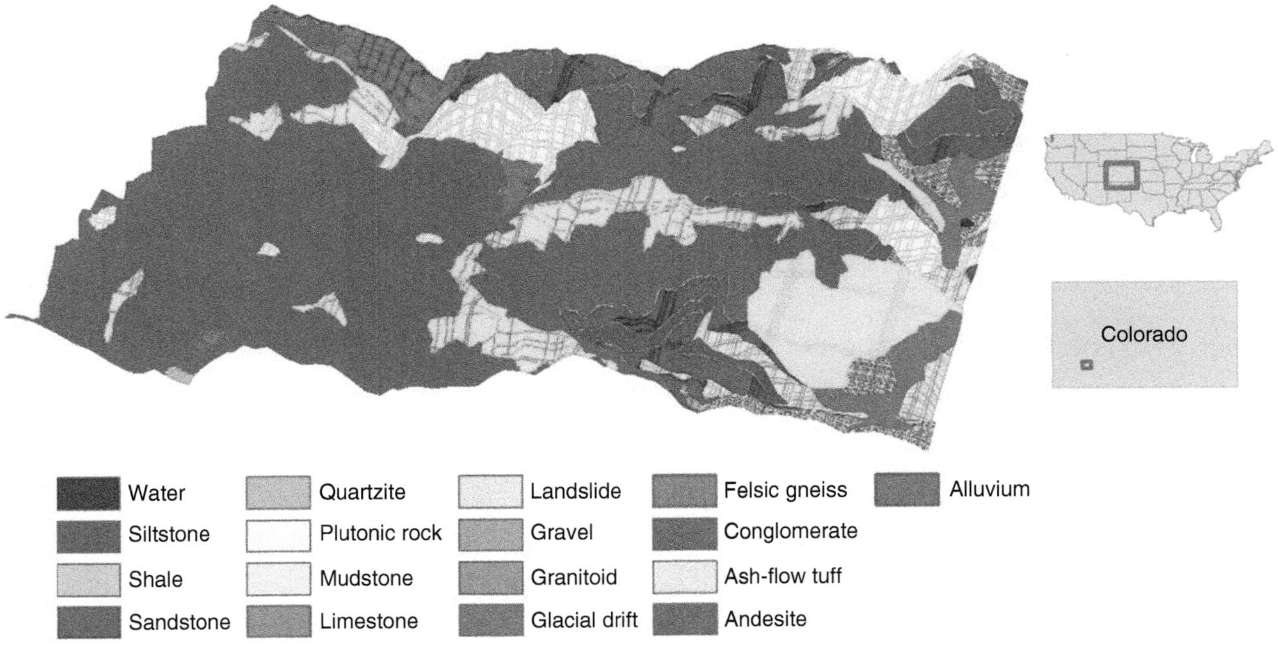

Figure 5.9 The lithology of the study area is represented in three dimensions. The San Juan has rugged topography dominated by ridges and valleys.

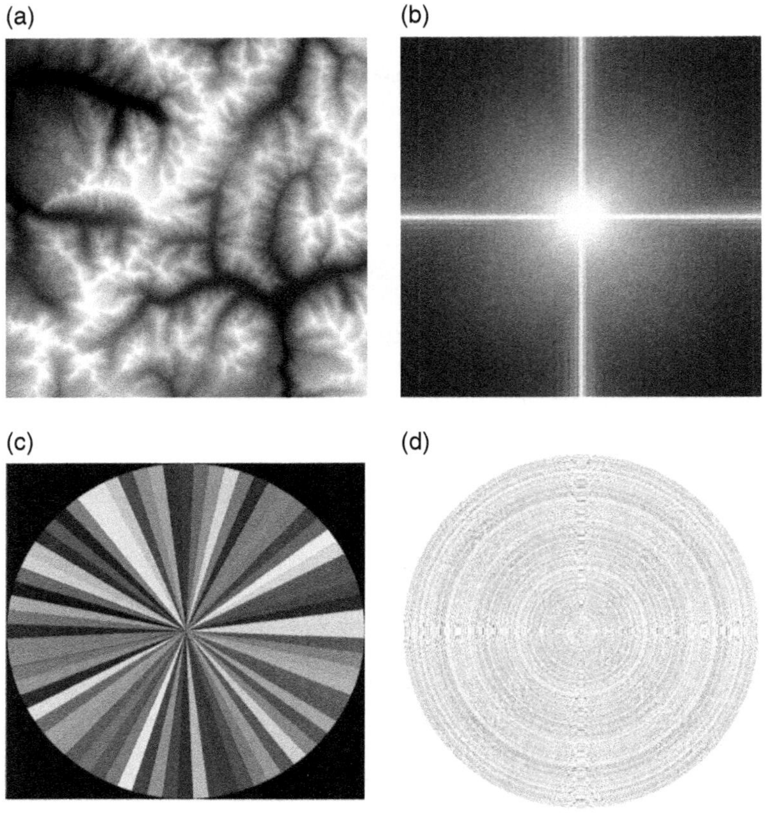

Figure 5.10 (a) The DEM of study area with 10-m spatial resolution. (b) The FFT image based on the DEM. A log transformation is applied for image enhancement. (c) A wedge-based filter is applied to the FFT image. The interval is every 5°. (d) A ring-based filter is applied to the FFT image. The interval is every 5 pixels.

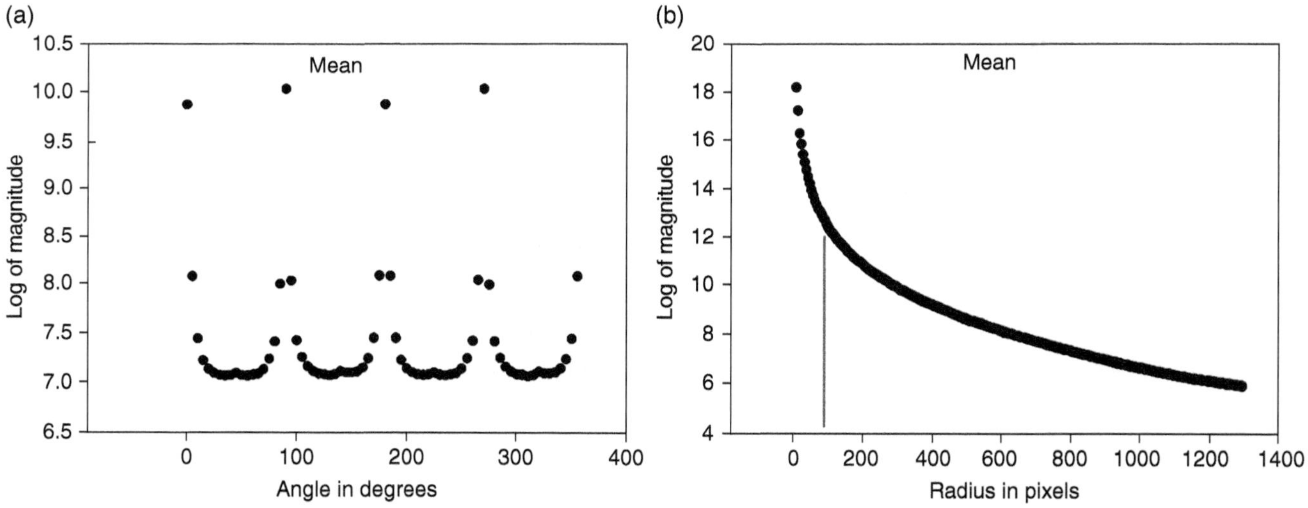

Figure 5.11 (a) The average magnitude within each wedge segment. (b) The average magnitude within each ring segment.

regarding orientation, a wedge-based filter with 5° interval (Figure 5.10c) is applied to the FFT image. To check scale regarding distance, a ring-based filter with 5-pixel interval (i.e. 50 m; Figure 5.10d) is also applied to the FFT image.

Within each filter segment (i.e. wedge and ring), FFT intensity is first summed and then averaged by a count of pixels in each segment. The final statistics are plotted as shown in Figure 5.11. Figure 5.11a illustrates no strong dependence based on surface orientation. Figure 5.11b illustrates that a strong scale dependence of 70 m is evident.

5.5.4.1.3 Terrain Analysis

Pennock et al. (1987) and Pennock (2003) have proposed a classification regime based on landform shapes, which

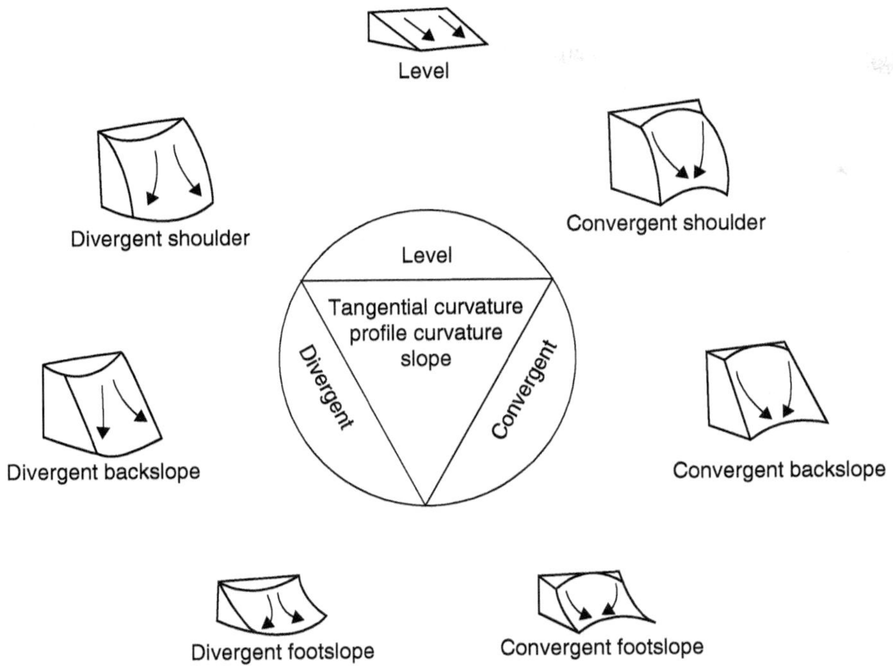

Figure 5.12 Illustration of original classification based on landform.

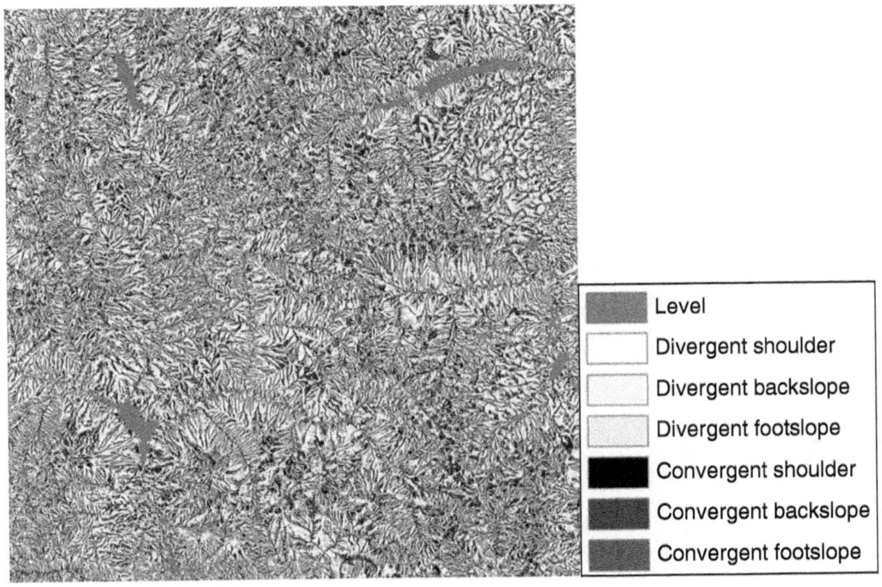

Figure 5.13 Landform classification map of the San Juan Mountains.

Giardino (1971) developed, including divergent shoulder, convergent shoulder, divergent backslope, convergent backslope, divergent footslope, convergent footslope, and level (Figure 5.12). We adopted this algorithm to reclassify the DEM of the San Juan Mountains. The result shows a unique spatial pattern of different landforms in the San Juan Mountains, which has not yet been applied in other research (Figure 5.13). Surface roughness can facilitate evaluating surface erosion rates, understanding drainage development and runoff pathways, and extracting geomorphology features (Hengl and Reuter, 2009; Florinsky, 2012). Hobson (1972) introduced the surface roughness factor (SRF) by incorporating slope and aspect of a local terrain. We adopted Hobson method to compute the surface roughness ratio over the San Juan Mountains (Figure 5.14). We are

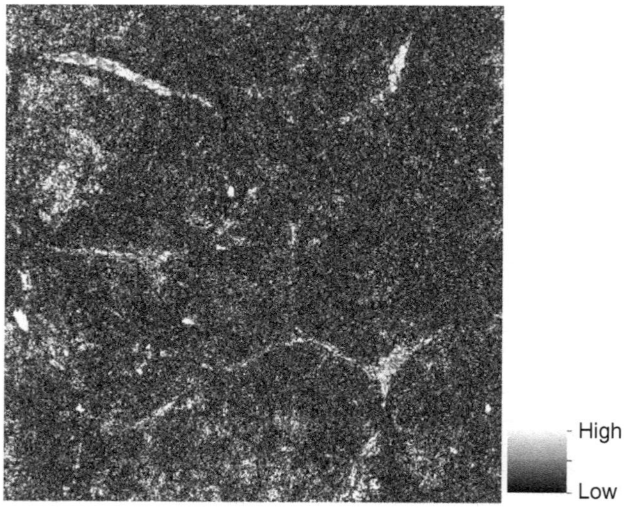

Figure 5.14 Surface roughness in the San Juan Mountains.

integrating landform, surface roughness, geomorphology, terrain convexity/concavity, and lithology to assess the erosion rate across the Western San Juan Mountains.

5.5.4.1.4 River Channel Extraction Based on New Topographic Index

Geomorphometry offers the efficiency and also flexibility for developing new algorithms for impact analysis. This case study illustrates how we developed a new topographic index for delineating river flow directions. This algorithm can be seen in Figure 5.15. For each pixel, if water flows toward all eight neighboring pixels, this pixel is classified as divergent. In contrast, if water flows into the center pixel from all eight neighboring pixels, this pixel is classified as convergent (Claps et al., 1994). Divergence and convergence images are computed based on this theory (Figure 5.16). Divergent features, such as mountain ridges, are highlighted in the divergent image (Figure 5.16a). Convergent features, such as river channel, are highlighted in the convergent image (Figure 5.16b).

To extract river channels from convergence images, two methods are tested and compared in this study, including a thresholding method and a fuzzy classification method. Because a divergence image is normalized between 0 and 1 (Figure 5.16b), based on the histogram, it is reasonable to establish several thresholds, which can separate a river channel from other low-intensity features. In this research, values at 0.8, 0.7, 0.6, 0.5, and 0.4 were established.

The thresholding method is a typical hard-classification rule to discriminate different classes. This rule omits the facts that in reality there is a transition between one class and another (Jensen, 2005). Fuzzy set theory, however, can compensate the omission because of its strong capacity of dealing with imprecise data (Wang, 1990; Zadeh, 1995). To apply a fuzzy classification, a fuzzy membership function needs to be selected first. In this research, we adopted an exponential decay function for fuzzy classification (Figure 5.17). The classification results are shown in Figures 5.18 and 5.19. By comparison, fuzzy classification approach is more reliable than thresholding approach.

5.5.4.1.5 Discussion

FFT are effective tools for characterizing scale-dependence issues in mountainous settings. This case study shows the dominate scale in the San Juan Mountains is 70 m and no strong scale associated with orientation exists. Landform classification and surface roughness can help better understand the geomorphology and subsequent weather and climatic conditions. Convergence index is a valuable resource for delineate drainage channels, and ongoing research in using rigorous statistics analysis can be used to validate the erosion rate in the San Juan Mountains. Drainage channel stratification can be implemented using topology and automation.

5.5.4.2 Modeling Glacier Change

5.5.4.2.1 Background

The second study area is in the Karakoram region of Pakistan. Good estimates of glacier-surface velocity can aid the understanding of processes related to glacier dynamics, including supraglacial mass transport, ice flow

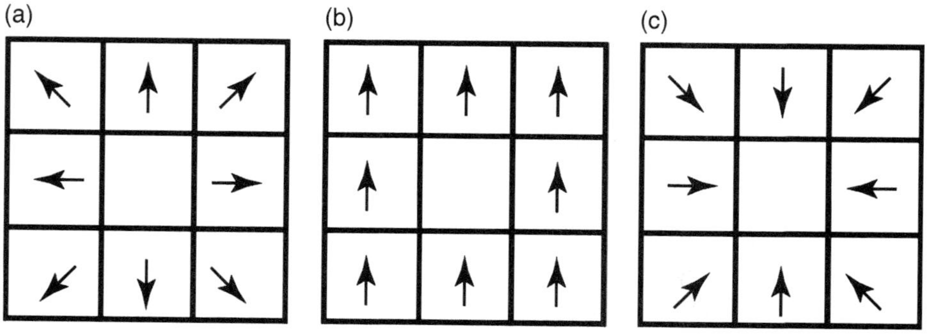

Figure 5.15 Illustration of the convergence and divergence concepts. (a) Divergence, (b) level, and (c) convergence.

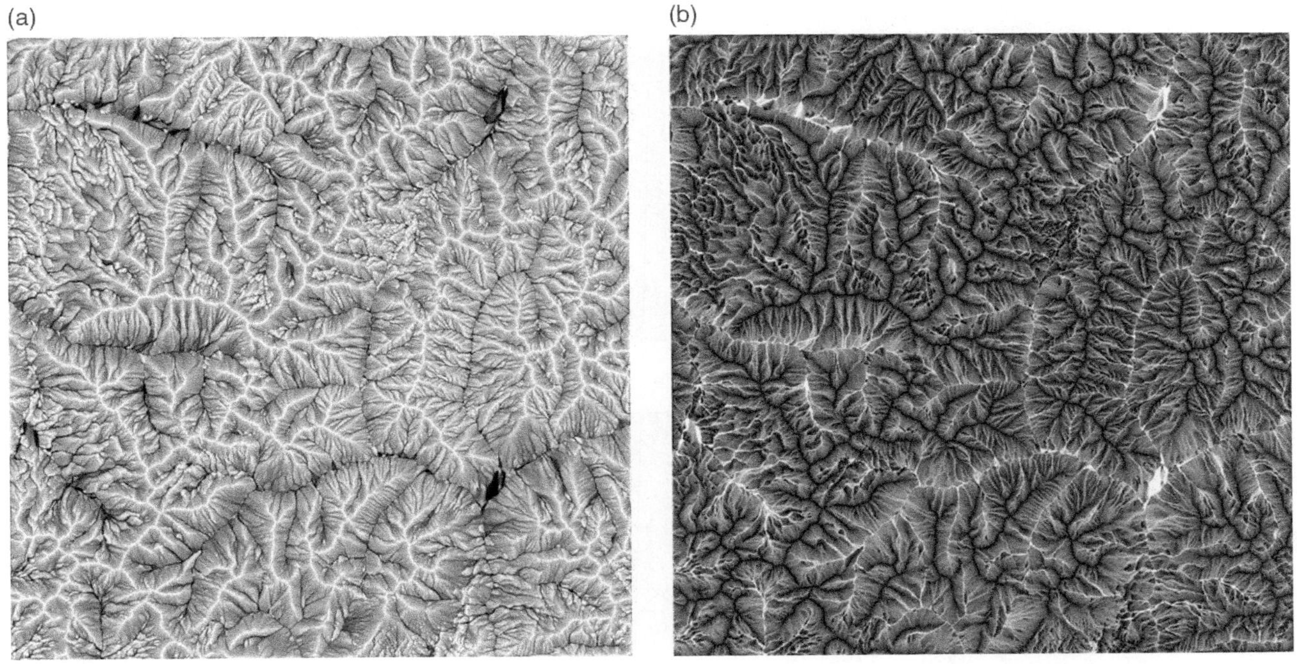

Figure 5.16 (a) The divergence index image, showing mountain ridges as highlighted. (b) The convergent index image, showing drainage channels as highlighted.

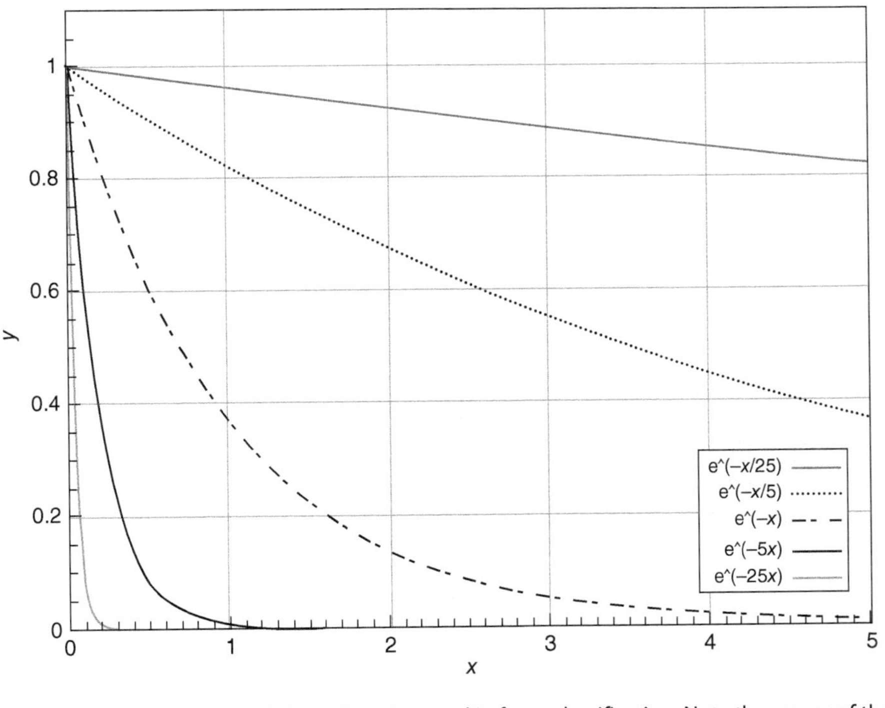

Figure 5.17 The exponential decay function used in fuzzy classification. Note the power of the function can dramatically alter the results of the classification.

instability (e.g. surging glacier), the supraglacier lake development, and ice flow direction and glacier boundary location (Paul et al., 2015). For optical satellite images, two major image-matching methods are widely used in glacier studies: normalized cross-correlation (NCC) in spatial domain and orientation correlation (CFF-O) in the frequency domain (Heid and Kaab, 2012). The NCC matching method was first proposed by Lewis (1995), and it has been used by research because of its suitability for measuring velocity fields in the regions of high visual

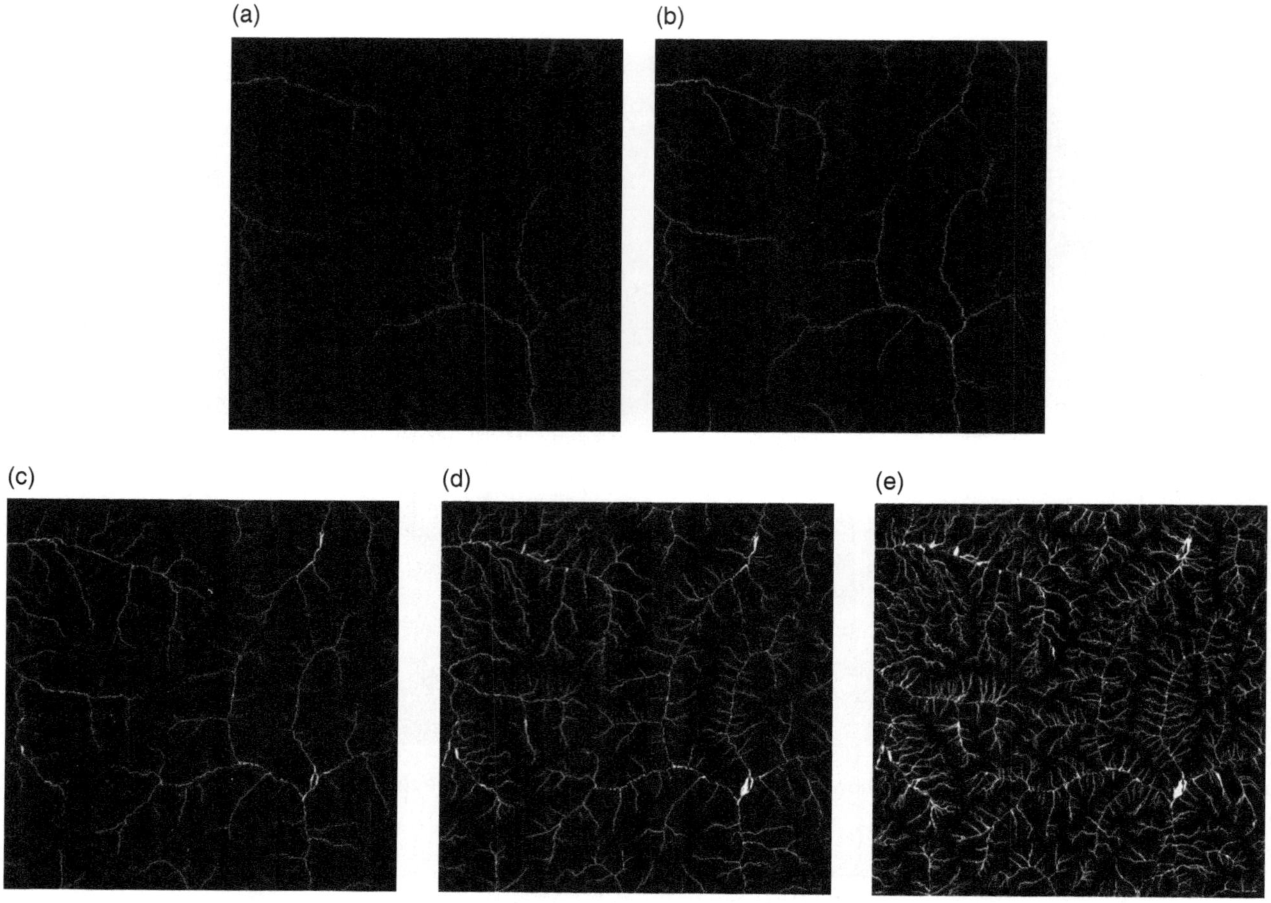

Figure 5.18 The results from the thresholding method at values of (a) 0.8, (b) 0.7, (c) 0.6, (d) 0.5, and (e) 0.4, respectively.

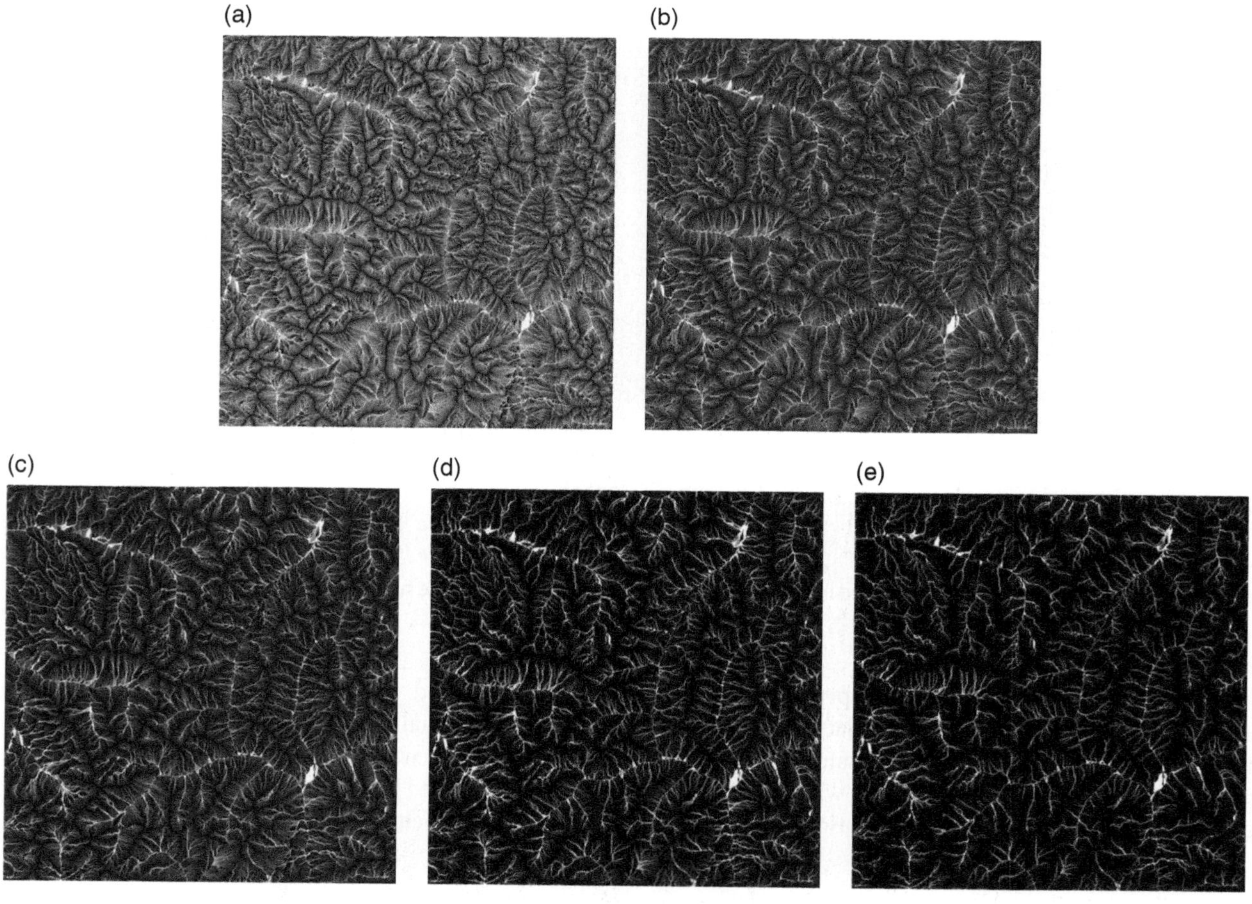

Figure 5.19 The results from the fuzzy classification at powers of (a) 50, (b) 25, (c) 10, (d) 5, and (e) 1, respectively.

contrast (i.e. ablation zone of debris cover; Paul et al., 2015). Currently we are investigating the velocity fields as derived by the CFF-O method, which is a part of the co-registration of optically sensed images and correlation (COSI-Corr) software module in ENVI (Exelis Visual Information Solutions; Leprince et al., 2007).

Extensive mountain glacier retreating is occurring in the Eastern Himalayas, which is a response to global warming (Quincey et al., 2009). However, glaciers in Western Himalayas show a unique pattern of glacier changes that stands in contradiction to global trends (Yadav et al., 2004). Thus advancing glaciers are particularly dominant in Karakoram, Pakistan. According to climate data (1961–2000), winter temperature increased, and the summer temperature decreased in this region, which accounts for such activity (Fowler and Archer, 2006). Quincey et al. (2009) used advanced synthetic-aperture radar (ASAR) data to examine the glacier velocity variations in Baltoro Glaciers covering a period from 1993 to 2008. However, because the footprint ranges from 20 to 30 m, and because Advanced Spaceborne Thermal Emission and Reflection Radiometer (ASTER) covers a finer footprint (15 m), it is fundamental to derive an updated glacier velocity profiles of Baltoro Glacier using ASTER images. In this case study, 3 years' velocities (2002–2004) of Baltoro Glacier are calculated by adopting CFF-O feature tracking technique.

Baltoro Glacier is located in Northeastern Pakistan (Figure 5.20), where it descends from the Western face of K2 (8611 m) via Godwin Austen Glacier. It terminates approximately 63 km down valley at an elevation of 3500 m. A number of small debris-covered glacier tributaries join the main glacier tongue along its northern and southern margins, some of which exhibit surge-type characteristics (Quincey et al., 2009).

5.5.4.2.2 Feature Tracking

In this case study, the COSI-Corr algorithm was used. This algorithm was developed by IDL and then integrated into ENVI, which is a remote sensing image processing software (Leprince et al., 2007). COSI-corr was developed to detect coseismic displacement but has been extended to other Earth science applications, such as geomorphometry change detections. Researches have

Figure 5.20 The Baltoro Glacier located at the Western Himalayas. The glacier itself is surrounded by a very complicated topographic condition. *Source:* From Quincey et al. (2009).

Figure 5.21 ASTER true color composite images for sequent 3 years.

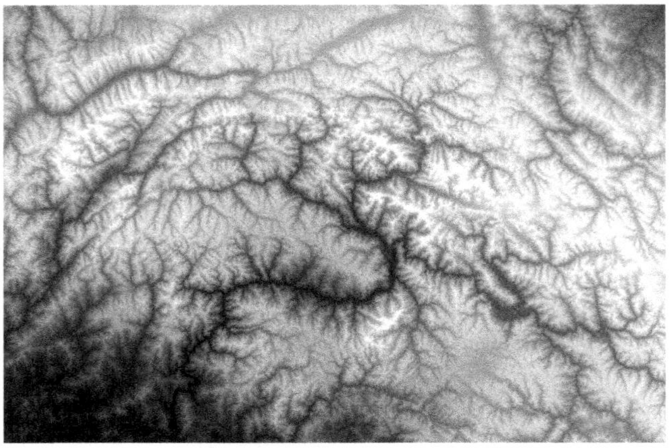

Figure 5.22 ASTER Global Digital Elevation Model (GDEM) over Karakoram.

demonstrated the utility of this algorithm, including glacier and dune movements (Scheidt and Lancaster, 2013).

In general, COSI-Corr utilizes a pair of images having clear temporal separations (i.e. before event and post-event). A series of processing steps need to be done within this algorithm: orthorectifications, displacement generations, velocity field calculations, and noise corrections (Leprince et al., 2007).

As at Baltoro, summer ablation could dramatically influence changes of the glacier, it would be wise to choose images that are captured at the end of summer, which will produce better results. In this research, 3 years' images were retrieved from ASTER Reverb website and these are 2002-10-03, 2003-09-20, and 2004-09-15 (Figure 5.21). To have an orthorectification output, ASTER Global Digital Elevation Model (GDEM) was also retrieved (Figure 5.22). To obtain velocity profiles, a centerline consisting of the whole glacier body was digitized and then used for extracting velocity profiles based upon the velocities generated.

The velocities generated from COSI-Corr are shown in Figure 5.23. The left side of Figure 5.23 shows the velocities between 2002 and 2003 ASTER images, whereas the right side is the velocities between 2003 and 2004. From visual detection, they both show a quite similar pattern of glacier velocities. For 2002–2003, the dominant velocity ranges from 50 to 100 (m year^{-1}), which is also the dominant velocity for 2003–2004. Also, from the up-glacier to terminus, they both show a descending trend of velocities. Especially for the terminus regions, the velocities decrease to below 20 (m year^{-1}).

By tradition, a velocity across the whole glacier body normally refers to the centerline velocity profile. Baltoro Glacier velocity profiles are extracted (Figure 5.23) from different pairs of velocity fields (i.e. 2002–2003 and 2003–2004 annual velocities). However, it should be noted that these profiles are extracted using exactly the same centerline, which reduces systematic errors. From Figure 5.24, both pairs of velocity profiles are decreasing consistently from up-glacier (around 70 m year^{-1}) to the terminus (around 20 m year^{-1}). The overall velocity of 2003–2004 is higher than the velocity in 2002–2003. Unique spikes in the profile are noise that has to be removed.

5.5.4.2.3 *Discussion*

Because the Karakoram glaciers are influenced by two significant driving factors, including the location with respect to the atmosphere circulation and air mass movement that climate change can have a marked effect. Another factor associated with mountain terrain is orthographic lifting. The regional climate varies seasonally and is related to global climate (Hewitt, 2005, 2013). Thus, it is fundamental to investigate the global climatic trend for past years, especially the decision of selection of 3 years' period under a globally climatic scenario. It is important to realize that this research can be further extended to investigate the correlation between climate forcing and glacier movement.

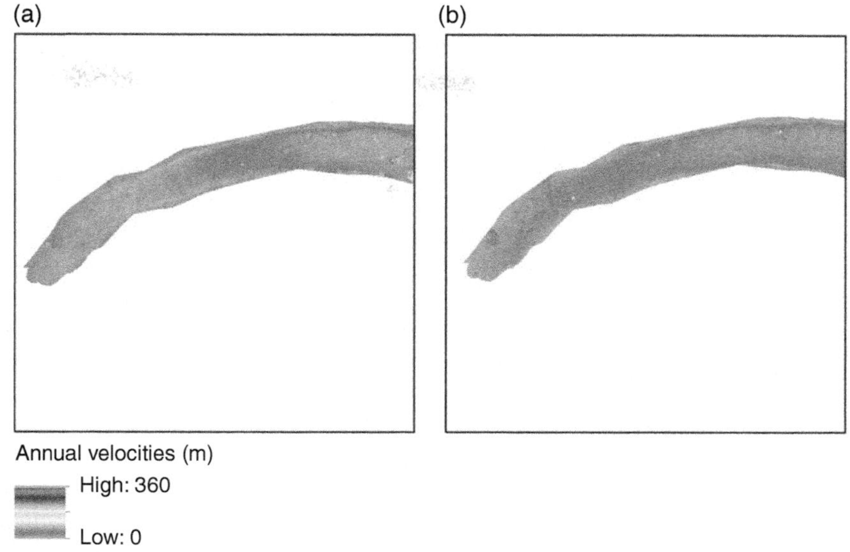

Figure 5.23 The velocity fields derived by COSI-Corr. (a) The annual velocity between 2002 and 2003; (b) the annual velocity between 2003 and 2004.

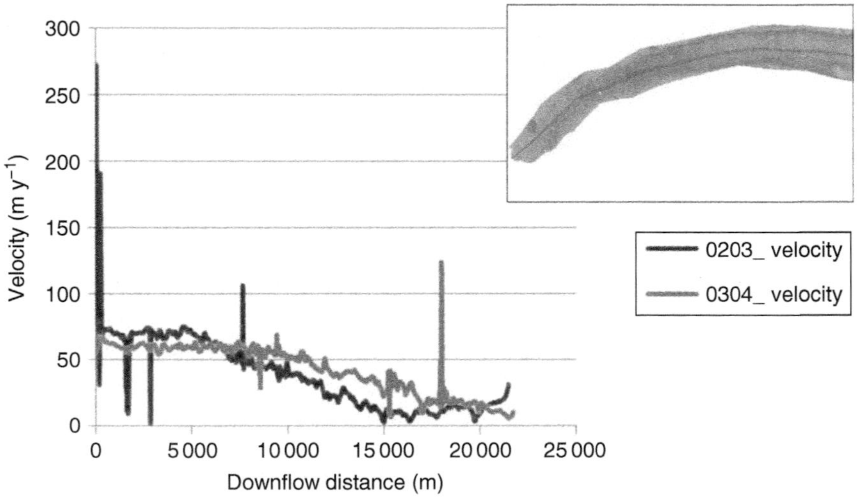

Figure 5.24 The velocity profile extracted from the centerline of the glacier (upper right shows the center line). The black line of the plot is the annual velocities between 2002 and 2003. The gray line represents the annual velocities between 2003 and 2004.

5.5.4.3 Automated Land Cover Stratification of Alpine Environments

5.5.4.3.1 Background

In Alpine glacier studies, there is a lack of glacier information regarding supraglacial lakes, debris load distribution, and ablation rates (Bajracharya and Mool, 2009; Benn and Lehmkuhl, 2000; Benn et al., 2001; Cogley et al., 2010). A better classification of basic land cover mapping of Alpine settings can facilitate the quantifying of the spatial distribution of different surfaces so as to enhance the understanding of glacier sensitivity of climate change. Unsupervised classifications such as Iterative Self-Organizing Data Analysis Technique (ISODATA) solely depend upon the statistical difference in multispectral feature space, which has limitations regarding characterizing spatial complexity, solving mix-pixel issues, and iteration efficiency (Dunn, 1973; Rees et al., 2003; Stow et al., 2003). Supervised classifications such as maximum likelihood or support vector machine (SVM) require a combination of a priori knowledge of the study area and field validations, which could potentially introduce bias and inaccuracies (Melgani and Bruzzone, 2004; Strahler, 1980). What is very problematic is a supervised classification demands intensive labor input (Hodgson et al., 2003). When considering the drawbacks of traditional classification regimes, we have successfully developed an automated decision-tree classification algorithm (Figure 5.25) based on MATLAB®, which is specifically suitable in Alpine settings, and the basic land cover can be classified into five categories: clouds, snow/ice, water, debris, and vegetation.

Cloud detection is the first step we use in this case study because clouds can hamper the reflectance of an Earth surface (Bolch et al., 2007). Although simple

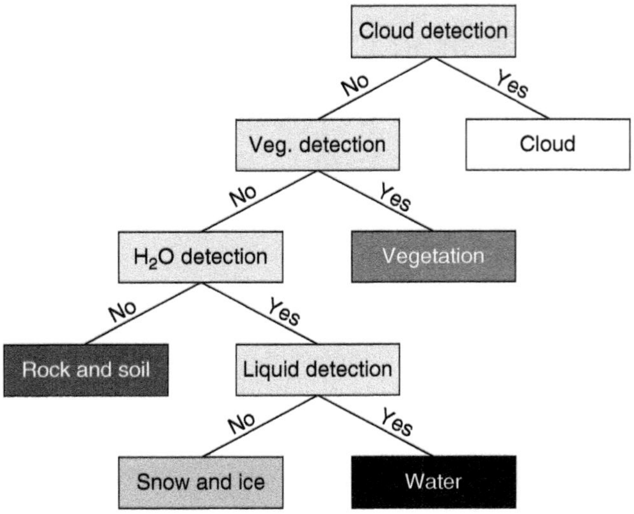

Figure 5.25 Decision-tree workflow of land cover stratification of an alpine environment.

selection of cloud-free images will significantly reduce the data availability, most geophysical parameters derived from ASTER, such as elevation, reflectance, and emissivity, are generated from individual scene without merging the data into a seamless dataset with respect to time and space. These raster-based products are important for many climate models such as GCM. For instance, a GCM may require land surface emissivity as an input to model SST evolution. In most cases, these geophysical parameters are not easily accessible in cloud-free conditions. Thus, it is a reason for development of a robust cloud detection algorithm.

Clouds are highly a reflective feature in visible–near-infrared (VNIR) and mid-infrared (MIR) bands but are low in the thermal bands as a result of low temperature. Besides, features like snow and highly reflective soil can also produce very similar spectral characteristics as clouds. Unfortunately ASTER cloud cover assessment algorithm (ACCAA) has limitations, including the inability to handle radiometric calibration, geographic registration, misdetection of cloud shadow, inaccuracy of snow elimination, etc. The applicability of other algorithms including Landsat-7 and MODIS are not suitable for Alpine settings because Landsat-7 algorithm will not account for cloud shadows and thin cirrus. In addition, MODIS has a 1-km footprint and also is not suitable for most Alpine studies (Jensen, 2005, 2007).

5.5.4.3.2 Stratification

In the stratification algorithm used in ASTER VNIR, SWIR, and thermal-infrared (TIR), bands are first converted to radiance and then planetary reflectance. Top of atmosphere (TOA) brightness temperatures are then calculated. The normalized difference vegetation index (NDVI), the normalized difference snow index (NDSI), and the normalized difference wetness index (NDWI) are all calculated and used as algorithm input. For each node in the decision tree, several thresholds are used for filtering purposes. The threshold values (Table 5.12) from Hulley and Hook (2008) are modified to enhance the suitability for an Alpine environment. In addition, spatial autocorrelation of spectral information is also considered in this decision tree. This study adopted the same study area of case study 1, which is Baltoro Glacier in Pakistan.

Preliminary result showed the classification worked well on Baltoro Glacier and its surrounding mountain area (Figure 5.26). The five basic land cover types, i.e. clouds, snow and ice, vegetation, water, and rock and soil, are well delineated (Figure 5.27). The stratification process is completely automated.

Table 5.12 Filters used in Hulley and Hook (2008).

Filer	Threshold test	Function
1) Brightness threshold	$r_2 > 0.08$	Eliminates low reflectance, dark pixels
2) Snow threshold	$\text{NDSI} = (r_1 - r_4)/(r_1 + r_4) < 0.7$	Eliminates snow
3) Temperature threshold	$T_{sat} < 300$	Eliminates warm surface features
4) Band 4/5 composite	$(1 - r_4)T_{sat} < 240 \rightarrow$ snow present	Eliminates cold surfaces – snow, tundra
	$(1 - r_4)T_{sat} < 250 \rightarrow$ snow absent	
5) Growing vegetation	$r_3/r_2 < 2$	Eliminates reflective growing vegetation
6) Senescing vegetation	$r_3/r_1 < 2.3$	Eliminates reflective senescing vegetation
7) Rocks and sand	$r_3/r_4 > 0.83$	Eliminates reflective rocks and sand
8) Warm/cold cloud	$(1 - r_4)T_{sat} > 235 \rightarrow$ warm cloud	Warm and cold cloud classification
	$(1 - r_4)T_{sat} < 235 \rightarrow$ cold cloud	
9) Cloud shadow	$r_3 < 0.05$ and $r_3/r_1 > 1.1$	Detects cloud shadow

Where r_i is the reflectance of band I and T_{sat} is temperature retrieved from thermal band.

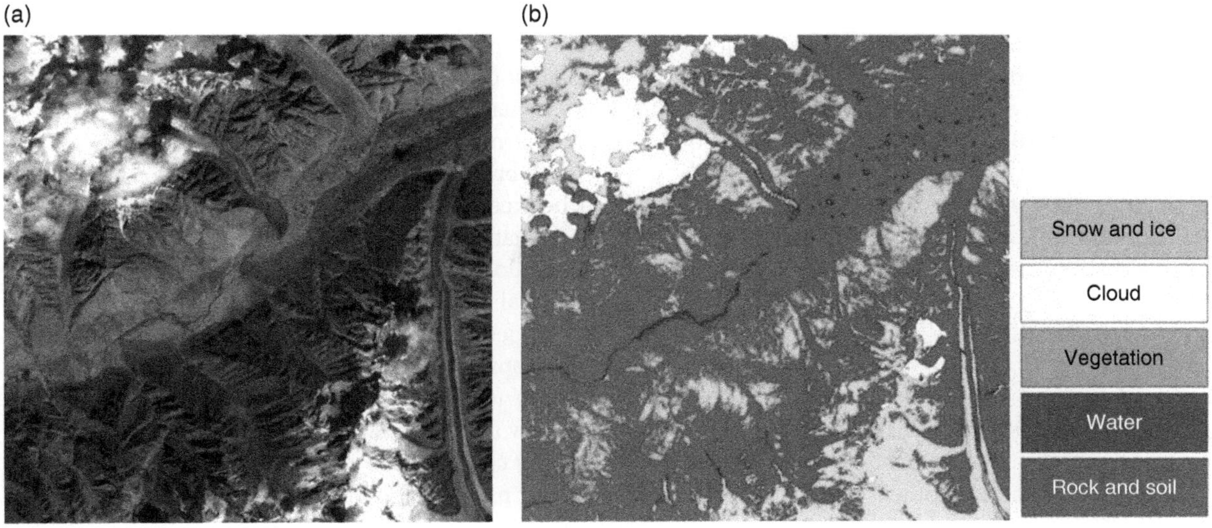

Figure 5.26 The RGB color composite of ASTER image over Baltoro Glacier and its surrounding area are shown in (a). Note there is cloud cover in the terminus area. The integrated stratification results are shown in (b). Note cloud is well picked up.

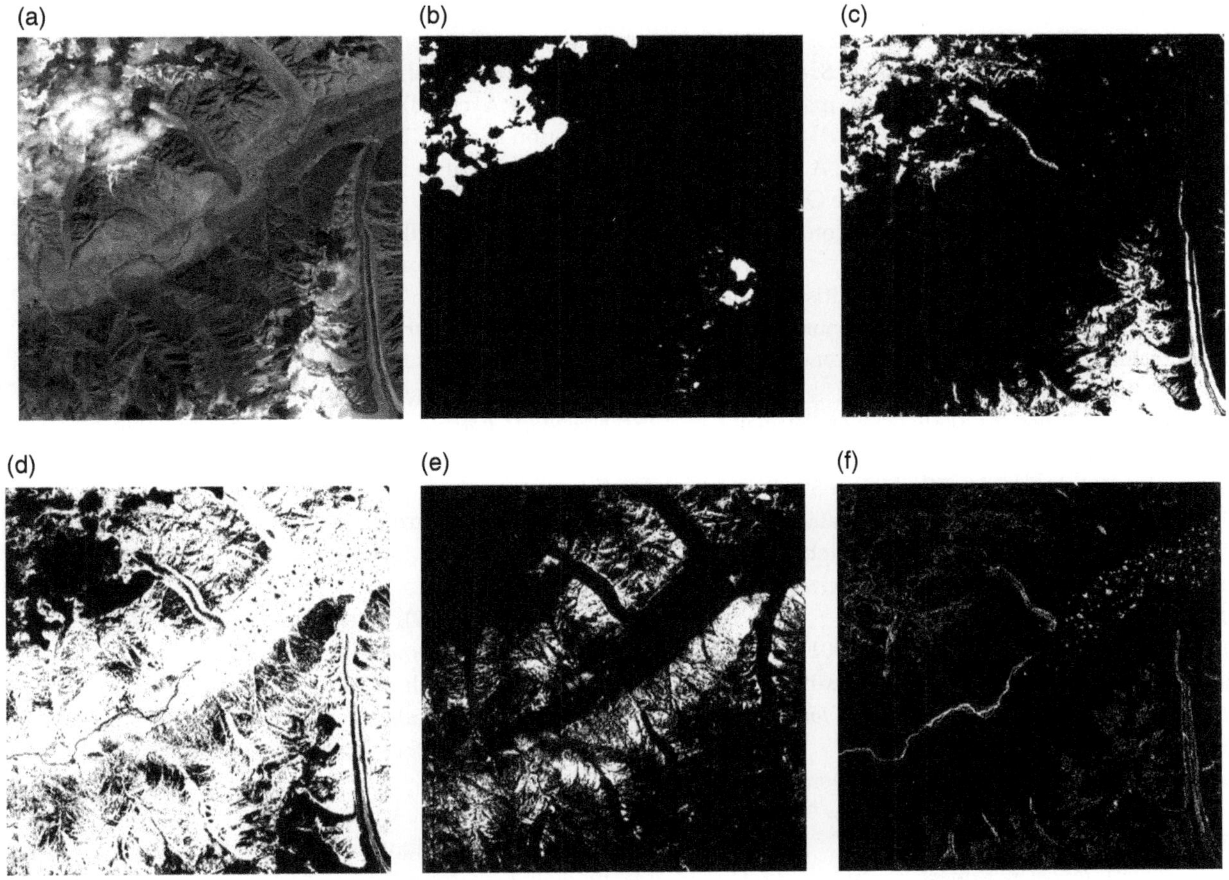

Figure 5.27 (a) The RGB color composite of an ASTER image over the study area. (b) Cloud mask delineated by this algorithm. (c) Snow and ice mask. (d) Rock and soil mask. (e) Vegetation mask. (f) Water mask.

5.5.4.3.3 Discussion

The ability to differentiate between clear and cloudy conditions using a cloud-filter algorithm is important for producing useful geophysical parameter products from ASTER for climate change studies (Bush and Philander, 1999). The algorithm used in this case study provides accurate land cover mapping in Alpine glacial environments. Accurate land cover stratification can also be used to understand glacier dynamics in the following ways: help detect cloud cover that interferes or masks interest in an alpine surface, help examine the elevational distribution of supraglacial debris loads, help evaluate the distribution and development of supraglacial lakes, mitigate GLOF, help estimate surface density for mass balance study, and assist in the delineation of glacier boundaries. This algorithm can also be extended to help study tree-line migrations and snowline fluctuations.

5.5.5 Summary

Three case studies in this section illustrate that geomorphometry is very powerful for climate change impact analysis. Many people have commented that a lot of researchers have been hampered by their narrow scope of tight coupling. However, as demonstrated in this section, a host of different software packages and programming languages are available for collaboration. A loose coupling among different technique platforms can significantly broaden the breath of understanding interactions between climate and the surface of the Earth. We suggest that engineers take advantage of open-source software along with programming, as compensation to commercial software. Besides, as noted, there are new trends in geomorphometry studies, including new sensor development producing higher resolution images, using UAV for data collection in rugged terrain, big data technology for information mining, super computers for enhancing computation capacity, computer vision technology for pattern recognition, and AI.

References

Ambaum, M.H.P. (2010). Significance tests in climate science. *American Meteorological Society* 23: 5927–5932.

Amundson, R., Richter, D.D., Humphreys, G.S. et al. (2007). Coupling between biota and earth materials in the critical zone. *Elements* 3 (5): 327–332.

Anderson, R.S., Anderson, S., Aufdenkampe, A.K. et al. (2010). Future Directions for Critical Zone Observatory (CZO) Science: National Science Foundation. CZO Community (29 December).

Anuta, P.E. (1970). Spatial registration of multispectral and multitemporal digital imagery using fast Fourier transform techniques. *IEEE Transactions on Geoscience Electronics* 8: 353–368.

Bajracharya, S.R. and Mool, P. (2009). Glaciers, glacial lakes and glacial lake outburst floods in the Mount Everest region, Nepal. *Annals of Glaciology* 50: 81–86.

Benn, D.I. and Lehmkuhl, F. (2000). Mass balance and equilibrium-line altitudes of glaciers in high-mountain environments. *Quaternary International* 65 (6): 15–29.

Benn, D.I., Wiseman, S., and Hands, K.A. (2001). Growth and drainage of supraglacial lakes on debris-mantled Ngozumpa glacier, Khumbu Himal, Nepal. *Journal of Glaciology* 47: 626–638.

Bjerknes, J. (1969). Atmospheric teleconnections from the equatorial Pacific. *Monthly Weather Review* 97: 163–172.

Bolch, T., Buchroithner, M.F., Kunert, A., and Kamp, U. (2007). Automated delineation of debris-covered glaciers based on ASTER data. In: *GeoInformation in Europe* (ed. M.A. Gomarasca), 403–410. Rotterdam: Millpress.

Brantley, S.L., Goldhaber, M.B., and Ragnarsdottir, K.V. (2007). Crossing disciplines and scales to understand the critical zone. *Elements* 3 (5): 307–314.

Bretherton, C.S., Smith, C., and Wallace, J.M. (1992). An intercomparison of methods for finding coupled patterns in climate data. *Journal of Climate* 5: 541–560.

Burrough, P.A. (1983). Multiscale sources of spatial variation in soil. I. The application of fractal concepts to nested levels of soil variation. *Journal of Soil Science* 34: 577–597.

Bush, A.B.G. and Philander, S.G.H. (1999). The climate of the last glacial maximum: results from a coupled atmosphere-ocean general circulation model. *Journal of Geophysical Research-Atmospheres* 104: 24509–24525.

Butler, K., Hammerschmidt, S., Steiner, F., and Zhang, M. (2009). *Reinventing the Texas Triangle: Solutions for Growing Challenges*, 29. Center for Sustainable Development, School of Architecture, University of Texas at Austin.

Carbonell, A. and Yaro, R.D. (2005). American spatial development and the new megalopolis. *Land Lines* 17 (2): 1–4.

Carleton, A.M. (1999). Methodology in climatology. *Annals of Association of American Geographers* 89: 713–735.

Carrara, A., Cardinali, M., Detti, R. et al. (1991). GIS techniques and statistical models in evaluation landslide hazard. *Earth Surface Processes and Landforms* 16: 427–445.

Chen, S.H., Jakeman, A.J., and Norton, J.P. (2008). Artificial intelligence techniques: an introduction to their use for modeling environmental systems. *Mathematics and Computers in Simulation* 78: 379–400.

Chiles, J.P. and Delfiner, P. (1999). *Geostatistics: Modeling Spatial Uncertainty*. New York: Wiley.

Claps, P., Fiorentino, M., and Oliveto, G. (1994). Informational entropy of fractal river networks. *Journal of Hydrology* 187 (1–2): 145–156.

Cogley, J.G., Kargel, J.S., Kaser, G., and Van der Veen, C.J. (2010). Tracking the source of glacier misinformation. *Science* 327: 522–522.

Cooley, J.W., Lewis, P.A., and Welch, P.D. (1969). The fast Fourier transform and its application. *IEEE Transactions on Education* 12: 27–34.

Cressie, N. (1993). *Statistics for Spatial Data*. New York: Wiley.

Crowell, J.C. (1978). Gondwanan glaciation, cyclothems, continental positioning, and climate change. *American Journal of Science* 278: 1345–1372.

Crutzen, P.J. (2002). Geology of mankind. *Nature* 415 (6867): 23–23.

Curran, P.J. (1988). The semivariogram in remote sensing: an introduction. *Remote Sensing of Environment* 24: 493–507.

Daubechies, I. (1992). *Ten Lectures on Wavelets (Regional Conference Series in Applied Mathematics)*. Philadelphia, PA: SIAM: Society for Industrial and Applied Mathematics.

Deng, Y. (2007). New trends in digital terrain analysis: landform definition, representation, and classification. *Progress in Physical Geography* 31 (4): 405–419.

Dewar, M. and Epstein, D. (2007). Planning for "Megaregions" in the United States. *Journal of Planning Literature* 22 (2): 108–124.

Dikau, R., Brunsden, D., Schrott, L., and Ibsen, M.L. (1996). *Landslide Recognition: Identification, Movement, and Causes*. Chichester: Wiley.

Doocy S., Daniels A., Murray S., and Kirsch T. D., (2013). The Human Impact of Floods: a Historical Review of Events 1980-2009 and Systematic Literature Review. PLOS Currents Disasters.

Dorfman, M.H., Mehta, M., Chou, B. et al. (2011). *Thirsty for Answers: Preparing for the Water-Related Impacts of Climate Change in American Cities*. New York: Natural Resources Defense Council.

Dunn, J.C. (1973). A fuzzy relative of the ISODATA process and its use in detecting compact well-separated clusters. *Journal of Cybernetic* 3: 32–57.

Feldstein, S.B. (2000). The timescale, power spectra, and climate noise properties of teleconnection patterns. *Journal of Climate* 13: 4430–4440.

Florinsky, I.V. (2012). *Digital Terrain Analysis in Soil Science and Geology*. Oxford: Elsevier.

Foufoula, E. and Kumar, P. (1994). *Wavelet in Geophysics*. London, England: Academic Press.

Fowler, H.J. and Archer, D.R. (2006). Conflicting signals of climatic change in the Upper Indus Basin. *Journal of Climate* 19: 4276.

Francis, P. and Oppenheimer, C. (2004). *Volcanoes*. Oxford: Oxford University Press.

Frohlich, C. (1998). The Sun's total irradiance: cycles, trends, and related climate change uncertainties since 1976. *Geophysical Research Letter* 25: 4377–4380.

Furfaro, R., Kargel, J.S., Lunine, J.I. et al. (2010). Identification of cryovolcanism on Titan using fuzzy cognitive maps. *Planetary and Space Science* 58: 761–779.

Gaffen, D.J. and Ross, R.J. (1999). Climatology and trends of U.S. surface humility and temperature range. *Journal of Climate* 12: 11–28.

Gallo, K.P., Easterling, D.R., and Peterson, T.C. (1996). The influence of land use/land cover on climatological values of the diurnal temperature range. *Journal of Climate* 9: 41–44.

Germann, U., Galli, G., Boscacci, M., and Bolliger, M. (2006). Radar precipitation measurement in a mountainous region. *Quarterly Journal of the Royal Meteorological Society* 132: 1669–1692.

Giardino, J.R. (1971). A comparative analysis of slope characteristics for the Colorado plateau. M.S. thesis. Arizona State University. Page 38 of 149.

Giardino, J.R. and Houser, C. (2015). *Principles and Dynamics of the Critical Zone*. Amsterdam: Elsevier.

Gonzalez, R.C., Woods, R.W., and Eddins, S.L. (2009). *Digital Image Processing Using Matlab*. Upper Saddle River, NJ: Pearson Prentice Hall.

Goodchild, M.F. (1992). Geographic information science. *International Journal of Geographic Information Systems* 6: 31–45.

Gornitz, V. (2009). Sea level change, post-glacial. In: *Encyclopedia of Paleoclimatology and Ancient Environments*. Dordrecht: Springer.

Goudie, A.S. (2006). Global warming and fluvial geomorphology. *Geomorphology* 79 (3–4): 384–394.

Graf, W.L. (2008). In the critical zone: Geography at the U.S. Geological Survey. *The Professional Geographer* 56 (1): 100–108.

Greiving, S., Fleischhauer, M., and Wanczura, S. (2006). Management of natural hazards in Europe: the role of spatial planning in selected EU member states. *Journal of Environmental Planning and Management* 49: 739–757.

Haykins, S. (1999). *Neural Networks: A Comprehensive Foundation*, 2e. Upper Saddle River, NJ: Prentice Hall.

Heid, T. and Kaab, A. (2012). Evaluation of existing image matching methods for deriving glacier surface displacements globally from optical satellite imagery. *Remote Sensing of Environment* 118: 339–355.

Hengl, T. and Reuter, H.I. ed. (2009). *Geomorphometry: Concepts, Software, Applications*, Developments in Soil Science, vol. 33. Amsterdam: Elsevier.

Hewitt, K. (2005). The Karakoram anomaly? Glacier expansion and the "elevation effect," Karakoram Himalya. *Journal of Glaciology* 13: 103.

Hewitt, K. (2013). *Glaciers of the Karakoram Himalaya*. Dordrecht: Springer.

Hobson, R.D. (1972). Surface roughness in topography: quantitative approach. In: *Spatial Analysis in Geomorphology* (ed. R.J. Chorley). New York: Harper & Row.

Hodgson, M.E., Jenson, J.R., Tullis, J.A. et al. (2003). Synergistic use of LIDAR and color aerial photography for mapping urban parcel imperviousness. *Photogrammetric Engineering & Remote Sensing* 69: 973–980.

Houser, C., Bishop, M.P., and Barrineau, P. (2015). Characterizing instability of aeolian environments using analytical reasoning. *Earth Surface Processes and Landforms* 40: 696–705.

Hsu, K., Gupta, H.V., and Sorroshian, S. (1995). Artificial neural network modeling of the rainfall-runoff process. *Water Resources Research* 31: 2517–2530.

Hubbard, B.B. (1998). *The World According to Wavelets*, 2e. Natick, MA: A. K. Peters Ltd.

Hulley, G.C. and Hook, S.J. (2008). A new methodology for cloud detection and classification with ASTER data. *Geophysical Research Letters* 35. doi: 10.1029/2008GL034644.

Jeffrey, S.J. (2001). Using spatial interpolation to construct a comprehensive archive of Australian climate data. *Environmental Modelling & Software* 16: 309–330.

Jensen, J.R. (2005). *Introductory Digital Image Processing.* Upper Saddle River, NJ: Pearson Prentice Hall.

Jensen, J.R. (2007). *Remote Sensing of the Environment.* Upper Saddle River, NJ: Pearson Prentice Hall.

Karafyllidis, I. and Thanilakis, A. (1997). A model for predicting forest fire spreading using cellular automata. *Ecological Modeling* 99: 87–97.

Kim, T.W. and Ahn, H. (2009). Spatial rainfall model using a pattern classifier for estimating missing daily rainfall data. *Stochastic Environmental Research and Risk Assessment* 23: 367–376.

Kosko, B. (1986). Fuzzy cognitive maps. *International Journal of Man-machine Studies* 24: 65–75.

Krige, D.G. (1951). A statistical approach to some mine valuations and allied problems at the Witwatersrand. Master's thesis. University of Witwatersrand.

Lau, K.M. and Weng, H.Y. (1995). Climate signal detection using wavelet transform and the scale analysis of the surface properties of sea ice. *IEEE Transactions on Geoscience and Remote Sensing* 34: 771–787.

Lemke, P., Ren, J., Alley, R.B. et al. (2007). Observations: changes in snow, ice and frozen ground. In: *Climate Change 2007: The Physical Science Basis. Contribution of Working Group I to the Fourth Assessment Report of the Intergovernmental Panel on Climate Change* (ed. S. Solomon, D. Qin, M. Manning, et al.). Cambridge, United Kingdom and New York, NY, USA: Cambridge University Press.

Leprince, S., Ayoub, F., Klinger, Y., and Avouac, J.P. (2007). Co-registration of optically sensed images and correlation (COSI-Corr): an operational methodology for ground deformation measurements. In: *IGARSS: 2007 IEEE International Geoscience and Remote Sensing Symposium*, vol. 1–12, 1943–1946. New York: IEEE.

Levin, R.B., Epstein, P.R., Ford, T.E. et al. (2002). U.S. drinking water challenges in the twenty-first century. *Environmental Health Perspect* 110 (Suppl 1): 43–52.

Lewis, J.P. (1995). Fast normalized cross-correlation. *Vision Interface* 10: 120–123.

Lin, H. (2010). Earth's critical zone and hydropedology: concepts, characteristics, and advances. *Hydrology and Earth System Sciences* 14: 25–45.

Ling, Y. and Mahadevan, S. (2013). Quantitative model validation techniques: new insight. *Reliability Engineering & System Safety* 111: 217–231.

Little, R.J.A. and Rubin, D.B. (2002). *Statistical Analysis with Missing Data.* New York: Wiley.

Longley, P., Goodchild, M.F., Maguire, D.J., and Rhind, D.W. (2004). *Geographic Information Systems and Science.* New York: Wiley.

Lorenz, E.N. (1956). *Empirical Orthogonal Functions and Statistical Weather Prediction*, Scientific Report No. 1, 49 pp. Cambridge, MA: Massachusetts Institute of Technology.

MacCollum, D. (2007). *Construction Safety Engineering Principles: Designing and Managing Safer Job Sites.* New York: McGraw-Hill.

Mace, R.E. and Wade, S.C. (2008). In hot water? How climate change may (or may not) affect the groundwater resources of Texas. *Proceedings 2008 Joint Meeting of the Geological Society of America, Soil Science Society of America, American Society of Agronomy, Crop Science Society of America, Gulf Coast Association of Geological Societies with the Gulf Coast Section of SEPM2008, GCAGS Transactions*, pp. 655–668.

Marden, J. (2000). Hypothesis testing: from *p* value to Bayes factors. *Journal of American Statistical Association* 95: 1316–1320.

Matheron, G. (1971). *The Theory of Regionalized Variables and Its Applications.* Paris: Mines Paris Tech.

Meentemeyer, V. (1989). Geographical Perspectives of Space, Time, and Scale. *Landscape Ecology* 3: 63–173.

Melgani, F. and Bruzzone, L. (2004). Classification of hyperspectral remote sensing images with support vector machines. *IEEE Transaction on Geoscience and Remote Sensing* 42: 1778–1790.

Mitchell, J.M. (1976). An overview of climatic variability and its causal mechanism. *Quaternary Research* 4: 481–494.

National Research Council (NRC) (2001). *Basic Research Opportunities in Earth Science.* Washington, DC: National Academy Press.

Ohnaka, M. (2013). *The Physics of Rock Failure and Earthquakes*. Cambridge: Cambridge University Press.

Padowski, J.C. and Jawitz, J.W. (2012). Water availability and vulnerability of 225 large cities in the United States. *Water Resources Research* 48 (12). doi: 10.1029/2012WR012335.

Palani, S., Liong, S., and Tkalich, P. (2008). An ANN application for water quality forecasting. *Marine Pollution Bulletin* 56: 1586–1597.

Papageorgiou, E.I. (2012). Learning algorithms for fuzzy cognitive maps – a review study. *IEEE Transaction on Systems, Man, and Cybernetics, Part C: Application and Reviews* 2: 150–163.

Papageorgiou, E.I. and Salmeron, J.L. (2013). A review of fuzzy cognitive maps research during the last decade. *IEEE Transaction on Fuzzy Systems* 21: 66–79.

Paul, F., Bolch, T., Kaab, A. et al. (2015). The glaciers climate change initiative: methods for creating glacier area, elevation change and velocity products. *Remote Sensing of Environment* 162: 408–426.

Pennock, D.J. (2003). Terrain attributes, landform segmentation, and soil redistribution. *Soil and Tillage Research* 69 (1–2): 15–26.

Pennock, D.J., Zebrath, B.J., and de Jong, E. (1987). Landform classification and soil distribution in hummocky terrain, Saskatchewan, Canada. *Geoderma* 40: 297–315.

Pichon, X.L. (1968). Sea-floor spreading and continental drift. *Journal of Geophysical Research* 73: 3661–3697.

Pierce, J.L., Meyer, G.A., and Timothy, J.A. (2004). Fire-introduced erosion and millennial-scale climate change in northern ponderosa pine forest. *Nature* 432: 87–90.

Pike, R.J. and Park, M. (1995). Geomorphometry – progress, practice, and prospect. *Zeitschrift fur Geomorphologie NF* 101: 221–238.

Quincey, D.J., Luckman, A., and Benn, D. (2009). Quantification of Everest region glacier velocities between 1992 and 2002, using satellite radar interferometry and feature tracking. *Journal of Glaciology* 192: 596.

Rees, W.G., Williams, M., and Vitebsky, P. (2003). Mapping land cover change in a reindeer herding area of the Russian Arctic using Landsat TM and ETM+ imagery and indigenous knowledge. *Remote Sensing of Environment* 85: 441–452.

Regmi, N.R., Giardino, J.R., and Vitek, J.D. (2013). Hazardness of a place. In: *Encyclopedia of Natural Hazards*. Dordrecht: Springer.

Richter, D.D. and Mobley, M. (2009). Monitoring earths critical one. *Science* 326: 1067–1067.

Rohli, R.V. and Vega, A.J. (2013). *Climatology*, 3e. Burlington, MA: Jones & Bartlett.

Scheidt, S. and Lancaster, N. (2013). The application of COSI-Corr to determine dune system dynamics in the southern Namib Desert using ASTER data. *Earth Surface Processes and Landforms* 38: 1004–1019.

Schneider, T. (2000). Analysis of incomplete climate data: estimation of mean values and covariance matrices and imputation of missing values. *American Meteorological Society* 14: 853–871.

Seaton, A., MacNee, W., Donaldson, K., and Godden, D. (1995). Particulate air pollution and acute health effects. *Lancet* 345: 176–178.

Short, D.A., Mengel, J.G., Crowley, T.J. et al. (1991). Filtering of Milankovitch cycles by Earth's geography. *Quaternary Research* 35: 157–173.

Smith-Spark, L. and H. Messia, 2017 Italy avalanche: hotel search ends with 29 dead, 11 rescued. *CNN News* (26 January).

Stow, D., Coulter, L., Kaiser, J. et al. (2003). Irrigated vegetating assessment for urban environment. *Photogrammetric Engineering & Remote Sensing* 69: 381–390.

Strahler, A.H. (1980). The use of prior probabilities in maximum likelihood classification of remotely sensed data. *Remote Sensing of Environment* 10: 135–163.

Sui, D.Z. and Giardino, J.R. (1995). Applications of GIS in environmental equity analysis: a multi-scale and multi-zoning scheme study for the city of Houston, Texas, USA. GIS/LIS'95, Nashville, TN (14–16 November), Vol. 2, pp. 950–959.

Thomas, D.S. (2004). Desertification. In: *Encyclopedia of Geomorphology*. London: Routledge.

Thuiller, W. (2004). Patterns and uncertainties of species' range shifts under climate change. *Global Change Biology* 10: 2020–2027.

Torrence, C. and Compo, G.P. (1997). A practical guide to wavelet analysis. *Bulletin of the American Metorological Society* 79: 61–78.

Trenberth, K.E. (1997). The definition of El Niño. *Bulletin of the American Meteorological Society* 78: 2771–2777.

USGS (2012). Ring of Fire. USGS. 2012-07-24.

Uyeda, S., Nagao, T., and Kamogawa, M. (2009). Short-term earthquake prediction: current status of seismo-electromagnetics. *Tectonophysics* 470: 205–213.

Varnes, D.J. (1984). *Landslide Hazard Zonation: A Review of Principles and Practice. IAEC Commission on Landslides and Other Mass Movement on Slopes*, 63. Paris: UNESCO.

Vaughan, D.G., Comiso, J.C., Allison, I. et al. (2013). Observations: Cryosphere. In: *Climate Change 2013: The Physical Science Basis. Contribution of Working Group I to the Fifth Assessment Report of the Intergovernmental Panel on Climate Change* (ed. T.F. Stocker, D. Qin, G.-K. Plattner, et al.). Cambridge, United Kingdom and New York, NY, USA: Cambridge University Press.

Wang, F. (1990). Fuzzy supervised classification of remote sensing images. *IEEE Transactions on Geoscience and Remote Sensing* 28: 194–201.

Wang, B., Wu, R., and Fu, X. (2000). Pacific-east Asian teleconnection: how does ENSO affect East Asian climate? *Journal of Climate* 13: 1517–1536.

Wilding, L.P. and Lin, H. (2006). Advancing the frontiers of soil science towards a geoscience. *Geoderma* 131 (3–4): 257–274.

Yadav, R.R., Park, W.K., Singh, J., and Dubey, B. (2004). Do the western Himalayas defy global warming? *Geophysical Research Letters* 31 (17).

Zadeh, L.A. (1995). Fuzzy sets. *Information and Control* 8: 338–353.

Zhang, M., Steiner, F., and Butler, K. (2007). Connecting the Texas triangle: economic integration and transportation coordination. *Proceedings of the Healdsburg Research Seminar on Megaregions*, Healdsburg, California, pp. 21–36.

Ziliak, S. and McCloskey, D. (2008). *The Cult of Statistical Significance*. Ann Arbor, MI: University of Michigan Press.

6

Adaptation Design to Sea Level Rise

Mujde Erten-Unal[1] and Mason Andrews[2]

[1] *Department of Civil and Environmental Engineering, Old Dominion University, Norfolk, VA, USA*
[2] *Department of Architecture, Hampton University, Hampton, VA, USA*

6.1 Introduction: Sea Level Rise

This chapter provides information on sea level rise and adaptation design strategies to alleviate the flooding impacts of this threat to the coastal communities of the United States. One of the impacts of climate change to coastal communities is the sea level rise. Sea level rise, an increase in the average surface level of oceans, is caused by thermal expansion and melting of polar ice caps due to climate change and temperature increases, thus adding to the existing volume of water in the oceans (IPCC, 2007; Warrick and Oerlemans, 1990; Withey et al., 2016). Climate change and sea level rise could lead to increased precipitation, storm surges, and higher occurrence of flooding to the residential and commercial infrastructure of many coastal cities throughout the world. This infrastructure must be adaptive to these conditions as our reliance on the coast is not something that will change.

6.1.1 Background

The Intergovernmental Panel on Climate Change (IPCC) estimated that sea level would rise by an additional 0.6–1.9 ft (0.18–0.59 m) by 2100 (Meehl et al., 2007; Solomon et al., 2007). However, the 2007 IPCC projections are conservative, and additional research showed higher estimates of sea level rise greater than 1 m (Nicholls, 2011; Nicholls et al., 2008). Even worse, sea level rise in itself will lead to an increase in the occurrence of what is presently considered extreme flooding.

In some areas throughout the Eastern United States, from Maine to Texas, just a one-meter rise in sea levels will result in immense amounts of coastal inundation. New Orleans will be 91% inundated with just another meter of sea level, and Miami and Tampa will be 18–15% inundated (America's Climate Choices, 2010). This does not include storm surge or higher than normal tides that may occur. On an average day, an immense amount of land in the Eastern United States will be inundated.

6.1.2 Causes of Sea Level Rise

The area that the authors work extensively on adaptation design to sea level rise is the Hampton Roads region located on the mid-Atlantic coast region. Hampton Roads relies on the sea heavily but is also dangerously affected by it as well. Sixteen cities and four rivers draining into the Chesapeake Bay make it one of the largest natural harbors in the world (Coastal Resilience Strategy). The Hampton Roads region is experiencing sea level rise at approximately twice the global rate (Boon, 2012; Ezer and Corlett, 2012a, b; Sallenger et al., 2012). This increased rate of sea level rise regionally is due to land subsidence and the slowing of the Gulf Stream (Boon et al., 2010; Ezer et al., 2013). This area, especially the city of Norfolk in Virginia, is at a further risk from inundation because the area is sinking due to several factors. The subsidence in Norfolk is caused by three factors including groundwater pumping, glacial isostatic adjustment, and that it is located at the edge of an ancient impact crater. The largest of these three factors is groundwater pumping. Groundwater pumping from industrial sites is what contributes most to the subsidence. When water is removed out of the aquifer, the clay layers compress, creating land subsidence across a large area (Leake, 2013).

The secondary factor causing subsidence is glacial isostatic adjustment. The last glacial maximum was about 16 000 years ago and much of North America was covered by glaciers several kilometers tall. The weight of the additional glaciers created a bulge on the edge of the glaciers, which is now sinking back. This takes thousands of years to occur and it will continue for several more thousand years. An estimate of how much the area of the

Handbook of Environmental Engineering, First Edition. Edited by Myer Kutz.

Chesapeake Bay will sink due to this is approximately 6 in. in the next 100 years (NOAA, 2015).

The last factor affecting subsidence in the area is the impact crater in the Chesapeake Bay. On the edge of the crater, older rock layers have already slid down into the crater; however newer sediments are constantly shifting and settling. These newer sediments also contribute to subsidence in Norfolk, which is on the edge of the crater. As the sediments settle, they compact effectively, causing subsidence (USGS, 2016).

6.1.3 Impacts of Sea Level Rise

Rising sea levels, storm surges, and floods may lead to other detrimental impacts along the coastal areas and may have ripple effect on a variety of infrastructure. Some of the impacts include but not limited to be the following:

- Increased storm surge and hence coastal erosion.
- Disruption of coasts, particularly along the lowlands and coastlines composed of soft sediments (Langis, 2013).
- Damage to infrastructure.
- Damage to the economy including businesses, tourism, and rising insurance rates.
- Saltwater intrusion to aquifer.

The objective of this chapter is to describe and examine various adaptation strategies for flooding caused by storm surge, sea level rise, and heavy storms. This chapter will help determine which strategies would be most effective in increasing the resiliency of coastal infrastructures against flooding while minimizing damage and cost with specific case studies from the Hampton Roads region.

6.2 Existing Structures and Adaptation Design to Sea Level Rise

6.2.1 Conventional Wisdom: Protect, Elevate, and Retreat

As individual structures have begun to experience more threat from flooding, three basic approaches have been most widely discussed. Each shall be examined briefly here. A challenge to each is the speculative understanding of how quickly sea levels will rise; uncertainty on the subject leads to uncertainty in planning the elevations of barriers, of first floors, and of where, in fact, safety lies. Each solution also involves fairly high costs.

6.2.1.1 Protection

The silver bullet of a barrier built to keep the sea off traditionally dry land is the most popular among building owners still hoping the life of a building and district can go on as it has. A dyke or some other armoring of a shoreline or some large-scale engineered barrier such as has been built on the Thames and is under construction around Venice, paid for, it is assumed, by a benign government entity, is particularly appealing to those just beginning to have concerns about flooding threats.

In the area of the authors' work, large-scale barriers pose a conflict with the very reason we fight to remain in place: the life of the area is rooted in maritime industry and naval defense. No sensible solutions for damming the Chesapeake Bay or the Hampton Roads harbor have been given serious consideration for these reasons, and it would be presumed, because the area is so low, that protection would also need to address the armoring of the Atlantic shore – a proposal too vast for reasoned consideration.

The Dutch are leaders in engineered solutions to keep seawater away from built areas. As a nation, they are vulnerable to sea level rise because 27% of the country's territory is below the sea level (Hunt and Mandia, 2012). The Dutch government spends around $1.3 billion per year to build and maintain flood protection structures and for water control projects. In addition, the local governments spend hundreds of millions to maintain their water control structures such as dikes and canals (Higgins, 2012). But as pioneers in massive engineered solutions to sea level rise, the Dutch have also been coming to terms with the complexities that a system of armoring entails. First, there has been loss of ecological diversity as tidal areas, and the ecosystems they support have been destroyed in pursuit of armoring, safety, and stability (Walker et al., 1994). Second, the waterways whose headlands lie outside the nation's borders but that flow into Holland can generate downstream flooding. With the global increase in the intensity of precipitation, swollen rivers beyond the control of Dutch engineers are causing flooding within the Netherlands (Klijn et al., 2012).

Along the Thames and in the Venetian Lagoon, massive deployable water barriers have been or are under construction. These represent enormous investments, and each is now seen to be not quite tall enough to have continued utility with the rapid rise of sea level.

6.2.1.2 Elevating Structures

Elevating buildings to a presumed level of safety is becoming more popular in the United States, in part because federal funding can sometimes be made available to property owners who have experienced repetitive losses from flooding. To elevate, a building is braced, detached from its existing foundation, and jacked to height deemed safe from potential floodwaters. While straightforward, this strategy, too, has several significant drawbacks.

First, there is the issue of the imprecision of prophesy: how high is high enough? The images below show buildings that have been raised 14 ft above grade. Each has experienced subsequent flooding, evidenced by the waterline visible on the building as shown in Figure 6.1. Given the cost of the house-jacking projects, frustration and expense seem likely to be cyclical.

Second, and most crucially to urban communities, the raising of individual buildings within an existing neighborhood can undercut the viability of that neighborhood. Urban districts "read" as wholes when there is alignment between buildings in relative height of rooflines and entry sequences. In addition to being visually jarring, the social fabric of life within a community in which the vital role of the porch as an intermediary zone for social activity and the provision of "eyes on the street" has an important role in public safety and civic life. For these benefits to be of value to a community, the relationship of porches to street cannot be successful much above 6 ft.

After the hurricane of 1900, almost all the structures in Galveston were raised, a case often cited as a success story for the procedure (Ramos, 1997). And, while not for reasons having to do with flooding, the buildings of Chicago also were systematically raised in 1855 (Young, 2017). In these two instances, the roadways were also elevated. With all buildings and streets raised, a viable urban district can be maintained. Yet marshaling funding for such large-scale interventions is difficult to imagine being possible for every coastal community. More piecemeal approaches, such as the current program to raise roadways in Miami-Dade, leave some curious urban conditions if the buildings that front them are not simultaneously raised (Flechas and Staletovich, 2015).

6.2.1.3 Retreat

Moving buildings to a higher elevation seems also a sensible solution, particularly for property parcels that have a higher elevation within the property lines. The solution, however, is expensive, and the numbers of parcels containing a mesa within them are few. Ultimately, structures of great and undisputed historical significance may be moved to higher ground. The relocation of one of Frank Lloyd Wright's Usonian Houses, the Pope-Leighey, out of the way of Interstate 66 in 1946[1] was a success. But more usually, the value of an architectural cultural resource includes the land and district in which it stands, upping the challenge – and the cost – of any potential relocation.

Eventually and, sadly, inevitably, discussions will begin about emigration from coastal areas that are too regularly flooded. But social and economic stability will require a long view – a multi-generational planning process. The good people of Tangier Island, which is sinking into the Chesapeake Bay, still hold out fierce hope that a barrier will be designed and constructed that will allow them to stay. Unhappily, this is unlikely.

6.2.2 Accommodation

To buy time for reasoned planning by individuals and communities, a fourth approach, adopted by the authors in their design studio, has been to learn strategies for living with occasional inundation. Our goal is to leave buildings and people in place into the next century by developing a plan for neighborhood-wide water management.

Developing strategies requires a series of solutions at many scales – the district, the street (public right of ways), public open space, and individual property parcels. This may sometimes create redundancy, which we believe a good thing for the goal of keeping communities safe and relatively dry. Designing a portfolio of solutions to let

Figure 6.1 Comparison of flood damage with amphibious house versus elevated house. Water line marks the water level reached the elevated house during a flood. *Source:* Courtesy of Dr. Elizabeth English, Buoyant Foundation Project.

communities live with water requires a multivalent approach, a shopping list of interventions, some of which can be realized by the efforts of the community and individual property owners and some of which can be selected for lobbying efforts in each municipal budget cycle.

As a first step, we look for opportunities to create protection from the incursion of seawater. In many locations now-undersized stormwater lines are actually acting as conduits into neighborhoods as outfalls are now submerged during storms. The insertion of check valves is the single most effective intervention we have modeled. Where possible, the creation of a living shoreline can provide both ecological benefits and, if allowed higher ground for retreat as sea level rise will literally drown wetlands, protection of land mass from erosion by rising seas. Armored barriers are sometimes necessary, although our program has worked to identify barriers that are not dependent on piping and pumping.

As a second step, it is important to look for places to store water until after storm tides subside. Public open space, if it lies over soil with good drainage with a low water table, can be made use of for deploying large-scale cisterns beneath the surface of park land. Ideally these cisterns re-discharge their water into the soil below. Where soil and water table conditions are less favorable, reshaping open space into dry swales is relatively economical and provides significant storage capability. Vacant lots can also be considered for the installation of cisterns, swales, or rain gardens.

Third, the areas of the public right of way have great potential utility. The area between the sidewalk and curb, the verge, can be rebuilt as a bioretention planting area. Permeable paving on sidewalks and the parking zone of the street itself can also play a role. The maintenance requirement of permeable paving entails persuasion of public works departments, but that it will soon be a necessary part of their operations seems inevitable.

Finally, rooftop disconnection programs are crucial. Reducing runoff from private property to zero leaves only the rainfall and tidal flooding amounts on the streets accumulating. The authors have found that most community members are willing and even thankful to have a role to play.

6.2.2.1 Letting the Water in

Engineers and architects have been trained to keep water away from the materials of habitable construction. Concerns with rot, mold, and the degradation of materials are valid guidance methods for new construction. But adaptation requires a reconsideration of these goals. Water has to go somewhere and is notoriously disrespectful of property lines. Our modeling shows, and anecdotal evidence from community engagement sessions confirms, that work waterproofing one basement will cause a neighbor to flood within a city district. Modeled, a block in an urban neighborhood in which basements are successfully waterproofed or filled per flood insurance recommendation, will be subject to flooding over first floor framing.[2] Clearly basements can have a role to play in water management. The city of Detroit also demolished blighted structures to create more stormwater retention basins (green infrastructure) and open space lands (Hester, 2016).

To ready a basement to take on this work, preparation is necessary. All utilities must be moved above the first floor level. The area of the basement may be used for the installation of cisterns, taking roof runoff piped directly to them. But as groundwater rises, the simplest solution is simply to let the water seep in through a pervious floor (many older structures have a sand-laid brick floor that works admirably). This also has the added benefit of relieving hydrostatic pressure from below. Similarly, flood vents protect the basement walls from the same pressures and resist racking of the structure from the foundation. To resist mold, batt insulation and dry wall should be replaced with closed cell foam insulation and plaster or some other antimicrobial finish material. Such preparations do, of course, reduce the homeowners' usable square footage, but in areas subject to recurrent flooding, this loss is a *fait accompli*.

Outside the house, there are effective strategies to diminish the dampness of basement/foundation walls. Swales dug along property lines and minimal regrading can lead water away from the exterior walls. Most importantly, a rooftop disconnection program in which all runoff from roofs and driveways is collected in cisterns and rain gardens can direct a significant amount of water away from foundations.

It is true that the average brick and mortar foundation was not designed to be regularly wet. But rising groundwater is making it so already. The intent of accommodation is to let buildings and communities buy time to plan for change and to live out useful lives in place. While ultimately foundations will experience some hastened degradation and while repointing scheduling will be increased to maintain mortar joints, the trade-off seems logical if far from the ideal of conventional construction.

Preparations of first floors can also be taken into consideration, although the need for this may not be immediate. The authors recommend to homeowners that as renovations occur, selections of waterproof, non-molding materials be made. Thus oak strip flooring, subject to buckling if made wet, might be replaced with epay, mahogany, teak, or other woods not so fragile, or linoleum over cementitious board. Each of these could take on water and then be hosed off after flood incursion. Similarly, replacing batt insulation and sheet rock with lath and plaster and closed cell foam insulation could leave a homeowner capable of returning home and hosing down after a flood. Raising utility boxes higher up on the wall is another sensible method to extend the life of a building and minimize the losses from flooding.

6.2.2.2 Rediscovered and New Technologies of Potential Utility

Several technologies have been identified that the authors believe have potential use in adaptation work. Patents exist or are pending on each, but only one is currently in large-scale use. Each represents an appropriate and promising potential in adaptation.

The first devices are rooted in harnessing buoyancy. The Belgian Aggeres company is manufacturing a water-actuated floor–wall system. A Kevlar-covered fiberglass panel is housed in a sleeve that admits rising waters; the wall then rises within the sleeve to the height required. This is an enormously appealing system in that there are no pumps or staging involved; the wall simply rises when needed. The Austrian company FLOODPROTECT is developing a similar product, distinguished by the ambition to use the housing to generate passive heating and/or energy.

The second, too, harness buoyancy. Dr. Elizabeth English of the University of Waterloo is leading the Buoyant Foundation Project (http://buoyantfoundation.org). Her studies involve raising existing structures and reattaching them to a pontoon foundation that rises on floodwaters. The building is held in place by telescoping tethers. Utilities are housed in a breakaway umbilical connection. One would not expect occupants to attempt to ride out a storm in such a structure, but in areas such as that in which the authors work with much light frame wood construction and only intermittent flooding events, it could buy time for a generation to plan for the future.

The third technology identified as having potential utility is Electro-Osmotic Pulse technology. It uses a very low electrical pulse, an anode, and a cathode, which in combination draw water from basements and foundations outside the building perimeter. Installed by the US Army at Fort Monroe, in Hampton Virginia, basements once sodden are, many years later, dry enough to be in use again. The utility for foundations now wetter than they were designed to be is enormous.

Finally, a student in the authors' design studio has a patent pending on an interlocking system of cistern/planters. Their variability, and their capacity to store and reutilize captured rain water for irrigation, demonstrates the potential for ingenuity to meet changing and challenging conditions (https://1701vb.com/meet-our-members-willie-parks-of-floark/).

6.2.2.3 New Construction within Flood Zones

Many municipalities in areas now subject to frequent inundation have revised building codes to require a base design elevation above the base flood elevations (BFE) specified by FEMA. Requiring that the bottom of first floor framing be some specified number of feet above the BFE anticipates rising sea levels. It also will require great design ingenuity to maintain the integrity of a community's street level identity, charm, and security.

Nonresidential uses are usually expected and may be built within the flood zone – as long as an engineer or architect certifies that the ground floor is either dry-flood proofed with flood barriers to keep the water out or wet-flood proofed, allowing water in. The specifics and/or standards of these designs has not yet made its way into zoning ordinances or building codes and is up to the professional to design. This is likely to change as failures occur.

As a matter of best practice, the selection of materials that will not mold and may be simply cleaned after inundation seems reasonable for all buildings in a flood zone.

6.2.2.4 Intersection between Adaptation Strategies and Community Resilience

That resilience has become the buzzword in adaptation planning speaks to the scale of the problems associated with climate change: the cost and mobilization for adaptation work is too large for single municipalities or for state or federal intervention. This being the case, educating and assisting the citizens of threatened municipalities prepare for more regular inundation is the focus of much energy among policymakers. In the business community, "Resilience was defined by most as the ability to recover from setbacks, adapt well to change, and keep going in the face of adversity," and the applicability for planning for the impacts of sea level rise for communities is apparent (Ovans, 2015).

As it happens, the community engagement process described within the case studies that follow has been an important first step in resilience planning. In Norfolk, city officials have been stymied working with communities because they are seen as those who should provide the solutions, although, sadly, the scale of the problem far outstrips available financial resources. The members of the design studio, however, by opening up discussions on the issues and possible solutions that involve participation by all members of the community, provide a crucial first step in both accepting the issue and understanding that all property owners have a role to play.

Much adaptation work in urban neighborhoods focuses on strategies within each block – creating drainage swales to route water from foundations, for example, requires the collaborative work of abutting property owners. The block, called *insula*, or island, by the Romans, is literally an island during flooding. Property owners who have begun to collaborate can then extend that effort in post-storm safety checks, as these abutting neighbors are those who are capable of reaching those in distress.

6.2.2.5 Intersection between Adaptation Strategies and Historic Preservation

The fine men and women who have organized efforts locally and nationally to preserve the buildings and districts that illuminate the physical culture of prior eras represent both a movement and a profession. The preservation of designated resources is the fundamental motivator and all owe a debt of gratitude for the efforts. But while development and business pressures, as well as changing tastes, have been their traditional adversaries, the sea will be much less tractable.

At a 2016 conference, *Keeping History above Water*, in Newport, Rhode Island, which gathered many members of the preservation community with scientists and design professionals, the threat was recognized, but the most spoken of solutions involved raising buildings. As mentioned previously, this solution may buy some time for certain buildings, but the destruction to the district fabric is severe, the costs are high, and the solution is likely temporary.

An alternate bitter pill to swallow is to leave buildings in place and alter them only to accept floodwaters and to cherish these cultural resources for the rest of their viable lives. Accepting that resources will be lost is the necessary reality of rising seas. Working to keep historic structures and districts intact as they were built is their only practical future. Expending energy and resources to work to protract their useful lives by managing floodwaters at a district scale and documentation of buildings and districts not destined for immortality should be the focus.

Many states have programs granting tax credits for work on historic structures. Examining and fine-tuning enabling legislation to include water management work aimed at keeping a structure as dry as possible in inevitable inundation is a reasonable focus of lobbying efforts.

6.2.2.6 Intersection between Adaptation Strategies and Policy Issues

In rapidly changing conditions, legislation governing building, public policy, and the federal flood insurance program can become out of sync with protracting a sustainable future. We believe retooling policy away from individual parcels toward goals of sustaining urban districts is crucial. For example, flood insurance rates and regulations look solely at individual properties without reference to neighboring structures. In our work, we have had repeated observation of "fixes" applied to a structure creating worsening flooding for the neighboring property. Further, homeowners are incentivized – with reduced insurance premiums – to fill in basements, a recipe for worsening flooding for all adjacent properties. Finally, the implementation of district management practices, such as those described in the subsequent case studies, under current regulations will not reduce premiums.

Policy implementation that incentivizes individual property owners to participate in water management practices must be pursued. In addition to historic tax credit programs including water management in covered expenses, programs such as reductions in stormwater fees for on-parcel retention installations should be included, and educational programs developed to help property owners design and install such systems will become crucial. Regulations covering development, such as zoning statutes, must require of the developer retention systems. And public works departments need to be armed with design standards for water retaining street improvements so that any time road surfaces are being replaced, they can be made to mitigate flooding.

6.3 Case Studies Reflecting Adaptation Design Solutions

The Hampton Roads region in the United States is recognized as being second only to New Orleans as the largest population center at greatest risk to sea level rise (IEN, 2011). This region is experiencing sea level rise at approximately twice the global rate (Boon, 2012; Ezer and Corlett, 2012a, b; Sallenger et al., 2012). This increased rate of sea level rise regionally is due to land subsidence and the slowing of the Gulf Stream (Boon et al., 2010; Ezer et al., 2013).

Approximately 40% of Hampton Roads has a seasonal high groundwater table less than 1 ft from the soil surface. The remaining 60% is within 2 ft (HRPDC, 2013). Throughout the year, depth to the water table can vary from less than a foot to between 4 and 8 ft. Seasonal highs can occur between January and May but are most common between April and May, with seasonal lows typically occurring between August and November (VSWCB, 1981). Groundwater levels can also experience a greater degree of variability on a daily basis closer to the coast due to the effects of ocean and estuarine tides.

Two separate neighborhoods in Norfolk, Chesterfield Heights and The Hague, were studied as a result of them being continuously inundated with sea level rise, tidal storm surge, heavy precipitation, and flooding. The results of adaptation design responses to sea level rise and flooding applied in the neighborhoods are discussed in the following sections.

6.3.1 Case Study 1: Chesterfield Heights

In the fall of 2014, a group of Hampton University students of architecture embarked on a project with local NGO Wetlands Watch to study possible design interventions in a local historic district with the aim of helping residents and their homes be better prepared to deal with floodwaters in a major storm. Although many projects had been undertaken on neighborhoods after catastrophes nationally, Wetlands Watch could find no such work seeking to safeguard a community prior to damage. Four communities within the cities of Southeastern Virginia were considered for the project, the one selected being chosen for the quality of the extant architecture and the activism of its community organizations.

Chesterfield Heights, listed on the National Register of Historic Places, was developed in the teens and twenties of the last century as a streetcar suburb intended to serve downtown Norfolk, Virginia. It fronts the Elizabeth River on the south and is now bounded on the north by an elevated interstate highway. To its west are industrial facilities, and to its east is a public housing property. The community is predominately low to moderate income and predominately African American. Houses are a handsome collection of four square and Arts and Crafts styles and most are well maintained (Figure 6.2).

6.3.1.1 Community Engagement

Because it is the only responsible way to begin any project involving urban design, project directors built on relationships established during the proceeding summer to set up a program of community engagement sessions. First, students and faculty introduced themselves and their project at a regularly scheduled meeting of the area's civic league. Second, they scheduled a work session at the community center to conduct interviews and map identified issues. Having established relationships in this way, they began a series of door-to-door interviews working alongside community leaders.

Establishing a working presence within the community was, and is, crucial to the work. The trust developed with students allows work to proceed in a way that reflects community concerns and aspirations as well as gathering real insight into conditions not necessarily apparent without it. For example, from interview sessions, the design team learned that more serious flooding was far inland from the river, that efforts by individual neighbors to waterproof basements were followed by worsening flooding in neighboring properties, and that the heart and soul of the community centered on a strong porch culture.

6.3.1.2 Observations of Conditions

Community members confirmed that the waterfront park shoreline was eroding in large part due to tug and barge traffic. It also had been observed to overtop more regularly. Also confirmed was that the two frequently submerged stormwater outfalls that served the community were acting as conduits, leading water back into the community. Some recent ADA compliant drains to the stormwater system were clogging badly, increasing flooding and requiring residents to wade out and clear them. The residents were proud of the remnants of the original brick street paving, which sent designers thinking hard about permeable unit pavers. The inland flooding was found to be along filled areas of former creek beds.

6.3.1.3 Cross-Disciplinary Inquiry

It was immediately apparent that an area of design work essentially uncharted involved seeking advice of professionals and policymakers from a number of different fields. Students worked with marine scientists, landscape architects, preservationists, and civil engineers to try to make sense of the interactions of potential design decisions. Indeed, by the second semester of work, students began a collaborative relationship with students from Old Dominion University's Department of Civil and Environmental Engineering. Collaboration presents its own challenges, but forging solutions in a new field requires it. The authors also believe it to be fundamental to the professional futures of the students as well as the most expeditious path to defining adaptation strategies.

6.3.1.4 Exploration of Full Block Raising/Storage

Initially, impressed by the residents' civic activities relating to porches and willing to explore standard solutions, designs were developed for raising full blocks of buildings, to keep the continuity of porch life intact, and using the zone of the

Figure 6.2 Chesterfield Heights housing styles.

Figure 6.3 Experimentation with raising structures by block over cisterns.

front yards to grade up over water storage cisterns to replicate the original relation between apparent grade and structure (Figure 6.3). From the beginning storing water under streets in the public right of way was explored as well.

6.3.1.5 Guiding Principles

As more insight was gained through consultation with preservationist and city planners, however, a set of principles for the design work began to emerge. The interest in being useful and the apparent lack of funding to execute complicated design strategies (such as raising full blocks) led the design team to a commitment to find a set of incremental interventions that were sustainable, simple, and, to whatever extent possible, could be worked upon by residents. Green infrastructure was given precedence following the Virginia Strormwater regulations over gray, and where possible powered pumps were avoided; the latter reflected concern for both cost and consequences of power outages during severe weather (VA DEQ, 2013).

6.3.1.6 Living Shoreline

With the help of consulting landscape architects and marine scientists, students developed designs for a living shoreline. An oyster reef and low rock sill were recommended to reduce wave action and a planting plan of marsh plants proposed. A small beach that had disappeared was proposed to be re-instituted, and a floating dock to extend residents' enjoyment of the waterfront (Figure 6.4).

6.3.1.7 Use of Existing Wetland Area for Storage

An inland wetland, a legacy of a filled creek, was reimagined for storage of runoff from inland areas. It was proposed that a roadway at its neck be raised and a tide gate installed to be deployed when both tidal water and significant precipitation were forecast.

6.3.1.8 Water Storage: Streetscape

Inspired by the residents' pride in their brick streets, students worked to make use of the public right of way for water storage. One of the challenges of adaptation work is the difficulty in obtaining geotechnical information. Eventually a city employee unearthed soil reports done prior to a public works project some decades ago. Happily, it indicated a high sand content, unusual in the region, and no appearance of groundwater to a 9′ depth. This allowed the team to propose a series of cisterns under sidewalks and parking areas on each side of the street, resurfaced with permeable paving. The geotechnical information suggested these cisterns could ultimately drain into and recharge groundwater, a desirable goal in an area suffering from subsidence. Further, the verge between these two previous zones was proposed to be reconfigured as a bioretention installation with water-tolerant plantings proposed over engineered soil and under drains (Figure 6.5) (VA DEQ, 2013).

6.3.1.9 Water Storage Parcel by Parcel

A highlight of the design proposal was focus on a parcel-by-parcel program rooftop disconnection program,

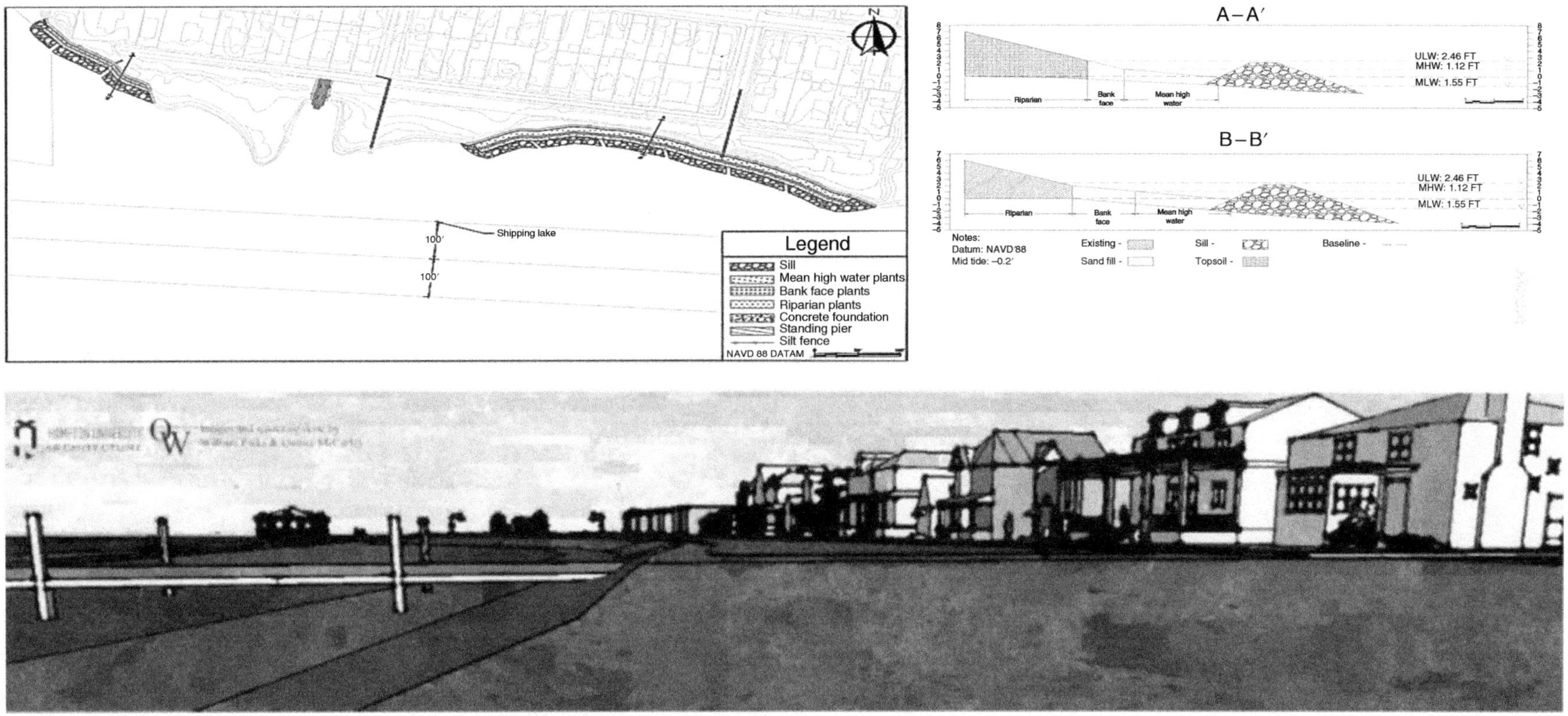

Figure 6.4 Living shoreline bank design in Chesterfield Heights.

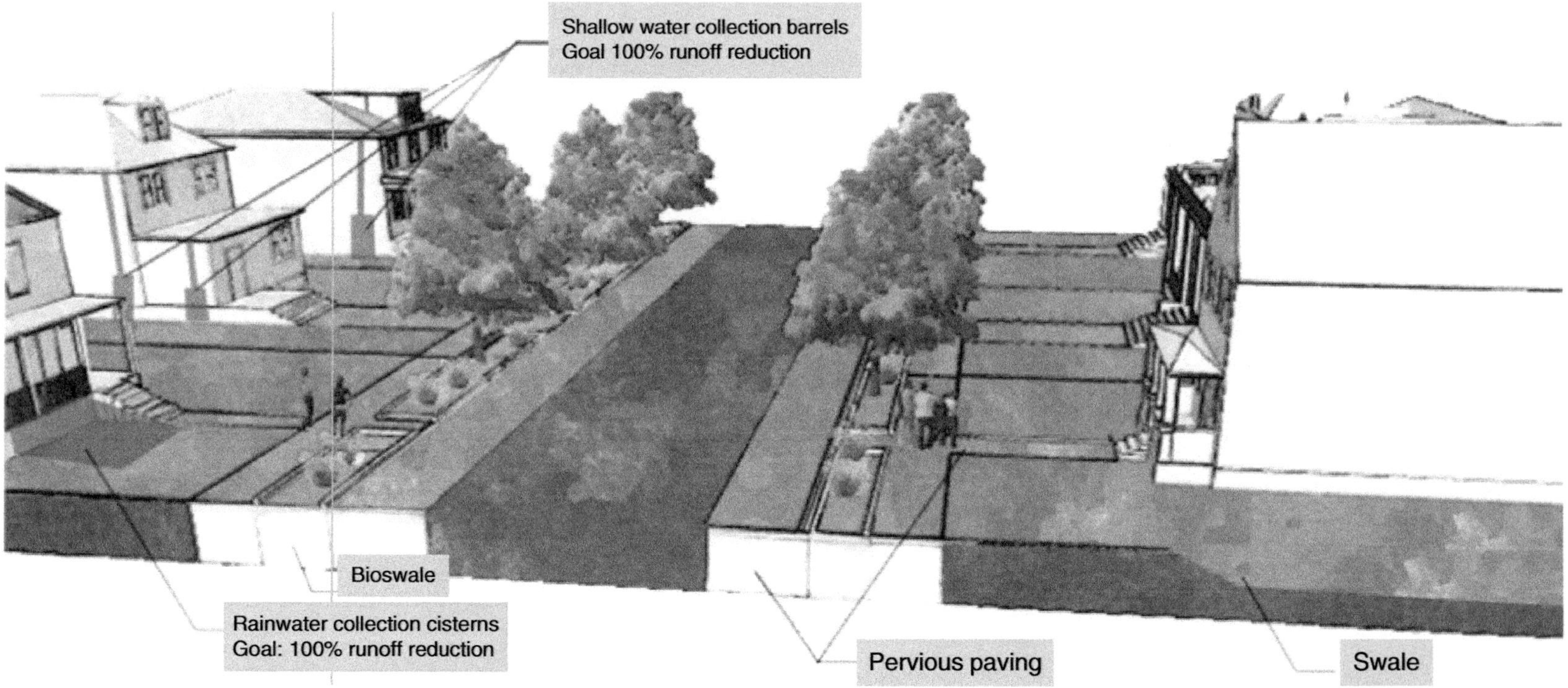

Hampton University Architecture Department

Figure 6.5 Combination of streetscapes and bioretention cells for water storage in Chesterfield Heights.

leading residents to deploy a combination of rain gardens and cisterns to hold water on site. Substantial reductions of runoff into streets from lots fronting them have significant impacts on flooding in the streets. Water retained in this program could be reused for irrigation, recharge of groundwater by allowing accumulated water to filter back into the soil, or discharge into stormwater system after the precipitating weather event has dissipated.

6.3.1.10 Tidal Check Valves and EPA SWMM Modeling

Students modeled the impact of combinations of their proposed interventions using the EPA's Storm Water Management Model (SWMM) program (US EPA, 2015). The single most effective proposal was the installation of check valves into the stormwater outfalls. But the suite of proposals was adequately strong not only to keep the buildings and residents free of significant flooding into the next century but also to do so without replacing the undersized (two 8″ outfall lines) stormwater system.

6.3.1.11 Epilogue: Dutch Dialogues

At a presentation of their project to the Regional Watershed Task Force, several city managers were in the audience, including the Rockefeller Foundation-funded chief resilience officer. So impressive did these officials find the proposed suite of relatively low-key interventions that they re-orchestrated an imminent international design charrette, The Dutch Dialogues Virginia: Life at Sea Level, to include the project area in the program. Students and faculty participated in the 5-day event. The Dutch guests proposed raising the bank of the waterfront park to create, in effect, a dyke at a height to accommodate a 100-year storm and 1 ft of sea level rise without overtopping. Students were enthusiastic about the berm allowing an area for the wetlands to retreat as sea levels rose. The tide gate proposal was endorsed and replicated in another similar wetland area (Figure 6.6).

6.3.1.12 Epilogue: NDRC

The city of Norfolk during the ensuing months prepared a proposal for funding in the US Department of Housing and Urban Development's National Disaster Resilience Competition. It was selected and awarded a $115 000 000.00 implementation grant and should be under construction by 2018.

6.3.2 Case Study 2: The Hague

In fall 2015, the collaborative adaptation design process continued between the Hampton University (HU) architecture students and Old Dominion University (ODU) civil and environmental engineering (CEE) students and faculty. The study area, The Hague, was another neighborhood that is inundated with flooding as a result of sea level rise, heavy precipitation, and subsidence.

Figure 6.6 Dutch Dialogues, tide gate proposal, and wetland area.

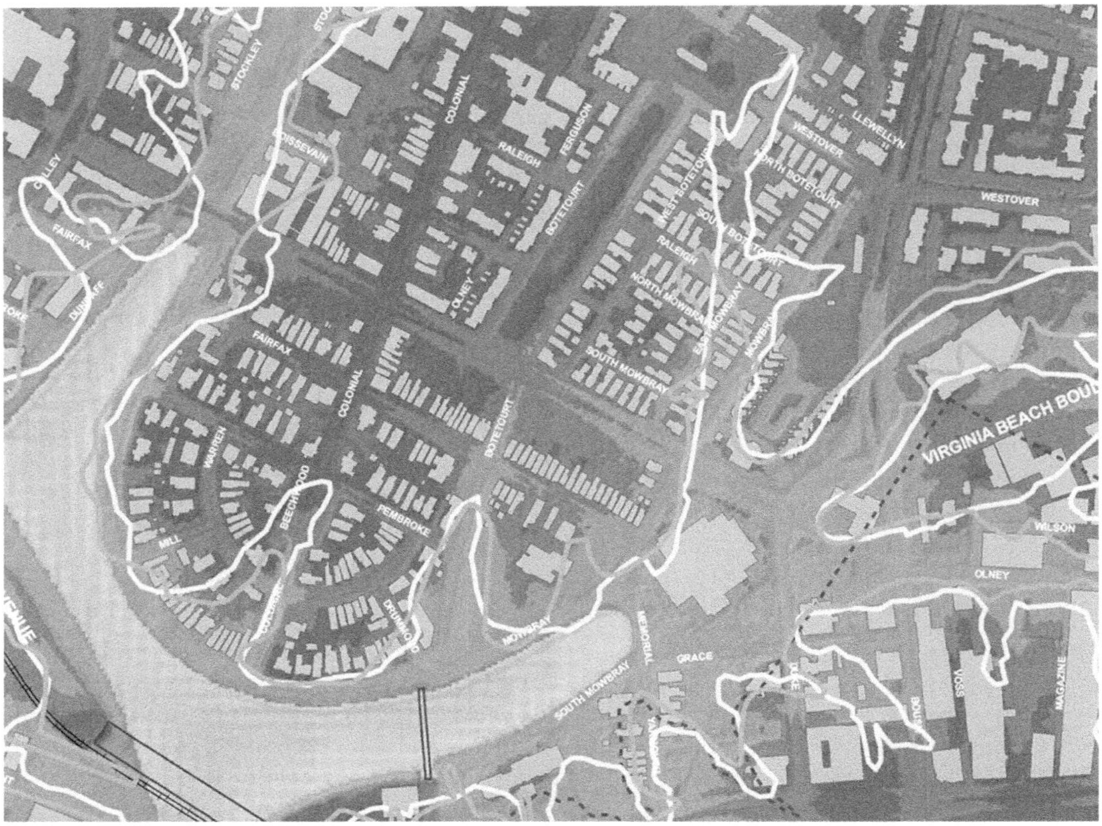

Figure 6.7 Location of The Hague watershed on Legacy Creek Bed.

The Hague is a bulk-headed estuary near downtown Norfolk, developed in the nineteenth century. Figure 6.7 shows an overlay of former shorelines; much, including the ground under the Chrysler Museum of Art, is Legacy creek bed, which regularly flooded. With land lower and soil poorer than in Chesterfield Heights, the challenge was significantly more difficult and more complicated. The shift of the shoreline and disappearance of it, combined with the extreme low-lying elevation of area itself, is the problem that Norfolk faces while developing viable solutions.

In another national register district, the buildings that line the basin are also threatened. A flood wall under a bridge at the estuary's mouth had been designed by Fugro but was thought to be beyond the cost to be borne by existing funding streams (Fugro Atlantic, 2012).

6.3.2.1 Observation of Conditions

6.3.2.1.1 Tidal Flooding

The Hague area is located in a low-lying area, has aging stormwater infrastructure, and does not have sufficient floodwater barriers. From a flood barrier point of view, the tailwater from tidal shifts combined with wind-driven water backs up into The Hague, even without stormwater runoff. Even at mean low low water (MLLW), roughly 90% of stormwater outfalls to The Hague are still completely submerged, with the tailwater reaching back into the streets and surrounding neighborhoods by way of the submerged outfalls. At high tides, the insufficient height of The Hague wall due to land subsidence and sea level rise allows water to spill over the crest of the wall and flood the surrounding lower elevation streets and property.

The wall of The Hague is insufficient to combat even a typical high tide cycle as well.

6.3.2.1.2 Stormwater Retention

Another aspect of the floodwater problem is the insufficient stormwater system surrounding The Hague. As Norfolk was developed, particularly throughout the last century, the initial stormwater system has become severely outdated. Another glaring factor is that ancient creek beds of Smith Creek extending on the north and south ends of The Hague that served as a natural stormwater retention areas were backfilled and developed upon. Much of those backfilled areas have also settled significantly. As the land settles and subsides, it creates structural issues with an already aged and outdated stormwater system. These problems are further compounded by a relatively high groundwater table.

6.3.2.2 Community Engagement

A principal first step was to talk to residents through a series of interview sessions arranged with the help of the community civic league. Following an introductory presentation to the Civic League, a resident on each block of

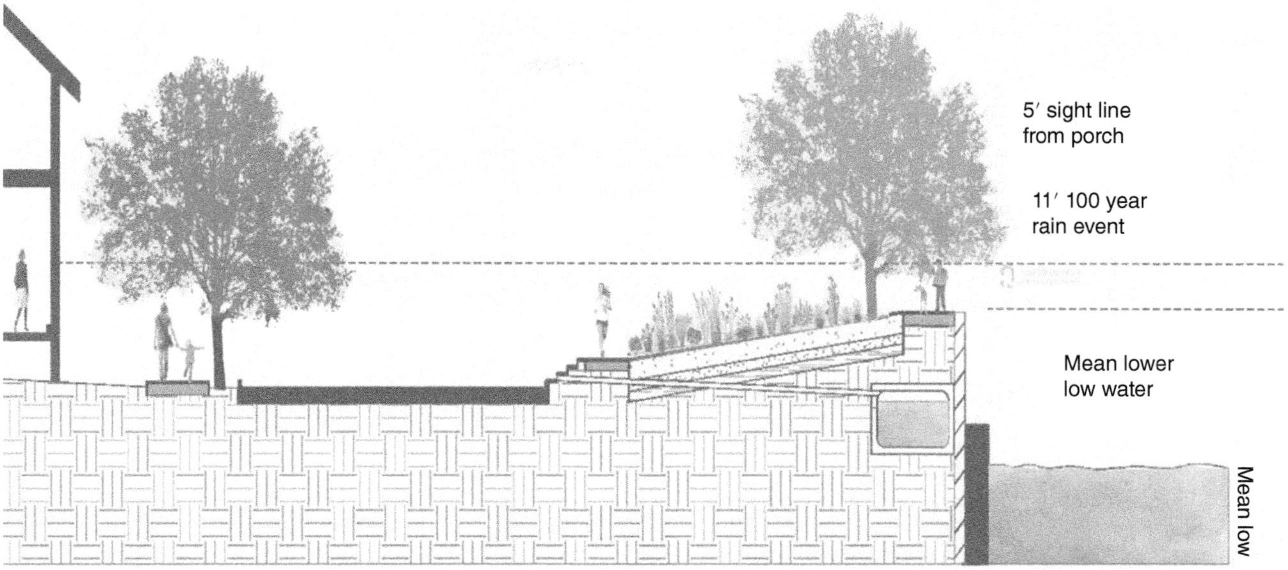

Hampton University Architecture Department

Figure 6.8 Barriers: perimeter basin sheet pile wall.

the primary subject area volunteered to organize and host a 2h interview session with their neighbors and the students. Interview materials were then analyzed, and informal responses were gathered about the appeal of various fairly obvious solutions. The quiet evenings with students allowed for a free and civil exchange of ideas and concepts were possible and productive. Area institutional stakeholders were also engaged during the design process.

6.3.2.3 Guiding Principles/Cross-Disciplinary Inquiry/Collaborators

With the multifaceted problem being faced, the collaborative team comprised of students from the CEE Department at ODU and the architecture students at HU separated to two groups to simultaneously create solutions: the *barrier* group, tasked with developing a solution to containing the incoming tidal surge and floodwaters, and the *sponge* group, tasked with stormwater retention and mitigation. Although the teams worked independently on their respective designs, the overall solution to the floodwater issue requires a cohesive solution since the stormwater outfalls of The Hague watershed empty into The Hague itself. The combination of these two groups maximized design potential for a dual pointed solution to floodwaters throughout The Hague watershed.

6.3.2.4 Barriers

6.3.2.4.1 Tidal Barriers

Those dubbed the barriers worked on solutions to tidal flooding. Solutions involved tidal check valves, as the stormwater outfalls are now regularly below the waterline and serve as quite efficient sluiceways bringing tidal water deep into the neighborhood. Barrier teamwork also involved the design of two alternate solutions to increase the height of the perimeter wall to a level established in the Dutch Dialogues, to offer protection at current levels for a 100-year storm, and to accommodate current sea level rise projections into the end of this century.

Figure 6.8 shows a design for a *sheet pile wall* graded with bioretention fill between a stressed and waterside set of running and walking paths. A toe drain system would capture water draining toward The Hague not captured upstream and hold until water levels receded sufficiently to allow outfall into Hague through sluice gates.

A second perimeter solution involved the use of a system in production by the Aggeres company in Holland, a *water-actuated fiberglass and Kevlar floor–wall system* relying on buoyancy rather than machinery shown in Figure 6.9. It is an elegant system, the use of which would allow an effective raised height with a much gentler modification of the slope of the subject area, as well as respond to the many floods that are allied to as yet unpredictable floods related not to moon phase or wind direction but, apparently, to naturally occurring fluctuations in the speed at which the Gulf Stream flows.

To accommodate existing outfalls through which the water-actuated wall could not rise, a series of short wall sections were proposed that would be used for public art and/or historical information and would be available as a fund-raising mechanism.

6.3.2.4.2 Check Valves

A crucial juncture of either barrier solution for the success of the overall project is the control of the tailwater or floodwater back into the stormwater system. This is accomplished through the use of backflow or *check valves*. The stormwater outfalls that deliver the runoff into The Hague must pass through the floodwall that is

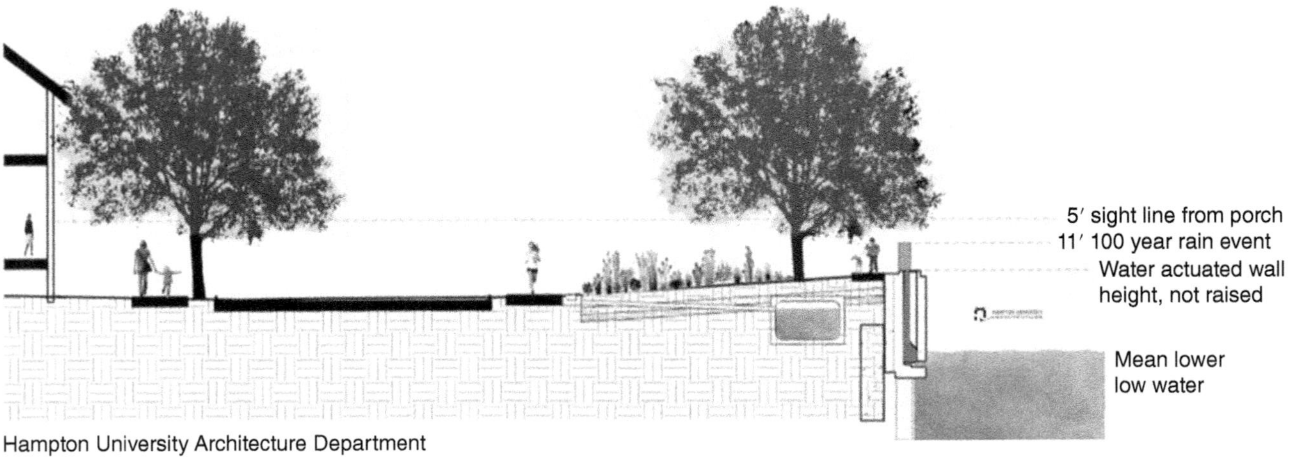

Figure 6.9 Barriers: water-actuated perimeter basin wall.

Figure 6.10 Tidal check valve. (Stormwater outflowing while the tidal water's entry is blocked.) https://stormrecovery.ny.gov/sites/default/files/uploads/nyrcr_seaford_wantagh_projectboards.pdf.

constructed. A typical example of a check vale is shown in Figure 6.10. The check valves that work on differential pressure will prevent the backflow of tidal water and storm surge into the stormwater delivery system. It also allows the stormwater to flow out.

6.3.2.5 Sponges

The sponges refer to stormwater retention systems designed for the neighborhood. The US Environmental Protection Agency Storm Water Management Model (EPA SWMM) was chosen to perform the hydrologic and hydraulic analysis (USEPA, 2015). This model is a dynamic rainfall–runoff simulator that can simulate short-term or long-term rainfall events. EPA SWWM is a computer program used to assess rainfall excess (runoff) and evaluate mitigation strategies. It is an engineering tool used to simulate precipitation–runoff events and receiving drainage systems for urban areas. The model results were used to determine the effectiveness of the proposed stormwater management alternatives.

6.3.2.5.1 Public Open Space

Dry swales are essentially bioretention cells that are shallower, configured as linear channels, and covered with turf or other surface material (other than mulch and ornamental plants) (VA DEQ, 2013, Stormwater Design Specifications No. 10). They may provide some reduction in nitrogen and phosphorus depending on the level of groundwater and serve as temporary storage. Having a dry grass swale also allows the residents to utilize the open spaces during non-rainfall events (Figure 6.11). However, the swales have a lower capacity for reducing runoff due to their limited storage capacity.

Below grade cisterns are large gray water holding tanks that can be installed above or below ground. They allow for collection of rainwater or other runoff to be used or disposed of at a later date (Figure 6.12). These tanks are designed with a multitude of applications and materials to maximize the efficiency of the particular system. Cisterns can be stand-alone tanks or a bank of multiple tanks to maximize volume. Rainwater can be diverted to cisterns for temporary storage for the duration of the rainfall event. The ability for cisterns to be constructed underground is crucial for densely populated areas such as Norfolk in order to maximize the volume of rainwater storage with minimal undeveloped area available.

Bioretention cells are landscaped depressions in the soil that are filled with a designed amount of specific soils. These soils act as filter beds as well as storage areas for runoff. The top layer is the surface soil that incorporates local vegetation, which then leads to a filter media underneath that is used to remove pollutants before the runoff is introduced to either the storage layer or underdrain (VA DEQ Manual). Bioretention cells provide more pervious space in highly impervious areas. These bioretention areas are usually in a landscaped area in a slight depression where the water is influenced and then allowed to infiltrate. Urban bioretention cells have the same ideas but are designed to fit into a specific area.

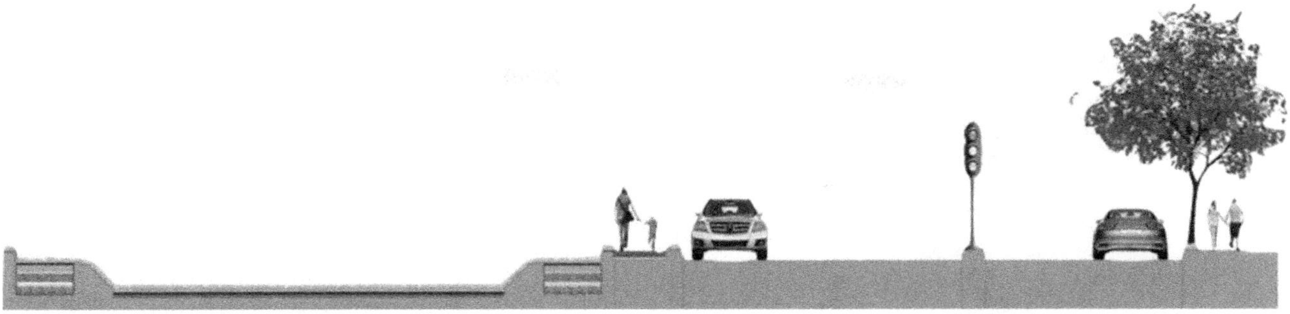

Hampton University Architecture Department

Figure 6.11 Vegetative dry swale.

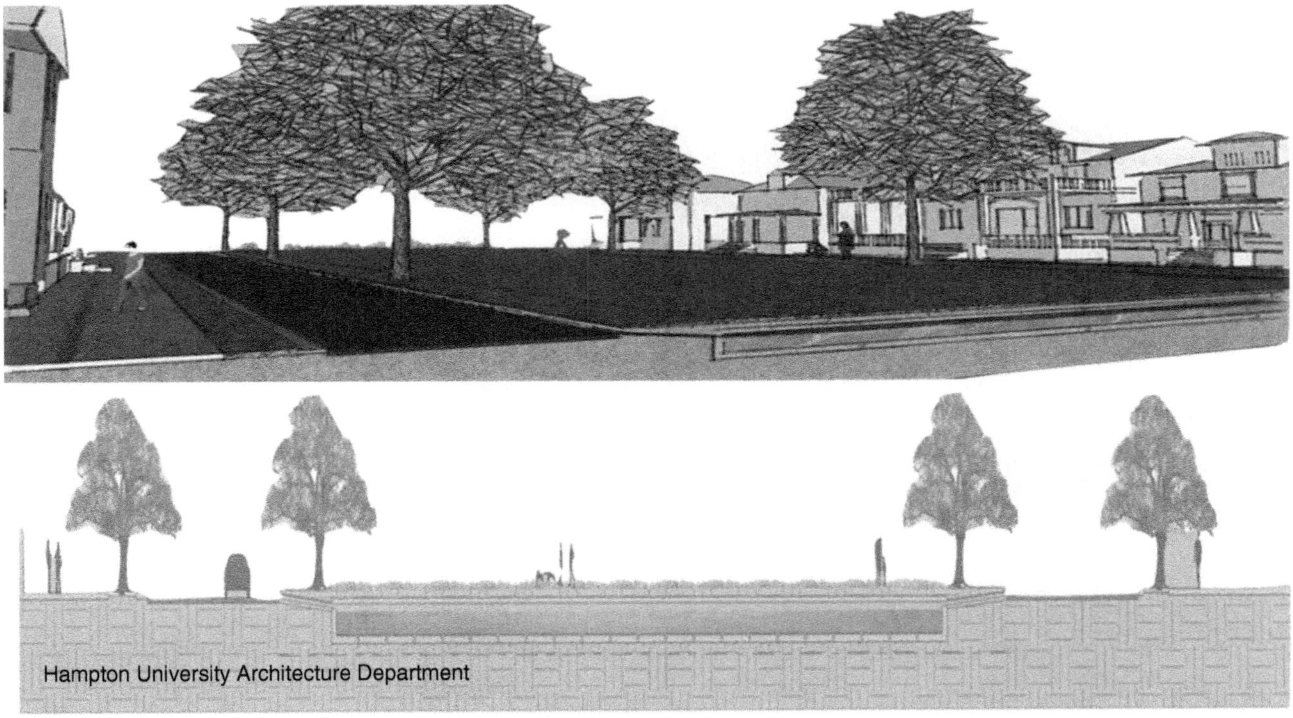

Figure 6.12 Buried cisterns.

These can be designed to fit an existing or upcoming streetscape in order to maximize land use and retention efficiency. Most bioretention systems have an underdrain, but that is not required in this neighborhood due to the well-drained soil. Bioretention systems are comprised of vegetation in a layer of mulch or top soil, a filtration layer made of mainly sand soil and organic material, an underlying stone aggregate reservoir layer, and a choking or small stone separation layer.

Permeable paving is a substitution for conventional concrete and asphalt solutions that help increase the amount of permeable area in a rain catchment. They can be composed of permeable concrete, permeable asphalt, and interlocking precast pavers. These systems include voids that allow water to infiltrate to an underlying reservoir or storage layer. This water can be either held or released at a controlled rate back into a storm drain system or directly into the ground, if the local soil conditions are suitable. The voids, however, must be maintained to ensure they are free from debris or any other clogging materials. These systems can be used in series in small, densely populated areas to maximize efficiency of the retention cell. Runoff will flow over the permeable pavement where some will be captured and the remaining water can then flow into the bioretention cell for treatment and storage. Figure 6.13 shows the proposed permeable paver and bioretention streetscape design introduced along the Colonial Avenue in The Hague.

6.3.2.5.2 Private Property

Rooftop disconnection program that utilizes rain barrels or small cisterns can be a very effective method of

Figure 6.13 Section of permeable paving with bioretention verge.

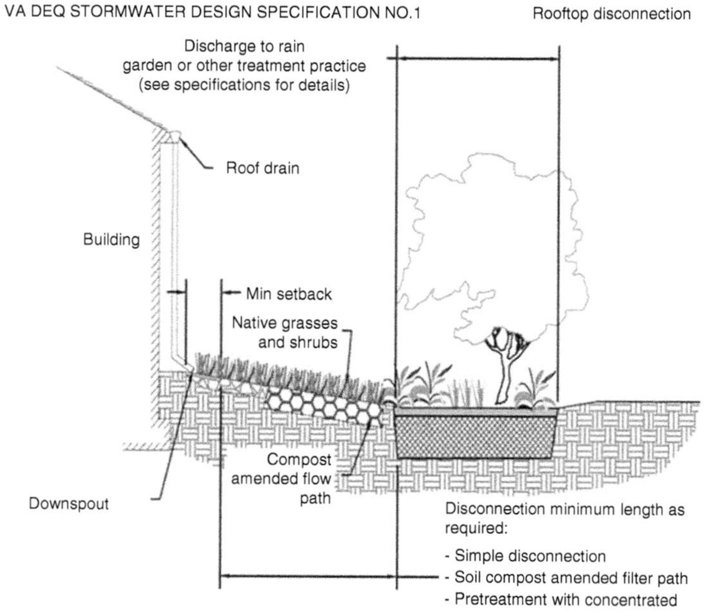

Figure 6.14 Rooftop disconnection. *Source*: VA DEQ Stormwater Design Specification No. 1.

alleviating runoff as well, and it is typically a very cost-effective solution. Rooftop disconnection (RD) is a sustainable best management practice (BMP) that reduces stormwater from residential lots. It directs runoff away from impervious streets, storm drains, and streams and redirects it to pervious landscaped areas. The redirected runoff gets infiltrated, filtered, treated, or reused prior to entering storm drains. This strategy reduces runoff volume and controls pollutants near their source. There are two types of rooftop disconnection. The first is simple rooftop disconnection, which is accomplished by cutting the downspout and redirecting it to a pervious area, such as a lawn, to infiltrate. The second type of rooftop disconnection is disconnection in which runoff is diverted into other BMP, such as bioretention or cisterns. Figure 6.14 shows a schematic of a rooftop disconnection options.

6.3.3 Summary

The proposed adaptation design responses for both neighborhoods would alleviate some of the peak flows and associated flooding, which would improve the existing capacity of the storm sewers in both neighborhoods. In addition, the proposed green infrastructure and low impact development solutions would assist in reducing the nitrogen and phosphorus loads discharging to the Chesapeake Bay and help with the nutrient reduction credits as part of the new stormwater regulations (Stiles, 2015).

The results of SWMM calculations showed that check valves decreased flooding volume and duration, vegetative swales are effective for low-intensity rainfall with high infiltration but with limited storage capacities, and rooftop disconnects provided localized runoff reduction practices. In Chesterfield Heights, the permeable pavers and bioretention streetscape reduced runoff approximately by 90%. The living shorelines helped with the protection of 2200 ft shoreline and dampened some of the tidal flooding. The Hague flooding would be controlled with the addition of perimeter walls and with addition of cisterns and green infrastructure measures. Finally, this work was the only adaptation design that was done before a hurricane or a major storm inundated and destroyed a community in the United States. According to our searches, no other design work existed before an area was devastated with high storm surge and excessive flooding.

Notes

1 http://www.woodlawnpopeleighey.org/aboutpope-leighey/.

2 http://www.citylab.com/cityfixer/2016/06/detroit-vacant-lots-gardens-stormwater/488342/

References

America's Climate Choices: Panel on Advancing the Science of Climate Change; Board on Atmospheric Sciences and Climate; Division on Earth and Life Studies; National Research Council (2010). Sea level rise and the coastal environment. In: *Advancing the Science of Climate Change*, 235–256. Washington, DC: National Academic Press.

Boon, J.D. (2012). Evidence of sea level acceleration at U.S. and Canadian tide stations, Atlantic Coast, North America. *Journal of Coastal Research* 28 (6): 1437–1445. doi: http://dx.doi.org/10.2112/JCOASTRES-D-12-00102.1.

Boon, J.D., Brubaker, J.M., and Forrest, D.R. (2010). Chesapeake Bay Land Subsidence and Sea Level Change, Report No. 425. Gloucester Point, VA: Virginia Institute of Marine Science.

Ezer, T., Atkinson, L.P., Corlett, W.B., and Blanco, J.L. (2013). Gulf Stream's induced sea level rise and variability along the U.S. mid-Atlantic coast. *Journal of Geophysical Research* 118: 685–697 http://dx.doi.org/10.1002/jgrc.20091.

Ezer, T. and Corlett, W.B. (2012a). Is sea level rise accelerating in the Chesapeake Bay? A demonstration of a novel new approach for analyzing sea level data. *Geophysical Research Letters* 39: L19605. doi: 10.1029/2012GL053435.

Ezer, T. and Corlett, W.B. (2012b). Analysis of relative sea level variations and trends in the Chesapeake Bay: Is there evidence for acceleration in sea level rise? *Proceedings of the Oceans'12 MTS/IEEE* (14–19 October). IEEE, Explore. doi:10.1109/OCEANS.2012.6404794.

Flechas, J. and Staletovich, J. (2015). Miami Beach's battle to stem rising tides. *Miami Herald* (5 October).

Fugro Atlantic (2012). Preliminary Engineering Feasibility Report. The Hague Watershed. Norfolk, VA: Fugro Atlantic.

Hampton Roads Planning District Commission (HRPDC) (2013). Land and Water Quality Protection in Hampton Roads. November. http://www.hrpdcva.gov/uploads/docs/11212013-PDC-E3A.pdf (accessed 15 February 2018).

Hester, J.L. (2016). Detroit is turning vacant lots into sponges for Stormwater. *The Atlantic City Lab* 24 June. https://www.citylab.com/solutions/2016/06/detroit-vacant-lots-gardens-stormwater/488342/ (accessed 15 February 2018).

Higgins, A. (2012). Lessons for US from a flood-prone land. *The New York Times* (14 November).

Hunt, J. and Mandia, S.A. (2012). A range of options to cope with sea level rise. In: *Excerpted from Rising Sea Levels: An Introduction to Cause and Impact*. Jefferson, NC: McFarland and Company.

Institute for Environmental Negotiation (IEN) (2011). Sea Level Rise in Hampton Roads: Findings from the Virginia Beach Listening Sessions. Final Report (30–31 March).

International Panel on Climate Change (IPCC) (2007). Coastal systems and low-lying areas. In: *Contribution of Working Group II to the Fourth Assessment Report of the Intergovernmental Panel on Climate Change, 2007*, Chapter 6 (ed. M.L. Parry, O.F. Canziani, J.P. Palutikof, et al.). Cambridge and New York: Cambridge University Press. http://www.ipcc.ch/publications_and_data/ar4/wg2/en/contents.html (accessed 9 January 2018).

Klijn, F., de Bruijn, K.M., Knoop, J., and Kwadijk, J. (2012). Assessment of the Netherlands' flood risk management policy under global change. *Ambio* 41 (2): 180–192 http://doi.org/10.1007/s13280-011-0193-x.

Langis, J. (2013). *Adaptation Measures for Floods, Storm Surges, and Sea Level Rise*. Report by Groupe Littoral et vie Université de Moncto. Canada: New Brunswick Environmental Trust Fund.

Leake, S.A. (2013). Land Subsidence from Groundwater Pumping. USGS. https://geochange.er.usgs.gov/sw/changes/anthropogenic/subside/ (accessed 12 December 2016).

Meehl, G.A., Stocker, T.F., Collins, W. et al. (2007). *Global Climate Projections*. Cambridge, UK, and New York, NY: Cambridge University Press.

National Ocean and Atmospheric Administration (NOAA) (2015). What is glacial isostatic adjustment. (6 August). http://oceanservice.noaa.gov/facts/glacial-adjustment.html (accessed 8 December 2016).

Nicholls, R.J. (2011). Planning for the impacts of sea level rise. *Oceanography* 24 (2): 144–157.

Nicholls, R.J., Hanson S., Herweijer C., et al. (2008). Ranking Port Cities with High Exposure and Vulnerability to Climate Extremes: Exposure Estimates. OECD Environment Working Papers, No. 1. Paris: OECD Publishing.

Ovans, A. (2015). What resilience means and why it matters. *Harvard Business Review* (5 January 2015).

Ramos, M.G. (1997). *Texas Almanac, 1998–1999*. Dallas, Texas: The Dallas Morning News. https://texashistory.unt.edu/ark:/67531/metapth162515/ (accessed 12 February 2018). University of North Texas Libraries, The Portal to Texas History, texashistory.unt.edu.

Sallenger, A.H., Doran, K.S., and Howd, P.A. (2012). Hotspot of accelerated sea-level rise on the Atlantic coast of North America. *Nature Climate Change* 2: 884–888. doi: 10.1038/nclimate1597.

Solomon, S., Qin, D., Manning, M. et al. eds. and IPCC (2007). *Climate Change 2007: The Physical Science Basis. Contribution of Working Group I to the Fourth Assessment Report of the Intergovernmental Panel on Climate Change*. Cambridge, UK and New York, NY: Cambridge University Press. https://www.ipcc.ch/publications_and_data/publications_ipcc_fourth_assessment_report_wg1_report_the_physical_science_basis.htm (accessed 9 January 2018).

Stiles, W. (2015). Tidewater Rising Resiliency Design Challenge: Conventional Solutions Won't Work. Interim Report to Virginia Sea Grant.

United States Environmental Protection Agency (US EPA) (2015). *Storm Water Management Model User's Manual Version 5.0*. USEPA.

United States Geological Survey (USGS) (2016). The Chesapeake Bay Bolide Impact: A New View of Coastal Plain Evolution. http://pubs.usgs.gov/fs/fs49-98/ (accessed 8 December 2016).

Virginia Department of Environmental Quality (VA DEQ) (2013). *Virginia Stormwater BMP Clearinghouse* Virginia Department of Environmental Quality and Virginia Water Resources Research Center (VWRRC). http://vwrrc.vt.edu/swc/index.html (accessed 15 February 2018).

Virginia State Water Control Board, Tidewater Regional Office (VSWCB) (1981). Ground Water Resources of the Four Cities Area, Virginia. Planning Bulletin 331, November. http://www.deq.state.va.us/Portals/0/DEQ/Water/GroundwaterCharacterization/GROUND_WATER_RESOURCES_FOUR_CITIES_AREA_VA.pdf (accessed 15 February 2018).

Walker, W., Abrahamse, A., Bolten, J. et al. (1994). A policy analysis of Dutch River dike improvements: trading off safety, cost, and environmental impacts. *Operations Research* 42 (5): 823–836 http://dx.doi.org/10.1287/opre.42.5.823.

Warrick, R. and Oerlemans, J. (1990). Sea level rise. In: *Climate Change – The IPCC Scientific Assessment* (ed. T. Houghton, G.J. Jenkins and J.J. Ephraums), 257–281. Cambridge, MA: Cambridge University Press.

Withey, P., Lantz, V.A., and Ochuodho, T.O. (2016). Economic costs and impacts of climate-induced sea-level rise and storm surge in Canadian coastal provinces: a CGE approach. *Applied Economics* 48 (1): 1–13.

Young, D. (2017). Raising the Chicago streets out of the mud. *Chicago Tribune* (January).

7

Soil Physical Properties and Processes

Morteza Sadeghi[1], *Ebrahim Babaeian*[2], *Emmanuel Arthur*[3], *Scott B. Jones*[1], *and Markus Tuller*[2]

[1] *Department of Plants, Soils and Climate, Utah State University, Logan, UT, USA*
[2] *Department of Soil, Water and Environmental Science, The University of Arizona, Tucson, AZ, USA*
[3] *Department of Agroecology, Aarhus University, Tjele, Denmark*

7.1 Introduction

Soil physics is concerned with the application of physical principles to characterize the soil system and mass and energy transport processes within the critical zone, which extends from the bedrock to the top of the plant canopy. The exploitation of soil physical principles is not only feasible for natural systems but also relevant for many industrial and engineering applications such as oil recovery, construction, filtration, powder technology, or production of ceramics and nanoporous materials. Soil physical processes are intimately intertwined with biological and chemical processes simultaneously occurring within the same soil volume.

This chapter introduces fundamental soil physical principles at the core of environmental engineering applications and provides a thorough overview of the current state of technology for measurement and monitoring of physical state variables and processes as well as techniques for characterization of basic soil properties. The chapter encompasses six sections, with basic soil properties presented in Section 7.2. Flow processes and applicable measurement methods are discussed in Section 7.3. Section 7.4 provides detailed information about the transport of solutes (contaminants). In Section 7.5, thermal soil properties and heat flow processes are presented, and a brief summary is provided in Section 7.6.

7.2 Basic Properties of Soils

Soils provide vital functions for society, directly or indirectly supplying an estimated 95% of our food. Besides food, feed, fiber, and wood production, soils sustain our terrestrial ecosystems, filter water, recycle waste, regulate the atmosphere, preserve our heritage, act as an aesthetic and cultural resource, and are utilized as building material and for numerous environmental engineering applications (Blum et al., 2006; Robinson et al., 2012). In contrast to man-made construction resources such as steel, plastic, glass, or concrete that have well-defined properties, soils are inherently more complex, exhibiting dynamic interactions of physical, chemical, and biological components and processes.

Soils are heterogeneous, polyphasic, particulate, and porous systems with often very large interfacial areas, which in conjunction with the intricate geometry of soil particles and pores give rise to phenomena such as adsorption of water and chemicals or capillarity. The major physicochemical soil constituents are mineral and organic particles, soil water with dissolved chemical substances, and air. In addition, myriad microbes occupy interfaces and pore spaces. The relative proportions of constituents that commonly exhibit great spatial and temporal variability impact virtually all engineering, environmental, and agricultural soil attributes.

7.2.1 Fundamental Mass–Volume Relationships

If we consider soils as a three-phase system made up of solid particles, water, and air, with each phase occupying a specific volume and exhibiting a specific mass, fundamental mass–volume relationships that provide valuable insights into soil behavior can be defined (Table 7.1). Note that because the mass of air is negligibly low, it is commonly not considered in the mass–volume relationship definitions. The mass of the solid phase is commonly determined after oven drying at 105 °C until mass consistency.

Handbook of Environmental Engineering, First Edition. Edited by Myer Kutz.

Table 7.1 Fundamental mass–volume relationships between soil phases.

Property	Definition[a]	Unit
Mean particle density	$\rho_s = M_s/V_s$	$kg\,m^{-3}$
Dry bulk density	$\rho_b = M_s/V_t = M_s/(V_s + V_a + V_w)$	$kg\,m^{-3}$
Porosity	$n = V_f/V_t = (V_a + V_w)/(V_s + V_a + V_w)$	—
Void ratio	$e = V_f/V_s = (V_a + V_w)/V_s$	—
Gravimetric water content	$\theta_m = M_w/M_s$	$kg\,kg^{-1}$
Volumetric water content	$\theta_v = V_w/V_t = V_w/(V_s + V_a + V_w)$	$m^3\,m^{-3}$
Degree of saturation	$S = V_w/V_f = V_w/(V_a + V_w)$	—

[a] M_s and M_w are the masses of the solid and water phases; V_a, V_s, and V_w are the volumes of the air, solid, and water phases, respectively; V_f is the volume of fluids (i.e. sum of the volumes of water and air); V_t is the total volume (i.e. sum of the volumes of air, solids, and water).

Mean particle density (ρ_s): The mean particle density is the mass of the oven-dry solid phase divided by its volume. Because of the abundance of quartz in many mineral soils, ρ_s often lies between 2600 and 2700 $kg\,m^{-3}$. For soils with significant amounts of organic matter, the mean particle density may be as low as 1300 $kg\,m^{-3}$ (Rühlmann et al., 2006). Soils with remnants of volcanic ash may exhibit mean particle densities of about 1000 $kg\,m^{-3}$. On the other hand, soils containing a significant amount of iron oxides or other heavy minerals may exhibit mean particle densities of 2900 $kg\,m^{-3}$ or higher. Water or gas pycnometer methods are commonly applied for direct determination of ρ_s (Flint and Flint, 2002). While gas displacement works well for coarse-grained soils, based on our experience, there are problems with capillary condensation of the probing gas (e.g. helium) in small pores of fine-grained soils, which may lead to vast overestimation of ρ_s.

Dry bulk density (ρ_d): The dry bulk density is defined as the ratio of the mass of oven-dry soil to the total soil volume. While the dry bulk density of sandy soils with low porosities may be as high as 1600 $kg\,m^{-3}$, loam or clay soils may exhibit bulk densities below 1200 $kg\,m^{-3}$. The bulk density in most natural soils increase with depth due increasing overburden and confinement pressures that tend to decrease the pore volume. The core, paraffin-coated clod, excavation, and radiation methods for determination of bulk density are well established and exhaustively discussed in literature (e.g. Grossman and Reinsch, 2002). More recently, Rossi et al. (2008) applied automated three-dimensional laser scanning to determine the dry bulk density of intact soil clods and rock fragments. They demonstrated close agreement between the laser scanning and the standard paraffin-coated clod method that is more labor intensive and complex, especially when coarse soils and rock fragments are involved.

Porosity (n): Porosity is defined as the volume of voids (liquid or air filled) per total sample volume. It is an important soil property that defines the pore volume available for water retention and storage and for water and gas flow and solute transport processes. Typical soil porosity values range from 0.3 to 0.6, with fine-textured soils tending to have more total pore space than coarse-textured soils. The porosity may be calculated from the soil's bulk and particle densities ($n = 1 - \rho_d/\rho_s$) or derived from water or gas desorption isotherms or from direct visualization of impregnated thin sections and image analysis (Flint and Flint, 2002). The recent advancements in X-ray micro-computed tomography (μCT) in conjunction with the development of sophisticated CT data segmentation algorithms provide new means for determination of porosity and pore topological attributes (Tuller et al., 2013; Vaz et al., 2014).

Void ratio (e): When active (i.e. swelling) clays are present, the total sample volume varies dependent on hydration state of the clay minerals. For such instances it is advantageous to define a void ratio that relates the volume of fluids (i.e. air and water) to the volume of solids, a fixed quantity, rather than porosity. The relationship between void ratio and porosity is defined as $e = n/(1 - n)$.

Gravimetric water content (θ_m): The ratio of the mass of water to the mass of oven-dry soil, the gravimetric water content, can be directly determined by oven-drying a wet bulk soil sample at 105 °C.

Volumetric water content (θ_v): The volumetric water content, defined as the volume of water within a given soil volume, may be expressed in terms of gravimetric water content ($\theta_v = \theta_m \rho_b/\rho_w$; with ρ_w as the density of water) or directly measured on core samples or with a variety of in situ sensors as discussed in Section 7.2.4. For some applications such as water balance calculations, it is convenient to express the volumetric water content in length units to be consistent with other variables such as precipitation or evapotranspiration. This so-called equivalent depth of wetting (D_e) is obtained by multiplying θ_v with the depth (D) of the soil layer of interest ($D_e = \theta_v D$).

Degree of saturation (S): In some instances, it is advantageous to express soil moisture (SM) in terms of degree of saturation, which is the volumetric water content normalized by the volume of fluids. In theory, S ranges from 0, when the soil is completely dry, to 1, when all soil pores are completely filled with water (see Section 7.2.4.). The degree of saturation may be expressed as a function of the volumetric water content as $S = \theta_v/n$.

7.2.2 The Size Distribution of Soil Particles and Soil Classification

The size distribution of primary particles that constitute the soil solid phase affects various other physicochemical and hydraulic soil properties (e.g. porosity, bulk density, specific surface area [SSA], cation exchange capacity, soil water characteristic [SWC], hydraulic conductivity) that virtually govern all critical zone processes. Dependent on their size, particles are commonly assigned to different fractions (i.e. gravel, sand, silt, and clay) based on various size scales such as instituted by the US Department of Agriculture (USDA), the American Society for Testing and Materials (ASTM), the US Public Roads Administration (USPRA), the American Association of State Highway and Transportation Officials (AASHTO), or the International Union of Soil Sciences (IUSS), only to name a few (Figure 7.1). A comprehensive review of particle size scales and classification schemes applied in soil and environmental science and engineering is provided in Blott and Pye (2012).

The size of sand particles commonly ranges from 0.02 or 0.05 mm to 2.00 mm. They can be discerned with the naked eye and feel gritty when rubbed between fingers. In many cases, sand grains appear spherical in shape and exhibit relatively small SSA (~0.01 to 10.0 $m^2\ g^{-1}$). The sand fraction is mainly composed of primary soil minerals such as quartz and feldspars.

Silt particle sizes range from 0.002 mm to 0.02 or 0.05 mm, dependent on the classification scheme. They are similar in shape as sand particles, but cannot be visually discriminated with the naked eye. Silt particles have silky feel when rubbed between fingers and have typical SSA larger than 1.0 $m^2\ g^{-1}$. In general, silt particles are the most susceptible to soil erosion among the three size classes. In some instances, silts exhibit properties typical for clays due to organic or clay mineral coatings.

The clay fraction comprises particles smaller than 0.002 mm and is considered the most reactive among the soil size classes. The major clay minerals kaolinite, montmorillonite, vermiculite, illite, and chlorite contribute many favorable properties to productive agricultural soils. Because of their large surface areas (e.g. up to 800 $m^2\ g^{-1}$ for pure montmorillonite) (Jury et al., 1991), they increase the cation exchange capacity and enhance the water storage and pH buffering capacities of soils. Clay minerals are composed of tetrahedral silica sheets and octahedral aluminum hydroxide and/or magnesium hydroxide sheets. Depending on the number and composition of sheets, they are classified as 1:1 or 2:1 minerals. While 1:1 clays (kaolinites) contain units of alternating tetrahedral and octahedral sheets, 2:1 clays (montmorillonites, vermiculites, illites) are composed of units with two tetrahedral and one octahedral sheet. In addition, some 2:1 clays (chlorites) have interlayer sites between successive 2:1 units, which are occupied by commonly hydrated interlayer cations.

Because of their extremely low hydraulic conductivity when hydrated, their capacity to retain radionuclides, and their abundance in large geological formations, active 2:1 clays are utilized for insulation of radioactive wastes (Clauer, 2005; Grambow, 2016) and as sealants or liners for landfills, sewage lagoons, or water retention ponds to prevent leaching of contaminants into the groundwater. Because of their expansive behavior – when pure montmorillonite is hydrated, its volume may expand to 30 times of its dry volume – active clays are responsible for significant damage to roads, bridges, buildings, and other infrastructure. Furthermore, dehydrated agricultural soils with appreciable amounts of active clays may develop large and deep desiccation cracks (Figure 7.2) that promote fast preferential transport of agrochemicals and other contaminants to deeper soil layers and the groundwater, causing public health concerns.

7.2.2.1 Methods for Determination of the Soil Particle Size Distribution

There are a number of techniques available to determine the particle size distribution (PSD) and particle fractions of soils. Because of significant technological advances

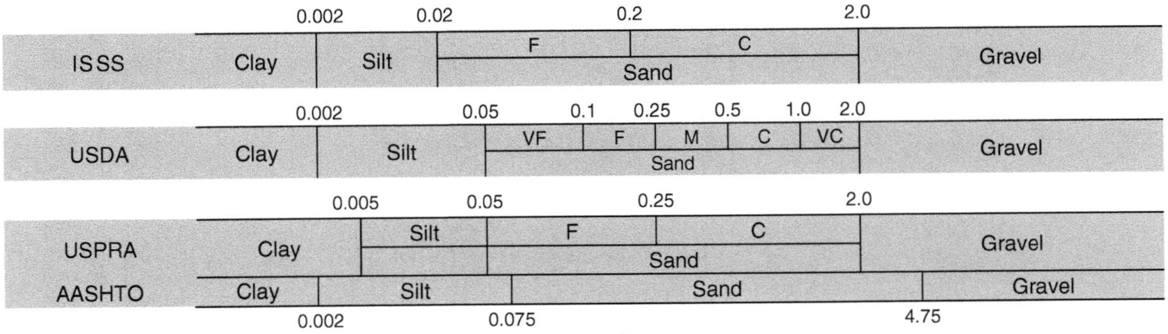

Figure 7.1 Selected soil classification schemes where F, C, VF, M, and VC stand for fine, coarse, very fine, medium, and very coarse, respectively.

Figure 7.2 Desiccation cracks in an agricultural soil with active clays (left) and damage to road pavement due to expansive clays (right).

over the last decade, there is an ongoing shift from very basic sieving and sedimentation methods to more advanced techniques that include light scattering and laser diffraction, particle counting, X-ray attenuation, and optical or transmission and scanning electron microscopy. Besides laboratory PSD determination, experienced individuals are able to approximate particle fraction percentages (i.e. sand, silt, and clay) in the field by rolling, squeezing, flattening, and pressing the soil between their fingers (Thien, 1979). Although this so-called "feel" method can be quite accurate when performed by a well-calibrated expert, quantitative laboratory methods are preferred to eliminate the human bias inherent to the "feel" technique. A thorough discussion of various PSD determination methods is provided in Gee and Or (2002).

7.2.2.1.1 *Dry and Wet Sieving Methods*

Sieving is probably the most widely applied PSD determination method that involves passing a fixed amount of soil through a stack of progressively smaller sieves (Figure 7.3) with a collection pan at the bottom (Gee and Or, 2002). The amount of soil remaining on each sieve is weighed, and the cumulative proportion of soil passing through each sieve is determined. Wet sieving with a water supply and distribution system mounted on top of the sieve stack is commonly more accurate than sieving soils in their oven-dry state. In many cases vibration sieve shakers with tunable shaking frequencies (Figure 7.3) are utilized for automation of the sieving step.

Because soil particles are generally not spherical, sieving results are reported as "effective diameters" that correspond to the sieve openings rather than to the actual particle sizes. Sieve analysis is commonly applied for particles larger than 0.053 mm (i.e. sand and gravel fractions).

Prior to the sieving process, soil samples are pretreated to remove particle binding agents such as organic matter, metal oxides, or carbonates. Organic matter is often

Standard sieve sizes

Sieve no.	Opening size (mm)
4	4.75
10	2.00
20	0.84
30	0.60
40	0.42
50	0.30
60	0.25
80	0.18
100	0.15
140	0.106
170	0.088
200	0.075
270	0.053

Figure 7.3 Standard sieve sizes commonly applied for PSD determination (left); digital wet sieving system with tunable frequency vibration shaker and water supply on top of the sieve stack (right).

removed with hydrogen peroxide, sodium hypochlorite, or disodium peroxodisulfate (Mikutta et al., 2005). Soils rich in iron oxides are pretreated with bicarbonate-buffered sodium dithionite-citrate, and carbonates are commonly removed by acidifying the sample with sodium acetate at pH 5 (Gee and Or, 2002). After removal of binding agents, the soil samples are commonly dispersed with sodium hexametaphosphate and physically agitated with blenders or sonicators to ensure that aggregates are broken up into primary soil particles.

7.2.2.1.2 *Sedimentation Methods*

Sedimentation techniques are based on the relationship between the settling velocity and the diameter of spherical and smooth particles suspended in a dilute dispersing agent solution. This relationship is derived from a force balance calculation, proposed by Stokes (1851), that considers the gravitational force of the settling particle, the buoyancy force of the displaced solution, and the drag force exerted on the particle by the surrounding liquid. From the force balance, the terminal settling velocity of silt- and clay-sized particles of a certain diameter is obtained and related to a specific settling depth to calculate the required time for these particles to settle below this depth (Gee and Or, 2002).

After pretreating the samples in the same fashion as described for the sieve method above, there are two techniques to determine the concentration of particles at a specified settling depth. The suspension may be directly extracted from the sedimentation cylinder (Figure 7.4) with a pipette from the specified settling depth at the time calculated for a specific particle diameter to have settled below this depth. For example, the time for clay particles with 2 μm diameter to settle below 10 cm depth in a 20 °C 0.5 g l^{-1} sodium hexametaphosphate solution is 8 h (Gee and Or, 2002). This means that after 8 h only particles less than or equal to 2 μm are in suspension at the 10 cm extraction depth. In other words, the suspension extracted with the pipette from 10 cm depth after 8 h only contains clay-sized particles (i.e. the larger silt-sized particles already settled below the 10 cm depth). The extracted suspension is transferred to an evaporation dish and oven-dried at 105 °C until mass consistency. The oven-dry mass is then related to the extracted suspension volume (e.g. 10 cm^{-3}) and the total initial soil mass and solution volume to calculate the mass percentage of the clay fraction.

Besides suspension extraction with the pipette method, the concentration (density) of the suspension may be directly measured with a calibrated hydrometer (e.g. ASTM 152 H type). When a hydrometer is applied, first the hydrometer settling depth that is dependent on the hydrometer design and the concentration of the dilute dispersant solution is calculated. For example, for a ASTM 152 H-type hydrometer (Figure 7.4), the settling depth, h^*, in cm is calculated as $h^* = -0.164C + 16.3$, with C as the concentration. Once the hydrometer settling depth is known, the settling times for silt- and clay-sized particles at which hydrometer readings need to be taken can be determined (Gee and Or, 2002). The measured concentrations are then related to the initial soil mass and dispersant solution volume to obtain the mass percentages of the fractions of interest.

7.2.2.1.3 *X-Ray Attenuation (SediGraph) Method*

The SediGraph (Micromeritics Instrument Corp., Norcross, GA, USA) is an analytical instrument that combines the sedimentation theory (i.e. Stokes' law; see

Figure 7.4 Sedimentation cylinders with soil suspensions extracted with a pipette (left); ASTM 152 H-type hydrometer (right).

sedimentation methods above) with Beer–Lambert's law for X-ray absorption. Beer–Lambert's law states that an X-ray beam passing through a medium is attenuated in proportion to the path length through the medium, its concentration, and the extinction coefficient of the medium. The method involves passing an X-ray beam through the sedimentation cell with settling particles and calculating the concentration values from the ratio of X-ray transmission through the soil suspension-filled cell and through a cell containing only dilute dispersant. The measurement range of the SediGraph is from 0.1 to 300 μm, and it requires less than 3 g of soil for analysis. Coates and Hulse (1985) compared the SediGraph technique with the standard pipette and hydrometer methods and concluded that the SediGraph yielded similar results for finer particles, but there were significant deviations for coarse particles. Buchan et al. (1993) revealed that the SediGraph consistently produced finer PSD than the pipette method, which led to the development of several equations for conversion of SediGraph to pipette method results (e.g. Andrenelli et al., 2013).

7.2.2.1.4 Coulter Counter Method

The core principle of Coulter counters (Beckman Coulter Life Sciences, Indianapolis, IN, USA) was developed by Wallace H. Coulter in the 1940s. It is based on measurable changes in electrical impedance produced by nonconductive particles suspended in a low-concentration electrolyte and passing through a small aperture. The instrument consists of a tube with a small aperture that is immersed in a container filled with a particle–electrolyte suspension. Two electrodes, one inside the aperture tube and one outside the tube, create a current path through the electrolyte when an electric field is applied. The impedance between the electrodes is measured. When a particle passes through the aperture, a volume of electrolyte equivalent to the immersed volume of the particle is displaced. This causes a short-term change in the impedance across the aperture that can be measured as a voltage or current pulse (Loveland and Whalley, 2001). The pulse height is proportional to the volume of the sensed particle. Count and pulse height analyzer circuits allow recording of the number and volume of each particle passing through the aperture. Equivalent spherical diameters are calculated from the recorded volumes and used in conjunction with associated count ratios to generate the PSD (Allen, 1997). The sizing range of Coulter counters is from 0.2 to 1600 μm.

7.2.2.1.5 Laser Diffractometry

The Fraunhofer or Mie diffraction theories are applied to determine the PSD from measurements of scattering intensity as a function of the scattering angle and the wavelength and polarization of light. Particles of a given size diffract light at a specific angle that increases with decreasing particle size (Agrawal et al., 1991). A parallel beam of monochromatic light is passed through a particle suspension and the diffracted light is focused onto a multielement ring detector. The detector measures the angular distribution of scattered light intensity. A lens with the detector at its focal point is positioned next to the illuminated sample and focuses the light to the detector center. This leaves only the surrounding diffraction pattern, which does not vary with particle movement. Thus, a stream of particles can be passed through the beam to generate a stable diffraction pattern.

There is some ambiguity about the capability of laser diffraction particle size analyzers to accurately size clay minerals that exhibit a platelet-shaped geometry. Fedotov et al. (2007) reported that traditional sedimentation methods yielded 1.5–5 times higher clay fractions than laser diffractometry (LD). Blott and Pye (2006) applied LD to evaluate its sensitivity to mixtures with vastly different grain fractions and differences in particle shape. They found that while their analyzer was highly sensitive to coarse particles when the fine fraction was dominating, it showed little sensitivity to finer particles when the coarse fraction was dominating. Their results revealed that the shape of individual soil particles affected LD measurements, which led to variations between LD and standard sieving. Similar findings are reported in Loizeau et al. (1994), Beuselinck et al. (1998), and Wen et al. (2002). The general consensus of reported comparisons of LD with classical methods is that there is a good correlation for individual size fractions with variable absolute agreement. LD commonly underestimates the clay fraction when compared with classical methods, which can be attributed to particle density, shape, and mineralogy. Based on our experience, results of such comparisons have to be analyzed with caution, as they seem to be highly dependent on the applied laser diffraction analyzer. Modern instruments such as the Beckman Coulter LS 13320 analyzer (Beckman Coulter Life Sciences, Indianapolis, IN, USA) employ two different laser systems: a standard system that covers the particle size range from 0.4 to 2000 μm and a polarization intensity differential scattering (PIDS) system that extends down 10-fold to 0.04 μm while still providing a continuous size distribution up to 2000 μm. To determine the PSD, a composite scattering pattern is measured by 126 detectors placed at angles up to approximately 35° from the optical axis. The light source is a 5-mW laser diode with a wavelength of 780 nm. For the PIDS system, a secondary tungsten-halogen light is projected through a set of filters that generates three wavelengths of 450, 600, and 900 nm (Beckman Coulter Inc., 2003) (Figure 7.5). Our own measurements with a Beckman Coulter LS 13320 analyzer for a wide range of soils show good agreement with standard methods.

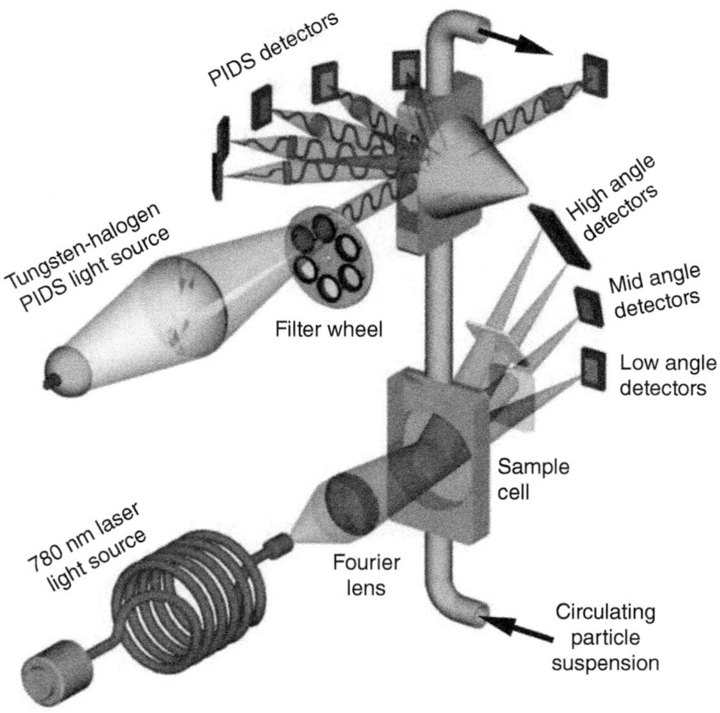

Figure 7.5 Sketch illustrating the standard laser and PIDS systems of the Beckman Coulter LS 13 320 analyzer.

7.2.2.1.6 Indirect Methods for Estimation of Particle Size Fractions

Over the last decade, several spectroscopic and water vapor sorption methods have been developed to indirectly estimate particle size fractions from other soil properties.

Spectroscopic techniques such as near-infrared (NIR) spectroscopy have been applied in conjunction with regression techniques for estimation of soil particle size fractions. The absorption of light of a specific wavelength by a soil sample or its transmission through the sample is measured. Because the magnitude of absorption or transmittance is dependent on soil composition, the estimation of particle size fractions may be influenced by spectrally active soil components such as water, clay minerals, and organic matter (Stenberg et al., 2010). Nevertheless, a comparison of particle size fractions obtained with NIR spectroscopy with results from the standard sieving/pipette method for Latosols from central Brazil (Figure 7.6) indicates good agreement (Vendrame et al., 2012).

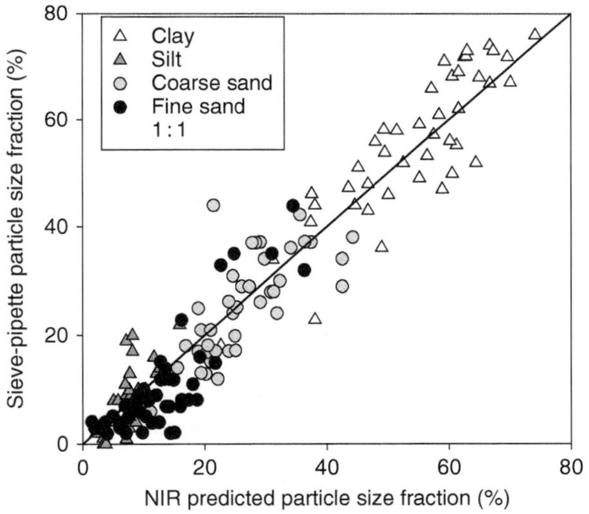

Figure 7.6 Comparison of particle size fractions obtained with NIR spectroscopy with results from standard sieving/pipette measurements for Latosols from central Brazil. *Source*: Adapted from Vendrame et al. (2012).

Another interesting method for estimation of the clay fraction is based on measurements of water vapor sorption. The amount of water retained in soils under very dry conditions is mainly in the form of thin films adsorbed onto particle surfaces. Because clay minerals exhibit a very large surface area (see Section 7.2.3), there is an intimate link between the clay fraction and the amount of adsorbed water (Wuddivira et al., 2012; Chen et al., 2014; Arthur et al., 2015). This method only requires water content measurements at a specific relative humidity (RH) that may be linked to the soil matric potential via the well-known Kelvin equation (see Section 7.2.4.2.1), which is much less time consuming than traditional measurements discussed above.

Chen et al. (2014) related the clay fraction to gravimetric soil water content (θ_m) and matric potential (ψ in cm H_2O): $\mathrm{Clay}(\%) = 6.2\,\theta_m/(6.8 - \log(-\psi)) + 1.57$.

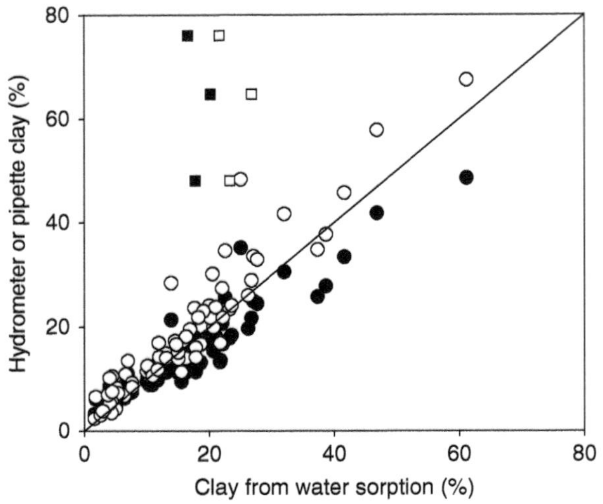

Figure 7.7 Comparison of clay fractions estimated for 90% RH based on the Chen et al. (2014) (open symbols) and Arthur et al. (2015) (closed symbols) relationships with pipette and hydrometer measurements. Square symbols represent kaolinite-rich soils. *Source*: Based on Chen et al. (2014) and Arthur et al. (2015).

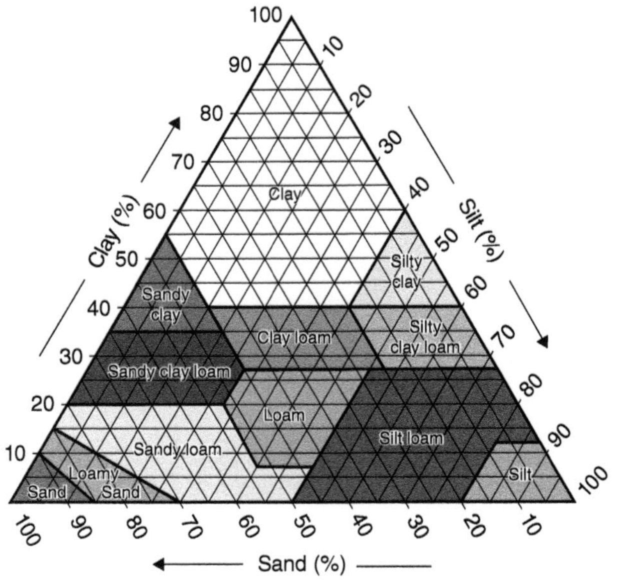

Figure 7.8 USDA textural triangle (Soil Survey Staff, 1975) depicting the boundaries between the 12 textural classes. For example, a soil composed of 35% sand, 35% silt, and 30% clay falls within the clay loam textural class.

A series of regression equations developed by Arthur et al. (2015) that consider effects of vapor desorption/adsorption hysteresis and a correction for soils with high organic matter contents provide reasonable results for soils with mixed clay mineralogy (i.e. the clay fraction is composed of a mixture of smectite, kaolinite, and illite) but underestimate the clay fractions measured with the hydrometer and pipette sedimentation methods for soils with appreciable amounts of nonexpandable clays (e.g. kaolinite) (Figure 7.7).

7.2.2.2 Application of PSD Information for Textural Classification of Soils

The information obtained from PSD analysis (i.e. sand, silt, and clay %) is regularly applied in soil and environmental science and engineering to group soils with anticipated similar hydraulic and mechanical behavior into different textural classes. The most widely applied textural classification scheme in soil and environmental science that dates back to the early twentieth century was developed by the USDA. The boundaries of the 12 textural classes are displayed in the USDA textural triangle (Figure 7.8). The size fractions (i.e. sand, silt, and clay) obtained from PSD analysis are used as input to determine the location (i.e. textural class) within the triangular space. Details about the USDA textural triangle and classification are provided by the Soil Survey Staff (1975). Several databases including the UNsaturated SOil hydraulic DAtabase (UNSODA) (Nemes et al., 2001) have been established based on the 12 USDA textural classes to provide class-specific soil hydraulic properties. In geotechnical engineering and construction, soils with similar engineering behavior are commonly grouped based on the ASTM unified soil classification system that requires PSD information in conjunction with engineering properties such as the Atterberg limits (ASTM-D4318-05, 2005; ASTM-D2487-11, 2011).

The utility of PSD information-based soil classification was recently questioned by Groenendyk et al. (2015). They proposed a process-based classification considering numerically modeled hydrologic response of soils. They demonstrated that maps generated via *k*-means clustering from modeled infiltration and drainage processes did not well correspond with the USDA textural classification.

7.2.3 The Specific Surface of Soils

The SSA of solid particles is a key soil parameter that governs numerous important soil characteristics and processes, including the retention of water, infiltration and drainage, ion exchange, adsorption and release of plant nutrients and contaminants, heat transport and storage, structural soil development, microbial processes, soil swelling, plasticity, cohesion, and soil strength. The SSA of the soil solid phase is commonly related to the dry mass of the solid particles and may range from 0.01 $m^2\ g^{-1}$ for coarse sands to greater than 800 $m^2\ g^{-1}$ for montmorillonite clay (Table 7.2).

The SSA of soils can be directly quantified via physical measurements of particle shapes and sizes (Borkovec et al., 1993). However, the most widely applied approach for SSA determination involves measurement of adsorption of

Table 7.2 Specific surface area of soil constituents.

	Specific surface area ($m^2 g^{-1}$)
Clay minerals	
Kaolinite	14–23
Vermiculite	100–300
Halloysite	~45
Allophane	260–300
Illite	76–100
Montmorillonite	280–830
Other soil constituents	
Organic matter	560–800
Calcite	0.047
Crystalline iron oxides	116–184
Amorphous iron oxides	305–412
Soil textural classes	
Sands	>10
Sandy loams and silt loams	5–20
Clay loams	15–40
Clays	>25

Source: From Diamond and Kinter (1958), Środoń and McCarty (2008), and Skopp (2011).

nonpolar probe molecules (e.g. N_2) from either gaseous or aqueous phases or the retention of polar liquids such as ethylene glycol monoethyl ether (EGME), ethylene glycol (EG), methylene blue (MB), or water (Hang and Brindley, 1970; Newman, 1983; Carter et al., 1986; Amali et al., 1994; Quirk and Murray, 1999; Pennell, 2002; Wuddivira et al., 2012). The application of polar liquids provides SSA estimates close to the actual surface area value due to their ability to penetrate the interlayer space of expandable clay minerals, unlike N_2, which only captures external surfaces (Pennell, 2002). However, application of polar liquids requires a difficult-to-establish measurement protocol and raises environmental concerns arising from their disposal. Studies by Newman (1983), Tuller and Or (2005a), Moiseev (2008), Resurreccion et al. (2011), and Leão and Tuller (2014) thus suggest to estimate SSA from water vapor adsorption isotherms.

7.2.3.1 Adsorption of Nonpolar Gases for SSA Estimation

Nonpolar gases such as nitrogen (N_2), argon (Ar), or krypton (Kr) are adsorbed onto solid particle surfaces due to weak molecular attractive forces. A sealed container at cryogenic temperature holding a small, dry, and outgassed soil sample is supplied with the probing gas (adsorbate) in sequential doses. After each dosing the pressure in the container is measured with a transducer. The adsorption of gas molecules onto the particle surfaces is associated with a pressure reduction. Considering constant cryogenic temperature, the volume of adsorbed gas can be determined from the measured pressure drop based on the ideal gas law. The resulting relationship between the volume of adsorbed gas and the relative pressure at constant cryogenic temperature, the adsorption isotherm, is analyzed based on the Brunauer, Emmett, and Teller (BET) equation (Brunauer et al., 1938) to determine the volume of gas required to form a monolayer of gas molecules on the particle surfaces. With known monolayer volume (V_m), the surface area is calculated as

$$\text{SSA} = \frac{V_m N_A A}{M_s V_g} \tag{7.1}$$

where N_A is Avogadro's constant ($6.022 \times 10^{23}\,\text{mol}^{-1}$), A is the area covered with one adsorbate molecule (e.g. $1.62 \times 10^{-19}\,\text{m}^2$ for N_2 and $1.95 \times 10^{-19}\,\text{m}^2$ for Kr), M_s is the mass (kg) of the oven-dry soil sample, and V_g is the volume occupied by 1 mol of adsorbate gas ($2.24 \times 10^{-2}\,\text{m}^3\,\text{mol}^{-1}$) under standard temperature and pressure conditions (i.e. 273.15 K and 1.013×10^5 Pa). Due to their size, N_2, Ar, and Kr gas molecules cannot penetrate the interlayer spaces of clay minerals. Hence the SSA obtained from adsorption of nonpolar gases, referred to as external surface area, is smaller than the total surface area that includes the surfaces of the clay interlayer spaces.

7.2.3.2 Adsorption of Polar Liquids for SSA Estimation

In contrast to nonpolar gases, polar liquids such as EGME, EG, MB, or water penetrate clay interlayers and thus provide estimates for the total surface area. This method involves saturating an initially oven-dry soil sample with the polar probing liquid and then evaporating the liquid in excess of monolayer coverage under high vacuum. Monolayer coverage is assumed when the mass of the evaporating sample remains constant for several consecutive weighings. The difference between monolayer-covered and dry sample masses is the mass of adsorbed EGME. From theoretical calculations for pure montmorillonite and experiments, it is known that 0.000 286 g of EGME is required to cover $1.0\,\text{m}^2$ with a monolayer of EGME molecules, which allows calculation of SSA from the dry sample mass and the mass of adsorbed EGME (Dyal and Hendricks, 1950). Uncertainties associated with the EGME method include the formation of multimolecular clusters around cation exchange sites prior to complete monolayer coverage (Orchiston, 1959) and the potential for cation solvation and dissolution into the organic phase for soils with appreciable amounts of organic matter (Tiller and Smith, 1990; Pennell et al., 1995). Despite these issues, EGME adsorption is widely considered as the reference method for estimation of the total SSA.

An alternative to EGME adsorption is the single-point water sorption method proposed by Newman (1983), who associated water molecule monolayer coverage of expansive soils with an RH of 47% and calculated SSA as

$$\mathrm{SSA} = \frac{\theta_{\mathrm{m}(47)} N_{\mathrm{A}} A}{M_{wa}} \tag{7.2}$$

where $\theta_{\mathrm{m}(47)}$ is the gravimetric water content at 47% RH, N_{A} is Avogadro's constant (6.022×10^{23} mol^{-1}), A is the area covered with one water molecule (1.06×10^{-19} m^2), and M_{wa} is the molar mass of water (0.018 kg mol^{-1}). Results obtained for 62 soils of different origins were in reasonable agreement with EGME SSA estimates (Newman, 1983). Later, Quirk and Murray (1999) and Arthur et al. (2013) associated monolayer water molecule coverage for nonexpansive soils with a RH of 20%.

Water vapor sorption isotherms rather than single-point measurements have been applied by Tuller and Or (2005a). They expressed the gravimetric water content (θ_{m}) as a function of adsorbed water film thickness (h) and SSA:

$$\theta_{\mathrm{m}} = \mathrm{SSA}\, h \rho_{\mathrm{w}} = \mathrm{SSA} \sqrt[3]{\frac{A_{svl}}{6\pi g \rho_{\mathrm{w}} \psi}}\, \rho_{\mathrm{w}} \tag{7.3}$$

where ρ_{w} (kg m^{-3}) is the density of water, A_{svl} (J) is the Hamaker constant for solid–vapor interactions through the intervening liquid, g (m s^{-2}) is the acceleration due to gravity, and ψ (J kg^{-1}) is the matric potential. The Hamaker constant represents average interactions between macro-objects such as mineral surfaces and liquid due to short-range van der Waals forces (Ackler et al., 1996; Bergström, 1997). An effective Hamaker constant of -6×10^{-20} J has been applied for soils (Tuller and Or, 2005a). The SSA is obtained from fitting Eq. (7.3) to measured sorption isotherms. SSA calculations based on the Tuller and Or (2005a) approach for 150 soils covering a wide range of textures are in very good agreement with results obtained with the EGME method (Figure 7.9).

7.2.4 Soil Water

Information about the energy state and the amount of water held within the soil pores is of significance for numerous engineering and earth and environmental science applications that include, but are not limited to, agricultural plant production, management and allocation of water resources, forecasting of weather and climate variability, prediction and monitoring of drought conditions, or monitoring of ecosystem response to climate change.

With regard to the energy state, kinetic energy due to motion of water within the soil porous system and potential energy that is dependent on the position of water within the soil profile and internal conditions are of primary interest. Because the flow velocity of water in soils is rather low (i.e. below 0.1 m h^{-1}), the kinetic energy is commonly neglected, unless preferential flow through macropores is considered. The driving force behind the

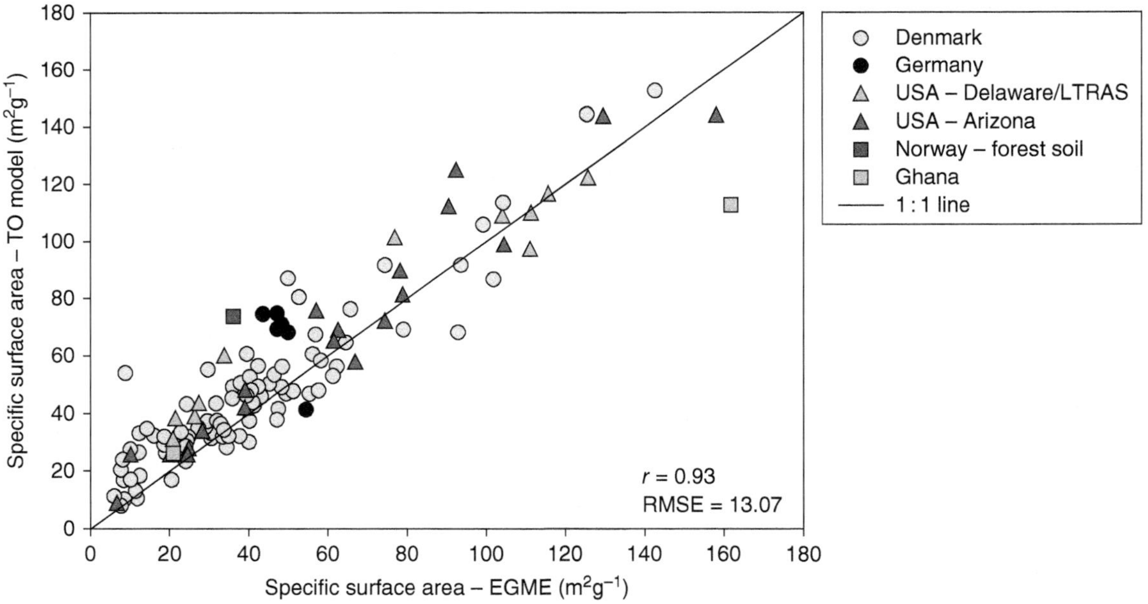

Figure 7.9 Comparison of the Tuller and Or (2005a) SSA estimation approach with standard EGME SSA measurements for 150 soils with varying textures.

movement of water in soils, the potential energy gradient, is the difference in potential energy between two points of interest within the soil divided by their Euclidean separation distance. Water always moves from locations of higher potential energy to locations of lower potential energy in pursuit of an equilibrium state.

The SM or soil water content represents the amount of water present in the soil at a given matric potential (Tuller and Or, 2005b). The matric potential is synonymous with the combined capillary and adsorptive surface forces that hold water within the solid soil matrix and is uniquely related to SM under hydrostatic conditions.

7.2.4.1 Soil Water Content

The soil water content may be expressed on a gravimetric basis or on a volumetric basis. The gravimetric water content, $\theta_m(\mathrm{kg\,kg^{-1}})$, is defined as the ratio of the mass of water within the soil sample and the mass of the oven-dry solid material. The volumetric SM content, $\theta_v(\mathrm{m^3\,m^{-3}})$, defined as the volume of water within a given soil volume may be expressed in terms of θ_m as

$$\theta_v = \theta_m \left(\frac{\rho_b}{\rho_w} \right) \tag{7.4}$$

where ρ_b is the dry bulk density ($\mathrm{kg\,m^{-3}}$) of the soil and ρ_w is the density of water ($\mathrm{kg\,m^{-3}}$). When all pores are filled with water, θ_v is termed saturated water content θ_s. In some instances, it is advantageous to express SM in terms of relative saturation, $S_e = \theta_v/\theta_s$, which is the volumetric SM content normalized to θ_s (i.e. pore volume). In theory, S_e ranges from 0, when the soil is completely dry, to 1, when all soil pores are completely filled with water. In practice however, it is not possible to attain complete dry or saturated conditions. There is always a residual moisture content, θ_r, present under dry conditions, and it is virtually impossible to completely de-air soil as air bubbles remain entrapped in dead-end pores and cavitation nuclei are held tightly in crevices of rough particle surfaces (Or and Tuller, 2002). To account for residual moisture, relative saturation is commonly defined as

$$S_e = \frac{\theta_v - \theta_r}{\theta_s - \theta_r} \tag{7.5}$$

Under wet conditions maximum attainable S_e ranges from about 0.94 to 0.97 for fine- (i.e. clay) and coarse-textured (i.e. sand) soils, respectively. For agricultural applications, a plant-available soil water content, θ_{PAW}, is often defined as the difference between the water content at field capacity, θ_{FC}, and the water content at the permanent wilting point, θ_{PWP} (Kabat et al., 1994). θ_{FC} is defined as the water content after internal redistribution of water within the soil matrix due to gravity (free drainage) and for practical purposes is often assumed to coincide with a matric potential of −330 cm. This definition is not entirely correct as θ_{FC} is dependent on the soil texture. For example, θ_{FC} for a sand soil more likely coincides with $\psi_m = -100$ cm. Below the permanent wilting point that is defined as the water content at −15 000 cm matric potential, water is so tightly bound within the soil matrix that plants are no longer able to recover their turgidity and irreversibly wilt. Again, this is only an approximation, as the permanent wilting point is dependent on plant physiology. Desert plants, for example, can withstand significantly lower matric potentials (dryer conditions) (Hupet et al., 2005). For water balance calculations it is convenient to express SM in length units to be consistent with other variables such as precipitation or evapotranspiration. This so-called equivalent depth of wetting is obtained by multiplying θ_v with the depth of the soil layer of interest.

7.2.4.1.1 Measurement of Soil Water Content

Besides direct measurement of θ_m or θ_v via oven drying of wet bulk soil or core soil samples at 105 °C, there are a plethora of techniques and sensors available for point-scale measurements of SM. In addition, several proximal and remote sensing methods amenable for larger-scale observations have been developed and refined over the last decades. Remote sensing techniques are discussed in Chapter 9 of this handbook.

Neutron Scattering The first portable neutron probes (NP) for in situ SM measurements have been introduced by Underwood and Swanson (1954) and Holmes (1956). The underlying neutron scattering technique is based on the tendency of hydrogen nuclei to slow (thermalize) fast, high-energy (2–4 MeV) neutrons to approach the characteristic speed of particles at ambient temperature with corresponding energies of about 0.03 eV. A typical NP consists of a radioactive neutron source (e.g. americium-241 and beryllium) and a detector to determine the flux of thermalized neutrons that form a cloud of nearly constant density near the probe. Neutrons lose different amounts of energy when colliding with various atomic nuclei. The greatest energy loss is due to collisions with particles of similar mass, such as hydrogen. While a few other elements such as chlorine, boron, and cadmium also tend to thermalize fast neutrons, the amount of hydrogen is nearly proportional to the number of slow neutrons. The quantity of hydrogen in the soil is largely dependent on the amount of water and to a lesser extent on the amount of organic matter and clay minerals. Therefore, a calibration function that relates the number of detected slow neutrons to soil water content can be developed. The sphere of influence (measurement volume) of NP depends on the SM content and chemical composition and on the

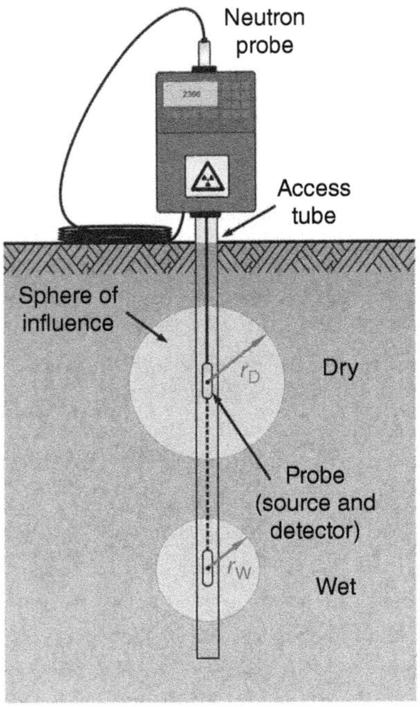

Figure 7.10 Illustration of a neutron probe.

strength of the radioactive source (Figure 7.10). Note that reliable measurements cannot be obtained close to the soil surface (i.e. top 20 cm) as neurons escape into the atmosphere; hence they are not captured by the detector.

Prior to measurements, an access tube needs to be installed by drilling a hole with slightly smaller diameter than the tube and then pushing the tube into the hole. A tight fit is crucial to prevent water from flowing down the outside of the access tube after irrigation or precipitation events. Tubes are typically made of steel, aluminum, or PVC. Access tube composition also affects the calibration. An advantage of plastic tubes is that it is possible to use a continuous roll of tubing for horizontal applications such as detection of leaks of landfill liners (Hanson and Dickey, 1993).

Ideally, the NP should be calibrated for each soil type and access tube material. The calibration consists of taking NP readings over a range of moisture contents in the soil of interest and then determining the moisture content volumetrically (core samples) in close vicinity of the NP access tube. To account for external conditions that might influence the NP measurements, it is common to normalize the actual count of thermalized neutrons by the shield count, i.e. an initial measurement conducted within the probe's lead shield. This so-called count ratio is plotted against the volumetric water content independently determined from the core samples and fitted with a regression line to obtain the calibration function. Calibrations are unique to each probe since the radioactive sources and the detectors may not be uniform from one probe to another, even when similar instruments are applied.

Electromagnetic Water Content Measurement Techniques Among the many options to determine volumetric soil water content in field settings and in the laboratory, electromagnetic (EM) techniques or sensors responding to soil dielectric permittivity are particularly advantageous because they do not use a radiation source and can be employed close to the soil surface (in contrast to neutron scattering and gamma ray attenuation techniques). Such techniques are considered noninvasive (in contrast to gravimetric techniques); they allow continuous monitoring and recording of SM from practically dry to saturated conditions, and they can be applied in most soil types and plant growth substrates (Vaz et al., 2013).

With the introduction of time-domain reflectometry (TDR), Topp et al. (1980) revolutionized EM soil water content measurement technology. While the first TDR instruments were very expensive and only amenable for laboratory measurements, costs have significantly decreased over the last decade, and a vast number of EM sensors with capacitance or impedance circuits (Table 7.3) that also capitalize on the strong correlation between SM and soil bulk dielectric permittivity have also been developed (Seyfried and Murdock, 2004; Jones et al., 2005; Bogena et al., 2007; Vaz et al., 2013). Permittivity (also called dielectric constant) is a measure of the electrical potential energy of a substance under the influence of an EM field. The EM sensor electric field is generally directed into the soil along 2-, 3-, or 4-parallel electrodes or an adjacent pair of rings. One sensor in particular, the HydraProbe (Stevens Water Monitoring Systems, Inc., Portland, OR, USA) that consists of one central and three outer electrodes (i.e. Hydra Probe, Stevens Water), is based on an impedance circuit operating at 50 MHz and turns out to be the only EM water content sensor offering both real and imaginary permittivity outputs (Figure 7.11a). While there is potential to extract additional information from complex permittivity, few have attempted to fully utilize these measurements (Peplinski et al., 1995; Kelleners et al., 2005). The HydraProbe is widely used by US federal agencies in weather and snow monitoring stations such as the USDA Natural Resources Conservation Service (NRCS) Soil Climate Analysis Network (SCAN) and Snow Telemetry (SNOTEL) monitoring networks.

The measurement of permittivity provides a highly accurate determination of water content in soil because the permittivity of water is about 80, while permittivities of solids and air are around 4 and 1, respectively. Interestingly, the most accurate water content determination method remains the original TDR method operating

Table 7.3 Off-the-shelf electromagnetic volumetric soil water content measurement systems and sensors.

Sensor	Company[a]	Measurement principle[b]	No. of electrodes	Electrode length (cm)	Sampling volume[c] (cm^3)	Sampling diameter[c] (cm)
TDR200[d]	Campbell	TDR	—	—	—	—
TDR-315	Acclima	TDR	3	15	—	—
TDT	Acclima	TDT	2	—	—	—
CS616	Campbell	TLO	2	30	3740	12.6
Theta P.	Delta-T	*I*	4	6	75	4
Hydra P.	Stevens	*I*	4	4.5	32	3
SM300	Delta-T	*I*	2	5.1	100	5
Wet2	Delta-T	*C*	3	6.8	500	9.7
5TE	METER	*C*	3	5.2	300	8.6
10HS	METER	*C*	2	10	1100	11.8

[a] Campbell Scientific, Inc., Logan, UT, USA; Acclima, Inc., Meridian, ID, USA; Delta-T Devices Ltd., Cambridge, UK; Stevens Water Monitoring Systems, Inc., Portland, OR, USA; METER Group, Inc., Pullman, WA, USA.
[b] TDR, time-domain reflectometry; TDT, time-domain transmission; TLO, transmission line oscillation; *I*, impedance; *C*, capacitance.
[c] From user manuals.
[d] The TDR200 system can be deployed with various off-the-shelf or customized probes.

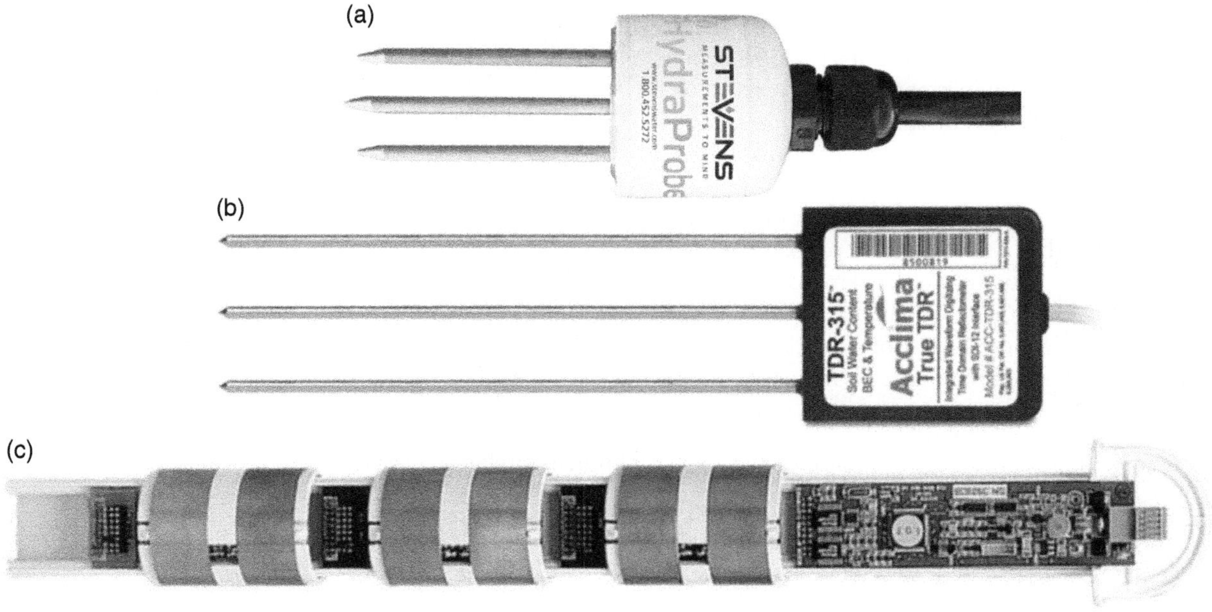

Figure 7.11 Examples for electromagnetic-based water content sensors: (a) The Stevens Water impedance-based HydraProbe measuring at 50 MHz. *Source*: Reproduced with permission of Stevens Water Monitoring Systems, Inc. (b) The Acclima True TDR-315 measuring in the range of 1000 MHz. *Source*: Photo courtesy of Realise Studios Australia. (c) The Sentek capacitance-based EnviroSCAN, sensor, which employs frequency-domain resonance, i.e. the water content determines the resonant frequency in the range from 100 to 150 MHz. Source: Reproduced with permission of Sentek Technologies.

at around 1 GHz frequency, despite a larger number of low-cost low-frequency (<100 MHz) water content sensors being marketed today (Blonquist et al., 2005; Vaz et al., 2013). In general, TDR does not require soil-specific calibrations and is less susceptible to secondary effects such as electrical conductivity (EC), temperature, and relaxation (e.g. from high clay or organic matter content). This immunity to secondary effects comes in part because of the high measurement frequency and the use of travel time as the measurement method. The travel time, t, of the EM signal is proportional to the relative permittivity, K_r, of the substance through which it travels according to

$$t = \frac{2l\sqrt{K_r}}{c} \quad (7.6)$$

where l is the length of the TDR electrodes and c is the speed of light (3×10^8 m s^{-1}) (Robinson et al., 2003). Once K_r is measured, the relationship between the volumetric water content (θ) of the soil and K_r is given from a general mixing model (Dobson et al., 1985; Jones et al., 2002), written as

$$\theta = \frac{K_r^\beta - (1-n)K_s^\beta - nK_a^\beta}{K_w^\beta - K_a^\beta} \quad (7.7)$$

where n is the soil porosity and subscripts s, a, and w refer to permittivities of constituent solids (minerals), air, and water, respectively. The exponent β summarizes the geometry of the medium and has a range $-1 < \beta < 1$, with $\beta = 0.5$ representing an isotropic mixed medium. Recently, GHz frequency TDR (Figure 7.11b) has been developed and reduced in size using low-cost cellular phone components, facilitating production of TDR sensors that are competitively priced with more common lower-frequency sensors.

Prior to the terrorist attacks on 9/11, NP were widely used for SM profile assessment as part of irrigation management owing to the ability to monitor water content in boreholes (Evett et al., 2002). The subsequent increase in regulations and paperwork related to radioactive devices has made NP impractical for most users. This has spurred an increase in EM-based borehole or downhole solutions, such as the EnviroSCAN probe (Sentek Technologies, Stepney, Australia), that employs parallel rings and capacitance circuitry to estimate water content in the vicinity of the cylindrical sensor (Figure 7.11c). A growing trend for environmental sensors in general is to capitalize on the SDI-12 communication protocol, which uses addressing to facilitate multiple sensors connected to a single digital port of a data logger. With options for addressing with single digit numbers and both upper- and lowercase letters, more than 50 sensors can be connected to each available port on a data logging device.

7.2.4.2 The Energy State of Soil Water

Water retained within the solid soil matrix is subjected to several forces that include matric forces due to interactions of the solid phase with the liquid and gaseous phases, osmotic forces due to differences in chemical composition of the soil solution, and gravitational body forces. These combined forces cause a deviation of potential energy relative to a reference state denoted as total soil water potential (ψ_T) that is defined as "The amount of work that an infinitesimal unit quantity of water at equilibrium is capable of doing when it moves (isothermally and reversibly) to a pool of water at similar standard (reference) state, i.e., similar pressure, elevation, temperature and chemical composition" (Or et al., 2005). Water always moves from locations exhibiting higher potential energy to locations of lower potential energy in pursuit of equilibrium. The driving force is the so-called potential energy gradient defined as the difference in potential energy between two points of interest within the soil profile divided by their separation distance. The total potential may be expressed as the sum of the potentials corresponding to the different force fields:

$$\psi_T = \psi_m + \psi_p + \psi_s + \psi_z \quad (7.8)$$

where ψ_m is the matric potential resulting from the combined effects of capillary and adsorptive surface forces within the soil matrix; ψ_p is the pressure potential defined as the hydrostatic pressure exerted by unsupported water on top of a point of interest in saturated soil; ψ_s is the solute or osmotic potential due to the presence of solutes in soil water, which lower its potential energy and its vapor pressure; and ψ_z is the gravitational potential, which is determined solely by the elevation of a point relative to some arbitrary reference point and is equal to the work needed to raise a body against the Earth's gravitational field from a reference level to its present position. Note that ψ_p and ψ_m are mutually exclusive, i.e. ψ_p is positive and ψ_m is zero when the soil is fully saturated or ψ_m is negative and ψ_p is zero when the soil is partially saturated. At the elevation of a free water table $\psi_p = \psi_m = 0$. A comprehensive depiction of potential energy theory of soil water is provided in Edlefsen and Anderson (1943), Tuller et al. (1999), and Or et al. (2005).

The potential components may be expressed as energy per unit of mass, energy per unit of volume, or energy per unit of weight (Table 7.4).

The different expressions for potential energy are interchangeable and can be simply converted with the following relationships:

$$\mu = \frac{\psi}{\rho_w} = gh \quad (7.9)$$

Table 7.4 Various expressions for potential energy of soil water.

Units	Symbol	Terminology	Dimensions	SI units
Energy/mass	μ	Chemical potential	L^2T^{-2}	J kg^{-1}
Energy/volume	ψ	Soil water potential, suction, or tension	$ML^{-1}T^{-2}$	N m^{-2} (Pa)
Energy/weight	H	Pressure head	L	m

where ρ_w is density of water (1000 kg m^{-3} at 20 °C) and g is acceleration due to gravity (9.81 m s^{-2}).

7.2.4.2.1 *Measurement of Potential Energy Components*

Matric Potential Several instruments and sensors are available to measure the matric potential, which may range from zero when the soil is completely saturated to very low negative values under dry conditions.

The most widely applied instrument, the tensiometer, consists of a porous cup with very fine pores made of porous stainless steel or ceramic, connected to a mechanical vacuum gauge or pressure transducer through a water-filled tube (Figure 7.12a). The porous cup is installed in close contact with the bulk soil at the depth of measurement. When the matric potential of the soil is lower (more negative) than the pressure inside the tensiometer, water moves from the tensiometer along a potential energy gradient to the soil through the saturated porous cup, thereby creating a negative pressure sensed by the gauge. Water flow into the soil continues until equilibrium is reached and the pressure inside the tensiometer equals the soil matric potential. Under dry conditions the water inside the tube needs to be replenished from time to time. When the soil is wetted in response to precipitation or irrigation events, flow may occur in the reverse direction, i.e. soil water enters the tensiometer until a new equilibrium is attained (Or et al., 2005). The gauge reading needs to be corrected to account for the weight of the water column overlaying the measurement point within the soil profile. If the matric potential is expressed as energy per unit of weight, the gauge or transducer reading is corrected as

$$\psi_{\text{m}} = \psi_{\text{gauge}} + \left(z_{\text{gauge}} - z_{cup}\right) \qquad (7.10)$$

with ψ_{gauge} representing the reading at the vacuum gauge location and z indicating depth. The vertical distance from the gauge plane to the cup, expressed as a negative quantity, must be added to the matric potential measured by the gauge (ψ_{gauge}) in order to obtain the matric potential at the depth of the cup. Tensiometer measurements may be fully automated to supply a continuous matric potential time series when calibrated pressure transducers connected to a data logger are employed. Modern tensiometers such as TS1 Smart Transducer Tensiometer (METER Group, Inc., Pullman, WA, USA) even incorporate an automated self-filling mechanism (Figure 7.12b). A limitation of the tensiometer is its narrow measurement range from 0 to about −10 m (i.e. −100 kPa) of pressure head. Hence, for lower matric potentials (i.e. drier soils), other measurement means such as heat dissipation sensors or psychrometers are required.

Heat dissipation sensors contain heating elements in line or point source configurations embedded in a rigid

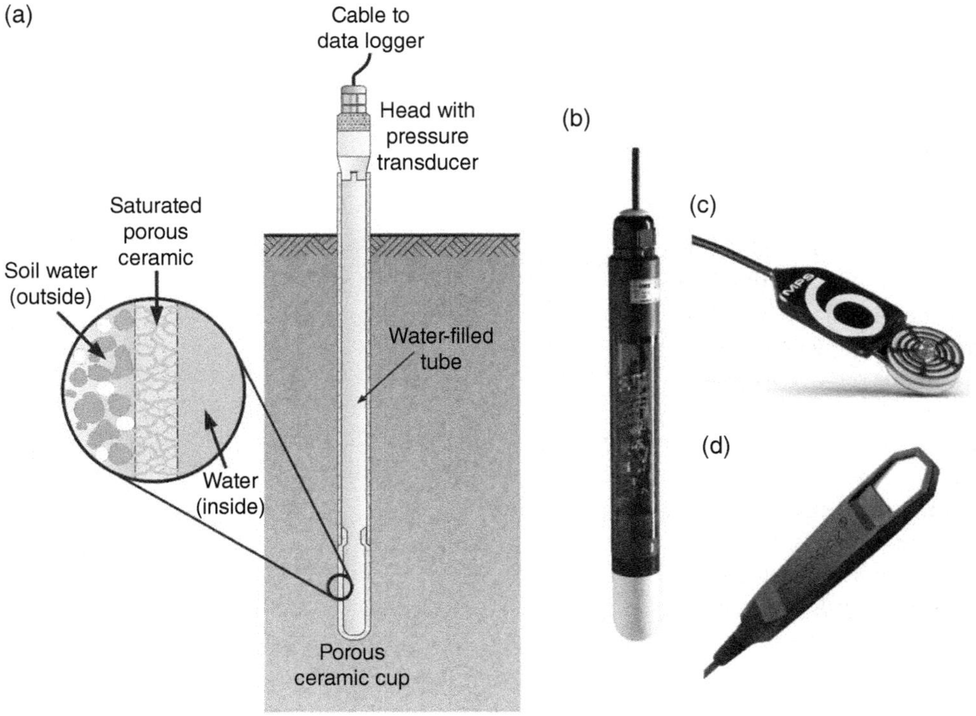

Figure 7.12 Illustration of a typical tensiometer with embedded pressure transducer (a) and self-refilling TS1 Smart Transducer Tensiometer (b). *Source*: Reproduced with permission of METER Group, Inc. USA. MPS-6 calibrated water potential sensor (c) and TensioMark matric potential sensor (d). *Source*: Reproduced with permission of Stevens Water Monitoring Systems, Inc.

porous matrix with fixed pore space. The measurement is based on application of a heat pulse by applying a constant current through the heating element for specified time period and analysis of the temperature response measured by a thermocouple placed at a fixed distance from the heating source (Phene et al., 1971; Bristow et al., 1993; Or et al., 2005). The rate of heat dissipation in soil is dependent on its specific heat capacity, thermal conductivity, and density. The heat capacity and thermal conductivity are affected by the water content. With the heat dissipation sensor buried in the soil, changes in soil water matric potential result in a gradient between the soil and the porous ceramic matrix, inducing water flow between the two materials until a new equilibrium is established. The water flow changes the water content of the ceramic matrix that, in turn, changes the thermal conductivity and heat capacity of the sensor and hence the measured temperature response to the applied heat pulse. A new-generation heat dissipation sensor, the TensioMark (Stevens Water Monitoring Systems, Inc., Portland, OR) with a measurement range from 0 to -1.0×10^6 kPa (distributor supplied information), is depicted in Figure 7.12d.

Another more recently developed technology encompasses a dielectric moisture sensor sandwiched between two ceramic disks with known relationship between the matric potential and volumetric water content. The sensor measures the dielectric permittivity of the ceramic disks, which are in equilibrium with the surrounding soil, and relates it to volumetric water content as outlined in Section 7.2.4.1.1. From the known water content–matric potential relationship of the disks (see Section 7.2.4.3), the matric potential is determined. The measurement range of the MPS-6 sensor (METER Group, Inc., Pullman, WA, USA) depicted in Figure 7.12c is from -9 to -1.0×10^5 kPa (distributor supplied information).

Another option for measurement of matric potential is psychrometric techniques. The potential of the soil solution (i.e. water and dissolved chemicals) is in thermodynamic equilibrium with its ambient water vapor. Taking the vapor pressure above pure water at reference state as e_0, the vapor pressure (e) over a salt solution or soil water held in soil pores by matric forces is depressed relative to the reference state, i.e. $e < e_0$. A convenient measure obtained by the psychrometer is the relative vapor pressure of the ambient soil atmosphere, which is related to the water potential ($\psi_w = \psi_m + \psi_s$) through the well-known Kelvin equation (Adamson and Gast, 1997; Or et al., 2005):

$$\mathrm{RH} = \frac{e}{e_0} = \exp^{(M_w \psi_w / \rho_w RT)} \tag{7.11}$$

where e is water vapor pressure (kPa), e_0 is saturated vapor pressure at the same temperature, M_w is the molecular weight of water ($0.018\,\mathrm{kg\,mol^{-1}}$), R is the ideal gas constant ($8.31\,\mathrm{J\,K^{-1}\,mol^{-1}}$ or $0.008\,314\,\mathrm{kPa\,m^3\,mol^{-1}\,K^{-1}}$), T is absolute temperature (K), and ρ_w is the density of water ($1000\,\mathrm{kg\,m^{-3}}$ at 20 °C). Rearranging and taking a log transformation of Eq. (7.11) yields an expression for water potential ψ_w:

$$\psi_w = \frac{RT\rho_w}{M_w} \ln\left(\frac{e}{e_0}\right) \tag{7.12}$$

Note that a psychrometer measures the combined matric and osmotic potentials ($\psi_w = \psi_m + \psi_s$). If the osmotic potential is negligible, the matric potential can be directly inferred from psychrometer measurements. If appreciable amounts of salts are present in the soil solution, the osmotic potential needs to be determined independently and subtracted from the psychrometer readings. A modern laboratory psychrometer, the WP4C (METER Group, Inc., Pullman, WA, USA) based on the chilled-mirror dew-point technique (Gee et al., 1992), and a thermocouple field psychrometer (Andraski and Scanlon, 2002) are depicted in Figure 7.13.

Pressure Potential The pressure potential, ψ_p, may be measured with a piezometer (Figure 7.14), which is simply a tube with a perforated bottom section that is flushed or pushed into predrilled holes to depths that extends below the unconfined water table. On the top the tube is open to the atmosphere. Soil water under positive hydrostatic pressure enters the tube through the perforations at the bottom and rises to a height equal to that of the unconfined water table. The elevation of the free water table is measured relative to the soil surface using a steel tape with bell sounder or other electro-optic devices that indicate water table depth. The value of pressure potential expressed as energy per weight is simply the vertical distance from a point of interest to the surface of the free water table. Note that pressure potentials above the unconfined water table surface are always zero (Or et al., 2005).

Osmotic Potential The soil water solutions contain varying quantities of dissolved salts. In dilute solutions the solute potential, also called the osmotic pressure, is proportional to the concentration and temperature according to

$$\psi_s = -RTC_s \tag{7.13}$$

where ψ_s is in kPa, R is the universal gas constant ($8.314 \times 10^{-3}\,\mathrm{kPa\,m^3\,mol^{-1}\,K^{-1}}$), T is absolute temperature (K), and C_s is solute concentration ($\mathrm{mol\,m^{-3}}$). A useful approximation that may be used to estimate ψ_s in kPa from the EC of the soil solution at saturation (ECs) in $\mathrm{dS\,m^{-1}}$ is

$$\psi_s \approx -36\,EC_s \tag{7.14}$$

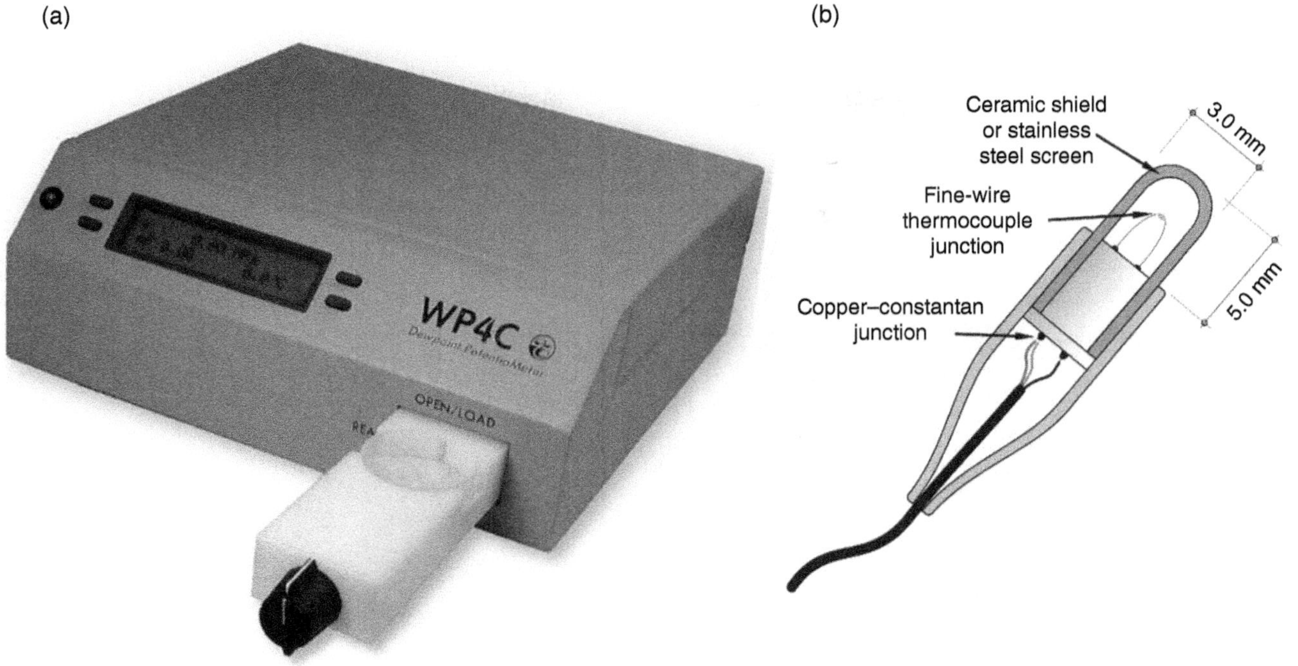

Figure 7.13 Modern WP4C Dewpoint Hygrometer (a). *Source*: Reproduced with permission of METER Group, Inc. USA. Thermocouple field psychrometer with porous ceramic shield (b). *Source*: Wescor Inc, Logan, UT, USA.

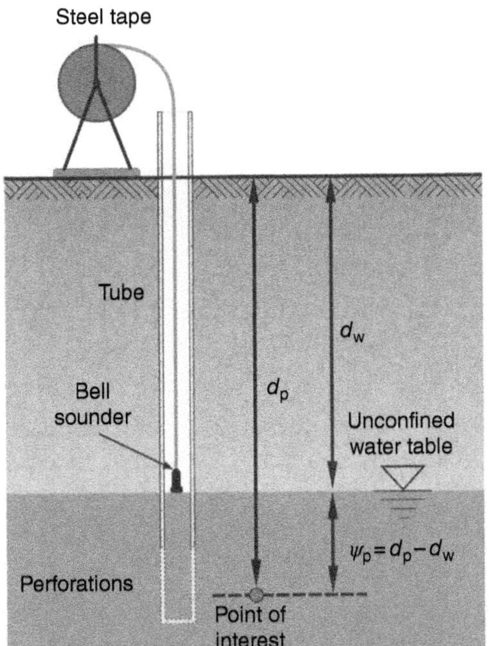

Figure 7.14 Sketch illustrating the concept of piezometer measurements.

Conventional measurement of soil solution EC involves solution extraction via centrifugation from saturated soil samples or via soil water samplers in the field and measuring the EC with an EC meter (Figure 7.15). Other laboratory or in situ methods to extract soil solution and measure EC are discussed in Tuller and Islam (2005).

7.2.4.3 The Soil Water Characteristic

The equilibrium relationship between soil water content and matric potential, ψ_m, is termed soil water characteristic and exhibits a very distinctive shape for each individual soil (Figure 7.16). The SWC is an important hydraulic property related to size and connectedness of pore spaces, hence strongly affected by soil texture and structure and by other constituents including organic matter. Modeling water distribution and flow in partially saturated soils requires knowledge of SWC; hence it plays a crucial role in water management and in prediction of solute and contaminant transport in the environment.

Typically, a SWC is highly nonlinear and is relatively difficult to obtain accurately. Because the matric potential extends over several orders of magnitude for the range of water contents commonly encountered in practical applications, the matric potential is often plotted on a logarithmic scale. Figure 7.16 depicts representative SWC curves for soils of different textures, demonstrating the effects of porosity (saturated water content) and the varied slopes of the relationships resulting from variable pore size distributions (Tuller and Or, 2005b).

Note that the water content and the matric potential are not uniquely related because the amount of water present at a given matric potential is dependent on the

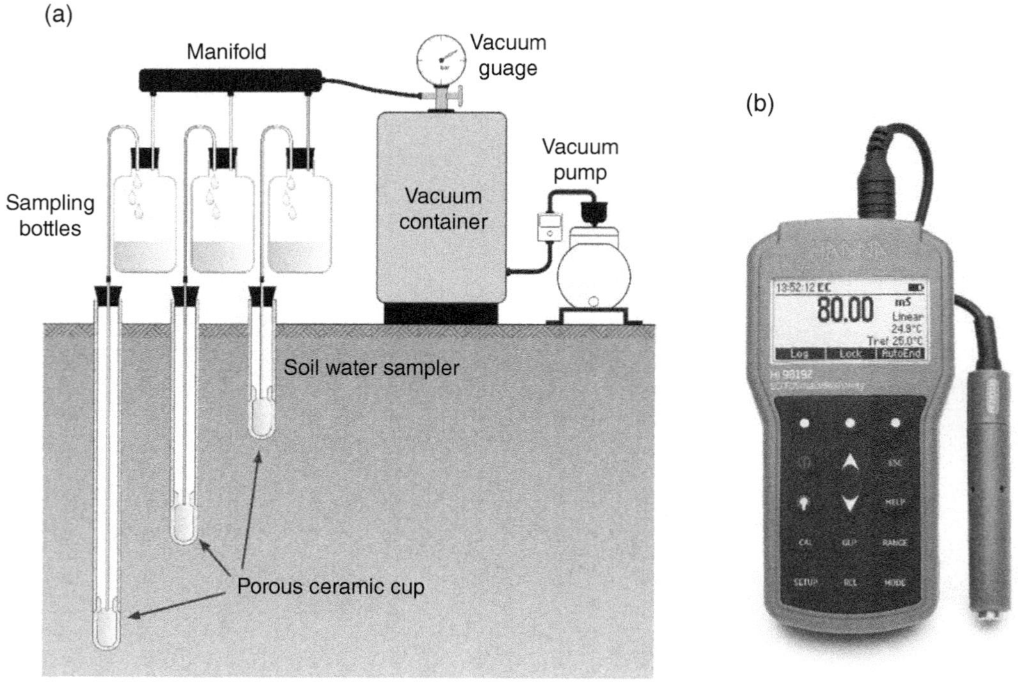

Figure 7.15 Automated field solution extraction setup (a); handheld electrical conductivity (EC) meter (b).

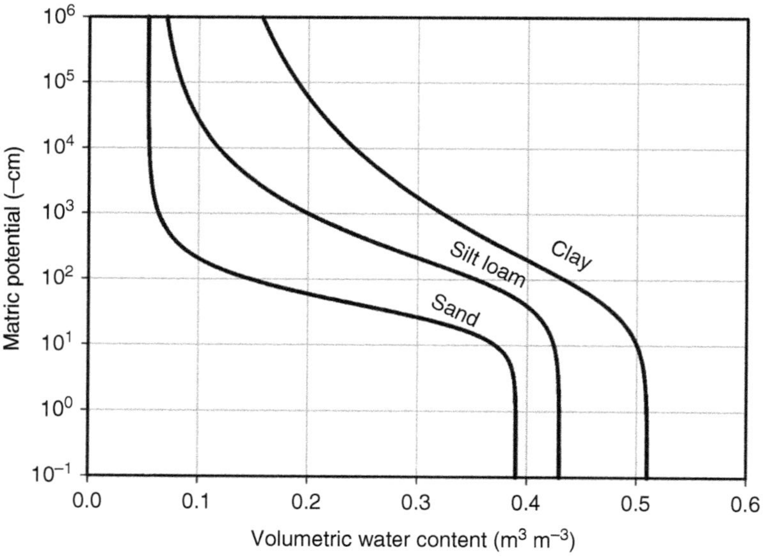

Figure 7.16 Soil water characteristics (SWC) for different soil textures.

pore size distribution and the properties of air–water–solid interfaces (Tuller and Or, 2005b). An SWC relationship may be obtained either by starting with an initially saturated sample and applying suction or pressure to desaturate it (desorption) or by gradually wetting an initially dry soil (sorption). These two pathways produce curves that in most cases are not identical; the water content of the "drying" curve is higher for a given matric potential than that of the "wetting" branch (Figure 7.17a). This is called "hysteresis," which can be related to several phenomena, including (i) the "ink bottle" effect resulting from nonuniformity in shape and sizes of interconnected pores (drainage is governed by the smaller pore radius r, whereas wetting is dependent on the large radius R (Figure 7.17c), (ii) different liquid–solid contact angles for advancing and receding water menisci (Figure 7.17b), (iii) entrapped air in a newly wetted soil, and (iv) swelling and shrinking of the soil under wetting and drying. Based on early observations of the phenomenon by Haines (1930) and on more recent theories (Mualem, 1984; Kool and Parker, 1987), the role of individual factors remains unclear and is subject to ongoing research. Part of the

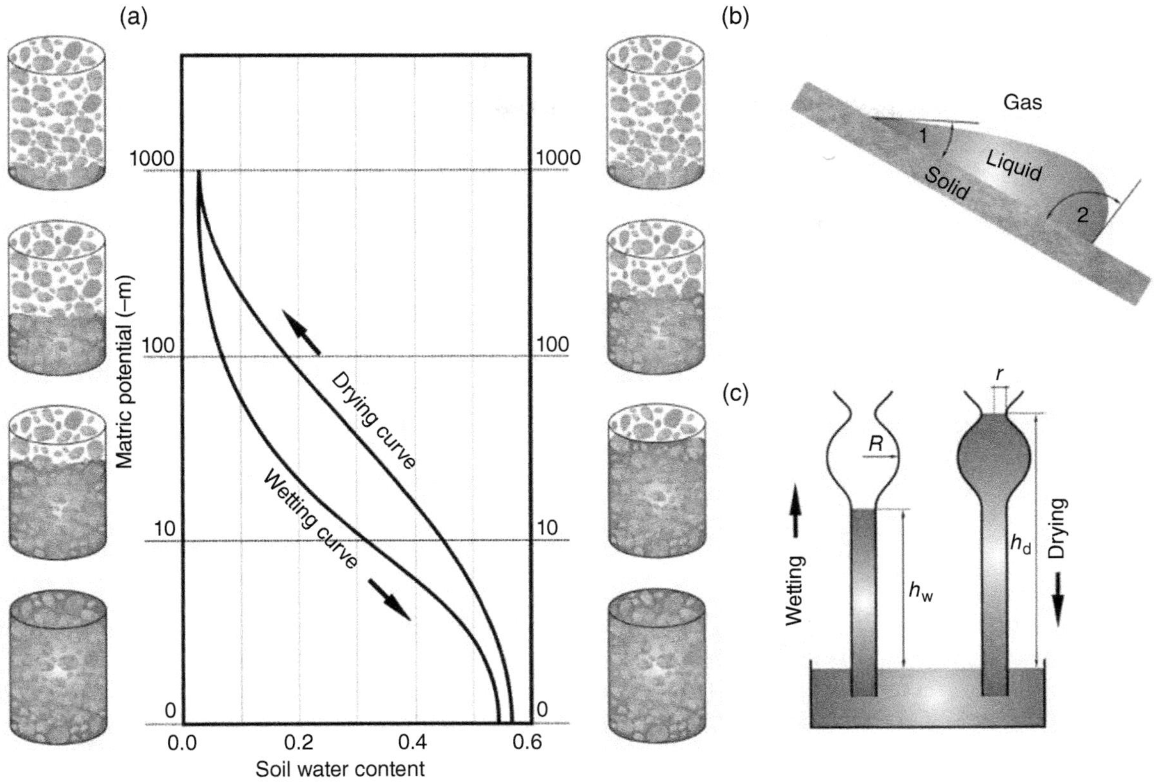

Figure 7.17 Conceptual sketch showing hysteresis of the SWC (a), the contact angle effect (b), and the ink bottle effect (c) as potential mechanisms for hysteresis. *Source*: Adapted from Or et al. (2005).

hysteresis phenomena may be attributed to measurement artifacts, for example, due to differences between tension- and pressure-induced desaturation (Chahal and Young, 1965). A potentially important aspect of desorption methods under tension is the possibility of liquid displacement (drainage) even in the absence of a continuous gaseous phase due to cavitation initiated by entrapped gas bubbles or the liquid's own vapor pressure (Or and Tuller, 2002). Surface heterogeneity and impurities in soil and rock water are conducive to lowering the cavitation tension threshold.

7.2.4.3.1 *Parametric Models for the Soil Water Characteristic*

Because measurement of the SWC is laborious and time consuming, experimentally determined θ–ψ_m pairs are in many cases fragmentary constituting only a few measurements over the applicable water content range. It is therefore beneficial to represent the SWC in a mathematically continuous form, which is required for numerical modeling of water flow and solute transport processes in soils. An important requirement for all parametric SWC models is a low degree of freedom (i.e. as few free model parameters as possible) to simplify parameter estimation and accurate description of SWC behavior at the wet and dry limits while closely matching the nonlinear shape of θ–ψ_m measurements (Tuller and Or, 2005b).

One of the most widely applied parametric SWC models for relating water content or effective saturation (S_e) to the matric potential (ψ_m) was proposed by van Genuchten (1980):

$$S_e = \frac{\theta - \theta_r}{\theta_s - \theta_r} = \left[\frac{1}{1 + \left(\alpha \psi_m\right)^n}\right]^m \tag{7.15}$$

where θ_r and θ_s are the residual and saturated water contents, respectively, and α, n, and m are parameters related to the shape of the SWC curve. A common simplification is to assume that $m = 1 - 1/n$. Thus the parameters required for estimation of the model are θ_r, θ_s, α, and n. θ_s is commonly measured or set to equal the soil's porosity, leaving only the three free model parameters to be determined via nonlinear regression analysis estimated from θ–ψ_m pairs.

Another widely applied parametric model was proposed by Brooks and Corey (1964):

$$\begin{aligned} S_e &= \frac{\theta - \theta_r}{\theta_s - \theta_r} = \left(\frac{\psi_b}{\psi_m}\right)^{\lambda} && \psi_m > \psi_b \\ S_e &= 1 && \psi_m \le \psi_b \end{aligned} \tag{7.16}$$

where ψ_b is a parameter related to the soil matric potential at air entry (i.e. pressure at which the largest pore in the soil systems starts draining) and λ is related to the soil pore size distribution. Matric potentials are expressed as positive quantities (i.e. in absolute values) in both van Genuchten (VG) (1980) and Brooks and Corey (BC) (1964) models. Other frequently employed parametric models are listed in Table 7.5. More complex physically based approaches to modeling of the SWC are discussed in Tuller et al. (1999), Or and Tuller (1999), and Tuller and Or (2005b).

Figure 7.18 depicts VG and BC models parameterized via nonlinear regression based on silt loam soil data from Or et al. (1991). The resulting best-fit parameters for the VG model are $\alpha = 0.417\,m^{-1}$, $n = 1.75$, $\theta_s = 0.513\,m^3\,m^{-3}$, and $\theta_r = 0.05\,m^3\,m^{-3}$ (with $r^2 = 0.99$). For the BC model the best-fit parameters are $\lambda = 0.54$, $\psi_b = 1.48\,m$, $\theta_s = 0.513\,m^3\,m^{-3}$, and $\theta_r = 0.03\,m^3\,m^{-3}$ (with $r^2 = 0.98$). Note that the most striking difference between the VG and the BC models is in the discontinuity at $\psi = \psi_b$ for BC.

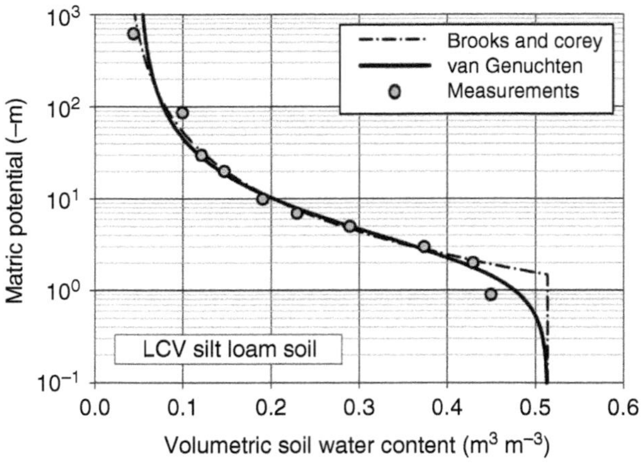

Figure 7.18 van Genuchten (VG) and Brooks and Corey (BC) models parameterized via nonlinear regression from measured data for silt loam soil.

7.2.4.3.2 Measurement of the Soil Water Characteristic

Pressure Flow Cells Tempe cells (i.e. pressure flow cells) in combination with psychrometric techniques (see Section 7.2.4.2.1 and Figure 7.13a) are the preferred methods to measure the SWC from saturation to very dry conditions. The design of a typical Tempe cell is depicted in Figure 7.19. It comprises a flow cell that holds a water-saturated porous plate, which allows water but not air to flow through its pores. The porous plate is at atmospheric pressure at the bottom, while the top surface is at the gas pressure applied to the top of the cell (Figure 7.19). Undisturbed core samples are placed in contact with the porous plate and initially saturated with water from the bottom. The cell is connected to a pressure manifold with regulator and precision manometer, and a known N_2 or air gas pressure is applied to force water out of the soil through the saturate ceramic plate. Water flows from the cell until the force exerted by the air pressure is in equilibrium with the force that retains water within the soil matrix (i.e. matric potential). Following equilibrium between soil matric potential and the applied air pressure (i.e. cessation of flow), the cells are disconnected from the

Table 7.5 Classical models for the soil water characteristic, $\theta(h)$, and hydraulic conductivity function, $K(h)$ or $K(\theta)$.

Model[a]	$S = \frac{\theta - \theta_r}{\theta_s - \theta_r} =$	$K_r = \frac{K}{K_s} =$
Gardner (1958)	—	$\left[1 + \left(\frac{h}{h_b}\right)^p\right]^{-1}$
Brooks and Corey (1964)	$\begin{cases} 1 & h \le h_b \\ (h/h_b)^{-\lambda} & h > h_b \end{cases}$	$\begin{cases} 1 & h \le h_b \\ (h/h_b)^{-P} & h > h_b \end{cases}$
van Genuchten (1980)	$[1 + (\alpha h)^n]^{-m}$	$S^{0.5}[1 - (1 - S^{1/m})^m]^2$
Gardner–Russo (Gardner, 1958; Russo, 1988)	$\left(\frac{1 + 0.5Ah}{e^{0.5Ah}}\right)^{2/(l+2)}$	e^{-Ah}
Kosugi (1994, 1996)	$0.5\,\mathrm{erfc}\left[\frac{\ln(h/h_m)}{\sigma\sqrt{2}}\right]$	$S^{0.5}\left\{0.5\,\mathrm{erfc}\left[\mathrm{erfc}^{-1}(2S) + \frac{\sigma}{\sqrt{2}}\right]\right\}^2$

Source: Modified from Sadeghi et al. (2017).

[a] θ, volumetric water content; θ_s and θ_r, saturated and residual volumetric water contents, respectively; K and K_s, unsaturated and saturated hydraulic conductivities, respectively; h, pressure head; h_b and h_m, air-entry and median pressure heads; σ, standard deviation of the log-transformed pore size; p, λ, P, α, n, m, A, and l, empirical parameters.

pressure manifold and weighed. This step is repeated several times with successively higher air pressures until the matric potential range of interest is covered. After the final weighing, the sample is oven-dried. From the known masses of the Tempe cell, the wet soil samples (corresponding to a specific matric potential), and the oven-dried sample, the gravimetric water contents can be calculated for each pressure step (matric potential). Finally, the gravimetric water contents are converted to volumetric water contents with the bulk density known from the dry sample mass and the volume of the sample cylinder. The water content of repacked soils at a given matric potential should not be used to infer θ of intact soils at the same ψ_m, due to modified pore sizes and pore geometry. The pressure range of Tempe cells is usually limited to 0–1 bar or 0.1 MPa. The porous ceramic plates used in pressure plates and flow cells must be completely water-saturated prior to use. The dry end of the SWC is commonly measured with a laboratory psychrometer such as the WP4C Dewpoint Hygrometer (METER Group, Inc., Pullman, WA, USA) (see Figure 7.13a).

Evaporation Method Another laboratory technique for measurement of the SWC and the unsaturated hydraulic conductivity, to be discussed later, is the so-called evaporation method originally proposed by Wind (1968). This method involves evaporation of water from the surface of an initially saturated soil column while concurrently measuring the change in matric potential and the water mass loss. Advanced benchtop evaporation systems such as the HYPROP (METER Group, Inc., Pullman, WA, USA) consist of two miniature transducer tensiometers for matric potential measurements, one installed close to the evaporating surface and the second close to the bottom of the column (Figure 7.20), and a highly accurate laboratory scale that continuously

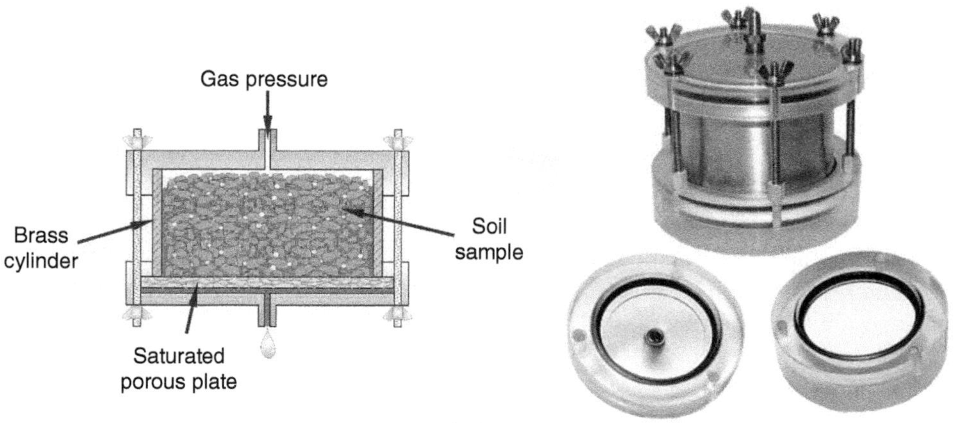

Figure 7.19 Tempe cell for wet-end SWC measurements.

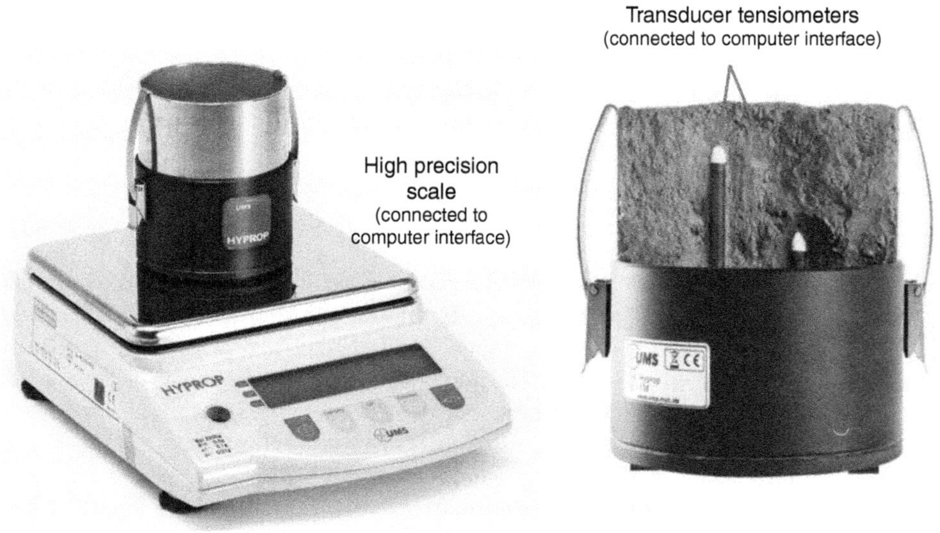

Figure 7.20 HYPROP evaporation measurement system. *Source*: Reproduced with permission of METER Group, Inc. USA.

measures the water mass loss. The HYPROP system is fully automated and interfaced with a computer for monitoring the experiment and recording the measurements. From the readings of the 2 tensiometers (kPa), the average matric potential and the hydraulic gradient are calculated. The mass difference, measured with the scale, is used to calculate the volumetric water content and the water flow rate. The experiment is terminated when either one of the tensiometers runs dry or the mass changes become marginal. Then, the remaining moisture content is determined by oven-drying the sample at 105 °C for 24 h. The analytic procedure is based on the assumption that the matric potential and water content distribute linearly through the column and that matric potential changes are linear between two evaluation points. The initial water content is determined from the total loss of water (i.e. evaporation and water loss by oven drying). Because HYPROP measurements are limited to the wet end of the SWC due to the functional range of the tensiometers, obtained data are commonly combined with WP4C Dewpoint Hygrometer data for dry conditions. A specifically developed software package allows combining wet- and dry-end measurements and provides means to parameterize the VG (1980) and BC (1964) SWC models.

Measurement of the SWC in the Field Despite of the importance of measuring SWCs in situ, few suitable methods are available for field or in situ application. Paired sensors may be used to measure θ and ψ_m in the same or closely adjacent soil volumes over a range of soil wetness. Examples include use of neutron moisture meter access tubes or TDR waveguides along with tensiometers. Application of paired sensors is often constrained by differences in soil volumes sampled by the respective sensors, different time constants required (e.g. many θ sensors obtain instantaneous measurements, while many ψ_m methods require equilibrium), and the fact that few matric potential techniques function over the entire wetness range of interest. This commonly provides limited overlap in soil water retention measured using different techniques, and measurements obtained using different methods may not be consistent in the ranges of overlap.

7.3 Water Flow in Soils

The flow of water in soils occurs under both saturated and unsaturated conditions. While saturated flow generally happens below the groundwater table where pressure heads (h) are positive, unsaturated conditions are prevalent in the vadose zone that extends from the water table to the soil surface. Unsaturated flow is governed by negative pressure heads. Water flow is a key process that directly impacts numerous other biogeochemical soil processes. For example, the transport of contaminants or colloids is intimately connected to water flow. Precipitation or irrigation water naturally enters through the soil surface and percolates to deeper depths, eventually replenishing the aquifer. Part of the water that enters the soil is returned to the atmosphere due to evaporation and plant transpiration. Governing equations for water flow can be derived for either steady or transient conditions. When discussing the total water potential ψ_T (Section 7.2.4.2), it was mentioned that whenever the potential energy of water differs from one location to another, a force acts to move water from locations of higher potential energy to locations of lower energy in pursuit of an equilibrium state. In this section we discuss the properties and laws governing water flow in soils under various conditions. We consider flow due to differences in hydraulic potential only, neglecting effects of solute potential (ψ_s) on water flow.

7.3.1 Basic Models for Water Flow in Soil

7.3.1.1 Darcy's Law

Water flows in soils from locations of higher energy to locations of lower energy. Considering that flow velocity in soils is generally low (i.e. the kinetic energy is negligible), energy for water movement in a saturated soil originates mainly from (i) the pressure potential (e.g. hydrostatic pressure exerted from a free water table) and (ii) the potential energy (i.e. mgz; mass × gravitational acceleration × elevation difference). Energy for water movement in soil is commonly expressed in terms of energy per unit of weight (mg), and termed "hydraulic potential" or "pressure head," and thus has the unit of length (Table 7.4).

Water flow in saturated soil was first formulated in 1856 by Henry Darcy who was a French engineer. Darcy's law states that water flux density (i.e. discharge per unit area) in saturated soil is proportional to the gradient of hydraulic potential, where the proportionality coefficient is called "saturated hydraulic conductivity":

$$q = -K_s \frac{\Delta H}{\Delta z} \tag{7.17}$$

In Eq. (7.17), q is the water flux density, H is the hydraulic potential, K_s is the saturated hydraulic conductivity, and z is the soil depth assumed positive downward from the soil surface. The minus sign in Eq. (7.17) appears because water flows from higher potentials toward lower potentials.

As mentioned above, it is commonly assumed that $H = h - z$, where h denotes either the pressure potential, ψ_p, or the matric potential, ψ_m (see Figure 7.21). Considering this assumption, Darcy's law in differential form can be written as

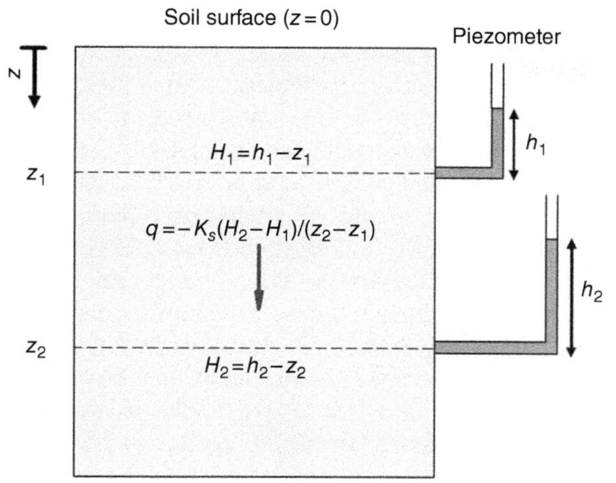

Figure 7.21 Sketch illustrating the Darcy flux (q from Eq. (7.17)) in a saturated soil with saturated hydraulic conductivity of K_s. Hydraulic potential for water flow (H) mainly originates from the pressure head (piezometric pressure, h) and the gravitational potential (considered to be zero at the surface and equal to $-z$ at the depth of z).

$$q = -K_s\left(\frac{\partial h}{\partial z} - 1\right) = -K_s\frac{\partial h}{\partial z} + K_s \tag{7.18}$$

As Eq. (7.18) indicates that the Darcy flux is composed of two terms, one describing flow due to the pressure head gradient and the other describing flow due to the gravity or gravitational potential.

7.3.1.2 Buckingham–Darcy Law

Buckingham (1907) modified Darcy's law for modeling flow in unsaturated soil by (i) replacing the pressure potential with the so-called matric potential, which is a negative potential for water movement due to the retention of water molecules by soil particles in unsaturated condition (see Section 7.2.4.2), and (ii) considering the fact that the hydraulic conductivity will highly decrease under unsaturated condition due to the reduction of the flow pathways (i.e. the saturated hydraulic conductivity is replaced by an unsaturated hydraulic conductivity function).

Calculation of the hydraulic potential for unsaturated soil is similar to saturated soil calculations illustrated in Figure 7.21 (i.e. $H = h - z$); only h now is a negative pressure that can be directly measured with a tensiometer (Section 7.2.4.2.1). Therefore, the Buckingham–Darcy (BD) law can be written as

$$q = -K(h)\left(\frac{\partial h}{\partial z} - 1\right) \tag{7.19}$$

where h is either the pressure potential or the matric potential (i.e. $h < 0$ for unsaturated parts and $h \geq 0$ for saturated parts of the soil) and K is the "unsaturated hydraulic conductivity," which is a function of h when $h < 0$.

7.3.1.3 Richards' Equation

Richards' Equation (RE) is a partial differential equation (PDE) governing water flow in unsaturated soil that is derived by combining the BD law (Eq. 7.19) and the continuity principle, $\partial\theta/\partial t = -\nabla q$. The one-dimensional (1D) form of RE is thus obtained as

$$\frac{\partial\theta}{\partial t} = \frac{\partial}{\partial z}\left[K(h)\frac{\partial h}{\partial z} - K(h)\right] \tag{7.20}$$

where θ is the soil water content and t is the time.

RE is basically a nonlinear convection–diffusion equation (CDE). The convection flux is due to gravity, while the diffusive flux is due to the moisture gradient. For solving RE, functional relationships between θ, h, and K need to be known. These functional relationships are termed "soil hydraulic functions" (SHFs). Having $h(\theta)$ and $K(h)$ [or $K(\theta)$] for a given soil and solving RE subject to specific initial and boundary conditions, one can simulate soil matric potential profile, $h(z)$, or SM profile, $\theta(z)$, at any time of interest, t.

7.3.1.3.1 Soil Hydraulic Functions

Various mathematical relationships exist for SHFs, $h(\theta)$ and $K(\theta)$. The most widely used unimodal SHFs are relationships introduced by Gardner (1958), Brooks and Corey (1964), van Genuchten (1980), Russo (1988), and Kosugi (1994, 1996), listed in Table 7.5.

In Table 7.5, the SWC curves are empirical relationships, and the conductivity models are derived from substitution of the retention curve into a bundle of cylindrical capillaries (BCC) model. The BCC model generally assumes soil pore space as a BCC. Then conductivity of all individual water-filled pores are calculated using Poiseuille's law, where each cylindrical pore of a known radius at a specific pressure head is assumed to be either completely water filled or completely empty based on the "Young–Laplace" equation.

The most widely applied BCC hydraulic conductivity model is that of Mualem (1976), given as

$$K = K_s S^{0.5}\left\{\frac{\int [h(S)]^{-1}\,\mathrm{d}S}{\int_0^1 [h(S)]^{-1}\,\mathrm{d}S}\right\}^2 \tag{7.21}$$

where S is the effective degree of saturation (i.e. $S = 0$ for fully dry soil and $S = 1$ for fully saturated soil).

The integral model of Eq. (7.21) can be solved in closed form if the $h(S)$ functional relationship (i.e. the SWC curve) is known. For example, van Genuchten (1980) assumed the following empirical $h(S)$ relationship:

$$S = \frac{\theta - \theta_r}{\theta_s - \theta_r} = \left[1 + (\alpha h)^n\right]^{-m} \tag{7.22}$$

where θ_s is the saturated water content, θ_r is called "residual water content" and is defined as θ when $h \to -\infty$, and

α, n, and m are empirical fitting parameters. It is commonly assumed that n and m are related (e.g. $m = 1 - 1/n$). By substituting Eq. (7.22) into Eq. (7.21) and solving the integral of Eq. (7.21), van Genuchten (1980) derived his popular closed-form hydraulic conductivity model, $K(S)$ or $K(h)$:

$$K = K_s S^{0.5}\left[1-\left(1-S^{1/m}\right)^m\right]^2 = K_s \frac{\left\{1-(\alpha h)^{n-1}\left[1+(\alpha h)^n\right]^{-m}\right\}^2}{\left[1+(\alpha h)^n\right]^{m/2}} \tag{7.23}$$

SHFs are commonly parameterized via nonlinear regression analysis of measured h–θ and K–θ data for a soil of interest. The VG model parameters for thousands of soils throughout the world have been characterized. Table 7.6 lists the averages of VG parameters for the 12 USDA soil textural classes, obtained by analyzing a big database of soils of various textures (Carsel and Parrish, 1988).

Several studies conducted within the past decade have provided evidence that in addition to capillary forces, adsorptive surface forces significantly impact the matric potential, especially in fine-textured soils with high SSA (Tuller et al., 1999; Tuller and Or, 2005a; Peters and Durner, 2008; Lebeau and Konrad, 2010). They pointed to the significant coexistence of liquid water films and capillary menisci that can be defined with a combination of the conventional Young–Laplace equation for the capillary pressure and the disjoining pressure concept for thin water films covering solid surfaces (Tuller et al., 1999; Bachmann and van der Ploeg, 2002).

Table 7.6 Average van Genuchten model parameters for the 12 soil textural classes of the USDA classification scheme.

Soil texture	θ_r	θ_s	α (cm^{-1})	n	K_s (cm d^{-1})
Sand	0.045	0.43	0.145	2.68	712.80
Loamy sand	0.057	0.41	0.124	2.28	350.20
Sandy loam	0.065	0.41	0.075	1.89	106.10
Loam	0.078	0.43	0.036	1.56	24.96
Silt	0.034	0.46	0.016	1.37	6.00
Silt loam	0.067	0.45	0.020	1.41	10.80
Sandy clay loam	0.100	0.39	0.059	1.48	31.44
Clay loam	0.095	0.41	0.019	1.31	6.24
Silty clay loam	0.089	0.43	0.010	1.23	1.68
Sandy clay	0.100	0.38	0.027	1.23	2.88
Silty clay	0.070	0.36	0.005	1.09	0.48
Clay	0.068	0.38	0.008	1.09	4.80

Source: From Carsel and Parrish (1988).

Direct measurements of SHPs mostly rely on inversion of Darcy's law for steady-state conditions or an approximate analytical or numerical solution of RE for transient conditions. Therefore, these methods are limited to the so-called Darcy scale, for which RE was derived. The Darcy scale encompasses the sample scale (Kool et al., 1985; Peters and Durner, 2008) or column scale (Shao and Horton, 1998; Sadeghi et al., 2014) for laboratory analyses and point scale (Šimůnek and Genuchten, 1996) and block scale (Hillel et al., 1972) for field tests, some of which will be discussed later in this section.

7.3.2 Mathematical Modeling of Soil Water Flow

Water flow in variably saturated soils can be simulated by solving the abovementioned governing equations. The solution can be performed either analytically or numerically. Because the flow equations are highly nonlinear, especially when considering complex mathematical relationship for SHFs (e.g. VG and Kosugi models), soil water flow simulation is generally performed with numerical codes such as HYDRUS (Šimůnek et al., 2005). Nonetheless, analytical solutions for the flow problems are still necessary in this field mainly for testing the accuracy of numerical schemes and to provide favorable means for inverse estimation of the hydraulic model parameters. Analytical solutions of the BD law and RE are commonly restricted to simplified cases such as simple hydraulic functions, steady-state flow conditions, and uniform and invariant initial and boundary conditions. In the following, some examples for analytical and numerical solutions to the soil water flow problems (BD law and RE) are presented. Note that the solutions presented here are only a few examples among the vast body of literature in this field.

7.3.2.1 Analytical Solution of Buckingham–Darcy Equation

7.3.2.1.1 *Steady-State Evaporation and Infiltration*

Assuming a steady-state flow rate through the soil column, e.g. evaporation rate from a constant water table, the matric potential distribution above the water table, $h(z)$, can be modeled by solving the BD law rearranged as

$$z = \int \frac{K(h)}{K(h)+q} dh \tag{7.24}$$

where z is considered here as the depth to water table, h is considered as the absolute value of the matric potential, and q is the steady-state flow rate being positive for upward flow and negative for downward flow.

The analytical solution of the integral appearing in Eq. (7.24) highly depends on the $K(h)$ function. For example, there exists exact or approximate analytical solution to Eq (7.24) for the Gardner and BC $K(h)$ functions (see Table 7.5), while no analytical solution exists for the more complex $K(h)$ functions (e.g. VG or Kosugi models). Some of the existing analytical solutions are referenced in Table 7.7.

As an example, here we present Sadeghi et al. (2012c) exact solution to $h(z)$ above a water table for the BC $K(h)$ function shown in Table 7.5:

$$z=\begin{cases}(1+r)^{-1}h & (h\le h_b)\\ (1+r)^{-1}h_b+h_e\sum_{i=0}^{\infty}\dfrac{(-1)^n(h/h_e)^{1+iP}}{1+iP}-h_e\sum_{i=0}^{\infty}\dfrac{(-1)^n(h_b/h_e)^{1+iP}}{1+iP} & (h_b<h\le h_e)\\ z(h_e)+h_e\sum_{i=1}^{\infty}\dfrac{(-1)^{i+1}(h/h_e)^{1-iP}}{1-iP}-h_e\sum_{i=1}^{\infty}\dfrac{(-1)^{i+1}}{1-iP} & (h_e<h)\end{cases} \tag{7.25}$$

where $r = q/K_s$ and h_e is the matric potential at which $K(h) = q$.

Sadeghi et al. (2012c) indicated that the complex exact solution of Eq. (7.25) can be approximated with the closed-form solution:

$$z=\begin{cases}(1+r)^{-1}h & (h\le h_b)\\ h_b\left[\dfrac{\ln(1+r)}{1+P}-\dfrac{r}{1+r}\right]+h-(1+P)^{-1}h\ln\left[1+r(h/h_b)^P\right] & (h_b<h\le h_e)\\ D_{max}+(1-P)^{-1}h\ln\left[1+r^{-1}(h/h_b)^{-P}\right] & (h_e<h)\end{cases} \tag{7.26}$$

where D_{max} is the depth at which $h\to\infty$ and is calculated as follows:

$$D_{max}=h_b\left\{\frac{\ln(1+r)}{1+P}-\frac{r}{1+r}-r^{-1/P}\left[\frac{\ln 2}{(1-P)}+\frac{\pi^2/12-\ln 2}{P(1-P)}-1+\frac{\ln 2}{1+P}\right]\right\} \tag{7.27}$$

In fact, D_{max} is the depth at which liquid water hydraulic conductivity becomes very low ($K\to 0$), and thus, according to Eq. (7.19), $\partial h/\partial z\to\infty$ to keep q constant. Therefore, a hydraulic disruption occurs at $z = D_{max}$, resulting in the formation of a "drying front" (DF) and a phase change (from liquid to vapor) at this depth.

Figure 7.22 depicts a schematic solution of $h(z)$ from Eq. (7.24) for two different cases: (i) both liquid and vapor flow are accounted for modeling the hydraulic conductivity (i.e. $K=K_{liquid}+K_{vapor}$), and (ii) vapor flow is neglected (i.e. $K=K_{liquid}$). Three different zones are distinguished along the profile: (i) a saturated zone known as the capillary fringe ($0<h<h_b$, where h_b is the air-entry pressure head), (ii) an intermediate, partially saturated zone between the capillary fringe and the DF ($h_b<h<h_{max}$, where h_{max} is the pressure head at which hydraulic liquid continuity is disrupted), and (iii) a dry zone (gas flow region) between the DF and the soil surface ($h_{max}<h<h_{sur}$, where h_{sur} is the pressure head at the soil surface). It is evident that the pressure head distribution within the dry zone cannot be modeled without accounting for the vapor flow contribution. However, the solution along the liquid flow domain (i.e. $0<z<D_{max}$) is the same for both cases.

Similar to other exact solutions listed in Table 7.7, the presented solution in Eqs. (7.25) and (7.26) is implicit regarding $h(z)$. In an attempt to derive an explicit solution to $h(z)$, Salvucci (1993) derived the following simple approximate analytical solution to the problem using the Gardner's power $K(h)$ function shown in Table 7.5:

$$h=h_b\left\{\left[\frac{z}{h_b}(1+r)\right]^{-p}-\frac{r}{1+r}\right\}^{-1/p}\quad (z<D_{max}) \tag{7.28}$$

Table 7.7 Existing analytical solutions to steady-state infiltration or evaporation from a water table based on the Buckingham–Darcy law, Eq. (7.19).

Process	$K(h)$	Solution type with respect to $h(z)$	Reference
Evaporation	G1	Exact in closed form (if $p = 1, 1.5, 2, 3,$ or 4)	Gardner (1958)
Evaporation	G2	Exact in closed form	Gardner (1958)
Evaporation	G1	Exact in integral form	Warrick (1988)
Evaporation	BC	Exact in integral form	Warrick (1988)
Infiltration, evaporation	G1	Approximate in closed form	Salvucci (1993)
Infiltration	G1, BC	Exact in series form	Zhu and Mohanty (2002)
Evaporation	BC	Exact in series form, approximate in closed form	Sadeghi et al. (2012c)

The $K(h)$ models used for the solutions are listed in Table 7.5 and include Gardner's (1958) power model (G1) and exponential model (G2) and Brooks–Corey model (BC).

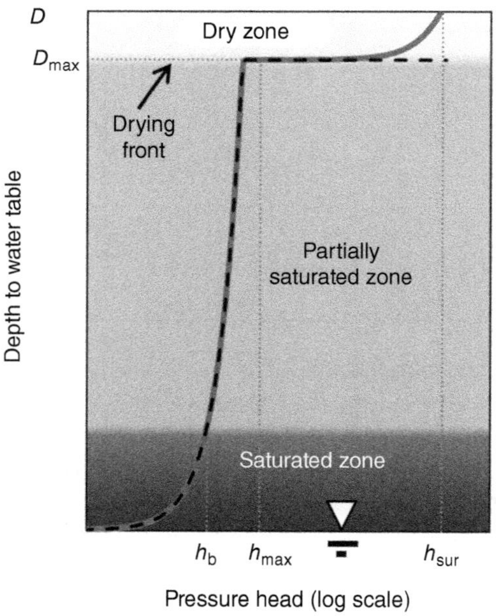

Figure 7.22 Sketch depicting the pressure head distribution above the water table during steady-state evaporation with (solid line) and without (dashed line) vapor flow contribution. In this figure, h_b denotes the air-entry pressure head, h_{max} is pressure head at the drying front where liquid continuity is disrupted, h_{sur} is pressure head at the surface, D is the water table depth, and D_{max} is the maximum extent of the hydraulically connected region above the water table. Pressure head is plotted on a logarithmic scale. *Source*: From Sadeghi et al. (2014).

where p is Gardner's model parameter and $D_{max} = z(\infty)$ is obtained as

$$D_{max} = h_b \left[\frac{1}{1+r} \left(\frac{r}{1+r} \right)^{-1/p} \right] \tag{7.29}$$

Although Salvucci's solution is easier to use when compared with the other analytical solutions, its accuracy declines for lower values of p and higher values of r as is the case for finer-textured soils (Figure 7.23).

Sadeghi et al. (2014) derived the following approximate analytical solution to Eq. (7.24) applied to any arbitrary $K(h)$ function that exhibits a steep slope (e.g. coarse- and medium-textured soils):

$$D_{max} = h_e \tag{7.30}$$

Eq. (7.30) states that pairs of q–D_{max} values coincide with the unsaturated hydraulic conductivity curve, $K(h)$, which means that the steady-state evaporation rate is approximately a measure for unsaturated hydraulic conductivity at the pressure head equal to D_{max}. In other words, when $h = D_{max}$, $K = q$. This solution offers a new approach for estimating profile-scale effective unsaturated hydraulic conductivity from evaporation rate and D_{max}, because these quantities are characteristics of the entire liquid flow domain (see Figure 7.22). Using this concept, Gohardoust et al. (2017) calculated the effective $K(h)$ of highly heterogeneous media at the profile scale as a function of all individual $K(h)$ curves of the constituent materials.

7.3.2.1.2 Transient Infiltration

Green and Ampt (1911) derived a solution to the BD law for transient infiltration into an initially nearly dry soil subject to a constant-head (h_0) surface boundary condition. A key assumption to this solution is that the SM profile during infiltration can be approximated with a step function as follows (see Figure 7.24):

$$\theta = \begin{cases} \theta_0 & (z \le z_f) \\ \theta_i & (z > z_f) \end{cases} \tag{7.31}$$

where θ_0 and θ_i are the wetted region and initial water contents, respectively, and z_f is the wetting front depth, which increases with time during infiltration.

The BD law can be rearranged as

$$\int_0^{z_f} (q - K)\,dz = \int_{-h_0}^{\infty} K\,dh \tag{7.32}$$

Note that the minus sign before h_0 in Eq. (7.32) was added because matric potential was considered as absolute value. Assuming the steplike SM profile, Eq. (7.32) can be solved as

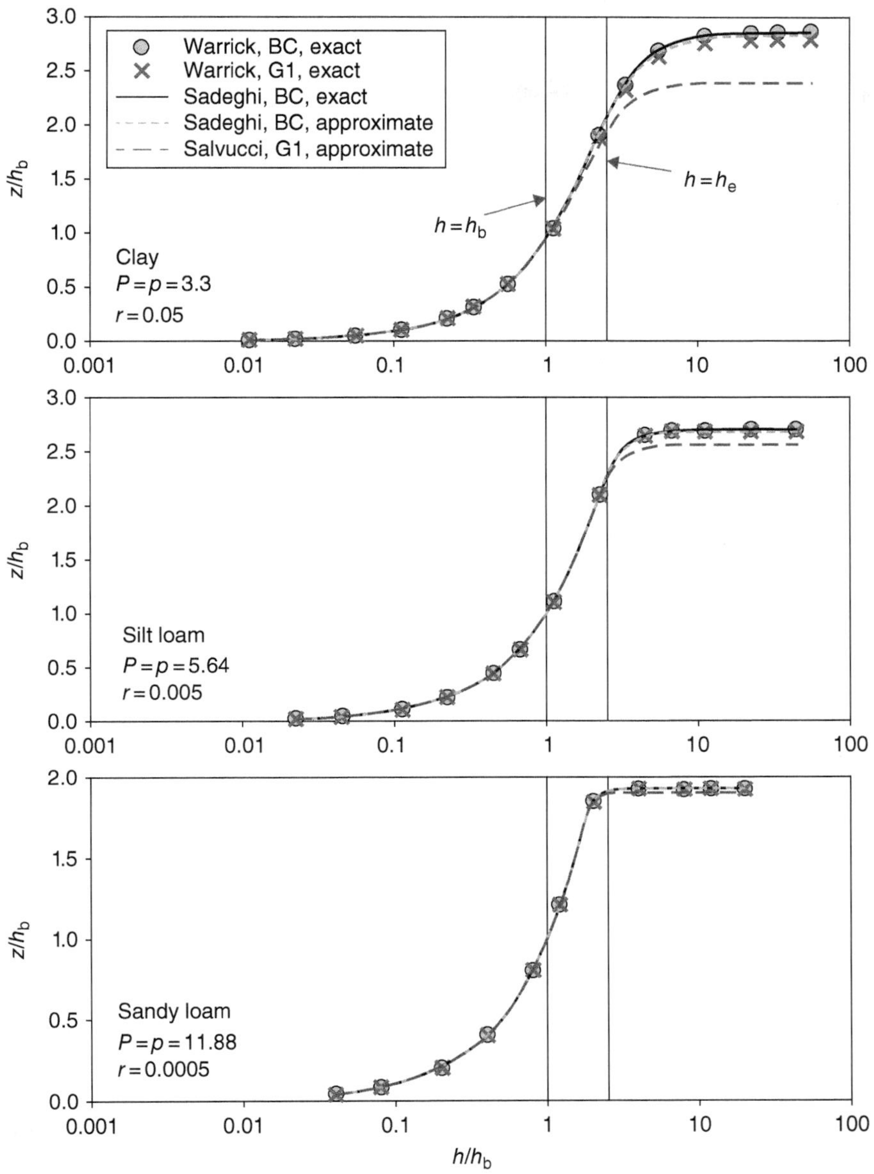

Figure 7.23 Comparison of different existing analytical solutions of Eq. (7.24) for steady-state evaporation from a water table (see Table 7.7). *Source*: From Sadeghi et al. (2012c).

$$i(t) = K_s + K_s \frac{h_0 + h_f}{z_f(t)} \tag{7.33}$$

where i is the surface flux or infiltration rate as a function of time t and h_f, named "wetting front suction (matric potential) head":

$$h_f = K_s^{-1} \int_0^{\infty} K \, dh \tag{7.34}$$

The parameter h_f is a constant soil property and, as Eq. (7.34) indicates, can be analytically or numerically estimated from the $K(h)$ function. For example, for the BC $K(h)$ function (shown in Table 7.5), we analytically obtain

$$h_f = K_s^{-1} \left[\int_0^{h_b} K_s dh + \int_{h_b}^{\infty} K_s \left(\frac{h}{h_b} \right)^{-P} dh \right] = h_b \left(\frac{P}{P-1} \right) \tag{7.35}$$

Considering that $z_f = I/(\theta_0 - \theta_i)$ and $i(t) = dI/dt$ (where I is the cumulative infiltration), Eq. (7.33) can be written as

$$\frac{dI}{dt} = K_s + K_s \frac{(h_0 + h_f)(\theta_0 - \theta_i)}{I} \tag{7.36}$$

Eq. (7.36) is an ordinary differential equation (ODE) that can be readily solved. Considering the initial condition of $I(0) = 0$, the solution yields the Green–Ampt cumulative infiltration model as a function of time in the following implicit form:

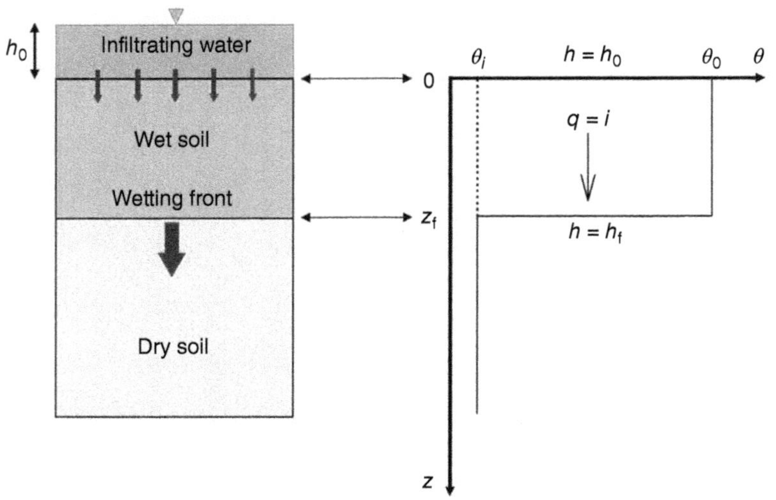

Figure 7.24 Sketch depicting a constant-head infiltration process into a dry soil (left) assuming a steplike soil moisture profile based on the Green and Ampt (1911) analysis (right). *Source:* Based on Green and Ampt (1911).

$$I(t) = K_s t + (h_0 + h_f)(\theta_0 - \theta_i)\ln\left[1 + \frac{I(t)}{(h_0 + h_f)(\theta_0 - \theta_i)}\right] \quad (7.37)$$

Explicit forms of the Green–Ampt model were also derived by several researchers, e.g. Salvucci and Entekhabi (1994) and Serrano (2001).

7.3.2.2 Analytical Solution of Richards' Equation

7.3.2.2.1 Diffusion Form of Richards' Equation

RE, Eq. (7.20), can be alternatively written in terms of water content as (Bruce and Klute, 1956)

$$\frac{\partial\theta}{\partial t} = \frac{\partial}{\partial z}\left[D(\theta)\frac{\partial\theta}{\partial z} - K(\theta)\right] \quad (7.38)$$

where D is the so-called soil water diffusivity, a function of θ similar to h and K. The definition of D comes from the equivalence of the two different forms of RE, Eqs. (7.20) and (7.38):

$$D(\theta) = K(\theta)\frac{\mathrm{d}h}{\mathrm{d}\theta} \quad (7.39)$$

For horizontal flow or for vertical flow in case that the gravity-driven flux, i.e. $K(\theta)$, is negligible compared with the diffusivity-driven flux, i.e. $D\partial\theta/\partial z$, RE in the form of Eq. (7.38) is reduced to a diffusion equation:

$$\frac{\partial\theta}{\partial t} = \frac{\partial}{\partial z}\left[D(\theta)\frac{\partial\theta}{\partial z}\right] \quad (7.40)$$

Eq. (7.40) may be solved analytically with some simple forms of the $D(\theta)$ function. In the simplest case, $D(\theta)$ may be assumed as constant, i.e. $D(\theta) = D_e$, reducing Eq. (7.40) to Fick's second law:

$$\frac{\partial\theta}{\partial t} = D_e\frac{\partial^2\theta}{\partial z^2} \quad (7.41)$$

Eq. (7.41) can be solved analytically for specific initial and boundary conditions. Solution to a similar problem is discussed later in Section 7.4 for the advection–dispersion equation (ADE) for solute transport in soils. For example, for infiltration of water into a horizontal soil column subject to conditions

$$\theta(z,0) = \theta_i, \quad \theta(0,t) = \theta_0 \quad (7.42)$$

Eq. (7.41) yields (Ostiachev, 1936; Kirkham and Feng, 1949)

$$\theta = \theta_i + (\theta_0 - \theta_i)\mathrm{erfc}\left(\frac{z}{\sqrt{4D_e t}}\right) \quad (7.43)$$

where z is defined as the distance from the wetted end of the soil column connected to the water source and D_0 can be considered as an effective diffusivity.

Because D in unsaturated soils is highly dependent on θ (e.g. Bruce and Klute, 1956), Eq. (7.41) cannot provide a realistic model for unsaturated soil water flow. Hence, solution of Eq. (7.40) has been of greater interest to soil physicists than that of Eq. (7.41).

A well-known approach to solve the horizontal infiltration problem through Eq. (7.40) is the "Boltzmann transformation":

$$\zeta = zt^{-1/2} \quad (7.44)$$

which converts the PDE of Eq. (7.40) to an ODE [i.e. $\theta(z, t) \rightarrow \theta(\zeta)$]:

$$-\frac{\zeta}{2}\frac{\mathrm{d}\theta}{\mathrm{d}\zeta} = \frac{\mathrm{d}}{\mathrm{d}\zeta}\left[D(\theta)\frac{\mathrm{d}\theta}{\mathrm{d}\zeta}\right] \quad (7.45)$$

The conversion of Eqs (7.40) to (7.45) using the Boltzmann transformation means that although the SM profile, $\theta(z)$, varies with time, the transformed SM profile, $\theta(\zeta)$, will be invariant for all times during the horizontal flow experiment. This concept was shown to be valid for real soils as well, indicating the adequacy of Eq. (7.40) for modeling horizontal flow. For example, experimental data of Nielsen et al. (1962) for horizontal infiltration into an initially dry Columbia silt loam are presented in Figure 7.25. As observed, the SM profile data were coalesced for all times after the distance was transformed.

The analytical solution of Eq. (7.45) is reviewed in Brutsaert (1976). Of course, the analytical solution of Eq. (7.45) is highly dependent on the functional relationship assumed for $D(\theta)$. For example, Brutsaert (1968) indicated that with a diffusivity function of the form

$$D = D_0 \frac{m}{2}\left(\frac{\theta-\theta_i}{\theta_0-\theta_i}\right)^m \left[1-\frac{1}{1+m}\left(\frac{\theta-\theta_i}{\theta_0-\theta_i}\right)^m\right] \quad (7.46)$$

solution of Eq. (7.45) becomes

$$\beta = \left[1-\left(\frac{\theta-\theta_i}{\theta_0-\theta_i}\right)^m\right]\left[2D_0\frac{(1+m)}{m^2}\right]^{1/2} \quad (7.47)$$

where D_0 and m are empirical soil parameters.

The cumulative infiltration during the horizontal infiltration process can be calculated as

$$I = \int_{\theta_i}^{\theta_0} z\,\mathrm{d}\theta = \int_{\theta_i}^{\theta_0} \zeta t^{1/2}\,\mathrm{d}\theta = st^{1/2} \quad (7.48)$$

where s, termed "sorptivity" by Philip (1957a), is given as

$$s = \int_{\theta_i}^{\theta_0} \zeta\,\mathrm{d}\theta \quad (7.49)$$

The sorptivity corresponding to Brutsaert's (1968) solution, Eq. (7.47), is given as

$$s = (\theta_0-\theta_i)\left(\frac{2D_0}{1+m}\right)^{1/2} \quad (7.50)$$

Bruce and Klute (1956) solved Eq. (7.45) for the diffusivity function:

$$D(\theta) = -0.5\frac{\mathrm{d}\zeta}{\mathrm{d}\theta}\int_{\theta_i}^{\theta} \zeta(\theta)\,\mathrm{d}\theta \quad (7.51)$$

From Eq. (7.51), the soil diffusivity function can be determined if the θ–ζ functional relationship is known. This relationship can be obtained for each soil

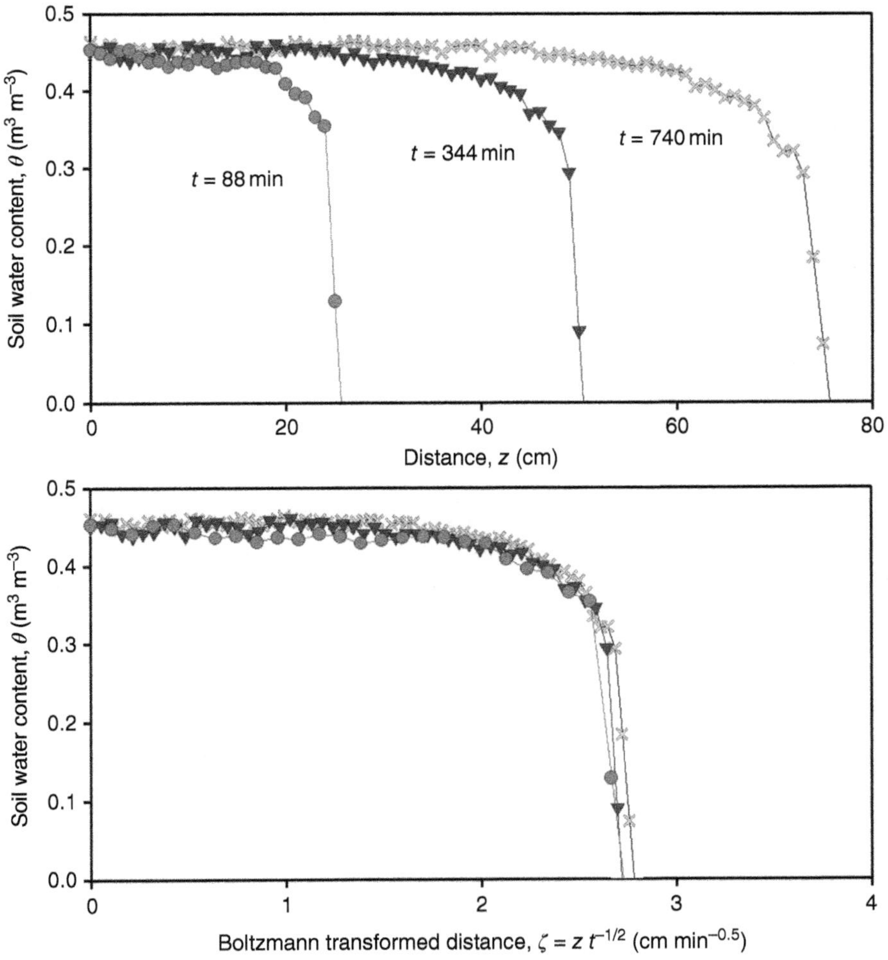

Figure 7.25 Soil water content distribution during infiltration into a horizontal Columbia silt loam soil column. *Source:* From Nielsen et al. (1962).

through a horizontal infiltration experiment, as shown in Figure 7.25.

The Boltzmann transformation approach for solving RE was indicated to be a special case of the "self-similar front solutions" (Caputo and Stepanyants, 2008) in the form

$$S(z,t)=t^{\alpha}\Phi(\zeta), \quad \zeta=zt^{-\beta} \tag{7.52}$$

where α and β are some unknown exponents to be solved ($\alpha=0$ and $\beta=0.5$ for the above-presented solution). Such solutions were sought by several researchers for infiltration and redistribution processes (e.g. Shao and Horton, 2000; Caputo and Stepanyants, 2008; Hayek, 2014). Although $\beta=0.5$ was shown to be valid in some experiments (e.g. Figure 7.25), some other experiments showed that the value of β is significantly less than 0.5 (Nielsen et al., 1962; Pachepsky et al., 2003). Assuming a generalized form of RE with fractional derivative of θ of the order γ with respect to time,

$$\frac{\partial^{\gamma}\theta}{\partial t^{\gamma}}=\frac{\partial}{\partial z}\left[D_{\gamma}(\theta)\frac{\partial\theta}{\partial z}\right] \tag{7.53}$$

Pachepsky et al. (2003) indicated that $\beta=\gamma/2$. The order of the fractional derivative γ can be less than or equal to 1, having $\gamma=1$ for the classical RE, i.e. Eq. (7.40).

For the vertical infiltration problem based on RE in the form of Eq. (7.38), Philip (1957b) showed that θ versus $zt^{-1/2}$ is no longer a unique functional relationship. He analytically derived the following expanded expression:

$$z=\zeta_1(\theta)t^{1/2}+\zeta_2(\theta)t+\zeta_3(\theta)t^{3/2}+\cdots+\zeta_j(\theta)t^{j/2}+\cdots \tag{7.54}$$

yielding the following solution to the cumulative infiltration:

$$I=\varphi_1(\theta)t^{1/2}+\varphi_2(\theta)t+\varphi_3(\theta)t^{3/2}+\cdots+\varphi_j(\theta)t^{j/2}+\cdots \tag{7.55}$$

where φ_j coefficients are functions of θ obtained with solving a set of ODEs.

7.3.2.2.2 The Linearized Richards' Equation

Another common approach to solve RE is to linearize it. The linearized form of RE is obtained assuming a constant value for the soil diffusivity, $D(\theta)=D_e$, and hydraulic conductivity K to be linearly dependent upon soil water content θ, that is, $V=\mathrm{d}K/\mathrm{d}\theta=\text{cons}$. The 1D linearized RE is obtained as

$$\frac{\partial\theta}{\partial t}=D_e\frac{\partial^2\theta}{\partial z^2}-V\frac{\partial\theta}{\partial z} \tag{7.56}$$

RE can also be linearized assuming Gardner's (1958) exponential $K(h)$ relationship, $K=K_s\exp(-Ah)$, and the K–θ linear relationship, $V=\partial K/\partial\theta=\text{cons.}$, yielding

$$\frac{\partial\theta}{\partial t}=\frac{V}{A}\frac{\partial^2\theta}{\partial z^2}-V\frac{\partial\theta}{\partial z} \tag{7.57}$$

Various authors presented analytical solutions for the linearized RE for different initial and boundary conditions (Warrick, 1975; Chen et al., 2001; Yuan and Lu, 2005; Menziani et al., 2007; Tracy, 2007). For example, assuming the following arbitrary conditions,

$$\theta=\begin{cases}\theta_i(z) & (z\geq 0,\ t=0)\\ \theta_0(t) & (z=0,\ t>0)\end{cases} \tag{7.58}$$

Menziani et al. (2007) solved the linearized RE, yielding the space and time evolution of the soil water content as

$$\theta(z,t)=\theta_1(z,t)+\theta_2(z,t) \tag{7.59}$$

where

$$\theta_1(z,t)=\frac{2}{\sqrt{\pi}}\int_0^t\theta_0(t')\mathrm{e}^{-\left[\left(V(t-t')-z\right)/\sqrt{4D_e(t-t')}\right]^2}\frac{4D_e z\,\mathrm{d}t'}{2\left[4D_e(t-t')\right]^{3/2}} \tag{7.60}$$

$$\theta_2(z,t)=\frac{\mathrm{e}^{Vz/2D_e-V^2t/4D_e}}{\sqrt{\pi}\sqrt{4D_e t}}\int_0^{\infty}\theta_i(z')\mathrm{e}^{-Vz'/2D_e}\left[\mathrm{e}^{-\left((z-z')/\sqrt{4D_e t}\right)^2}-\mathrm{e}^{-\left((z+z')/\sqrt{4D_e t}\right)^2}\right]\mathrm{d}z' \tag{7.61}$$

It can be shown that the linearized RE in Eq. (7.57) is obtained using the following SHFs provided that $P_\lambda=1$:

$$\theta=\theta_r+(\theta_s-\theta_r)\exp\left(-\frac{Ah}{P_\lambda}\right) \tag{7.62}$$

$$K=K_s\exp(-Ah) \tag{7.63}$$

$$K=K_s\left(\frac{\theta-\theta_r}{\theta_s-\theta_r}\right)^{P_\lambda} \tag{7.64}$$

where $P_\lambda=P/\lambda$ considering the BC functions (Table 7.5). The assumption of $P_\lambda=1$ is rarely met in reality, i.e. P_λ, for most natural soils is significantly larger than 1 (Sadeghi et al., 2012a). An approximate analytical solution of RE with the SHFs of (7.62)–(7.64) was derived by Sadeghi et al. (2017) considering $P_\lambda\neq 1$. They indicated that RE can be reduced to the following PDE:

$$\frac{\partial K^*}{\partial t^*}=P_\lambda K^*\left(\frac{\partial^2K^*}{\partial z^{*2}}-\frac{\partial K^*}{\partial z^*}\right) \tag{7.65}$$

where

$$K^*=\frac{K}{K_s},\quad z^*=Az,\quad t^*=\frac{AK_st}{\theta_s-\theta_r} \tag{7.66}$$

Eq. (7.65) is not linear, but is readily solvable. Using the separation of variables method, Sadeghi et al. (2017) solved Eq. (7.65) as follows:

$$\frac{K}{K_s} = \left(\frac{\theta - \theta_r}{\theta_s - \theta_r}\right)^{P_\lambda} = \frac{Az + c_1 \exp(Az) + c_2}{P_\lambda A K_s t / (\theta_s - \theta_r) + c_3} \tag{7.67}$$

where c_1, c_2, and c_3 are unknowns related to the initial/boundary conditions. This solution was derived for obtaining a simple analytical SM profile function, $\theta(z)$, having no more than three unknowns, for the NASA Airborne Microwave Observatory of Subcanopy and Subsurface (AirMOSS) P-band radar root zone SM retrieval algorithm (Tabatabaeenejad et al., 2015). Considering time as unknown and defining new unknowns, C_1, C_2, and C_3, the final SM profile was obtained as

$$\theta = \left[C_1 z + C_2 \exp(Az) + C_3\right]^{1/P_\lambda} \tag{7.68}$$

Sadeghi et al. (2017) left C_1, C_2, and C_3 as unknown to be retrieved by the radar, while they specified the soil hydraulic parameters, A and P_λ, for each soil texture. It was shown that the simple model of Eq. (7.68) well fits the numerical solution of RE with the VG hydraulic functions during soil drying (Figure 7.26).

7.3.2.2.3 The Nonlinear Richards' Equation

Despite the high nonlinearity of RE, there exist some analytical solutions to the problem for some specific initial/boundary conditions. For example, Broadbridge and White (1988) solved the nonlinear RE, Eq. (7.38), subject to the constant-flux infiltration and a uniform initial water content. They found the solution for specific mathematical forms of $D(\theta)$ and $K(\theta)$:

$$D = p_1 (p_2 - \theta)^{-2} \tag{7.69}$$

$$K = p_3 + p_4 (p_2 - \theta) + \frac{p_5}{2(p_2 - \theta)} \tag{7.70}$$

Warrick et al. (1990) extended the Broadbridge and White (1988) analysis for arbitrary initial conditions and constant surface flux, whether it be zero for drainage, positive for infiltration, or negative for evaporation.

The nonlinear RE can be solved for the infiltration problem based on the so-called front solutions. Self-similar front solutions introduced above, i.e. Eq. (7.52), are representative for such solutions. Another type is the "traveling wavefront solution" that may take the form (Caputo and Stepanyants, 2008):

$$S(z,t) = \Theta(\xi), \quad \xi = z - vt \tag{7.71}$$

where v is the velocity of the water content wave. An example of the traveling wavefront solution is the recent solution by Hayek (2016) who solved the nonlinear RE with the BC SHFs for infiltration through semi-infinite dry soils with prescribed water content or flux boundary conditions.

7.3.2.3 Numerical Solution of Richards' Equation

As discussed above, analytical solutions of RE are limited to simple cases, owing to the high degree of nonlinearity. Most of these simplified conditions are rarely met in nature. Consequently, numerical solutions without such restrictions have been developed. Numerical schemes mostly differ in the way of discretization or in the method of linearization used to solve RE (Haverkamp et al., 1977). Numerical approximations based on different

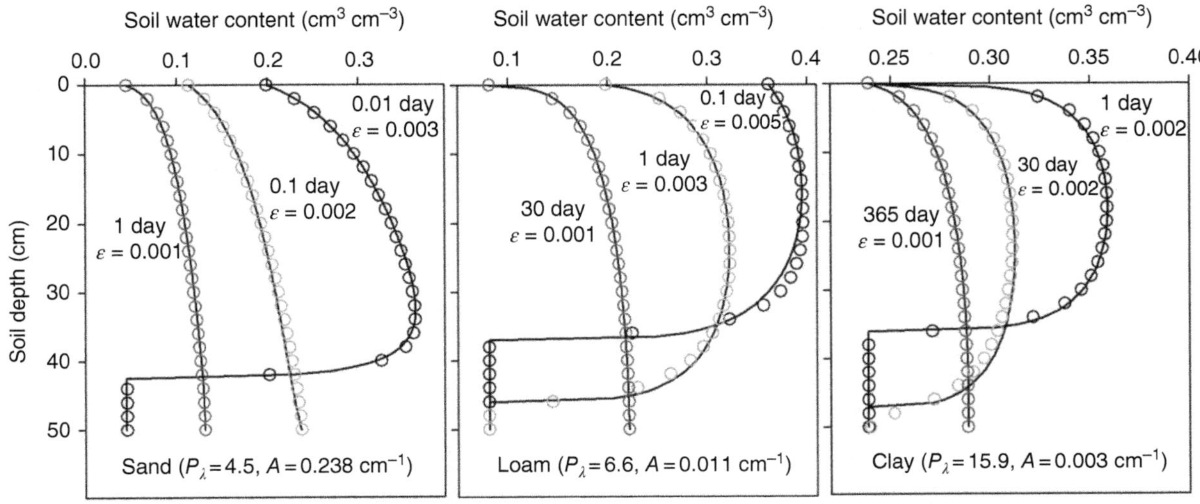

Figure 7.26 The best fit of Eq. (7.68) (solid lines) to numerical solutions of Richards' equation with the van Genuchten hydraulic functions (circles, HYDRUS-1D simulation results) for simultaneous evaporation and drainage in three different soils. The fitting accuracy is quantified with the mean absolute error, ε (cm^3 cm^{-3}). *Source*: From Sadeghi et al. (2017).

forms of RE (h based, θ based, mixed based) can lead to significantly different results for unsaturated flow problems (Celia et al., 1990). To provide an example for numerical solution of RE, we introduce a fully implicit finite difference scheme, adopted from the most widely applied numerical code for unsaturated flow in soils, HYDRUS-1D (Šimůnek et al., 2005).

In this scheme, the soil profile is discretized into $N-1$ layer, with the ends of the layers located at the nodal points and N being the number of nodes. RE is discretized as

$$\frac{\theta_i^{j+1,k+1}-\theta_i^j}{\Delta t}=\frac{1}{\Delta z}\left(K_{i+1/2}^{j+1,k}\frac{h_{i+1}^{j+1,k+1}-h_i^{j+1,k+1}}{\Delta z_i}-K_{i-1/2}^{j+1,k}\frac{h_i^{j+1,k+1}-h_{i-1}^{j+1,k+1}}{\Delta z_{i-1}}\right)+\frac{K_{i+1/2}^{j+1,k}-K_{i-1/2}^{j+1,k}}{\Delta z} \tag{7.72}$$

where

$$\Delta t=t^{j+1}-t^j,$$
$$\Delta z=\frac{z_{i+1}-z_{i-1}}{2},\quad \Delta z_i=z_{i+1}-z_i,\quad \Delta z_{i-1}=z_i-z_{i-1},$$
$$K_{i+1/2}^{j+1,k}=\frac{K_{i+1}^{j+1,k}-K_i^{j+1,k}}{2},\quad K_{i-1/2}^{j+1,k}=\frac{K_i^{j+1,k}-K_{i-1}^{j+1,k}}{2} \tag{7.73}$$

in which subscripts $i-1$, i, and $i+1$ indicate the position in the finite difference mesh; superscripts k and $k+1$ denote the previous and current iteration levels, respectively; and superscripts j and $j+1$ represent the previous and current time levels, respectively.

The mass-conservative method proposed by Celia et al. (1990), in which $\theta^{j+1,k+1}$ is expanded in a truncated Taylor series with respect to h about the expansion point $h^{j+1,k}$, is used in the time difference scheme of (7.72):

$$\frac{\theta_i^{j+1,k+1}-\theta_i^j}{\Delta t}=C_i^{j+1,k}\frac{h_i^{j+1,k+1}-h_i^{j+1,k}}{\Delta t}+\frac{\theta_i^{j+1,k}-\theta_i^j}{\Delta t} \tag{7.74}$$

where C_i represents the nodal value of the soil water capacity:

$$C_i^{j+1,k}=\left.\frac{d\theta}{dh}\right|^{j+1,k} \tag{7.75}$$

The presented scheme leads to the following matrix equation with matrix $[P_w]$ and vectors $\{h\}$ and $\{F_w\}$:

$$[P_w]^{j+1,k}\{h\}^{j+1,k+1}=\{F_w\} \tag{7.76}$$

The $[P_w]$ matrix in Eq. (7.76) is a symmetrical tridiagonal matrix of the form below:

$$[P_w]=\begin{bmatrix} d_1 & e_1 & 0 & 0 & 0 & 0 & 0\\ e_1 & d_2 & e_2 & . & . & . & 0\\ 0 & e_2 & . & . & . & . & 0\\ 0 & . & . & . & . & . & 0\\ 0 & . & . & . & . & e_{N-2} & 0\\ 0 & . & . & . & e_{N-2} & d_{N-1} & e_{N-1}\\ 0 & 0 & 0 & 0 & 0 & e_{N-1} & d_N \end{bmatrix} \tag{7.77}$$

where the diagonal entries d_i and above-diagonal entries e_i of the matrix $[P_w]$, and the entries f_i of vector $\{F_w\}$, are given by

$$d_i=\frac{\Delta z}{\Delta t}C_i^{j+1,k}+\frac{K_{i+1}^{j+1,k}+K_i^{j+1,k}}{2\Delta z_i}+\frac{K_i^{j+1,k}+K_{i-1}^{j+1,k}}{2\Delta z_{i-1}} \tag{7.78}$$

$$e_i=-\frac{K_i^{j+1,k}+K_{i+1}^{j+1,k}}{2\Delta z_i} \tag{7.79}$$

$$f_i=\frac{\Delta z}{\Delta t}C_i^{j+1,k}h_i^{j+1,k}-\frac{\Delta z}{\Delta t}\left(\theta_i^{j+1,k}-\theta_i^j\right)+\frac{K_{i+1}^{j+1,k}-K_{i-1}^{j+1,k}}{2} \tag{7.80}$$

The tridiagonal matrix [Pw] is symmetric and therefore the below-diagonal entries are equal to the above-diagonal entries. Eq. (7.76) can be solved using the tridiagonal matrix algorithms, such as the "Thomas algorithm." The entries d_1, e_1, and f_1 are dependent on the top boundary conditions, and the entries d_N, e_{N-1}, and f_N are dependent on the bottom boundary conditions. The head (Dirichlet) boundary condition can be specified using the following equations (Sadeghi, 2013):

$$\begin{aligned} &d_1=\mathrm{I},\quad e_1=0,\quad f_1=\mathrm{I}h_0,\\ &d_N=\mathrm{I},\quad e_{N-1}=0,\quad f_N=\mathrm{I}h_L \end{aligned} \tag{7.81}$$

where h_0 is the prescribed pressure head at the surface, h_L is the prescribed pressure head at the bottom, and I is an arbitrary large number (e.g. $I=1\mathrm{E}10$) that is considered to preserve the symmetry in $[P_w]$.

The flux (Neumann) boundary condition can also be specified using the following equations:

$$\begin{aligned} &d_1=\frac{K_1^{j+1,k}+K_2^{j+1,k}}{2\Delta z_1},\quad e_1=-\frac{K_1^{j+1,k}+K_2^{j+1,k}}{2\Delta z_1},\\ &f_1=\frac{K_1^{j+1,k}+K_2^{j+1,k}}{2}+q_0^{j+1},\\ &d_N=\frac{K_{N-1}^{j+1,k}+K_N^{j+1,k}}{2\Delta z_{N-1}},\quad e_{N-1}=-\frac{K_{N-1}^{j+1,k}+K_N^{j+1,k}}{2\Delta z_{N-1}},\\ &f_N=\frac{K_{N-1}^{j+1,k}+K_N^{j+1,k}}{2}+q_L^{j+1} \end{aligned} \tag{7.82}$$

where q_0 is the prescribed flux at the surface and q_L is the prescribed flux at the bottom.

7.3.2.4 Scaled Forms of Richards' Equation

It has been discussed that analytical solutions of RE are easy to implement but restricted to simple SHFs and initial/boundary conditions. In contrast to the analytical solutions, numerical methods are able to solve any complex problem; however, they require complicated calculations that should be repeated for any single problem (i.e. different soil, different initial/boundary conditions). Scaling of RE may be considered as a third approach, providing a favorable means to substantially reduce numerical calculations for specific problems (e.g. infiltration). Scaling in soil physics literature originates from the Miller and Miller (1956) "similar media" theory, reviewed in Sadeghi et al. (2016). Scaling of RE is basically an analytical approach to convert the original RE to a form that is invariant with respect to several scaling parameters.

To provide an example, let us assume the VG SHFs, Eqs. (7.22) and (7.23), and define the scaled variables, following Warrick et al. (1985):

$$\theta^* = \frac{\theta - \theta_r}{\theta_s - \theta_r}, \quad h^* = \alpha h, \quad K^* = \frac{K}{K_s}, \quad z^* = \alpha z, \quad t^* = \frac{\alpha K_s t}{\theta_s - \theta_r} \tag{7.83}$$

With these scaled variables, the scaled forms of RE and VG SHFs are obtained as follows:

$$\frac{\partial \theta^*}{\partial t^*} = \frac{\partial}{\partial z^*}\left(K^* \frac{\partial h^*}{\partial z^*} - K^*\right) \tag{7.84}$$

$$\theta^* = \left[1 + \left(h^*\right)^n\right]^{-m} \tag{7.85}$$

$$K^* = \theta^{*0.5}\left[1 - \left(1 - \theta^{*1/m}\right)^m\right]^2 \tag{7.86}$$

Let us consider an infiltration problem with the following initial/boundary conditions:

$$\theta(z,0) = 0, \quad \theta(0,t) = \theta_s, \quad \theta(\infty,t) = 0 \tag{7.87}$$

The scaled initial/boundary conditions are obtained as follows:

$$\theta^*\left(z^*,0\right) = 0, \quad \theta^*\left(0,t^*\right) = 1, \quad \theta^*\left(\infty,t^*\right) = 0 \tag{7.88}$$

As can be seen, Eqs. (7.84)–(7.86) subject to the conditions of Eq. (7.88) do not depend upon θ_s, θ_r, α, and K_s. The only remaining soil parameters are n and m, which are related (e.g. $m = 1 - 1/n$). It means that the solution of this problem will vary with n only and can be numerically obtained once forever for each n (e.g. for the 12 USDA textural classes). The scaled solutions then can be descaled using Eq. (7.83) for any soil of interest. This way scaling provides a simple approach to solving such a highly nonlinear problem, for which there exists no analytical solution.

The example provided above is the simplest method for scaling of RE that is merely based on SHFs. Warrick and Hussen (1993) further advanced this method by considering initial and boundary conditions in the scaled variables as well. They assumed the BC SHFs and scaled RE for infiltration and drainage problems. The scaled solutions of Warrick and Hussen (1993) were invariant with respect to θ_s, θ_r, h_b, K_s, and initial/boundary conditions, while the remaining parameters affecting the scaled solutions were the BC powers, λ and P. Sadeghi et al. (2011) developed a scaled form of RE for the soil water redistribution after wetting. The main advantage of this method over the previous methods is that it is not restricted to specific functional relationships for the hydraulic properties and any arbitrary model can be used in this scaling method.

As mentioned, the scaled equations of (7.84)–(7.86) or those of Warrick and Hussen (1993) are still dependent on the VG or BC powers. These parameters are generally called "shape parameters," and soils with the same shape parameters may be called "similar soils" (Sadeghi et al., 2016). Therefore, these scaling methods apply for "similar soils" only that is a restriction for these methods in practice. Toward releasing this restriction, assuming some exponential/power functions for soil hydraulic properties, Sadeghi et al. (2012a, b) developed some scaled forms of RE that showed to be invariant for a broad range of "dissimilar soils," for example, sand and clay (see Figure 7.27). They provided examples for the problems of drainage, constant-head infiltration, and constant-flux infiltration.

7.3.2.5 Preferential Flow

It should be acknowledged that theories presented above regarding the mathematical modeling of water flow in unsaturated soil are valid only for "matrix flow." Another type of soil water flow that occurs through soil macropores and fractures is "preferential flow." Preferential flow is an important phenomenon in soil hydrology that is receiving increased attention among soil physicists and hydrologists, who occasionally experienced failure of the classical RE in predicting flow and transport in field, e.g. infiltration faster than what RE would normally predict (Beven and Germann, 2013).

Equations governing the preferential flow are fundamentally different from the matrix flow governing Equations (BD law, RE). The standard models for the matrix flow cannot describe preferential flow in structured soil since they assume homogeneity and local-equilibrium conditions within a representative elementary

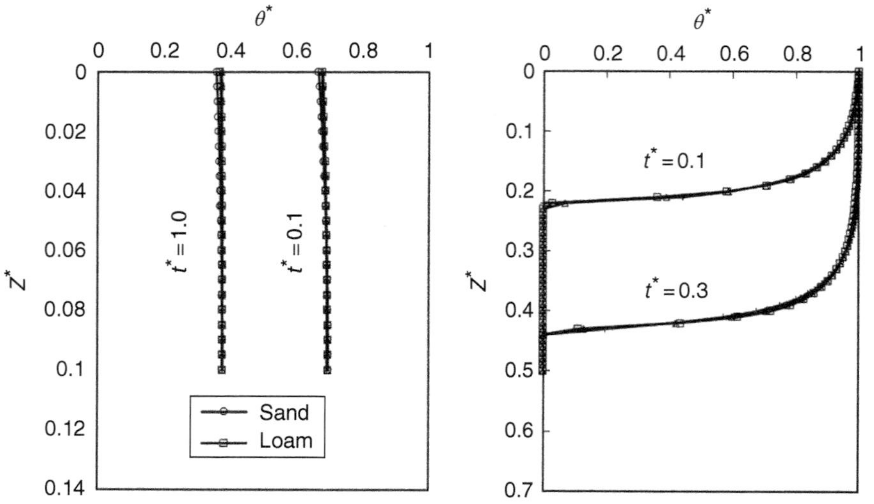

Figure 7.27 Scaled water content profile during drainage (left) and constant-head infiltration (right) obtained with numerical solution of the scaled Richards' equation. *Source*: From Sadeghi et al. (2012b).

volume among other simplifying assumptions, e.g. laminar flow, rigidity of the solid phase, and no air phase effects (Gerke, 2006).

Various soil physicists and hydrologists have attempted to incorporate preferential flow into the BD/RE framework. Beven and Germann (2013) categorized different preferential flow concepts as (i) the single-continuum approach in which preferential flow is treated as a modified relative conductivity curve close to saturation; (ii) the dual-continuum approach in which one domain is treated as immobile and one is a Darcy–Richards domain, with some exchange term between the two; (iii) the dual-permeability approach in which preferential flow is taken into account through high permeability inclusions embedded within a Darcy–Richards domain; and (iv) a dual-porosity approach in which simple volume filling or a kinematic equation is used for the preferential flows and exchanges with an immobile or Darcy–Richards matrix are treated separately.

Interested readers in mathematical modeling of preferential flow are referred to the review papers of Šimůnek et al. (2003), Gerke (2006), and Köhne et al. (2009).

7.3.3 Characterization of Soil Hydraulic Properties

Soil hydraulic properties, namely, the SWC and hydraulic conductivity function, can be either directly measured in laboratory or field relying on a simple analytical solution to the flow equations (Darcy, BD, or RE equations) or be estimated using inverse numerical solution of RE. In the former, parameters of any SHFs (e.g. those listed in Table 7.5) can be determined by least-squares fitting of the SHFs to the measured data. In the latter, parameters of the SHFs for a given soil are obtained by an optimization algorithm, yielding the best match between the RE solution and a set of measured h or θ. Direct measurement of the SWC curve has been discussed earlier in Section 7.2.4.3.1. Here, we briefly introduce some of the most common laboratory/field methods for direct measurement of saturated/unsaturated hydraulic conductivity. The introduced methods are only a few examples among abundant existing methods described in literature. Interested readers are referred to the more comprehensive reviews of Stolte et al. (1994), Benson and Gribb (1997), and Comegna et al. (2012). Also, readers interested in the inverse solution approach for estimating the soil hydraulic properties are referred to Šimůnek and De Vos (1999), Hopmans et al. (2002), and Vrugt et al. (2008).

7.3.3.1 Laboratory Methods

7.3.3.1.1 Constant-Head Flow Method

The constant-head flow experiment is a common laboratory method for measuring saturated hydraulic conductivity. Figure 7.28 depicts a sketch of the experimental setup. In this experiment, a constant pressure head (h) is applied at the bottom of a saturated soil column of known cross-sectional area (A) and length (L). The outflow from the top of the soil column is collected, and its volume (V) over a certain period of time (t) is measured. With these quantities, K_s can be readily calculated from Darcy's law:

$$q = \frac{V}{At} = K_s \frac{h-L}{L} \Rightarrow K_s = \frac{LV}{At(h-L)} \tag{7.89}$$

7.3.3.1.2 Falling-Head Flow Method

An alternative laboratory method for measurement of K_s is the falling-head flow experiment that requires neither the maintenance of a constant-head nor any outflow measurement. A potential experimental setup for measuring K_s with the falling-head method is illustrated in Figure 7.29. In this experiment, only the applied pressure head (i.e. height of ponding) needs to be recorded at the

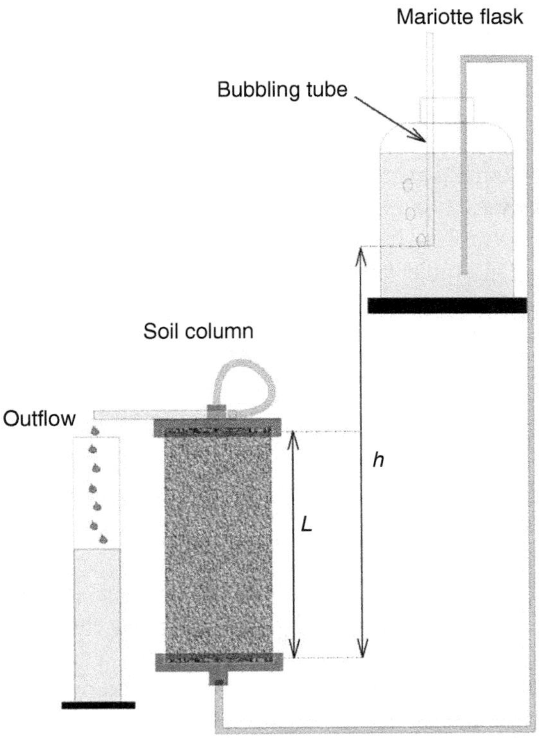

Figure 7.28 Sketch illustrating the setup of a constant-head flow experiment for measuring saturated hydraulic conductivity.

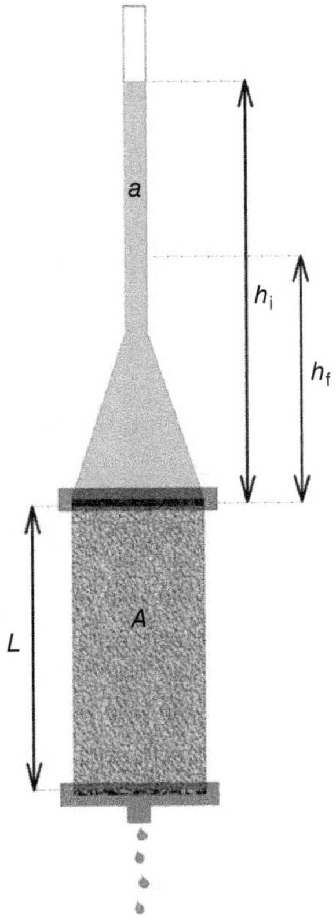

Figure 7.29 Sketch illustrating the setup of a falling-head flow experiment for measuring saturated hydraulic conductivity.

beginning ($t = 0$, $h = h_i$) and end of the experiment ($t = t_f$, $h = h_f$). The rate of decrease of the pressure head is equal to the flux density; hence, Darcy's law at each time during the experiment can be written as

$$\frac{a}{A}\frac{\mathrm{d}h}{\mathrm{d}t} = -K_s \frac{h+L}{L} \tag{7.90}$$

Rearrangement of Eq. (7.90) and integration of the resultant equation from $t = 0$ to t_f yields

$$\frac{a}{A}\int_{h_i}^{h_f} \frac{1}{h+L}\mathrm{d}h = -\int_0^{t_f} \frac{K_s}{L}\mathrm{d}t \tag{7.91}$$

Eq. (7.91) yields the following formula for calculating K_s:

$$K_s = \frac{aL}{At_f}\ln\left(\frac{h_i + L}{h_f + L}\right) \tag{7.92}$$

7.3.3.1.3 *Pressure Outflow Method*

The pressure outflow experiment was first proposed by Gardner (1956) for measuring unsaturated hydraulic conductivity over a wide range of pressure heads. Consider a sample of soil with bulk volume of V_b and height L, situated on a porous plate in a pressure plate apparatus (Tuller and Or, 2005b). Assume that the initial gauge pressure in the chamber is P_i, or in other words, the soil sample is initially in equilibrium with the matric potential of $h_i = P_i/\rho g$. The gauge pressure is then increased by an amount ΔP, forcing the matric potential of $h_f = h_i + \Delta P/\rho g$ onto the soil sample. This increase in the pressure causes water to flow out of the soil until equilibrium is attained once more. By choosing a sufficiently small ΔP, K can be assumed constant, and h–θ relationship can be assumed linear during the flow process. With these assumptions, Gardner (1956) indicated that RE is reduced to a linear diffusion equation similar to Eq. (7.41). He then derived an approximate analytical solution to the problem indicating a linear relationship between $\ln(Q - Q_0)$ and time, where Q is the outflow volume [L^3] at any time t and Q_0 is the total outflow volume. Gardner (1956) finally obtained the following relationship that can be used for calculation of K using the experimental data:

$$K(h_{avg}) = \frac{4S_{Qt}Q_0L^2}{\pi^2 V(h_f - h_i)} \tag{7.93}$$

where h_{avg} is the average of h_i and h_f and S_{Qt} [T^{-1}] is the slope of $\ln(Q - Q_0)$ against time. This experiment can be performed at various steps of h_i to h_f, yielding the unsaturated hydraulic conductivity at different matric potentials. This method is commonly called as "multistep pressure outflow," because it was modified later considering a "one-step pressure outflow" experiment (Gardner, 1962; Doering, 1965; Valiantzas et al., 1988).

7.3.3.1.4 Evaporation Method

Another laboratory technique for unsaturated hydraulic conductivity determination is the so-called evaporation method, originally proposed by Wind (1968) and discussed in detail in Section 7.2.4.3.1. Water is allowed to evaporate from the surface of an initially saturated soil column; all other boundaries are sealed. The weight change of the column is continuously recorded, and θ and h are measured in parallel a several depths. The water flux for different depths can be calculated from the changes in moisture content and weight. The h gradient can be derived from tensiometer readings, allowing calculation of K with the BD equation. Wind's evaporation method was later simplified assuming that distributions of θ and h along a short soil column during evaporation is nearly linear, so that measurements at two depths are sufficient for calculating K (Schindler, 1980; Schindler and Müller, 2006; Peters and Durner, 2008). In the simplified evaporation method, a 6-cm-tall soil sample with two miniature tensiometers installed in 1.5 and 4.5 cm depths is employed (see Figure 7.20). The soil column is first saturated and then exposed to the atmosphere for water to evaporate from the surface, while the change of mass (i.e. evaporation rate) is continuously recorded during the experiment. The unsaturated hydraulic conductivity at various matric potentials is then estimated from the relationship between evaporation rate and matric potential gradient based on the BD law. This method simultaneously provides the SWC (Section 7.2.4.3.1).

7.3.3.2 Field Methods

7.3.3.2.1 Double-Ring Infiltration Method

One of the most common techniques for measuring saturated hydraulic conductivity in field is the "double-ring infiltrometer" method. The infiltrometer consists of two thin-walled metal cylinders that are concentrically placed and driven into the soil to a depth of 5–10 cm (Figure 7.30). The idea behind such design is to establish 1D flow conditions for the inner ring by avoiding the lateral flow by means of the outer ring. The water levels within the inner and outer rings are maintained at the same level with two separate constant-head reservoirs (i.e. Mariotte devices). The water infiltration rate versus time from the inner ring is monitored by means of a Mariotte tower until it reaches a constant value. According to RE, the final infiltration rate for this condition is equal to K_s.

7.3.3.2.2 Single-Ring Infiltration Method

Based on Wooding's (1968) analysis, data for water infiltration from a shallow circular surface pond collected with a single-ring infiltrometer provide adequate information to estimate Gardner's (1958) exponential hydraulic conductivity model parameters, K_s and A, and hence the unsaturated hydraulic conductivity function, $K = K_s \exp(-Ah)$. Wooding's solution is given as

$$i_f = K_s + \frac{4K_s}{\pi r A} \tag{7.94}$$

where i_f is the final (steady-state) infiltration rate and r is the radius of the pond (i.e. radius of the ring in this case).

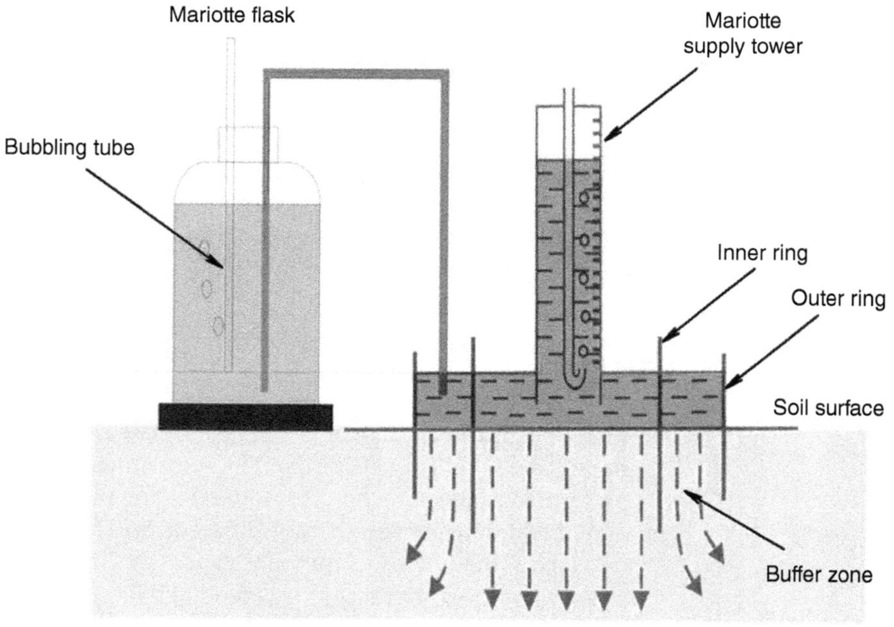

Figure 7.30 Sketch illustrating the setup of a double-ring infiltration experiment.

The ratio of K_s/A is related to the soil sorptivity s as follows (White and Sully, 1987):

$$\frac{K_s}{A} \approx \frac{bs^2}{\theta_0 - \theta_i} \quad (7.95)$$

where θ_0 and θ_i are the surface and initial water contents, respectively, and b is a shape parameter bounded between 1/2 and $\pi/4$. Assuming that Philip's series solution, Eq. (7.55), can be approximated by the first term at early times of infiltration (i.e. $I = st^{1/2}$, Eq. (7.48)), the sorptivity s can be approximated as the slope of the line fitting I versus $t^{1/2}$ data at early times of the infiltration experiment. This way K_s and A can be calculated with Eqs. (7.94) and (7.95).

Another approach to solve for K_s and A is to use a tension infiltrometer, where a negative pressure head is applied at the surface in the single-ring infiltration experiment. Wooding's solution can also be employed for unsaturated conditions (Reynolds and Elrick, 1991):

$$i_f = \frac{Q}{\pi r^2} = K(h)\left(1 + \frac{4}{\pi r A}\right) \quad (7.96)$$

where Q is the discharge of the infiltrated water. Considering Gardner's exponential model and performing the infiltration test for two different surface tensions, h_1 and h_2, we obtain

$$Q_1 = \pi r^2 K_s \exp(-Ah_1)\left(1 + \frac{4}{\pi r A}\right) \quad (7.97)$$

$$Q_2 = \pi r^2 K_s \exp(-Ah_2)\left(1 + \frac{4}{\pi r A}\right) \quad (7.98)$$

Combining Eqs. (7.97) and (7.98) yields the solution for A (Reynolds and Elrick, 1991):

$$A = \frac{\ln(Q_2/Q_1)}{|h_2 - h_1|} \quad (7.99)$$

Then, K_s can be calculated using either Eq. (7.97) or Eq. (7.98). Also K at additional values of h can be estimated through Eq. (7.96), once A is known.

7.3.3.2.3 *Dripper Method*

Another method for estimating soil hydraulic properties based on Wooding's solution, termed "dripper method," was developed by Shani et al. (1987). In this method, a series of pressure-compensated emitters with fixed discharge rates (Q) are used for a field infiltration test. During the test, the steady-state radii of the wetted regions on the soil surface (r) associated with various discharge rates are recorded. Knowledge of several values of r versus i_f $(=Q/\pi r^2)$ enables calculation of K_s and A based on Wooding's solution, Eq. (7.94). A practical way to do so is fitting a line to the data of i_f versus $1/r$ and calculating K_s and A as

$$K_s = I_{line}, \quad A = \frac{4K_s}{\pi S_{line}} \quad (7.100)$$

where I_{line} and S_{line} are the intercept and slope of the fitted line, respectively.

7.3.3.2.4 *Internal Drainage Method*

Another common filed method for the determination of unsaturated hydraulic conductivity is the internal drainage method, also referred to as "instantaneous profile method." An initially dry soil profile is saturated. At the end of the wetting phase, the soil surface is covered with a plastic sheet to prevent surface evaporation. The matric potentials and water contents are continuously recorded at different depths during internal drainage (i.e. water redistribution). Assuming that water flow occurs only in the vertical direction, the integration of RE yields the basic equation for estimation of $K(\theta)$ at any arbitrary depth L:

$$K_L(\theta) = -\frac{\int_0^L [\partial\theta/\partial t]_z \, dz}{[\partial h/\partial z]_{z=L} - 1} \quad (7.101)$$

Nielsen et al. (1964), Rose et al. (1965), Watson (1966), Davidson et al. (1969), and Hillel et al. (1972) were among the first to apply the internal drainage method. This method can be further simplified by assuming a unit hydraulic gradient during drainage (Davidson et al. 1969) that yields

$$K_L(\theta) \approx -\frac{\partial\theta_{avg}}{\partial t} L \approx -\frac{\partial\theta_{avg}}{\partial t} \frac{I}{\theta_{avg}} \quad (7.102)$$

where θ_{avg} is the average water content along the wet zone, θ_i is the initial water content before wetting, and I is the total infiltrated water during the wetting phase.

Eq. (7.102) may be less accurate than Eq. (7.101) (Reichardt, 1993). However, it highly simplifies the measurements by avoiding the need for measuring h as well as reducing the number of sensors required for measuring θ.

7.4 Solute Transport

Large amounts of solutes (soluble chemicals) are applied to the topsoil layer for various agricultural and industrial purposes. In many cases, these chemicals are transported to deeper layers, causing contamination of groundwater and other natural resources. Therefore, quantifying (i.e. measuring/modeling) solute transport rate in the vadose zone (i.e. partially saturated soil layer, extending from the soil surface to the groundwater level) can be of great importance for agricultural and environmental management purposes. Hence, we devote this section to introduce fundamentals of solute transport in partially saturated soils.

7.4.1 Solute Transport Processes

Solutes are transported in soils due to advection, molecular diffusion, and mechanical dispersion processes. As they pass through soils, solutes may interact with soil constituents leading to processes of solute adsorption, i.e. retarded transport, or solute exclusion, i.e. accelerated transport. A brief explanation of each of these processes follows.

7.4.1.1 Advection

Solutes migrate in soils via flow of the solvent (e.g. water), a mechanism of transport referred to as "advection" or "advective transport." Advective solute transport may be simply described as

$$j_{adv} = qc \tag{7.103}$$

where j_{adv} [$ML^{-2}T^{-1}$] is the advective solute flux density, q [LT^{-1}] is the water flux density, and c [ML^{-3}] is the solute concentration expressed as solute mass per solvent volume. The water flux density is related to mean water flow velocity, v [LT^{-1}], and can be described in unsaturated soils using the BD law (Buckingham, 1907), given for 1D vertical flow as

$$q = v\theta = -K(\theta)\frac{\partial h(\theta)}{\partial z} + K(\theta) \tag{7.104}$$

where θ [L^3T^{-3}] is the volumetric water content; h [L] is the hydraulic pressure head; K [LT^{-1}] is the unsaturated hydraulic conductivity, dependent on θ in unsaturated soils; and z [L] is the soil depth, positive downward.

7.4.1.2 Diffusion

Diffusion (also molecular diffusion) is basically the movement of substances due to random molecular motions. This process explains the net flux of molecules (e.g. nitrate) from a location of high concentration to surrounding regions of lower concentration. Diffusive flux density of solutes, j_{dif} [$ML^{-2}T^{-1}$], may be described by Fick's first law, where j_{dif} is proportional to the gradient of solute concentration (i.e. $\partial c/\partial z$) along a spatial coordinate of interest (e.g. soil depth, z) multiplied by a proportionality coefficient referred to as the "diffusion coefficient." The "Stokes–Einstein" equation indicates that the diffusion coefficient for a given solute in free solution depends on the fluid viscosity and temperature. In unsaturated soils, the bulk diffusion coefficient, D_{dif} [L^2T^{-1}], is commonly estimated from the diffusion coefficient in free water, D_0 [L^2T^{-1}], θ, to account for the reduced cross-sectional area and a saturation-dependent tortuosity factor, $\tau(\theta) = D_{dif}/D_0$, to account for the extended pathway. Therefore, 1D diffusive transport of solutes in unsaturated soils can be formulated as

$$j_{dif} = -D_{dif}\theta\frac{\partial c}{\partial z} = -D_0\tau(\theta)\theta\frac{\partial c}{\partial z} \tag{7.105}$$

Liquid diffusion coefficient values, D_0 (10^{-5} cm^2 s^{-1}), for common ions are given in Table 7.8. Various empirical and conceptual models have been suggested for describing the $\tau(\theta)$ functional relationship (e.g. Millington and Quirk, 1961; Bresler, 1973; Papendick and Campbell, 1980; Olesen et al., 1996), some of which are reviewed in Chou et al. (2012) and Ghanbarian et al. (2013).

7.4.1.3 Dispersion

In addition to diffusion and advection, another mechanism for solute transport arises from "mechanical dispersion," which occurs in soils due to local (microscale) variations of water flow. Figure 7.31 illustrates a few examples of these local variations, including (i) variation of flow velocity within an individual pore, (ii) variation of flow velocity in pores of different size, and (iii) variation of flow direction in differently shaped pores.

Although dispersion is a consequence of advective transport, its impact is similar to molecular diffusion. Taylor (1953) analytically indicated that mechanical dispersion in a single cylindrical pore can be modeled with a Fickian flux. Therefore, a formulation analogous to Eq. (7.105) is commonly applied for the flux associated with mechanical dispersion in unsaturated soils (Leij and van Genuchten, 2002):

$$j_{dis} = -D_{dis}\theta\frac{\partial c}{\partial z} \tag{7.106}$$

where j_{dis} [$ML^{-2}T^{-1}$] is the flux density of solutes due to mechanical dispersion and D_{dis} [L^2T^{-1}] is a virtual diffusion coefficient for mechanical dispersion, called the "coefficient of mechanical dispersion." For a single cylindrical pore of radius r, Taylor (1953) indicated that

$$D_{dis} = \frac{r^2v^2}{48D_0} \tag{7.107}$$

where v is the mean flow velocity in the pore. According to Eq. (7.107), similar to the advective transport, higher mechanical dispersion is expected for higher flow rates. In contrast, mechanical dispersion is inversely proportional to the molecular diffusion, because molecular diffusion will mitigate the transversal concentration gradient occurring due to the variation of flow velocity (i.e. velocity profile) within a single pore (see Figure 7.31a).

In unsaturated soils, it is assumed that D_{dis} is linearly proportional to the mean flow velocity, v, where the proportionality coefficient, D_L [L], is an empirical parameter known as the "longitudinal dispersivity" (Ogata, 1970):

$$D_{dis} = D_L|v| = D_L\frac{|q|}{\theta} \tag{7.108}$$

The initial belief was that the longitudinal dispersivity is a constant property of the porous materials (Ogata, 1970).

Table 7.8 Values of diffusion coefficient D_0 (10^{-5} cm^2 s^{-1}) for common ions in dilute aqueous solution at 25 °C.

Ion	D_0	Ion	D_0	Ion	D_0	Ion	D_0
Ag^+	1.648	Li^+	1.029	BrO_3^-	1.483	$1/2MoO_4^{2-}$	1.984
$1/3Al^{3+}$	0.541	$1/2Mg^{2+}$	0.706	CN^-	2.077	$N(CN)_2^-$	1.451
$1/2Ba^{2+}$	0.847	$1/2Mn^{2+}$	0.712	CNO^-	1.720	NO_2^-	1.912
$1/2Be^{2+}$	0.599	NH_4^+	1.957	$1/2CO_3^{2-}$	0.923	NO_3^-	1.902
$1/2Ca^{2+}$	0.792	$N_2H_5^+$	1.571	Cl^-	2.032	$NH_2SO_3^-$	1.286
$1/2Cd^{2+}$	0.719	Na^+	1.334	ClO_2^-	1.385	N_3^-	1.837
$1/3Ce^{3+}$	0.620	$1/3Nd^{3+}$	0.616	ClO_3^-	1.720	OCN^-	1.720
$1/2Co^{2+}$	0.732	$1/2Ni^{2+}$	0.661	ClO_4^-	1.792	OD^-	3.169
$1/3[Co(NH_3)_6]^{3+}$	0.904	$1/4[Ni_2(trien)_3]^{4+}$	0.346	$1/3[Co(CN)_6]^{3-}$	0.878	OH^-	5.273
$1/3[Co(en)_3]^{3+}$	0.663	$1/2Pb^{2+}$	0.945	$1/2CrO_4^{2-}$	1.132	PF_6^-	1.515
$1/6[Co_2(trien)_3]^{6+}$	0.306	$1/3Pr^{3+}$	0.617	F^-	1.475	$1/2PO_3F_2^-$	0.843
$1/3Cr^{3+}$	0.595	$1/2Ra^{2+}$	0.889	$1/4[Fe(CN)_6]^{4-}$	0.735	$1/3PO_4^{3-}$	0.824
Cs^+	2.056	Rb^+	2.072	$1/3[Fe(CN)_6]^{3-}$	0.896	$1/4P_2O_7^{4-}$	0.639
$1/2Cu^{2+}$	0.714	$1/3Sc^{3+}$	0.574	$H_2AsO_4^-$	0.905	$1/3P_3O_9^{3-}$	0.742
D^+	6.655	$1/3Sm^{3+}$	0.608	HCO_3^-	1.185	$1/5P_3O_{10}^{5-}$	0.581
$1/3Dy^{3+}$	0.582	$1/2Sr^{2+}$	0.791	HF_2^-	1.997	ReO_4^-	1.462
$1/3Er^{3+}$	0.585	Tl^+	1.989	$1/2HPO_4^{2-}$	0.759	SCN^-	1.758
$1/3Eu^{3+}$	0.602	$1/3Tm^{3+}$	0.581	$H_2PO_4^-$	0.959	$1/2SO_3^{2-}$	0.959
$1/2Fe^{2+}$	0.719	$1/2UO_2^{2+}$	0.426	$H_2PO_2^-$	1.225	$1/2SO_4^{2-}$	1.065
$1/3Fe^{3+}$	0.604	$1/3Y^{3+}$	0.550	HS^-	1.731	$1/2S_2O_3^{2-}$	1.132
$1/3Gd^{3+}$	0.597	$1/3Yb^{3+}$	0.582	HSO_3^-	1.545	$1/2S_2O_4^{2-}$	0.885
H^+	9.311	$1/2Zn^{2+}$	0.703	HSO_4^-	1.385	$1/2S_2O_6^{2-}$	1.238
$1/2Hg^{2+}$	0.913	$Au(CN)_2^-$	1.331	$H_2SbO_4^-$	0.825	$1/2S_2O_8^{2-}$	1.145
$1/2Hg^{2+}$	0.847	$Au(CN)_4^-$	0.959	I^-	2.045	$Sb(OH)_6^-$	0.849
$1/3Ho^{3+}$	0.589	$B(C_6H_5)_4^-$	0.559	IO_3^-	1.078	$SeCN^-$	1.723
K^+	1.957	Br^-	2.080	IO_4^-	1.451	$1/2SeO_4^{2-}$	1.008
$1/3La^{3+}$	0.619	Br_3^-	1.145	MnO_4^-	1.632	$1/2WO_4^{2-}$	0.919

Source: From Lide (2003).

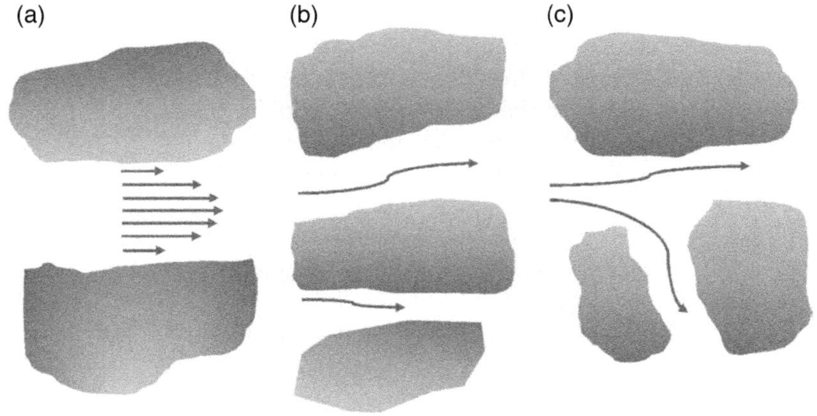

Figure 7.31 Sketch illustrating potential situations of local flow variations leading to mechanical dispersion of solutes in soils. (a) Variation of flow velocity within an individual pore. (b) Variation of flow velocity in pores of different sizes. (c) Variation of flow direction in differently-shaped pores. *Source*: After Leij and van Genuchten (2002).

More extensive studies (e.g. Gelhar et al., 1992; Vanderborght and Vereecken, 2007) indicated that D_L is mainly affected by the scale of transport (i.e. travel distance) rather than soil texture. A commonly accepted assumption based on dispersivity observations is that D_L is roughly about one-tenth of the mean travel distance (Gelhar et al., 1992).

In the literature, D_L values for porous aquifers can be found in the range of 0.1–500 m (Figure 7.32). There are

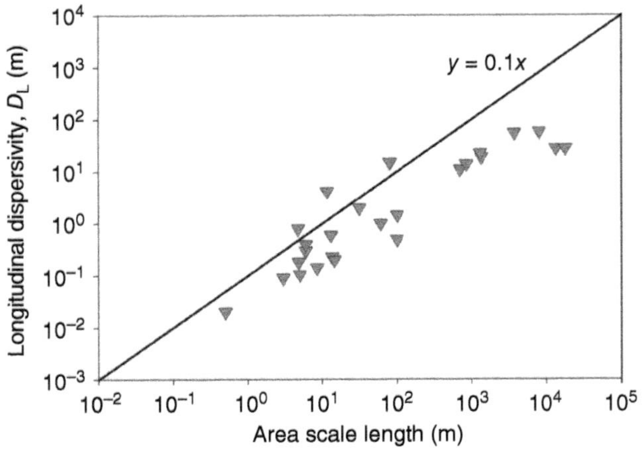

Figure 7.32 Scale dependence of longitudinal dispersivity. *Source*: Adapted from Kinzelbach (1986).

two main reasons for the observed scale dependence: (i) with growing size of a tracer cloud, larger and larger inhomogeneities may contribute to the dispersion, and (ii) the Fickian dispersion model is inadequate in situations that show too little randomness to resemble a diffusion process (Kinzelbach, 1986).

Because the mechanical dispersion and molecular diffusion cannot be distinguished at the Darcy scale, these two processes are often considered a combined process known as "hydrodynamic dispersion." Therefore, solute flux in soils, j, is determined as the sum of advection flux, j_{adv}, and hydrodynamic dispersion flux, j_{hd}:

$$j = j_{adv} + j_{hd} = qc - D\theta \frac{\partial c}{\partial z} \tag{7.109}$$

where

$$D = D_{dif} + D_{dis} = D_0 \tau(\theta) + D_L \frac{|q|}{\theta} \tag{7.110}$$

7.4.1.4 Adsorption

The clay and organic matter fraction of soils generally possess a net negative charge. Therefore, cations (i.e. positively charged solutes such as sodium and potassium) can be adsorbed by these negatively charged soil particles. In addition, anions (e.g. phosphorus) can be adsorbed by soil constituents, for example, hydrous iron oxides and aluminum hydroxides that are abundant in tropical and subtropical soils.

Adsorption in the solute transport governing equation is often considered by a functional relationship between the mass of adsorbed solutes per mass of dry soil, s [MM^{-1}], and solute concentration, c, assuming that s and c are in equilibrium. This functional relationship is called an "adsorption isotherm" and, in its simplest form, is given by the linear model $s = k_d c$, where k_d [L^3M^{-1}] is an empirical "distribution coefficient." Solute adsorption characterized with a linear adsorption isotherm is referred to as "linear adsorption."

Table 7.9 Common models used to describe solute adsorption isotherms, $s(c)$.

Model	Adsorption isotherm[a]
Linear	$s = k_d c$
Freundlich	$s = k_F c^p$
Langmuir	$s = s_{max} \frac{k_L c}{1 + k_L c}$
Dubinin–Radushkevitch	$s = s_{max} \exp\left\{-k_{DR}\left[\ln\left(1 + \frac{1}{c}\right)\right]^2\right\}$

Source: Adapted from Piccin et al. (2011).
[a] k_d, k_F, k_L, k_{DR}, and p are model parameters; s_{max} is the maximum adsorption (s when $c \to \infty$).

In reality, the $s(c)$ relationship is often nonlinear, reflecting the fact that the adsorption rate by the solid particles reduces at higher solute concentrations due to saturation of the soil particles. The most common models for the solute adsorption isotherm are presented in Table 7.9.

Adsorption of solutes by soil particles leads to transport retardation when compared with a similar nonreactive transport process. Hence, another way to account for the adsorption process is using a "retardation factor," R [dimensionless], such that $R = 1$ for nonreactive transport and $R > 1$ for retarded transport (i.e. solute movement is slower than water movement). The retardation factor is linked to the adsorption isotherm as follows:

$$R = 1 + \frac{\rho_b}{\theta} \frac{\partial s}{\partial c} \tag{7.111}$$

where ρ_b is the soil bulk density.

Eq. (7.111) yields $R = 1 + \rho_b \theta^{-1} k_d$ for the linear isotherm. For nonlinear isotherms, R will be a function of c and can be calculated based on Eq. (7.111) (e.g. $R = 1 + \rho_b \theta^{-1} k_F p c^{p-1}$ for the Freundlich model).

7.4.1.5 Exclusion

It is commonly observed that the average transport velocity of dissolved anions through soil is larger than that of accompanying water molecules or nonreactive solutes. This phenomenon, known as "anion exclusion," is attributed to electrostatic repulsion by negatively charged soil particles, forcing the anions into the center of pores where the flow velocity is faster (Figure 7.31a). The accelerated transport can be quantified with the retardation factor, R, being less than 1. In this case, R is given as

$$R = 1 - \frac{\theta_{ex}}{\theta} \tag{7.112}$$

where θ_{ex} is the anion exclusion volume (i.e. liquid phase not participating in the transport process).

A distinguished study regarding anion exclusion is the work of Gvirtzman and Gorelick (1991). Their field observations for a 27.5-m-deep soil profile, along with their calibrated numerical simulations for a 27-year period (1957–1984), uncovered the following insights: (i) Advection of anions (chloride and sulfate) was doubled compared with tritium (a nonreactive tracer). (ii) Dispersion coefficients associated with anion transport were approximately 30 times larger than those found for tritium transport. (iii) Attenuation of the anion concentration profiles was greatly enhanced. All of these results were explained by anion exclusion effects using two conceptual models, namely, the "dispersion model" (Figure 7.33a,b) and the "exchange model" (Figure 7.33c,d). The dispersion model suggests that, unlike tritium, anions cannot flow through small soil pores, leading to faster advective transport through the larger pores (Figure 7.33b). Gvirtzman and Gorelick (1991) suggest that anions necessarily follow a more tortuous pathway compared with that of tritium, leading to much larger dispersion of anions than the dispersion experienced by tritium, because anions effectively "see" a more heterogeneous porous medium. The enhanced attenuation of anion concentration profiles is justified by permitting the increased mechanical dispersion according to the dispersion model and by allowing for diffusional exchange between large pores and immobile pockets that are created during anion transport according to the exchange model. It was also observed that the divalent anion (sulfate) has a greater exclusion volume than the monovalent anion (chloride), leading to smaller dead-end pockets and smaller passageways through which mass transfer occurs.

7.4.2 Mathematical Modeling of Solute Transport

7.4.2.1 Transport in Unsaturated Soils

Solute transport in soils, considering the abovementioned processes, can be mathematically modeled based

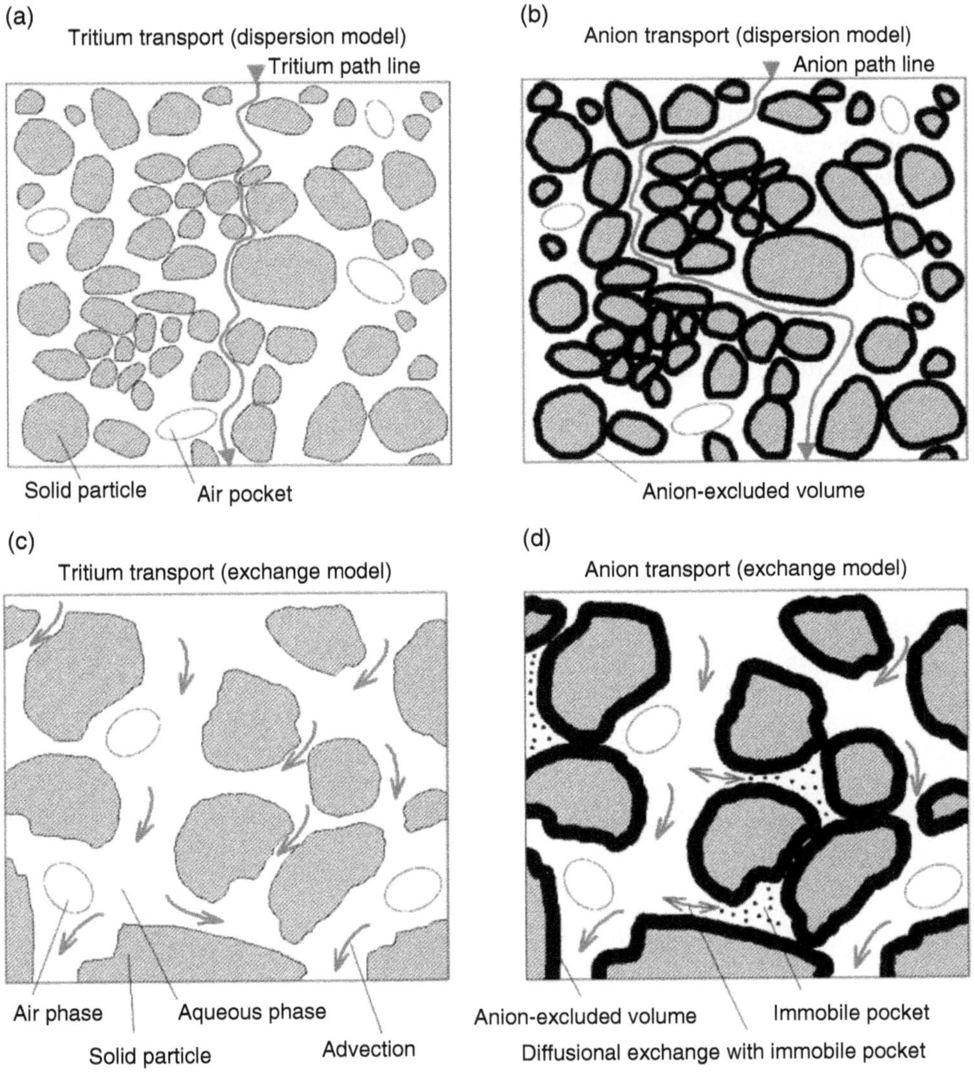

Figure 7.33 Two conceptual models, (a, b) "dispersion model" and (c, d) "exchange model," illustrating how anion exclusion substantially influences the active pore network. *Source*: From Gvirtzman and Gorelick (1991).

on the solute continuity equation, which in 1D form, can be written as follows:

$$\frac{\partial(c\theta R)}{\partial t}=-\frac{\partial j}{\partial z} \tag{7.113}$$

where t is time. Substituting the solute flux density, j, from Eq. (7.109) into Eq. (7.113) yields

$$\frac{\partial(c\theta R)}{\partial t}=-\frac{\partial}{\partial z}\left(qc-D\theta\frac{\partial c}{\partial z}\right) \tag{7.114}$$

Eq. (7.114) forms the governing equation of 1D solute transport in unsaturated soils and, in its current form, is highly nonlinear due to the dependence of c, θ, q, and R on each other, where q and θ themselves need to be solved using the highly nonlinear RE. Hence, analytical solution of Eq. (7.114) in unsaturated soils is very complicated, and it is solved usually via numerical models (e.g. HYDRUS-1D; Šimůnek et al., 2005). Nonetheless, there exist a few analytical solutions to the transport problem under unsaturated conditions, subject to limited cases of initial and boundary conditions as well as several simplifying assumptions. An example is the solution of Warrick et al. (1971) for 1D nonreactive transport (i.e. $R=1$) during infiltration, where D was assumed to be constant and q was calculated independently using a numerical solution. De Smedt and Wierenga (1978) also provided approximate analytical solutions for 1D nonreactive transport during infiltration and redistribution, assuming constant D and q. Another example is the analytical solution by Huang and van Genuchten (1995) for 1D solute transport in soil during ponded infiltration, where the Green–Ampt step-function moisture profile was assumed for the derivations.

For specific sets of initial and boundary conditions, the similar/dissimilar media scaling theories (Miller and Miller, 1956; Sadeghi et al., 2012a, b) may also provide means to reduce the complexity of the problem. This is achieved by transforming the flow and transport governing equations into generalized forms. An example is the solution of Sadeghi and Jones (2012) to 1D solute transport along with transient unsaturated flow during soil water redistribution. This is described using the coupled RE and Eq. (7.114), leading to nearly invariant solutions of the problem for a broad range of soils and conditions (Figure 7.34).

Despite the high nonlinearity of the solute transport governing equation in unsaturated soils, it can reduce to simple linear forms under steady-state saturated conditions that can be solved analytically, as presented briefly in the following.

7.4.2.2 Advection–Dispersion Equation and Its Analytical Solutions

In saturated soil, i.e. where $\theta=\theta_s$ (saturated water content) and $q=v\theta_s$, Eq. (7.114) reduces to the ADE when R is constant (i.e. linear isotherm):

$$R\frac{\partial c}{\partial t}=D\frac{\partial^2 c}{\partial z^2}-v\frac{\partial c}{\partial z} \tag{7.115}$$

If the advective transport is negligible compared with the dispersive transport, Eq. (7.115) can be written in the form of the classical diffusion equation:

$$\frac{\partial c}{\partial t}=D_R\frac{\partial^2 c}{\partial z^2} \tag{7.116}$$

where $D_R = D/R$. An analytical solution to Eq. (7.116), subject to the initial condition of $c(z, 0)=c_i$, surface boundary condition of $c(0, t)=c_0$, and bottom boundary condition of $c(\infty, t)=c_i$, is given as follows (Crank, 1975):

$$\frac{c(z,t)-c_i}{c_0-c_i}=0.5\,\mathrm{erfc}\left(\frac{z}{\sqrt{4D_R t}}\right)=0.5\,\mathrm{erfc}\left(\frac{Rz}{\sqrt{4RDt}}\right) \tag{7.117}$$

where erfc denotes the "complementary error function." By defining a moving coordinate,

$$\zeta=z-\frac{v}{R}t,\quad \tau=t \tag{7.118}$$

and writing the ADE (Eq. 7.115) in the new coordinate format using the chain rule,

$$R\left(\frac{\partial c}{\partial \tau}\frac{\partial \tau}{\partial t}+\frac{\partial c}{\partial \zeta}\frac{\partial \zeta}{\partial t}\right)=D\left(\frac{\partial}{\partial \zeta}\frac{\partial \zeta}{\partial z}+\frac{\partial}{\partial \tau}\frac{\partial \tau}{\partial z}\right)\left(\frac{\partial c}{\partial \zeta}\frac{\partial \zeta}{\partial z}+\frac{\partial c}{\partial \tau}\frac{\partial \tau}{\partial z}\right)-v\left(\frac{\partial c}{\partial \zeta}\frac{\partial \zeta}{\partial z}+\frac{\partial c}{\partial \tau}\frac{\partial \tau}{\partial z}\right) \tag{7.119}$$

the ADE is transformed to the form of a diffusion equation, similar to (7.116):

$$\frac{\partial c}{\partial \tau}=D_R\frac{\partial^2 c}{\partial \zeta^2} \tag{7.120}$$

Similarly, the ADE solution subject to the initial condition of $c(\zeta, 0)=c_i$ and boundary conditions of $c(0, \tau)=c_0$ and $c(\infty, \tau)=c_i$ is given as follows:

$$\frac{c(z,t)-c_i}{c_0-c_i}=0.5\,\mathrm{erfc}\left(\frac{\zeta}{\sqrt{4D_R \tau}}\right)=0.5\,\mathrm{erfc}\left(\frac{Rz-vt}{\sqrt{4RDt}}\right) \tag{7.121}$$

A more practical problem may be to consider a tracer pulse at the inlet boundary ($z=0$) for a finite period of time ($0<t<t_0$). An analytical solution to the ADE, subject to the conditions

$$c(z,0)=c_i,\quad c(0,t)=\begin{cases}c_0 & (0<t<t_0)\\ 0 & (t>t_0)\end{cases},\quad \frac{\partial c}{\partial z}(\infty,t)=0 \tag{7.122}$$

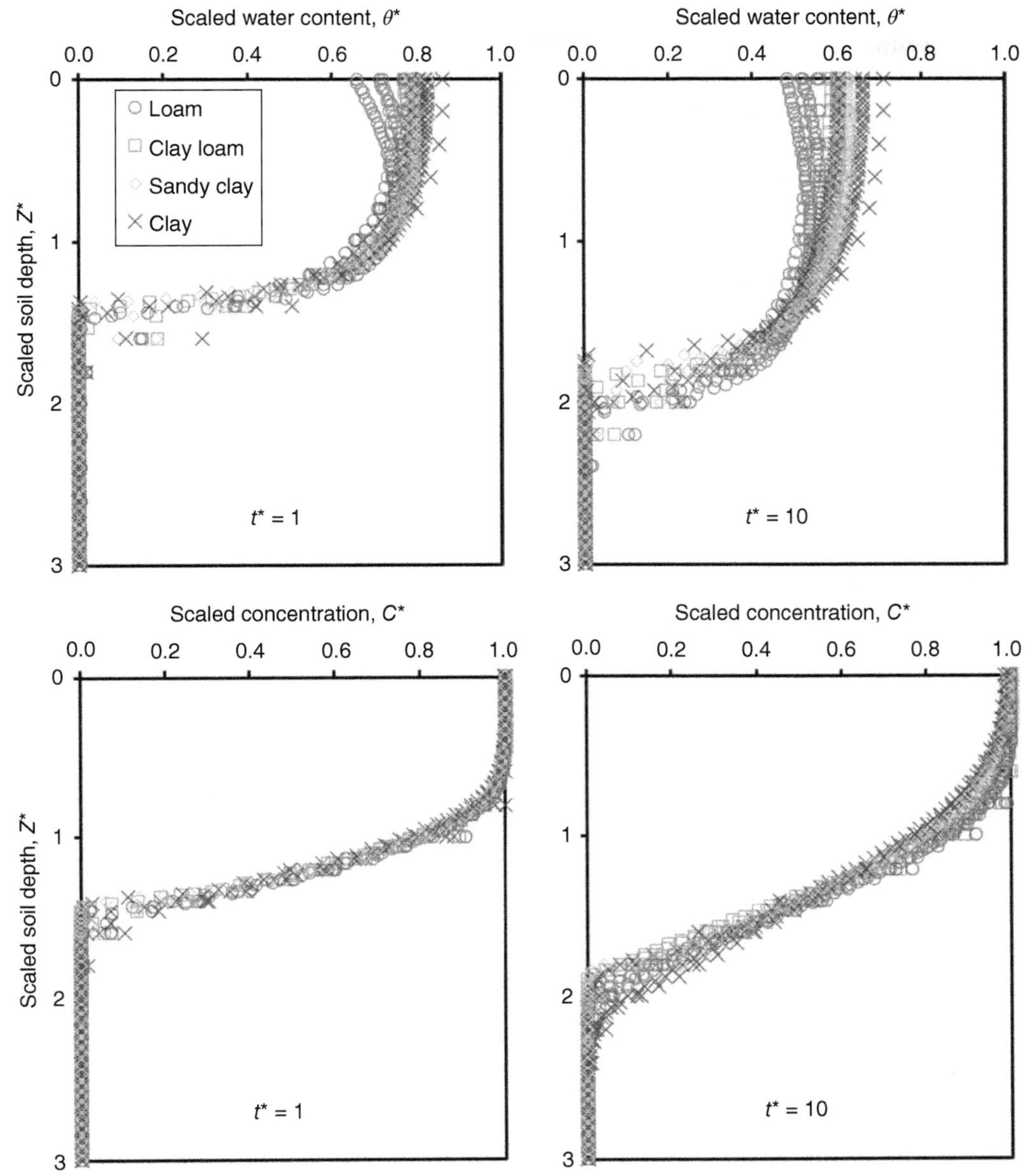

Figure 7.34 HYDRUS-1D numerical simulations of soil water content and concentration profiles during soil water redistribution for 20 different cases, varying in soil texture and initial conditions. The numerical data are presented in the form of scaled variables at two scaled times, $t^*=1$ and $t^*=10$. The solution exhibits nearly invariant scaled solutions for all cases. *Source*: From Sadeghi and Jones (2012).

is given as follows (after van Genuchten and Alves, 1982):

$$\frac{c(z,t)-c_i}{c_0-c_i}=\begin{cases}A(z,t) & (0<t<t_0)\\ A(z,t)-\dfrac{c_0}{c_0-c_i}A(z,t-t_0) & (t>t_0)\end{cases} \tag{7.123}$$

where

$$A(z,t)=0.5\,\mathrm{erfc}\left(\frac{Rz-vt}{\sqrt{4RDt}}\right)+0.5\exp\left(\frac{vz}{D}\right)\mathrm{erfc}\left(\frac{Rz+vt}{\sqrt{4RDt}}\right) \tag{7.124}$$

Interested readers are referred to the review of van Genuchten and Alves (1982) on analytical solutions of the ADE for various boundary conditions.

7.4.3 Characterizing Soil Transport Parameters

7.4.3.1 Breakthrough Curve

Soil parameters affecting solute transport (e.g. D and R) are commonly estimated from a soil column experiment involving injection of a solute with constant concentration, c_0, at the inlet of the column, followed by measuring the solute concentration at some specific depths or at the outlet (Figure 7.35, left). The transport parameters are then estimated by means of inverse solution of the transport equations, i.e. fitting an appropriate analytical or numerical solution to the observations. A convenient way for this analysis is using the so-called breakthrough curve (BTC) (Figure 7.35, right), which is the relationship between relative concentration, C, and a dimensionless time, T, equal to the number of displaced pore volumes. Defining the following dimensionless variables

$$C = \frac{c}{c_0}, \quad T = \frac{vt}{L}, \quad Z = \frac{z}{L} \tag{7.125}$$

leads to the following dimensionless and more functional form of the ADE for BTC experiments:

$$R\frac{\partial C}{\partial T} = \frac{1}{P_e}\frac{\partial^2 C}{\partial Z^2} - \frac{\partial C}{\partial Z} \tag{7.126}$$

where L is the observation depth or distance from the inlet (L is equal to the column's depth for the effluent BTC) and P_e is the so-called Peclet number, i.e. a dimensionless quantity that determines the relative strength of advection to hydrodynamic dispersion, written as

$$P_e = \frac{vL}{D} \tag{7.127}$$

The BTC can be modeled either numerically or with one of the analytical solutions of the ADE. For example, Eq. (7.121) can be written in the dimensionless form for $Z = 1$ ($z = L$):

$$C(T) = 0.5erfc\left[\sqrt{\frac{P_e}{4RT}}(R - T)\right] \tag{7.128}$$

The transport parameters, R and P_e, then can be estimated by fitting Eq. (7.128) to the measured BTC data. Average flow velocity, v, can be set during the experiment or be estimated from the observations. Then D can be determined from Eq. (7.127).

Figure 7.36 exhibits the BTC, modeled based on Eq. (7.128), for various values of P_e and R. In Figure 7.36a, P_e is varied while considering a nonreactive transport ($R = 1$). This figure indicates that extremely high P_e values (i.e. when hydrodynamic dispersion is negligible compared to advection) lead to a pistonlike BTC. Figure 7.36b indicates the effect of R on the BTC. As can be seen, $R = T$ at $C = 0.5$. This result is a property of Eq. (7.128) [i.e. erfc(0) = 1] and makes it possible to estimate R of experimental data even by visual inspection. Figure 7.36c includes experimental data for three different tracers regarding reaction with soil particles: Cl^- (an excluded anion), ^3H_2O (a nonreactive tracer), and Ca^{2+} (an adsorbed cation). Transport parameters, P_e and R, for these tracers were found by fitting Eq. (7.128) to the data. The obtained retardation factor, R, fairly shows the anion exclusion effect (i.e. acceleration in transport) for Cl^- and the adsorption effect (i.e. retardation in transport) for Ca^{2+}.

The BTC for a narrow pulse of solute [conditions of (7.122)] can also be modeled, for example, based on the solution of (7.123). Assuming $c_i = 0$, the following BTC (plotted in Figure 7.37) is obtained:

$$C(T) = \begin{cases} A(T) & (0 < T < T_0) \\ A(T) - A(T - T_0) & (T > T_0) \end{cases} \tag{7.129}$$

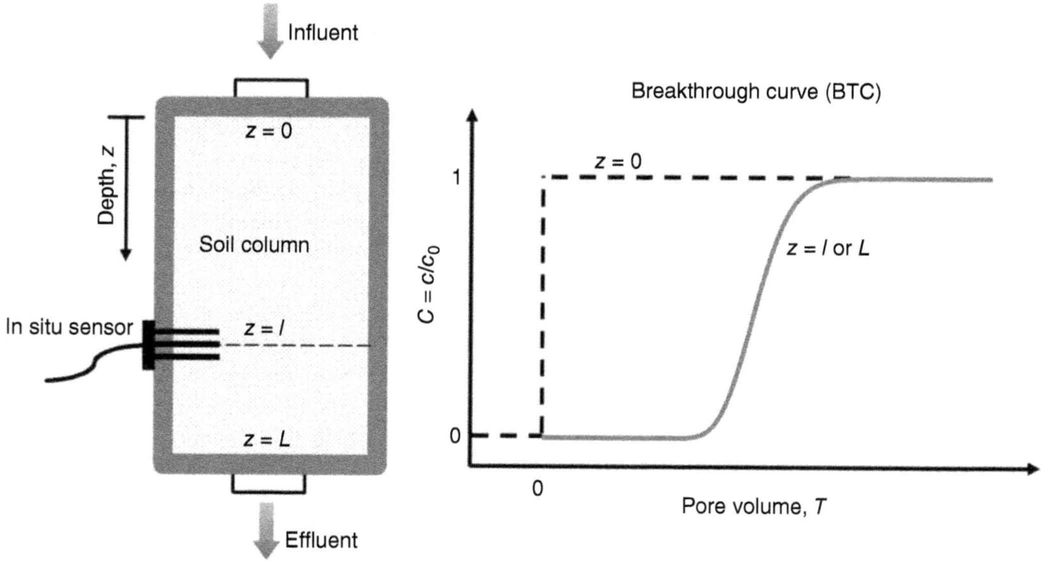

Figure 7.35 Sketch illustrating a laboratory experimental setup (left) for measuring the breakthrough curve (right) in order to estimate solute transport parameters. Solute concentration may be measured using an in situ sensor (e.g. TDR) for an intermediate depth ($z = l$) or by chemical analysis of the effluent ($z = L$). Note that because T is a dimensionless time, the BTC curves for both depths of l and L coincide (see Eq. (7.128)).

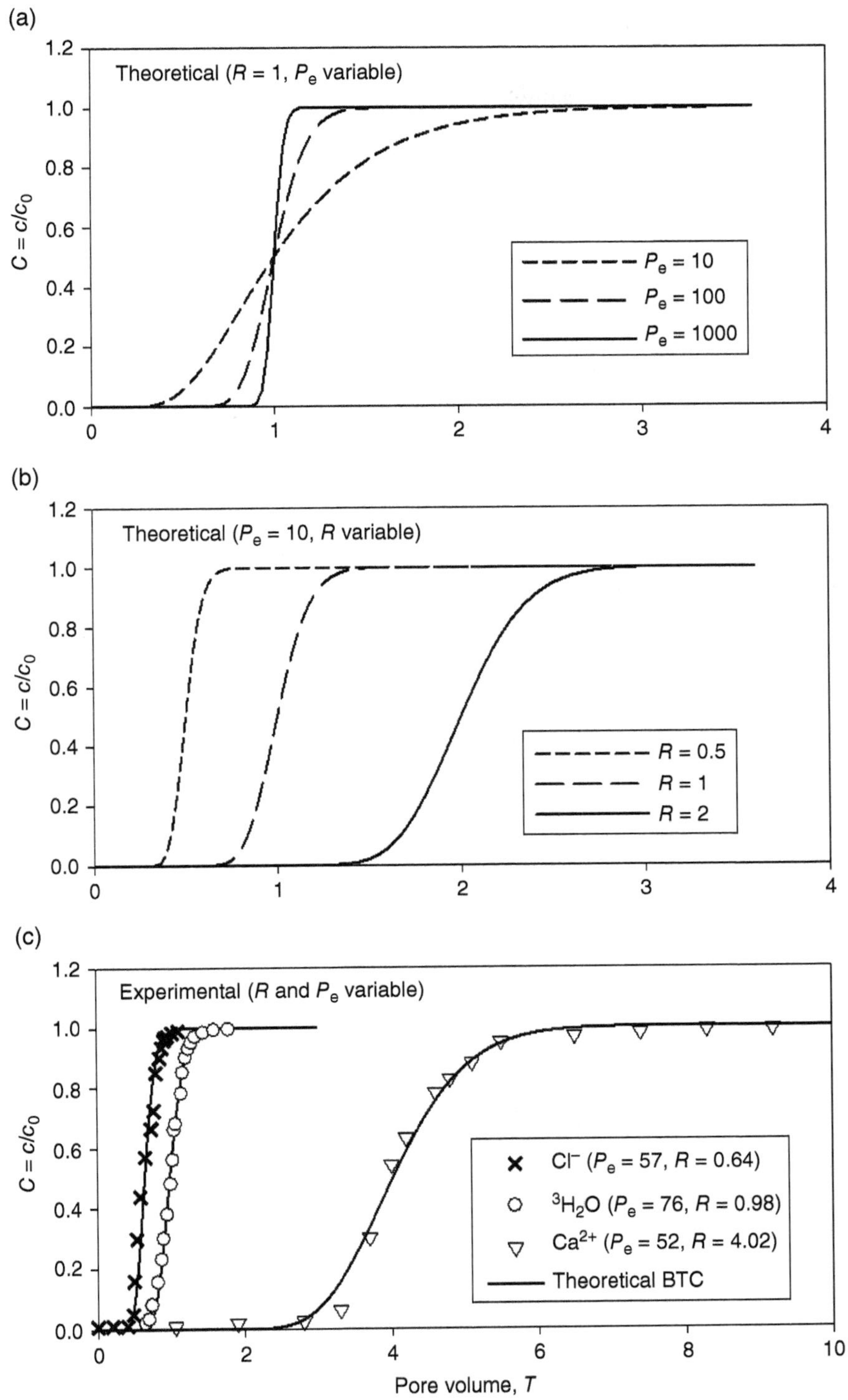

Figure 7.36 Sample breakthrough curves (BTCs) applying Eq. (7.128) (a) for $R = 1$ and different P_e values, (b) for $P_e = 10$ and different R values, and (c) fitted to actual experimental data from Leij and Sciortino (2011) for the transport of Cl^- (an excluded anion, $R = 0.64$) in a clay loam soil, ^3H_2O (a nonreactive tracer, $R = 0.98$) in the same clay loam soil, and Ca^{2+} (an adsorbed cation, $R = 4.02$) in a loam and fine loam soil.

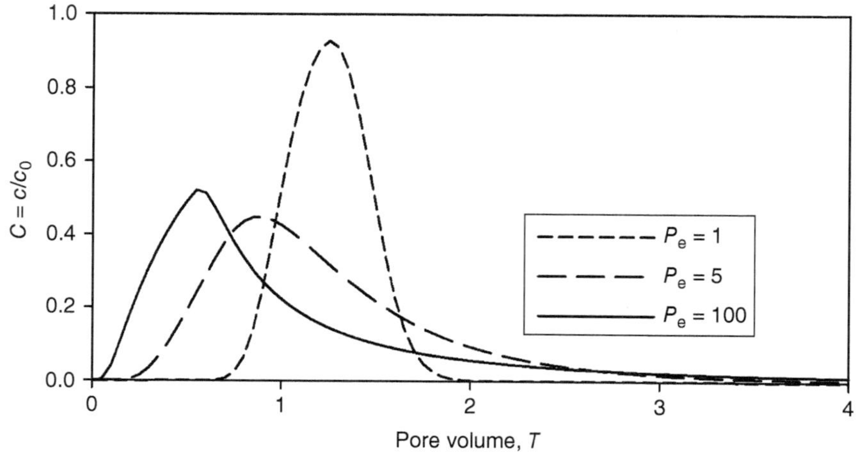

Figure 7.37 Theoretical breakthrough curves (BTCs) for a narrow pulse of solute modeled with Eq. (7.129) for $R = 1$, $T_0 = 0.5$ and different P_e values.

where

$$T_0 = \frac{vt_0}{L} \tag{7.130}$$

$$A(T) = 0.5\text{erfc}\left[\sqrt{\frac{P_e}{4RT}}(R-T)\right] + 0.5\exp(P_e)\,\text{erfc}\left[\sqrt{\frac{P_e}{4RT}}(R+T)\right] \tag{7.131}$$

7.4.3.2 Measurement of Solute Concentration

Methodologies for measuring solute concentration are an essential part of the BTC experiment or of field methods for characterizing transport parameters. As mentioned, one approach is the laboratory chemical analysis of the effluent collected at various times during the BTC experiment. For the case of field methods, in situ extraction of soil solution is required. A comprehensive review of in situ soil water extraction methods to monitor solute concentration was presented by Weihermüller et al. (2007). They described six different sampling devices: porous cups, porous plates, capillary wicks, pan lysimeters, resin boxes, and lysimeters. Or et al. (2006) also presented a detailed review of various methods for soil pore water extraction including suction cups, combined solution sampling–tensiometer probes, lysimeters, passive capillary samplers, and capillary absorbers.

Solute concentration is highly correlated to the EC of the soil water solution (see Cooper (2016) for the relationship between EC and solute concentration). Hence, another approach to determine the solute transport properties is to implement an in situ sensor that can measure soil EC. These sensors are based on three EM phenomena, namely, reflectometry, electromagnetic induction (EMI), and electrical resistivity (ER) (Visconti and de Paz, 2016). TDR has been widely employed for soil EC measurement, as reviewed by Noborio (2001). Frequency-domain reflectometry (FDR) is another reflectometry-based sensor that can be used for this purpose (Bouksila et al., 2008; Wilczek et al., 2012). Various researchers also studied EMI sensors (e.g. de Jong et al., 1979; Amakor, 2013) and ER sensors (e.g. Rhoades and van Schilfgaarde, 1976; Corwin and Hendrickx, 2002) for measuring EC. Interested readers are referred to Corwin and Lesch (2005) and Visconti and de Paz (2016) who reviewed in situ sensors for measuring soil EC.

7.5 Soil Temperature, Thermal Properties, and Heat Flow

Soil temperature is commonly monitored to understand environmental impacts on natural soils and engineered systems. Temperature affects most of the properties and processes within soil and porous media, and the thermal properties of these media often control the rates of change in temperature. Soil temperature is a critical factor in the germination of seed, microbial activity, chemical reactions, and the growth and maintenance of plant roots. There is growing interest on energy-related issues, and soil temperature plays a significant part in the efficiency of electrical systems (solar panels, power transmission lines), both above and below ground. Perhaps one of the most impactful processes is freezing of soil, which can cause significant damage to municipal infrastructure including roads, foundations, and structures, as well as water, sewer, and other pipelines. In addition, freezing soils impact agricultural practices such as plowing, planting, and harvesting. Understanding heat transport processes is key to addressing many of the problems associated with environmental temperature and global temperature change. Modeling these processes requires knowledge of soil thermal properties and the interdependent relationships involved in soil thermohydrodynamics. With advances in microelectronics and a resurgence of interest in the use of soil thermal sensors to detect soil properties and processes, application of heat pulse probes as well as hybrid thermal sensors (e.g. thermo-time domain reflectometry (TDR)) in research is evolving.

7.5.1 Soil Temperature

Soil temperature is a measure of the intensity of heat or, more specifically, the kinetic energy of a soil's combined molecules (Horton and Ochsner, 2011). Temperature increases as a result of an increase in kinetic energy locally. Soil heat is driven from locations of higher temperature to lower temperature locations. Temperature impacts most of the critical processes taking place in soil, including the germination of seed, microbial activity (Fierer et al., 2006), chemical reactions, carbon sequestration (Kirschbaum, 1995), gas production and emissions (Schaufler et al., 2010), the growth and maintenance of plant roots (Atkin et al., 2000), soil evaporation (Kalma et al., 2008), plant transpiration, and freeze–thaw cycling. Atmospheric conditions drive near-surface soil temperature, which drives heat transport in and out of soil on diel and annual cycles. The economic impact of soil temperature extremes (freezing, heat waves) can be enormous for municipal, state, and federal budgets absorbing costs for wide-scale damage to soil-based infrastructure (e.g. buildings, pipelines, roadways, utility lines, etc.) in addition to agricultural producers suffering from yield loss (Porter and Gawith, 1999).

Soil temperature is perhaps one of the most common soil measurements and reliable long-term monitoring stations (e.g. Rothamsted Research, 1931–present) provide valuable information on temperature trends associated with climate change and global warming. The World Meteorological Organization (WMO, 2006) recommends monitoring soil temperature measurement depths at 5, 10, 20, 50, and 100 cm. The USDA records hourly soil moisture and temperature measurements at these same depths in both of their national monitoring networks, SNOTEL and SCAN.

7.5.1.1 Measurement of Soil Temperature

Soil temperature is measured using both direct contact and noncontact methods. Thermometers (nonelectric), thermocouples, and thermistors all provide direct contact measurements, though thermometers that employ thermal expansion of liquids or solids are generally less accurate. Furthermore, measurement of the soil surface using contact thermometry is prone to serious errors (i.e. contact loss, surface disturbance, and thermal nonequilibrium).

7.5.1.1.1 Direct Soil Contact Temperature Measurement

Direct contact measurements of subsurface temperature are commonly measured using a thermocouple or a thermistor. Thermocouples employ a bimetal junction where the gradient in voltage developed is directly proportional to the gradient in temperature (Seebeck effect) and must be read using an advanced circuit common in multimeter and data logging devices. A thermistor is a type of negative coefficient resistor whose resistance is dependent on temperature that can be read by a simpler voltage reading circuit. Today's thermistors and thermocouples can resolve 0.01 °C, but absolute temperature calibration can be costly to achieve at this resolution. More accurate temperature measurements are available using a resistance temperature detector (RTD), which uses a more expensive metal-based temperature-dependent resistance circuit. Subsurface temperature sensors are generally epoxy-coated and may be embedded in stainless steel tubing for protection from corrosion and water damage. Advances in microelectronics have facilitated heat pulse probes (Campbell et al., 1991) that employ temperature rise measurements in soil to infer soil thermal properties as well as other processes of interest (e.g. soil heat and water flux density; Yang et al., 2013). Because of the low cost of temperature sensing circuits, most environmental sensors today include a temperature measurement in addition to other properties (e.g. soil moisture, electrical conductivity, etc.)

While most direct measurements are made at the point scale, the growing use of distributed temperature measurements with fiber-optic cable provides 0.1 °C precision at 1 m spatial resolution. These cables can be buried in soil and may extend across kilometers of distance (Selker et al., 2006), yielding unprecedented information on soil temperature with potential to provide other indirect properties such as soil moisture (Steele-Dunne et al., 2010).

7.5.1.1.2 Noncontact Soil Temperature Measurement

Noncontact measurements of soil temperature rely on surface-emitted infrared radiation sensed by a distant thermometer (infrared spot thermometer, infrared pyrometer, or infrared detector array). Early instruments employed fine-wire thermocouples, but advances in microelectronics have produced faster response and more accurate detectors for infrared thermometry applications. Infrared radiation emitted from a surface is a function of the surface emissivity as well as surface temperature. Although most surfaces including soils have high emissivity of 0.9–0.97 (Fuchs and Tanner, 1968), which is a key assumption of today's low-cost radiometers, measurement errors arise when lower emissivity surfaces (e.g. highly reflective) are detected without adjusting the assumed high emissivity value. Since most soils support plant growth naturally, it is important to understand the impact of plant canopies, which can insulate and alter temperature gradients (important in modeling) between the soil surface and the atmosphere.

This implies that both canopy and soil surface temperature measurements are important to consider.

7.5.2 Heat Transport

7.5.2.1 Modes of Energy Transfer in Soils

Energy transfer within porous media may occur by radiant, conductive, or convective heat transfer mechanisms (Figure 7.38). Within convective heat transfer, mechanisms may stem from flowing liquid water, from flowing air, or from latent heat transfer. Solar radiation provides the major source of heat to the Earth's surface with conduction and latent heat transfer comprising the major heat transport mechanisms within soil. Figure 7.38 depicts various modes of the heat transport process from the radiant-heated, sunny side of a soil ridge to the shady side. Within the pore domain, both convection resulting from expanding gas and conduction taking place primarily in the solid and liquid phases transport heat toward the shady (cool) side of the ridge.

Radiant heat transport provides the major source of Earth's surface warming from solar radiation. EM energy emitted from all bodies containing heat follows the Stefan–Boltzmann law, given as

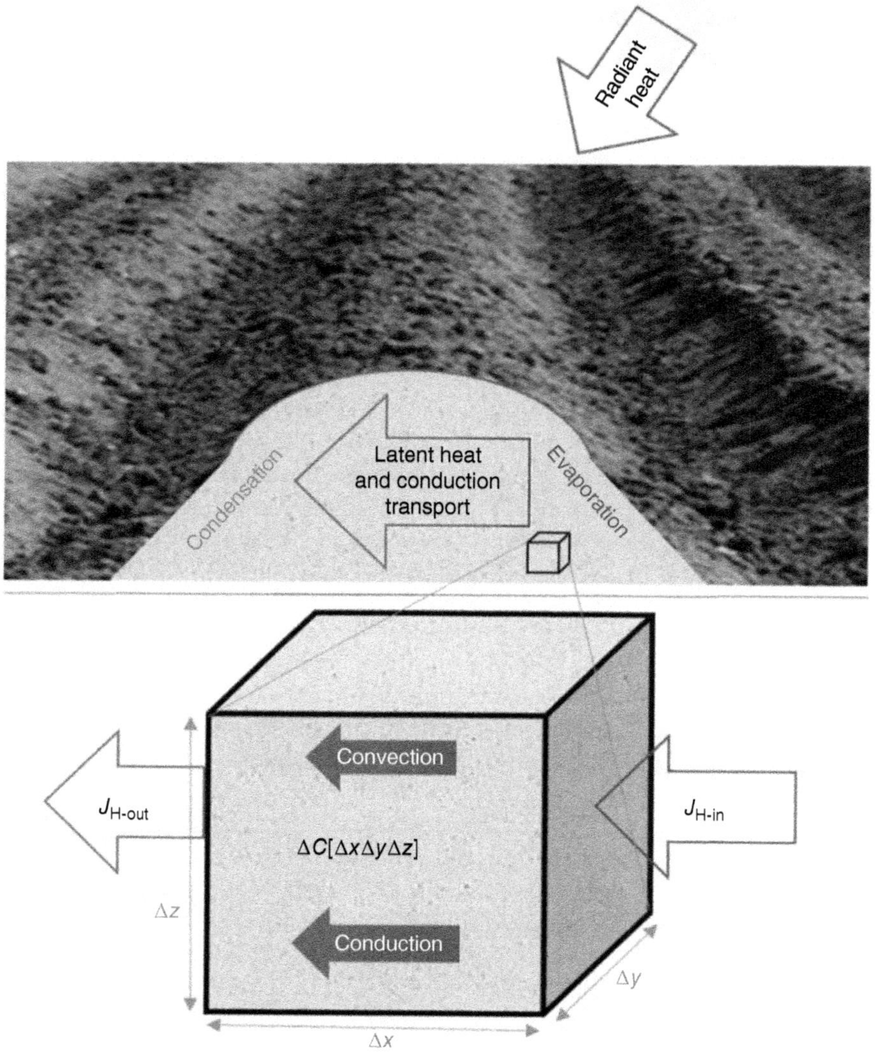

Figure 7.38 Illustration of radiant heat source warming of a soil ridge's sunny surface, causing evaporation and latent heat transport. At the pore scale a mixture of convective heat transport (gas expansion) in conjunction with heat conduction through liquid and solid phases takes place. Evaporation and condensation of water vapor occur at the pore scale driving latent and conductive heat to the cool, condensing side of the ridge, leading to a dry sunny side and a shady wet side ("Ridge and Furrow" photo 20 April 2009. Copyright by Derek Harper licensed for reuse under the Creative Commons Attribution-ShareAlike 2.0 license). *Source:* Courtesy of Derek Harper.

$$J_r = \varepsilon\sigma T^4 \tag{7.132}$$

where J_r is the total radiant energy of the emitter (body) integrated over all wavelengths, ε is the emissivity, T is the absolute surface temperature of the emitter, and σ is the Stefan–Boltzmann constant ($5.67037 \times 10^{-8}\,\mathrm{W\,m^{-2}\,K^{-4}}$) computed as

$$\sigma = \frac{2\pi^5 k^4}{15c^2h^3} \tag{7.133}$$

where k is the Boltzmann constant ($1.38064 \times 10^{-23}\,\mathrm{J\,K^{-1}}$), c is the speed of light, and h is Planck's constant ($6.62607 \times 10^{-34}\,\mathrm{J\,s}$). The emitter temperature dictates the wavelength distribution of the radiant energy as described by Wien's law, given as

$$\lambda_m = 2900T^{-1} \tag{7.134}$$

where λ_m is the maximum radiation intensity wavelength in micrometers. Planck's law provides further insight into the wavelength-dependent radiation intensity distribution, which is also a function of emitter temperature, written as

$$E_\lambda = \frac{2\pi hc^2}{\lambda_m^5\left[\exp\left(hc/k\lambda_m T\right)-1\right]} \tag{7.135}$$

where E_λ is the radiant energy flux emitted at a given wavelength. These relationships provide insight into the wavelength dependence of radiant energy transport from solar radiation as well as other sources of radiant heat.

Convective heat transport occurs as mass-carrying heat is transported within porous media as in the case of precipitation or irrigation water applied to the soil surface at a different temperature than the soil. Convective heat transport is generally described within the context of numerical modeling of soil water flow such as when heat transfer modeling is applied with RE calculations. Convective heat transport is generally short-lived and less significant than the daily cycles of radiant and conductive heat transport in and out of soil.

Conductive heat transport results from the internal energy or temperature driving molecular motion with the constituents of a porous medium (solids, water, and air). Heat conduction is analogous to a diffusion process where temperature gradients drive heat from high temperature regions to lower temperature locations. An example is where solar radiation warms the soil surface and heat is conducted downward to the subsurface during the day with heat conduction occurring in the opposite direction at night. Fourier's law describes steady-state conductive heat flux, J_H, as a function of the temperature gradient written as

$$J_H = -\lambda\nabla T \tag{7.136}$$

where λ is the porous medium thermal conductivity and ∇T is the spatial temperature gradient. The 1D heat flux can be expressed as

$$J_H = -\lambda_x\frac{dT}{dx} \tag{7.137}$$

where x is the 1D heat flux direction with corresponding thermal conductivity.

7.5.2.2 The Heat Conservation Equation

The differential equation for porous medium heat flow is derived using the Laplace equation similar to deriving the soil water flow equation. A heat energy balance can be performed, for example, on the unit soil volume shown in Figure 7.38, where heat flowing from the sunny to the shady side of the soil ridge is depicted as soil heat flux entering from right and exiting to the left. Applying the conservation of heat principle in the form of the continuity equation and assuming no source or sink of heat, the PDE is written as

$$C\frac{\partial T}{\partial t} = -\nabla J_H \tag{7.138}$$

where C is the soil volumetric heat capacity (soil bulk density, ρ_b, times specific heat capacity per unit mass, c_m) and t is time. Combining Eqs. (7.136) and (7.138), the second law of heat conduction is written as

$$C\frac{\partial T}{\partial t} = -\nabla(\lambda\nabla T) \tag{7.139}$$

which can be written for the 1D case of Figure 7.38 as

$$C\frac{\partial T}{\partial t} = -\frac{\partial}{\partial x}\left(\lambda\frac{\partial T}{\partial x}\right) \tag{7.140}$$

Eq. (7.140) can be simplified because it contains two of the three basic soil thermal properties, namely, C and λ, and we can divide these two properties by C and substitute the third property of soil diffusivity, κ, using

$$\kappa = \frac{\lambda}{C} \tag{7.141}$$

Assuming κ is constant in space, the resulting 1D heat transport equation is written as

$$\frac{\partial T}{\partial t} = -\kappa\frac{\partial^2 T}{\partial x^2} \tag{7.142}$$

The three basic soil thermal properties need to be estimated or measured in order to obtain accurate spatial and temporal solutions of heat transport in porous media.

7.5.2.3 Analytical Solutions to Heat Transport

The diel cycling of heat within the Earth's near-surface layer can be estimated analytically using specified or measured boundary conditions. A sin function has often been applied with a temperature boundary to estimate the temperature as a function of soil depth using daily or annual cycling. The general relationship for a constant average daily or annual temperature is given by

$$T(t) = T_A + A\sin(\omega t + \varphi) \qquad (7.143)$$

where T_A is the average daily or annual temperature at an arbitrary depth in the soil, A is the amplitude of temperature fluctuation taken as half of the maximum minus minimum temperature, ω is the angular frequency, and φ is the phase shift to ensure maximum and minimum temperatures are correctly aligned. Whether daily or annual cycling of temperature, the frequency is given in reciprocal time (e.g. 1 Hz = 1 cycle per second) and the angular frequency is written in terms of the wavelength or period of the cycle (1 day, 1 year), given as

$$\omega = \frac{2\pi}{\tau} \qquad (7.144)$$

where τ is the time required to complete one cycle, e.g. 24 h or 365 days. A more comprehensive solution to the temperature fluctuation at any given depth in the soil was derived by Carslaw and Jaeger (1959) and requires calculation of the damping depth, d, computed from the water content-dependent soil thermal diffusivity as

$$d = \sqrt{\frac{\kappa\tau}{\pi}} = \sqrt{\frac{2\kappa}{\omega}} \qquad (7.145)$$

The resulting analytical solution is given as

$$T(z,t) = T_A + A\exp\left(\frac{z}{d}\right)\sin\left(\omega t + \varphi + \frac{z}{d}\right) \qquad (7.146)$$

where z is a negative number indicating the position from the soil surface boundary downward in the profile. Figure 7.39 illustrates annual cycles of temperature at the soil surface and at three subsurface depths demonstrating the phase lag [$\Delta t = z/(\omega d)$] in peak and minimum temperatures between the surface and subsurface. While there is a lag in the occurrence of min and max temperatures, wavelengths for any given depth are the same ($\tau = 2\pi/\omega$). The mean temperature is the same for all depths (10 °C in Figure 7.39), and the depth-dependent temperature amplitude is scaled based on the surface temperature amplitude (12 °C at $z = 0$) using the function "exp (z/d)."

7.5.3 Soil Thermal Properties

The three fundamental soil thermal properties include (i) soil heat capacity, which may be expressed as soil-specific heat capacity (cm; $kJ\,kg^{-1}\,K^{-1}$) or as soil volumetric heat capacity (C; $MJ\,m^{-3}\,K^{-1}$); (ii) soil thermal conductivity (λ; $W\,m^{-1}\,K^{-1}$); and (iii) soil thermal diffusivity (κ; $m^2\,s^{-1}$). Analogies of these properties to soil hydraulic properties are found in soil hydraulic conductivity, soil hydraulic diffusivity, and soil-specific water capacity (slope of water retention curve). Unlike soil hydraulic conductivity, which is largely a function of the soil porosity and tortuosity, soil thermal conductivity mainly depends on the thermal conductivity of the soil solids and on the water content. In general, porous medium thermal properties range over approximately an order of magnitude from air-dry to saturated soil with some properties being linear functions, while others are nonlinear functions of soil water content and constituent makeup.

7.5.3.1 Soil Heat Capacity

Heat storage in soil is described through the thermal property referred to as soil heat capacity, which may be expressed on a mass basis or on a volume basis. The soil heat capacity characterizes the quantity of heat added to

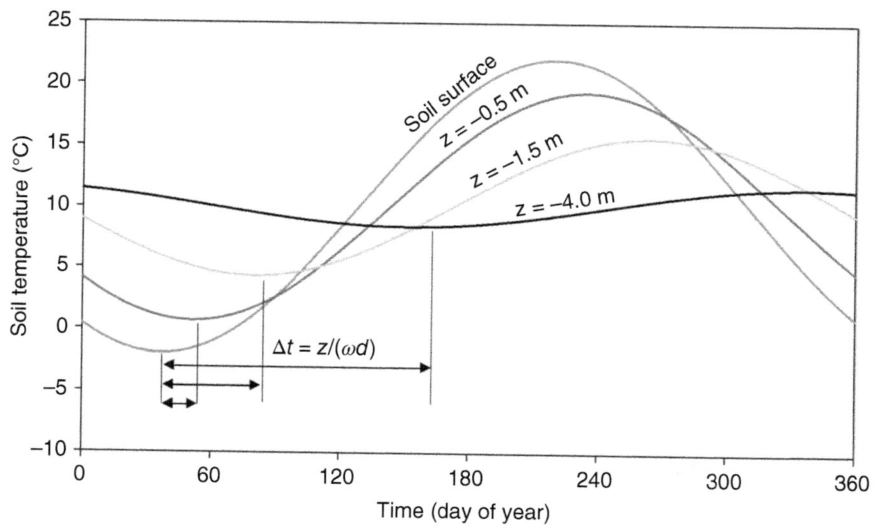

Figure 7.39 Analytically computed (Eq. 7.146) soil temperature at the surface and at −0.5, −1.5, and −4 m depths with mean annual temperature T_A = 10 °C, temperature amplitude, A = 12 °C, and damping depth d = 2 m. Note the phase or time lag [$\Delta t = z/(\omega d)$] from the surface minimum and maximum temperature increases with depth.

or removed from a unit volume of soil per degree Kelvin change in temperature. Soil with high heat capacity (e.g. wet soil) experiences lower temperature changes than soils with low heat capacity (e.g. dry soil) under similar thermal boundary conditions. Daily and annual soil temperature variations are therefore significantly influenced by the soil's heat capacity. The water content, bulk density, and constituent makeup of soil are major determinants for soil heat capacity as illustrated in Figure 7.40a. The relative proportions and individual heat capacities of constituents (i.e. mineral, water, ice, organic, and air constituents) dictate soil heat capacity. Soil constituent values of specific heat, cm, are listed in Table 7.10. These are multiplied by their corresponding densities, ρ, to obtain their volumetric heat capacity, C. While soil minerals and organic matter exhibit moderate volumetric heat capacities, water has nearly doubled the heat capacity of the solid phase and is the main variable affecting changes in C in large part because the volumetric heat capacity of air is negligible. A general volumetric mixing model written in terms of the density, volumetric fractions, and specific heat is used to estimate the volumetric heat capacity of soil, written as

$$C = \rho_m \theta_m c_m + \rho_w \theta_w c_w + \rho_a \theta_a c_a + \rho_{om} \theta_{om} c_{om} \quad (7.147)$$

where θ is the volume fraction of mineral, water, air, and organic matter phases denoted with subscripts m, w, a, and om, respectively. Representative soil constituent values from Table 7.10 or elsewhere may be used to estimate soil volumetric heat capacity. Additional details regarding C are available in Kluitenberg (2002). While the storage of heat is represented by the heat capacity, the rate of heat transport is described by the thermal conductivity term, which is also largely a function of soil water content.

7.5.3.2 Soil Thermal Conductivity

The thermal conductivity of soil (λ) is defined as the quantity of heat transported through a unit area over a unit of time (i.e. heat flux density) under the influence of a unit temperature gradient. As with heat capacity, λ is also largely a function of the soil bulk density and soil water content, both of which expand cross-sectional area for thermal transfer as they increase in volume. Soils containing significant fractions of quartz exhibit much higher thermal conductivity through the solid fraction, while λ for most other mineral constituents are about 4 times smaller and are 16 times smaller for water relative to quartz as illustrated in Table 7.11. Here the air phase is a significant barrier or insulator to heat transfer in soil with two orders of magnitude smaller λ than solid phases.

The bulk thermal conductivity of soil is described by the following generalized mixing model (Bristow, 2002):

$$\lambda = \frac{k_m \theta_m \lambda_m + k_w \theta_w \lambda_w + k_a \theta_a \lambda_a + k_{om} \theta_{om} \lambda_{om}}{k_m \theta_m + k_w \theta_w + k_a \theta_a + k_{om} \theta_{om}} \quad (7.148)$$

where k_i is the weighting factor depending on the shape of each constituent phase and θ and λ are volumetric fraction and thermal conductivity of each constituent

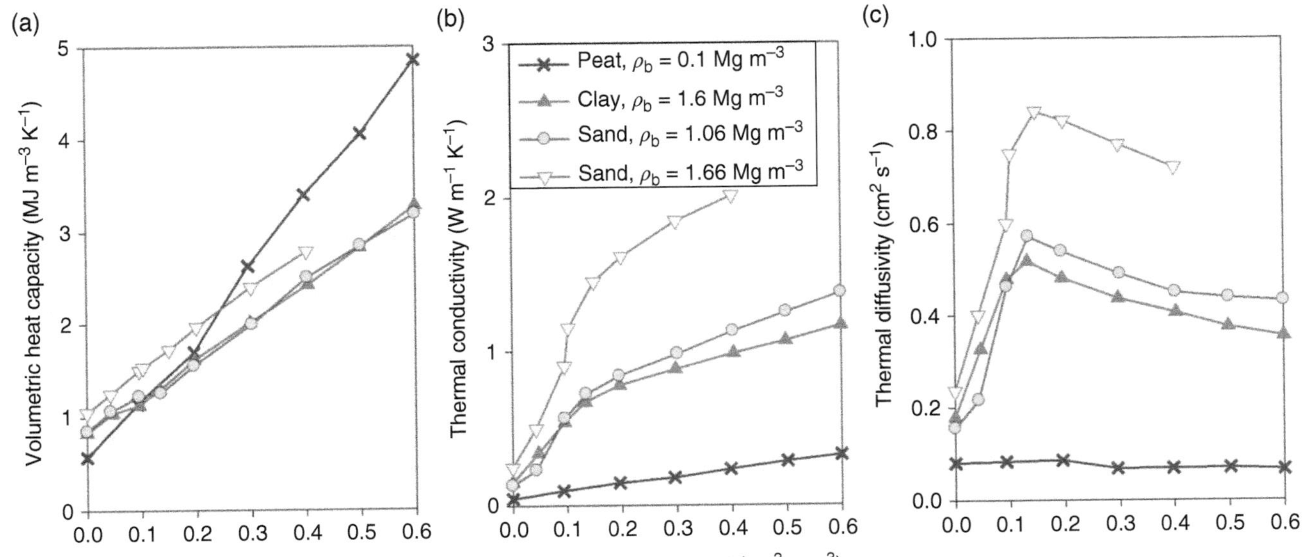

Figure 7.40 Thermal properties of sand, clay, and peat at various estimated bulk densities. (a) Volumetric heat capacity (C), (b) thermal conductivity (λ), and (c) thermal diffusivity (κ) of sand, clay, and peat at various estimated bulk densities (i.e. from porosities of sand-0.4, sand-0.6, clay-0.6, and peat-0.9 using mineral and organic matter densities of 2.65 and 1.0 Mg m^{-3}). Data from (b) and (c) were digitized from Jury and Horton (2004) after van Duin (1963), and values of C were approximated as λ/κ.

Table 7.10 Soil and constituent-specific heat, constituent density, and constituent volumetric heat capacity.

Soil constituent	Specific heat, c ($kJ\,kg^{-1}\,K^{-1}$)	Density, ρ ($Mg\,m^{-3}$)	Volume heat capacity, C ($MJ^{-3}\,K^{-1}$)
Soil minerals[a]	0.73	2.65	1.9
Soil organic matter[a]	1.9	1.3	2.5
Water[a]	4.18	1	4.18
Ice (0 °C)[a]	2	0.92	1.9
Air[a]	1	0.0012	0.0012
Quartz sand[b]	0.795–0.829	2.620[c]	2.084–2.172
Quartz powder[b]	0.791–0.875	2.620[c]	2.073–2.292
Kaolin[b]	0.975–1.021	2.600[c]	2.536–2.656
Feldspar (orthoclase)[b]	0.795–0.942	2.560[c]	2.036–2.411
Mica (muscovite)[b]	0.862–0.871	2.820[c]	2.432–2.455
Apatite[b]	0.766–0.921	3.190[c]	2.444–2.938
Dolomite[b]	0.929–0.963	2.840[c]	2.639–2.734
Al_2O_3[b]	0.870–0.940	4.000[d]	3.480–3.760
Fe_2O_3[b]	0.682–0.691	5.240[d]	3.575–3.619
Humus[b]	1.854–1.997	1.500[e]	2.782–2.995
Calcareous sand[b]	1.042		
Humus calcareous sand[b]	1.076		
Garden soil[b]	1.118		
Clay[b]	1.130		
Silty clay[b]	1.088		
Silt loam[b]	0.687–0.812		

[a] 20 °C and 0.101 MPa (after Van Wijk and deVries, 1963).
[b] Specific heat from table 5.5 in Jury and Horton (2004) who cited Bowers and Hanks (1962), Kersten (1949), Lang (1878), and Ulrich (1894).
[c] Mineral densities from engineeringtoolbox.com
[d] https://www.makeitfrom.com/
[e] Chesworth (2008).

Table 7.11 Soil mineral and soil constituent thermal conductivities.

Soil constituent	Thermal conductivity ($W\,m^{-1}\,K^{-1}$)
Quartz	8.8
Clay minerals	2.9
Basalt	2.2
Granite	2.0
Organic matter	0.25
Ice (0 °C)	2.18
Water	$0.552 + 2.34 \times 10^{-3}\,T - 1.10 \times 10^{-5}\,T^2$
Air	$0.0237 + 0.000064\,T$

Temperature (T)-dependent functions for water and air are in degrees Celsius.
After Buchan (2001), de Vries (1963), Hodgman et al. (1959), and Hopmans and Dane (1986).

phase. Also, mineral, water, air, and organic matter subscripts are defined as in Eq. (7.147). In addition to the physically based approach of Eq. (7.148), empirically derived estimates of thermal conductivity have been developed from measurements given in Campbell (1985):

$$\begin{aligned}
\lambda &= A + B\theta_w + (A - D)\exp\left[-(C\theta_w)^E\right] \\
A &= \frac{0.57 + 1.73\theta_q + 0.93\theta_m}{1 - 0.74\theta_q - 0.49\theta_m} - 2.8\theta_n(1 - \theta_n) \\
B &= 2.8\theta_n \\
C &= 1 + 2.6\theta_c^{-0.5} \\
D &= 0.03 + 0.7\theta_n^2 \\
E &= 4
\end{aligned} \tag{7.149}$$

where the subscripts w, n, q, c, and m refer to water, solid, quartz, clay, and mineral volume fractions.

Figure 7.41 illustrates the impact of soil bulk density on soil thermal conductivity as water saturation varies. Increases in both density and water content increase the surface areas through which heat can travel, or in other words, the thermal insulating capacity of air is reduced in both cases. Figure 7.40b demonstrates the impact of the solid constituency and solid volume fraction on soil thermal conductivity as a function of water content. There the solid constituent thermal conductivities presented in Table 7.11 show correlations to the resulting λ values at saturation in the figure. Bulk densities of 1.06 and 1.66 Mg m^{-3} for sand are shown to yield vastly different thermal conductivities as water content is increased. Furthermore, there is a change in the slope of the curve for the sand and clay soils between residual and field capacity water contents. This rapid increase in porous medium thermal conductivity with wetting from a dry state is caused by the transition from water films thickening on the solid surfaces to pendular water ring development between particles, which drastically increases the cross-sectional area for heat transport through the water phase with relatively small changes in volumetric water content.

7.5.3.3 Soil Thermal Diffusivity

Soil thermal diffusivity (κ) can be computed from the soil thermal conductivity divided by the soil heat capacity (Eq. (7.141)). Thermal diffusivity is not generally measured directly, but can be fitted to thermal measurements such as heat pulse sensor output when applied with

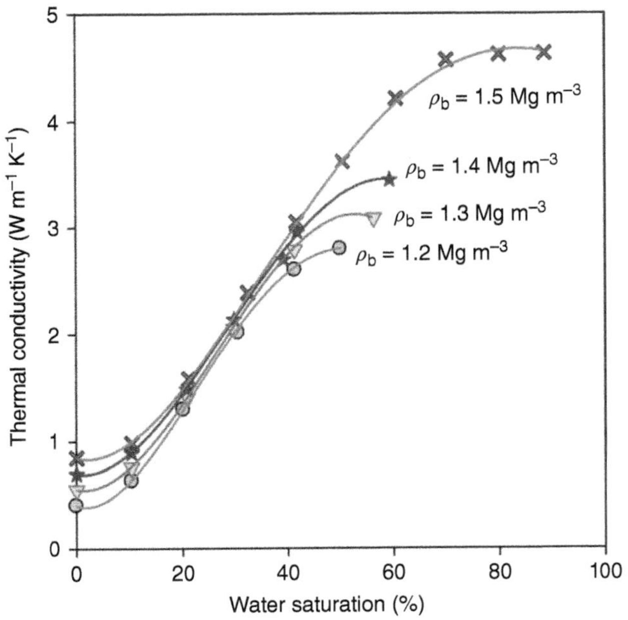

Figure 7.41 Saturation-dependent soil thermal conductivity as a function of soil bulk density. Data digitized from Jury and Horton (2004) after van Rooyen and Winterkorn (1959).

appropriate models such as the infinite line heat source theory (Kluitenberg et al., 1993; Bristow et al., 1994; Yang and Jones, 2009). Depending on the relationship between thermal conductivity and heat capacity, the thermal diffusivity may have nonunique values as a function of volumetric water content as illustrated in Figure 7.40c. The rapid rise and fall in thermal diffusivity illustrated in Figure 7.40c is explained by considering the shape of the thermal conductivity function (Figure 7.40b) relative to the linear volumetric heat capacity functions shown in Figure 7.40a. The rapid increase in thermal conductivity at low water contents and the subsequent reduction in the slope of λ with additional water content create a peak thermal diffusivity value at around one-quarter saturation in Figure 7.40c.

7.5.4 Heat Pulse Measurements

Several decades of research using heat pulse measurements have demonstrated a broad array of secondary soil properties or processes that can be inferred from the process of heat transport. These include determination of soil thermal properties and soil water content (Campbell et al., 1991; Bristow et al., 1993; Tarara and Ham, 1997; Ochsner et al., 2003; Ham and Benson, 2004). Heat pulse methods for measuring these physical properties employ a heat pulse applied to a line source with temperature increase about 6 mm from the source being monitored. A minimum of two parallel needles (Figure 7.42) are used for these measurements, one containing the heat source and the other containing a thermistor or thermocouple. With an additional thermistor or thermocouple needle (three-needle probe), soil water flux density can be determined (Ren et al., 2000; Ochsner et al., 2005; Kamai et al., 2008). These three-needle probes provide measurements of temperature rise both upstream and downstream from the heater needle.

7.5.4.1 Soil Thermal Property Fitting

Figure 7.43 illustrates thermal property determinations using the penta-needle heat pulse probe (PHPP) installed in a sand column under a four order of magnitude range of water flux density. Thermal properties were consistent for flow rates from around 5 up to around 5000 cm d^{-1}. Above that maximum flow rate, the increasing turbulence of water flow in the coarse sand likely alters the needle-to-needle heat transport and consequently adds increasing errors to the thermal property analysis. As of 2016 heat pulse probes were being used in research to infer soil water content (Campbell et al., 1991), soil heat flux density (Cobos and Baker, 2003), soil water flux density (Ochsner et al., 2003; Yang et al., 2013), and soil subsurface evaporation (Heitman et al., 2008a, b; Sakai et al., 2011) in addition

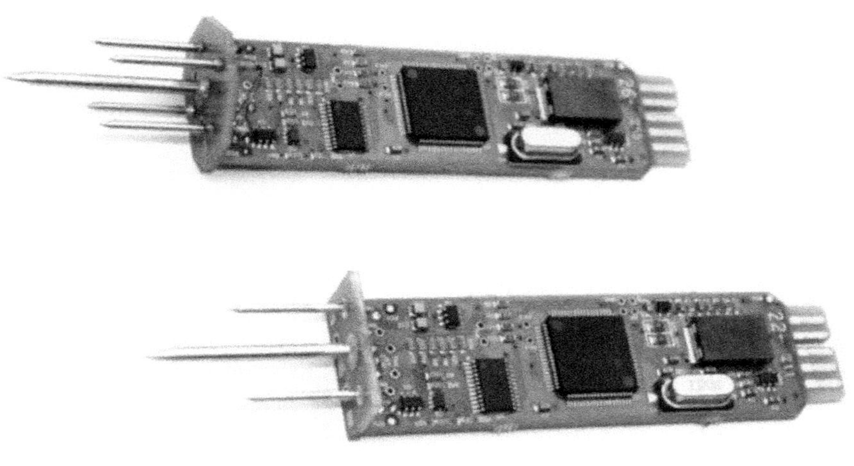

Figure 7.42 Penta- and tri-needle heat pulse probes developed at Utah State University employ onboard analysis of temperature rise data to optimize thermal diffusivity, thermal conductivity, and heat velocity between each thermistor needle and the central heater needle. *Source*: Yang et al. (2013). Reproduced with permission of John Wiley & Sons.

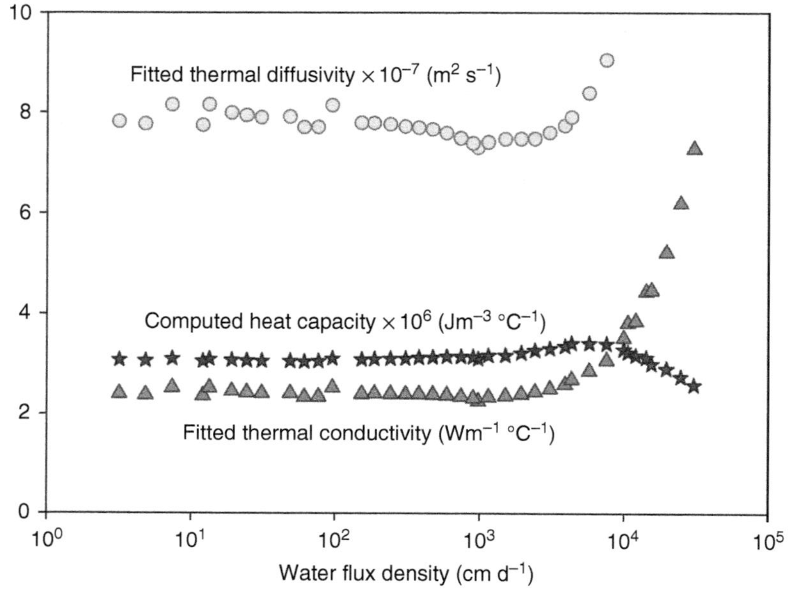

Figure 7.43 Averaged thermal properties either fitted or computed from triplicate PHPP measurements in a saturated coarse sand column with water flux density measured at the outlet. *Source*: Adapted from Yang et al. (2013). Reproduced with permission of John Wiley & Sons.

to simply estimating soil thermal properties. Beginning with the work of Campbell et al. (1991), expansion of the utility of heat pulse measurements has grown using 2-, 3- and 5-needle probes. Multifunctional probes have also been developed, which add soil EC (Mori et al., 2003) and soil dielectric (Ren et al., 1999; Ochsner et al., 2001; Sheng et al., 2016) measurements to the thermally based heat pulse measurements.

7.5.4.2 Soil Water Flux Density and Direction

Because three-needle heat pulse probes are only capable of quantifying the component of the soil water flux density in a direction defined by the orientation of the needles, estimating fluxes of arbitrary magnitude and direction require additional thermistors to resolve the components of water flux. To quantify flux magnitude and direction, Endo and Hara (2003) and a decade later Yang et al. (2013) developed analytical inverse methods to simultaneously estimate dual soil thermal properties in addition to the two components of the water flux density in the presence of flow. Figure 7.44 illustrates the concept of installing probes with needle orientation for three different PHPPs at different angles, namely, 15°, 30°, and 45°, relative to the direction of flow. The ability of the PHPP to estimate both water flux density magnitude and direction of water flow was demonstrated in this example. As seen, the range of water flux density is limited for both the low end at around $1\,\mathrm{cm\,d^{-1}}$, where there are signs of divergence, and the higher flow rate end, where again divergence of the data seems to occur at around $5000\,\mathrm{cm\,d^{-1}}$, likely due to transition from a laminar to turbulent hydraulic flow regime.

7.5.4.3 Soil Heat Balance and Subsurface Evaporation

Heat pulse probes are also being used to perform energy balance calculations in the near surface of soil. Heitman et al. (2008a, b) demonstrated sensible heat measurements

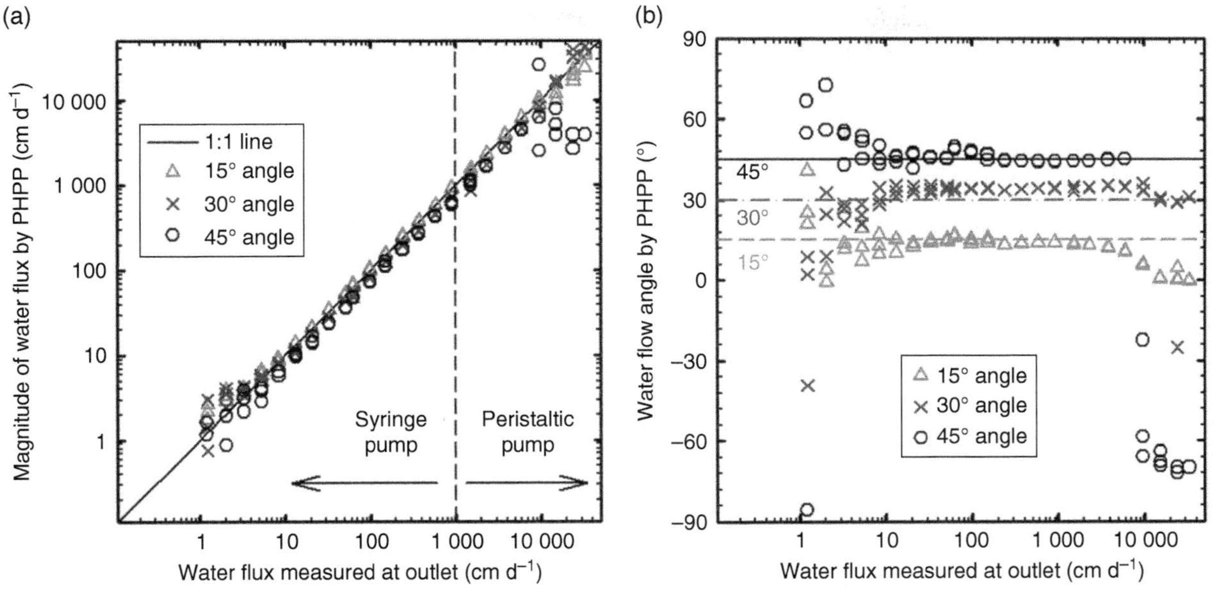

Figure 7.44 (a) Estimated magnitude of water flux density determined in a saturated sand column using three penta-needle heat pulse probes (PHPPs) (Figure 7.42). (b) Each PHPP yields flow direction estimates and were thus installed in the sand column with needles oriented at different directions relative to the direction of water flow, i.e. at 15°, 30°, and 45°. *Source*: From Yang et al. (2013). Reproduced with permission of John Wiley & Sons.

Figure 7.45 Numerically simulated subsurface evaporation for 3-mm soil layers beginning at 1.5 mm below the surface and extending to 19.5 mm. Each layer exhibits a normally distributed temporal contribution to the total subsurface evaporation computed by heat balance (dashed line) and water balance (dotted line) methods. Stage 1 evaporation and stage 2 between 0 and 1.5 mm are not detectable using the heat pulse probe method. *Source*: Sakai et al. (2011). Reproduced with permission of John Wiley & Sons.

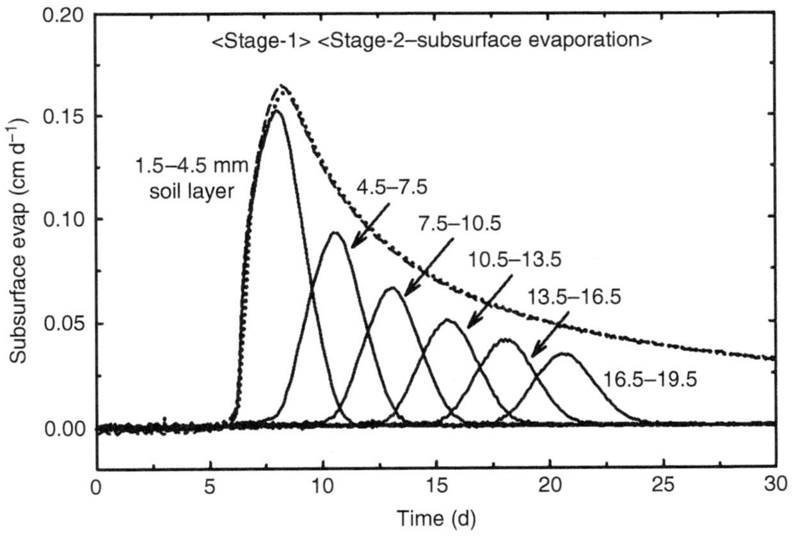

to infer depth and magnitude of subsurface soil water evaporation. Their calculations used three-needle heat pulse probe arrays in the soil profile. Each probe was used for determination of soil thermal conductivity, heat capacity, and thermal gradients in the soil profile to arrive at the sensible heat balance and subsurface evaporation estimate. This concept has sparked new opportunities to do energy and water balance estimation in the near surface of soil as evidenced by a number of studies since then (Xiao et al., 2011; Deol et al., 2012; Xiao et al., 2012; Trautz et al., 2014). Figure 7.45 shows numerical simulations of subsurface evaporation and the resulting calculations using both energy and water balance approaches. In addition, simulations and computed subsurface evaporation based on heat pulse probe measurement concept is shown for 3-mm soil layers that are integrated to arrive at the total subsurface evaporation values shown.

Soil heat flux is an important part of the energy balance comprised of net radiation, sensible heat, and latent heat in addition to soil heat flux (Heitman et al., 2007). The soil heat flux component is often challenging to determine since it requires buried sensors, and even then the measurements only represent a point in space where soil heat flux varies according to spatial variability in other environmental factors. Soil heat flux plates coupled with temperature arrays and water content sensors have traditionally been used to determine the soil heat flux (Sauer et al., 2007). The plates are generally buried between 5 and 10 cm deep with temperature and soil water content being determined between the surface and the plate depth in order to obtain an estimate of the heat storage, computed as

$$S = \frac{Cz\Delta T}{t} \tag{7.150}$$

where ΔT is the change in measured temperature over a time interval, t, and z is the heat flux plate burial depth. The term C is obtained from a combination of the specific heat of dry soil, c_d, and the specific heat of water, c_w, in the soil, given by

$$C = \rho_b c_d + \theta_v \rho_w c_w \tag{7.151}$$

where ρ_b is the bulk density of the soil, θ_v is the volumetric water content of the soil, and ρ_w is the density of water. The total soil heat flux, G, is the sum of S and the soil heat flux measured by the plate. A heat pulse probe can also be used to determine soil heat flux from the determination of the soil thermal conductivity and the soil temperature gradient, written as

$$G = \lambda \frac{\Delta T}{\Delta z} \tag{7.152}$$

In this case, if the heat pulse probe is placed near the surface, the storage term may be negligible, yielding the soil heat flux directly.

7.5.5 Temperature Effects on Biological Activity

Soils forming the Earth's crust contain at least three times as much organic carbon as the amount contained in the atmospheric carbon pool (Jobbágy and Jackson, 2000). Because of the global warming effect of increased CO_2 in the atmosphere, carbon sources and sinks have become a major focus of soil and atmospheric research. Soil temperature is among the most significant factors influencing the conversion of soil organic carbon to atmospheric CO_2 (Kirschbaum, 1995; Kätterer et al., 1998). Daily as well as annual variations in soil temperature (Figure 7.39) result in similar cycling in the magnitude of soil respiration where rates in summer may be an order of magnitude larger than those in winter. While all soils generally support some level of microbial activity and respiration, plant root respiration is limited to soils that support seed germination, plant growth, and maintenance. Plant roots and microbial respiration rates may be similarly impacted by variations in water, nutrients, and oxygen supply, and the presence of roots may double the respiration rate of soils compared with those without plants. Fierer et al. (2006) analyzed the temperature sensitivity of labile soil organic carbon decomposition from 70 US continental-scale samples classified into six different soil ecosystem types. That temperature dependence is illustrated in Figure 7.46 showing highest respiration rates in dry grassland and shrubland ecosystems compared with the lowest rates found in tropical forest and grassland systems. The mean respiration rate of all six ecosystems increased from about

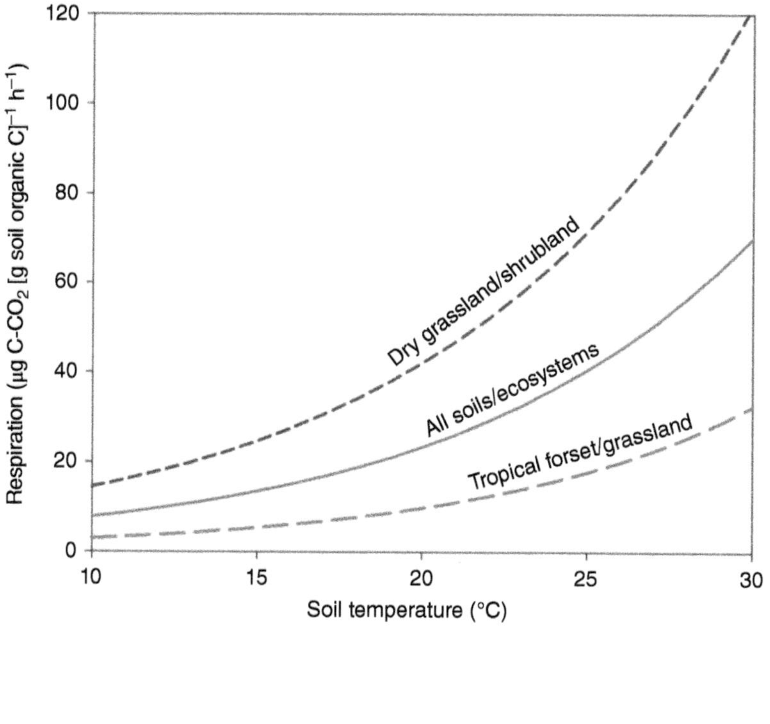

Figure 7.46 Temperature-dependent soil microbial respiration for high (dry grassland), low (tropical forest), and mean ecosystem rates. *Source:* Fierer et al. (2006). Reproduced with permission of John Wiley & Sons.

10 μg carbon as CO_2 per gram of soil organic carbon at 10 °C to a value of 70 at 30 °C.

Microorganisms thrive under different temperature regimes and in general are classified as psychrophiles, psychrotrophs, mesophiles, thermophiles, and hyperthermophiles. Each of these has a different temperature range for optimal activity as illustrated in Figure 7.47a where the minimum, maximum, and optimal temperature regimes are shown. Because of the broad distribution and heterogeneous mix of soil bacterial species, the optimal enzyme activity ranges in soil are difficult to pinpoint and generally vary both spatially and temporally. Figure 7.47b illustrates a hypothetical response for enzyme-controlled soil microbial activity where increasing temperature increases enzyme activity and respiration rate to an optimal value, which subsequently decreases with additional increase in temperature as a result of denaturation of the enzyme. The optimal respiration temperature illustrated in Figure 7.47b can vary considerably given the different ranges shown in Figure 7.47a combined with the likely broad distribution of bacteria types and conditions (i.e. nutrients, moisture, temperature, etc.) leading to numerous possible optimal temperature regimes in soils. The rate of increase in microbial respiration rate with temperature is another factor that provides insight into the CO_2 production potential of soils. The Q_{10} temperature coefficient is a measure of the increase in biological activity

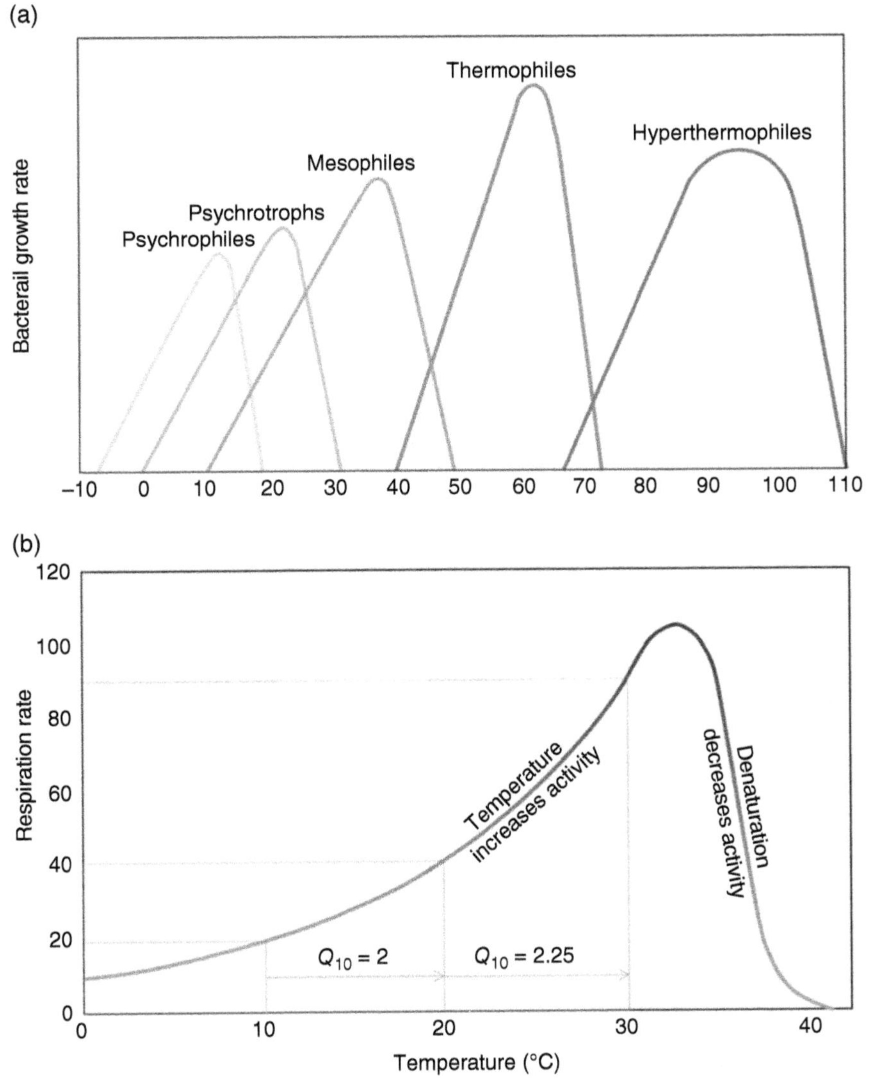

Figure 7.47 (a) Illustration of minimum, maximum, and optimal temperature for growth rates of bacteria within five different thermal classifications. (b) Enzyme-controlled microbial activity and resulting respiration rate as a function of enzyme activity that decreases rapidly above a given temperature threshold. Hypothetical values of Q_{10} are shown for temperature regime changes from 10 to 20 °C and from 20 to 30 °C using Eq. (7.153). *Source*: After Pommerville (2011).

(respiration rate here) with a 10 °C increase in temperature, computed as

$$Q_{10} = \left(\frac{R_2}{R_1}\right)^{10°/(T_2-T_1)} \qquad (7.153)$$

where R_1 and R_2 are the respiration rates at temperatures T_1 and T_2 (°C). The hypothetical temperature-dependent respiration function shown in Figure 7.47b provides an illustration of Q_{10} determination using Eq. (7.153) with temperature increases from 10 to 20 °C resulting in $Q_{10} = 2$ and from 20 to 30 °C giving $Q_{10} = 2.25$.

7.6 Summary

The physical properties and processes occurring within the Earth's thin surface layer are life sustaining and critical to maintenance of life on our planet. This chapter began by introducing basic soil properties that make up the texture and structure and how these hold water and maintain water's tug of war between the soil and the plant roots and microbial communities. The SWC is described as a key part of understanding soil hydrodynamic processes. The transport of water within soil is shown to be a function of the soil hydraulic conductivity, which typically occurs under unsaturated conditions. The saturated hydraulic conductivity, however, is easier to measure and serves as a key model parameter. Hydraulic transport is described using theories that attempt to explain saturated and unsaturated water flow in soil, and various measurement methods are presented. Because solutes (fertilizers, contaminants, etc.) are also transported with water in soil, we present concepts describing solute transport, retention, and release in soils. The CDE is presented and shown to account for key solute transport properties, such as the diffusion–dispersion transport term and the solute retardation coefficient. The CDE facilitates further explanation of solute transport using BTC. The use of various measurement instruments and sensors provide in situ determination of solute concentration change. Indirect measurements of soil solution EC are used to infer soil salinity when independent determination of in situ soil volumetric water content is also available. Soil temperature measurements and the importance of temperature on soil processes are illustrated. Modes of heat transport are described using various models, and soil thermal properties and their roles in heat transport are presented. Heat pulse probe measurements have received growing attention as tools to derive indirect estimates such as soil water content, soil thermal properties, soil water flux density, soil heat flux, soil heat balance, and subsurface evaporation, to name a few. The temperature of soil exerts significant control on the biological processes in soil and particularly on soil microorganisms, significantly influencing the types of microbes that are dominant in soils.

Acknowledgments

The authors gratefully acknowledge support from the Arizona and Utah Agricultural Experiment Stations.

Abbreviations

AASHTO	American Association of State Highway and Transportation Officials
ADE	Advection–dispersion equation
AirMOSS	Airborne Microwave Observatory of Subcanopy and Subsurface
ASTM	American Society for Testing and Materials
BC	Brooks–Corey
BCC	Bundle of cylindrical capillaries
BD	Buckingham–Darcy
BET	Brunauer, Emmett, and Teller
BTC	Breakthrough curve
EC	Electrical conductivity
EG	Ethylene glycol
EGME	Ethylene glycol monoethyl ether
EM	Electromagnetic
EMI	Electromagnetic induction
ER	Electrical resistivity
FDR	Frequency-domain reflectometry
IUSS	International Union of Soil Sciences
LD	Laser diffractometry
MB	Methylene blue
μCT	Micro-computed tomography
NIR	Near-infrared
NP	Neutron probe
NRCS	Natural Resources Conservation Service
ODE	Ordinary differential equation
PDE	Partial differential equation
PHPP	Penta-needle heat pulse probes
PIDS	Polarization intensity differential scattering
PSD	Particle size distribution
RE	Richards' equation
RH	Relative humidity
SCAN	Soil Climate Analysis Network
SHFs	Soil hydraulic functions
SM	Soil moisture
SNOTEL	Snow Telemetry
SSA	Specific surface area
SWC	Soil water characteristic

TDR	Time-domain reflectometry	USDA	U.S. Department of Agriculture
TDT	Time-domain transmission	USPRA	US Public Roads Administration
TLO	Transmission line oscillation	VG	van Genuchten
UNSODA	UNsaturated SOil hydraulic Database		

Physical Constants and Variables

Constant/variable	Symbol	Dimension	SI unit
Acceleration due to gravity	g	LT^{-2}	$m\,s^{-2}$
Adsorption	s	—	—
Advective solute flux density	j_{adv}	$ML^{-2}T^{-1}$	$kg\,m^{-2}\,s^{-1}$
Air-entry matric potential	ψ_b	—	kPa
Air-entry pressure head	h_b	—	kPa
Angular frequency	ω	T	s
Area	A	L^2	m^2
Avogadro's constant	N_A	—	mol^{-1}
Boltzmann constant	k	$ML^2T^{-2}K^{-1}$	$J\,K^{-1}$
Boltzmann transformation	ζ	—	—
Brooks–Corey function parameters	α, λ	—	—
Bulk diffusion coefficient	D_{dif}	L^2T^{-1}	$m^2\,s^{-1}$
Complementary error function	erfc	—	—
Concentration	C	ML^{-3}	$kg\,m^{-3}$
Damping depth	d	L	m
Darcian flux	q	LT^{-1}	$m\,s^{-1}$
Diffusion coefficient in water	D_0	L^2T^{-1}	$m^2\,s^{-1}$
Diffusive flux density of solutes	j_{dif}	$ML^{-2}T^{-1}$	$kg\,m^{-2}\,s^{-1}$
Dimensionless hydraulic conductivity	K^*	—	—
Discharge rate	Q	L^3	m^3
Distribution coefficient	k_d	L^3M^{-1}	$m^3\,kg^{-1}$
Dry bulk density	ρ_b	ML^{-3}	$kg\,m^{-3}$
Effective diffusivity	D_0	L^2T^{-1}	$m^2\,s^{-1}$
Electrical conductivity	EC	—	$dS\,m^{-1}$
Emissivity	ε	—	—
Empirical parameters in hydraulic functions	$p, \lambda, P, \alpha, n, A, l$	—	—
Equivalent depth of wetting	D_e	L	m
Field capacity	θ_{FC}	L^3L^{-3}	$m^3\,m^{-3}$
Final infiltration rate	i_f	LT^{-1}	$m\,s^{-1}$
Flux density	q	LT^{-1}	$m\,s^{-1}$
Gravimetric water content	θ_m	M^3M^{-3}	$kg\,kg^{-1}$
Gravimetric water content at 47% RH	$\theta_{m(47)}$	M^3M^{-3}	$kg\,kg^{-1}$
Gravitational potential	ψ_z	—	kPa
Hamaker constant	A_{svl}	ML^2	J

(*Continued*)

Constant/variable	Symbol	Dimension	SI unit
Heat flux	J_H	MT^{-3}	$W\,m^{-2}$
Heat storage	S	—	—
Hydraulic potential	H	L	m
Hydrodynamic dispersion flux	j_{hd}	$ML^{-2}T^{-1}$	$kg\,m^{-2}\,s^{-1}$
Ideal gas constant	R	—	$kPa\,m^3\,mol^{-1}\,K^{-1}$
Infiltration rate	i	LT^{-1}	$m\,s^{-1}$
Initial matric potential	h_i	L	m
Initial water content	θ_i	—	$m^3\,m^{-3}$
Length of TDR electrode	l	L	m
Longitudinal dispersivity	D_L	L	m
Mass of solid	M_s	M	kg
Mass of water	M_w	M	kg
Matric potential	ψ_m	—	kPa
Matric potential	ψ	—	$J\,kg^{-1}$
Matric potential of gauge	ψ_{gauge}	L	m
Matric potential profile	$h(z)$	—	—
Maximum adsorption	s_{max}	—	—
Maximum extent of the hydraulically connected region	D_{max}	L	m
Maximum pressure head	h_{max}	L	m
Maximum radiation intensity wavelength	λ_m	—	μm
Mechanical dispersion coefficient	D_{dis}	L^2T^{-1}	$m^2\,s^{-1}$
Molar mass of water	M_{wa} or M_w	—	$kg\,mol^{-1}$
Number of nodes	N	—	—
Outflow volume	Q	L^3	m^3
Particle density	ρ_s	ML^{-3}	$kg\,m^{-3}$
Peclet number	P_e	—	—
Permanent wilting point	θ_{PWP}	L^3L^{-3}	$m^3\,m^{-3}$
Planck's constant	h	ML^2T	J s
Pressure difference	ΔP	—	kPa
Pressure head	h	L	m
Pressure head at surface	h_{sur}	L	m
Pressure potential	ψ_p	—	kPa
Porosity	n	—	—
Radiant energy flux	E_λ	—	—
Radius	r	L	m
Relative humidity	RH	—	—
Relative saturation	S_e	—	—
Residual moisture content	θ_r	L^3L^{-3}	$m^3\,m^{-3}$
Relative permittivity	K_r	—	—
Retardation factor	R	—	—
Saturated hydraulic conductivity	K_s	LT^{-1}	$m\,s^{-1}$
Saturated vapor pressure	e_0	—	kPa
Saturated moisture content	θ_s	L^3L^{-3}	$m^3\,m^{-3}$
Saturation degree	S	—	—
Scaled depth	z^*	—	—
Scaled matric potential	h^*	—	—

Constant/variable	Symbol	Dimension	SI unit
Scaled time	t^*	—	—
Scaled water content	θ^*	—	—
Slope	S	—	—
Soil column height	L	L	m
Soil depth	z	L	m
Soil heat flux	G	MT^{-3}	$W m^{-2}$
Soil water diffusivity	D	L^2T^{-1}	$m^2 s^{-1}$
Solute or osmotic potential	ψ_s	—	kPa
Solute concentration	C_s	—	$mol m^{-3}$
Solute concentration	c	ML^{-3}	$kg m^{-3}$
Sorptivity	s	—	$m s^{-0.5}$
Specific heat	cm	—	$kJ kg^{-1} K^{-1}$
Specific heat of dry soil	c_d	—	$kJ kg^{-1} K^{-1}$
Specific heat of water	c_w	—	$kJ kg^{-1} K^{-1}$
Speed of light	c	LT^{-1}	$m s^{-1}$
Standard deviation	σ	—	—
Stefan–Boltzmann constant	σ	$MT^{-3}K^{-4}$	$W m^{-2} K^{-4}$
Settling depth	h^*	L	m
Surface area	A	L^2M^{-1}	$m^2 kg^{-1}$
Temperature	T	K	K
Temperature change	ΔT	K	K
Time	t	T	s
Time period	τ	—	—
Thermal conductivity	λ	—	$W m^{-1} K^{-1}$
Thermal diffusivity	κ	L^2T^{-1}	$m^2 s^{-1}$
Total outflow volume	Q_0	L^3	m^3
Total radiant energy of the emitter	J_r	—	—
Total soil water potential	ψ_T	—	kPa
Total volume	V_t	L^3	m^{-3}
Unsaturated hydraulic conductivity	K	LT^{-1}	$m s^{-1}$
van Genuchten function parameters	α, n, m	—	
Velocity of the water content wave	v	LT^{-1}	$m s^{-1}$
Void ratio	e	—	—
Volumetric water content	θ_v	L^3L^{-3}	$m^3 m^{-3}$
Volumetric heat capacity	C	—	$MJ m^{-3} K^{-1}$
Volumes of air	V_a	L^3	m^{-3}
Volume of fluids	V_f	L^3	m^{-3}
Volumes of solid	V_s	L^3	m^{-3}
Volumes of water	V_w	L^3	m^{-3}
Water density	ρ_w	ML^{-3}	$kg m^{-3}$
Water flux density	q	LT^{-1}	$m s^{-1}$
Water table depth	D	L	m
Water vapor pressure	e	—	kPa
Wet zone water content	θ_0	L^3L^{-3}	$m^3 m^{-3}$
Wetting front depth	z_f	L	m
Wetting front suction head	h_f	L	m
Water flow velocity	v	LT^{-1}	ms^{-1}

References

Ackler, H.D., French, R.H., and Chiang, Y.M. (1996). Comparisons of Hamaker constants for ceramic systems with intervening vacuum or water: from force laws and physical properties. *Journal of Colloid and Interface Science* 179 (2): 460–469.

Adamson, A.W. and Gast, A.P. (1997). *Physical Chemistry of Surfaces*, 6e. New York: Wiley. ISBN: 978-0-471-14873-9.

Agrawal, Y.C., McCave, I.N., and Riley, J.B. (1991). Laser diffraction size analysis. In: *Principles, Methods, and Application of Particle Size Analysis*, 119–128. New York: Cambridge University Press.

Allen, T. (1997). *Particle Size Measurement, Vol. 1. Powder Sampling and Particle Size Measurement*. New York: Springer.

Amakor, X.N. (2013). Using electromagnetic induction sensing to understand the dynamics and interacting factors controlling soil salinity. Ph.D. Dissertation. Utah State University, Logan, Utah.

Amali, S., Petersen, L.W., and Rolston, D.E. (1994). Modeling multicomponent volatile organic and water vapor adsorption on soils. *Journal of Hazardous Materials* 36 (1): 89–108.

Andraski, B.J. and Scanlon, B.R. (2002). Thermocouple psychrometry. In: *Methods of Soil Analysis: Part 4 Physical Methods*, 609–642. Madison, WI: Soil Science Society of America.

Andrenelli, M.C., Fiori, V., and Pellegrini, S. (2013). Soil particle-size analysis up to 250 μm by X-ray granulometer: device set-up and regressions for data conversion into pipette-equivalent values. *Geoderma* 192: 380–393.

Arthur, E., Tuller, M., Moldrup, P. et al. (2013). Soil specific surface area and non-singularity of soil-water retention at low saturations. *Soil Science Society of America Journal* 77 (1): 43–53.

Arthur, E., Tuller, M., Moldrup, P. et al. (2015). Prediction of clay content from water vapour sorption isotherms considering hysteresis and soil organic matter content. *European Journal of Soil Science* 66 (1): 206–217.

ASTM-D2487-11 (2011). *Standard Practice for Classification of Soils for Engineering Purposes (Unified Soil Classification System)*. West Conshohocken, PA: ASTM International.

ASTM-D4318-05 (2005). *Standard Test Methods for Liquid Limit, Plastic Limit, and Plasticity Index of Soils*. West Conshohocken, PA: ASTM International.

Atkin, O.K., Edwards, E.J., and Loveys, B.R. (2000). Response of root respiration to changes in temperature and its relevance to global warming. *The New Phytologist* 147 (1): 141–154.

Bachmann, J. and van der Ploeg, R.R. (2002). A review on recent developments in soil water retention theory: interfacial tension and temperature effects. *Journal of Plant Nutrition and Soil Science* 165 (4): 468–478.

Beckman Coulter Inc. (2003). *LS 13 320 Laser Diffraction Particle Size Analyzer Instrument Manual*. Miami, FL: Beckman Coulter Inc.

Benson, C. and Gribb, M. (1997). Measuring unsaturated hydraulic conductivity in the laboratory and the field. *Geotechnical Special Publication* 68: 113–168.

Bergström, L. (1997). Hamaker constants of inorganic materials. *Advances in Colloid and Interface Science* 70: 125–169.

Beuselinck, L., Govers, G., Poesen, J. et al. (1998). Grain-size analysis by laser diffractometry: comparison with the sieve-pipette method. *Catena* 32 (3): 193–208.

Beven, K. and Germann, P. (2013). Macropores and water flow in soils revisited. *Water Resources Research* 49 (6): 3071–3092.

Blonquist, J.M., Jones, S.B., and Robinson, D.A. (2005). Standardizing characterization of electromagnetic water content sensors. *Vadose Zone Journal* 4 (4): 1059–1069.

Blott, S.J. and Pye, K. (2006). Particle size distribution analysis of sand-sized particles by laser diffraction: an experimental investigation of instrument sensitivity and the effects of particle shape. *Sedimentology* 53 (3): 671–685.

Blott, S.J. and Pye, K. (2012). Particle size scales and classification of sediment types based on particle size distributions: review and recommended procedures. *Sedimentology* 59 (7): 2071–2096.

Blum, W.E.H., Warkentin, B.P., and Frossard, E. (2006). Soil, human society and the environment. *Geological Society, London, Special Publications* 266 (1): 1–8.

Bogena, H.R., Huisman, J.A., Oberdörster, C., and Vereecken, H. (2007). Evaluation of a low-cost soil water content sensor for wireless network applications. *Journal of Hydrology* 344 (1): 32–42.

Borkovec, M., Wu, Q., Degovics, G. et al. (1993). Surface area and size distributions of soil particles. *Colloids and Surfaces A: Physicochemical and Engineering Aspects* 73: 65–76.

Bouksila, F., Persson, M., Berndtsson, R., and Bahri, A. (2008). Soil water content and salinity determination using different dielectric methods in saline gypsiferous soil. *Hydrological Sciences Journal* 53 (1): 253–265.

Bowers, S.A. and Hanks, R.J. (1962). Specific heat capacity of soils and minerals as determined with a radiation calorimeter. *Soil Science* 94 (6): 392–396.

Bresler, E. (1973). Simultaneous transport of solutes and water under transient unsaturated flow conditions. *Water Resources Research* 9 (4): 975–986.

Bristow, K.L. (2002). Thermal conductivity. In: *Methods of Soil Analysis: Part 4 Physical Methods* (ed. C.G. Topp and J. Dane), 1209–1226. Madison, WI: Soil Science Society of America.

Bristow, K.L., Campbell, G.S., and Calissendorff, K. (1993). Test of a heat-pulse probe for measuring changes in soil water content. *Soil Science Society of America Journal* 57 (4): 930–934.

Bristow, K.L., Kluitenberg, G.J., and Horton, R. (1994). Measurement of soil thermal properties with a dual-probe heat-pulse technique. *Soil Science Society of America Journal* 58 (5): 1288–1294.

Broadbridge, P. and White, I. (1988). Constant rate rainfall infiltration: a versatile nonlinear model: 1. Analytic solution. *Water Resources Research* 24 (1): 145–154.

Brooks, R.H. and Corey, A.T. (1964). *Hydraulic Properties of Porous Media*. Hydrology Papers. Fort Collins, CO: Colorado State University.

Bruce, R.R. and Klute, A. (1956). The measurement of soil moisture diffusivity. *Soil Science Society of America Journal* 20 (4): 458–462.

Brunauer, S., Emmett, P.H., and Teller, E. (1938). Adsorption of gases in multimolecular layers. *Journal of the American Chemical Society* 60 (2): 309–319.

Brutsaert, W. (1968). The adaptability of an exact solution to horizontal infiltration. *Water Resources Research* 4 (4): 785–789.

Brutsaert, W. (1976). The concise formulation of diffusive sorption of water in a dry soil. *Water Resources Research* 12 (6): 1118–1124.

Buchan, G.D. (2001). Soil temperature regime. In: *Soil and Environmental Analysis: Physical Methods* (ed. K.A. Smith and C.R. Mullins), 539–594. New York: Marcel Dekker.

Buchan, G.D., Grewal, K.S., Claydon, J.J., and McPherson, R.J. (1993). A comparison of sedigraph and pipette methods for soil particle-size analysis. *Soil Research* 31 (4): 407–417.

Buckingham, E. (1907). *Studies on the Movement of Soil Moisture*. Bulletin 38. Washington, DC: USDA Bureau of Soils.

Campbell, G.S. (1985). *Soil Physics with BASIC*. New York: Elsevier.

Campbell, G.S., Calissendorff, C., and Williams, J.H. (1991). Probe for measuring soil specific heat using a heat-pulse method. *Soil Science Society of America Journal* 55 (1): 291–293.

Caputo, J.G. and Stepanyants, Y.A. (2008). Front solutions of Richards' equation. *Transport in Porous Media* 74 (1): 1–20.

Carsel, R.F. and Parrish, R.S. (1988). Developing joint probability distributions of soil water retention characteristics. *Water Resources Research* 24 (5): 755–769.

Carslaw, H.S. and Jaeger, J.C. (1959). *Heat in Solids*, vol. 1. Oxford: Clarendon Press.

Carter, D.L., Mortland, M.M., and Kemper, W.D. (1986). Specific surface. In: *Methods of Soil Analysis. Agronomy No. 9, Part 1. Physical and Mineralogical Methods*, 2e (ed. A. Klute), 413–423. Madison, WI: ASA, SSSA Publishers.

Celia, M.A., Bouloutas, E.T., and Zarba, R.L. (1990). A general mass-conservative numerical solution for the unsaturated flow equation. *Water Resources Research* 26 (7): 1483–1496.

Chahal, R.S. and Young, R.N. (1965). Validity of soil characteristics with the pressure apparatus. *Soil Science* 99: 98–103.

Chen, C., Ren, T., Hu, K. et al. (2014). Estimation of soil clay content using hygroscopic water content at an arbitrary humidity. *Soil Science Society of America Journal* 78 (1): 119–124.

Chen, J.M., Tan, Y.C., Chen, C.H., and Parlange, J.Y. (2001). Analytical solutions for linearized Richards equation with arbitrary time-dependent surface fluxes. *Water Resources Research* 37 (4): 1091–1093.

Chesworth, W. (2008). *Encyclopedia of Soil Science*. Dordrecht: Springer.

Chou, H., Wu, L., Zeng, L., and Chang, A. (2012). Evaluation of solute diffusion tortuosity factor models for variously saturated soils. *Water Resources Research* 48 (10). doi: 10.1029/2011WR011653.

Clauer, N. (2005). On the role of clay minerals in the disposal of radioactive waste in a clay geological formation. *Actualite Chimique* 285–286: 93–98.

Coates, G.F. and Hulse, C.A. (1985). A comparison of four methods of size analysis of fine-grained sediments. *New Zealand Journal of Geology and Geophysics* 28 (2): 369–380.

Cobos, D.R. and Baker, J.M. (2003). In situ measurement of soil heat flux with the gradient method. *Vadose Zone Journal* 2 (4): 589–594.

Comegna, V., Coppola, A., Basile, A., and Comegna, A. (2012). A review of approaches for measuring soil hydraulic properties and assessing the impacts of spatial dependence on the results. In: *Hydrogeology – a Global Perspective*. Rijeka, Croatia: InTech.

Cooper, J.D. (2016). Solute concentration estimation by electrical conductivity. In: *Soil Water Measurement: A Practical Handbook*, 292–300. Hoboken, NJ: Wiley.

Corwin, D.L. and Hendrickx, J.M.H. (2002). Solute content and concentration – indirect measurement of solute concentration – electrical resistivity: Wenner array. In: *Methods of Soil Analysis, Part 4: Physical Methods* (ed. J.H. Dane and G.C. Topp), 1282–1287. Madison, WI: Soil Science Society of America.

Corwin, D.L. and Lesch, S.M. (2005). Apparent soil electrical conductivity measurements in agriculture. *Computers and Electronics in Agriculture* 46 (1): 11–43.

Crank, J. (1975). *The Mathematics of Diffusion*, 2e. Oxford: Clarendon Press.

Davidson, J.M., Stone, L.R., Nielsen, D.R., and Larue, M.E. (1969). Field measurement and use of soil-water properties. *Water Resources Research* 5 (6): 1312–1321.

De Jong, E., Ballantyne, A.K., Cameron, D.R., and Read, D.W.L. (1979). Measurement of apparent electrical conductivity of soils by an electromagnetic induction probe to aid salinity surveys. *Soil Science Society of America Journal* 43 (4): 810–812.

De Smedt, F. and Wierenga, P.J. (1978). Approximate analytical solution for solute flow during infiltration and redistribution. *Soil Science Society of America Journal* 42 (3): 407–412.

Deol, P., Heitman, J., Amoozegar, A. et al. (2012). Quantifying nonisothermal subsurface soil water evaporation. *Water Resources Research* 48 (11). doi: 10.1029/2012WR012516.

DeVries, D.A. (1963). Thermal properties of soils. In: *Physics of Plant Environment* (ed. W.R. van Wijk), 210–235. Amsterdam: North-Holland.

Diamond, S. and Kinter, E.B. (1958). Surface areas of clay minerals as derived from measurements of glycerol retention. *Clays and Clay Minerals* 5: 334–347.

Dobson, M.C., Ulaby, F.T., Hallikainen, M.T., and El-Rayes, M.A. (1985). Microwave dielectric behavior of wet soil – part II: dielectric mixing models. *IEEE Transactions on Geoscience and Remote Sensing* 1: 35–46.

Doering, E.J. (1965). Soil water diffusivity by the one step method. *Soil Science* 99 (5): 322–326.

Dyal, R.S. and Hendricks, S.B. (1950). Total surface of clays in polar liquids as a characteristic index. *Soil Science* 69 (6): 503–509.

Edlefsen, N. and Anderson, A. (1943). Thermodynamics of soil moisture. *California Agriculture* 15 (2): 31–298.

Endo, A. and Hara, M. (2003). Simultaneous measurement of thermal front advection velocity vector and thermal properties of sandy soil under two-dimensional flow field with Quintuple-Probe Heat-Pulse technique and its application. *Transactions of the Society of Instrument and Control Engineers* 2: 88–95.

Evett, S., Laurent, J.P., Cepuder, P., and Hignett, C. (2002). Neutron scattering, capacitance, and TDR soil water content measurements compared on four continents. *17th World Congress of Soil Science* (August), pp. 14–21.

Fedotov, G.N., Shein, E.V., Putlyaev, V.I. et al. (2007). Physicochemical bases of differences between the sedimentometric and laser-diffraction techniques of soil particle-size analysis. *Eurasian Soil Science* 40 (3): 281–288.

Fierer, N., Colman, B.P., Schimel, J.P., and Jackson, R.B. (2006). Predicting the temperature dependence of microbial respiration in soil: a continental-scale analysis. *Global Biogeochemical Cycles* 20 (3).

Flint, A. and Flint, L. (2002). Particle density. In: *Methods of Soil Analysis, Part 4: Physical Methods* (ed. J.H. Dane and G.C. Topp), 241–253. Madison, WI: Soil Science Society of America.

Fuchs, M. and Tanner, C.B. (1968). Surface temperature measurements of bare soils. *Journal of Applied Meteorology* 7 (2): 303–305.

Gardner, W.R. (1956). Calculation of capillary conductivity from pressure plate outflow data. *Soil Science Society of America Journal* 20 (3): 317–320.

Gardner, W.R. (1958). Some steady-state solutions of the unsaturated moisture flow equation with application to evaporation from a water table. *Soil Science* 85 (4): 228–232.

Gardner, W.R. (1962). Note on the separation and solution of diffusion type equations. *Soil Science Society of America Journal* 26 (4): 404–404.

Gee, G.W., Campbell, M.D., Campbell, G.S., and Campbell, J.H. (1992). Rapid measurement of low soil water potentials using a water activity meter. *Soil Science Society of America Journal* 56 (4): 1068–1070.

Gee, G.W. and Or, D. (2002). Particle-size analysis. In: *Methods of Soil Analysis, Part 4: Physical Methods* (ed. J.H. Dane and G.C. Topp), 255–293. Madison, WI: Soil Science Society of America.

Gelhar, L.W., Welty, C., and Rehfeldt, K.R. (1992). A critical review of data on field-scale dispersion in aquifers. *Water Resources Research* 28 (7): 1955–1974.

Gerke, H.H. (2006). Preferential flow descriptions for structured soils. *Journal of Plant Nutrition and Soil Science* 169 (3): 382–400.

Ghanbarian, B., Hunt, A.G., Ewing, R.P., and Sahimi, M. (2013). Tortuosity in porous media: a critical review. *Soil Science Society of America Journal* 77 (5): 1461–1477.

Gohardoust, M.R., Sadeghi, M., Ahmadi, M.Z. et al. (2017). Hydraulic conductivity of stratified unsaturated soils: effects of random variability and layering. *Journal of Hydrology* 546: 81–89.

Grambow, B. (2016). Geological disposal of radioactive waste in clay. *Elements* 12: 239–245.

Green, W.H. and Ampt, G.A. (1911). Studies on soil physics. *The Journal of Agricultural Science* 4 (1): 1–24.

Groenendyk, D.G., Ferré, T.P., Thorp, K.R., and Rice, A.K. (2015). Hydrologic-process-based soil texture classifications for improved visualization of landscape function. *PLoS One* 10 (6): e0131299.

Grossman, R.B. and Reinsch, T.G. (2002). Bulk density and linear extensibility. In: *Methods of Soil Analysis: Part 4 Physical Methods*, 201–228. Madison, WI: Soil Science Society of America.

Gvirtzman, H. and Gorelick, S.M. (1991). Dispersion and advection in unsaturated porous media enhanced by anion exclusion. *Nature* 352 (6338): 793.

Haines, W.B. (1930). Studies in the physical properties of soil. V. The hysteresis effect in capillary properties, and the modes of moisture distribution associated therewith. *The Journal of Agricultural Science* 20 (1): 97–116.

Ham, J.M. and Benson, E.J. (2004). On the construction and calibration of dual-probe heat capacity sensors. *Soil Science Society of America Journal* 68 (4): 1185–1190.

Hang, P.T. and Brindley, G.W. (1970). Methylene blue absorption by clay minerals. Determination of surface areas and cation exchange capacities (clay-organic studies XVIII). *Clays and Clay Minerals* 18 (4): 203–212.

Hanson, B. and Dickey, G. (1993). When measuring soil water content, field practices affect neutron moisture meter accuracy. *California Agriculture* 47 (6): 29–31.

Haverkamp, R., Vauclin, M., Touma, J. et al. (1977). A comparison of numerical simulation models for one-dimensional infiltration. *Soil Science Society of America Journal* 41 (2): 285–294.

Hayek, M. (2014). On front solutions of the saturation equation of two-phase flow in porous media. *Applied Mathematical Modelling* 38 (19): 4694–4704.

Hayek, M. (2016). Analytical solution to transient Richards' equation with realistic water profiles for vertical infiltration and parameter estimation. *Water Resources Research* 52 (6): 4438–4457.

Heitman, J.L., Horton, R., Ren, T., and Ochsner, T.E. (2007). An improved approach for measurement of coupled heat and water transfer in soil cells. *Soil Science Society of America Journal* 71 (3): 872–880.

Heitman, J.L., Horton, R., Sauer, T.J., and DeSutter, T.M. (2008a). Sensible heat observations reveal soil-water evaporation dynamics. *Journal of Hydrometeorology* 9 (1): 165–171.

Heitman, J.L., Xiao, X., Horton, R., and Sauer, T.J. (2008b). Sensible heat measurements indicating depth and magnitude of subsurface soil water evaporation. *Water Resources Research* 44 (4).

Hillel, D., Krentos, V.D., and Stylianou, Y. (1972). Procedure and test of an internal drainage method for measuring soil hydraulic characteristics in situ. *Soil Science* 114 (5): 395–400.

Hodgman, C.D., Weast, R.C., and Shelby, S.M. (1959). *Handbook of Chemistry and Physics*, 41e. Cleveland, OH: Chemical Rubber Publishing Co.

Holmes, J.W. (1956). Calibration and field use of the neutron scattering method of measuring soil water content. *Australian Journal of Applied Science* 7: 45–58.

Hopmans, J.W. and Dane, J.H. (1986). Thermal conductivity of two porous media as a function of water content, temperature and density. *Soil Science* 142: 187–195.

Hopmans, J.W., Šimůnek, J., Romano, N., and Durner, W. (2002). Inverse methods. In: *Methods of Soil Analysis: Part 4 Physical Methods*, 963–1008. Madison, WI: Soil Science Society of America.

Horton, R. and Ochsner, T. (2011). Soil thermal regime. In: *Handbook of Soil Sciences: Resource Management and Environmental Impacts* (ed. P.M. Huang, Y. Li and M.E. Sumner). Boca Raton, FL: CRC Press.

Huang, K. and Van Genuchten, M.T. (1995). An analytical solution for predicting solute transport during ponded infiltration. *Soil Science* 159 (4): 217–223.

Hupet, F., Trought, M.C.T., Greven, M. et al. (2005). Data requirements for identifying macroscopic water stress parameters: a study on grapevines. *Water Resources Research* 41 (6).

Jobbágy, E.G. and Jackson, R.B. (2000). The vertical distribution of soil organic carbon and its relation to climate and vegetation. *Ecological Applications* 10 (2): 423–436.

Jones, S.B., Blonquist, J.M., Robinson, D.A. et al. (2005). Standardizing characterization of electromagnetic water content sensors. *Vadose Zone Journal* 4 (4): 1048–1058.

Jones, S.B., Wraith, J.M., and Or, D. (2002). Time domain reflectometry measurement principles and applications. *Hydrological Processes* 16 (1): 141–153.

Jury, W.A., Gardner, W.R., and Gardner, W.H. (1991). *Soil Physics*, 5e. New York: Wiley.

Jury, W.A. and Horton, R. (2004). *Soil Physics*. New York: Wiley.

Kabat, P., Beekma, J., and Ritzema, H. (1994). Water in the unsaturated zone. In: *Drainage Principles and Applications*, vol. 16, 383–434. Wageningen, the Netherlands: ILRI Publication.

Kalma, J.D., McVicar, T.R., and McCabe, M.F. (2008). Estimating land surface evaporation: a review of methods using remotely sensed surface temperature data. *Surveys in Geophysics* 29 (4–5): 421–469.

Kamai, T., Tuli, A., Kluitenberg, G.J., and Hopmans, J.W. (2008). Soil water flux density measurements near 1 cm d^{-1} using an improved heat pulse probe design. *Water Resources Research* 44 (4).

Kätterer, T., Reichstein, M., Andrén, O., and Lomander, A. (1998). Temperature dependence of organic matter decomposition: a critical review using literature data analyzed with different models. *Biology and Fertility of Soils* 27 (3): 258–262.

Kelleners, T.J., Seyfried, M.S., Blonquist, J.M. et al. (2005). Improved interpretation of water content reflectometer measurements in soils. *Soil Science Society of America Journal* 69 (6): 1684–1690.

Kersten, M.S. (1949). *Thermal Properties of Soils*. University of Minnesota. http://conservancy.umn.edu/bitstream/handle/11299/124271/eng_ex_bulletin_28.pdf?sequence=1 (accessed 15 February 2018).

Kinzelbach, W. (1986). *Groundwater Modelling: An Introduction with Sample Programs in BASIC*, vol. 25. Amsterdam: Elsevier.

Kirkham, D. and Feng, C.L. (1949). Some tests of the diffusion theory, and laws of capillary flow, in soils. *Soil Science* 67 (1): 29–40.

Kirschbaum, M.U. (1995). The temperature dependence of soil organic matter decomposition, and the effect of global warming on soil organic C storage. *Soil Biology and Biochemistry* 27 (6): 753–760.

Kluitenberg, G.J. (2002). Heat capacity and specific heat. In: *Methods of Soil Analysis: Part 4 Physical Methods*, 1201–1208. Madison, WI: Soil Science Society of America.

Kluitenberg, G.J., Ham, J.M., and Bristow, K.L. (1993). Error analysis of the heat pulse method for measuring soil volumetric heat capacity. *Soil Science Society of America Journal* 57 (6): 1444–1451.

Köhne, J.M., Köhne, S., and Šimůnek, J. (2009). A review of model applications for structured soils: a water flow and tracer transport. *Journal of Contaminant Hydrology* 104 (1): 4–35.

Kool, J.B. and Parker, J.C. (1987). Development and evaluation of closed-form expressions for hysteretic soil hydraulic properties. *Water Resources Research* 23 (1): 105–114.

Kool, J.B., Parker, J.C., and Van Genuchten, M.T. (1985). Determining soil hydraulic properties from one-step outflow experiments by parameter estimation: I. Theory and numerical studies. *Soil Science Society of America Journal* 49 (6): 1348–1354.

Kosugi, K.I. (1994). Three-parameter lognormal distribution model for soil water retention. *Water Resources Research* 30 (4): 891–901.

Kosugi, K.I. (1996). Lognormal distribution model for unsaturated soil hydraulic properties. *Water Resources Research* 32 (9): 2697–2703.

Lang, C. (1878). Ueber Wärmecapacität der Bodenconstituanten. *Forschungen auf dem Gebiete der Agricultur-Physik* 1: 109–145.

Leão, T.P. and Tuller, M. (2014). Relating soil specific surface area, water film thickness, and water vapor adsorption. *Water Resources Research* 50 (10): 7873–7885.

Lebeau, M. and Konrad, J.M. (2010). A new capillary and thin film flow model for predicting the hydraulic conductivity of unsaturated porous media. *Water Resources Research* 46 (12).

Leij, F.L. and Sciortino, A. (2011). Solute transport. In: *Handbook of Soil Sciences*, 2e, vol. 1 (ed. P.M. Huang, Y. Li and M.E. Sumner). Boca Raton, FL: CRC Press, Taylor and Francis Group. ISBN: 978-1-4398-0305-09.

Leij, F.J. and van Genuchten, M.T. (2002). Solute transport. In: *Soil Physics Companion*, 189. Boca Raton, FL: CRC Press.

Lide, D.R. (2003). *Handbook of Chemistry and Physics*, 84e. Boca Raton, FL: CRC Press.

Loizeau, J.L., Arbouille, D., Santiago, S., and Vernet, J.P. (1994). Evaluation of a wide-range laser diffraction grain-size analyzer for use with sediments. *Sedimentology* 41: 353–361.

Loveland, P.J. and Whalley, W.R. (2001). Particle size analysis. In: *Soil and Environmental Analysis, Physical Methods* (ed. K.A. Smith and C.E. Mullins), 281–314. New York: Marcel Dekker Inc.

Menziani, M., Pugnaghi, S., and Vincenzi, S. (2007). Analytical solutions of the linearized Richards equation for discrete arbitrary initial and boundary conditions. *Journal of Hydrology* 332 (1): 214–225.

Mikutta, R., Kleber, M., Kaiser, K., and Jahn, R. (2005). Review: organic matter removal from soils using hydrogen peroxide, sodium hypochlorite, and disodium peroxodisulfate. *Soil Science Society of America Journal* 69: 120–135.

Miller, E.E. and Miller, R.D. (1956). Physical theory for capillary flow phenomena. *Journal of Applied Physics* 27 (4): 324–332.

Millington, R.J. and Quirk, J.P. (1961). Permeability of porous solids. *Transactions of the Faraday Society* 57: 1200–1207.

Moiseev, K.G. (2008). Determination of the specific soil surface area from the hygroscopic water content. *Eurasian Soil Science* 41 (7): 744–748.

Mori, Y., Hopmans, J.W., Mortensen, A.P., and Kluitenberg, G.J. (2003). Multi-functional heat pulse probe for the simultaneous measurement of soil water content, solute concentration, and heat transport parameters. *Vadose Zone Journal* 2 (4): 561–571.

Mualem, Y. (1976). A new model for predicting the hydraulic conductivity of unsaturated porous media. *Water Resources Research* 12 (3): 513–522.

Mualem, Y. (1984). A modified dependent-domain theory of hysteresis. *Soil Science* 137 (5): 283–291.

Nemes, A., Schaap, M.G., Leij, F.J., and Wösten, J.H.M. (2001). Description of the unsaturated soil hydraulic database UNSODA version 2.0. *Journal of Hydrology* 251 (3): 151–162.

Newman, A.C.D. (1983). The specific surface of soils determined by water sorption. *European Journal of Soil Science* 34 (1): 23–32.

Nielsen, D.R., Biggar, J.W., and Davidson, J.M. (1962). Experimental consideration of diffusion analysis in unsaturated flow problems. *Soil Science Society of America Journal* 26 (2): 107–111.

Nielsen, D., Davidson, J., Biggar, J., and Miller, R. (1964). Water movement through Panoche clay loam soil. *California Agriculture* 35 (17): 491–506.

Noborio, K. (2001). Measurement of soil water content and electrical conductivity by time domain reflectometry: a review. *Computers and Electronics in Agriculture* 31 (3): 213–237.

Ochsner, T.E., Horton, R., Kluitenberg, G.J., and Wang, Q. (2005). Evaluation of the heat pulse ratio method for measuring soil water flux. *Soil Science Society of America Journal* 69 (3): 757–765.

Ochsner, T.E., Horton, R., and Ren, T. (2001). Simultaneous water content, air-filled porosity, and bulk density measurements with thermo-time domain reflectometry. *Soil Science Society of America Journal* 65 (6): 1618–1622.

Ochsner, T.E., Horton, R., and Ren, T. (2003). Use of the dual-probe heat-pulse technique to monitor soil water content in the vadose zone. *Vadose Zone Journal* 2 (4): 572–579.

Ogata, A. (1970). Theory of dispersion in a granular medium. In: *US Geological Survey Professional Paper 411-1*. Washington, DC: US Government Printing Office.

Olesen, T., Moldrup, P., Henriksen, K., and Petersen, L.W. (1996). Modeling diffusion and reaction in soils: IV. New models for predicting ion diffusivity. *Soil Science* 161 (10): 633–645.

Or, D., Groeneveld, D.P., Loague, K., and Rubin, Y. (1991). Evaluation of Single and Multi-parameter Methods for Estimating Soil-water Characteristic Curves (No. LASEP-REPORT-1991-001). University of California.

Or, D. and Tuller, M. (1999). Liquid retention and interfacial area in variably saturated porous media: upscaling from single-pore to sample-scale model. *Water Resources Research* 35 (12): 3591–3605.

Or, D. and Tuller, M. (2002). Cavitation during desaturation of porous media under tension. *Water Resources Research* 38 (5): 191–194.

Or, D., Tuller, M., and Stothoff, S. (2006). *Review of Vadose Zone Measurement and Monitoring Tools for Yucca Mountain Performance Confirmation Program*. San Antonio, TX: CNWRA.

Or, D., Tuller, M., and Wraith, J.M. (2005). Water potential. In: *Encyclopedia of Soils in the Environment* (ed. D. Hillel), 270–277. Oxford, UK: Elsevier Ltd.

Orchiston, H.D. (1959). Adsorption of water vapor: V. Homoionic illites at 25°C. *Soil Science* 87: 276–282.

Ostiachev, N.A. (1936). The law of distribution of moisture in soils and methods for study of the same. In: *International Congress on Soil Mechanics in Foundational Engineering Research*, Vol. 20, pp. 203–211.

Pachepsky, Y., Timlin, D., and Rawls, W. (2003). Generalized Richards' equation to simulate water transport in unsaturated soils. *Journal of Hydrology* 272 (1): 3–13.

Papendick, R.I. and Campbell, G.S. (1980). *Theory and Measurement of Water Potential, Water Potential Relations in Soil Microbiology*, 1–22. Madison, WI: Soil Science Society of America.

Pennell, K.D. (2002). Specific surface area. In: *Methods of Soil Analysis: Part 4 Physical Methods* (ed. J.H. Dane and C.G. Topp), 295–315. Madison, WI: Soil Science Society of America.

Pennell, K.D., Abriola, L.M., and Boyd, S.A. (1995). Surface area of soil organic matter reexamined. *Soil Science Society of America Journal* 59 (4): 1012–1018.

Peplinski, N.R., Ulaby, F.T., and Dobson, M.C. (1995). Dielectric properties of soils in the 0.3–1.3-GHz range. *IEEE Transactions on Geoscience and Remote Sensing* 33 (3): 803–807.

Peters, A. and Durner, W. (2008). Simplified evaporation method for determining soil hydraulic properties. *Journal of Hydrology* 356 (1): 147–162.

Phene, C.J., Hoffman, G.J., and Rawlins, S.L. (1971). Measuring soil matric potential in situ by sensing heat dissipation within a porous body: I. Theory and sensor construction. *Soil Science Society of America Journal* 35 (1): 27–33.

Philip, J.R. (1957a). The theory of infiltration: 4. Sorptivity and algebraic infiltration equations. *Soil Science* 84 (3): 257–264.

Philip, J.R. (1957b). Numerical solution of equations of the diffusion type with diffusivity concentration? Dependent. II. *Australian Journal of Physics* 10 (1): 29–42.

Piccin, J.S., Dotto, G.L., and Pinto, L.A.A. (2011). Adsorption isotherms and thermochemical data of FD&C Red n 40 binding by chitosan. *Brazilian Journal of Chemical Engineering* 28 (2): 295–304.

Pommerville, J.C. (2011). *Alcamo's Fundamentals of Microbiology*, 9e. Sudbury, MA: Jones and Bartlett.

Porter, J.R. and Gawith, M. (1999). Temperatures and the growth and development of wheat: a review. *European Journal of Agronomy* 10 (1): 23–36.

Quirk, J.P. and Murray, R.S. (1999). Appraisal of the ethylene glycol monoethyl ether method for measuring hydratable surface area of clays and soils. *Soil Science Society of America Journal* 63 (4): 839–849.

Reichardt, K. (1993). Unit gradient in internal drainage experiments for the determination of soil hydraulic conductivity. *Scientia Agricola* 50 (1): 151–153.

Ren, T., Kluitenberg, G.J., and Horton, R. (2000). Determining soil water flux and pore water velocity by a heat pulse technique. *Soil Science Society of America Journal* 64 (2): 552–560.

Ren, T., Noborio, K., and Horton, R. (1999). Measuring soil water content, electrical conductivity, and thermal properties with a thermo-time domain reflectometry probe. *Soil Science Society of America Journal* 63 (3): 450–457.

Resurreccion, A.C., Moldrup, P., Tuller, M. et al. (2011). Relationship between specific surface area and the dry end of the water retention curve for soils with varying clay and organic carbon contents. *Water Resources Research* 47 (6).

Reynolds, W.D. and Elrick, D.E. (1991). Determination of hydraulic conductivity using a tension infiltrometer. *Soil Science Society of America Journal* 55 (3): 633–639.

Rhoades, J.D. and van Schilfgaarde, J. (1976). An electrical conductivity probe for determining soil salinity. *Soil Science Society of America Journal* 40 (5): 647–651.

Robinson, D.A., Hockley, N., Dominati, E. et al. (2012). Natural capital, ecosystem services, and soil change: why soil science must embrace an ecosystems approach. *Vadose Zone Journal* 11 (1).

Robinson, D.A., Jones, S.B., Wraith, J.M. et al. (2003). A review of advances in dielectric and electrical conductivity measurement in soils using time domain reflectometry. *Vadose Zone Journal* 2 (4): 444–475.

Rose, C.W., Stern, W.R., and Drummond, J.E. (1965). Determination of hydraulic conductivity as a function of depth and water content for soil in situ. *Soil Research* 3 (1): 1–9.

Rossi, A.M., Hirmas, D.R., Graham, R.C., and Sternberg, P.D. (2008). Bulk density determination by automated three-dimensional laser scanning. *Soil Science Society of America Journal* 72 (6): 1591–1593.

Rothamstead Research (1931–present). e-RA: The Electronic Rothamsted Archive/Rothmet. http://www.era.rothamsted.ac.uk/Met/rothmet (accessed 20 January 2018).

Rühlmann, J., Körschens, M., and Graefe, J. (2006). A new approach to calculate the particle density of soils considering properties of the soil organic matter and the mineral matrix. *Geoderma* 130 (3): 272–283.

Russo, D. (1988). Determining soil hydraulic properties by parameter estimation: on the selection of a model for the hydraulic properties. *Water Resources Research* 24 (3): 453–459.

Sadeghi, M. (2013). Scaling Richards' equation for dissimilar soils. Ph.D. Dissertation. Ferdowsi University of Mashhad, Mashhad, Iran (In Farsi with English Abstract).

Sadeghi, M., Ghahraman, B., Davary, K. et al. (2011). Scaling to generalize a single solution of Richards' equation for soil water redistribution. *Scientia Agricola* 68 (5): 582–591.

Sadeghi, M., Ghahraman, B., Warrick, A.W. et al. (2016). A critical evaluation of the Miller and Miller similar media theory for application to natural soils. *Water Resources Research* 52 (5): 3829–3846.

Sadeghi, M., Ghahraman, B., Ziaei, A.N. et al. (2012a). Additional scaled solutions to Richards' equation for infiltration and drainage. *Soil and Tillage Research* 119: 60–69.

Sadeghi, M., Ghahraman, B., Ziaei, A.N. et al. (2012b). Invariant solutions of Richards' equation for water movement in dissimilar soils. *Soil Science Society of America Journal* 76 (1): 1–9.

Sadeghi, M. and Jones, S.B. (2012). Scaled solutions to coupled soil-water flow and solute transport during the redistribution process. *Vadose Zone Journal* 11 (4). doi: 10.2136/vzj2012.0023.

Sadeghi, M., Shokri, N., and Jones, S.B. (2012c). A novel analytical solution to steady-state evaporation from porous media. *Water Resources Research* 48 (9). doi: 10.1029/2012WR012060.

Sadeghi, M., Tabatabaeenejad, A., Tuller, M. et al. (2017). Advancing NASA's AirMOSS P-band radar root zone soil moisture retrieval algorithm via incorporation of Richards' equation. *Remote Sensing* 9 (1): 17.

Sadeghi, M., Tuller, M., Gohardoust, M.R., and Jones, S.B. (2014). Column-scale unsaturated hydraulic conductivity estimates in coarse-textured homogeneous and layered soils derived under steady-state evaporation from a water table. *Journal of Hydrology* 519: 1238–1248.

Sakai, M., Jones, S.B., and Tuller, M. (2011). Numerical evaluation of subsurface soil water evaporation derived from sensible heat balance. *Water Resources Research* 47 (2). doi: 10.1029/2010WR009866.

Salvucci, G.D. (1993). An approximate solution for steady vertical flux of moisture through an unsaturated homogeneous soil. *Water Resources Research* 29 (11): 3749–3753.

Salvucci, G.D. and Entekhabi, D. (1994). Explicit expressions for Green–Ampt (delta function diffusivity) infiltration rate and cumulative storage. *Water Resources Research* 30 (9): 2661–2663.

Sauer, T.J., Ochsner, T.E., and Horton, R. (2007). Soil heat flux plates. *Agronomy Journal* 99 (1): 304–310.

Schaufler, G., Kitzler, B., Schindlbacher, A. et al. (2010). Greenhouse gas emissions from European soils under different land use: effects of soil moisture and temperature. *European Journal of Soil Science* 61 (5): 683–696.

Schindler, U. (1980). Ein Schnellverfahren zur Messung der Wasserleitfähigkeit im teilgesättigten Boden an Stechzylinderproben. *Archiv für Acker- und Pflanzenbau und Bodenkunde* 24 (1): 1–7.

Schindler, U. and Müller, L. (2006). Simplifying the evaporation method for quantifying soil hydraulic properties. *Journal of Plant Nutrition and Soil Science* 169 (5): 623–629.

Selker, J.S., Thévenaz, L., Huwald, H. et al. (2006). Distributed fiber-optic temperature sensing for hydrologic systems. *Water Resources Research* 42 (12). doi: 10.1029/2006WR005326.

Serrano, S.E. (2001). Explicit solution to Green and Ampt infiltration equation. *Journal of Hydrologic Engineering* 6 (4): 336–340.

Seyfried, M.S. and Murdock, M.D. (2004). Measurement of soil water content with a 50-MHz soil dielectric sensor. *Soil Science Society of America Journal* 68 (2): 394–403.

Shani, U., Hanks, R.J., Bresler, E., and Oliveira, C.A.S. (1987). Field method for estimating hydraulic conductivity and matric potential-water content relations. *Soil Science Society of America Journal* 51 (2): 298–302.

Shao, M. and Horton, R. (1998). Integral method for estimating soil hydraulic properties. *Soil Science Society of America Journal* 62 (3): 585–592.

Shao, M. and Horton, R. (2000). Exact solution for horizontal water redistribution by general similarity. *Soil Science Society of America Journal* 64 (2): 561–564.

Sheng, W., Rumana, K., Sakai, M. et al. (2016). A multi-functional penta-needle thermo-dielectric sensor for porous media sensing. *IEEE Sensors Journal* 16 (10): 3670–3678.

Šimůnek, J. and De Vos, J.A. (1999). Inverse optimization, calibration and validation of simulation models at the field scale. In: *Modelling Transport Processes in Soils at Various Scales in Time and Space*, 431–445. Wageningen, the Netherlands: Wageningen Pers.

Šimůnek, J. and Genuchten, M.V. (1996). Estimating unsaturated soil hydraulic properties from tension disc infiltrometer data by numerical inversion. *Water Resources Research* 32 (9): 2683–2696.

Šimůnek, J., Jarvis, N.J., Van Genuchten, M.T., and Gärdenäs, A. (2003). Review and comparison of models for describing non-equilibrium and preferential flow and transport in the vadose zone. *Journal of Hydrology* 272 (1): 14–35.

Šimůnek, J., van Genuchten, M., and Sejna, M. (2005). The HYDRUS-1D software package for simulating the one-dimensional movement of water, heat, and multiple solutes in variably-saturated media. *University of California-Riverside Research Reports* 3: 1–240.

Skopp, J.M. (2011). Physical properties of primary particles. In: *Handbook of Soil Sciences: Properties and Processes*, 2e (ed. M. Huang, Y. Li and M.E. Sumner). Boca Raton, FL: CRC Press.

Soil Survey Staff (1975). Soil taxonomy: a basic system of soil classification for making and interpreting soil surveys. In: *Agriculture Handbook 436*. Washington, DC: U.S. Department of Agriculture, U.S. Government Printing Office.

Środoń, J. and MaCarty, D.K. (2008). Surface area and layer charge of smectite from CEC and EGME/H2O-retention measurements. *Clays and clay minerals* 56 (2): 155–174.

Steele-Dunne, S.C., Rutten, M.M., Krzeminska, D.M. et al. (2010). Feasibility of soil moisture estimation using passive distributed temperature sensing. *Water Resources Research* 46 (3). doi: 10.1029/2009WR008272.

Stenberg, B., Rossel, R.A.V., Mouazen, A.M., and Wetterlind, J. (2010). Chapter five – visible and near infrared spectroscopy in soil science. *Advances in Agronomy* 107: 163–215.

Stokes, G.G. (1851). On the effect of the internal friction of fluids on the motion of pendulums. *Transactions of the Cambridge Philosophical Society* 9: 8–106.

Stolte, J., Halbertsma, J.M., Veerman, G.J. et al. (1994). Comparison of six methods to determine unsaturated soil hydraulic conductivity. *Soil Science Society of America Journal* 58 (6): 1596–1603.

Tabatabaeenejad, A., Burgin, M., Duan, X., and Moghaddam, M. (2015). P-band radar retrieval of subsurface soil moisture profile as a second-order polynomial: first AirMOSS results. *IEEE Transactions on Geoscience and Remote Sensing* 53 (2): 645–658.

Tarara, J.M. and Ham, J.M. (1997). Measuring soil water content in the laboratory and field with dual-probe heat-capacity sensors. *Agronomy Journal* 89 (4): 535–542.

Taylor, G. (1953). Dispersion of soluble matter in solvent flowing slowly through a tube. *Proceedings of the Royal Society of London A: Mathematical, Physical and Engineering Sciences* 219 (1137): 186–203.

Thien, S.J. (1979). A flow diagram for teaching texture-by-feel analysis. *Journal of Agronomic Education* 8 (2): 54–55.

Tiller, K.G. and Smith, L.H. (1990). Limitations of EGME retention to estimate the surface area of soils. *Soil Research* 28 (1): 1–26.

Topp, G.C., Davis, J.L., and Annan, A.P. (1980). Electromagnetic determination of soil water content: measurements in coaxial transmission lines. *Water Resources Research* 16 (3): 574–582.

Tracy, F.T. (2007). Three-dimensional analytical solutions of Richards' equation for a box-shaped soil sample with piecewise-constant head boundary conditions on the top. *Journal of Hydrology* 336 (3): 391–400.

Trautz, A.C., Smits, K.M., Schulte, P., and Illangasekare, T.H. (2014). Sensible heat balance and heat-pulse method applicability to in situ soil-water evaporation. *Vadose Zone Journal* 13 (1). doi: 10.2136/vzj2012.0215.

Tuller, M. and Islam, M.R. (2005). *Field Methods for Monitoring Solute Transport*, 309–355. Boca Raton, FL: CRC Press LLC.

Tuller, M., Kulkarni, R., and Fink, W. (2013). Segmentation of X-ray CT data of porous materials: a review of global and locally adaptive algorithms. In: *Soil–Water–Root Processes: Advances in Tomography and Imaging*, 157–182. Madison, WI: SSSA Special Publication.

Tuller, M. and Or, D. (2005a). Water films and scaling of soil characteristic curves at low water contents. *Water Resources Research* 41 (9). doi: 10.1029/2005WR004142.

Tuller, M. and Or, D. (2005b). Water retention and characteristic curve. In: *Encyclopedia of Soils in the Environment* (ed. Hillel), 278–289. Oxford, UK: Elsevier Ltd.

Tuller, M., Or, D., and Dudley, L.M. (1999). Adsorption and capillary condensation in porous media: liquid retention and interfacial configurations in angular pores. *Water Resources Research* 35 (7): 1949–1964.

Ulrich, R. (1894). Untersuchungen über die Wärmekapazität der Bodenkonstituenten. *Forschungen auf dem Gebiete der Agrikulturphysik* 17: 1–31.

Underwood, N. and Swanson, R. (1954). A portable slow neutron flux meter for measuring soil moisture. *Soil Science* 77 (4): 339–340.

Valiantzas, J.D., Kerkides, P.G., and Poulovassilis, A. (1988). An improvement to the one-step outflow method for the determination of soil water diffusivities. *Water Resources Research* 24 (11): 1911–1920.

Van Duin, R.H.A. (1963). *The Influence of Soil Management on the Temperature Wave Near the Soil Surface. Technical Bulletin No. 29*. Wageningen, The Netherlands: Institute for Land and Water Management Research.

Van Genuchten, M.T. (1980). A closed-form equation for predicting the hydraulic conductivity of unsaturated soils. *Soil Science Society of America Journal* 44 (5): 892–898.

Van Genuchten, M.T. and Alves, W.J. (1982). Analytical solutions of the one-dimensional convective-dispersive solute transport equation (No. 157268). United States Department of Agriculture, Economic Research Service.

Van Rooyen, M. and Winterkorn, H.F. (1959). Structural and textural influence on the thermal conductivity of soils. *Highway Research Board Proceedings* 38: 576–621.

Van Wijk, W.R. and deVries, D.A. (1963). Periodic temperature variations in a homogeneous soil. In: *Physics of Plant Environment*, 103–143. Amsterdam: North Holland Publishing Company.

Vanderborght, J. and Vereecken, H. (2007). Review of dispersivities for transport modeling in soils. *Vadose Zone Journal* 6 (1): 29–52.

Vaz, C.M., Jones, S., Meding, M., and Tuller, M. (2013). Evaluation of standard calibration functions for eight electromagnetic soil moisture sensors. *Vadose Zone Journal* 12 (2). doi: 10.2136/vzj2012.0160.

Vaz, C.M.P., Tuller, M., Lasso, P.R.O., and Crestana, S. (2014). New perspectives for the application of high-resolution benchtop X-ray microCT for quantifying void, solid and liquid phases in soils. In: *Application of Soil Physics in Environmental Analyses*, 261–281. Switzerland: Springer International Publishing.

Vendrame, P.R.S., Marchão, R.L., Brunet, D., and Becquer, T. (2012). The potential of NIR spectroscopy to predict soil texture and mineralogy in Cerrado Latosols. *European Journal of Soil Science* 63 (5): 743–753.

Visconti, F. and de Paz, J.M. (2016). Electrical conductivity measurements in agriculture: the assessment of soil salinity. In: *New Trends and Developments in Metrology*. Rijeka, Croatia: InTech.

Vrugt, J.A., Stauffer, P.H., Wöhling, T. et al. (2008). Inverse modeling of subsurface flow and transport properties: a review with new developments. *Vadose Zone Journal* 7 (2): 843–864.

Warrick, A.W. (1975). Analytical solutions to the one-dimensional linearized moisture flow equation for arbitrary input. *Soil Science* 120 (2): 79–84.

Warrick, A.W. (1988). Additional solutions for steady-state evaporation from a shallow water table. *Soil Science* 146 (2): 63–66.

Warrick, A.W., Biggar, J.W., and Nielsen, D.R. (1971). Simultaneous solute and water transfer for an unsaturated soil. *Water Resources Research* 7 (5): 1216–1225.

Warrick, A.W. and Hussen, A.A. (1993). Scaling of Richards' equation for infiltration and drainage. *Soil Science Society of America Journal* 57 (1): 15–18.

Warrick, A.W., Lomen, D.O., and Islas, A. (1990). An analytical solution to Richards' equation for a draining soil profile. *Water Resources Research* 26 (2): 253–258.

Warrick, A.W., Lomen, D.O., and Yates, S.R. (1985). A generalized solution to infiltration. *Soil Science Society of America Journal* 49 (1): 34–38.

Watson, K.K. (1966). An instantaneous profile method for determining the hydraulic conductivity of unsaturated porous materials. *Water Resources Research* 2 (4): 709–715.

Weihermüller, L., Siemens, J., Deurer, M. et al. (2007). In situ soil water extraction: a review. *Journal of Environmental Quality* 36 (6): 1735–1748.

Wen, B., Aydin, A., and Duzgoren-Aydin, N.S. (2002). A comparative study of particle size analyses by sieve-hydrometer and laser diffraction methods. *Geotechnical Testing Journal* 25: 434–442.

White, I. and Sully, M.J. (1987). Macroscopic and microscopic capillary length and time scales from field infiltration. *Water Resources Research* 23 (8): 1514–1522.

Wilczek, A., Szypłowska, A., Skierucha, W. et al. (2012). Determination of soil pore water salinity using an FDR sensor working at various frequencies up to 500 MHz. *Sensors* 12 (8): 10890–10905.

Wind, G.P. (1968). Capillary conductivity data estimated by a simple method. In: *Wageningen Symposium on Water in the Unsaturated Zone* (ed. P.E. Rijtema and H. Wassink), 181–191. Gent, Brugge/Paris: International Association of Scientific Hydrologists/UNESCO.

WMO (2006). Measurement of temperature. WMO guide to meteorological instruments and methods of observation: WMO No. 8. Draft 7e. World Meteorological Organization, Geneva, Switzerland.

Wooding, R.A. (1968). Steady infiltration from a shallow circular pond. *Water Resources Research* 4 (6): 1259–1273.

Wuddivira, M.N., Robinson, D.A., Lebron, I. et al. (2012). Estimation of soil clay content from hygroscopic water content measurements. *Soil Science Society of America Journal* 76 (5): 1529–1535.

Xiao, X., Horton, R., Sauer, T.J. et al. (2011). Cumulative soil water evaporation as a function of depth and time. *Vadose Zone Journal* 10 (3): 1016–1022.

Xiao, Z., Lu, S., Heitman, J., and Horton, R. (2012). Measuring subsurface soil-water evaporation with an improved heat-pulse probe. *Soil Science Society of America Journal* 76 (3): 876–879.

Yang, C. and Jones, S.B. (2009). INV-WATFLX, a code for simultaneous estimation of soil properties and planar vector water flux from fully or partly functioning needles of a penta-needle heat-pulse probe. *Computers & Geosciences* 35 (11): 2250–2258.

Yang, C., Sakai, M., and Jones, S.B. (2013). Inverse method for simultaneous determination of soil water flux density and thermal properties with a penta-needle heat pulse probe. *Water Resources Research* 49 (9): 5851–5864.

Yuan, F. and Lu, Z. (2005). Analytical solutions for vertical flow in unsaturated, rooted soils with variable surface fluxes. *Vadose Zone Journal* 4 (4): 1210–1218.

Zhu, J. and Mohanty, B.P. (2002). Analytical solutions for steady state vertical infiltration. *Water Resources Research* 38 (8): 20-1–20-5.

8

In Situ Soil and Sediment Remediation

Electrokinetic and Electrochemical Methods

Sibel Pamukcu

Department of Civil and Environmental Engineering, Lehigh University, Bethlehem, PA, USA

8.1 Introduction and Background

Environmental sustainability relies on the health of natural soils and sediments, which are filters and conduits to surface and drinking waters and sources of nutrients for plant and animal life. Pollution is widely recognized as a significant challenge to soil and sediment and subsequently to groundwater resource management (Hadley and Ellis, 2009; Koch et al., 2013; McBratney et al., 2014). There are multiple pathways of soil–sediment pollution, a major one being point source pollution. Industrial leakage, improper discharge, leakage from energy production activities, urban runoff and construction activities, and improper management of waste (i.e. biowaste) are among some causes of point source pollution (Lima et al., 2012a; Mihelcic et al., 2017). When a plume of contaminants (i.e. organic molecular compounds, nano-substances, pharmaceuticals, immiscible liquids, heavy metals) permeating through soil encounters a low permeability layer, it initially sinks in this layer but can switch to being a source of long-term contamination over time. Pollution that remains as a potential point source in low permeability soils and sediments is a remediation challenge (Macdonald et al., 2000; Reddy, 2010).

In most field situations, the contaminants are found adsorbed onto soil surfaces, iron oxide coatings, soil colloids, and natural organic matter. Most contaminants are retained in clay interstices as hydroxycarbonate complexes or present in the form of immobile precipitates and products in soil pore throats and pore pockets that "lace" the vadose zone. This exacerbates cleanup efforts as the available technologies, such as in situ bioremediation, chemical treatment, or the traditional pump-and-treat method, may not be able to treat the entire site effectively in low permeability soils. Electrokinetic (EK) treatment, when designed to properly address the site-specific features, can potentially reduce the subsurface pollution by transporting and/or transforming contaminants and enhance resource recovery by extracting trapped materials (i.e. heavy metals, oils, and petroleum) that may not be extractable by other means. Electrochemically and electrokinetically assisted in situ remediation has been shown to be a viable and versatile technique for most soils and most contaminant types, particularly for low permeability clayey deposits containing soluble heavy metals (Ray and Ramsey, 1987; Pamukcu et al., 1990, 1997; Pamukcu and Wittle, 1992a, b, 1993; Acar and Alshawabkeh, 1993; Lageman, 1993; Ottosen et al., 1997; Reddy and Parupudi, 1997; Ribeiro and Mexia, 1997; Schultz, 1997; Gent et al., 2004; Nystroem et al., 2005; Jensen et al., 2007; Alshawabkeh, 2009; Elektorowicz, 2009; Traina, 2009; Yeung, 2011; Nieto Castillo et al., 2012; Merkx et al., 2013; Demir et al., 2014).

Applying a low direct current (DC) electrical field across a porous medium, such as wet soil, induces migration of ionic and surface-charged constituents and flow of the pore water to an electrode within the field. Commonly referred to as "EK," this process is a proven sustainable technology that can transport water and substances residing in or near the aqueous phase in clayey soil at a significantly higher rate than hydraulic methods. First used as early as 19th century for dewatering and stabilizing clays (Reuss, 1809; Casagrande, 1949), EK was intensively studied in the 1990s for removing metals from soils. EK is still a preferred method of in situ dewatering of heavily clay deposits and other hazardous slurries (Lockhart and Stickland, 1984; Yang et al., 2005, 2011; Huang et al., 2008). Since the 1990s EK and related electrochemical techniques have been shown to successfully treat in situ soils containing not only heavy metals but also mixed contaminants (Elektorowicz, 2009) including radioactive materials (Pamukcu and Wittle, 1992a; Acar et al., 1995; Maes et al., 1999; Roh et al., 2000; Kim et al., 2003; Valdovinos et al., 2016), organic compounds and immiscible fluids (Bruell et al., 1992; Pamukcu, 1994, 2009; Pamukcu et al., 1995, 2016;

Handbook of Environmental Engineering, First Edition. Edited by Myer Kutz.

Saichek and Reddy, 2003; Ribeiro et al., 2005, 2011; Elektorowicz et al., 2006; Korolev et al., 2008; Lima et al., 2011, 2012a, b; Reddy et al., 2011; Ghazanfari et al., 2012, 2014; Guedes et al., 2014a), and also wastewater sludges and biomass (Ray and Ramsey, 1987; Pulgarin et al., 1994; Bani-Melhem and Elektorowicz, 2010, 2011; Lei et al., 2012; Nieto Castillo et al., 2012; Guedes et al., 2014b; Hasan et al., 2014).

The process is particularly useful in applications where pump-and-treat methods are impractical due to the low permeability of the medium or in cases where the contaminants persist owing to their affinity to the solid phases in clay soils or the depth and extend of the contamination that do not lend "dig and dump" and landfilling options (Dresel et al., 2011). EK and electrochemical methods, engineered to site-specific requirements, have the potential to overcome traditional problems in soil remediation practices, including soil excavation, excess water usage, and disturbance to natural soil attributes such as porosity and biodiversity (Virkutyte and Sillanpää, 2002; Saichek and Reddy, 2005a; Yeung, 2011). The application of direct electric current in soils has been used successfully to transport reactive agents, providing the "driving force" necessary to deliver active reagents such as nanoparticle slurries to enhance in situ reactions that convert contaminants to their benign products (Roh et al., 2000; Chowdhury et al., 2012; Gomes et al., 2012, 2013, 2014, 2015a, b; Huang and Cheng, 2012; Kocur et al., 2013). Using EK, chemical surfactants, amendments, and chelating agents can be transported through low permeability soils to enhance solubilization of metals or desorption of organics from soil so they can be transported with water advection (Pamukcu, 1994; Yeung et al., 1996; Saichek and Reddy, 2004, 2005b; Roach and Reddy, 2006; Pazos et al., 2008; Chang et al., 2010; Chen et al., 2011; Ryu et al., 2011; Lima et al., 2012a; Ottosen et al., 2012). Additional benefits of EK transport have been demonstrated in stripping micropollutants, especially nonpolar compounds including heavy oils, from low permeability soils and sediments, where other treatment methods may fail (Pamukcu, 2009; Pazos et al., 2010; Alcántara et al., 2012; Cameselle and Reddy, 2012; Ghazanfari et al., 2012; Lima et al., 2012b).

Electrochemical and EK techniques have also been coupled successfully with conventional techniques, including *bioremediation and biotreatment* (Electorowicz and Boeva, 1996; Chilingar et al., 1997; DeFlaun and Condee, 1997; Wick et al., 2004; Luo et al., 2006; Niqui-Arroyo et al., 2006, Niqui-Arroyo and Ortega-Calvo, 2007; Elektorowicz et al., 2012; Mao et al., 2012; Riis et al., 2012; Ibeid et al., 2013; Gill et al.; 2014), *phytoremediation* (Aboughalma et al., 2008; Kubiak et al., 2012; Cameselle et al., 2013), *nanoscale zero-valent iron (nZVI) amendment* (Moon et al., 2005; Hannum and Pamukcu, 2007; Weng et al., 2007; Pamukcu et al., 2008; Chowdhury et al., 2012; Gomes et al., 2012, 2013, 2014, 2015a, b; Huang and Cheng, 2012), *enhanced oxidation–reduction and Fenton-like methods* (Pamukcu et al., 2004; Alshawabkeh and Sarahney, 2005; O'Mahony et al., 2006; Isosaari et al., 2007; Petri et al., 2011; Yukselen-Aksoy and Reddy, 2012; Brosky and Pamukcu, 2013; Innocenti et al., 2014; Sun et al., 2015), *permeable reactive walls* (Blowes et al., 1995; Chung and Lee, 2007; Ramírez et al., 2015; Mena et al., 2016; Zhou et al., 2016; Vieira dos Santos et al., 2017), *and resource recovery methods* (Amba et al., 1964; Fleureau and Dupeyart, 1988; Stevens, 1999; Wittle et al., 2008, 2011; Al Shalabi et al., 2012a; Ghosh et al., 2012; Lei et al., 2012; Haroun et al., 2013; Ghazanfari and Pamukcu, 2014; Ghazanfari et al., 2014; Guedes et al., 2014b).

Although the field-scale applications of the EK and electrochemical remediation methods are becoming more common at variety of scales (Banerjee et al., 1987; Ho et al., 1995, 1999a, 1999b; Schultz, 1997; Lindgren, 1998; Lageman and Godschalk, 2007; Wittle et al., 2009; Mao et al., 2012; Riis et al., 2012; Zhang et al., 2015; López-Vizcaíno et al., 2017), the technology has not yet advanced to a commonly accepted commercial status, still facing unique engineering challenges. These challenges stem mainly from short- and long-term side effects, such as pH fronts and hydrogen and chlorine gas generated at the electrodes and/or other unpredicted redox reactions (Probstein and Hicks, 1993; Eykholt and Daniel, 1994; Denisov et al., 1996; Dzenitis, 1997; Mattson et al., 2002; Shin et al., 2012; Peng et al., 2013). Other issues observed during EK treatment are temporally and spatially changing process parameters, such as the voltage gradients and pore water pressures (PWPs), migrating fronts of high ion concentrations in saline environments. Although some of these effects can have undesirable consequences on EK mass transport, they have also been shown to aid in degradation and/or extraction of some persistent substances (Pamukcu, 2009; Muraoka et al., 2011). For example, EK-assisted advection and mass displacement have been demonstrated as a stand-alone technique for oil transport and recovery from geological media including soils, sediments, and rock formations (Chilingar et al., 1968a, b; Pamukcu, 1994; Wittle et al., 2011; Ghazanfari et al., 2012, 2014; Ghosh et al., 2012; Ghazanfari and Pamukcu, 2014; Pamukcu et al., 2014, 2016). Other innovative uses of EK and electrochemistry for in situ cleanup of soils and sediments involve electrodialysis, which manages the electrode reaction products (Hansen et al., 1997, 2005; Nystroem et al., 2005; Jensen et al., 2007; Lima et al., 2012b; Ottosen et al., 1997, 2000, 2012; Sun et al., 2012; Guedes et al., 2014b). More recently, evidence was provided that shows how direct electric current can contribute to the success of the desired transformation reactions by lowering the energy for the redox reactions to occur. This enhancement of transformation reactions was attributed to the double layer polarization of the clay surfaces, leading to faradaic processes under the applied electric field (Pamukcu et al., 2004; Brosky and Pamukcu, 2013; Sun et al., 2015). Experimental evidence was also provided in effectiveness of low DC on separation and demulsification processes in oily sediments or sludges (Elektorowicz et al., 2006; Pamukcu et al., 2016).

This chapter first provides an overview of the use of direct electric current for environmental mitigation in the subsurface and discusses with examples the application of direct electric current to:

1) *Transport* of environmental contaminants for the purpose of extracting them out of subsurface soil and water.
2) *Transformation* of environmental contaminants for the purpose of their destruction in situ or conversion to environmentally less toxic/benign compounds.
3) *Recovery of resource* by extraction and accumulation of usable materials from subsurface soil and water.

8.2 Overview and Theory of Direct Electric Current in Soil and Sediment Remediation

Electrically induced migration of ions and water is a proven method of mass (i.e. water, ions, micelles, reagents, nutrients, and fine particles) transport in clay soils and sediments for environmental remediation purposes (Alshawabkeh and Acar, 1992, 1996; Pamukcu and Wittle, 1992a, b; Acar and Alshawabkeh, 1993, 1996; Lageman, 1993; Probstein and Hicks, 1993, Acar et al., 1994; Eykholt and Daniel, 1994; Hicks and Tondorf, 1994; Shapiro et al., 1995; Electorowicz and Boeva, 1996; Yeung et al., 1996; Dzenitis, 1997; Reddy and Parupudi, 1997; Paillat et al., 2000; Pamukcu and Huang, 2001; Elektorowicz, 2009). While electroosmosis (EO) is analogous to soil washing, ion migration is probably the primary mechanism of mass transport when the contaminants are ionic or surface charged. Relative contribution of EO and ion migration to the total mass transport varies according to soil type, water content, ion species, and their concentration. In silts and low-activity clays, EO flow reaches maximum in comparison with hydraulic flow. But the mass transported by ionic migration is always much higher than the mass transported by EO advection (Acar and Alshawabkeh, 1993). The effect of EO decreases significantly when pH and zeta potential drops in the later stages of a sustained EK process under a constant electric potential (Hamed et al., 1991; Pamukcu et al., 1997). When micelles (i.e. charged aggregate of molecules or particles) are formed with other species in the processing fluid, or when slurry masses are treated, electrophoresis may become significant (Pamukcu et al., 1995).

8.2.1 Transport of Charged Species: Electromigration

As in many electrochemical systems, flow of electric current through a network of a multiphase material, such as wet soil, occurs in different phases simultaneously: in the bulk liquid (electrolyte in the pores), on the surface of the solid (clay particles), and in the interface layer(s) between the solid and the liquid. Flow of the current can be achieved by ionic conduction through the liquid phase and electronic conduction through the solid phase and the interface layer(s). The electronic conduction orthogonal to and along the interface layer(s) takes place via charge transfer. In the classical treatise of "EK phenomenon" in colloidal systems (Hunter, 1981, 2001; Lyklema, 1995), it is this interface, known as the *electric double layer* (EDL) or *diffuse double layer* (DDL), that plays a critical role in the coupling between the ion motion and the fluid flow. The double layer intrinsically connects the solid and the liquid phase and mediates the relative motion between the liquid and solid phase through (i) accumulation of charge density, (ii) transport of charge and ions along the surface, and (iii) passage of charge to the surrounding electrolyte (Bard and Faulkner, 1980).

The bulk transport of ions in electrochemical systems without the contribution of advection by water is described by Poisson–Nernst–Planck (PNP) equations (Nernst, 1888; Planck, 1890; Rubinstein, 1990). The well-known Nernst–Planck equation describes the processes of *diffusion*, the process that drives the ions from regions of higher concentration to regions of lower concentration, and *electromigration* (also referred to as *ion migration*), the process that launches the ions in the direction of the electric field (Bard and Faulkner, 1980). Since the ions themselves contribute to the local electric potential, Poisson's equation that relates the electrostatic potential to local ion concentrations is solved simultaneously to describe this effect. The electroneutrality assumption simplifies the mathematical treatise of bulk transport in most electrochemical systems. Nevertheless, this "no charge density accumulation" assumption does not hold true at the interphase regions of *EDL* between the solid and liquid, hence the cause of most EK phenomenon in clay–electrolyte systems.

The analysis of mass transport by diffusion under chemical ($\partial C/\partial x$) gradient and by migration under electrical ($\partial \Phi/\partial x$) gradient in dilute solutions – for which the interactions between individual species can be neglected – is described by Nernst–Planck (Eq. (8.1)) and Poisson's (Eq. (8.2)) equations, together referred to as the PNP equations:

$$\frac{\partial C_i}{\partial t} = \nabla\left(D_i^* \nabla C_i + u_i z_i F C_i \nabla \Phi\right) \tag{8.1}$$

$$-\varepsilon_s \nabla^2 \Phi = \rho = \sum_i z_i F C_i \tag{8.2}$$

where C, D^*, u, and z are the concentration, diffusion coefficient, mobility, and charge number of a single species i, respectively, F is Faraday's constant (i.e. a mole of charge), Φ is the electrostatic potential, ε_s is the permittivity of the solvent, and ρ is the charge density.

For many electrochemical systems, the local electroneutrality condition is used, which sets the left hand side of Eq. (8.2) to zero for zero charge density. The mathematical implication is then that the electrical potential satisfies the Laplace Equation [$\nabla^2\Phi = 0$], hence uniform concentration distribution and uniform conductivity within the electrochemical cell. Yet in real electrochemical systems with concentration gradients, the current density of the system can be affected to cause current flow in the opposite direction of the electric field (Newman, 1991). Using the PNP equations and the electroneutrality condition, it can be shown mathematically that the concentration gradients give rise to spatial variation of conductivity, where a *diffusion potential* arises to ensure ion movement at the same speed to overcome the charge separation and violation of electrical neutrality (Newman, 1991). The charge density accumulation cannot be neglected for interphase layers, such as the electrical double layer. Using a non-dimensionalized form of Eq. (8.2), it was shown that the electroneutrality condition is a direct result of the Debye screening length, where the ratio of the Debye length, λ_D, to the field length, L, is described as follows in Eq. (8.3) (Chu, 2005):

$$-\varepsilon \nabla^2 \phi = \sum_i z_i c_i \tag{8.3}$$

where

$$\varepsilon = \frac{\lambda_D}{L},\quad \phi = \frac{\Phi}{RT/F},\quad c_i = \frac{C_i}{C_{eq}},\quad \lambda_D = \sqrt{\left(\frac{\varepsilon_s RT}{F^2 C_{eq}}\right)}$$

where

R = universal gas constant, 8.3144 J (K mol)$^{-1}$

T = absolute temperature, (K)

F = Faraday constant, 96 485 C mol^{-1} electrons

C_{eq} = equilibrium concentration in bulk

As shown in Eq. (8.3), unless the Laplace compliance holds true – which is not the case for most electrochemical systems owing to the presence of concentration gradients – the electroneutrality of the system can be satisfied when the Debye length λ_D is so small compared with L (s.t., $\varepsilon \ll$) that the left-hand side of the equation becomes zero. Asymptotic analysis of PNP equations numerically solved at several values of ε showed that for small values of ε – as would be in macroscopic systems – the charge density ρ is zero for majority of an electrochemical cell of length L, except at the boundaries where faradaic reactions and Stern layer capacitance are considered (Chu, 2005). As the Debye length approaches to the width of the EDL, nonzero charge density distribution appears in the entire electrochemical cell (Chu, 2005). Hence, the electroneutrality in an electrochemical system will hold when the charge density is small compared with total ion concentration, C_{eq}, of the bulk fluid. As the charge density approaches to total concentration within the electrochemical cell (e.g. EDL for which Debye length is the width of the layer), electroneutrality can be achieved considering Nernstian boundaries and faradaic reactions.

8.2.2 Transport of Water and Its Constituents: Electroosmosis

In 1809, Reuss observed the EK phenomena when a DC current was applied to a clay–water mixture (Reuss, 1809). Water moved through the capillary toward the cathode under the electric field. When the electric potential was removed, the flow of water immediately stopped. In 1861, Quinke found the electric potential difference across a membrane resulted from streaming potential (Quinke, 1861). Helmholtz first treated EO phenomena analytically in 1879 (Helmholtz, 1879) and provided a mathematical basis. Smoluchowski later in 1914 modified it to also apply to electrophoretic velocity, also known as the Helmholtz–Smoluchowski (H–S) theory. H–S theory describes the migration velocity of one phase of material dispersed in another phase under an applied electric potential. The EO velocity of a fluid of certain viscosity and dielectric constant through a surface-charged porous medium of zeta or EK potential (ζ), under an electric gradient, E, is given by the H–S equation as follows:

$$v_{eo} = \frac{\varepsilon_s \zeta}{\eta} \cdot \frac{\partial \Phi}{\partial x} = k_{eo} E \tag{8.4}$$

where

v_{eo} = EO (electrophoretic) velocity

ε_s = permittivity of the solvent or the pore fluid

η = viscosity of the solvent or the pore fluid

k_{eo} = coefficient of EO conductivity

$\partial\Phi/\partial x = E$ = electric field

The zeta (ζ) potential in Eq. (8.4) has been shown to vary with pH and ionic concentration of the pore fluid, as well as the electric field, and therefore is not constant during EO transport in clay media (Probstein and Hicks, 1993). This makes it difficult to assess a velocity term for temporal and spatial distribution predictions of the EO transport. The nonlinearity and nonuniformity associated with the EO velocity became apparent when experiments showed that v_e/E increased with E, hence rendering the other parameters of Eq. (8.4) to be functions of the electrical field (Ravina and Zaslavsky, 1967). In a given clay deposit in the field, pore cross sections, the electrical double layer properties, the fluid viscosity, the dielectric constant, solute distribution, and electrical field can vary significantly from one point to another, both along the flow path and in the orthogonal direction. Hence, averaging microscopic parameters (e.g. zeta potential) to use alongside with macroscopic parameters (e.g. E) in H–S equation can lead to the nonlinear effects.

The zeta, ζ, potential is the electrical potential at the junction between the fixed and mobile parts of the electrical double layer. The ζ potential is influenced by the type and concentration of the electrolytes added to the particle suspension (Kruyt, 1952; Hunter, 1981; Smith and Narimatsu, 1993). For clay soils, ζ potential is usually negative because of the net negative charge on clay particle surfaces. The magnitude and sign of this potential highly depends on pore fluid chemistry. As the hydrogen and hydroxyl ions are the potential determining ions, the lower pH will reduce ζ potential in magnitude for most clay. At low enough pH, ζ potential may become positive. Hunter and James (1992) observed that adsorption of partially hydrolyzed metal cations such as Co^{2+}, Cd^{2+}, and Cu^{2+} causes ζ potential reversals for kaolinite. As the concentration of hydrolyzable metal ions increases, ζ potential becomes more positive at low pH levels due to the accumulation of the cations in a compressed electrical double layer bearing a larger charge than is present on the solid surface (Kruyt, 1952). The effect is largest at an intermediate pH, slightly above the value at which precipitation of the metal hydroxide would be expected in the bulk solution. Due to the sorption of hydrolyzable metal ions, the sign reversal of the ζ potential can make the net EO flow insignificant in clay soils with high pore fluid electrolyte concentrations. In this case, and also for saline soils, electromigration becomes the dominant mechanism of EK transport.

Recognizing the inability to uncouple v_{eo} from E in the classical H–S equation for macroscopic systems, Khan (Khan and Pamukcu, 1989; Khan, 1991) proposed a modified theory of EO flow in clay soil. In Khan's work, the ionic conductance through the bulk fluid and the electronic conduction through the EDL at the solid–liquid interface were lumped in series, representing resistances in parallel or conductance in series. When modeling mass flow in the direction of the electric field, it is intuitive to line up the conducting elements in the direction of the electric field. The EK transport theory clearly shows that it is the electronic conductance at the EDL that induces EO transport and the ionic conductance in the bulk fluid that results in the ionic transport. Hence, the following relations are expected to hold:

$$i = i_s + i_b \tag{8.5}$$

$$E = \frac{R_s i_s}{L} = \frac{R_b i_b}{L} \tag{8.6}$$

$$v_{eo} = k_{eo}\frac{R_s i_s}{L} = \left[\frac{k_{eo}\tau}{\sigma_s n}\right]\cdot\frac{i_s}{A} \tag{8.7}$$

$$v_{eo}A = Q_{eo} = Ki_s \tag{8.8}$$

where k_{eo} is the H–S coefficient of EO conductivity; i, i_s, and i_b are the total, surface, and bulk currents (Amps or C/s), respectively; R_s and R_b (Ohm) are the surface and the bulk resistances; σ_s is the surface conductivity (l^{-1} Siemens or l^{-1} Ohm^{-1}; where l represents the unit of length); τ is the tortuosity; n is the porosity; and A is the total area ($A_{flow} = nA$). The relation between the resistance and conductivity is given as $R_s = L'/(\sigma_s A_{flow})$, and the tortuosity term τ is included to augment L ($L' = L\tau$) because the electric currents may follow a tortuous path through the length, L, of the electrochemical cell for which the pore walls are assumed lined up with connected EDLs. In this model, K emerges as a factor that relates volume rate of EO flow of fluid movement per electrical charge.

Shang et al. (1995) looked at the relation of conductivity and apparent polarization in several wet clay systems where they found that the surface conductivity and the apparent permittivity decreased significantly when the volume fraction of the particles was increased beyond 0.3. They also found that the conductivity of the bulk fluid was at least an order of magnitude higher than the surface conductivity (i.e. $R_b/R_s \cong 0.1$) and that the surface conductivity increased when the clay was remolded (i.e. $R_b/R_{sr} \cong 0.2$–0.5). These findings and other studies (Loch et al., 2005; Lima et al., 2010) support Khan's (1991) "parallel resistances" model when the flow direction is implicitly taken to be the direction of the electric current flow. Because of the EO velocity representation using the surface conductivity, σ_s would require fully developed EDLs with connectivity, increasing the dry clay content would inhibit the connectivity of the EDLs, and hence cause the reduction in the measurement of the surface conductivity for the overall system. Remolding of wet clay pastes has been shown to reduce the repulsion between particles and cause them to align in a face-to-face orientation (Quigley, 1980; Mitchell and Soga, 2005) and, therefore, possibly increase the connectivity of the electric layers in edge-to-edge correspondence. It follows that the preferential alignment of the particles in the direction of the electrical field helps increase the surface conductivity, as was shown by Shang et al. (1995).

Through laboratory experiments K has been shown to be a constant, independent of the electric field and soil mineralogy for similar major ion concentrations of the bulk fluid (Pamukcu and Wittle, 1992a, b; Wittle and Pamukcu, 1993). In these studies, the volume of EO flow (V, cc) per electric charge (either in Coulombs or moles of electrons), referred to as the *EO efficiency coefficient*, K_{ef} (cc/C), was determined from numerous EO tests of different clay and solute combinations. The relation between the cumulative electrical charge versus the cumulative volume of EO flow was found to be linear for a given clay type (e.g. kaolinite, montmorillonite, and sand–clay mixtures) at a given initial concentration of different ion species (As, Cd, Co, Cr, Cs, Hg, Ni, Pb, Sr, U, and Zn), all applied separately in the pore fluid. In these series of experiments (Wittle and Pamukcu, 1993), the clay matrixes used were pure kaolinite (KS), montmorillonite (MS), kaolinite with simulated groundwater electrolytes (KG), kaolinite with humic substances (KH), and a mixture of fine sand and 10% by weight montmorillonite (SS). In all cases the transport experiments were conducted over 48 h with a constant electric field of 4 V cm^{-1} applied across the electrodes. The clay specimens were sandwiched between two electrode chambers housing a set of graphite electrodes in tap water as shown in Figure 8.1.

The K_{ef} was demonstrated to be a constant factor for the soil type but changed with the initial concentration of the major ion. Figure 8.2 shows this variation for Cd^{+2}

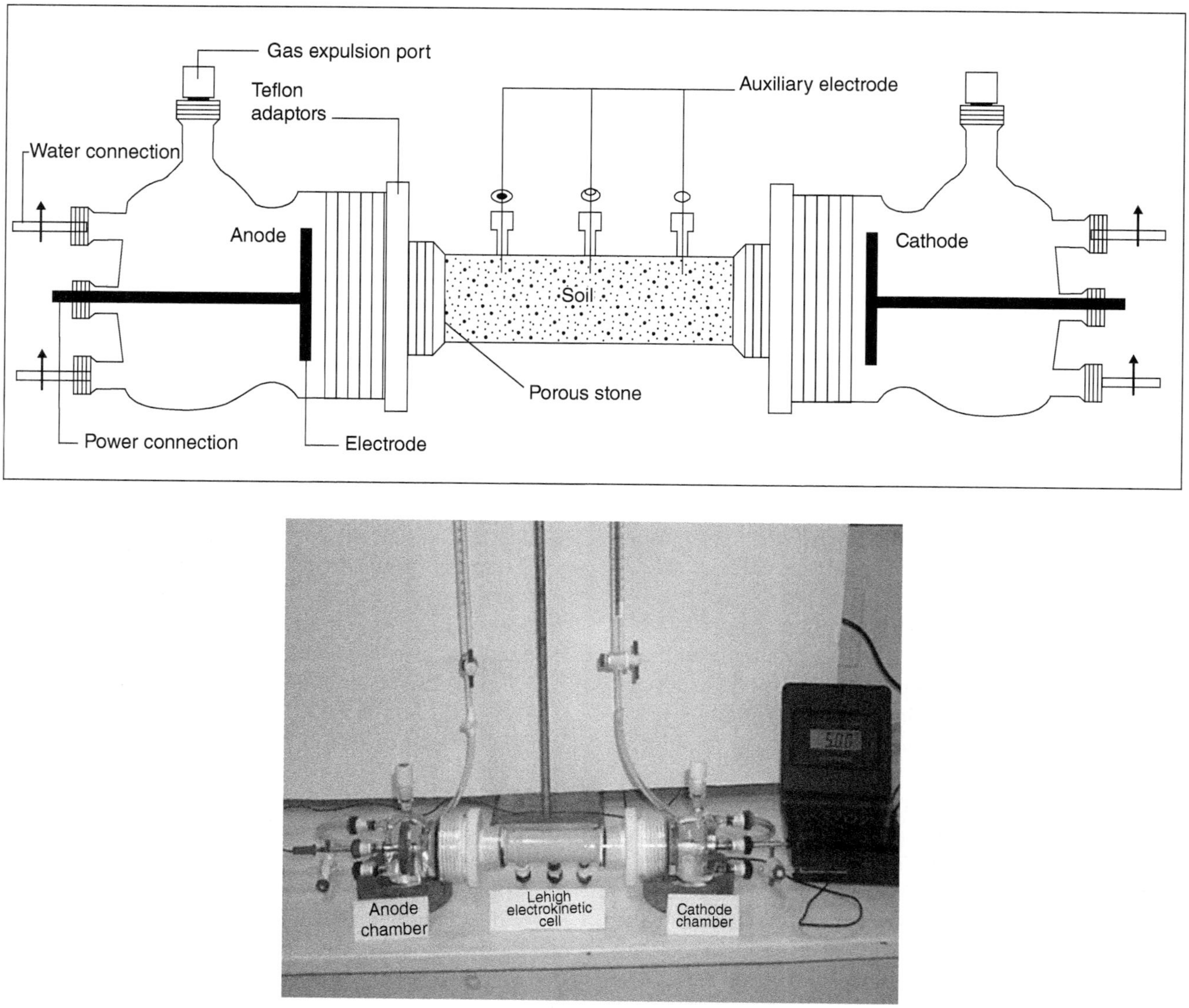

Figure 8.1 Schematic and photo of the Lehigh EK test cell assembly.

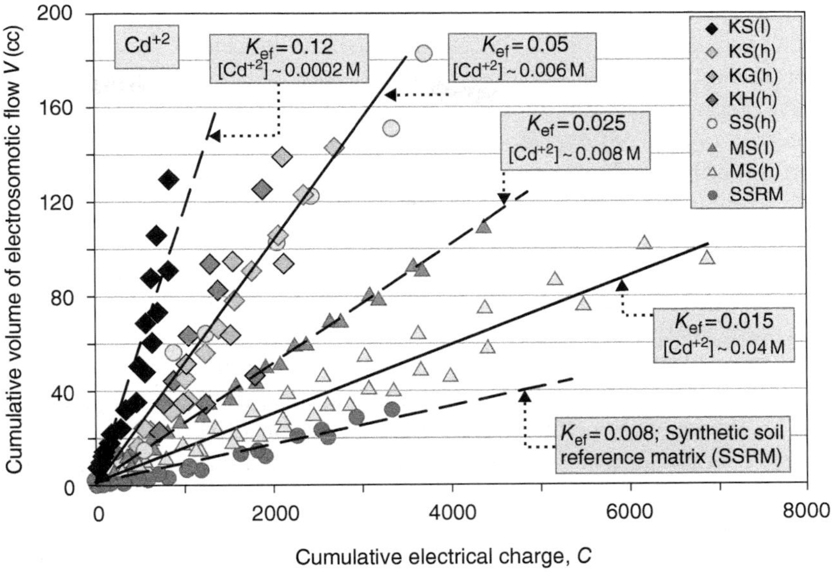

Figure 8.2 Variation of cumulative flow with electric charge. *Source:* Adapted from Pamukcu and Wittle (1992a) and Pamukcu (2009).

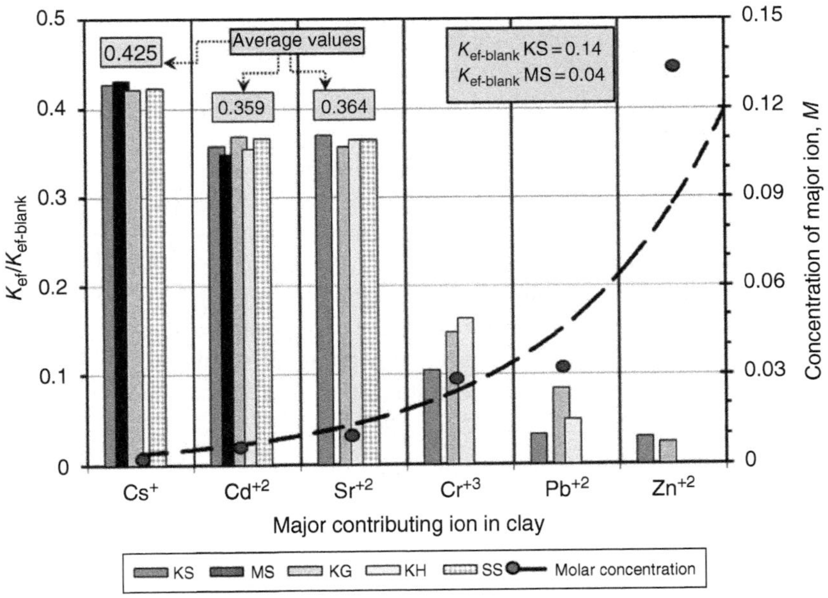

Figure 8.3 Variation of measured K_{ef} factors with ion and soil type at same concentration. *Source:* Adapted from Pamukcu and Wittle (1992a) and Pamukcu (2009).

as the major metal ion in clay pores. Also shown on the graph is data from a reference soil matrix (synthetic soil reference soil matrix [SSRM]) that represented a typical well-graded reference soil matrix (USEPA). The behavior of K_{ef} shown in Figure 8.2 is typical of all the other major contributing ions tested in these series of experiments (Pamukcu and Wittle, 1992a, b, 1993). The constancy of K_{ef} with soil type or clay mineral demonstrates that the assumptions made in developing Eq. (8.8) are valid. The following treatise strengthens further the validity of these assumptions. Figure 8.3 presents the variation of the measured K_{ef} for all soils and six selected major contributing ions (Cd, Cs, Cr, Pb, Sr, and Zn) in the bulk fluid, all applied separately. There is a clear trend of decreasing K_{ef} factor with increasing ion concentration, where the normalized K_{ef} of different soil types with similar molar concentration cohort of different ions are shown. The values are normalized by the measured K_{ef} values of the respective "blank" clays, those that were mixed and consolidated only with distilled water. At low concentrations (e.g. Cs, Cd, and Sr, all ≤0.01 M), the concentration and the type of the ion appear to control the K_{ef} factor, as all soil types show very close to constant normalized values of K_{ef} for the three ions involved (except MS with Sr). Subsequently, as the concentrations of the major contributing ions increase beyond 0.01 M, fewer soil types comply with the categorization, because the mineralogy of the soil and its unique chemical interaction with the dominant ion comes into effect. Because of the uniqueness of this interaction, the normalized values show greater variation for each ion involved at high concentration (e.g. Cr, Pb, Zn) then the low concentration (e.g. Cs, Cd, Sr) group.

K_{ef} is computed easily by measuring in the laboratory the total volume of flow through the clay matrix and the corresponding total current. In Eq. (8.5), the total current, i, is given as the summation of the bulk current, i_b, and the surface current, i_s. Unless the contribution of the surface current is uncoupled, the measured K_{ef} based on total current, i, will always display high dependency on soil composition and mineralogy and the type and concentration of the major contributing ion. During EK transport, as the bulk fluid chemistry changes, so does the EDL properties of the clay medium. Equation (8.7) derived earlier shows a ratio of the H–S coefficient of EO conductivity k_{eo} and the surface conductivity, σ_s, which should remain constant as they are uniquely related. Only when this constancy ($=k_{eo}/\sigma_s = \text{constant}$) is true, so is the expression given in Eq. (8.8). It is important to note here that the use of Eq. (8.8) necessitates the presence of distributed surface conductivity that enables the surface currents to travel within the electrochemical system for EO to occur. Yin et al. (1995) backed up Khan's theory, which relates EO mobility and surface conductivity. They found that there is no apparent relationship between EO mobility and the applied electric field. The term EO mobility, defined as the average velocity achieved by the pore water relative to the solid skeleton due to an externally applied electrical field of unit strength (cm^2 s-V^{-1}), appeared to be directly proportional to the specific conductance (mho cm^{-1}) of the specimens at different water saturations in independent tests (Yin et al., 1995).

8.2.3 Mathematical Modeling of Transport

In the past two decades, many researchers studied and proposed theoretical models of ion transport under electric field as it applied to contaminated clays. In most cases, dilute solutions, rapid dissociation–association chemical reactions, and small double layer thickness were considered. These models were all based on Nernst–Planck equation, and each was calibrated to the experimental findings (Mitchell and Yeung, 1991; Alshawabkeh and Acar, 1992, 1996; Shapiro and Probstein, 1993; Denisov et al., 1996; Cao, 1997). Cao in 1997 modeled the ion transport considering the effect of changing electrical field due to redistribution of charge concentration. The effect was later shown in a rigorous mathematical analysis by Chu (2005), where concentration gradients give rise to spatial variation of conductivity to overcome the violation of electroneutrality of the bulk fluid.

In the basic governing equation of advection diffusion, dispersion refers to the movement of species under the influence of gradient of chemical potential, whereas advection is the stirring or hydrodynamic transport caused by density gradient or forced convection. A general one-dimensional mass transfer to an electrode is governed by Nernst–Planck equation:

$$J_i(x) = -D_i^* \frac{\partial C_i(x)}{\partial x} - u_i z_i F C_i \frac{\partial \Phi(x)}{\partial x} - C_i v(x) \quad (8.9)$$

where

J_i = total flux of species i

$u_i = D_i^*/RT$ = mobility of species i (*Nernst–Einstein relation*)

$v(x)$ = advective velocity ($=v_{eo}(x)$, the EO velocity)

The mobility of an ion ui is defined as the limiting velocity of an ion moving in an electric field of unit strength. The minus sign arises because the direction of flux opposes the direction of increasing Ci. Applying Fick's second law, we arrive at another form of Nernst–Planck equation (also given earlier in Eq. (8.1)) with the added advection term:

$$\frac{\partial C_i}{\partial t} = \nabla\left(D_i^* \nabla C_i + u_i z_i F C_i \nabla \Phi\right) + \left(u_i z_i F \nabla \Phi + v\right) \nabla C_i \quad (8.10)$$

Equation (8.10) is the basic mass transfer equation for an electrochemical system under an electric field. In Eq. (8.10) the first parenthetical term is the diffusion, the second term (i.e. $u_i z_i F$ term) is the migration, and the third term (i.e. v term) is the advection contribution to the total mass transport of the species i in an electrochemical system. In clay soils where the hydraulic advection is negligible compared with EO advection, the velocity term v is simplified as the EO velocity (v_{eo}).

The two important system parameters that contribute to the ion flux, and hence the distribution of current density in the system, are the electromigration (v_{em}) and the EO (v_{eo}) velocities given in summation as the advective velocity (v_{adv}):

$$v_{adv} = v_{em} + v_{eo} \quad (8.11)$$

where

$$v_{em} = u_i z_i F \nabla \Phi$$

and, from Eqs. (8.4) to (8.7):

$$v_{eo} = \left[\frac{\varepsilon_s \varsigma \tau}{\eta \sigma_s n}\right] \cdot \frac{i_s}{A}$$

The electromigration velocity in Eq. (8.11) is the speed of ion movement in the pore water caused by an electric field in infinitely dilute solutions. In the pore fluids with finite ion concentrations, which more closely resemble the electrochemical systems of contaminated clays, influence of the interionic attraction should be considered. Generalized ion mobility that accounts for the possibility of interactions between the ionic species can be used to modify the electrochemical potentials of individual ions in the flux term (Eq. (8.9)) when strong chemical reactions between ionic species are considered (Newman, 1991). The resulting relative electric field due to the retardation effects of dissymmetrical ionic atmosphere during ion transport was considered in Cao's model (1997) as described below. The EO component of the EK transport is dependent on the EO velocity of the fluid flow. The classical H–S equation expresses this parameter as a function of the field gradient. Due to the tight coupling between the ion concentrations and electric potential – as the ions contribute to the local electric potential themselves – the use of H–S EO velocity in transport determination in clay soils can result in nonlinear predictions (Ravina and Zaslavsky, 1967; Chu, 2005). Hence, uncoupling this parameter from the electric potential using the surface conductivity σ_s and the resulting proportion of the current transferred over the solid–liquid interface i_s provides an intrinsic EO velocity dependent on clay surface properties only, as first introduced by Khan (1991) in Eq. (8.8).

The model predicted distribution and evolution of bulk conductivity, σ_b, and field strength, E, in kaolinite clay containing $Pb(NO_3)_2$ at the initial concentration of 0.05 M in its pore fluid are presented in Figures 8.4 and 8.5, respectively (Cao, 1997). In Figure 8.4, the normalized distribution of the conductivity (normalized by the initial conductivity of 0.28 S m^{-1}) shows a consistent decrease in the conductivity at the cathode region and a steady spread of the lower conductivity toward the anode area over time. Part of this result is attributed to the decrease of dissolved lead concentration, which prevails over the increase of H^+ concentration. The change of the conductivity is influenced by combination of three factors: (i) the initial concentrations of the species, (ii) the spatial and temporal distribution of ionic mobility of each species, and (iii) the production rate of H^+ at anode.

As the conductivity varies, the lower conductive area requires higher potential over it to keep a consistent current with the other parts of the soil column, resulting in a nonlinear electric potential profile across the cell. This tendency is observed in Figure 8.5 where the time evolution of the electric potential distribution is presented. As the electrochemical transport continues, the potential curve becomes less steep over the entire cell except in a narrow region adjacent to the cathode, where the electrical potential is likely to have the largest gradient. This tendency agrees well with the experimental data reported by Wittle and Pamukcu (1993) and Acar and Alshawabkeh (1996), among many others.

When charged ions transport under the influence of an externally applied electrical field, their concentration distributions change with time, which leads to a change in local electrical conductivity, as discussed previously. The change of local electric conductivity directly alters the value of potential gradient at that specific point. Hence, the changing electric conductivity and electrical field strength describe the transport process of the species implicitly. Furthermore, the mobility of the charged species decreases as they congregate in the direction of their migration. The increasing concentration of the ions leads to a decrease in equivalent conductance in their migration direction (Kortüm and Bockris, 1951). According to Kohlrausch's law of the independent migration of ions, the equivalent conductance of an electrolyte is additively composed of the ionic conductivities of the constituent ions. As ions move under an applied electric field at their terminal ionic velocity (u_i) of infinite dilution,

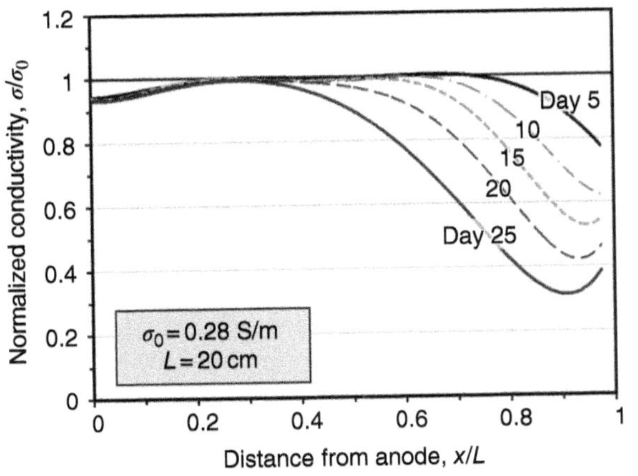

Figure 8.4 Model predicted spatial and temporal variation of conductivity during EK transport of Pb in kaolinite clay. *Source:* Adapted from Cao (1997) and Pamukcu (2009).

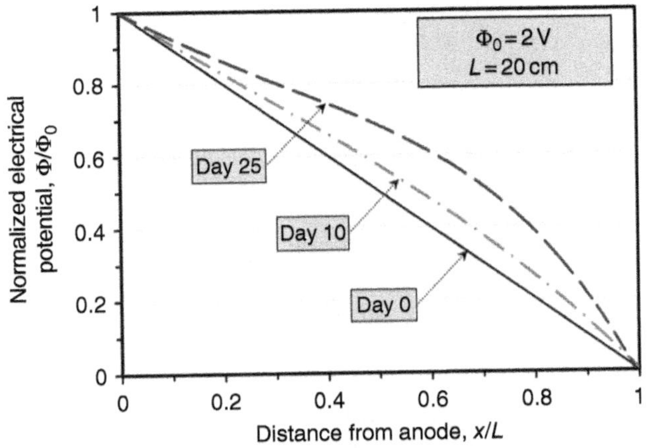

Figure 8.5 Model predicted spatial and temporal variation of electric potential during EK transport of Pb in clay. *Source:* Adapted from Cao (1997) and Pamukcu (2009).

two other phenomena occur that retard ion velocity and effectively reduce the bulk conductivity. These are called the "retardation" and "electrophoretic" effects. The retardation effect comes about when the ionic atmosphere becomes unsymmetrical about an ion in motion, as the charge density decreases in front of the ion and increases behind it. For a moving positive ion, a net negative charge trails behind it, exerting an electrostatic force in the opposite direction. The electrophoretic effect occurs as the negative ions in the ionic atmosphere around the positive ions migrate in the opposite direction, taking their solvent sheaths with them. The positive ion therefore travels against a medium moving in the opposite direction, hence a viscous drag. A quantitative formulation of ion mobility, relating the magnitudes of the electrostatic retardation force and the viscous drag force to the radius of the ionic atmosphere, directly provides a theoretical basis for the well-known empirical relation known as "Kohlrausch's square root law" (1900), in terms of equivalent conductance, Λ_v ($12\,\Omega^{-1}$) (Kortüm and Bockris, 1951). Λ_v is also the sum of individual ionic conductivities ($Fu+$ for positive ions and $Fu-$ for negative ions); hence the ionic mobility of a given ion, u_i, can be expressed in terms of the equivalent conductance (Bard and Faulkner, 1980):

$$u_i = \frac{\Lambda_v t_i'}{F} \tag{8.12}$$

where $t_i' = u_i/(u_+ + u_-)$ is the transference number of the of species i.

Eq. (8.12) effectively incorporates the retardation effects into the mobility determination for high-concentration solutions. As an example, the variation of the mobility of the positive ion (e.g. H^+, K^+, Na^+) of the 1,1 valence electrolytes HCl, KNO_3, and NaCl in aqueous solutions at room temperature was computed according to Equation (8.12), where Λ_v is described by "Kohlrausch's square root law." These relations are plotted in Figure 8.6. The variation of the transference numbers of the cations with concentration is also plotted to discern its effect on the mobility of each ion. As observed, the reduction of the mobility of each ion with increasing concentration appears to be mostly dependent on Λ_v.

Cao's model was used to predict the spatial and temporal distributions of four ionic species, Pb^{+2}, NO_3^-, H^+, and OH^-, in kaolinite clay. Four one-dimensional advection–dispersion equations were solved simultaneously, one for each ionic species. Changes in electrical potential, conductivity, and field strength were considered in the solutions. Figure 8.7 presents the spatial distribution of the total lead and the pH for 35-day runs of Cao's model. Superimposed on the numerical solutions are results from the laboratory tests of $Pb(NO_3)$-contaminated

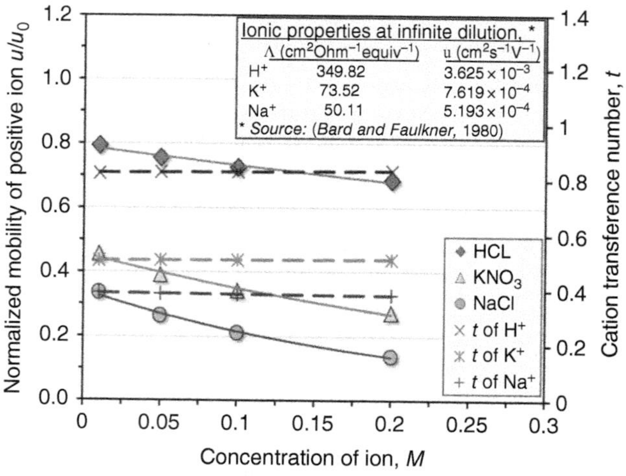

Figure 8.6 Variation of mobility and transference with concentration of selected ions. *Source:* Pamukcu (2009). Reproduced with permission of John Wiley & Sons.

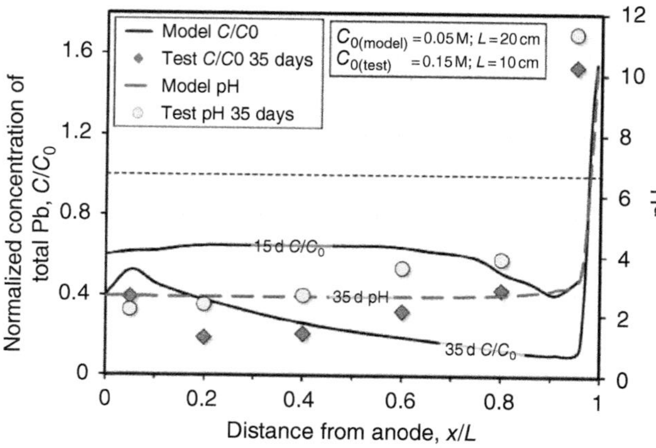

Figure 8.7 Model predicted and measured distribution of pH and total Pb concentration at 35-day EK transport. *Source:* Pamukcu (2009). Reproduced with permission of John Wiley & Sons.

kaolinite clay subjected to long-term EK treatment. The 35-day lead and pH distribution is predicted well by the model, including the accumulation of the metal in a *narrow zone adjacent to the cathode*. The model predictions of the 15- and 35-day distribution of lead concentration indicate that migration velocity of the ions becomes affected by the electrical field distribution over time. The reduction in electrical conductivity at the cathode zone results in higher electrical field in that region. The high electrical gradient hurls the ions to the narrow high pH zone to be accumulated by precipitation, hence evacuating the region immediately behind it. The progression of this effect is clearly seen by the evolution of lead distribution from 15 to 35 days, as the distribution toward the cathode regions continues to dip behind the high pH zone. The distribution of lead near the anode does not dip at the same rate because the conductivity remains high, and the electric potential distribution tends to flatten there. Consequently the ions slow down responding to a lower electric field. Factors such as sorption, retardation, and precipitation influence the model predictions, but their relative contributions may be small other than the narrow high pH zone near the cathode.

8.2.4 Electrochemical Transformations

Transport and extraction of both the inorganic and organic substances from clay pores frequently involve the addition of one or more chemical reagents to the system (Hansen et al., 2005; Pazos et al., 2008; Chang et al., 2010; Chen et al., 2011; Ryu et al., 2011; Liu and Wang, 2013), introducing with it the possibility of secondary pollution. Recently, "green technology" has been explored for in situ destruction of contaminants, enhancing beneficial electrochemical reactions by EK. Optimization parameters include changing electric field direction and intensity (Peng et al., 2013), applying a sinusoidal electric field with pulses and a polarity reversal (Rojo et al., 2012), and increasing potential gradient (Lei et al., 2012). DDL processes of clay have been suggested to influence in situ conversion of heavy metal soil contaminants to potentially less toxic forms when an external DC is applied to the clay–electrolyte system (Pamukcu et al., 2004; Pamukcu, 2009; Brosky and Pamukcu, 2013). The DDL shrinks, reducing the EK potential (i.e. zeta potential) of the system in the event of lowered pH and/or increased ionic concentration of the pore fluid due to accumulation of charges in DDL (Israelachvili and Adams, 1978). Therefore when the EK potential of a clay–electrolyte system changes under varying pH or electrolyte concentration, the system may become prone to spontaneous faradaic reactions as it tries to restore the equilibrium back. In EK treatment of clays, the transient acid front movement from anode to cathode and the regions of low and high pH developing near the electrode sites can set the stage for triggered faradaic reactions, resulting in beneficial transformation of the contaminants.

An electrode surface with its surrounding bulk solution is considered to explain the necessity of faradaic reactions for electroneutrality. Figure 8.8 shows a charged electrode surface with fully developed DDL, where the oxidants with a positive charge diffuse toward the negatively charged electrode, accept electrons from the electrode at the outer Helmholtz plane (OHP) interface, become the reductants, and diffuse toward the bulk solution. IHP and OHP are the inner and outer Helmholtz planes, respectively. An equivalent circuit representing each physical component at the interface and in the bulk solution is also shown. Here, Cddl is the double layer capacitor, Rp the polarization resistor represents the electrochemical processes in the region between IHP and OHP, the diffusion-controlled impedance ZW or Warburg resistor represents the processes in the diffused layer, and Rb is the bulk solution resistance. Similarly, Figure 8.9 depicts a schematic of a clay particle with negative surface charge with the fully developed DDL, where the capacitor and the diffuse layer structure mimic that of an electrode interface as shown in Figure 8.8. Under equilibrium, the electroneutrality of the particle is satisfied by the specifically adsorbed ions and those that move in and out of the diffused layer from the surrounding bulk solution. Electrochemical transformations do take place in natural system either through intervention of redox active

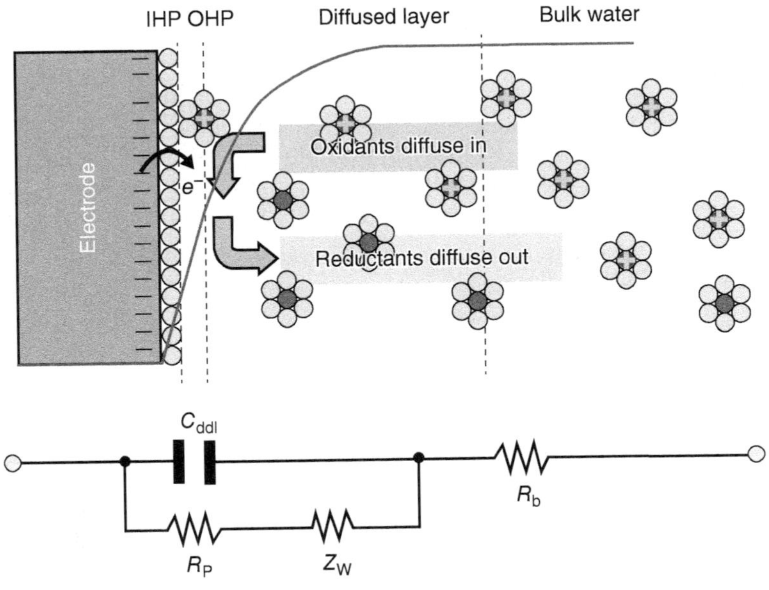

Figure 8.8 Schematic of a DDL interface of an electrode surface with equivalent circuit representation in the usual *Grahame* notation: C_{ddl}, R_p, Z_W, R_b. *Source:* Park and Yoo (2003). Reproduced with permission of American Chemical Society.

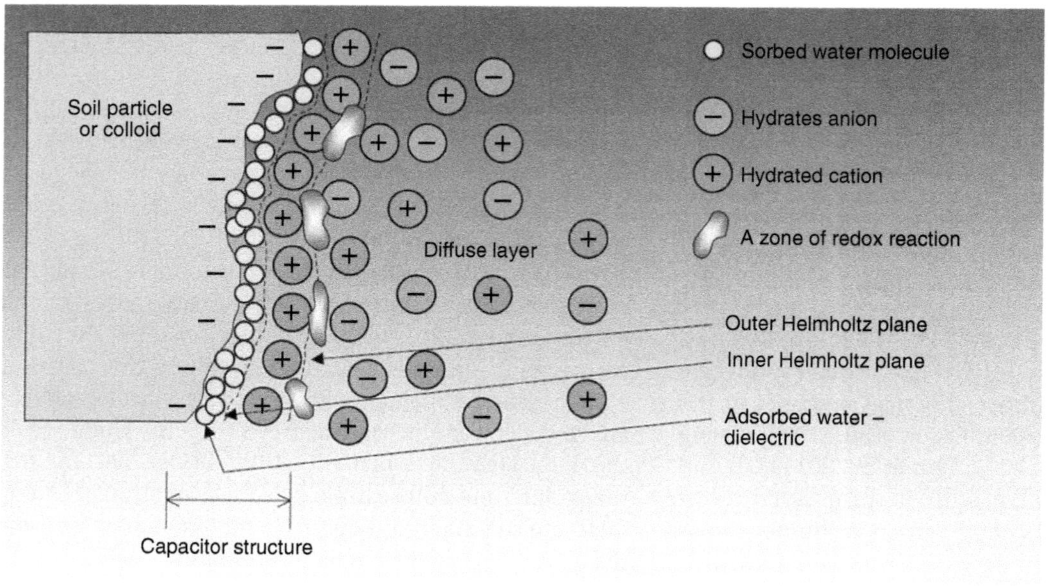

Figure 8.9 Conceptualized schematic of a clay particle with DDL structure that mimics the electric interface of an electrode with a "zone of redox reaction."

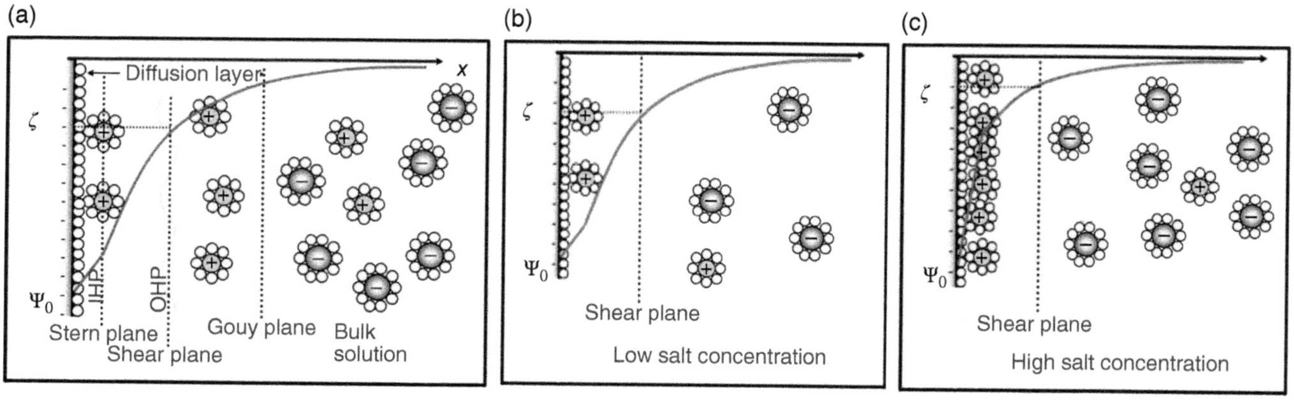

Figure 8.10 (a) The distribution of charges against charged clay surface in DDL formation. (b) The distribution of charges at low salt concentration. (c) The concept of electrolyte ions adsorbing on the surface and reducing the zeta potential without altering the surface potential or charge at high salt concentration. *Source:* Illustrations from PERC, http://perc.ufl.edu/.

compounds or by microbial activity (Kappler and Haderlein, 2003; Busalmen et al., 2008). Since most of these transformations take place in the bulk fluid, the clay DDL interface does not contribute to the process directly except for ion exchange and sorption reactions.

In soils, all particles with active interfaces (i.e. IHP and OHP [Hunter, 1981]) can be regarded to "act" as electrode interfaces when a current is supplied to them, adding up to a significantly large effective electrode surface area. Current flows across these interfaces via two pathways: the faradaic and non-faradaic. When an external electric field is applied to water-saturated clay of high ionic concentration in the pore fluid, given the incompatibility between the conductivity of two conducting layers in the mixture: (i) the DDL of clay particles with low conductivity (σ_s) and (ii) the surrounding electrolyte solution (bulk solution or pore fluid) with high conductivity (σ_b), a large electrical potential is induced across the DDL. This results in compression of the DDL, resulting in higher charge density in this region. This is similar to the effect when salt concentration of the bulk fluid is increased, causing the compression or shrinkage of DDL. The electric potential distribution shifts back (i.e. toward the solid) and down, lowering the EK potential at the shear plane (*ζ, zeta potential*) as shown in Figure 8.10a–c.

The potential distribution shift in this manner increases the intensity of the electric field within the Stern layer (between IHP and OHP) while decreasing the capacitance, *C*, of the DDL. The reduction in capacitance of the diffuse layer can trigger electron transfer across the diffuse layer toward the solution, giving rise to a faradaic "cathodic" current (Bard and Faulkner, 1980). The capacitance, *C*, of the double layer is directly related to the bulk electrolyte concentration and the surface potential of the particle. The capacitance reaches a minimum at the *point of zero charge* (PZC) and sharply increases on either side for the low electrolyte concentrations of the bulk solution. The surface charge or EK potential of most clay sharply varies with pH. For example, the PZC for most types of kaolinite clays have been reported to coincide between pH 2.0 and 4.0, depending on the electrolyte concentration of the pore solution (Bard and Faulkner, 1980). Accordingly, as the DDL compresses or expands, faradaic currents (cathodic or anodic) can generate across OHP (i.e. frequently assumed where the EK potential and shear surface develops), whereby the concentration gradients are overcome and the electrical equilibrium restored.

Brosky and Pamukcu (2013) proposed that a nonspontaneous change in the oxidation–reduction potential (ORP) (i.e. E_h) of a clay–electrolyte mixture represents the charge discharge due to the compression of the DDL. They demonstrated a shift in E_h values with increasing clay concentration of an aqueous slurry during electrically enhanced reduction of Cu(II) by Fe(II), as shown in the figure, "Temporal variation and shift of E_h with direct current in Cu(II)–Fe(II) mixtures of clay where maximum E_h shift up is 0.50 V" (Brosky and Pamukcu, 2013). The increased clay concentrations promoted electrochemical reduction of Cu^{2+} to more environmentally benign Cu^{+} or Cu^{0} in the system. This is highly relevant to bottom waters and sediments of surface water bodies, where many heavy metal contaminants are found. Electrically treated clay can switch its surface potential sign in time as pH lowers below an "apparent" PZC (Hunter, 1981). It can be seen from the figure, "Zeta Potential with pH in Cu(II)–Fe(II) mixtures of clay" (Brosky and Pamukcu, 2013), that the PZC is achieved for the kaolinite–Cu(II)–Fe(II) test slurries at around pH 2.4. The PZC signifies that the charged layers surrounding the clay particles become saturated with charges from ions in solution as well as electrons from current flow. At this state, the DDL capacitor is storing maximum charge and therefore has reached its minimum capacitance and maximum redox potential. Zeta potential data shows that electrically treated clay can switch in zeta potential sign in time as the pH decreases below the "apparent PZC" of pH 2.4. Observing the "Temporal variation..." figure, it appears that the clay system shift from base E_h is maximized (approximately between 4 and 6 h) when the pH of the system is recorded at around 2.4. This is consistent with the PZC of 2.4 shown in the "Zeta Potential..." figure.

Evidence of DDL-enhanced Cr(VI) reduction to less toxic Cr(III) by DC application has been reported previously (Banerjee et al., 1987; Pamukcu et al., 2004, 2008; Alshawabkeh and Sarahney, 2005; Hannum and Pamukcu, 2007). Electrolytic transformations of selected CHCs and PAHs have also been demonstrated successfully in water and wastewater (Pulgarin et al., 1994; Franz et al., 2002). There has also been recent evidence for the double layer-enhanced removal of Pb(II) (Ahn et al., 2010). Previously, controlled laboratory experiments of kaolinite clay (see test setup in 11) injected with Fe(II) showed that an externally applied electric field caused an additional "cathodic current" that drives forth the reduction of Cr(VI) in clay (Pamukcu et al., 2004). These transformations were characterized as to have resulted from the capacitive changes on the clay surfaces. The results in these experiments showed that the system ORP increased by a positive shift from the standard solution ORP in the presence of the clay medium and the induced electrical field. The ORP measurements were plotted against the reaction quotient of the Nernst relation, where the data was categorized by pH, as shown in Figure 8.11. The low pH range (pH range 2–3) data showed the best

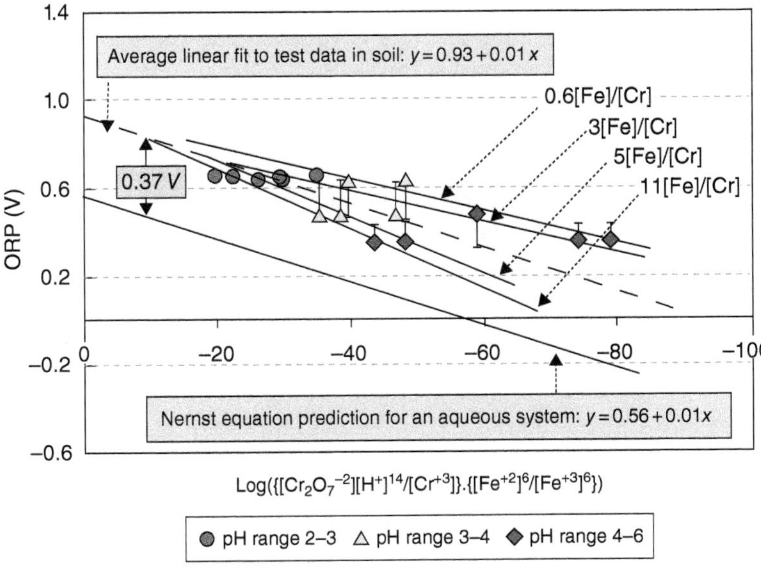

Figure 8.11 Measured and predicted redox potential variation with reaction quotient of species concentration in Nernst relation (average redox potential shift = 0.37 V). *Source:* Pamukcu et al. (2004). Reproduced with permission of American Chemical Society.

agreement with the linear fit, and the relative scatter of data at higher pH values was attributed to poor polarization of the electric layer. At low pH, the DDL is compressed with a higher ion concentration and higher electric field (Israelachvili and Adams, 1978), hence higher degree of polarization and lowered capacitance. At higher pH, the DDL is expanded with reduced degree of polarization, consequently higher capacitance. The ions in an expanded DDL are less restricted to move; therefore discernible fluctuations are likely to occur in the potential development across the diffuse layer.

8.3 Electrokinetically and Electrochemically Aided Soil and Sediment Remediation

The main and best-known advantage of electrokinetically aided environmental mitigation is its ability to move water, dissolved contaminants, and colloidal/micellar particles through low permeability soils and porous formations that are not amenable to hydraulic treatments (Acar and Alshawabkeh, 1993, 1996; Acar et al., 1994; Hamed et al., 1991; Pamukcu and Wittle, 1992a, b; Probstein and Hicks, 1993; Shapiro and Probstein, 1993; Wittle and Pamukcu, 1993; Pamukcu, 1994; Pamukcu et al., 1995, 1997). The electrokinetically aided transport is based on well-known EK processes (Casagrande, 1949; Hunter, 1981; Mitchell and Soga, 2005), primarily comprised of *EO*, *electrophoresis*, and *ion migration* in wet soil. The EK treatment involves applying a low DC (on the order of milliamps per unit cross-sectional area) or a low potential gradient (on the order of few volts per unit distance) between the power electrodes inserted in the ground (Wittle and Pamukcu, 1997; Wittle and Bell, 2002). As a result, the contaminants are transported toward the anode or cathode electrode sites by ionic or electrophoretic migration and EO advection. The contaminants are then removed at the electrode sites by one of several different methods: electroplating, adsorption, precipitation complexation, or pumping.

The EK treatment is a powerful soil cleanup technique explored in both the laboratory and the field for the last several decades, particularly for difficult site conditions. The field applications have frequently demonstrated that the success of the treatment is hinged on site-specific engineering design of the total system, which includes the enhancement of the transport (Lageman, 1993; Yeung et al., 1996; Acar et al., 1995; Lindgren, 1998). Some of the phenomena experienced with EK treatment of soils that interfere with the steady transport of fluid or charged particles to an electrode site are (i) the spatial and transient variation of soil pH, electric gradient and pore pressures, (ii) surface polarization effects on particles and electrodes, and (iii) redox reactions near electrodes.

The electrolysis reactions that occur at the electrode sites of the applied potential can produce substantial changes in pH of the pore solution. The reactions are oxidation at the anode generating hydrogen (hydronium) ion and oxygen gas and reduction at the cathode generating hydroxyl ions and hydrogen gas, as follows:

$$\text{Anode:}\quad H_2O \Leftrightarrow 2H^+ + \frac{1}{2}O_2 + 2e^- \quad (\text{Standard potential})\quad E^0 = +1.23\,\text{V}$$

$$\text{Cathode:}\quad 2H_2O + 2e^- \Leftrightarrow 2OH^- + H_2 \quad (\text{Standard potential})\quad E^0 = -0.83\,\text{V}$$

The rates at which the hydrogen and hydroxyl ions are produced depend on the current, as such the lower the current density, the lower the rate. The hydrogen and the hydroxide ions are transported into the soil by electromigration. The resulting pH fronts influence the adsorption/desorption and precipitation/dissolution reactions of ionic constituents in the soil pores, as well as the surface chemistry and electrochemistry of the mineral particles. The reduced pH is beneficial in desorption of most metals from soil, as well as dissolution of most metal precipitates. Natural soils with high buffering capacity and carbonate content, or those under the groundwater table, tend to neutralize the acid front and maintain a higher pH environment (Pamukcu, 1994). The base front generated at the cathode electrode penetrates the soil at a narrow distance, creating a zone of high pH and promoting precipitation of the inorganic species into insoluble hydroxide salts in the pore space and also their retention on the clay surfaces. These side effects may be overcome by continuous adjustment of the pH at the analyte and the catholyte and/or by adjusting the current density in the soil.

The mitigation outcomes of EK treatment can be categorized in three general processes, not exclusive of each other: *extraction and separation*, *immobilization*, and *transformation*, as discussed below.

8.3.1 Electrokinetically Aided Separation and Extraction

EK extraction is analogous to *soil washing* whereby the contaminant is *extracted* from the soil pore space and subsequently collected in aqueous phase in a collection well or deposited at the electrode site. The alkali metals, e.g. Na(I), K(I), and Cs(I), and alkali earth metals, e.g. Sr(II) and Ca(II), tend to remain ionic under a wide range of pH and redox potential values; therefore they are expected to electromigrate and be extracted from soils readily unless they become preferentially sorbed onto solid surfaces and clay interstices in soil. Under ideal conditions, the predominant cation and its accompanying anion may be caused to separate efficiently by electromigration only, for which little or no EO water advection may be necessary (Weinberger, 1993). The electroneutrality can be maintained by the hydrogen and the hydroxide ions produced by the electrolysis reactions at the electrodes and transported into soil by migration or by desorption of the counter ions from surfaces of soil constituents.

Described below are several selected laboratory experiments conducted using Lehigh EK test cell (Wittle and Pamukcu, 1993, 1997; Pamukcu, 1994) to exemplify the variety of EK extraction and separation processes that can be engineered to address site-specific situations. It is important to note that the soil samples used in all the experiments reported below were the actual contaminated field soil specimens retrieved from the specific sites.

8.3.1.1 NaCl Extraction/Separation

The first example of ionic specie separation and extraction is provided for NaCl removal from drilling mud obtained at an Eastern Pennsylvania site. Figures 8.12 and 8.13 show the separation of sodium (Na) and chloride (Cl) from drilling mud soil samples of different water saturations under constant electric potential. The final pH profiles attained at the end of the tests are superimposed on the graphs. As observed in the figure, "Post Electrokinetic distribution of *Na* in drilling mud of various initial water saturation degrees (S1, S2, S3), at various pore volumes of flow (PVF)" (Pamukcu and Wittle, 1993), close to 100% recovery of the Na is accomplished in the 81% saturated specimen, S1, at the termination of the test, while about 70% of Na is recovered for the 53% saturated specimen, S2. The specimen designated as S1 shows about 60% recovery of Cl (the figure, "Post Electrokinetic distribution of Cl in drilling mud of various initial water saturation degrees (S1, S2, S3), at various pore volumes of flow (PVF)" (Pamukcu and Wittle, 1993), although not as high a recovery as that of Na. The inability to account for all the Cl transported to the anode chamber is attributed to the formation of gaseous chlorine (at pH below 4) that would have escaped from the anode chamber without accumulation.

8.3.1.2 Perchlorate, ClO_4^-, Extraction

In another example, perchlorate-, ClO_4^-, contaminated soil samples retrieved from an aviation facility near Rancho Cordova, California, were treated to explore the extraction of this species electrokinetically. The soil contained little clay but mostly fine to medium sand and silt. The packed samples were saturated in the laboratory

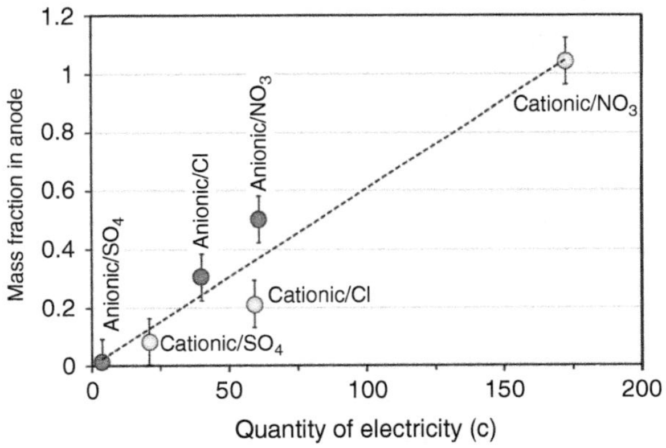

Figure 8.13 Mass fraction of anions extracted into anode per Coulomb of electricity.

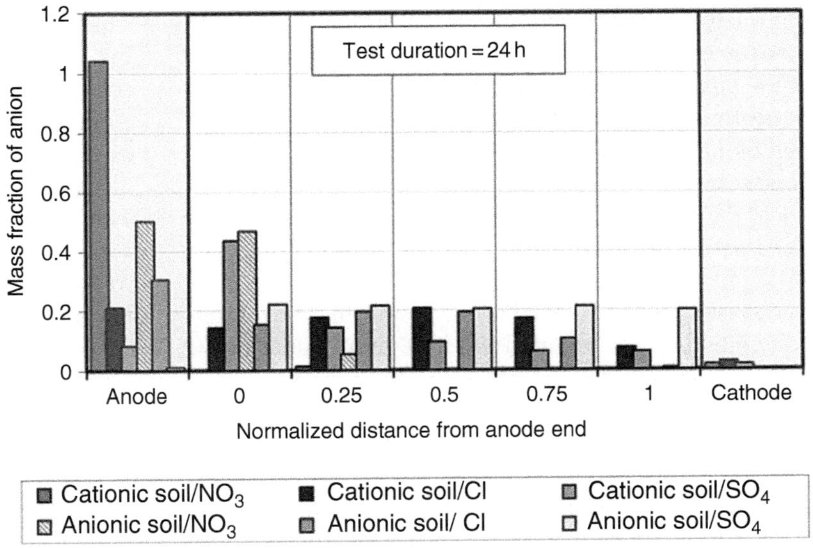

Figure 8.12 Post-EK distribution of mass fraction of anions of lead salts in cationic and anionic clay soils.

with the groundwater specimen obtained from the same location that contained high levels of perchlorate. The laboratory test specimens measured 25.7% water content by dry weight, bulk density of 2.8 g cm^{-3}, and void ratio of 0.6. Two tests were conducted in parallel, one of which served as a control test with no electrical current. In the treatment cell, the soil was subjected to constant 30 V across the electrodes residing in chambers filled with the local groundwater from site. The results of these tests are presented in the figures, "Electrokinetic removal of ClO_4^- from soil retrieved from an aviation facility in Rancho Cordova, Ca" and "Cumulative mass of ClO_4^- removed from soil into the electrode chambers during EK" (Pamukcu and Huang, 2001).

The current density varied from maximum of 1 mA cm^{-2} to minimum of 0.006 mA cm^{-2}, as shown in the figure, "Electrokinetic removal of ClO_4…." The efficiency of extraction (i.e. moles of perchlorate per quantity of electricity) is closely related to the current measured in the soil. The current peaked at 80% removal and diminished thereafter. Although the applied voltage remained at 30 V across the electrodes, the voltage potential measured in the soil was transient; decreasing thereafter the maximum removal efficiency was reached. A number of reasons can be offered for this occurrence, some of which are loss of electrical energy through the electrolysis of water, changes in soil–water chemistry, and transient pressure differentials in soil pores (Ray and Ramsey, 1987; Muraoka et al., 2011). This experiment exemplifies one of the typical situations where the majority of the current is carried by the ion flow in pore fluid and not through the double layers. Therefore, electromigration in pore fluid becomes the dominant mechanism of transport.

As observed in the figure, "Cumulative mass of ClO_4…," approximately 99% of the initial mass of the perchlorate (the initial perchlorate concentration was 840 mg kg^{-1}) were extracted out of the sample soil by the end of 7 days of EK treatment. The control specimen showed less than 30% removal by diffusion for the same duration. The rate of EK removal was significantly faster for the first 3 days of treatment, by the end of which 80% of the initial mass of perchlorate was removed. Although the soil remained conductive beyond the 3 days, most of the current carriers in the pore fluid were depleted during that time. The EO flow of water continued in the positive direction (from anode to cathode) for the first 8 days of treatment but ceased afterward. The average equivalent hydraulic conductivity was computed as 1×10^{-5} cm s^{-1} and the EO permeability as 1.7×10^{-5} cm^2 s^{-1} V^{-1}.

8.3.1.3 Ammonium, NH_4^+, Extraction

The processed municipal sludge by N-Viro Soil™ technique produces a Ca-rich material, a potentially inexpensive source of calcium for cement production. The fresh N-Viro soil gives out ammonia (NH_3(g)) as a by-product of the process, which needs to be mitigated prior to handling the material. EK extraction was proposed to reduce the excess ammonia in N-Viro Soil™. The visual and textural observation of the granular material indicated fine sand to silt size mixture with little plasticity. The leachate was a brown-colored liquid. Upon arrival, the leachate was acidified to arrest NH_4^+ in the liquid, thereby preserving its original concentration. The test specimens were prepared by mixing the N-Viro soil with the leachate at the reported liquid content of 65%. Two tests were conducted at constant potential of 30 V applied across the electrodes for the 200-h duration of treatment for each. The voltage gradient measured in the samples remained at approximately 0.3 V cm^{-1}. At the termination of each test, the soil was sampled at three points along its length and analyzed for moisture content, pH, and NH_4^+ concentration distribution. The figure, "Post Electrokinetic Distribution of NH_4 in Processed Municipal Sludge Soil" (Pamukcu and Huang, 2001), shows the post-PK distribution of the measured concentration of NH_4^+ and the pH along the soil sample and the anode and cathode liquid chambers. Overall, the concentration of ammonia in the soil samples was reduced from about 150 mg kg^{-1} to less than 6 mg kg^{-1} in both samples tested.

The mass balance determination showed that more than 50% of the ammonia was unaccounted. Larger concentrations of ammonia were found in the anode chamber in both tests, where pH was at 4. At this pH, the ammonia in the anode chamber would be in the form of ammonium ion. In the cathode chamber where the pH was well above 9.3, the ammonium would readily convert to uncharged ammonia (NH_3). The solubility of ammonia is fairly low in water (Henry's constant, $K_H = 57.6$ mol L^{-1} atm^{-1}). Therefore the ammonia removed to the cathode could have converted to vapor phase and escape into the head space of the chamber. Strong ammonia smell was detected when the trapped gasses were released at intervals from the cathode reservoir. The portion of ammonia that was collected in the anode chamber was due to colloidally enhanced transport whereby the positive ion NH_4^+ was expected to collect at the positive electrode by electromigrating colloids (i.e. bacterial, organic, or inorganic). The colloidal transport was observed visually as the anolyte clouded with suspended particles during the treatment. Charged colloids are known to strongly adsorb ions of opposite charge and enhance transport of these substances in porous media (Kretzschmar et al., 1999; Grolimund and Borkovic, 2001).

8.3.1.4 Simultaneous Extraction of Cyanide, CN^-, and Fluoride F^-

Spent pot-lining sludge from Columbia Falls Aluminum Company was tested to determine the efficacy of EK in removing cyanide and fluoride metals prior to its disposal. The sludge contained 0.2% cyanide and 9% fluoride by mass, initially. At the end of 7 days of EK treatment, about 8% of the cyanide and 2% of the fluoride by mass was removed from 118 g of a sample of the sludge. In these tests 15 V of constant potential was applied across the electrodes, and the anolyte and catholyte solutions were replenished with fresh solutions at 29 h of this constant potential. As shown in Figure 8.14, the electrical potential measured within the sample was small and changed with current when the electrolytes in the electrode chambers were replaced with fresh ones. The current remained high in the range of 15–20 mA in the first stage and dropped to about 7 mA in the second stage. The removal of metals over time, as shown in Figure 8.15, demonstrated that preferential removal of CN^- toward anode was prevalent, as the Fl^- transport picked up only after the 4th day against the control value of diffusion to the cathode.

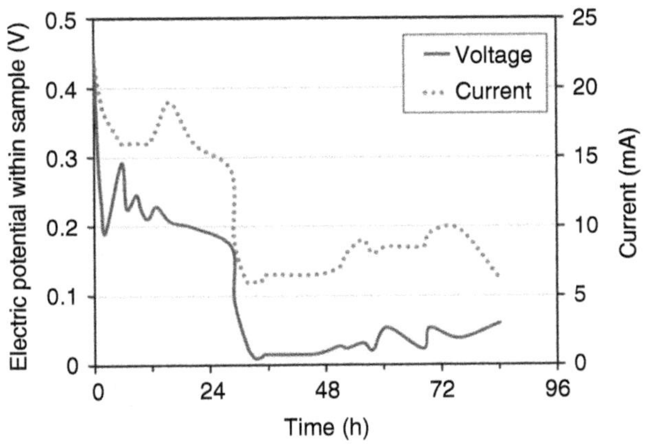

Figure 8.14 Voltage and current evolution in the pot-lining sludge sample during EK treatment under constant 15 V of potential. *Source:* From Pamukcu and Pervizpour (1998a).

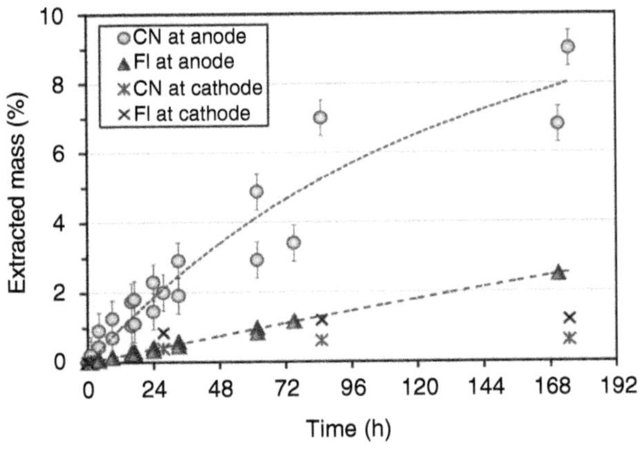

Figure 8.15 Cumulative mass of CN and Fl removed from the sample into the electrode chambers during EK treatment of sludge. *Source:* From Pamukcu and Pervizpour (1998a).

8.3.1.5 Hexavalent Chromium, Cr(IV), Extraction

An example of the influence of DDL reactions can be observed with hexavalent chromium in Cr(VI) treatment using EK. The Cr(VI), which is often found in anionic species of $Cr_2O_4^-$, is readily mobile in soil pore fluid but may not be readily extracted under an electrical gradient. The figure, "Variation of cumulative mass of Cr in anode chamber with EK treatment using constant current (Test I) and (Test II) tests of Aladdin Chrome Plating site soil" (Weeks and Pamukcu, 2001), shows typical results of EK extraction of Cr(VI) from a sample of hexavalent chromium-contaminated site in Eastern Pennsylvania (EPA SITE Demonstration Program at the Aladdin Plating Site in Chinchilla, PA). The soil sample was characterized as silty clay. As observed, less than 20% of the total Cr was removed to the anode chamber and very little to cathode chamber over 30 days despite the continuing electric field imparted at constant current of up to 1 mA. The figure, "Variation of the concentration of Cr, current and potential measured in test soil" (Weeks and Pamukcu, 2001), shows that the cumulative concentration increase over the treatment time flattens with each increase in applied current, unlike the previously observed direct correlation between current and removal efficiency. This result is attributable to tendency of the ensuing electrochemical reactions that reduce Cr(VI) to Cr(III), which subsequently is adsorbed onto the clay surfaces and no longer available for transport.

8.3.1.6 Sulfate, SO_4^{-2}, Nitrate, NO_3^-, and Chloride, Cl^- Extraction from Cationic/Anionic Soils

Simultaneous extraction of multiple anionic compounds introduced as lead salts was tested in anionic (positive surface charge) and cationic (negative surface charge) clayey soils. Mixed with medium sand, the clay substrates were predominantly kaolinite (cationic – PZC: 2–4.6) and gibbsite (anionic – PZC: 10). The test specimens were prepared by spiking with $Pb(SO_4)$, $Pb(NO_3)_2$, and $PbCl_2$ salt solutions at their respective solubility limits. EK tests were conducted over 24 h, at the end of which the distributions of the anions were measured in the electrode chamber solutions and across the soil specimens at five points. Figure 8.12 shows this distribution for the three anions in two different clay types. As observed, the extraction appears to be influenced by both the anion and the type of substrate. The anions of the soluble salts were extracted into the anode more effectively than the less soluble ones. The cationic soil appeared to release the anions more effectively than the anionic soil, as the anions would have higher affinity to the clay surface in the latter case. Figure 8.13 shows the mass fraction of anions extracted to anode per Coulomb of electricity used. As expected, a linear trend prevails between the extracted mass and the quantity of electricity required. The anionic soil appears to display slightly higher efficiency in energy requirement than the cationic soils for extraction of anions. This is in line with the assumption that the anionic soils will generate EO flow in the same direction as the anion migration hence will be enhanced for the same quantity of electricity when compared with that of the cationic soil case.

8.3.1.7 Electrokinetic Extraction of Contaminants from Unsaturated Soils

The application of EK in the vadose zone has been demonstrated by numerous studies in the past, most notably by Mattson et al. (2002) and by Banerjee et al. (1987). Unsaturated soils pose little difficulty when the EK transport occurs by ion migration mostly. The mobile ions residing in the liquid phase move to the respective electrodes, provided that there is continuity of the liquid phase over the diffuse layer interfaces. Since in most EK treatment schemes, clean water is supplied at the anode, the ionic constituents are eventually provided with adequate solution water to migrate in during the course of a treatment. The negative pore pressures that might generate in front or behind the water flux should have little effect on the electromigration of these charged species. This phenomenon is illustrated in the laboratory where kaolinite clay contaminated with strontium and sodium shows similar rates of transport per pore volume of liquid despite the different degrees of initial water saturation of the clay, as shown in the figure, "Production of Na and Sr at cathode with pore volume of EO flow in **unsaturated** clays" (S: degree of saturation) (Pamukcu and Wittle, 1993). In all three cases, majority of the contaminant is extracted into the cathode before the water advances one full pore distance between the anode and the cathode. However, the same may not hold true for situations where EO advection is the dominant mechanism of transport, as for nonaqueous phase liquids. In those circumstances, continuous supply of water at the influent electrode (anode for cationic soils, or soils with pH above PZC point, and cathode for anionic soils, or soils with pH below PZC point) is expected to progressively provide sufficient water concentration in the pores for the NAPLs to partition away from the solid surface.

8.3.1.8 Electrokinetic Extraction of Contaminants from Saturated Soils

The table, "Average Percent Reduction of Metal Concentration and Pore Volume Fraction of Water Transported Toward Cathode Chamber" (Pamukcu and Wittle, 1993), summarizes average metal concentration reduction (in percentage) in three replicate specimens of fully saturated synthetic soils after 24–48 h of EK treatment of four selected heavy metals. The measurements provided in the table were taken at the location of lowest concentration measurement along the specimens, as observed in the spatial distribution of strontium in the figure, "Post-EK distribution of mass fraction of Strontioum (Sr) in **saturated** clay soils" (Pamukcu et al., 1997). Also shown in the table is the average cumulative fraction of pore volume of water flow toward the cathode electrode compartment during each treatment. The variations in the concentration reduction percentages appear to have little or no correlation with the volume of flow achieved during treatment but depend on the soil matrix and the metal type mostly. For the cationic species (Cs, Sr, UO_2), the fraction of pore volume of flow ranged between 0.25 and nearly 1.00. The flow for the anionic form of Cr ($Cr_2O_7^{-2}$) was consistently lower than that of the cationic species. This is potentially due to the water dragging action of the hydrated anions as they migrate in the opposite direction of the net flow toward cathode.

Although significant removal of metal into the liquid phase was achieved in most cases, a clear trend of metal accumulation was observed at the discharge locations (cathode end for cations, anode end for anions) in all soil specimens. The accumulation of the metal is due to several conditions. One condition deduced from the experiments is that the time rate of migration of the metal reduces significantly as it approaches to the discharge end. This reduction is triggered by (i) concentration increase, (ii) precipitation of the metals, and (iii) increased retention capacity of the soil at the discharge ends due to the local pH levels. As the concentration of a particular ion increases near the discharge chamber (anode or cathode), the mobilities of the ions or the micelles decrease (Kohlrausch's law). A slight reduction of the contaminant mobility may trigger a progressive concentration buildup at that location if the discharge is inhibited by other mechanisms simultaneously. These mechanisms come about by the variation of soil pH from anode to the cathode end.

All the metals, both anionic and cationic, accumulated at their discharge locations, creating zones of high concentrations of metal (sometimes over that of the original). These zones are often narrow, residing at the interface between the soil and the electrode chamber, which is analogous of deposition of the metal on an electrode as if it was in contact with soil. In the case of cations, the production of hydroxide in the cathode chamber is one of the major causes of the phenomena. As the hydroxide ions (OH^-) migrate into the soil from the catholyte, they create a high pH zone in the soil before they can encounter the oncoming hydronium ions (H^+) and be neutralized. This zone may lessen in thickness as the acid front penetrates deeper toward the cathode side of the soil (Hamed et al., 1991). Subsequently, the high pH generated at the electrode site solution and the lower pH at the soil–water interface can create a large pH gradient across this interface, inhibiting the discharge of the metals. Another effect of high pH is on the adsorption capacity of the clay soils. Hydroxide ions are potential makers that subsequently cause the cation adsorption capacity of the soil to increase. Therefore, at high pH, the surface potential will increase, leading to thicker diffuse layers and increased surface conductivity. Since the bulk liquid conductivity would be reduced with precipitation, the combined effect should signal increased EO water transport and current efficiency. This suggests that with extended duration of treatment, the accumulated metal may eventually be transported into the electrode chamber by the steady but slower action of EO advection.

The figure, "Evolution of zeta-potential with pH in kaolinite clay of high salt concentration" (Pamukcu et al., 1997), shows the variation of zeta potential with pH for two clay–electrolyte systems of high ionic concentration. As observed, the PZC of either the Sr- or Pb-spiked kaolinite clay lie at about 4 pH units. The potential increases in negative values and then reverses at high pH. The changing zeta potential trends can also explain the interfacial accumulation of some compounds in response to reversed EO flow regimes. A typical example of post-EK metal accumulation near the soil–water interface at cathode side is displayed in the figure, "Post-EK distribution of mass fraction of Lead (Pb) in **saturated** clay soils" (Pamukcu et al., 1997).

8.3.1.9 Extraction of VOCs, PAHs, and Organic Acids

The application of EK phenomena to the transport of NAPL pollutants has not been widely investigated. Bruell et al. (1992) showed transport trichloroethylene (TCE) and several gasoline hydrocarbons by EO, assuming a constant rate of EO flux. Acar et al. (1992) showed that transport of phenol is possible by both EO and electromigration, owing to the amphoteric nature of the compound. Field experiments by Ho et al. (1995) showed that the dissolved fraction of TCE can be transported by EO.

Pamukcu (1994) showed that PAH compounds with higher water solubility and lower molecular weight (i.e. naphthalene) in coal tar-contaminated soils could be transported by EO without the aid of surface tension reducing agents. In these tests, extraction of coal tar PAHs, which included *naphthalene, acenaphthylene, acenaphthene, fluorene, phenanthrene, anthracene, fluoranthene, pyrene, chrysene, benz(a)anthracene, benzo(b)fluoranthene, benzo(k)fluoranthene, benzo(a) pyrene, dibenzo(a,h)anthracene, benzo(g,h,i)perylene,* and *indeno(1,2,3-c,d)pyrene*, was demonstrated using the EK process. Figure 8.16 displays the post-process distribution of the normalized mass of total PAHs in the coal tar-contaminated soil specimens retrieved from a manufactured gas plant (MGP) site in Champaign–Urbana, Illinois. As observed, the EO advection transported and accumulated the PAH mass in soil near the cathode interface. Use of an anionic surfactant (sodium dodecyl sulfate [SDS]) with or without a co-surfactant (butanol) appeared to reduce the PAH mass in the soil significantly, transporting the contaminants into the electrode chambers, mainly to the anode.

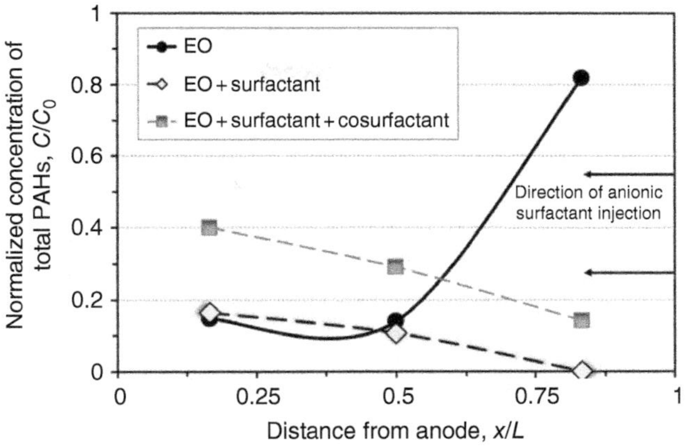

Figure 8.16 Post-EK distribution of the normalized concentration of total PAHs in coal tar-contaminated specimens from an MGP site in Illinois. *Source:* From Pamukcu (1994).

Pamukcu and Pervizpour (1998b), through well-controlled experiments, demonstrated the EK-aided transport of TCE in contaminated soil samples from Lawrence Livermore National Laboratory DOE complex site in California. Soil cores were injected with TCE using groundwater collected from the site and spiked with TCE below its solubility limit. The injection of TCE into the soil was accomplished using only electroosmotically aided transport of the solution from the anode (influent) to the cathode (effluent) chamber of the test cell. Figure 8.17 shows the trace and breakthrough of TCE in the cathode, while the concentration of TCE in anode was kept constant under its solubility limit. The breakthrough at the cathode occurred at approximately 1.5 pore volumes of flow. The hydraulic conductivity and the EO conductivity of the soil specimen were measured as $k_h = 10^{-8}$ cm s^{-1} and $k_{eo} = 2 \times 10^{-6}$ cm^2 s^{-1} V^{-1}, respectively. The equivalent hydraulic conductivity, k_{eh}, under EO attained an average value of 5×10^{-6} cm s^{-1}, producing steady flux of TCE into the cathode chamber. The tracer curve is indicative of a solute that undergoes partitioning between the solid and the liquid phase; therefore retardation of TCE on the solid phase was expected. Nevertheless, these results show clearly that the dissolved fraction of a chlorinated hydrocarbon could be transported by EO.

Potassium formate (potassium salt of formic acid), an organic acid used as a drilling fluid additive, was treated using EK. The goal of this treatment was to reduce the potassium content of the mixture by separating the potassium from its anion component (i.e. formic acid)

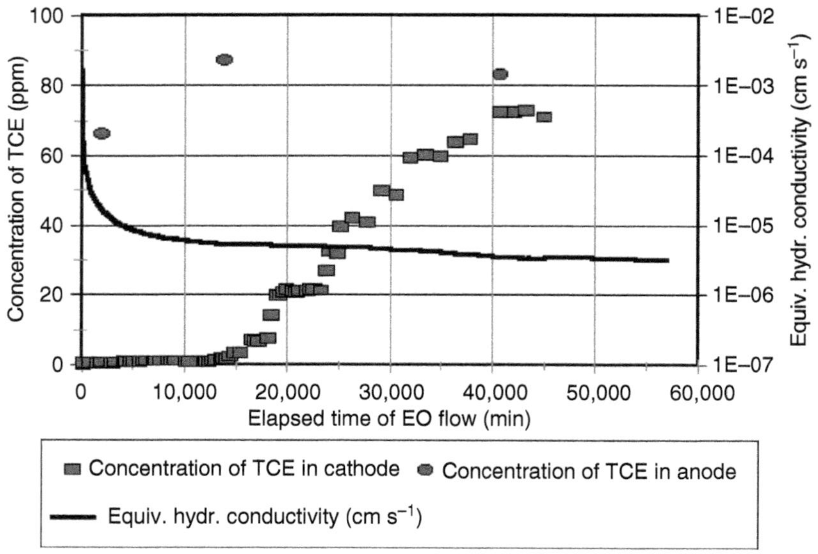

Figure 8.17 Cumulative mass transport of TCE from anode to cathode electrode chamber by EO in soil samples from LLNL DOE complex site. *Source:* From Pamukcu and Pervizpour (1998b).

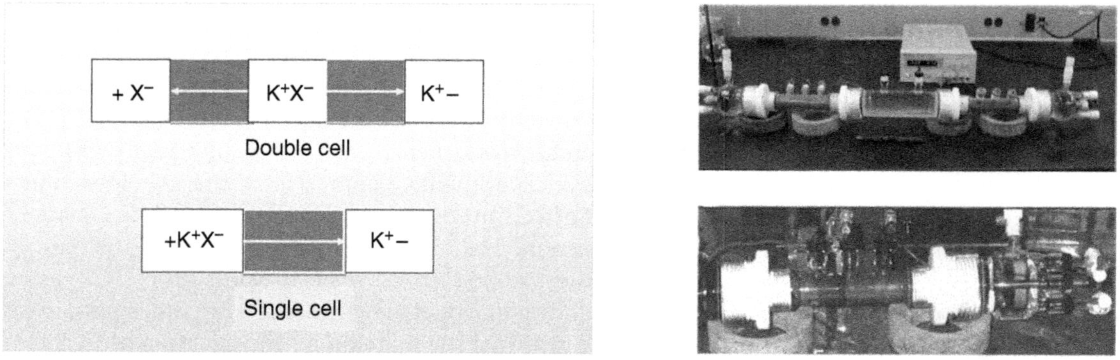

Figure 8.18 Single- and double-cell EK test configurations and photos of the test cells for extracting K$^+$ from drilling mud.

for recycling. Two types of tests, with single- and double-cell configurations and variable electric potential applications, were developed to assess the efficacy of EK. Figure 8.18 shows the conceptual scheme of the proposed separation and the photographs of the test cells. In the double-cell configuration, 5% potassium formate solution was placed in the center of anode and cathode liquid chambers whereby the potassium and the formic acid components were expected to filter through the soil into the oppositely charged electrode chambers. In the single-cell configuration, 5% potassium formate solution was placed only in the anode electrode chamber and potassium was expected to filter through the soil into the cathode electrode chamber. The results of these tests are presented in Figures 8.19 and 8.20. In both tests, the potassium concentration increased at the cathode and decreased in the anode and the center cell compartment. The concentration of the formic acid decreased in the anode chamber pursuant to migration of the compound into the soil as well, confirming potential in-field recycling of potassium from drilling fluid by EK.

8.3.2 Electrokinetically Aided Stabilization and Immobilization

In environmental restoration, *stabilization* is defined as fixing the toxic substance in place, thereby rendering it less likely to move elsewhere under the ambient hydrogeological conditions. In the subsurface, stabilization of a toxic substance can be accomplished by delivering an appropriate oxidizing or reducing agent that subsequently will (i) degrade the contaminant, (ii) change it to a nontoxic or immobile species, or (iii) enhance stable sorption and incorporation of the contaminant into the clay minerals. Zero-valent iron-enhanced degradation of TCE (Ho et al., 1995) and Fe(II) degradation of toxic Cr(VI) to less toxic and less mobile Cr(III) are examples of such processes (Haran et al., 1995; Pamukcu et al., 1997, 2008; Gomes et al., 2012, 2013, 2014, 2015b).

The metals that could be altered in such a manner are those that are least likely to be extracted by EK treatment. Owing to their complex electrochemistry, metals such as Cr, As, and Hg are possible candidates for EK

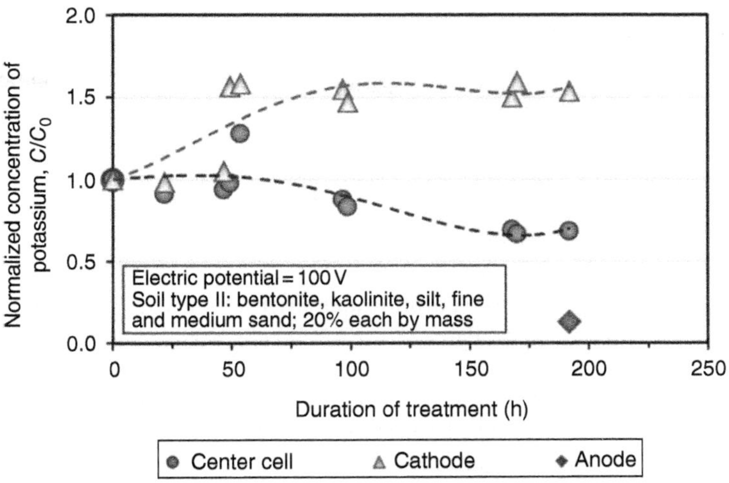

Figure 8.19 Evolution of potassium and concentrations in the electrode sites and center in *double-cell* EK test.

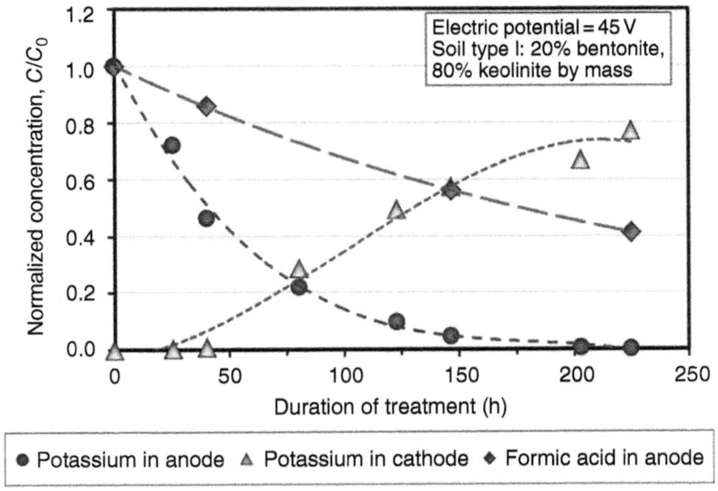

Figure 8.20 Evolution of potassium and formate concentrations in the electrode sites in *single-cell* EK test.

containment. A good example of electrochemical stabilization may be the relatively well-studied reduction of Cr(VI) to Cr(III) by delivering iron (Fe(0), Fe(II), or Fe(III)) with co-reagents in aqueous environments (Eary and Rai, 1991; Anderson et al., 1994; Powell, et al., 1995). Chromium exists in two possible oxidation states in soils: the trivalent Cr(III) and the hexavalent Cr(VI) chromium. At low pH conditions (2–6.5), the predominant form of the hexavalent chromium is chromate or dichromate ion. Due to their negative charge, they are not readily adsorbed or exchanged at clay surfaces and therefore remain in soil pore water and can be readily transported. At sufficiently low pH, the soil surface may become positively charged and tend to retain and accumulate these anions. Therefore complete removal may not be achieved unless precise control of pH is maintained at the anode during an EK process. As discussed previously, the aforementioned figure, "Variation of cumulative mass of Cr" (Weeks and Pamukcu, 2001), shows a good example where only a small mass of Cr is removed with increasing duration of treatment and applied current.

Hexavalent chromium can be reduced to Cr(III) under normal soil and pH conditions, for which soil organic matter acts as the electron donor (Rai et al., 1987). Bartlett (1991) reported that in natural soils, this reduction may be extremely slow, requiring years. In subsurface soils where there is less organic matter, the Fe(II)-containing minerals reduce Cr(VI) at pH less than 5 (Eary and Rai, 1991). Electrochemically injected Fe(II) into a matrix of soil containing hexavalent chromium should facilitate the reduction of Cr(VI) since electrochemical process produces low pH conditions. The delivery of Fe(II) has also been shown to enhance the formation of chromium–iron hydroxide solid solution [$(Cr_xFe_{1-x})(OH)_3(ss)$], which has a low equilibrium solution activity than pure solid phases (Powell, et al., 1995). The figure, "Post-EK formation of Cr,Fe solid solutions in soil at various Fe(II) concentration applications" (Weeks and Pamukcu, 2001), shows the formation of chromium–iron oxide hydroxide solutions when Fe(II) was delivered electrokinetically into Cr-contaminated soil where the original form of chromium was Cr(IV). As observed, most of the post-EK form of Cr coincided with the solid solution distribution where the Cr is transformed into nonaqueous Cr(III). The figure, "Post-EK distribution of Cr species in soil depicted on Pourbaix diagram."

Weeks and Pamukcu, (2001) confirms the post-EK abundance of Cr(III) per redox–pH measurements shown on Pourbaix diagram. These results confirm that the electrokinetically delivered Fe(II) had successfully transformed and arrested the Cr in soil where it is no longer available for leaching out into the groundwater environment, hence stabilized (Weeks and Pamukcu, 2001; Weeks, 2002).

As previously discussed, faradaic reactions may take place on clay particle surfaces when current passes in the pathways of the DDLs. These reactions can aid in electrically enhanced stabilization and/or immobilization of certain contaminants. Hence, external supply of electrical energy can help drive favorable oxidation–reduction reactions on contaminated clay surfaces where most of the contaminants tend to reside because of adsorption or exchange. Controlled laboratory experiments of kaolinite clay injected with Fe(II) showed that an externally applied electric field caused an additional "cathodic current" that drove forth the reduction of Cr(VI) in clay (Pamukcu et al., 2004). The results in these experiments showed that the system ORP increased by a positive shift from the standard solution ORP in the presence of the clay and the induced electrical field, as was shown earlier in Figure 8.11.

More evidence for electrokinetically assisted surface transformations in kaolinite clay were found when a clean specimen of the clay was permeated with polymer-coated dispersed nano-iron particles of positive surface charge (Sun et al., 2007) under an electrical gradient of 0.1 V cm^{-1} (Pamukcu et al., 2008). When there is no large concentration of a dominant ion in the clay matrix (i.e. contaminating heavy metals), the primary electron receptors are water and residual dissolved oxygen. Hence, the chemical reactions, which oxidize the nano-iron, are

$$Fe^0(s) + 2H_2O(aq) \rightarrow Fe^{+2}(aq) + H_2(g) + 2OH^-(aq) \tag{8.13}$$

$$2Fe^0(s) + 4H^+(aq) + O_2(aq) \rightarrow 2Fe^{+2}(aq) + 2H_2O(l) \tag{8.14}$$

Eq. (8.13) is the dominant redox process in presence of overwhelming concentration of water as the solvent. As hydrogen ions are utilized in Eq. (8.14), the pH would increase at corrosion locations, resulting in simultaneous decline in the solution ORP. The transient ORP measured across a 20-cm long, 2-mm thick bed of clay (see Figure 8.21) injected with nano-iron (nZVI) using EK are presented in the figure, "Variation of ORP in the kaolin medium in **a)** EK assisted transport of nZVI **b)** diffusion of nZVI and **c)** EK without nZVI; all using 1 mM NaCl as pore fluid electrolyte" (Gomes et al.,

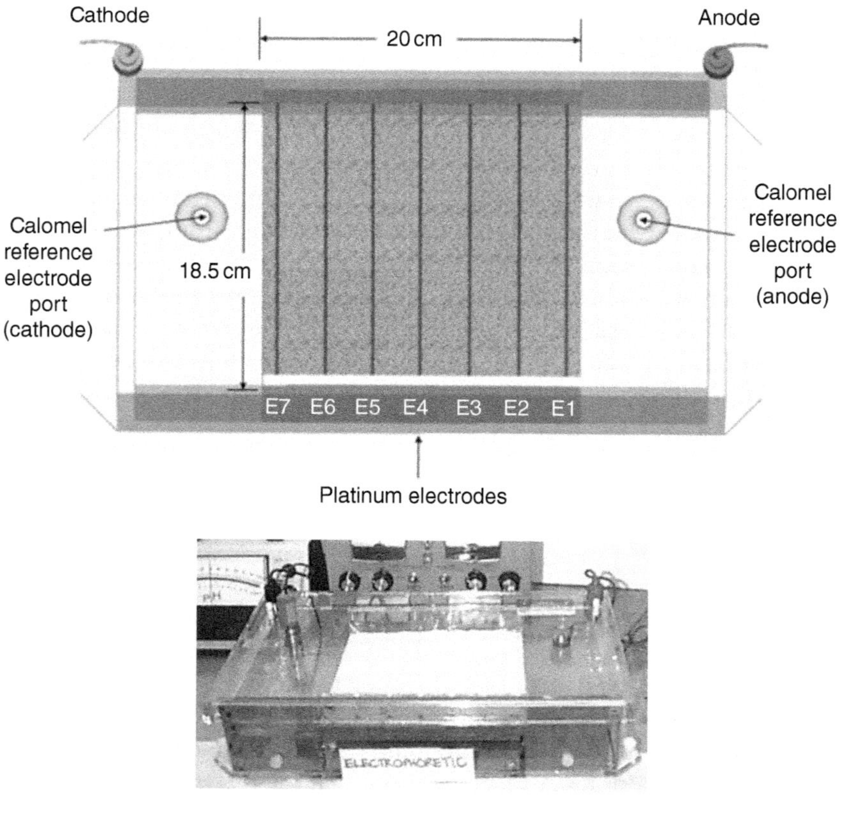

Figure 8.21 The electrophoretic setup to measure enhanced transformation of Cr(IV) to Cr(III) (Pamukcu et al., 2004, 2008; Hannum and Pamukcu, 2007).

2014). The ORP variations for control tests of diffusion and EK without the nano-iron are also presented. In these tests, the nZVI granules were placed in the clay inside a transverse groove at about 8 cm from the anode end of the clay bed, and the tests were run under oxygen-poor conditions using 5 V across the working electrodes. As observed, the ORP reduced across clay bed for the first 15 h of transport and increased to above the original values during the rest of the transport period. The ORP for the diffusion-only case remained mostly unchanged, and the EK-only case showed the typical pH-dependent ORP variation expected with EK. Based on the water chemistry of nano-iron (Sun et al., 2007), if the only contributor to the ORP was nano-iron, the change in the ORP of the system would be expected to vary within 50 mV between the pH units of 6.0–6.5. The ORP changes for the diffusion tests did remain between *+50* and *−50 mV* over the 48-h duration without the electric field application. The EK-assisted transport of nZVI, on the other hand, displayed larger changes in ORP across the clay bed indicating a synergistic effect of EK and nZVI activation within the first 15 h of the process. The longer-term ORP changes showed the contribution of clay surface interactions with Fe(II) and Fe(III) in presence of electrical potential as the ORP elevated over background values at the average low pH value of 3.

The ORP versus pH obtained from earlier experiments of similar configuration as above (Pamukcu et al., 2008) are plotted in the figure, "pH-ORP (Eh) variation of clay with and w/o nano-iron in electro- kinetic transport" (Pamukcu et al., 2008). The diagram depicts approximate regions of corrosion, passivity, and immunity for iron (Davis, 2000). The cluster of data that fell within the *immunity zone* corresponded to the first 6 h of EK transport where the ORP at the cathode end was below −1.0 V. This indicated that the nano-iron particles had already been transported to the cathode side of the clay bed by the end of 6 h of treatment or earlier. The evidence that corroborated with the visual observations of color change in clay (i.e. indicative of oxidation of iron) was the cluster of data that fell within the *passivity zone.* This data corresponded to measurements made later, from 6 to 46 h. By the end of the 46 h, the corrosion products with *higher* ORP appeared at the cathode end of the clay bed, as was evidenced by the brown tone color change in that region of the test bed.

The average voltage distributions across the clay bed with and without nZVI are plotted in the figure, "Average distribution of voltage in clay with and without nZVI transport over 48 h of EK" (Pamukcu et al., 2008). As observed, while the EK-only test displayed a constant electrical gradient of 0.12 V cm^{-1} (i.e. equation of the linearly fit line) for the 48-h duration of the test, the electrical gradient fluctuated between adjacent auxiliary electrodes inserted in soil bed for the nZVI case. The magnitude of these oscillations ceased attaining a constant profile in time. Variation of the electric gradient influences the velocity of the migrating particles from one station to the next, causing accumulation in some locations and evacuation in others. Theoretically, the particle velocity is expected to decrease at regions of low electrical fields, causing accumulation and higher charge density, but pick up at regions of higher electric field, hence dilution and low charge density. This process is likely to have a progressive nature that would allow for self-correction by diffusion and faradaic currents to maintain the electrical neutrality. Such a process could also result in the wave shape of the voltage distribution to remain stationary with only the magnitudes of the peaks changing. For example, if initially an accumulation were triggered by a physical obstruction (i.e. agglomeration of the particles in small clay pore throats), voltage gradient at that locale would increase because of the reduced conduction. The increased charge density would lead to surface capacitance and faradaic reactions at the location. At the same time, the increased voltage gradient would promote the particles to move and vacate the location, subsequently causing a drop in charge density and reversal of the surface reactions. It is difficult to capture such potential dynamics through sporadic voltage and ORP measurements in a typical EK transport testing in clay. Nevertheless, based on the data collected and evaluated, the empirical evidence appears to support that presence of clay and ensuing surface interactions enhance the ORP under DC electrical field.

The oscillating voltage gradients at high electrolyte concentration environments were observed again in a large-scale test designed to mimic potential application of electrically assisted dewatering of sea harbor sediments (Muraoka et al., 2011). The model experiment simulated parallel vertical electrodes that can be installed throughout the depth of unconsolidated sediments, allowing accelerated drainage laterally. Commercial kaolinite clay was used to mix a slurry at 35 : 65 solid : water ratio by mass. The mixing water and the electrode chamber waters were made up of 30 000 ppm electrolyte solution of instant ocean sea salt. Constant voltage, adjusted between 3.5 and 3.0 V, was applied across the electrodes to maintain a direct electric current of about 100 mA. The clay slurry and the electrode chamber waters were filled up to 40 cm from the bottom of the tank. The water levels in the electrolyte chambers were maintained at the mud line to alleviate any change in hydraulic potential. Five water pressure head (piezo tubes) and voltage (auxiliary electrodes) measurement probes were placed inside the clay slurry. The electrical gradients and the PWPs were measured across the dewatering/settling clay at time intervals.

The long-term experimental results (~7 days) showed that the settlement or the dewatering of clay slurry was hindered due to spatially nonlinear and temporally oscillatory voltage gradients, which give way to nonuniform distribution of stagnant PWPs within the clay slurry. These results are presented in the figures, "Spatial and temporal distribution of voltage gradient across adjacent probes at select times during EK" and "Spatial and temporal distribution of normalized pore water pressure (PWP) between adjacent probes at select times during EK" (Muraoka et al., 2011) for randomly selected time steps of the voltage gradient and PWP measurements, respectively. The significance of these observations is that the local voltage gradients appear to have direct influence on the development of local pressure gradients within the EO flow region of a high salt content medium. Fluctuating PWPs induce competing flow regimes, resulting in stagnant pore pressures that can impede EO mass transport. In the second "Spatial and temporal distribution" figure, mentioned above, the voltage gradient distribution between adjacent probe locations shows clearly zones of high and low gradients, above and below the self-manifesting threshold value of $0.015\,V\ cm^{-1}$. In final analysis, the nonlinear and oscillatory voltage gradients were attributed to streaming potential and migrating fronts of ions in the high salinity environment. The resistance to PWP dissipation was attributed to compression and polarization of particle DDL in the direction of the electric field. The combined consequence of these phenomena is the competing pressure gradients that keep the clay particles from settling and dewatering (Muraoka et al., 2011).

8.3.3 Electrokinetically Aided Containment

Containment in here is defined as causing controlled accumulation of a toxic substance by sorption in a small volume of substrate (i.e. permeable barrier), which can be flushed and regenerated for further use. EK containment may be accomplished by causing electromigration or EO transport of the contaminants to a reactive permeable barrier (Blowes et al., 1995; Chung and Lee, 2007; Ramírez et al., 2015; Mena et al., 2016; Zhou et al., 2016; Vieira dos Santos et al., 2017), strategically situated between the electrodes, where they are attenuated or transformed, and the filtered water is allowed to pass through (Hansen, 1995; Weeks and Pamukcu, 1996; Benner et al., 2002). In actual field applications, such permeable structures could be installed in multiples at various configurations throughout a contaminated site serving as primary and secondary treatment locations.

Pre-contaminated kaolinite samples with lead, Pb(II) (from $Pb(NO_3)_2$ salt), were tested to determine the effectiveness of using permeable reactive caps to contain heavy metals adjacent to the electrode chambers. The reactive permeable caps (GS caps) were comprised of 50% glauconite (green sand), 30% zeolite, and 20% bentonite clay by mass. The average particle size of green sand and zeolite was on the order of medium to fine sand. Bentonite was added to enhance the bonding of the matrix. The final water content of the cap mixture was 37.5%, and of the contaminated soil was 60% with initial Pb(II) concentration of 5000 ppm. A constant voltage of 15 V was applied to all the test samples. The tests were terminated when the current readings diminished to a constant low value.

The figures, "Post EK mass fraction (%) of total Pb remaining in soil and liquid tested without the GS Caps; pH superimposed" and "Post EK mass fraction (%) of total Pb remaining in soil and liquid tested with the GS Caps; pH superimposed" (Pamukcu et al., 1997), show the average pre- and post-EK mass fraction distributions of Pb(II) in the soil and the electrode chambers with and without the GS caps, respectively. The sample containing the GS caps showed larger concentrations of Pb(II) accumulated in the caps, and very little of the element passed into the electrode chambers. Overall, the results showed that in samples tested without the GS caps, more of the Pb(II) remained in the soil, while the samples containing the caps showed lesser concentrations of Pb(II) in the soil with larger concentrations in the areas of the GS caps for the same duration of EK treatment. The larger removal of the lead into the GS caps, particularly on the cathode side, is attributed to continued flow of higher current and EO advection, whereas precipitation of lead at the cathode interface of the soil without the caps inhibited the current and EO flow hindering the Pb(II) transport overall. Indeed, the steady low current measured with caps was three times larger than the one without the caps. The sorptive capacity of the cathode cap would have kept the Pb(II), and its hydroxide precipitates on the solid surfaces without causing blockage in the porous matrix. These results indicate that application of site- and contaminant-specific containment packs can work synergistically with EK remediation when the goal is to collect the contaminant in a small volume of substrate from where it can be recovered while regenerating the substrate for further use.

8.3.4 Recovery, Transformation, and Transport of Crude Oil

8.3.4.1 Laboratory Evidence of Oil Recovery by EK

Chilingar and coworkers investigated the use of EK to augment reservoir energy during petroleum production (Amba et al., 1964, 1965; Chilingar et al., 1968a, b). They found that application of direct electric current resulted in severalfold increase in the flow rate of oil and water (i.e. as much as 34-fold) in both synthetic and natural sandstone cores. Synthetic cores were prepared using various percentages of different types of clay admixed with 200-mesh size silica or $CaCO_3$ powder. Also, it was found that the effective permeability to kerosene of core samples containing connate water (i.e. entrapped water in the pores when the rocks were formed) was increased by 50% on temporary application of direct potential gradient. Successful field pilot tests have been reported on application of electrically enhanced oil recovery (EEOR) in heavy oil fields in the Santa Maria (California) basin and the Eastern Alberta Plains (Wittle et al., 2008). In these pilot tests, the EEOR was found to be more efficient than other conventional heavy oil recovery methods including steam flood or cyclic stream flood.

The results of DC electric field application in mobilizing the trapped oil through core-flood studies were reported (Haroun et al., 2010; Al Shalabi et al., 2012a, b). In the first set of experiments, DC was applied after ultimate waterflood recovery, and in the second set of experiments, the DC was applied concurrently with waterflood from the beginning of the test. Comparison of the two processes revealed that application of DC concurrently with waterflooding recovered approximately 1.5–4% more oil and reduced water requirement by 15–22% in comparison with the waterflooding process alone. Application of DC on a completely water-swept core was found to recover 3–9% additional oil. Also, up to 30% increase in permeability of cores was reported, which was attributed to clay dislodgement due to electrophoresis mechanism involved in EEOR.

Ghazanfari (2013) conducted a series of laboratory experiments on oil-bearing rock-core specimens retrieved from a shallow oil formation in Kentucky to investigate the oil extractability and change in the rock-core permeability under applied electric field. The physical properties of the three-core specimen (diameter: 36 mm, length: 82 mm) are given in Table 8.1. The physical properties of the natural oil in the rock cores and the brine solution used in the experiments as the surrogate formation water are given in Table 8.2. A current density of $1\,A\,m^{-2}$ was applied to the cores in accordance with the achievable current densities of $1.0–1.5\,A\,m^{-2}$ in practical applications in the field. The total flow generated in each phase (oil and water) was measured periodically for 120 h. The oil separation and analysis were conducted using standard techniques. As observed in Table 8.1, most of the oil recovered was produced at the catholyte, while the oil recovered at the anolyte was negligible. The total oil recovered was in the range of 5–11% by mass of the initial oil content in the cores. The oil recovery results suggest a direct relation between the initial oil content of the

Table 8.1 Average percent reduction of metal concentration and pore volume fraction of water transported toward cathode chamber.

	Soil type									
	KS		KH		KG		MS		SS	
Metal	% Rem	PV	% Rem	PV	% Rem	PV	% Rem	PV	% Rem	PV
Cr	93	0.18	97	0.15	95	0.23	95	0.12	97	0.12
Cs	72	0.65	74	0.30	77	0.64	55	0.96	89	0.30
Sr	98	0.41	96	0.88	99	0.44	92	0.53	99	0.63
U	79	0.35	70	0.69	85	0.25	44	0.77	33	0.64

Source: From Pamukcu and Wittle (1993).
KS, kaolinite; KH, kaolinite with humic substances; KG, kaolinite with groundwater salts; MS, montmorillonite; SS, clayey sand with 10% kaolinite; % Rem, percent removal at lowest concentration point; PV, pore volume of water to cathode.

Table 8.2 Physical properties and oil recovery in the natural sandstone cores from Kentucky.

Core	Porosity (%)	Bulk density ($g\,cm^{-3}$)	Permeability (mD)	Residual oil (%)	Residual water (%)	Recovered oil at the anolyte (ml)	Recovered oil at catholyte (ml)	Total recovered oil (%)
KY1	10.2	2.48	2.67	12.9	61	0.000	0.021	5.1
KY2	13.6	2.39	11.59	28.1	43.9	0.005	0.104	9.9
KY3	14.7	2.37	41.88	30.2	33.9	0.002	0.157	11.4

Source: From Pamukcu et al. (2014).

formation and the measured oil recovery. Modeling studies on two-phase EO flow revealed that the drag force exerted by the mobile water phase on the oil phase increases with water saturation of the pore up to about 50%. The oil production decreased when water saturation was above 50% and oil saturation dropped below 50% (Ghazanfari, 2013; Ghazanfari and Pamukcu; 2014). The higher oil recovery rate with higher initial oil/water (O/W) ratio was attributed to higher concentration of oil in the pore space.

In a separate set of experiments, sandstone core specimens retrieved from a deep oil formation in Pennsylvania were used to assess oil recovery, permeability change, and the effect of applied current density on oil extraction. Unlike the Kentucky cores discussed above, the Pennsylvania cores had very low initial oil content. In order to compute more accurately the posttest changes in oil and water saturation, these cores were impregnated with a known compositional content oil (API 47; dynamic viscosity = 38.7 cP) and surrogate formation water mixture (i.e. with the same properties as presented in Tables 8.2) prior to testing. First, the cores were dried and pores vacuumed. Then, oil and water were injected into the cores using a high-pressure injection pump (Quizix pump) to attain the desired initial O/W saturations. Once saturated, the cores were tested under a constant current density of 1 A m^{-2} for 130 h. The physical properties and test results of the four Pennsylvania cores used in these experiments are given in Table 8.3. As observed, most of the oil was produced at the catholyte, and the oil recovered at the anolyte was negligible. The total oil recovered was in the range of 0.5–6% by mass of the initial oil content of the cores. Similar to the Kentucky cores, the highest oil recovery corresponded to the core that had the highest initial oil content and highest hydraulic permeability (PA3).

8.3.4.2 Field Evidence of Oil Recovery by EK

A field application case in a Pennsylvania field was monitored over time to investigate the effectiveness of EEOR accompanied by water flooding (i.e. water pressure). The cumulative oil and water production was monitored at one of the production wells for 180 days. During this time, some amount of paraffin was observed to produce in the well also. The oil, water, and paraffin productions over time are plotted in Figure 8.22. The oil production was about 5% of the water production, and the paraffin production was about 4% of oil production by volume.

The oil production decreased with time in general, yet there was no immediate break in oil production when the power was terminated or reapplied or its magnitude changed. This is attributed to the initial buildup of the

Table 8.3 Physical properties of the Kentucky formation oil and surrogate water samples.

Fluid	Properties
Natural formation oil	API 22; dynamic viscosity = 66.5 cP; specific gravity = 0.92 at 20 °C
Electrolyte solution (surrogate formation water)	Salinity = 33 000 ppm; electrical conductivity = 45 000 μS; pH = 7.50; major elements: Na, Cl, Mg, S, K, Ca, Br

Source: From Pamukcu et al. (2014).

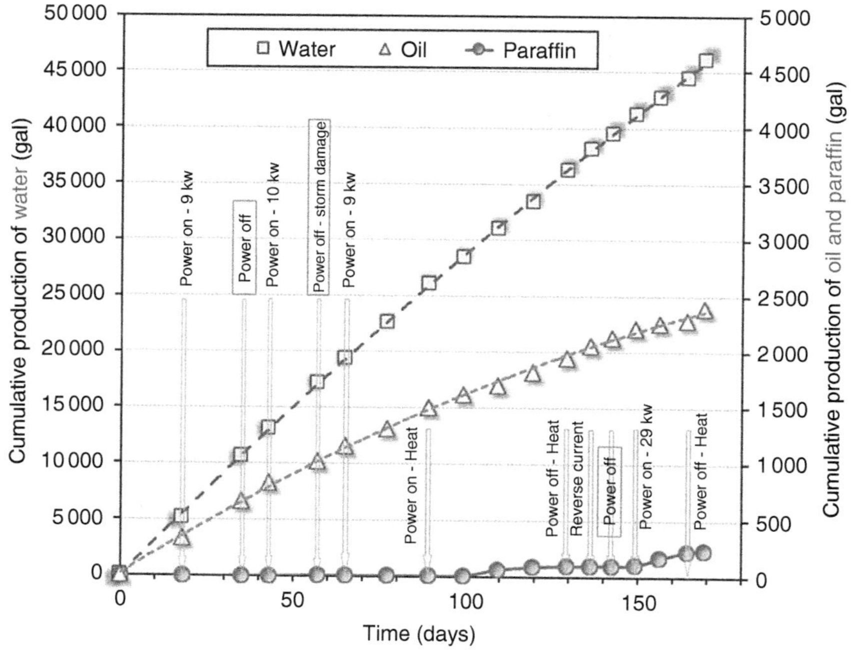

Figure 8.22 Water, oil, and paraffin production (Pennsylvania field). *Source:* From Ghazanfari (2013).

EO seepage and suction pressures that would require time to dissipate or adjust in low permeability formations as the flow regime subsides or changes (Muraoka et al., 2011). Figure 8.23 shows the oil and paraffin production over time. As observed, the paraffin production increased with power and elevated temperature. When the power was turned off, the paraffin production ceased. It is plausible that as pH, ionic concentration, and dissolved minerals (e.g. calcite) increase at the production well (cathode), it creates the ideal conditions for excess oil to transform to paraffin. A similar phenomenon was observed in lab experiments of core rock samples with Pennsylvania oil, where light-colored *foam*-like material was formed in the catholyte, later identified as paraffin.

Examining the daily production data, it was determined that when the applied current was reversed, the production of paraffin decreased. This is attributable to the counter current flow, resulting in decreased oil production and lowering of the pH at the well. Lower pH and less oil available at the location would potentially reverse the paraffin production. The increase in paraffin production corresponded to the periods of elevated temperature due to sustained power over long term (~ at 100 days) or when the maximum power of 29 kW was applied (~ at 150 days). As observed from Figure 8.23, at times of increased paraffin production, the oil production decreased or completely ceased. Among some of the plausible reasons for this exchange are the easier transport of existing paraffin in the pore space at elevated temperatures and the increased availability of oil and mineral components to form paraffin under elevated temperature and power.

8.3.4.3 Laboratory Evidence of Oil Transformation

The electrochemical reactions result in reduction of the interfacial tension between the formation water and crude oil, as well as reduction in the viscosity of the crude oil. It is postulated that both phenomena eventually lead to increased oil mobility and serve to enhance the oil recovery under applied DC electric field. The primary mechanism of interfacial tension reduction under applied electric field is reported as the reaction of electrolysis products of brine/formation water (hydroxyl ions, OH^-) on the acid impurities of oil (carboxylic acids) (Fleureau and Dupeyart, 1988). This results in the formation and accumulation of low solubility surface-active carboxylates analogous to that of directly adding a surfactant to reduce the interfacial tension (Sato et al., 1998). The amount of interfacial tension reduction is reported to be proportional to the applied potential (Tao and Xu, 2006).

Interactions between the pore fluids and the solids, which are normally affected by the ambient pH and redox conditions, will be enhanced upon application of DC electric field. These electrochemically enhanced reactions help oil transformation by cracking the heavy molecular structure of crude oil into its lighter components with lower viscosity (Wittle et al., 2008). As a result the functional group fractions of the crude oil, including the *s*aturates, *a*romatics, *r*esins, and *a*sphaltenes (SARA), will be affected. As the asphaltene and resin content of the formation oil decrease and the aromatic content increases, it is expected that the viscosity of the oil will decrease. The less viscous the crude oil, the higher is its mobility in the formation. Decrease in the oil viscosity under increased electric field intensity has been reported

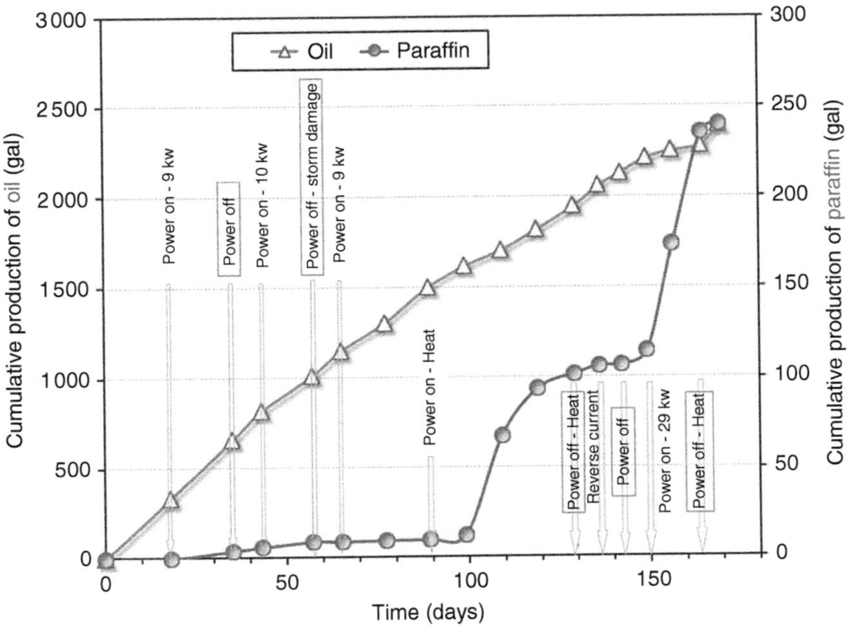

Figure 8.23 Oil and paraffin production (Pennsylvania field). *Source:* From Ghazanfari (2013).

in the literature (Sato et al., 1998; Tao and Xu, 2006; Wittle et al., 2008).

To investigate potential oil transformation under applied DC electric field, a series of synthetic cores were prepared and tested. Three of the test cells were run at low current density of $0.1\,A\,m^{-2}$, three at high current density of $1.0\,A\,m^{-2}$, and three control cells at no current. The cells were 100 cm in length and 16.2 cm in diameter as shown in Figure 8.24. Titanium electrodes and ion exchange resin packs were used at the two ends of each cell. The cells were designed in a manner that minimized transport of pH fronts into the cores, and they were hydraulically balanced to reduce flow through the cores. Liquid and solid samples were retrieved through the ports placed on the cell walls. Three different crude oils representing a wide range of dynamic viscosity but the same brine solution were used in preparation of the cores. The physical properties of the fluids used are presented in Table 8.4. The composition of the surrogate formation cores was 52% sand (mean particle size 0.23 mm), 40% silt (mean particle size 0.025 mm), and 8% Georgia kaolinite clay (mean particle size 0.001 1 mm). The initial porosity, water, and oil saturations were determined to be 30, 46, and 48%, respectively. The representative synthetic core permeability was measured as 510 mD (milliDarcy), typically at the high end of the permeability spectrum for oil formations. (Note: 1 darcy permits a flow of $1\,cm^3\,s^{-1}$ of a fluid with viscosity 1 cP [1 mPa s] under a pressure gradient of $1\,atm\,cm^{-1}$ acting across an area of $1\,cm^2$. One darcy is equivalent to $9.869\,233 \times 10{-}13\,m^2$ or $0.986\,923\,3\,[\mu m]^2$.)

The viscosity of the oil specimens extracted from the 107-day-treated samples under a constant current density of $1\,Amp\,m^{-2}$ showed a significant drop in all regions compared with their control value for the Pennsylvania and Canadian oil cores, as shown in Figures 8.25 and 8.26. The viscosity of the Pennsylvania oil sample dropped by 35, 23, and 13% in the anode, middle, and cathode

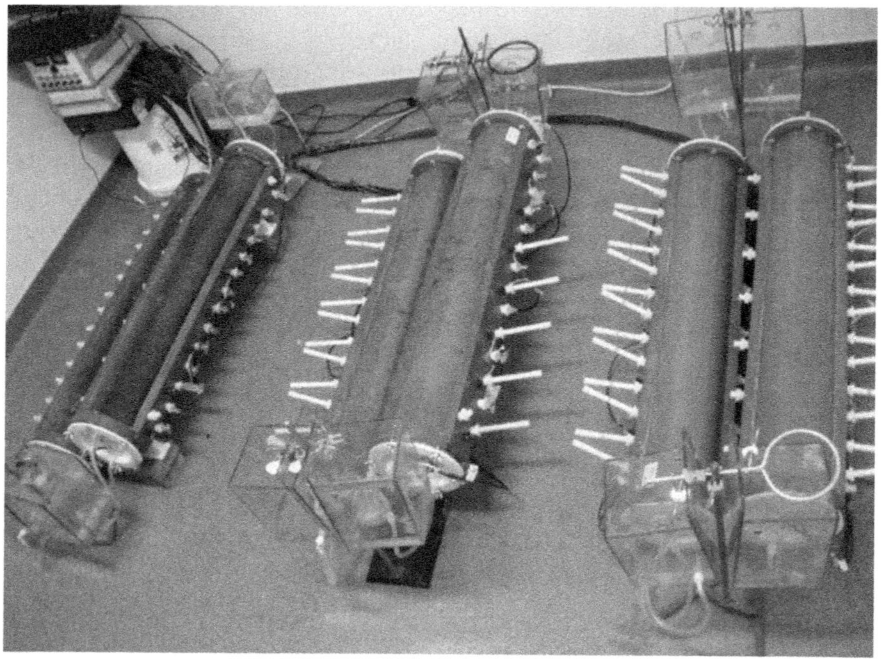

Figure 8.24 Floor scale surrogate core test setup.

Table 8.4 Physical properties and oil recovery in the oil-impregnated cores from Pennsylvania.

Core	Porosity (%)	Bulk density ($g\,cm^{-3}$)	Permeability (mD)	Oil saturation (%)	Water saturation (%)	Recovered oil at the anode side (ml)	Recovered oil at cathode side (ml)	Total recovered oil (%)
PA1	14.5	2.49	3.6	42	58	0.005 6	0.045 7	1.1
PA2	12.8	2.43	2.3	48	52	0.014	0.23	3.8
PA3	14.4	2.47	8.3	53	47	0.020	0.377	6.3
PA4	11.5	2.45	0.5	40	60	0.002 4	0.019 8	0.6

Source: From Pamukcu et al. (2014).

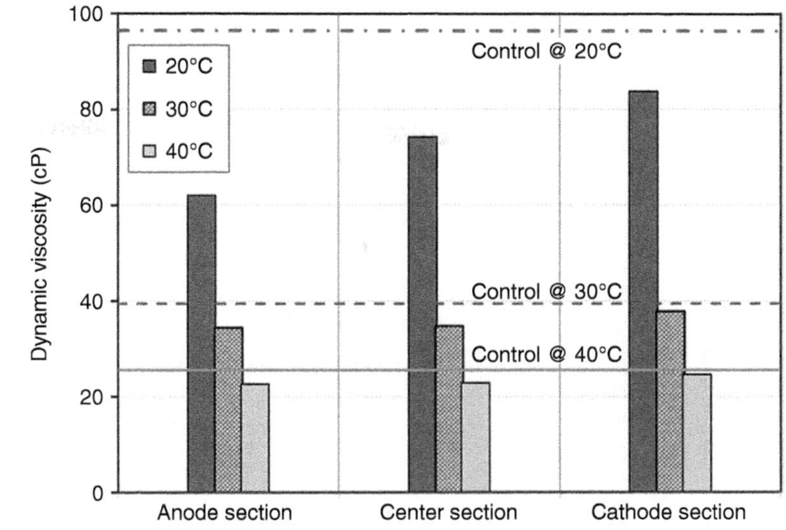

Figure 8.25 The posttest viscosity distribution of PA oil cell compared with its control. *Source:* From Ghazanfari (2013).

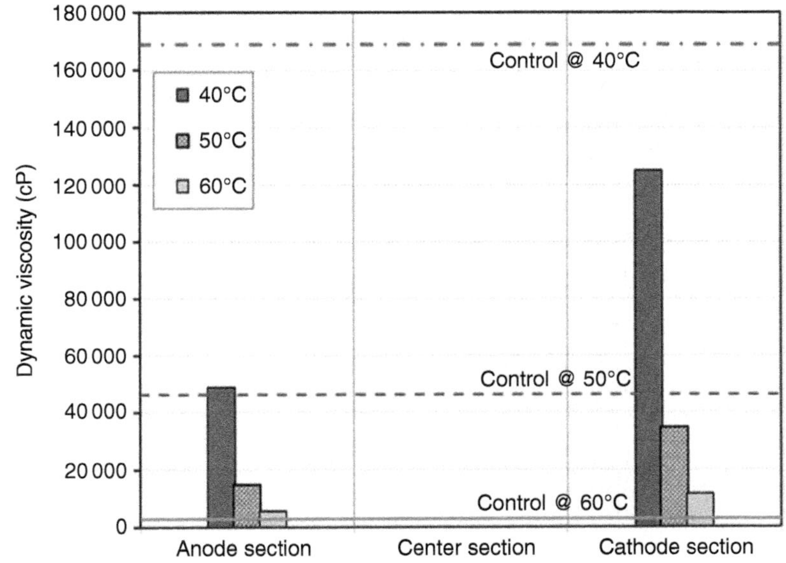

Figure 8.26 The posttest viscosity distribution of Canadian oil cell compared with its control. *Source:* From Ghazanfari (2013).

sections, respectively. For the Canadian oil sample, the viscosity change was more significant, dropping by 69% and 26% in the anode and cathode sections, respectively. The posttest SARA analysis (i.e. measures the contents of saturates, asphaltenes, resins, and aromatics in oil) of the California and Canadian oil samples are shown in Figures 8.27 and 8.28. For the Canadian oil specimens extracted from the anode and cathode regions, there was measurable decrease in the asphaltene content, while increase in saturate content, as shown in Figure 8.28. These trends confirm the viscosity changes observed with Canadian oil samples. The SARA analysis of the California oil sample, on the other hand, showed a decrease in saturates and aromatics, while a substantial increase of asphaltenes in the anode region. This was in concert with CA oil post-EK viscosity change trends, which are not presented here (Table 8.5).

8.3.4.4 Laboratory Evidence of Oil Transport by Displacement

Liquid hydrocarbons from offshore drilling and transportation can spill and seep deep into coastal sediments, in excess of several meters below the sediment surface, and persist as source of pollution over prolonged time. EK, particularly EO advection, has been applied to assist mass transport of hydrocarbon compounds in natural porous media where mobile water and its ionic constituents can induce hydrocarbon compounds and their micelles to move along with EO advection. *Viscous coupling* between the two liquid phases (i.e. pore water and oil) is one of the mechanisms that drag and drive the transport of hydrocarbon compounds in water wet porous media.

The dragging of oil requires some degree of coupling with pore water through an interface (i.e. similar to the

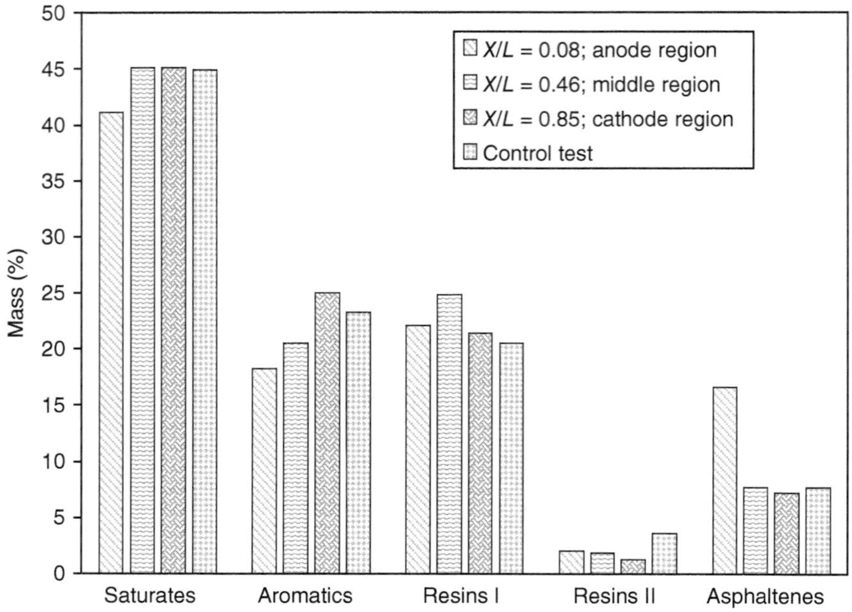

Figure 8.27 The posttest viscosity distribution of CA oil cell compared with its control. *Source:* From Ghazanfari (2013).

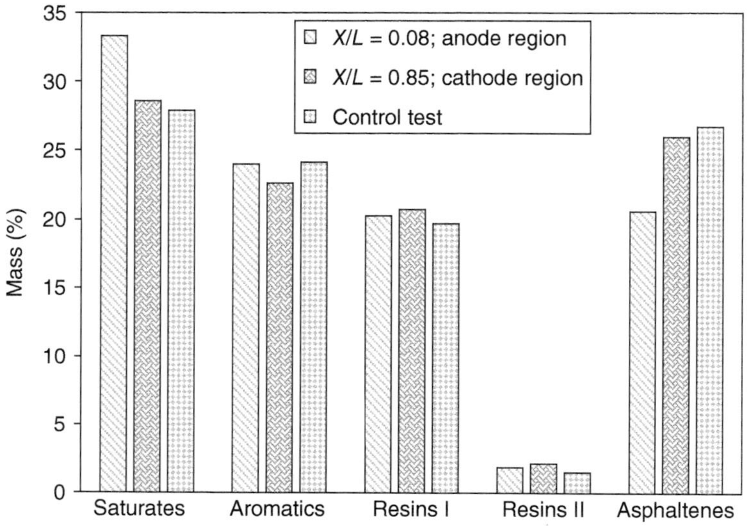

Figure 8.28 The posttest SARA distribution of Canadian core oil compared with its control. *Source:* From Ghazanfari (2013).

Table 8.5 Physical properties of the formation oils and surrogate water.

Fluid	Properties
PA field oil	API 47; dynamic viscosity = 38.7 cP; specific gravity = 0.79 at 20 °C
CA field oil	API 17.4; dynamic viscosity = 2782 cP; specific gravity = 0.95 at 20 °C
Canadian field oil	API 13; dynamic viscosity = 52 000 cP; specific gravity = 0.98 at 20 °C
Electrolyte solution (surrogate formation water)	Salinity = 33 000 ppm; electrical conductivity = 45 000 μS; pH = 7.50; major elements: Na, Cl, Mg, S, K, Ca, Br

DDL interface of clay and electrolyte) that generates an EK potential. This potential arises from the polarity of the two liquids (i.e. oil and water), the degree of salinity of the pore water, and the presence of functional groups and acidic impurities (i.e. O, N, S, compounds, carboxylic acids, amides) in the oil (Johansen, 1923). In presence of immiscible liquids, when one liquid is notably more conductive than the other and the in-between dipolar coupling force is weak (Warshavsky et al., 2001), the interface does not develop and viscous dragging of the oil phase may not take place. Instead, when put into motion by EO, the conductive liquid (i.e. saline water or pore water in soil and sediment) simply flows pass by the nonconductive mass (i.e. oil ganglion), as long as the pore space is available for flow. In compact clay sediments where the pore space is tight and water connectivity is poor and/or the chemistry of the two liquids does not support dispersion or dissolution of one in the other, the net result of a strong EO drive of the pore water may be the separation of two liquids by displacement of one with the other. As a result, as the water pushes into a confined space (i.e. tight pores of clay sediments) with a high EO driver (i.e. high electric field), it can tend to occupy the oil volume by forcing the oil mass to "flow" backward rather than by displacing it forward in the direction of flow (as in advective transport). As water moves forward in the direction of the electric field, it can progressively force the oil mass back in the opposite direction, where it may eventually accumulate in an oil bank (no longer able to move in the pores). This phenomenon, of aided oil transport in confined pore space of water wet sediments, is referred to as *electrically induced displacement*, particularly relevant for nonpolar and nonconductive oil phases residing within the polar conductive water phase in the pores.

The premise of electrically induced displacement of nonconducting nonpolar oil phase in saline sediments was examined in a large-scale experiment. In these tests the implications of spatially variable and transient voltage gradients and PWPs in electrically assisted transport of immiscible oils in high salinity coastal sediments were explored (Pamukcu et al., 2016). The figure, "Transient distribution of Pore Water Pressure (PWP) and corresponding Voltage at four measurement ports (intervals) of test cell P1: (E1-E2); P2: (E2-E3); P3: (E4-E5); P4: (E6-E7) & Photograph of the test cell at the end of 90 days; *note oil bank formation between E4-E5.*" (Pamukcu et al., 2016), shows the visual observation of posttest accumulation of clay and colloids, or "clay banks" – light tan-colored bed of fine material – at the anode end of the soil core, while accumulation of "oil bank" across interval E4–E5 and particularly under port P3 is clearly evident. Clay bank formation results from accumulation of clay and colloidal constituents in pore space or interfaces where prevalent electrophoretic migration transports and deposits them at locations where either the permeability is reduced by progressive accumulation or the enabling electrical potential is weakened significantly (Ghazanfari et al., 2012, 2014). Oil bank formation is similar to that of clay bank, except this time the pore space is occupied with agglomerated oil ganglia and/or oily micelles that may become too large in size to permeate or be mobilized through the soil pores. Water flow restriction at the regions of "oil banks" would normally cause a rise in PWP when the EO drive or the electric field is high. This trend is observed clearly from the voltage and PWP distribution at P3 graph (the figure, "Transient distribution of Pore Water Pressure (PWP)…"), which is unlike the trends observed at P1, P2, and P4 where flow is not restricted.

The figure, "Normalized Oil and Water saturation and Oil/Water ratio by volume distribution post 90 day test (normalized to initial values)" (Pamukcu et al., 2016), presents posttest distribution of oil and water saturation (i.e. pore volume fraction of either phase) along the specimen ports P1 to P4 (Pamukcu et al., 2016). These values are normalized by their initial values, as indicated on the graph. Superimposed on the graph is the distribution of O/W ratio by volume. The graph shows that the O/W ratio increased by about 50% at the anode half of the test sample (at P3 and P4), but it decreased by about 30% on average at the cathode half of the specimen (P1 and P2). The P3 location displayed the highest O/W increase, where restricted EO flow was postulated based on the voltage and PWP data and the visual observations of an oil bank. As the reduction of water content at P3 location was less than the maximum relative standard deviation of 8%, the significant increase in O/W is attributed to oil content increase at P3. The decrease in O/W ratio in the cathode half of the cell is explained by displacement of the resident oil volume by the EO water volume transported toward cathode, which emerges as most plausible mechanism in the absence of potential viscous coupling between the nonpolar oil phase and the pore fluid. In this case, the EO flow of water in one direction causes oil and water to exchange pore space locally as oil is forced to move in the opposite direction. The ultimate result appears to be the separation of the two immiscible fluids, which is best demonstrated by the formation of the oil bank under P3, confirming the postulated mechanism of *electrically induced displacement.*

8.4 Summary and Conclusions

The fundamental principles of electrochemical transport and transformations applied to EK processing of saturated clays and clayey formations were reviewed, and some findings were presented. The classical H–S equation expresses the EO velocity of the fluid as a function of the electric field and the EK potential of the clay. Both of these parameters vary during EK transport and result in a nonlinear process. It was shown analytically and experimentally that the EO velocity could be uncoupled from the applied electric field when surface conductivity σ_s and the resulting portion of the current transferred over the solid–liquid interface i_s are used as intrinsic properties of the clay to describe the velocity. The migration of the ions in the bulk fluid was modeled taking into consideration the changing electric field due to migration and other effects such as retardation and electrophoretic effects that reduce ion mobility. The model appeared to simulate well the long-term ion distribution in the soil as the conductivity and the electric field varied in time and space.

Finally, the experimental results of electrically enhanced extraction, separation, transformation, stabilization, containment, and displacement in clay and clay-rich porous media were presented and discussed in the framework of the theoretical electrochemical and EK models introduced earlier. The data appeared to support the hypothesis that faradaic current passage orthogonal to the planes in the EDL of clay particles may drive forth redox reactions on clay surfaces.

References

Aboughalma, H., Bi, R., and Schlaak, M. (2008). Electrokinetic enhancement on phytoremediation in Zn, Pb, Cu and Cd contaminated soil using potato plants. *Journal of Environmental Science and Health. Part A, Toxic/ Hazardous Substances and Environmental Engineering* 43: 926–933. doi: 10.1080/10934520801974459.

Acar, Y.B. and Alshawabkeh, A.N. (1993). Principles of electrokinetic remediation. *Environmental Science and Technology* 27 (13): 2638–2647.

Acar, Y.B. and Alshawabkeh, A.N. (1996). Electrokinetic remediation I: pilot scale tests with lead spiked kaolinite. *Geotechnical and Geoenvironmental Engineering* 122 (3): 173–185.

Acar, Y.B., Gale, R.J., Alshawabkeh, A.N., Marks, R.E., Puppala, S., Bricka, M., Parker, R. (1995). Electrokinetic remediation: basics and technology status. *Journal of Hazardous Materials* 40: 117–137. doi: 10.1016/0304-3894(94)00066-P.

Acar, Y.B., Hamed, J., Alshawabkeh, A.N., and Gale, R. (1994). Cd(II) removal from saturated kaolinite by application of electrical current. *Geotechnique* 44 (3): 239–254.

Acar, Y.B., Li, Heyi, and Gale, R.J., (1992). Phenol Removal From Kaolinite by Electrokinetics, Journal of the Geotechnical Engineering, *ASCE*, 118 (11): 837–1852.

Ahn, D.W., Kim, S.S., Han, S.J., and Kim, B.I. (2010). Characteristics of electrokinetic remediation of unsaturated soil contaminated by heavy metals-I: experimental study. *International Journal of Offshore and Polar Engineering* 20: 140–146.

Al Shalabi, E.W., Ghosh, B., Haroun, M., and Pamukcu, S. (2012a). Stimulation of sandstone reservoirs using DC potential. *Journal of Petroleum Science and Technology* 30 (20): 2137–2147.

Al Shalabi, E.W., Ghosh, B., Haroun, M., and Pamukcu, S. (2012b). The application of direct current potential to enhancing waterflood recovery efficiency. *Petroleum Science and Technology* 30 (20): 2160–2168. doi: 10.1080/10916466.2010.547902.

Alcántara, M.T., Gómez, J., Pazos, M., and Sanromán, M.A. (2012). Electrokinetic remediation of lead and phenanthrene polluted soils. *Geoderma* 173–174: 128–133. doi: 10.1016/j.geoderma.2011.12.009.

Alshawabkeh, A.N. (2009). Electrokinetic soil remediation: challenges and opportunities. *Separation Science and Technology* 44: 2171–2187.

Alshawabkeh, A.N. and Acar, Y.B. (1992). Removal of contaminants from soils by electrokinetics: a theoretical treatise. *Journal of Environmental Science and Health. Part A, Environmental Science and Engineering Toxicology* 27: 1835–1861. doi: 10.1080/10934529209375828.

Alshawabkeh, A.N. and Acar, Y.B. (1996). Electrokinetic remediation: II. Theoretical model. *Geotechnical and Geoenvironmental Engineering* 122 (3): 186–196.

Alshawabkeh, A.N. and Sarahney, H. (2005). Effect of current density on enhanced transformation of naphthalene. *Environmental Science and Technology* 39: 5837–5843.

Amba, S.A., Chilingar, G.V., and Beeson, C.M. (1964). Use of direct electrical current for increasing the flow rate of reservoir fluids during petroleum recovery. *Journal of Canadian Petroleum Technology* 3: 8–14. doi: 10.2118/64-01-02.

Amba, S.A., Chilingar, G.V., and Beeson, C.M. (1965). Use of direct electrical current for increasing the flow rate of oil and water in a porous medium. *Canadian Petroleum Technology* 4 (1): 81–88.

Anderson, L.D., Kent, D.B., and Davis, J.A. (1994). Batch experiments characterizing the reduction of Cr(VI) using suboxic material from a mildly reducing sand and gravel aquifer. *Environmental Science and Technology* 28 (1): 178–185.

Banerjee, S., Horng, J., Ferguson, J.F., and Nelson, P.O. (1987). *Field Scale Feasibility Study of Electrokinetic Remediation*, CR811762-01, 129pp. Cincinnati, OH: Risk Reduction Engineering Laboratory, Office of Research and Development, US EPA.

Bani-Melhem, K. and Elektorowicz, M. (2010). Development of a novel submerged membrane electro-bioreactor (SMEBR): performance for fouling reduction. *Environmental Science and Technology* 44: 3298–3304.

Bani-Melhem, K. and Elektorowicz, M. (2011). Performance of the submerged membrane electro-bioreactor (SMEBR) with iron electrodes for wastewater treatment and fouling reduction. *Journal of Membrane Science* 379 (1–2): 434–439.

Bard, A.J. and Faulkner, L.R. (1980). *Electrochemical Methods: Fundamentals and Applications*. New York: Wiley.

Bartlett, R.J. (1991). Chromium recycling in soils and water: links, gaps, and methods. *Environmental Health Perspective* 92: 17–24.

Benner, S.G., Blowes, D.W., Ptacek, C.J., and Mayer, K.U. (2002). Rates of sulfate reduction and metal sulfide precipitation in a permeable reactive barrier. *Applied Geochemistry* 17: 301–320. doi: 10.1016/S0883-2927(01)00084-1.

Blowes, D.W., Ptacek, C.J., Cherry, J.A. et al. (1995). Passive remediation of groundwater using in-situ treatment curtains. In: *Proceedings of the Specialty Conference on GeoEnvironment 2000*, 1588–1607. New Orleans, LA: ASCE.

Brosky, R.T. and Pamukcu, S. (2013). Role of DDL processes during electrolytic reduction of cu(II) in a low oxygen environment. *Journal of Hazardous Materials* 262: 878–882. doi: 10.1016/j.jhazmat.2013.09.032.

Bruell, C.J., Segal, B.A., and Walsh, M.T. (1992). Electroosmotic removal of gasoline hydrocarbons and TCE from clay. *Journal of Environmental Engineering* 118 (1): 68–83.

Busalmen, J.P., Esteve-Nuñez, A., and Feliu, J.M. (2008). Whole cell electrochemistry of electricity-producing microorganisms evidence an adaptation for optimal exocellular electron transport. *Environmental Science and Technology* 42 (7): 2445–2450.

Cameselle, C., Chirakkara, R.A., and Reddy, K.R. (2013). Chemosphere electrokinetic-enhanced phytoremediation of soils: status and opportunities. *Chemosphere* 93: 626–636. doi: 10.1016/j.chemosphere.2013.06.029.

Cameselle, C. and Reddy, K.R. (2012). Development and enhancement of electro-osmotic flow for the removal of contaminants from soils. *Electrochimica Acta* 86: 10–22. doi: 10.1016/j.electacta.2012.06.121.

Cao, X. (1997). Numerical modeling of electrokinetically enhanced transport mechanism in soils. Master of Science thesis, Department of CEE, Lehigh University, Bethlehem, PA.

Casagrande, L. (1949). Electro-osmosis in soils. *Geotechnique* 1: 159–177.

Chang, J.H., Shi, Y.H., and Tung, C.H. (2010). Stepwise addition of chemical reagents for enhancing electrokinetic removal of Cu from real site contaminated soil. *Journal of Applied Electrochemistry* 40: 1153–1160.

Chen, J.L., Yang, S.F., Wu, C.C., and Ton, S. (2011). Effect of ammonia as a complexing agent on electrokinetic remediation of copper-contaminated soil. *Separation and Purification Technology* 79: 157–163.

Chilingar, G.V., El-Nassir, A., and Stevens, R.G. (1968a). Effect of direct electrical current on permeability of sandstone cores. *Petroleum Technology* 22: 830–836. doi: 10.2118/2332-PA.

Chilingar, G.V., Loo, W.W., Khilyuk, L.F., and Katz, S.A. (1997). Electrobioremediation of soils contaminated with hydrocarbons and metals: progress report. *Energy Sources* 19: 129–146.

Chilingar, G.V., Sang, C.K., Davis, J.E. et al. (1968b). Possible use of direct electric current for augmenting reservoir energy during petroleum production. *Petroleum Technology* 4: 272–285.

Chowdhury, A.I.A., O'Carroll, D.M., Xu, Y., and Sleep, B.E. (2012). Electrophoresis enhanced transport of nano-scale zero valent iron. *Advances in Water Resources* 40: 71–82. doi: 10.1016/j.advwatres.2012.01.014.

Chu, K.T. (2005). Asymptotic analysis of extreme electrochemical transport. Ph.D. dissertation, Department of Mathematics, MIT, Cambridge, MA.

Chung, H.I. and Lee, M. (2007). A new method for remedial treatment of contaminated clayey soils by electrokinetics coupled with permeable reactive barriers. *Electrochimica Acta* 52: 3427–3431. doi: 10.1016/j.electacta.2006.08.074.

Davis, J.R. (2000). *Corrosion: Understanding the Basics*. Materials Park, OH: ASM International.

DeFlaun, M.F. and Condee, C.W. (1997). Electrokinetic transport of bacteria. *Journal of Hazardous Materials* 55: 263–277. doi: 10.1016/S0304-3894(97)00023-X.

Demir, A., Pamukcu, S., and Shrestha, R.A. (2014). Simultaneous removal of Pb, Cd and Zn from heavily contaminated mine tailing soil using enhanced electrochemical process. *Environmental Engineering Science* 32 (5): 416–424. doi: 10.1089/ees.2014.0384.

Denisov, G., Hicks, R.E., and Probstein, R.F. (1996). On the kinetics of charged contaminant removal from soils using electric fields. *Journal of Colloid Interface Science* 178: 309–323.

Dresel, P.E., Wellman, D.M., Cantrell, K.J., and Truex, M.J. (2011). Review: technical and policy challenges in deep vadose zone remediation of metals and radionuclides. *Environmental Science and Technology* 45: 4207–4216. doi: 10.1021/es101211t.

Dzenitis, J.M. (1997). Soil chemistry effects and flow prediction in electroremediation of soil. *Environmental Science and Technology* 31 (4): 1191–1197.

Eary, L.E. and Rai, D. (1991). Chromate reduction by surface soils under acidic conditions. *Journal of American Soil Science Society* 55: 676–683.

Electorowicz, M. and Boeva, V. (1996). Electrokinetic supply of nutrients in soil bioremediation. *Environmental Science and Technology* 17: 1339–1349.

Elektorowicz, M. (2009). Electrokinetic remediation of mixed metals and organic contaminants. In: *Electrochemical Remediation Technologies for Polluted Soils, Sediments and Groundwater* (ed. K.

Reddy and C. Cameselle) Chapter 15, 315–331. New York: Wiley.

Elektorowicz, M., Habibi, S., and Chifrina, R. (2006). Effect of electrical potential on the electro-demulsification of oily sludge. *Journal of Colloid and Interface Science* 295 (2): 535–541. doi: 10.1016/j.jcis.2005.08.042.

Elektorowicz, M., Hasan, S., and Oleszkiewicz, J. (2012). Charge it: a submerged membrane bioreactor achieves high removal efficiencies. *Water Environmental Technology* 24 (1): 60–63.

Eykholt, G.R. and Daniel, D.E. (1994). Impact of system chemistry on electroosmosis in contaminated soil. *Geotechnical and Geoenvironmental Engineering.* 120 (5): 797–815.

Fleureau, J.M. and Dupeyart, M. (1988). Influence of an electric field on the interfacial parameters of a water/oil/rock system: application to oil enhanced recovery. *Journal of Colloid and Interface Science* 123 (1): 249–258.

Franz, A.J., Rucker, J.W., and Flora, J.V. (2002). Electrolytic oxygen generation for subsurface delivery: effects of precipitation at the cathode and an assessment of side reactions. *Journal of Water Resources* 36: 2243–2254.

Gent, D.B., Bricka, R.M., Alshawabkeh, A.N. et al. (2004). Bench- and field-scale evaluation of chromium and cadmium extraction by electrokinetics. *Journal of Hazardous Materials* 110: 53–62. doi: 10.1016/j.jhazmat.2004.02.036.

Ghazanfari, E. (2013). Development of a mathematical model for electrically assisted oil transport in porous media. Ph.D. dissertation, Department of CEE, Lehigh University, Bethlehem, PA.

Ghazanfari, E. and Pamukcu, S. (2014). Mathematical modeling of electrokinetic transport and enhanced oil recovery in porous geo-media. In: *Electrokinetics for Petroleum and Environmental Engineers* (ed. G.V. Chilingar and M. Haroun), 177–236. Hoboken, NJ: Wiley. doi: 10.1002/9781118842805.

Ghazanfari, E., Pamukcu, S., Pervizpour, M., and Karpyn, Z. (2014). Investigation of generalized relative permeability coefficients for electrically assisted oil recovery in oil formations. *Transport in Porous Media* 105: 235–253. doi: 10.1007/s11242-014-0368-6.

Ghazanfari, E., Shrestha, R.A., Miroshnik, A., and Pamukcu, S. (2012). Electrically assisted liquid hydrocarbon transport in porous media. *Electrochimica Acta* 86: 185–191. doi: 10.1016/j.electacta.2012.04.077.

Ghosh, B., Al Shalabi, E.W., and Haroun, M. (2012). The effect of DC electrical potential on enhancing sandstone reservoir permeability and oil recovery. *Petroleum Science and Technology* 30: 2148–2159. doi: 10.1080/10916466.2010.551233.

Gill, R.T., Harbottle, M.J., Smith, J.W.N., and Thornton, S.F. (2014). Electrokinetic-enhanced bioremediation of organic contaminants: a review of processes and environmental applications. *Chemosphere* 107: 31–42. doi: 10.1016/j.chemosphere.2014.03.019.

Gomes, H.I., Dias-Ferreira, C., Ribeiro, A.B., and Pamukcu, S. (2012). Electrokinetic enhanced transport of zero-valent iron nanoparticles for chromium (VI) reduction in soils. *Chemical Engineering Transactions* 28: 139–144.

Gomes, H.I., Dias-Ferreira, C., Ribeiro, A.B., and Pamukcu, S. (2013). Enhanced transport and transformation of zerovalent nanoiron in clay using direct electric current. *Water, Air, & Soil Pollution* 224: 1710. doi: 10.1007/s11270-013-1710-2.

Gomes, H.I., Dias-Ferreira, C., Ribeiro, A.B., and Pamukcu, S. (2014). Influence of electrolyte and voltage on the direct current enhanced transport of iron nanoparticles in low permeability media. *Chemosphere* 99: 171–179. doi: 10.1016/j.chemosphere.2013.10.06.

Gomes, H.I., Ottosen, L.M., Ribeiro, A.B., and Dias-Ferreira, C. (2015b). Treatment of a suspension of PCB contaminated soil using iron nanoparticles and electric current. *Environmental Managment* 151: 550–555. doi: 10.1016/j.jenvman.2015.01.015.

Gomes, H.I., Rodríguez-Maroto, J.M., Dias-Ferreira, C. et al. (2015a). Numerical prediction of diffusion and electric field-induced iron nanoparticle transport. *Electrochimica Acta* 2015 (18): 5–12. doi: 10.1016/j.electacta.2014.11.157.

Grolimund, D. and Borkovic, M. (2001). Release and transport of colloidal particles in natural porous media: modeling. *Water Research* 37 (3): 559–570.

Guedes, P., Couto, N., Ottosen, L.M., and Ribeiro, A.B. (2014b). Phosphorus recovery from sewage sludge ash through an electrodialytic process. *Waste Management* 34: 886–892. doi: 10.1016/j.wasman.2014.02.021.

Guedes, P., Mateus, E.P., Couto, N. et al. (2014a). Electrokinetic remediation of six emerging organic contaminants from soil. *Chemosphere* 117: 124–131.

Hadley, P. and Ellis, D. (2009). Sustainable remediation white paper – integrating sustainable principles, practices, and metrics into remediation projects. *Remediation.* doi: 10.1002/rem.20210.

Hamed, J., Acar, Y.B., and Gale, R.J. (1991). Pb(II) removal from kaolinite by electrokinetics. *Journal of Geotechnical Engineering* 117 (2): 241–271.

Hannum, L. and Pamukcu, S. (2007). Aided transport of nano-iron in clay soils using direct electric field. In: *NSTI-Nanotech 2007*, Vol. 2, 635–638.

Hansen, H.K. (1995). Practical and theoretical aspects concerning the use of ion exchange membranes and resins in electrokinetic soil remediation. Ph.D. dissertation, Department of Physical Chemistry, The Technical University of Denmark.

Hansen, H.K., Ottosen, L.M., Kliem, B.K., and Villumsen, A. (1997). Electrodialytic remediation of soils polluted with Cu, Cr, Hg, Pb and Zn. *Journal Chemical Technology and Biotechnology* 70: 67–73.

Hansen, H.K., Rojo, A., and Ottosen, L.M. (2005). Electrodialytic remediation of copper mine tailings. *Journal of Hazardous Materials* 117: 179–183.

Haran, B.S., Zheng, G., Popov, B.N., and White, R.E. (1995). Electrochemical decontamination of soils: development of new electrochemical method for decontamination of hexavalent chromium from sand. *Electrochemical Society Proceedings* V95-12: 227–251.

Haroun, M., Chilingar, G.V., and Pamukcu, S. (2010). The efficacy of using electrokinetic transport in highly-contaminated offshore sediments. *Journal of Applied Electrochemistry* 40: 1131–1138.

Haroun, M.R., Chilingar, G. V., Pamukcu, S. et al. (2013). Optimizing electroosmotic flow potential for electrically enhanced oil recovery (EEORTM) in carbonate rock formations of Abu Dhabi based on rock properties and composition. *International Petroleum Technology Conference*, Doha, Qatar (7–9 December). International Petroleum Technology Conference. doi:10.2523/IPTC-13812-MS.

Hasan, S., Elektorowicz, M., and Oleszkiewicz, J. (2014). Start-up period investigation of pilot-scale submerged membrane electro-bioreactor (SMEBR) treating raw municipal wastewater. *Chemosphere* 97: 71–77. doi: 10.1016/j.chemosphere.2013.11.009.

Helmholtz, H. (1879). Studien uber electrische grenzschichten. *Annals of Physical Chemistry* 7 (3): 337–382.

Hicks, E.E. and Tondorf, S. (1994). Electrorestoration of metal contaminated soils. *Environmental Science and Technology* 28 (12): 2203–2210.

Ho, S.V., Athmer, C., Sheridan, P.W. et al. (1999a). The lasagna technology for in situ soil remediation. 1. Small field test. *Environmental Science and Technology* 33 (7): 1086–1091. doi: 10.1021/ES980332S.

Ho, S.V., Athmer, C., Sheridan, P.W. et al. (1999b). The lasagna technology for in situ soil remediation. 2. Large field test. *Environmental Science and Technology* 33 (7): 1092–1099. doi: 10.1021/ES980414G.

Ho, S.V., Sheridan, P.W., Athmer, C.J. et al. (1995). Integrated in situ soil remediation technology: the lasagna process. *Environmental Science and Technology* 29: 2528–2534.

Huang, Y.C. and Cheng, Y.W. (2012). Electrokinetic-enhanced nanoscale iron reactive barrier of trichloroethylene solubilized by Triton X-100 from groundwater. *Electrochimica Acta* 86: 177–184. doi: 10.1016/j.electacta.2012.03.048.

Huang, J., Elektorowicz, M., and Oleszkiewicz, J. (2008). Dewatering and disinfection of aerobic and anaerobic sludge using an electrokinetic (EK) system. *Water Science and Technology* 57 (2): 231–236. doi: 10.2166/wst.2008.009.

Hunter, R.J. (1981). *Zeta Potential in Colloidal Science Principles and Applications*. New York: Academic Press.

Hunter, R.J. (2001). *Foundations of Colloid Science*. Oxford: Oxford University Press.

Hunter, R.J. and James, M. (1992). Charge reversal of kaolinite by hydrolysable metal ions: an electroacoustic study. *Clay Mineralogy* 40 (6): 644–649.

Ibeid, S., Elektorowicz, M., and Oleszkiewicz, J. (2013). Modification of activated sludge properties caused by application of continuous and intermittent current. *Water Research* 47 (2): 903–910. doi: 10.1016/j.watres.2012.11.020.

Innocenti, I., Verginelli, I., Massetti, F. et al. (2014). Pilot-scale ISCO treatment of a MtBE contaminated site using a Fenton-like process. *Science of the Total Environment* 485–486: 726–738. doi: 10.1016/j.scitotenv.2014.01.062.

Isosaari, P., Piskonen, R., Ojala, P. et al. (2007). Integration of electrokinetics and chemical oxidation for the remediation of creosote-contaminated clay. *Journal of Hazardous Materials* 144: 538–548. doi: 10.1016/j.jhazmat.2006.10.068.

Israelachvili, J.N. and Adams, G.E. (1978). Measurements of forces between two silica surfaces in aqueous electrolyte solutions in the range of 0–100 nm. *Journal of the Chemical Society, Faraday Transactions 1: Physical Chemistry in Condensed Phases* 74 (4): 975–1001.

Jensen, P.E., Ottosen, L.M., and Ferreira, C. (2007). Electrodialytic remediation of soil fines (<63μm) in suspension – influence of current strength and L/S. *Electrochimica Acta* 52: 3412–3419. doi: 10.1016/j.electacta.2006.03.116.

Johansen, E.M. (1923). The interfacial tension between petroleum products and water. *Industrial and Engineering Chemistry* 16: 2.

Kappler, A. and Haderlein, S.B. (2003). Natural organic matter as reductant for chlorinated aliphatic pollutants. *Environmental Science and Technology* 37 (12): 2714–2719. doi: 10.1021/es0201808.

Khan, L.I. (1991). Study of electroosmosis in soil: a modified theory and its applications in soil decontamination. Ph.D. dissertation, Department of CEE, Lehigh University, Bethlehem, PA.

Khan, L.I. and Pamukcu, S. (1989). Validity of electroosmosis for groundwater decontamination. In: *Proceedings of Environmental Engineering Conference*, 563–570. Austin, TX: ASCE.

Kim, K.H., Kim, S.O., Lee, C.W. et al. (2003). Electrokinetic processing for the removal of radionuclides in soils. *Separation Science and Technology* 38: 2137–2163. doi: 10.1081/SS-120021617.

Koch, A., McBratney, A., Adams, M. et al. (2013). Soil security: solving the global soil crisis. *Global Policy* 4: 434–441. doi: 10.1111/1758-5899.12096.

Kocur, C.M., O'Carroll, D.M., and Sleep, B.E. (2013). Impact of nZVI stability on mobility in porous media. *Journal of Contaminant Hydrology* 145: 17–25. doi: 10.1016/j.jconhyd.2012.11.001.

Korolev, V.A., Romanyukha, O.V., and Abyzova, A.M. (2008). Electrokinetic remediation of oil-contaminated soils. *Journal of Environmental Science and Health. Part A, Toxic/Hazardous Substances and Environmental Engineering* 43 (8): 876–880.

Kortüm, G. and Bockris, J.M. (1951). *Textbook of Electrochemistry*, vol. 2. Amsterdam: Elsevier.

Kretzschmar, R., Borkovec, M., Grolimund, D., and Elimelech, M. (1999). Mobile subsurface colloids and their role in contaminant transport. *Advances in Agronomy* 66: 121–193.

Kruyt, H.R. ed. (1952). *Colloidal Science*. Amsterdam: Elsevier.

Kubiak, J.J., Khankhane, P.J., Kleingeld, P.J., and Lima, A.T. (2012). An attempt to electrically enhance phytoremediation of arsenic contaminated water. *Chemosphere* 87: 259–264.

Lageman, R. (1993). Electroreclamation: applications in the Netherlands. *Environmental Science and Technology* 27: 2648–2650.

Lageman, R. and Godschalk, M.S. (2007). Electro-bioreclamation. A combination of in situ remediation techniques proves successful at a site in Zeist, the Netherlands. *Electrochimica Acta* 52: 3449–3453.

Lei, H., Chen, K., and Li, Y. (2012). Electrokinetic recovery of copper, nickel and zinc from wastewater sludge: effects of electrical potentials. *Environmental Engineering Science* 29: 472–478.

Lima, A.T., Kleingeld, P.J., Heister, K., and Loch, J.P.G. (2011). Removal of PAHs from contaminated clayey soil by means of electro-osmosis. *Separation and Purification Technology* 79: 221–229.

Lima, A.T., Kleingeld, P.J., Heister, K., and Loch, J.P.G. (2012a). In situ electro-osmotic cleanup of tar contaminated soil – removal of PAHs. *Electrochimica Acta* 86: 142–147.

Lima, A.T., Loch, J.P.G., and Kleingeld, P.J. (2010). Bentonite electrical conductivity: a model based on series–parallel transport. *Journal of Applied Electrochemistry* 40: 1061–1068. doi: 10.1007/s10800-009-0060-7.

Lima, A.T., Ottosen, L.M., Heister, K., and Loch, J.P.G. (2012b). Assessing PAH removal from clayey soil by means of electro-osmosis and electrodialysis. *Science of the Total Environment* 435–436: 1–6. doi: 10.1016/j.scitotenv.2012.07.010.

Lindgren, E.R. (1998). *Electrokinetic Demonstration at the Unlined Chromic Acid Pit*, 144pp. Springfield, VA: Sandia National Laboratories, DOE-Office of Scientific and Technical Information.

Liu, S.H. and Wang, H.P. (2013). Electrochemical remediation of copper-contaminated soils enhanced by ethylenediaminetetraacetic acid: an in-situ x-ray absorption spectroscopic study. *International Journal of Electrochemical Science* 8: 4807–4817.

Loch, J.P., Gustav, K.R., and Thomas, J.S.K. (2005). Coupling between chemical and electrical osmosis in clays. *Physicochemical and Electrochemical Interactions* 2005: 283–288.

Lockhart, N.C. and Stickland, R.E. (1984). Dewatering coal washery tailings ponds by electroosmosis. *Powder Technology* 40: 215–221.

López-Vizcaíno, R., Risco, C., Isidro, J. et al. (2017). Scale-up of the electrokinetic fence technology for the removal of pesticides. Part II: does size matter for removal of herbicides? *Chemosphere* 166: 549–555. doi: 10.1016/j.chemosphere.2016.09.114.

Luo, Q., Wang, H., Zhang, X. et al. (2006). In situ bioelectrokinetic remediation of phenol-contaminated soil by use of an electrode matrix and a rotational operation mode. *Chemosphere* 64: 415–422. doi: 10.1016/j.chemosphere.2005.11.064.

Lyklema, J. (1995). *Fundamentals of Interface and Colloid Science. Volume II: Solid-Liquid Interfaces*. San Diego: Academic Press Limited.

Macdonald, R.W., Barrie, L.A., Bidleman, T.F. et al. (2000). Contaminants in the Canadian Arctic: 5 years of progress in understanding sources, occurrence and pathways. *Science of the Total Environment* 254: 93–234. doi: 10.1016/S0048-9697(00)00434-4.

Maes, N., Moors, H., Dierckx, A. et al. (1999). The assessment of electromigration as a new technique to study diffusion of radionuclides in clayey soils. *Journal of Contaminant Hydrology* 36: 231–247. doi: 10.1016/S0169-7722(98)00146-6.

Mao, X., Wang, J., Ciblak, A. et al. (2012). Electrokinetic-enhanced bioaugmentation for remediation of chlorinated solvents contaminated clay. *Journal of Hazardous Materials* 213–214: 311–317. doi: 10.1016/j.jhazmat.2012.02.001.

Mattson, E., Bowman, D., Robert, S., and Lindgren, E.R. (2002). Electrokinetic ion transport through unsaturated soil: 1. Theory, model development, and testing. *Contaminant Hydrology* 54 (1-2): 99–120.

McBratney, A., Field, D.J., and Koch, A. (2014). The dimensions of soil security. *Geoderma* 213: 203–213. doi: 10.1016/j.geoderma.2013.08.013.

Mena, E., Villaseñor, J., Rodrigo, M.A., and Cañizares, P. (2016). Electrokinetic remediation of soil polluted with insoluble organics using biological permeable reactive barriers: effect of periodic polarity reversal and voltage gradient. *Chemical Engineering Journal* 299: 30–36. doi: 10.1016/j.cej.2016.04.049.

Merkx, O.K., Loch, J.P.G., Lima, A.T. et al. (2013). The effectiveness of electro-remediation of aged, metal-contaminated sediment in relation to sequential extraction of metals. *Water, Air, & Soil Pollution* 224: 1–12. doi: 10.1007/s11270-013-1667-1.

Mihelcic, J.R., Naughton, C.C., Verbyla, M.E. et al. (2017). The grandest challenge of all: the role of environmental engineering to achieve sustainability in the world's developing regions. *Environmental Science and Engineering* 34 (1): 16–41. doi: 10.1089/ees.2015.0334.

Mitchell, J.K. and Yeung, T.C. (1991). Electro-kinetic flow barriers in compacted clay. *Transportation Research Records* 1288: 1–9.

Mitchell, J. and Soga, K. (2005). *Fundamentals of Soil Behavior*. Hoboken, NJ: Wiley.

Moon, J.W., Moon, H.S., Kim, H., and Roh, Y. (2005). Remediation of TCE-contaminated groundwater using zero valent iron and direct current: experimental results and electron competition model. *Environmental Geology* 48: 805–817. doi: 10.1007/s00254-005-0023-1.

Muraoka, T., Ghazanfari, E., Shrestha, R.A., and Pamukcu, S. (2011). Electrically induced pore pressures in high salt content clay slurry. *Separation and Purification Technology* 79: 133–138.

Nernst, W. (1888). Zur kinetik der in losung befindlichen korper. *Zeitschrift für Physikalische Chemie* 2: 613–637.

Newman, J. (1991). *Electrochemical Systems*, 2e. Englewood Cliffs, NJ: Prentice-Hall Inc.

Nieto Castillo, A., García-Delgado, R.A., and Cala Rivero, V. (2012). Electrokinetic treatment of soils contaminated by tannery waste. *Electrochimica Acta* 86: 110–114. doi: 10.1016/j.electacta.2012.04.132.

Niqui-Arroyo, J.-L., Bueno-Montes, M., Posada-Baquero, R., and Ortega-Calvo, J.-J. (2006). Electrokinetic enhancement of phenanthrene biodegradation in creosote-polluted clay soil. *Environmental Pollution* 142: 326–332. doi: 10.1016/j.envpol.2005.10.007.

Niqui-Arroyo, J.-L. and Ortega-Calvo, J.-J. (2007). Integrating biodegradation and electroosmosis for the enhanced removal of polycyclic aromatic hydrocarbons from creosote-polluted soils. *Journal of Environmental Quality* 36: 1444–1451.

Nystroem, G.M., Ottosen, L.M., and Villumsen, A. (2005). Electrodialytic removal of Cu, Zn, Pb, and Cd from harbor sediment: influence of changing experimental conditions. *Environmental Science and Technology* 39: 2906–2911.

O'Mahony, M.M., Dobson, A.D.W., Barnes, J.D., and Singleton, I. (2006). The use of ozone in the remediation of polycyclic aromatic hydrocarbon contaminated soil. *Chemosphere* 63: 307–314. doi: 10.1016/j.chemosphere.2005.07.018.

Ottosen, L.M., Hansen, H.K., and Hansen, C.B. (2000). Water splitting at ion-exchange membranes and potential differences in soil during electrodialytic soil remediation. *Journal of Applied Electrochemistry* 30: 1199–1207.

Ottosen, L.M., Hansen, H.K., Laursen, S., and Villumsen, A. (1997). Electrodialytic remediation of soil polluted with copper from wood preservation industry. *Environmental Science and Technology* 31: 1711–1715.

Ottosen, L.M., Jensen, P.E., Kirkelund, G.M. et al. (2012). Electrodialytic remediation of heavy metal polluted soil – treatment of water saturated or suspended soil. *Chemical Engineering Transactions* 28: 103–108. doi: 10.3303/CET1228018.

Paillat, T., Moreau, E., Grimaud, P.O., and Touchard, G. (2000). Electrokinetic phenomena in porous media applied to soil decontamination. *IEEE Transactions on Dielectrics and Electrical Insulation* 7: 693–704. doi: 10.1109/94.879363.

Pamukcu, S. (1994). *Electrokinetic Removal of Coal Tar Constituents from Contaminated Soils*, EPRI TR-103320. Palo Alto, CA: Electric Power Research Institute.

Pamukcu, S. (2009). Electrochemical transport and transformations. In: *Electrochemical Remediation Technologies for Polluted Soils, Sediments and Groundwater* (ed. K.R. Reddy and C. Cameselle), 29–65. Hoboken, NJ: Wiley. doi: 10.1002/9780470523650.

Pamukcu, S., Filipova, I., and Wittle, J.K. (1995). The role of electroosmosis in transporting PAH compounds in contaminated soils. In: *Proceedings of the Symposium on Electrochemical Technology Applied to Environmental Problems*, PV95-12, 252–266. Pennington, NJ: Electrochemical Society Transaction.

Pamukcu, S., Ghazanfari, E., and Wittle, J.K. (2014). Reduction of contaminants in soil and water by direct electric current. In: *Electrokinetics for Petroleum and Environmental Engineers* (ed. G.V. Chilingar and M. Haroun), 33–102. Beverly, MA: Wiley, Scrivener Publishing, LLC.

Pamukcu, S., Hannum, L., and Wittle, J.K. (2008). Delivery and activation of nano-iron by DC electric field. *Journal of Environmental Science and Health. Part A, Toxic/Hazardous Substances and Environmental Engineering* 43 (8): 934–944. doi: 10.1080/10934520801974483.

Pamukcu, S., and Huang, C.P. (2001). In-situ remediation of contaminated soils by electrokinetic processes. In: *Hazardous and Radioactive Waste Treatment Technologies Handbook* (ed. Oh C.H.), Chapter 3.1. Boca Raton, FL: CRC Press LLC.

Pamukcu, S., Khan, L.I., and Fang, H.Y. (1990). Zinc detoxification of soils by electro-osmosis. *Transportation Research Record* 1289: 41–46.

Pamukcu, S. and Pervizpour M. (1998a). Electrochemical Extraction of Cyanide and Fluoride from Conductive Porous Media. Final report to EPI, Wayne, PA (March 1998) (available from authors).

Pamukcu, S. and Pervizpour, M. (1998b). Electroosmotically Aided Restoration of TCE Contaminated Soil. Final Report to Lawrence Livermore National Laboratory, Environmental Restoration Division, Contract No. B3460123 (November), 52p. Livermore, CA: University of California.

Pamukcu, S., Shrestha, R.A., Ribeiro, A.B., and Mateus, E.P. (2016). Electrically induced displacement transport of immiscible oil in saline sediments. *Journal of Hazardous Materials* 313: 185–192. doi: 10.1016/j.jhazmat.2016.04.005.

Pamukcu, S., Weeks, A., and Wittle, J.K. (1997). Electrochemical extraction and stabilization of selected inorganic species in porous media. *Journal of Hazardous Materials* 55: 305–318 http://dx.doi.org/10.1016/S0304-3894(97)00025-3.

Pamukcu, S., Weeks, A., and Wittle, J.K. (2004). Enhanced reduction of Cr(VI) by direct electric current in a contaminated clay. *Environmental Science and Technology* 38 (4): 1236–1241. doi: http://dx.doi.org/10.1021/es034578v.

Pamukcu, S. and Wittle, J.K. (1992a). Electrokinetics for removal of low-level radioactivity from soil. *The 14th Annual DOE Low-Level Radioactive Waste Management Conference*, Phoenix, AZ (18–20 November), Conf-921137-Prcoeedings, pp. 256–278.

Pamukcu, S. and Wittle, J.K. (1992b). Electrokinetic removal of selected heavy metals from soil. *Environmental Progress* 11 (3): 241–250.

Pamukcu, S. and Wittle, J.K. (1993). Electrokinetically enhanced in-situ soil decontamination. In: *Remediation of Hazardous Waste Contaminated Soils* (ed. D.L. Wise and D.J. Trantolo), Chapter 13, 245–298. New York: Marcel Dekker Inc.

Park, S.-M. and Yoo, J.-S. (2003). Peer reviewed: electrochemical impedance spectroscopy for better electrochemical measurements. *Analytical Chemistry* 75 (21): 455-A–461-A.

Pazos, M., Gouveia, S., Sanroman, M.A., and Cameselle, C. (2008). Electromigration of Mn, Fe, Cu and Zn with citric acid in contaminated clay. *Journal of Environmental Science and Health. Part A, Toxic/Hazardous Substances and Environmental Engineering* 43 (8): 823–831.

Pazos, M., Rosales, E., Alcántara, T. et al. (2010). Decontamination of soils containing PAHs by electroremediation: a review. *Journal of Hazardous Materials* 177: 1–11. doi: 10.1016/j.jhazmat.2009.11.055.

Peng, C., Almeira, J.O., and Gu, Q. (2013). Effect of electrode configuration on pH distribution and heavy metal ions migration during soil electrokinetic remediation. *Environmental Earth Science* 69: 257–265.

Petri, B.G., Watts, R.J., Teel, A.L. et al. (2011). Fundamentals of ISCO using hydrogen peroxide. In: *In Situ Chemical Oxidation for Groundwater Remediation* (ed. R.L. Siegrist), 33–88. Berlin, Heidelberg: Springer. doi: 10.1007/978-1-4419-7826-4_2.

Planck, M. (1890). Ueber die erregung von elektricitat und warme in electrolyten. *Annals of Physical Chemistry* 39 (3): 161–186.

Powell, R.M., Puls, R.W., Hightower, S.K., and Sabatini, D.A. (1995). Coupled iron corrosion and chromate reduction: mechanisms for subsurface remediation. *Environmental Science and Technology* 29 (8): 1913–1922.

Probstein, R.F. and Hicks, R.E. (1993). Removal of contaminants from soils by electric fields. *Science* 260: 498–503.

Pulgarin, C., Adler, N., Peringer, P., and Comninellis, C. (1994). Electrochemical detoxification of a 1,4-benzoquinone solution in wastewater treatment. *Water Research* 28: 887–893.

Quigley, R.M. (1980). Geology, mineralogy and geochemistry of Canadian soft soils: a geotechnical perspective. *Canadian Geotechnical Journal* 20 (2): 288–298.

Quinke, G. (1861). Ueber die Fortfuhrung materieller Teilchen durch stromende Electrizitat. *Annalen der Physik und Chemie CXIII* 8: 33.

Rai, D., Sass, B., and Moore, M. (1987). Chromium (III) hydrolysis constants and solubility of chromium (III) hydroxide. *Journal of Chemistry* 26: 345–349.

Ramírez, E.M., Jiménez, C.S., Camacho, J.V. et al. (2015). Feasibility of coupling permeable bio-barriers and electrokinetics for the treatment of diesel hydrocarbons polluted soils. *Electrochimica Acta* 181: 192–199. doi: 10.1016/j.electacta.2015.02.201.

Ravina, I. and Zaslavsky, D. (1967). Non-linear electrokinetic phenomena I: review of literature. *Soil Science* 106 (1): 60–66.

Ray, C. and Ramsey, R.H. (1987). Removal of heavy metals in wastewater in a clay soil matrix using electro-osmosis. *Environmental Progress* 6 (3): 145–149.

Reddy, K.R. (2010). Technical challenges to in-situ remediation of polluted sites. *Geotechnical and Geological Engineering* 28: 211–221.

Reddy, K.R., Darko-Kagya, K., and Al-Hamdan, A.Z. (2011). Electrokinetic remediation of pentachlorophenol contaminated clay soil. *Water, Air & Soil Pollution* 221: 35–44.

Reddy, K.R. and Parupudi, U. (1997). Removal of chromium, nickel, and cadmium from clays by in-situ electrokinetic remediation. *Soil Contamination* 6 (4): 391–407.

Reuss, F.F. (1809). Sur un nouvel effet de l'électricité glavanique. *Mémoires de la Société impériale des naturalistes de Moscou* 2: 327–337.

Ribeiro, A.B., Mateus, E.P., and Rodríguez-Maroto, J.M. (2011). Removal of organic contaminants from soils by an electrokinetic process: the case of molinate and bentazone. Experimental and modeling. *Separation and Purification Technology* 79 (2): 193–203.

Ribeiro, A.B. and Mexia, J.T. (1997). A dynamic model for the electrokinetic removal of copper from a polluted soil. *Journal of Hazardous Materials* 56 (3): 257–271.

Ribeiro, A.B., Rodriíguez-Maroto, J.M., Mateus, E.P., and Gomes, H. (2005). Removal of organic contaminants from soils by an electrokinetic process: the case of atrazine. Experimental and modeling. *Chemosphere* 59: 1229–1239.

Riis, C., Bymose, M., Cox, E. et al. (2012). Successful pilot test of electrokinetic enhanced bioremediation (EK-BIO) as an innovative remedial approach for PCE DNAPL source area. *NORDROCS 2012: 4th Nordic Joint Meeting on Remediation of Contaminated Sites, International Conference*, Oslo, Norway, 4pp.

Roach, N. and Reddy, K.R. (2006). Electrokinetic delivery of permanganate into low-permeability soils. *International Journal of Environment and Waste Management* 1: 4–19.

Roh, Y., Lee, S.Y., Elless, M.P., and Cho, K.S. (2000). Electro-enhanced remediation of radionuclide-contaminated groundwater using zero-valent iron. *Journal of Environmental Science Health, Part A. Toxic/Hazardous Substances and Environmental Engineering* 35: 1043–1059.

Rojo, A., Hansen, H.K., and Cubillos, M. (2012). Electrokinetic remediation using pulsed sinusoidal electrical field. *Electrochimca Acta* 86: 124–129.

Rubinstein, I. (1990). *Electro-Diffusion of Ions. SIAM Studies in Applied Mathematics*. Philadelphia, PA: SIAM.

Ryu, B.G., Park, G.Y., Yang, J.W., and Baek, K. (2011). Electrolyte conditioning for electrokinetic remediation of As, Cu and Pb-contaminated soil. *Separation and Purification Technology* 79 (2): 170–176.

Saichek, R.E. and Reddy, K.R. (2003). Effect of pH control at the anode for the electrokinetic removal of phenanthrene from kaolin soil. *Chemosphere* 51: 273–287.

Saichek, R.E. and Reddy, K.R. (2004). Evaluation of surfactants/cosolvents for desorption/solubilization of phenanthrene in clayey soils. *International Journal of Environmental Studies* 61: 587–604.

Saichek, R.E. and Reddy, K.R. (2005a). Electrokinetically enhanced remediation of hydrophobic organic compounds in soils: a review. *Critical Reviews in Environmental Science and Technology.* 35: 115–192.

Saichek, R.E. and Reddy, K.R. (2005b). Surfactant-enhanced electrokinetic remediation of polycyclic aromatic hydrocarbons in heterogeneous subsurface environments. *Journal of Environmental Engineering and Science* 4 (5): 327–339.

Sato, M., Kudo, N., and Saito, M. (1998). Surface tension reduction of liquid by applied electric field using vibrating jet method. *IEEE Transactions on Industry Applications* 34 (2): 415–416.

Schultz, D.S. (1997). Electroosmosis technology for soil remediation: laboratory results, field trial, and economic modeling. *Journal of Hazardous Materials* 55: 81–91. doi: 10.1016/S0304-3894(97)00014-9.

Shang, J.Q., Lo, K.Y., and Inculet, I. (1995). Polarization and conduction of clay-water-electrolyte systems. *Geotechnical Engineering* 121 (3): 243–248.

Shapiro, A.P. and Probstein, R.F. (1993). Removal of contaminants from saturated clay by electroosmosis. *Environmental Science and Technology* 27: 283–291.

Shapiro, A.P., Probstein, R.F., and Hicks, R.E. (1995). Removal of cadmium (II) from saturated kaolinite by application of electric current. *Geotechnique* 45: 355–359.

Shin, S., Chilingar, G.V., Haroun, M. et al. (2012). The effect of generated chlorine gas on electroremediation of heavy metals from offshore muds. *Journal of Environmental Protection* 3: 363–373.

Smith, R.W. and Narimatsu, Y. (1993). Electrokinetic behavior of kaolinite in surfactant solutions as measured by both the microelectrophoresis and streaming potential methods. *Minerals Engineering* 6 (7): 753–763.

Stevens, S. (1999). *Enhanced Oil Recovery Scoping Study*, EPRI TR-113836. Palo Alto, CA: Electric Power Research Institute.

Sun, Y.-P., Li, X.-Q., Zhang, W.-X., and Wang, H.P. (2007). A method for the preparation of stable dispersion of zero-valent iron nanoparticles. *Colloid Surface A* 308: 60–66.

Sun, T.R., Ottosen, L.M., Jensen, P.E., and Kirkelund, G.M. (2012). Electrodialytic remediation of suspended soil – comparison of two different soil fractions. *Journal of Hazardous Materials* 203–204: 229–235. doi: 10.1016/j.jhazmat.2011.12.006.

Sun, T.R., Pamukcu, S., Ottosen, L.M., and Wang, F. (2015). Electrochemically enhanced reduction of hexavalent chromium in contaminated clay: kinetics, energy consumption, and application of pulse current. *Chemical Engineering Journal* 262: 1099–1107. doi: 10.1016/j.cej.2014.10.081.

Tao, R. and Xu, X. (2006). Reducing the viscosity of crude oil by pulsed electric or magnetic field. *Energy & Fuels* 2006 (20): 2046–2051.

Traina, G. (2009). Electrokinetic stabilization as a reclamation tool for waste materials polluted by both salts and heavy metals. *Chemosphere* 75: 819–824.

Valdovinos, V., Monroy-Guzmán, F., and Bustos, E. (2016). Electrokinetic removal of radionuclides contained in scintillation liquids absorbed in soil type phaeozem. *Journal of Environmental Radioactivity* 162: 80–86. doi: 10.1016/j.jenvrad.2016.05.017.

Vieira dos Santos, E., Sáez, C., Cañizares, P. et al. (2017). Reversible electrokinetic adsorption barriers for the removal of atrazine and oxyfluorfen from spiked soils. *Journal of Hazardous Materials* 322: 413–420. doi: 10.1016/j.jhazmat.2016.10.032.

Virkutyte, J. and Sillanpää, P.L. (2002). Electrokinetic soil remediation – critical overview. *Science of the Total Environment* 289: 97–121.

Warshavsky, V.B., Bykov, T.V., and Zeng, X.C. (2001). Effects of external electric field on the interfacial properties of weakly dipolar fluid. *The Journal of Chemical Physics* 114: 504–512.

Weeks, A. (2002). Electrokinetically enhanced reduction of Cr(VI) by aqueous Fe(II) in contaminated clays. Ph.D. dissertation, Lehigh University, Bethlehem, PA.

Weeks, A. and Pamukcu, S. (1996). Electrokinetic stabilization of selected heavy metals in contaminated soils. The 4th Great Lakes Geotechnical & Geoenvironmental Conference: In-Situ Remediation of Contaminated Sites, Chicago, IL (May), 151–167.

Weeks, A. and Pamukcu, S. (2001). Electrokinetically enhanced reduction of Cr (VI) in contaminated soils. In: *Proceedings of the XVth International Conference on Soil Mechanics and Geotechnical Engineering*, vol. 3, 1935–1938. Balkema Publishers.

Weinberger, R. (1993). *Practical Capillary Electrophoresis.* San Diego: Academic Press.

Weng, C.H., Lin, Y.T., Lin, T.Y., and Kao, C.M. (2007). Enhancement of electrokinetic remediation of hyper-Cr(VI) contaminated clay by zero-valent iron. *Journal of Hazardous Materials* 149: 292–302. doi: 10.1016/j.jhazmat.2007.03.076.

Wick, L.Y., Mattle, P.A., Wattiau, P., and Harms, H. (2004). Electrokinetic transport of PAH-degrading bacteria in model aquifers and soil. *Environmental Science and Technology* 38: 4596–4602. doi: 10.1021/es0354420.

Wittle, J.K. and Bell, C.W. (2002). Electrochemical system and method for rendering contaminated electrically conductive material nonhazardous. US Patent 6486376 B2.

Wittle, J.K., Hill, D.C., and Chilingar, G.V. (2008). Direct current electrical enhanced oil recovery in heavy-oil reservoirs to improve recovery, reduce water cut, and reduce H2S production while increasing API gravity, SPE 114012. Society of Petroleum Engineering.

Wittle, J.K., Hill, D.G., and Chilingar, G.V. (2011). Direct electric current oil recovery (EEOR) – a new approach to enhancing oil production. *Energy Sources, Part A Recovery Utilization and Environmental Effects* 33: 805–822. doi: 10.1080/15567036.2010.514843.

Wittle, J.K. and Pamukcu, S. (1993). Electrokinetic Treatment of Contaminated Soils, Sludges and Lagoons, DOE/CH-9206, No. 02112406, 65pp. Argonne, IL: Argonne National Laboratories, US Department of Energy.

Wittle, J.K. and Pamukcu, S. (1997). Electrochemical system and method for the removal of charged species from contaminated liquid and solid wastes. US Patent 5614077 A.

Wittle, J.K., Pamukcu, S., Bowman, D. et al. (2009). Field studies on sediment remediation. In: *Electrochemical Remediation Technologies for Polluted Soils, Sediments and Groundwater* (ed. K. Reddy and C. Camaselle), 661–696. Hoboken, NJ: Wiley. doi: 10.1002/9780470523650.ch32.

Yang, G.C.C., Chen, M.C., and Yeh, C.F. (2011). Dewatering of a biological industrial sludge by electrokinetics-assisted filter press. *Separation and Purification Technology* 79: 177–182. doi: 10.1016/j.seppur.2011.02.012.

Yang, L., Nakhla, G., and Bassi, A. (2005). Electro-kinetic dewatering of oily sludges. *Journal of Hazardous Materials* 125: 130–140.

Yeung, A.T. (2011). Milestone developments, myths, and future directions of electrokinetic remediation. *Separation and Purification Technology* 79: 124–132. doi: 10.1016/j.seppur.2011.01.022.

Yeung, A.T., Hsu, C.N., and Menon, R. (1996). EDTA-enhanced electrokinetic extraction of lead. *Geotechnical Engineering* 122 (8): 666–673.

Yin, J., Finno, R.J., and Feldkamp, J.R. 1995. Electro osmotic mobility measurement for kaolinite clay. *Proceedings of the Specialty Conference on Geotechnical Practice in Waste Disposal*, part 2 (of 2). New Orleans, LA: ASCE.

Yukselen-Aksoy, Y. and Reddy, K.R. (2012). Effect of soil composition on electrokinetically enhanced persulfate oxidation of polychlorobiphenyls. *Electrochimica Acta* 86: 164–169. doi: 10.1016/j.electacta.2012.03.049.

Zhang, S., Zhang, J., Cheng, X. et al. (2015). Electrokinetic remediation of soil containing Cr(VI) by photovoltaic solar panels and a DC-DC converter. *Journal of Chemical Technology and Biotechnology* 90: 693–700. doi: 10.1002/jctb.4359.

Zhou, M., Zhu, S., Yi, Y., and Zhang, T. (2016). An electrokinetic/activated alumina permeable reactive barrier-system for the treatment of fluorine-contaminated soil. *Clean Technologies and Environmental Policy* 18: 2691–2699. doi: 10.1007/s10098-016-1156-5.

9

Remote Sensing of Environmental Variables and Fluxes

Morteza Sadeghi[1], *Ebrahim Babaeian*[2], *Ardeshir M. Ebtehaj*[3], *Scott B. Jones*[1], *and Markus Tuller*[2]

[1] Department of Plants, Soils and Climate, Utah State University, Logan, UT, USA
[2] Department of Soil, Water and Environmental Science, The University of Arizona, Tucson, AZ, USA
[3] Department of Civil, Environmental and Geo-Engineering, Saint Anthony Falls Laboratory, University of Minnesota, Minneapolis, MN, USA

9.1 Introduction

The ability to accurately characterize and quantify large-scale environmental processes is at the core of numerous earth science and environmental engineering applications that directly impact the global environment and human society. Potential applications include forecasting of weather and climate variability, management and allocation of water resources, agricultural plant production, prediction and monitoring of drought conditions, remediation of contaminated natural resources, prevention of natural disasters, or monitoring of ecosystem response to climate change.

While direct quantification and monitoring of large-scale environmental processes are prohibitively expensive and laborious due to the large number of required sampling points, rapidly evolving remote sensing (RS) technology and retrieval techniques provide an exceedingly powerful means for determination of spatial highly heterogeneous environmental variables and fluxes at the regional to global scales.

RS is the science of acquiring, processing, and interpreting images that record the interaction between electromagnetic (EM) energy and matter, commonly applied from air and space for quantification and monitoring of land surface and atmospheric properties and processes. Because of a substantial intensification of interest in the application of RS technology over the last decade, considerable resources are currently devoted to advancement of sensors for air- and spaceborne measurement platforms as well as to the development of novel retrieval algorithms.

This chapter is intended to introduce the fundamental principles at the core of environmental RS applications and to provide a thorough overview of the current state of RS technology and associated methods for retrieval of environmental variables and fluxes in conjunction with a description of air- and spaceborne RS missions. The chapter encompasses seven sections, with radiative transfer theory that describes the propagation of EM waves in the environment presented in Section 9.2. Current remote sensors and associated space and airborne platforms are discussed in Section 9.3. Section 9.4 provides detailed information about RS of various static land surface properties. In Section 9.5, RS of relevant environmental state variables such as land surface temperature (LST) and soil moisture (SM) is presented. Section 9.6 is devoted to RS of environmental fluxes such as precipitation and evapotranspiration (ET), and a brief summary is provided in Section 9.7.

9.2 Radiative Transfer Theory

9.2.1 The Radiative Transfer Equation

EM radiation is at the core of RS technology. Because radiation can transport energy without a medium, it is the only means by which the earth interacts with the rest of the universe. The energy flux per unit time, unit frequency, unit solid angle, and unit area normal to the direction of propagation is called *spectral radiance* or *specific intensity of radiation* ($\mathrm{W\,m^{-2}\,sr^{-1}\,Hz^{-1}}$). In free space, the spectral radiance I at any given wavelength λ or frequency f (note that $f = c/\lambda$ where c is the speed of light) is conserved along a ray, i.e. $\mathrm{d}I(\lambda)/\mathrm{d}x = 0$, where x is the coordinate along the ray. When radiance interacts with intervening matter, it is modified. As illustrated in Figure 9.1, when a radiation beam intersects with a thin slab, radiance is modified due to *absorption*, *emission*, or *scattering* (directional change). The radiative transfer equation (RTE) generally states that as a beam of radiation

Handbook of Environmental Engineering, First Edition. Edited by Myer Kutz.

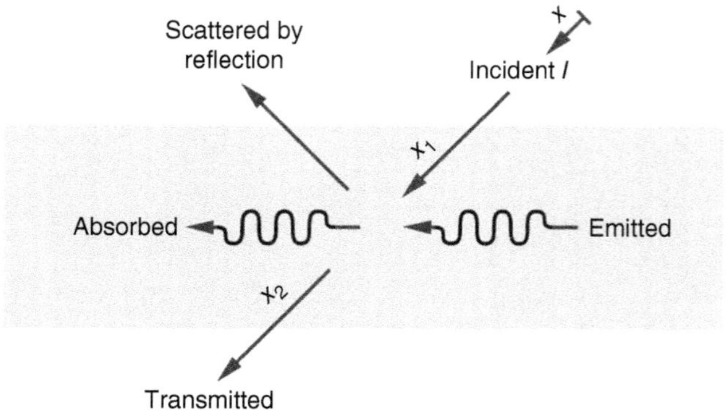

Figure 9.1 Sketch depicting absorption, emission, and scattering processes due to interaction of a radiation beam with spectral intensity *I*, propagating in direction *x*, within a thin slab.

travels, it loses energy due to absorption, gains energy by emission, and redistributes energy by scattering. Accounting for one or more of these processes, the RTE may be expressed in different forms that are amenable to analytical solutions for simple cases but require numerical solutions for more complex interactions between radiation and matter.

9.2.1.1 RTE for Absorption

If a radiation beam intersects with a thin slab, a reduction in the radiation intensity occurs due to absorption of photons. The reduction in spectral radiance is proportional to its intensity. Thus the RTE for absorption is given as

$$\frac{dI}{dx} = -kI \tag{9.1}$$

The wavelength-dependent proportionality coefficient k (m^{-1}), termed *linear absorption coefficient*, is expressed as absorptivity of the slab per unit length. The term "linear" implies that Eq. (9.1) is valid for optically thin layers (i.e. $dx \ll 1/k$). For a thick layer that absorbs a significant fraction of the photons, the probability of absorption increases nonlinearly with thickness. Note that all RTEs presented in this chapter are wavelength-dependent, while their dependence on wavelength is not shown for the sake of simplicity (i.e. $I = I(\lambda)$, $k = k(\lambda)$, etc.).

The dimensionless *absorption opacity* α of an object located at $x_1 < x < x_2$ is defined as

$$\alpha = \int_{x_1}^{x_2} k\,dx \tag{9.2}$$

The analytical solution of Eq. (9.1) yields the so-called *Beer–Lambert* law, which describes dimensionless *transmissivity* τ of an absorbing layer, i.e. the fraction of the radiation that is transmitted through the layer:

$$I(x_2) = I(x_1)\exp(-\alpha), \quad \text{i.e.} \quad \tau = \exp(-\alpha) \tag{9.3}$$

9.2.1.2 RTE for Absorption and Emission

It is well known that all objects with a temperature above absolute zero emit EM radiation. The spectral radiance (B) emitted by a *blackbody*, considered as a perfect absorber, is formulated by *Planck's* radiation law:

$$B = \frac{2hf^3}{c^2}\left[\exp\left(\frac{hf}{\kappa T_a}\right) - 1\right]^{-1} \tag{9.4}$$

where c is the speed of light, κ is the Boltzmann constant, h is the Planck constant, and T_a is the absolute temperature. For nonblackbody objects, a dimensionless parameter termed *emissivity* ε is defined. Emissivity is the ratio of energy emitted by a particular material and the energy emitted by a blackbody at the same temperature. Thus $\varepsilon = 1$ for blackbodies and $\varepsilon < 1$ for nonblackbodies. Integrating the blackbody spectral radiance over the entire EM spectrum, the *Stefan–Boltzmann* law states that the amount of energy per square meter per second that is emitted by a blackbody, β, is related to the fourth power of its absolute temperature

$$\beta = \int_0^\infty B\,df = \sigma T_a^4 \tag{9.5}$$

where $\sigma = 2\pi^5\kappa^4/(15c^2h^3)$ is the Stefan–Boltzmann constant. For a nonblackbody ($\varepsilon < 1$), the intensity of radiation integrated over all the EM frequencies is quantified as $\varepsilon\beta$.

In Eq. (9.1) emission was ignored, which is viable when the intervening material is not too hot (e.g. materials at room temperature). The RTE considering both absorption and emission processes is obtained by adding the emission term to Eq. (9.1):

$$\frac{dI}{dx} = -kI + eB \tag{9.6}$$

where e (m^{-1}) is the *spectral emission coefficient* (i.e. emissivity per unit length of the material at wavelength λ). Equation (9.6) is simplified when a local thermodynamic equilibrium (LTE) exists. *Kirchhoff's* law states that under an LTE condition $k = e$, a good absorber is a good emitter and vice versa. An analytical solution to Eq. (9.6) for the domain of $x_1 < x < x_2$ can be derived assuming LTE and homogeneity of the absorbing/emitting material (i.e. $k = e$ does not depend on x). The following solution is obtained by replacing e with k, multiplying both sides of Eq. (9.6) with $\exp[k(x - x_1)]$, and integrating the resultant equation from $x = x_1$ to $x = x_2$:

$$I(x_2) = I(x_1)\exp[-k(x_2 - x_1)] + B\{1 - \exp[-k(x_2 - x_1)]\} \tag{9.7}$$

9.2.1.3 RTE for Absorption, Emission, and Scattering

Scattering is generally defined as the redirection of radiation from its original path of propagation (x axis in our examples), commonly due to interactions with molecules and particles (e.g. aerosols and dust). Scattered radiation at an angle of less than $\pi/2$ from the incident radiation is called *backscatter*, otherwise it is termed *forward scatter*. Reflection, refraction, and diffraction are various forms of scattering.

There is a vast body of literature on radiation scattering theory. *Rayleigh* scattering and *Mie* scattering are two widely applied models in atmospheric science, describing light scattering from scattering particles such as air molecules and aerosols. These models formulate scattering for different domains of dimensionless size parameter, $\Delta = \pi D/\lambda$, where D is the diameter of scattering particles. *Rayleigh* scattering theory is applied to molecules and very small particles ($\Delta \ll 1$), while *Mie* scattering explains scattering of larger particles with diameters nearly the same as the wavelength ($\Delta \approx 1$). The *Rayleigh* scattering theory formulated for a single spherical particle with diameter D indicates that the scattered radiance is proportional to the inverse forth power of λ and to the sixth power of D. This theory explains why a clear sky looks blue because the sunlight scattering intensity is higher at shorter wavelengths of the visible spectrum (the blue end). In contrast to *Rayleigh* scattering, which is proportional to λ^{-4}, Mie scattering is linearly varying with wavelength. Nevertheless, it is critically dependent on the size parameter Δ and refractive index. Readers interested in scattering theory are referred to Platt et al. (2007).

In Eq. (9.6) scattering is ignored, which is viable whenever the intensity loss or gain due to scattering along a line of sight is negligible compared with losses due to absorption or gains due to thermal emission. However, scattering must be taken into account when the RTE is applied for environmental RS techniques, because remotely sensed data (e.g. reflectance or backscatter) are essentially scattered radiation. Scattering is commonly incorporated into the RTE by means of a *spectral scattering coefficient* (m^{-1}) that accounts for both energy loss and energy gain due to scattering. The RTE considering absorption, emission, and scattering is given as

$$\frac{\mathrm{d}I}{\mathrm{d}x} = -(k+s)I + eB + sJ \tag{9.8}$$

where J is the so-called *scattering source function*. The following analytical solution to Eq. (9.8) for the domain $x_1 < x < x_2$ can be derived assuming LTE and homogeneity (i.e. independence of k and s from x)

$$I(x_2) = I(x_1)\exp[-(k+s)(x_2 - x_1)] + \frac{kB + sJ}{k+s}\{1 - \exp[-(k+s)(x_2 - x_1)]\} \tag{9.9}$$

9.2.2 Radiative Transfer Modeling for the Optical Domain

9.2.2.1 Reflection and Refraction Processes

Remote sensors that operate within the optical EM frequency domain ($\lambda = 400$–$2500\,\mathrm{nm}$) determine *reflectance* of various surfaces at various frequencies or wavelengths. Reflectance is defined as the fraction of incident EM radiation that is reflected at an interface; hence, it is dimensionless and ranges between 0 and 1. The reflectance spectrum or spectral reflectance curve is the plot of the reflectance as a function of wavelength. In order to link optical RS data to surface properties (e.g. plant canopy chlorophyll content, soil water content, etc.), efforts have been made to formulate surface reflectance with radiative transfer models.

There is another material property termed *reflectivity*, which should be discerned from *reflectance*. Reflectance from the surface of compact and optically thick (semi-infinite) materials is called *reflectivity* (see Figure 9.2). Reflectivity is a constant property of a material and does not depend on its structure, whereas reflectance is influenced by infinite internal reflections or scattering of a structured medium. Scattering occurs due to the existence of several interfaces (e.g. solid–air or solid–liquid interfaces) within the structured medium. This is because of the modification of the so-called *refractive index* at interfaces.

Refraction is the change of the propagation direction of an EM wave or light when it passes from one medium into

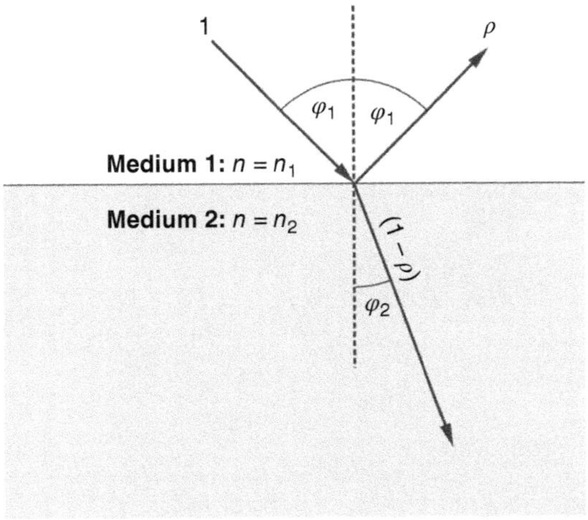

Figure 9.2 Sketch illustrating a single reflection from a semi-infinite layer with a smooth surface, termed reflectivity ρ.

another and changes its speed. The speed of light in a substance such as glass or water (c) is lower than the speed of light in vacuum or air (c_v). The refractive index of a substance, n, is defined as $n = c_v/c$. Figure 9.2 depicts a single reflection from a compact and thick material with a smooth surface. *Snell's* law or the law of refraction states that

$$n_1 \sin\varphi_1 = n_2 \sin\varphi_2 \tag{9.10}$$

where φ_1 is the incident angle and φ_2 is the refraction angle.

The reflectivity, ρ, of the interface in Figure 9.2 is calculated in accordance with *Fresnel's law* for vertical (V) and horizontal (H) polarizations:

$$\rho_V = \left(\frac{n_1\cos\varphi_1 - n_2\cos\varphi_2}{n_1\cos\varphi_1 + n_2\cos\varphi_2}\right)^2 = \left[\frac{\sin(\varphi_1 - \varphi_2)}{\sin(\varphi_1 + \varphi_2)}\right]^2 \tag{9.11}$$

$$\rho_H = \left(\frac{n_1\cos\varphi_2 - n_2\cos\varphi_1}{n_1\cos\varphi_2 + n_2\cos\varphi_1}\right)^2 = \left[\frac{\tan(\varphi_1 - \varphi_2)}{\tan(\varphi_1 + \varphi_2)}\right]^2 \tag{9.12}$$

Reflectivity for unpolarized incident radiation composed of equal contributions of V- and H-polarizations (e.g. typically at optical wavelengths) can be calculated as the average of the V- and H-polarization reflectivities:

$$\rho = 0.5\left[\frac{\sin(\varphi_1 - \varphi_2)}{\sin(\varphi_1 + \varphi_2)}\right]^2 + 0.5\left[\frac{\tan(\varphi_1 - \varphi_2)}{\tan(\varphi_1 + \varphi_2)}\right]^2 \tag{9.13}$$

While the reflectivity of a given material can be simply calculated according to *Fresnel's* law for an incidence angle φ_1 and a refractive index of the medium n_2, formulation of reflectance is not straightforward and needs to be solved for any presumed structure, leading to multiple internal reflections or scattering. Radiative transfer modeling approaches for calculating reflectance may be separated into several groups: (*i*) RTE-based models, (*ii*) plate models, (*iii*) multiple-flux models, (*iv*) compact spherical particle models, and (*v*) ray-tracing models as described in detail in Jacquemoud and Ustin (2008). To provide a brief overview of radiative transfer models applied for environmental RS, examples for groups (*ii*) and (*iii*) are presented below.

9.2.2.2 Reflectance from Plate Models

A widely applied properties spectra (PROSPECT) model for simulating plant canopy reflectance (Jacquemoud and Baret, 1990) that is based on the Allen et al. (1969) approach considers plant canopy as multiple plates (i.e. optically thin absorbing/transmitting layers). The Allen et al. (1969) model is an expansion of the pioneering solution of Stokes (1862), who first introduced a simple solution to reflectance and transmittance of a single plate and then extended the solution to a system of N grouped plates. Figure 9.3 illustrates *Stokes'* radiative transfer model for a single plate considering infinite reflections from and transmissions through the upper and lower interfaces. In this model, τ is the transmission coefficient of the plate that is related to its absorption coefficient via Eq. (9.3). This means that as a consequence of absorption, the light flux is reduced by a fraction of $1 - \tau$ when passing through the plate. Reflectance (R) and transmittance (T) of the plate (Figure 9.3) are calculated as

$$\begin{aligned} R &= \rho + (1-\rho)^2\rho\tau^2 + (1-\rho)^2\rho^3\tau^4 + (1-\rho)^2\rho^5\tau^6 + \cdots \\ &= \rho + (1-\rho)^2\rho\tau^2\left(1 + \rho^2\tau^2 + \rho^4\tau^4 + \cdots\right) \end{aligned} \tag{9.14}$$

$$\begin{aligned} T &= (1-\rho)^2\tau + (1-\rho)^2\rho^2\tau^3 + (1-\rho)^2\rho^4\tau^5 + \cdots \\ &= (1-\rho)^2\tau\left(1 + \rho^2\tau^2 + \rho^4\tau^4 + \cdots\right) \end{aligned} \tag{9.15}$$

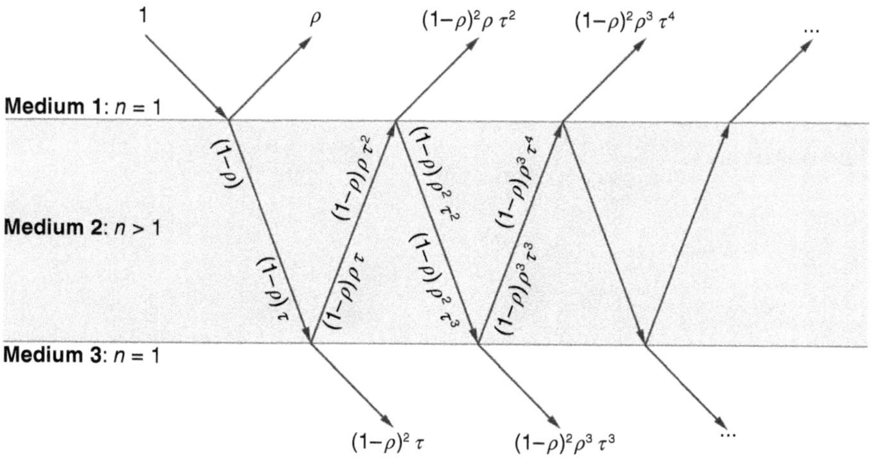

Figure 9.3 Sketch of the Stokes (1862) radiative transfer model for calculation of total reflectance from and transmittance through a single plate.

According to the Maclaurin series expansion of $(1-u)^{-1} = 1 + u + u^2 + \cdots$, reflectance and transmittance of the plate can be expressed in closed form as

$$R = \rho + \frac{(1-\rho)^2 \rho\tau^2}{1-\rho^2\tau^2} \tag{9.16}$$

$$T = \frac{(1-\rho)^2 \tau}{1-\rho^2\tau^2} \tag{9.17}$$

The total absorption of the plate (A) can also be derived from Eqs. (9.16) and (9.17) as

$$A = 1 - R - T = \frac{(1-\rho)(1-\tau)}{1-\rho\tau} \tag{9.18}$$

Stokes (1862) has provided mathematical proof that reflectance (R_N) and transmittance (T_N) for N grouped plates may be calculated as

$$R_N = \frac{b^N - b^{-N}}{ab^N - a^{-1}b^{-N}} \tag{9.19}$$

$$T_N = \frac{a - a^{-1}}{ab^N - a^{-1}b^{-N}} \tag{9.20}$$

where

$$a = \frac{1 + R^2 - T^2 + \Upsilon}{2R}, \quad b = \frac{1 - R^2 + T^2 + \Upsilon}{2T}, \\ \Upsilon = \left[\left(T^2 - R^2 - 1\right)^2 - 4R^2\right]^{0.5} \tag{9.21}$$

Allen et al. (1969) followed *Stokes'* approach for simulation of a compact plant leaf assumed as a single plate. In contrast to Stokes (1862), Allen et al. (1969) assumed that the transmissivity of the air-to-plate interface, τ_{ap}, differs from the transmissivity of the plate-to-air interface, τ_{pa} (note that the transmissivity of the interface should be distinguished from the transmissivity of the plate as the former is equal to $1-\rho$, while the latter is calculated from Eq. (9.3)). Considering *Stern's* (1964) relationship, $\tau_{pa} = n^{-2}\tau_{ap}$ (n: refractive index of the plate), Allen et al. (1969) solved reflectance and transmittance of a compact leaf as

$$R = \rho + \frac{(1-\rho)^2 \left(n^2 - 1 + \rho\right)\tau^2}{n^4 - \left(n^2 - 1 + \rho\right)^2 \tau^2} \tag{9.22}$$

$$T = \frac{(1-\rho)^2 \tau n^2}{n^4 - \left(n^2 - 1 + \rho\right)^2 \tau^2} \tag{9.23}$$

Jacquemoud and Baret (1990) developed the PROSPECT model based on the Allen et al. (1969) single leaf model linked to the Stokes N-layer system, i.e. Eqs. (9.19)–(9.23). The model is widely applied for simulation of canopy reflectance and transmittance, as discussed in Jacquemoud et al. (2009). The PROSPECT model assumes a stack of N homogeneous layers separated by $N-1$ air spaces and modifies the Stokes system by separating the first layer from the $N-1$ other layers. The first layer receives an incident beam with a variable angle of φ, while reflectivity and transmissivity of all other layers are calculated for an incidence angle, $\varphi = 90°$.

9.2.2.3 Reflectance from Multiple-Flux Models

A pioneering model that falls within this category is the Kubelka and Munk (KM) (1931) two-flux model. Kubelka and Munk (1931) developed a simple theory describing radiative transfer in an absorbing and scattering layer, considering a downward and an upward light propagation flux (I and J, respectively), perpendicular to the layer (Figure 9.4). The volume reflectance is modeled incorporating absorption and scattering processes (emission is neglected). The KM model consists of a pair of coupled ordinary differential equations describing I and J fluxes as functions of the layer's depth z:

$$\begin{cases} \dfrac{dI}{dz} = -(k+s)I + sJ \\ \dfrac{dJ}{dz} = -(k+s)J + sI \end{cases} \tag{9.24}$$

As is evident from Eq. (9.24), the KM model is basically the same as the RTE in Eq. (9.8), written only for the nadir direction. The derivation of the KM model from other forms of the RTE has been presented in Thennadil (2008) and Sandoval and Kim (2014).

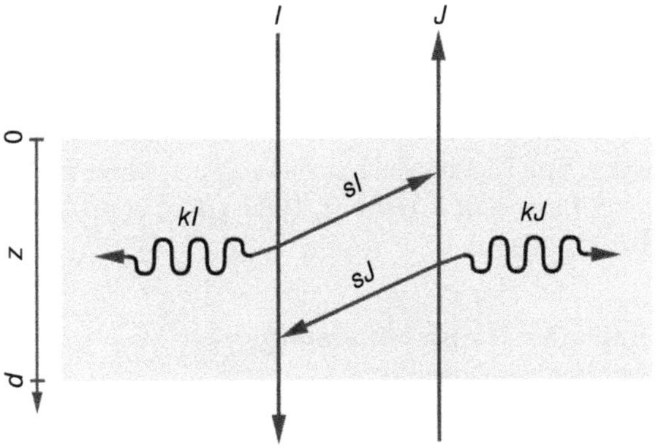

Figure 9.4 Sketch of the Kubelka and Munk (1931) model, where I and J are two opposite light fluxes and k and s are absorption and scattering coefficients, respectively. *Source*: Adapted from Ciani et al. (2005).

Reflectance and transmittance of the layer can be derived by solving Eq. (9.24) (Ciani et al., 2005):

$$R = \frac{(1-\chi^2)[\exp(\zeta d)-\exp(-\zeta d)]}{(1+\chi^2)\exp(\zeta d)-(1-\chi^2)\exp(-\zeta d)} \quad (9.25)$$

$$T = \frac{4\chi}{(1+\chi^2)\exp(\zeta d)-(1-\chi^2)\exp(-\zeta d)} \quad (9.26)$$

where $\chi = [k/(k+2s)]^{0.5}$, $\zeta = [k(k+2s)]^{0.5}$ and d is the layer thickness. In semi-infinite layers (i.e. $d \to \infty$), Eq. (9.25) is reduced to

$$R = 1+\frac{k}{s}-\sqrt{\left(\frac{k}{s}\right)^2+2\frac{k}{s}} \quad (9.27)$$

Several limitations of the KM model have been discussed, for example, dependencies between absorption and scattering coefficients k and s for translucent or strongly absorbing media. Nonetheless, the KM model is widely applied for multiple-scattering calculations due to its explicit form and simplicity (Edström, 2007). Yang and Kruse (2004) introduced a revised KM model that accounts for the effect of scattering on the path length of light propagation. However, Edström (2007) later showed that the revised KM model yields significant errors because of flawed theory. The more rigorous four-flux model of Maheu et al. (1984) represents radiation within the medium as two collimated and two diffuse fluxes propagating in opposite directions. Vargas and Niklasson (1997) indicated that the four-flux model compares well with numerical solutions of the RTE or with highly accurate Monte Carlo simulation results.

Both the KM and four-flux models have been successfully applied for environmental RS studies. For example, the KM model was applied by Allen and Richardson (1968) for calculation of plant canopy reflectance. Based on the KM model, Yamada and Fujimura (1991) proposed a mathematical model of reflectance and transmittance of plant leaves as a function of their chlorophyll pigment content. Fukshansky et al. (1991) applied the four-flux model to calculate the optical parameters of plant leaves. The KM model has been also applied for RS of soils (e.g. Barron and Torrent, 1986; Ciani et al., 2005; Sadeghi et al., 2015a).

9.2.3 Radiative Transfer Modeling for the Microwave Domain

Microwave RS techniques have shown great potential for large-scale mapping of SM because the *dielectric permittivity* ϵ of water (about 80 at frequencies below 5 GHz) significantly differs from ϵ of dry soil (about 3.5). As a result, the emissivity of variably-saturated soils varies over a wide range (Njoku and Entekhabi, 1996). Basic radiation laws for passive microwave RS of near-surface SM are briefly presented below. Interested readers are referred to Njoku and Kong (1977) and Njoku and Entekhabi (1996) for a more detailed discussion.

For low frequency microwave bands (f = 300 MHz to 300 GHz) ($hf/\kappa T_a$) in Eq. (9.4) is $\ll 1$; hence the Maclaurin series approximation of $\exp(u) \approx 1 + u$ may be applied. Therefore, Planck's law (Eq. (9.4)) reduces to the so-called *Rayleigh–Jeans* approximation for microwave frequencies:

$$B = \frac{2\kappa T_a f^2}{c^2} \quad (9.28)$$

For a nonblackbody ($I \neq B$), an equivalent blackbody temperature termed *brightness temperature* T_b can be defined based on Eq. (9.28):

$$T_b = \frac{Ic^2}{2\kappa f^2} = \frac{I\lambda^2}{2\kappa} \quad (9.29)$$

The brightness temperature T_b of a radiating body is related to its true temperature T_a via the general expression

$$T_b = \varepsilon T_a \quad (9.30)$$

Therefore, brightness temperature is equal to the true temperature of a blackbody and is lower than the true temperature of a nonblackbody. Equations (9.29) and (9.30) can be used to calculate emissivity ε of the land surface at any given location if the subsurface temperature profile $T_a(z)$ is uniform. The emissivity is related to the reflectivity ρ of the surface according to *Kirchhoff's* law:

$$\varepsilon = 1 - \rho \quad (9.31)$$

For a smooth surface and a medium of constant *complex dielectric constant* or *relative permittivity* ϵ, reflectivity for V- and H-polarizations is linked to the refractive index n and incident angle through the Fresnel equations, i.e. Eqs. (9.11) and (9.12). Therefore, ε is strongly correlated to soil water content θ since $\epsilon = n^2$. For a nonuniform temperature profile $T_a(z)$, T_a in Eq. (9.30) is replaced by an "effective profile temperature," which represents an average profile temperature weighted by the so-called *attenuation coefficient* that is dependent on soil depth (see eq. (6) in Njoku and Entekhabi (1996)). Equation (9.30) is valid for bare soils with a smooth surface. Incorporating vegetation and surface roughness leads to more complex modifications of Eq. (9.30) that are discussed in Njoku and Entekhabi (1996).

Wavelength-dependent attenuation of radiation through a medium can be quantified by means of the *penetration depth*, which is a measure of how deep EM radiation can penetrate into the medium. It is defined as the depth at which the intensity of the propagating radiation decreases to e^{-1} (≈37%) of its original magnitude received at the surface. In other words, the penetration depth is equivalent to the thickness of the medium layer from which 63% of the surface-emitted radiation originates. The penetration of microwave signals depends on soil properties and sensor characteristics including the moisture content, frequency (wavelength), polarization, and incidence angle. As shown in Figure 9.5, an increase in SM content leads to a decrease of penetration depth, while the penetration depth increases with a decrease in frequency (longer wavelengths). An important consequence of the wavelength dependence of penetration depth is that longer wavelengths can provide information for thicker layers (e.g. about 0.5 m for the P-band microwave radiation), whereas shorter wavelengths are useful for studying the layers close to the soil surface.

9.3 RS Technology

9.3.1 Spaceborne Sensors

Spaceborne or satellite RS techniques provide observations at various spatial scales, from a few meters to hundreds of kilometers. Satellite RS benefits from large, frequent, and repetitive coverage of an area of interest, quantitative measurement of ground features with radiometrically calibrated sensors, and relatively low costs per unit coverage area. With regard to the spectral frequencies applied for data acquisition, remote sensors can be classified into optical, thermal, and microwave imaging systems. In the following, properties of the most common optical, thermal, and microwave spaceborne sensors applied for environmental studies are discussed.

9.3.1.1 Optical/Thermal Sensors

Within the optical domain that includes visible, near-infrared (NIR), and short-wave infrared bands (λ from 350 to 2500 nm), sensors are used to image the Earth's surface by means of detecting the solar radiation reflected from objects on the ground. As described in Section 9.2, different materials absorb, reflect, and transmit different portions of EM radiation at different wavelengths. Thus, distinct materials can be identified in remotely sensed images due to their specific spectral signatures. Thermal RS techniques measure radiation emitted from ground objects, which, as discussed above, is related to their physical temperature. Optical/thermal spaceborne sensors are capable of providing much higher resolution data when compared with microwave spaceborne sensors and are classified as multispectral or hyperspectral depending on the number of spectral bands within which they operate. Multispectral sensors provide data at several discrete bands (e.g. Landsat Enhanced Thematic Mapper Plus [ETM$^+$] satellite measures reflectance for 6 optical bands and 1 thermal band), whereas hyperspectral sensors provide a near-continuous spectrum (e.g. Hyperion satellite is capable of resolving 220 spectral bands within the optical domain). The most widely applied satellite sensors are introduced in the following with basic features listed in Table 9.1.

9.3.1.1.1 Landsat

Landsat satellites, first launched in 1972, have gone through several stages of evolution. They were/are equipped with a number of RS instruments such as the Multispectral Scanner (MSS), the Thematic Mapper (TM), the ETM$^+$, and the Optical Land Imager (OLI). Landsat supplies moderate resolution visible and infrared imagery, with thermal imagery and a panchromatic image from the ETM$^+$ and OLI sensors. Landsat-8, launched in January 2013 with two optical and thermal sensors, namely, OLI and the Thermal Infrared Sensor (TIRS), is the latest of a continuous series of Landsat satellites. Landsat-8 collects data from the entire earth

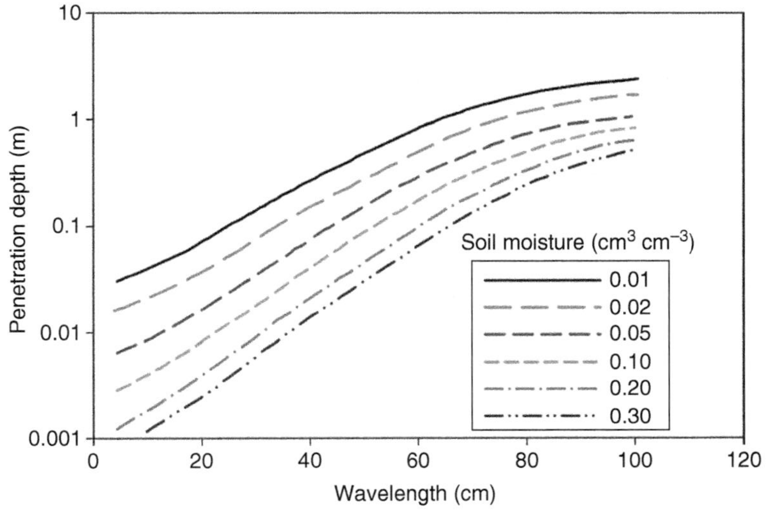

Figure 9.5 Soil penetration depth for microwave radiation as a function of frequency and moisture content. *Source*: Adapted from Njoku and Entekhabi (1996).

Table 9.1 Features of optical and thermal satellite remote sensors.

Instrument/satellite	No. of bands (wavelength in μm)	Spatial resolution	Time resolution	Data availability since
OLI, TIRS/Landsat-8	11 (0.43–12.51)	15-m (band 8)	16 d	April 2013
		30-m (bands 1–7, 9)		
		100-m (bands 10, 11)		
Sentinel-2A	13 (0.443–2.190)	10-m (bands 2, 3, 4, 8)	5–10 d	June 2015
		20-m (bands 5, 6, 7, 8a, 11, 12)		
		60-m (bands 1, 9, 10)		
MODIS/Aqua, Terra	36 (0.62–14.4)	250-m (bands 1–2)	Twice per day	1999 (Terra)
		500-m (bands 3–7)		2002 (Aqua)
		1000-m (bands 8–36)		

surface every 16 days within one panchromatic band (15-m resolution), eight optical bands (30-m resolution), and two thermal bands (100-m resolution, resampled to 30 m). The sensor readings obtained in digital number (DN) format can be converted to top-of-atmosphere radiance, I, and reflectance, R:

$$I = \iota_{I} + \sigma_{I}\mathrm{DN} \tag{9.32}$$

$$R = \iota_{R} + \sigma_{R}\mathrm{DN} \tag{9.33}$$

The parameters ι and σ are available from the recorded Landsat metadata.

Radiance obtained from TIRS band observations can be converted to brightness temperature according to

$$T_{b} = \frac{K_2}{\ln\left(1 + K_1/I\right)} \tag{9.34}$$

where K_1 and K_2 are constants (-) determined from the effective wavelength of a satellite sensor within the thermal region.

9.3.1.1.2 *MODIS*

The Moderate Resolution Imaging Spectroradiometer (MODIS) is a key instrument onboard of the National Aeronautics and Space Administration (NASA) Terra (EOS AM-1, launched in 1999) and Aqua (EOS PM-1, launched in 2002) satellites. Terra and Aqua MODIS are viewing the entire earth surface every 1–2 days, acquiring data within 36 optical and thermal bands with wavelengths ranging from 0.4 to 14.4 μm. The instrument collects data at three spatial resolutions of 250, 500, and 1000 m and provides global maps of several land surface characteristics, including land surface reflectance, albedo (% of shortwave solar radiation reflected back from the surface of the earth), LST, and several vegetation indices (VI) (e.g. the normalized difference vegetation index [NDVI]).

9.3.1.1.3 *Sentinel-2*

The Sentinel-2 mission, launched in June 2015 by the European Space Agency (ESA), is a land monitoring constellation of two satellites (2A, 2B) that provides high-resolution optical imagery. Sentinel-2 delivers global coverage of the land surface every 10 days with one satellite and every 5 days with 2 satellites. Sentinel-2 is a multispectral imager with 13 spectral bands from visible to shortwave infrared (SWIR) with 10–60 m spatial resolution. The image data support generation of basic land cover, land use and land change detection maps, and maps of geophysical variables for leaf area index (LAI), leaf chlorophyll content, and leaf water content. The instrument enables early detection of changes in plant health because of its high spatial/temporal resolution and three red-edge bands. This is particularly useful for agricultural management applications and for detection of plant stress conditions. The Sentinel-2A level 1C data are publicly available from the ESA web portal.

9.3.1.2 Microwave Sensors

Microwave sensors detect active and passive EM radiation at wavelengths ranging from 0.1 mm to more than 1 m. Active sensors (radars) transmit microwave pulses toward the Earth's surface and measure the backscattered radiation. The backscattered radiation is strongly correlated with the roughness and dielectric properties of the soil surface. Passive sensors (radiometers) measure the intensity of naturally emitted microwaves from the Earth's surface, expressed as brightness temperature. The energy detected by radiometers includes contributions from the atmosphere, reflected sky radiation, and land surface.

As indicated earlier, the most important application of microwave sensors (active and passive) is their suitability for SM estimation, because microwave radiation is mainly affected by the high dielectric permittivity of soil water. Currently operating satellite missions that provide

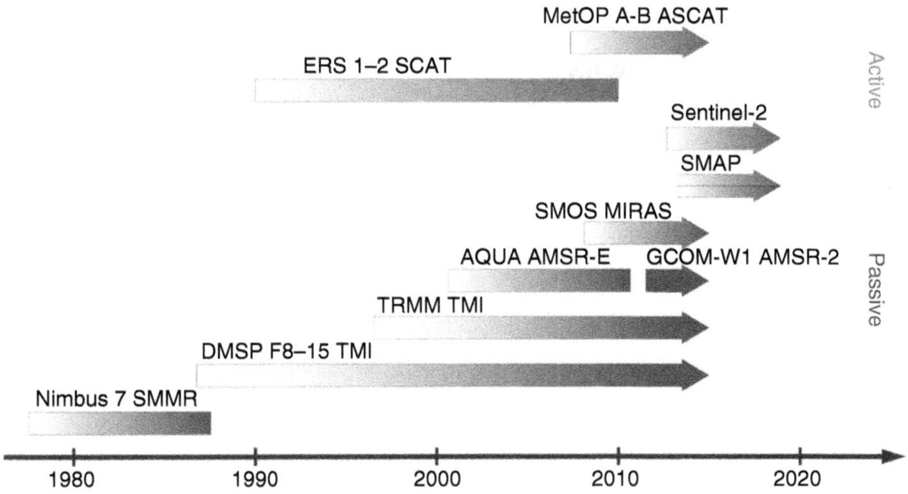

Figure 9.6 Active and passive microwave RS missions for soil moisture retrieval.

SM information include Soil Moisture and Ocean Salinity (SMOS), Advanced Microwave Scanning Radiometer 2 (AMSR-2), Advanced Scatterometer (ASCAT), Advanced Land Observation Satellite 2 (ALOS-2), Sentinel-1A, and Soil Moisture Active Passive (SMAP). The evolution of these missions for SM retrieval is illustrated in Figure 9.6. Precipitation is another important environmental variable that can be remotely sensed with microwave sensors. Current precipitation missions include the Tropical Rainfall Measuring Mission (TRMM) and Global Precipitation Measurement (GPM). An overview of microwave satellite missions is provided in the following with basic features of employed remote sensors listed in Table 9.2.

9.3.1.2.1 SMOS

The ESA SMOS mission, launched in November 2009, is the first satellite mission dedicated to the global measurement of near-surface SM (top 5 cm) over land and the salinity of oceans (Kerr et al., 2001, 2010). The SMOS satellite carries a novel interferometric radiometer that operates within the L-band (1.41 GHz, 21 cm) and senses brightness temperature T_b at several incidence angles. The radiometer exploits the interferometry principle, which measures the phase difference of incident radiation with 69 small receivers. The technique is based on cross-correlation of observations from all possible combinations of receiver pairs. As the satellite moves along its orbital path, the target area is observed under various viewing angles.

The sensitivity of the SMOS signal to SM is much higher than the sensitivity for atmospheric disturbances and other soil surface characteristics such as roughness, texture, topography, and bulk density. SMOS provides land surface and ocean observations at several levels. The SMOS level 2 product is generated via converting T_b (level 1) to SM with the L-MEB forward model proposed in Wigneron et al. (2007). In general, L-MEB is used to compute multiangular top-of-the-atmosphere brightness temperatures based on a set of soil (e.g. SM, temperature, texture), vegetation (e.g. water content, temperature), and radiative transfer parameters (e.g. single scattering albedo, polarization coupling factor, surface roughness). The retrieval algorithm is based on a Bayesian approach that retrieves SM by minimizing errors between L-MEB simulated brightness temperature and SMOS observed brightness temperature (Montzka et al., 2013). All additional higher-level products (e.g. daily or monthly SM) are calculated from the level 2 product.

The mission, with an SM retrieval error requirement of $0.04\,cm^3\ cm^{-3}$, has been experimentally validated in Europe (e.g. Dall'Amico et al., 2012; Dente et al., 2012; Sanchez et al., 2012; Rötzer et al., 2014; Pierdicca et al., 2015; Fascetti et al., 2016), the United States (e.g. Al Bitar et al., 2012; Jackson et al., 2012; Wagner et al., 2014), Africa (e.g. Louvet et al., 2015; Pierdicca et al., 2015; Fascetti et al., 2016), South America (e.g. Niclòs et al., 2016), and Australia (e.g. van der Schalie et al., 2015) for areas ranging from the local (e.g. Jackson et al., 2012; Louvet et al., 2015) to the continental (e.g. Al Bitar et al., 2012; Fascetti et al., 2016) scales. Validation experiments for SMOS SM product levels 2 and 3 have revealed varying retrieval accuracies ranging from 0.030 to $0.126\,cm^3\ cm^{-3}$ depending on soil type, seasonality, and vegetation density.

RS of SM is influenced by a large number of factors including atmospheric correction, vegetation factors, and soil surface roughness. Significant efforts have been devoted to the development of novel retrieval algorithms that provide reliable SM information for various land surface conditions (e.g. Piles et al., 2009; Montzka et al., 2013; McNairn et al., 2015; Fascetti et al., 2016).

Table 9.2 Satellite remote sensors and their features with regard to soil moisture and precipitation retrieval.

Features	SMOS	SMAP		Sentinel-1	GMP
		Radiometer	Radar		
Agency	ESA	NASA		ESA	JAXA, NASA
Sensor type (frequency)	L-band (1.4 GHz)	L-band (1.4 GHz)	L-band (1.22–1.3 GHz)	C-band (5.4 GHz)	Ka-band (35.5 GHz) Ku-band (13.6 GHz)
Spatial resolution	35 km	40 km	10 km	5, 20, 40 m	5 km
Time resolution	1–3 d	2–3 d	2–3 d	3–12 d	2–3 h
Operation time	2009–2017	2015–2020	January 2015–July 2015	2014–2030	2014–2019
Incidence angle	0–55°	35°–50°	35°–50°	20°–47°	52.8°
Polarization	HH, VV	V, H, 3rd and 4th stokes parameters	VV, HH, HV (or VH)	HH, HV, VH, VV	HV
Swath	1000 km	1000 km	1000 km	20–400 km	125, 245, 885 km
Altitude	758 km	685 km	685 km	700 km	407 km
Orbit	Polar	Near polar	Near polar	Near polar	Polar
Auxiliary data requirement	High	High	High	Medium	Low
Target property	Soil moisture	Soil moisture	Soil moisture	Soil moisture	Precipitation
Retrieval accuracy	$\pm 0.04\,cm^3\,cm^{-3}$	$\pm 0.04\,cm^3\,cm^{-3}$ for vegetation water content $\leq 5\,kg\,m^{-2}$	$\pm 0.04\,cm^3\,cm^{-3}$ for vegetation water content $\leq 5\,kg\,m^{-2}$	± 0.04–$0.08\,cm^3\,cm^{-3}$ for grassland and agricultural areas	$\pm 0.5\,mm\,h^{-1}$
Data latency	Within a few days after acquisition	Within a few days after acquisition	*Not operational*	Within a few hours after acquisition	Within a few days after acquisition

Because of the coarse spatial SMOS resolution (~40 km), observations are not well suited for small-scale (e.g. field and catchment scale) applications where SM commonly exhibits significant horizontal spatial variability. Recently, the spatial soil moisture representation has been improved via downscaling of SMOS soil moisture data with information supplied by high-resolution optical/thermal remote sensors such as MODIS LST/NDVI products (Piles et al., 2011, 2014; Merlin et al., 2012, 2013; Srivastava et al., 2013), CDF matching (Kornelsen and Coulibaly, 2014), and MODIS-based soil evaporative efficiency (Merlin et al., 2008; Djamai et al., 2015).

9.3.1.2.2 Sentinel-1

The ESA Sentinel-1 mission, launched in April 2014, is a polar-orbiting and day-and-night radar imaging system for land and ocean monitoring. The goal of the mission is to provide C-band (5.4 GHz) synthetic aperture radar (SAR) data continuity following the retirement of ERS-2 and the end of the ENVISAT mission. The SAR instrument on board of Sentinel-1 has four different modes. Over land it operates predominantly in interferometric wide swath (IWS) mode, which acquires images within a 250-km-wide swath at 5 × 20 m spatial and 3- to 12-day temporal resolutions. An advantage of the IWS mode when compared with the ENVISAT ASAR GM mode is the improved spatial, temporal, and radiometric resolutions, while both work at approximately the same frequencies (5.4 compared with 5.3 GHz). Another improvement is the application of cross-polarization to correct for seasonal vegetation effects in copolarized backscatter measurements. In general, due to better penetration of horizontally polarized (HH) waves through vertically oriented vegetation (grasses, crops), HH polarization is preferred (Brown et al., 2003). The data products from Sentinel-1 include level 1 georeferenced single-look complex and projected multilooked ground range detected products, and level 2 geolocated geophysical products that estimate wind speed and its direction over ocean surfaces.

Several attempts have been focused on the development of algorithms for SM retrieval from Sentinel-1A SAR data (e.g. Balenzano et al., 2012; Pierdicca et al., 2014). For example, Paloscia et al. (2013) presented an artificial neural network-based algorithm to retrieve SM from Sentinel-1 data and tested its applicability with ground truth SM measurements obtained at several test locations in Italy, Australia, and Spain. The obtained RMSEs ranged from 0.017 to 0.067 $cm^3\ cm^{-3}$ SM, which meets the imposed Sentinel-1 accuracy target of approximately 0.05 $cm^3\ cm^{-3}$ reasonably well.

In general, SM retrieval from SAR data is challenging due to the confounding effects of surface roughness and vegetation on the signal. Though the applicability of a number of backscatter models has been extensively studied, SM retrieval from Sentinel-1 observations remains an enigma due to large roughness changes at smaller scales (e.g. agricultural management operations) and higher complexity of small-scale characterization of vegetation (e.g. wheat versus sugar beet) (Barrett et al., 2009). Sentinel-1 is currently the only active spaceborne sensor with a meter-scale spatial resolution.

9.3.1.2.3 SMAP

The NASA Soil Moisture Active and Passive (SMAP) mission was launched in January 2015 and is designed to map the moisture content and state (i.e. frozen or thawed) of the top 2–5 cm soil layer. The major science objectives of SMAP are focused on the understanding of processes linking water, energy, and carbon cycles, land surface global water and energy fluxes, and improved prediction of weather and climate. This is accomplished with an instrument that combines a radar and a radiometer, operating within the L-band (1.4 GHz). Relative to each other, the SMAP radiometer brightness temperature measurements provide coarser spatial resolution (36 km) with higher SM sensitivity, and the SMAP radar backscatter cross-sectional measurements yield higher spatial resolution (3 km) with somewhat lower SM sensitivity, especially when applied for densely vegetated land surfaces. SMAP was designed to yield SM retrieval with reasonable accuracy and intermediate spatial resolution (9 km) from the merged measurements (Entekhabi et al., 2010).

SMAP generates 15 data products that represent 4 data processing and 3 SM levels. There are three L2, four L3, and two L4 SM products resulting from the radar and radiometer data streams. The levels 2 and 3 provide near-surface SM, while L4, which is based on the assimilation of SMAP data into the Catchment land surface model (Reichle and Koster, 2005), provides root zone (top 1-m layer of the soil profile) moisture products. A by-product of the data assimilation scheme provides error estimates for the L4 SM product.

While SMAP was intended to spend at least 3 years in orbit, the high-resolution radar instrument malfunctioned in July 2015 after less than 3 months in operation. Thus, only the radiometer instrument remains operational, though its level of detail is far more limited. Combining SMAP radiometer with other radar data (e.g. Sentinel-1 C-band radar) is considered as a potential solution to resolve the problem and fulfill the originally intended SMAP application of producing fine-resolution SM maps via combining passive and active microwave data.

9.3.1.2.4 GPM

The GPM mission is a collaboration between NASA and the Japanese Aerospace Exploration Agency (JAXA). The GPM core satellite was launched on 27 February 2014. Data collected from the core satellite serves as a reference standard that unifies precipitation measurements from research and operational satellites launched by a consortium of GPM partners in the United States, Japan, France, India, and Europe. The GPM constellation of satellites can observe precipitation over the entire globe every 2–3 h. The core satellite measures rain and snow with two instruments; the GPM Microwave Imager (GMI) and the Dual-Frequency Precipitation Radar (DPR). The GMI utilizes 13 different microwave channels to capture precipitation intensities and horizontal patterns, while the DPR provides three-dimensional information about precipitation particles derived from particle-reflected energy at different elevations within the cloud system. Together the GPM instruments provide base data that can be combined with microwave observations from partner satellites to yield a global precipitation dataset (Hou et al., 2014).

9.3.2 Airborne Sensors

Airborne RS is another avenue for imaging and mapping land surface features and properties. An aircraft with RS capabilities is deployed to provide data at very high spatial (e.g. cm-scale) resolutions, which is a major advantage over satellite RS. Airborne RS bridges the gap between in situ ground measurements and satellite observations. There are some drawbacks associated with airborne RS. In addition to significantly higher costs per unit area ground coverage relative to spaceborne sensors, airborne missions are often limited to single flights, while satellite instruments provide continuous monitoring capabilities.

Airborne platforms range from small, slow, and low-flying aircrafts to twin-engine turboprop jets flying at altitudes of up to 35 000 ft. Today, unmanned aerial vehicles (UAVs) are increasingly deployed for various airborne RS tasks. Flight altitude, aircraft speed, and range are critical determinants for selecting proper equipment for specific RS objectives. Features of airborne remote sensors are summarized in Table 9.3.

9.3.2.1 Manned Airborne RS Missions

9.3.2.1.1 HyMap

The Hyperspectral Mapper (HyMap^{TM}) is an aircraft-mounted commercial hyperspectral scanner manufactured by Integrated Spectronics Pty Ltd. (Castle Hill,

Table 9.3 Overview of airborne remote sensors and their features.

Feature	HyMap	AirMOSS	E-SAR	F-SAR	PLMR2
Range	Optical (hyperspectral)	Active microwave	Active microwave	Active microwave	Passive microwave
No. of bands (wavelength)	126 (0.45–2.45 μm)	1 (P; 0.42–0.44 GHz)	4 (X, C, L, P; 3–85 cm)	5 (X, C, S, L, P; 0.35–9.6 GHz)	1 (L; 1.401–1.425 GHz)
Spatial resolution	2–10 m	7 m slant range 0.8 m azimuth	2–4 m slant range 0.25–1.5 m azimuth	0.3–2.25 m slant range 0.2–1.5 m azimuth	~200–350 m
Incidence angle	45°	25°–45°	45°	25°–55°	38°
Polarization	—	HH, HV, VH, VV	HH, HV, VH, VV	HH, HV, VH, VV	VV, HH
Application	Vegetation, soil	Root zone soil moisture, carbon flux	Soil moisture, ice	Soil moisture	Soil moisture, soil salinity

NSW, Australia) that provides 126 spectral channels with bandwidths from 10 to 20 nm within the spectral range from 450 to 2450 nm and a spatial resolution of 2–10 m (Table 9.3). The HyMap has a signal-to-noise ratio of 500 : 1 or better, except near the major atmospheric water absorption bands (Cocks et al., 1998). Because of these features, the HyMap can be deployed for simulation of future spaceborne spectrometers with similar noise characteristics. HyMap data contains a wealth of information for a wide range of applications.

The HyMap has been initially designed to aid mineral exploration (e.g. Bedini et al., 2009) but has since been deployed for numerous environmental RS projects including the characterization of soil and water pollution (Wettle et al., 2009; Riaza et al., 2012), vegetation monitoring (Bachmann et al., 2002; Huang et al., 2004), and hydrological modeling (Getirana et al., 2012). For example, HyMap images from South Australia have been utilized for identification of kaolinite and iron oxides through their absorption features in the 600- to 800-nm and 1950- to 2300-nm spectral bands (Raggatt et al., 2004). HyMap data have also aided the characterization of the spatial distribution of desert pavements, calcareous soils, saline soils, and exposed gypsum, which was discriminated based on spectral angle mapper classification (Margate and Shrestha, 2001).

Some of the hyperspectral bands are affected by various atmospheric constituents such as water vapor, carbon dioxide, or aerosols. While these bands do not provide information for land surface characterization, they can be applied for atmospheric studies. For example, the vapor absorption band around 940 nm can be used for characterization of atmospheric water vapor (Halthore et al., 1997).

9.3.2.1.2 AirMOSS

The NASA Airborne Microwave Observatory of Subcanopy and Subsurface (AirMOSS) mission is aimed at estimation of root zone SM to aid the quantification of the overall carbon exchange between plants and the atmosphere. Because carbon dioxide affects the global climate, AirMOSS data are envisioned to improve the accuracy of climate projections for the next 50–100 years. Mounted to the fuselage of a Gulfstream-III jet (Figure 9.7), AirMOSS deploys a radar to collect SM information at nine North American sites representing different climate biomes to estimate sequestration and release of carbon (Chapin et al., 2012).

The AirMOSS radar is a fully polarimetric SAR operating within the P-band (0.42–0.44 GHz). Because low frequency microwave radiation penetrates dense vegetation canopies and the subsurface down to about 0.5–1 m depths depending on vegetation cover, soil type, and SM, AirMOSS provides excellent means for direct root zone SM determination (Moghaddam et al., 2000). The AirMOSS mission started with test flights in August 2012 and commenced science operations in October 2012. The surveys covered areas of about 100 × 25 km with FLUXNET towers in climate regions ranging from boreal forests in Saskatchewan, Canada, to tropical forests in La Selva, Costa Rica (Tabatabaeenejad et al., 2015). The flights provided high spatial resolution measurements (100 m) for nine regions representing the major North American climate biomes at subweekly, seasonal, and annual time scales. Tabatabaeenejad et al. (2015) indicated great potential of P-band radar for direct retrieval of root zone SM profiles. AirMOSS data has also provided direct means for validation of root zone SM products from the NASA SMAP mission and for assessing the impact of small-scale heterogeneities on coarse scale estimates.

Figure 9.7 Gulfstream-III jet (top) with the AirMOSS radar pod mounted to the fuselage (bottom).

9.3.2.1.3 E-SAR/F-SAR/PLMR2

The Experimental Synthetic Aperture Radar (E-SAR) system onboard the German Aerospace Center (DLR) DO228 aircraft simultaneously acquires data within the X-, C-, L-, and P-bands covering the wavelength range from 3 to 85 cm (Table 9.3). The E-SAR covers an area of 3–5 × 20 km with slant range resolutions from 2.3 to 4.5 m and azimuth resolutions from 0.7 to 2.5 m. It has been widely deployed for European campaigns such as AgriSAR, OPAQUE, SARTEO, and ICESAR for SM and ice studies (e.g. Parrella et al., 2012; Jagdhuber et al., 2015). Several studies demonstrated the utility of E-SAR data for SM retrieval (Jagdhuber et al., 2015; Zwieback et al., 2015). In 2012 the Microwaves and Radar Institute of the DLR developed a new E-SAR version, the F-SAR, which is designed to operate within the X-, C-, S-, L-, and P-bands. Figure 9.8 depicts examples for high spatial resolution F-SAR X-, C-, and S-band images from two test sites in Germany. The active F-SAR sensor has been operated in conjunction with the passive PLMR2 radiometer to acquire active–passive microwave data.

The PLMR2 is an advanced version of the Polarimetric L-band Microwave Radiometer (PLMR) developed in Australia for high-resolution mapping of near-surface SM (Summerell et al., 2009) and land surface salinity and temperature (Wang et al., 2007) based on measured brightness temperature data for low to moderate vegetation conditions. The PLMR2 has been deployed for validation of NASA SMAP SM products in an airborne campaign in the Rur Catchment in Germany (Montzka et al., 2014). Along with the multispectral RapidEye sensor for estimation of vegetation parameters, PLMR2 was applied for high-accuracy SM retrievals (RMSE $0.035\,cm^3\,cm^{-3}$) at the TERENO site in Germany (Hasan et al., 2014).

Figure 9.8 F-SAR X-band image with a spatial resolution of 0.4 m depicting Kaufbeuren in Germany (top); C-band image with a spatial resolution of 0.4 m depicting Wallerfing in Germany (center); and S-band image with a spatial resolution of 0.5 m for Kaufbeuren (bottom). *Source*: Adapted from Horn et al. (2009).

9.3.2.2 Unmanned Aerial Systems

An unmanned aerial system (UAS), also known as unmanned aerial vehicle (UAV) or drone, is a ground-controlled aircraft with camera or remote sensor for acquisition of aerial land surface images. Over the past decade, drones of various shapes, sizes, and capabilities have been developed for numerous applications for meteorology, precision agriculture, soil erosion, wildlife research, forestry, land management, infrastructure inspection, traffic monitoring, epidemic emergencies, natural disaster management, and wilderness search and rescue (Shahbazi et al., 2014; Perez et al., 2015). Benefits of drone-based RS applications are low operational costs and high spatial and operator-specified temporal resolutions. While the application of drone-based RS is still in a somewhat experimental stage, it is poised for increasing data acquisition efficiency (Ambrosia et al., 2011a; Wing et al., 2013; Shahbazi et al., 2014).

There are numerous classification schemes for UAVs (Watts et al., 2012). For example, they may be classified based on to the weight of RS instrument as large, medium, small/mini, or micro/nano (Anderson and Gaston, 2013). The NASA Ikhana Predator B and Sensor Integrated Environmental Remote Research Aircraft (SIERRA) are examples for large- and medium-sized drones, respectively (Fladeland, 2009). The Ikhana Predator B supports applications where manned aircraft deployment would be potentially unsafe for flight crews or inefficient such as real-time monitoring of wildfires in the western United States (Ambrosia et al., 2011b). The SIERRA is suitable for a wider range of applications, including ecological surveys. The main operational constraints of large and medium drones are the initial acquisition costs and operation expenses that render them impractical for most environmental research applications. In fact, to date most environmental RS campaigns have been conducted with small drones, typically weighing less than 20 kg and allow deployment for a few hours (Hardin and Jensen, 2011).

UAVs may be also classified based on their takeoff and landing modes. While fixed-wing UAVs take off and land in horizontal direction, rotary-wing UAVs take off and land in vertical direction. For RS applications, flight stability and range are the most critical considerations for drone design and development. Fixed-wing drones are commonly deployed to survey large areas, while rotary-wing drones are better suited for small fields where high spatial resolution measurements are desired. The novel stop-rotor UAV combines both technologies. It takes off and lands vertically, and once airborne transitions to fixed-wing mode. Different UAV types are depicted in Figure 9.9.

Small UAVs have been widely deployed for environmental research. For example, Jonsson et al. (1980) deployed a small UAV for measurement of air temperature profiles. Mayer et al. (2012) collected temperature, relative humidity, and wind profile data at altitudes as high as 3 km. Martin et al. (2011) and van den Kroonenberg et al. (2012) used the fixed-wing UAV M^2AV to measure the turbulent structure of the atmospheric boundary layer from the surface to 1500 m altitude.

UAVs are capable of carrying hyperspectral radiometers or optical and thermal cameras to provide high-resolution (cm-scale) maps of plant biophysical and biochemical variables. Hence, they can be deployed for precision agriculture applications, where high spatial resolution information and high revisit frequency are required for accurate assessment of crop productivity (Guillen-Climent et al., 2012). Within this context, Berni et al. (2009) characterized the thermal properties of crop canopies to estimate plant water stress based on thermal images acquired with a small UAV. Similarly, Zarco-Tejada et al. (2012) determined plant water stress from a combination of UAV-acquired thermal and fluorescence imagery of an orchard canopy. Hassan-Esfahani et al. (2015) used a UAV equipped with visual, NIR, and thermal cameras for high-resolution mapping of near-surface SM in an agricultural filed.

UAV RS has been also applied for monitoring soil erosion. d'Oleire-Oltmanns et al. (2012) deployed a fixed-wing UAV equipped with a digital camera to generate very high-resolution digital terrain models for quantifying and monitoring gully and badland erosion in Morocco. Perez et al. (2015) successfully applied this technology to evaluate soil erosion and control sediment transport at a 25-ha construction site in Alabama.

Figure 9.9 Geoscan 101 fixed-wing UAV (Geoscan Group, Saint-Petersburg, Russia) (top). *Source*: Reproduced with permission of © 2016–2017 Geoscan group of companies; R-Bat rotary-wing UAV from Northrop Grumman (center). *Source*: Courtesy of Northrop Grumman; SR800 StopRotor UAV prototype (Unmanned Aerial Systems Pty. Ltd, Sydney, Australia) (bottom). *Source*: Reproduced with permission of Small Unmanned Aerial Systems Lab at the Wake Forest Center for Energy, Environment, and Sustainability.

9.4 RS of Static Soil Properties

Detailed information about the static (time-invariant) properties of soils that include mineralogy, texture, organic matter content, and hydrostatic properties (i.e. soil water characteristic [SWC] and saturated hydraulic conductivity) is essential for characterization and simulation of hydrological processes as part of the water and energy cycles. Direct measurements of static properties for large-scale applications are prohibitively expensive and laborious due to the large number of sampling points required for detailed characterization of natural heterogeneity. Other issues associated with direct measurements are the invasive nature of sampling methods and the high uncertainty of specific measurement techniques. RS techniques provide a powerful and cost-efficient alternative means for large-scale determination of static soil properties. Spectral surface reflectance measured in the optical domain (i.e. visible, NIR, short-wave-infrared) is commonly applied for static soil property estimation. A major advantage of the application of spectral data for soil analysis is that multiple properties can be determined from a single spectrum with reasonable accuracy.

Soil spectroscopy was first introduced to soil and environmental research in the 1960s, when Bowers and Hanks (1965) were among the first to apply spectroscopy for estimation of the texture, organic matter content, and moisture of soils. Subsequently, Al-Abbas et al. (1972) applied soil spectroscopy to predict clay and organic matter contents. Since then, spectroscopy has been widely employed for determination of physicochemical soil properties, and it has been demonstrated that the mineral composition, color, texture, and organic matter and moisture content are the major determinants of soil surface reflectance (Viscarra Rossel et al., 2006; Stenberg et al., 2010).

9.4.1 Soil Minerals

Soils are heterogeneous, polyphasic, particulate, and porous systems with often very large interfacial areas, which in conjunction with the intricate geometry of soil particles and pores give rise to phenomena such as adsorption of water and chemicals or capillarity. The major physicochemical soil constituents are mineral and organic particles, soil water with dissolved chemical substances, and air. In addition, myriad microbes occupy interfaces and pore spaces. The relative proportions of constituents that commonly exhibit great spatial and temporal variability impact virtually all engineering, environmental, and agricultural soil attributes.

The mineral phase that mainly comprises of primary aluminosilicate and secondary clay silicate minerals reflects the parent material soils evolve from via fragmentation and weathering. In the following the most important minerals are briefly introduced.

The major *clay minerals* kaolinite, montmorillonite, vermiculite, illite, and chlorite contribute many favorable properties to productive soils. Because of their large surface area, they increase the cation exchange capacity and enhance the water storage and pH buffering capacities of soils. Clay minerals are composed of tetrahedral silica sheets and octahedral aluminum hydroxide and/or magnesium hydroxide sheets. Depending on the number and composition of sheets, they are classified as 1 : 1 or 2 : 1 minerals. While 1 : 1 clays (kaolinites) contain units of alternating tetrahedral and octahedral sheets, 2 : 1 clays (montmorillonites, vermiculites, illites) are composed of units with two tetrahedral and one octahedral sheet. In addition, some 2 : 1 clays (chlorites) have interlayer sites between successive 2 : 1 units, which are occupied by commonly hydrated interlayer cations.

The abundance of *carbonate minerals* (e.g. calcite, dolomite, sodium carbonate, siderite, etc.) in soils may vary from nearly zero to more than 50%. They constitute the primary mineral fraction of calcareous soils that are common in arid and semiarid regions of the world. The most abundant soil carbonates are calcite [$CaCO_3$] and dolomite [$CaMg(CO_3)_2$]; less common are sodium carbonate [Na_2CO_3] and siderite [$FeCO_3$] (Doner and Lynn, 1989). Carbonates buffer soil pH and are an indication for the relative abundance of bases. The presence of Ca^{+2} promotes clay flocculation. Besides these pedogenic implications, carbonate minerals play an important role in soil management (Doner and Lynn, 1989). The distribution and amount of carbonates affect water holding capacity, water flow and solute transport, carbon sequestration capacity, soil fertility, and erodibility. Powdery carbonate particles in high concentrations may lead to iron fixation and consequently inhibition of plant chlorophyll production.

While *sulfate minerals* (e.g. gypsum) that encompass a polyatomic sulfate anion [SO_4^{2-}] within their structure are rather rare in humid regions, they may significantly affect the physicochemical properties of arid region soils. Gypsum (hydrated calcium sulfate) [$CaSO_4^{2-}(H_2O)$] that forms due to reaction of calcium carbonate [$CaCO_3$] with sulfuric acid and easily dissolves and re-precipitates in soils is an important source of calcium and sulfur for plant nutrition, and it alleviates aluminum toxicity that often accompanies soil acidity. In addition, gypsum inhibits dispersion in sodic soils and promotes soil aggregation, which enhances water infiltration and salt leaching efficiency, while reducing surface runoff and erosion.

Iron minerals (e.g. hematite and goethite) in soils include mafic silicates, iron sulfides, iron carbonates, iron oxides, and smectite clay minerals. It is well documented that even a small amount of iron oxides may significantly alter soil color, and consequently soil surface reflectance. The red, brown, and yellow *hue* values, all associated with the presence of various iron oxides, have been widely used for soil classification. In addition to their pedological significance, iron oxides strongly affect a number of soil properties such as surface adsorption of numerous ions and molecules, soil aggregation, and redox behavior.

The direct characterization and quantification of the soil mineral composition is expensive and laborious, especially for investigation of large-scale heterogeneities that requires measurement of a large number of samples. As an alternative, reflectance spectroscopy has shown great potential for rapid identification of constituent soil minerals (Hunt, 1977; Clark et al., 1990), because of their distinct spectral reflectance patterns in the visible (Vis), NIR, and SWIR bands of the EM (Hunt, 1980). In the SWIR range (1300–2500 nm) soil spectra exhibit more features than observations at shorter wavelengths (Figure 9.10a). Particularly, frequencies from 1900 to 2500 nm encompass main spectral features of clay minerals, sulfates, and carbonates.

Figure 9.10 depicts reflectance spectra of common soil minerals for the Vis-NIR-SWIR range. Figure 9.10a specifies three major active spectral regions at SWIR wavelengths, including the 1300–1400, 1800–1900, and 2200–2500 nm frequency bands. The 1 : 1 clay mineral kaolinite exhibits strong absorption features near 1400 and 2200 nm. The absorption near 1400 nm is due to O–H stretching vibrational overtones, while absorption near 2200 nm is due to a combination of Al–OH bending and O–H stretching. Because of the relatively low surface area (i.e. less adsorbed water molecules) of kaolinite, the absorption band at 1900 nm is weak. Montmorillonite (i.e. 2 : 1 smectite clay) exhibits distinct absorptions features near 1400, 1900, and 2200 nm. The 1400 and 1900 nm bands are due to a combination of vibrations of water bound in the interlayer lattices and water adsorbed onto solid particle surfaces (Bishop et al., 1994). The band near 1400 nm can be attributed to the first overtone of structural O–H stretching in the octahedral layer. The band near 2200 nm is associated with the Al–OH (in highly enriched Al smectite), Fe–OH (in highly enriched Fe smectite), or Mg–OH (in highly enriched Mg smectite) bonds. Illite (2 : 1) also absorbs EM radiation near 1400, 1900, and 2200 nm, but generally the absorption is weaker than for montmorillonite.

Calcite exhibits absorption features near 2340 and 2450 nm, which are attributable to overtone and combination bands of the C–O fundamentals of carbonate minerals (Clark et al., 1990). The most distinct feature is close to 2350 nm, but some weaker absorption features are visible close to 1850, 2000, and 2150 nm (Gaffey, 1985) (Figure 9.10a). It is important to note that the locations of the absorption bands are dependent on the chemical soil composition (Hunt and Salisbury, 1970). Ben-Dor and Banin (1990) exploited the distinct mineral

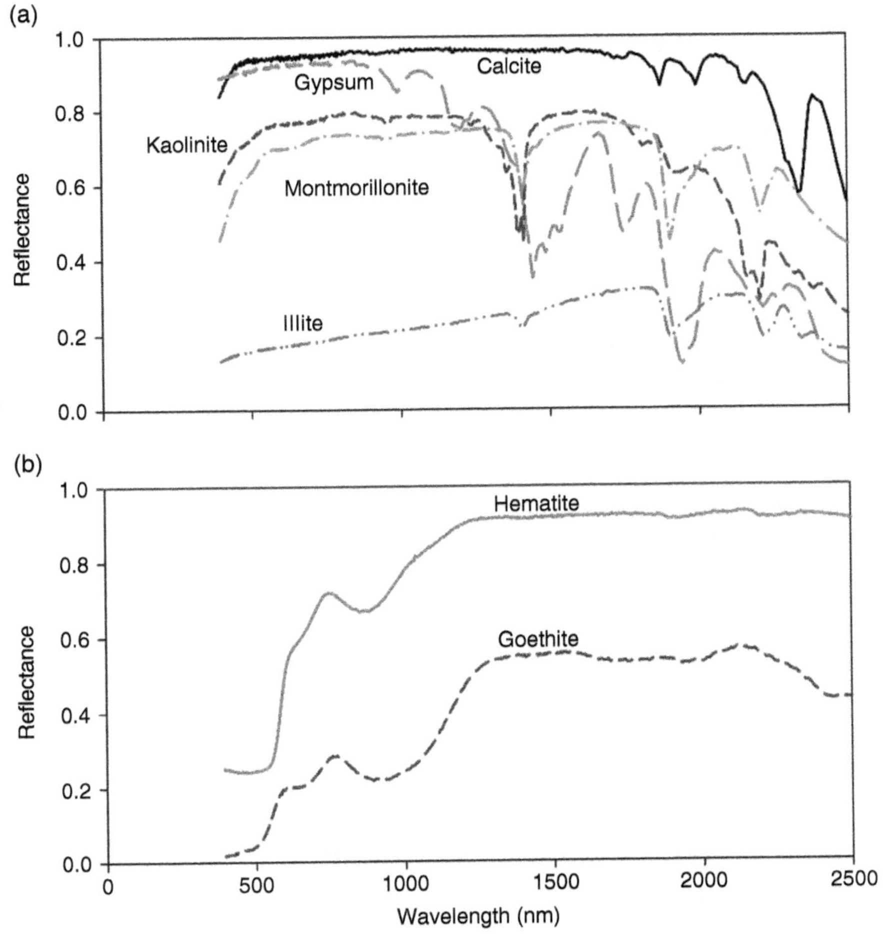

Figure 9.10 Reflectance spectra of common soil minerals exhibiting strong absorption features at SWIR (a) and Vis-NIR (b) wavelengths. *Source*: Adapted from the USGS spectral library.

absorption features to successfully predict carbonate content of arid soils from spectral reflectance observations at 1500, 2350, and 2360 nm.

Iron oxides (hematite and goethite) show distinct absorption features near 465, 650, and 900 nm (Figure 9.10b). The spectrum of goethite [α-FeO(OH)] shows weak absorption around 1700 nm, which is due to the first overtone of a stretching vibration of hydroxyl (OH) present in the crystal structure of goethite (Morris et al., 1985). The spectrum of hematite [α-Fe_2O_3] is characterized by high and nearly constant reflectivity at longer wavelengths within the NIR–SWIR spectral range, a local reflectivity minimum near 880 nm, a shoulder centered at 620 nm, and a band with very low reflectivity near 510 nm (Figure 9.10b) (Stenberg et al., 2010). For both hematite and goethite, the absorption features around 880 and 900–950 nm are attributable to ligand field transitions that involve excitations from a ground state to the first higher energy state (Sherman and Waite, 1985). In goethite, the shoulder near 600 nm and absorptions near 650 nm are due to charge transfer absorptions that lead to transition from a ground to a higher energy state. Absorption in the visible range (650 nm for hematite, 530 nm for goethite) causes the vivid colors of iron oxides (e.g. red hematite and yellow goethite). The shape or magnitude of iron oxide reflectance spectra may be altered when Fe is replaced with Al (Buckingham and Sommer, 1983) and when particle shape and particle size variations cause transopaque spectral effects in different parts of the spectrum (Cudahy and Ramanaidou, 1997).

9.4.2 Soil Organic Matter

Soil organic matter (SOM) comprises plant and animal tissues at various stages of decomposition, living microbial biomass, and substances synthesized by soil organisms. SOM is an essential soil constituent as it affects many biophysicochemical processes, and it is an indicator for soil health. It enhances the water holding capacity of soils, promotes aggregation and aggregate stability, improves water infiltration and soil aeration, reduces surface crusting, alleviates soil erosion, increases the cation exchange capacity (CEC), improves the pH buffering capacity, acts as a food source for soil organisms, and enhances microbial diversity and activity. The quantity of SOM in soils and its quality are intimately linked to soil formation processes that are governed by climate, topography, vegetation, time, and oxidation state.

The quantity and quality of SOM affect soil bulk reflectance. It has been reported that while organic matter in low quantities (<2%) only minimally impacts spectral reflectance (Baumgardner et al., 1970), high quantities commonly mask effects of other soil constituents. For example, the moisture effect on reflectance is significantly reduced in organic soils (Knadel et al., 2014; Sadeghi et al., 2015a).

The SOM decomposition stage (quality) significantly affects reflectance properties as shown for organic soils in Figure 9.11 (Baumgardner et al., 1985). Note that *fibric* refers to slightly decomposed organic material with rubbed fiber contents >40%, *hemic* is partially decomposed SOM with rubbed fiber contents >17 and <40%, and *sapric* refers to highly decomposed SOM. From Figure 9.11 it is evident that *sapric* SOM significantly increases soil reflectance between 800 and 1900 nm, while the effect is insignificant at 1950 nm.

A number of studies have demonstrated great potential of spectral observations for SOM determination. Schreier (1977) was among the first to apply multispectral airborne, ground, and laboratory measurements for SOM estimation. Vinogradov (1981) developed an exponential model to predict the humus content in the

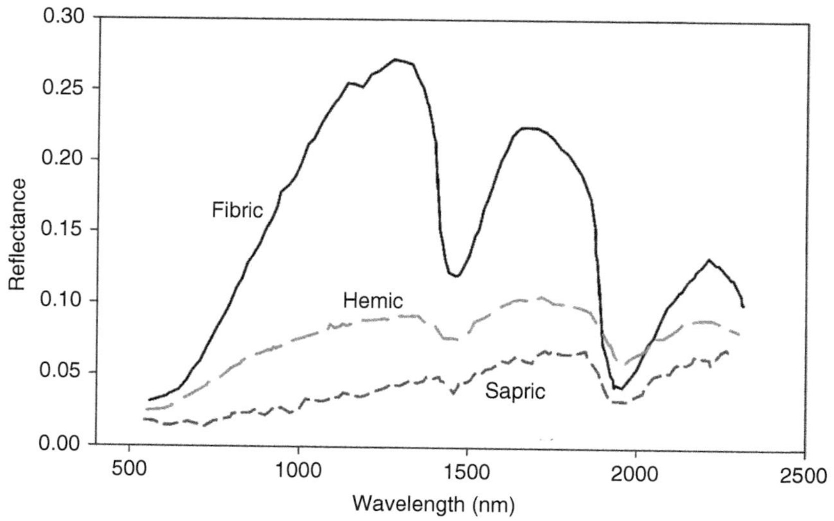

Figure 9.11 Reflectance spectra of three organic soils at various stages of decomposition. *Source*: Adapted from Baumgardner et al. (1985).

upper horizon of plowed forest soils based on spectral reflectance within the 600–700 nm frequency band. Similarly, Baumgardner et al. (1985) observed a curvilinear exponential relationship between SOM and reflectance. It should be noted that published SOM prediction models rely on analysis of absorption features spanning a wide range of frequencies including features near 1100, 1600, 1700–1800, 2000, and 2200–2400 nm (Dalal and Henry, 1986; Ben-Dor and Banin, 1995; Martin et al., 2002; Stenberg et al., 2010). It has also been reported that SOM is a strong absorber of visible EM radiation, which is attributable to the dark color of humic acid. The absorption features at specific wavelengths are due to combination and overtones of the fundamental vibrations of organic functional groups (i.e. O–H, C–H, and N–H) in SOM (Clark et al., 1990; Clark, 1999).

9.4.3 Soil Texture

Soil texture refers to the size distribution of primary particles that constitute the soil solid phase. Dependent on their size, particles are assigned to different fractions (i.e. gravel, sand, silt, and clay) that are subsequently applied to group soils with similar size distributions based on various classification schemes (e.g. US Department of Agriculture [USDA], American Society for Testing and Materials [ASTM], or International Union of Soil Sciences [IUSS]). Soil texture affects various other physicochemical and hydraulic properties (e.g. porosity, bulk density, specific surface area, cation exchange capacity, SWC, hydraulic conductivity) that virtually govern all critical zone processes.

The intimate link between the particle size distribution and spectral soil surface reflectance has been long recognized (Zwerman and Andrews, 1940; Bowers and Hanks, 1965; Salisbury and Hunt, 1968). In general, a decrease in particle sizes leads to an increase in reflectance, provided that the relative fractions of all constituents remain unchanged (Figure 9.12). This is obvious from a comparison of Figures 9.2 and 9.3. As the particle or aggregate sizes increase to approach infinity, reflectance approaches reflectivity (Figure 9.2) that is equivalent to the minimum reflectance of the considered material. This case corresponds to $\tau = 0$ in Eq. (9.16). Assuming a single plate-shaped particle similar to Figure 9.3, the effect of particle size (i.e. plate thickness) is manifested in τ, which may vary from 0 to 1. For the case $\tau = 1$, the plate thickness is 0 based on Eq. (9.3), which yields the maximum reflectance for the considered material (i.e. constant reflectivity, ρ) based on Eq. (9.16). Another reason for the close relationship between reflectance and particle size is self-shadowing of large

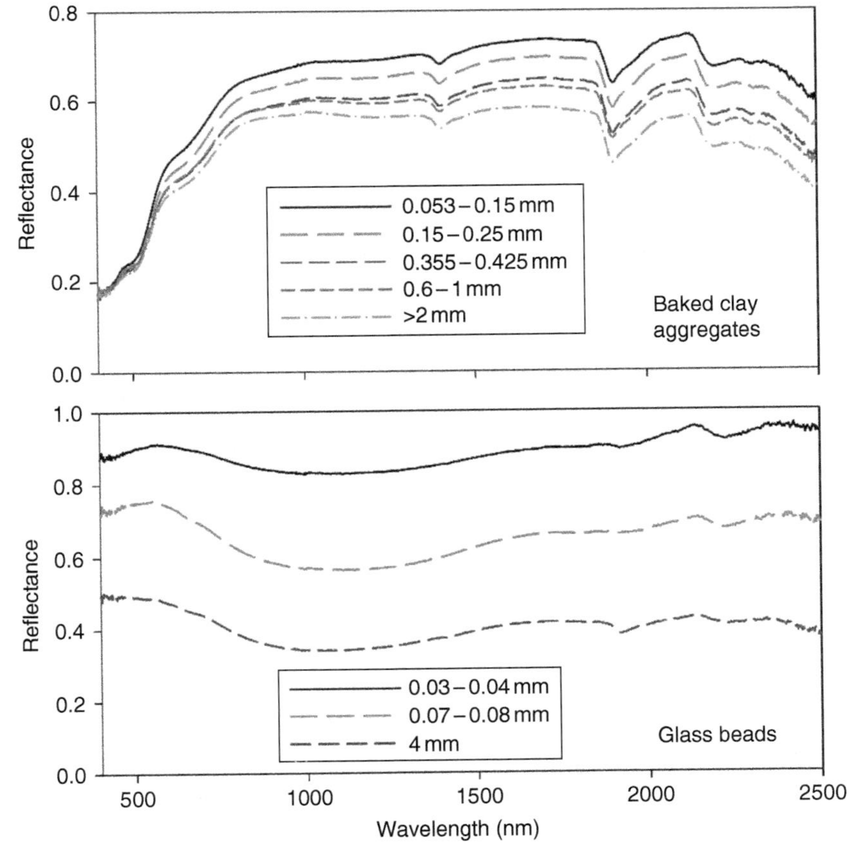

Figure 9.12 Reflectance spectra of differently sized baked clay aggregates (top) and glass beads (bottom).

particles and aggregates (Baumgardner et al., 1985). While small particles arrange to form smooth surfaces, the organization of irregular-shaped coarse particles yields a complex rough surface topology that reflects less EM radiation (Sinha, 1987). Laboratory observations for baked clay aggregates and glass beads of various sizes depicted in Figure 9.12 clearly show the inverse relationship between particle size and reflectance.

Reflectance spectra for differently textured sandy soils measured by Baumgardner et al. (1985) confirm the inverse relationship between particle size and reflectance (Figure 9.13), which was utilized in a number of studies to determine soil texture and particle size distributions from Vis-NIR-SWIR reflectance (e.g. Vendrame et al., 2012; Curcio et al., 2013; Sarathjith et al., 2014). Gerberman and Neher (1979) discovered a linear relationship between increasing sand fraction (0–100%) and reflectance at wavelengths of 440, 540, 720, and 860 nm. According to Sarathjith et al. (2014) and Babaeian et al. (2015a), frequencies around 1400, 1900, 2200, 2330 and 2450 nm are best suited for prediction of the soil sand, silt, and clay fractions. Stenberg et al. (2010) found that the position of above absorption bands slightly varies dependent soil composition.

The estimation errors for the sand, silt, and clay fractions vary in the order of clay < sand < silt (Chang et al., 2001; Islam et al., 2003; Babaeian et al., 2015b). The estimation error for clay is smallest because the spectral signatures of clay minerals are most pronounced within the SWIR frequency range (Stenberg et al., 2010). Sand particles, mainly quartz and feldspar, exhibit pronounced signatures within the Vis-NIR-SWIR frequency range (Hunt and Salisbury, 1970). Silt, on the other hand, is commonly a mixture of sand and clay particles; hence, it is challenging to distinguish in Vis-NIR-SWIR reflectance spectra (Wetterlind and Stenberg, 2010).

9.4.4 Soil Hydrostatic Properties

The SM (θ) is defined as the amount of water present in a soil at a given matric potential (h). The matric potential is synonymous with the combined capillary and adsorptive surface forces that hold water within the solid soil matrix and is uniquely related to θ under hydrostatic conditions. The highly nonlinear relationship between h and θ is termed soil water characteristic (SWC) and exhibits a very distinctive shape for each individual soil. A comprehensive discussion about the origin of h and the measurement and modeling of the SWC is provided in Tuller et al. (1999) and Tuller and Or (2005). The hydraulic conductivity (K) is a measure for the ability of a soil to transmit water when subjected to a hydraulic gradient. Both h and K are functions of θ that are termed *soil hydraulic properties* (SHPs). Both $K(\theta)$ and $h(\theta)$ are invariant for each considered soil and essential for parameterization of hydrological and land surface models.

Direct measurement of the spatially highly heterogeneous SHPs, especially at large scales, is prohibitively expensive and time consuming (Vereecken et al., 2010). Direct measurement methods primarily rely on inversion of the Buckingham–Darcy law (Buckingham, 1907) for steady state flow conditions or on approximate analytical or numerical solutions to Richards' (1931) equation (RE) for transient flow conditions. Because these methods are limited to the sample (Kool et al., 1985; Peters and Durner, 2008) or column scale (Shao and Horton, 1998; Sadeghi et al., 2014) when applied in laboratory settings, and to the point (Simunek and van Genuchten, 1996) and block scale (Hillel et al., 1972) in field applications, they are not suitable for large-scale (e.g. catchment, landscape, satellite footprint) modeling of water flow in soils. To overcome the scale issues,

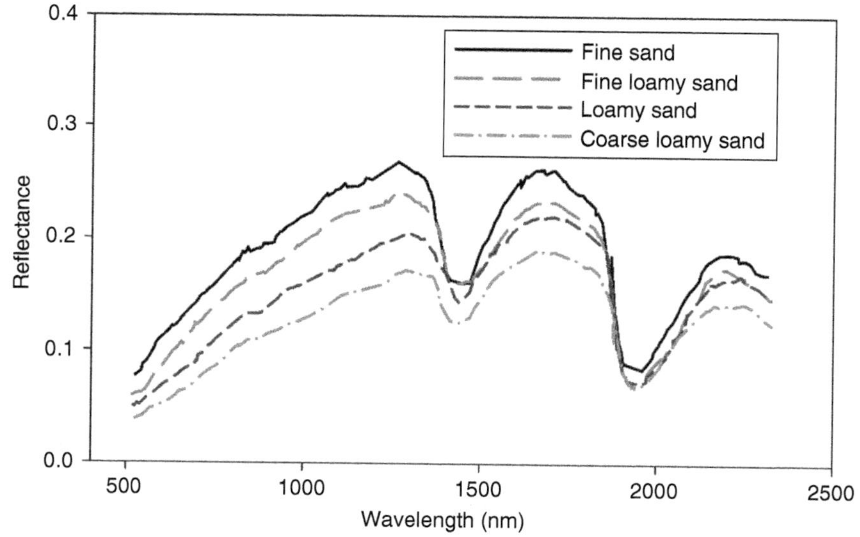

Figure 9.13 Reflectance spectra of differently textures sandy soils. *Source:* Adapted from Baumgardner et al. (1985).

effective SHPs for the satellite footprint scale have been recently determined with RS techniques.

The majority of existing RS-based SHP estimation methods applies the RE or its extended forms for inverse modeling of footprint scale effective SHPs so that model simulations best fit remotely sensed data. Camillo et al. (1986) were among the firsts to estimate SHPs via inversion of an extended form of the RE accounting for nonisothermal soil water flow and heat transfer to fit microwave-based retrievals of surface SM and soil temperature. They conducted a plot-scale experiment, where passive microwave data were acquired with a dual-polarized L-band radiometer mounted on a boom truck. More recently, incorporating time series of airborne and satellite (e.g. AMSR-E) RS-based near-surface SM, Ines and Mohanty (2008a, b, 2009) developed genetic-algorithm-based inverse modeling and data assimilation methods for estimating footprint-scale effective values of the van Genuchten (1980) soil hydraulic parameters. In a synthetic experiment, Montzka et al. (2011) explored the potential of applying near-surface SM retrieved from different satellites to estimate SM profiles and SHPs (e.g. van Genuchten parameters) with the RE and a particle filtering data assimilation method. A comprehensive review of inverse SHP modeling approaches based on RS data is provided in Mohanty (2013).

Hollenbeck et al. (1996) explored the potential of direct estimation of SHPs from RS data. They reported that soil hydraulic heterogeneity can be identified via quantification of relative changes of passive microwave RS observation time series and concluded that the variogram of relative change points to the existence of an effective correlation length much larger than that commonly observed in ground-based soil surveys. Their findings encourage application of passive microwave RS for assessing SHPs that are valid at a scale appropriate for hydrometeorological models. Mattikalli et al. (1998) observed a strong correlation between saturated hydraulic conductivity and passive microwave (L-band) brightness temperature. They demonstrated that regression models can accurately relate brightness temperature variations to saturated hydraulic conductivity across extensive regions.

In recent years, a number of laboratory studies have demonstrated the potential of hyperspectral RS for SHP estimation. The proposed methods primarily rely on predictive regression models termed *spectrotransfer functions*. Santra et al. (2009) and Babaeian et al. (2015a, b) relate soil reflectance at various optical wavelengths (400–2500 nm) to the van Genuchten (1980) soil hydraulic parameters. Although spectrotransfer functions were developed in laboratory settings with benchtop spectroradiometers, they are potentially applicable to airborne (e.g. HyMap) and satellite (e.g. Hyperion) hyperspectral RS for SHP estimation.

9.5 RS of State Variables

9.5.1 Land Surface Temperature

LST is an essential environmental state variable for determination of sensible and latent heat fluxes at the earth surface, and therefore one of the most critical elements in all land surface processes from the local to global scales (Anderson et al., 2008; Karnieli et al., 2010). The LST governs the partitioning of energy between soil and vegetation and affects near-surface air temperature (Zeng et al., 2015). It is a key variable for modeling various environmental processes such as soil evaporation, plant transpiration and growth, or climate change (Bastiaanssen et al., 1998; Kalma et al., 2008; Hansen et al., 2010).

The LST is defined as the temperature of the interface between the Earth's surface and the atmosphere or the radiative skin temperature of the ground. Because it depends on spatial and temporal highly varying surface properties such as albedo, vegetation cover, and SM (Neteler, 2010), the LST rapidly changes with space and time (Vauclin et al., 1982; Prata et al., 1995). In most cases, the LST represents a combination of vegetation and bare soil temperatures that rapidly respond to changes in solar radiation due to cloud cover and diurnal variation of illumination. Therefore, accurate quantification of the LST distribution and its temporal evolution necessitates measurements with detailed spatial and temporal sampling.

In situ LST measurements are commonly conducted with ground radiometers. Due to the complex heterogeneity of surface temperature, point measurements are not suitable for characterization of large areas. RS techniques offer a unique opportunity for measuring LST at various scales. Thermal infrared (TIR) satellite observations of the earth surface are commonly applied to retrieve LST because TIR data may be directly linked to the LST via radiative transfer models.

Despite the physical relationship between satellite TIR observations and LST, retrieval of LST is challenging because of difficulties with (i) implementing atmospheric corrections, (ii) removing effects of reflected solar radiation in the mid-infrared range during daytime, which potentially leads to significant retrieval errors; and (iii) conducting in situ measurements for validation of satellite-retrieved LST at a scale representative for the satellite sensor pixel. Besides these limitations, the accuracy of LST estimates also depends on the quality of TIR data (i.e. signal-to-noise ratio) and performance of the cloud mask used to isolate clear sky data (Li et al., 2013).

Over the past two decades, a number of algorithms have been proposed for LST retrieval from satellite TIR measurements. They can be classified as algorithms that require a priori knowledge of land surface emissivity (LSE), and algorithms where LSE is not known beforehand. While the first group encompasses single-channel (SC), multichannel (MC), and multiangle (MA) methods, the latter consists of stepwise retrieval methods, simultaneous LSE and LST retrieval methods with known atmospheric conditions, and simultaneous LSE and LST retrieval methods for unknown atmospheric conditions.

The SC methods require not only a priori LSE values for each pixel but also detailed information about the atmospheric profile (i.e. water vapor pressure) for the satellite swath, which is difficult to obtain for most practical applications. The MC methods do not require information about the atmospheric profile at the time of data acquisition but utilize the differential atmospheric absorption measured with two adjacent channels centered at 11 and 12 μm frequencies. The *split-window* (SW) algorithm, an MC method that has been first proposed by McMillin (1975), was later modified and extended to successfully retrieve LST (Kerr et al., 1992; Chrysoulakis and Cartalis, 2002; Yang and Yang, 2006; Zhao et al., 2009; Jiang et al., 2013; Du et al., 2015; Fan et al., 2015).

9.5.1.1 The Linear Split-Window Algorithm

At the core of the linear SW method is the stipulation that the radiance attenuation due to atmospheric absorption is proportional to the radiance difference between two simultaneous measurements at different wavelengths, each subjected to different quantities of atmospheric absorption (McMillin, 1975). The SW algorithm utilizes two TIR bands, usually located in the atmospheric window between 10 and 12.5 μm.

The linear split-window algorithm (LSWA) was derived via linearization of the RTEs for two adjacent TIR channels i and j, which yield LST as a linear combination of brightness temperatures sensed with channels i and j

$$\text{LST} = a_0 + a_1 T_{bi} + a_2 \left(T_{bi} - T_{bj}\right) \tag{9.35}$$

where T_{bi} and T_{bj} are the at-sensor (top of atmosphere) brightness temperatures (in Kelvin) for channels i and j (e.g. TIR bands 10 and 11 of Landsat-8) and the a_k values (k = 0, …, 2) are SW coefficients that are commonly determined from simulated data and primarily depend on (i) the spectral response function (the quantum efficiency of a sensor at specific wavelengths over the range of a spectral band) of the two channels (SRF_i and SRF_j), (ii) the emissivity of the two channels (ε_i and ε_j), (iii) the atmospheric water vapor content (w), and (iv) viewing zenith angle (φ)

$$a_k = F\left(\text{SRF}_i, \text{SRF}_j, \varepsilon_i, \varepsilon_j, w, \varphi\right) \tag{9.36}$$

To capitalize on the SW rather than on SC algorithms for more accurate LST estimates, the Landsat-8 satellite was designed to include two TIR bands. This feature was not incorporated in the TM and ETM$^+$ sensors of previous Landsat satellites. Because the single TIR band of the TM and ETM$^+$ sensors onboard Landsat-8 has been split into two bands, they are narrower than the TIR bands of Landsat-TM and Landsat-ETM$^+$, as depicted in Figure 9.14. A sample LST map obtained from Landsat-8 observations is shown in Figure 9.15.

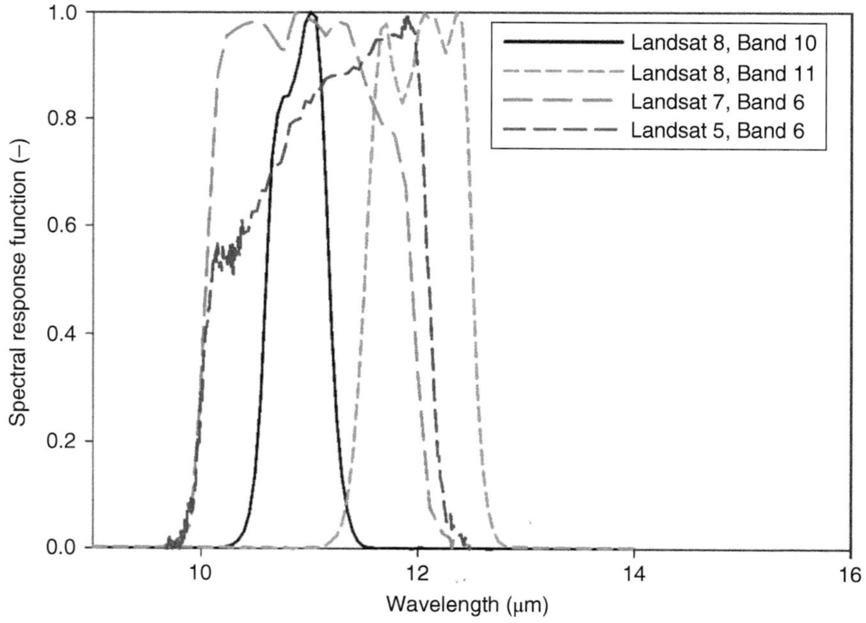

Figure 9.14 Spectral response functions for the thermal bands of Landsat 5 (TM), 7 (ETM+), and 8 (OLI/TIRS). Effective wavelengths are 11.457 μm for band 6 of Landsat 5, 11.269 μm for band 6 of Landsat 7, 10.904 μm for band 10 of Landsat 8, and 12.003 μm for band 11 of Landsat 8.

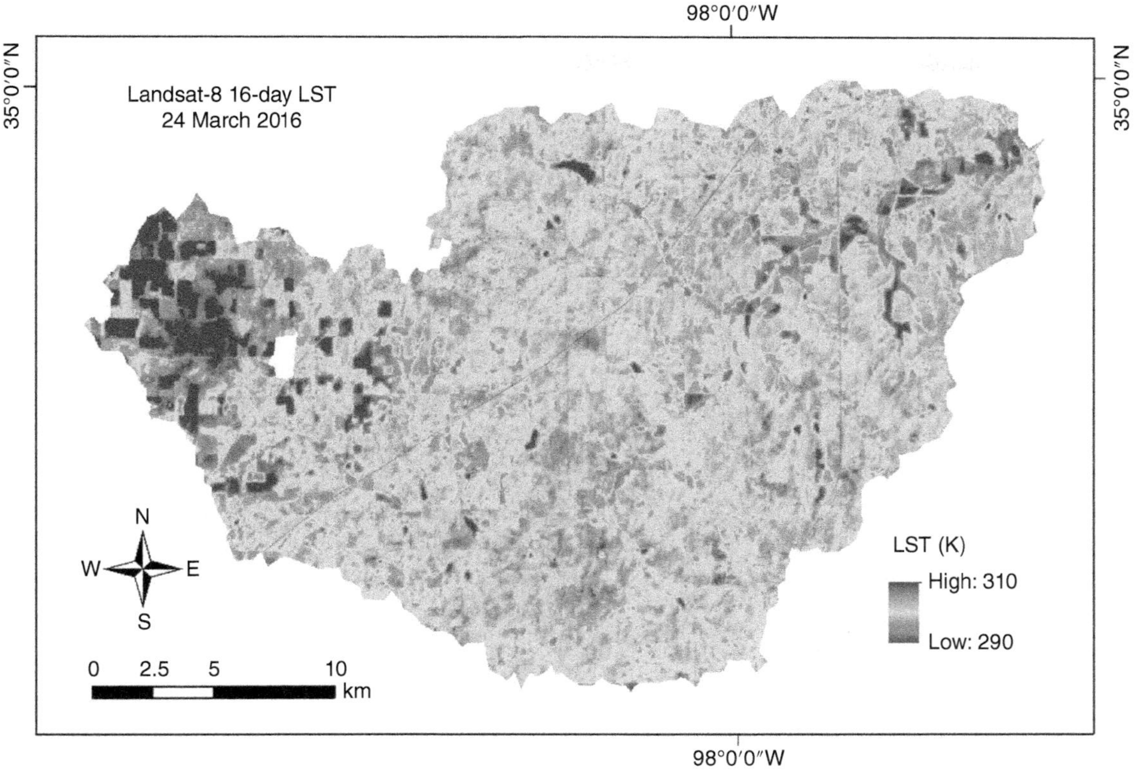

Figure 9.15 Land surface temperature map obtained from Landsat-8 observations of the Little Washita Watershed in Oklahoma on 24 March 2016.

It should be noted that the performance of the LSWA is dependent on the selection of appropriate SW coefficients (a_k), which may be determined either via regressing simulated satellite data to a set of atmospheric and surface parameters or by means of empirical relationships between satellite data and in situ ground LST measurements. Due to the high spatial variability of surface temperature, in situ ground LST measurements at the satellite pixel scale are not trivial. Therefore, it is more effective to regress simulated satellite data to preset LST values applied in the simulations (Li et al., 2013; Jiménez-Muñoz et al., 2014). A number of linear SW algorithms of similar form but with different coefficient parameterization schemes (i.e. linear and nonlinear functions) have been proposed.

Wan and Dozier (1996) developed a generalized SW algorithm for LST retrieval from MODIS TIR band 31 (channel i) and band 32 (channel j) and considered the emissivity (ε), atmospheric water vapor pressure (w), and viewing zenith angle (φ) for determination of a_k:

$$\mathrm{LST} = b_0 + \left(b_1 + b_2\frac{1-\varepsilon}{\varepsilon} + b_3\frac{\Delta\varepsilon}{\varepsilon^2}\right)\frac{T_{\mathrm{b}i}+T_{\mathrm{b}j}}{2} + \left(b_4 + b_5\frac{1-\varepsilon}{\varepsilon} + b_6\frac{\Delta\varepsilon}{\varepsilon^2}\right)\frac{T_{\mathrm{b}i}-T_{\mathrm{b}j}}{2} \tag{9.37}$$

where $\varepsilon = (\varepsilon_i + \varepsilon_j)/2$, $\Delta\varepsilon = \varepsilon_i - \varepsilon_j$, b_k ($k = 0, \ldots, 6$) are unknown coefficients for a given φ and for a given subrange of ε, w, and LST (Li et al., 2013). Expressing Eq. (9.37) in form of Eq. (9.35), the coefficients of Eqs. (9.35) and (9.37) can be linked:

$$\begin{aligned} a_0 &= b_0, \quad a_1 = b_1 + b_2\frac{1-\varepsilon}{\varepsilon} + b_3\frac{\Delta\varepsilon}{\varepsilon^2}, \\ a_2 &= \frac{b_4-b_1}{2} + \frac{b_5-b_2}{2}\frac{1-\varepsilon}{\varepsilon} + \frac{b_6-b_3}{2}\frac{\Delta\varepsilon}{\varepsilon^2} \end{aligned} \tag{9.38}$$

Validation for specific land covers under clear sky conditions yielded an accuracy of less than 1 K for LST calculated based on coefficients determined with Eq. (9.38) in the range of 263–322 K. A major advantage of the application of Eq. (9.38) is that LST retrieval is less sensitive to uncertainties related to LSE and atmospheric conditions. However, the errors may be significant for bare soils for cases where the LST is more than 16 K higher than the surface air temperature (Wan and Li, 2008; Li et al., 2013).

9.5.1.2 The Nonlinear Split-Window Algorithm

Because of uncertainties introduced due to linearization of the RTE and other applied approximations (e.g. approximation of transmittance as a linear function of water vapor), the LSWA yields significant LST retrieval errors under wet and hot atmospheric conditions. To overcome these issues, nonlinear split-window algorithms (NLSWA) have been developed. They may be expressed in general form as

$$\mathrm{LST} = c_0 + c_1 T_{bi} + c_2 \left(T_{bi} - T_{bj}\right) + c_3 \left(T_{bi} - T_{bj}\right)^2 \quad (9.39)$$

where c_k coefficients (k =0,...,3) are determined via regressing Eq. (9.39) to simulated satellite data analogs to the a_k coefficients in Eq. (9.35).

A number of NLSWAs of similar form have been proposed in the past (e.g. Sobrino et al., 1994, 1996; Sobrino and Raissouni, 2000; Atitar and Sobrino, 2009). Similar to LSWAs, some NLSWAs consider the LSE for derivation of c_k coefficients, some apply LSE and water vapor pressure, while others also incorporate the viewing zenith angle (φ). Considering effects of all three determinants, Atitar and Sobrino (2009) proposed the following parameterization for the Sobrino et al. (1996) algorithm

$$c_0 = d_0 + d_1(1-\varepsilon) + d_2\Delta\varepsilon + d_3 w(1-\varepsilon) + d_4 w\Delta\varepsilon,$$
$$c_1 = 1, \qquad c_2 = d_5, \qquad c_3 = d_6 \quad (9.40)$$

where d_k (k = 0, ..., 6) is a linear function of $\sec^2(\varphi)$

$$d_k = e_{0,k} + \frac{e_{1,k}}{\cos^2(\varphi)} \quad (9.41)$$

For sensors with three or more TIR channels, the LST may be estimated from linear and nonlinear combinations of at-sensor brightness temperatures of all channels. Sun and Pinker (2003, 2007) demonstrated that MC algorithms yield LST estimates with errors of less than 1 K. The Sun and Pinker (2007) four-channel algorithm has been specifically developed for the Spinning Enhanced Visible and Infrared Imager (SEVIRI) sensor with four TIR channels of i = 3.9, j = 8.7, k = 10.8, and l = 12.0 μm:

$$\mathrm{LST} = f_0 + f_1 T_{bk} + f_2\left(T_{bk} - T_{bl}\right) + f_3\left(T_{bi} - T_{bj}\right) + f_4\left(T_{bk} - T_{bl}\right)^2 + f_5\left[\sec(\varphi) - 1\right] + f_6 T_{bi}\cos(\varphi_s) \quad (9.42)$$

where the coefficients f_k (k = 0, ..., 6) depend on the land surface type that incorporates LSE effects, φ is the satellite viewing zenith angle, and φ_s is the solar zenith angle. Validation with in situ ground measurements revealed that the four-channel algorithm, Eq. (9.42), yielded better LST estimates than the SW algorithms.

9.5.1.3 LST-Based Estimation of Air Temperature

The near-surface air temperature (T_{air}), commonly measured at an elevation of 2 m above ground, is another important state variable that is essential numerous environmental applications that include weather forecasting, hydrological modeling, estimation of ET, prediction of crop yields, and prediction global climate change, only to name a few. The spatiotemporal patterns of T_{air} are highly variable and complex dependent on the variability of the environmental factors that determine the energy balance components. Therefore, point-scale T_{air} measurements at weather stations are not adequate for regional and larger-scale studies.

The advancement of RS techniques for LST retrieval has also provided a new approach to large-scale mapping of T_{air} based on LST maps. Three major classes for T_{air} estimation from remotely sensed LST data can be discerned (Janatian et al., 2016):

1) *Energy balance methods* (e.g. Sun et al., 2005; Zhang et al., 2015) with strong physical foundation that generally balance the sum of incoming net radiation and anthropogenic heat fluxes with the sum of the surface sensible and latent heat fluxes. A major limitation of energy balance methods is the complexity of required input parameters that are not readily available from RS observations.
2) *Temperature–vegetation index (TVX) method* (e.g. Stisen et al., 2007; Zhu et al., 2013) that is based on the assumption of a strong negative correlation between LST and VI such as the NDVI. However, this assumption is not always met for natural systems, and it may be affected by seasonality, ecosystem type, and SM variability, which leads to failure of the TVX method for certain cases (Sandholt et al., 2002; Vancutsem et al., 2010).
3) *Statistical methods* that are based on regression techniques (Benali et al., 2012; Chen et al., 2014; Good, 2015; Janatian et al., 2016) or more advanced machine learning approaches such as artificial neural networks (e.g. Jang et al., 2004) and support vector machine algorithms (e.g. Ho et al., 2014; Moser et al., 2015). Statistical methods empirically model T_{air} as a function of LST and additional auxiliary explanatory variables. Though statistical methods are site specific and require large datasets for calibration of the regression models, they generally provide accurate estimates of T_{air} for the spatial and time frames they were developed for. A recently developed statistical model based on MODIS LST data for eastern Iran yielded T_{air} mean absolute validation errors of 2.3 and 1.8 °C for daily and weekly temperatures, respectively (Janatian et al., 2016).

9.5.2 Soil Moisture

SM is a key hydrologic state variable that links land surface and atmospheric processes (Robinson et al., 2008).

Detailed knowledge about the state of SM and its spatial and temporal dynamics is of crucial importance for numerous meteorological, climatologic, and hydrologic applications and to improve our understanding of the water, energy, and carbon cycles, as well as for forecasting of extreme climate events (Vereecken et al., 2014).

9.5.2.1 Radiometric Signatures of Soil Moisture

As discussed in Sections 9.4.1 and 9.4.3, soils are heterogeneous, polyphasic, particulate, and porous systems that may be described as a three-component dielectric mixture of air, minerals, and water. The resulting EM soil properties for visible to microwave frequencies are dependent on temperature, solute concentration, particle size distribution, bulk density, surface roughness, and most importantly the volumetric soil water content.

Surfaces of soils with high water contents often appear darker than dry soils, which implies that soil surface reflectance generally decreases in the optical bands (0.4–2.5 μm) with increasing SM (Bowers and Hanks, 1965; Lobell and Asner, 2002). Spectral reflectance measurements in the SWIR frequency range by Sadeghi et al. (2015b) confirm this inverse relationship for various Arizona source soils (Jensen et al., 2015) that cover a wide range of mineral compositions and soil textures (Figure 9.16). The relationship is commonly expressed empirically with an exponential function (Lobell and Asner, 2002).

The surface reflectance–SM relationship may be also formulated based on the RTE. For example, Sadeghi et al. (2015a) have derived a simple physical relationship between soil reflectance and soil surface water content (θ) based on the KM radiative transfer model (Eq. (9.27)):

$$\frac{\theta}{\theta_s} = \frac{\sigma(r - r_d)}{r_s - r + \sigma(r - r_d)} \tag{9.43}$$

where θ_s is the saturated water content, $r = k/s$ is the KM transformation of reflectance that can be derived from inversion of Eq. (9.27) as $r = (1 - R)^2/2R$, r_s and r_d are the transformed reflectance of saturated and dry soils, respectively, and $\sigma = s_d/s_s$, where s_s and s_d are the scattering coefficients of saturated and dry soils, respectively. Experimental soil spectral reflectance data support the Sadeghi et al. (2015a) assumption that σ approaches 1 for SWIR frequencies (e.g. 1650 and 2210 nm) due to strong absorption of water molecules. This assumption has led to formulation of a linear physically based model for optical RS of SM of bare soils, which holds for a wide range of soil textures (Figure 9.17):

$$\frac{\theta}{\theta_s} = \frac{r - r_d}{r_s - r_d} \tag{9.44}$$

It is also well documented that the soil thermal inertia increases with increasing water content (Schmugge, 1978). Therefore, the surface skin temperature slower adjusts to diurnal variations of the ambient air temperature when the SM content is high (Friedl and Davis, 1994). This causality is often applied to infer SM from TIR (3.5–14 μm) RS observations. It has been also documented that while the skin temperature of a well-watered canopy surface is often cooler than the near-surface air temperature, this trend reverses for a canopy under considerable water stress (Moran et al., 1994). Based on this rationale, empirical methods have been developed to utilize both the optical and TIR bands to link SM to VI (e.g. NDVI) and surface canopy temperatures, which indirectly encode plant water stress levels (Carlson et al., 1994; Carlson, 2007).

For microwave radiation, typically 1–18 GHz, the dielectric constant of the soil–water mixture increases with increasing volumetric water content (Dobson et al., 1985; Hallikainen et al., 1985). This gives rise to an increase in soil reflectivity or a decrease in emissivity. As a result, moist soils are radiometrically colder than their dry counterparts. This cause and effect relationship is at the core of microwave RS of SM via physically based radiative transfer models (Basharinov and Shutko, 1975; Njoku and Kong, 1977; Jackson and Schmugge, 1991; Entekhabi and Njoku, 1994).

Optical methods are only applicable for clear sky and during daylight hours. EM radiation within the optical frequency range cannot even penetrate through scattered canopies, hence is not applicable for capturing SM variability for such conditions. Thermal methods that rely on VI are often site specific and prone to calibration errors. These limitations often impede the application of optical and thermal methods for reliable global-scale RS of SM. On the other hand, microwave RS techniques do not rely on reflected sunlight and therefore can operate during night time. More importantly, microwave radiation penetrates canopies with moderate vegetation density and can capture the SM beneath them. However, passive microwave RS methods provide substantially lower spatial resolutions (~40 km) when compared with optical and thermal RS techniques (~100–250 m). This low resolution often creates serious challenges for hydro-agronomic applications that content with processes at much smaller scales (i.e. field scale). Nonetheless, research over the past few decades has demonstrated that the microwave portion of the EM, especially the L-band frequencies (1–2 GHz), provides the most accurate information for RS of the surface (~0–5 cm) SM, provided that the vegetation water content is below 5 kg m^{-2} (Entekhabi et al., 2010).

9.5.2.2 Passive Microwave Soil Moisture Retrieval

In low microwave frequency domain, the brightness temperatures observed with a spaceborne radiometer

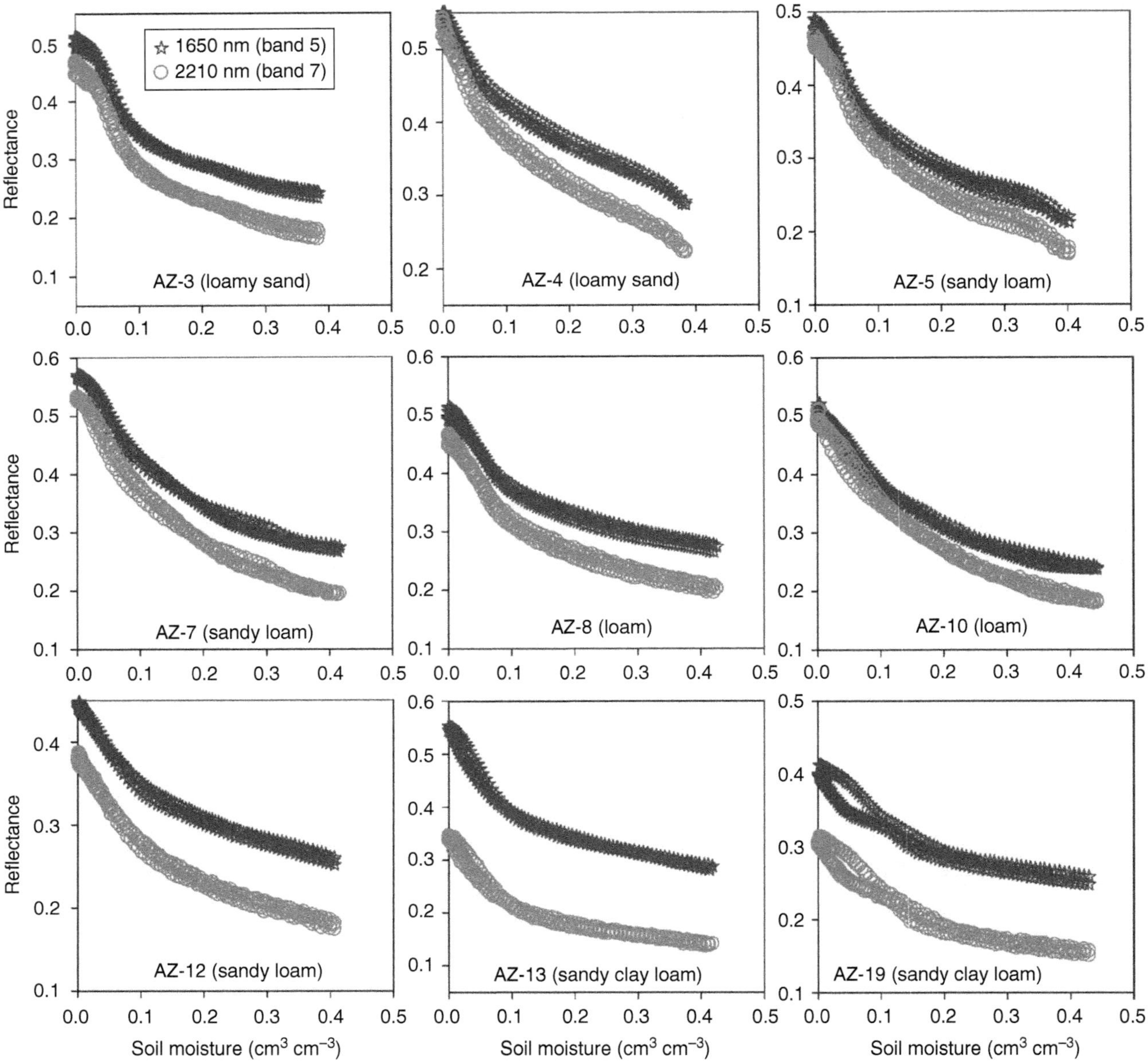

Figure 9.16 Relationship between surface reflectance and water content for Arizona (AZ) source soils covering a wide range of textures and mineral compositions.

consist of multiple components including (i) energy emitted from the soil and attenuated by overlying vegetation, (ii) energy emission from the vegetation cover, (iii) upwelling atmospheric emission, and (iv) reflected and attenuated downwelling atmospheric and cosmic emissions (Figure 9.18).

One of the most widely applied algorithms for passive SM retrieval, the *tau–omega* model, has been derived based on a zeroth-order approximation of the RTE (note that *tau* and *omega* refer to the original model parameters that are denoted differently below for consistency). The model assumes that the soil is covered by a homogenous canopy layer serving as a weakly scattering medium. At low frequencies, where the atmospheric emission is small and cosmic radiation is in the order of 2.7 K, the model can be formulated as

$$T_b = \varepsilon_s \tau_c T_s + (1 - \omega_c)(1 - \tau_c) T_c + \rho_s (1 - \omega_c)(1 - \tau_c) \tau_c T_c \quad (9.45)$$

where T_b is the total upward radiative flux expressed in terms of brightness temperature, T_s and T_c denote the soil surface and vegetation thermodynamic temperatures, ε_s is soil surface emissivity, $\rho_s = 1 - \varepsilon_s$ is the rough soil reflectivity, ω_c is the vegetation single scattering albedo, $\tau_c = \exp[-\alpha_c \sec(\varphi)]$ is the canopy layer transmissivity

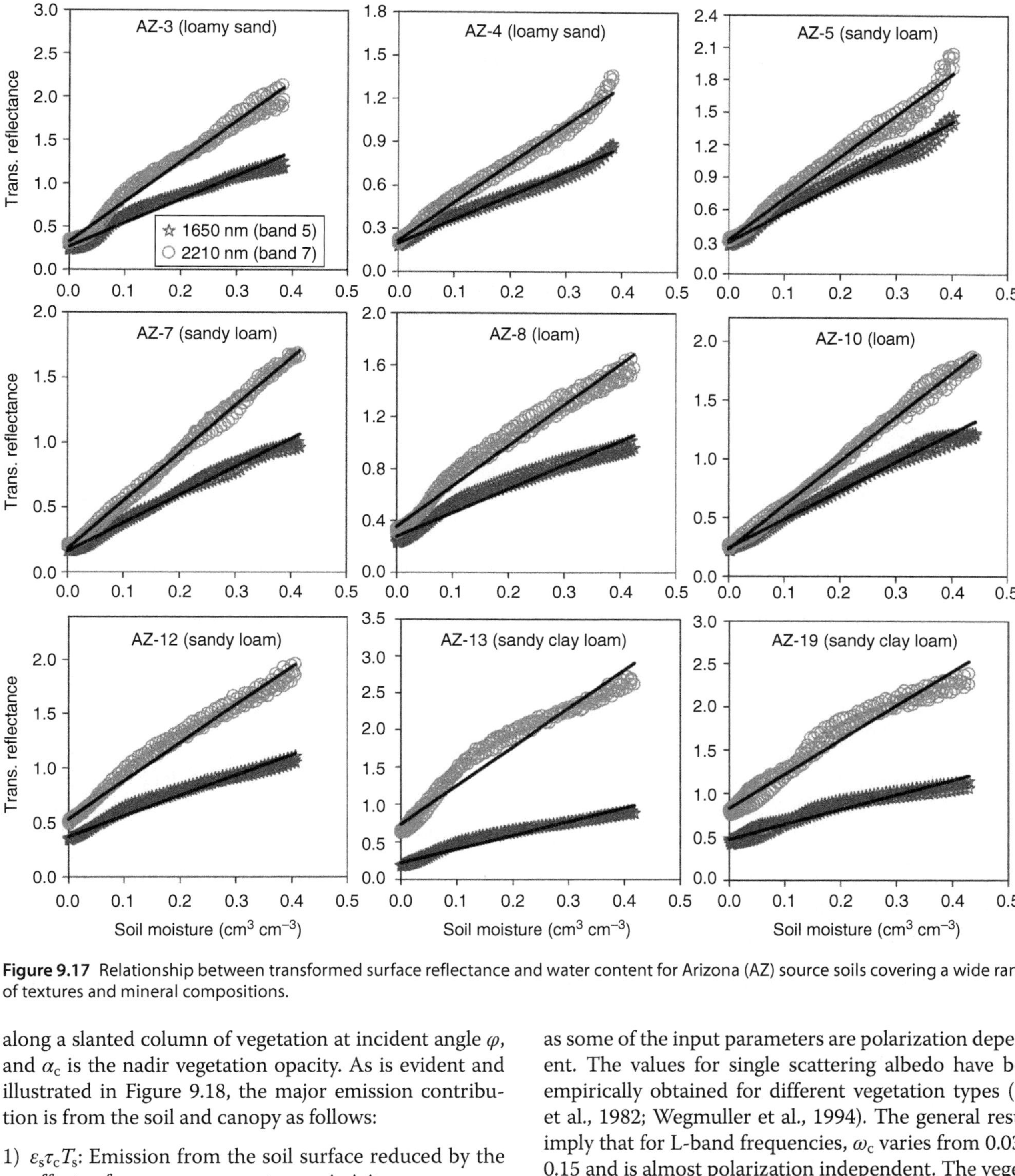

Figure 9.17 Relationship between transformed surface reflectance and water content for Arizona (AZ) source soils covering a wide range of textures and mineral compositions.

along a slanted column of vegetation at incident angle φ, and α_c is the nadir vegetation opacity. As is evident and illustrated in Figure 9.18, the major emission contribution is from the soil and canopy as follows:

1) $\varepsilon_s \tau_c T_s$: Emission from the soil surface reduced by the effects of one-way canopy transmissivity τ_c.
2) $(1-\omega_c)(1-\tau_c)T_c$: Upward slanted emission by the canopy.
3) $\rho_s(1-\omega_c)(1-\tau_c)\tau_c T_c$: Downward canopy emission followed by reflection at the soil surface and one-way canopy transmission.

Note that the *tau–omega* model can be used to obtain the total upward radiative flux for different polarizations, as some of the input parameters are polarization dependent. The values for single scattering albedo have been empirically obtained for different vegetation types (Mo et al., 1982; Wegmuller et al., 1994). The general results imply that for L-band frequencies, ω_c varies from 0.03 to 0.15 and is almost polarization independent. The vegetation opacity is often parameterized based on the vegetation water content C_w, which is the total water contained in a vertical column of a canopy per unit area. For example, $\alpha_c = 0.115 C_w$ is suggested for grass, corn, and soybeans for C_w in (kg m^{-2}) at a frequency of 1.4 GHz (Schmugge et al., 1986).

Considering the observed brightness temperatures T_b, ω_c, and α_c and assuming that the vegetation and the soil

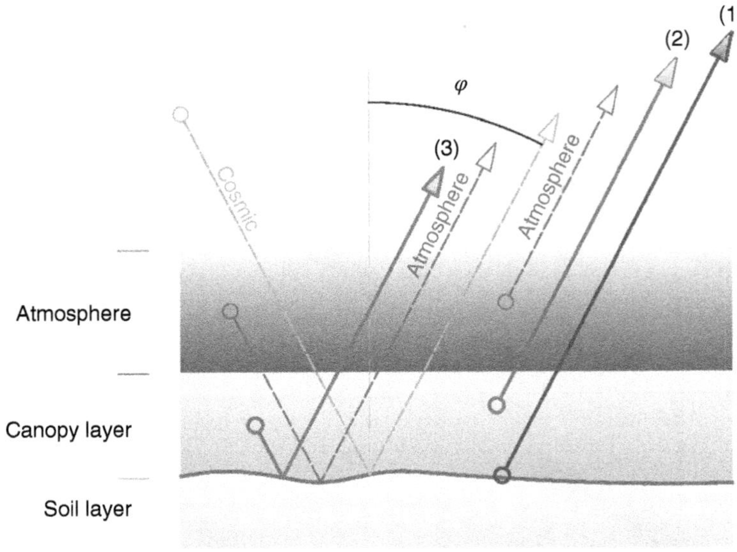

Figure 9.18 Emission contributions to observed low-frequency brightness temperatures for soil moisture retrievals in presence of a weakly scattering canopy. The solid/labeled arrows are major emission contributions associated with terms of Eq. (9.45). The dashed arrows refer to negligible components.

surface are in thermodynamic equilibrium, the model can be applied to estimate the rough soil surface reflectivity, ρ_s, which is related to the smooth surface reflectivity, ρ_{ss} (e.g. $\rho_s = \rho_{ss} \exp[-L \cos^2(\varphi)]$, where L is a linear function of the root mean squared surface roughness) (Choudhury et al., 1979). The obtained smooth surface reflectivity values allow estimation of the soil dielectric constant, $\epsilon = n^2$, through the Fresnel Eqs. (9.11) and (9.12). Then, in the final step, an empirical wet soil dielectric model is used to retrieve the SM content. For example, Dobson et al. (1985) suggested the following model for the real part of the soil complex dielectric constant (ϵ_{rs}):

$$\epsilon_{rs} = \left(2.122 + \theta^p \epsilon_{rw}^q - \theta\right)^{\frac{1}{p}} \tag{9.46}$$

where θ is the volumetric SM content, ϵ_{rw} denotes the real component of the water dielectric constant, $p = 0.65$, and $q = 1.27 - 0.519\ Sand\ - 0.152 Clay$, with $0 \leq Sand \leq 1$ and $0 \leq Clay \leq 1$ representing the mass fraction of sand and clay, respectively. Figure 9.19 depicts passive microwave retrievals based on the *tau–omega* model from data obtained with the radiometer onboard the SMAP satellite (Entekhabi et al., 2010).

9.5.3 Snow Cover and Water Equivalent

Snowfall and its accumulation on the ground, i.e. the snow cover, control the cryosphere mass balance, which is the most sensitive component of the hydrologic water cycle with regard to global warming. The snow cover regulates the temperature of permafrost during the winter, and its meltwater controls springtime SM and summertime streamflow. RS of snow cover is very important not only for understanding its climatic impacts but also for water resources management in agricultural and urban areas, where water supply relies on snowmelt and runoff from mountainous regions.

9.5.3.1 Radiometric Signatures of Snow Cover

The snow cover is a heterogeneous mixture of air, ice particles, and sometimes liquid water. The EM properties of snow are very elusive and change drastically dependent on its metamorphism. In the optical region, dry snow is highly reflective and that is why it appears white to the human eye. The reflectance is maximum between 0.4 and 0.6 μm and sharply declines for larger wavelengths (Figure 9.20). Snow reflectance in the optical region is largely a function of snow grain size and exhibits little dependence on snow depth and liquid water content. When the average snow grain size increases as the snowpack ages, the likelihood of light scattering declines as the number of air–ice interfaces is reduced; hence, reflectance typically decreases (Warren, 1982). In contrast, snow is not highly reflective in the thermal region and its emissivity ranges from 0.96 to 0.995.

In the microwave region, ground emission propagation through snow cover often generates more detailed information about the physical properties of snow than can be inferred from surface reflectance at optical and thermal frequencies. The snow cover attenuates the upwelling ground microwave emission through scattering and absorption. Attenuation of the ground emission throughout the snow layer alters the surface brightness temperature as a function of frequency, ice particle size and density, and liquid water content. The microwave propagation properties of dry snow are drastically different from the properties of wet snow. The dry snow ice particles strongly scatter the upwelling ground emission for frequencies above 20 GHz. The larger ice particles

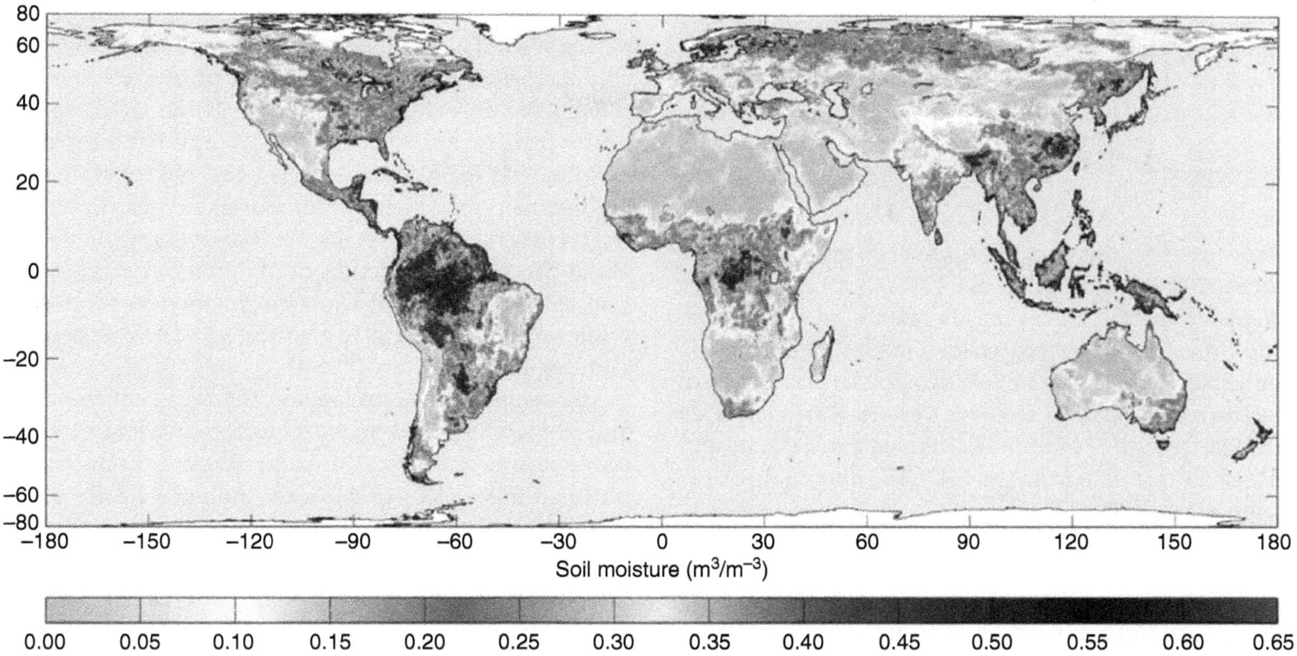

Figure 9.19 Passive soil moisture retrieval with the L-band radiometer onboard of the SMAP satellite over a one-week period from 1 June 2015 to 7 June 2015. *Source*: Adapted from Chan et al. (2016).

Figure 9.20 Spectral reflectance of a deep snowpack in the optical domain as a function of wavelength and grain size. *Source*: Adapted from Rees (2005) and Choudhury and Chang (1979).

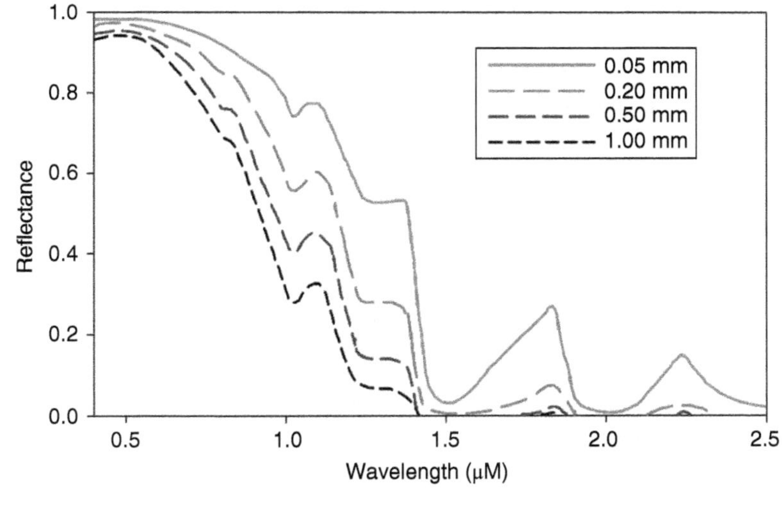

scatter microwaves more efficiently and attenuate them over a shorter traveling distance within the snow layer. In addition, nonspherical ice crystals in fresh snow often create strong polarization signals. It is obvious that the scattering efficiency of a thicker snow layer with higher density is higher than for a thin layer. Simulation and observational evidence indicate that the intensity of scattering increases with increasing frequency from 20 to 100 GHz. However, a small amount of liquid water (>1% volumetric) significantly alters the propagation properties of snow by changing its dielectric properties. Scattering markedly decreases with increasing wetness and the snowpack begins to behave similar to a black body. As a result, the radiometric temperatures of a wet snowpack are often much higher than for the dry counterpart. For example, Stiles and Ulaby (1980) showed that a 1% increase in volumetric water content can increase the observed brightness temperatures at 37 GHz by 70 K. It should be noted that due to the emission dominant wet snow regime, the observed brightness temperatures are almost independent of the snow particle size but are dependent on snow surface roughness.

9.5.3.2 Snow Cover Retrieval

As previously mentioned, snow reflectance is a function of snow grain size and sharply decreases from the visible to the NIR spectra. This sharp transition is often utilized for discrimination of snow via the normalized difference snow index (NDSI), which considers visible R_{Vis} (0.4-0.6 μm) and NIR R_{NIR} (1.55–1.75 μm) reflectance

$$\mathrm{NDSI} = \frac{R_{Vis} - R_{NIR}}{R_{Vis} + R_{NIR}} \tag{9.47}$$

Snow cover normally exhibits an NDSI greater than 0.4 (Dozier, 1984, 1989; Hall et al., 2002). In practice, other multispectral reflectance data are often used to eliminate cloud, forest, and surface water contamination effects (Rees, 2005, p. 144). Figure 9.21 depicts the percentage of time snow cover is detected over western North America obtained from the NDSI index derived from visible and NIR reflectance data from the MODIS onboard the Aqua satellite (Hall et al., 2002).

In addition to semiempirical methods such as the NDSI index, there are physically based retrieval approaches that rely on radiative transfer modeling of spectral reflectance of snow cover (Painter et al., 2009). This group of methods applies sophisticated radiative transfer models (e.g. Stamnes et al., 2007) to generate a library with a statistically representative number of snow physical properties (e.g. grain size) and their corresponding reflectance at wavelengths and solar zenith angles of interest. To provide tools for discrimination of snow cover from other land surface types, the library often contains ancillary data such as reflectance measured or simulated for different land surface types, e.g. vegetation,

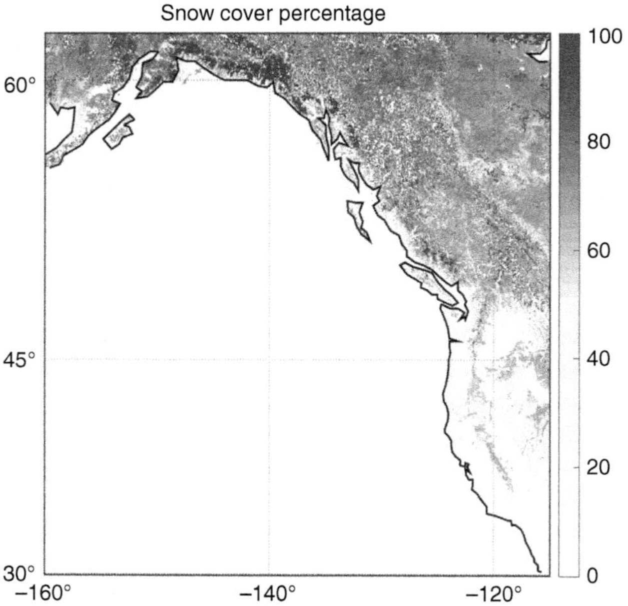

Figure 9.21 Annual snow cover time percentage over western North America in 2013. The snow cover retrievals are obtained from the normalized snow cover index applying reflectance data from the MODIS onboard the Aqua satellite.

rock, soil, or lake ice. The simulated spectral reflectance values of a homogenous land surface type are often called *endmember* (Figure 9.22). Given a pixel level spectral reflectance observation obtained with an air or space-borne instrument (e.g. MODIS), an optimal estimation technique is used to reconstruct the observation with endmembers of different land surface types. Based on the reconstruction error, the method not only can decide about the subpixel fraction of the land cover types but also allows retrieval of snow cover physical properties (Nolin and Dozier, 1993; Nolin et al., 1993; Rosenthal and Dozier, 1996).

Analogous to the optical region, there are semiempirical and physically based approaches for retrieval of snow cover extent and physical properties from microwave observations. As previously noted, fresh snow distinctively scatters surface emissions, with scattering becoming stronger at higher frequencies. Therefore, the snow surface brightness temperatures are lower at higher frequencies (37 GHz) than at lower frequencies (19 GHz). However, this trend is also observed for other land surface types such as deserts. Moreover, other intervening scattering media such as precipitating atmosphere may create similar scattering signatures in radiometric measurements from space. Based on extensive data analysis of the microwave observations from the special sensor microwave imager (SSM/I), Grody and Basist (1996) have suggested a series of thresholding schemes that allow discrimination of snow cover from other scattering media. In particular, a scattering medium is identified when the following condition holds:

$$T_b(22,V) - T_b(85,V) > 0 \quad \text{or} \quad T_b(19,V) - T_b(37,V) > 0 \tag{9.48}$$

where $T_b(f, P)$ denotes brightness temperature at frequency f (GHz) and polarization P. The scattering medium can then be identified as snow cover if it is not precipitation, cold desert, or frozen ground, which is detectable via specific threshold sets. For example, frozen surfaces exhibit a less distinct scattering signature but show a larger polarization difference at 19 GHz than snow cover and therefore can be detected based on the following condition:

$$\begin{aligned} &[T_b(22,V) - T_b(85,V) \leq 6\mathrm{K} \quad \text{or} \\ &T_b(19,V) - T_b(37,V) \leq 2\mathrm{K}] \\ &\text{and} \quad T_b(19,V) - T_b(19,H) \geq 8\mathrm{K} \end{aligned} \tag{9.49}$$

While such empirical methods are simple to implement, obtained results are often prone to large errors, especially when applied to identify snow cover below a precipitating cloud (Rees, 2005, p. 146). Assuming that

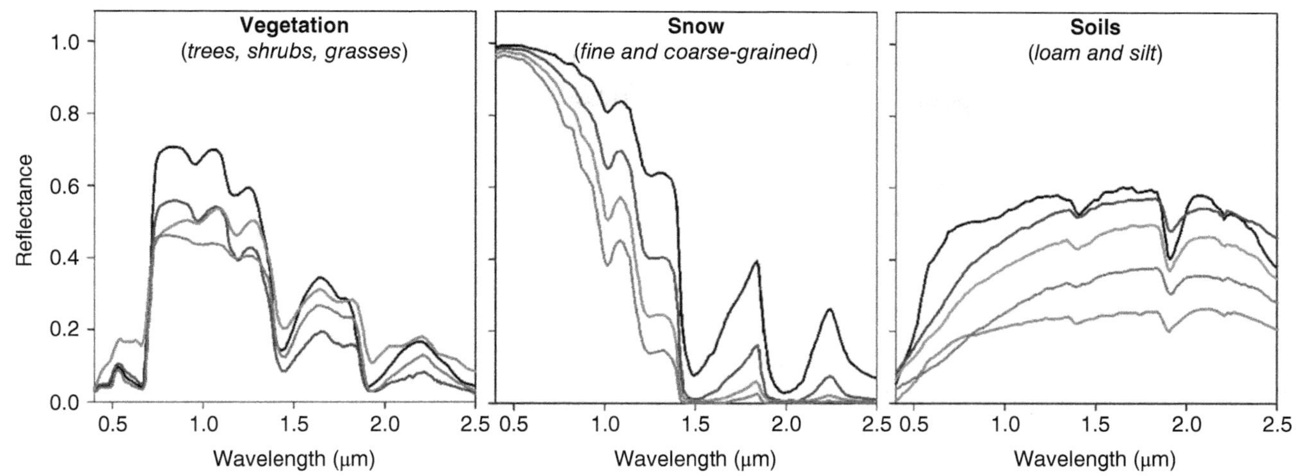

Figure 9.22 Examples for contrasting end-members of vegetation, snow, and soils *Source*: Adapted from Painter et al. (2009).

soil is covered with a homogenous snow layer, a zeroth-order approximation of the RTE can be formulated:

$$T_b = \varepsilon_s \tau_c T_s + (1-\omega_s)(1-\tau_s)T_{snow} + \rho_s(1-\omega_s)(1-\tau_s)\tau_s T_{snow} \tag{9.50}$$

The terms resemble the zeroth-order approximation of the SM radiative transfer Eq. (9.45), where T_b is snow surface brightness temperature, T_s and T_{snow} denote the soil surface and snow cover thermodynamic temperatures, ε_s is soil surface emissivity, $\rho_s = 1 - \varepsilon_s$ is the rough soil reflectivity, ω_s is the snow cover single scattering albedo, $\tau_s = \exp[-k_e d \sec(\varphi')]$ is the snow layer transmissivity along a column of snow of depth d in direction φ', which is related to the incident angle via Snell's law of refraction, i.e. Eq. (9.10), and k_e is the extinction coefficient due to both absorption and scattering. The extinction coefficient can be expressed in terms of the so-called mass extinction coefficient $k_{em} = k_e/\delta_{snow}$, where δ_{snow} is snow density. This equation relates the transmissivity to the snow water equivalent $W = \delta_{snow} d$ that yields $\tau_s = \exp[-k_{em} W \sec(\varphi')]$. It should be noted that ρ_s is often sufficiently small and the third term in Eq. (9.50) can be neglected for practical applications.

The zeroth-order approximation of the RTE cannot properly account for high-order scattering phenomena, which are intrinsic characteristics of the snow cover. Numerical solutions of the RTE show that the zeroth-order approximation is valid for media with a single scattering albedo of less than 0.2. It has been demonstrated that for a snow cover with a mean ice crystal size of 0.5 mm, the single scattering albedo is around 0.2 only for frequencies below 10 GHz (Ulaby and Long, 2014). Therefore, from a theoretical perspective, for larger ice crystals and higher frequencies, the zeroth-order approximation is not applicable. However, in practice, this approximation has been proven useful even for single scattering albedos greater than 0.2. The above approach is at the core of well-known semiempirical (Chang et al., 1987) and physically based models (Pulliainen and Hallikainen, 2001) for microwave snow depth and water equivalent retrievals. For example, Chang et al. (1987) applied Eq. (9.50) to compute the surface brightness temperature as a function of snow depth from observations obtained with the scanning multichannel microwave radiometer (SMMR) onboard of the NASA Nimbus satellite and suggested the following semiempirical model for snow depth (SD, mm) retrieval:

$$\mathrm{SD} = 15.9\left[T_b(19, H) - T_b(37, H)\right] \tag{9.51}$$

The performance of this model is depicted in Figure 9.23, based on observations from the radiometer onboard the GPM satellite.

9.5.4 Biophysicochemical Vegetation Variables

Vegetation is an important element that affects numerous land surface processes. Knowledge of biophysicochemical vegetation variables is essential for various environmental and agricultural applications. RS allows continuous mapping of several spatially heterogeneous vegetation variables such as fractional vegetation cover (FVC), LAI (defined as the projected area of leaves per unit of ground area), or chlorophyll and vegetation water contents. As shown in Figure 9.22, while vegetation exhibits relatively low reflectance in the VIS frequency range, it is much higher at NIR frequencies. This allows discrimination of vegetation from other surface components such as bare soil or water by means of the NDVI:

$$\mathrm{NDVI} = \frac{R_{NIR} - R_{red}}{R_{NIR} + R_{red}} \tag{9.52}$$

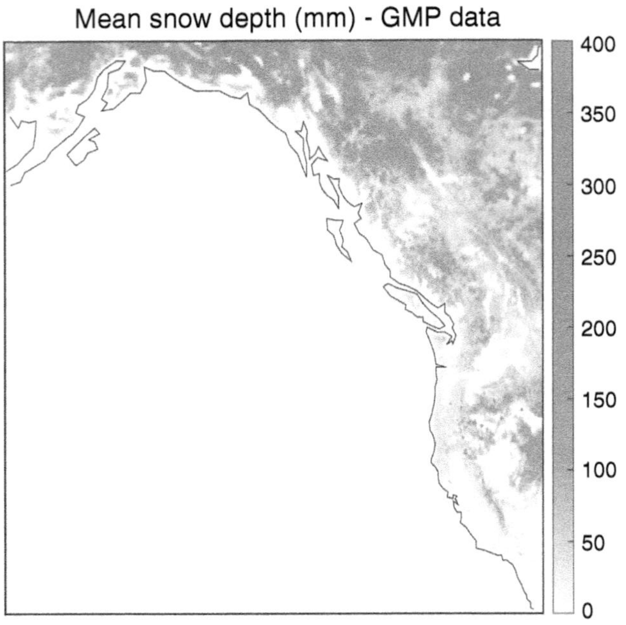

Figure 9.23 Snow depth map obtained with a semiempirical model (Chang et al., 1987) from mean brightness temperature values from all nonrainy overpasses of the radiometer onboard of the GPM satellite in 2013.

where R_{red} is reflectance in the red band. Sparse vegetation such as shrubs and grasslands commonly exhibits moderate NDVI values (~0.2 to 0.5), whereas dense vegetation, typical for temperate and tropical forests or agricultural crops at their peak growth stage, exhibits high-NDVI values (~0.6 to 0.9). Therefore, NDVI and other similar indices (Table 9.4) adjusted to minimize effects of soil background variations are strongly correlated with canopy biophysicochemical variables such as LAI, FVC, biomass, or chlorophyll content (Asrar et al., 1985; Baret and Guyot, 1991; Carlson and Ripley, 1997; Gilabert et al., 2002; Jiménez-Muñoz et al., 2009).

Several methods have been proposed to estimate vegetation variables from RS data via either empirically or physically based relationships that rely on inversion of a radiative transfer model. Empirical approaches are based on statistical relationships between the variables of interest and RS data. For example, a simple empirical model assumes that the FVC, a key parameter for many environmental and climate-related applications, is equal to the scaled NDVI* (Wittich and Hansing, 1995; Gutman and Ignatov, 1998):

$$\mathrm{FVC} = \mathrm{NDVI}^* = \frac{\mathrm{NDVI} - \mathrm{NDVI_s}}{\mathrm{NDVI_v} - \mathrm{NDVI_s}} \tag{9.53}$$

Table 9.4 Vegetation indices.

Vegetation index	Reference
$\mathrm{RVI} = \frac{R_{NIR}}{R_{red}}$	Pearson and Miller (1972)
$\mathrm{NDVI} = \frac{R_{NIR} - R_{red}}{R_{NIR} + R_{red}}$	Rouse et al. (1974)
$\mathrm{PVI} = \frac{R_{NIR} - bR_{red} - a}{\sqrt{1+b^2}}$	Richardson and Wiegand (1977)
$\mathrm{SAVI} = \frac{R_{NIR} - R_{red}}{R_{NIR} + R_{red} + L}(1+L)$	Huete (1988)
$\mathrm{TSAVI} = \frac{b(R_{NIR} - bR_{red} - a)}{R_{red} + b(R_{NIR} - a) + X(1+b^2)}$	Baret and Guyot (1991)
$\mathrm{MSAVI} = R_{NIR} + 0.5 - 0.5\sqrt{(2R_{NIR}+1)^2 - 8(R_{NIR} - R_{red})}$	Qi et al. (1994)
$\mathrm{OSAVI} = \frac{R_{NIR} - R_{red}}{R_{NIR} + R_{red} + Y}$	Rondeaux et al. (1996)
$\mathrm{GESAVI} = \frac{R_{NIR} - bR_{red} - a}{R_{red} + \frac{a' - a}{b' - b}}$	Gilabert et al. (2002)

Source: From Gilabert et al. (2002).
R_{red} and R_{NIR} are reflectance for red and near-infrared frequencies; a and b are parameters of the soil line in the R_{red}-R_{NIR} space (i.e. $R_{NIR} = a + bR_{red}$); a' and b' are parameters of the vegetation isoline in the R_{red}-R_{NIR} space (i.e. $R_{NIR} = a' + b'R_{red}$); L, X, and Y are empirical soil-adjustment parameters that depend on vegetation density ($L = 0.5$, $X = 0.08$, and $Y = 0.16$ have been previously proposed).

where $NDVI_s$ and $NDVI_v$ are the normalized difference vegetation indices of bare soil and dense vegetation, respectively.

While additional nonlinear relationships such as quadratic functions, i.e. $FVC = NDVI^{*2}$ (Carlson and Ripley, 1997), have been proposed, they did not significantly improve predictions based on Eq. (9.53) (Wittich and Hansing, 1995). Contrary to Gutman and Ignatov (1998), who assumed $NDVI_s$ and $NDVI_v$ as global constants independent of vegetation and soil type, Jiménez-Muñoz et al. (2009) postulated that correct identification of $NDVI_s$ and $NDVI_v$ values is critical due to their region- and season-specific nature. In contrast to the FVC, the LAI exhibits a highly nonlinear relationship such that the sensitivity of the NDVI decreases with increasing LAI (Gamon et al., 1995; Carlson and Ripley, 1997; Turner et al., 1999).

Other biophysicochemical variables such as vegetation biomass (Tucker et al., 1983; Gamon et al., 1995; Kogan et al., 2004; Kawamura et al., 2005), chlorophyll content (see table 1 in Hunt et al., 2013), water content (Ceccato et al., 2001; Chen et al., 2005; Yilmaz et al., 2008), and nitrogen or protein content (Gamon et al., 1995; Kawamura et al., 2005; Thulin et al., 2014) have been derived from optical reflectance data.

Inversion of a radiative transfer model (i.e. regressing model simulations to RS observations) provides a physically based mean for estimation of canopy biophysical and biochemical properties. An excellent example for such inversion is provided in Jacquemoud and Baret (1990). They have successfully estimated vegetation chlorophyll and water contents with the PROSPECT model that was developed based on Eqs. (9.19)–(9.23). As obvious from Eqs. (9.2) and (9.3), τ is linked to the absorption coefficient k. For a homogeneous plate of thickness d, Eqs. (9.2) and (9.3) yield $\tau = \exp[-kd\sec(\varphi_2)]$. Jacquemoud and Baret (1990) assumed the spectral absorption coefficient $k(\lambda)$ as a weighted average of leaf absorptive components:

$$k(\lambda) = \Sigma C_i k_i(\lambda) \tag{9.54}$$

where C_i is the content of the ith leaf component per unit leaf area and k_i is the absorption coefficient of the ith leaf component. Jacquemoud and Baret (1990) later discovered that at wavelengths between 400 and 800 nm, the coefficient k is affected by pigments such as chlorophyll a and b content (C_{a+b}), carotenoids and brown pigments (tannins), whereas k at 800–2500 nm is mainly affected by water content (C_w) due to strong absorption of water molecules within this frequency domain. They considered N (number of layers in Eqs. (9.19) and (9.20)), C_{a+b}, and C_w as main model input parameters and demonstrated that C_{a+b} and C_w can be accurately estimated via inversion of the PROSPECT model, i.e. fitting simulated reflectance and transmittance to measurements obtained with a spectrophotometer (see Figure 9.24).

The coupled PROSPECT leaf optical properties and scattering by arbitrary inclined leaves (SAIL) canopy bidirectional reflectance models (Verhoef, 1984), referred to as PROSAIL, is widely applied to study plant canopy spectral and directional reflectance in the solar domain and for simultaneous estimation of canopy biophysicochemical properties. Because of the ease of application, the computational efficiency, the small number of required input parameters, and the extensive validation based on laboratory, field, as well as satellite RS data in recent years, PROSAIL is currently the most popular radiative transfer modeling tool (Jacquemoud et al., 2009).

An example for the application of PROSAIL is provided in Thorp et al. (2012). They coupled PROSAIL with the cropping systems model, an ecophysiological model that simulates crop growth and development considering

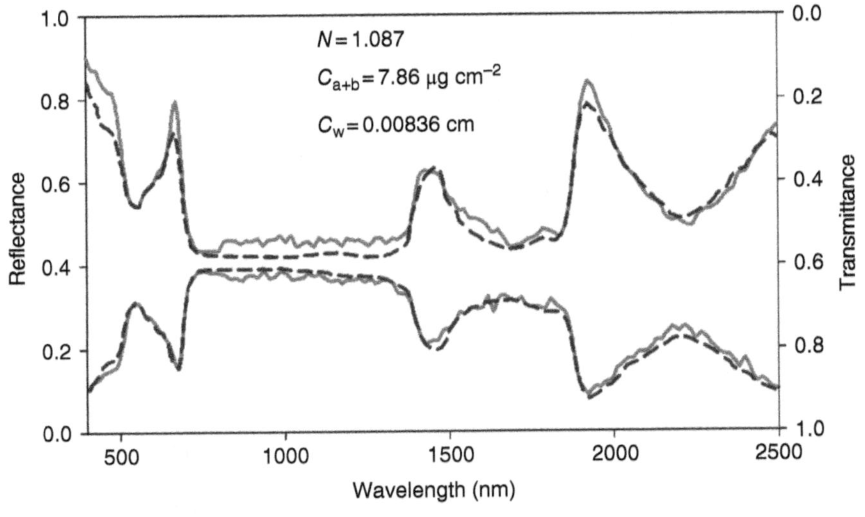

Figure 9.24 PROSPECT simulations of spectral reflectance and transmittance (dashed line) obtained via model inversion from measured (solid line) data for yellowing maize. *Source:* Adapted from Jacquemoud and Baret (1990).

effects of soil water and nutrient status on crop yield. Inversion of the coupled models and incorporation of measured field scale crop canopy spectral reflectance data yielded estimates LAI, canopy weight, plant nitrogen content, and yield for a durum wheat field in central Arizona (Thorp et al., 2012). This approach has been implemented at different scales based on data obtained with field spectroradiometers (e.g. Darvishzadeh et al., 2008; Thorp et al., 2015), airborne sensors (e.g. Darvishzadeh et al., 2011; Laurent et al., 2013), and satellite sensors (e.g. Weiss and Baret, 1999; Kallel et al., 2007; Boegh et al., 2013).

9.6 RS of Environmental Fluxes

9.6.1 Precipitation

Precipitation is the flux of either liquid or solid atmospheric water that connects the terrestrial and atmospheric water cycles. RS of precipitation is of paramount importance as ground-based gages are sporadically distributed and difficult to maintain at high altitudes and in regions with rough terrain. Because a significant portion of the precipitation falls over tropical oceans, where ground-based observations are not available, satellite RS plays an important role. RS of precipitation relies on active ground-based radars and active/passive satellite observations. While active precipitation radars often operate within the microwave band, optical to microwave EM frequencies have been applied for passive precipitation RS. Because the topic of precipitation RS is very vast and cannot be thoroughly covered in this chapter, discussions are confined to passive methods.

9.6.1.1 Radiometric Signatures of Precipitation

The top of a precipitating cloud with high liquid and vapor water content commonly exhibits higher reflectivity in the visible domain (Barrett, 1970) and appears cooler in the thermal region (Arkin, 1979) than a non-precipitating cloud. Therefore, higher reflectivity in the visible domain in conjunction with lower brightness temperatures in the thermal domain indicates higher surface precipitation rates. However, these causal links are not unique for precipitating regions of clouds and therefore cannot provide accurate information about the vertical structure of hydrometeors as the shortwaves are absorbed. Often, thermal brightness temperature signatures of high-elevation nonprecipitating cirrus clouds are similar to the signatures of thick precipitating convective clouds. Moreover, snow-covered surfaces also appear cool in the thermal domain similar to precipitating clouds. Ebert et al. (1996) and Kummerow et al. (1996) have shown that microwaves can penetrate clouds, interact with hydrometeors, and provide accurate information about the vertical structure and the phase of precipitation. Oceans are radiometrically cold in the low microwave frequency range (10–60 GHz) and less emissive than precipitating areas. Therefore, precipitation events over oceans give rise to radiometrically warm brightness temperatures across low-frequency channels. However, land surfaces are radiometrically warm, across these low frequencies, and their emissions mask any precipitation emission signal. In high-frequency channels (>80 GHz), atmospheric hydrometeors begin to scatter the upwelling land surface microwave emission when the wavelength approaches the particle size of the hydrometeors and thus reduce the observed brightness temperatures. These distinct microwave signatures allow more accurate inference of precipitation micro- and macrophysical properties than possible with infrared frequencies.

It should be noted that infrared observations from the geostationary operational environmental satellite (GOES) system are available for most of the globe in 30-min intervals, while microwave radiometers onboard of low earth orbiting satellites only revisit once or twice per day. To capitalize on both microwave and thermal observations, advanced machine learning algorithms have been developed to characterize and quantify relationships between microwave precipitation observations and associated thermal signatures for infrared precipitation retrieval (Hsu et al., 1997; Hong et al., 2004). Moreover, methods relying on GOES infrared imagery have been proposed for interpolating between sequential overpasses of spaceborne microwave radiometers (Joyce et al., 2004).

9.6.1.2 Precipitation Retrieval

While most of the passive retrieval models for visible and inferred frequencies are empirical, some microwave retrieval techniques implement the underlying physical processes, especially for retrieval over oceans where precipitation signatures are largely due to emissions from hydrometeors. Over land, where signatures are due to scattering, development of physically based models is still a complex task and most of the operation retrieval models are still empirical or semiempirical.

Radiative transfer modeling of precipitation is more complex than for SM and snow cover largely due to complicated and fast evolving space and time dynamics. Passive microwave precipitation retrievals often rely on an interrelated two-step approach that encompasses a "forward" and an "inverse" model. The forward model connects various atmospheric profiles to their brightness temperatures at the top of the atmosphere and consists of a cloud resolving model (Tao, 2003) coupled with an atmospheric radiative transfer model (Liu and Weng, 2013). The cloud resolving model is applied to generate a database of atmospheric profiles based on a statistically representative number of boundary and initial conditions that represent the climatological conditions of interest. Then the radiative transfer model is applied to obtain the brightness temperature of the generated atmospheric profiles at the incident angle and frequencies of the radiometer. For a given observed pixel-level vector of brightness temperatures, the inverse model then attempts to isolate a few atmospheric database profiles with similar brightness temperatures that are utilized to estimate the precipitation profile. This group of methods is referred to as *Bayesian retrieval algorithms*. In simple terms, for a given vector of brightness temperatures from N channels $\mathbf{y} \in (T_{b1}, T_{b2}, \ldots, T_{bN})^T \in \mathbb{R}^N$, Bayesian retrieval models assume that $\mathbf{y}$ can be expressed with a linear combination of simulated database brightness temperatures:

$$\mathbf{y} = \mathbf{B}\mathbf{c} + \mathbf{e} \tag{9.55}$$

where $\mathbf{B} \in \mathbb{R}^{N \times M}$ represents an N-by-M matrix that contains M simulated brightness temperatures from the database in its column space, $\mathbf{c} \in \mathbb{R}^M$ is the vector of linear representation coefficients, and $\mathbf{e} \in \mathbb{R}^N$ is the reconstruction error. Determination of optimal values of the representation coefficients requires robust regularization techniques as $M \gg N$, and the above linear system is underdetermined. Then these coefficients are used to linearly combine the simulated atmospheric profiles in the database and retrieve the precipitation profile. For additional information readers are referred to Petty and Li (2013), Ebtehaj et al. (2015), and Kummerow et al. (2015). Figure 9.25 depicts annual passive/active precipitation

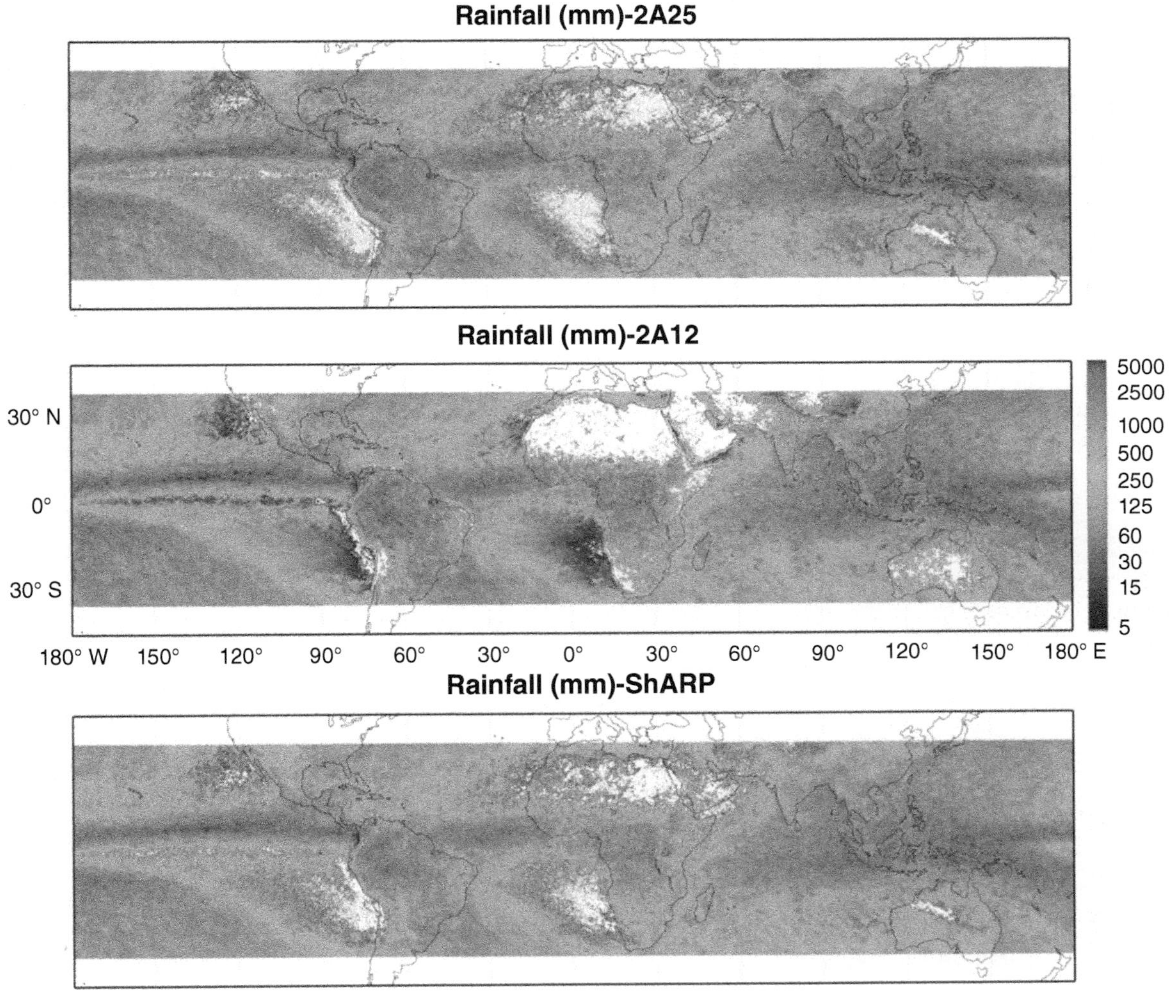

Figure 9.25 Annual total precipitation from all inner-swath overpasses of the TRMM satellite in 2013. Official active (2A25, top) and passive (2A12, center) precipitation retrievals (version 7) from the precipitation radar and radiometer onboard the TRMM satellite as well as from a new passive retrieval algorithm termed "Shrunken Locally Linear Embedding for Retrieval of Precipitation" (ShARP, Ebtehaj et al., 2015) (bottom).

retrievals from observations with the radiometer onboard the TRMM satellite.

9.6.2 Surface Water Storage and Fluxes

Storage of surface water in lakes, rivers, wetlands, and reservoirs is an essential element of all hydrologic systems and crucial for various managed and natural ecosystem processes.

9.6.2.1 Radiometric Signatures of Surface Water

Water reflects a small fraction of incident light in the visible frequency band dependent on water depth as well as on suspended and dissolved particulate matter. However, inundated areas strongly absorb solar radiation in the NIR region, which may be utilized to discern water bodies from surrounding soil and vegetated surfaces. While high-resolution optical sensors are able to accurately delineate the boundaries of inland water bodies, those capabilities are severely limited during cloudy conditions. At microwave frequencies, inundated areas are less emissive than the nearby soil and vegetated surfaces, and thus the surface brightness temperatures over water bodies are much cooler than the temperatures of surrounding areas (Figure 9.26). Moreover, water bodies are smooth within the microwave frequency band and thus their emission signal is strongly polarized when compared with soil and vegetated surfaces (Schmugge, 1987; Choudhury, 1989). Over land and at low frequencies (≤37 GHz), water vapor and precipitation emission/scattering are insignificant relative to land surface emissions. Therefore, unlike with optical observations, the delineation of water bodies is not limited to clear sky conditions, but for the microwave frequency range, the spatial resolution is typically coarser than 10 km for currently operational spaceborne radiometers. This resolution is often restrictive and limits the application of microwave RS to very large rivers and lakes.

9.6.2.2 Retrieval of Surface Water Storage and Fluxes

Flood mapping from space was first accomplished with optical observations (0.5–1.1 μm frequency range) obtained with the multispectral scanner system (MSS) sensors onboard of the Landsat-1 satellite (Rango and Anderson, 1974; Rango and Salmonson, 1974; McGinnis and Rango, 1975). In this pioneering work flooded areas were mapped by applying a surface reflectance threshold below which water absorption is strong in the NIR region. More recently, the NDVI has been applied as ancillary information to improve the thresholding scheme (Brakenridge and Anderson, 2006).

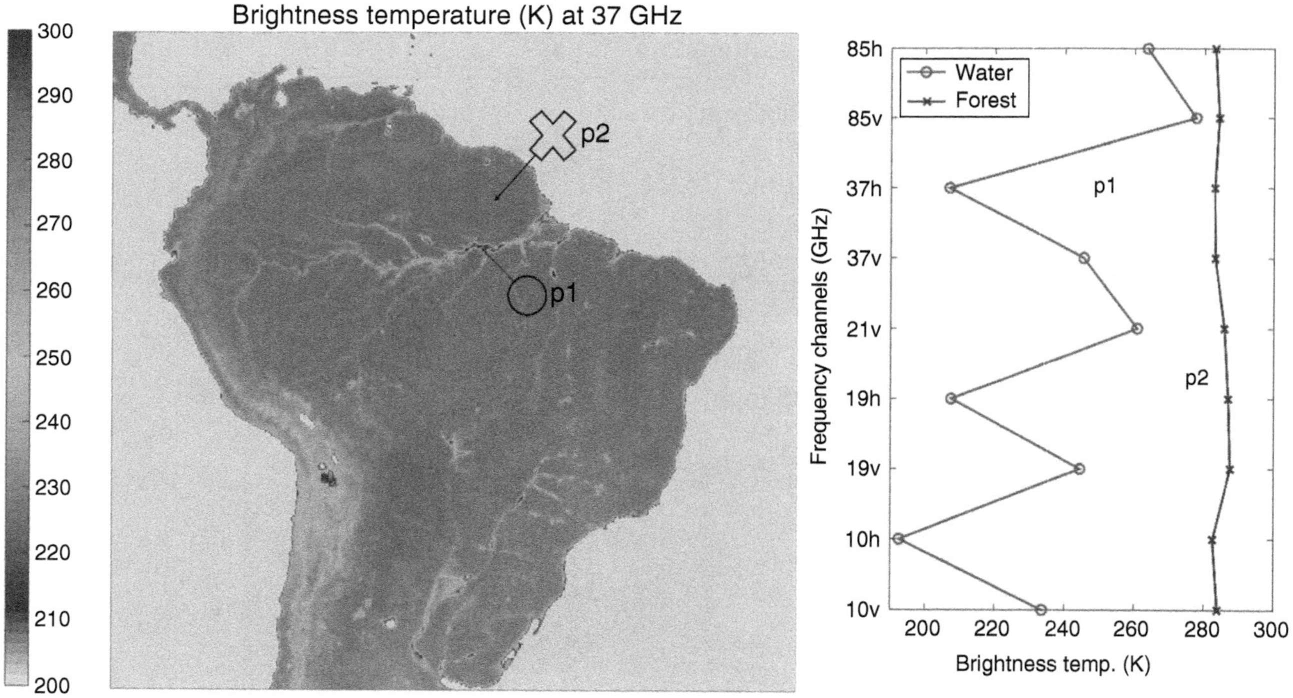

Figure 9.26 Water surfaces are radiometrically colder and more polarized within the microwave bands than bare soil and vegetated lands, allowing for delineation of surface water storage. Average brightness temperatures at 37 GHz obtained from all inner-swath overpasses of the radiometer onboard the TRMM satellite (left) and multifrequency brightness temperatures over water (p1) and forest (p2) (right). The rain scenes have been removed based on data from the onboard precipitation radar.

In the microwave frequency region, the temperature difference between vertical and horizontal polarizations at a single frequency is applied to detect inundated surfaces. For example, Giddings and Choudhury (1989) applied seasonal variability of the polarization signal at 37 GHz from the SMMR onboard the Nimbus-7 satellite for seasonal inundation mapping of rivers and wetlands in the Amazon basin.

While flood mapping techniques that rely on either optical or microwave observations are capable of capturing the approximate extent of inundated areas, estimation of depth and velocity of flowing water is at present not feasible with passive RS methods. This is why in situ measurements in conjunction with empirical or semiempirical relationships are applied to infer water fluxes from flow width. For example, regional and/or site-specific calibration of the well-known geomorphic relationship, $Q = aw^b$, has been widely applied to relate river discharge (Q) to remotely sensed width (w) (Gleason et al., 2014). However, river discharge is mainly a function of depth especially in valleys with steep banks. In other words, in many cases the river width does not substantially change in response to discharge variability and thus significant errors in the order of 40% have been reported for river discharge estimates. Incorporation of flow depth estimates can significantly reduce uncertainties of river discharge retrievals. To this end, Alsdorf and Lettenmaier (2003) suggested a topographic imager satellite mission with 100-m resolution to capture water depth variations in rivers and lakes. Alsdorf et al. (2007) introduced the idea of applying a Ka-band (35 GHz) radar interferometer to measure water surface elevation. Because of these proposals and the critical need for better characterization of space–time dynamics of surface water discharge at a global scale, in 2007 the National Research Council recommended that NASA designs a new Surface Water and Ocean Topography (SWOT) satellite mission that is scheduled to launch in 2020. A comprehensive review of SWOT objectives is provided in Biancamaria et al. (2016).

9.6.3 Evapotranspiration

ET combines evaporation from soil and water surfaces and plant transpiration. Evaporative fluxes from land and ocean surfaces to the atmosphere are among the most important components of the hydrologic cycle and play a crucial role in the climate system. To capture the high spatial and temporal variability of evapotranspirative fluxes, various monitoring and measurement techniques are applied. While ground-based measurement methods such as weighing lysimetry or eddy covariance systems provide relatively accurate estimates of ET, they are limited to small scales and require expensive and time-consuming maintenance. Because of this, large-scale RS methods that, as discussed below, may be classified into three main categories have received increased attention in recent years.

9.6.3.1 Physical Methods Based on the Surface Energy Balance

Physical methods commonly incorporate remotely sensed land surface parameters such as LST, surface albedo, or vegetation index into surface energy balance (SEB) models. For each pixel of a satellite image, the SEB models estimate ET or the latent heat flux LE (E is the ET rate and L is the latent heat of vaporization) as a residual component of the SEB by subtracting the ground heat (G) and sensible heat (H) fluxes from the net radiation (I_n) at the surface:

$$\mathrm{LE} = I_\mathrm{n} - G - H \tag{9.56}$$

The most widely applied SEB models include the Surface Energy Balance Algorithm for Land (SEBAL) (Bastiaanssen et al., 1998, Bastiaanssen, 2000), the Mapping EvapoTranspiration at high Resolution using Internalized Calibration (METRIC) algorithm (Allen et al., 2005, 2007, 2011), the Two-Source Energy Balance (TSEB) algorithm (Kustas et al., 2004), the Surface Energy Balance System (SEBS) (Su, 2002), the Simplified Surface Energy Balance Index (S-SEBI) (Roerink et al., 2000), the Simplified Surface Energy Balance (SSEB) method (Senay et al., 2007), and the operational Simplified Surface Energy Balance (SSEBop) method (Senay et al., 2013). These algorithms mainly differ in the parameterization of the terms in Eq. (9.56) with RS observations and meteorological data. A common approach is to apply the *hot* and *cold* pixel concept of Bastiaanssen et al. (1998). To estimate I_n, data for incoming and outgoing radiation, surface albedo, and emissivity fractions for shortwave and longwave bands are required. The G term can be estimated based on LST observations, albedo, and NDVI. The H term is intimately linked to the temperature gradient above the surface, surface roughness, and wind speed (Senay et al., 2007). To provide examples for physically based ET RS methods, the SEBS and S-SEBI algorithms are discussed below.

9.6.3.1.1 The SEBS Algorithm

The SEBS algorithm (Su, 2002) calculates the net radiation from incoming and outgoing short- and longwave radiation:

$$I_\mathrm{n} = (1-\omega) I_\mathrm{s} + \varepsilon (I_\mathrm{l} - \beta) \tag{9.57}$$

where I_s is the incoming shortwave (solar) radiation, ω is the land surface albedo, I_l is the incoming longwave radiation, β is the outgoing longwave land surface radiation (i.e. blackbody radiation; Eq. (9.5)), and ε is the LSE. The terms of Eq. (9.57) can be estimated from a combination of remotely sensed and local weather station data (Samani et al., 2007) or from RS data only (Bisht et al., 2005).

The ground heat flux in the SEBS algorithm is estimated as

$$G = I_n\left[\gamma_c + (1-\text{FVC})(\gamma_s - \gamma_c)\right] \tag{9.58}$$

It is assumed that the ratio of ground heat flux and net radiation for full vegetation canopy is $\gamma_c = 0.05$, and the ratio of ground heat flux and net radiation for bare soil is $\gamma_s = 0.315$. The FVC estimated with Eq. (9.53) is applied to interpolate between the two extremes.

The SEBS algorithm solves for actual ET (LE) in terms of the evaporative fraction (Λ) of a regional potential ET rate (LE_0) estimated with a model similar to the Penman–Monteith (PM) combination equation (Monteith, 1965):

$$\text{LE} = \Lambda \text{LE}_0 \tag{9.59}$$

The only unknown Λ is calculated for each pixel from comparison of LE (Eq. (9.56)) for dry and wet conditions. At the dry limit (maximum LST), LE in Eq. (9.56) is negligible and H is at its maximum (H_{dry}):

$$H_{\text{dry}} = I_n - G \tag{9.60}$$

At the wet limit (minimum LST), ET is at its potential rate (LE_0) and H is at its minimum (H_{wet}):

$$H_{\text{wet}} = I_n - G - \text{LE}_0 = H_{\text{dry}} - \text{LE}_0 \tag{9.61}$$

Combining Eqs. (9.56)–(9.58) yields

$$\Lambda = \frac{\text{LE}}{\text{LE}_0} = \frac{H_{\text{dry}} - H}{H_{\text{dry}} - H_{\text{wet}}} \tag{9.62}$$

The actual sensible heat flux (H) bounded by H_{dry} and H_{wet} is calculated as a function of LST as well as the near-surface air temperature and wind speed (i.e. eq. (5) in Su (2002)).

9.6.3.1.2 The S-SEBI Algorithm

In the Simplified-Surface Energy Balance Index (S-SEBI) algorithm developed by Roerink et al. (2000), the I_n term is calculated from the balance of all incoming and outgoing short- and longwave radiation (Eq. (9.57)). The G term is derived with an empirical relationship between vegetation and surface characteristics. The sensible and latent heat flux are not calculated as separate parameters, but as the evaporative fraction (Λ):

$$\Lambda = \frac{\text{LE}}{\text{LE} + H} = \frac{\text{LE}}{I_n - G} \tag{9.63}$$

The S-SEBI algorithm determines reflectance-dependent maximum and minimum LSTs for dry and wet conditions, respectively. For the dry condition LE is assumed to be zero; hence $\Lambda = 0$. For the wet condition H is assumed to be zero and $\text{LE}_0 = I_n - G$; hence $\Lambda = 1$. The evaporative fraction for each pixel is calculated based on the assumption of a linear relationship between Λ and LST:

$$\Lambda = \frac{\text{LST}_{\text{dry}} - \text{LST}}{\text{LST}_{\text{dry}} - \text{LST}_{\text{wet}}} \tag{9.64}$$

Therefore, the actual ET for each pixel can be estimated with the simplified relationship

$$\text{LE} = \Lambda \text{LE}_0 = \Lambda(I_n - G) \tag{9.65}$$

The simplified term in the S-SEBI model holds for the case where extreme temperatures can be determined from the image itself. This is feasible if the atmospheric conditions are constant for the imaged region of interest and a sufficient number of wet and dry pixels are captured. An advantage of the S-SEBI algorithm is that no ground-based meteorological measurements are required for parameterization. Furthermore, the extreme temperatures for wet and dry conditions vary with changing reflectance values, whereas other methods rely on single temperature limits for the entire image and/or land use class (Roerink et al., 2000).

The mathematical formulation of the SSEB model proposed by Senay et al. (2007) is similar to the S-SEBI algorithm. The SSEB implements a simplified version of the SEB approach to estimate actual ET while maintaining and extending the major assumptions in the SEBAL and METRIC algorithms. The SEBAL and METRIC methods apply a linear relationship between the near-surface temperature difference and the LST to estimate the sensible heat flux and assume that the *hot* pixel experiences no latent heat (i.e. LE = 0), while the *cold* pixel exhibits potential ET (LE_0). In addition, the SSEB algorithm assumes that the latent heat flux (actual ET) linearly varies between the *hot* and *cold* pixels. Similar to the S-SEBI algorithm, these assumptions lead to Eqs. (9.64) and (9.65). However, rather than applying the approximate solution $\text{LE}_0 = I_n - G$, Senay et al. (2007) used the global potential ET (LE_0) from the archives of USGS/FEWS NET operational model output to generate the spatial distribution of actual ET from MODIS LST data and Eqs. (9.64) and (9.65).

9.6.3.2 Empirical Methods Based on Vegetation Indices

Empirical methods, extensively reviewed in Glenn et al. (2010), are based on empirical relationships between ground-measured ET and vegetation indices (VI) such as the NDVI (Fisher et al., 2008; Zhang et al., 2009), the enhanced vegetation index (EVI) (Leuning et al., 2005; Mu et al., 2007; Yuan et al., 2010), the normalized difference water index (NDWI) (Lu and Zhuang, 2010), or other biophysical vegetation properties (e.g. the LAI) that are correlated with a VI (Cleugh et al., 2007; Mu et al., 2007; Leuning et al., 2008). Vegetation indices are applied either to directly retrieve ET from empirical relationships between ET ground measurements or the evaporative fraction (Λ) and vegetation indices (Nishida et al., 2003) or to parameterize the conductance term of the PM equation (Leuning et al., 2008; Glenn et al., 2010, 2011). Because most of the SEB algorithms are difficult to parameterize and require ground meteorological data in addition to RS observations, empirical methods based on vegetation indices have received increased attention in recent years (Hunsaker et al., 2005a, b; Campos et al., 2010; Padilla et al., 2011). Furthermore, satellite sensors with TIR observation capabilities are still limited in number (Glenn et al., 2010). Several studies have concluded that despite their simplicity, VI-based methods in many cases yield more accurate ET estimates than the more complex SEB models (Zhang et al., 2010; McVicar and Jupp, 1999; Cleugh et al., 2007; Mu et al., 2007; Guerschman et al., 2009; Glenn et al., 2011; King et al., 2011). For example, Cleugh et al. (2007) illustrated failures of SEB methods due to translation of small LST errors to large estimation errors for the H term in Eq. (9.56). In contrast, the VI model adequately estimated the magnitude and seasonality of ET.

9.6.3.3 Hybrid Methods Based on the FVC–LST Space

Another category of methods combines a vegetation index or the remotely sensed FVC with LST observations to estimate the actual ET. At the core of these methods is the pioneering work of Price (1990), Moran et al. (1994), and Carlson et al. (1994). It has been demonstrated that if a sufficiently large number of pixels are present after cloud and surface water outlier removal, the distribution of all pixel data in the FVC–LST space exhibits the shape of trapezoid or a triangle that can be applied for estimation of actual ET (Moran et al., 1994) or surface SM (Carlson et al., 1994; Carlson, 2007). With a combination of several physical and empirical relationships, Moran et al. (1994) demonstrated that for a partially vegetated surface, the evaporative fraction Λ is linearly related to the near-surface temperature gradient ($\Delta T = \text{LST} - T_{\text{air}}$). With the assumption that T_{air} variations within the area of interest are negligible, Λ can be expressed as:

$$\Lambda = \frac{\text{LE}}{\text{LE}_0} = \frac{\Delta T_{\text{dry}} - \Delta T}{\Delta T_{\text{dry}} - \Delta T_{\text{wet}}} = \frac{\text{LST}_{\text{dry}} - \text{LST}}{\text{LST}_{\text{dry}} - \text{LST}_{\text{wet}}} \quad (9.66)$$

Equation (9.66) is essentially the same as Eq. (9.64). However, in the trapezoid model, LST_{dry} and LST_{wet} are specified for every given FVC level, rather than for the entire image. For a specific area (i.e. satellite scene), LST_{dry} and LST_{wet} can be obtained from the LST–FVC space that commonly forms the trapezoid (Figure 9.27). The upper (dry) and lower (wet) edges of the trapezoid are used to solve for LST_{dry} and LST_{wet} at any given FVC:

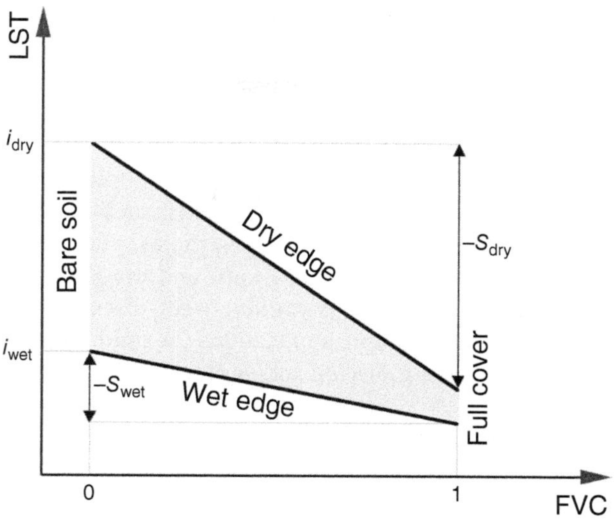

Figure 9.27 Sketch illustrating the trapezoid model parameters.

$$\text{LST}_{\text{dry}} = i_{\text{dry}} + s_{\text{dry}}\text{FVC} \quad (9.67)$$

$$\text{LST}_{\text{wet}} = i_{\text{wet}} + s_{\text{wet}}\text{FVC} \quad (9.68)$$

where i and s are the intercept and slope of the edges, respectively, and the subscripts *dry* and *wet* refer to the corresponding edge. Combining Eqs. (9.66)–(9.68) yields a simple model for estimation of Λ as a function of LST and FVC for each pixel:

$$\Lambda = \frac{i_{\text{dry}} + s_{\text{dry}}\text{FVC} - \text{LST}}{i_{\text{dry}} - i_{\text{wet}} + \left(s_{\text{dry}} - s_{\text{wet}}\right)\text{FVC}} \quad (9.69)$$

Surface SM can be also estimated from the trapezoid model when a linear relationship between Λ and surface SM (e.g. $\Lambda = \theta/\theta_s$) is assumed. The trapezoid model allows computationally efficient estimation of surface SM and ET as the pixel distribution itself is used for the solution and no ancillary atmospheric or surface data is required (Carlson, 2007). For uniform FVC Eq. (9.69) reduces to Eq. (9.64). Therefore, more accurate ET estimates are expected from the trapezoid model than from the SEB models that apply Eq. (9.64) without the FVC dependency of LST (e.g. Senay et al., 2007). Various recent studies have successfully applied the VI–LST concept in various forms for RS of ET (Tang et al., 2010; Long and Singh, 2012a, b). Interested readers are referred to the reviews of Carlson (2007) and Petropoulos et al. (2009).

9.7 Summary

The fundamentals of RS techniques for quantifying relevant environmental variables and fluxes at various scales were presented. Based on absorption, emission, and scattering processes, simple forms of the RTE were illustrated to elucidate links between remote sensor observations and land surface properties. It was shown that the terms of the RTE are based on fundamental radiation laws such as discovered and derived by Gustav Kirchhoff, Max Planck, Josef Stefan, Ludwig Boltzmann, Willebrord Snellius, and Augustin-Jean Fresnel. In addition, less complex approaches such as the *Plate Reflectance and Transmittance Model* and the *Kubelka-Munk Radiative Transfer Equation* were discussed in conjunction with example applications for estimation of soil and vegetation moisture and vegetation chlorophyll content.

RS hardware including air- and spaceborne sensors applicable for environmental observations were introduced, and underlying information about current optical and thermal satellite RS missions as well as about microwave satellite sensors for global mapping of SM and precipitation was provided. Furthermore, airborne sensors onboard manned aircrafts (i.e. HyMap, AirMOSS) and UAVs were discussed.

General relationships between land surface properties and RS observations were elucidated, and the current state of RS applications for estimation of static land surface properties such as mineralogical composition, soil texture, and hydraulic functions was illustrated. The "split-window" algorithm for LST retrieval from TIR observations was explained and current empirical and RTE-based techniques for estimation of SM and snow cover from remotely sensed optical and microwave data were introduced. It was illustrated how the inversion of a radiative transfer model (e.g. the "PROSPECT" model) can provide reliable physically based estimates for biophysicochemical properties of plant canopies.

Finally, the underlying concepts for RS of important components of the water balance, in specific the radiometric signatures and retrieval techniques for precipitation, ET, and surface water storage, were introduced.

Acknowledgments

The authors gratefully acknowledge support from the National Science Foundation (NSF) under grant no. 1521469 and from the Arizona and Utah Agricultural Experiment Stations.

Abbreviations

AirMOSS	Airborne Microwave Observatory of Subcanopy and Subsurface
ALOS-2	Advanced Land Observation Satellite 2
AMSR-2	Advanced Microwave Scanning Radiometer 2
ASCAT	Advanced scatterometer
ASTM	American Society for Testing and Materials
CEC	Cation exchange capacity
DN	Digital number
DPR	Dual-frequency precipitation radar
EM	Electromagnetic
ESA	European Space Agency
E-SAR	Experimental synthetic aperture radar
ET	Evapotranspiration
ETM^+	Enhanced Thematic Mapper Plus
FVC	Fractional vegetation cover
GOES	Geostationary Operational Environmental Satellite System
GPM	Global precipitation measurement
GMI	GPM Microwave Imager
HyMap	Hyperspectral Mapper
IUSS	International Union of Soil Sciences
IWS	Interferometric wide swath
JAXA	Japanese Aerospace Exploration Agency
KM	Kubelka–Munk
LAI	Leaf area index
Landsat	Land satellite
LSE	Land surface emissivity
LST	Land surface temperature
LSWA	Linear split-window algorithm
MA	Multiangle
MC	Multichannel
METRIC	Mapping EvapoTranspiration at high Resolution using Internalized Calibration
MODIS	Moderate Resolution Imaging Spectroradiometer
MSS	Multispectral scanner
NASA	National Aeronautics and Space Administration
NDSI	Normalized difference snow index
NDVI	Normalized difference vegetation index
NIR	Near infrared
NLSWA	Nonlinear split-window algorithms
OLI	Optical land imager
PLMR	Polarimetric L-band microwave radiometer
PROSAIL	PROSPECT + SAIL
PROSPECT	Properties spectra
RE	Richards' equation

RS	Remote sensing
RTE	Radiative transfer equation
SAIL	Scattering by arbitrary inclined leaves
SAR	Synthetic aperture radar
SC	Single-channel
SEB	Surface energy balance
SEBAL	Surface energy balance algorithm for land
SEBS	Surface energy balance system
SEVIRI	Spinning enhanced visible and infrared imager
SHPs	Soil hydraulic properties
SIERRA	Sensor Integrated Environmental Remote Research Aircraft
SMAP	Soil Moisture Active Passive
SMMR	Scanning multichannel microwave radiometer
SMOS	Soil Moisture and Ocean Salinity
SOM	Soil organic matter
SRF	Spectral response function
SSEB	Simplified surface energy balance
S-SEBI	Simplified surface energy balance index
SSEBop	Operational simplified surface energy balance
SSM/I	Special sensor microwave imager
SW	Split-window
SWIR	Shortwave infrared
SWOT	Surface water and ocean topography
TIR	Thermal infrared
TIRS	Thermal infrared sensor
TM	Thematic mapper
TRMM	Tropical Rainfall Measuring Mission
TSEB	Two-source energy balance
TVX	Temperature–vegetation index
UAS	Unmanned aerial system
UAV	Unmanned aerial vehicle
USDA	US Department of Agriculture
VI	Vegetation index
Vis	Visible

Physical Constants and Variables

Constant/variable	Symbol	Dimension	SI unit
Absorption coefficient	k	L^{-1}	m^{-1}
Absorption opacity	α	—	—
Absolute temperature	T_a	K	K
Air temperature	T_{air}	K	K
Angle of incidence/zenith angle	φ	—	rad
Blackbody radiation	β	MT^{-3}	$W\,m^{-2}\,sr^{-1}$
Boltzmann constant	κ	$ML^2T^{-2}K^{-1}$	$J\,K^{-1}$
Brightness temperature	T_b	K	K
Emission coefficient	e	L^{-1}	m^{-1}
Emissivity	ε	—	—
Evaporative fraction	Λ	—	—
Frequency	f	T^{-1}	Hz
Ground heat flux	G	MT^{-3}	$W\,m^{-2}$
Land surface temperature	LST	K	K
Latent heat flux	LE	MT^{-3}	$W\,m^{-2}$
Net radiation flux	I_n	MT^{-3}	$W\,m^{-2}$
Planck constant	h	ML^2T^{-1}	$J\,s^{-1}$
Reflectance	R	—	—
Reflectivity	ρ	—	—
Refractive index	n	—	—
Relative permittivity	ϵ	—	—
Scattering coefficient	s	L^{-1}	m^{-1}
Scattering source function	J	MT^{-2}	$W\,m^{-2}\,sr^{-1}\,Hz^{-1}$
Sensible heat flux	H	MT^{-3}	$W\,m^{-2}$
Spatial coordinate	x, z	L	m
Spectral radiance	I	MT^{-2}	$W\,m^{-2}\,sr^{-1}\,Hz^{-1}$
Spectral blackbody radiation	B	MT^{-2}	$W\,m^{-2}\,sr^{-1}\,Hz^{-1}$

(*Continued*)

Constant/variable	Symbol	Dimension	SI unit
Speed of light	c	LT^{-1}	$m\,s^{-1}$
Stefan–Boltzmann constant	σ	$MT^{-3}K^{-4}$	$W\,m^{-2}\,K^{-4}$
Surface albedo	ω	—	—
Transmittance	T	—	—
Transmissivity	τ	—	—
Transformed reflectance	r	—	—
Vegetation chlorophyll a and b content	C_{a+b}	ML^{-2}	$kg\,m^{-2}$
Vegetation water content	C_w	ML^{-2}	$kg\,m^{-2}$
Volumetric soil water content	θ	—	$m^3\,m^{-3}$
Wavelength	λ	L	m

References

Al Bitar, A., Leroux, D., Kerr, Y.H. et al. (2012). Evaluation of SMOS soil moisture products over continental US using the SCAN/SNOTEL network. *IEEE Transactions on Geoscience and Remote Sensing* 50 (5): 1572–1586.

Al-Abbas, A.H., Swain, P.H., and Baumgardner, M.F. (1972). Relating organic matter and clay content to the multispectral radiance of soils. *Soil Science* 114: 477–485.

Allen, W.A., Gausman, H.W., Richardson, A.J., and Thomas, J.R. (1969). Interaction of isotropic light with a compact plant leaf. *Journal of the Optical Society of America* 59 (10): 1376–1379.

Allen, R.G., Irmak, A., Trezza, R. et al. (2011). Satellite-based ET estimation in agriculture using SEBAL and METRIC. *Hydrological Processes* 25: 4011–4027.

Allen, W.A. and Richardson, A.J. (1968). Interaction of light with a plant canopy. *Journal of the Optical Society of America* 58 (8): 1023–1028.

Allen, R.G., Tasumi, M., Morse, A., and Trezza, R. (2005). A Landsat-based energy balance and evapotranspiration model in Western US water rights regulation and planning. *Irrigation and Drainage Systems* 19 (3–4): 251–268.

Allen, R.G., Tasumi, M., Morse, A. et al. (2007). Satellite-based energy balance for mapping evapotranspiration with internalized calibration (METRIC)-applications. *Journal of Irrigation and Drainage Engineering* 133: 395–406.

Alsdorf, D.E. and Lettenmaier, D.P. (2003). Tracking fresh water from space. *Science (New York, NY)* 301 (5639): 1491–1494.

Alsdorf, D.E., Rodriguez, E., and Lettenmaier, D.P. (2007). Measuring surface water from space. *Reviews of Geophysics* 45 (2): 1–24.

Ambrosia, V., Hutt, M., and Lulla, K. (2011b). Special issue: unmanned airborne systems (UAS) for remote sensing applications. *Geocarto International* 26 (2): 69–70.

Ambrosia, V.G., Wegener, S., Zajkowski, T. et al. (2011a). The Ikhana unmanned airborne system (UAS) western states fire imaging missions: from concept to reality (2006–2010). *Geocarto International* 26 (2): 85–101.

Anderson, K. and Gaston, K.J. (2013). Lightweight unmanned aerial vehicles will revolutionize spatial ecology. *Frontiers in Ecology and the Environment* 11 (3): 138–146.

Anderson, M.C., Norman, J.M., Kustas, W.P. et al. (2008). A thermal-based remote sensing technique for routine mapping of land-surface carbon, water and energy fluxes from field to regional scales. *Remote Sensing of Environment* 112 (12): 4227–4241.

Arkin, P.A. (1979). The relationship between fractional coverage of high cloud and rainfall accumulations during GATE over the B-scale array. *Monthly Weather Review* 107 (10): 1382–1387.

Asrar, G., Kanemasu, E.T., and Yoshida, M. (1985). Estimates of leaf area index from spectral reflectance of wheat under different cultural practices and solar angle. *Remote Sensing of Environment* 17: 1–11.

Atitar, M. and Sobrino, J.A. (2009). A split-window algorithm for estimating LST from Meteosat 9 data: test and comparison with in situ data and MODIS LSTs. *IEEE Geoscience and Remote Sensing Letters* 6 (1): 122–126.

Babaeian, E., Homaee, M., Montzka, C. et al. (2015b). Towards retrieving soil hydraulic properties by hyperspectral remote sensing. *Vadose Zone Journal* 14 (3). doi: 10.2136/vzj2014.07.0080.

Babaeian, E., Homaee, M., Vereecken, H. et al. (2015a). A comparative study of multiple approaches for predicting the soil–water retention curve: hyperspectral information vs. basic soil properties. *Soil Science Society of America Journal* 79: 1043–1058.

Bachmann, C.M., Donato, T.F., Lamela, G.M. et al. (2002). Automatic classification of land cover on Smith Island, VA, using HyMAP imagery. *IEEE Transactions on Geoscience and Remote Sensing* 40 (10): 2313–2330.

Balenzano, A., Mattia, F., Satalino, G. et al. (2012). SMOSAR algorithm for soil moisture retrieval using Sentinel-1 data. In: *2012 IEEE International Geoscience and Remote Sensing Symposium*, 1200–1203. Munich: IEEE.

Baret, F. and Guyot, G. (1991). Potentials and limits of vegetation indices for LAI and PAR assessment. *Remote Sensing of Environment* 35: 161–173.

Barrett, E.C. (1970). The estimation of monthly rainfall from satellite data. *Monthly Weather Review* 98 (4): 322–327.

Barrett, B.W., Dwyer, E., and Whelan, P. (2009). Soil moisture retrieval from active spaceborne microwave observations: an evaluation of current techniques. *Remote Sensing* 1 (3): 210–242.

Barron, V. and Torrent, J. (1986). Use of the Kubelka–Munk theory to study the influence of iron oxides on soil colour. *Journal of Soil Science* 37: 499–510.

Basharinov, A. and Shutko, A. (1975). Simulation studies of the SHF radiation characteristics of soils under moist conditions. NASA Transl. into ENGLISH of "Modelnyye Issledovaniya SVCh Radiatsionnykh Kharacteristik Pochvo-Gruntov v Usloviyakh Uvlazhneniya" (unpublished report) Moscow, Academy of Science of the USSR, Institute of Radio-Engineering and Electronics. 1–81.

Bastiaanssen, W.G.M. (2000). SEBAL-based sensible and latent heat fluxes in the irrigated Gediz Basin, Turkey. *Journal of Hydrology* 229: 87–100.

Bastiaanssen, W.G.M., Menenti, M., Feddes, R.A., and Holtslag, A.A.M. (1998). A remote sensing surface energy balance algorithm for land (SEBAL) 1. Formulation. *Journal of Hydrology* 212–213: 198–212.

Baumgardner, M.F., Kristof, S.J., Johannsen, C.J., and Zachary, A.L. (1970). Effects of organic matter on multispectral properties of soils. *Proc. Indiana Acad. Sci.* 79: 413–422.

Baumgardner, M.F., Silva, L.F., Biehl, L.L., and Stoner, E.R. (1985). Reflectance properties of soils. *Advances in Agronomy* 38: 2–44.

Bedini, E., van der Meer, F., and van Ruitenbeek, F. (2009). Use of HyMap imaging spectrometer data to map mineralogy in the Rodalquilar caldera, southeast Spain. *International Journal of Remote Sensing* 30 (2): 327–348.

Benali, A., Carvalho, A.C., Nunes, J.P. et al. (2012). Estimating air surface temperature in Portugal using MODIS LST data. *Remote Sensing of Environment* 124: 108–121.

Ben-Dor, E. and Banin, A. (1990). Near-infrared reflectance analysis of carbonate concentration in soils. *Applied Spectroscopy* 44 (6): 1064–1069.

Ben-Dor, E. and Banin, A. (1995). Near-infrared analysis as a rapid method to simultaneously evaluate several soil properties. *Soil Science Society of America Journal* 59: 364–372.

Berni, J.A.J., Zarco-Tejada, P.J., Suarez, L., and Fereres, E. (2009). Thermal and narrowband multispectral remote sensing for vegetation monitoring from an unmanned aerial vehicle. *IEEE Transactions on Geoscience and Remote Sensing* 47 (3): 722–738.

Biancamaria, S., Lettenmaier, D.P., and Pavelsky, T.M. (2016). The SWOT mission and its capabilities for land hydrology. In: *Remote Sensing and Water Resources* (ed. A. Cazenave, N. Champollion, J. Benveniste and J. Chen), 117–147. Cham: Springer International Publishing.

Bishop, J.L., Pieters, C.M., and Edwards, J.O. (1994). Infrared spectroscopic analyses on the nature of water in montmorillonite. *Clays and Clay Minerals* 42: 701–715.

Bisht, G., Venturini, V., Islam, S., and Jiang, L.E. (2005). Estimation of the net radiation using MODIS (moderate resolution imaging spectroradiometer) data for clear sky days. *Remote Sensing of Environment* 97 (1): 52–67.

Boegh, E., Houborg, R., Bienkowski, J. et al. (2013). Remote sensing of LAI, chlorophyll and leaf nitrogen pools of cropland grasslands in five European landscapes. *Biogeosciences* 10: 6279–6307.

Bowers, S.A. and Hanks, R.J. (1965). Reflection of radiant energy from soils. *Soil Science* 100 (2): 130–138.

Brakenridge, R. and Anderson, E. (2006). MODIS-based flood detection, mapping and measurement: the potential for operational hydrological applications. In: *Transboundary Floods: Reducing Risks Through Flood Management* (ed. J. Marsalek, G. Stancalie and G. Balint), 1–12. Dordrecht: Springer.

Brown, S.C., Quegan, S., Morrison, K. et al. (2003). High-resolution measurements of scattering in wheat canopies-implications for crop parameter retrieval. *IEEE Transactions on Geoscience and Remote Sensing* 41: 1602–1610.

Buckingham, E. (1907). *Studies on the movement of soil moisture*, USDA Bureau of Soils Bulletin, vol. 38. Washington, DC: US Government Printing Office.

Buckingham, W.F. and Sommer, S.E. (1983). Mineralogical characterization of rock surfaces formed by hydrothermal alteration and weathering – application to remote-sensing. *Economic Geology* 78 (4): 664–674.

Camillo, P.J., O'Neill, P.E., and Gurney, R.J. (1986). Estimating soil hydraulic parameters using passive microwave data. *IEEE Transactions on Geoscience and Remote Sensing* 24: 930–936.

Campos, I., Neale, C.M.U., Calera, A. et al. (2010). Assessing satellite-based basal crop coefficients for irrigated grapes (*Vitis vinifera L.*). *Agricultural Water Management* 98: 45–54.

Carlson, T. (2007). An overview of the triangle method for estimating surface evapotranspiration and soil moisture from satellite imagery. *Sensors* 7 (8): 1612–1629.

Carlson, T.N., Gillies, R.R., and Perry, E.M. (1994). A method to make use of thermal infrared temperature and NDVI measurements to infer surface soil water content and fractional vegetation cover. *Remote Sensing Reviews* 9: 161–173.

Carlson, T.N. and Ripley, D.A. (1997). On the relation between NDVI, fractional vegetation cover, and leaf area index. *Remote Sensing of Environment* 62 (3): 241–252.

Ceccato, P., Flasse, S., Tarantola, S. et al. (2001). Detecting vegetation leaf water content using reflectance in the optical domain. *Remote Sensing of Environment* 77 (1): 22–33.

Chan, S.K., Bindlish, R., O'Neill, P.E. et al. (2016). Assessment of the SMAP passive soil moisture product. *IEEE Transactions on Geoscience and Remote Sensing* 54 (8): 4994–5007.

Chang, A., Foster, J.L., and Hall, D.K. (1987). Nimbus-7 SMMR derived global snow cover parameters. *Annals of Glaciology* 9: 39–44.

Chang, C.W., Laird, D.A., Mausbach, M.J., and Hurburgh, C.R. (2001). Near-infrared reflectance spectroscopy-principal components regression analysis of soil properties. *Soil Science Society of America Journal* 65: 480–490.

Chapin, E., Chau, A., Chen, J. et al. (2012). AirMOSS: an airborne P-band SAR to measure root-zone soil moisture. In: *2012 IEEE Radar Conference*, 693–698. Atlanta, GA: IEEE.

Chen, D., Huang, J., and Jackson, T.J. (2005). Vegetation water content estimation for corn and soybeans using spectral indices derived from MODIS near- and short-wave infrared bands. *Remote Sensing of Environment* 98: 225–236.

Chen, F., Liu, Y., Liu, Q., and Qin, F. (2014). A statistical method based on remote sensing for the estimation of air temperature in China. *International Journal of Climatology* 35 (8): 2131–2143.

Choudhury, B.J. (1989). Monitoring global land surface using Nimbus-7 37 GHz data – theory and examples. *International Journal of Remote Sensing* 10 (10): 1579–1605.

Choudhury, B.J. and Chang, A.T.C. (1979). The solar reflectance of a snow field. *Cold Regions Science and Technology* 1 (2): 121–128.

Choudhury, B.J., Schmugge, T.J., Chang, A., and Newton, R.W. (1979). Effect of surface roughness on the microwave emission from soils. *Journal of Geophysical Research, Oceans* 84 (C9): 5699–5706.

Chrysoulakis, N. and Cartalis, C. (2002). Improving the estimation of land surface temperature for the region of Greece: adjustment of a split window algorithm to account for the distribution of perceptible water. *International Journal of Remote Sensing* 23 (5): 871–880.

Ciani, A., Goss, K.U., and Schwarzenbach, R.P. (2005). Light penetration in soil and particulate minerals. *European Journal of Soil Science* 56: 561–574.

Clark, R.N. (1999). *Spectroscopy of Rocks and Minerals, and Principles of Spectroscopy*. New York: Wiley.

Clark, R.N., King, T.V.V., Klejwa, M., and Swayze, G.A. (1990). High spectral resolution reflectance spectroscopy of minerals. *Journal of Geophysical Research* 12: 653–680.

Cleugh, H.A., Leuning, R., Mu, Q., and Running, S.W. (2007). Regional evaporation estimates from flux tower and MODIS satellite data. *Remote Sensing of Environment* 106: 285–304.

Cocks, T., Jenssen, R., Stewart, A. et al. (1998). The HyMap^{TM} airborne hyperspectral sensor: the system, calibration and performance. *Proceedings of the First EARSeL Workshop on Imaging Spectrometry*, Zurich (6–8 October), 37–42.

Cudahy, T.J. and Ramanaidou, E.R. (1997). Measurement of the hematite: goethite ratio using field visible and near-infrared reflectance spectrometry in channel iron deposits, Western Australia. *Australian Journal of Earth Sciences* 44 (4): 411–420.

Curcio, D., Ciraolo, G., D'Asaro, F., and Minacapilli, M. (2013). Prediction of soil texture distributions using VNIR–SWIR reflectance spectroscopy. *Procedia Environmental Sciences* 19: 494–503.

Dalal, R.C. and Henry, R.J. (1986). Simultaneous determination of moisture, organic carbon and total nitrogen by near infrared reflectance spectrophotometry. *Soil Science Society of America Journal* 50: 120–123.

Dall'Amico, J.T., Schlenz, F., Loew, A., and Mauser, W. (2012). First results of SMOS soil moisture validation in the upper Danube catchment. *IEEE Transactions on Geoscience and Remote Sensing* 50: 1507–1516.

Darvishzadeh, R., Atzberger, C., Skidmore, A., and Schlerf, M. (2011). Mapping grassland leaf area index with airborne hyperspectral imagery: a comparison study of statistical approaches and inversion of radiative transfer models. *ISPRS Journal of Photogrammetry and Remote Sensing* 66 (6): 894–906.

Darvishzadeh, R., Skidmore, A., Schlerf, M., and Atzberger, C. (2008). Inversion of a radiative transfer model for estimating vegetation LAI and chlorophyll in a heterogeneous grassland. *Remote Sensing of Environment* 112 (5): 2592–2604.

Dente, L., Su, Z., and Wen, J. (2012). Validation of SMOS soil moisture products over the Maqu and Twente regions. *Sensors* 12: 9965–9986.

Djamai, N., Magagi, R., Goita, K. et al. (2015). Evaluation of SMOS soil moisture products over the CanEx-SM10 area. *Journal of Hydrology* 520: 254–267.

Dobson, M.C., Ulaby, F.T., Hallikainen, M.T., and El-Rayes, M.A. (1985). Microwave dielectric behavior of wet soil-part II: dielectric mixing models. *IEEE Transactions on Geoscience and Remote Sensing* 23 (1): 35–46.

d'Oleire-Oltmanns, S., Marzolff, I., Peter, K.D., and Ries, J.B. (2012). Unmanned aerial vehicle (UAV) for monitoring soil erosion in Morocco. *Remote Sensing* 4 (11): 3390–3416.

Doner, H.E. and Lynn, W.C. (1989). Carbonate, halide, sulfate, and sulfide minerals. In: *Minerals in the Soil Environment*, Soil Science Society of America Book Series No. 1, 2e (ed. J.B. Dixon and S.B. Weed). Madison, WI: Soil Science Society of America.

Dozier, J. (1984). Snow reflectance from LANDSAT-4 thematic mapper. *IEEE Transactions on Geoscience and Remote Sensing* 22 (3): 323–328.

Dozier, J. (1989). Spectral signature of alpine snow cover from the Landsat Thematic Mapper. *Remote Sensing of Environment* 28: 9–22.

Du, C., Ren, H.Z., Qin, Q.M. et al. (2015). A practical split-window algorithm for estimating land surface temperature from Landsat 8 data. *Remote Sensing* 7 (1): 647–665.

Ebert, E.E., Manton, M.J., Arkin, P.A. et al. (1996). Results from the GPCP algorithm intercomparison programme. *Bulletin of the American Meteorological Society* 77 (12): 2875–2887.

Ebtehaj, A.M., Bras, R.L., and Foufoula-Georgiou, E. (2015). Shrunken locally linear embedding for passive microwave retrieval of precipitation. *IEEE Transactions on Geoscience and Remote Sensing* 53 (7): 3720–3736.

Edström, P. (2007). Examination of the revised Kubelka-Munk theory: considerations of modeling strategies. *Journal of the Optical Society of America* 24 (2): 548–556.

Entekhabi, D. and Njoku, E.G. (1994). Solving the inverse problem for soil moisture and temperature profiles by sequential assimilation of multifrequency remotely sensed observations. *IEEE Transactions on Geoscience and Remote Sensing* 32 (2): 438–448.

Entekhabi, D., Njoku, E.G., O'Neill, P.E. et al. (2010). The soil moisture active passive (SMAP) mission. *Proceedings of the IEEE* 98 (5): 704–716.

Fan, X.W., Tang, B.H., Wu, H. et al. (2015). Extension of the generalized split-window algorithm for land surface temperature retrieval to atmospheres with heavy dust aerosol loading. *IEEE Journal of Selected Topics in Applied Earth Observations and Remote Sensing* 8 (2): 825–834.

Fascetti, F., Pierdicca, N., Pulvirenti, L. et al. (2016). A comparison of ASCAT and SMOS soil moisture retrievals over Europe and Northern Africa from 2010 to 2013. *International Journal of Applied Earth Observation and Geoinformation* 45: 135–142.

Fisher, J.B., Tu, K.P., and Baldocchi, D.D. (2008). Global estimates of the land-atmosphere water flux based on monthly AVHRR and ISLSCP-II data, validated at 16 FLUXNET sites. *Remote Sensing of Environment* 112: 901–919.

Fladeland, M. (2009). *SIERRA: A New Unmanned Aircraft for Earth Science Investigations*. Moffett Field, CA: NASA Ames Research Center.

Friedl, M.A. and Davis, F.W. (1994). Sources of variation in radiometric surface temperature over a tallgrass prairie. *Remote Sensing of Environment* 48 (1): 1–17.

Fukshansky, L., Fukshansky-Kazarinova, N., Martinez, V., and Remisowsky, A. (1991). Estimation of optical parameters in a living tissue by solving the inverse problem of the multiflux radiative transfer. *Applied Optics* 30 (22): 3145–3153.

Gaffey, S.J. (1985). Reflectance spectroscopy in the visible and near infrared (0.35–2.55 μm): applications in carbonate petrology. *Geology* 13: 270–273.

Gamon, J.A., Field, C.B., Goulden, M.L. et al. (1995). Relationships between NDVI, canopy structure, and photosynthesis in three Californian vegetation types. *Ecological Applications* 5 (1): 28–41.

Gerberman, A.H. and Neher, D.D. (1979). Reflectance of varying mixtures of a clay soil and sand. *Photogrammetric Engineering and Remote Sensing* 45: 1145–1151.

Getirana, A.C.V., Boone, A., Yamazaki, D. et al. (2012). The hydrological modeling and analysis platform (HyMAP): evaluation in the Amazon basin. *Journal of Hydrometeorology* 13 (6): 1641–1665.

Giddings, L. and Choudhury, B.J. (1989). Observation of hydrological features with Nimbus-7 37 GHz data, applied to South America. *International Journal of Remote Sensing* 10 (10): 1673–1686.

Gilabert, M.A., González-Piqueras, J., Garcıa-Haro, F.J., and Meliá, J. (2002). A generalized soil-adjusted vegetation index. *Remote Sensing of Environment* 82 (2): 303–310.

Gleason, C.J., Smith, L.C., and Lee, J. (2014). Retrieval of river discharge solely from satellite imagery and at-many-stations hydraulic geometry: sensitivity to river form and optimization parameters. *Water Resources Research* 50 (12): 9604–9619.

Glenn, E.P., Doody, T.M., Guerschman, J.P. et al. (2011). Actual evapotranspiration estimation by ground and remote sensing methods: the Australian experience. *Hydrological Processes* 25: 4103–4116.

Glenn, E., Nagler, P., and Huete, A. (2010). Vegetation index methods for estimating evapotranspiration by remote sensing. *Surveys in Geophysics* 31: 531–555.

Good, E. (2015). Daily minimum and maximum surface air temperatures from geostationary satellite data. *Journal of Geophysical Research – Atmospheres* 120 (6): 2306–2324.

Grody, N.C. and Basist, A.N. (1996). Global identification of snowcover using SSM/I measurements. *IEEE Transactions on Geoscience and Remote Sensing* 34 (1): 237–249.

Guerschman, J.P., Van Dijk, A.I., Mattersdorf, G. et al. (2009). Scaling of potential evapotranspiration with MODIS data reproduces flux observations and catchment water balance observations across Australia. *Journal of Hydrology* 369: 107–119.

Guillen-Climent, M., Zarco-Tejada, P.J., Berni, J. et al. (2012). Mapping radiation interception in row-structured orchards using 3D simulation and high-resolution airborne imagery acquired from a UAV. *Precision Agriculture* 13 (4): 473–500.

Gutman, G. and Ignatov, A. (1998). The derivation of the green vegetation fraction from NOAA/AVHRR data for use in numerical weather prediction models. *International Journal of Remote Sensing* 19 (8): 1533–1543.

Hall, D.K., Riggs, G.A., Salomonson, V.V. et al. (2002). MODIS snow-cover products. *Remote Sensing of Environment* 83 (1): 181–194.

Hallikainen, M.T., Ulaby, F.T., Dobson, M.C. et al. (1985). Microwave dielectric behavior of wet soil – Part I: empirical models and experimental observations. *IEEE Transactions on Geoscience and Remote Sensing* 23 (1): 25–34.

Halthore, R.N., Eck, T.F., Holben, B.N., and Markham, B.L. (1997). Sun photometric measurements of atmospheric water vapor column abundance in the 940-nm band. *Journal of Geophysical Research* 102: 4343–4352.

Hansen, J., Ruedy, R., Sato, M., and Lo, K. (2010). Global surface temperature change. *Reviews of Geophysics* 48 (4).

Hardin, P.J. and Jensen, R.R. (2011). Small-scale unmanned aerial vehicles in environmental remote sensing: challenges and opportunities. *GIScience & Remote Sensing* 48: 99–111.

Hasan, S., Montzka, C., Rüdiger, C. et al. (2014). Soil moisture retrieval from airborne L-band passive microwave using high resolution multispectral data. *ISPRS Journal of Photogrammetry and Remote Sensing* 91: 59–71.

Hassan-Esfahani, L., Torres-Rua, A., Jensen, A., and McKee, M. (2015). Assessment of surface soil moisture using high-resolution multi-spectral imagery and artificial neural networks. *Remote Sensing* 7 (3): 2627–2646.

Hillel, D., Krentos, V.D., and Stylianou, Y. (1972). Procedure and test of an internal drainage method for measuring soil hydraulic characteristics in situ. *Soil Science* 114: 395–400.

Ho, H.C., Knudby, A., Sirovyak, P. et al. (2014). Mapping maximum urban air temperature on hot summer days. *Remote Sensing of Environment* 154: 38–45.

Hollenbeck, K.J., Schmugge, T.J., Hornberger, G.M., and Wang, J.R. (1996). Identifying soil hydraulic heterogeneity by detection of relative change in passive microwave remote sensing observations. *Water Resources Research* 32: 139–148.

Hong, Y., Hsu, K.L., Sorooshian, S., and Gao, X. (2004). Precipitation estimation from remotely sensed imagery using an artificial neural network cloud classification system. *Journal of Applied Meteorology* 43 (12): 1834–1853.

Horn, R., Nottensteiner, A., Reigber, A. et al. (2009). F-SAR – DLR's new multi-frequency polarimetric airborne SAR. In: *2009 IEEE International Geoscience and Remote Sensing Symposium*, University of Cape Town, Cape Town, South Africa, (12–17 July), 902–905.

Hou, A.Y., Kakar, R.K., Neeck, S. et al. (2014). The global precipitation measurement (GPM) mission. *American Meteorological Society* 95: 701–722.

Hsu, K.L., Gao, X., Sorooshian, S., and Gupta, H.V. (1997). Precipitation estimation from remotely sensed information using artificial neural networks. *Journal of Applied Meteorology* 36 (9): 1176–1190.

Huang, Z., Turner, B.J., Dury, S.J. et al. (2004). Estimating foliage nitrogen concentration from HYMAP data using continuum removal analysis. *Remote Sensing of Environment* 93 (1-2): 18–29.

Huete, A.R. (1988). A soil-adjusted vegetation index (SAVI). *Remote Sensing of Environment* 25: 295–309.

Hunsaker, D.J., Barnes, E.M., Clarke, T.R. et al. (2005b). Cotton irrigation scheduling using remotely sensed and FAO-56 basal crop coefficients. *Transactions of ASAE* 48: 1395–1407.

Hunsaker, D.J., Pinter, P.R., and Kimball, B.A. (2005a). Wheat basal crop coefficients determined by normalized difference vegetation index. *Irrigation Science* 24: 1–14.

Hunt, G.R. (1977). Spectral signatures of particulate minerals in the visible and near-infrared. *Geophysics* 42: 501–513.

Hunt, G.R. (1980). Spectroscopic properties of rock and minerals. In: *Handbook of Physical Properties Rocks* (ed. C.R. Stewart), 295. Boca Raton, FL: CRC Press.

Hunt, E.R. Jr., Doraiswamy, P.C., McMurtrey, J.E. et al. (2013). A visible band index for remote sensing leaf chlorophyll content at the canopy scale. *International Journal of Applied Earth Observation and Geoinformation* 21: 103–112.

Hunt, G.R. and Salisbury, J.W. (1970). Visible and near-infrared spectra of minerals and rocks. I. Silicate minerals. *Modern Geology* 1: 283–300.

Ines, A.V.M. and Mohanty, B.P. (2008a). Near-surface soil moisture assimilation for quantifying effective soil hydraulic properties under different hydroclimatic conditions. *Vadose Zone Journal* 7: 39–52.

Ines, A.V.M. and Mohanty, B.P. (2008b). Near-surface soil moisture assimilation to quantify effective soil hydraulic properties using genetic algorithm. 1. Conceptual modeling. *Water Resources Research* 44: W06422.

Ines, A.V.M. and Mohanty, B.P. (2009). Near-surface soil moisture assimilation to quantify effective soil hydraulic properties using genetic algorithm. 2. With air-borne remote sensing during SGP97 and SMEX02. *Water Resources Research* 45 (1): W01408.

Islam, K., Singh, B., and McBratney, A.B. (2003). Simultaneous estimation of various soil properties by ultra-violet, visible and near-infrared reflectance spectroscopy. *Australian Journal of Soil Research* 41: 1101–1114.

Jackson, J., Bindlish, R., Cosh, M.H. et al. (2012). Validation of soil moisture and ocean salinity (SMOS) soil moisture over watershed networks in the U.S. *IEEE Transactions on Geoscience and Remote Sensing* 50: 1530–1543.

Jackson, T.J. and Schmugge, T.J. (1991). Vegetation effects on the microwave emission of soils. *Remote Sensing of Environment* 36 (3): 203–212.

Jacquemoud, S. and Baret, F. (1990). PROSPECT: a model of leaf optical properties spectra. *Remote Sensing of Environment* 34: 75–91.

Jacquemoud, S. and Ustin, S. (2008). Environmental photobiology: Modelingleaf optical properties. *Photobiological Sciences Online*. http://www.photobiology.info/#Environ (accessed 13 January 2018).

Jacquemoud, S., Verhoef, W., Baret, F. et al. (2009). PROSPECT+SAIL models: a review of use for vegetation characterization. *Remote Sensing of Environment* 113: S56–S66.

Jagdhuber, T., Hajnsek, I., and Papathanassiou, K.P. (2015). An iterative generalized hybrid decomposition for soil moisture retrieval under vegetation cover using fully polarimetric SAR. *IEEE Journal of Selected Topics in Applied Earth Observations and Remote Sensing* 8 (8): 3911–3922.

Janatian, N., Sadeghi, M., Sanaeinejad, S.H. et al. (2016). A statistical framework for estimating air temperature using MODIS land surface temperature data. *International Journal of Climatology*. doi: 10.1002/joc.4766.

Jang, J.D., Viau, A.A., and Anctil, F. (2004). Neural network estimation of air temperatures from AVHRR data. *International Journal of Remote Sensing* 25 (21): 4541–4554.

Jensen, D.K., Tuller, M., de Jonge, L.W. et al. (2015). A new two-stage approach to predicting the soil water characteristic from saturation to oven-dryness. *Journal of Hydrology* 521: 498–507.

Jiang, G.M., Zhou, W., and Liu, R.G. (2013). Development of split-window algorithm for land surface temperature estimation from the VIRR/FY-3A measurements. *IEEE Geoscience and Remote Sensing Letters* 10 (4): 952–956.

Jiménez-Muñoz, J.C., Sobrino, J.A., Plaza, A. et al. (2009). Comparison between fractional vegetation cover retrievals from vegetation indices and spectral mixture analysis: case study of PROBA/CHRIS data over an agricultural area. *Sensors* 9 (2): 768–793.

Jiménez-Muñoz, J.C., Sobrino, J.A., Skoković, D. et al. (2014). Land surface temperature retrieval methods from Landsat-8 thermal infrared sensor data. *IEEE Geoscience and Remote Sensing Letters* 11 (10): 1840–1843.

Jonsson, I., Mattson, J.O., Okla, L., and Stridsberg, S. (1980). Photography and temperature measurements from a remotely piloted vehicle. *Oikos* 35: 120–125.

Joyce, R.J., Janowiak, J.E., Arkin, P.A., and Xie, P. (2004). CMORPH: a method that produces global precipitation estimates from passive microwave and infrared data at high spatial and temporal resolution. *Journal of Hydrometeorology* 5 (3): 487–503.

Kallel, A., Le Hégarat-Mascle, S., Ottlé, C., and Hubert-Moy, L. (2007). Determination of vegetation cover fraction by inversion of a four-parameter model based on isoline parametrization. *Remote Sensing of Environment* 111 (4): 553–566.

Kalma, J.D., McVicar, T.R., and McCabe, M.F. (2008). Estimating land surface evaporation: a review of methods using remotely sensed surface temperature data. *Surveys in Geophysics* 29: 421–469.

Karnieli, A., Agam, N., Pinker, R.T. et al. (2010). Use of NDVI and land surface temperature for drought assessment: merits and limitations. *Journal of Climate* 23 (3): 618–633.

Kawamura, K., Akiyama, T., Yokota, H.O. et al. (2005). Quantifying grazing intensities using geographic information systems and satellite remote sensing in the Xilingol steppe region, Inner Mongolia, China. *Agriculture, Ecosystems and Environment* 107 (1): 83–93.

Kerr, Y.H., Lagouarde, J.P., and Imbernon, J. (1992). Accurate land surface-temperature retrieval from AVHRR data with use of an improved split window algorithm. *Remote Sensing of Environment* 41 (2-3): 197–209.

Kerr, Y.H., Waldteufel, P., Wigneron, J.P. et al. (2001). Soil moisture retrieval from space: the soil moisture and ocean salinity (SMOS) mission. *IEEE Transactions on Geoscience and Remote Sensing* 39: 1729–1735.

Kerr, Y.H., Waldteufel, P., Wigneron, J.P. et al. (2010). The SMOS mission: new tool for monitoring key elements of the global water cycle. *Proceedings of the IEEE* 98 (5): 666–687.

King, E., van Niel, T., van Dijk, A. et al. (2011). *Actual Evapotranspiration Estimates for Australia*. Canberra, Australia: Intercomparison and evaluation.

Knadel, M., Deng, F., Alinejadian, A. et al. (2014). The effects of moisture conditions – from wet to hyper dry – on visible near-infrared spectra of Danish reference soils. *Soil Science Society of America Journal* 78: 422–433.

Kogan, F., Stark, R., Gitelson, A. et al. (2004). Derivation of pasture biomass in Mongolia from AVHRR-based vegetation health indices. *International Journal of Remote Sensing* 25 (14): 2889–2896.

Kool, J.B., Parker, J.C., and van Genuchten, M.T. (1985). Determining soil hydraulic properties from one-step outflow experiments by parameter estimation: I. Theory and numerical studies. *Soil Science Society of America Journal* 49: 1348–1354.

Kornelsen, K.C. and Coulibaly, P. (2014). Data-based disaggregation of SMOS soil moisture. In: *2014 IEEE Geoscience and Remote Sensing Symposium*, 3334–3337. IEEE: Quebec city, QC.

van den Kroonenberg, A., Martin, S., Beyrich, F., and Bange, J. (2012). Spatially averaged temperature structure parameter over a heterogeneous surface measured by an unmanned aerial vehicle. *Boundary-Layer Meteorology* 142 (1): 55–77.

Kubelka, P. and Munk, F. (1931). Ein Beitrag zur Optik der Farbanstriche. *Zeitschrift für Technische Physik* 12: 593–601.

Kummerow, C., Oison, W.S., and Giglio, L. (1996). A simplified scheme for obtaining precipitation and vertical hydrometeor profiles from passive microwave sensors. *IEEE Transactions on Geoscience and Remote Sensing* 34 (5): 1213–1232.

Kummerow, C.D., Randel, D.L., Kulie, M. et al. (2015). The evolution of the Goddard profiling algorithm to a fully parametric scheme. *Journal of Atmospheric and Oceanic Technology* 32 (12): 2265–2280.

Kustas, W.P., Norman, J.M., Schmugge, T.J., and Anderson, M.C. (2004). Mapping surface energy fluxes with radiometric temperature. In: *Thermal Remote Sensing in Land Surface Processes* (ed. D. Quattrochi and J. Luvall), Chapter 7, 205–253. Boca Raton, FL: CRC Press.

Laurent, V.C., Verhoef, W., Damm, A. et al. (2013). A Bayesian object-based approach for estimating vegetation biophysical and biochemical variables from APEX at-sensor radiance data. *Remote Sensing of Environment* 139: 6–17.

Leuning, R., Cleugh, H.A., Zegelin, S.J., and Hughes, D. (2005). Carbon and water fluxes over a temperate Eucalyptus forest and a tropical wet/dry savanna in Australia: measurements and comparison with MODIS remote sensing estimates. *Agricultural and Forest Meteorology* 129: 151–173.

Leuning, R., Zhang, Y., Rajaud, A. et al. (2008). A simple surface conductance model to estimate regional evaporation using MODIS leaf area index and the Penman–Monteith equation. *Water Resources Research* 44 (17): W10419.

Li, Z.L., Tang, B.H., Wu, H. et al. (2013). Satellite-derived land surface temperature: current status and perspectives. *Remote Sensing of Environment* 131: 14–37.

Liu, Q. and Weng, F. (2013). Using advanced matrix operator (AMOM) in community radiative transfer model. *IEEE Journal of Selected Topics in Applied Earth Observations and Remote Sensing* 6 (3): 1211–1218.

Lobell, D.B. and Asner, G.P. (2002). Moisture effects on soil reflectance. *Soil Science Society of America Journal* 66 (3): 722–727.

Long, D. and Singh, V.P. (2012a). A two-source trapezoid model for evapotranspiration (TTME) from satellite imagery. *Remote Sensing of Environment* 121: 370–388.

Long, D. and Singh, V.P. (2012b). A modified surface energy balance algorithm for land (M-SEBAL) based on a trapezoidal framework. *Water Resources Research* 48: W02528.

Louvet, S., Pellarin, T., al Bitar, A. et al. (2015). SMOS soil moisture product evaluation over West-Africa from local to regional scale. *Remote Sensing of Environment* 156: 383–394.

Lu, X. and Zhuang, Q. (2010). Evaluating evapotranspiration and water-use efficiency of terrestrial ecosystems in the conterminous United States using MODIS and AmeriFlux data. *Remote Sensing of Environment* 114: 1924–1939.

Maheu, B., Letoulouzan, J.N., and Gouesbet, G. (1984). Four-flux models to solve the scattering transfer equation in terms of Lorenz–Mie parameters. *Applied Optics* 23: 3353–3362.

Margate, D.E. and Shrestha, D.P. (2001). The use of hyperspectral data in identifying "desert-like" soil surface features in Tabernas area, southeast Spain. *Paper Presented at the 22nd Asian Conference on Remote Sensing.* 5: 9.

Martin, S., Bange, J., and Beyrich, F. (2011). Meteorological profiling of the lower troposphere using the research UAV M2AV Carolo. *Atmospheric Measurement Techniques* 4: 705–716.

Martin, P.D., Malley, D.F., Manning, G., and Fuller, L. (2002). Determination of soil organic carbon and nitrogen at the field level using near-infrared spectroscopy. *Canadian Journal of Soil Science* 82: 413–422.

Mattikalli, N.M., Engman, E.T., Jackson, T.J., and Ahuja, L.R. (1998). Microwave remote sensing of temporal variations of brightness temperature and near surface soil water content during a watershed-scale field experiment, and its application to the estimation of soil physical properties. *Water Resources Research* 34: 2289–2299.

Mayer, S., Sandvik, A., Jonassen, M.O., and Reuder, J. (2012). Atmospheric profiling with the UAS SUMO: a new perspective for the evaluation of fine-scale atmospheric models. *Meteorology and Atmospheric Physics* 116 (1-2): 15–26.

McGinnis, D.F. and Rango, A. (1975). Earth resources satellite systems for flood monitoring. *Geophysical Research Letters* 2 (4): 132–135.

McMillin, L.M. (1975). Estimation of sea surface temperatures from two infrared window measurements with different absorption. *Journal of Geophysical Research* 80 (36): 5113–5117.

McNairn, H., Jackson, T.J., Wiseman, G. et al. (2015). The soil moisture active passive validation experiment 2012 (SMAPVEX12): prelaunch calibration and validation of the SMAP soil moisture algorithms. *IEEE Transactions on Geoscience and Remote Sensing* 53: 2784–2801.

McVicar, T.R. and Jupp, D.L.B. (1999). Estimating one-time-of-day meteorological data from standard daily data as inputs to thermal remote sensing based energy balance models. *Agricultural and Forest Meteorology* 96: 219–238.

Merlin, O., Escorihuela, M.J., Mayoral, M.A. et al. (2013). Self-calibrated evaporation-based disaggregation of SMOS soil moisture: an evaluation study at 3 km and 100 m resolution in Catalunya, Spain. *Remote Sensing of Environment* 130: 25–38.

Merlin, O., Rüdiger, C., Al Bitar, A. et al. (2012). Disaggregation of SMOS soil moisture in southeastern Australia. *IEEE Transactions on Geoscience and Remote Sensing* 99: 1–16.

Merlin, O., Walker, J.P., Chehbouni, A., and Kerr, Y. (2008). Towards deterministic downscaling of SMOS soil moisture using MODIS derived soil evaporative efficiency. *Remote Sensing of Environment* 112 (10): 3935–3946.

Mo, T., Choudhury, B.J., Schmugge, T.J. et al. (1982). A model for microwave emission from vegetation-covered fields. *Journal of Geophysical Research* 87 (1): 11229–11237.

Moghaddam, M., Saatchi, S., and Cuenca, R.H. (2000). Estimating subcanopy soil moisture with radar. *Journal of Geophysical Research – Atmospheres* 105 (D11): 14899–14911.

Mohanty, B.P. (2013). Soil hydraulic property estimation using remote sensing: a review. *Vadose Zone Journal* 12 (4). doi: https://doi.org/10.2136/vzj2013.06.0100.

Monteith, J.L. (1965). Evaporation and environment. *Symposium of the Society of Experimental Biology* 19: 205–234.

Montzka, C., Bogena, H., Jagdhuber, T. et al. (2014). Active and passive L-band microwave remote sensing for soil moisture; a test-bed for SMAP fusion algorithms. In: *2014 IEEE Geoscience and Remote Sensing Symposium*, 2427–2430. IEEE: Quebec city, QC.

Montzka, C., Grant, J.P., Moradkhani, H. et al. (2013). Estimation of radiative transfer parameters from L-band passive microwave brightness temperatures using advanced data assimilation. *Vadose Zone Journal* 12 (3).

Montzka, C., Moradkhani, H., Weihermüller, L. et al. (2011). Hydraulic parameter estimation by remotely-sensed top soil moisture observations with the particle filter. *Journal of Hydrology* 399 (3-4): 410–421.

Moran, M.S., Clarke, T.R., Inoue, Y., and Vidal, A. (1994). Estimating crop water deficit using the relation between surface-air temperature and spectral vegetation index. *Remote Sensing of Environment* 49 (3): 246–263.

Morris, R.V., Lauer, H.V., Lawson, C.A. et al. (1985). Spectral and other physicochemical properties of submicron powders of hematite (alpha-Fe_2O_3), maghemite (gamma-Fe_2O_3), magnetite (Fe_3O_4), goethite (alpha-FeOOH), and lepidocrocite (gamma-FeOOH). *Journal of Geophysical Research* 90: 3126–3144.

Moser, G., De Martino, M., and Serpico, S.B. (2015). Estimation of air surface temperature from remote sensing images and pixelwise modeling of the estimation uncertainty through support vector machines. *IEEE Journal of Selected Topics in Applied Earth Observations and Remote Sensing* 8: 332–349.

Mu, Q., Heinsch, F.A., Zhao, M., and Running, S.W. (2007). Development of a global evapotranspiration algorithm based on MODIS and global meteorology data. *Remote Sensing of Environment* 111: 519–536.

Neteler, M. (2010). Estimating daily land surface temperatures in mountainous environments by reconstructed MODIS LST data. *Remote Sensing* 2 (1): 333–351.

Niclòs, R., Rivas, R., García-Santos, V. et al. (2016). SMOS Level-2 soil moisture product evaluation in rain-fed croplands of the Pampean region of Argentina. *IEEE Transactions on Geoscience and Remote Sensing* 54 (1): 499–512.

Nishida, K., Nemani, R.R., Glassy, J.M., and Running, S.W. (2003). Development of an evapotranspiration index from aqua/MODIS for monitoring surface moisture status. *IEEE Transactions on Geoscience and Remote Sensing* 41: 493–501.

Njoku, E.G. and Entekhabi, D. (1996). Passive microwave remote sensing of soil moisture. *Journal of Hydrology* 184: 101–129.

Njoku, E.G. and Kong, J.A. (1977). Theory for passive microwave remote sensing of near-surface soil moisture. *Journal of Geophysical Research* 82 (20): 3108–3118.

Nolin, A.W. and Dozier, J. (1993). Estimating snow grain size using AVIRIS data. *Remote Sensing of Environment* 44 (2-3): 231–238.

Nolin, A.W., Dozier, J., and Mertes, L.A.K. (1993). Mapping alpine snow using a spectral mixture modelling technique. *Annals of Glaciology* 17: 121–124.

Padilla, F.L.M., González-Dugo, M.P., Gavilán, P., and Domínguez, J. (2011). Integration of vegetation indices into a water balance model to estimate evapotranspiration of wheat and corn. *Hydrology and Earth System Sciences* 15: 1213–1225.

Painter, T.H., Rittger, K., McKenzie, C. et al. (2009). Retrieval of subpixel snow covered area, grain size, and albedo from MODIS. *Remote Sensing of Environment* 113 (4): 868–879.

Paloscia, S., Pettinato, S., Santi, E. et al. (2013). Soil moisture mapping using Sentinel-1 images: algorithm and preliminary validation. *Remote Sensing of Environment* 134: 234–248.

Parrella, G., Al-Kahachi, N., Jagdhuber, T. et al. (2012). Ice volume characterization using long-wavelength airborne PolSAR data. In: *2012 IEEE International Geoscience and Remote Sensing Symposium*, 3245–3248. Munich: Germany.

Pearson, R.L. and Miller, L.D. (1972). Remote mapping of standing crop biomass for estimation of the productivity of the shortgrass prairie, Pawnee National Grasslands, Colorado. *Proceedings of the Eighth International Symposium on Remote Sensing of the Environment* (2–6 October), 1355–1379. Ann Arbor, MI: Environmental Research Institute of Michigan.

Perez, M.A., Zech, W.C., and Donald, W.N. (2015). Using unmanned aerial vehicles to conduct site inspections of erosion and sediment control practices and track project progression. *Transportation Research Record* 2528: 38–48.

Peters, A. and Durner, W. (2008). A simple model for describing hydraulic conductivity in unsaturated porous media accounting for film and capillary flow. *Water Resources Research* 44: W11417.

Petropoulos, G., Carlson, T.N., Wooster, M.J., and Islam, S. (2009). A review of T-s/VI remote sensing based methods for the retrieval of land surface energy fluxes and soil surface moisture. *Progress in Physical Geography* 33 (2): 224–250.

Petty, G.W. and Li, K. (2013). Improved passive microwave retrievals of rain rate over land and ocean. Part II: validation and Intercomparison. *Journal of Atmospheric and Oceanic Technology* 30 (11): 2509–2526.

Pierdicca, N., Fascetti, F., Pulvirenti, L. et al. (2015). Analysis of ASCAT, SMOS, in-situ and land model soil moisture as a regionalized variable over Europe and North Africa. *Remote Sensing of Environment* 170: 280–289.

Pierdicca, N., Pulvirenti, L., and Pace, G. (2014). A prototype software package to retrieve soil moisture from sentinel-1 data by using a Bayesian multitemporal algorithm. *IEEE Journal of Selected Topics in Applied Earth Observations and Remote Sensing* 7: 153–166.

Piles, M., Camps, A., Vall-Llossera, M. et al. (2011). Downscaling SMOS-derived soil moisture using MODIS visible/infrared data. *IEEE Transactions on Geoscience and Remote Sensing* 49 (9): 3156–3166.

Piles, M., Entekhabi, D., and Camps, A. (2009). A change detection algorithm for retrieving high-resolution soil moisture from SMAP radar and radiometer observations. *IEEE Transactions on Geoscience and Remote Sensing* 47: 4125–4131.

Piles, M., Sánchez, N., Vall-llossera, M. et al. (2014). A downscaling approach for SMOS land observations: evaluation of high-resolution soil moisture maps over the Iberian peninsula. *IEEE Journal of Selected Topics in Applied Earth Observations and Remote Sensing* 7 (9): 3845–3857.

Platt, U., Pfeilsticker, K., and Vollmer, M. (2007). Radiation and optics in the atmosphere. In: *Handbook of Lasers and Optics* (ed. F. Träger). Berlin: Springer.

Prata, A.J., Caselles, V., Coll, C. et al. (1995). Thermal remote sensing of land surface temperature from satellites: current status and future prospects. *Remote Sensing Reviews* 12 (3-4): 175–224.

Price, J.C. (1990). Using spatial context in satellite data to infer regional scale evapotranspiration. *IEEE Transactions on Geoscience and Remote Sensing* 28 (5): 940–948.

Pulliainen, J. and Hallikainen, M. (2001). Retrieval of regional snow water equivalent from space-borne passive microwave observations. *Remote Sensing of Environment* 75 (1): 76–85.

Qi, J., Chehbouni, A., Huete, A.R. et al. (1994). A modified soil adjusted vegetation index (MSAVI). *Remote Sensing of Environment* 48: 119–126.

Raggatt, T.J., Lewis, M., and Fitzpatrick, R.W. (2004). Spectral discrimination of soil and regolith attributes within Herrmanns Catchment, Mount Lofty Ranges, SA. In: *Regolith*, 287–291. CRC LEME.

Rango, A. and Anderson, A.T. (1974). Flood hazard studies in the Mississippi river basin using remote sensing. *Journal of the American Water Resources Association* 10 (5): 1060–1081.

Rango, A. and Salmonson, V. (1974). Regional flood mapping from space. *Water Resources Research* 10 (3): 1944–7973.

Rees, W.G. (2005). *Remote Sensing of Snow and Ice*. Boca Raton, FL: CRC Press, Taylor and Francis Group.

Reichle, R.H. and Koster, R.D. (2005). Global assimilation of satellite surface soil moisture retrievals into the NASA catchment land surface model. *Geophysical Research Letters* 32 (2).

Riaza, A., Buzzi, J., Garcia-Melendez, E. et al. (2012). River acid mine drainage: sediment and water mapping through hyperspectral HyMap data. *International Journal of Remote Sensing* 33 (19): 6163–6185.

Richards, L.A. (1931). Capillary conduction of liquids through porous mediums. *Journal of Applied Physics* 1 (5): 318–333.

Richardson, A.J. and Wiegand, C.L. (1977). Distinguishing vegetation from soil background information. *Photogrammetric Engineering and Remote Sensing* 43: 1541–1552.

Robinson, D.A., Campbell, C.S., Hopmans, J.W. et al. (2008). Soil moisture measurement for ecological and hydrological watershed-scale observatories: a review. *Vadose Zone Journal* 7 (1): 358–389.

Roerink, G.J., Su, Z., and Menenti, M. (2000). A simple remote sensing algorithm to estimate the surface energy balance. *Physics and Chemistry of the Earth* 25: 147–157.

Rondeaux, G., Steven, M., and Baret, F. (1996). Optimization of soil-adjusted vegetation indices. *Remote Sensing of Environment* 55: 95–107.

Rosenthal, W. and Dozier, J. (1996). Automated mapping of montane snow cover at subpixel resolution from the Landsat thematic mapper. *Water Resources Research* 32 (1): 115–130.

Rötzer, K., Montzka, C., Bogena, H. et al. (2014). Catchment scale validation of SMOS and ASCAT soil moisture products using hydrological modeling and temporal stability analysis. *Journal of Hydrology* 519: 934–946.

Rouse, J.W., Haas, R.H., Schell, J.A., and Deering, D.W. (1974). Monitoring vegetation systems in the Great Plains with ERTS. In: *Third Earth Resources Technology Satellite – 1 Syposium. Volume I: Technical Presentations* (eds S.C. Freden, E.P. Mercanti, and M. Becker), NASA SP-351, pp. 309–317. Washington, DC: NASA.

Sadeghi, M., Jones, S.B., and Philpot, W.D. (2015a). A linear physically-based model for remote sensing of soil moisture using short wave infrared bands. *Remote Sensing of Environment* 164: 66–76.

Sadeghi, M., Jones, S.B., and Tuller, M. (2015b). Toward a calibration-free model for optical remote sensing of soil moisture. *AGU Fall Meeting*, San Francisco, CA (14–18 December). doi: 10.13140/RG.2.1.1963.4004.

Sadeghi, M., Tuller, M., Gohardoust, M.R., and Jones, S.B. (2014). Column-scale unsaturated hydraulic conductivity estimates in coarse textured homogeneous and layered soils derived under steady-state evaporation from a water table. *Journal of Hydrology* 519: 1238–1248.

Salisbury, J.W. and Hunt, G.R. (1968). Martian surface materials: effect of particle size on spectral behavior. *Science* 161: 365–366.

Samani, Z., Bawazir, A.S., Bleiweiss, M. et al. (2007). Estimating daily net radiation over vegetation canopy through remote sensing and climatic data. *Journal of Irrigation and Drainage Engineering* 133 (4): 291–297.

Sanchez, N., Martínez-Fernández, J., Scaini, A., and Perez-Gutierrez, C. (2012). Validation of the SMOS L2 soil moisture data in the REMEDHUS network (Spain). *IEEE Transactions on Geoscience and Remote Sensing* 50 (5): 1602–1611.

Sandholt, I., Rasmussen, K., and Andersen, J. (2002). A simple interpretation of the surface temperature/vegetation index space for assessment of surface moisture status. *Remote Sensing of Environment* 79 (2–3): 213–224.

Sandoval, C. and Kim, A.D. (2014). Deriving Kubelka-Munk theory from radiative transport. *Journal of the Optical Society of America A* 31: 628–636.

Santra, P., Sahoo, R.N., Das, B.S. et al. (2009). Estimation of soil hydraulic properties using proximal spectral reflectance in visible, near-infrared, and shortwave-infrared (VIS–NIR–SWIR) region. *Geoderma* 152: 338–349.

Sarathjith, M.C., Das, B.S., Vasava, H.B. et al. (2014). Diffuse reflectance spectroscopic approach for the characterization of soil aggregate size distribution. *Soil Science Society of America Journal* 78: 369–376.

Schmugge, T. (1978). Remote sensing of surface soil moisture. *Journal of Applied Meteorology* 17 (10): 1549–1557.

Schmugge, T. (1987). Remote sensing applications in hydrology. *Reviews of Geophysics* 25 (2): 148–152.

Schmugge, T., O'Neill, P.E., and Wang, J.R. (1986). Passive microwave soil moisture research. *IEEE Transactions on Geoscience and Remote Sensing* 24 (1): 12–22.

Schreier, H. (1977). Quantitative predictions of chemical soil conditions from multispectral airborne ground and laboratory measurements. In: *Proceedings of the fourth Canadian Symposium on Remote Sensing*, 106–112. Ottawa, QC: Aeronautics and Space Institute.

Senay, G.B., Bohms, S., Singh, R.K. et al. (2013). Operational evapotranspiration mapping using remote sensing and weather datasets: a new parameterization for the SSEB approach. *Journal of the American Water Resources Association* 49: 577–591.

Senay, G.B., Budde, M., Verdin, J.P., and Melesse, A.M. (2007). A coupled remote sensing and simplified surface energy balance approach to estimate actual evapotranspiration from irrigated fields. *Sensors* 7 (6): 979–1000.

Shahbazi, M., Theau, J., and Menard, P. (2014). Recent applications of unmanned aerial imagery in natural resource management. *GIScience & Remote Sensing* 51 (4): 339–365.

Shao, M. and Horton, R. (1998). Integral method for estimating soil hydraulic properties. *Soil Science Society of America Journal* 62: 585–592.

Sherman, D.M. and Waite, T.D. (1985). Electronic spectra of Fe oxides and oxyhydroxides in the near infrared to ultraviolet. *American Mineralogist* 70: 1262–1269.

Simunek, J. and van Genuchten, M.T. (1996). Estimating unsaturated soil hydraulic properties from tension disc infiltrometer data by numerical inversion. *Water Resources Research* 32: 2683–2696.

Sinha, A.K. (1987). Variations in soil spectral reflectance related to soil moisture, organic matter and particle size. *Journal of the Indian Society of Remote Sensing* 15 (2): 7–11.

Sobrino, J.A., Li, Z.L., Stoll, M.P., and Becker, F. (1994). Improvements in the split-window technique for land surface temperature determination. *IEEE Transactions on Geoscience and Remote Sensing* 32 (2): 243–253.

Sobrino, J.A., Li, Z.-L., Stoll, M.P., and Becker, F. (1996). Multi-channel and multi-angle algorithms for estimating sea and land surface temperature with ATSR data. *International Journal of Remote Sensing* 17 (11): 2089–2114.

Sobrino, J.A. and Raissouni, N. (2000). Toward remote sensing methods for land cover dynamic monitoring: application to Morocco. *International Journal of Remote Sensing* 21 (2): 353–366.

Srivastava, P.K., Han, D., Ramirez, M.R., and Islam, T. (2013). Machine learning techniques for downscaling SMOS satellite soil moisture using MODIS land surface temperature for hydrological application. *Water Resources Management* 27 (8): 3127–3144.

Stamnes, K., Li, W., Eide, H. et al. (2007). ADEOS-II/GLI snow/ice products – part I: scientific basis. *Remote Sensing of Environment* 111 (2): 258–273.

Stenberg, B., Rossel, R.A.V., Mouazen, A.M., and Wetterlind, J. (2010). Visible and near infrared spectroscopy in soil science. In: *Advances in Agronomy*, vol. 107 (ed. D.L. Sparks), 163–215. Burlington, MA: Academic Press.

Stern, F. (1964). Transmission of isotropic radiation across an interface between two dielectrics. *Applied Optics* 3 (1): 111–113.

Stiles, W.H. and Ulaby, F.T. (1980). The active and passive microwave response to snow parameters: 1. Wetness. *Journal of Geophysical Research* 85: 1037–1044.

Stisen, S., Sandholt, I., Norgaard, A. et al. (2007). Estimation of diurnal air temperature using MSG SEVIRI data in West Africa. *Remote Sensing of Environment* 10 (2): 262–274.

Stokes, G.G. (1862). On the intensity of the light reflected from or transmitted through a pile of plates. *Proceedings of the Royal Society of London Series B* 11: 545–556.

Su, Z. (2002). The surface energy balance system (SEBS) for estimation of the turbulent heat fluxes. *Hydrology and Earth System Science* 6 (1): 85–99.

Summerell, G., Wilford, J., Shoemark, V. et al. (2009). High resolution passive microwave response to landscape controls influencing soil moisture patterns: a case study for the Livingstone Creek Catchment. *MODSIM Congress*, Cairns, Australia (13–17 July).

Sun, D. and Pinker, R.T. (2003). Estimation of land surface temperature from a geostationary operational environmental satellite (GOES-8). *Journal of Geophysical Research – Atmospheres* 108 (D11).

Sun, D. and Pinker, R.T. (2007). Retrieval of surface temperature from the MSG-SEVIRI observations: part I. Methodology. *International Journal of Remote Sensing* 28 (23): 5255–5272.

Sun, Y.J., Wang, J.F., Zhang, R.H. et al. (2005). Air temperature retrieval from RS data based on thermodynamics. *Theoretical and Applied Climatology* 80 (1): 37–48.

Tabatabaeenejad, A., Burgin, M., Duan, X., and Moghaddam, M. (2015). P-band radar retrieval of subsurface soil moisture profile as a second-order polynomial: first AirMOSS results. *IEEE Transactions on Geoscience and Remote Sensing* 53 (2): 645–658.

Tang, R., Li, Z.-L., and Tang, B. (2010). An application of the Ts–VI triangle method with enhanced edges determination for evapotranspiration estimation from MODIS data in arid and semi-arid regions: implementation and validation. *Remote Sensing of Environment* 114: 540–551.

Tao, W.K. (2003). Goddard cumulus ensemble (GCE) model: application for understanding precipitation processes. In: *Cloud Systems, Hurricanes, and the Tropical Rainfall Measuring Mission (TRMM): A Tribute to Dr. Joanne Simpson* (ed. W.-K. Tao and R. Adler), 107–138. Boston, MA: American Meteorological Society.

Thennadil, S.N. (2008). Relationship between the Kubelka–Munk scattering and radiative transfer coefficients. *Journal of the Optical Society of America. A* 25: 1480–1485.

Thorp, K.R., Wang, G., West, A.L. et al. (2012). Estimating crop biophysical properties from remote sensing data by inverting linked radiative transfer and ecophysiological models. *Remote Sensing of Environment* 124: 224–233.

Thorp, K.R., Gore, M.A., Andrade-Sanchez, P. et al. (2015). Proximal hyperspectral sensing and data analysis approaches for field-based plant phenomics. *Computers and Electronics in Agriculture* 118: 225–236.

Thulin, S., Hill, M.J., Held, A. et al. (2014). Predicting levels of crude protein, digestibility, lignin and cellulose in temperate pastures using hyperspectral image data. *American Journal of Plant Sciences* 5: 997–1019.

Tucker, C.J., Vanpraet, C.L., Boerwinkel, E., and Gaston, A. (1983). Satellite remote sensing of total dry matter production in the Senegalese Sahel. *Remote Sensing of Environment* 13: 461–474.

Tuller, M. and Or, D. (2005). Water retention and characteristic curve. In: *Encyclopedia of Soils in the Environment* (ed. D. Hillel). Oxford: Elsevier.

Tuller, M., Or, D., and Dudley, L.M. (1999). Adsorption and capillary condensation in porous media: liquid retention and interfacial configurations in angular pores. *Water Resources Research* 35 (7): 1949–1964.

Turner, D.P., Cohen, W.B., Kennedy, R.E. et al. (1999). Relationships between leaf area index and Landsat TM spectral vegetation indices across three temperate zone sites. *Remote Sensing of Environment* 70 (1): 52–68.

Ulaby, F. and Long, D. (2014). *Microwave Radar and Radiometric Remote Sensing*. Ann Arbor: University of Michigan Press.

van der Schalie, R., Parinussa, R.M., de Jeu, R.A.M. et al. (2015). SMOS soil moisture retrievals using the land parameter retrieval model: evaluation over the Murrumbidgee Catchment, southeast Australia. *Remote Sensing of Environment* 163: 70–79.

van Genuchten, M.T. (1980). A closed-form equation for predicting the hydraulic conductivity of unsaturated soils. *Soil Science Society of America Journal* 44: 892–898.

Vancutsem, C., Ceccato, P., Dinku, T., and Connor, S.J. (2010). Evaluation of MODIS land surface temperature data to estimate air temperature in different ecosystems over Africa. *Remote Sensing of Environment* 114 (2): 449–465.

Vargas, W.E. and Niklasson, G.A. (1997). Applicability conditions of the Kubelka–Munk theory. *Applied Optics* 36: 5580–5586.

Vauclin, M., Vieira, S.R., Bernard, R., and Hatfield, J.L. (1982). Spatial variability of surface temperature along two transects of a bare soil. *Water Resources Research* 18 (6): 1677–1686.

Vendrame, P.R.S., Marchão, R.L., Brunet, D., and Becquer, T. (2012). The potential of NIR spectroscopy to predict soil texture and mineralogy in Cerrado Latosols. *European Journal of Soil Science* 63 (5): 743–753.

Vereecken, H., Huisman, J.A., Pachepsky, Y. et al. (2014). On the spatio-temporal dynamics of soil moisture at the field scale. *Journal of Hydrology* 516: 76–96.

Vereecken, H., Weynants, M., Javaux, M. et al. (2010). Using pedotransfer functions to estimate the van Genuchten–Mualem soil hydraulic properties: a review. *Vadose Zone Journal* 9: 795–820.

Verhoef, W. (1984). Light scattering by leaf layers with application to canopy reflectance modeling: the SAIL model. *Remote Sensing of Environment* 16 (2): 125–141.

Vinogradov, B.V. (1981). Remote sensing of the humus content of soils. *Soil Science* 11: 114–123.

Viscarra Rossel, R.A., Walvoort, D.J.J., McBratney, A.B. et al. (2006). Visible, near infrared, mid infrared or combined diffuse reflectance spectroscopy for simultaneous assessment of various soil properties. *Geoderma* 131: 59–75.

Wagner, W., Brocca, L., Naeimi, V. et al. (2014). Clarifications on the comparison between SMOS, VUA, ASCAT, and ECMWF soil moisture products over four watersheds in US. *IEEE Transactions on Geoscience and Remote Sensing* 52 (3): 1901–1906.

Wan, Z. and Dozier, J. (1996). A generalized split-window algorithm for retrieving land-surface temperature from space. *IEEE Transactions on Geoscience and Remote Sensing* 34 (4): 892–905.

Wan, Z. and Li, Z.L. (2008). Radiance-based validation of the V5 MODIS land-surface temperature product. *International Journal of Remote Sensing* 29 (17-18): 5373–5395.

Wang, Y., Heron, M.L., Prytz, A. et al. (2007). Evaluation of a new airborne microwave remote sensing radiometer by measuring the salinity gradients across the shelf of the great barrier reef lagoon. *IEEE Transactions on Geoscience and Remote Sensing* 45 (11): 3701–3709.

Warren, S.G. (1982). Optical properties of snow. *Reviews of Geophysics* 20 (1): 67–89.

Watts, A.C., Ambrosia, V.G., and Hinkley, E.A. (2012). Unmanned aircraft systems in remote sensing and scientific research: classification and considerations of use. *Remote Sensing* 4 (6): 1671–1692.

Wegmuller, U., Mätzler, C., and Njoku, E.G. (1994). Canopy opacity model. In: *Passive Microwave Remote Sensing of Land-Atmosphere Interactions* (ed. B.J. Choudhury, Y.H. Kerr, E.G. Njoku and P. Pampaloni), 375–387. Utrecht, Netherland: VSP BV.

Weiss, M. and Baret, F. (1999). Evaluation of canopy biophysical variable retrieval performances from the accumulation of large swath satellite data. *Remote Sensing of Environment* 70 (3): 293–306.

Wetterlind, J. and Stenberg, B. (2010). Near-infrared spectroscopy for within field soil characterization: small local calibrations compared with national libraries spiked with local samples. *European Journal of Soil Science* 61: 823–843.

Wettle, M., Daniel, P.J., Logan, G.A., and Thankappan, M. (2009). Assessing the effect of hydrocarbon oil type and thickness on a remote sensing signal: a sensitivity study based on the optical properties of two different oil types and the HYMAP and Quickbird sensors. *Remote Sensing of Environment* 113 (9): 2000–2010.

Wigneron, J.P., Kerr, Y., Waldteufel, P. et al. (2007). L-band microwave emission of the biosphere (L-MEB) model: description and calibration against experimental data sets over crop fields. *Remote Sensing of Environment* 107 (4): 639–655.

Wing, M.G., Burnett, J., Sessions, J. et al. (2013). Eyes in the sky: remote sensing technology development using small unmanned aircraft systems. *Journal of Forestry* 111 (5): 341–347.

Wittich, K.P. and Hansing, O. (1995). Area-averaged vegetative cover fraction estimated from satellite data. *International Journal of Biometeorology* 38 (4): 209–215.

Yamada, N. and Fujimura, S. (1991). Nondestructive measurement of chlorophyll pigment content in plant leaves from three-color reflectance and transmittance. *Applied Optics* 30 (27): 3964–3973.

Yang, L. and Kruse, B. (2004). Revised Kubelka–Munk theory. I. Theory and applications. *Journal of the Optical Society of America A* 21: 1933–1941.

Yang, H. and Yang, Z.D. (2006). A modified land surface temperature split window retrieval algorithm and its applications over China. *Global and Planetary Change* 52 (1-4): 207–215.

Yilmaz, M.T., Hunt, E.R. Jr., and Jackson, T.J. (2008). Remote sensing of vegetation water content from equivalent water thickness using satellite imagery. *Remote Sensing of Environment* 112 (5): 2514–2522.

Yuan, W.P., Liu, S.G., Yu, G.R. et al. (2010). Global estimates of evapotranspiration and gross primary production based on MODIS and global meteorology data. *Remote Sensing of Environment* 114: 1416–1431.

Zarco-Tejada, P.J., Gonzalez-Dugo, V., and Berni, J.A.J. (2012). Fluorescence, temperature and narrow-band indices acquired from a UAV platform for water stress detection using a micro-hyperspectral imager and a thermal camera. *Remote Sensing of Environment* 117: 322–337.

Zeng, L.L., Wardlow, B.D., Tadesse, T. et al. (2015). Estimation of daily air temperature based on MODIS land surface temperature products over the corn belt in the US. *Remote Sensing* 7 (1): 951–970.

Zhang, K., Kimball, J.S., Mu, Q. et al. (2009). Satellite based analysis of northern ET trends and associated changes in the regional water balance from 1983 to 2005. *Journal of Hydrology* 379: 92–110.

Zhang, K., Kimball, J.S., Nemani, R.R., and Running, S.W. (2010). A continuous satellite derived global record of land surface evapotranspiration from 1983–2006. *Water Resources Research* 46: W09522.

Zhang, R., Rong, Y., Tian, J. et al. (2015). A remote sensing method for estimating surface air temperature and surface vapor pressure on a regional scale. *Remote Sensing* 7: 6005–6025.

Zhao, S.H., Qin, Q.M., Yang, Y.H. et al. (2009). Comparison of two split-window methods for retrieving land surface temperature from MODIS data. *Journal of Earth System Science* 118 (4): 345–353.

Zhu, W., Lu, A., and Jia, S. (2013). Estimation of daily maximum and minimum air temperature using MODIS land surface temperature products. *Remote Sensing of Environment* 130: 62–73.

Zwerman, C.H. and Andrews, A.I. (1940). Relation of particle size and characteristic of light reflected from porcelain enamel surfaces. *Journal of the American Ceramic Society* 23: 93–102.

Zwieback, S., Hensley, S., and Hajnsek, I. (2015). Assessment of soil moisture effects on L-band radar interferometry. *Remote Sensing of Environment* 164: 77–89.

10

Environmental Fluid Mechanics

Nigel B. Kaye[1], *Abdul A. Khan*[1], *and Firat Y. Testik*[2]

[1] *Glenn Department of Civil Engineering, Clemson University, Clemson, SC, USA*
[2] *Civil and Environmental Engineering Department, University of Texas at San Antonio, San Antonio, TX, USA*

Environmental fluid mechanics is an enormous field covering a vast range of scales from mixing of oceans and lakes, heat trapping in urban areas (urban heat island effect), pollutant dispersion in rivers and streams, heat flows through buildings, and all the way down to pollutant transport through rock and soil pores. Fluid mechanics is ubiquitous in the field of environmental engineering as the majority of environmental heat and pollutant transport processes are driven by fluid flows. There are many books entirely devoted to various topics in the field of environmental fluid mechanics (Boeker and van Grondelle, 1999; Rubin and Atkinson, 2001; Fernando, 2013; Imberger, 2013), and the reader is referred to these books for more details on the field as a whole or with regard to particular topics. Given the enormous scope of the field (the handbook of environmental fluid dynamics (Fernando, 2013) has 81 chapters and over 1100 pages in 2 volumes), this chapter can only provide a brief overview of the dominant processes that control heat, mass, and pollutant transport in the environment. The material presented in this chapter is quite theoretical and, at times, abstract, with the goal of providing the mathematical underpinnings of some of the models presented in the more applied chapters. The chapter assumes a solid knowledge at the undergraduate level of fluid mechanics and calculus including fluid properties, control volume analysis of mass, energy, and momentum, scalar conservation laws, dimensional analysis, open-channel flow, and ordinary and partial differential equations. All sections in this chapter include extensive reference lists for further reading.

The first three sections below review free surface and subsurface flows. In Section 10.1 Dr. Abdul Khan reviews the modeling of open-channel flow including reviews of steady uniform flow, sub- and supercritical flow, flow control, and gradually varied flow (GVF). Surface waves are described by Dr. Firat Y. Testik in Section 10.2, while in Section 10.3 Dr. Khan reviews the mechanics of flow in porous media and its application to infiltration and groundwater flow. Dr. Nigel B. Kaye covers the basics of pollutant transport through advection and diffusion in Section 10.4. This section presents the basic advection diffusion equations and the solution to a set of idealized problems that illustrate basic behavior. The chapter concludes with detailed discussions of turbulent shear flows including turbulent jets (Section 10.5, Dr. Khan), turbulent plumes (Section 10.6, Dr. Kaye), and gravity currents (Section 10.7, Dr. Testik).

10.1 Open-Channel Flow

Open-channel flow is a gravity-driven flow with a free surface that is exposed to atmospheric pressure. Open-channel flow can occur in rivers, canals, culverts, and sanitary sewers. Open-channel flow can be identified as steady flow, unsteady flow, uniform flow, and nonuniform flow. The flow in an open channel is unsteady if the flow properties such as depth, velocity, and/or discharge are changing with time at a specific channel cross section. In some instances an unsteady flow can be transformed into a steady flow by coordinate transformation, i.e., having a coordinate system that is based on a moving frame of reference. A more general definition can be construed by realizing that for steady flow the local acceleration is zero. Uniform and nonuniform flows are determined based on the variation of flow properties along the length of the channel or along the flow direction at a given instance in time. The variations of velocity within a channel cross section are ignored. If the flow properties do not change along the channel length or the convective acceleration is zero, the flow is deemed uniform (Henderson, 1966; Chin, 2006; Chaudhry, 2008).

Handbook of Environmental Engineering, First Edition. Edited by Myer Kutz.

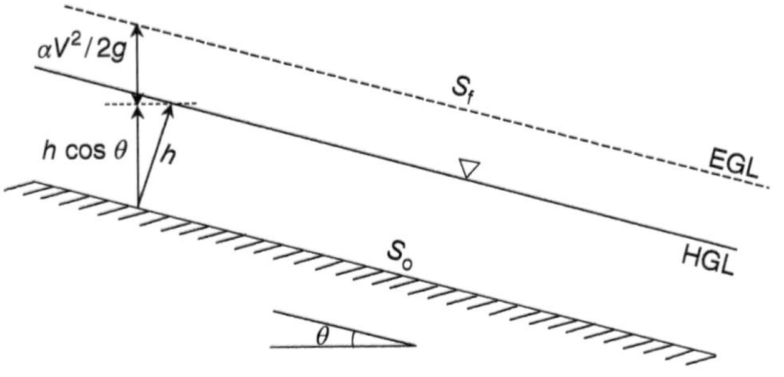

Figure 10.1 Uniform flow in open channels.

10.1.1 Uniform Flow

Uniform flows occur in a prismatic channel (channels with constant geometric shape and slope) with constant discharge. Under uniform flow conditions the water depth and velocity remain the same along the length of the channel. As shown in Figure 10.1, the bed slope (S_o), the slope of the hydraulic grade line (HGL, represented by water surface), and the slope of the energy grade line (EGL), S_f, are the same. The bed and energy slopes are considered positive when the bed elevation and energy is decreasing in the direction of flow. In uniform flow, the gravitational force is balanced by the friction force that leads to Eq. (10.1), where τ_o is the bed shear stress, P_w is the wetted perimeter, L is the length along the channel, A is the flow area, θ is the inclination of the bed with the horizontal axis, R_h is the hydraulic radius ($=A/P_w$), $S_o=\tan\theta$ is the bed slope, and $\cos\theta \simeq 1.0$ for small bed slopes ($\theta < 10°$):

$$\tau_o P_w L = \rho g A L \sin\theta = \rho g A L S_o \cos\theta \Rightarrow \tau_o = \rho g R_h S_o \tag{10.1}$$

The friction loss in the channel is given by Eq. (10.2), where S_f is the slope of the EGL. The Chezy equation and the Manning equation (in SI units) are shown in Eq. (10.3), where S_f is equal to S_o for uniform flow and C is known as Chezy constant. Manning (1890) showed that C was not constant and is given by $C = R_h^{1/6}/n$, where n is the Manning coefficient. It is obvious that n is not constant and is a function of friction factor, f, and should vary with Reynolds number and channel bed material. For a given channel surface, the Manning coefficient can be assumed constant only for rough turbulent flow (friction factor is independent of Reynolds number) and for $4 \le R_h/k_s \le 500$, where k_s is the equivalent sand roughness (Yen, 1991). The Manning coefficient is related to bed sediment and equivalent sand roughness (Chin, 2006; Chaudhry, 2008). Typical values for Manning coefficient for various surfaces are provided by French (1985) and Chow (1959):

$$h_f = \frac{fLQ^2}{8gR_hA^2} \Rightarrow Q = A\sqrt{\frac{8g}{f}}\sqrt{R_hS_f\cos\theta} \tag{10.2}$$

$$Q = CA\sqrt{R_hS_o} \Rightarrow Q = \frac{1}{n}AR_h^{2/3}S_o^{1/2} \tag{10.3}$$

The Manning equation relates discharge to flow depth for a given channel geometry, roughness coefficient, and channel bed slope. The flow depth determined for a given discharge and bed slope using Manning equation is called normal depth or uniform flow depth h_o. In a prismatic channel, away from imposed boundary conditions at the upstream and/or downstream end, the flow depth will approach the normal depth, which guarantees the balance between the gravitational and friction forces. Thus, for prismatic channels, the normal depth is the flow depth that the flow will achieve away from any imposed disturbances.

10.1.2 Specific Energy

The specific energy (E) comprises kinetic energy and work done by the pressure force as given by Eq. (10.4), where h is the flow depth normal to the bed, V is the cross-sectional averaged flow velocity, and α is the kinetic energy flux coefficient. The pressure distribution is hydrostatic normal to bed as the flow is assumed parallel to the bed (refer to Figure 10.1). For a rectangular channel of width B, the total discharge Q can be replaced by specific discharge or discharge per unit width q ($=Q/B$), and for small bed slope the equation can be approximated as shown in Eq. (10.5):

$$E = h\cos\theta + \alpha\frac{V^2}{2g} = h\cos\theta + \alpha\frac{Q^2}{2gA^2} \tag{10.4}$$

$$E = h + \alpha\frac{q^2}{2gh^2} = h + \alpha\frac{V^2}{2g} \tag{10.5}$$

Equation (10.5) represents a cubic equation in flow depth, and for a given discharge and specific energy, the equation will yield three roots for the depth. One of the roots is negative and can be ignored, while the other two are positive values and represent possible solutions. Figure 10.2 shows graphical representation of Eq. (10.5) for a given q. The curve lies below the $E=h$ (45°) line as E is always higher than h. The $E-h$ curve has three distinct parts: the minimum specific energy point is represented

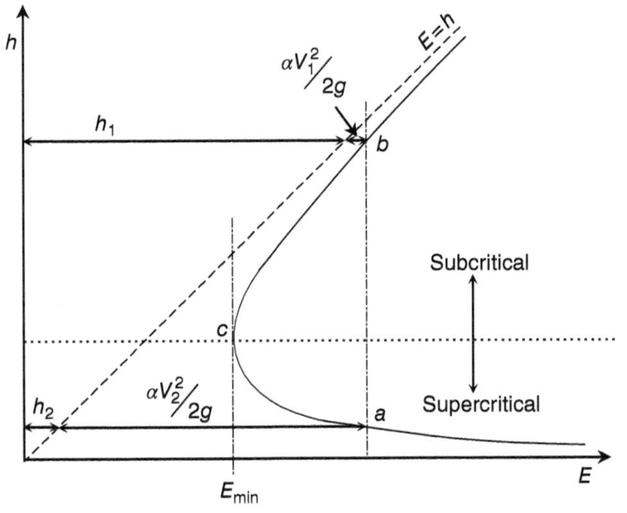

Figure 10.2 Specific energy diagram.

by "c"; the portion of the curve that lies above "c," where the depth is relatively large with small velocity head; and the lower portion, where the depth is relatively small compared with the velocity head. For a given q as the depth increases the velocity will decrease (based on $q = Vh$), and the $E - h$ curve will approach the $E = h$ line, while the curve will approach the horizontal axis as the depth decreases and velocity increases.

The condition for minimum specific energy can be obtained by differentiating Eq. (10.4) with respect to the depth and setting it equal to zero as shown in Eq. (10.6), where T is the top width or width of the channel section at the water surface. Equation (10.6) can be expressed as Eq. (10.7), where F_r is the Froude number. Thus, at point "c," which represents the minimum specific energy, the Froude number is equal to one and is called the critical flow condition or critical flow. The minimum specific energy (E_{min}) is given by Eq. (10.9), where the subscript "c" signifies that the depth, area, and top width are for critical flow condition, i.e. $F_r = 1.0$. For small bed slopes $\cos\theta \simeq 1.0$ and Eqs. (10.8) and (10.9) can be simplified. Furthermore, for $\alpha \simeq 1.0$ and a rectangular channel, these equations can be written as $F_r = V/\sqrt{gh}$ and $E_{min} = 1.5y_c$, respectively:

$$\frac{dE}{dh} = 0 = \cos\theta - \frac{\alpha Q^2}{gA^3}\frac{dA}{dh} = \cos\theta - \frac{\alpha Q^2}{gA^3}T \quad (10.6)$$

$$\frac{\alpha Q^2 T}{gA^3\cos\theta} = 1.0 \Rightarrow F_r^2 = 1.0 \Rightarrow F_r = 1.0 \quad (10.7)$$

$$F_r = \frac{Q}{A\sqrt{(g/\alpha)(A/T)\cos\theta}} \quad (10.8)$$

$$E_{min} = h_c\cos\theta + \frac{1}{2}\frac{A_c}{T_c}\cos\theta \quad (10.9)$$

As shown in Figure 10.2, for $E > E_{min}$ there exists two depths for a given discharge and specific energy and are represented as "a" and "b." At point "b" the depth is larger, and velocity is smaller compared with the critical condition, thus the Froude number is less than one, and the flow is termed as subcritical. At point "a" the depth is smaller, and the velocity is higher than at the critical condition resulting in a Froude number higher than one, and the flow regime is considered as supercritical. In subcritical flow any disturbance caused in the channel can move both upstream and downstream along the channel, while in supercritical flow information can travel only in the downstream direction. The concept of specific energy is important for rapidly varied flow (RVF) analysis in open channel. In RVF the energy equation is applied over relatively short distance, and friction losses are often neglected.

10.1.3 Gradually Varied Flow

As the name suggests, GVF occurs over long distances, and hence friction/energy losses cannot be ignored. As opposed to uniform flow, GVF falls under the class of nonuniform steady flow. As shown in Figure 10.3, the bed, the water surface (HGL), and the EGL are not parallel. In the Manning equation, given by Eq. (10.3), the bed slope term should be replaced with the slope of the EGL. The total energy, assuming small bed slope, at any given section in an open-channel flow, is given by Eq. (10.10), where H is the total energy and z_b is the bed elevation from a horizontal datum. Equation (10.11) results from differentiating Eq. (10.10) with respect to s (distance along the channel), where $dA/ds = T dh/ds$ for prismatic channels. Applying the same procedure to the specific energy form of Eq. (10.10) results in Eq. (10.13):

$$H = z_b + h + \alpha\frac{Q^2}{2gA^2} = z_b + E \quad (10.10)$$

$$\frac{dH}{ds} = -S_f = -S_o + \frac{dh}{ds} - \alpha\frac{Q^2}{gA^3}T\frac{dh}{ds} \quad (10.11)$$

$$\frac{dh}{ds} = \frac{S_o - S_f}{1 - F_r^2} \quad (10.12)$$

$$\frac{dE}{ds} = S_o - S_f \quad (10.13)$$

In open-channel flows, for a given channel geometry, discharge, and bed slope, one can easily determine the

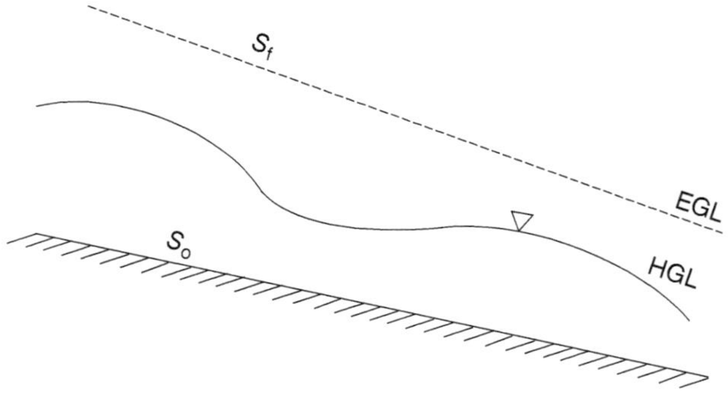

Figure 10.3 Nonuniform flow in open channels.

normal flow depth h_o and critical flow depth h_c and thus identify a channel slope as mild, steep, or critical: for mild slope $h_o > h_c$, for steep slope $h_o < h_c$, and for critical slope $h_o = h_c$. In addition, if the flow depth in a channel is higher than normal depth, $S_f < S_o$, and for a flow depth lower than normal depth, $S_f > S_o$. Based on Eq. (10.12), it is clear that at the critical condition the water surface slope is vertical. As mentioned before, for prismatic channel, away from the imposed depth (control), the flow will adjust its depth so as to approach the normal depth; this approach to the normal depth is asymptotic in nature.

It should be noted that if a dam is built across the river at a certain point, the water level in the channel, upstream of the dam, will rise slowly to reach the dam/spillway height. Also, if a channel terminates in a free overfall and upstream flow is subcritical, the flow in the vicinity of the fall is critical. Furthermore, if the slope of a channel changes such that the upstream section is mild and the downstream section is steep, the flow must pass through the critical point, which is assumed to be located at the junction. Installation of sluice gate in a channel may change flow conditions both upstream and downstream as the sluice gate may act as a dam for the upstream section and forces the flow to exit sluice gate at a certain depth in the downstream section determined by the sluice gate opening. A dam, critical flow point, or sluice gate in the channel forces the channel to adjust its depth upstream and/or downstream. These points are called controls and act as boundary conditions for the flow upstream and/or downstream. The subcritical flow is affected upstream of the control, and the supercritical flow is impacted downstream of the control, as the flow tends to achieve normal depth away from the control. Anytime the flow has to transition from a supercritical flow to subcritical flow, a hydraulic jump (HJ) forms and is associated with an abrupt change in the water level.

In general, the flow depth in a channel may lie in three different zones. Zone 1 is above the normal depth (for mild slope) or critical depth (for steep or critical slope).

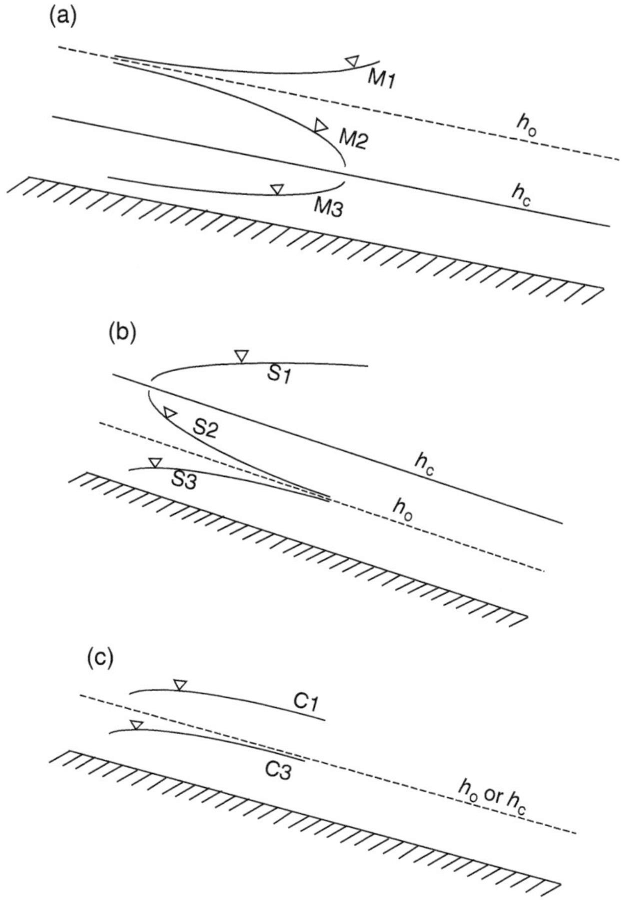

Figure 10.4 Gradually varied flow profiles: (a) mild slope, (b) steep slope, and (c) critical slope.

The higher flow depth in this zone may be caused by a dam. The flow in this zone is always subcritical; hence flow upstream is affected by the control. Zone 2 lies between the normal depth and the critical depth. For a mild slope the flow is subcritical, and for a steep slope the flow is supercritical in this zone. Hence, the control is downstream for the mild slope and upstream for the steep slope. The control in this zone may be caused by a free overfall or sluice gate. This zone is absent for the critical slope. Zone 3 occurs between the bed and critical depth (for mild slope) or normal depth (for steep slope). The flow in this zone is supercritical and the control should always be upstream.

The GVF profiles in different zones are shown in Figure 10.4. For mild slopes the water surface profiles are labeled as M1, M2, and M3 in zones 1, 2, and 3, respectively. The change in water depth along the channel in each zone can be ascertained using Eq. (10.12), e.g., in zone 1 for mild slope $S_f < S_o$ and $F_r < 1.0$, resulting in positive value for the right-hand side. That is, the water depth is increasing along the channel. Similar reasoning can be followed for all other profiles. For steep slopes the water

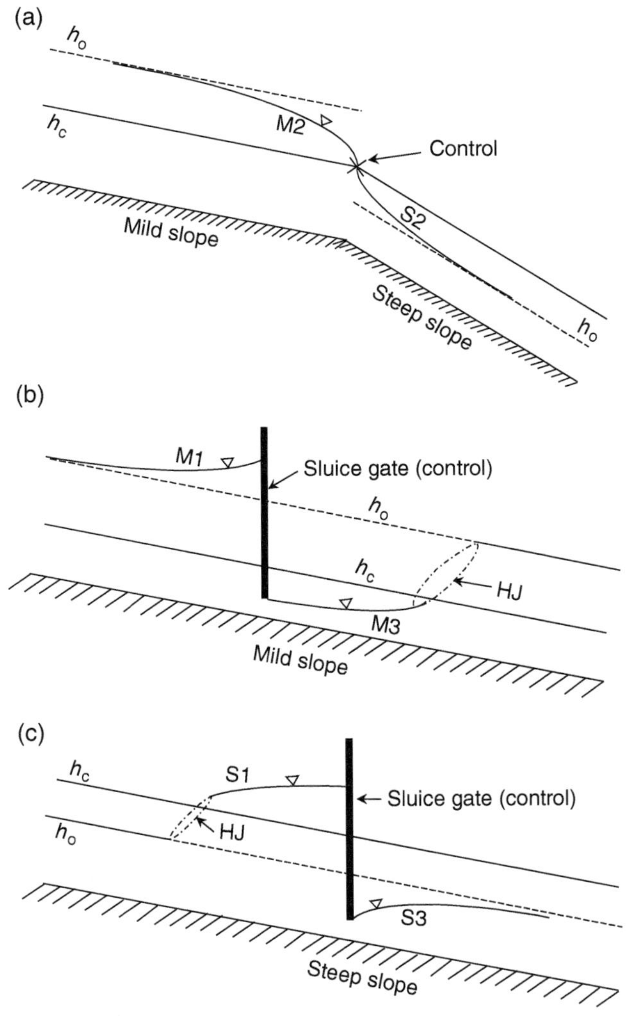

Figure 10.5 Examples of gradually varied flow profiles.

surface profiles are labeled as S1, S2, and S3. For critical slopes the curves are C1 and C3 as zone 2 is absent.

Examples of GVF are shown in Figure 10.5. In Figure 10.5a the slope changes from mild to steep slope. The flow must pass through the critical point, which acts as a control for the upstream subcritical flow and downstream supercritical flow with M2 and S2 profiles, respectively. A sluice gate in the middle of a mild channel is shown in Figure 10.5b. The sluice gate acts as a dam for the upstream flow with M1 profile and generates a flow depth in zone 3 downstream. In the downstream channel an M3 profile is generated in zone 3. Since the flow eventually has to achieve normal depth, it must cross the critical depth through a HJ. The critical depth cannot act as a control for the downstream reach in zone 2 as the flow in subcritical and control must be downstream. In Figure 10.5c, a sluice gate is placed in the middle of a steep channel with the sluice gate opening in zone 3. The sluice gate acts as a control for the upstream and downstream section of the channel. In the downstream channel an S3 profile develops to achieve normal depth. In the upstream section the flow must pass critical flow to reach zone 1, which is achieved through a HJ followed by an S1 profile.

The computation of GVF profiles is achieved using Eq. (10.12) or (10.13) for prismatic channels using a direct step method, where starting from the control the distance is calculated for a given depth. Let us assume the control is at section 1 that means V_1 and h_1 are known at section 1. Based on the type of the curve, the h_2 value is selected, and the distance from section 1 where the depth will exist is found using Eq. (10.14). The slopes S_{f1} and S_{f2} are calculated using Manning equation. In Eq. (10.14) arithmetic mean is used to approximate the energy slope between sections 1 and 2 and can be replaced with geometric or harmonic mean. For natural rivers, where cross-sectional geometry data are available at particular locations and bed slope may change from section to section, the standard step method is adopted. The standard step method is illustrated in Eq. (10.15), where $Z = h + z_b$ and h_e represents eddy losses within the section arising due to contraction, expansion, etc. Notice that the discharge at section 1 may be different than that at section 2 due to inflow or outflow that may take place within the section. The equation has to be solved iteratively for water surface level at section 2 knowing the water surface level at section 1 and discharges at sections 1 and 2:

$$s_2 = s_1 + \frac{E_2 - E_1}{S_o - 0.5\left(S_{f1} + S_{f2}\right)} \tag{10.14}$$

$$Z_1 + \alpha_1 \frac{Q_1^2}{2gA_1^2} = Z_2 + \alpha_1 \frac{Q_2^2}{2gA_2^2} + \Delta s \frac{\left(S_{f1} + S_{f2}\right)}{2} + h_e \tag{10.15}$$

10.1.4 Unsteady Flow Equations

The unsteady flow in open channels is three-dimensional (3D) in nature and, for turbulent flows, is represented by 3D Reynolds-averaged Navier–Stokes (RANS) equations. These equations represent mass and momentum conservation equations for turbulent flows. Turbulent closure schemes are required to model turbulent normal and shear stresses that arise due to Reynolds averaging. Furthermore, the location of the free surface must be determined as part of the solution. Various methods are available to locate the free surface such as marker-and-cell approach (de Paulo et al., 2007) or volume of fluid methods (Gopala and van Wachem, 2008). The turbulent closure schemes for determining turbulent normal and shear stresses are needed to close the system of equations. Several turbulent closure schemes are available and are discussed in detail by Rodi (1984). RANS equations with turbulence closure schemes must be solved using numerical modeling techniques as direct solution is generally not possible. Even with the advent of ever-faster computing facilities, the use of 3D computational models is limited to small river reaches, i.e., to study local flow phenomena in detail. One-dimensional (1D) and two-dimensional (2D) computational models are more prevalent in modeling river flows for relatively long river reaches.

The simplest unsteady river flow models are based on the 1D unsteady mass and momentum flow equations. In these models, the flow is assumed to be predominantly along the length of the river or longitudinal direction and the pressure distribution hydrostatic. For natural channels, the 1D equations are given by Eqs. (10.16) and (10.17) representing mass and momentum equations (called Saint-Venant equations), respectively, where x is the distance along the channel and S_f is given by Eq. (10.18). As shown by Khan and Lai (2014, 2012a), the eigenvalues for the Saint-Venant equations are given by Eq. (10.19). The eigenvalues represent the characteristic directions and determine the application of boundary conditions in subcritical and supercritical flows and are used for obtaining stable solutions for numerical schemes. Various numerical techniques such as finite difference, finite volume, and finite element can be used to solve these equations (Khan and Lai, 2014):

$$\frac{\partial A}{\partial t}+\frac{\partial Q}{\partial x}=0 \tag{10.16}$$

$$\frac{\partial Q}{\partial t}+\frac{\partial\left(Q^2/A\right)}{\partial x}=-gA\frac{\partial Z}{\partial x}-gAS_f \tag{10.17}$$

$$S_f=\frac{n^2Q|Q|}{R_h^{4/3}A^2} \tag{10.18}$$

$$\lambda_1=\frac{Q}{A}-\sqrt{\frac{gA}{T}}=u-c;\quad \lambda_2=\frac{Q}{A}+\sqrt{\frac{gA}{T}}=u+c \tag{10.19}$$

The 2D depth-averaged unsteady open-channel flow equations are given by Eqs. (10.20), (10.21), and (10.22), representing mass conservation equation, momentum equation in x direction, and momentum equation in y direction, respectively. In these equations, x, y are coordinate directions in the longitudinal and lateral directions, respectively; and q_x, q_y are discharges per unit width in the x, y directions, respectively. The eigenvalues and various numerical techniques employed to solve these equations are discussed in detail by Khan and Lai (2014, 2012b):

$$\frac{\partial h}{\partial t}+\frac{\partial q_x}{\partial x}+\frac{\partial q_y}{\partial y}=0 \tag{10.20}$$

$$\frac{\partial q_x}{\partial t}+\frac{\partial}{\partial x}\left(\frac{q_x^2}{h}\right)+\frac{\partial}{\partial y}\left(\frac{q_xq_y}{h}\right)=-gh\frac{\partial Z}{\partial x}-g\frac{n^2q_x\sqrt{q_x^2+q_y^2}}{h^{7/3}} \tag{10.21}$$

$$\frac{\partial q_y}{\partial t}+\frac{\partial}{\partial x}\left(\frac{q_xq_y}{h}\right)+\frac{\partial}{\partial y}\left(\frac{q_y^2}{h}\right)=-gh\frac{\partial Z}{\partial y}-g\frac{n^2q_y\sqrt{q_x^2+q_y^2}}{h^{7/3}} \tag{10.22}$$

10.2 Surface Waves

Surface waves, undulations of a water surface, are a very important phenomenon for various engineering and scientific applications, especially in large bodies of water such as oceans and lakes. Interactions of waves and coastal structures (Young and Testik, 2011), wave-driven sediment transport, morphodynamics, beach erosion, and scour around structures (Voropayev et al., 2003a, b; Sumer and Fredsoe, 2005; Testik et al., 2005, 2006; Young and Testik, 2009; Johnson et al., 2013) are only few examples of such applications. Given the importance and applications of this subject, there is a very large body of literature on this topic. A few of the fundamental sources for an overview of the essentials of waves and related processes are Mei (1989), Mei et al. (2005), Dean and Dalrymple (1991, 2004), and Fredsoe and Deigaard (2005).

Wave phenomenon covers a wide range of spatial and temporal scales, ranging from an order of milliseconds of wave periods and millimeters of wavelengths for capillary waves to an order of minutes to hours of wave periods and hundreds to thousands of kilometers of wavelengths for tsunamis and tidal waves. As such, waves are classified into a number of different types. Wave classifications based upon wave period can be found in, for example, Dean and Dalrymple (2004). There are various mechanisms for wave generation such as the wind blowing over the water surface, gravitational attraction by the sun and moon, landslides, and vessel motion. Surface gravity waves are wind-generated waves and have a significant practical importance in various applications such as design/construction/maintenance of coastal structures and modeling of coastal sediment transport and morphodynamics. Therefore, here we focus our attention to surface gravity waves. Wind velocity, duration, fetch, and the state of the water surface affect the characteristics of these waves. Simplified analyses of these waves typically consider 2D regular surface gravity waves, which have single frequency, length, and height. In nature, waves are typically superposed with different frequencies and amplitudes, which are referred to as irregular waves. For the purposes of this section, we restrict our discussion to 2D regular waves.

Surface gravity waves have relatively short periods, T, typically from 1 to 30 s. Their effect is limited to a relatively shallow layer of the water column in deep waters, and it is felt all the way to the bottom only in rather shallow waters as discussed later in this section. There have been various wave theories to describe surface gravity waves, the most common one being the Stokes wave theory. This theory is based upon the solutions of equations of motion and continuity equations with appropriate boundary conditions for an inviscid fluid. The simplest form of the Stokes wave theory solution is referred to as the small amplitude wave theory (often referred to as linear wave theory or Airy wave theory). A major assumption of the linear wave theory is that the wave height is infinitesimally small compared with the wavelength ($a/L \ll 1$) and water depth ($a/h \ll 1$). Due to the inviscid fluid and small wave height assumptions, linear wave theory would be most suitable for small waves in deep water. Nevertheless, its simplicity and the design calculation requirements for many coastal applications make the linear wave theory a popular tool for preliminary calculations for surface gravity waves in shallow waters as well. Therefore, among different wave theories available, linear wave theory is discussed next.

10.2.1 Linear Wave Theory

A simplified schematic of a surface gravity wave is given in Figure 10.6. As depicted in this schematic, the linear wave theory predicts a symmetric sinusoidal wave profile as described in (10.23):

$$\eta(x,t) = \frac{H}{2}\cos k(x - Ct) = \frac{H}{2}\cos(kx - \sigma t) \tag{10.23}$$

Here, η is the water surface elevation; $k = 2\pi/L$ is the wave number, and L is the wavelength; $\sigma = 2\pi/T$ is the wave frequency, and T is the wave period; $C = L/T = \sigma/k$ is the wave celerity; H is the wave height; x is the coordinate along the wave propagation; and t is time.

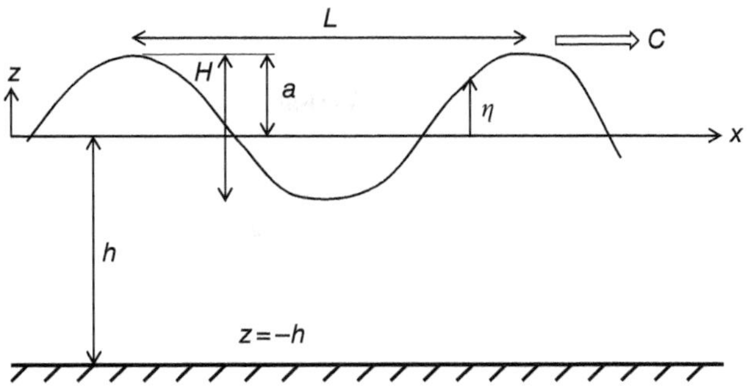

Figure 10.6 Simplified schematic of a surface gravity wave and the relevant nomenclature.

The wavelength, wave period, and water depth (h) are related through the frequency dispersion relationship, which is given in (10.24) for the linear wave theory as follows:

$$\sigma^2 = gk\tanh(kh) \tag{10.24}$$

Manipulation of (10.24) yields an implicit parameterization for the wavelength as follows:

$$L = \frac{g}{2\pi}T^2\tanh(kh) \tag{10.25}$$

As the waves propagate, while the wave energy and form are transmitted, the water mass does not propagate with the wave. Water particles under a passing wave are subject to closed orbital motions with the following horizontal, u, and vertical, w, velocities:

$$u(x,z,t) = \frac{H\sigma}{2}\frac{\cosh k(h+z)}{\sinh(kh)}\cos(kx - \sigma t) \tag{10.26}$$

$$w(x,z,t) = \frac{H\sigma}{2}\frac{\sinh k(h+z)}{\sinh(kh)}\sin(kx - \sigma t) \tag{10.27}$$

Here, z is the vertical coordinate with an origin at the free surface, positive being upward as illustrated in Figure 10.6. These water particle velocity solutions of the linear wave theory show that particle velocities depend on the wave height and frequency, position in the water column (i.e. reduce with increasing vertical distance from the surface), and wave phase. The horizontal, ε, and vertical, ξ, amplitudes of the orbital motions of the water particles can be calculated as follows:

$$\varepsilon = \frac{H}{2}\frac{\cosh k(h+z)}{\sinh kh}\sin(kx - \sigma t) \tag{10.28}$$

$$\xi = \frac{H}{2}\frac{\sinh k(h+z)}{\sinh kh}\cos(kx - \sigma t) \tag{10.29}$$

One can show that as water depth h goes to infinity, ε and ξ become identical and the orbital trajectories are circular. In practice, waves that are in water depths larger than $L/2$ are considered to be deep water waves, and wave-bottom interactions are considered to be negligible. For deep water waves, water particle trajectories can be considered as circular. Waves that are in shallower depths have elliptical orbits.

As the waves propagate from deep to shallow waters, wave transformation processes take place. Two such processes are wave shoaling and refraction, which are caused by the interactions of the waves and the underlying bathymetry. Wave shoaling is the process of wavelength decrease and wave height increase, which results in the increase of the wave asymmetry (i.e. increased deviations from the sinusoidal profiles of the linear wave theory solution). Wave refraction is the process of wave direction change, i.e. turning of wave crests to propagate toward regions of shallow water. During the change of direction, wave heights are also altered. Wave interactions with a man-made (e.g. breakwater) or natural (e.g. coral reef) structure may yield reflection of some portion of the wave energy back toward the incident waves (i.e. wave reflection process) and/or turning of the waves into the shadow zone that forms behind the structure (i.e. wave diffraction process). The modified waves due to wave transformation processes eventually break when they reach a limiting steepness, which is a function of the water depth. Onshore of the wave breaker is characterized by increased turbulence and reformation of waves. This region is very active in sediment transport and shaping the shoreline.

10.3 Groundwater Flow

Groundwater is water below the ground surface sometimes called subsurface water. Groundwater is an important resource and commonly used for water supply in many parts of the world. Groundwater may exist in several different forms such as solid, liquid, or vapors and may be chemically bound to the soil particles (Driscoll, 1986). In general, groundwater refers to water in liquid form moving through a soil or rock medium. Groundwater is referred to water in saturated or unsaturated zones below the ground surface. Unsaturated zone consists of water in the root zone just below the ground surface, gravitational water (water in transit, moving from root zone to a deeper zone), and capillary zone (water above the water table). The saturated zone is where the voids in the soil are filled with water.

Another term that is commonly used in the context of groundwater is aquifer. An aquifer is a saturated zone below the ground surface that can yield water. Aquifers are classified as confined or unconfined. In the case of the unconfined aquifer (Figure 10.7), the saturated soil layer has a water surface that is exposed to atmospheric pressure and has an underlying impervious or highly impermeable layer. This means that the water level in a well dug in an unconfined aquifer will rise up to the water table. Unconfined aquifers are replenished through infiltration throughout its surface area and possibly from up seepage from a pressurized aquifer underneath. Confined aquifers (Figure 10.8) have impervious layers both at the top and bottom and hence represent pressurized flow. The water level in a well dug in a confined aquifer will rise above the top of the aquifer. The recharge area of the confined aquifer is where the confining surfaces crop out. The following discussion is limited to groundwater flow in aquifers.

10.3.1 Hydraulic Head and Hydraulic Conductivity

The driving force in groundwater flow is the hydraulic head difference between two points and is defined as the sum of pressure head and elevation head. A piezometer (which can be considered as an observation well) is

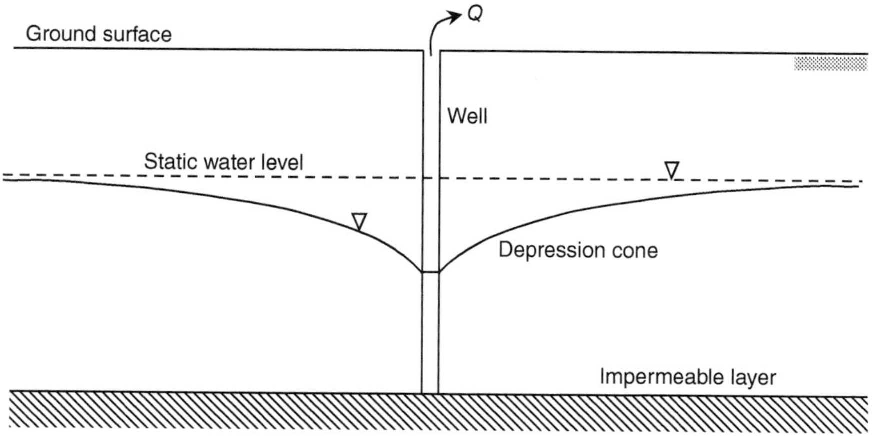

Figure 10.7 Unconfined aquifer layout with a well.

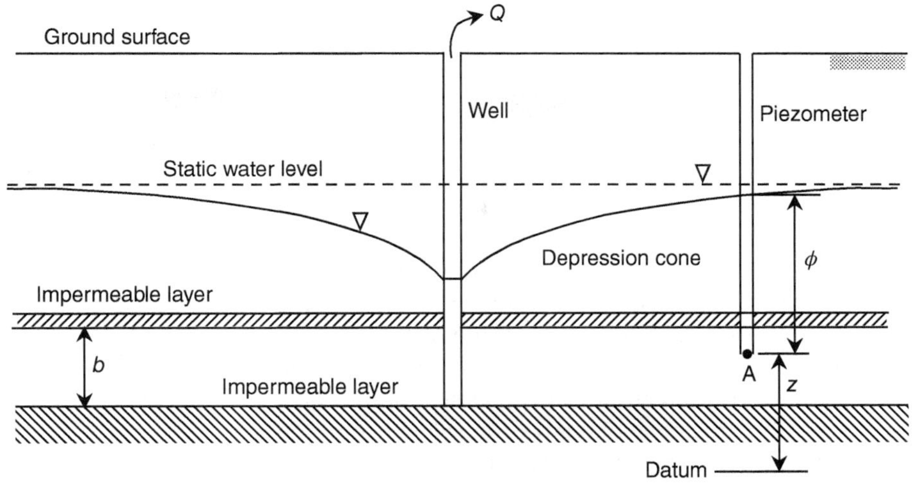

Figure 10.8 Confined aquifer layout with a well and piezometer.

shown in Figure 10.8. At point "A" the pressure head is ϕ and the elevation head from a selected datum is z. The hydraulic head at point "A" is given by Eq. (10.30), where p is the pressure at point "A," ρ is the density of the fluid, and g is the gravitational acceleration. The water surface elevation in a piezometer or a well from a datum is equal to the hydraulic head:

$$h = \frac{p}{\rho g} + z = \phi + z \tag{10.30}$$

Hydraulic conductivity (K) is a measure of ease with which a given fluid flows through a given soil medium. It has a unit of velocity and is generally measured using a permeameter in a laboratory. Hydraulic conductivity reflects fluid and soil properties. Intrinsic or specific permeability (k) is independent of fluid properties and is related to the properties of soil only in terms of allowing any fluid to flow through it. The relation between the hydraulic conductivity and intrinsic permeability is given by Eq. (10.31), where μ is the viscosity of the fluid. Once the intrinsic permeability of a soil sample is known, the hydraulic conductivity for any fluid or the same fluid at different temperatures can be computed. The intrinsic permeability has units of L^2:

$$K = k\frac{\rho g}{\mu} \tag{10.31}$$

10.3.2 Darcy's Law

The flow through soil is governed by Darcy's law. For 1D flow through the soil, the discharge through the soil is described by Darcy equation as shown in Eq. (10.32), where A is the cross-sectional area of the soil normal to the flow direction, s is the distance along the flow direction, h_1 is the hydraulic head at the upstream section, h_2 is the hydraulic head at the downstream section, and L is the length along the flow direction between the upstream and downstream sections. It should be noted that the whole cross-sectional area of the soil is used as opposed to the pore area and as such hydraulic conductivity or intrinsic permeability are determined and reported based on the whole soil area. The negative sign indicates that the flow is in the direction of decreasing hydraulic head. The velocity through the pores of the soil, which represents the real velocity through the soil matrix, involves the porosity (ψ) of the soil and is given by Eq. (10.33). This velocity can be used to determine the time for any fluid or solute particle to travel a certain distance within the soil matrix:

$$Q = -KA\frac{\mathrm{d}h}{\mathrm{d}s} = KA\frac{h_1 - h_2}{L} \tag{10.32}$$

$$V = \frac{Q}{\psi A} \tag{10.33}$$

10.3.3 Wells in Aquifers

Wells are commonly dug in the confined or unconfined aquifer to draw water. Typical layouts of wells in unconfined and confined aquifers are shown in Figures 10.7 and 10.8, respectively. Before pumping water out of the well, the water level in the aquifer is identified as the static water level; in the case of a confined aquifer, the static water level is above the top of the aquifer. As the well is pumped at a constant flow rate, Q, the water level in the vicinity of the well will drop and form a cone around the well that is known as cone of depression. The radius of the depression cone increases with time and eventually stabilizes (steady state) for a given flow rate. The radius of the depression cone after stabilization is called radius of influence (R_o). The radius of influence varies directly with the discharge from the well. The difference between the static water level and the depression cone at any point is called the drawdown. The drawdown is zero at the radius of influence. As the well is pumped, the flow in the aquifer moves radially toward the well.

The flow through the well in confined and unconfined aquifers under steady state is given by Eqs. (10.34) and (10.35), respectively. In these equation h_1 is the hydraulic head at a radius r_1 from the center of the well, and h_2 is hydraulic head at a distance r_2 from the center of the well, where $r_2 > r_1$. The maximum value of r_2 is R_o and the minimum value of r_1 is the radius of the well. In the case of an unconfined aquifer, the thickness of the aquifer varies along the cone of depression. The unconfined aquifer thickness is given by $\bar{b}$ and represents an average thickness of the thicknesses of aquifer at r_1 and r_2 away from the center of the well. If the bottom of the unconfined aquifer is used as datum (only if it is horizontal), then the hydraulic heads will be equal to the thickness of the aquifer, and Eq. (10.35) can be written as Eq. (10.36). The time for a fluid or solute particle to travel from r_2 to r_1 is given in Eq. (10.37). This equation is valid for a confined aquifer; for an unconfined aquifer b should be replaced by $\bar{b}$:

$$Q = 2\pi b K \frac{h_2 - h_1}{\ln(r_2/r_1)} \tag{10.34}$$

$$Q = 2\pi \bar{b} K \frac{h_2 - h_1}{\ln(r_2/r_1)} \tag{10.35}$$

$$Q = 2\pi\left(\frac{h_2 + h_1}{2}\right) K \frac{h_2 - h_1}{\ln(r_2/r_1)} = \pi K \frac{h_2^2 - h_1^2}{\ln(r_2/r_1)} \tag{10.36}$$

$$t = \frac{\pi b \psi}{Q}\left(r_2^2 - r_1^2\right) \tag{10.37}$$

The equations given above are valid for a single well. In addition, the depression cone developed by a single well must be unhindered by a river/lake (where hydraulic head is fixed) or a no-flow boundary. For multiple wells and where no-flow or prescribed head boundary is nearby, the method of superposition may be used to determine steady-state solutions. Details of the solution procedures in these cases are provided by Gupta (2008).

10.3.4 Unsteady Flow Equations in Aquifers

Mass conservation equations for unsteady flow in confined and unconfined aquifers are given by Eqs. (10.38) and (10.39), respectively. In these equations K_x, K_y, and K_z are hydraulic conductivities in the Cartesian coordinate system along the x, y, and z directions, respectively; x is the longitudinal direction, y is the transverse direction, and z is the vertical direction; S_s is the specific storage and is the volume of water released from a unit volume of aquifer under a unit drop in head; ω is the effective volume of water yielded per unit volume of aquifer under gravity; q is the outflow discharge per unit volume of the aquifer, and q^* is the outflow discharge per unit surface area of the aquifer; and T is the transmissivity and is the product of hydraulic conductivity and thickness of the aquifer:

$$S_s \frac{\partial h}{\partial t} + q = \frac{\partial}{\partial x}\left(K_x \frac{\partial h}{\partial x}\right) + \frac{\partial}{\partial y}\left(K_y \frac{\partial h}{\partial y}\right) + \frac{\partial}{\partial x}\left(K_z \frac{\partial h}{\partial z}\right) \tag{10.38}$$

$$\omega \frac{\partial h}{\partial t} + q^* = \frac{\partial}{\partial x}\left(T_x \frac{\partial h}{\partial x}\right) + \frac{\partial}{\partial y}\left(T_y \frac{\partial h}{\partial y}\right) \tag{10.39}$$

10.3.5 Solute Transport in Aquifers

Solute transport in aquifers is defined by the advection–diffusion equation as shown in Eq. (10.40), where C is the pollutant concentration; u, v, and w are the velocity components along x, y, and z directions, respectively; and D_x, D_y, and D_z are the dispersion coefficients (also called coefficient of hydrodynamic dispersion) in x, y, and z directions, respectively, and have unit of L^2/T (see Section 5.4 for more details on the advection–diffusion equation). The dispersion coefficient consists of molecular diffusion and dispersivity (Freeze and Cheery, 1979). The dispersion coefficient is given by Eq. (10.41), where τ is the tortuosity of the soil medium and α is the dispersivity (Gupta, 2008). Generally, the transverse dispersion coefficients (D_y, D_z) are small and may be neglected (Gupta, 2008):

$$\frac{\partial C}{\partial t} + \frac{\partial uC}{\partial x} + \frac{\partial vC}{\partial y} + \frac{\partial wC}{\partial z} = \frac{\partial}{\partial x}\left(D_x \frac{\partial C}{\partial x}\right) + \frac{\partial}{\partial y}\left(D_y \frac{\partial C}{\partial y}\right) + \frac{\partial}{\partial z}\left(D_z \frac{\partial C}{\partial z}\right) \tag{10.40}$$

$$D_x = \tau D_m + u\alpha \tag{10.41}$$

The 1D advection–diffusion equation, where velocity and dispersion coefficient distributions are homogeneous, is given by Eq. (10.42). For the boundary conditions given by Eq. (10.43), the analytical solution of Eq. (10.42) is given by Eq. (10.44), where "erfc" is the complementary error function (Chanson, 2004). The boundary conditions is related to the continuous release of pollutant in the yz-plane at $x = 0$. A variety of analytical solutions for solute transport in aquifers under 1D, 2D, and 3D scenarios with different initial and boundary conditions are presented by (Wexler, 1992). The groundwater flow and transport equations must be solved numerically for nonhomogeneous and nonisotropic soil characteristics:

$$\frac{\partial C}{\partial t}+u\frac{\partial C}{\partial x}=D_x\frac{\partial}{\partial x}\left(\frac{\partial C}{\partial x}\right) \tag{10.42}$$

$$\left.\begin{aligned} C\left(0<x<\infty, t=0\right)&=0 \\ C\left(x=0, t\geq 0\right)&=C_o \end{aligned}\right\} \tag{10.43}$$

$$C(x,t)=\frac{C_o}{2}\left\{\operatorname{erfc}\left(\frac{x-ut}{\sqrt{4D_x t}}\right)+\exp\left(\frac{ux}{D_x}\right)\operatorname{erfc}\left(\frac{x+ut}{\sqrt{4D_x t}}\right)\right\} \tag{10.44}$$

10.4 Advection and Diffusion

Advection and diffusion of a tracer such as a chemical pollutant, heat, or fine sediment are ubiquitous transport processes in the environment. This section reviews the basic equations and standard solutions for some idealized problems. The section begins by developing the full advection–diffusion equations. This is followed by analysis of pure diffusion problems and then advection–diffusion problems. As with the rest of the chapter, practical application of these models is left to other, more applied chapters.

10.4.1 Advection–Diffusion Equations

Fickian diffusion is a molecular scale process whereby a tracer is transported down spatial concentration gradients. For a tracer characterized by its concentration C (kg m^{-3}), the diffusive flux f_x in a given direction is proportional to the local gradient of tracer concentration in that direction. That is,

$$f_x=-D\frac{\partial C}{\partial x} \tag{10.45}$$

where D is the diffusion coefficient (with units of m^2 s^{-1}) for the particular combination of tracer and ambient fluid. The negative sign on the right-hand side indicates that the tracer moves from regions of high to low concentration. Typically these coefficients are very small and diffusion is a very slow process. For example, the coefficients of diffusion for carbon dioxide in air and water are $D=1.6\times10^{-5}$ m^2 s^{-1} and $D=1.6\times10^{-9}$ m^2 s^{-1}, respectively (Boeker and van Grondelle, 1999).

For an infinitesimal control volume of stationary fluid (of length dx, height dy, and unit width, see Figure 10.9a) with tracer concentration only varying in the x direction, the Reynolds transport equation simplifies to

$$\frac{d}{dt}C\,dx\,dy+\left(f_x+\frac{1}{2}df_x\right)dy-\left(f_x-\frac{1}{2}df_x\right)dy=0 \tag{10.46}$$

which can be simplified to

$$\frac{\partial C}{\partial t}+\frac{\partial f_x}{\partial x}=0 \tag{10.47}$$

or, for constant D,

$$\frac{\partial C}{\partial t}=D\frac{\partial^2 C}{\partial x^2}. \tag{10.48}$$

For diffusion in three dimensions, (10.48) becomes

$$\frac{\partial C}{\partial t}=D\nabla^2 C. \tag{10.49}$$

If the fluid is flowing, then the tracer will be advected as well as diffused. In this case, the advective flux is the product of the local flow velocity U and concentration C (see Figure 10.9b). Following the same analysis as above and assuming a uniform flow field, the full 1D advection–diffusion equation is given by

$$\frac{\partial C}{\partial t}+U\frac{\partial C}{\partial x}=D\frac{\partial^2 C}{\partial x^2} \tag{10.50}$$

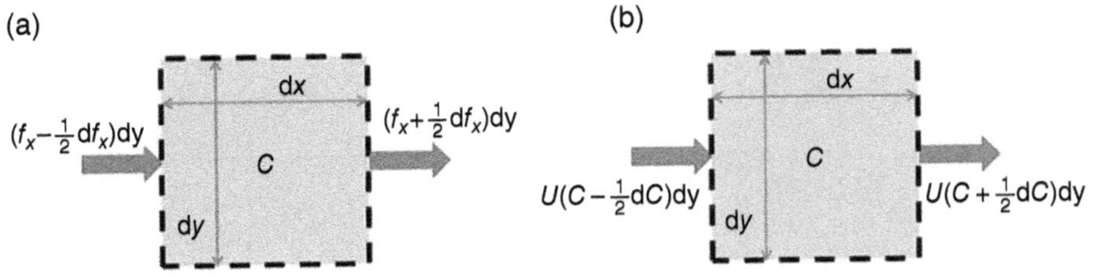

Figure 10.9 Infinitesimal control volumes for calculating the concentration change due to (a) diffusion and (b) convection.

which can be extended to three dimension to give

$$\frac{\partial C}{\partial t} + \vec{U}.\nabla C = D\nabla^2 C \qquad (10.51)$$

To understand the relative importance of these two process, the equations can be nondimensionalized using a length scale appropriate to the flow L, a velocity scale U_0, and a reference concentration C_0 to give the nondimensional length, time, velocity, and concentration scales $\chi = x/L$, $\tau = tU_0/L$, $u^* = U/U_0$, and $\kappa = C/C_0$, respectively. The nondimensional 1D equation can then be written as

$$\frac{\partial \kappa}{\partial \tau} + u^* \frac{\partial \kappa}{\partial \chi} = \frac{1}{\mathrm{Pe}} \frac{\partial^2 \kappa}{\partial \chi^2} \qquad (10.52)$$

where $\mathrm{Pe} = U_0 L/D$ is the Peclet number that quantifies the relative role of advection over diffusion. For very high Peclet numbers, tracer transport is dominated by advection. To get a sense of scale, consider the advection and diffusion of dissolved salt in a slow moving river of depth $L = 2$ m and uniform velocity of $U_0 = 0.3$ m s^{-1}. The diffusivity of salt in water is $D = 1.3 \times 10^{-9}$ m^2 s^{-1} (Boeker and van Grondelle, 1999) that leads to a Peclet number of $\mathrm{Pe} = 4.6 \times 10^8$. The Peclet number is analogous to the Reynolds number in which the kinematic viscosity quantifies the molecular diffusivity of momentum.

10.4.2 Pure Diffusion

There are a number of idealized diffusion problems that can be solved analytically. For ease of presentation of the data, we present the 1D cases of these problems to provide insight into pure diffusion problems. These solutions can be extended to two or three dimensions; see Boeker and van Grondelle (1999), Chanson (2004), and Fischer et al. (1979). We briefly review finite releases, infinite releases, and how various boundary conditions influence the concentration field.

A finite release of a mass M of pollutant over a very small length can be approximated as a point release of infinite concentration. The development of the concentration field in space is given by

$$C(x,t) = \frac{M}{\sqrt{4\pi Dt}} \exp\left(-\frac{x^2}{4Dt}\right). \qquad (10.53)$$

The concentration has a Gaussian shape with width growing in time and peak concentration decaying over time. A plot of this curve for unit mass and diffusion coefficient is shown in Figure 10.10a for a range of times after release. As the problem is treated as 1D the concentration is given in units of kg m^{-1}. If the pollutant is released over some finite length L, then the concentration can be calculated by treating the initial pollutant distribution as an infinite set of point source releases. The solution is calculated by adding up the contributions of all these infinitesimal contributions by integrating (10.53) over the release area of $-L/2$ to $L/2$. This leads to

$$C(x,t) = \frac{C_0}{2}\left(\mathrm{erf}\left(\frac{L/2 + x}{\sqrt{4Dt}}\right) + \mathrm{erf}\left(\frac{L/2 - x}{\sqrt{4Dt}}\right)\right) \qquad (10.54)$$

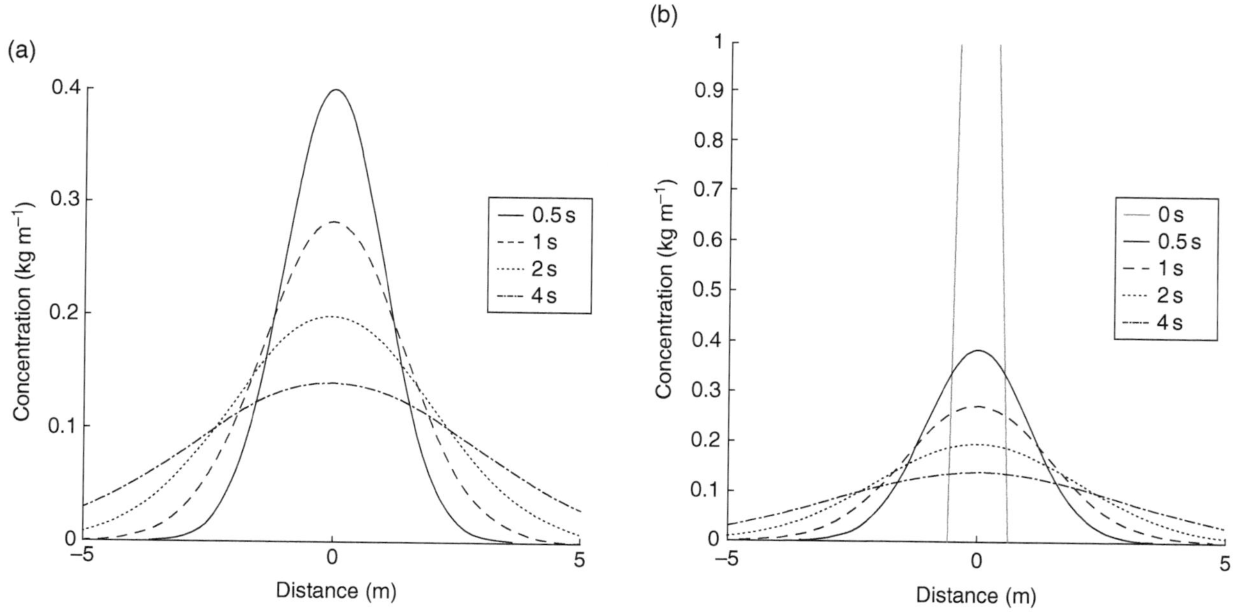

Figure 10.10 Plot of the concentration field over time for (a) a point release of mass $M = 1$ kg and with diffusion coefficient $D = 1$ m^2 s^{-1} and (b) a 1 kg release spread over 1 m (i.e. over $-0.5 < x < 0.5$) and $D = 1$ m^2 s^{-1}.

where C_0 is the initial concentration of the release. A plot of the concentration field at a range of times for $C_0 = 1$ over a 1 m release centered on the origin is shown in Figure 10.10b. This example is the distributed release equivalent of the point 1 kg mass release shown in Figure 10.10a, and, after some time, the concentration field has diffused laterally enough that the difference between the point and finite area release is negligible.

For a very large spatial release of pollutant, the diffusion of the sharp interface between the release and the surrounding ambient is similar to the diffusion of a finite release only for very large L. In this case the solution is given by integration of (10.53) over half the domain. Therefore, for a release of initial concentration C_0 over the range $x > 0$, the concentration is given by

$$C(x,t) = \frac{C_0}{2}\left(1 + \operatorname{erf}\left(\frac{x}{\sqrt{4Dt}}\right)\right). \tag{10.55}$$

This is plotted in Figure 10.11a for an initial concentration of $C_0 = 1\ \mathrm{kg\,m^{-1}}$. If the tracer/pollutant maintains a constant concentration at a boundary, such as water vapor above a liquid of constant temperature, then the solution is given by

$$C(x,t) = C_0\left(1 - \operatorname{erf}\left(\frac{x}{\sqrt{4Dt}}\right)\right) \tag{10.56}$$

which is plotted in Figure 10.11b.

The previous results all assume an infinite domain that would be appropriate for small releases in large fluid bodies such as the atmosphere, oceans, or lakes. However, there are many cases in which the fluid domain is limited such as in a river or building. In this case boundary conditions away from the source need to be imposed. Boundary conditions are typically complex but can be broadly categorized as some combination of fixed concentration or fixed flux (fixed gradient). Two simplified boundary conditions that can be analyzed analytically are a zero-flux boundary or a perfectly absorbing boundary in which the diffusion continues across the boundary unaffected by the boundary's presence. More complex boundary conditions require detailed computational evaluation.

For a perfectly absorbing boundary, the concentration field is unchanged as the boundary absorbs the flux of concentration as if the boundary was not there. This is shown in Figure 10.12a for the case considered in Figure 10.2a but with a perfectly absorbing boundary at $x = -2$ m.

For a solid boundary across which there is zero flux of pollutant, the concentration gradient must be zero at the boundary (see Eq. (10.45)). This can be achieved mathematically by adding a second mirrored concentration field on the other side of the boundary. For example, the diffusion of an instantaneous point release (10.53) that is located at distance x_B from a solid wall is given by

$$C(x,t) = \frac{M}{\sqrt{4\pi Dt}}\left(\exp\left(-\frac{x^2}{4Dt}\right) + \exp\left(-\frac{(x+2x_B)^2}{4Dt}\right)\right) \tag{10.57}$$

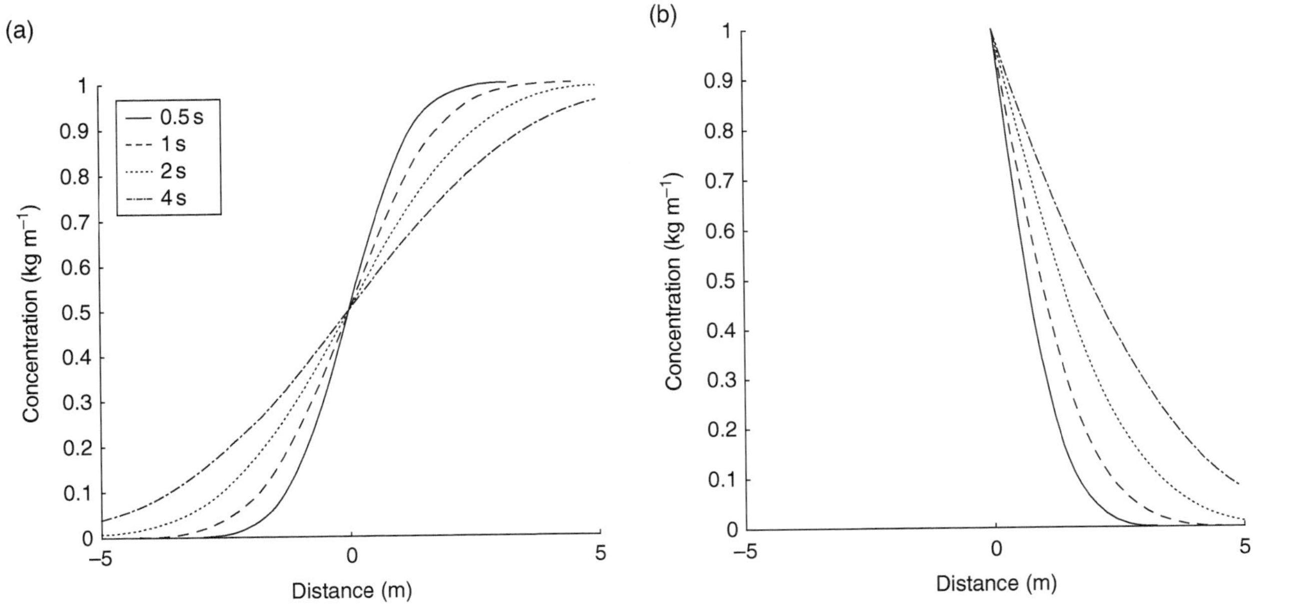

Figure 10.11 Plots of the concentration field over time for (a) an infinitely large release of initial concentration $C_0 = 1\ \mathrm{kg\,m^{-1}}$ over $x > 0$ and with diffusion coefficient $D = 1\ \mathrm{m^2\,s^{-1}}$ and (b) the diffusion of a constant concentration boundary over time for the same initial conditions as (a).

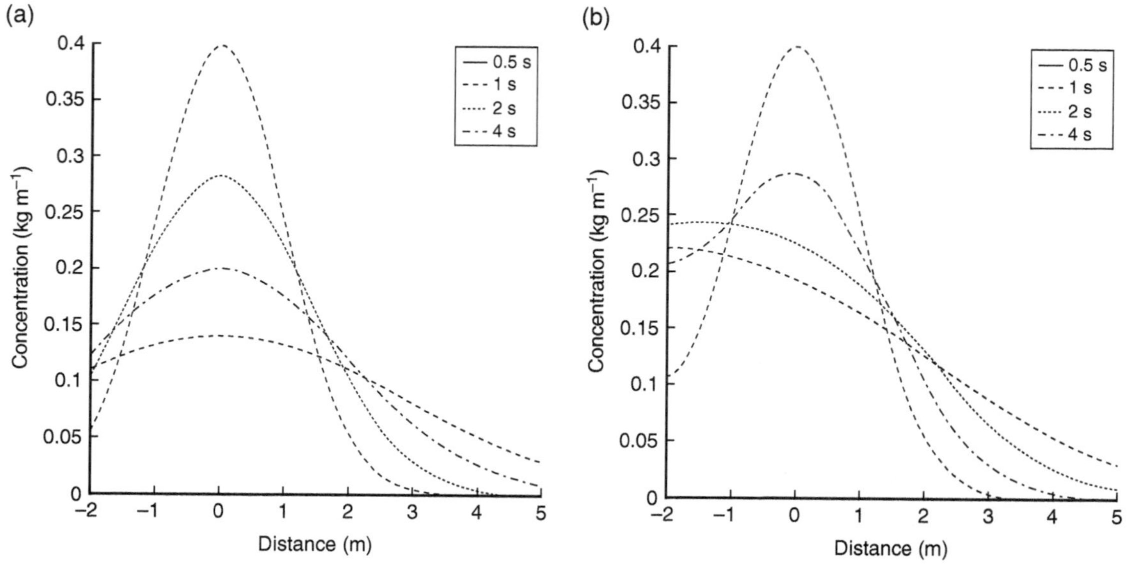

Figure 10.12 Plots of the concentration field over time for (a) a point release of mass $M = 1$ kg over at $x = 0$ with $D = 1\ \mathrm{m^2\,s^{-1}}$ and a perfectly absorbing boundary at $x = -2$ m and (b), the same case as (a) but with a zero flux boundary at $x = -2$ m.

The first term in brackets is the point-source diffusion solution for an infinite ambient (10.53). The second term in brackets is the mirrored point source located at $x = -2x_B$. The resulting concentration field is shown in Figure 10.12b for the same case as Figure 10.12a and illustrates the zero concentration gradient at the boundary created by the inclusion of the mirrored virtual point release.

In all of the examples in the figures above, the diffusion coefficient ($D = 1\ \mathrm{m^2\,s^{-1}}$) is several orders of magnitude higher than typical molecular diffusion coefficients, and the processes plotted occur more rapidly than typical diffusive processes. However, in all these examples the diffusion coefficient and time always appear as a product, Dt. Therefore, if one were to consider a release in which say $D = 1 \times 10^{-9}\ \mathrm{m^2\,s^{-1}}$, then the same plots could be used but with time scaled up by a factor of 10^9 such that the 1 s line would actually occur after 10^9 s.

This section has only considered molecular diffusion. However, turbulent mixing in the environment is often parameterized in terms of a turbulent diffusivity. Turbulent diffusion coefficients are typically several orders of magnitude greater than molecular values and are parameterized in terms of mean flow velocity and length scales. A detailed discussion of turbulent diffusion is beyond the scope of this chapter as the parameterizations are very application specific. See Chaudhry (2008) and Chin (2006) for more details on turbulent diffusion.

10.4.3 Advection–Diffusion

For constant velocity flows (most rivers have a constant average velocity on the time scale of diffusive events), the flow is equivalent to a purely diffusive flow attached to a coordinate system translating with the flow velocity. Therefore, the solutions for the diffusion equation shown above can be adjusted to be a solution to the constant velocity advection–diffusion equation by replacing the spatial coordinate x with $x - Ut$. So, for example, the concentration field that results from the instantaneous release of a point source of mass Mat $x = 0$ is given by

$$C(x, t) = \frac{M}{\sqrt{4\pi Dt}} \exp\left(-\frac{(x - Ut)^2}{4Dt}\right) \tag{10.58}$$

This is identical to (10.53) but with the peak concentration moving with the flow and located at $x = Ut$. A plot of this concentration distribution is shown in Figure 10.13a. In this case it is not possible to rescale time for different diffusion coefficients, as discussed above, because of the second time scale introduced by the velocity.

An extension of the standard advection–diffusion equation is to include the possibility of a chemically reacting pollutant. In this case the mass of pollutant decreases over time due to the chemical reaction. The simplest reaction form assumes that the reaction rate is proportional to the local concentration of the pollutant. The advection–diffusion equation therefore needs an additional sink term to give

$$\frac{\partial C}{\partial t} + U\frac{\partial C}{\partial x} = D\frac{\partial^2 C}{\partial x^2} - RC \tag{10.59}$$

where R is a reaction rate constant with units of 1 s. This is a very simplistic model and assumes an endless supply of other reactants such that the reaction rate is

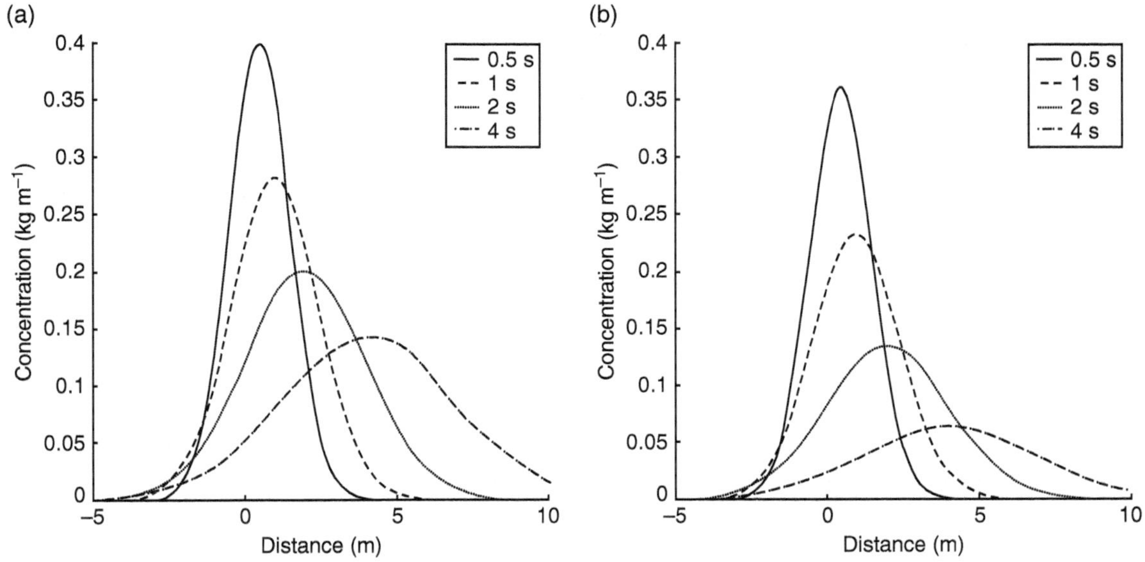

Figure 10.13 Plot of the concentration field over time for (a) a point release of mass $M = 1$ kg and with diffusion coefficient $D = 1\ \mathrm{m^2\,s^{-1}}$ advected by a flow velocity of $1\ \mathrm{m\,s^{-1}}$ and (b) the same case but with a chemical reaction ($R = 0.2$ 1 s) leading to a decay of pollutant over time.

only controlled by the concentration of the pollutant. This results in an exponential decay in pollutant over time and (10.58) becomes

$$C(x,t) = \frac{M}{A\sqrt{4\pi Dt}} \exp\left(-\frac{(x-Ut)^2}{4Dt} - Rt\right) \qquad (10.60)$$

A plot of this concentration field for the same conditions as the previous example is shown in Figure 10.13b for $R = 0.2$ 1 s.

While the previous examples are unsteady, there are advection–diffusion problems for which the concentration field is steady. For example, the steady merger of two streams of different tracer concentration, such as when two rivers merge, results in a steady downstream concentration field. In this case advection is the dominant process in the stream-wise (x) direction, and diffusion is the dominant process in the transverse (y) direction. For this case the full 2D advection–diffusion equation can be approximated by

$$U\frac{\partial C}{\partial x} = D\frac{\partial^2 C}{\partial y^2} \qquad (10.61)$$

This is analogous to the diffusion of a sharp interface (10.55) in the y direction with the concentration field advected downstream (x direction) with a velocity U. Therefore, for a given transverse slice, the diffusion time is the time since the two streams merged $t = x/U$, and the solution is given by (10.55) with t replaced by x/U:

$$C(x,y) = \frac{C_0}{2}\left(1 + \mathrm{erf}\left(\frac{y}{\sqrt{4Dx/U}}\right)\right)s \qquad (10.62)$$

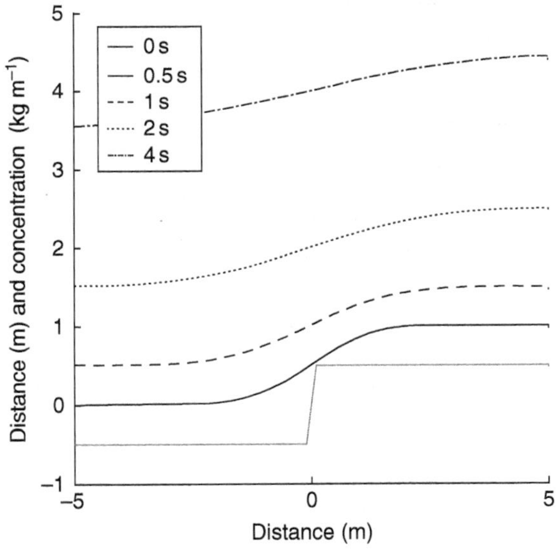

Figure 10.14 Concentration field for the merger of two streams with velocity $1\ \mathrm{m\,s^{-1}}$, with a concentration difference of 1 and a diffusion coefficient of $D = 1\ \mathrm{m^2\,s^{-1}}$. The concentration contours are displaced upward to represent the different distances downstream from the merger point at $x = 0$.

This simplified solution assumes there is negligible stream-wise diffusion compared with transverse diffusion and stream-wise advection. This is often a reasonable assumption as the stream-wise gradient is quite small. A plot of this concentration field is shown in Figure 10.14.

There are clearly many other applications of advection and diffusion in the environment that go beyond the simplified cases discussed above. In many cases the velocity field is nonuniform, and turbulent mixing is also important.

In these cases numerical solutions are required to account for nonstandard velocity and boundary condition cases. These topics are left to the application chapters.

10.5 Turbulent Jets

Turbulent jets are commonly used to discharge and dilute polluted effluent into an ambient environment. Jets entrain large volumes of ambient fluid and provide an efficient mechanism for dilution, thus minimizing the environmental impact of the effluents. Jets are normally discharged into rivers, lakes, reservoirs, or coastal environments. The ambient fluid may be still (lakes or reservoirs) or moving (rivers, streams, etc.). In general, jets are assumed to be nonbuoyant (density of the fluid is same as the ambient fluid), and pollutants, if any, are conservative. The outlet geometry and the jet velocity at the outlet affect the spread and mixing with the ambient fluid. Most common outlet geometries are circular and rectangular. Rectangular jets where the width to height ratio is large (also known as slot jets or plane jets) are considered as 2D jets, whereas circular jets are axisymmetric and hence can be considered as 2D using a longitudinal and radial coordinate system. Jets can be released in an ambient fluid of infinite extent, i.e. the bed, surface, and side boundaries are far way and do not influence the behavior of the jet and are termed as free jets. Jets that are discharged in an environment where one or more boundaries are in close proximity are designated as confined jets (Raiford and Khan, 2009).

10.5.1 Free Turbulent Jets in Quiescent Flow

The two most studied turbulent free jets are circular and slot jets (Abramovich, 1963; Rajaratnam, 1976; Aziz et al., 2008). A typical jet issuing out of a nozzle in an ambient quiescent fluid of infinite extent is shown in Figure 10.15. The origin of the coordinate system is at the center of the nozzle. The coordinate direction normal to the x-axis is y for the slot jet and r (radial distance) for the circular jet. The velocity at the jet nozzle is U_o and is assumed uniform. As the jet exits out of the nozzle, it interacts with the surrounding fluid developing a shear layer at the interface. The shear layer penetrates inward and reduces the zone where the velocity is uniform and at the same time expanding the jet by entraining the ambient fluid into the jet at the outer boundary of the shear layer. At the end of the potential core, the shear layer has penetrated to the center of the jet. Within the potential core the velocity at the center of the jet is U_o; outside this zone the centerline line velocity at any point along x-axis is given by u_m.

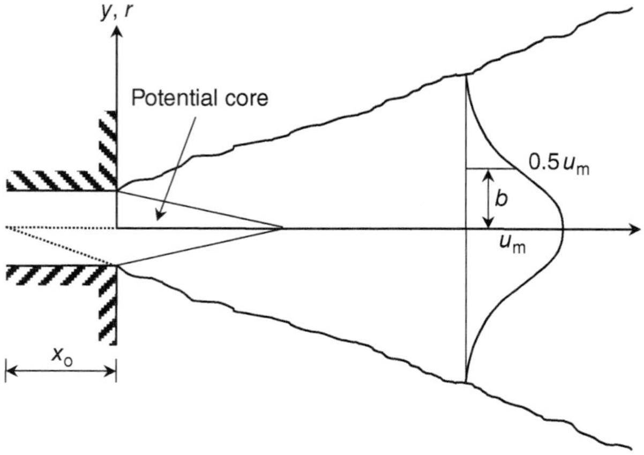

Figure 10.15 Spreading of a jet in an ambient environment.

The distance "b" is measured from the x-axis to a point across the jet where the velocity is $0.5u_m$. The jet grows and slows down in the downstream direction; however the momentum flux is conserved. Conservation of momentum within the jet leads to the similarity of velocity profiles across the jet when scaled appropriately.

The main properties of the free jet have been explored in depth through experimental and computational studies. The virtual origin is a point x_o behind the nozzle and is the location from which an equivalent jet issuing from an idealized point source of pure momentum would be located. The location of the virtual origin and the length of the potential core depends on the velocity profile and turbulence level at the exit point. Some salient features of the free circular and slot jets are discussed below.

The jet growth, as determined by the increase in the length b along the jet, is given by Eqs. (10.63) and (10.64) for the slot and circular jets, respectively, where d_o is half the height of a slot jet and d is the diameter of the circular jet. The constants B_1 and B_2 are found to vary based on the experimental data, but an average value of 0.097 may be used as recommended by Abramovich (1963) and Rajaratnam (1976). The virtual origin is found to range from 0 to $2.4d_o$ behind the nozzle for plane turbulent jets and $0.6d$–$2.2d$ behind the nozzle for circular turbulent jets. For practical use, the effect of virtual origin may be ignored:

$$\frac{b}{d_o} = \frac{B_1(x + x_o)}{d_o} \tag{10.63}$$

$$\frac{b}{d} = \frac{B_2(x + x_o)}{d} \tag{10.64}$$

The decay of the centerline velocity beyond the potential core is given by Eqs. (10.65) and (10.66) for plane and circular turbulent jets, respectively. Average values for constants B_3 and B_4 of 3.5 and 6.3 are recommended by

Rajaratnam (1976). As before, for practical use, the effect of virtual origin may be ignored. The velocity profiles outside the potential core are found to be similar and may be defined by Eq. (10.67), where u is velocity at a point y or r from the x-axis and α is equal to y/b for plane turbulent jets and r/b for circular turbulent jets:

$$\frac{u_m}{U_o} = \frac{B_3}{\sqrt{(x+x_o)/d_o}} \tag{10.65}$$

$$\frac{u_m}{U_o} = \frac{B_4}{(x+x_o)/d} \tag{10.66}$$

$$\frac{u}{u_m} = \exp\left(-0.693\alpha^2\right) \tag{10.67}$$

The velocity in the vertical or radial direction (v) is also found to be similar (Aziz et al., 2008). The turbulent normal and shear stresses within the jet are also explored experimentally as well as through computational models (Rajaratnam, 1976; Aziz et al., 2008; Aziz and Khan, 2011). Jets with 3D exit geometry are also common. 3D jets and intersecting jets are discussed in detail by Rajaratnam (1976) and Rajaratnam and Khan (1992).

10.5.2 Jets in a Crossflow

Figure 10.16 shows a nonbuoyant circular turbulent jet issuing into a crossflow. The crossflow is moving left to right with a uniform velocity U_s, and the jet is issuing into the crossflow normal to the flow at velocity U_o. Due to the crossflow, the jet bends in the direction of the crossflow while penetrating into the ambient fluid. The jet develops two counterrotating vortices, and the cross section deforms into a kidney shape (Rajaratnam, 1976). The potential core persists in the crossflow as the jet bends. For velocity ratio, $\beta(=U_o/U_s)$, greater than 4, the potential core (represented by zone 1) is within the straight portion of the jet. That is, for β greater than 4, zone 1 stays normal to the crossflow (Keffer and Baines, 1963). For a velocity ratio less than 4, the jet starts to bend within zone 1. In zone 2, the jet experiences a large deflection in the direction of crossflow, while zone 3 is the final stage of the jet where the jet moves mainly in the direction of the crossflow. In Figure 10.16, the curves C1, C2, C3, and C4 are identified as outer boundary that shows the penetration of the jet in the crossflow, the inner boundary, the jet axis that is the trace of the path where maximum velocity occurs in the cross section, and the geometric center of the jet, respectively.

The profiles of C1, C2, and C4 were investigated in detail by Pratte and Baines (1967) for velocity ratios ranging from 5 to 35. Based on the experimental results, the profiles for these curves are given by Eqs. (10.68), (10.69), and (10.70), respectively. The origin for the coordinate system is at the center of the nozzle, x-axis is in the horizontal direction and y-axis is in the vertical direction. The distances y_1, y_2, and y_4 represent vertical locations of C1, C2, and C4 curves as a function of x, respectively:

$$\frac{y_1}{\beta d} = 2.63\left(\frac{x}{\beta d}\right)^{0.28} \tag{10.68}$$

$$\frac{y_2}{\beta d} = 1.35\left(\frac{x}{\beta d}\right)^{0.28} \tag{10.69}$$

$$\frac{y_4}{\beta d} = 2.05\left(\frac{x}{\beta d}\right)^{0.28} \tag{10.70}$$

The center profile (C4) of the jet in terms of a curvilinear coordinate η was also investigated by Pratte and Baines (1967). The profiles for zone 1 and zone 3 are given by Eqs. (10.71) and (10.72), respectively. The first zone extends up to $\eta/\beta d \simeq 5$ and represents the potential core (Pratte and Baines, 1967). The region between these two zones is represented by zone 2:

$$\frac{y_4}{\beta d} = \frac{\eta}{\beta d} \tag{10.71}$$

$$\frac{y_4}{\beta d} = 1.63\left(\frac{\eta}{\beta d}\right)^{1/3} \tag{10.72}$$

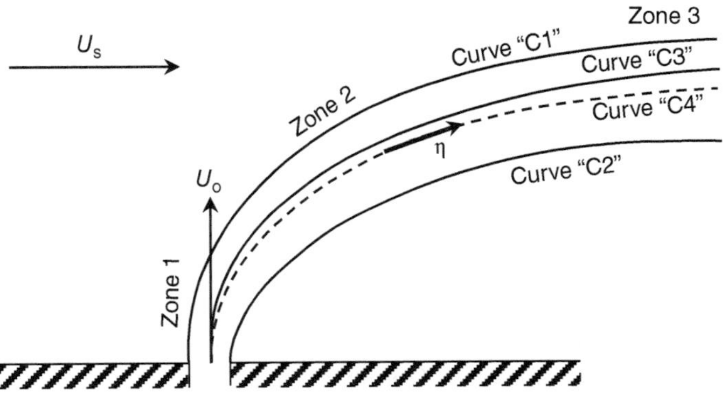

Figure 10.16 Layout of circular jet in a crossflow (Rajaratnam, 1976).

Jets issuing into a crossflow at an angle less than 90° have been studied by Margason (1968) and Platten and Keffer (1971). In addition, Abramovich (1963) reported other studies for jets issuing at different angles in the crossflow. These studies suggest that the profiles are strongly dependent on the angle at which a jet issues into the crossflow.

Plane turbulent jets in a crossflow have received limited attention. Ramaprian and Haniu (1983) and Haniu and Ramaprian (1989) experimentally investigated the mean and turbulent flow properties of plane turbulent jets in crossflow. Kalita et al. (2002) used 3D RANS equations with a $k-\varepsilon$ turbulence closure model to simulate plane turbulent jets in a crossflow. The computational model was verified using the two experimental studies mentioned earlier.

10.5.3 Pollutant Transport in Jets

A jet issuing into an ambient fluid diffuses after exiting. If a jet has a well-mixed pollutant with uniform concentration C_o at the exit, then, as the jet entrains ambient fluid, it expands, and the pollutant concentration will decrease in a manner similar to the velocity. Out of the potential core, the maximum concentration for a free jet will occur along the longitudinal axis of the jet (x-axis) and is denoted by c_m, and the concentration will decrease along the y or r direction. The growth or length scale b_c is similar to the velocity scale b, i.e., b_c is the distance along y or r direction where $c = 0.5c_m$. The concentration profiles along and across plane and circular turbulent jets are provided by (Zhang et al., 2012). The concentration growth is $b_c = 1.17b$, where b is the growth of plane or circular turbulent jets. The decay of maximum concentration along the centerline (x-axis) of plane and circular turbulent jets is given by Eqs. (10.73) and (10.74), respectively. The concentration profile for both types of jets is given by Eq. (10.75), where α is as defined before. The characteristics of pollutant transport in circular turbulent jets in crossflow have been studied by several authors, and details are provided by Zhang et al. (2012):

$$\frac{c_m}{C_o} = \frac{3.45}{\sqrt{x/d_o}} \tag{10.73}$$

$$\frac{c_m}{C_o} = \frac{5.34}{x/d_o} \tag{10.74}$$

$$\frac{c}{c_m} = \exp\left[-0.693\left(\frac{\alpha}{1.17}\right)^2\right] \tag{10.75}$$

10.6 Turbulent Buoyant Plumes

Turbulent plumes, like turbulent jets discussed in the previous section, are ubiquitous in environmental flows. Like a round jet, round turbulent plumes have a broad range of flow scales, are relatively long and thin, and have an approximately linear radial growth rate with distance from the source. The significant difference between a jet and a plume is that plumes are buoyant. That is, they have a different density from their surroundings. There is, therefore, a buoyancy force acting on the flow and a resulting continuous change in momentum flux along the axis of the plume. This is in contrast to turbulent jets that have a constant momentum flux. The density difference is typically parameterized in terms of a reduced gravity given by

$$g' = g\left(\frac{\rho_a - \rho_p}{\rho_a}\right) \tag{10.76}$$

where g is gravitational acceleration, ρ_a is the ambient fluid density, and ρ_p is the local density of the plume fluid. The plume is driven by the rate at which reduced gravity is released at the source that is parameterized in terms of a buoyancy flux F. For a thermal plume, that is, a plume where the density difference is generated by a temperature difference, the buoyancy flux depends on the source power output and is given by

$$F = \frac{Pg\varepsilon_V}{C_P \rho_a} \tag{10.77}$$

where P is the power output of the source in Watts, ε_V is the coefficient of thermal expansion in K^{-1}, and C_P is the specific heat of the plume fluid in Joules $kg^{-1} K^{-1}$ (Batchelor, 1954). For a gas, the coefficient of thermal expansion is typically given by $\varepsilon_V = 1/T_O$, where T_O is the absolute temperature of the ambient gas in Kelvin. In the case of a plume with a density difference generated by some constituent of the plume, such as dissolved salt or suspended sediment, the buoyancy flux is given by the product of the source-reduced gravity and source volume flux:

$$F = g'_0 Q_0 \tag{10.78}$$

where Q is the volume flux of the plume in $m^3 s^{-1}$ at any height and F has units of $m^4 s^{-3}$. The subscript "0" indicates a source condition.

Turbulent plumes occur in the atmosphere (e.g. smokestacks [Carson and Moses, 1969], cooling towers [Rothman and Ledbetter, 1975], volcanic eruptions [Woods, 2010], and wild fires [Bhutia et al., 2010]), lakes, rivers, and oceans (e.g. thermal power plant cooling water discharge [Lee and Jirka, 1981], fresh wastewater discharges into the salty ocean [Rawn et al., 1960], or downward flowing sediment plumes from dredge waste disposal [Chowdhury and Testik, 2014a]). They also occur within the built environment (e.g. smoke plumes in building fires [Jaluria and Kapoor, 1992] or thermal plumes from heat sources in rooms [Linden et al., 1990]).

This section will begin with a review of the basic equations of motion for steady round (point source) and slot (line source) turbulent plumes. This will be followed by discussion of how plume dynamics are altered by unsteady source conditions, stratified ambient conditions, and chemical reactions. The section will conclude with a description of plume flows in confined regions.

10.6.1 Round Turbulent Plumes

The classic steady integral from of the turbulent plume equations was originally derived by integrating the Navier–Stokes equations across a horizontal plane assuming a consistent shape for the profiles of vertical velocity and density difference (see Morton et al., 1956). Herein the equations are derived using a simplified control volume approach which, while less mathematically rigorous, is easier to understand physically. The model assumes that the plume is relatively long and thin, that is, the plume radius (b) is small compared with distance from the source (z). It is also assumed that, at any given height, the velocity within the plume (U) is constant across the width of the plume and that the pressure is hydrostatic. Finally, the plume is assumed to develop in a uniform density quiescent environment and that there are no chemical or thermodynamic processes that alter the local density of the fluid (such as endo- or exothermic reactions or phase changes).

Conservation of volume and the momentum equation can then be applied to a control volume of height dz at any arbitrary distance above the plume source (see Figure 10.17a). For a steady flow the sum of the inflows must balance the sum of the outflows such that

$$Q + u_e 2\pi b dz = Q + \mathrm{d}Q \tag{10.79}$$

In (10.79) the second term on the left-hand side represents the entrainment of ambient fluid into the plume envelope. This inflow is characterized in terms of a mean entrainment velocity u_e. The momentum equation applied to the control volume balances the sum of the forces in the vertical direction with the change in momentum flux (see Figure 10.17b). That is,

$$\rho_a g \mathrm{d}\forall - \rho_p g \mathrm{d}\forall = M + \mathrm{d}M - M \tag{10.80}$$

where the forces on the left-hand side are the buoyancy force acting up and the weight force acting down. Equations (10.79) and (10.80) can be simplified to read

$$\frac{\mathrm{d}Q}{\mathrm{d}z} = 2\pi b u_e \text{ and } \frac{\mathrm{d}M}{\mathrm{d}z} = \left(\rho_a - \rho_p\right) g \pi b^2 \tag{10.81}$$

However, the pair of Eq. (10.81) have three unknowns (Q, M, and u_e). To close the problem (Morton et al., 1956) employed the entrainment assumption that states that the entrainment velocity scales linearly on the characteristic plume velocity at any given height. That is,

$$u_e = \alpha U \tag{10.82}$$

This assumption requires the plume to be fully developed and self-similar. The limitations of the entrainment assumption for plumes are discussed later in this section. Equations (10.81) and (10.82) can be combined to form a

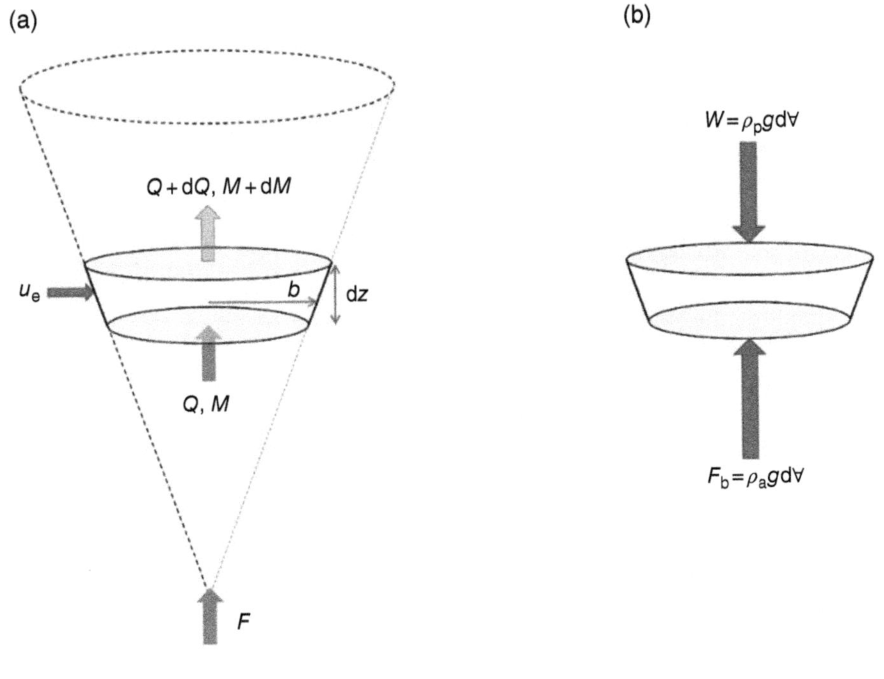

Figure 10.17 (a) Schematic diagram of a round plume with source buoyancy flux F showing the control volume of radius b and height $\mathrm{d}z$ at an arbitrary vertical distance z. Also shown are the inflowing and outflowing fluxes of volume (Q) and momentum (M) and the entrainment velocity (u_e). (b) Force diagram for the control volume showing the weight acting down and the buoyancy force acting up.

set of two first-order separable ordinary differential equations for the fluxes of volume and momentum. However, the variables are often rescaled for simplicity to give

$$\hat{Q} = Ub^2\left(=\frac{Q}{\pi}\right), \hat{M} = U^2b^2\left(=\frac{M}{\pi\rho_a}\right), \text{and } \hat{F} = g'\hat{Q} \tag{10.83}$$

where it is assumed that the velocity is uniform across the plume width at any height (the so-called top-hat profile). The final pair of ODEs is then

$$\frac{d\hat{Q}}{dz} = 2\alpha\sqrt{\hat{M}} \tag{10.84a}$$

and

$$\frac{d\hat{M}}{dz} = \frac{\hat{Q}\hat{F}}{\hat{M}} \tag{10.84b}$$

For an idealized point source of buoyancy flux F with no source volume or momentum flux, the ODEs can be integrated to give the rescaled volume flux, momentum flux, and plume radius as

$$\hat{Q} = \left(\frac{5\hat{F}}{8\alpha}\right)^{1/3}\left(\frac{6\alpha z}{5}\right)^{5/3}, \hat{M} = \left(\frac{5\hat{F}}{8\alpha}\right)^{2/3}\left(\frac{6\alpha z}{5}\right)^{4/3}, \text{ and } b = \frac{6\alpha z}{5} \tag{10.85}$$

It can be seen from (10.85) that the volume flux increases with height raised to the power of 5/3, which is more rapid than for a jet where the volume flux increases linearly with distance from the source. This is due to the buoyancy force continuously increasing the momentum flux with height in contrast to the fixed momentum flux in a jet. The same modeling approach used for plumes can be applied to jets simply by setting $F = 0$ in (10.84). In this case the momentum flux is constant (10.84b), and the volume flux scales on the square root of the source momentum flux and linearly with distance from the source (10.84a).

In general the primary parameter of interest is the volume flux Q as it is a measure of the dilution of any pollutant transported by the plume and is given by

$$Q = CF^{1/3}z^{5/3} \tag{10.86}$$

where $C = 2.5\alpha^{4/3}$. The local concentration of any conserved pollutant (nonreacting) can be calculated given the pollutant loading at the source and the local volume flux. For a source pollutant flux of K (kg s^{-1}), the local concentration (k kg m^{-3}) is simply the flux divided by the local volume flux. That is,

$$k = 6.3KF^{-1/3}z^{-5/3} \tag{10.87}$$

As such, plumes are much more efficient at diluting pollutants due to their significantly larger entrainment of ambient fluid compared with a jet.

The above analysis assumes that the plume originates from an idealized source of buoyancy only. However, almost all real-world plumes have finite area sources with finite fluxes of volume, momentum, and buoyancy. Far from the source the plume behaves as if it originated from an idealized point source though the location of virtual point source differs from the actual physical source location. In this case, the vertical coordinate, z, in (10.85) is replaced with $z + z_0$ where z_0 depends on the source conditions (Hunt and Kaye, 2001). The near-field dynamics depends on the balance of source fluxes. At any height above the plume source, it is possible to calculate a number of characteristic length scales from the local fluxes (Q, M, F). Two of particular interest are the volume flux length scale and momentum jet length given by

$$L_Q = \frac{6\hat{Q}}{5\alpha\sqrt{\hat{M}}} \text{ and } L_M = \sqrt{\frac{5\hat{M}^{3/2}}{9\alpha\hat{F}}}, \tag{10.88}$$

respectively. The first length scale represents the vertical distance a pure plume (based on Eq. (10.85)) would need to reach in order to have a radius equal to the local radius. The momentum jet length is the height a pure plume would have to reach in order to achieve the local momentum flux. Far from the source the two length scales become equal; however, near the source, the plume dynamics depend on their relative size. For example, if $L_Q > L_M$, then the plume has a larger diameter than the equivalent pure plume, and the plume may initially contract before developing into a pure plume. If $L_M > L_Q$, then there is an excess of momentum compared with a pure plume, and the plume initially behaves like a jet before adjusting to pure plume behavior further from the source. This balance is often parameterized in terms of the square of the length scales

$$\Gamma = \frac{L_Q^2}{L_M^2} = \frac{5\hat{Q}^2\hat{F}}{4\alpha\hat{M}^{5/2}}\left(= \frac{5}{4\alpha}\frac{bg'}{U^2} = \frac{5}{4\alpha}\frac{1}{Fr^2}\right) \tag{10.89}$$

sometimes referred to as the plume Richardson number. Plume behavior is determined by the source value of Γ (Morton and Middleton, 1973). If $\Gamma_0 = 1$, and the plume is a pure plume at all heights. If $\Gamma_0 = 0$, the flow is a jet at all heights. For $0 < \Gamma_0 < 1$, the plume is described as forced and will behave as a jet over some adjustment distance prior to developing into a pure plume (Morton, 1959). For $\Gamma_0 > 1$, the plume is described as lazy, and, if Γ_0 is large enough, the plume will accelerate and contract as it moves away from the source before developing in to a pure plume (Hunt and Kaye, 2005).

Measurements of entrainment into jets and plumes indicate that the top-hat profile entrainment coefficient for a jet ($\alpha_{\text{jet}} = 0.076$) is less than that of a plume ($\alpha_{\text{plume}} = 0.12$) (Kaye, 2008). There have been a number of attempts to develop variable entrainment models for jets and plumes with the entrainment coefficient typically being written as a function of the local plume Richardson number, that is, $\alpha = \alpha(\Gamma)$ (Fischer et al., 1979). Experimental evidence suggests that in the range between pure jet and pure plume flow ($0 < \Gamma < 1$), the entrainment coefficient is a linear function of Γ (Wang and Law, 2002).

10.6.2 Turbulent Line Plumes

Plume sources can have many different shapes; however, the most commonly modeled are round plumes discussed above and line plumes. A line plume can form above a linear heat source such as strip-heating elements in a room or from the merger of a line of round plumes such as a line of diffusers (Jirka, 2006). Other examples include salt plumes that form under cracks in the arctic ice cap (Ching et al., 1996) or from the merger of thermal plumes rising above a line of people sitting in an auditorium.

Line plumes (see schematic diagram in Figure 10.18) are modeled in the same manner as round plumes though they are typically characterized in terms of their buoyancy flux per unit length of source, i.e. $F^* = F/L$ (with units of $m^3\ s^{-3}$). The pair of ODEs analogous to (10.84) is given by

$$\frac{dQ^*}{dz} = 2\alpha \frac{M^*}{Q^*} \text{ and } \frac{dM^*}{dz} = \frac{Q^* F^*}{M^*} \tag{10.90}$$

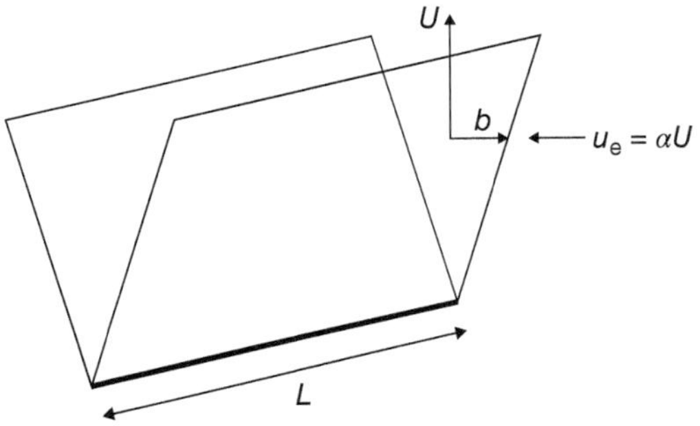

Figure 10.18 Schematic diagram of a line plume of source length L showing the plume half width b, vertical velocity U, and entrainment velocity u_e.

where Q^* and M^* are the fluxes of volume and momentum per unit length. The solution for an idealized line source of buoyancy only is

$$Q^* = \left(\frac{F^*}{2\alpha}\right)^{1/3} 2\alpha z, M^* = \left(\frac{F^*}{2\alpha}\right)^{2/3} 2\alpha z, U = \left(\frac{F^*}{2\alpha}\right)^{1/3}, \text{ and}$$
$$b = \alpha z \tag{10.91}$$

The entrainment coefficient for a line plume is $\alpha = 0.15$ (Lee and Chu, 2003), which is larger than that for a round plume though their dilution rate is smaller as the volume flux only increases linearly with distance from the source. The adjustment of a line plume formed by a finite width source is similar to that of a finite diameter round plume as discussed above.

10.6.3 Bent-over Plumes

The above analysis assumes that the ambient fluid surrounding the plume is quiescent. However, many plumes discharge vertically into a horizontally flowing ambient (such as smokestack plumes on a windy day). In this case, entrainment of horizontal momentum from the flowing ambient into the plume will cause the plume to bend over in the direction of the ambient velocity. Therefore, a third equation is required to account for the horizontal component of the plume's momentum flux (Hoult and Weil, 1972; Ooms and Mahieu, 1981). For a round plume in a crossflow, equations are often written in terms of the volume flux

$$\frac{d\hat{Q}}{ds} = 2\frac{\hat{Q}}{\sqrt{\hat{M}}}\left(\alpha\left|\frac{\hat{M}}{\hat{Q}} - U\cos\theta\right| + \beta\left|U\sin\theta\right|\right), \tag{10.92}$$

momentum flux

$$\frac{d\hat{M}}{ds} = \frac{\hat{Q}\hat{F}}{\hat{M}}\sin\theta + 2\frac{\hat{Q}}{\sqrt{\hat{M}}}\left(\alpha\left|\frac{\hat{M}}{\hat{Q}} - U\cos\theta\right| + \beta\left|U\sin\theta\right|\right)U\cos\theta, \tag{10.93}$$

and trajectory angle

$$\hat{M}\frac{d\theta}{ds} = \frac{\hat{Q}\hat{F}}{\hat{M}}\cos\theta + 2\frac{\hat{Q}}{\sqrt{\hat{M}}}\left(\alpha\left|\frac{\hat{M}}{\hat{Q}} - U\cos\theta\right| + \beta\left|U\sin\theta\right|\right)U\sin\theta. \tag{10.94}$$

Written in this form the coordinate s is the distance along the plume centerline, $\hat{Q}$ is the scaled volume flux (see Eq. (10.83) for definition), $\hat{F}$ is the scaled buoyancy flux (assumed constant with height), $\hat{M}$ is the scaled momentum flux, θ is the angle of the plume axis to the horizontal (see Figure 10.19), U is the horizontal ambient velocity, and α and β are entrainment coefficients.

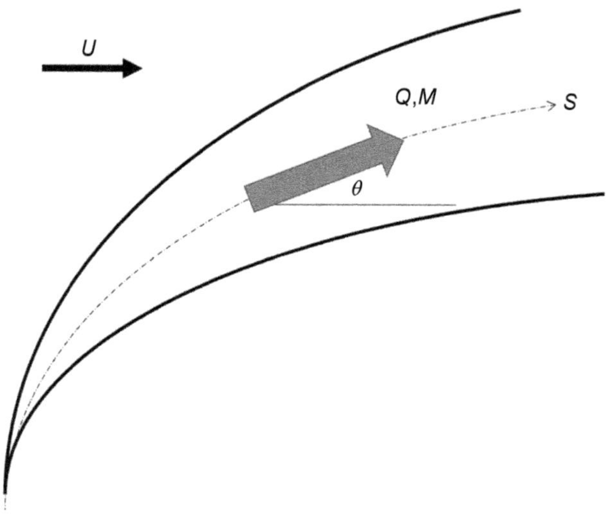

Figure 10.19 Schematic diagram of a bent-over plume showing the centerline trajectory coordinate *s*, local trajectory angle *θ*, the volume (*Q*) and momentum (*M*) fluxes, and the ambient wind speed *U*.

Entrainment due to longitudinal shear, that is shear parallel to the plume centerline, is parameterized by α and is directly analogous to the vertical round plume entrainment coefficient except that the entrainment velocity is given by the entrainment coefficient multiplied by the difference between the plume axial velocity and the component of the ambient velocity parallel to the plume axis (first term in brackets on the right-hand side of Eq. (10.92)). Entrainment due to transverse shear, that is shear due to the component of the ambient velocity normal to the plume centerline, is parameterized by β. This entrainment is analogous to the mixing between the ambient and a line of buoyant fluid rising in a quiescent ambient (Turner, 1973).

There is no simple analytic solution to these equations, and they are typically solved numerically for a given set of plume source and crossflow conditions. The introduction of the ambient crossflow adds an additional nondimensional parameter to the problem, namely, the ratio of the plume exit velocity to the ambient characteristic horizontal velocity. See Tohidi and Kaye (2016) for a more detailed discussion of bent-over plume models and the impact of source conditions and the ambient velocity profile on plume dynamics.

Bent-over plume models have also been used to model the flow of a plume with a source momentum flux that has a horizontal component (Lane-Serff et al., 1993). This may occur when a diffuser used to disperse a buoyant fluid into a lake issues horizontally or when cold air vented from an air-conditioning system blows horizontally out of a wall.

10.6.4 Plumes in a Stratified Ambient

The models presented above are for plumes in a uniform density environment. However, many plumes form in stratified environments such as the atmosphere and oceans. In this case, the density difference between the plume and its surroundings changes due to both mixing and the change in ambient density with height. In this case an additional equation is needed to calculate the rate of change of buoyancy flux with height and is given by

$$\frac{\mathrm{d}\hat{F}}{\mathrm{d}z} = -N^2\hat{Q} \tag{10.95}$$

where

$$N^2 = -\frac{\mathrm{d}g'_{\mathrm{a}}}{\mathrm{d}z} \tag{10.96}$$

N is known as the Brunt–Vaisala frequency, and g'_{a} is the local ambient reduced gravity relative to the ambient density at a reference height. The stratification is stable provided the ambient density decreases with increasing height ($N^2 > 0$).

A plume in a stable linearly stratified environment (constant N) will entrain ambient fluid and get denser with height while the environment density decreases with height. The plume will eventually reach a height at which the ambient and plume densities are the same, the so-called neutral height. Upon reaching this height the plume will continue to rise due to its momentum but lose momentum with height due to being surrounded by less dense fluid. It will eventually collapse back toward the neutral height and spread out laterally.

The neutral height will depend on the strength of the stratification and the source buoyancy flux of the plume. Dimensional analysis suggests that the neutral height will scale on the stratification length scale

$$L_{\mathrm{N}} = F^{1/4}N^{-3/4} \tag{10.97}$$

provided that length is significantly greater than the source length scales defined in (10.88). Laboratory and field data suggests that the neutral height is approximately $3.8L_{\mathrm{N}}$ (Briggs, 1969), while the maximum rise height is approximately $4.5L_{\mathrm{N}}$ (Wong and Wright, 1988).

10.6.5 Chemically Reacting Plumes

Another mechanism by which the buoyancy flux of a plume can change is through heat addition or extraction due to an exothermic or endothermic chemical reaction. The case typically considered is a plume containing one reactant issuing into an environment containing the second reactant. As the plume entrains ambient fluid, the reactants are mixed together, and the reaction occurs. If the reaction is exothermic, the buoyancy flux of the plume increases, while it will decrease if the reaction is endothermic. For an endothermic reaction the buoyancy flux could be reduced to zero and then become negative.

Here the situation is more complex than the simple addition of an equation for the rate of change of buoyancy flux with height. Transport equations are required for each of the reactants and reactant products within the plume, a reaction rate equation, and a heat transport equation that can be converted into a buoyancy flux equation using (10.77). It is also possible for the reaction products to be of a different density compared with the reactants that also must be accounted for in the plume buoyancy flux equation. The plume behavior depends on the relative speed of the entrainment process and the reaction rate. For a fast reaction rate, the heat release (or extraction) occurs at the height at which each fluid packet is entrained into the plume, and the local change in buoyancy flux is a function of the local entrainment rate. For slow reaction rates the reaction (and heat release or extraction) continues as the entrained fluid is transported vertically up the plume. See Conroy and Llewellyn Smith (2008), Campbell and Cardoso (2010), Cardoso and McHugh (2010), and Domingos and Cardoso (2013) for more detailed discussions of chemically reacting plumes.

10.6.6 Plumes with Unsteady Source Conditions

All the prior analysis assumes that the plume source conditions are constant. This is often a reasonable assumption for man-made plumes such as smoke plumes from power plants or wastewater outfalls. However, there are cases, such as plumes formed by convection above land heated by the sun (where the heat release rate varies considerably over time) or plumes from volcanic eruptions. In these cases the steady plume model is inappropriate, and the steady plume equations should be extended to account for time-varying source conditions (Scase et al., 2006). This results in a set of coupled partial differential equations that, for a round plume in a uniform quiescent environment, are given by

$$\frac{\partial}{\partial t}\left(\frac{\hat{Q}^2}{\hat{M}}\right)+\frac{\partial \hat{Q}}{\partial z}=2\alpha\sqrt{\hat{M}},\ \frac{\partial \hat{Q}}{\partial t}+\frac{\partial \hat{M}}{\partial z}=\frac{\hat{Q}\hat{F}}{\hat{M}},$$
$$\text{and}\ \frac{\partial}{\partial t}\left(\frac{\hat{Q}\hat{F}}{\hat{M}}\right)+\frac{\partial \hat{F}}{\partial z}=0 \tag{10.98}$$

The solution to these equations has been investigated for step changes in source buoyancy flux (Scase et al., 2008) though more general solutions to the equations will require numerical calculation.

10.6.7 Turbulent Fountains

Turbulent plume theory has also been applied to the case of a plume with source momentum that is opposed by the source buoyancy flux (a dense plume injected upward or a light plume injected downward). Such flows are called fountains and have a wide variety of applications in environmental and industrial flows (Hunt and Burridge, 2015). Fountains will flow in the direction of their source momentum until the retarding action of the buoyancy force reduces the flow momentum to zero, and after which the fountain reverses direction and flows back toward its source. Their behavior depends on source value of Γ (10.89). For small values of Γ, that is, flows where the momentum jet length is relatively large, the fountain behaves like a jet for a substantial distance before reversing direction. In this case the rise height scales on the momentum jet length (so-called forced fountains (Kaye and Hunt, 2006)). For larger values of Γ, the flow reversal distance is relatively low and scales on the source diameter (so-called weak fountains).

10.6.8 Confined and Semiconfined Plumes

The preceding discussion assumes that the plume is in an infinite volume ambient. However, there are many cases in which plumes occur in finite enclosures such as thermal plume rising from a person in a room or a dense mud plume dumped in a lake. In the discussion that follows, the plume is assumed to be positively buoyant and rising vertically up in a finite enclosure. There are two main cases to consider, a sealed enclosure and a ventilated enclosure, in which fluid can drain in to and out of the enclosure. These cases are illustrated in Figure 10.20.

The flow of a plume in a finite enclosure is known as a filling box flow (Baines and Turner, 1969) (see Figure 10.20a). When the plume is first turned on, the plume rises to the top of the box and spreads out laterally forming a buoyant layer of plume fluid. Over time more plume fluid is added to the buoyant upper layer, and the layer thickens, and the interface between the buoyant layer and the ambient layer, called the first front, moves down toward the plume source. Assuming that there is no mixing of the upper buoyant layer down into the ambient fluid below (a reasonable assumption as the density difference suppresses mixing), the thickness of the upper buoyant layer $(H-h)$ can be calculated using conservation of volume

$$A_E\frac{\mathrm{d}}{\mathrm{d}t}(H-h)=CF^{1/3}h^{5/3} \tag{10.99}$$

where A_E is the cross-sectional area of the enclosure and the right-hand side is the plume volume flux at the height

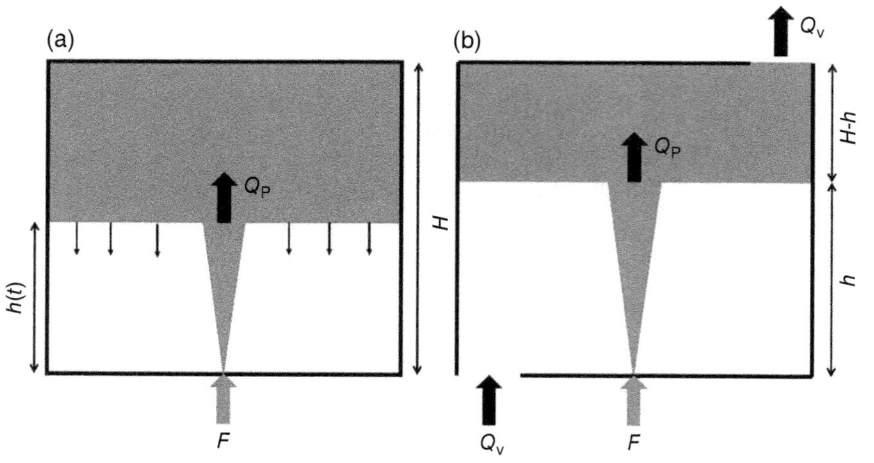

Figure 10.20 Schematic diagram of (a) a filling box flow and (b) a ventilated filling box.

of the first front above the plume source. For a uniform cross-sectional area, this can be solved for the front position over time as

$$h^{-2/3} - H^{-2/3} = \frac{2CF^{1/3}}{3A_E}t \tag{10.100}$$

The stratification in the buoyant layer above the first front can also be calculated (Worster and Huppert, 1983). The filling box model has been used to investigate a number of environmental flows including the stratification of the oceans and the stability of liquid natural gas tanks (Germeles, 1975).

The classic filling box model has been extended to consider an enclosure that is connected to an external ambient through vents at the top and bottom of the enclosure (Linden et al., 1990). As a result, the buoyant upper layer drives a flow out the top vent and induces a flow in through the bottom vent (see Figure 10.20b). The flow rate can be calculated by balancing the hydrostatic pressure head with the head loss through the vents to give

$$g'(H-h) = Q^2\left(\frac{C_B}{2A_B^2} + \frac{C_T}{2A_T^2}\right) \tag{10.101}$$

where C_B and C_T are the loss coefficients for the bottom and top vents, respectively. Equation (10.101) is usually simplified to

$$Q = A^*\sqrt{g'(H-h)} \tag{10.102}$$

where

$$A^* = \left(\frac{C_B}{2A_B^2} + \frac{C_T}{2A_T^2}\right)^{-1/2} \tag{10.103}$$

is the effective area of the vents. After an initial transient adjustment (Kaye and Hunt, 2004), a steady-state flow is established in which the flow rate through the enclosure (10.102) is matched by the flow rate in the plume at the height of the interface separating the upper buoyant layer from the lower ambient layer. In this steady state the upper layer is well mixed and has a reduced gravity equal to that of the plume at the height of the interface. The upper layer buoyancy and interface height are given by

$$g' = \frac{F^{2/3}}{Ch^{5/3}} \text{ and } \frac{A^*}{C^{3/2}H^2} = \frac{\xi^{5/2}}{\sqrt{1-\xi}} \tag{10.104}$$

respectively, where $\xi = h/H$ is the nondimensional interface height and C is the prefactor in the plume volume flux Eq. (10.86).

The so-called emptying filling box flow was initially developed to model flow in a naturally ventilated building containing localized heat sources (Linden et al., 1990) though it has been extended to more complex enclosure geometries (Flynn and Caulfield, 2006; Kaye and Hunt, 2007), the modeling of indoor air quality (Hunt and Kaye, 2006; Bolster and Linden, 2009a, b), and even plume flow in confined porous media (Roes et al., 2014).

10.7 Gravity Currents

A gravity current, often referred to as a density current, is the flow of a fluid into another fluid due to the horizontal pressure gradient created by the density difference between the two fluids (Chowdhury and Testik, 2014b; Testik and Ungarish, 2016). A typical gravity current created in a laboratory experiment is shown in Figure 10.21 (from Chowdhury et al., 2009). The density difference between the fluids can result from, for example, dissolved or suspended materials and/or temperature differences (Ungarish, 2009). Gravity currents are ubiquitous in the environment. A few examples of environmental gravity currents are turbidity currents, mud slides, sea breezes, katabatic flows, and oil slicks. Simpson (1997) discusses

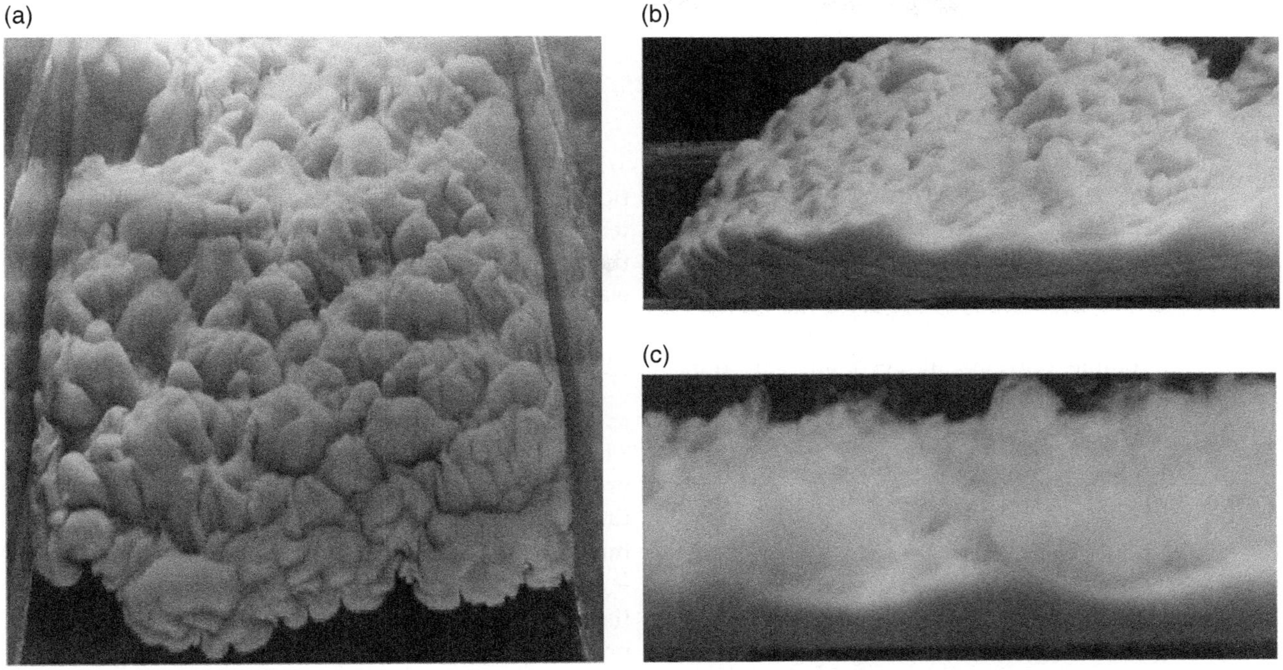

Figure 10.21 Photographs of a typical gravity current from a laboratory experiment: (a) top view showing lobe-cleft instability patterns at the leading edge of the current, (b and c) side view showing the current nose and Kelvin–Helmholtz billows and their decay. *Source:* Reproduced from Chowdhury et al. (2009) with permission of Springer.

many examples of such gravity currents. Given the widespread occurrence of these environmental flows, there have been a large number of investigations of these flows leading to a wide body of literature including two monographs (Simpson, 1997; Ungarish, 2009) and a special issue of a journal dedicated to environmental gravity currents (Testik, 2014).

Gravity currents are classified into various types depending on different factors (e.g. Ungarish, 2009). Based upon the source release conditions, the currents can be considered as constant/fixed- and nonconstant-volume currents. As the names imply, a constant-volume current has a fixed volume of current fluid, whereas a nonconstant-volume current is supplied by a continuous flux of current fluid (henceforth, referred to as continuous-flux current). The currents can be further grouped as compositional and particle-driven, Boussinesq and non-Boussinesq, and 2D and axisymmetric geometries, among other classifications. Compositional currents refer to the case when the density difference between the current fluid and the ambient fluid is due to a dissolved material such as salt or temperature difference, while particle-driven currents are those with density difference due to suspended particles. Boussinesq current refers to the case when the density difference between the current fluid and the ambient fluid is relatively small, enabling simplifications in the theoretical treatment of the problem. Since many gravity currents in the environment are of Boussinesq type, we will only consider Boussinesq currents in this chapter. Furthermore, given the intended scope of this chapter, we will only consider 2D currents over a horizontal surface in a homogenous ambient fluid. A large body of research on different types of gravity currents can be found in the literature (see the reviews by Chowdhury and Testik (2014b), Hoult (1972), Middleton (1993), Kneller and Buckee (2000), Monaghan (2007), and Meiburg and Kneller (2010)).

10.7.1 Gravity Current Anatomy

In a typical gravity current over a horizontal surface, the flow is driven by the buoyancy force and resisted by the inertia and the viscous forces. After creation, a gravity current undergoes an adjustment phase, which is the jet propagation phase in the case of continuous-flux currents and the slumping phase in the case of constant-volume currents. This adjustment phase is typically followed first by the inertia–buoyancy phase and then by the viscous–buoyancy phase. During the inertia–buoyancy phase, the driving buoyancy forces and resisting inertia forces govern the current flow and viscous forces are negligible compared with these forces. Eventually viscous forces may dominate over the inertia force as the length of the propagating current increases, yielding the viscous–buoyancy propagation phase. In some cases, viscous forces may dominate over the inertia force before the adjustment phase terminates. In those cases, the gravity current may bypass the inertia–buoyancy phase and experience the viscous–buoyancy phase. During each propagation phase, a gravity current has a distinct anatomy and propagation characteristics, which are discussed in the following subsections. Our discussions are for constant-volume gravity currents in deep ambient fluid that can be analyzed using one-layer models (see Ungarish, 2009). Gravity currents in shallow ambient fluids have different anatomy and propagation characteristics than the ones in deep ambient fluids and require a more complex analysis using two-layer models, which are not discussed here. Nevertheless, insights from the 2D constant-volume currents in deep ambient fluid would provide the necessary background information to the readers for further readings on different types of gravity currents and ambient conditions.

10.7.2 Inertial Propagation

A typical constant-volume gravity current has two main parts: the frontal zone/head and the body. A distinct dividing line between the ambient fluid and the current fluid forms in the frontal zone. The foremost point of the frontal zone, referred to as the nose, is raised above the bed due to the no-slip boundary condition (Parsons and Garcia, 1998). At the frontal zone, there are two types of instabilities at the interface between the current fluid and the ambient fluid. These instabilities are (i) the lobe and cleft instability that are protrusion cusp features at the leading edge and (ii) Kelvin–Helmholtz billows that form at the current head due to the velocity shear and decay in the current body (see Figure 10.21 and Chowdhury et al., 2009). The Kelvin–Helmholtz billows and the elevated nose are responsible for the entrainment of ambient fluid into the current, diluting the current (Jacobson and Testik, 2013, 2014). As such, the body of the current has a dense underlying layer and a layer with diluted less dense fluid.

The dynamics of the current head and body differ significantly (Middleton, 1993). The current head possesses a larger thickness than the body, which provides the necessary drive to displace the stationary ambient fluid and propagate. There are various theoretical models to analyze the gravity current propagation. Three main mathematical modeling approaches for gravity current propagation are force balance models, box models, and shallow water models for inertial currents (lubrication theory models for viscous currents). These models typically assume negligible entrainment of the ambient fluid into the current for ease of the analytical treatment. It is important to note that entrainment may be significant during slumping and inertial propagation phases (e.g. Jacobson and Testik, 2014) and may have an effect on the predictive capabilities of these models. All of these three mathematical modeling approaches admit the following general form of parameterization for the front position of a compositional 2D constant-volume gravity current during inertial propagation:

$$x_{\mathrm{N}} = K_{\mathrm{I}}\left(g'_{\mathrm{a}}q\right)^{1/3} t^{2/3} \tag{10.105}$$

Here, x_{N} is the current front position, K_{I} is the proportionality constant function, g'_{a} is the reduced gravity in terms of ambient fluid density (see Eq. (10.76)), q is the current volume released per unit width, and t is the elapsed propagation time. K_{I} differs depending upon the modeling approaches and governs the difference among the predictions of different mathematical models.

As an example, propagation of an inertial gravity current as a function of elapsed time is presented in Figure 10.22. This figure is from Chowdhury and Testik (2011) and presents the current front position measurements with time from a number of select experiments with fluid mud gravity currents. In this figure, predictions of $x_{\mathrm{N}} = K_{\mathrm{I}}\left(g'_{\mathrm{a}}q\right)^{1/3} t^{2/3}$, the force balance model solution (i.e. Eq. (10.105) with the associated K_{I} values), are also provided for each of the experimental currents. A detailed comparison for the predictive capabilities of different modeling approaches is provided by Chowdhury and Testik (2011). It is important to note that comparing model predictions with the observations is not a straightforward task. The model predictions may need to be adjusted using the additional distance correction procedure that is described by Chowdhury and Testik (2012) in detail.

10.7.3 Viscous Propagation

During the inertial propagation of the current, viscous forces continuously increase. After some propagation distance/time, viscous forces become the dominant resisting force and the current transitions into the viscous–buoyancy propagation phase. Since viscous forces depend on the rheological properties of the current fluid,

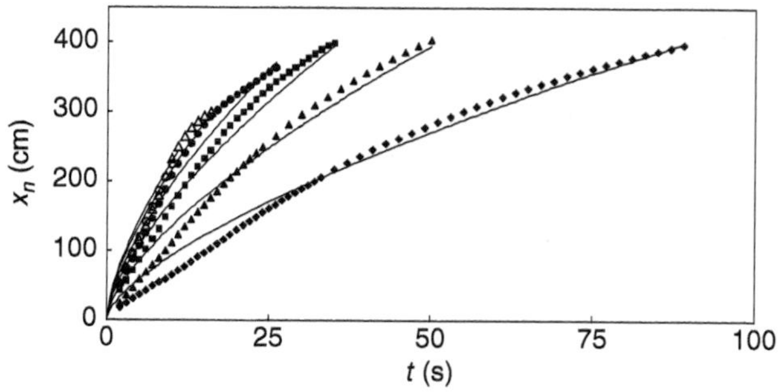

Figure 10.22 Comparison of the fluid mud gravity current front position as a function of time as predicted by the force balance model (solid lines) and measured in different laboratory experiments with different concentrations of fluid mud mixtures (symbols). *Source:* Reproduced from Chowdhury and Testik (2011) with permission of Elsevier.

gravity current dynamics during and after the propagation phase transition depends heavily on the current fluid rheology. Chowdhury and Testik (2011) developed the following parameterization for the inertia–buoyancy to viscous–buoyancy transition time, t^{**}:

$$t^{**} = c\left(\frac{\rho_c^3 q_0^{n+3}}{m^3 g_c'^{2n}}\right)^{1/(n+6)} \tag{10.106}$$

Here, m and n are Ostwald power-law rheological parameters (m is the consistency index and n is the flow behavior index), and c is an empirical coefficient. This parameterization is applicable to both Newtonian currents and non-Newtonian currents that can be modeled using Ostwald power-law model. For Newtonian currents, $n = 1$ and m should be replaced with the dynamic viscosity μ of the current fluid. The value of the empirical constant c can be obtained by experimental observations, and for the case of fluid mud gravity currents studied by Chowdhury and Testik (2011), the c value was determined as 6.

The aforementioned mathematical modeling approaches that are used for inertial propagation modeling are used also for viscous propagation modeling. These modeling approaches admit a general parameterization form for the viscous propagation of 2D constant-volume gravity currents as follows:

$$x_N = K_V q^{(2+n)/(2n+3)}\left(\frac{\rho_c g_c'}{m}\right)^{1/(2n+3)} t^{n/(2n+3)} \tag{10.107}$$

Here, K_V is the proportionality constant function, which differs for different viscous propagation models and depends upon n. This equation is also applicable to both Newtonian currents and non-Newtonian currents of Ostwald power-law fluids. It is clear from Eq. (10.107) that while current front position is a function of a constant power of time for all Newtonian currents, it is a function of a variable power of time that depends on n for non-Newtonian currents. Therefore, for example, while front positions of Newtonian saline gravity currents with different concentrations propagate with the same time dependency, front positions of non-Newtonian fluid mud gravity currents with different concentrations propagate with different time dependencies

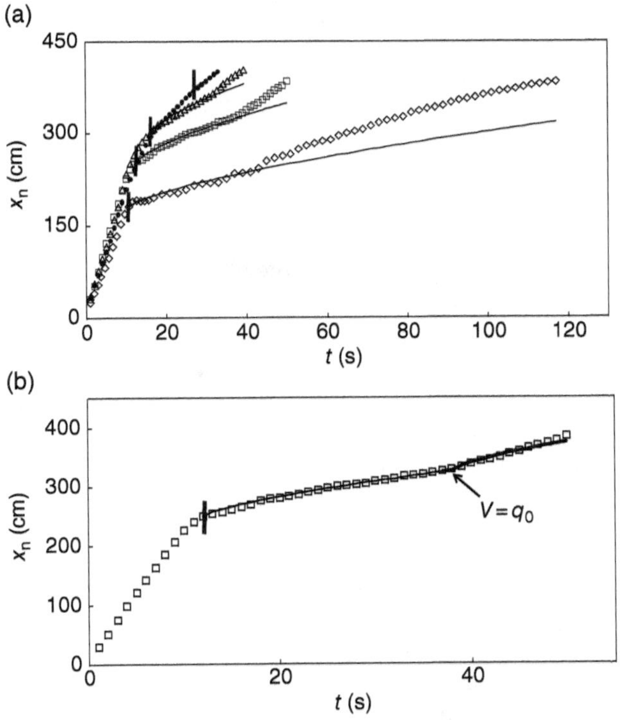

Figure 10.23 Comparison of the viscous force balance model predictions (solid lines) and the laboratory experimental measurements (symbols) for the front position of the fluid mud gravity currents in the viscous–buoyancy propagation phase. In (a), a fixed source location (x_N at t^{**}, shown with thick vertical solid lines) is used for predictions, and in (b), a shifting source location (shown in the graph as $V = q_0$) is used for predictions for one of the experimental currents. *Source:* Reproduced from Chowdhury and Testik (2011) with permission of Elsevier, and the details can be found in that paper.

(e.g. see Figure 10.23a). In Figure 10.23, a comparison of the force balance model predictions and experimental observations of fluid mud gravity currents from Chowdhury and Testik (2011) are presented. For model prediction calculations, the current fluid release source is considered to be stationary at the viscous transition position for Figure 10.23a, and a correction to the source location is introduced in Figure 10.23b for one of the experimental currents as an example. Details of such corrections for model comparisons as well as intermodel comparisons can be found in Chowdhury and Testik (2011, 2012).

References

Abramovich, G.N. (1963). *The Theory of Turbulent Jets*. Massachusetts: MIT Press (English Translation).

Aziz, T.N. and Khan, A.A. (2011). Simulation of vertical plane turbulent jet in shallow water. *Advances in Civil Engineering* 2011: 10pp. doi: 10.1155/2011/292904.

Aziz, T.N., Raiford, J.P., and Khan, A.A. (2008). Numerical simulation of turbulent jets. *Engineering Applications of Computational Fluid Mechanics* 2 (2): 234–243.

Baines, W.D. and Turner, J.S. (1969). Turbulent buoyant convection from a source in a confined region. *Journal of Fluid Mechanics* 37: 51–80.

Batchelor, G.K. (1954). Heat convection and buoyancy effects in fluids. *Quarterly Journal of the Royal Meteorological Society* 80: 339–358.

Bhutia, S., Jenkins, M.A., and Sun, R. (2010). Comparison of firebrand propagation prediction by a plume model and a coupled-fire/atmosphere large-eddy simulator. *Journal of Advances in Modeling Earth Systems* 2 (4): 1–5.

Boeker, E. and van Grondelle, R. (1999). *Environmental Physics*, 2e. Chichester, UK: Wiley.

Bolster, D.T. and Linden, P.F. (2009a). Particle transport in low energy ventilation systems. Part 1: theory of steady states. *Indoor Air* 19: 122–129.

Bolster, D.T. and Linden, P.F. (2009b). Particle transport in low energy ventilation systems. Part 2: transients and experiments. *Indoor Air* 19: 130–144.

Briggs, G. A. (1969). "Plume Rise" U.S. Atomic Energy Commission Critical Review, 81pp.

Campbell, A. and Cardoso, S. (2010). Turbulent plumes with internal generation of buoyancy by chemical reaction. *Journal of Fluid Mechanics* 655: 122–151.

Cardoso, S.S. and McHugh, S.T. (2010). Turbulent plumes with heterogeneous chemical reaction on the surface of small buoyant droplets. *Journal of Fluid Mechanics* 642: 49–77.

Carson, J.E. and Moses, H. (1969). The validity of several plume rise formulas. *Journal of the Air Pollution Control Association* 19: 862–866.

Chanson, H. (2004). *Environmental Hydraulics of Open Channel Flows*. Oxford, UK: Elsevier Butterworth Heinemann.

Chaudhry, M.H. (2008). *Open-Channel Flow*, 2e. New York: Springer.

Chin, D.A. (2006). *Water-Resources Engineering*, 2e. Pearson Prentice Hall.

Ching, C.Y., Fernando, H.J.S., Mofor, L.A., and Davies, P.A. (1996). Interaction between multiple line plumes: a model study with applications to leads. *Journal of Physical Oceanography* 26: 525–540.

Chow, V.T. (1959). *Open-Channel Hydraulics*. New York: McGraw-Hill.

Chowdhury, M.R. and Testik, F.Y. (2011). Laboratory testing of mathematical models for high-concentration fluid-mud turbidity currents. *Ocean Engineering* 38 (1): 256–270.

Chowdhury, M.R. and Testik, F.Y. (2012). Viscous propagation of two-dimensional non-Newtonian gravity currents. *Fluid Dynamics Research* 44: 045502.

Chowdhury, M. and Testik, F. (2014a). Axisymmetric underflows from impinging buoyant jets of dense cohesive particle-laden fluids. *Journal of Hydraulic Engineering* 141: 04014079. doi: 10.1061/(ASCE)HY.1943-7900.0000969.

Chowdhury, M.R. and Testik, F.Y. (2014b). A review of gravity currents formed by submerged single-port discharges in inland and coastal waters. *Environmental Fluid Mechanics* 14 (2): 265–293.

Chowdhury, M.R., Testik, F.Y., and Khan, A.A. (2009). Three-dimensional flow structure at the frontal zone of a gravity-driven fluid mud flow. *Journal of Visualization* 12 (4): 287–288.

Conroy, D.T. and Llewellyn Smith, S.G. (2008). Endothermic and exothermic chemically reacting plumes. *Journal of Fluid Mechanics* 612: 291–310.

Dean, R.G. and Dalrymple, R.A. (1991). *Water Waves Mechanics for Engineers and Scientists*. Singapore: World Scientific.

Dean, R.G. and Dalrymple, R.A. (2004). *Coastal Processes with Engineering Applications*. New York: Cambridge University Press.

Domingos, M.G. and Cardoso, S.S.S. (2013). Turbulent two-phase plumes with bubble-size reduction owing to dissolution or chemical reaction. *Journal of Fluid Mechanics* 2013 (716): 120–136.

Driscoll, F.G. (1986). *Groundwater and Wells*, 2e. St. Paul, MN: Johnson Division.

Fernando, H.J.S. ed. (2013). *Handbook of Environmental Fluid Dynamics*. Boca Raton, FL: CRC Press.

Fischer, H.B., List, E.J., Koh, R.C.Y. et al. (1979). *Mixing in Inland and Coastal Waters*. San Diego, CA: Academic Press.

Flynn, M.R. and Caulfield, C.P. (2006). Natural ventilation of interconnected chambers. *Journal of Fluid Mechanics* 564: 139–158.

Fredsoe, J. and Deigaard, R. (2005). *Mechanics of Coastal Sediment Transport*. Singapore: World Scientific Press.

Freeze, R.A. and Cheery, J.A. (1979). *Groundwater*. Englewood Cliffs, NJ: Prentice Hall.

French, R.H. (1985). *Open-Channel Hydraulics*. New York: McGraw-Hill.

Germeles, A.E. (1975). Forced plumes and mixing of liquids in tanks. *Journal of Fluid Mechanics* 71: 601–623.

Gopala, V.R. and van Wachem, B.G.M. (2008). Volume of fluid methods for immiscible-fluid and free-surface flows. *Chemical Engineering Journal* 141 (1–3): 2014–2221.

Gupta, R.S. (2008). *Hydrology and Hydraulic Systems*, 3e. Long Grove, IL: Waveland Press, Inc.

Haniu, H. and Ramaprian, B.R. (1989). Studies on two-dimensional curved nonbuoyant jets in cross flow. *ASME Journal of Fluids Engineering* 111: 78–86.

Henderson, F.M. (1966). *Open Channel Flow*. Macmillan Publishing.

Hoult, D. (1972). Oil propagation on the sea. *Annual Review of Fluid Mechanics* 4: 341–368.

Hoult, D.P. and Weil, J.C. (1972). Turbulent plume in a laminar cross flow. *Atmospheric Environment* 6: 513–531.

Hunt, G. and Burridge, H. (2015). Fountains in industry and nature. *Annual Review of Fluid Mechanics* 47: 195–220.

Hunt, G.R. and Kaye, N.G. (2001). Virtual origin correction for lazy turbulent plumes. *Journal of Fluid Mechanics* 435: 377–396.

Hunt, G.R. and Kaye, N.B. (2005). Lazy plumes. *Journal of Fluid Mechanics* 533: 329–338.

Hunt, G.R. and Kaye, N.B. (2006). Pollutant flushing with natural displacement ventilation. *Building and Environment* 41: 1190–1197.

Imberger, J. (2013). *Environmental Fluid Dynamics – Flow Processes, Scaling, Equations of Motion, and Solutions to Environmental Flows*. Oxford, UK: Academic Press.

Jacobson, M.R. and Testik, F.Y. (2013). On the concentration structure of high-concentration constant-volume fluid mud gravity currents. *Physics of Fluids* 25: 016602.

Jacobson, M.R. and Testik, F.Y. (2014). Turbulent entrainment into fluid mud gravity currents. *Environmental Fluid Mechanics* 14 (2): 541–563.

Jaluria, Y. and Kapoor, K. (1992). Wall and corner flows driven by a ceiling jet in an enclosure fire. *Combustion Science and Technology* 18: 311–326.

Jirka, G.H. (2006). Integral model for turbulent buoyant jets in unbounded stratified flows part 2: plane jet dynamics resulting from multiport diffuser jets. *Environmental Fluid Mechanics* 6: 43–100.

Johnson, E.B., Testik, F.Y., Ravichandran, N., and Schooler, J. (2013). Levee scour from overtopping storm waves and scour countermeasures. *Ocean Engineering* 57: 72–82.

Kalita, K., Dewan, A., and Dass, A.K. (2002). Prediction of turbulent plane jet in crossflow. *Numerical Heat Transfer, Part A* 41: 101–111.

Kaye, N.B. (2008). Turbulent plumes in stratified environments: a review of recent work. *Atmosphere-Ocean* 46: 433–441.

Kaye, N.B. and Hunt, G.R. (2004). Time-dependent flows in an emptying filling box. *Journal of Fluid Mechanics* 520: 135–156.

Kaye, N.B. and Hunt, G.R. (2006). Weak fountains. *Journal of Fluid Mechanics* 558: 319–328.

Kaye, N.B. and Hunt, G.R. (2007). Heat source modelling and natural ventilation efficiency. *Building and Environment* 42: 1624–1631.

Keffer, J.F. and Baines, W.D. (1963). The round turbulent jet in a cross-wind. *Journal of Fluid Mechanics* 8: 481–496.

Khan, A.A. and Lai, W. (2012a). Discontinuous Galerkin method for 1d shallow water flow in nonrectangular and nonprismatic channels. *Journal of Hydraulic Engineering ASCE* 138 (3): 285–296.

Khan, A.A. and Lai, W. (2012b). A discontinuous Galerkin method for two-dimensional shallow water flows. *International Journal of Numerical Methods in Fluids* 70 (8): 939–960.

Khan, A.A. and Lai, W. (2014). *Modeling Shallow Water Flow Using the Discontinuous Galerkin Method*. New York: CRC Press, Taylor & Francis Group.

Kneller, B. and Buckee, C. (2000). The structure and fluid mechanics of turbidity currents: a review of some recent studies and their geological implications. *Sedimentology* 47: 62–94.

Lane-Serff, G.F., Linden, P.F., and Hillel, M. (1993). Forced, angled plumes. *Journal of Hazardous Materials* 33: 75–99.

Lee, J.H.W. and Chu, V.H. (2003). *Turbulent Jets and Plumes a Lagrangian Approach*. Norwell, MA: Kluwer Academic Publishers.

Lee, J.H.W. and Jirka, G.H. (1981). Vertical round buoyant jet in shallow water. *Journal of Hydraulics Division, Proceedings of ASCE* 107: 1651–1975.

Linden, P.F., Lane-Serff, G.F., and Smeed, D.A. (1990). Emptying filling boxes, the fluid mechanics of natural ventilation. *Journal of Fluid Mechanics* 212: 309–335.

Manning, R. (1890). Flow of water in open channels and pipes. *Transaction of the Institute of Civil Engineers (Ireland)* 20: 161–207.

Margason, R.J. (1968). The Path of a Jet Directed at Large Angles to a Subsonic Free Stream. Technical Report NASA TN D-4919.

Mei, C.C. (1989). *The Applied Dynamics of Ocean Surface Waves*. Singapore: World Scientific.

Mei, C.C., Stiassnie, M., and Yue, S.K.P. (2005). *Theory and Applications of Ocean Surface Waves: Part 1, Linear Aspects; Part 2, Nonlinear Aspects*. Singapore: World Scientific Publishers.

Meiburg, E. and Kneller, B. (2010). Turbidity currents and their deposits. *Annual Review of Fluid Mechanics* 42: 135–156.

Middleton, G. (1993). Sediment deposition from turbidity currents. *Annual Review of Earth and Planetary Sciences* 21: 89–114.

Monaghan, J.J. (2007). Gravity current interaction with interfaces. *Annual Review of Fluid Mechanics* 39: 245–261.

Morton, B.R. (1959). Forced plumes. *Journal of Fluid Mechanics* 5: 151–163.

Morton, B.R. and Middleton, J. (1973). Scale diagrams for forced plumes. *Journal of Fluid Mechanics* 58: 165–176.

Morton, B.R., Taylor, G.I., and Turner, J.S. (1956). Turbulent gravitational convection from maintained and instantaneous sources. *Proceeding of the Royal Society of London* 234: 1–23.

Ooms, G. and Mahieu, A. (1981). A comparison between a plume path model and a virtual point source model for a stack plume. *Applied Scientific Research* 36: 339–356.

Parsons, J.D. and Garcia, M.H. (1998). Similarity of gravity current fronts. *Physics of Fluids* 10: 3209–3213.

de Paulo, G.S., Tome, M.F., and Mckee, S. (2007). A marker-and-cell approach to viscoelastic free surface flows using the PTT model. *Journal of Non-Newtonian Fluid Mechanics* 147 (3): 149–174.

Platten, J.L. and Keffer, J.F. (1971). Deflected turbulent jet flows. *Transactions of the ASME, Journal of Applied Mechanics* 38: 756–758.

Pratte, B.D. and Baines, W.D. (1967). Profiles of the round turbulent jets in cross flow. *Journal of Hydraulic Division, ASCE* 92: 53–64.

Raiford, J.P. and Khan, A.A. (2009). Investigation of circular jets in shallow water. *Journal of Hydraulic Research, IAHR* 47 (5): 611–618.

Rajaratnam, N. (1976). *Turbulent Jets*. New York: Elsevier Scientific.

Rajaratnam, N. and Khan, A.A. (1992). Intersecting circular turbulent jets. *Journal of Hydraulic Research, IAHR* 30 (3): 373–387.

Ramaprian, B. R. and Haniu, H. (1983). Turbulence Measurements in Plane Jets and Plumes in Cross Flow. Technical Report No. 266, IIHR, University of Iowa, Iowa City, IA.

Rawn, A., Bowerman, F., and Brooks, N. (1960). Diffusers for disposal of sewage in sea water. *Journal of the Sanitary Engineering Division: Proceedings of ASCE* 86: 65–105.

Rodi, W. (1984). Turbulence Models and Their Application in Hydraulics. *International Association for Hydro-Environment Engineering and Research (IAHR)*, Madrid, Spain, IAHR Monograph.

Roes, M.A., Bolster, D.T., and Flynn, M.R. (2014). Buoyant convection from a discrete source in a leaky porous medium. *Journal of Fluid Mechanics* 755: 204–229.

Rothman, T. and Ledbetter, J.O. (1975). Droplet size of cooling tower fog. *Environmental Letters* 10 (3): 191–203.

Rubin, H. and Atkinson, J. (2001). *Environmental Fluid Mechanics*. New York: CRC Press.

Scase, M.M., Caulfield, C.P., and Dalziel, S.B. (2008). Temporal variation of non-ideal plumes with sudden reductions in buoyancy flux. *Journal of Fluid Mechanics* 600: 181–199.

Scase, M.M., Caulfield, C.P., Dalziel, S.B., and Hunt, J.C.R. (2006). Time-dependent plumes and jets with decreasing source strengths. *Journal of Fluid Mechanics* 563: 443–461.

Simpson, J. (1997). *Gravity Currents: In the Environment and the Laboratory*. Cambridge, MA: Cambridge University Press.

Sumer, B.M. and Fredsoe, J. (2005). *The Mechanics of Scour in the Marine Environment*. Singapore: World Scientific Publishing.

Testik, F.Y. (2014). Preface: gravity currents in the environment. *Environmental Fluid Mechanics* 14 (2): 263–264.

Testik, F.Y. and Ungarish, M. (2016). On the self-similar propagation of gravity currents through an array of emergent vegetation-like obstacles. *Physics of Fluids* 28: 056605. doi: 10.1063/1.4947251.

Testik, F.Y., Voropayev, S.I., Balasubramanian, S., and Fernando, H.J.S. (2006). Self-similarity of asymmetric sand-ripple profiles formed under nonlinear shoaling waves. *Physics of Fluids* 18: 108101.

Testik, F.Y., Voropayev, S.I., and Fernando, H.J.S. (2005). Adjustment of sand ripples under changing water waves. *Physics of Fluids* 17: 072104.

Tohidi, A. and Kaye, N.B. (2016). Highly buoyant bent-over plumes in a boundary layer. *Atmospheric Environment* 131: 97–114.

Turner, J.S. (1973). *Buoyancy Effects in Fluids*. Cambridge, UK: Cambridge University Press.

Ungarish, M. (2009). *An Introduction to Gravity Currents and Intrusions*. Boca Raton, FL: Chapman and Hall/CRC Press.

Voropayev, S.I., Testik, F.Y., Fernando, H.J.S., and Boyer, D.L. (2003a). Morphodynamics and cobbles behavior in and near the surf zone. *Ocean Engineering* 30 (14): 1741–1764.

Voropayev, S.I., Testik, F.Y., Fernando, H.J.S., and Boyer, D.L. (2003b). Burial and scour around short cylinder under progressive shoaling waves. *Ocean Engineering* 30 (13): 1647–1667.

Wang, H. and Law, A.W.-K. (2002). Second-order integral model for a round turbulent buoyant jet. *Journal of Fluid Mechanics* 459: 397–428.

Wexler, E.J. (1992). *Analytical Solutions for One-, Two-, and Three-Dimensional Solute Transport in Ground-Water Systems*. Denver, CO: USGS.

Wong, D.R. and Wright, S.J. (1988). Submerged turbulent jets in stagnant linearly stratified fluids. *Journal of Hydraulic Research* 26 (1): 199–223.

Woods, A.W. (2010). Turbulent plumes in nature. *Annual Review of Fluid Mechanics* 42: 391–412.

Worster, M.G. and Huppert, H.E. (1983). Time-dependent density profiles in a filling box. *Journal of Fluid Mechanics* 132: 457–466.

Yen, B.C. (1991). Hydraulic resistance in open channels. In: *Channel Flow Resistance: Centennial of Manning's Formula* (ed. B.C. Yen), 1–35. Water Resources Publications: Highlands Ranch, CO.

Young, D.M. and Testik, F.Y. (2009). Onshore scour characteristics around submerged vertical and semicircular breakwaters. *Coastal Engineering* 56: 868–875.

Young, D.M. and Testik, F.Y. (2011). Wave reflection by submerged vertical and semicircular breakwaters. *Ocean Engineering* 38 (10): 1269–1276.

Zhang, W., Rajaratnam, N., and Zhu, D.Z. (2012). Transport with jets and plumes of chemicals in the environment. In: *Transport and Fate of Chemicals in the Environment* (ed. J.S. Gulliver). New York: Springer.

11

Water Quality

Steven C. Chapra

Department of Civil & Environmental Engineering, Tufts University, Medford, MA, USA

Water, water, everywhere, nor any drop to drink
"The Rime of the Ancient Mariner"
Samuel Taylor Coleridge (1772–1834)

11.1 Introduction

Life as we know it on this planet would not be possible without the remarkable compound: water. In this regard, both the quantity and quality of water are critical. Anyone who has ever lived through a drought knows that having a sufficient water supply is of paramount importance to both humans and terrestrial animals as well as to the myriad organisms living in the aquatic environment. Similarly, anyone who has suffered through a flood recognizes that excessive uncontrolled water flow can be highly destructive.

Although an adequate and moderate water quantity is certainly essential, of equal importance is its quality. The quality of water can be conceptualized as its suitability (i) for particular human uses and (ii) for the support and health of organisms living within and in proximity to aquatic ecosystems. As expressed poetically by Coleridge, you can have plenty of water, but if it is unsafe to drink, you can die of thirst. Hence, ensuring adequate water quantity is futile without also considering water quality.

The quality of water is typically based on the water's physical, chemical, and biological characteristics. Although the relationship of quality to these characteristics can be complex and multifaceted, water quality deterioration can be grouped into three major categories: rubbish, stink, and death (Chapra, 2011).

The term "rubbish" reflects the aesthetic importance of water in that humans tend to devalue water when it is visually unappealing. This aversion may have originally been connected with health concerns in that colored, murky, or particle-laden water may be unpotable and/or unpalatable. But regardless of its basis, there is no question that humans place higher value on pure, clear water. This is attested to by the phenomena ranging from the billion-dollar bottled water industry to the premium on residential and vacation real estate located on the shoreline of clear, clean natural waters.

At first hearing, the term "stink" suggests another aesthetic quality, as humans have an obvious distaste for malodorous water. However, in the present context, the term is meant to represent ecosystem health. To explain, the "health" of most aquatic ecosystems experienced directly by humans (i.e. while at the beach, fishing, or boating) is often related to the water's dissolved oxygen level. When excessive quantities of sewage and other sources of organic matter are discharged into such systems, bacterial oxidation can reduce dissolved oxygen concentrations to low levels. Such cases are formally referred to as hypoxia. In the extreme, concentrations can approach zero, which is referred to as anoxia. This, of course, is an ecological disaster for organisms such as fish that depend on dissolved oxygen for their survival. However, along with direct biological health impacts such as fish kills, anoxic waters also lead to significant chemical changes. In particular, sulfate in the water will be converted to hydrogen sulfide gas, creating an intense "rotten egg" smell. Thus, the water's bad odor provides humans with an olfactory signal that the aquatic ecosystem is "sick."

Of the three terms, "death" is the most obvious. Humans clearly care whether water is safe to consume without the risk of immediate or long-term harm. But beyond such public health concerns, the two other less dramatic characteristics, rubbish and stink, also contribute significantly to how humans value water.

While this triad of symptoms informs humans that water quality is threatened, when was this awareness translated into remedial action? That is, when did society

Handbook of Environmental Engineering, First Edition. Edited by Myer Kutz.

collectively marshal its resources to consider and then solve the water quality problem? A historical perspective is useful in answering this question.

11.2 Historical Background

In Koln, a town of monks and bones,
And pavements fanged with murderous stones,
And rags, and hags, and hideous wenches,
I counted two-and-seventy stenches.
All well defined, and separate stinks!
Ye nymphs that reign o'er sewers and sinks,
The river Rhine, it is well known,
Doth wash your city of Cologne;
But tell me, nymphs, What power divine
Shall henceforth wash the river Rhine?

"Cologne"
Samuel Taylor Coleridge (1772–1834)

The following review is intended to provide historical perspective to our subsequent description of current water quality issues and problems. Because the modern conception of water quality was primarily developed in Europe and the Americas, this historical review primarily focuses on Western history.

11.2.1 Ancient Water Sanitation

For most of their 200 000-year history, *Homo sapiens* were primarily widely dispersed hunter-gatherers. Although their sparse distribution undoubtedly mitigated the threat of waterborne disease, the general aversion to water that stank, tasted awful, and/or looked filthy must have developed quite early during humankind's evolution.

However, with the advent of agriculture 10 000 years ago and the subsequent concentration of humans and animals in permanent settlements in about 7500 BCE, pathogens carried by contaminated water became a very serious health risk. Although the ancients had no scientific concept of the connection between contaminated water and disease, certain historical practices may have been originally based on an empirically derived recognition of the connection. For example, the Hebrew stricture against consuming shellfish may have been related to the fact that such bottom-dwelling organisms when situated near human settlements often carry waterborne diseases (Lebeau, 1983). Further, the addition of salt was an old remedy for contaminated water as recorded in the Bible (2 Kings 2:19–21):

> The men of the city said to Elisha,..."this town is well situated,..., but the water is bad and the land is unproductive"...Then he went out to the spring [of Jericho] and threw the salt into it, saying, "This is what the Lord says: I have healed this water. Never again will it cause death or make the land unproductive."

By the second millennium BCE, the Mycenaean culture of Crete had developed the first purposeful construction of a Western water supply/sanitation system. Greek philosophical and medical thinkers also recognized the importance of water for public health. For example, Hippocrates, the "Father of Western Medicine," invented the first bag filter (called the "Hippocratic sleeve") to trap sediments that caused taste and odor problems in about 500 BCE.

Starting in about the third century BCE, the Romans began constructing massive aqueducts to carry clean water long distances from sparsely settled high-elevation locations to their cities and settlements. Vitruvius, a great Roman author, architect, and engineer during the first century BCE, expressed his own view on pure water, which provides a window into Roman consciousness of water quality (Vitruvius, 1934):

> The trial and proof of water are made as follows. If it be of an open and running stream,...the shape of the limbs of the inhabitants of the neighborhood should be looked to and considered. If they are strongly formed, of fresh colour, with sound legs, and without blear eyes, the supply is of good quality. Also, if digging to a fresh spring, a drop of it be thrown into a Corinthian vessel made of good brass, and leave no stain thereon, it will be found excellent... Moreover, if the water itself, when in the spring is limpid and transparent, and the places over which it runs do not generate moss, nor reeds, nor other filth be near it, everything about it having a clean appearance, it will be manifest by these signs, that such water is light and exceedingly wholesome.

In addition to aqueducts, the Romans also built extensive drainage and wastewater sewer systems to collect urban runoff as well as wastewater from latrines and public baths. Although the sewers effectively removed the waste from the city proper, they ultimately emptied it into the adjacent Tiber River, making it a heavily polluted, foul water body.

After the fall of Rome at the end of the fifth century CE, Europe entered the Dark Ages and most of the Roman infrastructure including their water quality innovations fell into disrepair. Consequently, poor water quality once again became ubiquitous, particularly in urban environments. Along with other types of epidemics, such as the Black Death, waterborne diseases were a common contributing factor to the extremely high mortality rates and associated low populations in Europe during this period.

Nevertheless, there were still some customs that reflected at least an empirically derived awareness of the danger of consuming contaminated water. In fact, it was common practice throughout medieval Europe to avoid water as the major source of liquid sustenance. Rather, because of its low pH and the disinfecting action of ethyl alcohol, people of all ages in the West consumed wine and beer as preferred major thirst quenchers. Even when these alcoholic beverages were dear or in short supply, they were mixed with water to accrue the same benefit.

Interestingly, a different empirically derived approach was taken in the East. Because roughly half of all Asians lack a key enzyme, they cannot completely metabolize alcohol, which results in some unpleasant side effects. Hence, they created a potable nonalcoholic beverage by consuming tea that had been brewed in boiling water!

However, drinking some water was often unavoidable, and waterborne diseases plagued the West well into the nineteenth century. Although much of the disease outbreaks occurred in cities, other high-density population areas also suffered. A notable example occurred in temporary wartime military encampments where crowding, poor hygiene, and the absence of clean water sources or adequate sanitation facilities made the typical camp a breeding ground for waterborne disease. Consequently, disease claimed more lives than battles in many wars. For example, typhus claimed more soldiers than were killed by the Russians during Napoleon's 1812 retreat from Moscow. Bad sanitary practices were to blame for the spread of disease during the American Civil War (1860–1865) where twice as many soldiers died of disease (400 000) than of wounds (200 000).

So when did society collectively marshal its resources to understand and begin to effectively solve the water quality problem? To answer this question, we must turn to nineteenth-century England.

11.2.2 London and the Birth of Rational Water Quality Control

The roots of modern water quality control can be traced to London in the mid-nineteenth century. Due primarily to the economic and demographic impacts of the industrial revolution coupled with the great wealth generated by the British Empire, the population of London grew almost an order of magnitude during the nineteenth century (Figure 11.1). As a result, it was the world's largest city from 1830 to 1925.

As is currently the case in many parts of the world, a great deal of this population growth consisted primarily of poor displaced peasants migrating from the countryside to urban centers like London in search of employment. These settlers tended to congregate in densely populated slums with inadequate water and sanitation infrastructure.

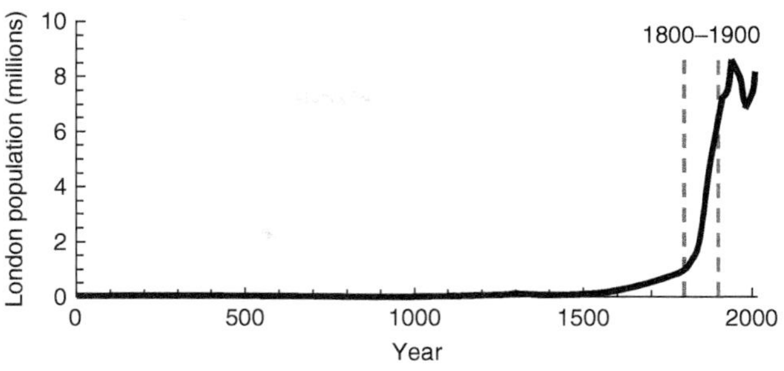

Figure 11.1 Population of London showing the rapid rise during the nineteenth century, as people migrated from the countryside to the city.

Figure 11.2 A drawing from British magazine, *Punch*, depicting the state of the Thames River in London circa 1858, the "Year of the Great Stink."

Much of their drinking water was drawn from the groundwater via shallow wells, and their sewage was disposed into cesspits or to storm sewers that ultimately fed into the adjacent Thames River.

In 1858, the problem reached an environmental tipping point, as the large quantities of sewage combined with a particularly hot summer created a perfect storm of rubbish, stink, and death. As depicted in Figure 11.2, "rubbish," in the form of dead animals and debris, floated in the Thames River. That year was dubbed "The Great Stink of 1858" due to the terrible stench emanating from the river. Doctor William Budd, a pioneer in the study of infectious diseases, stated: "Never before, at least, had a stink risen to the height of an historic event." The smell was so overwhelming that it threatened the functioning of the House of Commons and the law courts. As public awareness and political pressure grew, a government committee was appointed to develop a solution of the problem.

But what about death? During the same period, cholera became widespread in London. At first, the science of the

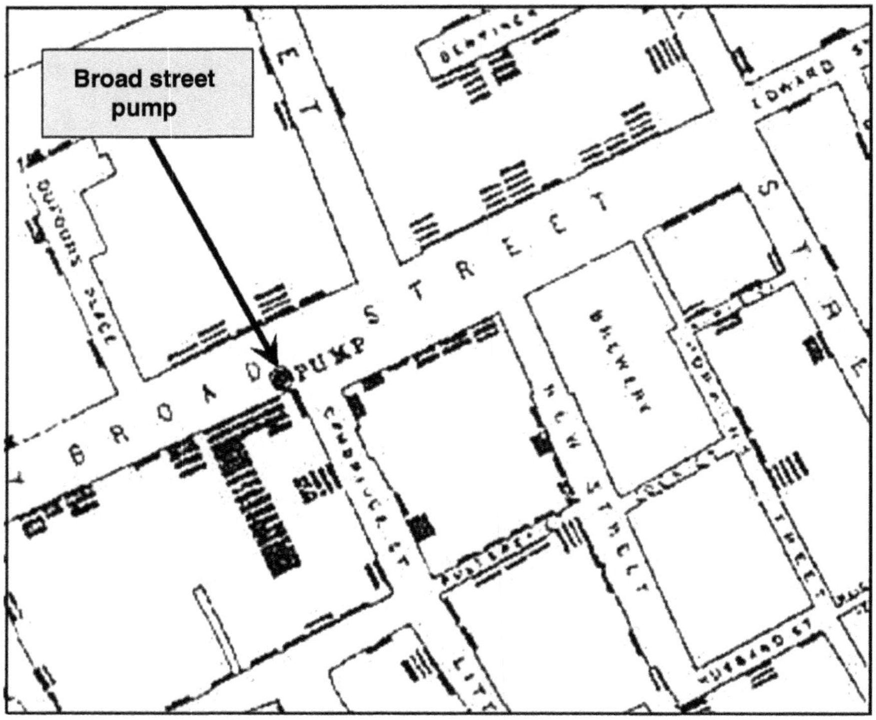

Figure 11.3 A portion of the so-called Ghost Map developed by John Snow. Each hash mark signified a fatality due to a cholera outbreak in 1854. The fact that the deaths centered on the Broad Street pump suggested that the disease was waterborne. In addition, no deaths occurred at the nearby brewery where workers were provided with beer.

day erroneously posited that the disease was caused by an airborne "miasma." In 1854, the London physician John Snow developed a counter hypothesis that the cholera epidemics were due to contaminated drinking water. A primary line of evidence supporting this hypothesis was a map he developed on which he plotted the deaths for a particular outbreak (Figure 11.3). A precursor of GIS, the so-called "Ghost Map," provided strong circumstantial evidence that the disease was waterborne (Johnson, 2006).

Unfortunately, Snow's hypothesis was not widely accepted until Robert Koch isolated the cholera organism in the 1880s. Nevertheless, the political awareness and pressure created by the Big Stink was sufficient to stimulate a number of remedial measures, including a major expansion of the London sewer system. Designed by chief engineer Joseph Bazalgette, the sewer expansion was designed to carry most of London's wastewater to a discharge point well downstream of the city. Although the principle motivation for the new sewer system was to alleviate the stench, ancillary benefits were to improve appearance of the Thames River and to greatly diminish the cholera outbreaks.

This huge and expensive sewer project can be considered the birth of modern environmental engineering. That is, it was the first major modern example of the application of the engineering approach to solve an environmental problem. From this point forward, other cities around the world began to consider how infrastructure development could be used to alleviate the rubbish, stink, and death created by urban water pollution.

11.3 Overview of Modern Water Quality

Because of its complexity and multifaceted nature, describing and defining "water quality" is a daunting task. For one thing, it involves many, often subjective, judgments from a variety of different perspectives. Before reviewing the current threats to water quality, it is useful to describe the state of unpolluted surface water.

11.3.1 The "Natural" State of Clean Water

From the human perspective, water from uncontaminated sources is clear, colorless, and odorless. It is typically low in dissolved solids such as salts and nutrients and suspended solids such as silts and clays. Dissolved gases like dissolved oxygen and carbon dioxide should be at or near equilibrium with the atmosphere (i.e. at saturation). It should also be free of pathogens and toxic substances such as solvents, pesticides, flame retardants, etc.

However, from a deeper and broader perspective, as depicted in Figure 11.4, the aquatic biosphere can be viewed as a cycle of life (left) and death (right). Powered by the sun, autotrophic[1] organisms (primarily plants) convert simple inorganic nutrients into more complex organic molecules. In this photosynthesis process, solar energy is stored as chemical energy in the organic molecules. In addition, oxygen is liberated and carbon dioxide is consumed.

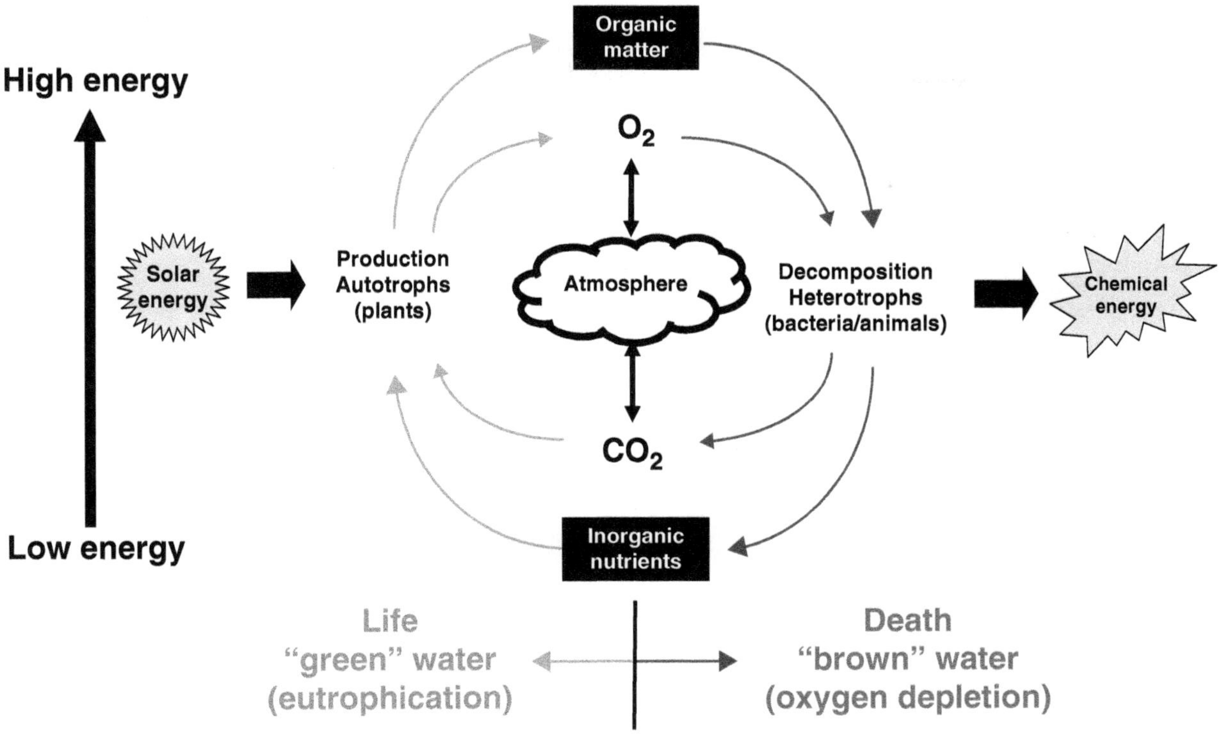

Figure 11.4 The natural cycle of organic production and decomposition (Chapra, 1997).

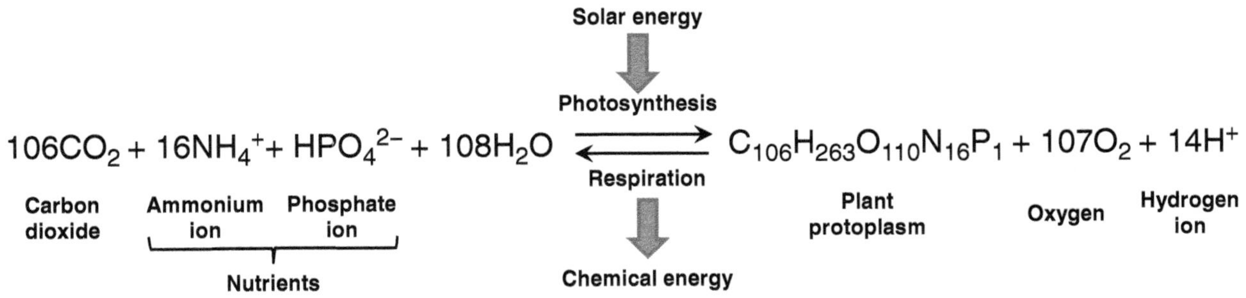

Figure 11.5 An equation expressing the major chemical transformations of the photosynthesis and respiration reactions that are at the heart of the life/death cycle in nature.

In the reverse processes of respiration and decomposition, the organic matter then serves as an energy source for heterotrophic organisms (bacteria and animals). This process returns the organic matter to the simpler inorganic nutrient state. During breakdown, oxygen is consumed and carbon dioxide is liberated.

The equation in Figure 11.5 expresses both parts of the cycle in chemical terms. Thus, we can see that the forward reaction (corresponding to the left side of Figure 11.4) transforms the dead inorganic ingredients into living plant biomass (AKA food) and generates the life-supporting gas, oxygen. Conversely, the reverse reaction (the right side of Figure 11.4) consumes the food liberating energy for the benefit of the animal and bacterial consumers and returning the nutrients to their original dead state.

This wheel of life and death, which has been cycling continuously for millions of years, offers an alternative basis for defining water quality that transcends the purely human perspective. That is, that there is intrinsic importance in keeping the cycle going smoothly as well as maintaining a balance between the two processes. As we will see in the remainder of this chapter, humans can interfere with the cycle in two fundamental ways. Thus, as summarized in Table 11.1, we can divide water quality problems into two broad categories:

- *Natural or "conventional" pollution.* In what we collectively dub "natural" pollution, humans overload the system with contaminants that in themselves and in moderation are not intrinsically harmful. These represent

Table 11.1 Major water quality categories.

Major category	Subcategory	Components	Impacts
Natural pollutants	Dissolved oxygen depletion (anoxia)	• Biochemical oxygen demand • Dissolved oxygen • Hydrogen sulfide	• Aesthetics • Ecosystem health • Death/disease
	Eutrophication	• Phosphorus • Nitrogen	• Aesthetics • Ecosystem health • Death/disease
	Dissolved inorganic solids	• Salinity • Hardness and rust • Groundwater • Saltwater intrusion • Industrial waste • Road salt	• Aesthetics • Ecosystem health • Death/disease
	Suspended solids	• Soil erosion • Resuspension • Turbidity • Floatables/plastics	• Aesthetics • Death/disease
	Pathogens	• Bacteria • Viruses • Protozoa • Helminths	• Death/disease
Modern pollutants	Toxicants	• Toxic organics • Metals • Radionuclides	• Death/disease
	Acids and bases	• Acid rain • Mine drainage • Biotic effects	• Death/disease
	Spills	• Oil • DNAPLs	• Aesthetics • Ecosystem health • Death/disease
	Heat and temperature	• Cooling water • Riparian vegetation	• Death/disease
	Emerging pollutants	• Pharmaceuticals • Nanoparticles • Greenhouse gases	• Ecosystem health • Death/disease

pollutants that would have occurred in the preindustrial period as a natural by-product of human and animal existence. However, when discharged in excessive quantities, one and/or the other side of the life/death cycle can be accelerated. If they discharge excess organic carbon primarily in the form of untreated wastewater, the right or "brown" side of the cycle will reduce the receiving water's dissolved oxygen to critically low levels, resulting in disastrous outcomes ranging from massive fish kills to the generation of noxious odors. Conversely, adding excess nutrients from land runoff and/or sewage can overstimulate the left or "green" side. In such cases, the water body would exhibit a number of symptoms including excessive plant growth, unsightly scums, and toxic algae blooms.

- *Modern or industrial/technological pollution.* Second, in what we loosely term as industry-related pollution, humans can discharge contaminants that interfere with the life/death cycle. Hence, in contrast with the overstimulation of the conventional pollutants, the industry-related contaminants can poison the cycle. This can have direct health impact on aquatic organisms, as well as threatening humans and animals that use it as a source of food and drinking water.

Notice that beyond sewage, nutrients, and industry-derived pollutants, we have added other "natural" and "modern" contaminants to the table. For example, beyond nutrients, agriculture can increase the introduction of both dissolved and particulate solids into natural waters. Similarly, industrial cooling water and emerging pollutants such as pharmaceuticals are clustered in the modern category. Although the division is not perfect, it is useful in distinguishing between the natural contaminants that are mostly the result of high densities of humans and livestock and the modern pollutants that emanate in large part from advanced industry and technology.

11.4 Natural or "Conventional" Water Quality Problems

As stated previously, by conventional pollutants, I mean water quality degradation that results from too many people and/or animals concentrated in close proximity to a water body. As such, this type of pollution has occurred ever since humans evolved from widely dispersed nomadic hunter-gatherers into sedentary farmers and city dwellers.

11.4.1 Dissolved Oxygen Depletion

High-density populations of humans, and animals generate a steady stream of sewage, which is high in organic carbon and the essential nutrients: nitrogen and phosphorus. Out of convenience, these wastes were typically discharged directly to adjacent waters. In the absence of any waste treatment, the concentrated organic carbon would be broken down by bacteria that would deplete the dissolved oxygen in the water.

Although the focus on the breakdown of organic carbon is certainly important, Figure 11.4 does not encompass all of the features of the dissolved oxygen problem. In order to gain a more complete understanding, we must describe two other processes that underlie the problem: nitrification and the redox sequence.

Nitrification. Another important oxidation process is nitrification, the conversion of the ammonium ion (NH_4^+) to nitrate ion (NO_3^-). The process actually occurs in several steps (Figure 11.6). Any nonliving organic nitrogen in the water is first converted into ammonia. Depending on pH, the total ammonia occurs in two forms: ammonium ion (low to normal pH) and unionized ammonia (high pH). Nitrification of the ammonium then occurs in two steps: ammonium to nitrite (NO_2^-) and nitrite to nitrate. These reactions are catalyzed by two separate genera of bacteria, *Nitrosomonas* and *Nitrobacter*, each utilizing oxygen to extract energy from the process.

Thus, nitrification creates a significant oxygen sink over and above organic carbon oxidation. Further, beyond the oxygen problem, two of the components in Figure 11.6 are water pollutants in their own right. In high pH systems, unionized ammonia can rise to levels that are harmful to fish in the phenomenon called "ammonia toxicity." The end product of the process in oxic systems is nitrate, which at sufficiently high concentrations ($>10\,mg\,L^{-1}$) can cause methemoglobinemia (AKA "blue baby syndrome").

The Redox Sequence. To this point, we have limited our description of oxygen depletion to the oxidation of organic carbon and ammonium by aerobic organisms. When water is depleted of oxygen, anaerobic bacteria take over, which utilize oxidants other than O_2. In addition, even when the overlying water is oxygenated, bottom

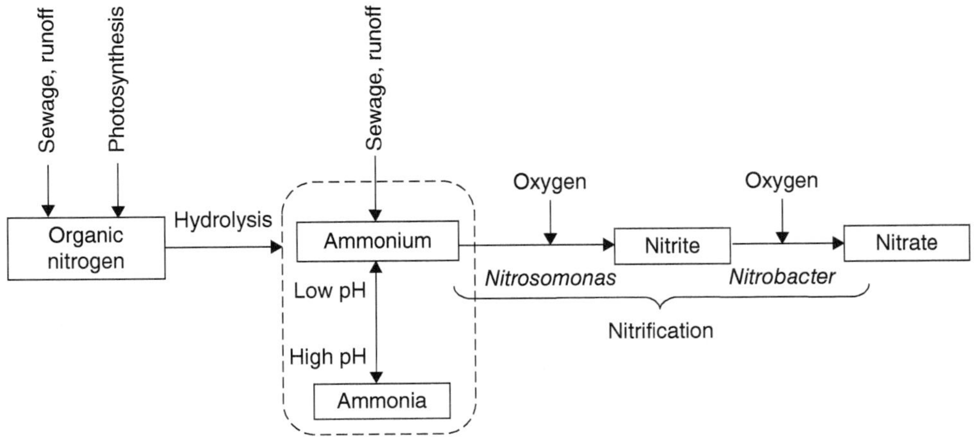

Figure 11.6 The nitrification process that occurs in oxic waters.

sediments are typically rich in organic carbon and, aside from a very thin oxygenated layer at the sediment–water interface, the bottom sediments invariably will be anaerobic. In fact, the sediments are where much of the breakdown of organic matter takes place.

In either the water or the sediments, the organic matter oxidation is primarily controlled by five compounds. In decreasing order of energy produced, these are nitrate (NO_3^-), manganese dioxide (MnO_2), ferric hydroxide ($Fe(OH)_3$), sulfate (SO_4^{2-}), and organic carbon (CH_2O) itself. The microbes first use the oxidant that produces the most energy until it is depleted; only then does another agent become dominant. This sequence of oxidation processes (aka the redox sequence) is depicted in Figure 11.7.

Aside from providing a means to break down organic matter in the absence of oxygen, the anoxic redox sequence has several additional environmental impacts:

- Note that all steps in the process generate carbon dioxide. This tends to lower the pH of the sediments and bottom waters. However, if it is released to the air, CO_2 can have effects on the atmosphere as it is a greenhouse gas.
- The products Mn^{2+}, Fe^{2+}, HS^-, and CH_4 are soluble and hence mobile in the sediment pore water. Thus, they can diffuse upward until they reach the upper sediment layers or the overlying water where they can be oxidized by oxygen or nitrate.
- Although, as depicted in Figure 11.7, denitrification usually generates a harmless end product, in some cases the process does not go to completion, and an intermediate product, nitrous oxide, results. Although this is inconsequential from a water quality perspective, if released to the atmosphere, nitrous oxide is a potent greenhouse gas.
- Sulfate is at sufficiently low levels in most freshwaters that the sulfate step in the redox sequence is small enough to be negligible. In such cases, given enough organic carbon, methanogenesis occurs with carbon dioxide and methane (CH_4) as the end products. Because methane is relatively insoluble, methane bubbles can form (aka ebullition), which rise rapidly out of the sediments, up through the water column and into the atmosphere. This loss of oxidizable carbon is good for water quality (i.e. it does not oxidize in the water and consume oxygen). However, this is bad for the atmosphere as gaseous methane is a potent greenhouse gas.
- In all marine systems as well as certain freshwaters (particularly those with drainage basins with significant sulfur minerals such as gypsum), the sulfate step will become prominent, and large amounts of hydrogen sulfide (HS^-) can be generated. Because HS^- is much more soluble than CH_4, most will be subject to oxidation. However, in cases where oxidation is not significant, it can impart a rotten egg smell and taste to the water.

Wastewater treatment is the primary remedy for remedying the oxygen problem. Conventional wastewater treatment consists of three stages:

- Primary (or physical) treatment. Typically sedimentation to remove both organic and inorganic particles. This reduces suspended solids with a small removal of organic carbon.
- Secondary (or biological) treatment. This involves optimizing the breakdown of organic carbon compounds by enhanced growth of bacteria. This goes a long way toward removing organic carbon and further reducing solids. In addition, it can be extended to begin to foster some nitrification to convert ammonium to nitrate.
- Tertiary (or advanced) treatment. This is a broad term for additional treatment processes to achieve a number of different ends. From the standpoint of oxygen, one goal would be to foster further nitrification to convert most of the ammonium to nitrate. In addition, it can also represent additional processes to remove nutrients from the wastewater effluent for cases where control of both oxygen and eutrophication is necessary.

In addition to wastewater treatment, older cities with aging infrastructure may have leaky sewage collection systems (pipes, pumps, valves), which can cause sanitary sewer overflows. Some cities also have combined sewers, which carry both wastewater and rainwater in the same conduits. Although these are often adequate in conveying wastewater to treatment facilities during dry periods,

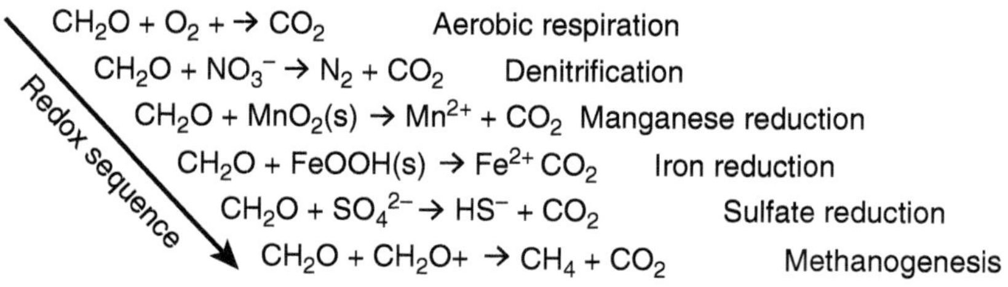

Figure 11.7 Simplified representation of the redox sequence.

a significant portion of settleable particulate sewage sludge tends to build up in the sewer bottoms. This sludge can then be flushed in huge quantities out into receiving waters during subsequent rain storms. Because these combined sewer systems are extremely expensive to retrofit or replace, alternatives to sewage treatment such as injection of oxygen into the receiving water during periods immediately following storms can often represent a cost-effective alternative.

In summary, the oxygen problem is complex and multifaceted. As should be clear from the foregoing, at the extreme it can drive an aquatic ecosystem to anoxia with the resulting destruction of oxygen-dependent organisms as well as the generation of smells and tastes that humans find objectionable. We now turn to the flip side of the life/death cycle, eutrophication. Whereas the oxygen problem represents overfeeding bacteria with energy-yielding organic carbon and ammonium, eutrophication involves overstimulating plant life with excessive inputs of nutrients.

11.4.2 Eutrophication

Derived from Greek, eutrophication literally means "well nourished." As it applies to natural waters, it represents the response of a natural water to the addition of large quantities of nutrients (most commonly phosphorus and nitrogen), which leads to excessive plant growth. This enhanced growth of aquatic vegetation or phytoplankton disrupts normal ecosystem functions in myriad ways. Most directly, the water surface can become dominated by unsightly plant scums that can decrease the water body's aesthetic and recreational value.

Whereas excessive plants represents the overstimulation of the life side (left) of life/death cycle (Figure 11.4), the plants in the surface water will eventually die and decompose. Hence, this stimulates the death side (right) of the cycle. The fraction of plant biomass that settles can lead to anoxic conditions of the bottom water, which can lead to the death of fish and shellfish. Some can also wash up and accumulate on shorelines where they can rot and emit noxious odors. These and a variety of other problems are listed in Table 11.2.

It should be noted that eutrophication occurs in both freshwaters and salt waters. With regard to the latter, because they are often at the terminus of nutrient-rich river systems, estuaries and coastal embayments are especially prone to eutrophication. In contrast with freshwater systems, nitrogen is more likely to be the limiting nutrient of marine waters. Upwelling in coastal systems also promotes increased productivity by conveying deep, nutrient-rich waters to the surface, where the nutrients can be assimilated by algae.

Table 11.2 Major water quality impacts of eutrophication.

Quantity
- Increased biomass of microscopic floating plants (phytoplankton)
  - Loss of water clarity
  - Unsightly scums
  - Increase in the frequency of algal blooms
  - Filter clogging in water treatment
  - Shading of bottom plants
- Increased biomass of attached algae (periphyton, filamentous algae, macrophytes)
  - Hinder navigation
  - Interfere with recreation
  - After death, wash up on beaches
  - Addition of toxic chemicals to control weeds (e.g. copper, diquat, 2,4-D, etc.)

Chemistry
- Anoxia in bottom waters
  - Low DO harms fish and macroinvertebrates
  - Sediment release of nutrients and toxicants
- pH changes
  - High pH can harm fish
  - Ammonia toxicity
- Drinking water impacts
  - Taste and odor
  - Disinfection by-products

Biology
- Inedible phytoplankton species
- Harmful algal blooms (HABs)
- Loss of desirable fish species
- Species diversity decreases and the dominant biota changes

Historically, the sources of excess phosphorus were detergents, domestic sewage, and land runoff. With the phasing out of phosphate detergents in the 1970s, domestic sewage, agriculture, and urban runoff have emerged as the dominant drivers of eutrophication. For domestic sewage, untreated or inadequately treated sewage discharges are often the case in less-developed countries. Even when sewers and wastewater treatment are installed, the initial focus is typically on removing organic carbon and sometimes fostering nitrification (converting ammonium to nitrate) in order to mitigate oxygen depletion. Because such treatment has minimal impact on nutrients, the eutrophication problem remains. Hence, costly nutrient removal (called "tertiary treatment") is often necessary if both low oxygen and excessive plant growth are to be remedied.

11.4.3 Pathogens

Disease-causing microorganisms are referred to as pathogens. As summarized in Table 11.3, they include certain types of bacteria, viruses, protozoans, parasitic worms, and algae.

Table 11.3 Some common waterborne pathogens.

Category	Description	Species and groups
Bacteria	Microscopic, unicellular organisms that lack a fully defined nucleus and contain no chlorophyll	*Vibrio cholera, Salmonella, Shigella, Legionella*
Viruses	A large group of submicroscopic (10–25 nm) infectious agents. They are composed of a protein sheath surrounding a nucleic acid core and, thus, contain all the information required for their own reproduction. However, they require a host in which to live	Hepatitis A, Enteroviruses, Polioviruses, Echoviruses, Coxsackie viruses, Rotaviruses
Protozoa	Unicellular animals that reproduce by fission	*Giardia, Entamoeba histolytica, Cryptosporidium*
Helminths	Intestinal worms and wormlike parasites	Nematodes, *Schistosoma haematobium*
Algae	Large group of nonvascular plants. Certain species produce toxins that if consumed in large quantities, may be harmful	*Anabaena, Microcystis, Aphanizomenon*

High levels of pathogens may result from seepage from septic tanks and pit latrines or untreated or inadequately treated sewage discharges as is often the case in less-developed countries. As was the case with oxygen and eutrophication, older cities with aging infrastructure may have leaky sewage collection systems (pipes, pumps, valves), which can cause sanitary sewer overflows. Some cities also have combined sewers, which may discharge untreated pathogen-laden sewage during rain storms.

11.4.4 Salts

Like so many terms, "salt" has both a common meaning and a scientific definition. In common parlance, the term "salt" denotes a white crystalline substance that gives seawater its characteristic taste and is used for seasoning or preserving food. Most commonly this is sodium chloride (NaCl) with potassium chloride (KCl) sometimes used by individuals with low sodium tolerance.

From a scientific perspective, a salt is an ionic compound formed from the reaction of an acid and a base, with all or part of the acid's hydrogen replaced by a metal or other cation. Salts are composed of related numbers of cations (positive ions) and anions (negative ions) so that the product is electrically neutral. These component ions can be inorganic, such as chloride (Cl^-), or organic. Here, we will focus on the inorganic types because of their more dominant impacts on water quality.

As an essential nutrient that humans and animals cannot produce for themselves, a little salt is a good thing. Among other things, it helps regulate many bodily functions including the transport of oxygen and nutrients as well as maintaining the body's overall fluid balance. Hence, saltiness is one of the basic human tastes and our cravings for salty foods reflect an expression of our fundamental need.

Historical evidence of the essential nature of salt is provided by the inclusion of the term "lick" in many town names. For example, particularly in the US Appalachian region, you will commonly come upon towns with names like French Lick (Indiana), Grants Lick (Kentucky), Licking (Missouri), and Salt Lick (Kentucky). These town names stemmed from their proximity to rich surface salt deposits. Among other things, the "licks" drew game as well as providing a commodity that the original settlers could refine for direct consumption, food preservation, or trade. In England, "wich" and "wych" are names often associated with brine springs or salt production. Examples include Middlewich, Nantwich, Northwich, and Leftwich in Cheshire. On a larger scale, major deposits could be mined providing the original economic basis for several major cities such as Syracuse, New York ("The Salt City"), and the eponymously named Salzburg, Austria.

Whereas a little salt in your diet is necessary and beneficial, excess consumption can lead to increased blood pressure, osteoporosis, asthma, stomach cancer, and weight gain. Thus, too much salt can make a drinking water supply nonpotable. In addition, high salt concentrations can interfere with other water uses beyond human and animal consumption. For example, too much dissolved salt can diminish or prevent the growing of crops as well as interfering with many industrial uses.

Salt Buildup Due to Irrigation. Rivers flowing through heavily agriculturalized watersheds are commonly used as the source of water to irrigate adjacent fields. As the water flows through and over the soil, it can dissolve salts that are then discharged back to the river as return flow. Thus, as the water flows downstream, salt concentrations in the river can rise. For arid region, the problem is greatly magnified because the added dissolved salts are concentrated as permanent water losses (evaporation, evapotranspiration, and water incorporated into plant tissue) are much higher than the meager precipitation.

Road Salt. In regions subject to significant snowfall, salts are often applied as deicing agents on roads, parking

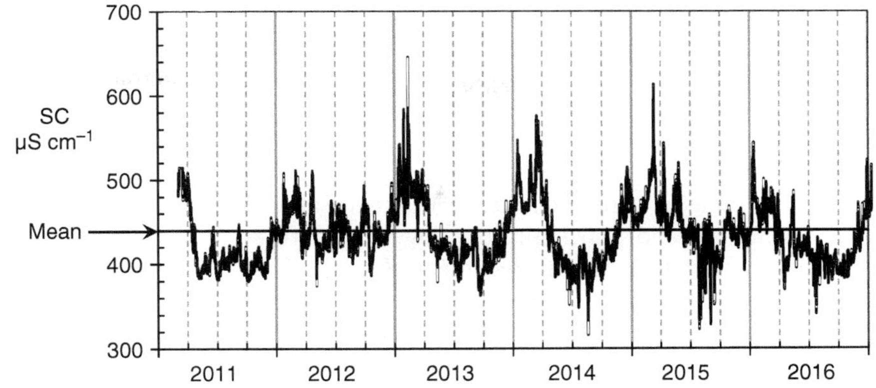

Figure 11.8 The specific conductance (SC) of the Fox River (Wisconsin) just above its outlet into Green Bay. The elevated levels in late winter are due to runoff of road salt applied to roads, parking lots, and sidewalks during and after heavy snowstorms.

lots, and sidewalks. Hence, salt levels in runoff are elevated during thawing periods in winter (Figure 11.8).

Saltwater Intrusion and Well Water Contamination. Saltwater intrusion is the movement of saline water (for the ocean, primarily sodium and chloride ions) into freshwater aquifers, which can lead to contamination of drinking water sources. Saltwater intrusion occurs naturally to some degree in most coastal aquifers, owing to the hydraulic connection between groundwater and seawater. Because saline water has a higher mineral content than freshwater, it is denser and has a higher water pressure. As a result, salt water can push inland beneath the freshwater. Certain human activities, especially groundwater pumping from coastal freshwater wells, have increased saltwater intrusion in many coastal areas. Water extraction drops the level of fresh groundwater, reducing its water pressure and allowing salt water to flow further inland. Other contributors to saltwater intrusion include navigation channels or agricultural and drainage channels, which provide conduits for salt water to move inland. Saltwater intrusion can also be worsened by extreme events like hurricane storm surges. In addition, sea level rise due to global warming could also exacerbate the problem by increasing the oceanic head.

Well Water Contamination. As with saltwater intrusion, salt contamination of inland well waters typically involves two independent processes related to supply and demand. On the supply side, the most common example relates to the leaching of salts from agricultural land. In contrast with oceanic sodium and chloride, the principle pollutant in such cases is nitrate. Farmers typically apply large quantities of nitrogen-rich commercial fertilizer and manure to increase crop yields. Much of the excess nitrogen not taken up by the plants is converted by nitrification to dissolved nitrate that flows freely through soil. Consequently the aquifers under heavily farmed land are often contaminated with high NO_3^- concentrations. As with saltwater intrusion, groundwater pumping for drinking water supply then provides a route to deliver the contaminated water to consumers.

11.4.5 Heat and Temperature

When we think of water quality, we typically think of dissolved substances and particles. Of equal importance are anthropogenic impacts on receiving waters' energy budgets. In particular, a variety of human activities can induce both local and global water temperature rises.

The earliest recognized local temperature effects were due to heated cooling water discharges from power generation. More recently, it has been recognized that the removal of vegetation from watersheds can lead to temperature increases. This is particularly true for the shade reductions caused by the removal of riparian or streamside vegetation canopies. Finally, rising temperature due to climate change could have serious global impacts.

The actual impacts of rising water temperatures are numerous. These include:

Biochemical reactions. All biochemical reactions are strongly temperature dependent. As a rule of thumb, biochemical reaction rates approximately double for each 10 °C rise in temperature. Consequently, rising water temperatures speed up biochemical reactions, which can exacerbate and localize problems such as oxygen depletion and eutrophication.

Dissolved gases. As temperatures rise, less dissolved gas can be held in solution. The critically important gas, dissolved oxygen, is of particular significance in this regard. The saturation concentration of oxygen in a natural unpolluted 0 °C freshwater at sea level is about 14 $mg\,l^{-1}$. As summarized in Figure 11.9a, this value drops almost 50% (7.6 $mg\,l^{-1}$) at temperatures (~30 °C) common in tropical sea level freshwaters. Further, the presence of salts compounds the problem with the 0 and 30 °C saturations dropping to 11.5 and 6.2 $mg\,l^{-1}$, respectively. Finally, elevation above sea level decreases the partial pressure of atmospheric oxygen and consequently reduces oxygen saturation.

As in Figure 11.9b, 0 and 30 °C saturations at high elevations (4.8 km or 3 miles as occur in the Andes) drop to 6.6 and 3.4 mg l^{-1}, respectively.

Species shifts and survivability. All aquatic organisms have temperature tolerances above which they cannot exist. Such is the case for "cold water" fish such as salmonids (i.e. trout, char, salmon, whitefish, and grayling). So rising temperatures can obviously be to the detriment of such organisms. In particular, the young of these species are typically more sensitive than adults to elevated temperatures.

11.4.6 Suspended Matter

As outlined in Table 11.4, suspended solids are often divided as to their sources, with allochthonous solids entering the system from the watershed, point discharges (e.g. sewage outfalls), and the atmosphere, and autochthonous solids generated within the system by organisms and chemical reactions. Within both these categories, they can be further divided into inorganic and organic fractions.

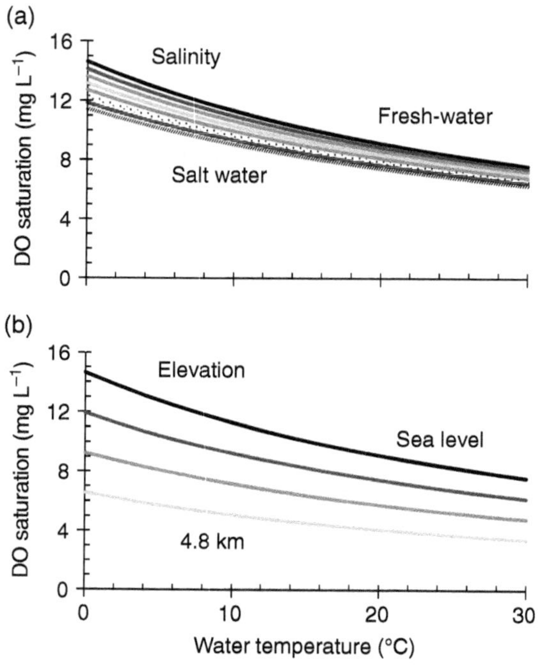

Figure 11.9 Relationship of dissolved oxygen (DO) saturation concentration versus temperature as a function of (a) salinity and (b) elevation.

All types of suspended solids contribute to water clarity. Waters with high sediment loads are very obvious because of their "muddy" appearance. This is especially evident in rivers, where the force of moving water keeps the sediment particles suspended.

The geology and vegetation of a watershed affect the amount of suspended solids. If the watershed has steep slopes with little plant life, topsoil will wash into the waterway with every rain. On the other hand, if the watershed has firmly rooted vegetation, it will trap water and soil and thereby diminish most erosion. Most suspended solids come from accelerated erosion from agricultural land, logging operations (especially clear-cutting), surface mining, and construction. Another source of suspended solids is the resuspension of sediments, which accompanies dredging that is undertaken to keep channels navigable.

There are both direct and indirect environmental impacts of solids. Suspended solids can clog fish gills and decrease light penetration. This reduces the ability of algae to photosynthesize. When water velocities are low (e.g. when they enter a reservoir), the suspended sediment settles to the bottom in a process called siltation. This causes the water to clear, but as the sediment settles, it may smother bottom-dwelling organisms and fish eggs and cover breeding areas.

Indirectly, the suspended solids affect other pollution problems such as eutrophication, pathogens, and toxic substances. For eutrophication, suspended particles can both absorb and scatter light. This reduces the penetration of photosynthetically active radiation. Hence, beyond nutrients, the growth of plants in turbid systems can be light limited. Similarly, suspended particles can limit the penetration of the parts of solar radiation that can kill pathogens (ultraviolet radiation) or that destroys (photolyzes) certain toxics. In addition, both toxics and

Table 11.4 Major types of suspended solids.

	Organic	Inorganic
Allochthonous (from outside the water body)	Sewage sludge Leaves Pollen	Soil particles (clay, silt, fine-grained sand) Floatables (plastic bottles, garbage)
Autochthonous (generated within the water body)	Living organic matter (algae, zooplankton, bacteria) Dead organic matter (detritus)	Chemical precipitates (e.g. calcite, iron precipitates)

bacteria can sorb or attach to settling particles, which effectively reduces levels by transporting or "scavenging" the pollutants out of the water column and into the bottom sediments.

Suspended solids interfere with effective drinking water treatment. High sediment loads interfere with coagulation, filtration, and disinfection. More chlorine is required to effectively disinfect turbid water. They also cause problems for industrial users. Suspended sediments also interfere with recreational use and aesthetic enjoyment of water. Poor visibility can be dangerous for swimming and diving. Siltation, or sediment deposition, eventually may close up channels or fill up the water body, converting it into a wetland. A positive effect of the presence of suspended solids in water is that toxic chemicals such as pesticides and metals tend to adsorb to them or become complexed with them, which makes the toxics less available to be absorbed by living organisms.

11.4.7 Acids and pH

Just as with oxygen and temperature, all organisms have pH ranges where they thrive and values where they suffer. The pH of natural, undisturbed aquatic systems will be primarily governed by the geology of its drainage basin. With the exception of special cases (e.g. saline lakes, bogs, and other high tannin "blackwaters"), most will be in the neutral pH range of about 6.5–8.5. As mentioned previously, overstimulation of the life/death cycle (Figure 11.4) can have a profound effect on pH, particularly for poorly buffered waters. When productivity dominates (eutrophication), pH will rise, and, conversely, when respiration dominates, pH falls.

Beyond the direct effect on organisms, pH has subtler indirect effects due to the fact that the speciation of many chemicals is highly pH dependent. As mentioned previously, the unionized form of ammonia that is toxic to fish predominates at high pH. Similarly, the toxic species of some metals is also pH dependent.

In addition to pH changes connected with overstimulation of the life/death cycle as well as certain industrial discharges, the primary causes of extreme pH are connected with mining and acid rain.

Mining and Acid Mine Drainage. Acid mine drainage refers to the outflow of acidic water from metal and coal mines. Although acid drainage occurs naturally in some watersheds as part of rock weathering, it is greatly magnified by the large-scale earth disturbances characteristic of mining operations, especially those containing large quantities of sulfide minerals. After being exposed to air and water, oxidation of metal sulfides (often iron sulfide) generates acidity. In addition, acid drains from coal mining and processing as well as from tailing piles or ponds, or other mining disposal areas.

When the pH of acid mine drainage rises above about 3, through contact with either freshwater or neutralizing minerals, previously soluble iron(II) ions precipitate as iron(III) hydroxide, a yellow-orange solid commonly dubbed "yellow boy." Other types of iron precipitates include iron oxides and oxyhydroxides, which can discolor water and disrupt stream ecosystems. The process also produces additional hydrogen ions, which can further decrease pH.

Acid Rain. The main cause of acid rain is the burning of coal and other fossil fuels with coal-burning energy plants and internal combustion engines widely considered to be the primary causes. In addition, manufacturing, oil refineries, and other industries also can contribute waste gases containing sulfur dioxide and nitrogen oxides. These can rise very high into the atmosphere, where they mix and react with water, oxygen, and other chemicals to form acid rain.

The ecological impacts of acid rain are most clearly manifest in poorly buffered aquatic environments. For example, mountainous parts of the Northeast US, Scandinavia, and Japan have soils that cannot adequately neutralize the acid in the rainwater. As a result, these areas are particularly vulnerable. As it flows through the soil, acidic rainwater leaches aluminum from soil clay particles and then flows into streams and lakes. Although some plants and animals are able to tolerate acidic waters and moderate amounts of aluminum, many are threatened or destroyed as the pH declines. As was the case with temperature, the young of most species are especially vulnerable to low pH.

11.5 Toxic Substances

To this point the models described have almost exclusively dealt with what are called conventional pollutants, that is, pollutants that are a natural by-product of human and animal metabolism and that overstimulate the natural production/decomposition cycle. Thus, they are not innately harmful, particularly at the levels in undisturbed, natural aquatic ecosystems. In contrast, toxic substances are directly harmful to organisms and in the worst cases can be considered poisonous. This along with three other factors distinguishes toxics from conventional wastes:

Natural versus alien. The conventional pollution problem typically deals with the natural cycle of organic production and decomposition. The discharge of sewage adds both organic matter and inorganic nutrients to a water body. The decomposition of the organic matter by bacteria can result in oxygen depletion. The inorganic nutrients can stimulate excess plant growth. In both cases the problem involves an overstimulation of the natural processes governing the water body. In contrast many toxics do not occur naturally. Prime examples are herbicides and pesticides, which are expressly designed to kill flora and fauna. For such "alien" pollutants, the problem is one of poisoning or interference with natural processes. Thus, rather than accelerating the life/death cycle as is the case for natural organic compounds and nutrients, toxics threaten the processes governing its healthy rotation.

Aesthetics versus health. Although it would be an overstatement to contend that conventional pollution deals solely with aesthetics, a strong case can be made that the mitigation of "visual" pollution has been a prime motivation for conventional waste treatment. In contrast the toxic substance issue is almost totally dominated by health concerns as much toxic remediation focuses on preventing the contamination of drinking water and aquatic food stuffs.

Few versus many. Conventional water quality management deals with on the order of about 20 "pollutants." In contrast there are tens of thousands of organic chemicals that could potentially be introduced into our natural waters. Further, a large fraction of these are synthetic and are increasing every year (Schwarzenbach et al., 1993). If even a fraction of these proved toxic, the sheer numbers of potential toxicants will have a profound effect on the resulting control strategies. Further it is costly to obtain detailed information on the factors governing their transport and fate in the environment. The study of problems such as dissolved oxygen depletion and eutrophication is facilitated by the fact that they involve a few chemicals. In contrast toxicant control is complicated by the vast number of contaminants that are involved.

Speciation. The conventional paradigm usually treats the pollutant as a single species. Consequently its strength in the water body is measured by a single concentration. In contrast the transport, fate, and ecosystem impact of toxicants are intimately connected with how they partition into species. In the context of this chapter, speciation refers to the different chemical forms that a toxicant can occur in an aqueous solution. For example, as depicted in Figure 11.10, the total mercury concentration in a water body is composed of several species, some of which are toxic, whereas other species are not. In addition to chemical speciation, another type of speciation relates to sorption, that is, the tendency of dissolved toxicants to associate with solid matter in and below the water body (i.e. in the bottom sediments). Thus, toxic substance analysis must distinguish between dissolved and particulate forms. This distinction has an impact on transport and fate in the sense that certain mechanisms differentially impact the two forms. For example, volatilization acts only on the dissolved component. The distinction also has importance from toxicity. For example, larger organisms can be exposed in different ways depending on the form of the toxicant. That is, only dissolved contaminants can be taken up via their gills, whereas only particulate forms will be consumed via their guts.

11.5.1 Types of Toxic Substances

Toxic substances can be roughly divided into organic and inorganic forms.

11.5.1.1 Toxic Organics

Although some can occur naturally, toxic organics are overwhelmingly synthetic organic compounds designed for primarily industrial, agricultural, and pharmaceutical applications. In addition, many hydrocarbons, called "polynuclear aromatic hydrocarbons" (PAHs), are generated by combustion of fossil fuels (coal, oil, gasoline, tobacco, wood, etc.) or other organic substances such as charcoal-broiled meat. Table 11.5 summarizes major categories of toxic organics. Many of these man-made organic compounds are resistant to environmental degradation. Sometimes referred to as persistent organic pollutants (POPs), many POPs are toxic and can bioaccumulate in the food chain.

Disinfection by-products (DBPs) are created as an unintended consequence of disinfection of drinking water that has high concentrations of dissolved organic carbon. For example, chlorine can react with the drinking water to form toxic compounds such as trihalomethanes (THMs), haloacetic acids (HAAs), and chlorite. In cases

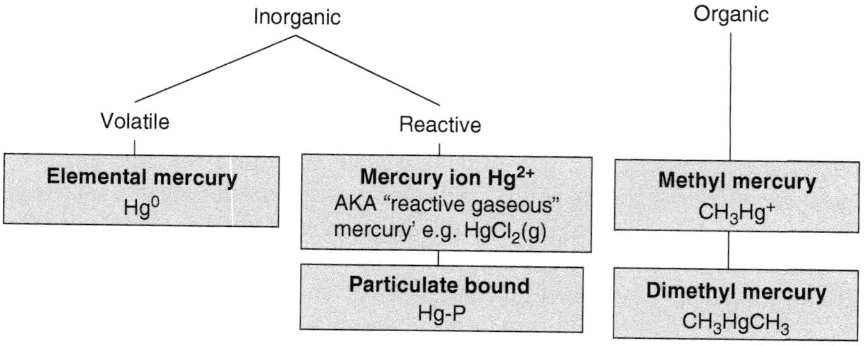

Figure 11.10 Total mercury in natural waters can take on different forms or "species." These species have differing toxicities and are subject to different transport and fate behavior.

Table 11.5 The categories of the organic priority pollutants (CEQ 1978).

Pollutant	Characteristics	Sources and remarks
Pesticides: Generally chlorinated hydrocarbons	Readily assimilated by aquatic animals, fat soluble, concentrated through the food chain (biomagnified), persistent in soil and sediments	Direct application to farm and forest lands, runoff from lawns and gardens, urban runoff, discharge in industrial wastewater. Several chlorinated hydrocarbon pesticides already restricted by EPA: aldrin, dieldrin, DDT, DDD, endrin, heptachlor, lindane, chlordane
Polychlorinated biphenyls (PCBs): Used in electrical capacitors and transformers, paints, plastics, insecticides, other industrial products	Readily assimilated by aquatic animals, fat soluble, subject to biomagnification, persistent, chemically similar to the chlorinated hydrocarbons	Municipal and industrial discharges disposed of in dumps and landfills. TOSCA ban on production after 1 June 1979 but will persist in sediments; restrictions in many freshwater fisheries as a result of PCB pollution (e.g. lower Hudson, upper Housatonic, parts of Lake Michigan)
Halogenated aliphatics (HAHs): Used in fire extinguishers, refrigerants, propellants, pesticides, solvents for oils and greases, and dry cleaning	Largest single class of "priority pollutants"; can cause damage to the central nervous system and liver, not very persistent	Produced by chlorination of water, vaporization during use. Large-volume industrial chemicals, widely dispersed but less threat to the environment than persistent chemicals
Ethers: Used mainly as solvents for polymer plastics	Potent carcinogens; aquatic toxicity, and fate not well understood	Escape during production and use. Although some are volatile, ethers have been identified in some natural waters
Phthalate esters: Used chiefly in the production of polyvinyl chloride and thermoplastics as plasticizers	Common aquatic pollutant, moderately toxic but teratogenic and mutagenic properties in low concentration; aquatic invertebrates are particularly sensitive to toxic effects; persistent and can be biomagnified	Waste disposal vaporization during use (in nonplastics)
Monocyclic aromatics (MAHs) (excluding phenols, cresols, and phthalates): Used in the manufacture of other chemicals, explosives, dyes, and pigments and in solvents, fungicides, and herbicides	Central nervous system depressant; can damage the liver and kidneys	Enter environment during production and by-product production states by direct volatilization, wastewater
Phenols: Large-volume industrial compounds used chiefly as chemical intermediates in the production of synthetic polymers, dyestuffs, pigments, herbicides, and pesticides	Toxicity increases with degree of chlorination of the phenolic molecule; very low concentrations can taint fish flesh and impart objectionable odor and taste to drinking water; difficult to remove from water by conventional treatment; carcinogenic in mice	Occur naturally in fossil fuels; wastewater from coke ovens, oil refineries, tar distillation plants, herbicide, and plastic manufacturing can all contain phenolic compounds
Polycyclic aromatic hydrocarbons (PAHs): Used as dyestuffs, chemical intermediates, pesticides, herbicides, motor oils, and fuels	Carcinogenic in animals and indirectly linked to cancer in humans; most work done on air pollution; more is needed on the aquatic toxicity of these compounds; not persistent and are biodegradable, although bioaccumulation can occur	Fossil fuels (use, spills, and production), incomplete combustion of hydrocarbons
Nitrosamines: Used in the production of organic chemicals and rubber; patents exist on processes using these compounds	Tests on laboratory animals have shown the nitrosamines to be some of the most potent carcinogens	Production and use can occur spontaneously in food cooking operations

where ozone is employed to kill organisms, ketones, carboxylic acids, and aldehydes (formaldehyde) can be formed. Further, for source waters with significant bromide levels, ozonation can lead to the creation of the carcinogen bromate as well as other brominated DBPs.

11.5.1.2 Inorganic Toxicants

These are not only primarily metals and radionuclides but also some special inorganic compounds such as cyanide.

Heavy Metals. A toxic heavy metal is any relatively dense metal or metalloid that is noted for its potential toxicity, especially in environmental contexts (Table 11.6). Many are toxic at relative low concentrations and can have direct physiological effects or can bioconcentrate.

Metals differ from toxic organics in some key ways. First, they are persistent in that they are never destroyed. Second, they can exist in a much wider range of physical and chemical forms and can change between forms in the environment or within organisms. Third, in many cases, some of their species are toxic, whereas others are harmless.

Although they can occur naturally often at low levels, anthropogenic processes can elevate their concentrations to high, toxic levels. These include certain industrial processes (smelting), mining, fossil fuel combustion, and road runoff.

Finally, humans can be exposed to high concentrations through drinking water via old plumbing systems. For example, high levels of lead can leach into drinking water through corrosion of pipes, solder, and plumbing fixtures

11.5.2 Radionuclides

Radionuclides are atoms that are radioactive (Table 11.7). That is, they are atoms with an unstable nucleus that, to become more stable, emits energy in the form of rays or high-speed particles. The most common radionuclides in drinking water are radium, radon, and uranium.

Table 11.6 Metals found in natural waters.

Name	Symbol	Name	Symbol	Name	Symbol
Aluminum	Al	Chromium	Cr	Mercury	Hg
Antimony	Sb	Cobalt	Co	Nickel	Ni
Arsenic	As	Copper	Cu	Selenium	Se
Barium	Ba	Iron	Fe	Silver	Ag
Cadmium	Cd	Lead	Pb	Thallium	Tl
Cesium	Cs	Manganese	Mn	Zinc	Zn

Table 11.7 Radionuclides found in natural waters.

Name	Symbol
Cesium-137	^{137}Cs
Lead-210	^{210}Pb
Plutonium-239, 240	$^{239,240}Pu$
Radium-226, 228	$^{226,228}Ra$
Strontium-90	^{90}Sr
Uranium	U

Certain radionuclides (or radioactive substances) occur naturally. They are created in the upper atmosphere and are found in certain types of rocks containing trace amounts of the radioactive forms (isotopes) of uranium, thorium, or actinium. As these rocks weather, the resulting runoff can enter surface and groundwaters.

In addition to natural sources, man-made inputs can result from nuclear energy generation, nuclear weapons development, and some industrial applications. Many devices and processes result in radioactivity including color television, medical instruments (X-rays and chemotherapy), and coal power plants. Finally, nuclear testing and accidental spills can release large amounts of radioactivity to the aqueous environment. The latter could become more prominent in the future as the first generation of nuclear power plants approach or exceed their design lifetimes. Further, more nuclear power plants could be developed as an alternative to greenhouse gas-emitting fossil fuel plants.

People who are exposed to relatively high levels of radionuclides in drinking water for long periods may develop a variety of serious health problems, such as cancer, anemia, osteoporosis, kidney disease, liver disease, and impaired immune systems. And obviously, the exposure to high levels of radiation connected with accidents can be deadly.

11.6 Emerging Water Pollutants

Emerging water contaminants are any synthetic or naturally occurring chemical or any microorganism that is not commonly monitored or regulated in the environment but has the potential to enter surface and groundwaters and cause adverse human and/or ecological health effects. These include but are not limited to prescription drugs, personal care products, and new chemicals used in agriculture and industry.

Of particular interest are personal care products and pharmaceuticals, such as birth control pills, painkillers, and antibiotics. Little is known about their long-term human or ecosystem impacts, although some are believed to mimic natural hormones in humans and other species.

Finally, new chemicals and substances such as flame retardants, nanoparticles, and new pesticides all can find their way into natural waters.

11.7 Back to the Future

In this chapter, I have outlined how water quality has been defined in the United States and Western Europe since the 1860s. As I hope is clear, a century and a half ago, those parts of our planet suffered from comparable levels of "rubbish, stink, and death" that plague many developing countries today. At first, economic development exacerbated and focused these problems as populations moved from the country to the city. Then, the subsequent growth of stable middle classes coupled with democratic political systems led to societal actions to ameliorate these symptoms.

Figure 11.11 provides a depiction of this evolution. As societies move from agrarian to developed economies, the emphasis on water quality control moves from survival to sustainability. While past concerns for water quality were driven by "rubbish, stink, and death," future efforts will be increasingly dictated by higher-order goals such as quality of life, tourism, and trade. Today, developing countries around the world are beginning to recognize that environmental protection must be coupled with economic development. For these countries, cost-effective control strategies could provide a means to control water pollution while maintaining economic growth. In this sense, the tools and approaches developed over the past century by environmental engineers and scientists could greatly contribute to environmentally sustainable development across the planet.

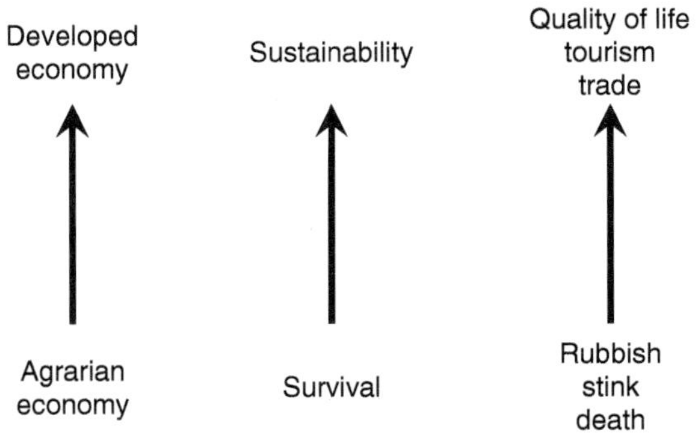

Figure 11.11 As societies move from primitive to developed economies, the emphasis of water quality control moves from survival to sustainability. Whereas past concern for water quality was driven by "rubbish, stink, and death," future efforts will be dictated by higher-order characteristics such as quality of life, tourism, and trade.

Note

1 The term autotrophic refers to organisms like plants that do not depend on other organisms for nutrition. In contrast, heterotrophic organisms consist of animals and most bacteria that subsist on organic matter.

References

CEQ (1978). *Environmental Quality: The Ninth Annual Report of the Council on Environmental Quality.* Washington, DC: U.S. Government Printing Office.

Chapra, S.C. (1997). *Surface Water-quality Modeling.* Long Grove, IL: Waveland Press.

Chapra, S.C. (2011). Rubbish, stink, and death: The historical evolution, present state, and future direction of water-quality management and modeling. *Environmental Engineering Research* 16 (3): 113–119.

Johnson, S. (2006). *The Ghost Map: The Story of London's Most Terrifying Epidemic – And How It Changed Science, Cities, and The Modern World.* New York, NY: Riverhead Books.

Lebeau, J.M. (1983). *The Jewish Dietary Laws Sanctify Life* (ed. S. Garfinkel). New York, NY: United Synagogue of Conservative Judaism, Department of Youth Activities.

Schwarzenbach, R.P., Gschwend, P.M., and Imboden, D.M. (1993). *Environmental Organic Chemistry.* New York: Wiley.

Vitruvius (1934). *On Architecture.* Cambridge: Harvard University Press.

12

Wastewater Engineering

Say Kee Ong

Department of Civil, Construction, and Environmental Engineering, Iowa State University, Ames, IA, USA

12.1 Introduction

There are more than 15 400 municipal wastewater treatment plants in the United States treating more than 32 billion gallons per day of municipal and industrial wastewaters (US EPA, 2012). To minimize environmental impact and to protect human health, these municipal and industrial wastewaters must be treated to remove various chemical constituents and toxic/hazardous compounds before they are released to the environment. As required by the Clean Water Act (1987), municipal wastewater treatment plants are required to obtain a National Pollutant Discharge Elimination System (NPDES) permit to discharge treated effluent to a receiving body of water. Effluent limits in an NPDES permit are determined using technology-based effluent limits, i.e. based on available technology to control the pollutants, and water quality-based effluent limits, i.e. based on limits that are protective of the water quality standards of the receiving water (US EPA, 2010). Typical pollutant limitations for municipal wastewater treatment plants include biochemical oxygen demand (BOD), nutrients such as nitrogen (ammonia and total nitrogen) and phosphorus, total coliforms, and suspended solids (SS) (US EPA, 2004). Similarly, industries discharging treated wastewater directly to a receiving body of water are required to obtain an NPDES permit. The effluent quality to be discharged is based on several factors including specific pollutants found in the industrial wastewaters and the water quality and assimilative capacity of the receiving water body. Industries that are major contributors of industrial wastewaters include chemical, petrochemical, petroleum refining, food and consumer product, metal, and pulp and paper industries. To avoid obtaining an NPDES permit, many industries discharge their wastewaters into sanitary sewers and are treated in municipal wastewater treatment plants along with municipal wastewaters. Under these circumstances, municipal wastewater treatment plants require the industries to comply with the National Pretreatment Program as outlined in the 40 Code of Federal Regulations (CFR) 403 (US EPA, 2011).

The National Pretreatment Program controls the discharge of 126 priority pollutants from industries into sewer systems as described in the Clean Water Act. These priority pollutants fall into two categories:

- Metals – Examples include arsenic, mercury, lead, chromium, and cadmium. Metals cannot be degraded or broken down through treatment.
- Organic compounds – Examples include volatile organics, semivolatile organics, and persistent organic compounds (POPs). Examples include solvents such as trichloroethylene, aromatics such as toluene and benzene, pesticides, and polychlorinated biphenyls (PCBs). Many can be degraded to innocuous compounds or to carbon dioxide and water. However, some compounds such as PCBs can be recalcitrant.

Other pollutants include:

- Acid/alkalis – Examples include nitric acid, sulfuric acid, and caustic soda. They have low or high pH and are corrosive but can be neutralized.
- Inorganic salts – Examples include innocuous ions such as sodium, potassium, chloride, and sulfates that are harmless, but they increase the total dissolved solids of the receiving water.
- Suspended solids – May come in various forms; inorganic or organic solids from processing of raw materials such as ores or raw products.

12.2 Wastewater Characteristics and Treatment Requirements

Municipal (sanitary) wastewaters from homes and institutions such as schools and airports are the most common and ubiquitous source of water pollution. Concentrations

Handbook of Environmental Engineering, First Edition. Edited by Myer Kutz.

of various pollutants in a medium strength municipal wastewater are presented in Table 12.1. Municipal wastewater treatment plants are required to reduce the 5-day biochemical oxygen demand (BOD_5), SS, ammonia, and bacteria to levels acceptable by the receiving water body. BOD_5 is generally regulated to prevent further depletion of dissolved oxygen in the receiving water, while ammonia is regulated for its toxicity to aquatic organisms. For water bodies that are heavily impaired, other requirements such as total phosphorus and total nitrogen may be imposed to minimize further eutrophication of the receiving water body.

Wastewater from an industrial facility may be categorized as (i) process wastewaters from the manufacturing processes, (ii) utility wastewaters from blowdowns of boilers and cooling towers, and (iii) wastewaters from sanitary activities. All three categories of wastewaters are significantly different in their quality, with varying wastewater characteristics. Table 12.1 shows the wide range of concentrations of some of the common pollutants found in wastewaters of several typical industries. One of the treatment strategies for large industrial facilities is to separate the different wastewater streams from each process and treat the wastewater streams individually before discharging into a sanitary sewer or directly discharging into a body of water. There are instances where it is more economical to combine the process and utility wastewaters and treat them at a central wastewater treatment plant before it is discharged. In some industrial complexes, several industries may combine their wastewaters and treat them together at a central wastewater facility. Some of the disadvantages of a central wastewater treatment plant treating wastewaters from several industries include designing and operating various unit treatment processes to remove the pollutants sequentially along with careful monitoring of wastewater pollutants from different industries that may upset the treatment plant. Sanitary wastewaters from industrial facilities are typically not mixed with the process wastewaters and are discharged separately into sanitary sewers.

If the wastewater is to be pretreated before it discharges into the sanitary sewer, the effluent will be regulated by National Pretreatment Program. The National Pretreatment Program identifies specific requirements that apply to different categories of industrial users (see 40 CFR 403.6 and 40 CFR 403.3(i)) that discharge industrial wastewaters to municipal wastewater treatment plants (see Table 12.2). For each of the different categories of users, three types of discharge standards are applied and enforced. They are:

- Prohibited discharge standards – General, national standards applicable to all industrial users to a municipal wastewater treatment plant, designed to protect against pass-through and interference, protect the sewer collection system, and promote worker safety

Table 12.1 Characteristics of municipal and several industrial wastewaters.

Contaminants	Municipal wastewater (medium strength)[a]	Salad dressing manufacturing wastewater[b]	Steel plating mill wastewater[c]	Thermochemical pulping wastewater[d]
Total solids ($mg\,l^{-1}$)	720	—	—	—
Total dissolved solids ($mg\,l^{-1}$)	500	—	2100	—
Total suspended solids ($mg\,l^{-1}$)	210	742	291	383–810
BOD_5 ($mg\,l^{-1}$)	190	2286	20	2800
COD ($mg\,l^{-1}$)	430	4260	175	5600–7210
Ammonia ($mg\,l^{-1}$ as N)	25	—	80	—
Total phosphorus ($mg\,l^{-1}$ as P)	7	—	—	2.3
Oil and grease ($mg\,l^{-1}$)	90	382	5	—
Volatile organic compounds ($mg\,l^{-1}$)	100–400	—	—	235[e]
Metals ($mg\,l^{-1}$)	<1	—	See below[f]	—
Total coliform (No./100 ml)	10^7–10^9	—	—	—
pH	6.5–8	5.1–12.3	≅2	4.2

[a] Metcalf and Eddy (2003).
[b] Ersu et al. (2008a).
[c] Matheson et al. (2015).
[d] Chromium, $1\,mg\,l^{-1}$; copper, $0.3\,mg\,l^{-1}$; iron, $70\,mg\,l^{-1}$; lead, $0.6\,mg\,l^{-1}$; mercury, $0.190\,mg\,l^{-1}$; zinc, $70\,mg\,l^{-1}$; nickel, $140\,mg\,l^{-1}$.
[e] Pokhrel and Viraraghavan (2004).
[f] Acetic acid.

Table 12.2 Effluent guidelines and standard title 403–471.

Category	40 CFR Part	Pollution of concern (partial listing only)
Aluminum forming	467	
Asbestos manufacturing	427	Suspended solids, pH, organic loading (COD)
Battery manufacturing	461	
Canned and preserved fruits and vegetables processing	407	Organic loading, suspended solids, pH, oil and grease
Canned and preserved seafood processing	408	Organic loading, suspended solids, pH, oil and grease
Carbon black manufacturing	458	
Coil coating	465	
Copper forming	468	
Cement manufacturing	411	Suspended solids, temperature, pH
Centralized waste treatment	437	
Concentrated aquatic animal producing	451	
Concentrated animal feeding operations (CAFO)	412	Organic loading (BOD_5), pH, solids, nitrogen, and phosphorus
Dairy products processing	405	Organic loading, suspended solids, pH
Electrical and electronic components	469	
Electroplating	413	Cyanide, heavy metals (copper, nickel, chromium, zinc, lead, cadmium, silver), total solids, pH, total toxic organics (e.g. acrolein, benzene, chlorinated compounds), chelating agents
Explosives manufacturing	457	
Ferroalloy manufacturing	424	Suspended solids, chromium, manganese, pH, cyanide, phenols, ammonia
Fertilizer manufacturing	418	Ammonia, organic nitrogen, nitrate, phosphorus, fluoride, suspended solids, pH, organic loading
Glass manufacturing	426	Organic loading (BOD_5, COD), phenol, suspended solids, pH, oil (mineral, animal, and vegetable), phosphorus, lead, fluoride, ammonia
Grain mills	406	Organic loading, suspended solids, pH
Gum and wood chemical manufacturing	454	
Hospital	460	
Ink formulating	447	
Inorganic chemical manufacturing	415	Organic carbon, suspended solids, pH, metals (antimony, arsenic, barium, cobalt, mercury, lead, copper, nickel, chromium, iron, selenium, silver, zinc), fluoride, residual chlorine, cyanide, sulfide, ammonia, oil and grease
Iron and steel manufacturing	420	Suspended solids, oil and grease, ammonia, cyanide, phenols, pH, aromatic (benzo(a) pyrene, naphthalene, 2,3,7,8-TCDF, tetrachloroethylene), lead, zinc, chromium, nickel
Landfill	445	
Leather tanning and finishing	425	Organic loading (BOD_5), suspended solids, oil and grease, chromium, pH, sulfide
Meat poultry products	432	
Metal finishing	433	
Metal molding and casting	464	
Mineral mining and processing	436	
Metals products and machinery	438	

(*Continued*)

Table 12.2 (Continued)

Category	40 CFR Part	Pollution of concern (partial listing only)
Nonferrous metals forming and metal powders	471	
Nonferrous metal manufacturing	421	Fluoride, suspended solids, pH, benzo(a)pyrene, metals (arsenic, antimony, beryllium, cobalt, nickel, aluminum, indium, molybdenum, mercury, copper, cadmium, lead, zinc, selenium, silver, tantalum, tin, titanium, tungsten), ammonia, cyanide, oil and grease, organic loading (COD), total phenolics, fluoride, hexachlorobenzene
Oil and gas extraction	435	
Ore mining and dressing	440	
Organic chemicals, plastics, and synthetic fibers	414	Organic loading, suspended solids, pH, specific organic compounds (benzene, anthracene, chlorobenzene, phenols, etc.), metals (chromium, copper, lead, nickel, zinc), cyanide
Paint formulating	446	
Paving and roofing materials (tars and asphalt)	443	
Pesticide chemicals	455	
Petroleum refining	419	Organic loading (BOD_5, COD), suspended solids, oil and grease, phenolic compounds, ammonia, sulfide, chromium, pH
Pharmaceutical manufacturing	439	
Photographic	459	
Plastics molding and forming	463	
Porcelain enameling	466	
Phosphate manufacturing	422	Total phosphorus, fluoride, suspended solids, pH
Pulp, paper, and paperboard	430	
Rubber manufacturing	428	Organic loading, oil and grease, suspended solids, pH, lead, chromium, zinc
Soap and detergent manufacturing	417	Organic loading (BOD_5), suspended solids, oil and grease, pH, surfactants
Steam electric power generating	423	Suspended solids, oil and grease, copper, iron, free available chlorine, 126 priority pollutants (cooling tower), chromium and zinc
Sugar processing	409	Organic loading (BOD_5), suspended solids, coliform, temperature, pH
Textile mills	410	Organic loading (BOD_5), suspended solids, oil and grease, sulfide, phenol, chromium, pH
Timber products processing	429	
Transportation equipment cleaning	442	
Waste combustors	444	

and beneficial biosolids use (40 CFR 403.5). A listing of the effluent guidelines and standards is presented in Table 12.3.

- Categorical pretreatment standards – These are national, technology-based standards that limit pollutant discharges for specific process wastewaters of particular industrial categories called categorical industrial users. The standards applicable to industrial discharges are designated in the Effluent Guidelines and Limitations (40 CFR Parts 405–471) (Table 12.2 provides a list of pollutants of concern regulated for certain selected industries).
- Local limits – Developed to reflect specific needs and capabilities at individual municipal wastewater treatment plants and are designed to protect the publicly owned treatment work (POTW) receiving waters (40 CFR 403.8(f)(4)).

Within each industry category, there are subparts in the 40 CFR that refer to specific manufacturing process within the industry. For example, the iron and steel manufacturing category has a total of 13 subparts for different processes. Subpart A of this category addresses

Table 12.3 Prohibited discharge standards.

General	• Any pollutant(s) that cause pass-through or interference
Specific	• Pollutants that create a fire or explosion hazard, waste streams with a closed cup flashpoint of less than 60 °C or 140 °F
	• Pollutants that will cause corrosive structural damage, in no case discharges with pH lower than 5.0, unless the works are specifically designed to accommodate such discharges
	• Solid or viscous pollutants that will cause obstruction to the flow in the POTW
	• Any pollutant, including oxygen demanding pollutants (BOD) released that will cause interference with the POTW
	• Heat that will inhibit biological activity in the POTW, in no case heat discharges in such quantities that the temperature at the POTW treatment plant exceeds 40 °C (104 °F) unless approved
	• Petroleum oil, nonbiodegradable cutting oil, or products of mineral oil origin that will cause interference or pass-through
	• Pollutants that result in the presence of toxic gases, vapors, or fumes within the POTW in a quantity that may cause acute worker health and safety problems
	• Any trucked or hauled pollutants, except at discharge points designated by the POTW

wastewater from the coke making process, while subpart F addresses wastewater from continuous casting process. The effluent guidelines and limits for an industrial facility are based on whether it is an existing source or a new source and the use of best available technology (BAT) or best practicable control technology (BPT). Effluent limits are production based with daily maximums and monthly averages, concentration-based with daily maximums and monthly averages, or mass-based (concentration-based standards multiplied by process flow) with daily maximums and monthly averages. In some cases, the guidelines will prohibit discharges of process wastewater at a flow rate or mass loading rate, which are excessive over any time period during the peak period. For example, for the continuous casting process, the pretreatment effluent limits are production based with maximums for any day equal to 0.00730, 0.00313, 0.0000939, and 0.000141 lb/1000 lb. of product for total SS, oil and grease, lead, and zinc, respectively. pH of the wastewater is permissible between 6 and 9. For the metal finishing industries (Part 433), the pretreatment effluent limits for a new source are concentration based that include maximum concentrations for any day for cyanide (1.2 $mg\,l^{-1}$), oil and grease (52 $mg\,l^{-1}$), total SS (60 $mg\,l^{-1}$), pH (6–9), and seven heavy metals such as cadmium (0.11 $mg\,l^{-1}$), copper (3.38 $mg\,l^{-1}$), and lead (0.69 $mg\,l^{-1}$).

12.3 Treatment Technologies

Wastewater treatment facilities consist of unit processes placed in series to remove one or more constituents from the wastewater. As the wastewater passes through the unit processes, pollutants are removed to the concentrations needed before the treated effluent is discharged. Unit processes can be broadly categorized into:

- Physical unit processes – Use of physical forces (e.g. mixing, screening, flocculation, sedimentation, filtration).
- Chemical unit processes – Use of chemical agents to remove or convert pollutants to less harmful or innocuous compounds (e.g. chemical oxidation, precipitation, adsorption).
- Biological unit processes – Use of microbial or biological agents to remove or degrade pollutants to innocuous compounds (e.g. microbial degradation, nitrification).

Table 12.4 provides a list of pollutants and possible unit processes that can be used to remove the pollutant. Since a unit process may affect the performance of another unit process or a certain pollutant must be removed before a unit process can be applied, each unit process must be carefully selected and placed sequentially for optimal removal of pollutants from the wastewater. A typical flow diagram consisting of various unit processes for the removal of SS, carbonaceous compounds (BOD), ammonia, and bacteria in municipal wastewater is presented in Figure 12.1. The treatment train can be divided into preliminary, primary, secondary, and tertiary treatment. Preliminary treatment consists of screens to physically remove large debris and grit sedimentation tanks to gravitationally settle sand and grit from the wastewater. The primary clarifier in the primary treatment uses gravity to settle out organic solids. In the activated sludge tank of the secondary treatment, soluble and particulate organics are biologically assimilated by microorganisms and mineralized to carbon dioxide and water, while ammonia is biologically converted to nitrate. Depending on the treatment system, a tertiary treatment system may be included consisting of sand filters for further removal of SS, activated carbon filters to adsorb any metals and/or trace anthropogenic organic compounds, or a nutrient removal system to remove nitrogen and phosphorus. Municipal treatment systems typically have a disinfection system such as a chlorination system or an ultraviolet (UV) light system to remove coliforms before the effluent is discharged into the body of water.

An example flow diagram for the treatment of industrial wastewater such as in food processing industries with high concentrations of carbonaceous waste (between 1000 and 10 000 $mg\,l^{-1}$ BOD_5) but with minimal inorganic

Table 12.4 Major pollutants and unit processes.

Pollutant	Specific pollutants	Physical processes	Chemical processes	Biological processes
Inorganic solids	Coarse solids (>1 cm)	• Screening • Sedimentation	—	—
	Suspended solids (>1 μm)	• Sedimentation • Flotation • Filtration	• Precipitation • Hydrolysis	• Biological hydrolysis
	Dissolved solids (<1 μm)	• Adsorption • Membrane separation	• Precipitation	• Bacterial assimilation
Organic solids and matter	BOD in suspension (particulate BOD >1 μm)	• Sedimentation • Absorption	• Hydrolysis	• Biological degradation (aerobic and anaerobic)
	Soluble BOD	• Adsorption • Absorption	• Chemical oxidation	• Biological degradation (aerobic and anaerobic)
Nitrogen	Organic nitrogen	—	—	• Ammonification
	Ammonia	• Air or steam stripping	• Chemical oxidation (chlorination)	• Nitrification • Bacterial assimilation
	Nitrate	• Membrane separation	• Ion exchange • Chemical reduction	• Denitrification
Phosphorus	Organic phosphorus	• Membrane separation	• Chemical precipitation	• Bacterial assimilation • Biological accumulation
	Phosphates	• Membrane separation	• Chemical precipitation	• Bacterial assimilation • Biological accumulation
Inorganic pollutants	Heavy metals	• Membrane separation	• Chemical precipitation • Adsorption • Electrodialysis	• Biological assimilation
Organic pollutants	Volatile and semi-volatile organic compounds	• Volatilization	• Adsorption • Chemical oxidation	• Biological degradation
Oils	Oil and grease	• Flotation	• Adsorption	—
Microorganisms	Bacteria and virus	• Heat	• Ultraviolet light • Chlorination	—
	Protozoa cysts and helminth eggs	• Sedimentation • Filtration	—	—
Residual chlorine		• Volatilization	• Dechlorination • Chemical reduction	—

wastes such as metal ions is presented in Figure 12.2. Screens are used to remove large debris in the wastewater. This is followed by a balancing or equalization tank to even the wastewater flow pumped into the treatment plant. Because of the high carbonaceous waste, the wastewater is typically treated with an anaerobic process where between 70 and 90% of the BOD are removed. If the treated wastewater is to be discharged into the sewer system, SS in the effluent of the anaerobic digester are removed and pH adjusted before the effluent is discharged into the sewers. If the wastewater is to be discharged into a body of water, the effluent from the anaerobic digester is further treated using an aerobic process such as an activated sludge process to reduce the BOD_5 to concentrations suitable for discharge. The wastewater may be further treated using sand filters or activated carbon filters to remove SS and other pollutants such as metals or micropollutants.

12.3.1 Physical Unit Process

12.3.1.1 Air Stripping

Air stripping is the process of transferring volatile organic compounds (VOCs) and semi-VOCs from the wastewater (liquid phase) into vapor phase by passing a high volume of stripping medium through the wastewater.

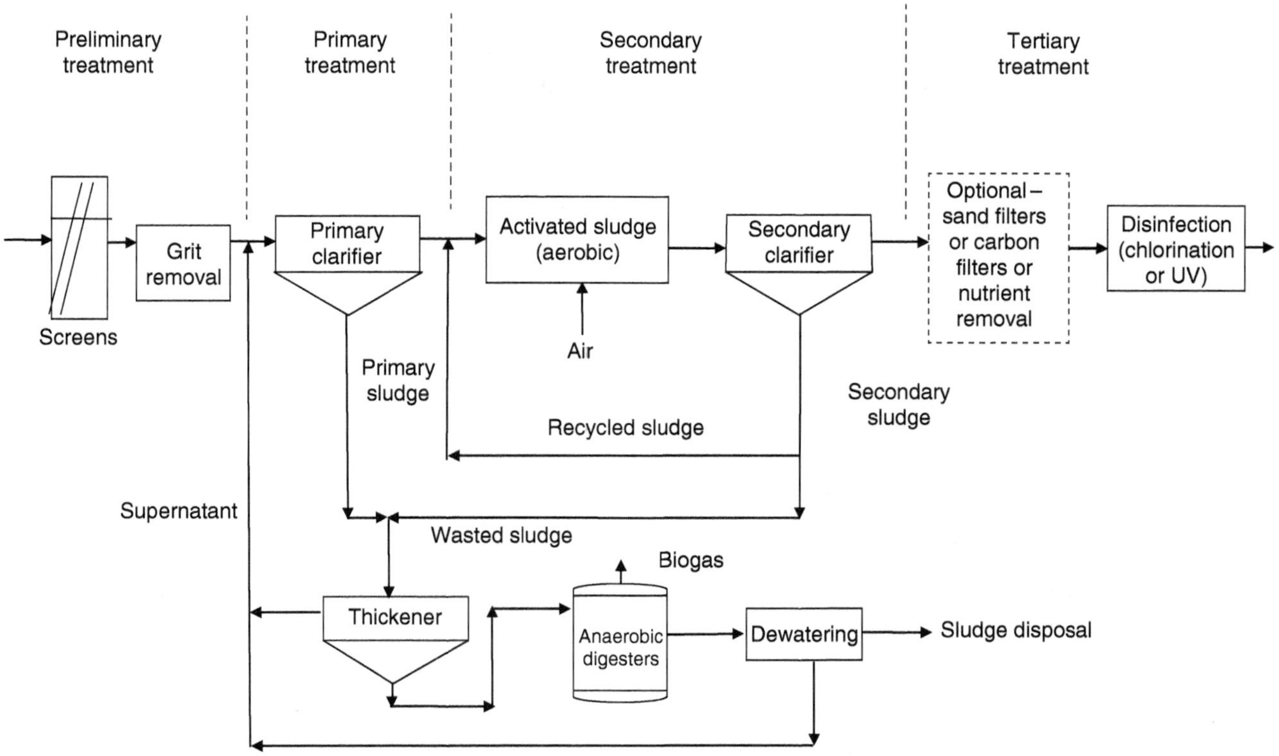

Figure 12.1 Generalized flow diagram of an activated sludge treatment plant treating municipal wastewater.

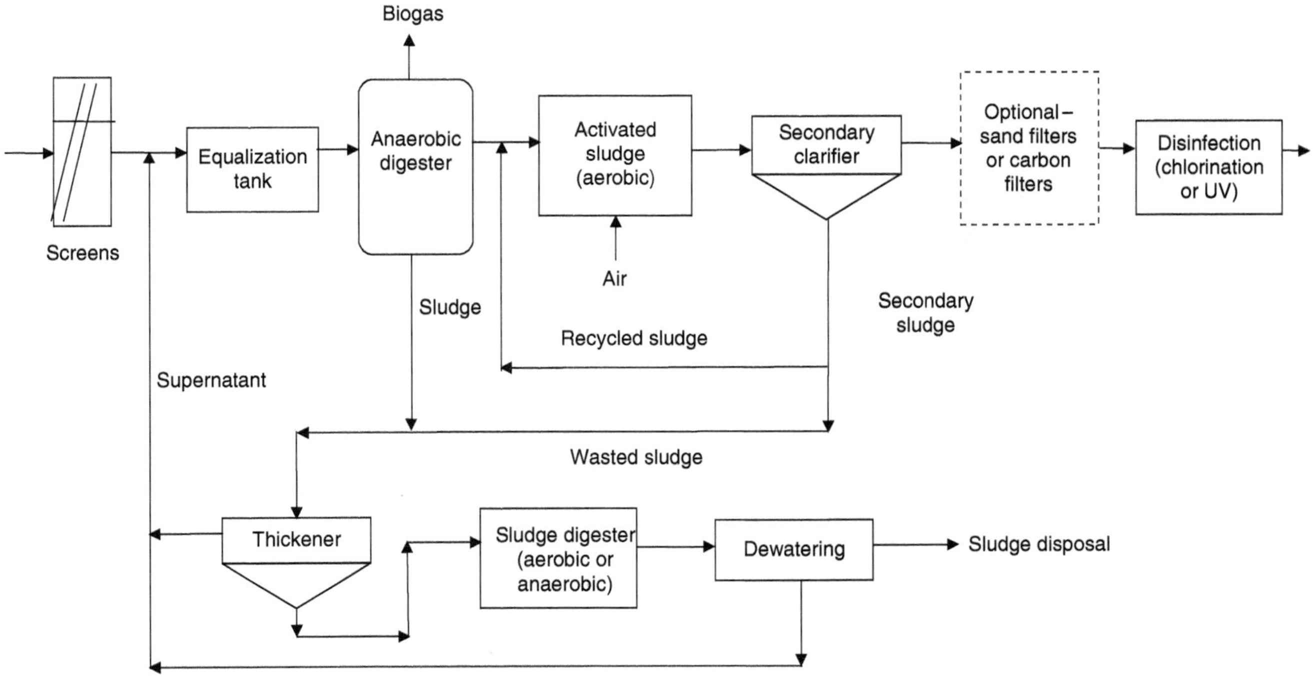

Figure 12.2 Example flow diagram for the treatment of high BOD industrial wastewater.

Two common stripping media used to remove VOCs are air and steam. Air stripping is cost effective for treatment of waste streams with low concentrations of the hazardous VOCs especially for constituents that can be volatilized at room temperature. When semivolatile compounds are present, steam is often used to increase the temperature of the liquid phase to enhance volatilization of the organic compounds. Stripping can be performed by using packed towers, tray towers, spray systems, diffused aeration systems, or mechanical aeration. In a typical packed tower system, water is distributed evenly at the top of the tower through a packing material with a high specific surface area. Water is introduced into the tower as fine water droplets and is made to spread over the surface of the packing material resulting in a high surface area for the volatilization of the volatile organics from the liquid phase into air or steam. Air is introduced from the bottom of the tower and leaves from the top of the tower. Factors affecting the volatilization efficiency of an air stripping tower include volatility of the compounds as measured by Henry's law constant, air flow rate, water loading rate, mass transfer at the air–water interface, packing material, and depth of the packing material (McCarthy, 1983; Gavaskar et al., 1995).

The height of the packing tower needed for the removal of a pollutant to a certain effluent concentration can be estimated using the following equations:

Tower packing height, $Z = \text{HTU} \times \text{NTU}$.

NTU is defined as the number of transfer units and is given by

$$\text{NTU} = \left(\frac{S}{S-1}\right)\ln\left[\frac{(S-1)}{S}\frac{C_{\text{L,in}}}{C_{\text{L,out}}} + \frac{1}{S}\right]$$

S is called the stripping factor and is given by

$$S = K'_{\text{H}}\frac{Q_{\text{G}}}{Q_{\text{w}}}$$

HTU is defined as the height of transfer units and is given by

$$\text{HTU} = \frac{Q_{\text{w}}}{K_{\text{L}}aA} = \frac{L_{\text{M}}}{\rho_{\text{L}}K_{\text{L}}a}$$

K'_{H} is Henry's law constant (dimensionless), $K_{\text{L}}a$ is the overall volumetric mass transfer coefficient ($\text{m}^3\ \text{s}^{-1}$-$\text{m}^2$), a is the interfacial surface area for mass transfer ($\text{m}^2\ \text{m}^{-3}$), Q_{G} is the air flow rate ($\text{m}^3\ \text{s}^{-1}$), L_{M} is the liquid mass flux ($\text{kg}\,\text{m}^{-2}$-s), Q_{w} is the liquid flow rate ($\text{m}^3\ \text{s}^{-1}$), $C_{\text{L,in}}$ is the influent liquid solute concentration ($\text{mg}\,\text{l}^{-1}$), $C_{\text{L,out}}$ is the effluent liquid solute concentration ($\text{mg}\,\text{l}^{-1}$), and ρ_{L} is the liquid density ($\text{kg}\,\text{m}^{-3}$).

12.3.1.2 Solids Removal: Clarification

Settling of solids in wastewaters by gravity is a common and inexpensive technology for separation and removal of solids (Metcalf and Eddy, 2003). The process is termed as sedimentation or clarification. Sedimentation is carried out in a clarifier or a sedimentation tank. The clarified liquid that is low in SS may be reused as process water or discharged, while the concentrated suspension or sludge is concentrated further to produce a drier product before disposal. In the design of a clarifier, two processes are occurring: clarification as demonstrated by hindered settling and thickening as demonstrated by compression settling at the bottom of the clarifier. Surface overflow rates are typically used in the design of sedimentation tanks (see Table 12.5 for design surface overflow rates).

Conventional settling tanks have surface overflow rates ranging from 1.0 to 2.5 m h^{-1}. To increase the sedimentation rates, lamella tube settlers have been used giving surface overflow rates in the range of 2.5–7.5 m h^{-1} (Crittenden et al., 2005). In recent years, sedimentation technologies with high processing rates and, consequently, a smaller equipment footprint are becoming more attractive than conventional sedimentation tanks. There are a variety of high rate clarification technologies for the separation of solids. High rate clarification systems may rely on the addition of ballast (such as fine sand) or recycling of flocculated solids to enhance the formation of microflocs and settling of the flocs. Most high rate clarification systems use tube settlers to maximize the settling surface area. Examples of ballasted flocculation system are the Actiflo® and Microsep® process. A ballasted system consists of a mixing zone, a maturation zone for the formation of the flocs, and a settling zone using lamella plate settling. Sand and flocs are removed from the clarifier and pumped into a cyclone separator for sand removal. The clean sand is returned to the injection tank, and solids from the cyclone are sent to

Table 12.5 Typical surface overflow rates for clarifier design.

Type of waste	Specific gravity	Surface overflow rates ($\text{m}^3\ \text{m}^{-2}$-h)	Average hydraulic retention time (h)
Precipitation treatment			
Aluminum and iron floc	1.002	1.0–3.1	2–8
Calcium carbonate precipitates	1.2	1.0–4.4	1–4
Metals precipitation (using NaOH)	1.1	0.87–2.6	1–4
Biological solids (activated sludge)	1.005	0.7–2.0	1–4

Table 12.6 High rate sedimentation processes from various vendors.

Technology name	Surface overflow rates (gpm ft^{-2})	(m h^{-1})	Description	Vendors
Lamella settler	Up to 3.0	Up to 7.5	Inclined tubes or parallel plates to increase sedimentation	Various
Actiflo®	20–70	50–175	Microsand ballasted flocculation and lamella clarification	Kruger
Microsep®	20–40	50–100	Microsand ballasted flocculation/solids contact and clarification	Veolia Water
Densadeg®	10–50	25–125	Two-stage flocculation with chemically conditioned recycled sludge followed by lamella clarification	Infilco Degremont
Trident® HS	5–25	12–65	Adsorption of flocs onto floating media followed by clarification	US filter
CONTRAFAST™	>6	>15	Flocculation enhanced with chemically conditioned recycled sludge, followed by lamella clarification, within one tank	US filter

a solids handling system for disposal. Surface overflow rates ranging from 50 to 150 m h^{-1} (20–60 gpm ft^{-2}) have been reported. Examples of some of the more recent clarification technologies are presented in Table 12.6.

12.3.1.3 Filtration

Filtration removes solid particles in fluid by passing the fluid through a filtering medium on which the solids are deposited or trapped. Concentrations of solids that can be removed may vary from trace concentrations to very high percentages, depending on the filtration medium. Removal of solids by filtration may be enhanced by adding a filter aid such as an organic polymer. Filters are divided into three main categories: cake filters, clarifying filters, and cross flow filters.

Cake filter separates relatively large amounts of solids by forming a cake on the surface of the filtration medium. Cake filter may be operated by applying pressure on the upstream section of the filter medium or a vacuum on the downstream section of the filter medium. The operation of the filter may be in batches or continuous. However, most cake filters are batch operated since the operation of the filter under positive pressure needs to be stopped to facilitate the removal and discharge of solids. Examples of cake filters include filter press, vacuum filter, and centrifugal separator.

In a typical clarifying filter, solids are removed within the filter media through straining, inertial impact of the solids to the media, and adhesion of solids to the media. Removal efficiency is fairly even throughout the filter run before the solids "breakthrough," resulting in a sharp increase in the effluent solid concentrations. The pores of a clarifying filter media are typically larger than that of a cake filter, allowing the solids to penetrate into and be removed by the filtering medium. An example of clarifying filter is the rapid sand filter for water treatment.

In crossflow filtration, the wastewater flows tangentially to the surface of the filter medium at a fairly high velocity. A thin layer of solids forms on the surface of the medium, but the high liquid velocity keeps the layer from building up. At the same time, the water permeates the membrane producing a clear filtrate. Filter media may be ceramic, metal (e.g. sintered stainless steel or porous alumina), or a polymer membrane (cellulose acetate, polyamide, and polyacrylonitrile) with pores small enough to exclude most suspended particles. Examples of crossflow filtration are microfiltration with pore sizes ranging from 0.1 to 5 μm and ultrafiltration with pore sizes from 1 down to about 0.001 μm.

12.3.1.4 Flotation

Dissolved air flotation (DAF) is a proven, robust technology for removal of oil and grease, fibers, low density materials, and SS and is used for the thickening of wastewater sludge or chemical sludge (Officer et al., 2000). Fine gas bubbles are attached to the SS, resulting in an increase in their buoyancy. SS then rise to the surface and are removed using scrapers or overflow weirs. Most air flotation systems operate with a recycle where a portion of the clarified liquid is pumped to a retention tank and is pressurized (5–10 atm) with air. The recycled clarified liquid containing dissolved air is mixed with fresh wastewater at the entrance of the flotation unit. The reduction in pressure at the entrance results in the release of the dissolved air as fine air bubbles. Figure 12.3 shows a typical air flotation unit. Recycle ratio varies between 20% and above. Bubbles sizes are between 30 and 120 μm. Surface overflow rates that can be achieved by DAF are between 10 and 20 m h^{-1}, and solids loading rates are in the range of 5–20 kg m^{-2} h^{-1}. Solids removal can be as high as 98% and the concentrated sludge can be as high as 5%. Some of the advantages of DAF include

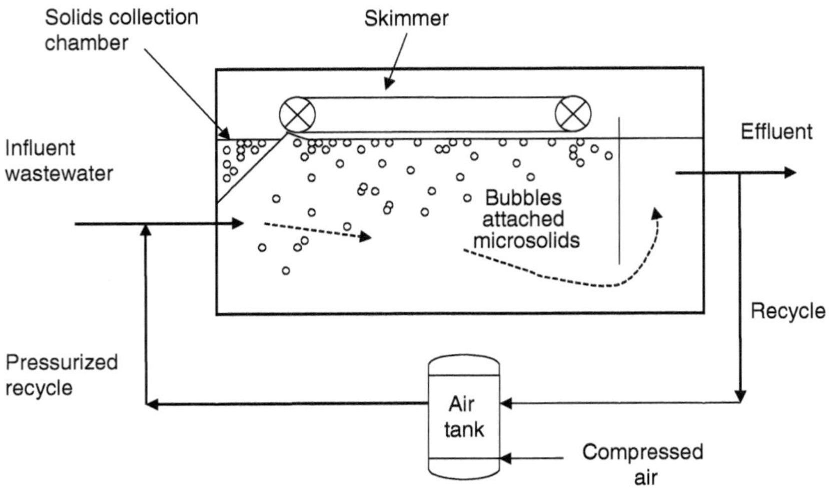

Figure 12.3 Schematic of a dissolved air flotation unit.

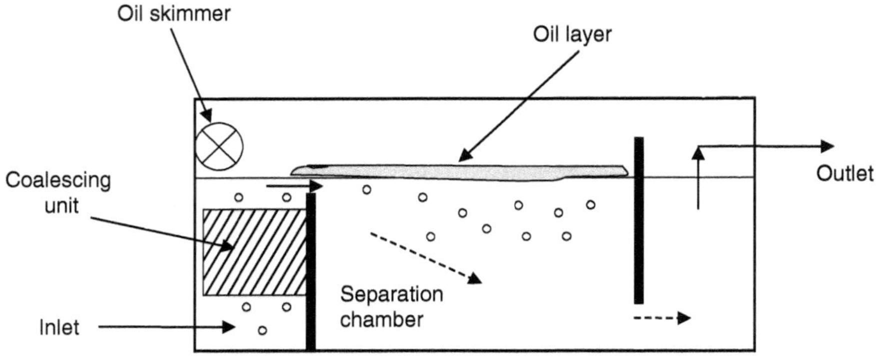

Figure 12.4 Schematic diagram of an oil/water separator.

(i) rapid start-up (in minutes) with flexibility of withstanding periodic stoppages, (ii) high solids capture, particularly of finer solids (80–90%), (iii) reduced chemical usage, and (iv) mechanical float removal, which can produce a relatively thick sludge (2–5% sludge dry solids).

12.3.1.5 Oil and Grease Removal

Oil/water separators employ various separation methods to separate the oil from the aqueous phase. The most common oil/water separators employ the principle of flotation with or without physical coalescing of the oil droplets for the removal of oils and greases from industrial wastewaters (US Army, 1996). The application of a particular separation process depends on the properties of the oil in the oil/water mixture. Figure 12.4 shows a simplified diagram of a typical oil/water separator system. The oil layer at the surface of the water is then skimmed to an oil holding tank where it is recycled or disposed of accordingly. The treated wastewater is passed under a baffle to the outlet chamber and is discharged for further treatment or into the sewer.

Depending on the size of the oil droplets in the wastewater, a simple oil/water flotation separator may not be sufficient to remove oil to the regulatory discharge standards. Under such situations, removal of oil may be enhanced by coalescing the oil droplets to form larger oil droplets, making them more buoyant and causing them to rise faster. For example, the time needed for a 100-μm diameter oil droplet to rise by 15 cm in water is approximately 10 min, while the time needed for a 20-μm diameter oil droplet to travel the same distance is approximately 2 h. In a typical oil/water separator, the minimum water depth of the separator may be between 1.2 and 1.5 m, meaning that oil droplets of certain diameters may pass through the oil/water separator uncollected.

Placing inclined plates within the separation chamber of the oil/water separators is one of the approaches used to make the wastewater with the oil droplets to travel short vertical distances (~0.6 cm). As the oil droplets encounter the fixed surface, they coalesce and rise along the plates to the water surface. Another approach is to use a filter media made of fine oleophilic (oil "loving") fibers such as polypropylene. As wastewater flows through the filter, fine oil droplets are attached to the fibers. The attached droplets get larger with time and become buoyant and detach from the fibers and rise to the water surface. The use of detergents and soaps for the removal of oil and grease from equipment surfaces can adversely affect the operation of an oil/water separator. These detergents and soaps, known as emulsifying agents, are specifically formulated to increase the dispersal of oil into tiny drops in water. The parameters affecting the effectiveness of an oil/water separator are the residence time of the wastewater in the oil/water separator and the surface area of the chamber for the accumulation of oil on the surface. If too much oil is accumulated and not removed in a timely manner, oil may flow out of the oil/water separator with the treated wastewater.

12.3.1.6 Evaporation

Evaporation is used to concentrate a particular wastewater by evaporating the solvent or the aqueous phase. The targeted/valuable product is the concentrated solution of the solute while the vapor is condensed and reused or discarded. If the targeted/valuable product is the vapor or the condensed solvent, then the process is known as distillation. An example of distillation is the production of pure water (solute-free solvent) from seawater. Single- and multiple-effect evaporators are used for evaporation or distillation. Steam is typically used to heat the liquid waste to the required boiling point. Factors to be considered for evaporators include:

1) Concentration – As concentration increases, the density and viscosity of the solution may become saturated or viscous, resulting in crystallization/precipitation/scaling, which may clogged the heat transfer tubes or heat transfer may not be adequate because the properties of the viscous solution are different from the starting material.
2) Foaming – This may result in entrainment and carryover of the valuable product in the evaporation process or contamination of the valuable product as in distillation.

12.3.2 Chemical Treatment

12.3.2.1 Neutralization

Neutralization is a process where acid reagents are added to an alkaline wastewater or alkaline reagents are added to an acidic wastewater to adjust the pH of the wastewater to a more acceptable pH for subsequent treatment or disposal into the sanitary sewers. Typical acceptable pH range before the wastewater can be discharged into the sewer is between 6.0 and 9.0. Typical acid reagents used in neutralization are sulfuric acid and nitric acid and in some cases waste acid streams from the manufacturing process. Alkaline reagents used are sodium hydroxide, potassium hydroxide, or alkaline waste streams from the manufacturing process. The process for neutralization is fairly simple and is accomplished in a mixing tank with a pH sensor. The pH sensor monitors the pH of the treated wastewater and sends a signal to a pump where the needed amount of acid or base is added accordingly to neutralize the wastewater. The amount of reagent needed is usually determined by conducting an acid- or base-titration curve.

12.3.2.2 Precipitation

Precipitation is a common treatment method for metal finishing waste streams. The physical state of dissolved metals is altered by adding precipitating chemicals such that the solubility of a metal compound is exceeded, resulting in an insoluble solid termed as the precipitate. The precipitation reaction can be generalized as follows:

$$A_Z B_Y(s) \Leftrightarrow zA^{Y+} + yB^{Z-}$$

where A is the metal or cation, B is the anion, and z and y are number of molecular units in the compound. The product of the activities of the species involved in the precipitation is represented by the solubility product, K_{sp}, which provides an indication of the extent of the solubility of the compound, i.e.

$$K_{sp} = \{A^{y+}\}^z \{B^{z-}\}^y$$

The process can be reversed with the precipitate dissolving in the aqueous phase when the activities of the precipitate species in the aqueous phase are less than the solubility of the precipitate or when the environmental conditions such as pH and redox are changed. For example, typical K_{sp} values of various lead compounds (for example, $Pb(OH)_2 = 10^{-14.3}$, $PbSO_4 = 10^{-7.8}$, $PbS = 10^{-27.0}$, and $PbCO_3 = 10^{-13.1}$) indicate that the least soluble of the four compounds is PbS, while the most soluble is $PbSO_4$. Although the K_{sp}s provide an indication of the solubility of the compound, the interactions of metal ions in wastewater with other ions or molecules to form complex ions or coordination compounds may have an impact on their solubilities. Examples of complexes formed include hydro-, cyano-, and ammonium complexes when cyanide and ammonium ions are present.

A common approach in precipitating metals from wastewater is to precipitate the metals as metal hydro-complexes. Lime ($Ca(OH)_2$) and sodium hydroxide (NaOH) are typically used. Lime is widely used and is the less expensive of the two. The formation of various metal hydro-complexes is dependent on the solution pH. Figure 12.5 shows the solubilities of several metal hydroxides at different pH in equilibrium with the metal precipitate. Hydro-complexes of metals are amphoteric, resulting in a minimum solubility between pH 9 and 11

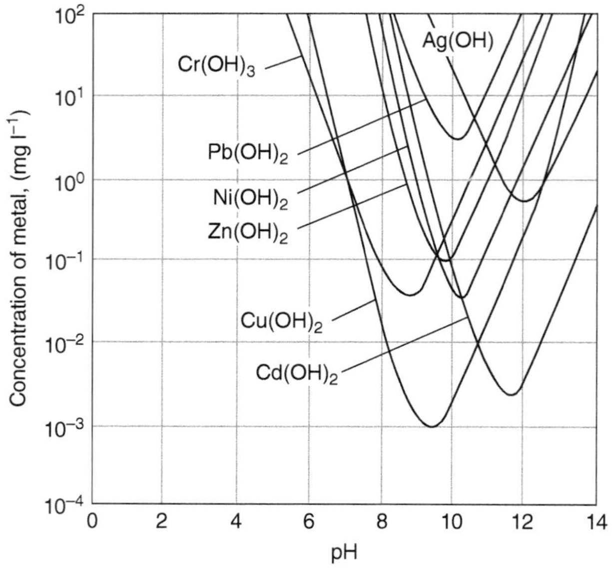

Figure 12.5 Solubilities of metal hydroxides as a function of pH (US EPA, 1983).

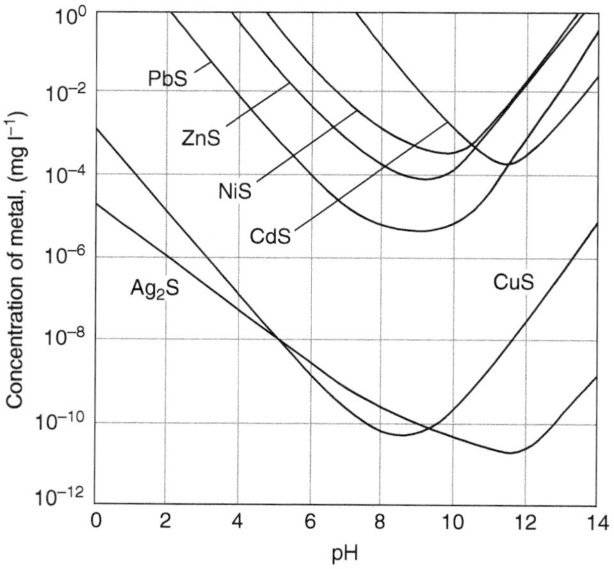

Figure 12.6 Solubilities of metal sulfides as a function of pH (US EPA, 1983).

but are fairly soluble at elevated pH (>11) and at low pH or acidic conditions (<6). The pH range of the minimum solubility depends on the metal and other constituents present in the wastewater.

As indicated earlier (using lead as an example), metal removal from wastewater by precipitation can be maximized by precipitating the metals in the form of metal sulfide. Figure 12.6 shows the solubilities of several metal sulfide compounds for different pH values in equilibrium with the metal precipitate. Chemicals used for sulfide precipitation are sodium sulfide or bisulfide (Na_2S or NaHS). Extremely low concentrations of the metals can be obtained using sulfide precipitation. Since it is possible that hydrogen sulfide (H_2S) gas may be generated during treatment, sulfide precipitation is generally operated under alkaline conditions.

A typical metal precipitation system is shown in Figure 12.7. Some systems have equalization basins to equalize flow into the treatment system along with a pre-reaction tank for pH adjustment. The central part of the treatment plant is the rapid mixing tank with a detention time of between 1 and 5 min where coagulating chemicals are added and a slow mix tank or flocculation tank with a detention time of 20–30 min for the agglomeration of metal precipitates or flocs. Organic polymers may be added to aid in the flocculation. The precipitates are settled out in a settling tank or clarification tank with typical overflow rates ranging between 1 and 5 m h^{-1}. The precipitates are either reprocessed to recover the metals or dewatered and stabilized before disposal.

12.3.2.3 Chemical Oxidation and Reduction

Chemical oxidation is a well-established technology for the destruction of a wide range of compounds to harmless or less toxic pollutants by an oxidizing agent. Compounds that can be degraded include chlorinated hydrocarbons, aromatic compounds, and inorganics such as cyanide and ammonia. Chemical reduction, however, has limited applications but is effective in the destruction of certain halogenated compounds such as tetrachloroethene, trichloroethene, and carbon tetrachloride and is used to alter the ionic charge of metal ions to facilitate precipitation. An example is the reduction of dichromate anions (Cr(VI)) to chromium ion (Cr(III), which can be precipitated as a hydroxide.

Common oxidizing agents used in the oxidation of hazardous waste include ozone (O_3), hydrogen peroxide (H_2O_2), chlorine, permanganate, and persulfate. In recent years, the application of chemical oxidation has advanced by using a combination of oxidants to enhance oxidation and the use of various techniques such as UV light, ultrasound (US), and inorganic catalysts to increase the oxidizing power of the oxidants by generating oxidizing radicals such as hydroxyl radicals (OH•) (Hoigne and Bader, 1976). The oxidizing power of hydroxyl radicals is several orders of magnitude more powerful than ozone and hydrogen peroxide, but the hydroxyl radicals have extremely short half-lives. Systems that are engineered to produce hydroxyl radicals are termed as advanced oxidation processes (AOPs). The more common AOP applications are Fenton's reagent (use of hydrogen peroxide and Fe(II) catalyst), ozone/UV, hydrogen peroxide/ozone/UV, and ozone or hydrogen peroxide in the presence of a metal catalyst such as titanium dioxide (TiO_2).

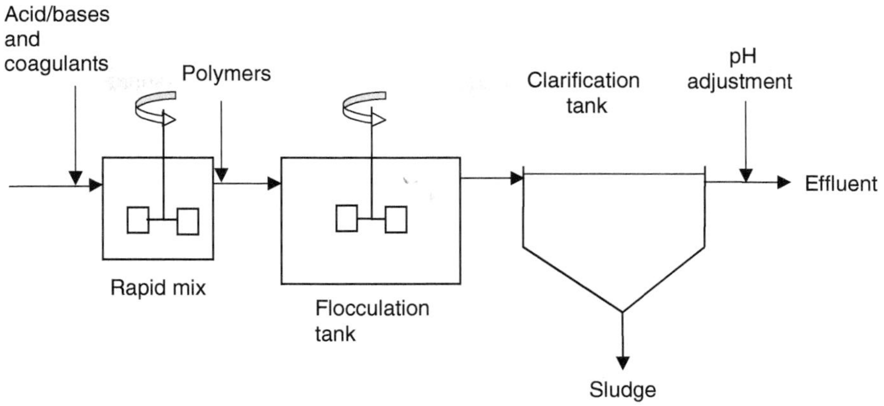

Figuer 12.7 Schematic diagram for chemical precipitation system.

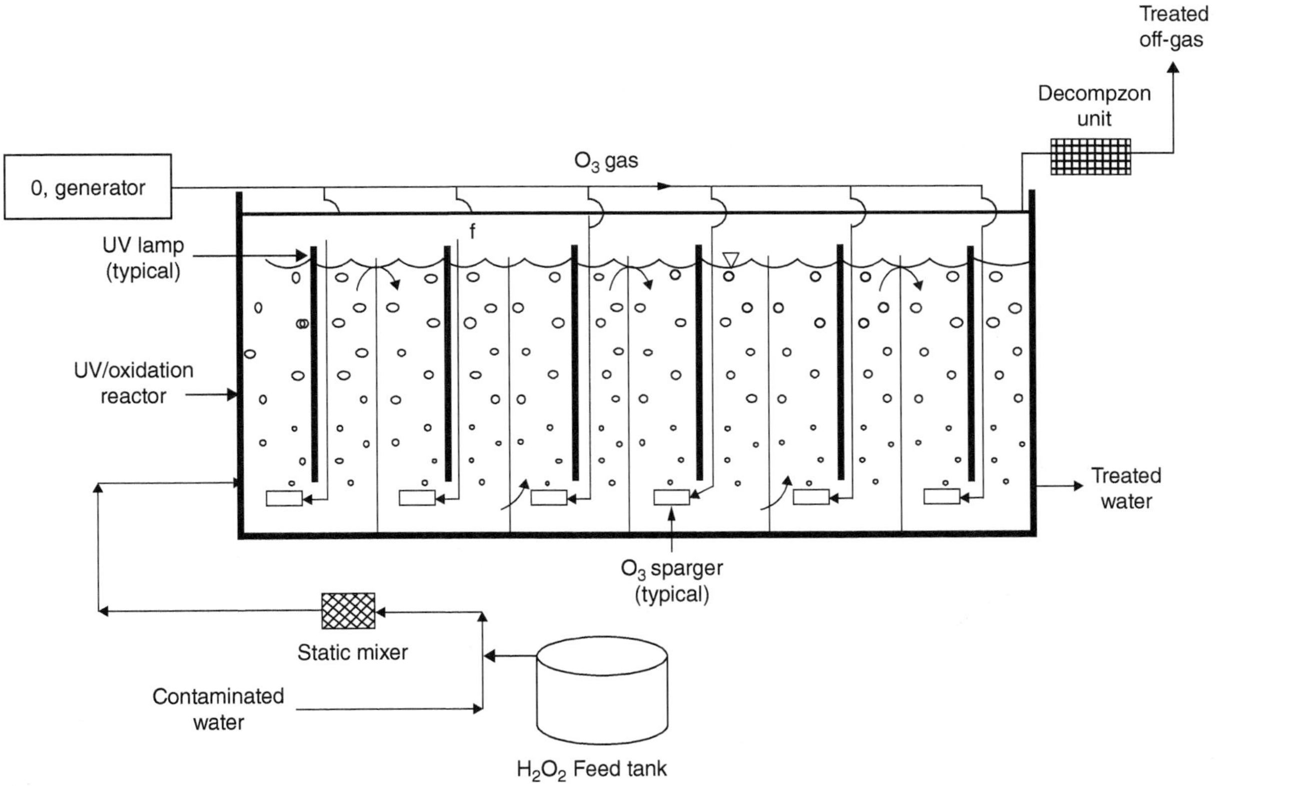

Figure 12.8 Flow diagram of US Filter ozone/H_2O_2/UV system (US EPA, 1998).

AOPs using a combination of O_3 and/or H_2O_2 with UV irradiation have been successfully applied for the oxidation of organic compounds. Commercially available systems include Calgon perox-pure UV/H_2O_2, US Filter UV/O_2/H_2O_2, Calgon Rayox® UV/H_2O_2, Magnum CAV-OX® UV/H_2O_2, Wedeco UV/H_2O_2, Wedeco UV/O_3, and Matrix UV/TiO_2. The flow diagram for the US Filter UV/O_2/H_2O_2 is presented in Figure 12.8. The hydraulic retention time of the system is about 5 min. More than 99% destruction of chlorinated compounds have been reported as summarized in Table 12.7 for a pilot study treating water contaminated with chlorinated compounds. Reactions for a UV/O_2/H_2O_2 system are assumed to proceed by H_2O_2 initiating the decomposition of O_3 to form hydroperoxide ions (HO_2^-), which in turn react with ozone to produce ozonide ions (O_3^-) and hydroxyl radicals (HO•).

A classical example of the application of chlorine in industrial waste treatment is the alkaline destruction of cyanide wastes. The wastewater is maintained at a pH greater than 10, and the cyanide is destroyed in a two-step process using separate tanks. The reactions where cyanide is oxidized to cyanate (CNO^-) is written as follows:

$$NaCN + Cl_2 + 2NaOH \Rightarrow NaCNO + 2NaCl + H_2O$$

Table 12.7 Pilot study results for Calgon UV/H_2O_2 and US Filter O_3/H_2O_2/UV systems (US EPA, 1998).

	Compounds	Initial concentration removal ($\mu g\,l^{-1}$)	Percent (%)
Calgon UV/H_2O_2	Benzene	52	>96
	Chlorobenzene	3100	>99.9
	Chloroform	41–240	93.6 to >97
	1,1-Dichloroethane	120–400	>95.8 to >99.5
	1,2-DCA	22	>92
	Tetrachloroethylene	63–2500	>98.7 to >99.9
US Filter O3/H_2O_2/UV	1, 1-Dichloroethane	9.5–13	65
	1,1,1-Trichloroethane	2–4.5	87
	Trichloroethylene	50–520	>99

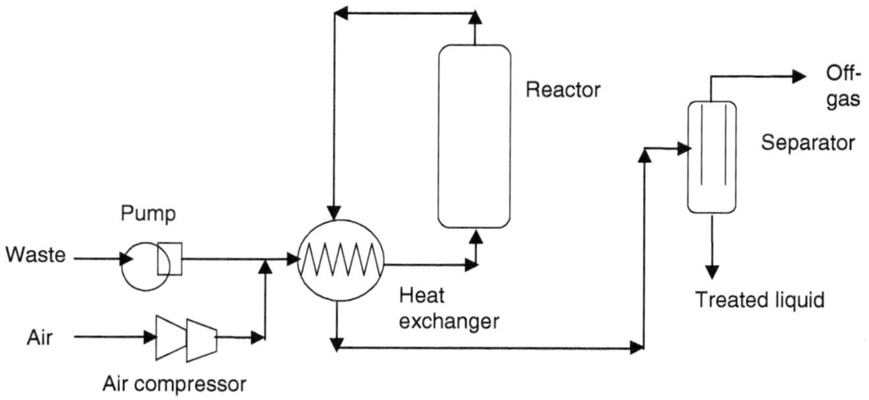

Figure 12.9 Schematic diagram of a wet air oxidation system.

Cyanate is subsequently oxidized to carbon dioxide and nitrogen as follows:

$$2NaCNO + 3Cl_2 + 4NaOH \Rightarrow 2CO_2 + N_2 + 6NaCl + 2H_2O$$

However, in a typical wastewater, cyanides are complexed with copper, nickel, and precious metals, which slow the destruction of cyanide as compared with free cyanide. Excess chlorine is needed to oxidize these cyanide complexes. Reaction times for the complete destruction of cyanide ranged from 60 to 120 min.

12.3.2.4 Wet Air Oxidation and Supercritical Water Oxidation

For wastewater containing high organic carbon content in the range of 10 000–200 000 $mg\,l^{-1}$ and with refractory content, the use of chemical oxidation or biological treatment may not be cost effective. Under such circumstances, the organic wastes may be oxidized in the liquid phase using wet air oxidation. For wet air oxidation systems, the wastewater is oxidized in the presence of air at elevated pressures and temperatures but below the critical point of water (374 °C and 218 atm) (Copa and Gitchel, 1988). Temperatures and pressures of wet air oxidation systems are in the range of 150–325 °C and 100–200 atm, respectively. When the system is operated at temperatures and pressures above the critical point, the system is called supercritical water oxidation (Modell, 1989). Beyond the critical point, the liquid and gas phases exist as a single phase fluid where the solubility of organics is enhanced while the solubility of inorganics in the fluid is decreased by three to four orders of magnitude. The gas-like properties of the fluid enhance the contact between the target organics and the oxidizing radicals, maximizing the degradation of the target organics. A typical wet air oxidation system is shown in Figure 12.9.

In wet air oxidation, COD removal between 75 and 90% can be achieved. The end products consist of simpler forms of biodegradable compounds such as acetic acid and inorganic salts along with the formation of carbon dioxide and water. Depending on the pollutants in the wastewater, further treatment of the waste stream may be needed. Residence time of reactor is typically between 15 and 120 min. The system is adaptable to a wide variety of oxidizable materials, and water acts as a heat sink assisting in the control of the temperature within the reactor. Special alloy materials are needed for the reactor due to the high corrosivity of the reaction by-products. A disadvantage of wet air oxidation is the high maintenance costs of the system.

Residence time for supercritical water oxidation systems may be as short as several minutes at temperatures of 600–650 °C. More than 99.9% conversion of EPA priority pollutants such as chlorinated solvents has been achieved in a pilot-scale plant with retention time less than 5 min (Gloyna and Li, 1995). The system is limited to the treatment of liquid wastes or solids less than 200 μm in diameter. Formation of char during the reaction may impact the oxidation time of the organics, while separation of inorganic salts followed by precipitation during the process may be a problem. Typical materials for the reactor are Hastelloy C-276 and Inconel 625 (high nickel alloys), which can withstand high temperatures and pressures and the corrosive conditions.

12.3.2.5 Electrochemical Process

In an electrochemical process, pollutants in the liquid wastes are chemically oxidized and reduced by applying electricity across appropriate electrodes to create the oxidation and reduction potential. However, reactions in an electrochemical cell may be enhanced by adding oxidizing chemicals. Electrodes used are of special materials, allowing for selectivity in pollutants removed and may, at the same time, prevent the production of unwanted by-products (Juttner et al., 2000). In some processes, a separation membrane may be used in the electrochemical cell to improve the removal of specific pollutants. Chemical reactions in an electrochemical cell can be controlled by controlling the electrode potential and the environment at the surface of the electrodes. Electrochemical processes can be viewed as reactions due to direct electrolysis (reactions at the cathode or the anode) or reactions due to indirect electrolysis. Figure 12.10 illustrates the two processes.

Examples of direct electrolysis reactions include removal of specific metals by cathodic deposition where a metal ion is reduced by accepting electrons at the cathode. To minimize side reactions and to provide a certain level of selectivity, ion selective membranes of thin polymeric materials may be used within the electrochemical cell. Indirect electrolysis electrochemically generates redox reagents as a chemical reactant to convert pollutants to less harmful products. The redox reagent acts as an intermediary for shuttling electrons between the pollutant substrate and the electrode. An example is the generation of chlorine (Cl_2) from chloride (Cl^-) in the waste solution at the anode, which in turn is used as an oxidizing agent to oxidize the pollutants.

A commercial application of electrochemical cells is the CerOx process (a mediated electrochemical oxidation (MEO) or catalyzed electrochemical oxidation (CEO)), which has been shown to successfully destroy PCBs at a concentration of 2 mg l^{-1} in alcohol in a patented electrochemical cell (called the T-cell). The process uses cerium metal ion, Ce^{4+}, placed in contact with an organic compound that reduces the cerium ion to Ce^{3+}. The process operates at low temperature (90–95 °C) and near atmospheric pressure. The $CerO_x$ process may also treat wastes containing refractory pesticide compounds such as DDT, silvex, and chlordane and pharmaceuticals wastewaters (Anonymous, 2000). Advantages of electrochemical processes include versatility in treatment – treating small to large volumes, oxidation and reduction of pollutants directly or indirectly, low energy use in comparison with thermal processes, and control of the reactions with a certain level of selectivity. Some of the disadvantages include stability of the electrodes, electrodes fouling, mass transfer limitations due to the size of electrode area, and reactions that are dependent on the conducting medium, electrolyte.

12.3.2.6 Adsorption and Ion Exchange

Waste pollutants can be removed from the waste stream by preferential accumulation of the pollutants at the surface of a solid phase or adsorbent. Adsorption is one of the more widely applied technologies for the treatment

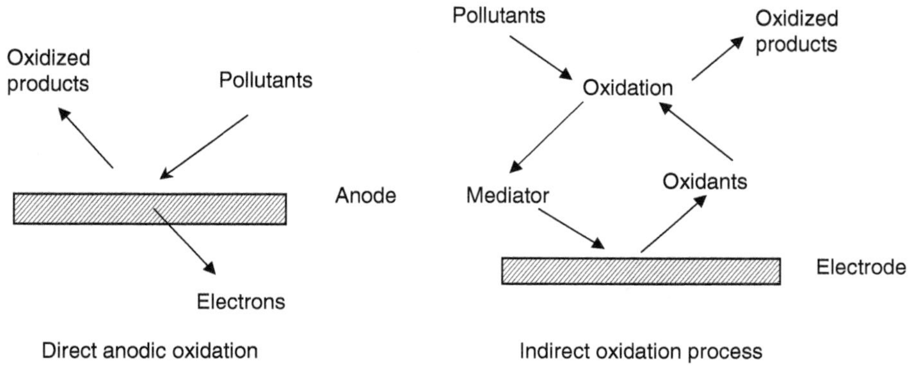

Figure 12.10 Direct and indirect reactions of electrochemical process.

of industrial wastewater. Ion exchange is a form of adsorption whereby an ion in the solid phase is replaced by another ion in a solution in contact with the solid. Since replacement takes place at the interface, this process may be classified as adsorption. Adsorption is used to remove a wide range of pollutants from synthetic organic chemicals such as pesticides and petroleum hydrocarbons to inorganic compounds such as heavy metals and anions such as perchlorate. Common adsorbents used for wastewater are activated carbon and synthetic ion exchange resins. Other adsorbents include activated alumina, forage sponge, and clays.

Adsorption isotherms are typically used to describe the equilibrium relationship between the bulk aqueous phase activity (concentration) of the adsorbate and the amount adsorbed on the interface at a given temperature. The more common models used are empirical models such as the linear model and the Freundlich model. First principle models such as the Langmuir model are also used.

Activated carbon comes in two forms: granular activated carbon (GAC) (average diameters ranging from 0.42 to 2.38 mm) and powdered activated carbon (PAC) (average diameter about 44 μm). GAC and PAC are made from wood, peat, lignite, bituminous coal, or coconut shells. The surface of GAC consists of different functional groups such as OH– and COO– that can adsorb a range of different compounds (Crittenden et al., 2005). Most GAC systems are used as a tertiary process and are preceded with solids removal and filtration to minimize fouling of the GAC.

Synthetic ion exchange materials are made of cross-linked polymer matrix with charged functional groups attached by covalent bonding. The base material is polystyrene and is cross-linked for structural stability with 3–8% divinylbenzene. Functional groups on the base material include sulfonate (SO_3^-) groups for strong acid ion exchange resins and quaternary amine ($N(CH_3)_3^+$) groups for strong basic ion exchange resins. Cation resins are regenerated with strong acids such as sulfuric acid, nitric acid, and hydrochloric acid, while anion resins are regenerated by a strong base such as sodium hydroxide.

The reactor configuration, typically used for carbon adsorption and ion exchange systems, is the fixed bed system. Important design parameters for GAC and ion exchange fixed bed systems include the type of GAC or ion exchange resin used, surface loading rate, GAC or ion exchange resin usage rate in terms of mass of GAC or ion exchange resin used per volume of water treated, GAC or ion exchange resin depth or volume based on required breakthrough of a pollutant, and empty bed contact time. Methods used in sizing of GAC fixed bed systems include the use of pilot- and laboratory-scale column tests such as the rapid small-scale column tests (RSSCTs) (Crittenden et al., 2005), the bed depth service time method (BDST) (Droste, 1997), and the kinetic approach (Reynolds and Richards, 1996). Typical GAC surface loading rates are between 2 and 30 m h^{-1} with empty bed contact times between 10 and 30 min. The average GAC depth is approximately 1 m. In the case of ion exchange systems, the volume of resin needed to treat a wastewater is based on pilot studies or estimated based on the data provided by resin suppliers (Dow, 2002). Typical surface loading rates range from 5 to 60 m h^{-1} with resin bed depths for cocurrent regeneration and countercurrent packed bed systems of approximately 1.2 and 2 m, respectively.

12.3.3 Biological Waste Treatment

Biological processes harness the metabolic abilities of microorganisms to degrade organic materials (both dissolved and suspended organics) into stable and simple end products with the production of energy for growth and maintenance and the synthesis of new microbial cells. Several factors impact the microbial processes. These factors may be broadly divided into substrate related, microorganism related, and environment related. Substrate-related factors include the impact of the concentration and the physical–chemical properties of the substrate (e.g. structure of organic compounds, solubility, sorption). Microorganism-related variables include selection and acclimatization of the microorganism to the compound, presence of certain species of microorganism, and required enzyme system. Environment-related conditions include the presence of electron acceptors, pH of the wastewater, presence of nutrients, temperature, and total dissolved solids. Two of the important environmentally related conditions in wastewater are the availability of electron acceptors and nutrients. To facilitate the microbial oxidation–reduction reaction, oxygen is used as an electron acceptor for aerobic metabolism, while nitrate, Mn(IV) and Fe(III), sulfate, and carbon dioxide are used for anaerobic metabolism. The electron acceptor that derives the maximum free energy by the microorganisms will be used first. In a closed environment with a fixed amount of oxygen, oxygen will be used first by the dominant aerobic heterotrophs. When the oxygen is used up, denitrifiers will use nitrate as electron acceptors followed by sulfate reducers, fermenters, and finally methanogens. The two nutrients that are most likely to be limiting are nitrogen and phosphorus. Nutrients must be provided at a minimum level in order to sustain microbial growth. The approximate formula for a bacteria cell is $C_5H_7O_2NP_{0.074}$, indicating that the ratio of C:N for cell synthesis and as an energy source is approximately 10 : 1. In typical wastewater systems, a ratio of C:N:P of 100 : 10 : 1 is used. Micronutrients such as sulfur, potassium, calcium, magnesium, iron, cobalt, and molybdenum are also needed.

12.3.3.1 Types of Biological Treatment System

Biological processes can be divided into aerobic (oxygen as the electron acceptors) or anaerobic (nitrate, sulfate, Fe(III) and carbon dioxide as the electron acceptors). For each of the biological process using different electron acceptors, treatment technologies may be classified as suspended growth, fixed film (biofilm), or hybrid (combination of biofilm and suspended growth). Examples of aerobic suspended growth technologies include the activated sludge process and the sequencing batch reactor, while anaerobic systems include the conventional high rate anaerobic digesters. In recent years, membrane bioreactors (MBRs) have emerged as a very strong contender to replace the activated sludge system for suspended growth technologies. Examples of aerobic fixed film technologies include trickling filters and rotating biological reactors while anaerobic fixed film technologies include anaerobic biofilters. Some of the newer fixed film technologies include the biological aerated filters (BAFs) or the combined oxic–anoxic biological filters. Biological treatment is widely used for wastewater from municipal wastewaters, pulp and paper, food processing, oil and gas, and petrochemical industries.

12.3.3.2 Activated Sludge Process

SS consisting of microbial cells and inert materials are kept in suspension in the aeration basin of the activated sludge process by continuous aeration of the basin with air or oxygen. The mixed liquor (SS and treated effluent) is transferred to a sedimentation tank where the SS are separated from the treated effluent by gravity settling (see Figure 12.11). Part of the SS at the bottom of the sedimentation (called sludge) is returned to the aeration tank to maintain an SS concentration in the aeration tank. The rest of the sludge is wasted and treated accordingly. The efficiency of the activated sludge process and the quality of the final effluent are highly dependent on maintaining the active biomass in the aeration basin, the settling characteristics of biomass produced, and the performance of the settling tanks (Metcalf and Eddy, 2003). Operational characteristics such as maintaining the appropriate solids residence time (SRT) play an important role in avoiding poor biomass settleability and sludge bulking (Benefield and Randall, 1980).

The hydraulic retention times (HRTs) of an activated sludge plant typically range from 4 to as high as 24 h. With recirculation of sludge and wasting in the sedimentation tank, the SS in the aeration basin are maintained at a concentration of 2500–4000 mg l^{-1}, while the SRTs are kept between 10 and 30 days. Removal of micropollutants such as estrogenic compounds in activated sludge plants is very much a function of the SRT of the treatment plant where longer SRTs provide sufficient time for the micropollutants to be degraded (Khanal et al., 2006; Limpiyakorn et al., 2011).

12.3.3.3 Sequencing Batch Reactors

Sequencing batch bioreactors (SBRs), as indicated by its name, operate in a batch mode with two or more bioreactors in a typical system (Demoulin et al., 2001). Treatment in an SBR is accomplished by operating the reactor in a sequence of events within a cycle (see Figure 12.12). For a treatment plant with two SBRs, wastewater is added, in the "fill" sequence, to the first SBR that contains the biomass for the biochemical reactions from the previous cycle. When the first SBR is filled, the influent wastewater is diverted to the second SBR. The wastewater in the first SBR is aerated and treatment of the wastewater occurs. This is called the "react" sequence. After the "react" sequence, the air is turned off and the biomass is allowed to settle in the "settle" sequence. The settled supernatant is then decanted in the "decant" sequence and discharged, leaving behind the biomass for the next cycle, and the sequences are repeated. Typically about 1/3 of the reactor is decanted. The time period for a cycle varies from 4 to as much as

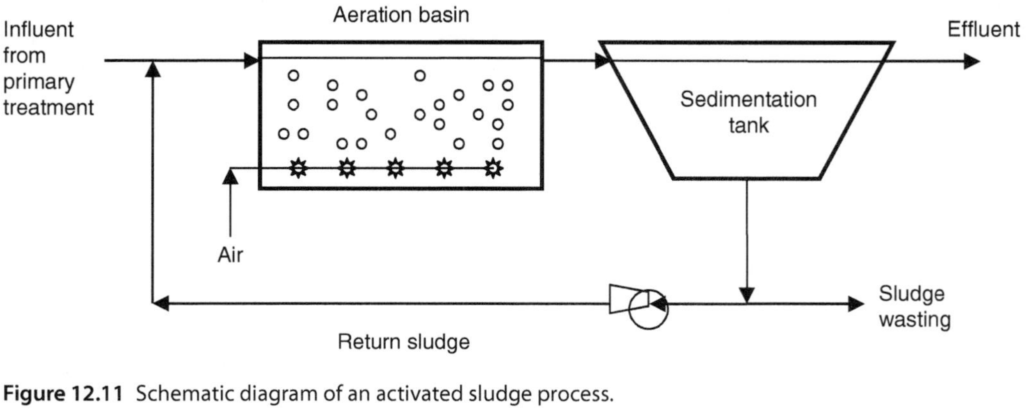

Figure 12.11 Schematic diagram of an activated sludge process.

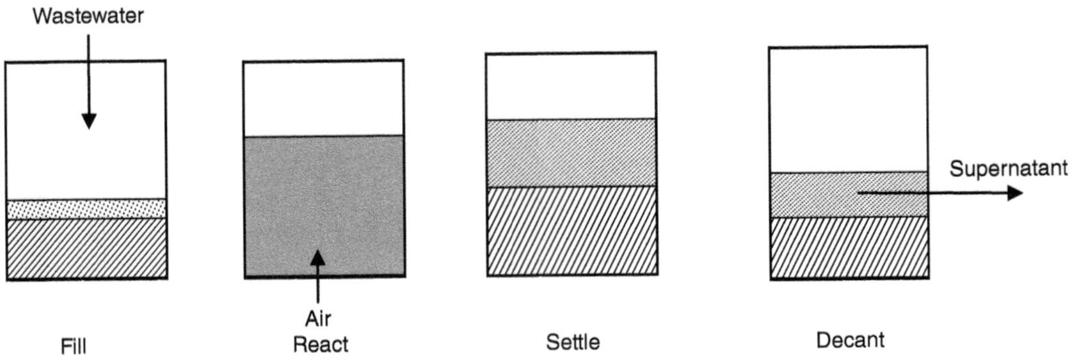

Figure 12.12 Different sequences of a sequencing batch bioreactor.

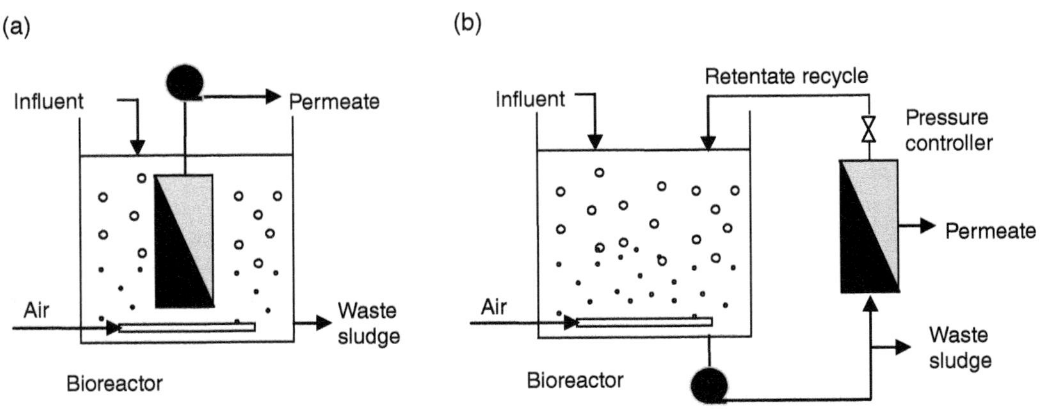

Figure 12.13 Schematic diagram of membrane bioreactors: (a) submerged and (b) sidestream.

24 h, depending on the wastewater. The biomass in the SBR is typically maintained between 2000 and 4000 mg l^{-1}.

The SBR is highly flexible in that the sequences can be manipulated for different redox conditions (anaerobic, anoxic, and oxic conditions), allowing for the treatment of organic compounds that are more amenable to anaerobic degradation followed by oxic degradation or for the removal of nitrogen and phosphorus from the wastewater (Ersu et al., 2008a). A novel variation of the SBR is the hybrid system called the sequencing batch biofilm bioreactor (SBBR). The SBBR has a support media that allows a biofilm to grow along with maintaining suspended growth within the reactor (Protzman et al., 1999). This allows the SBBR to retain a higher concentration of biomass within the bioreactor, resulting in higher rate of biochemical reactions.

12.3.3.4 Membrane Bioreactors

Poor settling of sludge and the requirement of large surface area of sedimentation tanks are some of the disadvantages with activated sludge systems especially in places where space availability is limited. To overcome some of these disadvantages, membranes such as microfiltration and ultrafiltration instead of gravity settling as in sedimentation tanks can be used to separate the SS from the treated effluent (Stephenson et al., 2000). These reactors are called membrane bioreactors (MBR). Membranes can be installed either inside the reactor (submerged) or outside the reactor (sidestream) with the submerged configuration being the more common and cost-effective system used (Figure 12.13, Knoblock et al., 1998). Tubular membrane modules are usually installed for sidestream configuration, while plate-frame membranes or hollow fiber membrane modules are generally used in submerged MBR systems. Sidestream tubular modules are typically used for the treatment of harsh industrial wastewaters such as low pH or high pH wastewaters where ceramic membranes are needed (Ersu and Ong, 2008).

Some of the advantages of MBRs over conventional activated sludge systems include excellent effluent quality, smaller plant size, lower sludge production, high operational flexibility, high decomposition rate of organics, process reliability, and excellent microbial separation and odor control (Zhang et al., 1997; Crawford et al., 2000). Disadvantages include fouling of the membranes, higher capital cost and energy consumption, and higher aeration requirements than the activated sludge process.

MBR allows the control of SRT independently from HRT, resulting in much longer SRTs (typically 15–50 days) as compared with activated sludge (10–30 days) (Ersu et al., 2008a, b, 2010). MBRs operate at mixed liquor suspended solids (MLSS) concentrations of between 4000 and 10 000 mg l^{-1} and at HRTs as low as 4 h, resulting in a smaller bioreactor volume, which can be one-fourth of the size of an activated sludge plant. Sludge produced in MBRs may be as low as 0.22 kg MLSS kg^{-1} BOD_5 at 50 days SRT (Takeuchi et al., 1990) as compared with 0.7–1 kg MLSS kg^{-1} BOD_5 at 10–20 days SRT for activated sludge (Hsu and Wilson, 1992). In addition, by adding an anaerobic tank and an anoxic tank prior to the MBR, both phosphorus and nitrogen can be effectively removed from municipal wastewater (Brown et al., 2011). A recent study by Chung et al. (2014) showed that a four-stage Bardenpho® MBR system can effectively remove 97 and 93% of COD and total nitrogen in the semiconductor manufacturing wastewater.

12.3.3.5 Biological Aerated Filters

BAFs are biofilm bioreactors with a support medium for the growth of a biofilm. BAFs are fully submerged with air injected into the bottom of the bioreactor (Mendoza-Espinosa and Stephenson, 1999; Ha et al., 2010a). BAFs are well suited as a secondary treatment process and for upgrading existing treatment process as an add-on treatment process (M'Coy, 1997). BAFs are operated either in an upflow or downflow mode (Figure 12.14a and b). Downflow systems with countercurrent air flow have the advantage of efficient mass transfer of oxygen to the biofilm, while upflow systems with cocurrent air and wastewater flow can handle higher influent flow rates than downflow systems (Figure 12.14b). In upflow systems, odor problems are reduced since the wastewater is fed from the bottom. Upflow systems can be modified to include different electron acceptor zones – with an anoxic (nitrate as the electron acceptor) or anaerobic zone created at the bottom of the BAF with an oxic (oxygen as the electron acceptor) zone in the upper half of the BAF by injecting air at a certain depth within the filter (see Figure 12.14c) (Ha et al., 2010b). For example, the Biostyr® technology marketed by Veolia Water has a combination of oxic–anoxic zones in a single reactor (Borregaard, 1997). BAFs can be alternatively operated in an anaerobic and aerobic mode to treat municipal wastewater and, at the same time, accumulate and recover phosphorus from the wastewater (Tian et al., 2015).

The media used for BAFs can be sunken type for downflow and upflow configurations or floating type for upflow systems (Mendoza-Espinosa and Stephenson, 1999). The materials used include proprietary materials such as 3- to 5-mm fired clay material, 2- to 4-mm polystyrene beads, a 60 : 40 mixture of polypropylene and calcium carbonate, and common materials such as 5-mm diameter sand (Metcalf and Eddy, 2003; Ha et al., 2005; Ha and Ong, 2007).

Operation of BAFs with an organic loading of 2.5 kg BOD_5 m^{-3}-day has been reported as compared with 0.06 kg BOD_5 m^{-3}-day for activated sludge plants (Smith et al., 1990). HRTs of BAF are typically in the range of 1–4 h although Pujol et al. (1998) reported that HRT as low as 10 min did not seem to influence the reactor performance treating municipal wastewater. Depths of media may range from 1.6 to as much as 2.5 m.

12.3.3.6 Anaerobic Processes

Anaerobic metabolism uses electron acceptors such as nitrate, Mn(IV) and Fe(III), sulfate, or carbon dioxide other than oxygen to facilitate microbial oxidation–reduction

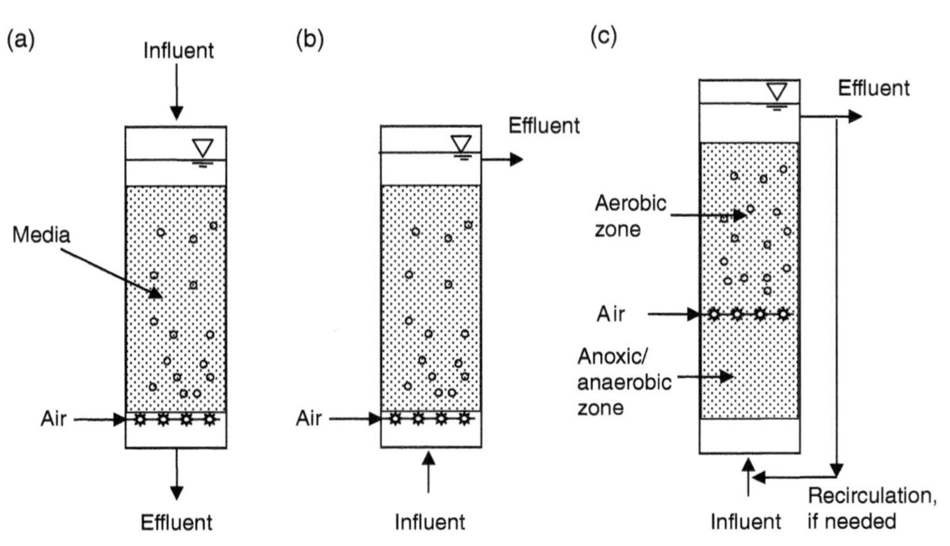

Figure 12.14 (a) Downflow biological aerated filters (BAFs), (b) upflow BAFs, and (c) combined anoxic/oxic upflow BAFs.

reactions. Anaerobic digestion can be viewed as consisting of four reaction steps occurring simultaneously: hydrolysis, acidogenesis, acetogenesis, and methanogenesis. In the hydrolysis step, large complex organic molecules such as proteins and lipids are hydrolyzed and converted to smaller molecules such as amino acids and fatty acids by exoenzymes excreted by anaerobic microorganisms. The acidogenesis step results in the conversion of the hydrolyzed products into low molecular weight volatile fatty acids such as acetic acid, propionic acid, butyric acid, alcohols, aldehydes, and gases such as hydrogen and carbon dioxide. In the acetogenesis step, the products of the acidogenesis phase are converted into acetic acid, hydrogen, and carbon dioxide by acetogenic bacteria. In the final methanogenesis step, acetic acid is converted to methane and carbon dioxide, while hydrogen and carbon dioxide are converted to methane by methanogens.

There are many different anaerobic digestion systems. Some of the more common systems include:

- Standard rate (mesophilic phase) – Typically a single reactor system consisting of three zones within the reactor: clarification zone, active digestion zone, and sludge collection zone. The contents are typically not mixed, and untreated waste is introduced into the zone where the sludge is actively digested. Gas produced rises to the surface and is collected while digested sludge is removed from the bottom.
- Single-stage high rate (mesophilic) – Typically a single-stage reactor where digested contents are mixed by gas recirculation, mechanical mixers, pumping, or draft tube mixers. In addition, the contents are heated to achieve optimal digestion rates.
- Two-stage high rate (mesophilic) – The reactor is similar to the single-stage high rate reactor except that a second stage is used as settling tank to separate the digested sludge from the supernatant.
- Upflow anaerobic sludge blanket (UASB) – A single reactor where wastewater is introduced into the bottom of the reactor to fluidize a suspended sludge blanket that filters the solids and treats the wastewater. The upflow pattern of the reactor allows mixing of the wastewater without mechanical devices and carries the biogas formed to the top of the reactor.
- Anaerobic biofilter – Consists of a single reactor or a two-reactor system where the anaerobic microorganisms are attached to a supporting medium. Wastewater is typically introduced into the bottom of the biofilter.
- Two-phase anaerobic digestion – Consists of two reactors in series with the first reactor operated with a short retention time in the acid phase and with the second reactor operated in the methanogenic phase.
- Temperature-phased anaerobic digestion (TPAD) – Consists of two completely mixed reactors in series with the first reactor operated under thermophilic conditions (~55 °C) and the second reactor operated under mesophilic conditions (~35 °C).

The typical composition of biogas in municipal anaerobic systems is 60–70% methane, 30–40% carbon dioxide, and with small percentages by volume of hydrogen sulfide, ammonia, and water vapor. The energy content of the biogas ranged from 22 to 26 MJ m^{-3}. The maximum theoretical yield of methane, based on a stoichiometric utilization of oxygen for the complete conversion of methane to carbon dioxide and water, is approximately 0.35 m^3 of methane for 1 kg COD. In a full-scale anaerobic reactor, the yield is lower and is usually in 0.20 m^3 of methane for 1 kg COD.

Two of the more important operating parameters for anaerobic systems are SRT and organic loading rate (OLR). SRTs for standard rate digesters (no mixing and heating) treating municipal sludge range from 30 to 60 days, while high rate digesters (with heating and mixing) have shorter SRTs ranging from 10 to 20 days. For anaerobic systems treating municipal sludge, the recommended OLR for standard rate digesters is approximately 0.65 kg of volatile solids m^{-3} of active digestion volume per day (40 lb of volatile solids/1000 ft^3 day^{-1}), and for high rate digesters, the OLR is about 1.3 kg of volatile solids m^{-3} of active digestion volume per day (80 lb of volatile solids/1000 ft^3 day^{-1}) (Great Lakes – Mississippi River Board of State and Provincial Public Health and Environmental Managers, 2014). OLR for USABs are dependent on the wastewater type, the concentration of the organics (COD) in the wastewater and the temperature of the treatment. OLR can range from 1 to 15 kg COD m^{-3} day^{-1}, while the HRT can be as low as 6 h with SRTs ranging from 30 to 50 days (Lettinga et al., 1983; Metcalf and Eddy, 2003).

Optimal operations of anaerobic reactors are affected by various operational and environmental factors. These factors include:

- Oxygen – Methanogenic bacteria are strict anaerobes. Environmental conditions should be completely free of oxygen for their growth. Redox potential needed is less than 330 mV.
- pH – Optimal pH range for methanogenic bacteria is between 6.5 and 7.5 with a typical pH of 7.
- Hydrogen – Conversion of fatty acids and alcohols by acetogenic bacteria can be limited by the hydrogen present. Low hydrogen concentrations in the anaerobic process are maintained by converting carbon dioxide and hydrogen to methane by methanogenic bacteria.
- Inhibitors – Methanogenic bacteria are sensitive to various inhibitors such as common ions (ammonium, sodium, magnesium, metal ions) and organic compounds.

12.4 Summary

The Clean Water Act and the National Pretreatment Program (40 CFR 403) regulate the discharge of wastewaters into bodies of water and into municipal wastewater treatment plants via the sewers. Compliance of these regulations is enforced by US EPA and by state authorities. Industries can draw on the different physical, chemical, and biological treatment technologies to treat industrial wastewaters to stay in line with the regulations and to protect the environment from pollution. Selection of treatment technologies is dependent on the quantity and quality of the wastewater, the final effluent concentrations requirement, capital and operating costs, and on-site constraints such as land availability, labor availability, and labor skill levels. A thorough understanding of the physical principles and the chemical or biological reactions within each treatment process is essential in the selection of the right treatment process for municipal and industrial wastewater. By combining a series of unit treatment processes, most wastewaters can be treated to levels where it can be safely discharged. In areas where water is scarce, many municipal wastewater treatment plants and industries are reusing treated wastewater for utility use or for noncontact water such as flushing of toilets and lawn irrigation. As more and more municipalities and industries explore different ways of implementing "sustainability" in their operations, conservation, recovery of useful products, and reuse of water have become important factors in their operational and business philosophy.

References

Anonymous (2000). Electrochemical process oxidizes PCBs. *Chemical Engineering Progress* 96 (12): 17.

Benefield, L.D. and Randall, C.W. (1980). *Biological Process Design for Wastewater Treatment.* Englewood Cliffs, NJ: Prentice-Hall.

Borregaard, V.R. (1997). Experience with nutrient removal in a fixed-film system at full-scale wastewater treatment plants. *Water Science and Technology* 36 (1): 129–137.

Brown, P., Ong, S.K., and Lee, Y.W. (2011). Influence of anoxic and anaerobic hydraulic retention time on biological nitrogen and phosphorus removal in a membrane bioreactor. *Desalination* 270 (1–3): 227–232.

Chung, J., Fleege, D., Ong, S.K., and Lee, Y.W. (2014). Organic semiconductor wastewater treatment using a four stage Bardenpho® with membrane system. *Environmental Technology* 35 (22): 2837–2845.

Copa, W.M. and Gitchel, W.B. (1988). *Wet Air Oxidation in Standard Handbook of Hazardous Waste Treatment and Disposal* (ed. H.M. Freeman). New York: McGraw-Hill.

Crawford, G., Thompson, D., Lozier, J. et al. (2000). Membrane bioreactors: a designer's perspective. *CD-ROM Proceedings of WEFTEC 73rd Annual Conference and Exposition*, Anaheim, CA (14–18 October 2000).

Crittenden, J.C., Trussell, R.R., Hand, D.W. et al. (2005). *Water Treatment: Principles and Design*, 2e. Hoboken, NJ: Wiley.

Demoulin, G., Rudiger, A., and Goronszy, M.C. (2001). Cyclic activated sludge technology: recent operating experience. *Water Science and Technology* 43 (3): 331–337.

Dow (2002). *Dowex Marathon C.* Midland, MI: Ion Exchange Resins Engineering Information, Dow Chemical Company.

Droste, R.L. (1997). *Theory and Practice of Water and Wastewater Treatment.* New York: Wiley.

Ersu, C.B. and Ong, S.K. (2008). Treatment of phenolic wastewater using a tubular ceramic membrane bioreactor. *Environmental Technology* 29 (2): 225–234.

Ersu, C.B., Ong, S.K., Arslankaya, E. et al. (2008a). Modification of a full-scale sequencing batch reactor operational mode for biological nutrient removal. *Water Environment Research* 80 (3): 257–266.

Ersu, C.B., Ong, S.K., Arslankaya, E., and Brown, P. (2008b). Comparison of recirculation configurations for biological nutrient removal in a membrane bioreactor. *Water Research* 42 (6–7): 1651–1663.

Ersu, C.B., Ong, S.K., Arslankaya, E., and Brown, P. (2010). Impact of solids residence time on biological nutrient removal performance of membrane bioreactor. *Water Research* 44 (10): 3192–3202.

Gavaskar, A., Kim, B.C., Rosansky, S. et al. (1995). Crossflow air stripping and catalytic incineration for remediation of volatile organics in groundwater. *AIChE Environ Progress* 14 (1): 33–40.

Gloyna, E.F. and Li, L. (1995). Progress in supercritical water oxidation: research and development. *Fifth International Chemical Oxidation Symposium and Principles and Practices Workshop*, Nashville, TN.

Great Lakes – Mississippi River Board of State and Provincial Public Health and Environmental Managers (2014). *Recommended Standards for Wastewater Facilities*, 2014e. Albany, NY: Health Research, Inc., Health Education Services Division.

Ha, J.H. and Ong, S.K. (2007). Nitrification and denitrification in partially aerated biologica aerated filter (BAF) with dual size sand media. *Water Science and Technology* 55 (1–2): 9–171.

Ha, J.H., Ong, S. K., and Surampalli, R. (2005). Nitrification and denitrification in partially aerated biological aerated filter (BAF) with dual size sand media. *IWA Specialty Conference, Wastewater Reclamation and Reuse for Sustainability*, Jeju, Korea (8–11 November 2005).

Ha, J.H., Ong, S.K., and Surampalli, R. (2010a). Impact of media type and various operating conditions on nitrification in polishing biological aerated filters. *Environmental Engineering Research* 15 (2): 79–84.

Ha, J.H., Ong, S.K., and Surampalli, R. (2010b). Nitrification in polishing biological aerated filters (BAFs) – impact of temperature. *Environmental Technology* 31 (6): 671–680.

Hoigne, J. and Bader, H. (1976). The role of hydroxyl radical reactions in ozonation processes in aqueous solutions. *Water Research* 10: 377.

Hsu, M. and Wilson, T.E. (1992). Activated sludge treatment of municipal wastewater – USA practice. In: *Activated Sludge Process Design and Control: Theory and Practice*, vol. 1 (ed. W.W. Eckenfelder and P. Grau). Lancaster: Technomic Publishing Co., Inc.

Juttner, K., Galla, U., and Schmieder, H. (2000). Electrochemical approaches to environmental problems in the process industry. *Electrochimica Acta* 45: 2575–2594.

Khanal, S.K., Xie, B., Thompson, M. et al. (2006). Fate, transport and biodegradation of natural estrogens in the environment and engineered systems. *Environmental Science and Technology* 40 (21): 6537–6546.

Knoblock, M.D., Sutton, P.M., and Mishra, P.N. (1998). Lessons learned from operation of membrane bioreactors. *CD-ROM Proceedings of WEFTEC 71st Annual Conference and Exposition*, Orlando, FL (3–7 October 1998).

Lettinga, G., Roersma, R., and Grin, P. (1983). Anaerobic treatment of raw domestic sewage at ambient temperatures using a granular bed UASB reactor. *Biotechnology and Bioengineering* 25: 1701–1723.

Limpiyakorn, T., Homklin, S., and Ong, S.K. (2011). Fate of estrogens and estrogenic potential in sewerage systems. *Critical Reviews in Environmental Science and Technology* 41 (13): 1231–1270.

M'Coy, W.S. (1997). Biological aerated filters: a new alternative. *Water Environment and Technology* 42: 39–42.

Matheson, W., Morton, C.A., and Titus, J. (2015). Wastewater recycling at a steel plating mill. http://www.duraflow.biz/pdfs/literature/Wastewater-Recycling-at-a-Steel-Plating-Mill.pdf (accessed December 2015).

McCarthy, P.L. (1983). Removal of organic substances from water by air. In: *Stripping in Control of Organic Substances in Water and Wastewater* (ed. B.B. Berger). Office of Research and Development EPA-600/8-83-011.

Mendoza-Espinosa, L.G. and Stephenson, T. (1999). A review of biological aerated filters (BAFs) for wastewater treatment. *Environmental Engineering Science* 16 (3): 201–216.

Metcalf and Eddy (2003). *Wastewater Engineering, Treatment and Reuse*, 4e. New York: McGraw Hill Inc.

Modell, M. (1989). Supercritical water oxidation. In: *Standard Handbook of Hazardous Waste Treatment and Disposal* (ed. H.M. Freeman). New York: McGraw-Hill.

Officer, J., Ostrowski, J.A., and Woollard, P.J. (2000). The design and operation of conventional and novel flotation systems on a number of impounded water types: particle removal from reservoirs and other surface waters. *Water Science and Technology Water Supply* 1 (1): 63–69.

Pokhrel, D. and Viraraghavan, T. (2004). Treatment of pulp and paper mill wastewater - a review. *Science of the Total Environment* 333 (1–3): 37–58.

Protzman, R.S., Lee, P.H., Ong, S.K., and Moorman, T.B. (1999). Treatment of formulated atrazine Rinsate by *Agrobacterium radiobacter* strain J14a in a sequencing batch biofilm reactor. *Water Research* 33 (6): 1399–1404.

Pujol, R., Lemmel, H., and Gousailles, M. (1998). A Keypoint of nitrification in an upflow biofiltration reactor. *Water Science and Technology* 38 (3): 43–49.

Reynolds, T.D. and Richards, P.A. (1996). *Unit Operations and Processes in Environmental Engineering*. Boston, MA: PWS Publishing Company.

Smith, A.J., Quinn, J.J., and Hardy P.J. (1990). The development of an aerated filter package plant. *1st International Conference on Advances in Water Treatment and Environmental Management*, Lyon, France (27–29 June 1990).

Stephenson, T., Judd, S., Jefferson, B., and Brindle, K. (2000). *Membrane Bioreactors for Wastewater Treatment*. London: IWA Publishing.

Takeuchi, K., Futamura, O., and Kojima, R. (1990). *Integrated Type Membrane Separation Activated Sludge Process for Small Scale Sewage Treatment Plants*. Tokyo, Japan: Ebara Infilco Ltd.

Tian, Q., Ong, S.K., Wang, K.W. et al. (2015). Enhanced phosphorus recovery and biofilm microbial community changes in an alternating anaerobic/aerobic biofilter. *Chemosphere* 144: 1797–1806.

US Army (1996). Oil/Water Separator Installation and Maintenance: Lessons Learned. U.S. Army Center for Public Works Technical Note (31 October 1996), Aberdeen, MD.

US EPA (1983). *Development Document for Effluent Limitations Guidelines and Standards for the Metal Finishing Point Source Category*. Washington, DC: Environmental Protection Agency EPA 440/1-83-091.

US EPA (1998). *Advanced Photochemical Oxidation Processes*. Washington DC: Office of Research and Development, Environmental Protection Agency EPA l6251R-981004.

US EPA (2004). *Primer for Municipal Wastewater Treatment Systems*. Washington DC: Environmental Protection Agency EPA 832-R-04-001.

US EPA (2010). *NPDES Permit Writer's Manual*. Washington DC: Environmental Protection Agency EPA-833-K-10-001.

US EPA (2011). *Introduction to the National Pretreatment Program*. Washington DC: Environmental Protection Agency EPA-833-B-11-001.

US EPA (2012). *Clean Water Needs Survey 2008, Report to Congress*. Washington, DC: Environmental Protection Agency EPA-832-R-10-002.

Zhang, B., Yamamoto, K., Ohgaki, S., and Kamiko, N. (1997). Floc size distribution and bacterial activities in membrane separation activated sludge processes for small-scale wastewater treatment/reclamation. *Water Science and Technology* 35 (6): 37–44.

13

Wastewater Recycling

Judith L. Sims[1] *and Kirsten M. Sims*[2]

[1] *Utah Water Research Laboratory, Utah State University, Logan, UT, USA*
[2] *WesTech Engineering, Inc., Salt Lake City, UT, USA*

13.1 Introduction

Clean water is essential to the development of nations; however, according to a 2006 United Nations report, over one billion people have no access to clean water, and 2.6 billion people lack access to adequate sanitation (UNDP, 2006). To provide this access, existing water resources must be used more efficiently, and new sources of freshwater need to be found, especially in urban areas where most of the world's population now lives. New sources are essential for the world's population due to:

- Population growth.
- Contamination and deterioration of surface and groundwaters.
- Uneven distribution of water resources throughout the world.
- Frequent droughts.

Desalination of ocean water, brackish groundwater, and other salty water sources is one potential new source of water. Desalination is usually used in areas without an adequate supply of water, as it is an expensive and energy-intensive process.

Reclaiming and reusing wastewater is another source of additional clean water, especially as large amounts of wastewater are generated near urban population centers. The terms "reclaimed water" and "recycled water" are both commonly used to refer to water that, as a result of wastewater treatment, is suitable for a direct beneficial or controlled use. *De facto* reuse has been practiced for many years but has not been officially recognized (e.g. a drinking water supply intake located downstream from a wastewater treatment plant). However, planned water reuse can be used to conserve and extend existing water supplies, serve as an alternate wastewater disposal method, and provide pollutant abatement by diverting effluent discharges away from sensitive receiving waters. Reusing wastewater allows communities to maintain more green spaces, which aids in public health protection by leading to cleaner air, which in turn reduces respiratory diseases. Green spaces also allow people to be more active, thus reducing obesity and cardiovascular disease.

Water reclamation and nonpotable reuse requires typical conventional wastewater treatment technologies that are widely used and available throughout the world. With the use of advanced treatment technologies, any desired level of water quality can be achieved. The governing principle of wastewater reuse is that the quality of the wastewater should be appropriate for the intended use. Higher levels of use, such as irrigation of vegetables, require a higher degree of wastewater treatment and reliability than lower level uses, such as irrigation of forage crops and pastures.

Nonpotable reclaimed wastewater may be substituted for an existing potable water source, possibly providing a more economical available source for many different types of activities that do not need high quality sources. Examples include irrigation of lawns, parks, roadway medians and roadsides, air conditioning, industrial cooling towers, stack gas scrubbing, industrial processing, toilet flushing, dust control and construction, vehicle washing, cleaning and maintenance activities, scenic waters and fountains, and environmental and recreational purposes (US EPA, 2004).

Although water reclamation and recycling is generally fairly simple, only a few countries have widespread use of the reclaimed water. Israel recycles 100% of its wastewater using 70% for agriculture and 30% used as gray water or for industrial purposes. In the United States, water recycling is often used in the southwestern states, an area with deserts and often underwater stress. In 2006, California reported reuse of 580 million gallons per day, which in terms of reuse per capita per day is 16 gallons.

Handbook of Environmental Engineering, First Edition. Edited by Myer Kutz.

However, Florida in the southeastern part of the United States has led the United States in recycling for over 20 years. Florida Statutes 373 and 403 defines the reuse of reclaimed water and water conservation as critical state objectives. In 2016, Florida reused approximately 760 million gallons per day for beneficial purposes, which represents an average reuse per capita of 37.7 gallons per day per person. This reuse is estimated by the Florida Department of Environmental Protection to have avoided the use of over 148 billion gallons of potable water while adding more than 89 billion gallons to available groundwater supplies. The 2016 reuse capacity represents about 65% of the total permitted domestic wastewater treatment capacity in Florida (FDEP, 2018). The use of reclaimed wastewater in 2015 was 57% for irrigation of public access areas, 17% for industrial uses, 13% for groundwater recharge, 9% for agricultural irrigation, and 4% for wetlands and other uses (FDEP, 2016).

13.2 Uses of Reclaimed Wastewater

Water reclamation and recycling can supplement the water requirements for many needs of society, including the following categories as defined by the US Environmental Protection Agency (US EPA, 2012) and shown in Table 13.1.

13.2.1 Urban Reuse of Reclaimed Wastewater

Increasing water demand in urban areas is driven by population growth, industrial development, and the increasing integration of irrigated peri-urban agriculture into urban life. In fact, in the United States, urban reuse is one of the highest volume uses for reclaimed water. Urban reclaimed water is the third largest reuse category in California in 2009 (after agricultural and natural systems) and first largest in Florida in 2010 (the two states with the highest reuse rates) (US EPA, 2012).

Reuse of wastewater is applicable to offset growing urban water consumption demand and contributes to a reduction in volumetric and constituent loadings on urban wastewater treatment plants, delaying or reducing the need for plant expansion as population increases.

The term "urban reuse" is typically applied where reclaimed domestic wastewater that has been subjected to post-secondary treatment is used for nonpotable purposes in an urban environment. The term may also be applied to treated industrial wastewater that is used for the same purposes.

Reused wastewater in urban applications may be categorized as accessible to the public (unrestricted) or as public access limited (restricted). Currently, 32 US states or territories have rules, regulations, or guidelines addressing treatment of wastewater for unrestricted urban use, while 40 US states or territories have rules, regulations, or guidelines addressing restricted urban wastewater reuse. It is important for practitioners to understand the rules, regulations, guidelines, and restrictions that a potential reuse project will be subjected to based on state and local jurisdictions.

Examples of unrestricted uses include beneficial irrigation of areas that are intended to be accessible to the public, such as irrigation of golf courses, parks, athletic fields, roadway medians, residential and commercial landscaping, and cemeteries. Expanded application of unrestricted urban reuse may include nonirrigation applications such as toilet flushing, dust control, and fountain water. Examples of restricted water reuse applications include irrigation of areas with controlled access, fire protection water sources, and construction water.

13.2.1.1 Constituents of Concern in Urban Reuse

While typically not ingested as a drinking water source, the potential for direct human contact is anticipated with urban wastewater reuse. Therefore, urban wastewater reuse typically necessitates sophisticated levels of treatment to reduce or remove constituents of concern that cause a pathological or toxicological threat to human and environmental health.

When considering application of urban water reuse, it is important to understand what constituents are of concern and in what concentrations. Untreated domestic wastewater may contain a wide range of organic and inorganic constituents, some of which are hazardous to environmental and human health at given concentrations.

All reuse systems require a minimum of primary and secondary treatment, wherein the level of conventional wastewater constituents including biochemical oxygen demand (BOD), total suspended solids (TSS), pH, and pathogens is reduced. However, the level of treatment achieved by conventional secondary treatment is not typically considered adequate for the protection of human health when the water is reused in an urban environment.

13.2.1.1.1 Microorganisms

While a fraction of microorganisms present in wastewater may be benign or even beneficial to human health, raw domestic water can contain a large variety of pathogenic (disease-causing) microorganisms that are dangerous to humans. Pathogenic constituents associated with waterborne diseases, typically referred to as "microorganisms" or "microbes," may be classified into three broad groups. These groups include bacteria, viruses, and parasites.

Table 13.1 Water reuse categories and number of states with rules, regulations, or guidelines addressing these reuse categories (US EPA, 2012).

Category of reuse		Description	Number of US states or territories with rules, regulations, or guidelines addressing reuse category
Urban reuse	Unrestricted	The use of reclaimed water for nonpotable applications in municipal settings where public access is not restricted	32
	Restricted	The use of reclaimed water for nonpotable applications in municipal settings where public access is controlled or restricted by physical or institutional barriers, such as fencing, advisory signage, or temporal access restriction	40
Agricultural reuse	Food crops	The use of reclaimed water to irrigate food crops that are intended for human consumption	27
	Processed food crops and nonfood crops	The use of reclaimed water to irrigate crops that are either processed before human consumption or not consumed by humans	43
Impoundments	Unrestricted	The use of reclaimed water in an impoundment in which no limitations are imposed on body-contact water recreation activities (some states categorize snowmaking in this category)	13
	Restricted	The use of reclaimed water in an impoundment where body contact is restricted (some states include fishing and boating in this category)	17
Environmental reuse		The use of reclaimed water to create, enhance, sustain, or augment water bodies, including wetlands, aquatic habitats, or stream flow	17
Industrial reuse		The use of reclaimed water in industrial applications and facilities, power production, and extraction of fossil fuels	31
Groundwater recharge – nonpotable reuse		The use of reclaimed water to recharge aquifers that are not used as a potable water source	16
Potable reuse	Indirect potable reuse (IPR)	Augmentation of a drinking water source (surface or groundwater) with reclaimed water followed by an environmental buffer that precedes normal drinking water treatment	9
	Direct potable reuse (DPR)	The introduction of reclaimed water (with or without retention in an engineered storage buffer) directly into a water treatment plant, either collocated or remote from the advanced wastewater treatment system	0

Individual state reuse programs often incorporate different terminology, so the reader should exercise caution in comparing the categories in these tables directly to state regulatory definitions.

Municipal wastewater typically contains a wide variety of bacteria, viruses, and parasite species, only some of which pose a pathological threat to humans. However, the potential for transmission of infectious diseases by pathogenic constituents in wastewater is the most common concern associated with reuse of treated municipal wastewater in urban applications. In raw, untreated wastewater, the concentrations of pathogenic constituents may vary according to the general health and habits of the contributing communities. Concentrations may also vary seasonally.

As urban water reuse often involves spraying or drip irrigation of reused wastewater, it is also important to consider threats posed by pathogenic agents that are contained in aerosols. Aerosols are particles less than 50 μm in diameter that are suspended in the air. Inhalation of aerosols is a potential direct route for human infections. The infective dose of some pathogenic agents is lower for respiratory transmission than for ingestion and transmission through the digestive tract.

Many pathogens are associated with particulate matter in wastewater, and the association with particulate matter can often reduce the effect of disinfectants such as chlorine and ultraviolet (UV) light due to a shielding effect. Further, organic matter reduces the availability of chlorine for disinfection of pathogenic constituents, as organic matter consumes chlorine present in the wastewater. Reduction of particulate matter to levels around 5 $mg l^{-1}$ of TSS is generally agreed to be sufficient for reliable destruction of pathogenic microorganisms during disinfection process.

Concentrations of microorganisms are typically reported in logarithmic scales due to the large number of microbes present. For example, a 1-log removal indicates 90% removal, 2-log removal indicates 99% removal, 3-log removal indicates 99.9% removal, and so forth.

"Indicator organisms" are used to measure treatment efficiency with respect to reduction of pathogenic organisms that originate from fecal contamination. Indicator organisms are not dangerous to human health but rather serve as a proxy indication of the level of health risk posed for a given wastewater. In some states, total coliform bacteria are used as an appropriate indicator. However, in many states with specific regulations, the pathogenic risk of a reclaimed water source must be evaluated by daily monitoring of fecal coliform bacteria in disinfected effluent based on a single, 100-ml grab sample.

13.2.1.1.2 *Organic Constituents*

Carbonaceous compounds, both dissolved and particulate, are found in raw wastewater. These organic compounds may be relatively simple, comprising a backbone of only a few carbon atoms, or they may be complex, comprising a long-chain carbon backbone, aromatic rings, and strong covalent bonds with inorganic side groups or radical (free electron) components. The degree of recalcitrance is based on the types of organics present. Naturally occurring humic substances, oils, grease, industrial products, consumer products, pharmaceuticals, food waste, detergents, and fecal matter that are present in wastewater may contribute to the organic content.

For urban reuse, concern with the presence of residual organics is attributable to several adverse effects associated with elevated levels of organics in water. These may include the proliferation of microorganisms, which use the organics as a food source, as well as a potential malodorous effect. Particulate organic matter may accumulate in soil, causing a decrease in permeability. Further, particulate organic matter may contribute to clogging of nozzles, sprinklers, and other discharge points.

Of particular concern to human health is the potential for interaction between certain organic compounds with chlorine during a disinfection process prior to urban water reuse. Organic compounds may be converted into disinfection by-products (DBPs), which have been linked to the development of bladder, kidney, and liver cancer among genetically prone individuals. These include trihalomethane (THM) compounds and haloacetic acid (HAA) compounds. THM include chloroform and other chlorine or bromine-substituted forms of methane. Chloroform has been indicated as a potential cause for liver and kidney cancer. HAA have been linked to similar adverse health effects. There is also a growing concern over the presence of the carcinogenic compound *N*-nitrosodimethylamine (NDMA), which is commonly present after disinfection (US EPA, 2012).

One method for addressing concerns associated with DBPs is to incorporate advanced oxidation processes (AOPs) such as UV radiation as a posttreatment to reverse osmosis (RO) to address NDMA and other trace organics.

The level of organics in water is typically described by proxy indicators, including BOD, chemical oxygen demand (COD), and total organic carbon (TOC). TOC may further be categorized as dissolved organic carbon (DOC), which is the level of TOC that passes through a 0.45-um pore-size filter, and particulate organic carbon (POC), which is the level that is retained on the 0.45-um pore-size filter.

13.2.1.1.3 *Inorganic Chemical Constituents*

Inorganic constituents of concern in wastewater vary according to the wastewater source and on the type of treatment applied for urban reuse. These include salts, nutrients, metals, and oxyhalides. Levels of inorganic constituents present in water are often represented through an aggregate measurement by assessing total dissolved solids (TDS) and conductivity. These may also account for some organic material present.

Typically, the concentration of toxic metals and salt is far below the level of concern for human health in treated wastewater. Some metals that remain present at low levels after treatment, such as boron (present in detergents), may still have detrimental effects on plant growth and therefore should be monitored for urban reuse applications that involve providing water for irrigation or landscaping.

Salts may also remain in reclaimed wastewater. Though not of concern to human health, salinity may reduce the permeability of soils with a significant clay component. Salts may also affect plant health directly by causing leaf burn.

The poorest quality of reclaimed wastewater with respect to TDS in the United States is typically found in the southwestern regions, where the greatest golf course reuse occurs. One strategy for coping with the elevated TDS is to choose salt-tolerant grasses such as alkali, Bermuda, fine leaf, salt grass, etc. Additional practices may include applying extra water to leach excess salts and provide adequate drainage.

Further, salinity may cause issues such as scaling and corrosion in distribution systems. Considerations for salinity removal may be undertaken for urban reuse applications, although processes for salinity treatment are often energy and cost intensive.

Depending on the intended urban reuse application, nutrients (nitrogen and phosphorus) may pose concerns for environmental and human health due to the facilitation of overgrowth of biological agents such as algae. However, for urban reuse applications including landscaping and irrigation, nutrients enhance reuse effectiveness. Reclaimed water that has not been nitrified or denitrified may contain greater than 20 $mg\,l^{-1}$ of ammonia nitrogen, which may exert over 100 $mg\,l^{-1}$ of nitrogenous oxygen demand.

While the nutrients contained in reuse water may be advantageous for enhancing plant growth, it is important to monitor oversupply of nutrients. For golf courses with excessive runoff from watering, excess nitrogen is often supplied at loadings much higher than the turfgrass needs (>5.1 lb of water-soluble nitrogen per 1000 ft^3) (US EPA, 2012).

13.2.1.1.4 Trace Chemical Constituents

Many trace chemical constituents, which are sometimes referred to as "emerging contaminants of concern," have been found to be present even in treated wastewater as analytical techniques have become more sophisticated and capable of detecting very low levels of organic and inorganic constituents that have not previously been regulated but may pose a risk to human and environmental health.

Trace chemical constituents may include degradation products from various industrial and domestic chemicals, including pesticides, biocides, herbicides, pharmaceuticals, personal care products, household chemicals, food additives, and industrial chemicals.

Pharmaceuticals and personal care products in traditional wastewater treatment are not significantly removed during the treatment process, although some removal or chemical conversion can be achieved via oxidation disinfection processes. While direct ingestion of reuse water is not anticipated in urban reuse, inhalation of aerosolized components and skin contact may be encountered. Therefore, it is important to understand the risks these constituents may pose on human health and the environment. Public exposure to trace chemical constituents in urban reuse for nonpotable applications is considered negligible; however, the scientific community is continuing to investigate the potential for human harm and establish a consensus on upper limits that may become the basis for design and operation of urban reuse facilities in the future.

Of greater concern is the potential for negative impact of trace chemical constituents on ecological health of urban communities. Where urban reuse water is applied and potentially collected as runoff, there is a potential for causing altered reproduction physiologies and elevated incidences of hermaphrodism (Sumpter and Johnson, 2008) to feral fish in receiving waters.

Endocrine disrupting compounds (EDCs) of concern in reclaimed waters include hormones and detergent residues. Primarily, the natural estrogenic hormone 17-β estradiol and the synthetic estrogenic hormone ethinylestradiol contribute to the estrogenic activity in reclaimed waters (Metcalf and Eddy, 2007).

Of further concern is the release of antibiotics that are not removed during treatment for urban reclaimed water, which may lead to increased antibiotic resistance in pathogenic and nonpathogenic environmental organisms (Pauwels and Verstraete, 2006). A majority of antibiotics (up to 75%) that are ingested by humans and animals are excreted unaltered or as metabolites (Bockelmann et al., 2009). Ecological health should be considered when designing an urban reuse system.

13.2.1.2 Decentralization of Urban Wastewater Treatment and Recycling

Presently, urban water reuse typically involves large-scale applications that are based on existing centralized sewage treatment plants. However, given the existing collection networks that precede centralized treatment, a cost advantage may be realized for decentralized water reuse applications wherein wastewater is collected en route to the centralized treatment facility and treated in small package-type systems and then used for irrigation or other nonpotable urban uses.

This semicentralized approach to wastewater treatment and reuse may minimize the growing discrepancy between urban growth and the provision of treatment and application infrastructure. Further, this approach offers added flexibility in energy expenditures for treatment and through reduced need to transport treated wastewater to the site of reuse.

When considering a design for an urban water reuse project, consideration should be given to the feasibility of incorporating decentralized or semidecentralized components to reduce loading on centralized urban wastewater treatment plants and deliver treated wastewater for use near the point of treatment.

13.2.1.3 Treatment Objectives: Urban Reuse

As the potential for human contact of reused wastewater increases, higher levels of treatment processes beyond conventional secondary treatment may be required. Generally, two disinfection threshold levels are recommended depending on the probability of human contact for nonpotable urban reuse.

The behavior and fate of contaminants in wastewater treatment have been fairly well modeled and understood with the exception of some emerging contaminants of concern. Maintaining protection of environmental and human health has led several states to develop regulations and criteria for various reuse applications, including urban reuse. Title 22 of the California Code of Regulations for Water Recycling Criteria (California Department of Health Services, 2014) and Chapter 62-610 of the Florida Administrative Code for Reuse of Reclaimed Water and Land Application both require a multibarrier approach for treatment of wastewater for reuse.

13.2.1.3.1 Unrestricted Urban Reuse

Public exposure is highly likely in unrestricted urban reuse. Further, high levels of treatment are generally required. All states that specify regulations regarding this type of reuse require a minimum of disinfection following secondary treatment.

For uses where direct or indirect contact with reclaimed water is likely and for dual water systems where there is any potential for cross-connection contamination with potable waterlines, disinfection to produce reclaimed water with no detectable fecal coliform indicator organisms per 100 ml is recommended as a minimum treatment goal (US EPA, 2012). Additional levels of treatment, including oxidation, coagulation, and filtration, are additional unit processes that are sometimes required as well for reduction in additional constituents, such as emerging contaminants of concern. The level of treatment required for unrestricted urban reuse in 10 states is shown in Tables 13.2 and 13.3.

Many states base their regulatory and design criteria on the removal of indicator organisms to account for the removal of bacterial, viral, or protozoa pathogens for public health protection. Total and fecal coliform counts are generally used as the relevant indicator organisms. Regulations vary from state to state and may be based on either removal efficiencies or treatment technologies with specified performance requirements.

North Carolina, for example, has developed reuse specifications for reclaimed wastewater with direct human contact potential that requires 6-log removal (99.9999%), removal for *E. coli*, 5-log removal for coliphage, and 4-log removal for *Clostridium perfringens*. Alternatively, California reclaimed water must be subjected to oxidation, sedimentation, coagulation, filtration, and disinfection, leading to undetectable or very low microbe levels in reclaimed water for urban applications (California Department of Health Services, 2001).

A critical component of wastewater treatment for reuse is also dependent on effective industrial source control through the National Pretreatment Program established under the Clean Water Act (US EPA, 2011). This requires nondomestic entities that discharge to publicly owned treatment works (POTW) to implement treatment and management of their wastewater to reduce or eliminate the discharge of harmful pollutants to sanitary sewers. The US EPA has established numeric effluent guidelines for 56 categories of industry. These guidelines are technology based, and the Clean Water Act requires that the US EPA must annually review its effluent guidelines and pretreatment standards. Further, the US EPA must identify potential new categories for pretreatment standards. The recommendations are presented in a preliminary effluent guidelines program plan.

Due to the significant reduction in chemical discharge to POTWs achieved through industrial pretreatment programs, specific objectives for reduction or removal of emerging chemicals of concern and inorganic constituents are not considered mandatory for urban reuse applications. Of course, if the application of urban reclaimed water requires specific water quality criteria (such as low salinity for residential/landscape irrigation), then targeted treatment of these constituents may need to be implemented as post-secondary processes.

Regarding emerging contaminants of concern, there are few regulations in place regarding treatment requirements. As noted before, industrial pretreatment requirements may reduce loadings of many of these chemicals on downstream treatment facilities. Without federal guidelines, states may choose to regulate specific chemical constituents. The threat of these constituents to human health in urban reuse applications is generally thought to be negligible. However, concerns about proliferating antibiotic resistance in naturally occurring microbial populations, including pathogens, may be of growing concerns. More research is required before specific treatment objectives are agreed upon.

13.2.1.3.2 Restricted Urban Reuse

Public exposure is minimized or controlled in restricted urban reuse applications. Reclaimed water used for applications where no direct public or worker contact with the water is expected should be disinfected to achieve an average fecal coliform concentration not exceeding 200/100 ml. At this indicator bacterial concentration, disinfection of secondary effluent is readily achievable with well-established and relatively low-cost technologies (US EPA, 2012).

While some states, such as Florida, impose the same requirements on both unrestricted and restricted urban access reuse, generally treatment requirements for controlled access may be less than those associated with unrestricted urban reuse. Generally, states require a minimum of secondary treatment followed by disinfection prior to restricted urban reuse. The level of treatment required for restricted urban reuse in 10 states is shown in Tables 13.4 and 13.5.

Table 13.2 Selected state standards for urban reuse – unrestricted – Arizona, California, Florida, Hawaii, and Nevada (US EPA, 2012).

	Arizona class A	California disinfected tertiary	Florida	Hawaii R1 water	Nevada category A
Unit processes	Secondary treatment, filtration, disinfection	Oxidized, coagulated, filtered, disinfected	Secondary treatment, filtration, high-level disinfection	Oxidized, filtered, disinfected	Secondary treatment, disinfection
UV dose, if UV disinfection used	*NS*	NWRI UV Guidelines	NWRI UV Guidelines enforced, variance allowed	NWRI UV Guidelines	*NS*
Chlorine disinfection requirements, if used	*NS*	CrT > 450 mg min l^{-1}; 90 min modal contact time at peak dry weather flow	TRC > 1 mg l^{-1}; 15 min contact time at peak hour flow[a]	Min residual >5 mg l^{-1}; 90 min modal contact time	*NS*
BOD_5 (or $CBOD_5$)	*NS*	*NS*	$CBOD_5$: • 20 mg l^{-1} (ann avg) • 30 mg l^{-1} (mo avg) • 45 mg l^{-1} (wk avg) • 60 mg l^{-1} (max)	30 or 60 mg l^{-1} depending on design flow	30 mg l^{-1} (30-d avg)
TSS	*NS*	*NS*	5 mg l^{-1} (max)	30 or 60 mg l^{-1} depending on design flow	30 mg l^{-1} (30-d avg)
Turbidity	• 2 NTU (24-h avg) • 5 NTU (max)	• 2 NTU (avg) for media filters • 10 NTU (max) for media filters • 0.2 NTU (avg) for membrane filters • 0.5 NTU (max) for membrane filters	Case by case (generally 2–2.5 NTU) Florida requires continuous online monitoring of turbidity as indicator for TSS	• 2 NTU (95 percentile) • 0.5 NTU (max)	*NS*
Bacterial indicators	Fecal coliform: • None detectable in the last 4 of 7 samples • 23/100 ml (max)	Total coliform: • 2.2/100 ml (7-d med) • 23/100 ml (not more than one sample exceeds this value in 30 d) • 240/100 ml (max)	Fecal coliform: • 75% of samples below detection • 25/100 ml (max)	Fecal coliform: • 2.2/100 ml (7-d med) • 23/100 ml (not more than one sample exceeds this value in 30 d) • 200/100 ml (max)	Total coliform: • 2.2/100 ml (30-d geom) • 23/100 ml (max)
Pathogens	*NS*	*NS*	*Giardia* and *Cryptosporidium* sampling once each 2-yr period for plants ≥1 mgd; once each 5-yr period for plants ≤1 mgd	*TR*	*TR*
Other	If nitrogen >10 mg l^{-1}, special requirements may be mandated to protect groundwater	—	—	—	—

NS, not specified by the state's reuse regulation; *TR*, monitoring is not required but virus removal rates are prescribed by treatment requirements.

[a] In Florida when chlorine disinfection is used, the product of the total chlorine residual and contact time (CrT) at peak hour flow is specified for three levels of fecal coliform as measured prior to disinfection. If the concentration of fecal coliform prior to disinfection is ≤1000 cfu per 100 ml, the CrT shall be 25 mg min l^{-1}; is 1000–10 000 cfu per 100 ml, the CrT shall be 40 mg min l^{-1}; and is ≥10 000 cfu per 100 ml, the CrT shall be 120 mg min l^{-1}.

Table 13.3 State standards for urban reuse – unrestricted – New Jersey, North Carolina, Texas, Virginia, and Washington (US EPA, 2012).

	New Jersey type I RWBR	North Carolina type 1	Texas type I	Virginia level 1	Washington class A
Unit processes	Filtration, high-level disinfection	Filtration (or equivalent)	*NS*	Secondary treatment, filtration, high-level disinfection	Oxidized, coagulated, filtered, disinfected
UV dose, if UV disinfection used	100 mJ cm^{-2} at max day flow	*NS*	*NS*	*NS*	NWRI UV Guidelines
Chlorine disinfection requirements, if used	Min residual >1 mg l^{-1}; 15 min contact time at peak hour flow	*NS*	*NS*	TRC CAT <1 mg l^{-1}; 30 min contact time at avg. flow or 20 min at peak flow	Chlorine residual >1 mg l^{-1}; 30 min contact time (CrT > 30 may be required)
BOD_5 (or $CBOD_5$)	NS	• 10 mg l^{-1} (mo avg) • 15 mg l^{-1} (daily max)	5 mg l^{-1}	• 10 mg l^{-1} (mo avg) or $CBOD_5$: • 8 mg l^{-1} (mo avg)	30 mg l^{-1}
TSS	5 mg l^{-1}	• 5 mg l^{-1} (mo avg) • 10 mg l^{-1} (daily max)	*NS*	*NS*	30 mg l^{-1}; this limit is superseded by turbidity
Turbidity	2 NTU (max) for UV	10 NTU (max)	3 NTU	• 2 NTU (daily avg), CAT >5 NTU	• 2 NTU (avg) • 5 NTU (max)
Bacterial indicators	Fecal coliform: • 2.2/100 ml (wk med) • 14/100 ml (max)	Fecal coliform or *E. coli*: • 14/100 ml (mo mean) • 25/100 ml (max)	Fecal coliform or *E. coli:* • 20/100 ml (30-d geom) • 75/100 ml (max) *Enterococci*: • 4/100 ml (30-d geom) • 9/100 ml (max)	Fecal coliform: • 14/100 ml (mo geom), CAT >49/100 ml *E. coli*: • 11/100 ml (mo geom), CAT >35/100 ml *Enterococci:* • 11/100 ml (mo geom), CAT >24/100 ml	Total coliform: • 2.2/100 ml (7-d med) • 23/100 ml (max)
Pathogens	*NS*	*NS*	*NS*	*NS*	*NS*
Other	(NH_3-N + NO_3-N) <10 mg l^{-1} (max)	Ammonia as NH_3-N: • 4 mg l^{-1} (mo avg) • 6 mg l^{-1} (daily max)	—	—	Specific reliability or redundancy requirements based on formal reliability assessment

NS, not specified by the state's reuse regulation; *TR*, monitoring is not required but virus removal rates are prescribed by treatment requirements.

Table 13.4 Selected state standards for urban reuse – restricted – Arizona, California, Florida, Hawaii, and Nevada (US EPA, 2012).

	Arizona class B	California disinfected secondary-23	Florida[a]	Hawaii R2 water	Nevada category B
Unit processes	Secondary treatment, disinfection	Oxidized, disinfected	*NS*	Oxidized, disinfected	Secondary treatment, disinfection
UV dose, if UV disinfection used	*NS*	*NS*	*NS*	*NS*	*NS*
Chlorine disinfection requirements, if used	*NS*	*NS*	*NS*	Chlorine residual >5 $mg l^{-1}$, actual modal contact time of 10 min	*NS*
BOD_5 (CBOD for Florida)	*NS*	*NS*	*NS*	30 or 60 $mg l^{-1}$ depending on design flow	30 $mg l^{-1}$ (30-d avg)
TSS	*NS*	*NS*	*NS*	30 or 60 $mg l^{-1}$ depending on design flow	30 $mg l^{-1}$ (30-d avg)
Turbidity	*NS*	*NS*	*NS*	*NS*	NS
Bacterial indicators	Fecal coliform: • Less than 200/100 ml in the last four of seven samples • 800/100 ml (max)	Total coliform: • 23/100 ml (7-d med) • 240/100 (not more than one sample exceeds this value in 30 d)	*NS*	Fecal coliform: • 23/100 ml (7-d med) • 200/100 ml (not more than one sample exceeds this value in 30 d)	Fecal coliform: • 2.2/100 ml (30-d geom) • 23/100 ml (max)
Other	If nitrogen >10 $mg l^{-1}$, special requirements may be mandated to protect groundwater	—	—	—	—

NS, not specified by the state reuse regulation.
[a] Florida does not specifically include urban reuses in its regulations for restricted public access under F.A.C. 62-610-400; requirements for restricted public access reuse are provided in Agricultural Reuse – Nonfood Crops, Tables 13.6–13.15.

13.2.1.4 Treatment Technologies and Urban Reuse

Objectives for reuse of wastewater are based on intended end-use application. Further, the appropriate treatment for urban reuse will vary depending on state-specific requirements. In order to determine the appropriate treatment system for a given urban reuse application, it is important to consider the relevant regulations or recommendations, the community's wastewater composition including industrial contributions, as well as the climate, topography, and socioeconomic factors.

A majority of states that allow and permit urban reuse applications do not specifically require certain treatment technologies but rather have regulations based on reuse water quality criteria, while some require both. Other aspects to consider are requirements or recommendations for reliability and resilience to process upsets, equipment failure, or power outages.

Removal of pathogenic constituents is primarily achieved through physical separation of particulate matter from wastewater. Microbiological constituents often absorb to particulate matter and floc particles, making them more difficult to deactivate through typical disinfection processes. This may take place in primary screening steps, primary and secondary sedimentation processes, dissolved air flotation (DAF) processes, or other physical/chemical separation steps. Advanced treatment processes such as coagulation and flocculation, media filtration, and membrane filtration may further reduce the particulate matter from the wastewater if necessary.

Advanced treatment technologies that can be used in addition to conventional primary and secondary treatment technologies include filtration, depth filtration, surface filtration, membrane filtration, granular activated carbon (GAC), alternative disinfection methods, advanced oxidation, and natural systems.

Table 13.5 Selected state standards for urban reuse – restricted – New Jersey, North Carolina, Texas, Virginia, and Washington (US EPA, 2012).

	New Jersey type II RWBR	North Carolina[a] type 1	Texas type II	Virginia level 2	Washington class C
Unit processes	Case by case	Filtration (or equivalent)	*NS*	Secondary treatment, disinfection	Oxidized, disinfected
UV dose, if UV disinfection used	75 mJ cm^{-2} at max day flow	*NS*	*NS*	NS	NWRI UV Guidelines
Chlorine disinfection requirements, if used	Chlorine residual >1 mg l^{-1}; 15 min contact time at peak hour flow	*NS*	*NS*	TRC CAT <1 mg l^{-1} 30 min contact time at avg. flow or 20 min at peak flow	Chlorine residual >1 mg l^{-1}; 30 min contact time
BOD_5 (CBOD for Florida)	*NS*	• 10 mg l^{-1} (mo avg) • 15 mg l^{-1} (daily max)	Without pond system: 20 mg l^{-1} (or CBOD 15 mg l^{-1}) With pond: 30 mg l^{-1}	• 30 mg l^{-1} (mo avg) • 45 mg l^{-1} (max wk) Or $CBOD_5$ • 25 mg l^{-1} (mo avg) • 40 mg l^{-1} (max wk)	30 mg l^{-1}
TSS	30 mg l^{-1}	• 5 mg l^{-1} (mo avg) • 10 mg l^{-1} (daily max)	*NS*	• 30 mg l^{-1} (mo avg) • 45 mg l^{-1} (max wk)	30 mg l^{-1}
Turbidity	*NS*	10 NTU (max)	*NS*	*NS*	*NS*
Bacterial indicators	Fecal coliform: • 200/100 ml (mo geom) • 400/100 ml (wk geom)	Fecal coliform or *E. coli*: • 14/100 ml (mo mean) • 25/100 ml (daily max)	Fecal coliform or *E. coli*: • 200/100 ml (30-d geom) • 800/100 ml (max) *Enterococci*: • 35/100 ml (30-d geom) • 89/100 ml (max)	Fecal coliform: • 200/100 ml (mo geom), CAT >800/100 ml *E. coli*: • 126/100 ml (mo geom), CAT >235/100 ml *Enterococci*: • 35/100 ml (mo geom), CAT >104/100 ml	Total coliform: • 23/100 ml (7-d med) • 240/100 ml (max)
Other	(NH_3-N + NO_3-N): <10 mg l^{-1} (max)	Ammonia as NH_3-N: • 4 mg l^{-1} (mo avg) • 6 mg l^{-1} (daily max)	—	—	—

NS, not specified by the state reuse regulation.
[a] There is no expressed designation between unrestricted and restricted urban reuse in North Carolina regulations.

13.2.1.4.1 Filtration

Filtration technologies are applied for the removal of particulates, suspended solids, and some dissolved constituents, depending on the pore size of the media. Most types of filtration will remove large pathogens such as protozoan cysts. Smaller pathogens, such as bacteria and viruses, may be absorbed to larger particulate matter and thus also removed with the larger particulate particles. Those that are not absorbed to particulate matter may also be removed by size exclusion filtration, wherein the media is comprised of very small pore sizes. Removal of particulates during filtration may result in more effective disinfection processes downstream. Chemical addition to coagulate smaller particles into larger, more easily filterable particles may also be recommended.

State-specific regulatory factors should be taken into account for the design of filtration processes for urban water reuse activities. For example, California stipulates that the filtration technology applied must be conditionally accepted by the California Department of Public Health for treatment of recycled water. The performance of the filtration system must also meet strict turbidity limits. Several types of filtration, including membrane filtration, depth filtration, and surface filtration, have received approval in California, and the loading rate at which the accepted filter technology can be operated is also specified.

13.2.1.4.2 Depth Filtration

Depth filters include a bed of media, which may be compressible or noncompressible. In a noncompressible media filter, several feet of sand or anthracite are packed into columns and are backwashed (may be continuous, semicontinuous, or batch backwash process). Constituents targeted by the application of noncompressible media filters include TSS, turbidity, and some protozoan oocysts and cysts. The nominal pores size typically ranges from 60 to 300 μm. Further, granular media beds may support biological activity that enhances filtration treatment by removing additional biodegradable constituents such as TOC. These granular media beds may aid in the treatment or reduction in pesticide residuals and other emerging contaminants of concern.

Compressible media depth filters are a more recent technology, wherein a synthetic media has a high porosity (around 88%), allowing for higher hydraulic loading rates. During filtration, the synthetic media is compressed 15–40%. Backwash is then applied in a batch process as the media is uncompressed and then cleaned with air scour and a hydraulic wash (Figure 13.1).

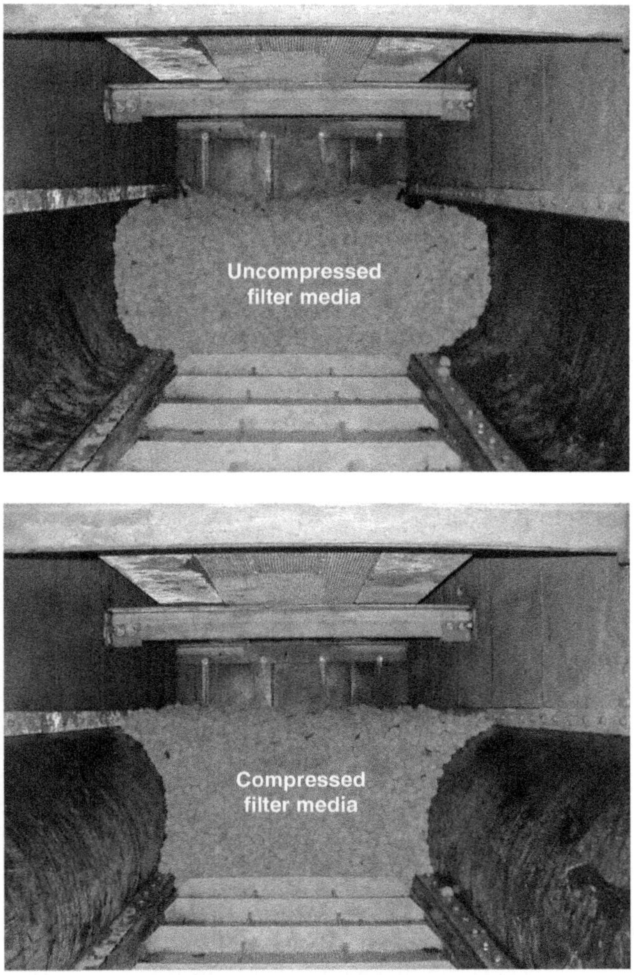

Figure 13.1 Compressible media depth filter. *Source:* From Fitzpatrick et al. (2015).

13.2.1.4.3 Surface Filtration

Surface filters typically consist of screens or fabric. The material of construction may be nylon, acrylic, nylon, or even stainless steel fibers. Most are gravity fed and operate primarily in a semicontinuous backwash mode. Disk-type filters have been granted regulatory approval in California, with improvements in design allowing for increasing loading rates (referred to as "high rate" disk filters). Like noncompressible media filters, the target constituents for removal include TSS, turbidity, and some protozoan oocysts and cysts (US EPA, 2012).

13.2.1.4.4 Membrane Filtration

Membranes may be applied as a selective barrier to the transport of matter from one side to another. Membrane processes are typically pressure driven, and the filter effluent quality that may be achieved is higher than the quality that may be achieved in the surface of depth filters. Equipment and energy costs, however, tend to be significantly higher than with surface or depth filters. Membrane filtration pore sizes range from 0.05 μm (microfiltration) to <0.002 μm (nanofiltration and RO).

Microfiltration is typically applied for the removal of TSS, turbidity, some protozoan oocysts and cysts, and some bacteria and viruses. Ultrafiltration (pore sizes of 0.002–0.050 μm) may remove macromolecules, colloids, most bacteria, and some viruses and proteins. Nanofiltration may remove molecules, some hardness, and viruses. RO will remove very small molecules, color, hardness, sulfates, nitrates, sodium, and other ions. Depending on the urban reuse application, removal of hardness (to reduce potential for scaling) or salts (to reduce impact to plant life) may necessitate the application of one or more membrane filtration steps. For example, ultrafiltration may be applied upstream of RO or nanofiltration.

13.2.1.4.5 Granular Activated Carbon

In addition to size exclusion filtration, GAC also has adsorption properties, which may achieve high levels of particulate removal. Also, like granular media filters, GAC may also support biological activity, which can enhance removal of biodegradable constituents in wastewater. Due to the absorptive properties, GAC must be regenerated periodically.

13.2.1.4.6 Disinfection

Due to the potential for human contact in urban wastewater reuse, disinfection for inactivation of bacteria, protozoan oocysts and cysts, helminthes (worms), and viruses is often required following traditional primary and secondary wastewater treatment. However, disinfection processes can make up about one-third of the total energy used at a wastewater treatment plant, excluding the energy required for pumping (WEF, 2009). Therefore, understanding the level of treatment required for urban reuse applications may prevent “overdesign” and the addition of unnecessary cost.

The most common process for disinfection is chlorination (using free chlorine or chloramines). Contact time and mixing are often the governing design parameters; however, additional considerations include the water temperature, pH, presence of interfering substances, characteristics and concentration of pathogens, and form of chlorinating species. Monitoring requirements usually include monitoring of total chlorine remaining in the treated water after a designated contact time (residual chlorine). The efficacy of disinfection with chlorine is typically measured as CrT, which is the product of the concentration of total chlorine residual and the contact time with the water being disinfected. It is typically expressed in units of mg-min l^{-1}.

UV disinfection is also widely applied as an alternative to chlorine. Advances in energy efficiency and reduction in required capital investment have made UV increasingly attractive as an alternative to chlorination. However, UV disinfection performance may be affected by the presence of particle-associated microorganisms and UV transmittance of the wastewater. Particle sizes of larger than 10 μm can shield microorganisms from disinfection by UV light. Therefore, upstream filtration of particulate matter is recommended.

Additional disinfection technologies available for implementation include ozone, ferrate, and pasteurization. Ozone disinfection is attractive in its ability to both disinfect and degrade residual complex organics that may originate from pharmaceutically active compounds and other emerging contaminants of concern. Ozone disinfection may not be useful in many urban reuse applications as emerging contaminants of concern are not currently a high priority for reduction or removal.

13.2.1.4.7 Advanced Oxidation

Advanced oxidation processes (AOPs) are water treatment processes that may be placed on the end of a treatment train and have a broad range of applications including reduction of emerging contaminants of concerns and toxicity of industrial effluent. Examples of AOP processes include UV/hydrogen peroxide, ozone/hydrogen peroxide, ozone/UV, UV/titanium dioxide, and a variety of Fenton reactions (including iron [Fe]/hydrogen peroxide, and Fe/ozone). There are likely few urban reuse applications that would require this level of treatment, but it may be considered where direct human contact is anticipated or the wastewater includes complex/toxic components that need to be removed to protect environmental health.

13.2.1.4.8 Natural Systems

Natural treatment systems may be employed for the reduction or removal of pathogens, organic carbon, and nutrients, and even some contaminants of concern. Natural systems may include wetlands and soil aquifer treatment (SAT) systems. Natural treatment of reclaimed water may be particularly attractive for urban reuse applications, as the public perception is often positive and received as in-line with objectives of environmental stewardship and conservation. Incorporation of natural systems may reduce the energy footprint of the overall reuse treatment system and may provide secondary environmental benefits such as enhanced stream flows and wildlife habitat. Additional description of natural treatment systems is provided in the groundwater recharge section.

With appropriate design, construction, and maintenance, wetlands have been found to have the ability to provide consistent water quality improvement. Wetlands may be natural or constructed and support vegetation and microbes that assimilate constituents of concern. Constructed wetlands are effective in treatment of BOD, TSS, nutrients, metals, sulfates, and pathogens.

If optimum conditions are maintained, assimilation of nitrogen and BOD will occur indefinitely by being consumed by microorganisms in the wetlands. Higher loadings of BOD or nitrogen will stimulate additional microorganisms for treatment. Phosphorus assimilation, however, is finite, as phosphorus is adsorbed by the soil. Therefore, phosphorus removal from the water is related to the adsorption capacity of soil, and long-term storage within the system is experienced. Most metals and sulfates behave similarly to phosphorus, so the adsorption capacity of the system for these constituents should be measured.

SAT systems involve applying reclaimed wastewater in spreading basins where the reclaimed water percolates into the soil. The soil profile typically includes loam, sand, gravel, silt, and clay layers. Each layer allows the waster to undergo increasing physical, biological, and chemical treatment as the water flows downward from layer to layer.

13.2.1.5 Governing Design Considerations for Urban Reuse

13.2.1.5.1 Development of a Temporal Water Balance

One key to implementing urban reuse is to determine how much wastewater will be generated and how much urban reuse demand there will be for the reclaimed water. This may change seasonally. For example, during the winter, there may be a decrease in demand for residential and commercial irrigation water relative to the amount of wastewater that is produced. Therefore, a National Pollutant Discharge Elimination System (NPDES) discharge permit, or additional storage, may be necessary to achieve consistent delivery of reusable water for urban applications. If insufficient reuse demand exists, the reclaimed water may be land applied in some states at a dedicated site. The land application must be applied at application rates based on soil characteristics, which is defined as a reclaimed water land application system (RLAS). This is considered disposal rather than reuse, unless the land application system is being used for agricultural purposes (US EPA, 2012).

13.2.1.5.2 Economic and Technical Viability

Urban water reuse is not suitable for all wastewater treatment facilities. Reuse applications require additional manpower, increased operator sophistication, an increase in permit reporting frequency and complexity, and appropriate proximity to the points of reuse to make reuse feasible.

13.2.1.5.3 Constituent Application

The constituents of concern were described in detail in Section 13.2.1.1. Residual constituents present in the reclaimed water may be of negligible concern for certain urban reuse applications but of considerable concern for others. For example, salinity present in reclaimed water may have a detrimental effect on cover crops that will be irrigated on residential or commercial properties. For specialized users such as nurseries, a detailed evaluation of the reclaimed water constituents should be undertaken as part of the process design.

13.2.1.5.4 Reliability

Reliability is the ability of a unit process or system to perform its designated function without failure in the event of an anomaly event, such as a power outage, flow change, or physical impediment. Since human contact is anticipated in urban reuse applications, reliability is critical in maintaining continuous operation and can be achieved through sophisticated application of process design, including unit process redundancy.

The ability to isolate and bypass all unit processes without losing treatment integrity must be included in the design. Further, the system must be capable of handling significant surges or peaks in flow without allowing dangerous constituents to pass through untreated to the point of urban reuse. This may be achieved through flow equalization and oversized and/or redundant parallel treatment trains.

Spare parts must be maintained so that repairs may be achieved quickly and the system may be returned to operation before a reduction in treatment efficiency is realized. A system for maintaining the appropriate spare parts inventory should be explicitly communicated and implemented.

Another option may be to identify an appropriate land application site for events in which treatment efficiency has been compromised temporarily and disposal is necessary while treatment effectiveness is restored to urban reuse standards.

13.2.1.5.5 Opportunity for Decentralized Treatment and Application

As urban populations continue to increase, POTWs are experiencing increasing hydraulic and constituent loadings. Substantial upgrades are often required to cope with the increased loadings, increasing utility rates, and power consumption associated with large-scale wastewater treatment. As described in Section 13.2.1.2, examples of decentralized and colocalized collection, treatment, and reuse are becoming increasingly common. A designated volume of domestic wastewater may be extracted from the common collection system, treated (including disinfection and perhaps advanced treatment) in on-site wastewater treatment systems, and then is applied for the irrigation of golf courses or residential developments.

13.2.2 Agricultural Reuse of Reclaimed Wastewater

Agriculture still represents the largest reuse volume, and continued growth is driven particularly by agricultural development in developing countries. As agriculture requires freshwater for irrigation, the availability of freshwater for agricultural application is critical for global food security and economic growth. As of 2012, nearly 60% of global freshwater is applied for irrigation uses. In the United States, approximately 37% of all freshwater is applied for agricultural irrigation. The U.S. Department of Agriculture (USDA) and the National Institute of Food and Agriculture have made funding for agricultural water reuse projects one of their key priorities (US EPA, 2012).

Advantages of applying reused wastewater for agricultural purposes include a reduction in cost compared with drawing potable water from unconventional water sources (such as desalination), the supply of reclaimed water is usually reliable and grows with population increases, and municipalities often prefer to allocate reclaimed water for irrigation from a management perspective. The nutrient content in reused wastewater may also provide a low-cost supplementation of fertilizer components.

13.2.2.1 Constituents of Concern in Agricultural Reuse

Several factors, including soil–plant–water interactions, are important in successful crop production. Plant sensitivity is generally determined by the plant's tolerance to constituents encountered in the root zone or deposited on the foliage. Reclaimed water may have higher concentrations of several constituents of concern than groundwater or surface water.

US EPA guidelines for the interpretation of water quality for irrigation purposes are given in Table 13.6.

Table 13.6 Guidelines for the interpretation of water quality for irrigation (US EPA, 2012).

Potential irrigation problem		Units	None	Slight to moderate	Severe
Salinity (affects crop water availability)[a]					
	EC_w	$dS\,m^{-1}$	<0.7	0.7–3.0	>3.0
	TDS	$mg\,l^{-1}$	<450	450–2000	>2000
Infiltration (affects infiltration rate of water into the soil; evaluate using EC_w and SAR together)[b]					
SAR	0–3	and EC_w=	>0.7	0.7–0.2	<0.2
	3–6		>1.2	1.2–0.3	<0.3
	6–12		>1.9	1.9–0.5	<0.5
	12–20		>2.9	2.9–1.3	<1.3
	20–40		>5.0	5.0–2.9	<2.9
Specific ion toxicity (affects sensitive crops)					
	Sodium (Na)[c]				
	Surface irrigation	SAR	<3	3–9	>9
	Sprinkler irrigation	$meq\,l^{-1}$	<3	>3	
	Chloride (Cl)[c]				
	Surface irrigation	$meq\,l^{-1}$	<4	4–10	>10
	Sprinkler irrigation	$meq\,l^{-1}$	<3	>3	
	Boron(B)	$mg\,l^{-1}$	<0.7	0.7–3.0	>3.0
Miscellaneous effects (affects susceptible crops)					
	Nitrate (NO_3-N)	$mg\,l^{-1}$	<5	5–30	>30
	Bicarbonate (HCO_3)	$meq\,l^{-1}$	<1.5	1.5–8.5	>8.5
	pH		**Normal range 6.5–8.4**		

Source: Adapted from FAO (1985).

[a] EC_w means electrical conductivity, a measure of the water salinity, reported in deciSiemens per meter at 25 °C ($dS\,m^{-1}$) or in millimhos per centimeter ($mmho\,cm^{-1}$); both are equivalent.

[b] SAR is the sodium adsorption ratio; at a given SAR, infiltration rate increases as water salinity increases.

[c] For surface irrigation, most tree crops and woody plants are sensitive to sodium and chloride; most annual crops are not sensitive. With overhead sprinkler irrigation and low humidity (<30%), sodium and chloride may be absorbed through the leaves of sensitive crops.

13.2.2.1.1 Salinity

Salinity is one of the key constituents of concern for agricultural reuse applications. It is important to keep in mind that irrigation management practices may play a significant role in determining the load of salt that an agricultural reuse site can tolerate before detrimental effects are experienced. For example, low salinity (and high quality) reuse water may be applied to agricultural land with poor drainage conditions, leading to a buildup of salinity and crop failure. In contrast, well-drained soils with proper leaching fractions may tolerate higher levels of salinity in the reclaimed irrigation water. Salts (both cations and anions) will build up in the soil without proper drainage.

Using TDS as a proxy measure for salinity, no detrimental effects on crop yield are generally noticed below reuse water concentrations of $500\,mg\,l^{-1}$. Between 500 and $1000\,mg\,l^{-1}$, sensitive plants may be affected. Above $1000\,mg\,l^{-1}$, many crop types may experience detrimental effects, such as leaf burning. At TDs levels greater than $2000\,mg\,l^{-1}$, reuse water may be used regularly for only designated salt-tolerant plants (US EPA, 2012).

Chlorine residuals (a component of salinity) may be a by-product of disinfection. Free chlorine at concentrations less than $1\,mg\,l^{-1}$ usually poses no problem to plants, and up to $5\,mg\,l^{-1}$ may be tolerable. Above chlorine residual levels of $5\,mg\,l^{-1}$, severe damage may be caused to most plants.

13.2.2.1.2 Macronutrients and Micronutrients

Fertilizers are applied to soils that contain inadequate levels of the mineral nutrients required for plant growth. These mineral nutrients may be classified as macronutrients and micronutrients.

Macronutrients Macronutrients may further be classified into primary macronutrients (nitrogen, phosphorus, and potassium, which are all required in large amounts for plant survival) and secondary macronutrients (calcium, magnesium, and sulfur).

Nitrogen is important for fostering rapid plant growth, increasing fruit production, and improving crop quality. It is especially useful for replacing the use of commercial fertilizer during early to midseason crop-growing periods. However, nitrogen application during late-season periods may be detrimental to many crops, causing excessive vegetative growth rather than the desirable fruit or vegetable plant portions, delayed or uneven maturity, or reduced crop quality. If low-nitrogen water is available, water supplies can be switched or blended with other water supplies during late-season irrigation.

Phosphorus is important for facilitating blooming and root proliferation. Potassium is absorbed by plants in larger quantities than most other mineral elements (besides nitrogen and sometimes calcium) and enhances fruit quality and disease prevention. Therefore, elevated levels of these in reuse water may be beneficial for agricultural application.

Micronutrients Micronutrients include boron, copper, iron, manganese, zinc, chloride, and molybdenum and are sometimes referred to as trace elements. They are necessary for plant growth in small amounts. Excessive concentrations of trace elements are toxic for plant growth. Accumulation of micronutrients in the soil is the primary concern for agricultural reuse. Most of these elements are readily absorbed by soil and therefore will accumulate with time such that repeated application in excess of suggested levels is likely to induce phytotoxicity. Further, acidity (pH) can affect the influence of metal toxicity on plants and therefore the acceptable loading levels. A pH range of 6–8 is recommended for reclaimed water use for irrigation. The maximum application concentrations of these constituents and additional metals are provided in Table 13.7.

Irrigation of food crops with reclaimed water is common in the United States and globally. Therefore, in order to avoid risks to human health, pathogens are also a significant constituent of concern. Pathogen reduction is not of a significant concern for irrigation of nonfood crops (seed crops, processed food crops, etc.).

13.2.2.1.3 Emerging Constituents of Concern

Recent research has indicated that there may be indirect health effects resulting from the use of reclaimed wastewater in agricultural applications (Knapp et al., 2010), which showed evidence of growing resistance of soil bacteria to antibiotics. This resistance may lead to the contamination of human and animal feedstocks with pathogens that are more difficult to remove using conventional vector control methods.

13.2.2.2 Treatment Objectives: Agricultural Reuse

As stated before, treatment objectives for water intended for reuse differ depending on the intended application. Reuse water application on food crops that will be directly ingested by humans is regulated to a higher level of treatment than reclaimed water for application on nonfood crops.

13.2.2.2.1 Food Crops

Guidelines for agricultural reclaimed water use were developed by the World Health Organization (WHO) in World Health Organization (2006). These guidelines specify treatment processes, water quality standards, and monitoring processes that should be undertaken to minimize risks for the use of reclaimed water for irrigation of crops that are ingested by humans. In the United States, the use of reclaimed water for irrigation of food crops is prohibited in some states, while others

Table 13.7 Recommended water quality criteria for irrigation (US EPA, 2012).

Constituent	Maximum concentrations for irrigation ($mg\,l^{-1}$)	Remarks
Aluminum	5.0	Can cause nonproductiveness in acid soils, but soils at pH 5.5–8.0 will precipitate the ion and eliminate toxicity
Arsenic	0.10	Toxicity to plants varies widely, ranging from 12 $mg\,l^{-1}$ for Sudan grass to <0.05 $mg\,l^{-1}$ for rice
Beryllium	0.10	Toxicity to plants varies widely, ranging from 5 $mg\,l^{-1}$ for kale to 0.5 $mg\,l^{-1}$ for bush beans
Boron	0.75	Essential to plant growth; sufficient quantities in reclaimed water to correct soil deficiencies. Optimum yields obtained at few-tenths $mg\,l^{-1}$; toxic to sensitive plants (e.g. citrus) at 1 $mg\,l^{-1}$. Most grasses are tolerant at 2.0–10 $mg\,l^{-1}$
Cadmium	0.01	Toxic to beans, beets, and turnips at concentrations as low as 0.1 $mg\,l^{-1}$; conservative limits are recommended
Chromium	0.1	Not generally recognized as an essential element; due to lack of toxicity data, conservative limits are recommended
Cobalt	0.05	Toxic to tomatoes at 0.1 $mg\,l^{-1}$; tends to be inactivated by neutral and alkaline soils
Copper	0.2	Toxic to a number of plants at 0.1–1.0 $mg\,l^{-1}$
Fluoride	1.0	Inactivated by neutral and alkaline soils
Iron	5.0	Not toxic in aerated soils, but can contribute to soil acidification and loss of phosphorus and molybdenum
Lead	5.0	Can inhibit plant cell growth at very high concentrations
Lithium	2.5	Tolerated by most crops up to 5 $mg\,l^{-1}$; mobile in soil. Toxic to citrus at low doses – recommended limit is 0.075 $mg\,l^{-1}$
Manganese	0.2	Toxic to a number of crops at few-tenths to few $mg\,l^{-1}$ in acidic soils
Molybdenum	0.01	Nontoxic to plants; can be toxic to livestock if forage is grown in soils with high molybdenum
Nickel	0.2	Toxic to a number of plants at 0.5–1.0 $mg\,l^{-1}$; reduced toxicity at neutral or alkaline pH
Selenium	0.02	Toxic to plants at low concentrations and to livestock if forage is grown in soils with low levels of selenium
Tin, tungsten, and titanium	—	Excluded by plants; specific tolerance levels unknown
Vanadium	0.1	Toxic to many plants at relatively low concentrations
Zinc	2.0	Toxic to many plants at widely varying concentrations; reduced toxicity at increased pH (6 or above) and in fine-textured or organic soils

limit this application to crops that will be peeled or thermally processed prior to being consumed. California's regulatory code requires treatment to near-potable quality for agricultural reuse (United States Environmental Protection Agency, 2012).

Depending on the type of crop or type of irrigation system, treatment requirements range from secondary treatment and disinfection to oxidation, coagulation, filtration, and advanced oxidation/high-level disinfection.

The level of treatment required for reuse on food crops in 10 states is given in Tables 13.8 and 13.9.

13.2.2.2.2 Nonfood Crops

In the United States, many states have adopted regulations for the use of reclaimed water for nonfood and processed food crop production that are generally less stringent than for food crop applications. In some cases, direct application of disinfected secondary effluent may be allowed.

Some of these states have implemented regulatory levels for pathogens based on the risk of exposure of farm workers to the recycled water. For example, California (under Title 22 (California Department of Health Services, 2014)) requires total coliform bacteria to be less than 23 MPN/100 ml for the irrigation of nonfood crops. This level is generally achieved with secondary treatment processes and disinfection.

The level of treatment required for reuse on nonfood crops in 10 states is shown in Tables 13.10 and 13.11.

13.2.2.3 Treatment Technologies and Agricultural Reuse

Treatment technologies applied for agricultural reuse range from simple post-secondary treatment, such as

Table 13.8 Selected state standards for agricultural reuse – food crops – Arizona, California, Florida, Hawaii, Nevada, and New Jersey (US EPA, 2012).

	Arizona class A	California disinfected tertiary	Florida[a]	Hawaii R1 water	Nevada	New Jersey type III RWBR
Unit processes	Secondary treatment, filtration, disinfection	Oxidized, coagulated, filtered, disinfected	Secondary treatment, filtration, high-level disinfection	Oxidized, filtered, disinfected	*NP*	Filtration, high-level disinfection
UV dose, if UV disinfection used	*NS*	NWRI UV Guidelines	NWRI UV Guidelines enforced, variance allowed	NWRI UV Guidelines	*NP*	100 mJ cm^{-2} at max day flow
Chlorine disinfection requirements, if used	*NS*	CrT > 450 mg min l^{-1}; 90 min modal contact time at peak dry weather flow	TRC > 1 mg l^{-1}; 15 min contact time at peak hour flow[b]	Min residual >5 mg l^{-1}, actual modal contact time of 90 min	*NP*	Min residual >1 mg l^{-1}; 15 min contact at peak hour flow
BOD_5 (or $CBOD_5$)	*NS*	*NS*	CBOD5: • 20 mg l^{-1} (ann avg) • 30 m l^{-1} (mo avg) • 45 mg l^{-1} (wk avg) • 60 mg l^{-1} (max)	30 or 60 mg l^{-1} depending on design flow	*NP*	*NS*
TSS	*NS*	*NS*	5 mg l^{-1} (max)	30 or 60 mg l^{-1} depending on design flow	*NP*	5 mg l^{-1}
Turbidity	• 2 NTU (24-h avg) • 5 NTU (max)	• 2 NTU (avg) for media filters • 10 NTU (max) for media filters • 0.2 NTU (avg) for membrane filters • 0.5 NTU (max) for membrane filters	Case by case (generally 2–2.5 NTU) Florida requires continuous online monitoring of turbidity as indicator for TSS	• 2 NTU (95 percentile) • 0.5 NTU (max)	*NP*	2 NTU (max) for UV
Bacterial indicators	Fecal coliform: • None detectable in the last 4 of 7 samples • 23/100 ml (max)	Total coliform: • 2.2/100 ml (7-d med) • 23/100 ml (not more than one sample exceeds this value in 30 d) • 240/100 ml (max)	Fecal coliform: • 75% of samples below detection • 25/100 ml (max)	Fecal coliform: • 2.2/100 ml (7-d med) • 23/100 ml (not more than one sample exceeds this value in 30 d) • 200/100 ml (max)	*NP*	Fecal coliform: • 2.2/100 ml (wk med) • 14/100 ml (max)
Viral indicators	*NS*	*NS*	*NS*	*TR*	*NP*	*NS*
Pathogens	*NS*	*NS*	*Giardia*, *Cryptosporidium* sampling once per 2-yr period for plants ≥1 mgd; once per 5-yr period for plants ≤1 mgd	—	*NP*	*NS*
Other	If nitrogen >10 mg l^{-1}, special requirements may be mandated to protect groundwater	—	—	Oxidized, filtered, disinfected	—	(NH_3-N + NO_3-N): <10 mg l^{-1} (max) Special information, crop tests may be required

NP, not permitted by the state; *NS*, not specified by the state's reuse regulation; *TR*, monitoring is not required but virus removal rates are prescribed by treatment requirement.

[a] In Texas and Florida, spray irrigation (i.e. direct contact) is not permitted on foods that may be consumed raw (except Florida makes an exception for citrus and tobacco), and only irrigation types that avoid reclaimed water contact with edible portions of food crops (such as drip irrigation) are acceptable.

[b] In Florida when chlorine disinfection is used, the product of the total chlorine residual and contact time (CrT) at peak hour flow is specified for three levels of fecal coliform as measured prior to disinfection. If the concentration of fecal coliform prior to disinfection is ≤1000 cfu per 100 ml, the CrT shall be 25 mg min l^{-1}; is 1 000–10 000 cfu per 100 ml, the CrT shall be 40 mg min l^{-1}; and is ≥10 000 cfu per 100 ml, the CrT shall be 120 mg min l^{-1}.

Table 13.9 Selected state standards for agricultural reuse – food crops – North Carolina, Texas, Virginia, and Washington (US EPA, 2012).

	North Carolina				
	Processed type 1	**NOT processed type 2**	**Texas[a] type I reclaimed water**	**Virginia[b] level 1**	**Washington class A**
Unit processes	Filtration (or equivalent)	Filtration, dual UV/chlorination (or equivalent)	*NS*	Secondary treatment, filtration, high-level disinfection	Oxidized, coagulated, filtered, disinfected
UV dose, if UV disinfection used	*NS*	Dual UV/chlorination (or equivalent)	*NS*	*NS*	NWRI UV Guidelines
Chlorine disinfection requirements, if used	*NS*	Dual UV/chlorination (or equivalent)	*NS*	TRC CAT >1 mg l^{-1} 0 min contact time at avg. flow or 20 min at peak flow	Chlorine residual >1; 30 min contact time
BOD_5 (or $CBOD_5$)	• 10 mg l^{-1} (mo avg) • 15 mg l^{-1} (daily max)	• 5 mg l^{-1} (mo avg) • 10 mg l^{-1} (daily max)	5 mg l^{-1}	• 10 mg l^{-1} (mo avg) or $CBOD_5$ • 8 mg l^{-1} (mo avg)	30 mg l^{-1}
TSS	• 5 mg l^{-1} (mo avg) • 10 mg l^{-1} (daily max)	• 5 mg l^{-1} (mo avg) • 10 mg l^{-1} (daily max)	*NS*	*NS*	30 mg l^{-1}
Turbidity	10 NTU (max)	5 NTU (max)	3 NTU	2 NTU (daily avg) CAT >5 NTU	• 2 NTU (avg) • 5 NTU (max)
Bacterial indicators	Fecal coliform or *E. coli*: • 14/100 ml (mo mean) • 25/100 ml (daily max)	Fecal coliform or *E. coli*: • 3/100 ml (mo mean) • 25/100 ml (mo mean)	Fecal coliform or *E. coli:* • 20/100 ml (30-d geom) • 75/100 ml (max) *Enterococci:* • 4/100 ml (30-d geom) • 9/100 ml (max)	Fecal coliform: • 14/100 ml (mo geom), CAT >49/100 ml *E. coli*: • 11/100 ml (mo geom), CAT > 35/100 ml *Enterococci:* • 11/100 ml (mo geom), CAT >24/100 ml	Total coliform: • 2.2/100 ml (7-d med) • 23/100 ml (max)
Viral indicators	*NS*	Coliphage: • 5/100 ml (mo mean) • 25/100 ml (daily max)	*NS*	*NS*	*NS*
Pathogens	*NS*	Clostridium: • 5/100 ml (mo mean) • 25/100 ml (daily max)	*NS*	*NS*	*NS*
Other	Ammonia as NH_3-N: • 4 mg l^{-1} (mo avg) • 6 mg l^{-1} (daily max)	Ammonia as NH_3-N: • 1 mg l^{-1} (mo avg) • 2 mg l^{-1} (daily max)	—	—	Specific reliability and redundancy requirements based on formal assessment

NP, not permitted by the state; *NS*, not specified by the state's reuse regulation; *TR*, monitoring is not required but virus removal rates are prescribed by treatment requirement.
[a] In Texas and Florida, spray irrigation (i.e. direct contact) is not permitted on foods that may be consumed raw (except Florida makes an exception for citrus and tobacco), and only irrigation types that avoid reclaimed water contact with edible portions of food crops (such as drip irrigation) are acceptable.
[b] The requirements presented for Virginia are for food crops eaten raw.

Table 13.10 Selected state standards for agricultural reuse – nonfood crops and processed food crops (where permitted) – Arizona, California, Florida, Hawaii, and Nevada (US EPA, 2012).

	Arizona		California			Nevada[a]
	Class B	Class C	California disinfected tertiary	Florida	Hawaii R2 water	Nevada[a] category E
Unit processes	Secondary treatment, disinfection	Secondary treatment, with or without disinfection	Oxidized	Secondary treatment, basic disinfection	Secondary-23: oxidized, disinfected	Secondary treatment[b]
UV dose, if UV disinfection used	*NS*	*NS*	*NS*	*NS*	*NS*	*NS*
Chlorine disinfection requirements, if used	*NS*	*NS*	*NS*	TRC > 0.5 mg l^{-1}; 15 min contact time at peak hour flow[b]	Chlorine residual >5 mg l^{-1}; 10 min actual modal contact time	*NS*
BOD_5 (or $CBOD_5$)	*NS*	*NS*	*NS*	$CBOD_5$: • 20 mg l^{-1} (ann avg) • 30 mg l^{-1} (mo avg) • 45 mg l^{-1} (wk avg) • 60 mg l^{-1} (max)	30 or 60 mg l^{-1} depending on design flow	30 mg l^{-1} (30-d avg)
TSS	*NS*	*NS*	*NS*	• 20 mg l^{-1} (ann avg) • 30 mg l^{-1} (mo avg) • 45 mg l^{-1} (wk avg) • 60 mg l^{-1} (max)	30 or 60 mg l^{-1} depending on design flow	30 mg l^{-1} (30-d avg)
Turbidity	*NS*	*NS*	*NS*	*NS*	*NS*	*NS*
Bacterial indicators	Fecal coliform: • 200/100 ml in the last four of seven samples • 800/100 ml (max)	Fecal coliform: • 1000/100 ml in last four of seven samples • 4000/100 ml (max)	*NS*	Fecal coliform: • 200/100 ml (avg) • 800/100 ml (max)	Fecal coliform: • 23/100 ml (7-d med) • 200/100 ml (not more than one sample exceeds this value in 30 d)	*NS*
Other	If nitrogen >10 mg l^{-1}, special requirements may be mandated to protect groundwater	If nitrogen >10 mg l^{-1}, special requirements may be mandated to protect groundwater	—	—	—	—

NS, not specified by the state's reuse regulation.

[a] Nevada prohibits public access and requires a minimum buffer zone of 800 ft for spray irrigation of nonfood crops.

[b] In Florida when chlorine disinfection is used, the product of the total chlorine residual and contact time (CrT) at peak hour flow is specified for three levels of fecal coliform as measured prior to disinfection. If the concentration of fecal coliform prior to disinfection is ≤1000 cfu per 100 ml, the CrT shall be 25 mg min l^{-1}; is 1 000–10 000 cfu per 100 ml, the CrT shall be 40 mg min l^{-1}; and is ≥10 000 cfu per 100 ml, the CrT shall be 120 mg min l^{-1}.

Table 13.11 Selected state standards for agricultural reuse – nonfood crops and processed food crops (where permitted) – New Jersey, North Carolina, Texas, Virginia, and Washington (US EPA, 2012).

	New Jersey type II RWBR	North Carolina type 1	Texas type II	Virginia level 2	Washington class C
Unit processes	Case by case	Filtration (or equivalent)	*NS*	Secondary treatment, disinfection	Oxidized, disinfected
UV dose, if UV disinfection used	75 mJ cm^{-2} at max day flow	*NS*	*NS*	*NS*	NWRI UV Guidelines
Chlorine disinfection requirements, if used	Chlorine residual >1 mg l^{-1}; 15 min contact time at peak hour flow	*NS*	*NS*	TRC CAT <1 mg l^{-1}; 30 min contact time at avg. flow or 20 min at peak flow	Chlorine residual >1 mg l^{-1}; 30 min contact time
BOD_5 (or $CBOD_5$)	*NS*	• 10 mg l^{-1} (mo avg) • 15 mg l^{-1} (daily max)	Without pond: 20 mg l^{-1} (or $CBOD_5$ 15 mg l^{-1}) With pond: 30 mg l^{-1}	• 30 mg l^{-1} (mo avg) • 45 mg l^{-1} (max wk) or $CBOD_5$ • 25 mg l^{-1} (mo avg) • 40 mg l^{-1} (max wk)	30 mg l^{-1}
TSS	30 mg l^{-1}	• 5 mg l^{-1} (mo avg) • 10 mg l^{-1} (daily max)	*NS*	• 30 mg l^{-1} (mo avg) • 45 mg l^{-1} (max wk)	30 mg l^{-1}
Turbidity	*NS*	10 NTU (max)	*NS*	*NS*	*NS*
Bacterial indicators	Fecal coliform: • 200/100 ml (mo geom) • 400/100 ml (wk geom)	Fecal coliform or *E. coli*: • 14/100 ml (mo mean) • 25/100 ml (daily max)	Fecal coliform or *E. coli*: • 200/100 ml (30-d geom) • 800/100 ml (max) *Enterococci*: • 35/100 ml (30-d geom) • 89/100 ml (max)	Fecal coliform: • 200/100 ml (mo geom), CAT >800/100 ml *E. coli*: • 126/100 ml (mo geom), CAT >235/100 ml *Enterococci*: • 35/100 ml (mo geom), CAT >104/100 ml	Total coliform: • 23/100 ml (7-d med) • 240/100 ml (max)
Other	(NH_3-N + NO_3-N): <10 mg l^{-1} (max)	Ammonia as NH_3-N: • 4 mg l^{-1} (mo avg) • 6 mg l^{-1} (daily max)	—	—	—

NS, not specified by the state's reuse regulation.

disinfection and oxidation, to advanced treatment methods including coagulation, filtration, and advanced oxidation/high-level disinfection.

13.2.2.3.1 Disinfection

Chlorination is the most commonly applied disinfection method. However, higher chlorine residual and/or a longer contact time may be necessary to ensure that viruses are effectively inactivated or destroyed. However, it must be taken into account that many plant types are highly sensitive to chlorine residuals. Therefore, alternative methods for disinfection, such as UV radiation or ozone, may be considered.

Several states (California, Washington, Hawaii) have regulations on UV dosages required for disinfection application. In these cases, UV systems must be either prevalidated or undergo extensive on-site validation after construction that includes detailed third-party research of the reactors over a range of potential operating conditions.

Additional disinfection technologies are described in the urban reuse technology treatment section.

13.2.2.3.2 Advanced Treatment

While chemical addition prior to filtration may be necessary to meet water quality recommendations, analysis of plant tolerance/sensitivity to chemical constituents (e.g. coagulants) should also be taken into consideration when designing a reuse system for agriculture application.

For further description of advanced treatment technologies, refer to the urban reuse technology treatment section.

13.2.2.4 Governing Design Considerations for Agricultural Reuse

13.2.2.4.1 Variations in Seasonal Demand

The demand for irrigation water will vary throughout the year due to seasonal agricultural applications and precipitation levels. Storage, alternative treated wastewater discharge/disposal, or treatment plant adjustments may be employed to react to the variations in demand.

13.2.2.4.2 Compatibility and Reliability

While regulatory requirements may not require treatment beyond secondary processes and disinfection, compatibility issues may be encountered between the wastewater quality and irrigation application method. For example, reclaimed water treated to secondary standards may not be suitable for use in drip irrigation systems due to suspended solids that may increase clogging.

It may also be problematic to apply secondary quality reclaimed water where the method of delivery causes creation of aerosols to form where human contact may be anticipated.

13.2.2.4.3 Salinity

To assess the applicability of reclaimed water for irrigation based on salinity, the factors listed below should be taken into account when considering the application of degraded water with elevated salinity levels (US EPA, 2012). Recommendations are based on the Food and Agriculture Organization (FAO) report issued in 1985:

1) Irrigation method: Recommended irrigation methods include normal surface or sprinkler irrigation methods that supply water infrequently (as needed). At least 15% of the applied water should percolate below the root zone, and the crop should utilize at least 50% of the applied water prior to the next irrigation. This may be modified for drip irrigation type systems.
2) Soil conditions: Drainage and plant uptake are affected significantly by the soil profile. Soil texture may range from sandy loam to clay loam for good internal drainage. The climate should be semiarid to arid and rainfall should not be excessive.
3) Yield potential of the crops: Specific utilization of water constituents is associated with different crop types.

13.2.3 Reuse of Reclaimed Wastewater for Impoundments

Impoundments are typically earthen basins that collect water for a variety of uses. Examples of impoundments that may receive at least a portion of reuse water include aesthetic impoundments (such as golf course hazards) or recreational impoundments (such as water bodies for boating, fishing, and swimming).

Recreational impoundments may be further classified as contact or noncontact applications. In contact-type impoundments, human body contact is allowed, such as swimming. Noncontact-type impoundments allow human access wherein only incidental contact (rather than intentional contact) with human bodies is anticipated, such as experience with boating or fishing.

Artificial snowmaking may also be considered a type of impoundment. Snowmaking with reclaimed water is being done in the United States, Canada, and Australia. In addition to supplementing low snow levels to enhance and/or prolong ski seasons, using reuse water for snowmaking also provides an application for generated reuse water in winter months when demand for other reuse applications experiences a decrease.

13.2.3.1 Constituents of Concern for Reuse for Impoundments

13.2.3.1.1 Pathogens

For contact-type impoundments, pathogens (viruses, bacteria, and protozoa, described in additional detail in the section on urban reuse) are the primary constituents of concern due to the high risk of ingestion and absorption of the water on the skin. A relationship between gastrointestinal illness and estimates of fecal indicator organisms has been demonstrated, particularly in children less than eleven years old.

Even for noncontact-type impoundments where incidental contact with the reuse water is anticipated, pathogens are of concern. Particular concern must also be given to aerosols due to splashing or other agitation of the water.

13.2.3.1.2 Nutrients

Nutrients are of particular concern in snowmaking applications. Care must be taken to ensure that sensitive water bodies do not receive large surges in snowmelt from frozen reclaimed water that contains relatively high levels of phosphorus and nitrogen.

Further, excess nutrient loading on impoundment water bodies where fish or plant life is abundant may cause environmental harm due to the potential for algae blooms and subsequent eutrophication, possibly leading to depletion of oxygen.

13.2.3.1.3 Other Constituents of Concern

Both contact- and noncontact-type impoundment reuse applications where any level of human contact is expected should not contain chemical constituents that may pose a toxicological risk following inadvertent ingestion or that may be irritating to the skin or eyes. The pH and temperature should be maintained at levels that will not cause harm.

Maintenance of low turbidity is important in impoundments where visual assessments need to be made for safety reasons (for example, visual indication of physical hazards such as rocks in a swimming hole).

Also, where fishing is allowed on an impoundment that receives reused wastewater, it is important to maintain low levels of metals that accumulate in fish and plants to levels that present health risks to consumers of these foods.

13.2.3.2 Treatment Objectives: Impoundments

The level of treatment required for the application of reclaimed water in impoundments is dependent on several factors, including water demand and designated level of human contact allowed.

For contact-type impoundments, complete removal of pathogenic constituents is recommended and often required. Treatment requirements for contact-type impoundments often reflect requirements required for potable reuse.

The level of treatment required for contact impoundments in 10 states is indicated in Tables 13.12 and 13.13.

The level of treatment required for noncontact impoundments is often less stringent, although care should be taken to remove chemicals and pathogens that may cause irritation when contacted or ingested. Further, metals should be removed that may accumulate in fish and or plants that may be consumed as a human food source. In some states, such as Arizona, reclaimed water that is being used for recreational impoundments where boating or fishing is an intended use of the impoundment must meet class A requirements, which include secondary treatment, filtration, and disinfection so that no detectable fecal coliform organisms are present in four of the last seven daily reclaimed water samples taken and no single sample maximum concentration of fecal coliform indicator organisms exceeds 23/100 ml (US EPA, 2012).

The level of treatment required for noncontact impoundments in 10 states is shown in Tables 13.14 and 13.15

13.2.3.3 Treatment Technologies and Reuse for Impoundments

Even for noncontact-type impoundments where incidental contact with the reuse water is anticipated, such as boating, some states require that the reclaimed water be subjected to secondary treatment, filtration, and disinfection so that no detectable fecal coliform organisms are present. Membrane filtration for removal of metals, turbidity, and salinity may also be required, depending on the type of impoundment receiving reuse wastewater.

Regarding snowmaking applications for impoundment reuse, while various states do not yet have in place regulations for reuse water, some states require that reclaimed water must be filtered with site-specific nutrient removal depending on snowmelt and runoff to surface streams. Treatment beyond secondary quality is commonly achieved using a variety of biological nutrient removal technologies. The processed wastewater may be further filtered using membrane filtration (ultrafiltration) to achieve 4-log reduction of viral pathogens. Disinfection is then applied as the final treatment process.

More detailed description of these unit processes is provided in Section 13.2.1.1.

Table 13.12 Selected state standards for impoundments – restricted – Arizona, California, Florida, Hawaii, and Nevada (US EPA, 2012).

	Arizona class B	California disinfected secondary-2.2	Florida	Hawaii R-2 water	Nevada category A
Unit processes	Secondary treatment, disinfection	Oxidized, disinfected	*NR*	Oxidized, disinfected	Secondary treatment, disinfection
UV dose, if UV disinfection used	*NS*	*NS*	*NR*	*NS*	*NS*
Chlorine disinfection requirements, if used	*NS*	*NS*	*NR*	Chlorine residual >5 mg l^{-1}; actual modal contact time of 10 min	*NS*
BOD_5	*NS*	*NS*	*NR*	30 mg l^{-1} or 60 mg l^{-1} depending on design flow	30 mg l^{-1} (30-d avg)
TSS	*NS*	*NS*	*NR*	30 mg l^{-1} or 60 mg l^{-1} depending on design flow	30 mg l^{-1} (30-d avg)
Turbidity	*NS*	*NS*	*NR*	*NS*	*NS*
Bacterial indicators	Fecal coliform: • 200/100 ml in the last four of seven samples • 800/100 ml (max)	Total coliform: • 2.2/100 ml (7-d med) • 23/100 (not more than one sample exceeds this value in 30 d)	*NR*	Fecal coliform: • 23/100 ml (7-d med) • 200/100 ml (not more than one sample exceeds this value in 30 d)	Total coliform: • 2.2/100 ml (30-d geom) • 23/100 ml (max)
Other	If nitrogen >10 mg l^{-1}, special requirements may be mandated to protect groundwater	—	—	—	—

NR, not regulated by the state under the reuse program; *NS*, not specified by the state's reuse regulation; *TR*, monitoring is not required but virus removal rates are prescribed by treatment requirements.

13.2.3.4 Governing Design Considerations Reuse for Impoundments

Intended level of human contact is the governing design consideration for determining the appropriate treatment of reclaimed wastewater in impoundment applications.

Also, potential impact to environmental elements, such as wildlife and plants in the water, must be considered, particularly those that will be used for human consumption.

For aesthetic impoundments, such as golf course hazards, the cost of treatment must be considered with respect to desired levels of color or odor removal.

13.2.4 Environmental Reuse of Reclaimed Wastewater

Environmental reuse primarily includes the use of reclaimed water to support or restore wetlands and to provide supplemental stream and river flows. Aquifer recharge is also considered a type of environmental reuse.

Many natural wetlands have been drained or altered for purposes associated with agriculture, mining, forestry, and urbanization. Therefore, reclaimed water is advantageous for mitigating the effects of urbanization and alterations and for restoration or augmentation of wetlands.

In some arid regions, reclaimed water may serve as the primary source of water to maintain a base flow in the rivers. Reclaimed water is a reliable water source that can be supplied constantly for aquatic and riparian habitat enhancement. River or stream flow augmentation may provide an economical method of ensuring water quality, as well as protecting and enhancing aquatic and riparian habitats along the water body.

According to the US Army Corps of Engineers and the US EPA, wetlands are defined as areas that are saturated by surface water or groundwater at a frequency and duration sufficient to support a prevalence of wetland vegetation and a diverse ecological community.

Table 13.13 Selected state standards for impoundments – restricted – New Jersey, North Carolina, Texas, Virginia, and Washington (US EPA, 2012).

	New Jersey	North Carolina	Texas type II	Virginia level 2	Washington class B
Unit processes	*NR*	*NS*	*NS*	Secondary treatment, disinfection	Oxidized, disinfected
UV dose, if UV disinfection used	*NR*	*NS*	*NS*	*NS*	NWRI UV Guidelines
Chlorine disinfection requirements, if used	*NR*	*NS*	*NS*	TRC CAT <1 mg l^{-1} after minimum contact time of 30 min at avg. flow or 20 min at peak flow	Chlorine residual >1 mg l^{-1}; 30 min contact time
BOD_5	*NR*	*NS*	Without pond: 20 mg l^{-1} (or $CBOD_5$ 15 mg l^{-1}) With pond: 30 mg l^{-1}	30 mg l^{-1} (mo avg) 45 mg l^{-1} (max wk) Or $CBOD_5$: 25 mg l^{-1} (mo avg) 40 mg l^{-1} (max wk)	30 mg l^{-1}
TSS	*NR*	*NS*	*NS*	30 mg l^{-1} (mo avg) 45 mg l^{-1} (max wk)	30 mg l^{-1}
Turbidity	*NR*	*NS*	*NS*	*NS*	NS
Bacterial indicators	*NR*	*NS*	Fecal coliform or *E. coli:* • 200/100 ml (30-d geom) • 800/100 ml (max) *Enterococci*: • 35/100 ml (30-d geom) • 89/100 ml (max)	Fecal coliform: • 200/100 ml (mo geom), CAT >800/100 ml *E. coli*: • 126/100 ml (mo geom), CAT >235/100 ml *Enterococci*: • 35/100 ml (mo geom), CAT >104/100 ml	Total coliform: • 2.2/100 ml (7-d med) • 23/100 ml (max)
Other	*NR*	—	—	—	Specific reliability and redundancy requirements based on formal assessment

NR, not regulated by the state under the reuse program; *NS*, not specified by the state's reuse regulation; *TR*, monitoring is not required but virus removal rates are prescribed by treatment requirements.

13.2.4.1 Constituents of Concern for Environmental Reuse

13.2.4.1.1 Chlorine Residual

Following disinfection, the presence of elevated levels of chlorine residuals may pose toxicity risks to many fish species. Typically 0.1–1 mgl^{-1} may result in toxicity. Therefore, UV may be considered as an alternative option to chlorination for disinfection (US EPA, 2012).

13.2.4.1.2 Total Dissolved Solids

TDS may pose a toxicity risk to some aquatic species found in both rivers/streams and wetlands.

13.2.4.1.3 Organic Matter

Dissolved oxygen (DO) may be depleted during the process of degradation of organic matter by oxygen-consuming microorganisms present in the water. A BOD level of less than 20 mgl^{-1} is desirable to avoid potential issues with oxygen depletion.

13.2.4.1.4 Nutrients

In order to prevent eutrophication, nitrogen levels of less than 3 mgl^{-1} and phosphorus levels of less than 1 mgl^{-1} are desirable. An increase in flow rate may be applied to reduce the amount of stagnation in the water and reduce the potential for algae growth.

13.2.4.2 Treatment Objectives: Environmental Reuse

13.2.4.2.1 Wetlands

Some states, including Florida, South Dakota, and Washington, provide regulations specifically for the use of reclaimed water in wetland systems. Further, natural wetlands are considered waters of the United States and

Table 13.14 Selected state standards for impoundments – unrestricted – Arizona, California, Florida, Hawaii, and Nevada (US EPA, 2012).

	Arizona[a] class A	California disinfected tertiary	Florida	Hawaii	Nevada
Unit processes	Secondary treatment, disinfection	Oxidized, coagulated, filtered, disinfected[b]	*NR*	*NR*	*NP*
UV dose, if UV disinfection used	*NS*	NWRI UV Guidelines	*NR*	*NR*	*NP*
Chlorine disinfection requirements, if used	*NS*	CrT > 450 mg min l^{-1}; 90 min modal contact time at peak dry weather flow	*NR*	*NR*	*NP*
BOD_5	*NS*	*NS*	*NR*	*NR*	*NP*
TSS	*NS*	*NS*	*NR*	*NR*	*NP*
Turbidity	*NS*	• 2 NTU (avg) for media filters • 10 NTU (max) for media filters • 0.2 NTU (avg) for membrane filters • 0.5 NTU (max) for membrane filters	*NR*	*NR*	*NP*
Bacterial indicators	Fecal coliform: • None detectable in the last four of seven samples • 23/100 ml (max)	Total coliform: • 2.2/100 ml (7-d med) • 23/100 ml (not more than one sample exceeds this value in 30 d) • 240/100 ml (max)	*NR*	*NR*	*NP*
Other	If nitrogen >10 mg l^{-1}, special requirements may be mandated to protect groundwater	Supplemental pathogen monitoring	—	—	*NP*

NP, not permitted by the state; *NR*, not regulated by the state under the reuse program; *NS*, not specified by the state's reuse regulation.
[a] Arizona does not allow reuse for swimming or "other full-immersion water activity with a potential of ingestion." Arizona also allows "class A" and "A+" waters to be used for snowmaking, which is included in this definition.
[b] Disinfected tertiary recycled water that has not received conventional treatment shall be sampled/analyzed monthly for *Giardia*, enteric viruses, and *Cryptosporidium* during the first 12 months of operation and use. Following the first 12 months, samples will be collected quarterly, and ongoing monitoring may be discontinued after the first 2 years, with approval.

thus are protected under the US EPA's NPDES permit and water quality standard programs. The quality of reclaimed water entering natural wetlands is regulated by federal, state, and local agencies and must be treated to secondary treatment levels or greater. Constructed wetlands, however, are not considered waters of the United States and, if nondischarging, do not require an NPDES permit.

13.2.4.2.2 Stream Augmentation

Similar to impoundments, water quality for stream or river augmentation will be governed by the designated use of the waterway and to enhance an acceptable appearance. Unlike the regulations that some states have adopted for wetland reuse water, requirements for reclaimed water quality and monitoring for augmentation of rivers or streams are often covered under a discharge permit.

For both wetlands and stream augmentation, residual chlorine levels of less than 0.1 mg l^{-1} are desirable to avoid potential toxicity to fish species. DO levels of at least 5 mg l^{-1} are required for the sustainability of aquatic species. Reduced concentrations of DO may cause stress and death of certain sensitive species of fish. The target DO concentration should be based on the sensitivity of the most sensitive fish species present in the aquatic environment. Also, water temperatures should be kept near ambient temperatures to prevent stress to temperature-sensitive fish species.

Selected state standards for environmental reuse are given in Tables 13.16 and 13.17.

13.2.4.3 Treatment Technologies and Environmental Reuse

Unit processes to achieve the treatment objectives for environmental reuse are described in Section 13.2.1.1

With regard to disinfection, it is recommended that UV radiation be applied in place of chlorination in order to reduce the level of chlorine residual applied onto wetlands or streams.

Table 13.15 Selected state standards for impoundments – unrestricted – New Jersey, North Carolina, Texas, Virginia, and Washington (US EPA, 2012).

	New Jersey	North Carolina	Texas	Virginia level 1	Washington class A
Unit processes	*NR*	*NS*	*NS*	Secondary treatment, filtration, high-level disinfection	Oxidized, coagulated, filtered, and disinfected
UV dose, if UV disinfection used	*NR*	*NS*	*NS*	*NS*	NWRI UV Guidelines
Chlorine disinfection requirements, if used	*NR*	*NS*	*NS*	TRC CAT <1 mg l^{-1} after minimum contact time of 30 min at avg. flow or 20 min at peak flow	Chlorine residual >1 mg l^{-1}; 30 min contact time
BOD_5	*NR*	*NS*	5 mg l^{-1}	10 mg l^{-1} (mo avg) or $CBOD_5$: 8 mg l^{-1} (mo avg)	30 mg l^{-1}
TSS	*NR*	*NS*	*NS*	*NS*	30 mg l^{-1}
Turbidity	*NR*	*NS*	3 NTU	2 NTU (daily avg), CAT >5 NTU	• 2 NTU (avg) • 5 NTU (max)
Bacterial indicators	*NR*	*NS*	Fecal coliform or *E. coli*: • 20/100 ml (avg) • 75/100 ml (max) *Enterococci*: • 4/100 ml (avg) • 9/100 ml (max)	Fecal coliform: • 14/100 ml (mo geom), CAT >49/100 ml *E. coli*: • 11/100 ml (mo geom), CAT >35/100 ml *Enterococci*: • 11/100 ml (mo geom), CAT >24/100 ml	Total coliform: • 2.2/100 ml (7-d med) • 23/100 ml (max)
Other	*NR*	—	—	—	Specific reliability and redundancy requirements based on formal assessment

NP, not permitted by the state; *NR*, not regulated by the state under the reuse program; *NS*, not specified by the state's reuse regulation.

13.2.4.4 Governing Design Considerations for Environmental Reuse

13.2.4.4.1 Wetlands

When designing a wetland application for reclaimed wastewater, the hydrology of the wetlands must be considered. For example, the flow rate of reclaimed water may need to be adjusted to accommodate natural seasonal changes that affect the growth and life cycle of some species. Inlet and outlet structures should be designed so that water levels and flow rates may be adjusted as necessary. Bypass or transfer structures should also be included to redistribute water to different areas of the wetlands as needed. Water may also be lost due to evapotranspiration; therefore, the wetlands may experience an accumulation or concentration of some constituents near the outflow of the wetland.

Habitat and fishery protection are also important design considerations. As wetlands are vital to fish and animal health (and therefore the multibillion dollar fishing industry), reuse water must be treated to protect target species from harm, including toxic metal removal to prevent accumulation in fish (Metcalf and Eddy, 2007).

13.2.4.4.2 Stream Flow Augmentation

Understanding the baseline water quality and sensitivity of the receiving water body is critical for ensuring the ecotoxicological consequences are minimized when the stream or river is augmented with reclaimed water.

13.2.5 Industrial Reuse of Reclaimed Wastewater

Historically, the traditional industrial application of reclaimed water has been for cooling water makeup for industries such as pulp and paper facilities, textile production facilities, etc. In these applications, wastewater was typically treated and reused on-site.

Table 13.16 Selected state standards for environmental reuse – Arizona, California, Florida, Hawaii, and Nevada (US EPA, 2012).

	Arizona[a]	California	Florida[b]	Hawaii	Nevada category C
Unit processes	*NR*	*NR*	Secondary treatment, nitrification, basic disinfection	*NR*	Secondary treatment, disinfection
UV dose, if UV disinfection used	*NR*	*NR*	*NS*	*NR*	*NS*
Chlorine disinfection requirements, if used	*NR*	*NR*	TRC > 0.5 mg l^{-1}; 15 min contact time at peak hour flow[c]	*NR*	*NS*
BOD_5 (or $CBOD_5$)	*NR*	*NR*	$CBOD_5$: • 5 mg l^{-1} (ann avg) • 6.25 mg l^{-1} (mo avg) • 7.5 mg l^{-1} (wk avg) • 10 mg l^{-1} (max)	*NR*	30 mg l^{-1} (30-d avg)
TSS	*NR*	*NR*	• 5 mg l^{-1} (ann avg) • 6.25 mg l^{-1} (mo avg) • 7.5 mg l^{-1} (wk avg) • 10 mg l^{-1} (max)	*NR*	30 mg l^{-1} (30-d avg)
Bacterial indicators	*NR*	*NR*	Fecal coliform: • 200/100 ml (avg) • 800/100 ml (max)	*NR*	Fecal coliform: • 23/100 ml (30-d geom) • 240/100 ml (max)
Total ammonia	*NR*	*NR*	• 2 mg l^{-1} (ann avg) • 2 mg l^{-1} (mo avg) • 3 mg l^{-1} (wk avg) • 4 mg l^{-1} (max)	*NR*	*NS*
Nutrients	*NR*	*NR*	Phosphorus: • 1 mg l^{-1} (ann avg) • 1.25 mg l^{-1} (mo avg) • 1.5 mg l^{-1} (wk avg) • 2 mg l^{-1} (max) Nitrogen: • 3 mg l^{-1} (ann avg) • 3.75 mg l^{-1} (mo avg) • 4.5 mg l^{-1} (wk avg) • 6 mg l^{-1} (max)	*NR*	*NS*

NR = not regulated by the state under the reuse program; *NS* = not specified by the state's reuse regulation.

[a] Though Arizona reuse regulations do not specifically cover environmental reuse, treated wastewater effluent meeting Arizona's reclaimed water classes is discharged to waters of the United States and creates incidental environmental benefits. Arizona's NPDES surface water quality standards include a designation for this type of water, "effluent-dependent waters."

[b] Florida requirements are for a natural receiving wetland regulated under Florida Administrative Code Chapter 62-611 for wetland's application.

[c] In Florida when chlorine disinfection is used, the product of the total chlorine residual and contact time (CrT) at peak hour flow is specified for three levels of fecal coliform as measured prior to disinfection. If the concentration of fecal coliform prior to disinfection is ≤1000 cfu per 100 ml, the CrT shall be 25 mg min l^{-1}; is 1 000–10 000 cfu per 100 ml, the CrT shall be 40 mg min l^{-1}; and is ≥10 000 cfu per 100 ml, the CrT shall be 120 mg min l^{-1}.

Recently, the industrial use of reclaimed water has grown in a wider variety of industries and applications including food processing, power generation, manufacturing, and electronics. In addition to cooling water makeup, reclaimed water now applies for purposes ranging from boiler feedwater to process water and may also include toilet flushing and on-site irrigation. Also, municipal facilities have begun to produce reclaimed

Table 13.17 Selected state standards for environmental reuse – New Jersey, North Carolina, Virginia, and Washington (US EPA, 2012).

	New Jersey	North Carolina type 1	Texas	Virginia[a]	Washington class A
Unit processes	*NR*	Filtration (or equivalent)	*NR*	*NS*	Oxidized, coagulated, filtered, disinfected
UV dose, if UV disinfection used	*NR*	*NS*	*NR*	*NS*	NWRI UV Guidelines
Chlorine disinfection requirements, if used	*NR*	*NS*	*NR*	*NS*	Chlorine residual >1 $mg l^{-1}$; 30 min contact time
BOD_5 (or $CBOD_5$)	*NR*	• 10 $mg l^{-1}$ (mo avg) • 15 $mg l^{-1}$ (daily max)	*NR*	*NS*	20 $mg l^{-1}$
TSS	*NR*	• 5 $mg l^{-1}$ (mo avg) • 10 $mg l^{-1}$ (daily max)	*NR*	*NS*	20 $mg l^{-1}$
Bacterial indicators	*NR*	Fecal coliform or *E. coli*: • 14/100 ml (mo mean) • 25/100 ml (daily max)	*NR*	*NS*	Total coliform: • 2.2/100 ml (7-d med) • 23/100 ml (max)
Total ammonia	*NR*	Ammonia as NH_3-N: • 4 $mg l^{-1}$ (mo avg) • 6 $mg l^{-1}$ (daily max)	*NR*	*NS*	Not to exceed chronic standards for freshwater
Nutrients	*NR*	Phosphorus: 1 $mg l^{-1}$ (max)[b] Nitrogen: 4 $mg l^{-1}$ (max)[b]	*NR*	*NS*	Phosphorus: 1 $mg l^{-1}$ (ann avg)[c]

NR, not regulated by the state under the reuse program; *NS*, not specified by the state's reuse regulation.
[a] Wetlands in Virginia, whether natural or created as mitigation for impacts to existing wetlands, are considered state surface waters; release of reclaimed water into a wetland is regulated as a point source discharge and subject to applicable surface water quality standards of the state.
[b] These limits are not to be exceeded unless net environmental benefits are provided by exceeding these limits.
[c] The phosphorus limit is as an annual average for wetland augmentation/restoration, while for stream flow augmentation is the same as that required to NPDES discharge limits, or in other words variable.

water for industrial and power company users so that the cost of on-site treatment at the industrial facilities for reuse is reduced.

As reclaimed water is becoming increasingly attractive as an economical alternative to the acquisition of potable water for industrial processes, the variety of reclaimed industrial water applications is predicted to expand.

13.2.5.1 Constituents of Concern for Industrial Reuse

13.2.5.1.1 Cooling Water Makeup

Inorganic constituents in reclaimed water affect the operation of cooling water systems, as they may cause corrosion or scaling to occur, increase chemical costs, and decrease the number of cycles of concentration (which are determined by calculating the ratio of the concentration of dissolved solids in the blowdown water compared to the make-up water). When the cycles of concentration are in the range of 3–7, some dissolved solids in the circulating water (including calcium, phosphorus, and silica) may exceed their solubility limits and precipitate as calcium phosphate, silica, and calcium sulfate, causing scale formation in coolers and pipe systems (Tchobanoglous et al., 2003). Fluoride and magnesium may also come out of solution to cause scaling issues as calcium fluoride and magnesium silicate.

The presence of nutrients may encourage biological growth, which may produce undesirable biofilm deposits and, which may also interfere with heat transfer and cause microbiologically induced corrosion from acid that is produced. Further, biological films grow rapidly and may plug heat exchangers or plug cooling tower water distribution nozzles/sprays. Ammonia levels are also of concern, as elevated levels of ammonia reduce the effectiveness of copper and brass heat exchanger surfaces.

Water may exit cooling systems through wind action, causing aerosols to form containing resilient pathogenic constituents and potentially come into contact with humans. While reclaimed water has typically been subjected to disinfection prior to cooling tower reuse, one organism that has been specifically linked to cooling tower water is the heterotrophic bacterium *Legionella,* which causes Legionellosis, a severe and sometimes fatal form of pneumonia. Management to reduce the incidence of this bacterium includes an increase in the temperature to over 60 °C (Metcalf and Eddy, 2007).

13.2.5.1.2 Boiler Water Makeup

Chemicals that contribute to water hardness are of particular concern for boiler water makeup. Insoluble scales of calcium and magnesium, silica, and alumina in boiler feedwater will cause significant detrimental issues with boiler operation.

Alkalinity of the reclaimed water (as determined by bicarbonate, carbonate, and hydroxyl content) is also of concern, as alkalinity may contribute to foaming carryover. This may result in deposits in the superheater, reheater, and turbine units. Localized corrosion in steam-using equipment and condensate return systems may be caused by the breakdown of bicarbonate alkalinity under the influence of boiler heat to produce carbon dioxide.

Organics in reclaimed water may also cause foaming in boiler systems, leading to carryover of boiler water where only steam should be present and contributing to corrosion.

Additional constituents of concern and their harmful effects include the following:

- Hydrogen sulfide: Corrosive to most metals; may be removed via aeration and filtration.
- Dissolved oxygen (DO): Corrosion and pitting of boiler tubes; may be treated using deaeration and chemical treatment with sodium sulfite or hydrazine.
- Iron (Fe) and manganese (Mn): Deposits in boiler may inhibit heat transfer; may be removed in aeration, filtration, and ion exchange.
- Oil and grease: May cause foaming and carryover; may be removed via coagulation and filtration
- Sulfate: Forms hard scale if calcium is present; may be removed in deionization (Metcalf and Eddy, 2007).

13.2.5.1.3 Other Uses

Reclaimed water in high-technology manufacturing is becoming more common. For example, reclaimed water is used in rinse operations for circuit board manufacturing. Similar to treatment for boiler feedwater, circuit board rinsing water requires extensive treatment.

Because of public perception concerns, the food and beverage industry was initially reluctant to apply reclaimed water reuse. However, the growing "green" movement has made reuse more feasible from a perception standpoint. Like many other reuse applications, constituents of concern and required treatment levels are heavily dependent on the risk associated with human contact/ingestion of the water or products produced with the water and protection of the environment. Reuse water may be employed for washing, plant process water, flume water transport, etc.

Reuse water applied in pulp and paper mill and paper production processes is also common. Reuse water may be applied to aid in heating and cooling systems, as well as for direct production of the paper products. Constituents of concern, depending on the grade of paper products produced, may include the following: iron, manganese, microbial contamination, and suspended solids, which may all affect the brightness of the paper, while phosphates, surfactants, and metal ions may affect the efficiency of resins in the stock preparation process.

Opportunities for application of reclaimed water in textile industries also exist for several production processes including cotton fabrication and carpet dyeing. Turbidity, color, iron, and manganese all may cause staining of fabric during production and therefore should be managed. Also, hardness in the reclaimed water may cause precipitation of some dyes and damage certain material fibers (like silk).

13.2.5.2 Treatment Objectives: Industrial Reuse

The level of treatment required for industrial reuse in 10 states is indicated in Tables 13.18 and 13.19.

13.2.5.2.1 Cooling Water Makeup

Management of cooling water systems should be designed to control corrosion, scale, fouling, and microbiological growth.

13.2.5.2.2 Boiler Water Makeup

Required levels of treatment of reclaimed water for boiler feedwater are typically much higher than what is required for cooling water.

The primary concern for boiler water makeup is scale buildup and corrosion. Control or removal of hardness from reclaimed water is required for use as boiler makeup water. Further, control of insoluble scales of calcium and magnesium as well as control of silica and alumina is also required. For steam generation, TDS levels are recommended to be less than 0.2 ppm and less than 0.05 ppm for once-through steam generation (Metcalf and Eddy, 2007).

Specific treatment targets for boiler water limits may vary according to the design operating pressure of the drum.

13.2.5.3 Treatment Technologies and Industrial Reuse

13.2.5.3.1 Cooling Water Makeup

Constituents with the potential to form scale may be evaluated and controlled by chemical treatment and/or adjusting the cycles of concentration. Specifically, pretreatment of reclaimed water to lower concentrations of calcium and phosphate as well as the application of scale inhibition chemicals is recommended.

Table 13.18 Selected state standards for industrial reuse[a] – Arizona, California, Florida, Hawaii, and Nevada (US EPA, 2012).

	Arizona[b]	**California**[c] **disinfected tertiary**	**Florida**[c]	**Hawaii**[a] **R-2 water**	**Nevada category E**
Unit processes	Individual reclaimed water permit, case specific[b]	Oxidized, coagulated, filtered, disinfected	Secondary treatment, filtration, high-level disinfection	Oxidized, disinfected	Secondary treatment, disinfection
UV dose, if UV disinfection used	*NS*	NWRI UV Guidelines	NWRI UV Guidelines enforced, variance allowed	*NS*	*NS*
Chlorine disinfection requirements, if used	*NS*	CrT > 450 mg min l^{-1}; 90 min modal contact time at peak dry weather flow	TRC > 1 mg l^{-1}; 15 min contact time at peak hour flow[d]	Chlorine residual >5 mg l^{-1}, actual modal contact time of 10 min	*NS*
BOD_5 (or $CBOD_5$)	*NS*	*NS*	$CBOD_5$: • 20 mg l^{-1} (ann avg) • 30 mg l^{-1} (mo avg) • 45 mg l^{-1} (wk avg) • 60 mg l^{-1} (max)	30 or 60 mg l^{-1} depending on design flow	30 mg l^{-1} (30-d avg)
TSS	*NS*	*NS*	5 mg l^{-1} (max)	30 or 60 mg l^{-1} depending on design flow	30 mg l^{-1} (30-d avg)
Turbidity	*NS*	• 2 NTU (avg) for media filters • 10 NTU (max) for media filters • 0.2 NTU (avg) for membrane filters • 0.5 NTU (max) for membrane filters	Case by case (generally 2–2.5 NTU) Florida requires continuous online monitoring of turbidity as indicator for TSS	*NS*	NS
Bacterial indicators	*NS*	Total coliform: • 2.2/100 ml (7-d med) • 23/100 ml (not more than one sample exceeds this value in 30 d) • 240/100 ml (max)	Fecal coliform: • 75% of samples below detection • 25/100 ml (max)	Fecal coliform: • 23/100 ml (7-d med) • 200/100 ml (not more than one sample exceeds this value in 30 d)	Fecal coliform: • 2.2/100 ml (30-d geom) • 23/100 ml (max)
Pathogens	*NS*	*NS*	*Giardia, Cryptosporidium* sampling once each 2-yr period if high-level disinfection is required	*NS*	*TR*

NR, not regulated by the state under the reuse program; *NS*, not specified by the state's reuse regulation; *TR*, monitoring is not required but virus removal rates are prescribed by treatment requirements.
[a] All state requirements are for cooling water that creates a mist or with exposure to workers, except for Hawaii. Hawaii includes industrial processes that do not generate mist, do not involve facial contact with recycled water, and do not involve incorporation into food or drink for humans or contact with anything that will contact food or drink for humans.
[b] Arizona regulates industrial reuse through issuance of an individual reclaimed water permit, which provides case-specific reporting, monitoring, record keeping, and water quality requirements.
[c] For industrial uses in Florida, such as once-through cooling, open cooling towers with minimal aerosol drift and at least a 300-ft setback to the property line, wash water at wastewater treatment plants, or process water at industrial facilities that does not involve incorporation of reclaimed water into food or drink for humans or contact with anything that will contact food or drink for humans, which do not create a mist or have potential for worker exposure, less stringent requirements, such as basic disinfection (e.g. TRC > 0.5 mg l^{-1}, no continuous online monitoring of turbidity, fecal coliform <200/100 ml, etc.), secondary treatment standards (e.g. TSS < 20 mg l^{-1} annual average, etc.), no sampling for pathogens (except in the case of open cooling towers regardless of setbacks), may apply.
[d] In Florida when chlorine disinfection is used, the product of the total chlorine residual and contact time (CrT) at peak hour flow is specified for three levels of fecal coliform as measured prior to disinfection. If the concentration of fecal coliform prior to disinfection is ≤1000 cfu per 100 ml, the CrT shall be 25 mg min l^{-1}; is 1000–10 000 cfu per 100 ml, the CrT shall be 40 mg min l^{-1}; and is ≥10 000 cfu per 100 ml, the CrT shall be 120 mg min l^{-1}.

Table 13.19 Selected state standards for industrial reuse[a] – New Jersey, North Carolina, Texas, Virginia, and Washington (US EPA, 2012).

	New Jersey type IV RWBR	North Carolina type 1	Texas[b] type II	Virginia[c] level 2	Washington class A
Unit processes	Case by case	Filtration (or equivalent), unless there is no public access or employee exposure	*NS*	Secondary treatment, disinfection	Oxidized, coagulated, filtered, and disinfected
UV dose, if UV disinfection used	*NS*	*NS*	*NS*	*NS*	NWRI UV Guidelines
Chlorine disinfection requirements, if used	*NS*	*NS*	*NS*	TRC CAT <1 $mg\,l^{-1}$; 30 min contact time at avg. flow or 20 min at peak flow	Chlorine residual >1 $mg\,l^{-1}$; 30 min contact time
BOD_5 (or $CBOD_5$)	*NS*	• 10 $mg\,l^{-1}$ (mo avg) • 15 $mg\,l^{-1}$ (daily max)	Without pond: 20 $mg\,l^{-1}$ (or $CBOD_5$ 15 $mg\,l^{-1}$) With pond: 30 $mg\,l^{-1}$	• 30 $mg\,l^{-1}$ (mo avg) • 45 $mg\,l^{-1}$ (max wk) Or $CBOD_5$ • 25 $mg\,l^{-1}$ (mo avg) • 40 $mg\,l^{-1}$ (max wk)	30 $mg\,l^{-1}$
TSS	Case by case	• 5 $mg\,l^{-1}$ (mo avg) • 10 $mg\,l^{-1}$ (daily max)	*NS*	• 30 $mg\,l^{-1}$ (mo avg) • 45 $mg\,l^{-1}$ (max wk)	30 $mg\,l^{-1}$
Turbidity	*NS*	10 NTU (max)	*NS*	*NS*	• 2 NTU (avg) • 5 NTU (max)
Bacterial indicators	*NS*	Fecal coliform or *E. coli*: • 14/100 ml (mo mean) • 25/100 ml (daily max)	Fecal coliform or *E. coli*: • 200/100 ml (30-d geom) • 800/100 ml (max) *Enterococci*: • 35/100 ml (30-d geom) • 89/100 ml (max)	Fecal coliform: • 200/100 ml (mo goem), CAT >800/100 ml *E. coli*: • 126/100 ml (mo geom), CAT >235/100 ml *Enterococci*: • 35/100 ml (mo geom), CAT >104/100 ml	Total coliform: • 2.2/100 ml (7-d med) • 23/100 ml (max)
Pathogens	*NS*	*NS*	*NS*	*NS*	*NS*

NR, not regulated by the state under the reuse program; *NS*, not specified by the state's reuse regulation; *TR*, monitoring is not required but virus removal rates are prescribed by treatment requirements.
[a] All state requirements are for cooling water that creates a mist or with exposure to workers, except for Texas. Texas requirements are for cooling tower makeup water.
[b] For industrial uses that do not create a mist or have potential for worker exposure, less stringent requirements may apply.
[c] In Virginia, these are the minimum reclaimed water standards for most industrial reuses of reclaimed water; more stringent standards may apply as specified in the regulation. For industrial reuses not listed in the regulation, reclaimed water standards may be developed on a case-by-case basis relative to the proposed industrial reuse.

For management of *Legionella*, routine disinfection of cooling systems may be applied using thermal treatment of chemical disinfectants, including ozone (typical dose of 0.3 $mg l^{-1}$) and chlorination (may be applied continually or intermittently).

Antibiologics, such as sodium hypochlorite or bromine chloride, are commonly used to prevent or reduce the growth of biofilms within the cooling towers. Biofilm growth may also be prevented by the upstream removal of nutrients from reclaimed water.

13.2.5.3.2 Boiler Water Makeup

A typical flow scheme for the treatment of reclaimed water for boiler feedwater includes clarification, filtration, low pressure membrane filtration (ultrafiltration), RO (often including first- and second-pass RO), and demineralization. For boiler feedwater treatment, reclaimed water must be subject to demineralization to remove minerals that will deposit on the hot metal surfaces.

An example flow sheet (Figure 13.2) illustrates a "zero-liquid discharge" power-station reclaimed water treatment process developed by WesTech Engineering in Salt Lake City, Utah, for power generation reuse water, including cooling tower and boiler feed makeup water treatment and reuse. The blue lines represent the forward flow path of the liquid portion of the reclaimed water, while the brown lines represent the forward flow path of the solids constituents. As illustrated, pretreatment may be applied to the raw incoming reclaimed water for general water needs in the plant, and then a demineralization train may be applied to produce boiler feedwater. The process may also include treatment of a slip stream from the cooling tower for internal reuse.

13.2.5.4 Governing Design Considerations for Industrial Reuse

For evaluating the feasibility of applying reclaimed wastewater in industrial uses, the proximity of the site of reuse to the site of reclaimed water treatment must be considered. In some cases, the cost of constructing and maintaining reclaimed water collection and transport systems is more expensive than using other water sources. If a central treatment system is located too far away, considerations of on-site treatment systems may be evaluated.

13.2.6 Groundwater Recharge of Reclaimed Wastewater

Groundwater recharge refers to the application of reclaimed wastewater to recharge aquifers, usually for use as a nonpotable water source. One of the major

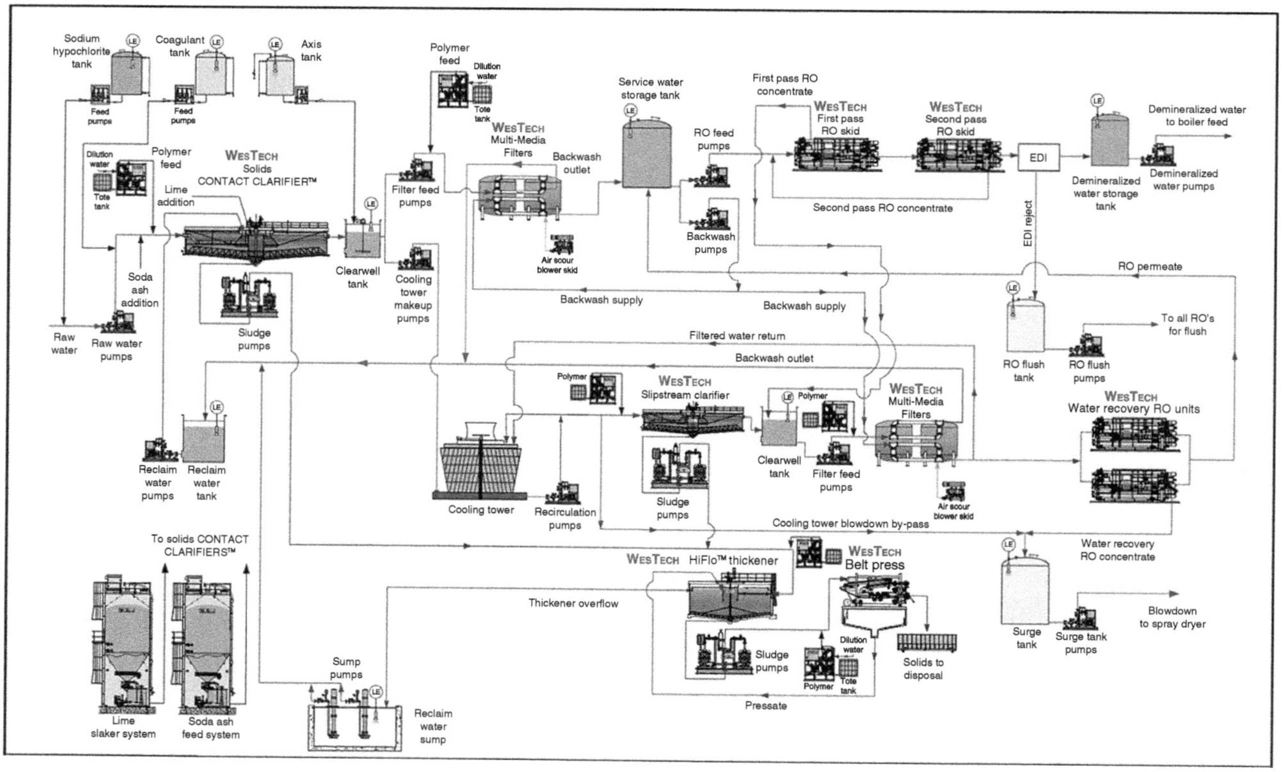

Figure 13.2 A typical flow scheme for treatment of reclaimed water for boiler feedwater. *Source:* WesTech (2016a). Reproduced with permission of WesTech Engineering, Inc.

purposes of groundwater recharge has typically been to provide long-term water storage capacity.

Compared with surface water storage, groundwater recharge is advantageous due to the additional treatment that is achieved as the reclaimed water percolates through the soil. Groundwater recharge also provides protection from evaporation and reduces the likelihood of algae blooms that can carry and release potential toxins into the environment. Additional advantages of groundwater recharge include the replenishment of depleted aquifers, reduction in energy costs associated with pumping from deeper aquifers, the avoidance of environmental impacts associated with the construction of surface-level storage facilities, and the prevention of seawater intrusion into aquifers (Metcalf and Eddy, 2007).

Groundwater recharge is primarily achieved by surface spreading due to its high loading rates and relatively low maintenance requirements. Surface spreading may be applied on unconfined aquifers and typically requires only primary or secondary wastewater treatment levels prior to application.

At the point of application, referred to as a spreading basin, the reclaimed water percolates through the vadose (unsaturated) layers of the soil, including of loam, sand, gravel, silt, and clay layers. Excavation is typically necessary to remove surface soils of low permeability. The excavated soil may be used to construct berms around recharge basins.

Operational requirements of recharge basins using reclaimed water include the application of wetting and drying periods to maintain adequate infiltration rates.

As the water travels through the soil profile, it undergoes further treatment via physical, chemical, and biological mechanisms. For example, following traditional secondary or advanced wastewater treatment methods, residual constituents of concern are subject to filtration, adsorption, hydrolysis, and biotransformation.

An additional advantage of surface spreading for groundwater recharge may include the colocation of the groundwater recharge site with the site at which the reclaimed water may be applied, such as metropolitan and agricultural areas where groundwater overdraft is a concern. However, the requirement of relatively large land area for surface spreading may create economical barriers in urbanized areas where the cost of land is increasing. Large costs may also be associated with the distribution system necessary to deliver water to the recharge basins.Another method for achieving groundwater recharge is via injection, wherein treated wastewater is conveyed and placed directly into an aquifer. Injection is typically applied where groundwater is deep, land availability is not suitable for surface spreading, or the subsurface hydrogeological conditions make surface spreading impractical. Injection may also be applied as a method for creating a freshwater barrier in coastal aquifers to protect against the intrusion of salt water. This may be accomplished via vadose zone injection and direct injection wells.

The relatively recent development of vadose zone injection well technology is a result of the increasing cost of land in urbanized areas. These are essentially a variation of dry wells that are designed to inject water continuously into the vadose zone. Like surface spreading, vadose zone injection wells are applied for unconfined aquifer types. Advantages of vadose zone injection when compared with surface spreading include a reduction in land requirements and reduced potential for evaporation.

In direct injection, highly treated reclaimed water is pumped directly into the groundwater zone, usually into a contained aquifer. Direct injection systems may be used for both injection and extraction of reclaimed water and may achieve a high rate of reclaimed water injection. Unlike surface spreading or vadose injection systems, direct injection may be used in both saturated and unsaturated aquifers, and the flow may be reversed, thereby allowing for periodic maintenance and cleaning. However, direct injection systems are relatively expensive to construct and operate due to energy-intensive high-pressure pumping that is required for injection, and, similar to vadose injection wells, the major operational concern is clogging that occurs at the edge of the borehole.

13.2.6.1 Constituents of Concern for Groundwater Recharge

13.2.6.1.1 Suspended Solids

The removal of suspended solids is critical in order to prevent clogging. In surface spreading applications, reduced suspended solids enhance infiltration rates in the soil. In both vadose and direct injection systems, suspended solids will cause clogging and decreased infiltration rates at the borehole/soil interface.

Because the flow is irreversible in vadose zone injection wells, clogging may be irreversible, and the life span of vadose zone injection wells will then be severely diminished.

13.2.6.1.2 Organic Carbon and Nutrients

While for some groundwater recharge methods (surface spreading), only primary or secondary treatment levels are required prior to land application, excessive carbon and nutrient loadings may cause biological growth that leads to clogging of the water flow path. Particularly in areas of high solar incidence, algae growth may be a major fact in contributing to a reduction in infiltration rates in surface spreading methods. Therefore, drying periods must be employed wherein the spreading basin is drained and allowed to dry to desiccate the organic and nutrient matter on the surface.

Regarding nitrogen, nitrified effluents with nitrate concentrations in excess of 10 mg NO_3-N l^{-1} should not be used for groundwater recharge unless the recharge basin is coupled to a wetland where plants may provide a source of organic carbon to stimulate denitrification (Metcalf and Eddy, 2007).

13.2.6.1.3 Redox Potential

Oxygen in the water is consumed near the soil/water interface as easily biodegradable organics are oxidized at the top of the soil profile. Therefore, the water entering the aquifer may be depleted of oxygen, creating an anoxic plume of reclaimed water. While the potential for adverse interactions of this anoxic plume on the native aquifer materials is generally low, it is important to keep in mind that the low redox potential may lead to the solubilization of reduced iron, manganese, and arsenic from native aquifer materials.

13.2.6.1.4 Pathogens

Concerns over pathogens during the recharge of groundwater using reclaimed wastewater include the fate and transport of parasites, bacteria, and viruses. Studies have been conducted on many types of pathogens during subsurface transport. There have been no demonstrated hosts for pathogenic microorganisms in the subsurface. Further, bacteria and parasites are too large to be transported effectively during subsurface flow.

Therefore, survival of viruses is the primary concern during subsurface transport. Soil type, pH, moisture content, and virus type all affect the adsorptive capacity and virus reduction potential in soil. Global regulatory efforts therefore have focused on the ability of a virus to survive in the environment. For example, the Netherlands and Germany have indicated detention times of 70 and 50 days, respectively, for bank filtration systems. If public access to recharge is expected, then extensive disinfection may be required.

13.2.6.2 Treatment Objectives: Groundwater Recharge

Pretreatment requirements vary considerably for groundwater recharge applications for reclaimed water, depending on the recharge methods, location, source of reclaimed water, and final use.

Application of groundwater recharge and reuse that employ surface spreading requires typically only primary or secondary levels of wastewater treatment prior to discharge at the spreading basin, as surface spreading method of groundwater recharge is in itself an effective form of wastewater treatment. The primary treatment objective for surface spreading is to maintain optimized infiltration rates during groundwater recharge.

To prevent or minimize clogging, well protection may be accomplished by using a screen and filling the well with sand or highly permeable backfill material. Primary treatment to reduce solids that may affect infiltration rates is typically required, and often secondary treatment is required to reduce the likelihood that algae blooms will occur near the surface where solar incidence is high.

The pretreatment objectives for vadose zone injection wells are similar to those for surface spreading, where the primary goal is to maintain infiltration rates. However, a minimum requirement of tertiary treated and disinfected effluent may be necessary to limit the accumulation of suspended solids at the borehole/soil interface. A membrane bioreactor (MBR) may be employed for this purpose. An MBR combines a membrane process like ultrafiltration with a suspended growth biological reactor. The elevated retention of biomass allows the system to maintain elevated levels of mixed liquor suspended solids compared with traditional activated sludge-type processes, thus reducing the required reactor volume to achieve an equivalent loading rate.

If elevated levels of biodegradable materials are present in the reclaimed water, a high degree of treatment such as RO may be applied to create a biologically stable water that will not result in microbial clogging. Further, biological growth in vadose zone injection wells may be inhibited by the addition of a disinfectant agent such as chlorine.

The water quality requirements associated with direct injection are often greater than surface spreading of vadose injection and may include the need for RO and advanced oxidation to eliminate water quality concerns prior to injection. Antidegradation laws regarding water quality may apply, wherein injected water must have equivalent or better quality than the existing quality of the aquifer to comply.

Also, the level of treatment necessary may also be contingent on the surface area available to support biofiltration reactions for the injected water. If water is injected directly into fractured geological formations, the potential for water quality improvement decreases as filtration through soil with beneficial treatment properties is reduced.

Direct injection is often associated with indirect potable reuse (IPR) of reclaimed wastewater. Therefore, water quality requirements associated with direct injection are often much greater than surface spreading of vadose injection.

After storage, water is recovered for use using recovery wells or dual-purpose storage and recovery wells.

Posttreatment of the recovered water may be required prior to final use.

The level of treatment required for groundwater recharge in 10 states is shown in Tables 13.20 and 13.21.

13.2.6.3 Treatment Technologies and Groundwater Recharge

Treatment technologies applied for pretreatment of wastewater prior to groundwater recharge may range from simple primary treatment methods of solids/liquid separation only (sedimentation or flotation), secondary biological treatment for soluble organic material and nutrient reduction, tertiary treatment including filtration, and even to advanced oxidation treatment for pathogen removal and microbial growth prevention. These tertiary and posttertiary treatment technologies have been discussed in greater detail in urban reuse.

Table 13.20 Selected state standards for groundwater recharge – nonpotable reuse[a] – Arizona, California, Florida, Hawaii, and Nevada (US EPA, 2012).

	Arizona[b]	California	Florida[c]	Hawaii	Nevada
Unit processes	Regulated by Aquifer Protection Permit[b]	Case by case	Secondary treatment, basic disinfection	Case by case	*ND*
UV dose, if UV disinfection used	*NS*	*NS*	*NS*	*NS*	*ND*
Chlorine disinfection requirements, if used	*NS*	*NS*	TRC > 0.5 $mg\,l^{-1}$; 15 min contact time at peak hour flow[d]	*NS*	*ND*
BOD_5 (or $CBOD_5$)	*NS*	*NS*	$CBOD_5$: • 20 $mg\,l^{-1}$ (ann avg) • 30 $mg\,l^{-1}$ (mo avg) • 45 $mg\,l^{-1}$ (wk avg) • 60 $mg\,l^{-1}$ (max)	*NS*	*ND*
TSS	*NS*	*NS*	• 20 $mg\,l^{-1}$ (ann avg) • 30 $mg\,l^{-1}$ (mo avg) • 45 $mg\,l^{-1}$ (wk avg) • 60 $mg\,l^{-1}$ (max)	*NS*	*ND*
Turbidity	*NS*	*NS*	*NS*	*NS*	*ND*
Bacterial indicators	*NS*	*NS*	Fecal coliform: • 200/100 ml (avg) • 800/100 ml (max)	*NS*	*ND*
Total nitrogen	*NS*	*NS*	NS (nitrate <12 $mg\,l^{-1}$)	*NS*	*ND*
TOC	*NS*	*NS*	*NS*	*NS*	*ND*
Primary and secondary drinking water standards	*NS*	*NS*	*NS*	*NS*	*ND*

ND, regulations have not been developed for this type of reuse; *NR*, not regulated by the state under the reuse program; *NS*, not specified by the state's reuse regulation.

[a] All state requirements are for groundwater recharge of a nonpotable aquifer.

[b] Groundwater recharge using reclaimed water is pervasive in Arizona but is not considered part of the reclaimed water program; Arizona Department of Environmental Quality (ADEQ) regulates quality under the Department's Aquifer Protection Permit Program (which governs all discharges that might impact groundwater). The Arizona Department of Water Resources (ADWR) oversees a program to limit withdrawals of groundwater to prevent groundwater depletion; municipalities and other entities can offset these pumping limitations by recharging reclaimed water through detailed permits under its recharge program.

[c] Higher treatment standards may be required, such as filtration, high-level disinfection, total nitrogen below 10 $mg\,l^{-1}$, and meeting primary and secondary drinking water standards, if there may be a connection to a potable aquifer or other conditions such as groundwater recharge overlying the Biscayne aquifer in Southeast Florida.

[d] In Florida when chlorine disinfection is used, the product of the total chlorine residual and contact time (CrT) at peak hour flow is specified for three levels of fecal coliform as measured prior to disinfection. If the concentration of fecal coliform prior to disinfection is ≤1000 cfu per 100 ml, the CrT shall be 25 $mg\,min\,l^{-1}$; is 1 000–10 000 cfu per 100 ml, the CrT shall be 40 $mg\,min\,l^{-1}$; and is ≥10 000 cfu per 100 ml, the CrT shall be 120 $mg\,min\,l^{-1}$.

Table 13.21 Selected state standards for groundwater recharge – nonpotable reuse[a] – New Jersey, North Carolina, Texas, Virginia, and Washington (US EPA, 2012).

	New Jersey[b]	North Carolina	Texas	Virginia[c]	Washington class A
Unit processes	*NR*	Aquifer storage and recovery in accordance with GS 143-214.2	*NR*	*NS*	Oxidized, coagulated, filtered, nitrogen reduced, disinfected
UV dose, if UV disinfection used	*NR*	*NR*	*NR*	*NS*	NWRI UV Guidelines
Chlorine disinfection requirements, if used	*NR*	*NR*	*NR*	*NS*	Chlorine residual >1 $mg l^{-1}$ 30 min contact time at peak hour flow
BOD_5 (or $CBOD_5$)	*NR*	*NR*	*NR*	*NS*	5 $mg l^{-1}$
TSS	*NR*	*NR*	*NR*	*NS*	5 $mg l^{-1}$
Turbidity	*NR*	*NR*	*NR*	*NS*	• 2 NTU (avg) • 5 NTU (max)
Bacterial indicators	*NR*	*NR*	*NR*	*NS*	Total coliform: • 2.2/100 ml (7-d med) • 23/100 ml (max day)
Total nitrogen	*NR*	*NR*	*NR*	*NS*	Case by case
TOC	*NR*	*NR*	*NR*	*NS*	Case by case
Primary and secondary drinking water Standards	*NR*	*NR*	*NR*	*NS*	Case by case

ND, regulations have not been developed for this type of reuse; *NR*, not regulated by the state under the reuse program; *NS*, not specified by the state's reuse regulation.
[a] All state requirements are for groundwater recharge of a nonpotable aquifer.
[b] All discharges to groundwater for nonpotable reuse are regulated via a New Jersey Pollutant Discharge Elimination System permit in accordance with N.J.A.C. 7:14A-1 et seq. and must comply with applicable groundwater quality standards (N.J.A.C. 7:9C).
[c] In Virginia, groundwater recharge of a nonpotable aquifer may be regulated in accordance with regulations unrelated to the Water Reclamation and Reuse Regulation (9VAC25-740).

13.2.6.4 Governing Design Considerations for Groundwater Recharge

Two critical considerations in choosing a groundwater recharge method include the type of aquifer that is present and the availability of land. If a vadose zone (nonsaturated conditions) is not sufficient or not present, then direct injection is required. Where a confined aquifer with a vadose zone is present and adequate land space is available, then surface spreading is often the most economically attractive and technically feasible choice.

Hydrogeological conditions often govern which groundwater recharge method is suitable for any given application. A soil characterization must be undertaken to understand the type of flow experienced in the subsurface environment.

In both surface spreading and injection groundwater recharge applications, the placement of extraction wells in locations that are far away from the spreading basin or injection well increases the flow path length and residence time of the recharged water. This contributes to the advantageous mixing of the reclaimed water and the aquifer contents.

13.2.7 Potable Reuse of Reclaimed Wastewater

Reuse of reclaimed wastewater for potable reuse applications may be classified as either indirect potable reuse (IPR, or may also be referred to as *de facto* potable reuse) or direct potable reuse (DPR). The practice of discharging treated wastewater effluent to a natural environmental buffer, such as a stream or aquifer, has long occurred as an example of IPR. However, it has been demonstrated that well-engineered advanced water treatment plants can perform equally or better than natural systems in attenuating constituents of concern. Both natural and engineered systems are considered IPR.

It can be argued that treatment of wastewater to effluent quality higher than drinking water standards followed by discharge to aquifers or lakes is counterproductive. Therefore, DPR, i.e. the introduction of highly treated reclaimed water either directly into the potable water supply distribution system downstream of a water treatment plant or into the raw water supply immediately upstream of a water treatment plant, is of increasing interest.

13.2.7.1 Constituents of Concern for Potable Reuse

As the intended use of reclaimed water for potable applications is directly suited for high levels of human contact, all constituents that have been identified previously as urban reuse constituents of concern that pose a health risk to humans and the environment must be considered when designing a potable reuse system. Most chemical constituents found in treated municipal wastewater are present at concentrations that may be of concern only with chronic exposure. However, those constituents found in treated wastewater at concentrations that are higher than those considered safe for potable use may present a health risk due to acute exposure.

Of particular concern is the presence of organics in treated wastewater that have not yet been studied thoroughly enough to determine the potential health risk. While typical advanced wastewater treatment is effective for sufficient pathogen reduction, some experts express concern that DBPs and other organics that have not been expressly classified as "nonhealth risk associated" may need to be addressed.

For treatment design purposes, constituents of concern for both IPR and DPR may be categorized into those that are aesthetic (have no direct established link to detrimental human health effects, such a turbidity and color), microbiological constituents, inorganic salts and metals with potential for toxicological effects, and others without health risk (such as calcium carbonate).

13.2.7.2 Treatment Objectives: Potable Reuse

13.2.7.2.1 Indirect Potable Reuse

Whether IPR is achieved by spreading treated wastewater via spreading basins into potable aquifers, direct injection into potable aquifers, or by augmentation of surface water supply reservoirs, suggested guidelines for water reuse indicate that there should be no detectable total coliform/100 ml. Minimum chlorine residual levels should be $1\,\mathrm{mg\,l^{-1}}$ Cl_2. Color of less than 2 NTU and TOC levels of less than $2\,\mathrm{mg\,l^{-1}}$ are also recommended. The reclaimed water should meet drinking water standards after the wastewater has percolated through the vadose zone.

Further, a minimum of 2-month retention time of the treated wastewater in the underground prior to extraction should determine the setback distance of the point at which the wastewater enters the potable water source to where the point of extraction is used.

The level of treatment required for IPR in 10 states is indicated in Tables 13.22 and 13.23.

13.2.7.2.2 Direct Potable Reuse

Nearly complete removal of all constituents of concern to levels well below even drinking water standards is required for treatment of wastewater for DPR.

Regarding barrier design, for aesthetic constituents of concern, two barriers are typically required for treatment. For microbiological pollutants, three barriers are often required, and for other parameters where no direct human health risk has been demonstrated, only one barrier may be required.

13.2.7.3 Treatment Technologies and Potable Reuse

13.2.7.3.1 Indirect Potable Reuse

For groundwater recharge by spreading into potable aquifers, typical post-secondary treatment technologies include filtration, disinfection, and SAT. For additional description of SAT, refer to the ground water recharge section.

Where indirect potable reclamation of wastewater is achieved by injection into potable aquifers or by augmentation of surface water supply reservoirs, secondary treatment is typically followed by filtration, disinfection, and advanced wastewater treatment.

13.2.7.3.2 Direct Potable Reuse

In addition to advanced treatment methods described previously to treat urban reuse constituents of concern, including disinfection, membrane filtration, and advanced oxidation, advances in technologies including enhanced membrane filtration coupled with AOPs are capable of producing potable reuse water with quality that has achieved full removal of trace constituents and far exceeds drinking water standards.

As multiple treatment barriers are often required for DPR, additional unit processes and technologies may be employed to mitigate against the risk of process upset or equipment failure. For example, a barrier for turbidity may include a combination of processes such as flocculation, DAF, and dual media filtration.

Additional unit processes that may be considered for designing sophisticated barrier systems and for removal of constituents of concern may include incorporation of high pH lime treatment, sedimentation (with ferric chloride addition), recarbonation, filtration, UV irradiation, rapid sand filtration, carbon adsorption (including GAC and biological activated carbon systems), RO, air stripping (to remove carbon dioxide and volatile organic chemicals), ozonation, and chloramination.

13.2.7.4 Governing Design Considerations for Potable Reuse

13.2.7.4.1 Indirect Potable Reuse

Water rights, permits, and storage contracts must exist in order to ensure beneficial withdrawal of the additional yield of the augmented water supply. Public education and acceptance of IPR is also critical.

Table 13.22 Selected state standards for indirect potable reuse (IPR) – Arizona, California, Florida, Hawaii, Nevada, and New Jersey (US EPA, 2012).

	Arizona[a]	California	Florida[b]	Hawaii	Nevada	New Jersey[c]
Unit processes	*NR*	Oxidized, coagulated, filtered, disinfected, multiple barriers for pathogen and organics removal	Secondary treatment, filtration, high-level disinfection, multiple barriers for pathogen and organics removal	Case by case	*ND*	*NR*
UV dose, if UV disinfection used	*NR*	NWRI Guidelines[b]	NWRI UV Guidelines enforced, variance allowed	*NS*	*ND*	*NR*
Chlorine disinfection requirements, if used	*NR*	CrT > 450 mg min l^{-1}; 90 min modal contact time at peak dry weather flow[d]	TRC > 1 mg l^{-1}; 15 min contact time at peak hour flow[e]	*NS*	*ND*	*NR*
BOD_5 (or $CBOD_5$)	*NR*	*NS*	$CBOD_5$: • 20 mg l^{-1} (ann avg) • 30 mg l^{-1} (mo avg) • 45 mg l^{-1} (wk avg) • 60 mg l^{-1} (max)	*NS*	*ND*	*NR*
TSS	*NR*	*NS*	5 mg l^{-1} (max)	*NS*	*ND*	*NR*
Turbidity	*NR*	• 2 NTU (avg) for media filters • 10 NTU (max) for media filters • 0.2 NTU (avg) for membrane filters • 0.5 NTU (max) for membrane filters	Case by case (generally 2–2.5 NTU) Florida requires continuous online monitoring of turbidity as indicator for TSS	*NS*	*ND*	*NR*
Bacterial indicators	*NR*	Total coliform: • 2.2/100 ml (7-d med) • 23/100 ml (not more than one sample exceeds this value in 30 d) • 240/100 ml (max)	Total coliform: • 4/100 ml (max)	*NS*	*ND*	*NR*
Total nitrogen	*NR*	10 mg l^{-1} (avg of four consecutive samples)	10 mg l^{-1} (ann avg)	*NS*	*ND*	*NR*
TOC	*NR*	0.5 mg l^{-1}	• 3 mg l^{-1} (mo avg) • 5 mg l^{-1} (max); TOX[f]: <0.2 (mo avg) or 0.3 mg l^{-1} (max); alternate limits allowed	*NS*	*ND*	*NR*
Primary and secondary drinking water standards	*NR*	Compliance with most primary and secondary	Compliance with most primary and secondary	*NS*	*ND*	*NR*
Pathogens	*NR*	*TR*	*Giardia, Cryptosporidium* sampling quarterly	*NS*	*ND*	*NR*

ND, regulations have not been developed for this type of reuse; *NR*, not regulated by the state under the reuse program; *NS*, not specified by the state's reuse regulation; *TR*, monitoring is not required, but virus removal rates are prescribed by treatment requirements.

[a] Arizona currently does not have IPR regulations; however, ADEQ regulates recharge facilities where mixed groundwater-reclaimed water may be recovered by a drinking water well through its Aquifer Protection Permit program. The Governor's Blue Ribbon Panel on Water Sustainability issued a report including a recommendation to develop a more robust regulatory/policy program to address IPR.

[b] Florida requirements are for the planned use of reclaimed water to augment classes F-I, G-I, or G-II groundwaters (US drinking water sources) with a background TDS of 3000 mg l^{-1} or less. For G-II groundwaters greater than 3000 mg l^{-1} TDS, the TOC and TOX limits do not apply. Florida also includes discharges to class I surface waters (public water supplies) or discharges less than 24 h travel time upstream from class I surface waters as IPR. For discharge to class I surface waters or water contiguous to or tributary to class I waters (defined as a discharge located less than or equal to 4 h travel time from the point of discharge to arrival at the boundary of the class I water), secondary treatment with filtration, high-level disinfection, and any additional treatment required to meet TOC and applicable surface water quality limits is required. The reclaimed water must meet primary and secondary drinking water standards, except for asbestos, prior to discharge. The TOX limit does not apply and a total nitrogen limit is based on the surface water quality. Outfalls for surface water discharges are not to be located within 500 ft (150 m) of existing or approved potable water intakes within class I surface waters. Pathogen monitoring for class I surface water augmentation is the same, except that if discharge is 24–48-h travel time from domestic water supply, *Giardia, Cryptosporidium* sampling is once every 2 years.

[c] Since groundwater recharge is on a case-by-case basis, State Groundwater Quality Standards must be met.

[d] Additional pathogen removal is required for groundwater recharge through other treatment processes in order to achieve 12-log enteric virus reduction, 10-log *Giardia* cyst reduction, and 10-log *Cryptosporidium* oocysts reduction.

[e] In Florida when chlorine disinfection is used, the product of the total chlorine residual and contact time (CrT) at peak hour flow is specified for three levels of fecal coliform as measured prior to disinfection. If the concentration of fecal coliform prior to disinfection is ≤1000 cfu per 100 ml, the CrT shall be 25 mg min l^{-1}; is 1000–10 000 cfu per 100 ml, the CrT shall be 40 mg min l^{-1}; and is ≥10 000 cfu per 100 ml, the CrT shall be 120 mg min l^{-1}.

[f] Total organic halides (TOX) are regulated in Florida.

Table 13.23 Selected state standards for indirect potable reuse (IPR) – North Carolina, Texas, Virginia, and Washington (US EPA, 2012).

	North Carolina	Texas	Virginia	Washington		
				Surface percolation class A	Direct groundwater recharge[a] class A	Stream flow augmentation case by case
Unit processes	*NR*	Case by case	Case by case	Oxidized with nitrogen reduction, filtered, disinfected	Oxidized, coagulated, filtered, RO-treated, disinfected	Oxidized, clarified, disinfected
UV dose, if UV disinfection used	*NR*	*NS*	*NS*	NWRI Guidelines	NWRI Guidelines	NWRI Guidelines
Chlorine disinfection requirements, if used	*NR*	*NS*	*NS*	Chlorine residual >1 $mg l^{-1}$; 30 min contact time at peak hour flow	Chlorine residua l >1 $mg l^{-1}$; 30 min contact time at peak hour flow	Chlorine residual to comply with NPDES permit
BOD_5 (or $CBOD_5$)	*NR*	5 $mg l^{-1}$	*NS*	30 $mg l^{-1}$	5 $mg l^{-1}$	30 $mg l^{-1}$
TSS	*NR*	*NS*	*NS*	30 $mg l^{-1}$	5 $mg l^{-1}$	30 $mg l^{-1}$
Turbidity	*NR*	3 NTU	*NS*	• 2 NTU (avg) • 5 NTU (max)	• 0.1 NTU (avg) • 0.5 NTU (max)	*NS*
Bacterial indicators	*NR*	Fecal coliform or *E. coli:* • 20/100 ml (30-d geom) • 75/100 ml (max) *Enterococci:* • 4/100 ml (30-d geom) • 9/100 ml (max)	*NS*	Total coliform: • 2.2/100 (7-d med) • 23/100 (max)	Total coliform: • 1/100 ml (avg) • 5/100 ml (max)	Fecal coliform: • 200/100 ml (avg) • 400/100 ml (max wk)
Total nitrogen	*NR*	*NS*	*NS*	*NA*	10 $mg l^{-1}$	NPDES requirements to receiving stream
TOC	*NR*	*NS*	*NS*	*NA*	1 $mg l^{-1}$	*NS*
Primary and secondary drinking water standards	*NR*	*NS*	*NS*	Compliance with SDWA MCLs	Compliance with most primary and secondary	NPDES requirements to receiving stream
Pathogens	*NR*	*NS*	*NS*	*NS*	*NS*	*NS*

ND, regulations have not been developed for this type of reuse; *NR*, not regulated by the state under the reuse program; *NS*, not specified by the state's reuse regulation; *TR*, monitoring is not required but virus removal rates are prescribed by treatment requirements.
[a] Washington requires the minimum horizontal separation distance between the point of direct recharge and point of withdrawal as being withdrawn as a drinking water supply.

13.2.7.4.2 Direct Potable Reuse

Development of public trust is key for the implementation of DPR. Clear communication of all risk mitigation methods must be established with the community.

Multiple barrier systems that include sequential and redundant processes should be applied to remove constituents of concern reliably and consistently, even in the face of a power outage or other anomaly. Barriers may be applied to treatment processes (ensuring the processes are continually present to reduce undesired substances in the water to an acceptable level) or may also be applied to monitoring systems, ensuring complete and redundant monitoring at the inlet and outlet, allowing corrective action to be undertaken before deleterious effects of process upsets are experienced.

Introduction methods of the reclaimed wastewater into the potable water supply must also be considered and may include holding the purified water in an engineered storage buffer before being blended with a water supply prior to water treatment. Alternatively the water may be blended into the distribution system for delivery to water consumers following water treatment.

13.3 Reliability Requirements for Wastewater Reclamation and Recycling Systems

Due to the need for wastewater reclamation plants to deliver reclaimed wastewater of adequate quantity and quality for the intended use or uses, reclamation plants must have a high standard of reliability. If improperly treated reclaimed wastewater is provided, there is a potential for harm to the receiving environment. Reliability features must be considered during the design, construction, and operation of the wastewater reclamation plants.

The Federal Water Quality Administration, the precursor to the US EPA, issued guidelines for treatment reliability in 1970: *Federal Guidelines: Design, Operation, and Maintenance of Wastewater Treatment Facilities* (Federal Water Quality Administration, 1970). In 1974, the US EPA issued an additional guidance: *Design Requirements for Mechanical, Electric, and Fluid Systems and Component Reliability* (US EPA, 1974).

The Federal Water Quality Administration (1970) and the US EPA (1974) identified the following elements as especially critical for reliability:

- Power supply
  - Duplicate dual sources of electrical power as well as standby on-site power for essential plant processes are recommended.
- Flexibility of piping and control
  - Rerouting of flows under emergency conditions to emergency storage facilities or to approve nonreuse areas may be necessary. Installation points of pipes and pumps should not allow inadequately treated effluent to enter reclaimed water distribution systems.

  Dual distribution systems (i.e. reclaimed water distribution systems paralleling a potable water system) must include safeguards to prevent cross-connections of the systems and misuse of the reclaimed water. Piping, valves, and hydrants should be color coded (purple) and marked. Backflow prevention devices should be used, and hose bibs on reclaimed waterlines should be prohibited to prevent misuse at the point of delivery. Periodic use of tracer dyes may be used to detect the occurrence of possible cross-contamination into potable supply lines.
- Monitoring
  - Monitoring of treatment systems has three different purposes:
    - Validation: used to prove the system is meeting its design requirements; used when a new system is developed or when new processes are added to test or prove that the system is capable of meeting specified targets.
    - Operational: used on a routine basis to indicate that processes are working as expected; relies on simple measurements that can be read quickly so decisions can be made in time to remedy a problem; online monitoring systems (e.g. turbidimeters, chlorine residual analyzers, chemical feed facilities) may be used to analyze appropriate parameters in real time.
    - Verification: used to show that the end product, i.e. the reclaimed water, meets treatment targets; collected periodically so information is too late for decision making for problems to prevent a hazard breakthrough; used to monitor trends through time to determine efficiency changes through time.
- Quality assurance program for sampling and laboratory analyses
  - A quality assurance/quality control (QA/QC) plan contains defined protocols with data quality objectives and procedures to develop quality control data, including precision, bias, accuracy, and other reliability factors. QA/QC programs are required to ensure that systems and procedures are maintained and calibrated and produce accurate results.
    - QA/QC procedures may be dictated by regulatory agencies and represent necessary operating overhead costs, which may equal the costs of wastewater reclamation itself.

- Individual treatment units
  - Multiple treatment units and backup equipment may be required to provide redundancy in the case of breakdowns or failures.
- Maintenance program
  - A strict preventive maintenance schedule as well as guidelines for troubleshooting problems and breakdowns is essential.
- Operating personnel
  - The plant operator is the most critical reliability factor in the wastewater reclamation system. If the operator is not conscientious and capable, even well-designed and well-constructed systems will not perform adequately. Operator attendance, operator competence, and operator training are all essential to ensure reliability. Most regulatory agencies require operator certification to ensure that facilities are operated by qualified personnel. Frequent continuing education can enhance operator competence. Special training and certification should be considered for operators of wastewater reclamation facilities.

Comprehensive operating protocols should be followed that define the responsibilities and duties of the operators to ensure reliable production and delivery of reclaimed water.

- Alarm systems
  - Alarm systems are required at all water reclamation facilities, especially those without full-time operators. Alarms should be placed at critical treatment locations to alert operators to malfunctions. Supervisory control and data acquisition (SCADA) may be used when the information is made available to locations that are staffed when operators are not on-site at the reclamation facility. Examples of when alarms may be used include indications of loss of power, high water levels, failures of pumps or blowers, loss of DO, loss of coagulant feed, high head loss on filters, high effluent turbidity, or loss of disinfection.
- Storage requirements
  - Reclaimed wastewater is usually continuously generated, and if all of it cannot be used immediately, it must be stored. Depending on the volume and pattern of reuse demands, seasonal storage requirements may be a significant design and capital cost consideration. If during the storage the quality of the water is degraded by algal growth and requires retreatment to meet reclaimed water use standards, the systems might also impact the operational costs. Many local and state regulations specify required storage volumes.
  - Storage may be required during periods of low demand and subsequent use during periods of peak demand. Alternatively storage may be required to reduce or eliminate discharge of excess reclaimed water into surface or groundwaters. The use of storage methods with finite capacity, such as tanks, ponds, or reservoirs, must be large in comparison with the design flows in order to provide complete equalization of seasonal supplies and demands.
  - Aquifer storage and recovery (ASR) of reclaimed water, involving the injection of reclaimed wastewater into a subsurface formation for storage and recovery for use at a later time, can be an environmentally sound and effective approach for storage. The potential storage of an ASR system is unlimited.
- Industrial Pretreatment Program
  - Designing a wastewater treatment plant for reuse with industrial flows requires identification and monitoring of constituents that may interfere with the potential reuse application. If hazardous constituents are not removed during pretreatment or in the wastewater treatment plant, the number and type of reuse applications may be limited.
  - Industrial waste streams may contain high levels of toxic compounds and elements that may adversely affect wastewater treatment plant performance and reclaimed water quality. A rigorous pretreatment program is required for a water reclamation facility that receives industrial wastes to ensure the reliability of the biological treatment processes by preventing entry of potentially toxic pollutants to the sewer system.
  - In the United States, a national pretreatment program, which is a component of the NPDES permit program, requires nondomestic dischargers to comply with pretreatment standards to ensure the goals of the Clean Water Act are attained. The pretreatment program identifies specific discharge standards and requirements that apply to sources of nondomestic wastewater discharged to a publicly owned wastewater treatment (POWT) plant. By reducing or eliminating waste at the industries (i.e. source reduction), fewer toxic pollutants are discharged to and treated by the POTWs, providing benefits to both the POTWs and the industrial users.
  - The objectives of the National Pretreatment Program are to:
    - Prevent the introduction of pollutants into a POTW that will interfere with its operation, including interference with its use or disposal of municipal sludge.

- Prevent the introduction of pollutants into a POTW that will pass through the treatment works or otherwise be incompatible with it.
- Improve opportunities to recycle and reclaim municipal and industrial wastewaters and sludges.

13.4 Planning and Funding for Wastewater Reclamation and Reuse

Planning for water reclamation and reuse projects is more complex than planning for conventional wastewater treatment facilities, where planning is required only for conveyance, treatment, and disposal of wastewater. Wastewater reclamation program plans should include (Metcalf and Eddy, 2003; US EPA, 2004):

- *Objectives of the project*: Objectives including (i) assessment of wastewater treatment and disposal needs, (ii) assessment of water supply and demand, and (iii) assessment of water supply benefit based on water reuse potential.
- *Project study areas*: Study areas including (i) collection system area to be served by the wastewater treatment plant and (ii) the area that potentially could be served and benefit from the reclaimed water, which may be a community not served by the wastewater treatment plant.
- *Market assessment*: Identification of (i) potential uses of reclaimed water, (ii) survey of potential customers capable and willing to use reclaimed water, and (iii) community support.
- *Technical issues*: Treatment requirements for producing safe and reliable reclaimed water suitable for intended use(s), storage facilities required to balance fluctuations in supply with fluctuations in demand, supplemental facilities required to operate a water reuse system, including conveyance and distribution networks, and operational storage facilities.
- *Environmental impacts*: Potential impacts of implementing water reclamation.
- Identification of knowledge skills and abilities necessary to operate and maintain water reclamation system.
- Regulatory requirements.
- *Monetary analyses and cost-effectiveness*: Often the measure of feasibility of a wastewater reuse project but more emphasis on environmental considerations, public acceptance, and public policy issues recommended; basic questions include (i) based on perceived costs and benefits of a project, *should* a water reuse project be constructed and (ii) *can* a water reuse project be constructed? Comparison required on costs (i) to construct new freshwater facilities versus (ii) to operate and maintain the reclamation facilities.

Key elements of a water reuse program as suggested by the US EPA (2012) are given in Table 13.24.

There are several mechanisms for funding reclamation water systems. Typical sources include (US EPA, 2012):

- *Internal funding*: Revenue generated from customers, including individual large volume users or a broad network of users; large volume users can fund significant portions of a project; large volume users typically include industrial users, large-scale agricultural operations, or golf courses.
- *Revenue bonds and low interest loans*: Typically long term, with funding received up front from bondholders.

In some areas, grant programs may be available to underwrite portions of capital requirements, and state subsidies may be available to aid in annual operating costs. However, federal funds cannot be used to cover operating costs. Once a project is established, a reuse project should work to achieve self-sufficiency as soon as possible.

Examples of US federal funding sources include several from the USDA, including the Rural Development Agency, the Rural Utilities Service through the water and waste programs, the Rural Housing Service, and the Rural Business-Cooperative Service through the Rural Business Enterprise Grant program.

The US Bureau of Reclamation also can fund water reclamation and reuse projects after Congressional approval. This funding is limited to projects in the 17 western states, unless authorized by Congress.

Comprehensive information about US federal funding sources is available in the Catalog of Federal Domestic Assistance (https://www.cfda.gov), prepared by the Federal Office of Management and Budget.

13.5 Legal and Regulatory Issues

Most states have laws, policies, rules, and regulations that support and challenge the use of reclamation projects. During project planning, water rights laws, water use, wastewater discharge regulations, land use restrictions, and environmental rules may all affect project development. When implementing projects, policies for reclaimed water rates, agreements among reclaimed water producers, wholesalers, retailers, and customers and rules affecting system construction and liability for water reuse should be considered.

Water right laws are especially important because in the United States, the states generally retain ownership of natural or public waters within its borders, and state statutes, regulations, and case law govern the allocation and administration of private parties and government entities to use such water. Depending on particular state laws, water right laws can either promote reuse or pose an obstacle.

Table 13.24 Key elements of a water reuse program (US EPA, 2012).

1	Establish the objectives	Objectives that encourage and promote reuse should be clear and concise
2	Commit to the long run	A water reuse program should be considered a permanent, high-priority program within the state
3	Identify the lead agency or agencies	The lead agencies should be able to issue permits for the production, distribution, and use of the reclaimed water. These permits are issued under state authority and are separate from the federal requirements for wastewater discharges to surface waters under the NPDES permit program. Preference to the lead agency determination should be given to the public health agency since the intent of the use of reclaimed water is for public contact and/or consumption following adequate and reliable treatment
4	Identify water reuse leader	A knowledgeable and dedicated leader of the water reuse program who develops and maintains relationships with all water programs and other agencies should be designated
5	Enact needed legislation	Initial legislation generally should be limited to a clear statement of the state objectives, a clear statement of authorization for the program, and other authorizations needed for implementation of specific program components. States also will want to review and evaluate existing state water law to determine what constraints, if any, it will impose on water reuse and what statutory refinements may be needed
6	Adopt and implement rules or guidelines governing water reuse	With stakeholder involvement, a comprehensive and detailed set of reuse regulations or guidelines that are fully protective of environmental quality and public health should be developed and adopted in one location of the regulations. Formal regulations are not a necessity – they may be difficult and costly to develop and change and therefore overly rigid. Frameworks that have an ability to adapt to industry changes are most effective
7	Be proactive	The water reuse program leader should be visible within the state and water reuse community, while permitting staff of the lead agency must have a positive attitude in reviewing and permitting quality water reuse projects
8	Develop and cultivate needed partnerships	Partnerships between the agency responsible for permitting the reclaimed water facilities (usually the lead agency) and the agency(ies) responsible for permitting water resources as well as the agency responsible for protection of public health are critical. Other agency partnerships, such as with potential major users of reclaimed water such as the department of transportation, are also helpful in fostering statewide coordination and promotion of water reclamation
9	Ensure the safety of water reuse	Ensuring the protection of public health and safety can be accomplished by placing reliance on production of high quality reclaimed water with minimal end-use controls, or allowing lower levels of treatment with additional controls on the use of reclaimed water (setback distances, time of day restrictions, limits on types of use, etc.), or by a combination of both types of regulations. A formal reliability assessment to assure a minimum level or redundancy and reliability to review and detail operating standards, maintainability, critical operating conditions, spare parts requirements and availability, and other issues that affect the ability of the plant to continuously produce reclaimed water. A critical component to ensuring the safety of reclaimed water for public access and contact-type reuse is defining requirements for achieving a high level of disinfection and the monitoring program necessary to ensure compliance (this is described further in Chapter 6)
10	Develop specific program components	Program components are going to differ from state to state and maturity of the reuse program
11	Focus on quality, integrity, and service	Not only should the reclaimed water utilities implement high quality reuse systems that are operated effectively, but the lead agency should also model this commitment to quality and prompt service to the regulated and general public regarding reuse inquiries and permitting issues. In effect, the lead agency should focus on building the same level of trust public potable water systems develop and re-establish daily
12	Be consistent	A comprehensive and detailed set of state regulations, as well as having a lead reuse role, help keep the permitting of reuse systems consistent. If there are multiple branches around the state involved in permitting, training, and other measures of retaining, consistency must be taken
13	Promote a water reuse community	The lead agency should be proactive in developing and maintaining the state's water reuse community – reuse utilities, consulting engineers, state agencies, water managers, health departments, universities, researchers, users of reclaimed water, and others – in an effort to disseminate information and obtain feedback related to possible impediments, issues, and future needs. Active participation in the national and local reuse organizations is valuable
14	Maintain a reuse inventory	Maintenance of a periodical (e.g. annual) reuse inventory is essential in tracking success of a state's water reuse program. Facilities in Florida that provide reclaimed water are required by their permits to submit an annual reuse report form every year. That data not only is used in the states annual reuse inventory report and reuse statistics but is also shared with the WateReuse Association's National Reuse Database
15	Address cross-connection control issues	Coordination and joint activity between agencies and within agencies (drinking water program, wastewater program, water reuse program, etc.) must be taken to address cross-connection control issues (this is described further in Chapter 2)

Source: Adapted from WateReuse Association (2009).

Water supply and use regulations affect water reuse projects by determining how an agency with water rights decides to distribute the water supply to various parts. US states through mandates by the US Clean Water Act of 1972 must set water quality standards, thus allowing the states to control the pollution from wastewater treatment plants. Strict regulations on the discharge of pollutants and limits on treatment effluent discharged to a receiving body are powerful tools to encourage the reuse of treated effluent.

There are many state and federal environmental regulations that affect the development of a water reclamation program. A guide to the development of a water reuse regulatory program for states is given in Table 13.25.

13.6 Public Involvement and Participation

Public involvement, trust, and support are critical to the success of a water reclamation and reuse program. Fears, concerns, and lack of knowledge of the community can result in the failure of a potential reclamation program. The key to success is two-way communications involving the reclamation agency and the public, customers, media, internal staff, regulators, other government agencies, and anyone else who may be affected or involved. Acceptance by the public policy that values water conservation and reuse rather than relying on the development of additional water resources is key to a successful reuse program.

The WateReuse Foundation (2006) has developed a guidebook, *Marketing Nonpotable Recycled Water, A Guidebook for Successful Public Outreach & Customer Marketing*, to aid in the implementation of a recycling wastewater program by providing tools, approaches, strategies, and tasks that an agency can use in their outreach program. Depending on the reclamation project, some projects may only require contact with a number of specific uses, and some may require an extensive campaign of public engagement. The public may be concerned with public health issues, but growth, economic, and political issues may also be involved. All issues of concern must be addressed truthfully through the outreach process and to each stakeholder group. With more understanding of recycled water usually comes a higher rate of support.

The WateReuse Foundation (2006) *Guidebook* presents recommendations on the development of a strategic outreach plan. The elements of the plan include:

- *Introduction*: Purpose of strategic plan.
- *Situational analysis*: Description of agency and community's need for a recycled wastewater project; possible needs may include a growing population, drought conditions, limited potable supply, or some other factors.
- *Budget*: Costs of implementing strategies, completing tasks, and developing communication tools and advertising.
- *Public outreach and marketing goals*: Possible goals include public information to secure project support, gain consensus on how a project will be implemented, market to potential customers to obtain support, or support agency's efforts to design and construct recycled water facilities within the time frame and budget.
- *Challenges and opportunities*: Identify potential problems and issues so a response or plan can be developed rather than waiting for obstacles to occur.
- *Key outreach messages*: Establishment of consistent messages presented in informational materials, advertisements, press releases, presentations, websites, and stakeholder meetings to prevent confusion.
- *Stakeholder groups*: Identification of stakeholders (individuals, groups, or organizations) that have a real or perceived interest in the project outcome; stakeholders may include general public, elected officials, media, internal staff, business community, government agencies, and recycled water customers.
- *Strategies for stakeholder groups*: Development of strategies for reaching the stakeholder groups throughout the planning, construction, and development of the reclamation program.
- *Development of communication materials and advertising*: In addition to presentations and face-to-face communications, development of materials that can be distributed, including the name of project and logo, brochures, fact sheets, newsletters, radio ads, magazine and newspaper advertisements, websites, feature articles and bill inserts, press kits, and display booths.
- *Timeline*: Timeline for outreach to each individual stakeholder group and should be coordinated with timeline for the overall project.

13.7 Additional Considerations for Wastewater Recycling and Reclamation: Integrated Resource Recovery

In recent decades, the energy, transportation, water, wastewater, and solid waste treatment sectors have become increasingly interconnected. Efforts to reduce consumption, recover and reuse resources, and develop innovative energy-efficient technologies have revealed an interdependent relationship on resource flow across these sectors.

Table 13.25 Fundamental components of a water reuse regulatory framework for states (US EPA, 2012).

Category	Comment
Purpose and/or goal statement	Frame the state's purpose for developing the rule or regulation (e.g. to satisfy a need or fulfill a statutory requirement), and describe the ultimate vision for the water reuse program. The process to authorize, develop, and implement rules or changes to rules is time consuming and costly. After adoption, rules are difficult to change, which limits the ability to accommodate new technologies and information
Definitions	Define type of use and other water reuse-related terms used within the body of the rule or regulation
Scope and applicability	Define the scope and applicability of the rules or regulations that delineates what facilities, systems, and activities are subject to the requirements of the rules or regulations
	Include grandfathering or transitioning provisions for existing facilities, systems, or activities not regulated prior to the adoption of the rules or regulations
Exclusions and prohibitions	Describe facilities, systems, and activities that are (i) not subject to the requirements of the rules or regulations and (ii) specifically prohibited by the rules or regulations
Variances	Describe procedures for variances to design, construction, operation, and/or maintenance requirements of the regulation for hardships that outweigh the benefit of a project, and the variance, if granted, would not adversely impact human health, other beneficial uses, or the environment. These variance procedures give regulators flexibility to consider projects that may deviate only minimally from the requirements with no significant adverse impact or opportunities that are not anticipated during initial development of a regulation. Since variances need to be based on sound, justifiable reasons for change, regulatory programs should develop guidance on how to develop adequate justification that can be relied upon as precedence setting for future regulatory decisions and actions
Permitting requirements	Describe the permitting framework for water reuse. Indicate whether the water reuse rule or regulation will serve as the permitting mechanism for water reuse projects or identify other regulations through which the water reuse rule or regulation will be implemented and projects permitted
	Describe if or how end users of reclaimed water will be permitted and rights of end user to refuse reclaimed water if not demanded
	Describe permit application requirements and procedures. Specify all information that the applicant must provide in order to appropriately evaluate and permit the water reuse projects
Define or refine control and access to reclaimed water	Determine the rights to and limits of access and control over reclaimed water for subsequent use and the relationship between the underlying water right, wastewater collection system ownership, reclamation plant ownership, and downstream water users who have demonstrated good-faith reliance on the return of the wastewater effluent into a receiving stream within the limits and requirements of the state's water rights statutory and regulatory requirements
Relationship to other rules	Describe relationship between water reuse rule or regulation and, for example, water and wastewater regulations, environmental flow requirements, solid waste or hazardous waste rules, groundwater protection, required water management plans, and relevant health and safety codes for housing, plumbing, and building
Relationship to stakeholders	Identify regulatory or nonregulatory stakeholders from various sectors (e.g. water, wastewater, housing, planning, irrigation, parks, ecology, public health, etc.) that have a role or duty in the statewide reuse program
Relationship to regulations or guidelines for uses of other nonconventional water sources	Describe other rules or regulations that exist for gray water recycle and storm water or rainwater harvesting and use
	Some states may choose to develop a more comprehensive approach that encompasses rules or regulations for all nonconventional water sources, including water reuse, within one set of rules or regulations
Reclaimed water standards	Include a provision to evaluate and allow standards to be developed on a case-by-case basis for less common uses of reclaimed water that are not listed
	Require points of compliance to be established to verify compliance with standards
	Describe response and corrective action for occurrence of substandard reclaimed water (a component of the contingency plan, below)

(*Continued*)

Table 13.25 (Continued)

Category	Comment
Treatment technology requirements	In addition to reclaimed water standards, some states specify treatment technologies for specific reuse applications
Monitoring requirements	Describe methods and frequency for monitoring all standards listed in the rules or regulations
Criteria or standards for design, siting, and construction	Describe criteria or standards of engineering design, siting, and construction for water reuse facilities and systems that typically include, but are not limited to, facilities or systems to treat/reclaim, distribute, and store water for reuse
	Develop requirements for dual-plumbed distributions systems (separate distribution of potable and nonpotable water) that are colocated
	Describe requirements for the transfer of reclaimed water and its alternative disposal if unsuitable or not required by target user (e.g. during wet seasons)
Construction requirements	Describe requirements for engineering reports, pilot studies, and certificates required to construct and to operate
Operations and maintenance (O&M)	Describe minimum requirements for the submission and content of O&M manual. The scope and content of an O&M manual will be determined by the type and complexity of the system(s) described by the manual
Management of pollutants from significant industrial users as source water protection	Where facilities or systems with inputs from significant industrial users are proposing to generate reclaimed water suitable for human contact or potable reuse, describe programs that must be implemented to manage pollutant of concern from significant industrial users
	Pretreatment programs of combined publicly owned treatment works and reclamation systems may satisfy program requirements
	Develop program requirements for satellite reclamation systems also affected by inputs from significant industrial users
	Such pretreatment programs should develop discharge limits that are intended to protect source water rather than wastewater treatment and sewer system integrity
Access control and use area requirements	Describe requirements to control access to sites where reclaimed water will be generated or, in some cases, stored or utilized
	Describe requirements for advisory sign placement, message, and size
	Describe requirements for proper use of reclaimed water by end users to ensure protection of the environment and human health (e.g. setbacks, physical barriers or practices to prevent reclaimed water from leaving the site of use, etc.)
Education and notification	Include requirements for generators or providers of reclaimed water to educate end users of appropriate handling and use of the water and to provide notification to end users regarding the discharges of substandard water to reuse and loss of service for planned or unplanned cause
Operational flow requirements	Requirements for maintaining flow within design capacity of treatment system or planning for additional treatment capacity as needed
Contingency plan	Include a requirement for a contingency plan that describes how system failures, unauthorized discharges, or upsets will be remedied or addressed
Recordkeeping	Describe what operating records must be maintained, the location where they are retained, and the minimum period of retention
Reporting	Describe what items must be reported, the frequency of reporting, and to whom they are reported
Stakeholder participation	Requirements on public notice, involvement, and decision making. This will apply where the water reuse rule or regulation is used as the vehicle to permit water reuse projects
Financial assistance	Describe state, local, or federal funding or financing sources

Indicative of this trend is a profound paradigm shift that has occurred in wastewater treatment. Historical treatment of wastewater has focused on remediation of the water for the protection of human health and the environment. The focus has recently been redirected toward the perception of wastewater as a rich mixture of resources (including water itself) that may be recovered and reused. Integrated resource recovery efforts have led to a "rebranding" of wastewater treatment plants as "resource recovery" facilities where waste is viewed as a potential resource.

According to a report issued jointly by the US EPA, the U.S. Department of Energy, and the National Science Foundation in 2015 (National Science Foundation, U.S.

Department of Energy, and U.S. Environmental Protection Agency, 2015), the aging US water and wastewater infrastructure will require an investment of $600 billion over the next 20 years. In recognizing the inherent value of resources that are contained in both municipal and industrial wastewater, water reclamation and reuse considerations should be coupled to efforts to recover additional resources such as energy, nutrients, and metals. This may provide a reduction in capital and operational expenses associated with the necessary replacement and expansion of water and wastewater infrastructure.

13.7.1 Energy

Wastewater treatment plants are often the single largest consumer of energy in a community. The integration of water resource recovery with power systems may allow utilities to reduce energy consumption, reduce infrastructure capital and operating costs, and increase energy generation. As illustrated in Figure 13.3, energy contained in wastewater may be further characterized as thermal energy, chemical energy, and hydraulic energy.

Regarding thermal energy, the average temperature of wastewater is typically several degrees warmer than ambient temperatures. This low-grade heat may be captured to drive other heat intensive processes in the wastewater treatment plant, such as anaerobic digestion preheating.

Biogas, a by-product of anaerobic treatment of organics in wastewater, is the primary source of chemical energy that may be extracted from wastewater. Typical composition of biogas includes approximately 60% methane and 40% carbon dioxide. The energy-dense methane fraction of biogas should be recognized as an in situ resource as energy to be recovered rather than flared. Advancements in biogas upgrading and conversion technologies have allowed biogas utilization to be economically feasible for wastewater treatment facilities of all sizes.

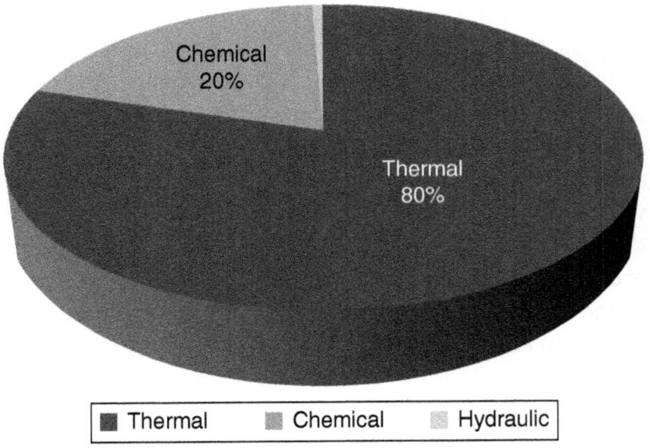

Figure 13.3 Energy sources associated with wastewaters (NSF, US DOE, and US EPA, 2015).

Secondary constituents contained in biogas may include water, siloxanes, and hydrogen sulfide (H_2S). These secondary constituents have provided some challenges to the economical conversion of biogas to heat and/or electricity, as they need to be removed prior to energy generation in traditional internal combustion engines or turbine-type systems. However, advancements in both cleaning and conversion technologies have made biogas energy more accessible to wastewater treatment plants of all sizes. For example, Stirling engine technology has recently been applied for combined heat and power production from biogas, as the Stirling engine relies on external combustion technology that reduces or eliminates the need to "scrub" the biogas of H_2S and siloxanes prior to conversion.

Also, the application of anaerobic pretreatment toward high-strength organics-rich industrial wastewater can facilitate an elevated level of energy recovery from these types of wastewater, as well as a reduction in the cost, footprint, and energy associated with downstream treatment of the pretreated wastewater toward discharge or reuse quality.

While only a small fraction of the energy may be recoverable from wastewater, hydraulic energy may be captured as water flowing downhill that can provide energy used to drive turbines or other mechanical systems.

13.7.1.1 Example of Energy Recovery and Water Reuse Treatment in Industrial Wastewater Treatment

An example flow sheet illustrating the coupling of wastewater treatment for an industrial high-strength/high-BOD reuse application to a resource recovery system for water reuse and chemical energy recovery is provided in Figure 13.4.

Referring to Figure 13.4, the wastewater may first be passed through a screening step to remove some of the larger particulate matter. The screened effluent may then be sent to an equalization/preacidification basin, wherein complex organics are partially fermented and converted to easily degradable volatile fatty acids. This process will increase treatment efficiency and reduce the energy consumption of downstream biological unit processes.

Further reduction in fats, oils, and greases (FOG) as well as TSS may then be achieved using a primary liquid–solid separation process, such as sedimentation or DAF. A portion of the insoluble organics is removed in this step.

The wastewater may then be fed into a high rate anaerobic reactor wherein influent wastewater is evenly distributed beneath a bed of granulated active biological sludge and flows upward through the sludge layer. Dissolved organic material in the wastewater is degraded

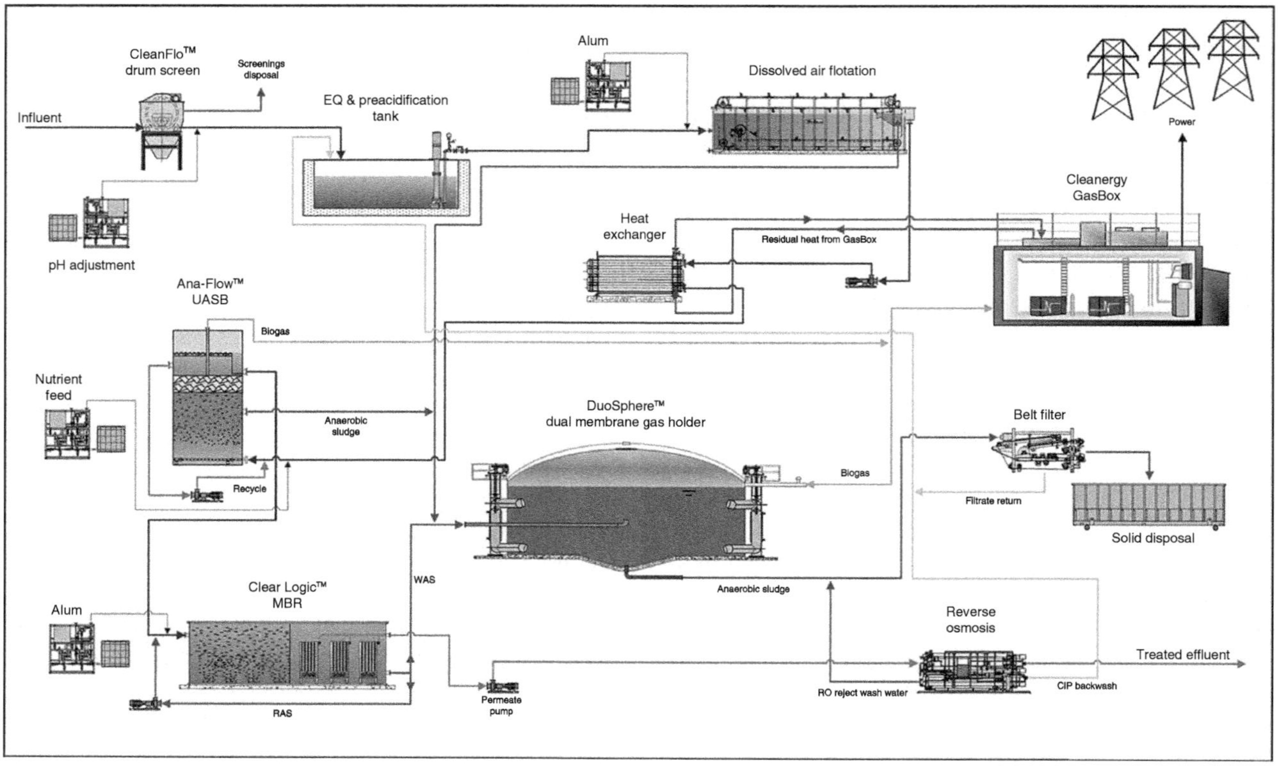

Figure 13.4 Reuse treatment schematic for high-strength waste. *Source:* WesTech (2016b). Reproduced with permission of WesTech Engineering, Inc.

in the absence of oxygen and converted to biogas, which may be captured for heat and/or electricity generation. A reduction in BOD of 70–90% may be achieved.

The overflow from the anaerobic treatment may be further treated in an aerobic polishing process to further reduce the organics and to treat/recover nutrients that may be in the waste stream. A variety of aerobic processes may be employed depending on the volume and strength of the waste stream, as well as the reuse application.

Wasted solid material from the DAF, anaerobic stage, and aerobic polishing stage may be fed into a conventional completely stirred anaerobic digester to produce additional biogas that is combined with the biogas from the anaerobic pretreatment step and converted to heat and electricity.

A final filtration step may be employed to further reduce dissolved solids depending on reuse quality requirements. This may be accomplished using ultrafiltration, RO, ion exchange, or nanofiltration (or a combination of these).

13.7.2 Nutrients

Increasing attention is being given to technologies that may be applied to recover nutrients rather than remove nitrogen and phosphorus from wastewater. The usual treatment approach has been to treat them by converting them into nitrogen gas and phosphate salts. However, both nitrogen and phosphorus are critical components in fertilizer and thus carry value as a potential revenue stream for a resource recovery facility. Further, phosphorus supplies on the planet are finite, and current mining methods for phosphorus are unsustainable long term. Therefore, recycling phosphorus may help long-term sustainability of global fertilizer supplies.

Ammonia/ammonium, typically found in anaerobic digestates, may be converted to ammonium fertilizers, such as ammonium sulfate, rather than oxidizing then reducing ammonium/ammonia to nitrites/nitrates and then to nitrogen gas. When ammonia is produced as a synthetic chemical, large energy and cost expenditures are incurred. Therefore, recovery of ammonia from wastewater and other organic sources may help to offset the global energy demand associated with new ammonia production.

Phosphate's precipitation as a constituent of struvite was once considered a nuisance for anaerobic digestion systems. However, the value of struvite as a source of phosphorus and magnesium for fertilizer applications has recently inspired the development of technology that intentionally precipitates and recovers struvite prior to anaerobic digestion as a method for both decreasing the adverse effects on anaerobic digesters and as a method for recovering and selling a valuable fertilizer constituent.

Other potential methods for nutrient recovery are in various stages of development, including the use of microalgae and other microorganisms as a biological uptake method of nitrogen and phosphorus (and in some cases, soluble carbon). The microalgae may be collected and then used as higher value products such as protein production for animal feed and fertilizer components.

13.7.3 Future of Wastewater Treatment, Reuse, and Resource Recovery

These systems may provide economic, environmental, and social benefits, including reduction in greenhouse gas emissions associated with energy consumption and biogas flaring, production of carbon-neutral forms of energy, flexible infrastructure, production of new revenue streams to offset infrastructure cost, and a reduction of costs when compared with managing each waste stream individually.

With regard to water reuse, the implementation and integration of various forms of resource recovery into water reclamation facilities may provide for on-site generation of energy that is used to drive more energy-intensive advanced wastewater treatment for reuse applications, as well as additional economical drivers for enhanced wastewater treatment, such as more effective nutrient recovery.

Public perception of the beneficial impacts of a holistic resource recovery initiative that includes water reuse may be extremely beneficial in the development of a robust water reuse program with regard to human health protection, environmental sustainability, innovation, and long-term global economic stability.

13.8 Additional Sources of Information

The US EPA has developed two documents addressing water reuse in both the United States and throughout the world: *2004 Guidelines for Water Reuse* (US EPA, 2004) and *2012 Guidelines for Water Reuse* (US EPA, 2012). The guidelines were updated in 2012 to address new applications and advances in technology and to update information concerning the US state regulations.

The WHO has also developed comprehensive guidelines for wastewater recycling, *Guidelines for the Safe Use of Wastewater, Excreta, and Greywater* (WHO, 2006).

References

Bockelmann, U., Dorries, H., Ayuso-Gabella, M.N. et al. (2009). Quantitative PCR monitoring of antibiotic resistance genes and bacterial pathogens in three European artificial groundwater recharge systems. *Applied and Environmental Microbiology* 75: 154.

California Department of Health Services (2014). Water recycling criteria. In: *California Code of Regulations, Title 22, Division 4*, Chapter 3. Sacramento, CA: California Department of Health Services.

Federal Water Quality Administration (1970). *Federal Guidelines: Design, Operation, and Maintenance of Waste Water Treatment Facilities*. Washington, DC: Federal Water Quality Administration, U.S. Department of the Interior.

FAO (1985). FAO Irrigation and Drainage Paper, 29 Rev. 1. Food and Agriculture (FAO) of the United Nations, Rome, Italy.

Fitzpatrick, J., Jerry Ussher, J., Tim Weaver, T. et al. (2015). Compressible media filtration helps increase peak wet-weather flow treatment capacity and decrease untreated overflows. *Water Environment & Technology Magazine* 27 (6): 46–49.

Florida Department of Environmental Protection (FDEP) (2016). *Florida's Reuse Activities*. Florida DEP, Tallahassee, FL. https://floridadep.gov/water/domestic-wastewater/content/floridas-reuse-activities (1 July 2016).

Florida Department of Environmental Protection (FDEP) (2018). *Florida's Reuse Activities*. Florida DEP, Tallahassee, FL. https://floridadep.gov/water/domestic-wastewater/content/floridas-reuse-activities (25 January 2018).

Knapp, W., Dolfing, J., Ehlert, I., and Graham, W. (2010). Evidence of increasing 429 antibiotic resistance gene abundances in archived soils since 1940. *Environmental Science and Technology* 44: 580–587.

Metcalf and Eddy (2003). *Wastewater Engineering, Treatment and Reuse*, 4th edition. New York, NY: McGraw-Hill.

Metcalf and Eddy (2007). *Water Reuse: Issues, Technologies, and Applications*. New York, NY: McGraw-Hill.

National Science Foundation, U.S. Department of Energy, and U.S. Environmental Protection Agency (2015). *Energy-Positive Water Resource Recovery Workshop*

Report (28–29 April). Arlington, VA: National Science Foundation, U.S. Department of Energy, and U.S. Environmental Protection Agency.

Pauwels, B. and Verstraete, W. (2006). The treatment of hospital wastewater: an appraisal. *Journal of Water and Health* 4: 405.

Sumpter, J.P. and Johnson, A.C. (2008). Reflections on endocrine disruption in the aquatic environment: From known unknowns to unknown unknowns (and many things in between). *Journal of Environmental Monitoring* 10: 1476–1485.

Tchobanoglous, G., Burton, F.L., and Stensel, D.H. (2003). *Wastewater Engineering: Treatment and Reuse*. New York, NY: McGraw-Hill.

United Nations Development Programme (2006). *Beyond Scarcity: Power, Poverty, and the Global Water Crisis*. New York, NY: Human Development Report, UNDP.

United States Environmental Protection Agency (1974). *Design Requirements for Mechanical, Electric, and Fluid Systems and Component Reliability, Supplement to Federal Guidelines: Design, Operation, and Maintenance of Wastewater Treatment Facilities*, EPA-430-99-74-001. Washington, DC: Office of Water Program Operations, U.S. Environmental Protection Agency.

United States Environmental Protection Agency (2004). *2004 Guidelines for Water Reuse*, EPA/625/R-04/108. Washington, DC: Office of Wastewater Management, U.S. Environmental Protection Agency; Cincinnati, OH: National Risk Management Research Laboratory, Office of Research and Development; and Washington, DC: U.S. Agency for International Development.

United States Environmental Protection Agency (2011). *Introduction to the National Pretreatment Program*, EPA-833-B-11-001. Washington, DC: Office of Wastewater Management, U.S. Environmental Protection Agency.

United States Environmental Protection Agency (2012). *2012 Guidelines for Water Reuse*, EPA/625/R-04/108. Washington, DC: Office of Wastewater Management, U.S. Environmental Protection Agency; Cincinnati, OH: National Risk Management Research Laboratory, Office of Research and Development; and Washington, DColumbia: U.S. Agency for International Development.

WateReuse Association. (2009). How to Develop a Water Reuse Program: Manual Of Practice. WRA-105. WateReuse Association, Alexandria, VA.

WateReuse Foundation (2006). Marketing Nonpotable Recycled Water, A Guidebook for Successful Public Outreach & Customer Marketing. WateReuse Foundation, Alexandria, Virginia.

WesTech (2016a). A typical flow scheme for treatment of reclaimed water for boiler feed water. In: *WesTech Process Flow Sheet Manual*. Salt Lake City, UT: WesTech Engineering, Inc.

WesTech (2016b). Re-use schematic for high strength waste. In: *WesTech Process Flow Sheet Manual*. Salt Lake City, UT: WesTech Engineering, Inc.

World Health Organization (2006). *WHO Guidelines for the Safe Use of Wastewater, Excreta and Greywater*. Geneva, Switzerland: World Health Organization and United Nations Environment Program.

14

Design of Porous Pavements for Improved Water Quality and Reduced Runoff

Will Martin[1], Milani Sumanasooriya[2], Nigel B. Kaye[3], and Brad Putman[3]

[1] General Engineering Department, Clemson University, Clemson, SC, USA
[2] Department of Civil & Environmental Engineering, Clarkson University, Potsdam, NY, USA
[3] Glenn Department of Civil Engineering, Clemson University, Clemson, SC, USA

Clean water regulations, flood risk in urban areas, and other environmental concerns have led to the creation and adoption of low impact development (LID) technologies and best management practices (BMP) for land development. LID/BMP technologies include infiltration trenches, rain gardens, green roof systems, bioretention cells, cisterns, rain barrels, and porous pavements (Field et al., 2004). The primary goals of many of these LID systems are to retain the water quality volume for a given land development (Akan and Houghtalen, 2003) and reduce downstream sediment loads through retention or filtration (Urbonas and Stahre, 1993). The secondary goals of these systems are to reduce the total runoff and peak discharge from a watershed for a given rainfall event and to increase stormwater infiltration and groundwater recharge compared with traditional land development stormwater management practices. The focus of this chapter is on the design and installation of porous pavement systems as an LID/BMP technology to improve water quality and reduce runoff.

14.1 Introduction

Porous pavements are essentially regular pavements with significantly less (and at times no) fine aggregate. In fact, pervious concrete is sometimes referred to as "no fines concrete." Leaving the fines out of the mix makes the resulting pavement highly porous. Laboratory tests have shown porosity values as high as 40% (Martin et al., 2013). The network of large pores (see Figure 14.1) gives the pavement a high hydraulic conductivity. Laboratory measurements of pervious concrete have exhibited hydraulic conductivity values as high as $10\,\text{cm}\,\text{s}^{-1}$ (West et al., 2016a).

The high permeability exhibited by correctly installed and well-maintained porous pavements allows them to infiltrate rainfall at a rate significantly greater than the peak rainfall intensity of even the most intense rainfall events. This leads to significantly reduced surface runoff and negligible surface water on the pavement. This is clearly seen in Figure 14.2, which shows adjacent porous and standard pavements. The foreground porous pavement has no surface water, while the background standard pavement has surface water that runs off onto the porous pavement where it percolates down into the pavement subbase.

There are a number of different porous pavement systems that can be used depending on the design requirements (discussed in more detail later). The three pavement types of relevance to this chapter are pervious concrete, porous asphalt, and permeable interlocking concrete pavers (PICP) (see Figure 14.3). All three pavement systems have a high permeability pavement layer above a porous subbase that also acts as a rainfall storage layer. Another widely used type of porous pavement called open-graded friction course (OGFC) has a porous surface layer above an impermeable pavement layer. Rain falling on OGFC percolates down to the impermeable pavement layer and then drains to the side of the roadway and then into traditional storm sewer systems. OGFC has a minimal filtration capability and does not retain water or enhance infiltration and groundwater recharge and is not discussed further in the chapter.

During a rain event, the precipitation will rapidly percolate through the pavement layer into the subbase and then infiltrate into the soil. In the event that the rainfall intensity exceeds the infiltration capacity of the soil, only some will be infiltrated, and the remainder will begin to pond in the subbase. For a large enough storm, the water

Handbook of Environmental Engineering, First Edition. Edited by Myer Kutz.

could fill the entire pavement structure and back up onto the pavement surface and runoff laterally. A schematic of a typical pavement system showing the various component dimensions and material properties is shown in Figure 14.4. This fill then overflow behavior is similar to a stormwater detention pond and, therefore, must be modeled as a pond rather than as sub-basin as would be the case for an impermeable pavement. More detail on the hydraulic and hydrologic behavior and modeling of porous pavement systems is presented in Section 14.4.

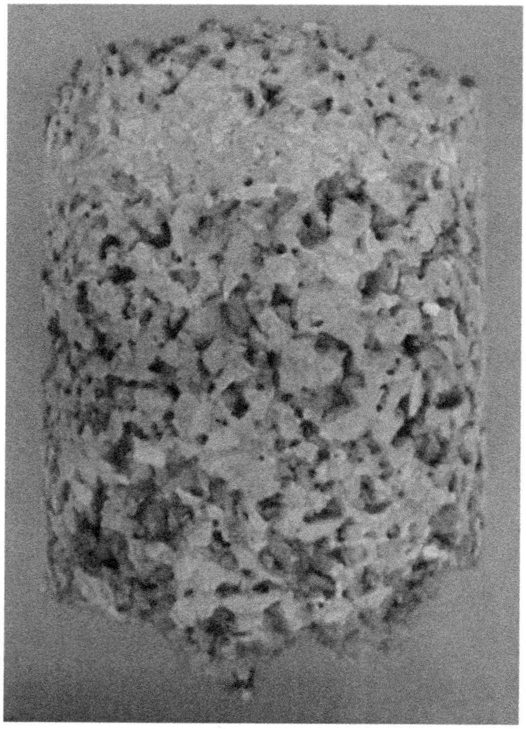

Figure 14.1 Image of a pervious concrete cylinder showing the extensive pore network that allows water to percolate down through the pavement into the subbase.

Installation of porous pavement systems has many environmental advantages over traditional pavements. The major advantage is the reduction in stormwater runoff, which reduces downstream flooding and increases infiltration of rainwater into the soil, recharging groundwater aquifers (Dietz, 2007). Rainfall retention rates can be very high, particularly for small storms (Pratt et al., 1989; Dreelin et al., 2006; Collins et al., 2008). The reduction in runoff has the more immediate economic benefit of reducing the size of downstream stormwater management infrastructure (Dietz, 2007). The prevention of runoff, particularly early in a rainfall event, reduces the impact of the first flush removal of surface pollutants into runoff. Porous pavement has, as a result, been observed to significantly reduce pollutant loads in storm sewer systems. For example, field study measurements have indicated significant reductions in ammonia, phosphorous, and zinc (Brattebo and Booth, 2003; Bean et al., 2007). Porous pavement has also been observed to aid in the reduction of urban heat island air temperature anomalies, though the exact mechanism is quite complex (Stempiha et al., 2014). A more detailed review of the environmental benefits of porous pavements is discussed in Section 14.2.

These benefits are not without cost. Porous pavements are harder to install and, therefore, more expensive than traditional pavements. Poor installation can result in

Figure 14.2 Photograph of Centennial Blvd. on the campus of Clemson University with adjacent permeable (foreground) and impermeable (background) pavements showing the lack of surface runoff from the permeable pavement. *Source:* Photo courtesy of Brad Putman.

Figure 14.3 Images of various porous pavements. From left to right: Pervious concrete, porous asphalt, and permeable interlocking concrete pavers. *Source:* Photos courtesy of Brad Putman.

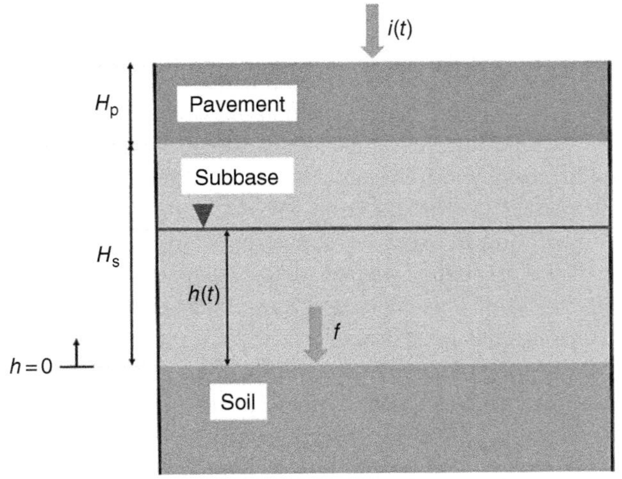

Figure 14.4 Schematic diagram of a typical undrained porous pavement cross section showing the pavement layer of thickness H_p, the subbase of thickness H_s, and the soil layer. Also shown is the rainfall intensity $i(t)$, the depth of water in the pavement system $h(t)$ measured from the top of the soil, and the infiltration rate into the soil layer $f(t)$.

surface sealing and poor performance. They can only be used for certain applications such as low traffic volume streets and parking lots. Once installed, porous pavements require regular maintenance to maintain performance. Sediment and debris can clog the pavement pores reducing performance. Frequent sweeping or vacuuming is needed, and sediment control systems may be needed to protect against sediment run-on. These and other materials and construction issues are discussed in Section 14.5.

The goals associated with porous pavement installation vary considerably, leading to a range of different design criteria. Some pavements are installed purely to eliminate surface waters in which case pavement hydraulic conductivity and subsurface drainage are key. Pavements designed to retain the first flush or water quality volume are designed to store that volume in the subbase and then provide adequate drainage so that the subbase draws down within a prescribed time. Pavements can be designed to reduce the peak runoff from a design storm event in which case modified retention pond design tools are appropriate. More recently, models have been developed to characterize porous pavements using an effective curve number (ECN) (Schwartz, 2010), which has led to the development of tools for designing a pavement to achieve a desired ECN (Martin and Kaye, 2014, 2016).

The remainder of this chapter is structured as follows. Section 14.2 reviews field data on the environmental benefits of porous pavement systems including improved water quality, pollutant removal, sediment filtering, reduced runoff, and enhanced on-site infiltration of rainfall. The hydraulic properties of porous pavements are discussed in Section 14.3. This section reviews standard test techniques for characterizing pavement porosity and permeability as well as the latest research on advanced measurement techniques. Methods for characterizing the hydraulic and hydrologic behavior of porous pavements are reviewed in Section 14.4 including standard flow routing models and available preliminary hydrologic design tools that allow porous pavement systems to be incorporated into stormwater management plans. Finally, Section 14.5 reviews the mechanical design, installation, and maintenance of porous pavement systems including pavement material mix designs, placement, on-site quality control, and maintenance planning.

14.2 Benefits

Porous pavement systems have many benefits, but the primary motivators behind the use of porous pavements are their ability to improve the quality and reduce quantity of stormwater runoff. This section looks at these main benefits as well as other minor ones.

14.2.1 Water Quality

Pollutants in stormwater can be either physical particles, measured as suspended solids, or dissolved chemical pollutants like heavy metals, hydrocarbons, and organic pollutants like nitrogen and phosphorus. Porous pavements have the ability to reduce pollutant loads of both physical and chemical pollutants through a combination of filtration, sorption, chemical degradation, and biological activity. The magnitude of impact each of these processes has depends on many factors such as whether the pavement system has an underdrain or not or the type of soil under the pavement, but all these processes are present to some degree in most pavement systems.

The first process by which water quality is improved is through physical filtration, which is a natural by-product of the porous pavement material properties. Because the stormwater has to flow through the pore structure of the surface pavement and the aggregate subbase, larger sediment particles are trapped as in a filter. Most commonly, the lowest porosity layer in the pavement structure is the surface pavement where the sediments can then be removed through regular maintenance. The exact size of particles trapped by the surface pavement will depend on the mix design, construction methods, and the final pore distribution after construction. Preliminary research looking at the pore distribution of compacted pavements (porous asphalt and pervious concrete) found that the porosity of the surface pavement was not constant from top to bottom but rather tended to have a minimum near the top and gradually increased further down the pavement due to surface compaction (Martin et al., 2014). See Figure 14.5. This means that most of the larger trapped sediment will be located very near the top of the surface pavement, which is conducive to its removal if maintenance is conducted in a timely manner.

Due to the subbase typically having higher porosity and larger pores, the finer sediment that passes through the surface pavement will likely pass through the entire pavement structure and collect at the bottom of the pavement if there is no underdrain or on a geotextile if present (Chopra et al., 2010; Lucke and Beecham, 2011; Mata and Leming, 2012). Pavements with underdrains can see some of the fine material washed out, but this can be limited by having the underdrain slightly above the bottom of the subbase to create a space for the sediments to settle and not be washed out of the underdrain. Clearly any additional space below the underdrain would be a finite capacity; therefore it is still important that the pavement is well maintained and designed as to reduce the likelihood of excessive sediment loading (like excessive sediment being washed onto the pavement during construction). This is discussed more in Section 14.5.

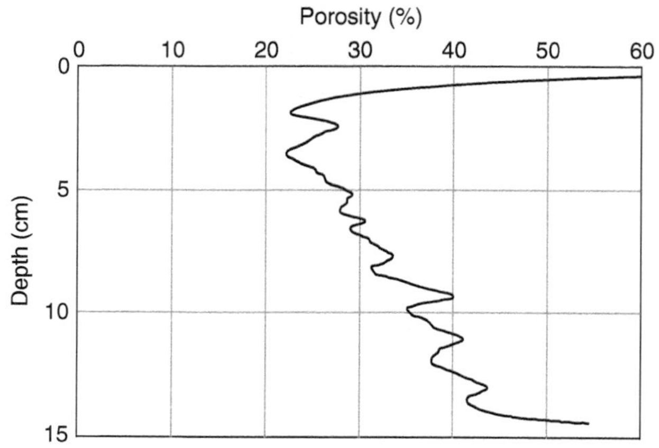

Figure 14.5 An example of the vertical porosity distribution that can be present in surface-compacted porous surface pavements. *Source:* From Martin et al. (2014).

Many studies have looked at the total suspended solid (TSS) removal rate of pavements, and most report a reduction in the range of 67–99% (Rushton, 2001; Gilbert and Clausen, 2006; Van Seters et al., 2006; Bean et al., 2007; UNHSC (University of New Hampshire Stormwater Center), 2009).

While the physical filtration reduces the TSS in the runoff, it has the added benefit of reducing some of the chemical pollutants adsorbed to these sediment particles. Similarly, some soil types, notably clay, have the ability to adsorb heavy metals, which can further reduce their presence in outflow from the pavement system (Debo and Reese, 2002). Of the studies which measured heavy metals, the range of reduction was 13–97% with the median value around 75% (Rushton, 2001; Barrett et al., 2006; Gilbert and Clausen, 2006; Van Seters et al., 2006; Bean et al., 2007; UNHSC (University of New Hampshire Stormwater Center), 2009).

Other common organic pollutants such as hydrocarbons, nitrogen, and phosphorous compounds can be removed from the stormwater by providing an environment conducive to the natural breakdown of pollutants through degradation or biological activity. Early on Pratt et al. (1999) demonstrated that porous pavement systems could be used as an effective in situ aerobic bioreactor for hydrocarbons. This ability comes from the presence of microbial communities that survive in the subbase due to the availability of moisture, air, and organic compounds (Fan et al., 2013). The availability of hydrocarbons actually increases the microbial population, which in turn increases the pavement's ability to degrade organic material (Mbanaso et al., 2013). When a microbial community from a four-year-old pavement was compared with a commercially available oil-degrading microbial mixture, there was no significant difference in their ability to degrade oil (Newman et al., 2002).

This same microbiological community helps remove nitrogen and phosphorus from the stormwater. Reported reductions for nitrogen compounds fall in the range of 35–67%, and the reductions of phosphorus are reported at 34–65% (Rushton, 2001; Gilbert and Clausen, 2006; Van Seters et al., 2006; Bean et al., 2007; UNHSC (University of New Hampshire Stormwater Center), 2009). The Interlocking Concrete Pavement Institute suggests designing the subbase to detain water longer to encourage denitrification (Smith, 2011), which could improve the removal efficiencies.

14.2.2 Water Quantity

As important as the water quality benefits porous pavements provide, they also provide significant water quantity benefits. This includes reducing the total volume of runoff and peak runoff rate and delaying the runoff. These benefits, if accounted for in design, can reduce or eliminate the need for conventional stormwater infrastructure.

Porous pavement systems reduce the runoff volume for a storm event by allowing infiltration of the stormwater into the soil below the pavement. This infiltration occurs both during the storm and after if water is retained in the pavements' subbase. However, the overall volume reduction depends on a number of factors including if the pavement system has an underdrain, the depth of the pavement system, and even the rainfall intensity. Clearly pavement systems without an underdrain will tend to have a much larger runoff reduction as all the rainfall that enters the pavement will be infiltrated. Thus only large storms that completely fill the pavement will have any runoff. Because of this behavior, the majority case studies that look at undrained pavements report a high percent reduction of 100% since most studies are only a year or two in duration and likely experience smaller, more frequent storms (Booth and Leavitt, 1999; Pratt et al., 1999; Rushton, 2001; Brattebo and Booth, 2003).

While pavements without underdrains will generally have a larger rainfall retention, underdrained pavements still have the capacity to retain a significant runoff depth. The reported percent reductions in runoff from underdrained pavements range from 25 to 66% (Booth and Leavitt, 1999; Roseen et al., 2009; Fassman and Blackbourn, 2010). As mentioned before, this large range is dependent on a number of variables, including the infiltration rate of the underlying soil, the rainfall depths experienced during the case study, and the size and placement of the underdrain. For example, instead of placing the underdrain at the bottom of the subbase layer, it can be raised, leaving a region below the drain where all water will have to infiltrate into the soil. This will significantly improve the runoff reduction for smaller depth storms. Section 14.4 will address the design of porous pavement systems and show exactly how the design variables impact the overall performance.

It is important to note while looking at the results of case studies that even though a simple overall percent reduction in runoff is easy to report and understand, those values do not provide a good understanding of the actual hydrologic behavior of these pavements. Moreover, they cannot be accurately used in hydrologic design or modeling. Because of this, a number of researchers have worked on better quantifying the volume reduction in runoff for porous pavements, and those methods and models are discussed in detail in Section 14.4.

By reducing the runoff volume by use of infiltration, porous pavements have the additional benefit of recharging the groundwater system. This mimics the natural water cycle more than traditional stormwater infrastructure (Klein, 1979; Finkenbine et al., 2000) and can play a part in restoring aquifers that have been depleted by years of overuse and limited recharge.

Peak runoff and timing control are two other significant benefits for porous pavement systems. For underdrained systems, the timing and rate of the discharge will be attenuated similar to a retention pond. In fact, some people view porous pavements as a type of underground retention pond. This analogy is actually well suited for underdrained pavement systems as the porosity provides the storage volume, the underdrain acts as the outlet control, and the pavement system filling and overflowing would be like runoff from an emergency spillway.

One noticeable exception to the idea of a porous pavement as an underground retention pond is undrained pavements. It should be rare for an undrained pavement to completely fill and overflow, but if it does, there will be virtually no attenuation of the runoff that does occur. At that time there will be little difference than runoff from a traditional impervious pavement. This underscores the importance in the design of undrained porous pavements that they be sized to store the design storm volume of interest.

14.2.3 Other Benefits

Beyond water quality and quantity, porous pavements have other less well-known benefits. These include a potential decrease in project cost, increase in road safety (improved traction and visibility during rain events), noise reduction, and reduced heat island effect.

Cost savings have been reported (US EPA, 2007; Smith, 2011) when using porous pavements on projects. This is not due to the porous pavement material or installation being less expensive than conventional pavements (typically porous pavements are more expensive). However, the overall cost savings come from the elimination or significant decrease in the size of stormwater infrastructure as part of the project. This means that cost savings will only be realized on a project if the hydrologic performance of the pavement can be accurately determined and regulators approve it replacing the traditional stormwater infrastructure. Therefore, the exact cost–benefit of porous pavements is very difficult to quantify because different permitting agencies evaluate and credit porous pavements differently.

Driver safety is improved in two ways by porous pavements. The first is improved visibility due to reduced splash and spray during rain events (Rungruangvirojn and Kanitpong, 2010). Because the water is removed from the surface of the pavement almost instantaneously, there is less water to be thrown up by a passing car. Similarly, because the water is removed from the pavement surface, there is less likelihood of hydroplaning during rainstorms. Additionally, the texture porous pavements have been shown to increase surface friction and therefore decrease driver's stopping distance even when the pavement is not wet (Kowalski et al., 2009).

Porous pavement has also been found to produce reductions in road noise. Because of the interconnected voids in the pavement, the tire–pavement interaction produces less noise, and the pore structure absorbs frequencies of sound that are in the typical range associated with traffic noise (Crocker et al., 2004).

In urban areas, the use of conventional pavements and presence of buildings have contributed to a buildup of heat known as the urban heat island effect. Typical pavements absorb heat during the day and then release the heat during the night, significantly warming the evening temperatures and causing heat to build up in the area over a longer period of time. Porous pavements have the potential to combat this problem as it has a more open structure, which transfers less heat to the ground below. Therefore, there is less heat to be released at night (Haselbach and Gaither, 2008; Stempiha et al., 2014). Limiting the heat gain to the ground not only decreases air temperature overall, but it also reduces the temperature increase of runoff.

14.3 Hydraulic Characterization

Porous pavements should be designed primarily to achieve desired hydraulic benefits. Thus, permeability is the most important hydraulic property of porous pavements from a functional perspective, and undeniably porosity is one of the most important pore structure features that ensure the desired hydraulic performance. This section reviews porosity and permeability of porous pavements and several test methods available to examine these properties. It should be noted that the test methods discussed in this section are pertinent to any porous pavement materials in general including porous asphalt, even though some of the results presented here are based on pervious concrete studies.

14.3.1 Porosity

Porosity or air void fraction is one of the most definitive features of the material structure of any porous material. Porous pavements are primarily designed for water transport through its material structure while maintaining sufficient mechanical properties. Therefore, porosity is to be considered as its most important physical property, and a certain nonminimal porosity should be maintained in the pavement through its life span to assure desired performance requirements. It has been reported that (Meininger, 1988) a minimum of 15% porosity is needed to assure flow through pervious concretes. The porosity of pervious concrete is typically in the range of 15–30% (Meininger, 1988; Marolf et al., 2004; Tennis et al., 2004a; Neithalath et al., 2006; American Concrete Institute (ACI), 2013). Proper mixture proportioning methods should be adopted to produce porous pavement materials having porosities sufficient to transport water at desired rates. Material contents in the mixture and the compaction efforts have to be carefully controlled in order to prevent the paste from flowing off through the aggregates and closing the open pore structure, thus reducing the effective porosity and connectivity of the pores in a porous pavement.

Pore structure properties of porous materials is highly dependent on factors such as aggregate size and gradation, cementitious material or asphalt content, water content, water-to-cement ratio, and compaction efforts. It has been reported that the porosity is a function of compactive effort, particle shape and texture of the aggregate, and uniformity coefficient of the aggregate for a constant paste content for pervious concretes (American Concrete Institute (ACI), 2013). Some studies have reported influence of aggregate size and gradation on the porosity of hardened pervious concretes (Marolf et al., 2004; Neithalath, 2004). The variation in porosity

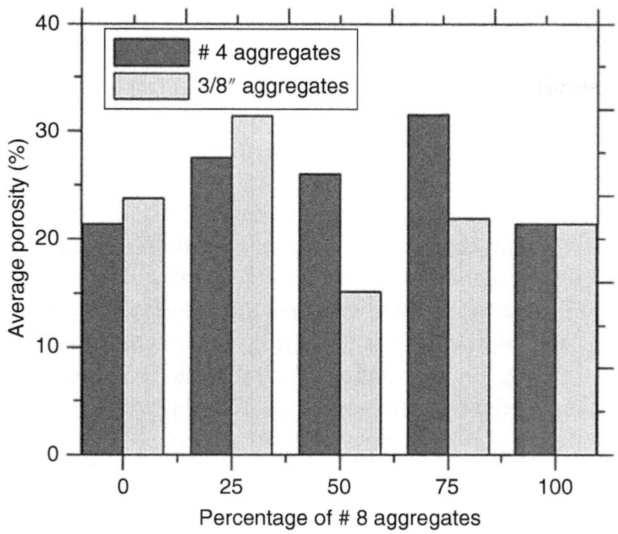

Figure 14.6 Variation in porosity of pervious concretes made with binary blends of #8 and either #4 or 3/8″ aggregates. *Source:* From Neithalath (2004).

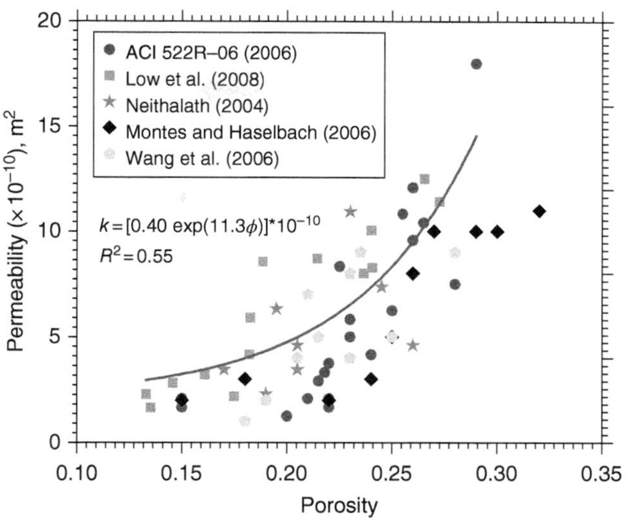

Figure 14.7 Porosity–permeability relationships for several pervious concretes mixtures (Neithalath et al., 2010). *Source:* Reproduced with permission of Elsevier.

with percentage of #8 aggregates for pervious concrete mixtures with binary blends of #8 and either #4 or 3/8″ aggregates is shown in Figure 14.6. The difference in the porosities of the pervious concrete mixtures made with single-sized aggregates is insignificant as seen from this figure. However, it is observed that porosities of pervious concrete made with binary blends of aggregates of different sizes typically result in a higher porosity as compared with the single-sized aggregates as a result of loosening effect exerted by the smaller aggregates (de Larrard, 1999). ACI 522R-10 (American Concrete Institute (ACI), 2010) has reported that blending aggregates should be controlled such that the ratio of the diameter of the larger aggregate to that of the smaller one in the blend should not exceed 2.5 in order to prevent pore clogging from smaller aggregates, as it will result in reducing the porosity and consequently lowering functional performance such as reducing the permeability.

14.3.2 Permeability

Permeability or hydraulic conductivity is the most important performance characteristic of porous pavements, as it defines the flow through the material structure. It is evident that for any porous material, water transport properties are inherently dependent on several pore structure features such as porosity, pore sizes and distribution of pores, pore connectivity, tortuosity, and the specific surface area of the pores. Although, porosity alone does not determine the permeability of porous pavement materials, it has been common to relate the permeability to its porosity, primarily because of the ease with which porosity can be measured in such highly porous materials. The influence of mixture proportioning on the porosity and permeability of pervious concretes has been experimentally investigated (Chopra et al., 2006; Montes and Haselbach, 2006; Neithalath et al., 2006; Wang et al., 2006). Figure 14.7 shows the porosity–permeability relationships of pervious concretes from a few reported studies (Neithalath, 2004; ACI 522R – 06, 2006; Montes and Haselbach, 2006; Wang et al., 2006; Low et al., 2008). A general trend of increasing permeability with increasing porosity can be observed. However, it is obvious from this figure that representing the permeability as a function of porosity alone is not adequate. Martin et al. (2014) has shown that the porosity distribution across the depth of a pavement must be considered for more accurate permeability predictions. Figure 14.5 shows an example of a typical vertical porosity distribution. Empirical or semiempirical relationships such as the Kozeny–Carman equation (Berryman and Blair, 1987) and Katz–Thompson equation (Katz and Thompson, 1986) use other features of the pore structure such as the characteristic length scale that is defined by the pore structure of the material and pore connectivity (or tortuosity) to predict permeability of porous materials. Application of such methods to predict permeability of pervious concretes has been discussed in Neithalath et al. (2010).

14.3.3 Test Methods to Examine Porosity and Permeability

There are numerous methods to quantify the porosity and permeability of porous pavement materials.

This section reviews standard and commonly used test methods to examine the porosity and permeability of porous paving materials.

14.3.3.1 Determination of Porosity

Determination of the hardened state porosity of porous paving materials is relatively easy as compared with conventional paving materials due to the highly porous nature of its material structure. This section reviews determination of hardened porosity by the ASTM D7063/D7063M-11 standard test method (ASTM D7063/D7063M-11, 2011), ASTM C1754/C1754M-12 standard test method (ASTM C1754/C1754M-12 ASTM C1754/C1754M-12, 2012), the volumetric method, and the image analysis method. Determination of fresh porosities of pervious concrete using the ASTM C1688/C1688M standard test method (ASTM C1688/C1688M-14a, 2014) is also reviewed.

14.3.3.1.1 Determination of Porosity by ASTM D7063/D7063M-11 Standard Test Method

The effective porosity (accessible porosity) of porous paving materials including concrete cylinders can be determined using the ASTM D7063/D7063M-11 standard test method for effective porosity and effective air voids of compacted bituminous paving mixture samples (ASTM D7063/D7063M-11, 2011), although it is originally developed for compacted field and laboratory bituminous paving samples. This method measures the porosity of compacted samples by the use of a vacuum sealing method. The sample is first vacuum sealed inside a plastic bag and submerged under the water, and the bulk density (specific gravity), SG_1, is calculated using a water displacement method and the dry weight of the sample. The sample is then unsealed under the water, and the apparent maximum density, SG_2, is calculated knowing the saturated weight of the sample. The effective porosity of the sample is then calculated as

$$\phi = \frac{SG_2 - SG_1}{SG_2} \times 100 \tag{14.1}$$

This method has been used to determine the porosity of several porous paving material specimens including pervious concrete and porous asphalt using the CoreLok® vacuum sealer as detailed in Martin and Putman (2016).

14.3.3.1.2 Determination of Porosity by ASTM C1754/C1754M-12 Standard Test Method

The effective porosity of pervious concretes can be determined using the ASTM C1754/C1754M-12 standard test method for density and void content of hardened pervious concrete (ASTM C1754/C1754M-12 ASTM C1754 / C1754M-12, 2012) as summarized in this section. In this method, the dimensions of the hardened cylindrical pervious concrete specimen are determined using a jaw caliper. The average length (L) and the average diameter (D) of the specimen are recorded. The constant dry mass (M_D) of the oven-dried specimen is determined. The specimen is then completely submerged in a water bath (with appropriate dimensions to allow a specimen to soak completely and facilitate determining the submerged mass of the specimen without being removed from the water) and allowed to sit upright for 30 ± 5 min. The side of the fully submerged specimen is tapped 10 times using a rubber mallet while rotating the specimen slightly after each tap. The specimen is then inverted, and the submerged mass (M_S) of the specimen is determined. The density of the water (ρ_w) at temperature of the water bath is determined, and the porosity of the hardened pervious concrete is calculated as

$$\phi = \left[1 - \left(\frac{C_P \times (M_D - M_S)}{\rho_w \times D^2 \times L}\right)\right] \times 100 \tag{14.2}$$

where factor $C_P = 1\,273\,240$ in SI units or 2200 in (inch-pound) units.

14.3.3.1.3 Determination of Porosity by Volumetric Method

The volumetric method of porosity determination is another commonly adopted method to determine the effective porosity of hardened pervious concretes (Neithalath, 2004; Neithalath et al., 2006). In this method, the mass of water required to fill a sealed test is measured and converted into an equivalent volume of voids to determine the porosity. Although this method has been primarily developed for pervious concretes, it can be used to determine effective porosity of compacted samples of highly porous materials, as it uses direct measurement of the specimen dimensions to calculate the total specimen volume and the added mass of water to determine the volume of voids. The volumetric method uses cylindrical test specimens, and the top and bottom of the specimen are removed to avoid finishing effects. Several studies (Neithalath, 2004; Neithalath et al., 2006; Sumanasooriya and Neithalath, 2009; Deo et al., 2010) have used 100-mm × 200-mm-diameter-long test cylinders, and 25-mm-thick slices from the top and bottom of the cylinder are removed to obtain 150-mm-long test specimens. The test specimen is then immersed in water for 24 h to allow the pores in the cement paste to be completely saturated. After removing the specimen from water, the specimen tightly enclosed in a latex membrane. The bottom of the specimen is sealed to a stainless steel plate, and the mass of the unit, M_1, is determined. Water is added to the top of the specimen until it is filled, and the total mass of the water-filled specimen unit is determined. The mass of the added water, M_2, is calculated by subtracting the mass of the unit from the total mass of the water-filled specimen unit. The volume of water needed to fill the specimen that is equal to the volume of voids is then determined knowing the density of water, ρ_w. The volume of the voids expressed as a percentage of the total volume of the specimen (V) is the effective porosity as given in Eq. (14.3):

$$\phi = \frac{(M_1 - M_2)/\rho_w}{V} \times 100 \qquad (14.3)$$

14.3.3.1.4 Determination of Porosity by Image Analysis Method

Image analysis techniques can be used to extract several pore structure features of porous materials including porosity using stereological and morphological techniques on two-dimensional images of parent materials. Such method is also beneficial in ascertaining the variation in porosity with depth of a porous material specimen or layer. Image analysis technique has been well established as a tool to characterize the pore structure of cement-based materials (Dequiedt et al., 2001; Soroushian and Elzafraney, 2005; Hu and Stroeven, 2006). This method has been used to determine the porosity and porosity distribution of porous paving materials including pervious concrete and porous asphalt (Neithalath et al., 2010; Sumanasooriya and Neithalath, 2011; Martin et al., 2013). In this method, test specimens are sectioned into several slices (horizontally or vertically), and the cut sectioned are processed to obtain flat and smooth surfaces. The solid phase of these surfaces can be painted using an appropriate color to enhance contrast between the solid and pore phases. The processed slices are then scanned in grayscale mode. These images can be further processed and analyzed using an image analysis software. Figure 14.8 shows the typical steps in image processing and analysis using horizontal slices. As seen from this figure, the outer circumference of the image is cropped to avoid edge effects. The grayscale image is then converted to a binary image by thresholding to differentiate the pore and solid phases by carefully analyzing the gray level histogram of the image. In the resultant binary images shown in Figure 14.8, the pore phase is in black, and the solid phase is in white. The binary image is further cleaned by carefully removing the noise in the image. Individual pore structure features such as area can be extracted using an image analysis software after setting the scale of the image by knowing the actual size of the scanned surface. The individual pore areas are summed up and divided by the total area of the image to determine the area fraction of pores. The results from several random images are averaged to determine the total porosity of the specimen. It should be noted that a representative area of the sample must be imaged, and random samples must be considered in order to obtain statistically meaningful results from image analysis method of porous materials.

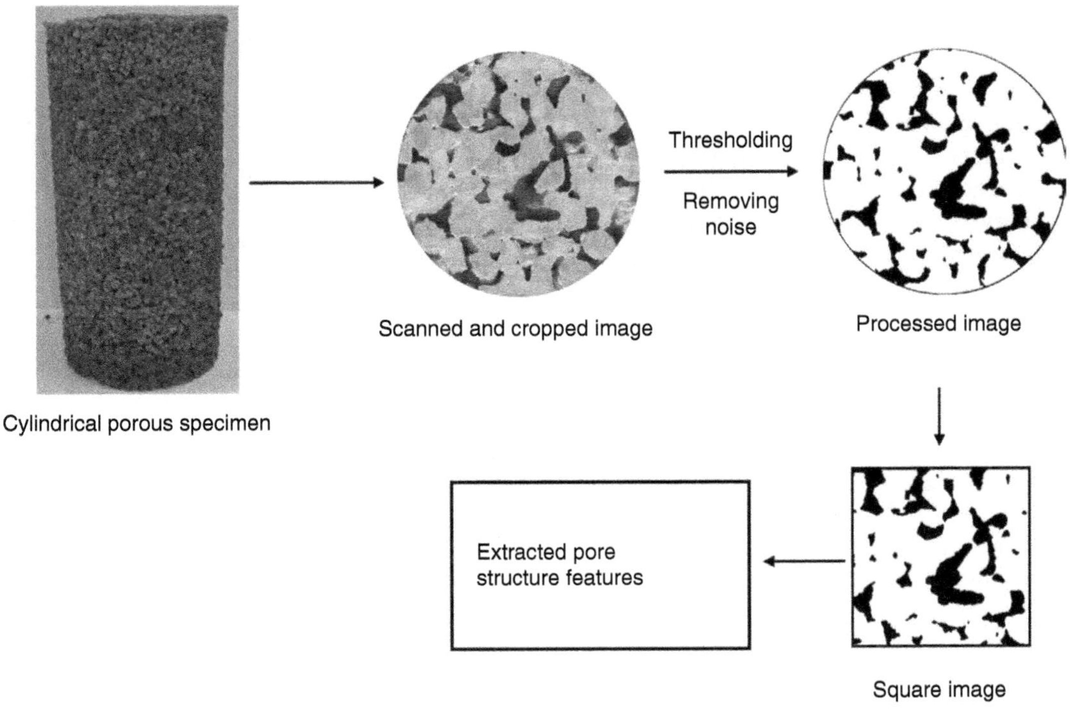

Figure 14.8 Steps in image processing and analysis.

14.3.3.1.5 Determination of Fresh Porosity by ASTM C1688 Standard Test Method

ASTM C1688/C1688M standard test method can be used to determine the density and void content (porosity) of freshly mixed pervious concrete (ASTM C1688/C1688M-14a, 2014) as summarized in this section. The masses of the component materials (cement, course aggregate, fine aggregate (if used), water, and any other materials used) are measured prior to mixing, and the total mass is calculated as the sum of masses of all the component materials. The theoretical density of the concrete (T) is then calculated as the ratio of the total mass and the sum of the absolute volumes of the component materials in the batch. The mass of a cylindrical container made of steel or other suitable metal with a capacity of $7.0 \pm 0.6L$ ($0.25 \pm 0.02\,\text{ft}^3$) and a diameter equal to 0.75–1.25 times the height is measured, and the freshly mixed pervious concrete is placed in two layers of approximately equal depth using a scoop. For each layer, 20 drops of standard proctor hammer is used at its full 305-mm (12 in.) drop height. For each layer, the tamper is positioned such that the entire surface area of the pervious concrete in the container is consolidated equally. The container is filled to overflowing before consolidating the final layer. A small quantity of fresh concrete is added after 10 proctor hammer drops to the final layer, if it appeared that there is insufficient concrete in the container or a small quantity is removed if there is too much concrete. After consolidation of the final layer, the top surface of the concrete is struck off and finished flat with a strike-off plate. After strike-off, the excess concrete is cleaned from the exterior of the container, and the mass of the concrete-filled container is determined. The net mass of the fresh concrete is calculated by subtracting the mass of the container from the mass of the concrete-filled container. The density (D_{fresh}) is calculated as the ratio of the net mass of the fresh concrete and the volume of the container. The fresh porosity of the pervious concrete mixture (ϕ_{fresh}) is then calculated as

$$\phi_{\text{fresh}} = \frac{T - D_{\text{fresh}}}{T} \times 100\% \tag{14.4}$$

Table 14.1 lists the design porosity, average fresh porosity using the ASTM C1688/C1688M method, average effective porosity using the volumetric method, and average porosity (area fraction of pores) using the image analysis method for selected pervious concrete mixtures. It can be observed that the porosity values determined using different values are fairly close. However, in general, porosity values of hardened porous paving mixtures using different methods discussed above have been found to be statically different primarily because of the manner in which they are determined as detailed in Martin and Putman (2016).

Table 14.1 The fresh and hardened porosities for pervious concrete mixtures designed for desired porosities using #8, #4, and 3/8″ aggregates.

Aggregate composition	Design porosity (%)	Fresh porosity (%)	Volumetric porosity (%)	Area fraction of pores (%)
100% #8	22	23.5	20.2	21.6
100% #4	19	19.6	19.5	18.9
	22	23.3	24.2	23.1
	27	27.2	28.9	27.7
100% 3/8″	19	19.5	17.8	20.2
	22	21.1	24.2	23.0
	27	27.1	26.4	26.2

Source: Adapted from Neithalath et al. (2010) and Sumanasooriya and Neithalath (2011).

14.3.3.2 Determination of Permeability

This section reviews determination of permeability of porous materials by the falling head test and constant head test using the standard test methods for determining hydraulic conductivity of porous materials detailed in ASTM D5856-15 (2015) and the use of ASTM C1701/C1701M-09 (2009) standard test method to determine the permeability of in-place pervious concretes.

14.3.3.2.1 Determination of Permeability by Falling Head Test

This test method uses a falling head permeameter to determine the hydraulic conductivity of porous materials by measuring the time taken for water to pass through a test specimen under an applied head loss as detailed in ASTM D5856-15 standard test method for the measurement of hydraulic conductivity of porous material using a rigid-wall, compaction-mold permeameter (ASTM D5856-15, 2015). In this method water is allowed to flow downward through the test specimen placed in the permeameter cell, and the elapsed time (t) required for water to fall from a head loss of h_1 to h_2 across the specimen is measured. Knowing the final length of the test specimen along the path flow (L), cross-sectional area of the reservoir containing water (A_1), and the cross-sectional area of the test specimen (A_2), the hydraulic conductivity K can be calculated as

$$K = \frac{A_1 L}{A_2 t} \ln\left(\frac{h_1}{h_2}\right) \tag{14.5}$$

The hydraulic conductivity, K, can be converted to intrinsic permeability (often referred as permeability) using the specific weight (γ) and dynamic viscosity (μ) of water as

$$k = \frac{\mu}{\gamma} K \tag{14.6}$$

It should be noted that the hydraulic conductivity is calculated in this test method by applying Darcy's law, assuming a laminar flow through the pervious concrete test specimen. Falling head test has been extensively used to measure the permeability of pervious concrete as detailed in (Neithalath et al., 2006; Sumanasooriya and Neithalath, 2011; American Concrete Institute (ACI), 2013; West et al., 2016a).

14.3.3.2.2 Determination of Permeability by Constant Head Test

This test method measures the quantity of flow through a porous material test specimen within a certain interval of time while keeping the head loss across the test specimen constant. The complete test procedure is detailed in ASTM D5856-15 standard test method for measurement of hydraulic conductivity of porous material using a rigid-wall, compaction-mold permeameter (ASTM D5856-15, 2015). The test specimen is placed inside the permeameter cell, and the water is allowed to pass through the specimen in the downward direction. The quantity of inflow and the quantity of outflow are measured within a time interval, Δt, driven by a difference in hydraulic head across the specimen, Δh. The quantity of flow through the specimen (ΔQ) is then calculated as the average of inflow and outflow. Knowing the final length of the test specimen along the path flow (L) and the cross-sectional area of the test specimen (A_2), the hydraulic conductivity K can be calculated as

$$K = \frac{\Delta QL}{A_2 \Delta t \Delta h} \tag{14.7}$$

The permeability of the porous specimen is then calculated from Eq. (14.6).

14.3.3.2.3 Determination of Permeability by Infiltration Test

The permeability of an in-place pervious concrete can be estimated by the ASTM C1701/C1701M-09 standard test method for infiltration rate of in-place pervious concrete (ASTM C1701/C1701M-09, 2009). In this method, a rigid cylindrical infiltration ring with a 300 ± 10 mm (12 ± 0.5 in.) diameter and a minimum height of 50 mm (2.0 in.) is sealed tightly to a clean surface of the pavement test location. After prewetting the test area by pouring a total of 3.6 ± 0.05 kg (8.0 ± 0.1 lb) water into the ring at a sufficient rate, a certain mass of water (based on the prewetting time) is poured into the ring. The time taken (t) for the water to completely pass through the pavement is recorded, and the infiltration rate, I (mm h^{-1} or in. h^{-1}), is calculated using the equation.

$$I = \frac{KM}{D^2 t} \tag{14.8}$$

where M is the mass of infiltrated water, D is the inside diameter of the infiltration ring, and the factor $K = 4\,583\,666\,000$ in SI units or 126 870 in [inch-pound] units. This test should be conducted in multiple locations (three locations for areas up to 25 000 m^2 or 25 000 ft^2) of the pavement, and results are averaged to determine the infiltration rate of the entire pavement. The infiltration test method can be viewed as a combination of the falling head test and constant head test for permeability determination (West et al., 2016b). However, a vertical flow in the downward direction through the porous medium cannot be maintained as compared with those test methods. This method can be utilized to determine the infiltration rate of new pavements and also to determine the reduction of infiltration rate of the pavement over the time due to pore clogging. However, it is not recommended to use this method as an acceptance criterion for a pervious concrete pavement as there may be a wide variation in the test results (Brown and Sparkman, 2012).

14.4 Hydraulic and Hydrologic Behavior

This section examines the hydraulic behavior of porous pavement systems and models for predicting their response to design storm events.

14.4.1 Flow Routing through Porous Pavement Systems

A typical porous pavement system consists of an aggregate subbase with a porous pavement laid on top. The whole system is placed on top of the underlying soil and may or may not include a drain within the subbase. A drain is typically required if the rate of infiltration into the soil is too slow to meet local drawdown regulations or the system needs to be drained to prevent freezing. A schematic diagram of a generic porous pavement system is shown in Figure 14.9.

The typical pavement layer has a thickness H_p between 10 and 15 cm (Martin et al., 2013) with a permeability k_p of order 10^{-9} to 10^{-8} m^2 (equivalent to a hydraulic conductivity of 1–10 cm s^{-1}) (West et al., 2016a) and porosity ϕ_p between 0.15 and 0.3 (Martin et al., 2013). This sits on an aggregate layer with thickness, permeability, and porosity (denoted H_p, k_p, and ϕ_p, respectively) all typically

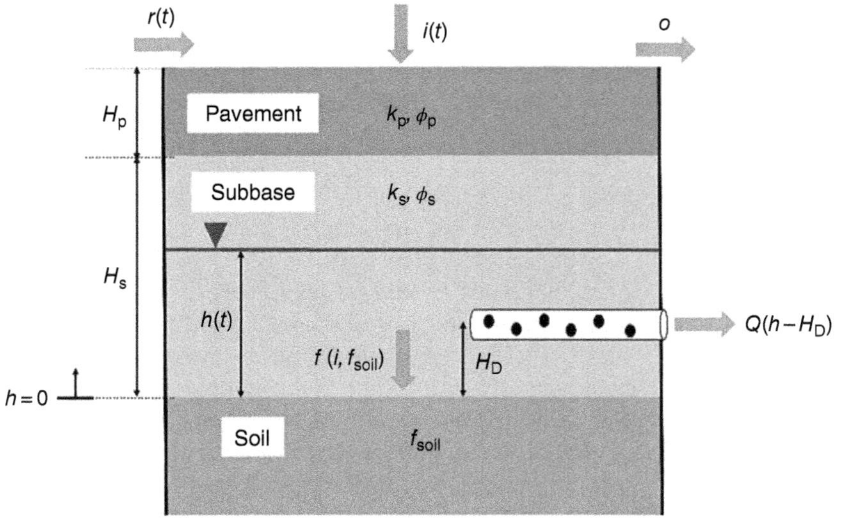

Figure 14.9 Schematic diagram of flow through a drained porous pavement layer of thickness H_p, permeability κ_p, and porosity ϕ_p, the subbase of thickness H_s, permeability κ_s, and porosity ϕ_s, and the soil layer with infiltration capacity f_{soil}. The routing parameters are the rainfall intensity $i(t)$, the depth of water in the pavement system $h(t)$ measured from the top of the soil, the infiltration rate into the soil layer $f(i, f_{soil})$, and a perforated pipe underdrain located a distance H_D above the soil.

slightly larger than the pavement layer values. Most hydraulic models treat the soil as an outlet to the system, so the only soil parameter of significance is the soil infiltration capacity (f_{soil}), which is often taken to be the soils saturated infiltration capacity. The underdrain, if needed, is typically a perforated pipe that outlets to a stormwater system typically assumed to be at atmospheric pressure. The main parameters that control the behavior of the underdrain are its height above the soil layer H_D and the stage-discharge behavior, $Q(h - H_D)$,which is discussed in more detail below.

During a rainstorm the porous pavement behaves in a manner similar to that of a detention pond. Rain will fall on the surface at a rate $i(t)$ m s^{-1} and water will run on to the pavement at a per unit area rate $r(t)$ m s^{-1} (though in order to prevent clogging, porous pavement systems are often designed so that there is no run-on to prevent clogging from suspended sediment in the run-on water). The rainfall and run-on will percolate down through the pavement and into the subbase. If the total inflow rate is greater than the infiltration rate into the soil, then the subbase will begin to fill. The depth of water is denoted herein as $h(t)$. The water accumulating in the pavement system only occupies the voids in the subbase and pavement layer, and, therefore, the storage per unit area per unit depth is the local porosity $\phi(h)$. The storage capacity of the system per unit area is, therefore, $\phi_p H_p + \phi_s H_s$. If this storage is exceeded during a rainfall event, then water will run off the pavement surface at a rate per unit area of $o(t)$ and will need to be captured and managed downstream.

During all these processes there are a range of time scales including the duration of the rainfall event (of the order of hours), the time taken to fill the total system storage (also of the order of hours), the time taken for the stored water to infiltrate into the soil layer (of the order of hours to days), and the time taken for the rainfall and run-on to percolate down into the subbase (of the order of seconds to minutes). As percolation is so much faster than all the other processes, it is reasonable to approximate the percolation process as instantaneous.

The following discussion of flow routing through porous pavement systems is based on the model of (Schwartz, 2010). The flow routing equation that allows calculation of the outflow through the underdrain and surface runoff is conservation of volume for a control volume of unit pavement area bounded by the soil and pavement surface. This can be written as

$$\phi(\mathrm{h})\frac{dh}{dt} + f + \frac{Q}{A_p} + o + a - i - r = 0. \tag{14.9}$$

where the first term is the rate of change of storage over time. The remaining positive terms are the system outflows and the negative terms are the system inflows. The local pavement porosity $\Phi(h)$ is the porosity of the pavement system at a height h above the soil and is given by

$$\phi(h) = \begin{cases} \Phi_s & 0 < h < H_s \\ \Phi_p & H_s < h < H_s + H_p \end{cases}. \tag{14.10}$$

The outflow terms are the rate of infiltration into the soil f, the flow rate out of the underdrain Q expressed per unit area of pavement A_p, the surface overflow runoff per unit area o, and an abstraction term a to account for pavement wetting losses. It is assumed that evaporation is negligible during a storm event. The system inflows are the rainfall hyetograph $i(t)$, and the run-on hydrograph per unit area $r(t)$. It is assumed that the inflow is a known function of time. The rainfall intensity hyetograph will be location and event specific, and the run-on hydrograph will depend on the rainfall hyetograph and the hydrology of the surrounding terrain.

Schwartz (2010) assumed that the infiltration capacity of the soil was constant and that the presence of the aggregate layer masked some of the areas available for infiltration, thus reducing the infiltration capacity to $\Phi_s f_{\text{soil}}$. This approach is commonly used in the literature though recent research (Martin et al., 2015) suggests that this will underestimate the true infiltration capacity of the soil. When there is no water stored in the subbase, the infiltration rate is controlled by the smaller of the rainfall intensity or the effective soil infiltration capacity. Once there is water stored in the subbase, the infiltration is limited only by the effective soil infiltration capacity. The infiltration rate into the soil is, therefore, given by

$$f = \begin{cases} \min\left(i, \Phi_s f_{\text{soil}}\right) & h = 0 \\ \Phi_s f_{\text{soil}} & h > 0 \end{cases} \tag{14.11}$$

In the original model of Schwartz (2010), a perforated pipe underdrain was treated as an orifice. That is,

$$Q = C_D A_0 \sqrt{2g\left(h - H_D\right)} \tag{14.12}$$

where $C_D$is the effective pipe discharge coefficient and A_0 is the pipe cross-sectional area. This is supported by full-scale experimental (Murphy et al., 2014) and computational (Afrin et al., 2016b) data. In general, the underdrain discharge coefficient is a function of the pipe length to diameter ratio and the perforation area per unit pipe surface area (Afrin et al., 2016b). However, all of these datasets only apply to the case where the underdrain is fully submerged and the pipe is running full at the outlet. It is, therefore, unclear how to calculate the discharge through the underdrain for a partially full pipe. See (Hager, 1999) for more details on partially full pipe outlets.

The calculation of surface runoff requires information about the flow control structures into which the runoff will flow and the local pavement topography. Schwartz (2010) and Martin and Kaye (2014) assumed that the runoff was instantaneous and that there was no significant storage of water above the pavement surface. This is a reasonable assumption for a large area parking lot; however, for smaller installations such as porous pavement bike lanes next to impervious travel lanes (Tosomeen and Lu, 2008), the depth of water above the pavement surface next to a curb may have significant storage.

For a pavement of width W, length L (parallel to the curb), and cross slope S_B ending in a curb that controls the surface runoff, the storage in the pavement is given by

$$\frac{S(h)}{WL} = \phi_s h + \left(\phi_p - \phi_s\right)\left(h - H_s\right)\mathcal{H}\left(h - H_s\right) + \frac{1-\phi_p}{2S_B}\left(h - H_s - H_p\right)^2 \mathcal{H}\left(h - H_s - H_p\right) \tag{14.13}$$

where $\mathcal{H}$ is the Heaviside step function. The first term on the right hand side is the storage in the subbase, the second term is the difference in storage in the pavement layer due to the different porosities, and the third term is the storage above the surface of the pavement. A schematic of such a pavement is shown in Figure 14.10 for an undrained pavement system. Eq. (14.9) then becomes

$$\frac{1}{WL}\frac{dS}{dt} + f + \frac{Q}{A} + o + a - i - r = 0 \tag{14.14}$$

where the first term is a more generalized rate of change of storage term (Eq. (14.13)) and the other terms are the same as for Eq. (14.9). However, in this case, the surface runoff o is now controlled by flow control structures along the curb and will, therefore, be a function of the depth of water above the curb $h - H_s - H_p$. The flow control structures along the curb can be sized based on the local spread regulations and the results of flow routing through the pavement (West et al., 2016b).

Detailed calculation of the abstraction pavement wetting loss term requires knowledge of the aggregate properties, the surface evaporation rate within the pavement system, and the antecedent moisture conditions. For an event-based stormwater model, this is overly complex. To overcome this problem (Schwartz, 2010),

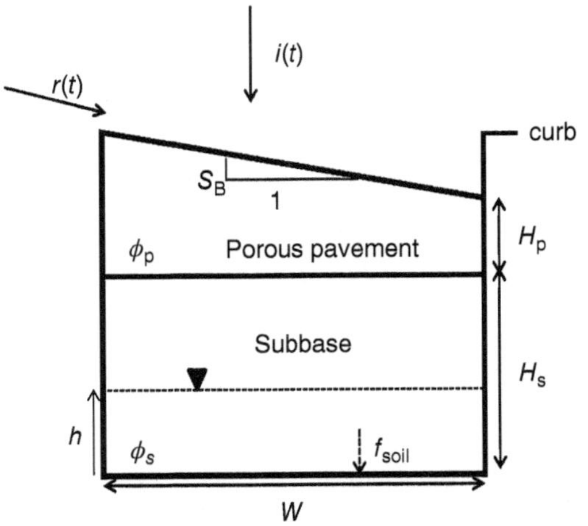

Figure 14.10 Schematic diagram of a sloped porous pavement of width W and slope S_B and a curb to manage surface runoff once the pavement is full.

approximate the abstraction by routing the rainfall hyetograph through a virtual sub-basin of the same area as the pavement, a time of concentration of zero and a runoff curve number (RCN) of 98 (the same RCN used for paved surfaces). This will typically be a conservative approach (in that it will underestimate the abstraction) as the aggregate surface area of the pavement system will be substantially greater than the plan area of the pavement (A_p).

14.4.2 Undrained Porous Pavement Hydraulics and Hydrology

Porous pavement systems can be installed without an underdrain in places where the soil permeability is high enough that drawdown will be relatively fast and where there is no risk of ground freeze preventing infiltration. The hydraulic behavior of such a system can be broken down into four distinct periods, infiltration, accumulation (or fill), runoff, and drawdown. Initially the rainfall rate is less than the effective infiltration capacity of the soil ($\Phi_s f_{soil}$); all the rainfall is infiltrated and the subbase water depth is effectively zero. Once the rainfall intensity exceeds the effective infiltration capacity ($i > \Phi_s f_{soil}$), the excess rainfall ($i_e = i - \Phi_s f_{soil}$) is stored in the pavement voids, and the subbase begins to fill (accumulation phase). Provided the storm depth is large enough, the pavement will eventually fill completely ($h = H_s + H_p$), and any further excess rainfall will run off. Finally, as the rainfall intensity decreases toward the end of the storm, the excess rainfall intensity reduces to zero, and the water level in the pavement begins to draw down.

This is more easily illustrated through a numerical example. The example considered is that of a porous pavement system with a pavement layer that is 5 in. thick with a porosity of 20% placed over a 6-in. subbase with an aggregate porosity of 30%. The underlying soil has a saturated infiltration capacity of 0.33 in. h^{-1}. For this example the total storage for the system is $\phi_p H_p + \phi_s H_s = 2.8$ in., and the effective infiltration capacity is $\Phi_s f_{soil} = 0.1$ in. h^{-1}. The hydraulic behavior of the system is illustrated by exposing the pavement system to a 6-in. SCS type II, 24-h rainfall event. The cumulative and instantaneous rainfall and runoff hydrographs for this example are shown in Figure 14.11.

The infiltration phase is illustrated in Figure 14.11a. The cumulative infiltration hydrograph (dashed line) overlies the cumulative hyetograph (solid line) until the slope of the rainfall curve exceeds the effective infiltration capacity of the soil. At this point (5.3 h into this example storm), water begins to accumulate in the storage layer. The depth of water in the storage area is the vertical distance between the dashed and solid lines. The effective storage line (dot-dash line) is the sum of the pavement total storage ($\phi_p H_p + \phi_s H_s = 2.8$ in.) and the total infiltration depth. When the total rainfall depth exceeds the effective storage, the pavement is full and

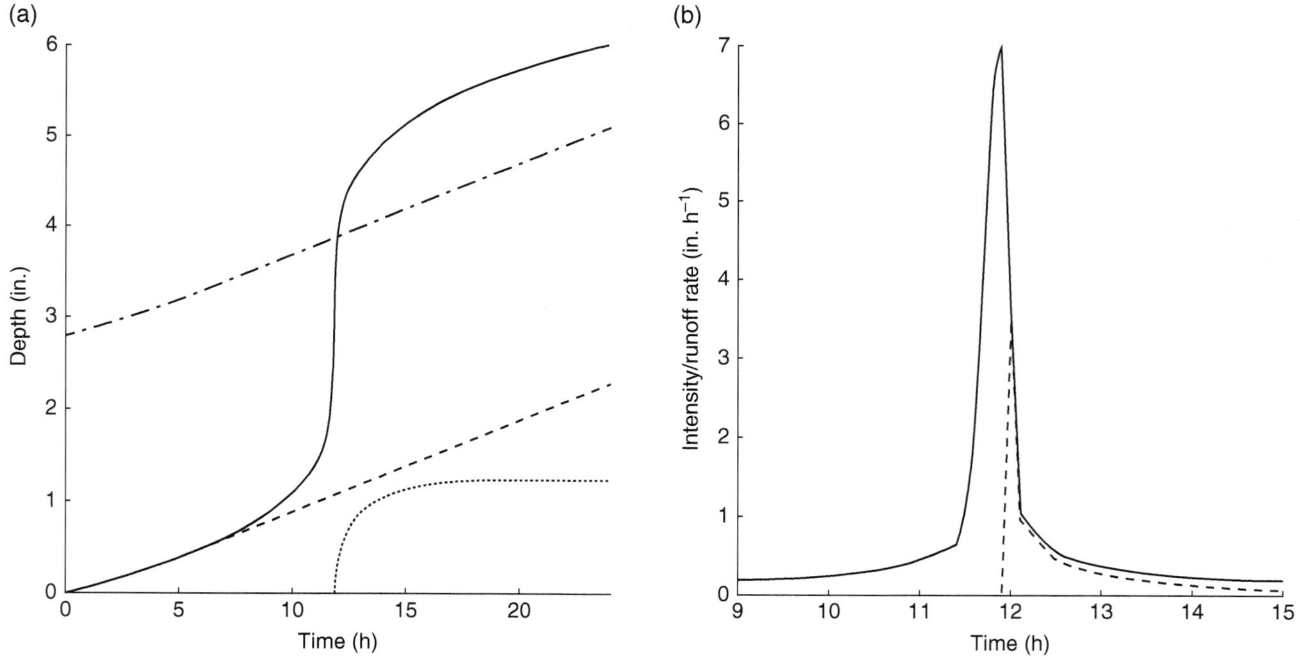

Figure 14.11 Example hydrographs for 5″ pavement ($\phi = 0.2$) and 6″ subbase ($\phi = 0.3$) placed over the soil with a saturated infiltration capacity of 0.33 in. h^{-1} for a 6-in. SCS type II, 24-h rainfall event. (a) The cumulative rainfall hyetograph (solid line), cumulative infiltration hydrograph (dashed line), the total effective storage line (dot-dash line), and the cumulative runoff (dotted line). (b) The middle 6 h of the instantaneous rainfall hydrograph (solid line) and surface runoff hydrograph (dashed line).

surface runoff begins (12 h into this example). After this time, the total infiltration and effective storage continue to increase as water still infiltrates into the underlying soil. The cumulative runoff is the vertical distance between the effective storage and total rainfall. The cumulative runoff is shown as the dotted line in Figure 14.11a. Runoff continues until the rainfall intensity drops below the effective infiltration rate (18.5 h into this example). At this point the cumulative runoff hydrograph flattens out, and the water level in the pavement system begins to draw down. Drawdown will continue for some time after the rainfall ceases (for about 27 h in this example).

The example illustrates a number of aspects of porous pavement hydraulic behavior that differ from standard pavements and undeveloped sub-basins. The first is that the runoff hydrograph (dashed line in Figure 14.11b) has a step jump in the runoff rate. There is no smooth growth in the runoff rate as the rainfall intensity increases. There is what is sometimes referred to as fill and spill behavior. That is, there is no runoff for a long time and then a sudden jump in the runoff rate. In fact, if the runoff does not start until after the peak rainfall, then the initial runoff rate is also the peak rate (see Figure 14.11b). Second, the rainfall depth required to fill the system storage can be greater than the storage capacity ($\phi_p H_p + \phi_s H_s$) due to infiltration into the subsoil during the rainfall. In the example above, the system storage is 2.8 in.; however, the pavement does not fill until almost 4 in. has fallen. Therefore, to fully characterize the hydraulic behavior of a porous pavement, one needs to know the storage capacity ($\phi_p H_p + \phi_s H_s$) and the effective infiltration rate that is sometimes quantified as the 24-h infiltration depth $24\phi_s f_{soil}$.

This simple modeling approach has been extended to cases in which there is significant run-on, potentially significant storage above the pavement surface, and flow control structures limiting the rate of surface runoff (West et al., 2016b).

The fill and spill behavior is seen temporally in individual events (Figure 14.11b) and in the pavements' hydrologic behavior for a range of storm depths. A plot of runoff depth versus rainfall depth will show zero runoff up to some depth beyond which the vast majority of any additional rainfall will run off. This is shown in Figure 14.12, which is a plot of runoff versus rainfall for the example pavement described above. Two approaches have been taken characterizing this behavior: the ECN method of (Schwartz, 2010) and the broken line model of (Martin and Kaye, 2015).

The ECN method proposed by Schwartz (2010) uses the storm routing method described above to generate a rainfall–runoff paired dataset for a range of rainfall depths and then fit a curve through the data of the form

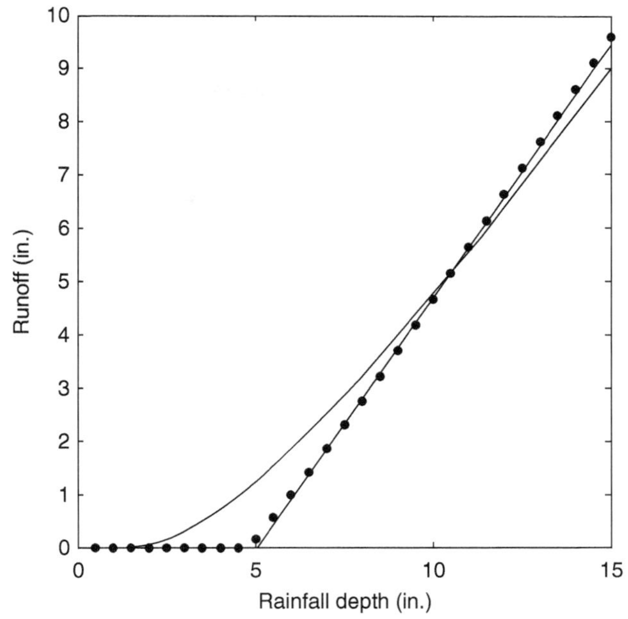

Figure 14.12 Plot of calculated runoff versus rainfall (points) for the example pavement described above. Also shown are the fitted ECN (solid line) and the broken line model (dashed line).

specified by the SCS RCN model to establish an ECN that best represents the rainfall–runoff behavior. Applying this approach to the example pavement gave an ECN of 59.

A more formal process that defined the set of rainfall depths and hyetographs to be used was presented by (Martin and Kaye, 2014). The ECN is a function of the storage capacity, the 24-h infiltration depth, and the local design storm depths (as the curve fit was shown to be sensitive to the choice of rainfall depths used). The ECN is, therefore, location specific due to soil properties, SCS storm type, and various return period storm depths. This is different to the original RCN model that depends only on the site conditions and soil type. The method does, however, give a simple method for initial calculation of the likely runoff from a given storm that can be used in preliminary design calculations.

The ECN for all possible pavement systems for a given location can be summarized in a contour plot of ECN as a function of storage capacity ($\phi_p H_p + \phi_s H_s$) and 24-h infiltration depth ($24\phi_s f_{soil}$). An example plot for Atlanta, GA, USA, is shown in Figure 14.13. The greater the 24-h infiltration depth or storage capacity, the lower the ECN. The figure also shows lines of constant drawdown time. The application of this type of plot to preliminary design is discussed later. At the time of printing, there is a web-based tool (at wmartin.people.clemson.edu/ECN.html) that takes as input the 1-, 2-, 5-, 10-, 25-, 50-, and 100-year 24-h storm depths and SCS storm type and produces an ECN plot similar to that of Figure 14.13.

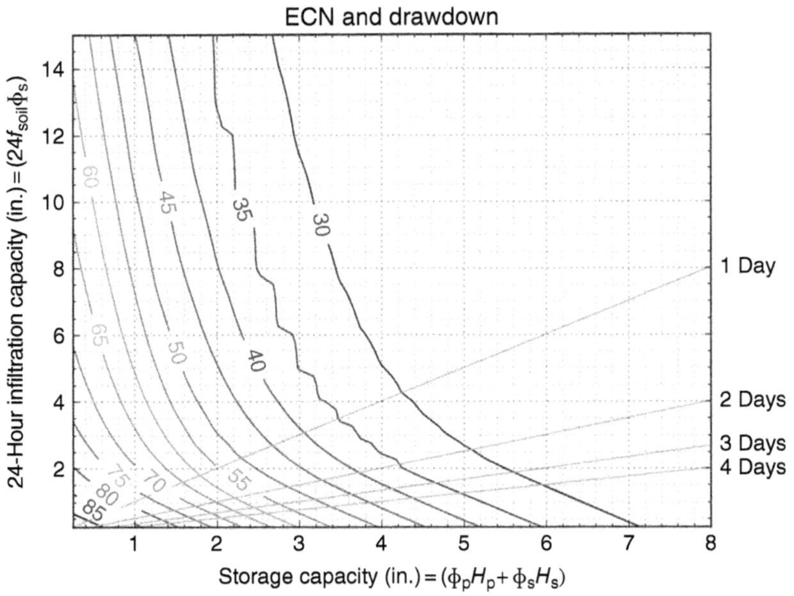

Figure 14.13 ECN contour plot for Atlanta, GA., USA, showing contours of constant ECN as a function of effective storage and 24-h infiltration capacity. Also shown are lines of constant drawdown.

An alternate method for characterizing the behavior of a porous pavement system is the broken line model (Martin and Kaye, 2015). In this model the pavement is described in terms of an initial abstraction I_a and a runoff coefficient C. Any rainfall depths (I) less than the initial abstraction will result in no runoff. The runoff that results from a rainfall event of depth greater than the initial abstraction is the rainfall depth in excess of the initial abstraction multiplied by the runoff coefficient:

$$R = \begin{cases} 0 & I < I_a \\ C(I - I_a) & I > I_a \end{cases} \tag{14.15}$$

For the example pavement used previously, the initial abstraction is 5.0″ and the runoff coefficient is 0.95. These two parameters are functions of the rainfall hyetograph type, storage capacity, and 24-h infiltration depth. The model is insensitive to the choice of rainfall depths used in the curve fitting and can be summarized in contour plots similar to those for the ECN (see Martin and Kaye, 2015). A plot of both the ECN and broken line models is shown in Figure 14.12 for the example pavement described above.

Many stormwater regulations limit the time allowed for a stormwater management system to draw down following a rainfall event. For an undrained porous pavement system, the drawdown is via infiltration into the underlying soil. The drawdown time T_D in days for a full pavement ($h = H_s + H_p$) is the 24-h infiltration depth divided by the pavement storage capacity

$$T_D = \frac{24\phi_s f_{soil}}{\phi_p H_p + \phi_s H_s}. \tag{14.16}$$

14.4.3 Undrained Porous Pavement Hydraulic Design Considerations

Design engineers may have many design goals and/or constraints when designing a porous pavement system. Structural and maintenance considerations are discussed later in this chapter. Herein the focus is on hydraulic and hydrologic design considerations and preliminary design calculations. The main design goals considered are design for storage/capture, design for peak discharge, and design for hydrologic performance similar to predevelopment conditions.

Design for capture and storage can be based on either water quality or water quantity. For water quality the goal will likely be to retain the first flush (or water quality volume). Water quantity-based design will have a goal of retaining some design storm depth (say, a 2-year return period storm) within the pavement system. In both cases the conservative approach is to ensure that the storage capacity of the pavement ($\phi_p H_p + \phi_s H_s$) is greater than the design depth being considered.

Peak discharge design requires that the peak surface runoff be limited to some specified rate, often equal to the predevelopment peak runoff. As porous pavements are effectively retention ponds, standard retention pond analysis techniques can be applied. For example, the hydrograph truncation method (Akan and Houghtalen, 2003) can be used. In this method the time of the peak runoff rate in in/h is located on the falling limb of the rainfall hyetograph. The total effective storage required is then equal to the total rainfall depth prior to that point in time. The total effective storage is the sum of the pavement effective storage and the depth infiltrated prior to the time calculated from the hydrograph truncation.

To calculate this requires a full storm routing through the pavement. However, an approximate depth could be calculated by ignoring the infiltration component and only using the storage capacity $\phi_P H_P + \phi_s H_s$.

Design to comply with predevelopment hydrologic conditions is complex and will require detailed flow routing and iterative design. Fortunately flow routing is easily conducted in most hydrologic modeling programs such as HEC-HMS and EPA SWMM. The main difficulty in comparing a predevelopment sub-basin with a porous pavement system is that the system behaves as a retention pond. As such, the time of concentration of the pavement is not clearly defined. It is, however, possible to design a porous pavement system to match the predevelopment RCN with a postdevelopment ECN using charts similar to the one in Figure 14.13. The design constraints would be a desired ECN, the local regulatory drawdown time, and the sites' soil infiltration capacity. It is also likely that the porosity of the aggregate to be used in the subbase is constrained by what is locally available and/or what is required for the structural needs of the system. Therefore, the 24-h infiltration depth is predetermined. It is then a simple matter of tracing a horizontal line at the height of the calculated 24-h infiltration depth in Figure 14.13 (or a local equivalent) and noting the storage capacities where the line crosses the desired ECN contour. However, if the intersection of this horizontal line and the desired ECN contour is below and to the right of the required drawdown timeline, then the storage capacity will be too large to comply. In this case, the minimum ECN achievable will be the ECN where the horizontal line intersects the appropriate drawdown line. If this ECN is too large, then an underdrain may be needed (see below).

All the preliminary design calculations described above provide a starting point for detailed hydrologic modeling of the system as part of the full site model. The preliminary calculations can also form the basis for a preliminary cost–benefit analysis for the use of the porous pavement system. For example, a site with low soil infiltration capacity may well require a large storage capacity to replicate the predevelopment RCN. In such a case, the porous pavement may be better used for retention of the water quality volume rather than total runoff reduction or peak flow control.

14.4.4 Porous Pavements with Underdrains

There are cases where an undrained pavement is not appropriate such as when the infiltration capacity of the underlying soil is very low such that the pavement system does not draw down rapidly enough or for locations where there is a significant risk of ground freeze and the pavement system must be drained. In these cases a perforated pipe underdrain or some other subsurface outlet is required. Stormwater manuals and LID/BMP guides offer little guidance on either the design of underdrains or the routing of flow through drained porous pavement systems.

Full-scale experiments of flow through a perforated pipe underdrain show that when the pipe is running full at its outlet, the pipe behaves like an orifice, that is,

$$Q = A_D\sqrt{2gH} \tag{14.17}$$

where A_D is the cross-sectional area of the pipe and H is the height of water above the pipe centerline. Two types of drainage pipe were tested, namely, "leached" and "perforated" (the names used are the terms commonly used by manufacturers). Leached pipes have round holes punched in the side walls on the lower half of the pipe, whereas "perforated" pipes have slits cut in the side walls around the whole cross section.

For tests using standard drainage stone as the surrounding aggregate (Murphy et al., 2014), found that the discharge coefficient for standard plastic corrugated pipe depended on the pipe diameter (D) and the percentage of the side wall area that was perforated: $\varphi = A_i/\pi DL$ (where A_i is the total wall inlet area and L is the pipe length). The results are summarized in Table 14.2. A subsequent computational study (Afrin et al., 2016b) supported these results, and a more detailed parametric study (Afrin et al., 2016a) showed that the pipe discharge was not significantly altered by changes in the size or shape of the side wall perforations, only φ and the pipe length. The parametric study also showed that the pipe discharge coefficient will reach a maximum when the wall inlet area is more than three times the pipe cross-sectional area, that is, $L > 3D/4\varphi$. This suggests that the experiments of (Murphy et al., 2014) underestimate the potential discharge from a given pipe due to the relatively short pipe lengths tested. However, as they are the only full-scale test results available in the literature, they can be used to provide a conservative (under) estimate of the drain outflow.

The data in Table 14.2 can be used to size and locate a perforated pipe underdrain in order to prevent surface

Table 14.2 Discharge coefficients for 10-ft underdrain pipes of different wall opening types and wall inlet area percentages (φ) based on the measurements of Murphy et al. (2014).

Diameter (D) (in.)	Wall inlet type	φ	C_D (exp)
4	Perforated	0.023	0.49
4	Leached	0.021	0.41
6	Leached	0.018	0.34

Source: Reproduced with permission of American Society of Civil Engineers.

runoff for a given storm. For these calculations the peak discharge will be the pavement area multiplied by the peak rainfall intensity ($Q_{max} = A_p i_{max}$). In order to prevent surface runoff, the pipe centerline must be located a distance H below the pavement surface and have a diameter large enough such that $C_D \pi D^2 \sqrt{2gH}/4 > A_p i_{max}$. This will provide a worst-case design as it assumes that the pavement system is exactly full at the time of the peak rainfall.

The results of (Murphy et al., 2014) for the discharge from a perforated pipe underdrain are only valid when the pipe is running full at the outlet. When the pipe is not running full, the flow is generally controlled by the pipe outlet brink depth (Hager, 1999). There is some experimental data on partially full underdrains (Murphy, 2013) that suggests (though not conclusively) that, under these circumstances, the pipe outflow may behave like a weir.

A fit to the data of Murphy (2013) by Martin and Kaye (2016) was used to investigate the response of a drained porous pavement system to a design storm event. The routing procedure used was the same as for the undrained pavement. The total runoff was taken to be the sum of the drain outflow and surface runoff. The behavior of such a system is quite different to that of an undrained pavement system. Discharge through the underdrain occurs well before the pavement is full, which means that the fill and spill behavior of undrained pavements is no longer observed and there is a gradual buildup of discharge over the course of the storm. This difference in behavior is also observed in plots of runoff depth versus rainfall depth (see Figure 14.14 for a sample plot), which does not have the sharp transition from no runoff to runoff. As such, the behavior of a drained system is much closer to the standard RCN model than undrained pavements (Figure 14.12). However, the introduction of an underdrain adds three additional design parameters, namely, the drain height above the soil, the drain size, and the pavement area per drainage pipe. As such, while it is possible to summarize the behavior of a given system in terms of an ECN, it is not possible to summarize the ECN efficiently in a single chart such as can be done for undrained systems (e.g. Figure 14.13). See (Martin and Kaye, 2016) for more details on drained porous pavement systems.

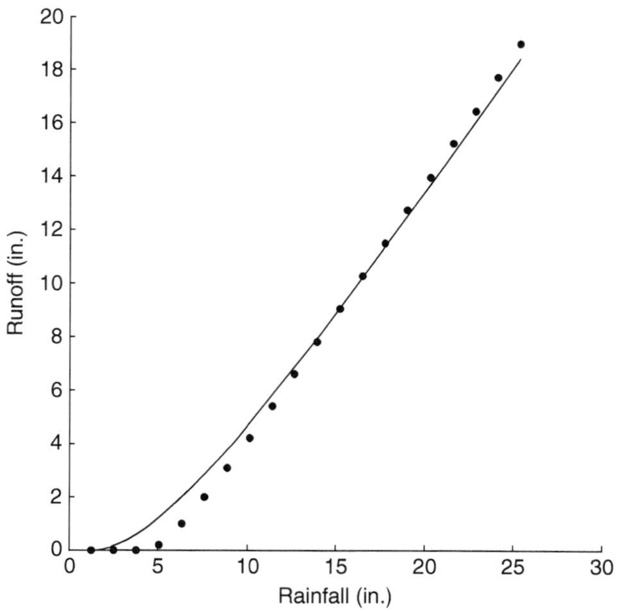

Figure 14.14 Plot of runoff depth versus rainfall depth for the same pavement as used in the previous example (see Figure 14.13) with a 4-in. perforated pipe underdrain located 3 in. above the soil. Also shown is the equivalent RCN curve based on the pavement ECN = 78.

14.5 Design, Construction, and Maintenance

14.5.1 Structural Design

In addition to accommodating hydrologic demands, porous pavements must also withstand vehicular loading. Therefore, the structural design of the pavement is equally important as the hydrologic design. The role of the pavement structure (i.e. all layers above the subgrade) is to reduce the stresses applied by traffic loads to a magnitude that the subgrade can withstand without resulting in excessive deformation. The stresses are absorbed by each individual layer (e.g. surface course, choker course, reservoir course, etc.), and the ability for each layer to absorb the applied load is a function of the strength of the material and thickness of the particular layer.

There are two main types of pavement surfaces: flexible and rigid (Figure 14.15). Flexible pavements include asphalt pavements and interlocking concrete pavements, while concrete pavements are considered rigid. This designation (flexible or rigid) is based on the behavior of the surface material. For example, asphalt is a mixture of aggregate "glued" together with an asphalt binder that remains flexible, thus allowing the pavement surface to flex (to a degree) under loading. Concrete is a mixture of aggregates bound with a cementitious binder that becomes rigid after curing. Interlocking concrete pavements, although made of rigid concrete units, are flexible in nature because the units are not bound together. Figure 14.15 illustrates how loads are distributed throughout rigid and flexible pavement structures.

Porous pavements, like conventional pavements, should be designed to support the expected amount of traffic over the design life of the pavement, which is commonly accomplished using the procedures outlined by the American Association of State Highway and Transportation Officials (AASHTO) (1993). This design methodology, as with others, requires the user to have an

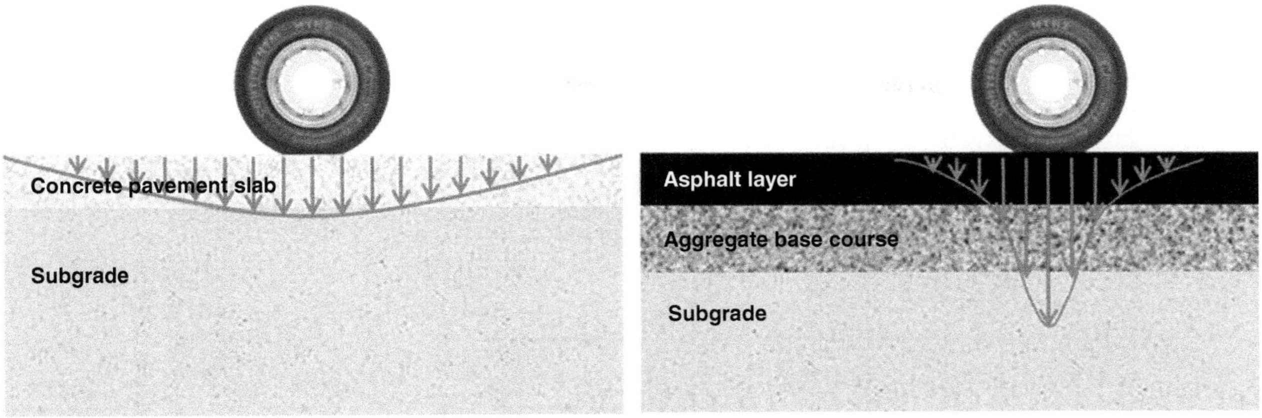

Figure 14.15 Comparison of traffic stress distribution in rigid and flexible pavements.

understanding of certain factors that will significantly impact the design: traffic and subgrade quality. A pavement must be designed to sustain traffic loading over its design life.

14.5.1.1 Traffic

Pavements are designed to carry many different types of vehicles in the traffic stream including automobiles, light trucks, buses, freight trucks, construction equipment, and sanitation trucks, among other vehicle types and loads. Although the main component of most traffic streams is passenger vehicles, the primary consideration in pavement design is heavy trucks. This is because heavy trucks impart far more stress on pavements compared with automobiles and thus are the primary contributors to pavement damage. Based on the axle load factors provided in (American Association of State Highway and Transportation Officials (AASHTO), 1993), a loaded 5 axle tractor trailer imparts more than 1600 times more damage than a typical passenger car and more than 200 times greater than a large sport utility vehicle (SUV).

Because different vehicles have different loads and wheel configurations, the AASHTO method normalizes all axle loads to an 18 000-lb equivalent single axle load (ESAL) using load equivalency factors (American Association of State Highway and Transportation Officials (AASHTO), 1993) and then the total number of ESALs is summed for the design life. It should be noted, however, that porous pavements are typically not recommended for pavements exposed in heavy vehicle loading because the porous materials (e.g. concrete, asphalt, aggregate base, etc.) are typically weaker due to the high void content required to achieve the desired hydrologic properties (porosity and hydraulic conductivity).

14.5.1.2 Subgrade

The success of any structure, including a pavement structure, is highly dependent on the quality of the foundation upon which it is built. In the case of a pavement structure, the foundation refers to the soil (or subgrade) that the pavement is constructed upon. A higher quality (or stronger) subgrade can withstand greater stresses, which means that the thickness of the pavement structure can be reduced compared with that needed for a weaker subgrade. For this reason, it is important that the subgrade soil be thoroughly examined and understood before developing a pavement design. The strength of a subgrade is typically characterized using the resilient modulus for flexible pavement design (M_r) and modulus of subgrade reaction (k) for rigid pavement design. If it is not feasible to directly measure these values, they can be estimated based on other soil strength tests such as the California bearing ratio (CBR), R-value, and others using the relationships illustrated in Figure 14.16 (ARA, Inc., 2001) and

$$k(\text{pci}) = \frac{M_r}{19.4} \tag{14.18}$$

where M_r is in psi or

$$k\left(\text{MPa m}^{-1}\right) = 2.03 M_r \tag{14.19}$$

where M_r is in MPa (National Ready Mixed Concrete Association (NRMCA), n.d.).

14.5.1.3 Thickness Design

There are separate methods used to design the thickness of the pavement structure for both flexible and rigid pavements using the AASHTO design process (Federal Highway Administration (FHWA), 2008, 2012,). Specific details and design guidance can be found in (American Association of State Highway and Transportation Officials (AASHTO), 1993) as well as (Tennis et al., 2004b; Smith, 2011; Al-Rubaei et al., 2013). Porous pavements should always be designed for the specific site conditions (traffic, subgrade, and materials), but Table 14.3 provides typical layer thickness used for porous pavements.

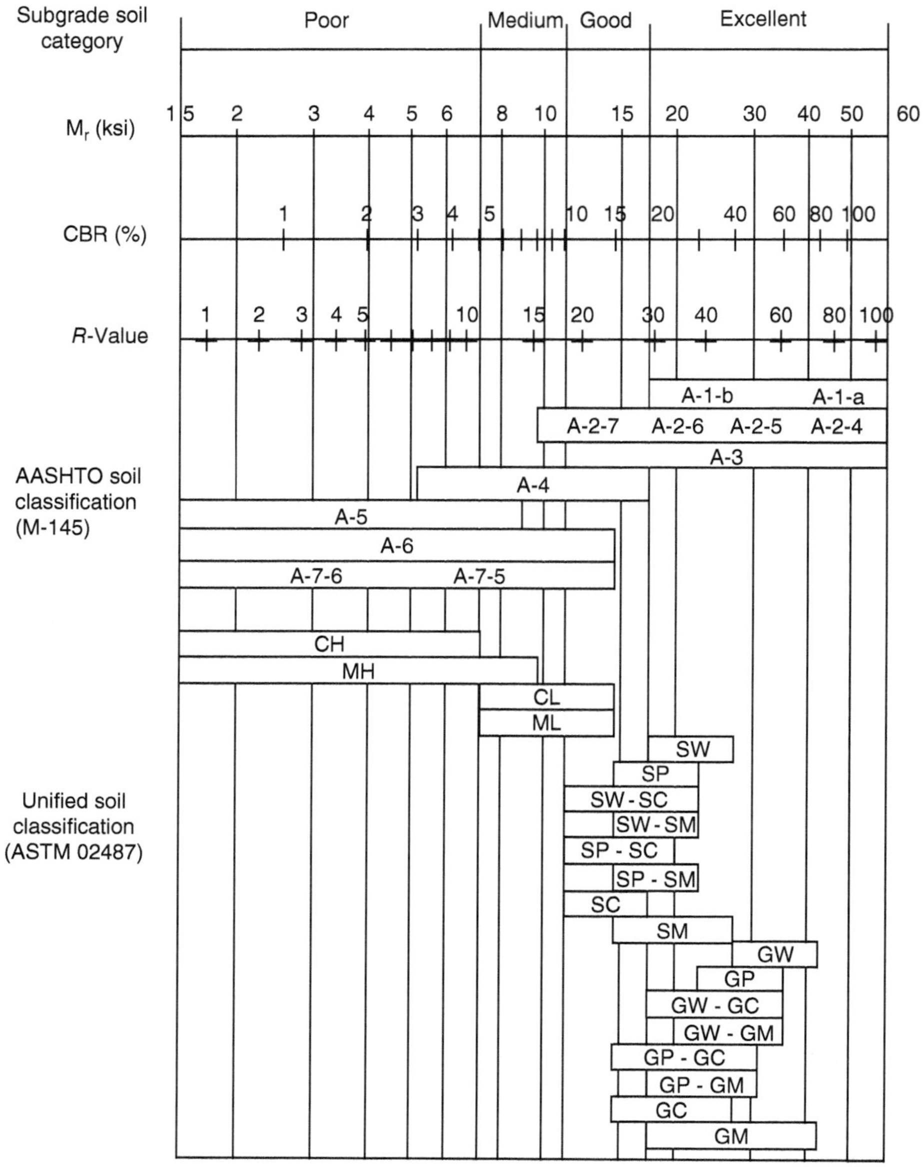

Figure 14.16 Typical resilient modulus correlations to empirical soil properties and classification categories. *Source:* From ARA, Inc. (2001). Reproduced with permission of the Transportation Research Board.

14.5.2 Construction Practices

14.5.2.1 Subgrade Preparation

In many cases, once excavation has been completed, the existing subgrade will be left uncompacted to maximize the infiltration capacity. There are times when the subbase will need to be compacted to meet structural requirements, but compaction should only be used when it is specifically called for in the construction documents. If compaction is not required, construction vehicle traffic should be limited over the excavated subgrade to minimize compaction. In cases where a subgrade is accidentally compacted or if the soil is naturally compacted, it can be scarified to improve infiltration.

Use of geotextiles between the subgrade and subbase is not consistent at this time. Advantages of using geotextiles are that they limit soil migration into the porous subbase and can provide some additional structural benefits. Disadvantages include the additional cost and the potential for fine particles in the stormwater to build up on the geotextile and therefore clog the system. There are different recommendations from different agencies; for example, the National Asphalt Pavement Association recommends using a geotextile below porous asphalt

Table 14.3 Typical pavement layer thicknesses for porous pavements.

Pavement layer	Porous asphalt	Pervious concrete	PICP
Surface	Porous asphalt (2–6 in.)	Pervious concrete (6–12 in.)	Permeable paver (3⅛ in. standard)
Bedding layer	No. 8 stone (2 in.)[a]	N/A	No. 8 stone (2 in.)
Base	No. 57 stone (4+ in.)	No. 57 stone (min 4 in. if used)[a]	No. 57 stone (4 in.)
Subbase	No. 2 stone (6+ in.)	N/A	No. 2 stone (6+ in.)
Subgrade			

[a] Denotes that the layer is optional.

installations, while the University of New Hampshire Stormwater Center recommends eliminating geotextiles from the design (UNHSC (University of New Hampshire Stormwater Center), 2009).

14.5.2.2 Porous Subbase

Many different gradations of aggregate are used as material for the subbase, but all must be open graded and washed. It is very important that all the fines have been removed from the aggregate, because if any remain they will be washed down and accumulate on top of the subgrade and potentially reduce the soil's infiltration rate.

The exact aggregate size and number of layers used for the subbase varies between different applications. For pervious concrete and porous asphalt, a single subbase layer can be used, though many different subbase configurations have been used, including a choker course with a smaller aggregate size above the subbase and sand filter layers for water quality purposes. The Interlocking Concrete Pavement Institute (ICPI) (Smith, 2011) recommends three subbase layers for PICP: a minimum of six inches of ASTM No. 2 stone as the subbase, a minimum of four inches of ASTM No. 57 stone as a base, and then a 2-in. bedding layer of ASTM No. 8 stone as seen in Table 14.3.

The aggregate subbase should be compacted with either a steel vibratory roller or plate compactor. Each lift should be compacted until no more movement of the aggregate is visible. The density can be checked with a nuclear gauge or a base stiffness gauge (Smith, 2011).

In preparation for paving, the top of the subbase should be relatively smooth and even. Any ruts left by construction equipment should be removed to limit a large variation in surface pavement depth. While preparing the surface for porous asphalt or pervious concrete is important, there is some margin of error as the paving material can conform to the subbase surface. However, surface preparation of the subbase is critical to the placement of PICP as they rest directly on the bedding material, and any deviations on the bedding surface will be visible on the surface. Because of this, the ICPI has very stringent guidelines for the placement and preparation of the bedding layer, which includes screeding the entire surface (Smith, 2011).

14.5.2.3 Surface Pavement

14.5.2.3.1 Porous Asphalt

Porous asphalt is placed using the same technique and equipment as conventional asphalt. A paver is fed the asphalt mixture from a truck and is placed in 2–4-in. lifts, depending on the project specifications. The asphalt is then compacted with two to four passes of a static roller. However, careful attention should be paid to the compaction specifications (including temperature restrictions and compaction limits) because overcompaction can decrease the overall porosity of the asphalt to the point where the hydraulic conductivity may be impacted.

14.5.2.3.2 Pervious Concrete

Before the placement of pervious concrete, the subbase should be wetted so the aggregate does not draw water out of the pervious concrete mix and change the water-to-cement ratio. Ready-mix concrete trucks then can deposit the pervious concrete mix directly onto the subbase. It is important to monitor and test the mixture frequently while it is being placed as it is much easier to reject a load if needed than to remove and replace the concrete later. The water-to-cement ratio is especially critical as if it is too high, the cement paste can seal pores and a ratio that is too low can cause the pavement to fail due to raveling (surface aggregates becoming dislodged).

The concrete should be spread evenly between forms and struck off with a screed to a height slightly above the desired pavement height. The pavement can then be compacted down to the final pavement height. The amount of compaction will be specified as part of the pavement design. When placing the concrete and finishing it, no vibrators, trowels, or power finishing equipment should be used as all of these can cause sealing of the pores in the pavement. Once the mixture has been poured, curing should begin within 20 min (American Concrete Institute (ACI), 2013). To cure, the pavement should be covered with a minimum 6-mil thick polyethylene sheet, and the pavement should remain covered for a minimum of 7 days.

The pavement can be jointed immediately after placement using a joint tool, or the joint can be cut using a concrete saw after the pavement has cured.

14.5.2.3.3 Permeable Interlocking Concrete Pavers

PICP is relatively simple to install once the bedding is properly prepared. The pavers can be laid by hand for small pavements or with the use of mechanized installation for larger areas. The placement of the PICP pavers is exactly the same as placing conventional pavers as joint spacing is created by offsets built into the pavers.

After the pavers have been installed, the paver joints are filled with ASTM No. 8, 89, or 9 stone. The aggregate is spread on the pavement and then swept into the joints, either by hand or with a mechanized sweeper. Once the surface has been swept clean, the pavers are compacted with two passes of a plate compactor, and additional filler aggregate is added if necessary.

One advantage of PICP is that the pavers can be removed and then reinstalled if access is needed under the pavement (for example, to a utility), whereas porous asphalt and pervious concrete would need to be cut, removed, and then replaced.

14.5.2.4 Postinstallation

Often in construction with typical impervious pavements, the pavements are one of the first things installed on-site and during construction are exposed to sediment laden runoff and construction equipment with muddy wheels/tracks and are even used as staging locations for soils and/or mulch for landscaping. With porous pavements, once it has been placed (subbase or surface pavement), it is critical to limit sediment on the pavement. If a porous pavement is exposed to a large load of sediment as is common with construction, it can easily become severely clogged past the point of remediation by common maintenance practices and would need to be replaced.

Suggestions for options to prevent contamination of porous pavement installations during construction if they must be installed before all soil-disturbing construction is complete include (Smith, 2011):

- Construct the subbase layers as designed, but install a geotextile and two inches of additional base material (or an impervious asphalt pavement) to serve as a temporary surface. Once construction is finished, the temporary surface material and geotextile can be removed, and the final surface course can be installed.
- Install the entire pavement system including surface pavement, but then cover it with geotextile and two inches of open-graded aggregate.
- Restrict access to the porous pavement areas by constructing temporary construction access roads.
- Set up washing stations to remove sediment from construction equipment accessing the paved area.

14.5.3 Other Considerations

14.5.3.1 Layout and Siting

Applications of porous pavements typically include parking lots and low volume roads; however, they have also been used in many other configurations depending on the site or project. Some installations have porous pavement covering the entire parking lot or road, while others use the porous pavements in key areas such as parking stalls, pedestrian areas, or even as shoulder or bike lanes (West et al., 2016b). These configurations have different advantages and disadvantages including costs, ability to use as a retrofit (such as adding porous bike lanes to an existing impervious road), and available storage volume.

While porous pavements can be used to handle runoff from adjoining sites with proper hydrologic design, it is important to protect the pavement from runoff containing excessive sediment, like during construction as previously mentioned. Runoff from other impervious areas like adjacent building roofs will likely not pose much of an issue, but runoff from landscaped areas with loose soil or mulch or poorly established lawns should be limited. Similarly, porous pavements should be avoided in areas where hazardous materials are loaded, unloaded, or stored to prevent the possibility of groundwater contamination.

A common recommendation is that the porous pavements be sited on relatively level sites (up to 5%). This is due to two concerns. The first is that if a site is sloped, the water will pond in the low end of the pavement and any storage capacity of the pavement located above the lowest pavement surface will be unused. The second is that excessive water flow down the soil aggregate interface could cause soil erosion under the subbase, jeopardizing pavement stability in some locations and clogging the subbase pore space in others.

A common solution that addresses both of these concerns is the use of check dams or berms in the subbase, which prevents the free flow of water along the length of the pavement. This breaks the area up into much smaller infiltration areas and therefore retains the water over a greater total area of the pavement, which increases overall infiltration. In cases where the concern is soil erosion of the subgrade, a geotextile could be used to lessen the impact.

14.5.3.2 Maintenance

All the benefits discussed in Section 14.2 will only be realized if the porous pavement system is functioning as designed. Because clogging is one of the biggest problems all types of porous pavement face, regular maintenance is critical (Bean et al., 2004; Briggs, 2006; Chai et al., 2012; Drake and Bradford, 2013). The ICPI recommends inspection and cleaning of a pavement once or twice during the first year and then adjusting the frequency of cleaning based on the specific site conditions (Smith, 2011).

Sediments can be introduced to a porous pavement from a number of sources. The most common are:

- Sediment tracked onto the pavement by vehicles from other areas.
- Rubber particles from tire wear.
- Organic plant material such as leaves dropped from a nearby tree.
- Sediment laden run-on from surrounding areas (as discussed in Section 14.5.2.4).
- Sand applied for snow or ice conditions. This practice should ideally be limited for porous pavements.
- Breakdown of pavement structure. This will be limited in a well-constructed pavement, but for poorly designed or installed pervious concrete and porous asphalt, raveling can be an issue. The structural considerations behind raveling are of a larger concern than the clogging aspect, and a pavement experiencing this should ideally be replaced.

As part of inspection, it is important to note visible sediments (leaves or other large particles) or signs of sediments (such as dirt stains from muddy runoff from a landscaped area). Not only must the sediments be removed, but the sources of the sediments be identified and proactively addressed if possible. To assess the functionality of the pavement qualitatively, look for a ponding or surface runoff during rain events. For a more accurate assessment, the infiltration capacity of the pavement can be measured using the ASTM C1701 test procedure as described in Section 14.3 (ASTM C1701/C1701M-09, 2009).

Maintenance should be scheduled based on the primary cause of clogging. For example, if a pavement is used in northern climates where snow and ice are common and sand is used, maintenance should be scheduled in the early spring to limit the sand from migrating into the pavement. If a porous pavement is located near trees that drop leaves on the pavement, then maintenance should be scheduled in the late fall to remove the leaves before they are broken down by traffic and the small pieces can enter the pavement.

There are three commonly used maintenance practices used with porous pavements in sediment removal: sweeping, vacuuming, and pressure washing (listed in order of increasing effectiveness) (Golroo and Tighe, 2010; Henderson and Tighe, 2011; Winston et al., 2016).

14.5.3.2.1 Sweeping

Mechanical sweeping has been shown to be the least effective method of maintenance (Henderson and Tighe, 2011). It is not recommended by the ICPI (Smith, 2011) because no sediment is being removed from the pavement surface; it is simply being relocated. Combination sweeper–vacuum trucks, where the brushes direct the material to the vacuum inlets, can effectively be used for larger debris such as leaves but will likely be less effective for finer sediment if the sediment is in the pore structure of the pavement surface.

14.5.3.2.2 Vacuuming

Vacuuming is preferred to sweeping as the sediments are removed from the surface and from a certain depth within the pavement surface. Depending on the sediment type, vacuuming can sometimes be improved by wetting the sediment and pavement before vacuuming.

For porous asphalt and pervious concrete, vacuuming removes the loose sediment from the pore spaces of the pavement. However, since PICP has aggregate material in the joints, often vacuuming removes the sediment and a certain amount of aggregate, which should then be replaced by clean aggregate. If it is not replaced, future sediment will build up deeper in the joints and will be harder to remove. While the replacement of aggregate after maintenance for PICP is a reoccurring cost to the owner, PICP has been reported to respond better to maintenance after it has experienced heavy clogging (Smith, 2011).

14.5.3.2.3 Power Washing

In many cases, power washing has been shown to be more effective than vacuuming at restoring infiltration rates for clogged pavements (Chopra et al., 2010; Henderson and Tighe, 2011). This is likely due to the water being able to enter into the pores and dislodge the sediment better than a vacuum alone. However, a common concern is that while pressure washing dislodges the sediment, it either drives the sediment deeper into the pavement or relocates it to a different area similar to sweeping (Chopra et al., 2010; Hein et al., 2013). Therefore, when using pressure washing, it is recommended to spray the pavement at a low angle to prevent driving the sediment deeper, and pressure washing in conjunction with vacuuming that was shown to be more effective at restoring the infiltration rate than either method was alone (Hein et al., 2013).

If all three maintenance practices fail to improve infiltration rates to the required level, the pavement can either be replaced, or if clogging is solely confined to the top of the surface pavement, then milling can be used to remove the clogged portion of the pavement (Henderson and Tighe, 2011).

References

ACI 522R-06 (2006). *Pervious Concrete*. Farmington Hills: American Concrete Institute Committee.

ACI 522R-10 (2010). *Report on Pervious Concrete*. Farmington Hills: American Concrete Institute Committee.

Afrin, T., Kaye, N.B., Khan, A.A., and Testik, F.Y. (2016a). A parametric study of perforated pipe underdrains surrounded by loose aggregate. *ASCE Journal of Hydraulic Engineering* 142 (12).

Afrin, T., Khan, A.A., Kaye, N.B., and Testik, F.Y. (2016b). Numerical model for the hydraulic performance of perforated pipe underdrains surrounded by loose aggregate. *ASCE Journal of Hydraulic Engineering* 142 (8).

Akan, A.O. and Houghtalen, R.J. (2003). *Urban Hydrology, Hydraulics, and Stormwater Quality: Engineering Applications and Computer Modeling*. Hoboken, NJ: Wiley.

Al-Rubaei, A., Stenglein, A., Viklander, M., and Blecken, G. (2013). Long-term hydraulic performance of porous asphalt pavements in northern Sweden. *Journal of Irrigation and Drainage Engineering* 139 (6): 499–505.

American Association of State Highway and Transportation Officials (AASHTO) (1993). *Guide for Design of Pavement Structures*. Washington, DC: AASHTO.

American Concrete Institute (ACI) (2013). Specification for Pervious Concrete Pavements. ACI Standard 522.1-13, Farmington Hills, MI.

ARA, Inc. (2001). Guide for Mechanistic-Empirical Design of New and Rehabilitated Pavement Structures, Appendix CC: Correlation of CBR Values with Soil Index Properties. *NCHRP Rep. 1-37A*.

ASTM C1688/C1688M-14a (2014). *Standard Test Method for Density and Void Content of Freshly Mixed Pervious Concrete*. ASTM International, West Conshohocken, PA. www.astm.org (27 January 2018).

ASTM C1701/C1701M-09 (2009). *Standard Test Method for Infiltration Rate of in-Place Pervious Concrete*. ASTM International, West Conshohocken, PA. www.astm.org (27 January 2018).

ASTM C1754/C1754M-12 ASTM C1754/C1754M-12 (2012). *Standard Test Method for Density and Void Content of Hardened Pervious Concrete*. ASTM International, West Conshohocken, PA. www.astm.org (27 January 2018).

ASTM D5856-15 (2015). *Standard Test Method for Measurement of Hydraulic Conductivity of Porous Material Using a Rigid-Wall Compaction-Mold Permeameter*. West Conshohocken, PA: ASTM International www.astm.org.

ASTM D7063/D7063M-11 (2011). *Standard Test Method for Effective Porosity and Effective Air Voids of Compacted Bituminous Paving Mixture Samples*. ASTM International, West Conshohocken, PA. www.astm.org (27 January 2018).

Barrett, M.E., Kearfott, P., and Malina, J.F. (2006). Stormwater quality benefits of a porous friction course and its effect on pollutant removal by roadside shoulders. *Water Environment Research* 78 (11): 2177–2185.

Bean, E.Z., Hunt, W.F., and Bidelspach, D.A. (2004). Study on the surface infiltration rate of permeable pavements. Critical Transitions in Water and Environmental Resources Management, American Society of Civil Engineers, 1–10.

Bean, E.Z., Hunt, W.F., and Bidelspach, D.A. (2007). Evaluation of four permeable pavement sites in eastern North Carolina for runoff reduction and water quality impacts. *Journal of Irrigation and Drainage Engineering* 6 (583): 583–592. doi: 10.1061/(ASCE)0733-9437(2007)133.

Berryman, J.G. and Blair, S.C. (1987). Kozeny–carman relations and image processing methods for estimating Darcy's constant. *Journal of Applied Physics* 62: 2221–2228.

Booth, D. and Leavitt, J. (1999). Field evaluation of permeable pavement systems for improved stormwater management. *Journal of the American Planning Association* 65 (3): 314–325.

Brattebo, B.O. and Booth, D.B. (2003). Long term stormwater quantity and quality performance of permeable pavement systems. *Water Research* 37 (18): 4369–4376.

Briggs, J.F. (2006). Performance assessment of porous asphalt for stormwater treatment. M.S. thesis. University of New Hampshire, Durham, NH.

Brown, H.J. and Sparkman, A. (2012). The development, implementation, and use of ASTM C1701 field infiltration of in place pervious concrete. *Pervious Concrete, STP* 1551: 69–79.

Chai, L., Kayhanian, M., Givens, B. et al. (2012). Hydraulic performance of fully permeable highway shoulder for storm water runoff management. *Journal of Environmental Engineering* 138 (7): 711–722.

Chopra, M., Kakuturu, S., Ballock, C. et al. (2010). Effect of rejuvenation methods on the infiltration rates of pervious concrete pavements. *Journal of Hydrologic Engineering* 15 (6): 426.

Chopra, M., Wanielista, M., Spence, J. et al. (2006). Hydraulic performance of pervious concrete pavements. *Proceedings in CD of the 2006 Concrete Technology Forum*. National Ready Mix Concrete Association, Nashville.

Collins, K.A., Hunt, W.F., and Hathaway, J.M. (2008). Hydrologic comparison of four types of permeable pavement and standard asphalt in eastern North

Carolina. *Journal of Hydrologic Engineering* 1146–1157. doi: 10.1061/(ASCE)1084-0699(2008)13:12(1146).

Crocker, M., Hanson, D., Li, Z. et al. (2004). Measurement of acoustical and mechanical properties of porous road surfaces and tire and road noise. *Transportation Research Record: Journal of the Transportation Research Board* 1891 (1): 16–22.

Debo, T.N. and Reese, A. (2002). *Municipal Stormwater Management*. Boca Raton, FL: CRC Press.

Deo, O., Sumanasooriya, M., and Neithalath, N. (2010). Permeability reduction in pervious concretes due to clogging: experiments and modeling. *Journal of Materials in Civil Engineering* 22: 741–751.

Dequiedt, A.-S., Coster, M., Chermant, L., and Chermant, J.-L. (2001). Study of phase dispersion in concrete by image analysis. *Cement and Concrete Composites* 23: 215–226.

Dietz, M.E. (2007). Low impact development practices: a review of current research and recommendations for future directions. *Water, Air, and Soil Pollution* 186 (1–4): 351–363.

Drake, J. and Bradford, A. (2013). Assessing the potential for restoration of surface permeability for permeable pavements through maintenance. *Water Science and Technology* 68 (9): 1950–1958.

Dreelin, E., Fowler, L., and Ronaldcarroll, C. (2006). A test of porous pavement effectiveness on clay soils during natural storm events. *Water Research* 40 (4): 799–805.

Fan, L.-F., Wang, S.-F., Chen, C.-P. et al. (2013). Microbial community structure and activity under various pervious pavements. *Journal of Environmental Engineering* 140 (3): 4013012.

Fassman, E.A. and Blackbourn, S. (2010). Urban runoff mitigation by a permeable pavement system over impermeable soils. *Journal of Hydrologic Engineering* 15 (6): 475–485.

Federal Highway Administration (FHWA) (2008). Porous Asphalt Pavements with Stone Reservoirs. Tech Brief, FHWA-HIF-15-009, 2015. Hanson, K., Porous Asphalt Pavements for Stormwater Management, IS-131, National Asphalt Pavement Association (NAPA).

Federal Highway Administration (FHWA) (2012). Pervious Concrete. Tech Brief, FHWA-HIF-13-006.

Field, R., Tafuri, A. N., Muthukrishnan, S.. et al. (2004). The Use of Best Management Practices (BMPs) in Urban Watersheds. National Risk Management Research Laboratory, Office of Research and Development, U.S. Environmental Protection Agency, Washington, DC.

Finkenbine, J.K., Atwater, J.W., and Mavinic, D.S. (2000). Stream health after urbanization. *JAWRA Journal of the American Water Resources Association* 36 (5): 1149–1160.

Gilbert, J. and Clausen, J. (2006). Stormwater runoff quality and quantity from asphalt, paver, and crushed stone driveways in Connecticut. *Water Research* 40 (4): 826–832.

Golroo, A. and Tighe, S.L. (2010). Developing an overall combined condition index for pervious concrete pavements using a specific panel rating method. *Transportation Research Record: Journal of the Transportation Research Board* 2153 (1): 40–48.

Hager, W.H. (1999). Cavity outflow from a nearly horizontal pipe. *International Journal of Multiphase Flow* 25 (2): 349–364.

Haselbach, L.M. and Gaither, A. (2008). Preliminary Field Testing: Urban Heat Island Impacts and Pervious Concrete. Concrete Technology Forum. Focus on Sustainable Development.

Hein, M.F., Dougherty, M., and Hobbs, T. (2013). Cleaning methods for pervious concrete pavements. *International Journal of Construction Education and Research* 9 (2): 102–116.

Henderson, V. and Tighe, S.L. (2011). Evaluation of pervious concrete pavement permeability renewal maintenance methods at field sites in Canada. *Canadian Journal of Civil Engineering* 38 (12): 1404–1413.

Hu, J. and Stroeven, P. (2006). Proper characterization of pore size distribution in cementitious materials. *Key Engineering Materials* 302–303: 479–485.

Katz, A.J. and Thompson, A.H. (1986). Quantitative prediction of permeability in porous rock. *Physical Review B* 34 (11): 8179–8181.

Klein, R.D. (1979). Urbanization and stream quality impairment. *JAWRA Journal of the American Water Resources Association* 15 (4): 948–963.

Kowalski, K., McDaniel, R., Shah, A., and Olek, J. (2009). Long-term monitoring of noise and frictional properties of three pavements. *Transportation Research Record: Journal of the Transportation Research Board* 2127 (1): 12–19.

de Larrard, F. (1999). *Concrete Mixture Proportioning: A Scientific Approach*, Modern Concrete Technology, 421. London-New York: E & FN Spon, An imprint of Routledge.

Low, K., Harz, D., and Neithalath, N. (2008). Statistical characterization of the pore structure of enhanced porosity concrete. *Proceedings in CD of the 2008 Concrete Technology Forum*. National Ready Mix Concrete Association, Denver.

Lucke, T. and Beecham, S. (2011). Field investigation of clogging in a permeable pavement system. *Building Research and Information* 39 (6): 603–615.

Marolf, A., Neithalath, N., Sell, E. et al. (2004). Influence of aggregate size and gradation on acoustic absorption of enhanced porosity concrete. *ACI Materials Journal* 101 (1): 82–91.

Martin, W.D. III, Kaye, N.B., and Putman, B.J. (2014). Impact of vertical porosity distribution on the permeability of pervious concrete. *Construction and Building Materials* 59: 78–84.

Martin, W. and Kaye, N. (2014). Hydrologic characterization of undrained porous pavements. *ASCE Journal of Hydrologic Engineering* 19 (6): 1069–1079.

Martin, W. and Kaye, N.B. (2015). Characterization of undrained porous pavements using a broken-line model. *ASCE Journal of Hydrologic Engineering* 20 (2): 04014043-1-8.

Martin, W. and Kaye, N.B. (2016). Hydrologic characterization of an underdrained porous pavement. *ASCE Journal of Hydrologic Engineering* 20 (2).

Martin, W., Kaye, N.B., and Putman, B. (2015). Effects of aggregate masking on soil infiltration under an aggregate bed. *ASCE Journal of Irrigation & Drainage Engineering* 141 (9).

Martin, W.D. and Putman, B.J. (2016). Comparison of methods for measuring porosity of porous paving mixtures. *Construction and Building Materials* 125: 229–305.

Martin, W., Putman, B., and Kaye, N.B. (2013). Using image analysis to measure the porosity distribution of a porous pavement. *Construction and Building Materials* 48: 210–217.

Mata, L.A. and Leming, M.L. (2012). Vertical distribution of sediments in pervious concrete pavement systems. *ACI Materials Journal* 109 (2): 149–155.

Mbanaso, F.U., Coupe, S.J., Charlesworth, S.M., and Nnadi, E.O. (2013). Laboratory-based experiments to investigate the impact of glyphosate-containing herbicide on pollution attenuation and biodegradation in a model pervious paving system. *Chemosphere* 90 (2): 737–746.

Meininger, R.C. (1988). No-fines pervious concrete for paving. *Concrete International* 10 (8): 20–27.

Montes, F. and Haselbach, L. (2006). Measuring hydraulic conductivity in pervious concrete. *Environmental Engineering Science* 23 (6): 960–969.

Murphy, P. (2013). The hydraulic performance of perforated pipe underdrains surrounded by loose aggregate. M.S. thesis. Clemson University, Clemson, SC.

Murphy, P., Kaye, N.B., and Khan, A.A. (2014). Hydraulic performance of aggregate beds with perforated pipe underdrains flowing full. *ASCE Journal of Irrigation & Drainage Engineering* 140 (8): 04014023-1-7.

National Ready Mixed Concrete Association (NRMCA) (n.d.). Pervious Concrete Pavements Structural Design Considerations. www.perviouspavement.org/design/structural.html (27 January 2018).

Neithalath, N. (2004). Development and characterization of acoustically efficient cementitious materials. Ph.D. thesis. Purdue University, West Lafayette, IN.

Neithalath, N., Sumanasooriya, M.S., and Deo, O. (2010). Characterizing pore volume, sizes, and connectivity in pervious concretes towards permeability prediction. *Materials Characterization* 61: 802–813.

Neithalath, N., Weiss, J., and Olek, J. (2006). Characterizing enhanced porosity concrete using electrical impedance to predict acoustic and hydraulic performance. *Cement and Concrete Research* 36 (11): 2074–2085.

Newman, A., Pratt, C., Coupe, S., and Cresswell, N. (2002). Oil bio-degradation in permeable pavements by microbial communities. *Water Science and Technology* 45 (7): 51–56.

Pratt, C., Mantle, J., and Schofield, P. (1989). Urban stormwater reduction and quality improvement through the use of permeable pavements. *Water Science and Technology* 21 (8–9): 769–778.

Pratt, C., Newman, A., and Bond, P. (1999). Mineral oil bio-degradation within a permeable pavement: long term observations. *Water Science and Technology* 39 (2): 103–109.

Roseen, R.M., Ballestero, T.P., Houle, J.J. et al. (2009). Seasonal performance variations for storm-water management systems in cold climate conditions. *Journal of Environmental Engineering* 135 (3): 128–137.

Rungruangvirojn, P. and Kanitpong, K. (2010). Measurement of visibility loss due to splash and spray: porous, SMA and conventional asphalt pavements. *International Journal of Pavement Engineering* 11 (6): 499–510.

Rushton, B. (2001). Low-impact parking lot design reduces runoff and pollutant loads. *Journal of Water Resources Planning and Management* 127 (3): 172–179.

Schwartz, S. (2010). Effective curve number and hydrologic design of pervious concrete storm-water systems. *ASCE Journal of Hydrologic Engineering* 15 (6): 465–474.

Smith, D.R. (2011). *Permeable Interlocking Concrete Pavements Manual: Design, Specification, Construction Maintenance*. American Concrete Institute.

Soroushian, P. and Elzafraney, M. (2005). Morphological operations, planar mathematical formulations, and stereological interpretations for automated image analysis of concrete microstructure. *Cement and Concrete Composites* 27: 823–833.

Stempiha, J., Pourshams-Manzouri, T., Kaloush, K., and Rodezno, M. (2014). Porous asphalt pavement temperature effects for urban heat island analysis. *Transportation Research Record: Journal of the Transportation Research Board* 2293. doi: 10.3141/2293-15.

Sumanasooriya, M.S. and Neithalath, N. (2009). Stereology and morphology based pore structure descriptors of enhanced porosity (pervious) concretes. *ACI Materials Journal* 106: 429–438.

Sumanasooriya, M.S. and Neithalath, N. (2011). Pore structure features of pervious concretes proportioned for desired porosities and their performance prediction. *Cement and Concrete Composites* 33: 778–787.

Tennis, P.D., Lemin, M.L., and Akers, D.J. (2004b). *Pervious Concrete Pavements*. Skokie, IL: Portland Cement Association and National Ready Mixed Concrete Association.

Tennis, P.D., Leming, M.L., and Akers, D.J. (2004a). *Pervious Concrete Pavements*. Skokie, IL: Portland Cement Association.

Tosomeen, C. and Lu, Z. (2008). Pervious concrete bicycle lanes: roadway stormwater mitigation within the right-of-way. *Low Impact Development for Urban Ecosystem and Habitat Protection, International Low Impact Development Conference 2008*, Seattle.

UNHSC (University of New Hampshire Stormwater Center) (2009). UNHSC Design specifications for porous asphalt pavement and infiltration beds. University of New Hampshire, Durham, NH, USA.

Urbonas, B. and Stahre, P. (1993). *Stormwater Best Management Practices and Detention for Water Quality, Drainage, and CSO Management*. Englewood Cliffs, NJ: Princeton Hall.

US EPA (2007). *Reducing Stormwater Costs through Low Impact Development (LID) Strategies and Practices*, vol. 37. Washington, DC: US EPA.

Van Seters, T., Smith, D., and MacMillan, G. (2006). Performance evaluation of permeable pavement and a bioretention swale. *Proceedings 8th International Conference on Concrete Block Paving*.

Wang, K., Schaefer, V. R., and Suleiman, M. T. (May 2006). Development of mix proportion for functional and durable pervious concrete. *Proceeding of the NRMCA Concrete Technology Forum: Focus on Pervious Concrete*, Nashville, TN.

West, D., Kaye, N.B., Putman, B.J., and Clark, R. (2016a). Quantifying the non-linear hydraulic behavior of pervious concrete. *ASTM Journal of Testing and Evaluation* 44 (6).

West, D.W., Kaye, N.B., and Putman, B. (2016b). Surface flow and spread calculations for the preliminary design of porous pavement bike lanes. *ASCE Journal of Irrigation & Drainage Engineering* 142 (2).

Winston, R.J., Al-Rubaei, A.M., Blecken, G.T. et al. (2016). Maintenance measures for preservation and recovery of permeable pavement surface infiltration rate – the effects of street sweeping, vacuum cleaning, high pressure washing, and milling. *Journal of Environmental Management* 169: 132–144.

15

Air Pollution Control Engineering

Kumar Ganesan[1] *and Louis Theodore*[2]

[1] *Department of Environmental Engineering, Montana Tech, Butte, MT, USA*
[2] *Manhattan College, New York, NY, USA*

15.1 Overview of Air Quality

The Industrial Revolution changed the landscape of lifestyle and along came the pollution problems. Smoke control ordinance was in place even around 1881 in large cities like Chicago. Coal was used as the main source of energy. Switching from coal to oil and gas in the 1950s significantly reduced the air pollution problem. However, after World War II, industrial growth increased the fossil fuel use, and therefore air pollution problems were getting severe. In late 1940, a California Professor, Haagen-Smit, indicated that photochemical reactions of pollutants in the atmosphere are the cause of severe smog problems in the Los Angeles area. In 1948, Donora, Pennsylvania, experienced an air pollution episode, resulting in 20 deaths, and about 7000 people got sick. In 1952, the London smog caused 4000 deaths related to air pollution, and it disproportionally affected the younger and the elderly population.

The United States took notice of the growing air pollution problem, and the Air Pollution Control Act was formulated in 1955 by the Congress, making the way to fund federal agencies to conduct research in air pollution. In 1963 the Clean Air Act replaced the 1955 Act and added funding and grants to nonfederal agencies. The Air Quality Act of 1967 articulated that the federal government has the right and duty to enforce control measures for air pollution. In 1970 one of the most powerful environmental legislation, Clean Air Act Amendment (CAAA), was enacted. The National Environmental Policy Act created the Environmental Protection Agency (EPA). The US EPA became operational on 2 December 1970, and it was charged with the responsibility to protect the land, air, and water systems of the United States.

The EPA developed and implemented comprehensive environmental programs in coordination with the state and local agencies. There were subsequent amendments to the Clean Air Act including the 1990 CAAA, which specifically identified seven titles: Title I applies to urban air quality, Title II deals with mobile sources, Title III deals with air toxics, Title IV deals with acid deposition, Title V deals with operating permit, Title VI deals with stratospheric ozone protection, and Title VII focused on enforcement activities.

The National Ambient Air Quality Standards (NAAQS) were established for outdoor air by the 1970 CAAA. Also, the New Source Performance Standards (NSPS) were established for controlling emissions from industrial sources. The NAAQS were changed periodically to reflect the new scientific information on health effects related to air pollution. The original NAAQS included suspended particulate matter (SPM), and subsequently the SPM standard was replaced by PM-10 and PM-2.5 particle size standards. The NAAQS include PM-10, PM-2.5, SO_2, NO_2, CO, lead, and ozone. The NAAQS and NSPS are given in Tables 15.1 and 15.2.

The following sections focus on control techniques for particulate and gaseous pollutants. The gaseous pollutants included are SO_2, NO_x, and volatile organic compound (VOC).

15.2 Emissions of Particulates

The solid and liquid particles suspended in air are generally termed as "particulate matter" (PM). The chemical and physical characteristics of these airborne particulates vary significantly. The physical size of the particulates can vary from coarse to very fine particulates that may need a powerful microscope to see them. The larger particles tend to settle down more readily close to the emission source and not readily transported to longer distances. Consequently, the spatial impact of large

Handbook of Environmental Engineering, First Edition. Edited by Myer Kutz.

Table 15.1 National Ambient Air Quality Standards (40 CFR 50, 2017).

Pollutant		Primary/secondary	Averaging time	Level	Form
Carbon monoxide (CO)		Primary	8 h	9 ppm	Not to be exceeded more than once per year
			1 h	35 ppm	
Lead (Pb)		Primary and secondary	Rolling 3 mo average	$0.15\,\mu g\,m^{-3}$ [a]	Not to be exceeded
Nitrogen dioxide (NO_2)		Primary	1 h	100 ppb	98 percentile of 1-h daily maximum concentrations, averaged over 3 yr
		Primary and secondary	1 yr	53 ppb [b]	Annual mean
Ozone (O_3)		Primary and secondary	8 h	0.070 ppm [c]	Annual fourth highest daily maximum 8-h concentration, averaged over 3 yr
Particle pollution (PM)	PM-2.5	Primary	1 yr	$12.0\,\mu g\,m^{-3}$	Annual mean, averaged over 3 yr
		Secondary	1 yr	$15.0\,\mu g\,m^{-3}$	Annual mean, averaged over 3 yr
		Primary and secondary	24 h	$35\,\mu g\,m^{-3}$	98 percentile, averaged over 3 yr
	PM-10	Primary and secondary	24 h	$150\,\mu g\,m^{-3}$	Not to be exceeded more than once per year on average over 3 yr
Sulfur dioxide (SO_2)		Primary	1 h	75 ppb [d]	99 percentile of 1-h daily maximum concentrations, averaged over 3 yr
		Secondary	3 h	0.5 ppm	Not to be exceeded more than once per year

[a] In areas designated nonattainment for the Pb standards prior to the promulgation of the current (2008) standards and for which implementation plans to attain or maintain the current (2008) standards have not been submitted and approved, the previous standards ($1.5\,\mu g/m^3$ as a calendar quarter average) also remain in effect.

[b] The level of the annual NO_2 standard is 0.053 ppm. It is shown here in terms of ppb for the purposes of clearer comparison with the 1-h standard level.

[c] Final rule signed on 1 October 2015 and effective on 28 December 2015. The previous (2008) O_3 standards additionally remain in effect in some areas. Revocation of the previous (2008) O_3 standards and transitioning to the current (2015) standards will be addressed in the implementation rule for the current standards.

[d] The previous SO_2 standards (0.14 ppm 24 h and 0.03 ppm annual) will additionally remain in effect in certain areas: (i) any area for which it is not yet 1 year since the effective date of designation under the current (2010) standards, and (ii) any area for which an implementation plan providing for attainment of the current (2010) standard has not been submitted and approved and which is designated nonattainment under the previous SO_2 standards or is not meeting the requirements of a SIP call under the previous SO_2 standards (40 CFR 50.4(3)). A SIP call is an EPA action requiring a state to resubmit all or part of its State Implementation Plan to demonstrate attainment of the required NAAQS.

particles is limited to the local areas. The two-size range of particulates that are commonly addressed in ambient air are particulate matter with aerodynamic diameter less than or equal to 10 μm (PM-10) and particulate matter with aerodynamic diameter less than or equal to 2.5 μm (PM-2.5). These two particle sizes behave differently in the atmosphere. The PM-2.5 tends to remain suspended in air for much longer periods of time. This results in transport of such particulates to longer distances, over hundreds of miles, from its emission sources. Therefore, the spatial distribution of PM-2.5 is much larger. The PM-10 particulates travel shorter distances from its emission sources and have relatively lower spatial impact compared with the PM-2.5 particulates.

The "primary" particulates are emitted directly from a source into the atmosphere, while the "secondary" particulates are formed in the atmosphere due to chemical reactions. In general, primary particles are course in size, and secondary particles are very fine in size. Sulfate and nitrate particles in the atmosphere are mainly due to sulfur dioxide and nitrogen oxide emissions that undergo chemical reaction in the atmosphere. In most cases the PM-2.5 contains large portion of secondary particulates. There are multiple sources of primary PM-2.5 sources such as emissions from motorized vehicles, emissions from wood-burning stoves, forest fires, and many more.

The National Emissions Inventory (NEI) tracks the emissions data of particulates. The NEI obtains the data from actual emission measurement as well as estimated emissions for various sources. Because it is almost impossible to measure all emission sources, emissions are estimated using methodologies such as mass balance and modeling approaches. Some of the primary sources of particulate emissions include power plants and industrial, commercial, residential, institutional, and mobile sources. The NEI data primarily comes from the US EPA

Table 15.2 Selected examples of National New Source Performance Standards (NSPSs) (40 CFR 60, 2017).

1) Steam electric power plants (coal-fired)
   a) Particulate matter: 0.015 lb per million Btu heat input, or 0.03 lb per million Btu heat input and 99.9% reduction
   b) NO_x: 1.0 lb $(MWh)^{-1}$ gross energy output
   c) SO_2: 1.4 lb $(MWh)^{-1}$ gross energy output or 95% reduction
   d) Hg: 0.020 lb $(GWh)^{-1}$ gross energy output

2) Large (>250 tons d^{-1}) municipal solid waste (MSW) combustors: There are individual standards for dioxins/furans, cadmium, lead, mercury, HCl, particulate matter, NO_x, and SO_2. Three examples are:
   a) PM: 20 mg $(dscm^{a})^{-1}$ corrected to 7% O_2
   b) HCl: 25 ppm dry volume, corrected to 7% O_2
   c) Hg: 50 µg $(dscm)^{-1}$ corrected to 7% O_2

3) Nitric acid plants: The standard is a maximum 3-h average NO_x emission of 1.5 kg per metric ton of 100% acid produced. All NO_x emissions are to be expressed as 100% NO_2. Also, the stack gases must meet 10% opacity (where 0% opacity represents perfectly clear stack gas and 100% opacity means completely opaque)

4) Sulfuric acid plants: The standard is a maximum 3-h average emission of SO_2 of 2 kg per metric ton of 100% acid produced. An acid mist is a maximum 3-h emission of 0.075 kg SO_2 per metric ton of acid produced. Also, the stack gases must meet 10% opacity

5) Primary copper smelters: The particulate emission standard is 50 mg $(dscm)^{-1}$, the SO_2 standard is 0.065% by volume, and the opacity is limited to 20%

6) Wet-process phosphoric acid plants: The total fluorides emission standard is 10.0 g per metric ton of P_2O_5 feed

7) Iron and steel plants: Particulate discharges may not exceed 50 mg $(dscm)^{-1}$, and the opacity must be 10% or less except for 2 min in any hour

8) Sewage sludge incinerators: The particulate emission standard is 0.65 g kg^{-1} sludge input (dry basis). The opacity standard is 20%

9) Hospital/medical/infectious waste incinerators: large (>500 lb h^{-1} of waste feed). There are individual standards for PM, CO, dioxins/furans, HCl, SO_2, NO_x, and several metals, all corrected to 7% O_2. Examples include:
   a) PM: 34 mg $(dscm)^{-1}$
   b) CO: 40 ppmv
   c) Dioxins/furans: 25 ng $(dscm)^{-1}$ total CDD/CDF or 0.6 ng $(dscm)^{-1}$ TEQ
   d) HCl: 15 ppmv or 99% reduction
   e) Selected metals: Pb – 0.07 mg $(dscm)^{-1}$, Cd – 0.04 mg $(dscm)^{-1}$, Hg – 0.55 mg $(dscm)^{-1}$

[a] dscm means dry standard cubic meter.

and state, tribal, and local air quality management agencies. Wildfire emissions are estimated based on fire activity and location on satellite detection system. The on-road and off-road vehicle emissions are determined based on the input provided by state agencies to models to estimate the vehicle emissions.

The NEI data have been available since 1990 for all states and counties in the United States and US territories of Puerto Rico and Virgin Islands. Between 1990 and 2011, the primary PM-10 emissions decreased by 40%. The US EPA has divided the United States into 10 regions for administrative purposes. All regions showed decrease in PM-10 primary emissions except Region 9. Region 9 includes Arizona, Nevada, California, Hawaii, the Pacific Islands, and 148 tribal nations. Between 1990 and 2011, the PM-10 emission was reduced by 67% from the fuel combustion category, which was the highest reduction. Fugitive dust from roads contributed to the majority of the PM-10 emissions.

The PM-2.5 emission trends during the years between 1990 and 2011 showed a decrease of 53% from anthropogenic sources. Most of the reduction was due to decrease in PM-2.5 emissions from fuel combustion sources. Sixty percent of the PM-2.5 emissions were from fugitive, natural, and miscellaneous sources. All EPA regions showed reduction in PM-2.5 emissions ranging from 33% in Region 4 to about 82% in Region 1. The anthropogenic PM-2.5 emissions for the United States by source category are given in Figure 15.1. The relative amounts of PM-2.5 from anthropogenic and other sources for 2011 are given in Figure 15.2.

15.2.1 Control Technologies for Particulate Matter

Particulates discharged into the atmosphere can be controlled by treating the airstream to remove the particulates or preventing the formation of particulates in the first place. Although prevention is the most economically and environmentally desirable approach, opportunities for reduction of emission at the source are limited.

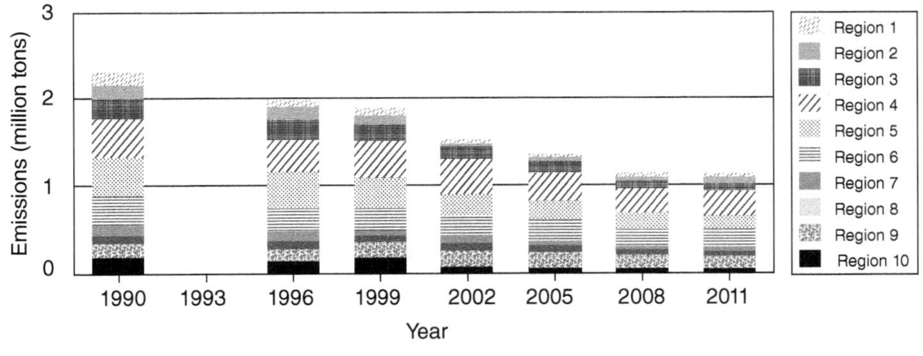

Figure 15.1 Anthropogenic PM-2.5 emissions in the United States by EPA Region, 1990–2011 (U.S. EPA, 2017).

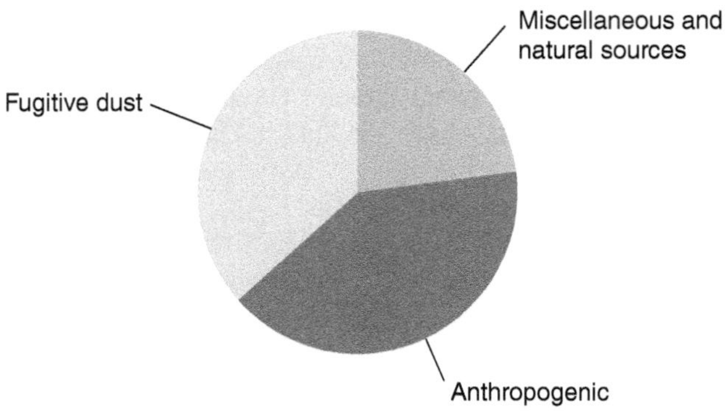

Figure 15.2 Relative amounts of US PM-2.5 emissions from anthropogenic and other sources, 2011 (U.S. EPA, 2017).

Process modification and optimization strategies at times are effective in reducing the particulate emissions. Few examples are optimizing the process, increasing the size of particulates generated, and reducing the quantity formed or mass loading. Stable process operations can be combined to reduce particulate emissions from a specific process. In many cases switching of fuel, for example, from coal to natural gas, will considerably prevent the discharge of particulates into the atmosphere. That is, switching of coal-fired power plants to natural gas-fired power plants will significantly reduce the particulate emissions. But there are more regulatory, economic, and engineering implications to overcome to accomplish this. Installation of particulate control systems of various types is the conventional strategy to reduce particulate emissions from process and industrial operations into the atmosphere.

It is clear from Table 15.3 that particulate emissions from coal are significantly higher than the natural gas. Development of syngas from coal and using an integrated gasification combined cycle (IGCC) will improve the efficiency of coal-fired power plants in reducing particulate emissions. Syngas is produced in a gasifier. If oxygen is used instead of air in the gasifier, then the amount of flue gas to be treated is significantly reduced in volume and reducing the cost of gas cleaning. However, the initial capital cost for separating oxygen from air needs to be considered in the economic equation. In this process the syngas is cleaned to reduce the particulates and sulfur gases before it is burned in a gas turbine. Syngas burns differently than natural gas, because syngas is composed mainly of hydrogen and carbon monoxide. The gas turbine and steam turbines generate energy in the IGCC process instead of steam turbines only in a conventional coal-fired power plant.

15.2.2 Process Optimization

Process optimization involves modification of feed material, process unit functions, and process variables to minimize particulate emissions. This can also change the chemical and physical characteristics of the particulates. For example, it may influence the particulate size distribution and even reduce the volume of gas generated, thus decreasing the air pollution control equipment cost and its size. A change in particulate size distribution may favor a low-cost control system instead of a relatively more expensive control system. In addition, the change in size distribution, for example, reduces the emission of PM-2.5 size particulates. This helps in the air quality compliance relative to having flue gas with large amounts of PM-2.5 particulates.

The properties of raw materials fed into a process such as particle size, moisture, and chemical composition can have significant effect on the emissions. For example, prescreening of raw phosphate rock before drying can reduce fine materials in the feedstock, resulting in reduced fine particulate emissions. Manufacturing steps involve multiple unit operations. A careful analysis of the process flow, the number, order, and types of process steps can reduce the number of emission points, especially fugitive particulate emissions, and save wasteful products. In many cases, consideration of enclosed

Table 15.3 Emissions from fossil fuel combustion (EIA, 2008).

	Emissions from fossil fuel combustion				
	Sources (lb per billion Btu)			Percent emissions in natural gas	
Pollutant	Natural gas	Oil	Coal	Percent oil	Percent coal
Carbon dioxide	117 000	164 000	208 000	71	56
Carbon monoxide	40	33	208	21	19
Nitrogen oxides	92	448	457	21	20
Sulfur dioxides	0.6	1122	2591	0.05	0.02
Particulates	7	84	2744	8.33	0.26

conveyer systems reduces emissions and saves money to the plant in the long run. Wetting or agglomeration of materials can have direct effect on the particulate emissions. For example, in a wood fiber plant, partial polymerization of the heat-setting resin that coats the fibers before its transfer can significantly reduce the emissions and save product materials. Optimizing each process and industrial operations using pollution prevention concepts will benefit the environment and save money for the company.

15.2.3 Gas Cleaning for Particulates

The process optimization can prove to be very effective in reducing emissions, but in many cases further reduction is needed to meet the regulatory requirements. In these cases, gas cleaning techniques are employed to remove particulates from gas streams. The traditional devices available for gas cleaning are cyclones, multicyclones, fabric filters, electrostatic precipitators (ESPs), and wet scrubbers. Each of these devices operates under specific physical and chemical principles. Selection of a control equipment uses a complex set of variables, including regulatory limitations, physical and chemical characteristics of the emissions, the removal efficiency of each equipment, long-term reliability, consistency in performance, ease of maintenance and operation, space requirements, safety issues, and the cost of the system.

In most cases, the control devices must perform to meet the regulatory requirements of emission limits or ambient air quality standards. There are federal and local regulations that are applicable for industrial sources. The NSPS, National Emission Standards for Hazardous Air Pollutants (NESHAP), NAAQS, and Prevention of Significant Deterioration (PSD) are some of the federal regulations that will affect the decision on type of control device selection for a specific plant in a specific location. In most cases, federal authorities like EPA delegate authorities to individual states for implementation and enforcement of the regulations. The first step in determining regulatory requirements is to determine which local, state, and federal agencies have jurisdiction over a particular industrial operation. It is best to consult the involved agencies up front to avoid any delays in the permitting process.

15.2.4 Cost of Particulate Control Devices

The selection of a control device that will provide efficient and reliable service over the life of the device at the least possible cost is what most companies want. The cost in general includes the installed cost of the device, direct and indirect operating costs. The installed cost includes the cost of engineering and design of the equipment, cost of materials for the construction, cost of fabrication and manufacturing, cost of transportation, taxes, cost of monitoring equipment, and the cost of labor for installation. The direct operating costs involves the cost of electricity, labor to operate the devices, the maintenance cost, waste disposal cost, monitoring and performance evaluation cost, inventory of spare parts for replacement, cost of routine preventive maintenance, and cost related to safety and personnel protection and training. Indirect cost involves, overhead costs, insurance, taxes, and capital recovery of the investment. An accurate cost estimates will help proper budgeting and avoid any future financial issues to maintain and operate the equipment through its lifetime.

15.2.5 Characteristics of Particulates

In designing a control device for particulates, the most critical physical property to consider is the particle size characteristics. The particle size provides the aerodynamic behavior of the particles in a pollution control device. Particulates are of varying shapes including spherical and fibrous. They are of different densities and chemical compositions. They can exist as an agglomeration of smaller particles, and these agglomerated particles at times are fragile and may break down during their transport in airstreams. Asbestos is one of the fibrous particles. Mechanical processes such as grinding, milling, polishing, sawing, and hammer mills can generate particulates having varying irregular shapes. In general, aerosols formed by the condensation of organic compounds are spherical in shape and much finer than the particles generated from mechanical processes. Liquid particles formed in the process are also spherical in shape. Compounds in vapor form can condense on solid particles and provide spherical shape to the original particle. The particle size also determines the light scattering properties of the particles, which is directly related to the plume opacity. The particle size between 0.1 and 1 μm size scatters light in the visible light range and is responsible for plume opacity. Particles in these size ranges are very difficult to control. The particle size range of interest is between 0.01 and 1.00 μm for designing particulate control devices. One micrometer is one millionth of a meter ($1\,\mu m = 1 \times 10^{-6}$ m). The most common representation of particle size is by "aerodynamic diameter." An aerodynamic diameter is the diameter of an imaginary spherical particle of unit density having the same aerodynamic characteristics as the actual particle. This will further be explained using Stokes' law in the following sections. The aerodynamic diameter is used to understand the particle behavior in the particulate control devices. A cascade impactor separates particles based on their aerodynamic particle size. It cascades particulate-laden air through a series of slotted plates and impaction plates. The size of the slots or the diameter of the slots decreases as the air passes through different stages to separate finer particles. The finest particles will be captured in the last impaction plate, while the first impaction plate will capture the coarse particles. The amount of particles collected in each stage will be used to determine the size distribution of those particulates.

Particle size distribution can be characterized by its number concentration, surface area, and mass. For purposes of air pollution control device design, a mass distribution with respect to size range of particles is considered. The mass and size distribution of a typical particle size distribution follows a lognormal distribution. In a lognormal distribution, log of the diameter of the particles is plotted against the cumulative mass percentage of particulates. When plotted, this curve will look like an "S" curve. However, same data when plotted on a log-probability paper will yield a straight line. The slope of the line represents the geometric standard deviation, and the geometric mean diameter is the particle diameter corresponding to 50% cumulative distribution. By plotting on a log-probability paper, the particle size distribution data become more versatile and useful to obtain mass percentages of various particle sizes of interest. For example, PM-2.5 and PM-10 mass percentage of particles can be easily obtained from such graphs.

15.2.6 Aerodynamic Behavior of Particles

Particulate control devices are designed based on one or more basic mechanisms to separate the particles from the airstream. The three most common mechanisms are impaction, interception, and diffusion. The inertial impaction is based on the relative difference between the inertia of moving gas molecules and that of the moving particles. By having greater mass than the gas, the particles are unable to change their direction of motion, while gas can easily change its direction of flow. Therefore, if the gas changes its direction of flow due to an obstacle, then the particles get separated from the airstream. These separated particles can be collected and removed from the airstreams. This inertial impaction mechanism to collect particle is being used in several particulate control devices.

While the particles are moving in air, there exists a drag force that is exerted by the fluid on the particle opposing the particle motion. This drag force is proportional to the viscosity of the gas, diameter of the particle, and relative velocity of the particle with respect to the airstream for spherical particles with Reynolds number less than 1. Under this condition, the drag force is termed as the "Stokes' law":

$$F_{\mathrm{D}} = 3\pi\mu v_{\mathrm{r}} d_{\mathrm{p}}$$

where F_{D} is the drag force,μ is the viscosity of the fluid (gas stream), v_{r} is the relative velocity between the gas stream and the particle, and d_{p} is the particle diameter.

15.2.7 Terminal Settling Velocity of Particle

When there is an external force like the gravitational force acting on a particle, the gravitational force will tend to pull the particle toward the ground, while the drag force will oppose its downward motion in a still air. At some point, within a fraction of seconds, the gravitational force and the drag forces become equal, and the particle ceases to accelerate and attains a constant velocity. This constant velocity of particle traveling under gravitational force or an external force is called the terminal velocity, or Stokes' velocity or settling velocity of particles. There could be more than one force acting on the particle.

15.2.7.1 Particle Settling by Gravity

When a hypothetical particle is released in a still air, the particle gains acceleration due to the gravitational force that pulls the particle toward the ground. As the particle starts its motion, there is an opposing force, the drag force, that comes into play opposing its downward motion. Within a very short distance (μm) from its initial acceleration, the particle attains a constant velocity or gravitational settling velocity when the gravitational force is equal and opposite to the drag force. In this equilibrium status, the particle has no more acceleration and will travel at a constant velocity. This velocity of particles is the gravitation settling velocity of the particles and called Stokes' velocity. Therefore, particles in a room will settle down to the floor based on its settling velocity. The settling velocity is directly proportional to the square of the diameter of the particle. Thus, larger particles will settle relatively faster than smaller particles. In general, gravitational settling is effective for particles greater than 5 μm in diameter, and as such it is not the prominent technique to control particles in industrial operations.

Stokes' settling velocity is given as follows:

$$v_{\mathrm{t}} = \frac{d_{\mathrm{p}}^2 \left(\rho_{\mathrm{p}} - \rho_{\mathrm{g}}\right) g}{18\mu}$$

where

d_{p} is particle diameter, μm
v_{t} is settling velocity, $\mathrm{cm\,s^{-1}}$
ρ_{p} is particle density, $\mathrm{g\,cm^{-3}}$
ρ_{g} is gas density, $\mathrm{g\,cm^{-3}}$
g is 9.81 $\mathrm{m\,s^{-2}}$, acceleration of gravity
μ is gas viscosity, $\mathrm{kg\,m\text{-}s^{-1}}$

15.3 Control of Particulates

As a gas stream approaches an obstacle, the gas stream will pass around the object while the particles by its inertia propel itself toward the object. If the particles are too small, they tend to follow the gas stream lines and will not be captured on the object. Therefore, inertial impaction can be used to separate particles from gas stream effectively for a range of particle sizes. As the particle size gets too small, below a micron size, the impaction energy needed to separate the particles becomes larger and gets to be very energy intensive. The impaction parameter and the ratio of drag force to viscous force are indicators of the efficiency of impaction:

$$K_{\mathrm{I}} = \frac{C d_{\mathrm{p}}^2 \rho_{\mathrm{p}} v}{18 \mu D_{\mathrm{c}}}$$

where
- K_{I} is Stokes' number
- C is Cunningham slip factor, dimensionless
- d_{p} is particle diameter, μm
- ρ_{p} is particle density, g cm^{-3}
- v is particle velocity, cm s^{-1}
- D_{c} is diameter of collector, cm
- μ is gas viscosity, kg m-s^{-1}

Based on the above equation, impaction is directly proportional to the square of the particle diameter. High-density particles have relatively high impaction parameters, thus being separated from gas stream much more easily than the smaller particles. Interception is another mechanism by which particles get separated from gas stream. An obstacle can intercept the particle if the radius of the particle is equal or larger than the streamline displacement. Interception is more applicable for particles greater than micron size and adds to the impaction mechanism in separating the particles from gas streams. Diffusion is another mechanism that helps to separate particles from the gas stream. Particles sized similarly to gas molecules (0.001 μm) are subjected to random movement due to collisions with gas molecules. This random motion effectively can have some of those particles reaching the obstacle and be collected by the obstacle. The effectiveness of particle collection by diffusion can be mathematically explained by the Stokes–Einstein equation:

$$D = \frac{CKT}{3\pi \mu d_{\mathrm{p}}}$$

where
- D is diffusion coefficient (diffusivity)
- C is Cunningham correction factor, dimensionless
- K is the Boltzmann constant
- T is absolute temperature, K
- μ is absolute viscosity, kg m-s^{-1}
- d_{p} is particle diameter, m

The higher the diffusivity of the particle, the higher the potential for those particles to be captured by the obstacle. The diffusivity of the particles increases as the particle diameter decreases or it is inversely proportional to the diameter of the particle. Therefore, diffusion mechanism is mainly viable for fine particles of less than one micron size. On a mass basis, the diffusion mechanism plays a smaller role in collecting particles relative to particle collection by impaction. The diffusivity of particles larger than micron size is significantly smaller and thus does not affect the collection of particles greater than micron size. The impaction, interception, and diffusion of particles are represented pictorially in Figures 15.3–15.5.

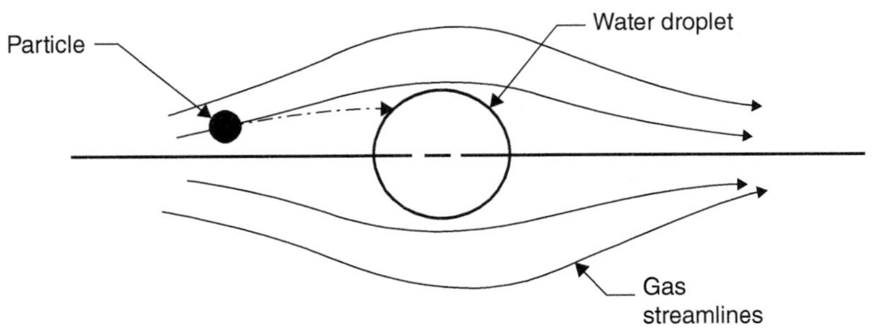

Figure 15.3 Impaction of particles on a target in a moving gas stream (U.S. EPA, 1982).

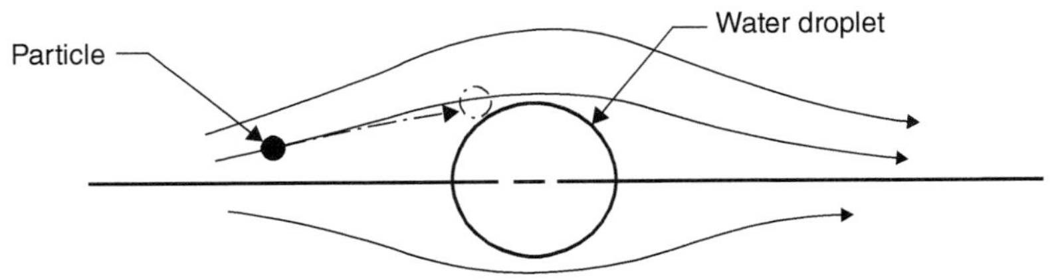

Figure 15.4 Interception of a particle on a target in a moving gas stream (U.S. EPA, 1982).

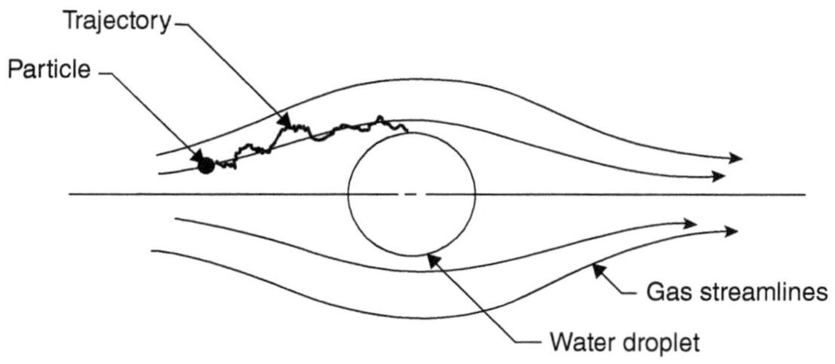

Figure 15.5 Diffusion of a particle to a target in a moving gas stream (U.S. EPA, 1982).

15.3.1 Efficiency of Control Devices

Particulate control systems are evaluated based on their efficiency in removing particles. Efficiency can be represented as a fraction or in percentages. It is estimated by the ratio of the difference in inlet and outlet particulate mass to the inlet particulate mass. The performance of a control system can also be represented by penetration, which is the ratio of outlet particulate mass to the inlet particulate mass loading, indicating how much of the incoming mass is escaping the control device (Penetration = 1 − Efficiency).

$$\text{Efficiency} = \frac{\text{Incoming particle mass} - \text{Outgoing particle mass}}{\text{Incoming particle mass}}$$

$$\text{Penetration} = \frac{\text{Outgoing particle mass}}{\text{Incoming particle mass}}$$

In many cases selection of a specific particulate control device is determined by the regulatory requirements. Mechanical collectors are favored for coarse particle to achieve 90% or less efficiency. When the particle removal efficiency requirement is above 95%, then bag filters, ESPs, or high energy wet scrubbers are favored. In addition, particle size distribution also plays a role in selecting the appropriate type of particulate control system. If control system is required for an existing plant, then particle size characterization by stack testing method following the EPA methods will provide the necessary data. For a new plant, data can be obtained from stack testing results from similar plants, or from the EPA AP-42 emission factor document. The AP-42 document provides information on the particulate emission rate of the size distribution of particles. The AP-42 document also provides space requirements, which limit the selection of control devices that can be utilized. A plant located in an urban complex with limited space will choose a system much different from a plant having additional space. Finally, cost of the device will be a major factor in selecting the type of device for the plant. Table 15.4 shows various particle capture mechanisms in various particulate control devices.

The control system is connected by ductwork and fans to move the gas through the control devices. For proper operation of the device, a well-designed ductwork with right types of fans is critical. Materials for construction of duct must factor in the corrosiveness of gas stream, abrasiveness and toxicity of particulates, and fire hazards, if any. The selection of fan depends on the amount and type of gas flow rates, the pressure drop requirement, the particulate load, and the corrosive, toxic, and explosive nature of the particulates. Maintaining proper velocity in the ductwork and the proper angles for smooth material flow will avoid unnecessary maintenance issues in the future. Different varieties of fans are available; the most rugged type is the radial blade, with centrifugal fans that can withstand high dust loadings with minimum vibration and desirable efficiencies. Fan can be located on the inlet side of the system for handling dirty gas streams. The backward curved fans are relatively more efficient than the radial blades, but handle cleaner gas streams, which are more applicable in the outlet side of the control systems. The forward curved fans, though, are more efficient and can handle only clean gas streams – therefore not common in air pollution control systems.

15.3.2 Mechanical Collectors

Mechanical collectors use gravitational, inertial, or centrifugal forces to separate particles from the gas stream. They are suitable to remove coarser particles. A properly designed mechanical particulate control system could be very effective as a precollector to reduce the particulate loading on more expensive control systems in the downstream. Mechanical collectors are simple to operate and low cost, have low maintenance and low pressure drop, and can handle varying particulate loads.

Settling chambers are designed based on gravitational settling of particulates and meant to remove large particles, mainly to protect the downstream equipment from abrasion and/or excessive particulate loading. A low velocity of 0.3–3.0 m s^{-1} superficial gas velocity is maintained essentially to minimize any flow turbulence. The gravity settling chambers can be as simple as a single chamber to a chamber with multiple horizontal trays. Particles get settled on the trays. Another type of particle separator, an elutriator, has one or more tubes through which gas passes upward while particles settle at the bottom if the settling velocity of

Table 15.4 Particle capture mechanisms normally active in conventional particulate control devices (U.S. EPA, 1982).

Control device	Principal particle capture mechanism	Particle size dependence[a]
Settling chamber	Gravity settling	d_p^2
Momentum separator	Gravity settling	d_p^2
	Inertial separation	d_p^2
Large-diameter single cyclone	Inertial separation	d_p^2
Small-diameter multiple cyclone	Inertial separation	d_p^2
Fabric filters	Impaction on dry surfaces	d_p^2
	Interception	d_p
	Diffusion to dry surfaces	$1/d_p$
Electrostatic precipitator	Electrostatic attraction	d_p^2 and $1/d_p$
	Gravity settling	d_p^2
Wet scrubber	Impaction on surfaces	d_p^2
	Impaction on liquid droplets	d_p^2
	Diffusion to wetted surfaces	$1/d_p$
	Diffusion to liquid droplets	$1/d_p$
Incinerator	Particle oxidation	$1/d_p$

[a] Based on particle capture mechanism.

the particle is greater than the upward velocity of the gas. Finer particles with lower settling velocities will be carried along with the airstream and escape. Multiple-tube elutriators with varying diameter tubes can effectively be used to separate different sizes of particles, which are commonly used in mineral processing industries, petrochemical operations, and agricultural industries.

A momentum separator uses gravity and momentum to separate the particles from airstream. A drastic change in the flow direction of airstream allows the particles to cross streamlines and settle into a collection chamber. Baffles can also be added to increase particle separation and have multiple chambers to further improve particle collection. In some cases, inclined louvers are used such that the gas stream changes its flow angle while particles get settled on the louvered shutters. Mechanically aided separators provide external power to increase the momentum of particles, enabling the collection of smaller particles.

15.3.3 Cyclones

Cyclones work based on a continuously spiraling flow where particles get separated from the gas stream and impinge on the walls of the cyclone (Figure 15.6). The spiraling motion of the air can be initiated by having a tangential inlet or the flow pass through a fixed vane, creating the spiral motion of gas flow. The tangential inlet or the vane-axial flow cyclones are commonly used in industries due to their low cost, them being simple to operate, their low maintenance, their low pressure drop, and their high efficiency for coarse particles. There are no moving parts in a typical cyclone, and thus it is low maintenance. Cyclones can also tolerate varying particulate loads and gas flows under harsh conditions. In a cyclone, the efficiency increases as the gas flow increases to a certain extent. Higher gas flow increases inlet velocity of the cyclone that improves the efficiency of the cyclone.

In vane-axial cyclones, the gas streams enter through a fixed vane, which initiates a swirling motion to the gas stream. The gas spirals down the cyclone and turns 180° moving upward to the top of the cyclone and leaves through a concentric outlet. A multicyclone system consists of hundreds of smaller-diameter vane-axial cyclones operating in parallel (Figure 15.7). The particulate-laden airstream enters the cyclone through a fixed vane and escapes through an extended outlet. The particles get

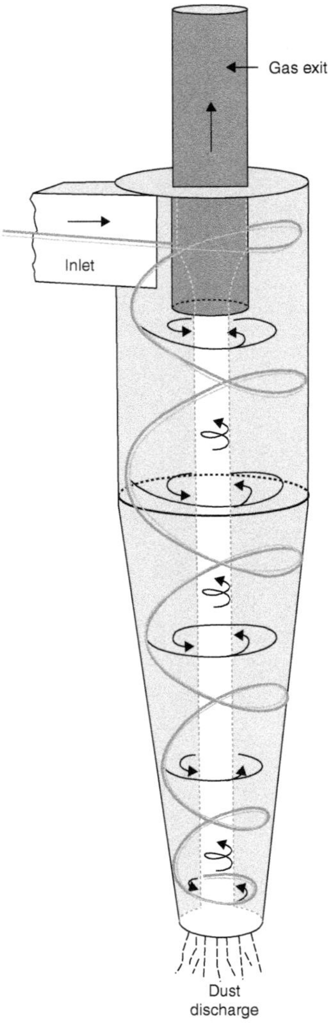

Figure 15.6 Image of a cyclone to remove particulates (U.S. EPA, 1982).

separated from the outer core of the airstream as it spirals downward. The airstreams turn 180 degrees and travel upward toward the outlet. The inner core of the airstream moves upward with relatively cleaner air and exits axially. The inlet and outlet gas streams are physically separated by enclosures. The multicyclones have a common inlet and common outlet for the gas stream. The individual cyclones are of 15–60 cm in diameter, which maintains high efficiency. The pressure drop typically ranges from 0.5 to 1.5 kPa. The number of cyclones in a multicyclone system is limited by the availability of space and the amount of gas to be treated. Multicyclones are a popular particulate precleaning system, especially preceding an ESP.

15.3.3.1 Performance of Cyclone Separators

The performance of cyclones heavily depends on the particle size. Cyclones can achieve up to 90% efficiency for particles of 10 μm diameter and up to 40% for particles of 2.5 μm size. The performance of a cyclone can be theoretically evaluated using a "50% cut size diameter," which is defined as the diameter of the particles collected with 50% efficiency. This approach was developed by several authors: Lapple, Leith, Licht, and Kalen and Zenz. The cyclone efficiency is a function of several parameters including:

- The number of effective turns a spiraling airflow makes before it turns back.
- The width of the inlet that determines the distance a particle should travel to reach the cyclone body for its collection.
- The inlet velocity of the cyclone.
- The density of particle and the viscosity of the gas.

The lower the cut diameter of the particle for a specific cyclone, the higher its efficiency would be. Therefore, to determine the efficiency of a cyclone, first its 50% cut diameter will be determined. Based on the cut diameter, the following equation adopted from Theodore and DePaola (1980) can be used to determine the efficiency of cyclone for various sizes of particles. Since particle sizes are given in size ranges, mid diameter of the size range is used as the particle diameter.

The equations represent cyclone efficiency:

$$\eta_j = \frac{1}{1+\left(\frac{d_{\mathrm{pc}}}{d_{\mathrm{p}}}\right)^2}$$

where

η_j is collection efficiency for the jth particle size range

d_{p} is characteristic diameter of the jth particle size range

$$d_{\mathrm{pc}} = \left[\frac{9\mu W}{2\pi N_{\mathrm{e}} V_i \left(\rho_{\mathrm{p}} - \rho_{\mathrm{g}}\right)}\right]^{1/2}$$

where

W is width of inlet in m

d_{pc} is diameter of a particle size collected with 50% efficiency, m

V_i is terminal velocity, m s^{-1}

ρ_{p} is particle density, kg m^{-3}

μ is gas viscosity, kg m-s^{-1}

ρ_{g} is gas density, kg^3

N_{e} is number of effective turns, dimensionless

The number of turns the outer dirty airstream achieves before it turns upward influences the size of the d_{pc}. Increasing the number of spirals will theoretically decrease the diameter of the cut size equation.

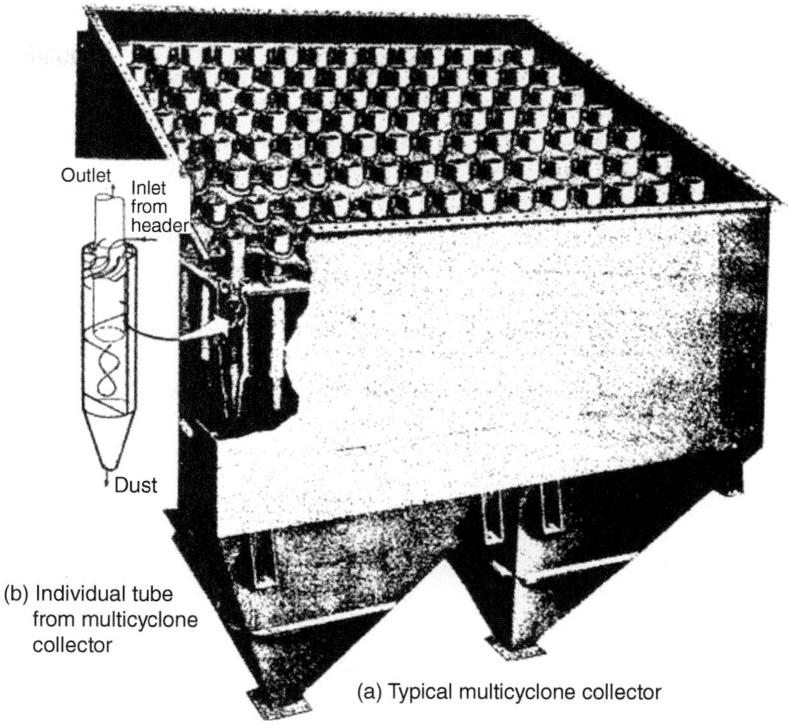

Figure 15.7 Image of a multicyclone collector (U.S. EPA, 1982).

The fractional efficiency equation provides efficiency for a mid-size particle of a specific size range. The overall efficiency for the set of particles range can be calculated based on the efficiency for each size range and the mass in each size range. The sum of the efficiency times the mass fraction will give the overall efficiency of the cyclone. Therefore, overall efficiency, η_0, is

$$\eta_0 = \Sigma \eta_i m_i$$

where

η_i is efficiency for particles of a specific diameter
m_i is mass fraction of a specific particle size

The overall efficiency of a cyclone is affected by its dimensions. The inlet of a cyclone for a specific flow determines the inlet velocity of the gas. If the inlet area is smaller, the velocity is higher, and consequently the efficiency of the cyclone is higher. The increased inlet velocity will also increase the pressure drop of the cyclone that increases the electricity cost for the daily operation of the cyclone. There is always a balance between efficiency and pressure drop to achieve the desired design results. For this reason, the cyclones are classified into three categories: conventional, medium, and high efficiency cyclones. The three categories have specific dimensional guidelines to fall under each of these categories. Table 15.5 and Figure 15.8 provide the dimensional requirements for each of the three cyclones. The dimensions are given in relation to the diameter of the cyclone.

Table 15.5 Dimensionless design ratios for tangential entry cones.

		Efficiency		
Symbol	Nomenclature	High	Medium	Conventional
D_c	Body diameter	1.0	1.0	1.0
H_c	Inlet height	0.5	0.75	0.5
B_c	Inlet width	0.2	0.375	0.25
S_c	Outlet length	0.5	0.875	0.625
D_c	Outlet diameter	0.5	0.75	0.5
L_c	Cylinder length	1.5	1.5	2.0
Z_c	Cone length	2.5	2.5	2.0

Source: From Theodore (2008).

While designing a cyclone, the gas flow rate and efficiency required must be known. Also the particle size distribution is needed to determine the overall efficiency of the cyclone. Medium efficiency cyclones are used to treat airstreams with coarse particulates, and the high efficiency cyclones can handle smaller-sized particles with up to 90% efficiency for PM-10 particles. High efficiency cyclones will have a smaller body diameter and a relatively smaller inlet and outlet area of cross section.

The vendors who supply cyclones will be very valuable to obtain information for specific industrial applications. It is essential to verify the vendor claims through cyclone efficiency calculations to determine the best suited cyclone acceptable for the particular operation. Specifically, vendors' claims of overall efficiency of cyclones may not give the full picture. Obtaining the cyclone efficiency from the vendors for varying sizes of particles is critical for the plant to make decisions whether that cyclone will meet expected emissions reduction at particle size ranges.

Cyclones can also be used in series when particulate loading is high. The initial cyclone can be of a medium efficiency cyclone to remove larger-sized particles and that can be followed by a high efficiency cyclone to remove smaller-sized particles. Cyclones with longer bodies are expected to have a higher efficiency than the ones with shorter body length. There are other operating conditions that affect the performance of cyclones (Table 15.6).

The ratio of efficiency change of two cyclones with respect to gas flow rate, particle density, and gas viscosity depends on the square root of the ratio between the new and original conditions. The particulate loading ratio between the two cyclones has a lower effect, as shown in Table 15.7.

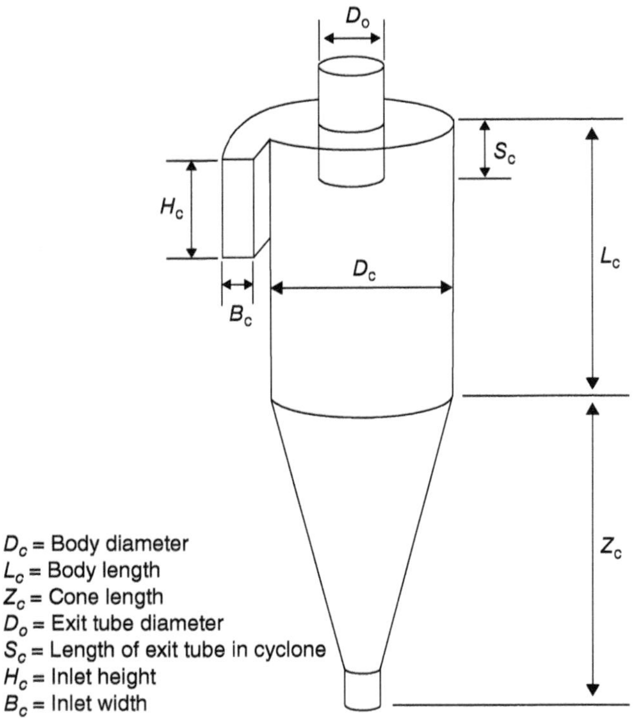

Figure 15.8 Nomenclature for a tangential entry cyclone. *Source:* U.S. EPA, 1998.

15.3.3.2 Pressure Drop in a Cyclone

Several factors affect the pressure drop in a cyclone: the kinetic energy losses in the cyclone, frictional losses at the cylinder walls, and the losses due to inlet and outlet ductwork. The predominant loss is due to the kinetic energy loss. The pressure drop of a cyclone can be estimated using the following equation, where it is directly proportional to the density of gas, the square of the inlet velocity, and the number of inlet velocity heads. The number of velocity heads is calculated based on the size of the inlet (width and height) divided by the diameter of the outlet. A cyclone configuration constant of 16 is used in most cases:

$$H_v = K\frac{HW}{D_o^2}$$

where

H_v is the pressure drop, expressed in number of inlet velocity heads

K is a constant that depends on cyclone configuration and operating conditions

W is width of inlet

H is height of inlet

D_o is exit tube diameter

$$\Delta P = \frac{1}{2}\rho_g V_i^2 H_v$$

Table 15.6 Changes in performance characteristics (U.S. EPA, 1998).

Cyclone and process design changes	Pressure drop	Efficiency	Cost
Increase cyclone size (D_c)	Decreases	Decreases	Increases
Lengthen cylinder (L_c)	Decreases slightly	Decreases	Increases
Lengthen cone (Z_c)	Decreases slightly	Increases	Increases
Increase exit tube diameter (D_c)	Decreases	Decreases	Increases
Increase inlet area (maintaining velocity)	Increases	Decreases	Decreases
Increase velocity	Increases	Increases	Operating costs higher
Increase temperature (maintaining velocity)	Decreases	Decreases	No change
Increased dust concentration	Decreases for large increases	Increases	No change
Increasing particle size and/or density	No change	Increases	No change

Source: From Theodore (2008).

Table 15.7 Effects of operating conditions on cyclone performance (U.S. EPA, 1982).

Variable	Relationship	Reference
Gas flow rate	$\frac{P_1}{P_2}=\left(\frac{Q_2}{Q_1}\right)^{0.5}$	Licht, Theodore, and Buonicore
Particle density	$\frac{P_1}{P_2}=\left[\frac{\rho_{p_2}-\rho_{g_2}}{\rho_{p_1}-\rho_{g_1}}\right]^{0.5}$	Licht, Theodore, and Buonicore
Gas viscosity	$\frac{P_1}{P_2}=\left(\frac{\mu_1}{\mu_2}\right)^{0.5}$	Licht, Theodore, and Buonicore
Dust loading	$\frac{P_1}{P_2}=\left(\frac{C_2}{C_1}\right)^{0.182}$	Baxter

where

ΔP is the pressure drop, N m^{-2} or Pa

ρ_g is the gas density, kg m^{-3}

V_i is the inlet gas velocity, m s^{-1}.

In general, cyclones have low pressure drops ranging from 2 to 10 in. of water. Higher pressure drops correspond to high efficiency of the cyclones. Pressure drops greater than 10 in. may not be an economically desirable choice for the cyclone.

The entrainment of particles can be avoided by designing a proper particle collection hopper. The top particulate level in the hopper always should be below the lowest vortex point in the cyclone. Design of valves that periodically or continuously remove particles will minimize the potential of particle entrainment that affects the efficiency of the cyclone.

15.3.4 Electrostatic Precipitators (ESP)

ESPs are very effective in removing particulates from variety of source categories (Figures 15.9 and 15.10). In ESPs energy is spend directly on the particles to be separated rather than on the whole gas stream. The pressure drop across an ESP, therefore, is among the lowest relative to other particulate separators. The energy is spent ionizing the particles which then are removed in a collection plate having an opposing charge. An efficiency of over 99.9% can be accomplished in an ESP.

ESPs can be classified into dry and wet systems. The dry types collect particles in the dry form on the collection

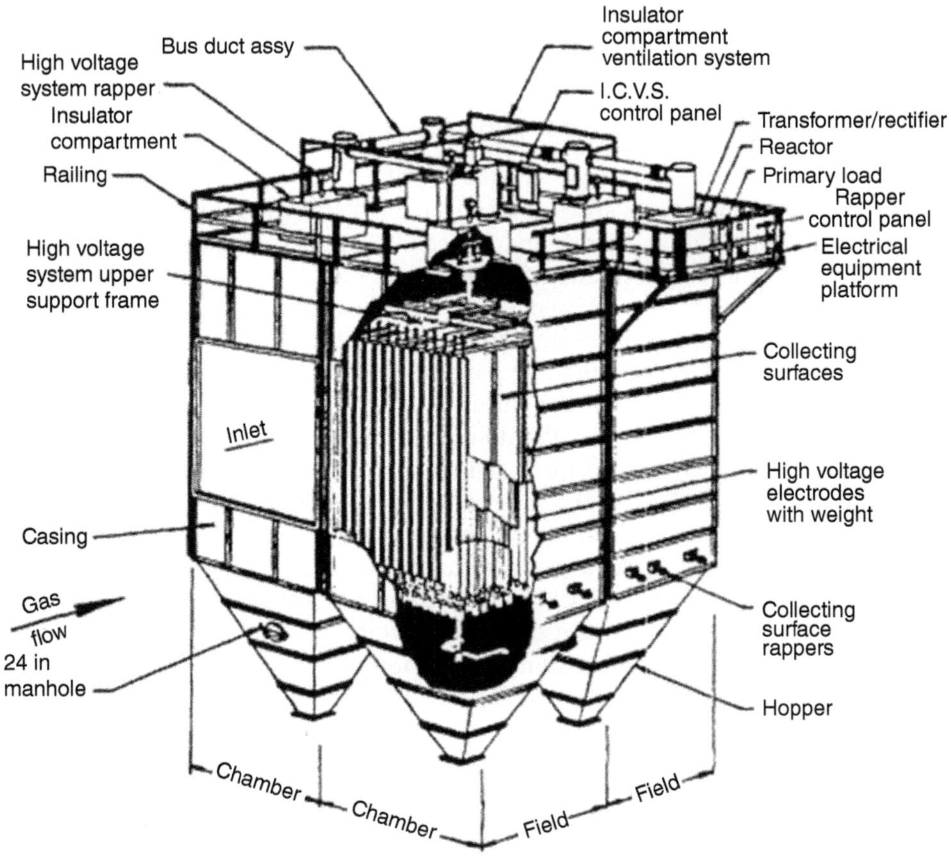

Figure 15.9 Typical ESP with insulator components (U.S. EPA, 1982).

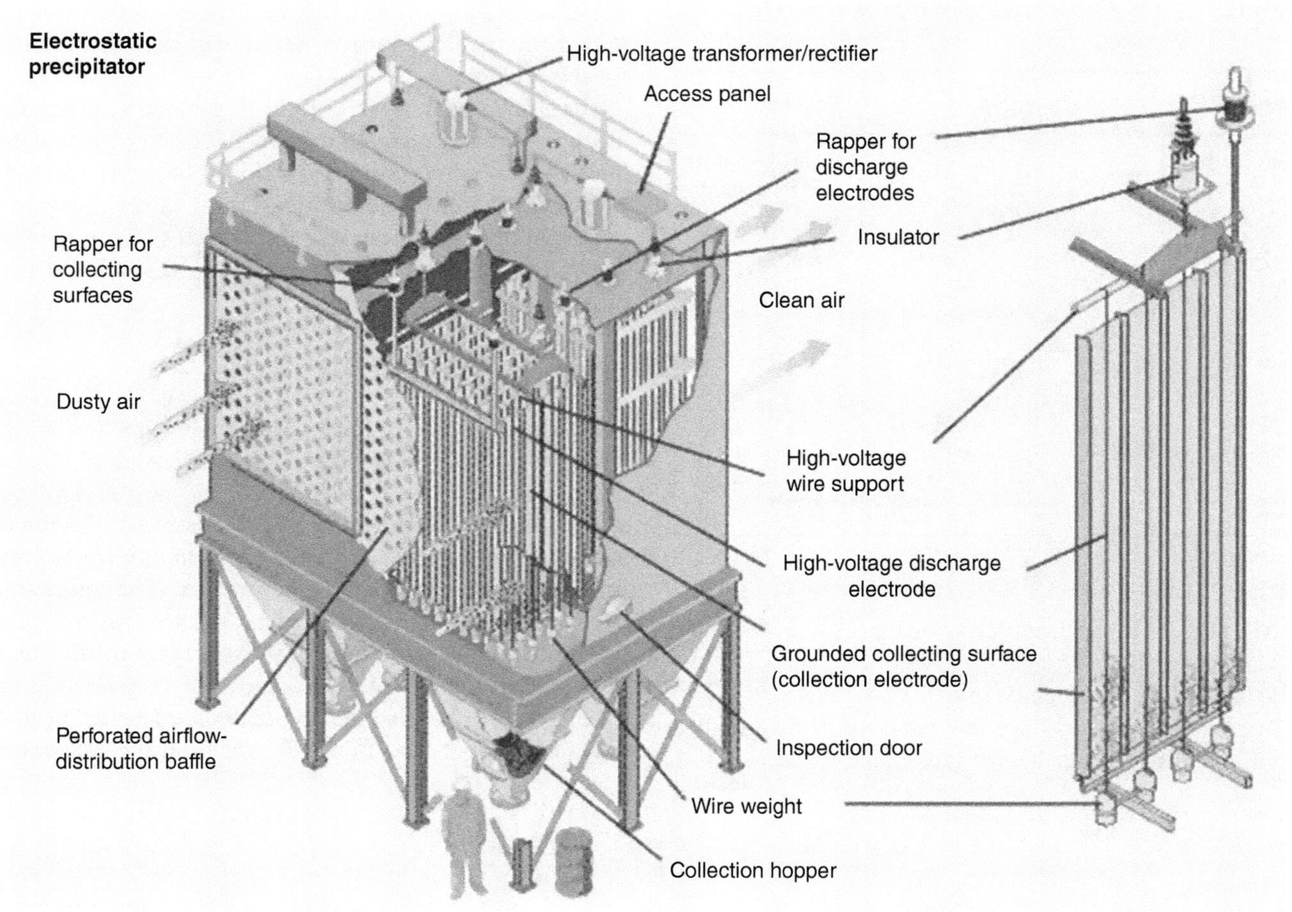

Figure 15.10 Electrostatic precipitator, a common particle collection device at a fossil fuel power-generating station. *Source:* U.S. EPA, 1982.

plates, which are then being deposited on the hoppers. In wet ESP water spray is being used to enhance the particle capture on the collection plate. Each vendor has variations of these two systems to make it attractive for specific application. Utility boilers, cement kilns, Kraft pulp recovery boilers, and metallurgical furnaces prefer dry ESP.

The ESP was initially designed to remove acid mist particles from a sulfuric acid plant. In 1907, a University of California, Berkeley, professor, Dr. Cottrell, developed a small ESP to handle about 100–200 acfm of gas flow. Based on its successful operation, ESP became a very prominent control device for removing particulates, especially from coal-fired power plants, cement plants, minerals industries, and others.

15.3.4.1 Particle Charging

The ESPs work based on charging the particles and collecting them in an oppositely charged collection plate (Figure 15.11). A high voltage is applied on a discharge electrode. This creates a highly nonuniform electric field around the discharge electrode and generates corona. Particle charging and collection takes place between the corona and the collection plate. The corona discharges tremendous number of electrons that in turn ionize gas molecules. These ions then migrate toward the collection plate, and during their migration toward the collection plates, the particles get charged. Charging is done by field and diffusion charging. The size of particles influences the dominant mechanism. In field charging, the ions are driven onto the particles due to the electric field. As the ions continue to impinge on the particles, the particles get charged up to a saturation point. The time required for saturation charge is a function of ion density that exists around the particles. Under normal conditions it takes only few milliseconds for particles to gain saturation charge. However, particle resistivity and size of particle can lengthen the charging time and cause the particle to travel several meters before saturation charge is reached.

Diffusion charging is the dominant mechanism occurring for particles of less than 0.2 μm in size. The ion movements are governed by the electric field, as well as the diffusional forces. Both diffusional forces and electric

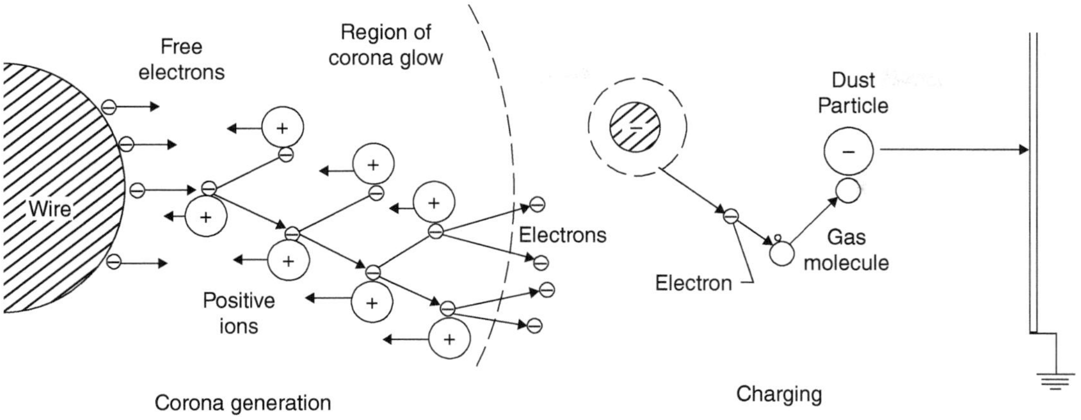

Figure 15.11 Basic processes involved in electrostatic precipitation (U.S. EPA, 1982).

field charging mechanisms occur for particles ranging between 0.2 and 0.5 μm in size. The charged particles then are attracted toward the collection plate that has an opposing charge. The particles collected onto the plates are dislodged by periodically rapping the plates. The particles then are collected onto a hopper and discharged either in a dry or wet form. The charged particles under the electric field create the force to travel toward the collection plate. The magnitude of this force is a function of the amount of charge on the particle and the strength of the electric field. The number of particles collected onto the plate is based on the mechanical, electrical, and molecular forces. If the forces are too strong, then dislodging the particles from the collection plate becomes difficult. If the forces are too weak, then the particles will not cling to the plates and will be re-entrained into the gas stream, resulting in a lower efficiency. The particles collected on the plates ideally should fall as a coherent particulate sheet into the collection hopper avoiding or minimizing the potential for re-entrainment. A properly designed system will avoid such problems and provide consistently high particulate collection efficiency.

The ESPs can be of plate and wire type or of cylindrical tubes with discharge electrode running at the center of the cylinder. Most of the large commercial ESPs are of plate and wire type. The cylindrical ESPs are used for smaller operations. ESPs differ in many ways, based on their configuration, based on the charging methods, and based on being wet or dry.

The plate- and wire-type ESP consists of several plates as collection electrodes in parallel to each other at 9–15 in. apart. The discharge electrode is generally a rigid wire of few millimeters (2–3 mm) in diameter. Some of the discharge electrode uses barbs or serrated strips instead of round wires. The discharge electrodes are rigid enough that it maintains equidistance between the two parallel collection plates. The potential for sparking is higher when the distance between the discharge electrode and the parallel plates are not maintained along the height of the plate. The discharge electrodes are also subject to acid condensation creating corrosion, thus weakening the electrode.

The collection plates are preferably perfectly flat, hung straight, and parallel. The spacing between the plates as mentioned earlier is uniform within a tolerance of few millimeters. The plates can be of steel or any other metals that resist corrosion. The plates are of 6–12 m high and 1–4 m long along the direction of the flow. The total height of an ESP includes height of the hopper, the control and electrical systems on the top of the plates, and the rapping mechanisms. The ESP housing is about 2–3 times the height of the plate.

In most cases charging of particles and collection are done at the same stage, while in few cases the particles are charged upstream stage and then be collected by collection plates at the downstream. The two-stage systems are of low voltage (12–13 kV), while the single-stage commercial systems are of 60–100 kV. The two-stage ESPs were originally developed as "electronic air filters" for cleaning gas streams with low particulate concentration such as in air-conditioning systems. These two-stage systems are also used to control acid mists and other aerosol particles. The ionizing stage consists of positively charged thin wires of about 0.2 mm, spaced about 3–5 cm apart with parallel ground tubes to ionize the particles. The second stage consists of parallel plates about 2.5 cm apart and negatively charged. In the second stage the particles get collected and removed from the plates. If liquid particles are captured, then the liquid will drain down the plates to a collection system.

Another version of the ESP is the tubular systems, where the discharge electrode is located at the center of the cylindrical tube. The tube acts as the collection electrode where particles get collected and removed. The tubes are about 15–30 cm in diameter with a height of 200–500 cm. This is a single-stage system where both ionization and collection of particles take place in the same region. These types of systems are mainly used for collecting liquid particles.

One of the drawbacks of ESP is its sensitivity to particle resistivity. Since particle resistivity varies for different types of particles, the collection efficiency can be affected by this particle character. It is critical to understand the characteristics of particle the ESP is meant to collect and design the system accordingly. Changes in resistivity characteristics of particles can even fail to meet the expected efficiency. Particle resistivity is a function of many factors. For example, the resistivity of fly ash from coal-fired power plants depends on the sulfur content of coal and the temperature of flue gas.

When the resistivity of particles is too high, above 10^{11} Ω-cm, the particles resist the current to pass across the collected particulate, thus making the particles difficult to dislodge during the rapping process. If the resistivity is too low, then the current passes through the dust cake and it loses its ability to stick to the collection plates, causing re-entrainment. The particle with high resistivity can also induce sparking between the electrodes, thus lowering the operating voltage, resulting in reduced collection efficiency. A particle resistivity below 10^{11} Ω-cm is preferable. Particle resistivity can also affect the migration or drift velocity of the particles. The migration velocity is the velocity by which particles move toward the plate from the gas stream due to ionic charges. The fly ash migration velocity reduces to half the velocity between 10^{11} and 10^{12} Ω-cm.

The amount of sulfur in the coal affects the resistivity of the particles. The resistivity of the particles increases as the sulfur content decreases. Burning low-sulfur coal increases the resistivity of the particles, consequently decreasing the efficiency of the ESP. The temperature of the flue gas also has influence on the particle resistivity. At a temperature of about 300 F, the resistivity for fly ash from coal containing 0.5–1%, 1.5–2%, and 2.5–3% sulfur is about $10^{11.5}$, $10^{10.5}$, and $10^{9.5}$ Ω-cm, respectively.

Increase in particle resistivity to a certain extent can be controlled by conditioning the flue gas before it enters the ESP (Table 15.8). Injection of SO_3 or ammonia-based salts or water into the flue gas is one method used to reduce resistivity. In cement plants, adding steam or injecting water has positive effects on the efficiency of the system. Locating the ESP on the cold side and treating the flue gas with chemical seems to work better in a coal-fired power plant compared with locating the ESP on the hot side, that is, before the air preheater or heat exchanger.

Table 15.8 Reaction mechanisms of major conditioning agents (U.S. EPA, 1982).

Conditioning agent	Mechanism(s) of action
Sulfur trioxide and sulfuric acid	Condensation or adsorption on fly ash surfaces; may also increase cohesiveness of fly ash. Reduce resistivity
Ammonia	Mechanism is not clear; various ones proposed
	Resistivity modification
	Increase in cash cohesiveness
	Enhances space charge effect
Ammonium sulfate	Little known about the actual mechanisms; claims made about the following:
	Resistivity modification
	Increase in ash cohesiveness
	Enhances space charge effect
	Experimental data lacking to substantiate which of the following is predominant
Triethylamine	Particle agglomeration claimed; no supporting data
Sodium compounds	Natural conditioner if added with coal
	Resistivity modifier injected into gas stream
Compounds of transition metals	Postulated that they catalyze oxidation of SO_2 to SO_3; no definitive tests with fly ash verify this postulation
Potassium sulfate and sodium chloride	In cement and lime kiln ESPs:
	Resistivity modifiers in the gas stream
	NaCl: natural conditioner when mixed with coal

15.3.4.2 Wet ESP

Wet ESP is more suitable for particles that are sticky, flammable, moist, or explosive or even have high resistivity. Water is either used continuously or intermittently to remove the particles from the collection plate. Wet ESPs also reduce the potential for particle re-entrainment. Wet ESP also is more efficient for fine particulates and for controlling acid mists. It also potentially can reduce the mercury emissions by collecting fine particles on which mercury may be condensed and the cooler wet ESP can also condense and capture some of the vapor forms of mercury. Also, if the vapor form of mercury is converted to soluble form, then the wet ESP is likely to remove more mercury than the dry ESP.

ESPs are preferred because it has high efficiency even for fine particles at a low pressure drop. They can handle large volume flows with varying temperature and dust loads. Collection of particulates in a dry form helps to recover valuable materials. However, the capital cost of ESP is high and it also needs lot of space, which could be a limitation in some locations. Also, ESPs do not work well with particles having high resistivity.

The performance of a typical ESP depends on the uniform gas flow conditions through the precipitator, particle resistivity, and the number of electrical sections. Maintaining a uniform gas flow through the ESP is critical to maintain the high efficiency. Perforated inlet plates and proper duct design are important for proper flow distribution. Very low or very high resistivity is not desirable for attaining the expected efficiency of ESP (Figure 15.12). For example, black carbon particles are hard to collect by ESP due to their very low resistivity. Particles with high resistivity are difficult to charge (Table 15.9). Once they are charged, they will not give up the charges as they reach the collection plate. A high potential field is created as the dust deposits onto the grounded collection plate. The outer layer of collected particle will have a negative charge, while the interior of the particle is neutral, and the collection plate is grounded. This creates "back corona," which generates positive ions that can accelerate toward the discharge electrode. These positive ions will counteract with the charging of particles with a corresponding reduction in efficiency. These situations could be avoided during the designing stages by including relevant factors.

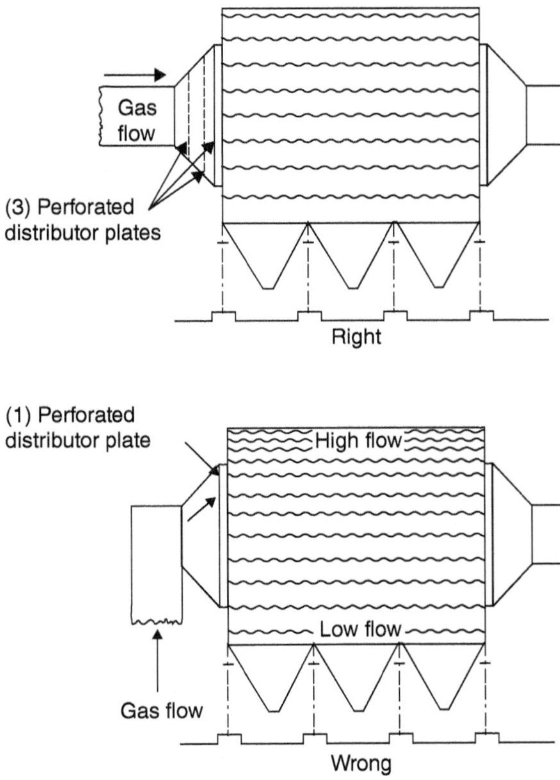

Figure 15.12 Effect of two different methods of gas distribution on flue characteristics in an ESP (U.S. EPA, 1982).

Table 15.9 Design power density (U.S. EPA, 1982).

Resistivity (ohm-cm)	**Power density (w m^{-2} of collecting plate)**
10^{4-7}	40
10^{7-8}	30
10^{9-10}	25
10^{11}	20
10^{12}	15
$>10^{12}$	10

The performance of ESP is evaluated by monitoring particulate loading at the inlet and outlet of the system, either by continuous particulate monitors or periodic stack testing. The author has conducted stack tests to evaluate the performance of an ESP that claimed to have over 99% efficiency by the plant managers. After conducting the stack test, the ESP's performance was estimated to be around 97.5%. That low efficiency surprised the plant operators and the managers. The stack test was repeated to confirm the previous results. As the results were remained the same, the plant decided to shut down for early maintenance. The plant noticed several issues: broken or corroded discharge electrode and particle deposition on the ductwork due to improper velocity distribution. The plant took the initiative to correct the issues, and another stack testing was performed. The efficiency dramatically increased over the expected 99% efficiency. This testing was conducted in a coal-fired power plant in India in the mid-1970s, and it was one of the cleanest coal-fired power plants operated by a reputable private company. In many cases, developing countries required to have the most current control technologies for their new plants to obtain the permit. However, after its commissioning, its maintenance takes the back seat due to poor enforcement by the governing agencies and the lack of interest among the plant operators to allocate proper maintenance budget. However, these practices are becoming less and less due to better government/agency oversight and awareness of plant operators in keeping up with the maintenance.

15.3.4.3 Design Parameters for ESP

A mathematical relationship of ESP for its efficiency is known as the Deutsch equation. This equation relates the ESP efficiency to the total plate area, the drift or migration velocity of particles, and the gas flow rate as follows:

$$\eta = 1 - e^{-\frac{wA}{Q}}$$

where

η is the fractional collection efficiency
w is the terminal drift velocity, m s^{-1}
A is the total collection area, m^2
Q is the volumetric airflow rate, m^3 s^{-1}

The equation assumes a uniform flow across the ESP. Another limitation of this equation is that it does not involve particle size distribution, but it uses single migration velocity. As larger particles get collected, the finer particles travel farther in the ESP and will have a lower migration velocity than the average design velocity. Thus, the efficiency prediction may not be as accurate especially for high efficiency systems. Also, particle re-entrainment is not accounted in the theoretical equation. An effective migration velocity that is more representative of actual performance of the system must be used for designing the system. The effective migration velocity can be estimated using complex empirical relationships. The capital cost of a typical commercial ESP is expensive; therefore a pilot-scale test performed on gas streams with particles may be cost effective to obtain a more realistic migration velocity for designing the actual system. However, one should be aware that the pilot scale tests tend to overpredict the performance and this factor should be considered in designing full scale system.

The vendors use complex computer models to design ESP Systems in deciding on the number of electrical sections and the amount of applied voltage in each section (Figure 15.13). Gas velocities in the range of 1–1.2 m s^{-1} are used in the ESP. The ratio of length to the height of the gas passage is known as the aspect ratio kept in the range of 1–1.5. Plate spacing varies from 15 to 40 cm. Plate heights are determined by the amount of flow rate to be handled and the aspect ratio. Unrealistically tall plates will be structurally unstable; therefore a plate height less than 12 m is preferred. The ESP can be sectionalized along the direction of flow based on the number of electrical sections and several chambers running across the direction of flow. Thus, a 3-chamber ESP with 4 electrical sections will have 12 mechanical sections.

Rapper system is an integral part of the ESP for effective removal of particles from the collection plates. Electromagnetic or pneumatic impulse-type rappers are more common. Another type uses vibration to dislodge the collected particles. Systems with vibration technique are more suitable for particles that are less difficult to dislodge. A direct current pulse in the rapper coil supplies the energy, and the steel plunger is raised by the magnetic field and is allowed to fall and strike a rapper bar that is connected to the collection plates. The shock transmitted by the rapper to the collection electrodes dislodges the collected particles. The numbers of rappers, the size of the rapper, the weight of the rapper, and the rapping frequency are design parameters to be considered for proper design of an ESP.

15.3.5 Baghouse

The fabric filters can be classified based on the types of fabric being used, the geometry of the equipment, and the mode of operation. Most common classification is based on the type of cleaning mechanisms. The fabric filters can be operated continuously and intermittently or batch process. The batch systems are generally used for smaller operations, where the fabric filters are cleaned as needed. Commercial and industrial carpentry shops and small wood working shops have batch systems that are lower in cost compared with the continuous systems. In continuous systems, the bags are compartmentalized such that the dirtiest compartment will be isolated and cleaned. That compartment will be in operation after its cleaning. In a pulse-jet cleaning system, however, the filtration is simultaneous with the cleaning process.

The ability to collect very fine particles at very high efficiency separates the fabric filters from other particulate control devices. Because of this reason, the fabric filters have become more desirable particulate control system. A particulate control efficiency of higher than 99% can be expected from a well-designed system. The pressure drop across the filter system is one of the limiting parameters. The pressure drop across a fabric filter is directly related to the cost of electricity to run the operation.

15.3.5.1 Shaker-Type Fabric Filters

Tubular fabric filters are of 6–9 in. in diameter and they are hung from the top with opening at the bottom where the dirty air enters (Figure 15.14). The dirty air enters the bottom of the bags and filtered through the inside of the bag surface to the outside surface. As filtration continues,

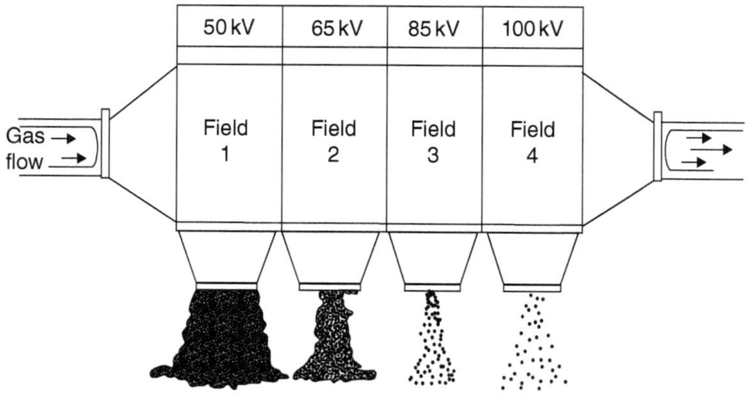

Figure 15.13 Stage or field sectionalization. *Source:* U.S. EPA, 1992.

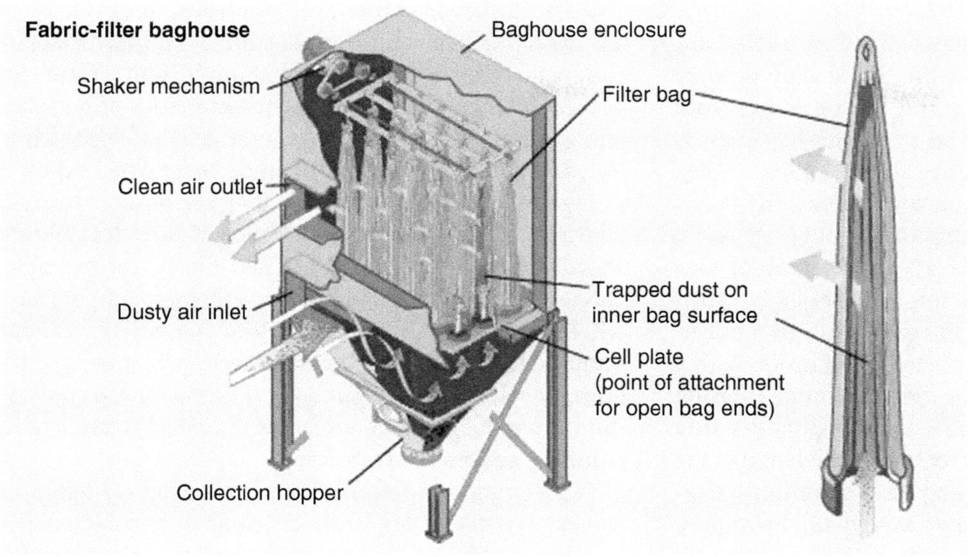

Figure 15.14 Baghouse employing an array of fabric bags for filtering the airstream. *Source*: U.S. EPA, 1998.

dust cake builds up on the inside of the bag that acts as additional filtering medium. The dust cakes sticking to the bags will resist flow through the bags and creates the pressure drop. When the pressure drop reaches a preset value, the compartment containing these bags will be isolated from other compartments for cleaning. Mechanical systems are used to clean the bags in large systems, while smaller systems will use manual operation. In mechanical cleaning system, the top of bag is attached to a shaker bar. A rapid horizontal motion of the shaker bar induced by a mechanical force causes flexing of the bags and causes the dust cake to dislodge from the bag into the collecting hopper at the bottom of the bag. The cleaning intensity is a function of the tension of the hanging bags, the frequency, duration, and the amplitude of shaking.

The cleaned fabric filters still have some dust attached to it; however, the resistance to flow has significantly reduced and is ready for filtration again. The compartment with cleaned fabric filters will join the other compartments in the filtration process. Thus, if there are N number of compartments in the system, during the normal filtration cycle, all the gas flow is going through all these N compartments at an average filtration velocity (flow/total filter area), V. When one of these compartments is removed for cleaning, the amount of filter area for the same gas flow rate is reduced; consequently, the filtration velocity is increased. This average velocity when one of the compartments is under cleaning is the design velocity. The total number of filtration area required is calculated based on this design velocity.

The low design filtration velocity of about 1 m min^{-1} requires a greater filtration area, or more fabric filter bags relative to higher design velocities. The filtration velocity is also called the air to cloth ratio (A/C), represented as $m^3\ m^{-2}$-min^{-1} (m min^{-1}). There are industry- or vendor-dictated A/C ratios that are in common practice for use in designing shaker-type systems. Using higher than the design A/C values will reduce the life of bags significantly.

15.3.5.2 Reverse Air Cleaning

In a reverse air-cleaning system, the collected dust on the fabric filter bags is dislodged by reversing the airflow for a very short period, thus making the bags to flux. The filtration will take place from inside to outside of the bags, resulting in particulate being collected on the inside of the bag. In the reverse air system, an airflow flowing from outside the bag toward the inside of the bag makes the particles dislodge from the filter bags. The design requires compartments similar to shaker type. All the bags with in a compartment are cleaned by reversing the airflow when the bags become dirty. There are also designs where each bag is cleaned individually. The bags in this case are radially aligned such that the bags will be cleaned by a reverse-air manifold rotating around the bag units and stop at each bag to induce the required reverse flow. This radial reverse-flow cleaning system is more applicable to smaller systems. For larger systems compartmentalization and cleaning individual compartment containing multiple bags at the same time are common. The reverse air is supplied by the cleaned air or a separate intake from the ambient air based on the climatic conditions.

The filter bags may collapse during the reverse-air process; therefore, noncollapsible rings or shelter is provided to avoid such bag collapse during cleaning. The filter bags can be of woven or felt type. The design filtration velocity for reverse-air process is like the shaker type of around 1 m min^{-1}.

15.3.5.3 Pulse-Jet Baghouse

Pulse-jet baghouses are popular because of their continuous operation compared with the shaker type or reverse-air. Bag filters are hung and the air passes from outside bag to the inside of the bag. The bag then is cleaned by a blast of high pressure air jets of 90–100 Psi, and this pulse creates a wave on the bag and the particles are dislodged during flexing of the bags. Pulse-jet lasts for 30–100 milliseconds at an interval of few minutes. The design velocity for a pulse-jet is much higher than the shaker type and reverse-air. In a pulse-jet baghouse, the bags are supported from inside with a supporting cage. For example, the maximum filtering velocity varies from 5 to 14 ft min^{-1}. The higher filtering velocity reduces the number of bags needed for the same gas flow to treat the same amount of air in a pulse-jet system compared with other systems. The capital cost of the system significantly reduced due to the high filtering velocity. It requires only less than half of the footprint of a reverse-air.

The pulse-jet system however needs a compressor to provide the high pressure jet for cleaning the bag. About 0.2–0.8% of the flow of filtered air is needed for the cleaning application. The cost of the compressor adds to the capital cost of the pulse-jet system. Use of pulse-jet baghouse to remove particulates from large industries such as coal-fired power plants is designed with number of compartments for the purposes of maintenance and easier installation.

15.3.5.4 Selection of Fabric for Baghouse

The selection of an appropriate fabric for a specific application includes average and maximum temperature of gas stream, cleaning method, characteristics of particles and gas stream, pressure drop, cost, and safe operation. Polymer and natural fabrics degrade faster when exposed to higher than the temperature they are designed for. In addition, the moisture and stickiness of particles can also shorten the life of bags. Fiberglass filters, for example, can tolerate acid gases and reasonably high temperature, while natural fibers will not withstand acid or higher temperature. The filter manufactures will have guidelines for selecting the right type of fabric for the application in hand. See Table 15.10 for chemical resistance of common commercial fabrics.

Two common types of filter fabrics used are felted or woven. Felt filters are more efficient in particle collection compared with the woven fabrics. Felted fabrics consist of thick randomly oriented fibers, resulting in a strongly bonded fabric. The thickness of the felt enhances the particle impingement, thus improving particle capture. The pressure drop will be higher as well. Felt-type fabrics are normally used in pulse-jet systems that are used at a higher filtration velocity. On the other hand, woven fabrics are preferred in shaker-type or reverse-air systems with relatively lower filtration velocity. Woven fabrics are made of spun yarns or filament in different patterns with specific spacing and finish. For example, plain weave fabric is least expensive

Table 15.10 Chemical resistance of common commercial fabrics (U.S. EPA, 1982).

Fabric	Generic name	Type of yarn	Acid resistance	Fluoride resistance	Alkali resistance	Flex and abrasion resistance
Cotton	Natural fiber cellulose	Staple	Poor	Poor	Fair to good	Fair to good
Wool	Natural fiber protein	Staple	Very good	Poor to fair	Poor to fair	Fair
Nylon	Nylon polyamide	Filament spun	Fair	Poor	Very good to excellent	Very good to excellent
Dynel®	Modacrylic	Filament spun	Good to very good	Poor	Good to very good	Fair to good
Polypropylene	Polyolefin	Filament spun	Excellent	Poor	Excellent	Very good to excellent
Orlon®	Acrylic	Spun	Good to excellent	Poor to fair	Fair	Fair
Dacron®	Polyester	Filament spun	Good	Poor to fair	Fair to good	Very good
Nomex®	Nylon aromatic	Filament spun	Fair	Good	Excellent	Very good to excellent
Teflon®	Fluorocarbon	Filament spun	Excellent	Poor to fair	Excellent	Fair
Fiberglass	Glass	Filament spun bulked	Fair to good	Poor	Fair	Poor
Polyethylene	Polyolefin	Filament spun	Very good to excellent	Poor to fair	Very good to excellent	Good
Stainless steel (type 304)			Excellent		Excellent	

with reasonable efficiency for particle collection but has high blinding potential. The other common weave patterns are twill and sateen. The permeability of woven fabric in general depends on the type of fiber used, the tightness of the twist, yarn size, and the tightness of weave.

Cotton fabrics need to be preshrunk to maintain the dimension of the bag, while synthetic fabrics need to be heat-set to maintain the dimensional stability. Bags can also be treated with silicon to improve dust cake release and abrasion resistance.

15.3.5.5 Wet Collectors

In wet scrubber, liquid is used to contact with the contaminated gas stream to remove the particulates, and to a smaller extent it also removes soluble gaseous contaminants (Figures 15.15 and 15.16). Wet scrubbers can condition the gas stream such as adding moisture and cooling the gas stream. They can also be operated at a range of efficiencies determined effectively by the amount of input energy into the system. High energy venturi scrubbers can achieve very high efficiency for fine particulates with pressure drops reaching as high as 60 in. of water. Spray

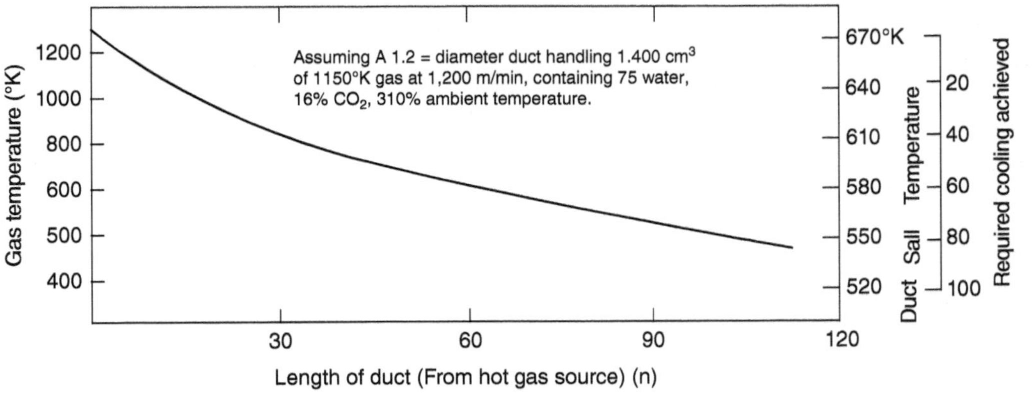

Figure 15.15 Radiation effectiveness in cooling hot gases (U.S. EPA, 1982).

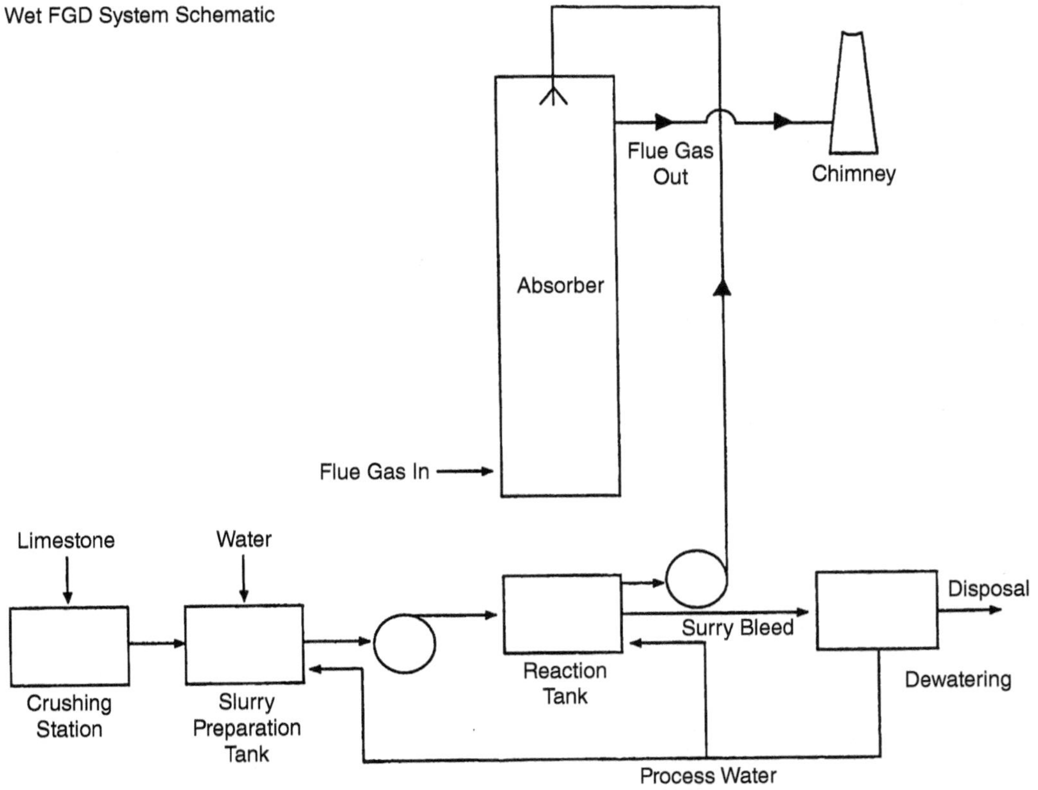

Figure 15.16 Wet scrubber using a limestone slurry to remove sulfur dioxide gas from flue. *Source:* U.S. EPA, 2000.

towers with a pressure drop of few inches of water have low efficiency for particles of less than 5 μm size. They can be effectively used for cooling gas streams and act as a precleaner for more expensive control system.

The major mechanisms by which particulates are removed in wet scrubber are impaction, interception, and diffusion. There are different types of wet scrubbers such as preformed spray, packed bed, tray-type, mechanically aided, venturi, and orifice scrubbers. The wet scrubbers generate different sizes of droplets and the droplets capture particulates. The droplets that captured the particles can then be separated from the gas stream and disposed. Particles are also captured on water layers surrounding a packing material directing the gas streams between a tight space between the packing materials.

15.3.5.6 Preformed Spray Towers

Preformed spray towers contain array of spray nozzles. The droplets travel downward by gravity and the gas flows upward. The particles from the gas stream encounter the droplets on its travel upward and get captured by the droplet by impaction, interception, and diffusion. The spray tower has low efficiency especially for particles less than 5 μm in size. Therefore, spray towers are appropriate for treating gas streams with large particles and high mass loadings.

In cyclonic spray towers, the gas stream is given a cyclonic motion by which the gas spirals upward and additional particles get collected in addition to the removal by droplets. Therefore, the cyclonic spray towers have slightly higher efficiency than the simple spray towers. Some of the cyclonic spray towers where the inlet velocity into the tower can be adjusted such that it can maintain a constant velocity for varying gas flow rates by an adjustable valve. The droplets travel crosscurrent to the gas flow. The cyclonic scrubbers can capture particle up to 2 μm in size. The pressure drop in a cyclonic spray tower is between 4 and 6 in. of water.

In an impingement plate tower, the dominant particle collection mechanism is inertial impaction on droplets while the gas stream passes through the impingement holes. The aerodynamic particle size, relative velocity between particles and droplets, and the liquid to gas ratio are some of the key parameters that control the performance of impingement plate towers. Up to 70% efficiency can be expected for one-micron-size particles in an impingement plate tower.

The preformed spray scrubbers can have countercurrent, cocurrent, or crosscurrent systems (Figure 15.17). The spray is delivered from a set of nozzles from the top of the tower while the gas is passing upward in a countercurrent system. The particles get impacted onto the droplets as they fall through the height of the spray tower. The primary mechanism by which particles are captured is inertial impaction. The droplet size, the liquid to gas flow, and the gas velocity are the main factors that determine the efficiency of spray towers. For a countercurrent system, Calvert et al. developed an equation to estimate the penetration of particles using the impaction parameters. The higher the impaction parameter, the higher the fractional efficiency of single droplets. The pressure drop is relatively lower in a spray tower, less than 3 in. of water, but it is inefficient in removing fine particles. Spray towers can also be used to cool down hot flue gases and to remove small amounts of contaminants that are soluble in water like sulfur dioxide.

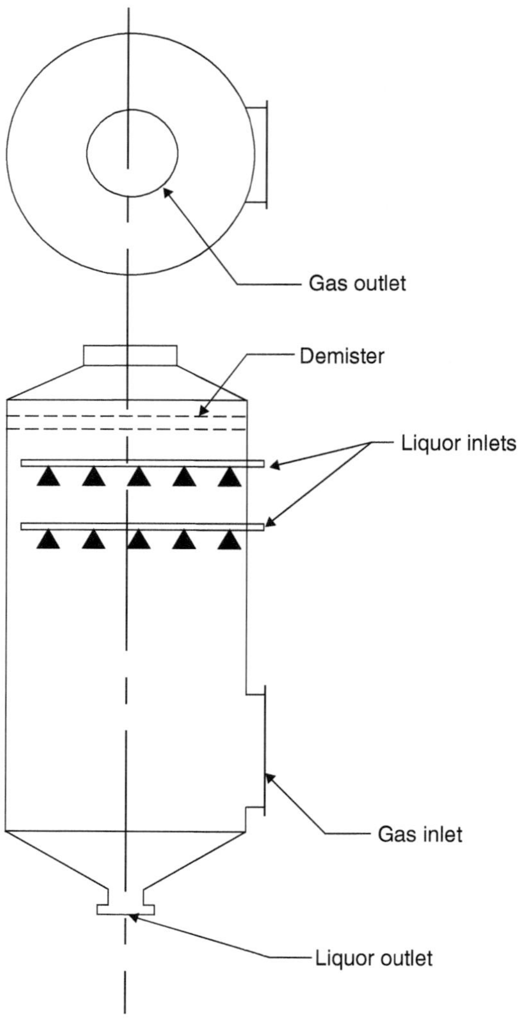

Figure 15.17 Spray tower scrubber (U.S. EPA, 1982).

15.3.5.7 Venturi Scrubbers

Venturi scrubbers are mainly used to remove particles from gas streams. A venturi scrubber consists of a converging section, a throat section, and a diverging section. The liquid is atomized by the airstream that is passing through the sections at a very high velocity of 100–150 fps. In many cases the liquid is sprayed before the converging section and the liquid gets atomized to droplets of less than 100 μm in size. These droplets sweep the particles as they travel along with the airstream cocurrently, especially at the throat section. The droplets that capture the particles then are separated from the gas stream in a cyclonic scrubber, and the cleaner gas passes through the top of the cyclonic scrubber while the liquid is collected at the bottom of the cyclonic scrubber.

Venturi scrubbers can be classified into low, medium, and high energy scrubbers. High energy scrubbers can achieve very high efficiencies of 99% for fine particulates. They demand a higher pressure drop of up to 60 in. of water. The pressure drop is to maintain a very high gas velocity and to atomize the liquid into fine droplets.

The design of venturi scrubber involves the determination of throat size for the gas flow and the velocity to be maintained to achieve the design efficiency. The liquid to gas ratio is also critical to generate the required amount of fine droplets sufficient to collect the particles. A simple approach by Lapple and Kamack relates the amount of energy spent to the efficiency, otherwise known as contacting power theory. Contacting power is the amount of energy spent to treat a unit volume of gas. In general, the contacting power includes the amount of energy supplied by the gas, the liquid, and any added mechanical energy into the system. The coefficient and the exponent for the equation can be determined by conducting pilot studies measuring the efficiency of the system and corresponding energy input for two or more conditions. This simple contact power theory will be valid if the amount of energy spent on the system can be determined more accurately:

$$N_t = \ln\left(\frac{1}{1-E}\right) \tag{15.1}$$

where

N_t is the number of transfer units (dimensionless)

E = fractional collection efficiency

$$N_t = \alpha p_t^{\beta} \tag{15.2}$$

where α, β are parameters for the types of particulates being collected.

The penetration equation to evaluate the performance of venturi scrubber is originally developed by Calvert et al. It is based on inertial impaction of particles on fine water droplets. In this equation penetration of a specific aerodynamic diameter particle is calculated. For example, if the goal is to capture 90% (10% penetration) of 1-µm-diameter particles, this equation will help to calculate that efficiency using appropriate correction factors. The equation requires the determination of droplet diameter size and nonuniformity correction factor. The Sauter mean diameter of droplet is calculated based on the Nukiyama and Takanawa equation. Another parameter needed is the impaction parameter, which is a dimensionless parameter calculated using the equation as given below. A theoretical penetration curve for different throat velocities and for different particle sizes is presented in Figure 15.18. For a one-micron-size particle, the penetration is 0.02 at a throat velocity of 15 000 cm s^{-1}, the penetration is 0.15 at 10000 cm s^{-1}, and the penetration is 0.6 at 5000 cm s^{-1}. Increasing the throat velocity will yield higher efficiency, but the pressure drop of the system also will be higher. The ratio of liquid to gas flow also influences the penetration to a smaller degree – the higher the liquid to gas ratio, the higher the removal efficiency. A typical liquid to gas ratio in common practice for venturi scrubbers is between 0.5 and 1 l m^{-3}:

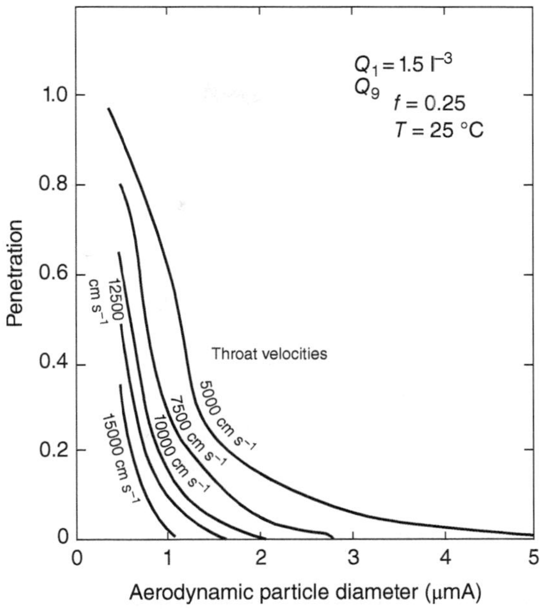

Figure 15.18 Theoretical penetration curve for a venture scrubber illustrating the effect of throat velocity (U.S. EPA, 1982).

$$P_i = e - \left[\left(\frac{0.036\rho_l d_d V_g}{\mu}\right)\left(\frac{Q_l}{Q_g}\right)\right]\left[-0.7 - k_i f + 1.4\ln\left(\frac{k_i f + 0.7}{0.7}\right) + \left(\frac{0.49}{k_i f + 0.7}\right)\right]\left(\frac{1}{k_i}\right) \tag{15.3}$$

where

P_i is the penetration value for particles with aerodynamic diameters of i

ρ_p is the droplet density, kg m^{-3}

d_d is the droplet diameter, m

V_g is the superficial gas velocity in venturi throat, m s^{-1}

μ is the gas viscosity, kg m^{-1} s^{-1}

Q_l/Q_g is the liquid to gas ratio, dimensionless

k_i is the impaction parameter, dimensionless

f is the nonuniformity correction factor, dimensionless

$$d_d = \left(\frac{50}{V_g}\right) + 91.8\left(\frac{Q_l}{Q_g}\right)^{1.5}$$

where

d_d is the Sauter mean droplet size, cm
V_g is the gas velocity in venturi throat, $cm\,s^{-1}$
Q_l/Q_g is the liquid to gas ratio, dimensionless

$$K_i = \frac{V_R (d_i)^2 \rho_w}{9 \mu d_d}$$

where

K_i is the impaction parameter, dimensionless
V_R is the particle-droplet relative velocity, $cm\,s^{-1}$
di is the aerodynamic particle diameter i, cm
μ is the gas viscosity at actual temperature, $g\,cm^{-1}\,s^{-1}$
d_d is the Sauter mean droplet size, cm
ρ_w is density of water, $g\,cm^{-3}$

15.4 Control of Gaseous Compounds

15.4.1 Control of Oxides of Nitrogen (NO_x)

Oxides of nitrogen (NO_x) are formed during high temperature fuel combustion and is called thermal NO_x. In addition, fuel like coal contains small amounts of nitrogen. A portion of this fuel nitrogen is also converted to NO_x during the combustion called the fuel NO_x. The NO_x in the flue gas consists of 95% nitric oxide and 5% nitrogen dioxide. Thermal NO_x is formed when temperatures are as high as 2000 °C, while the formation of fuel NO_x depends on the amount of nitrogen in the fuel and the equivalent ratio, which is the inverse of air–fuel ratio.

There are several combustion modification techniques that can reduce the formation of NO_x. Some of the specific techniques used to reduce the NO_x formation are reducing the peak temperature of the flame zone, reducing the gas residence time in the hot flame zone, and reducing the oxygen levels at the hot combustion zones. Low NO_x burners, flue gas recirculation, gas reburning low excess air, and off-stoichiometric combustion are few of the methods used in combustion systems to reduce NO_x. Cooling the flame by injecting water on the flame is another technique used to reduce the flame temperature to further reduce the formation of NO_x. The above combustion modification techniques reduce NO_x for about 40–60%. In many cases the regulatory requirements demand even higher NO_x reduction. Flue gas treatment (FGT) for NO_x reduction is necessary to meet the regulatory requirements.

The two major techniques for FGT of NO_X are selective catalytic reduction (SCR) and selective noncatalytic reduction (SCNR). These technologies are based on the chemical reactions of ammonia or urea to reduce NO_x into molecular nitrogen and water.

The SCR systems are one of the most popular FGT systems to reduce NO_x emissions across the world. In the United States, they are used mainly to treat flue gas from coal-fired power plants and gas-fired combustion systems. They are preferred when the NO_x reduction required is greater than 90%. They also use catalyst, which are titanium and vanadium oxides, in a pellet form or honeycombed shape. Honeycombed catalysts can tolerate small amounts of particulate in the gas stream and more suitable for treating flue gas from coal-fired power plants. Anhydrous ammonia and urea are the preferred chemicals used in the NO_x reduction technique. Although ammonia is more effective, it is a toxic chemical and requires additional safety precautions in transporting and handling as well as its storage on-site. Handling of ammonia on-site also involves strict safety measures. The use of catalyst in an SCR system reduces the reaction temperature significantly and has a broader temperature window than the SNCR system. The SCR systems are more expensive to install and operate.

In a SCR system ammonia is injected through a set of injection grid into the hot flue gas stream mounted in the ductwork. The ductwork is expanded to accommodate the catalyst as well as to maintain the required gas velocity through the catalyst. The ammonia is diluted and injected into the flue gas where it gets mixed with the gas stream. A simplified flow diagram for an SCR system is given in Figure 15.19.

Ammonia can penetrate through the catalyst pores, providing a better reaction environment compared with urea. Ammonia, either in anhydrous or aqueous form, passes through a vaporizer before its injection into the flue gas. The ammonia then reacts with NO_x as per the following reactions to reduce NO_x into water and molecular nitrogen.

15.4.2 Reactions

The equations indicate that one mole of NH_3 will react with one mole of NO_x. In reality higher than one mol is needed per mole of NO_x for proper mixing and completion of the reaction. When the NO_x reduction efficiency needed is greater than 85%, additional ammonia may be required due to the NO_2 reaction rates. In specific cases an extra amount of catalyst is needed to achieve the expected high efficiency for NO_x reduction:

$$4NO + 4NH_3 + O_2 \rightarrow 4N_2 + 6H_2O$$

$$2NO_2 + 4NH_3 + O_2 \rightarrow 3N_2 + 6H_2O$$

The performance of the SCR system depends on several factors including catalyst performance, proper mixing of reactants, the pressure drop, and the efficiency of NO_x reduction. The SNCR system varies from the SCR system in many ways. The reaction is accomplished in SNCR without the presence of catalyst. The optimum temperature for SCR systems ranges from 370 to 400 °C, while for SNCR system it varies from 870 to 1100 °C, as shown in Figure 15.20.

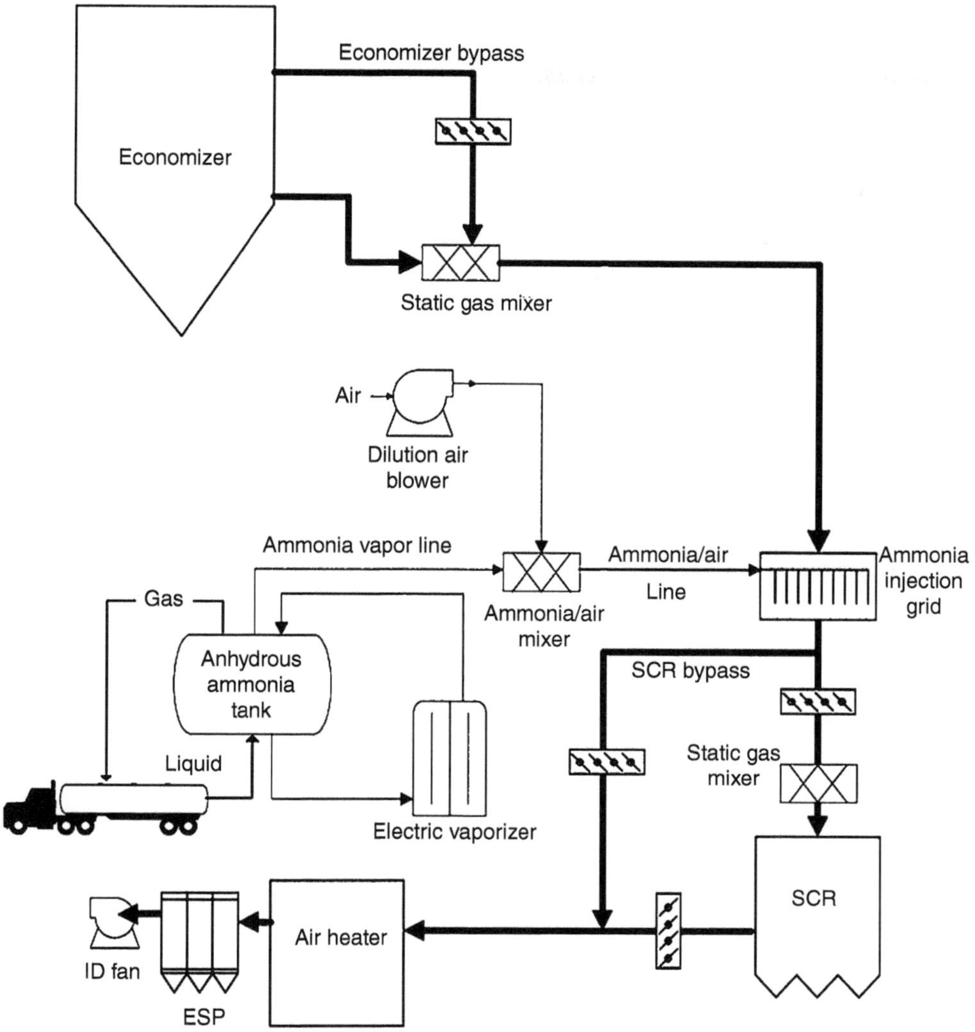

Figure 15.19 SCR process flow diagram (U.S. EPA, 1981).

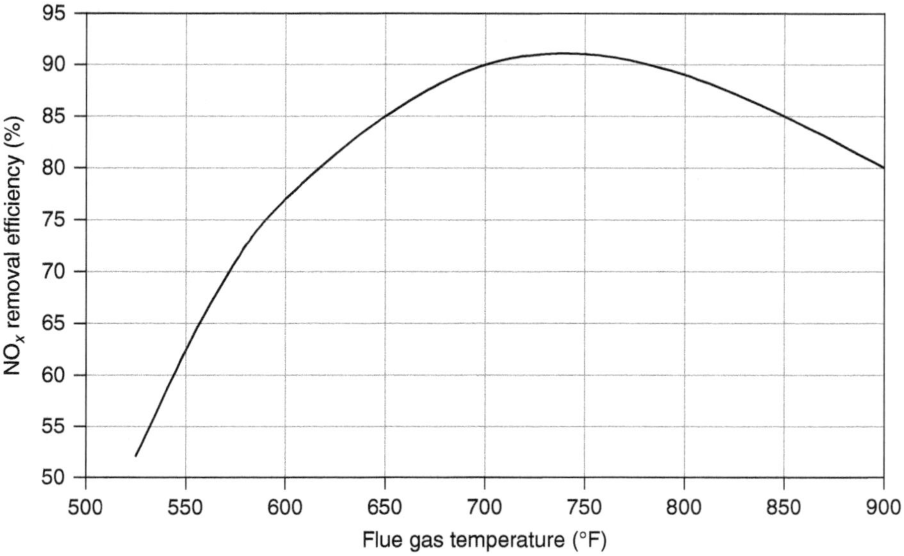

Figure 15.20 NO_x removal versus temperature (U.S. EPA, 1982).

The residence time for an SCR system depends on the gas flow rate and the amount of surface area of catalyst available for the reaction to take place. The specific surface area of a typical catalyst ranges from 300 to 1200 $m^2\ m^{-3}$ of catalyst. The increase of catalyst specific surface area will increase the NO_x reduction for a given gas flow rate. Therefore, during the selection of catalysts, factors including higher specific surface area, low pressure drop, and the least catalyst to poisoning potential. The catalysts generally last 5–10 years depending on the quality of catalysts and the type of flue gas to be treated.

In a honeycombed catalyst, the area of cross section of the cells determines the velocity of flow. For a given flow rate, channels with higher cross-sectional area will result in lower interstitial gas velocity, improving the NO_x reduction. Also, the size of the pitch is important to minimize any plugging of the channels by the deposition of particulates from gas stream, which reduces the available surface area and increases the pressure drop. Catalysts can be deactivated by poisoning, thermal sintering and plugging, fouling, and aging. Proper selection of catalyst and proper operation and maintenance of the SCR system are paramount to achieve the expected NO_x reduction. Major equipment required for an SCR application is given in Table 15.11.

The SNCR for NO_x control converts the NO_x to elemental nitrogen and water. This is accomplished without having a catalyst unlike the SCR system. The SNCR system is less expensive and simple to operate. However, the NO_x reduction is in the range of 40–60% in comparison with SCR system where the NO_x reduction is greater

Table 15.11 Major equipment list for an SCR application (U.S. EPA, 1981).

Item	Description/size
SCR reactors (1–2)	Vertical flow type, 805 000 acfm capacity, 44 ft × 44 ft × 31 ft high (excluding outlet duct and hoppers), equipped with 9604 ft^3 of ceramic honeycomb catalyst, insulated casing, sootblowers, hoppers, and hoisting mechanism for catalyst replacement
Anhydrous ammonia tank (1 or more)	Horizontal tank, 250 psig design pressure, storage tanks 15 000 gal, 34-ton storage capacity
Air compressor (2)	Centrifugal type, rated at 3200 acfm and 30 hp motor
Vaporizers (2)	Electrical type, rated at 80 kW
Mixing chamber	Carbon steel vessel for mixing or air and ammonia
Ammonia injection grid	Stainless steel construction, piping, valves, and nozzles
Ammonia supply piping diameter, with valves and fittings	Piping for ammonia unloading and supply, carbon steel pipe: 1.0 in.
Sootblowing steam	Steam supply piping for the reactor soot-piping blowers, 2-in.-diameter pipe with an on–off control valve and drain and vent valved connections
Air ductwork	Ductwork between air blowers, mixing chamber, and ammonia injection grid, carbon steel, 14 in diameter, with two isolation butterfly dampers and expansion joints
Flue gas ductwork	Ductwork modifications to install the SCR modifications reactors, consisting of insulated duct, static mixers, turning vanes, and expansion joints
Economizer bypass	Ductwork addition to increase flue gas temperature during low loads consisting of insulated duct, flow control dampers, static mixers, turning vanes, expansion joints, and an opening in the boiler casing
Ash handling	Extension of the existing fly ash handling modifications system: modifications consisting of 12 slide gate valves, 12 material handling valves, 1 segregating valve, and ash conveyor piping
Induced draft fans	Centrifugal type, 650 000 acfm at 34 in. wg and 4000 hp motor
Controls and Instrumentation	Stand-alone, microprocessor-based controls for the SCR system with feedback from the plant controls for the unit load, NO_x emissions, etc. including NO_x analyzers, air and ammonia flow monitoring devices, ammonia sensing and alarming devices at the tank area, and other miscellaneous instrumentation
Electrical supply	Electrical wiring, raceway, and conduit to connect the new equipment and controls to the existing plant supply systems
Electrical equipment	System service transformer OA/FA/-60 Hz, 1000/1250 kVA (65 °C)
Foundations	Foundations for the equipment and ductwork/piping, as required
Structural steel	Steel for access to and support of the SCR reactors and other equipment, ductwork, and piping

than 90%. The SNCR system can use ammonia or urea as the reducing agent.

The reaction between NO_x and ammonia occurs at a higher temperature range than in an SCR system. At lower temperatures, the reaction is slow and ammonia slip occurs. At higher temperatures, ammonia can be oxidized to make additional NO_x. The narrow temperature window is one of the disadvantages of the system especially when the process gas temperature is variable. The optimum temperature range is from 870 to 1100 °C. The temperature dependence of NO_x reduction in an SNCR system is shown in Figure 15.21.

Urea can be injected directly into the boiler in a location where the temperature is appropriate for the reaction. This technique may also be used to directly reduce NO_x emissions from the boiler. The chemical residence time for the reaction between ammonia and NO_x should be enough for the mixing of the ammonia and initiate the reaction with NO_x. Once well mixed the reaction time is very short as 100–500 milliseconds. The residence time can vary from 0.001 to 1 s. When urea is used in liquid form, it requires sufficient time for the evaporation of water, decomposition of urea into ammonia, and reaction of ammonia with NO_x. Therefore, urea injection system requires additional residence time for the reaction to occur. Increasing the residence time especially at the lower temperature window is necessary for the completion of the reaction. Residence time can be increased up to 10 s, but the increased benefit is minimal after 0.5 s. Figure 15.22 shows the relationship between the temperature and residence time on NO_x reduction.

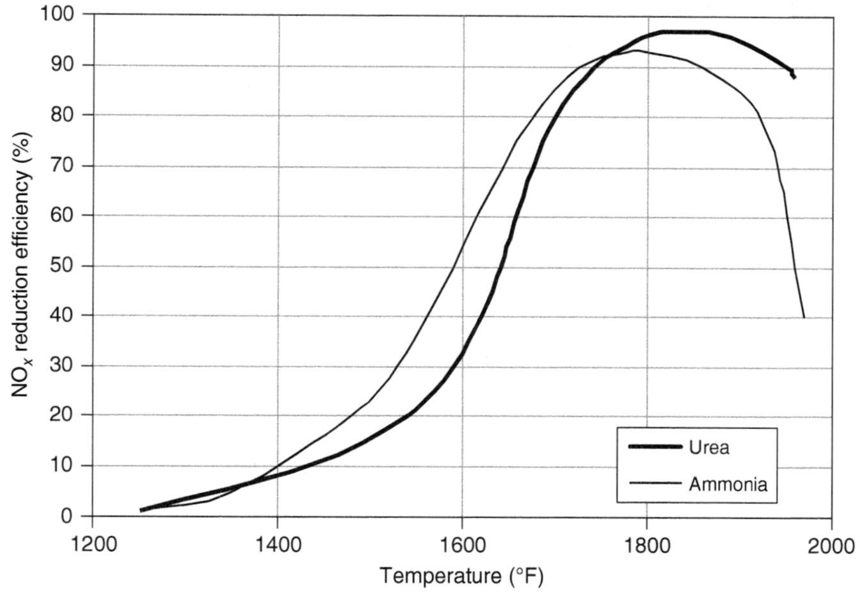

Figure 15.21 Effect of temperature on NO_x reduction (U.S. EPA, 1982).

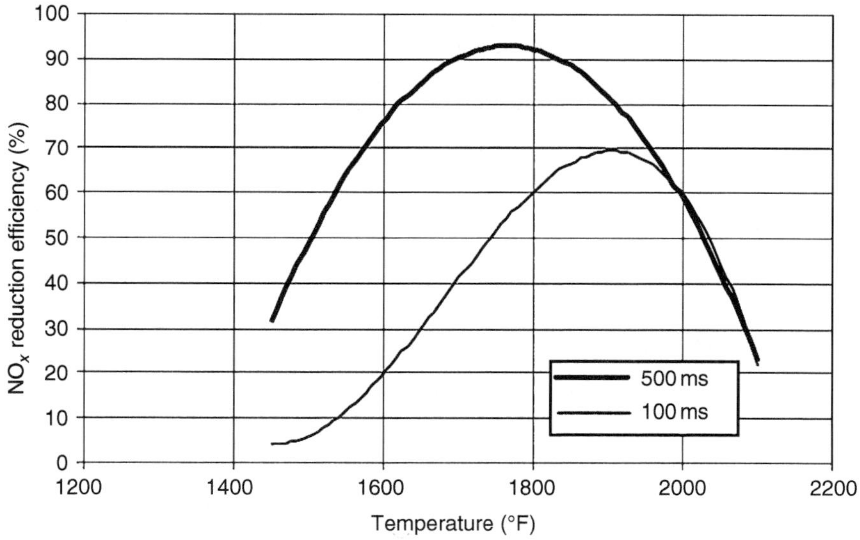

Figure 15.22 Effect of residence time on NO_x reduction (U.S. EPA, 1982).

The aqueous urea is injected through nozzles, creating optimum size droplets for easy distribution and to enhance water evaporation. Larger-sized droplets although can penetrate longer distance requires longer time for water to evaporate, while too small a droplet size may not penetrate for proper distribution. Inadequate mixing and distribution will result in insufficient NO_x reduction.

Ammonia that escapes, the ammonia slip, in the flue gas poses health concerns. The concentration of ammonia slip is regulated due to its negative impacts. Also, it can make a visible stack emission due to the formation of ammonium chloride. Excess ammonia can also help to form other salts that can plug and corrode downstream equipment. As a result, excess amounts of ammonia are restricted in many of the NO_x reduction systems. Ammonia slip is restricted to 5–10 ppm in NO_x reduction system that uses ammonia or urea. The urea-based system can employ modular design where required components are put in a skid module for easy transport to the site. This modular skids are installed on-site with very little additional work. A typical skid module system flow diagram is shown in Figure 15.23 and Table 15.12.

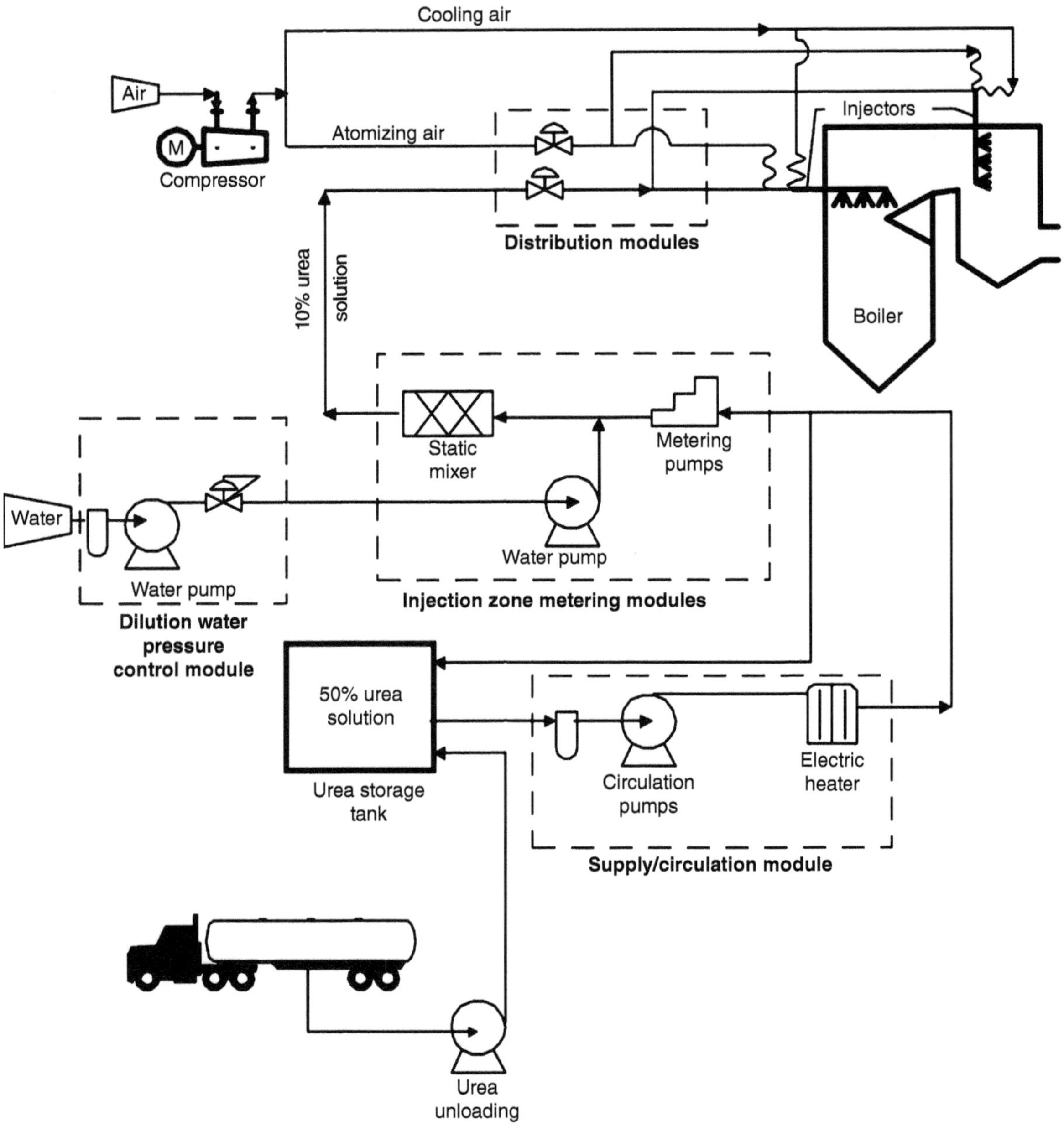

Figure 15.23 Urea SNCR process flow diagram (U.S. EPA, 1981).

Table 15.12 Urea-based SNCR system equipment (U.S. EPA, 1981).

Item	Description/size
Urea unloading skid	Centrifugal pumps with hoses to connect to rail tank car or truck
Urea storage tanks	Vertical, insulated fiberglass-reinforced plastic (one or more tanks) (vinyl ester resin) tank, atmospheric pressure design, and equipped with a vent, caged ladder, manway, and heating pads
Circulation module	Skid-mounted circulation module consisting of: • Circulation pumps • Electric heaters • Insulated/heat traced piping • Isolation valves for pumps and heaters • Instrumentation for flow, pressure, temperature, and a control panel
Injection zone metering (IAM) modules (1–5 modules)	Skid mounted metering modules consisting of: • Metering pumps, hydraulic diaphragm type equipped with a variable speed motor drive • Water booster pumps, turbine type • Insulated/heat traced piping • Isolation and control valves for pumps • Instrumentation for flow, pressure, temperature, and a control panel
Air compressor	Rotary type
Distribution modules (1–5 modules)	Urea solution distribution module consisting of: • Valved connections for urea and atomizing air • Isolation valve and a pressure control valve for the air/urea supply to each injector • Pressure indicator for air/urea supply to each injector • Flow indicator for urea supply to each injector
Injectors (4–12 per distribution)	Wall type: Dual-fluid-type wall injector, with modules, furnace wall panels, and hoses for air and urea supplies Lance type: Dual-fluid-type lance injector, with furnace wall panels and hoses for air and urea supplies
Piping	Between urea unloading skid and urea tank; urea tank and circulation module; and circulation module and IZM modules(s). Insulate/heat traced piping, stainless steel
Piping	Between IZM module(s) and distribution modules. Insulated/heat traced tubing, stainless steel
Tubing	Between distribution modules and injectors. Insulated/heat traced tubing, stainless steel
Dilution water piping	Insulated/heat traced piping, carbon steel, with isolation and pressure reducing valves
Miscellaneous piping	Piping/tubing and valves for flushing water, atomizing air, and control air
Piping supports	Structural support steel, including a pipe bridge, for supporting all piping
Economizer outlet emission monitors	Monitor NO and O in the flue gas and provide a feedback signal for urea injection control
Instrumentation and controls	Instrumentation and stand-alone, microprocessor-based controls for the SNCR system with feedback from the plant controls for the unit load, NO emissions, etc.
Enclosures	Preengineered, heated, and ventilated enclosure for the circulation and metering skids
Foundations	Foundations and containment walls for the tank and equipment skids, enclosure, and piping support steel, as required

15.4.3 Control of Sulfur Dioxide

Sulfur dioxide is one of the major primary pollutants emitted by fossil fuel combustion. In 2008, coal-fired power plants contributed over 70% of the SO_2 emissions in the United States. One of the major environmental impacts due to SO_2 emission is acidic deposition. Acidic deposition through rain, snow, and other forms of precipitation including dry deposition has affected several lakes and water bodies in the United States, Canada, and Scandinavian countries. The 1990 CAAA clearly identified acidic deposition as a major issue and called for SO_2 emission reductions under Title IV. The CAAA requires a reduction of 10 million tons of sulfur dioxide from 1980 levels, especially from power plants. Sulfur dioxide also can be responsible for fine particulate formation in the atmosphere and can reduce visibility and contribute to PM-2.5 particulates. The US EPA has been very effective in reducing sulfur dioxide emissions from 31 million tons of SO_2 in 1970 to about 11 million tons in 2008.

Sulfur dioxide emissions into the atmosphere can be reduced by either removing the sulfur from the fuel before it is burned or capturing it from the flue gas after its combustion. The clean coal technologies developed over the years helped to reduce the sulfur content of the coal before it is being burned. The cost of cleaning the coal is not economical and the industries prefer flue gas treatment (FGT) systems. There are several technologies available to treat the flue gas to reduce SO_2 emissions. They can be classified as wet and dry process and throwaway and regenerative process. In addition, there are processes developed to remove sulfur from natural gas. The "Claus" process, for example, converts hydrogen sulfide from the produced gas to elemental sulfur as a by-product. Many gas wells contain hydrogen sulfide as high as 40–50%. Canada leads in recovering sulfur from the hydrogen sulfide in the natural gas, and many other countries are successful in reducing SO_2 emissions using similar technologies.

Sulfur is contained in coal as pyrites or mineral sulfates. This sulfur can be removed by clean coal technology, mainly by washing. Gasification of coal can also remove sulfur from coal, which is more expensive. There are new gasification processes in development that may reduce the cost of sulfur reduction in gasification process. One of the major emitters of sulfur dioxide is smelters. The sulfur-containing minerals during its processing, especially during drying, release sulfur compounds. The gas containing sulfur compounds after cleaning is sent to a sulfuric acid manufacturing process to make sulfuric acid. This approach is not only reducing the emissions but also creating a salable product.

Coal-fired power plants use coal with sulfur content between 0.5 and 5%. The sulfur then released as mainly sulfur dioxide along with the flue gas. There are several flue gas desulfurization (FGD) techniques available to control sulfur dioxide emissions from gas streams. Limestone, lime, dual alkali, citrate, magnesium oxide, and Wellman–Lord (W-L) are some of the processes. They are summarized in Table 15.13 and Figures 15.24–15.29.

In a limestone process the limestone slurry is contacted in a spray tower. The limestone droplets capture sulfur dioxide from the flue gas and forms calcium sulfite and calcium sulfate. The limestone is inexpensive, and, in general, the source of limestone is closure to the power plants. Although the cost is lower compared with other processes, there will be additional maintenance problems like plugging and scaling of the system. The lime processes use calcium oxide instead of limestone that forms calcium hydroxide with water. Calcium hydroxide in turn absorbs sulfur dioxide and forms calcium sulfite. The dual-alkali system prevents the scale formation due to its solubility in water. Sodium sulfite and sodium hydroxide solution is used to convert sulfur dioxide to a soluble sodium sulfate solution. The sodium hydroxide then is regenerated in a separate vessel adding lime or limestone. The effluent rich in sodium causes a water pollution problem and needs to be disposed properly. Also, the dual-alkali system needs a prescrubber to remove any acid, especially hydrochloric acid, from the flue gas to minimize the consumption of sodium hydroxide.

In a citrate process, citrate ion reacts with hydronium ion that is generated by the reaction of sulfur dioxide. Reaction of citrate ion with hydronium ion promotes the reaction of sulfur dioxide to form sulfurous acid, which dissociates bisulfite ion and hydronium ion, as shown in the equation below. The SO_2-loaded citrate solution is

Table 15.13 Summary of SO_2 removal processes (U.S. EPA, 1981).

Process	Description	Illustration
Limestone	Throwaway process uses limestone slurry to absorb SO_2 and forms calcium sulfite and calcium sulfate	Figure 15.24
Lime	Throwaway process uses lime slurry to absorb SO_2 and forms calcium sulfite and calcium sulfate	Figure 15.25
Dual alkali	Regenerative process use sodium solution, which is regenerated using lime or limestone	Figure 15.26
Citrate	Regenerative process where citric acid helps to enhance the SO_2 removal by reacting with hydronium ions. Pure SO_2 or elemental sulfur can be recovered	Figure 15.27
Magnesium oxide	Regenerative process that uses magnesium hydroxide slurry removes SO_2. Regeneration by calcination produces SO_2 gas and the chemical is recovered for reuse	Figure 15.28
Wellman–Lord	Regenerative process where sodium sulfite reacts with SO_2 to form sodium bisulfite. When sodium sulfite is heated, it releases SO_2 and sodium sulfite also is regenerated	Figure 15.29

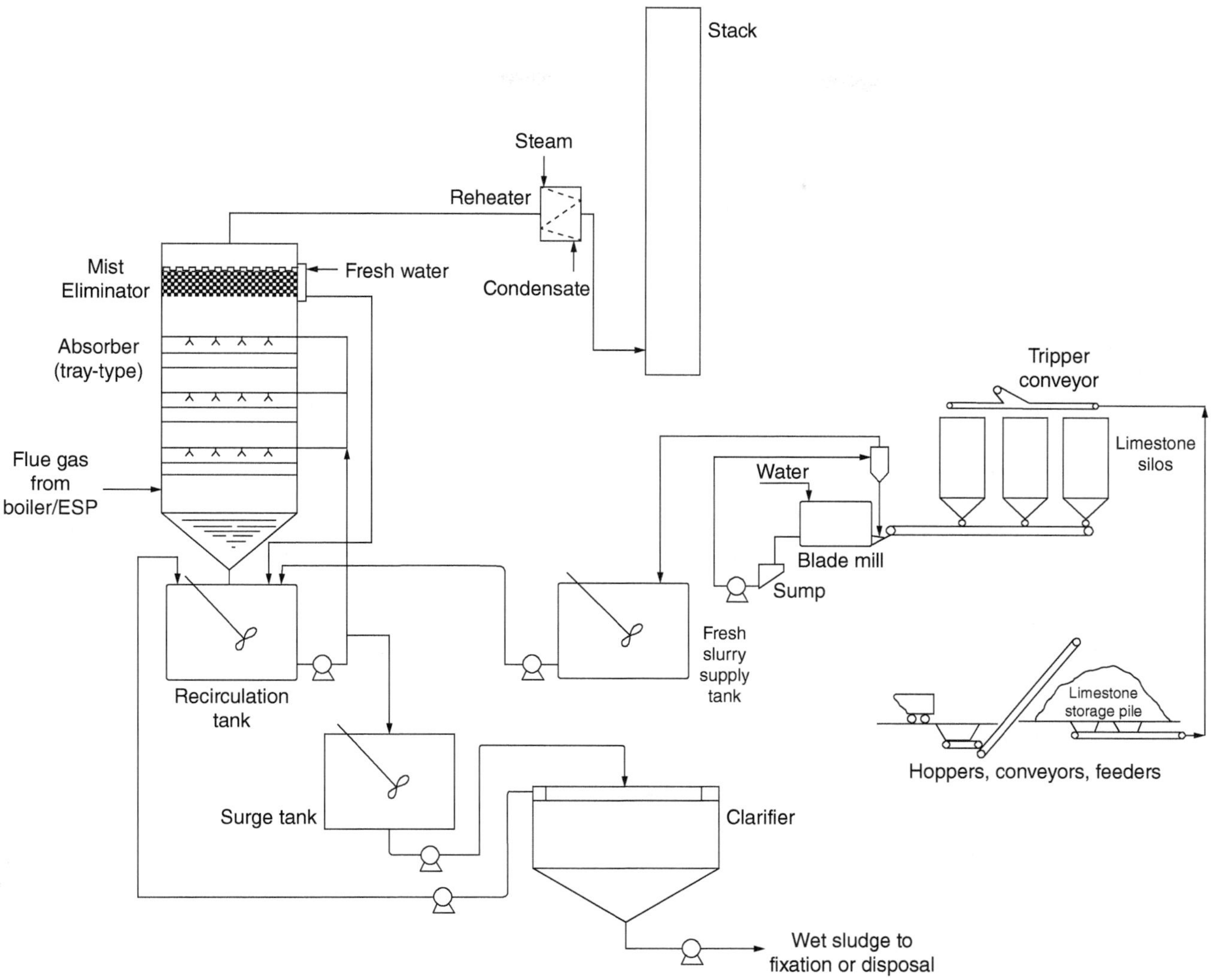

Figure 15.24 Diagram of a typical limestone FGD system (U.S. EPA, 1981).

further treated with H_2S to form elemental sulfur or SO_2 and stripped from the solution as a marketable product.

$$SO_2(g) + H_2O \rightleftharpoons H_2SO_3$$

$$H_2SO_3 \rightleftharpoons H^+ + HSO_3^-$$

$$Ci^{-3} + H^+ \rightleftharpoons HCi^{-2}$$

$$HCi^{-2} + H^+ \rightleftharpoons H_2Ci^-$$

In a magnesium oxide process, magnesium hydroxide slurry is used to capture SO_2, producing magnesium sulfite and sulfate solids. The solids then undergo calcination in a calciner, releasing SO_2 gas and magnesium oxide. Therefore, this process needs a calciner, adding cost to the process, but does not produce huge amounts of solid waste as in the lime or limestone process:

The W-L process is a true regenerative process where the chemicals are regenerated and a salable sulfur product as SO_2 gas or elemental sulfur can be recovered. A sodium sulfite solution is used to absorb sulfur dioxide to form sodium bisulfite. In a separate heated evaporator vessel, the sodium bisulfite solution is heated to release concentrated SO_2 gas and to form sodium sulfite crystals. The sodium sulfite is then reused in the process. There will be sodium loss in the process that will be compensated by adding sodium carbonate. It requires approximately 1 mol of sodium carbonate per 42 mol of SO_2 removed. The sodium carbonate readily reacts with sulfur dioxide and forms sodium sulfite. Thus W-L process needs a prescrubber to remove any hydrochloric acid from the flue gas to minimize the use of more expensive sodium sulfite chemicals and to remove any particulates from the flue gas.

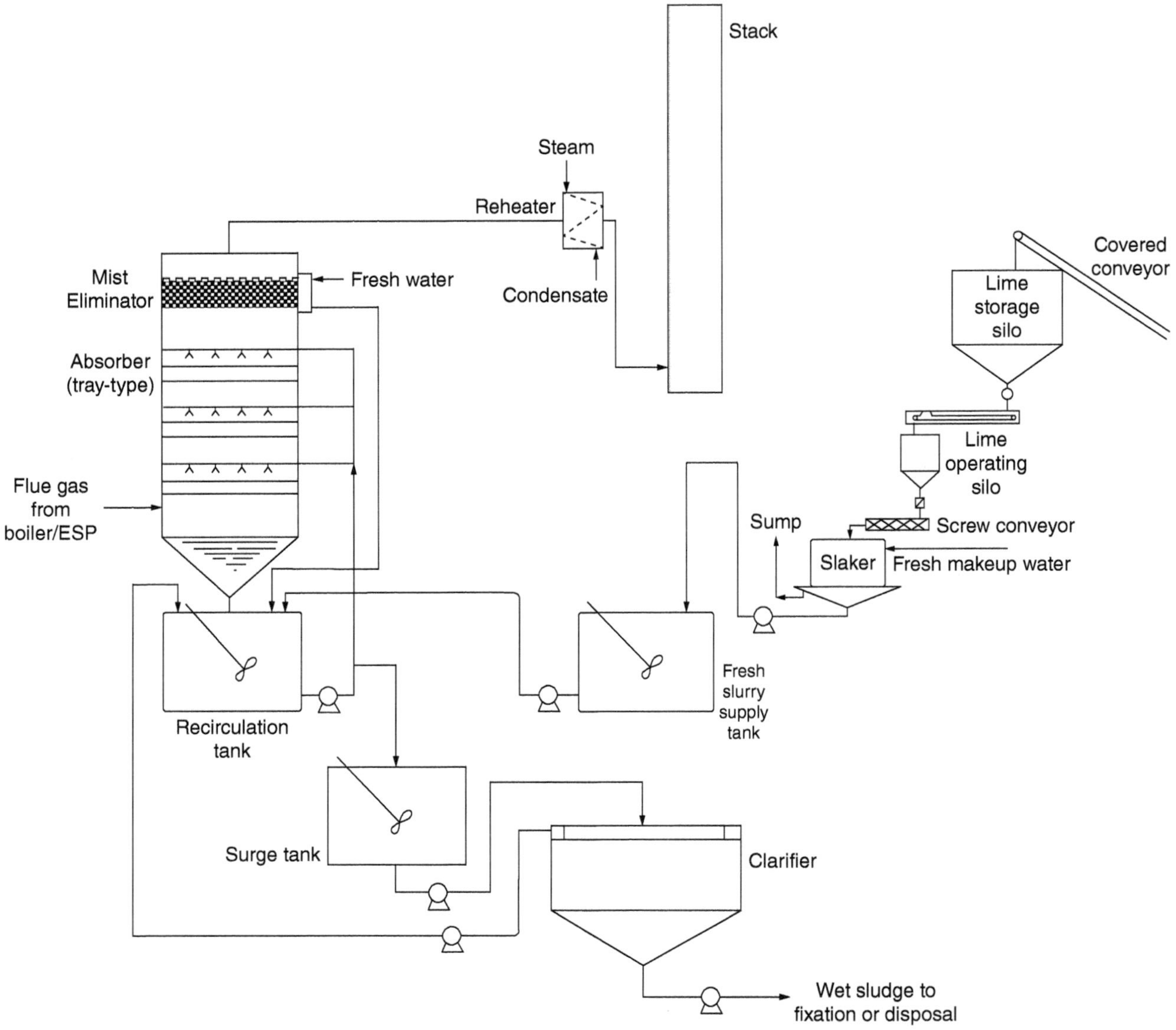

Figure 15.25 Typical lime FGD system (U.S. EPA, 1981).

15.4.3.1 Wet Scrubbers for Gases

In chemical industry, wet scrubbing or absorption technique is used to recover products from the process stream. In air pollution control, wet scrubbers are used to remove contaminants from the gas streams and are effective in reducing water-soluble gases such as sulfur dioxide and hydrogen fluoride. To reduce contaminants, wet scrubbers often use chemicals in the scrubbing solution to initiate chemical absorption between the contaminant and the solution and are effective for pollutant concentration of 250–10 000 ppm. Commonly the scrubber solutions include water, aqueous solutions, mineral oils, and heavy hydrocarbon oils. The contaminants are removed by simple absorption into the scrubbing solution or by more complex chemical reactions. For example, wet scrubbers add alkaline compounds in water to enhance the removal of acid gases such as sulfur dioxide.

For wet scrubbers, the removal efficiencies vary for each pollutant–solvent system and with the type of absorber used. Most absorbers have removal efficiencies higher than 90%. Packed bed absorbers can achieve efficiencies as high as 99.0% for some pollutant–solvent systems. Packed towers are packed with lightweight packing materials to increase the mass transfer between the gas stream and the scrubbing solution. A thin layer of solution on the surface of the packing material provides the surface area needed for the mass transfer.

The choice of using a wet scrubber depends on several factors including the availability of solvent at a reasonable cost, the amount of contaminant to be removed,

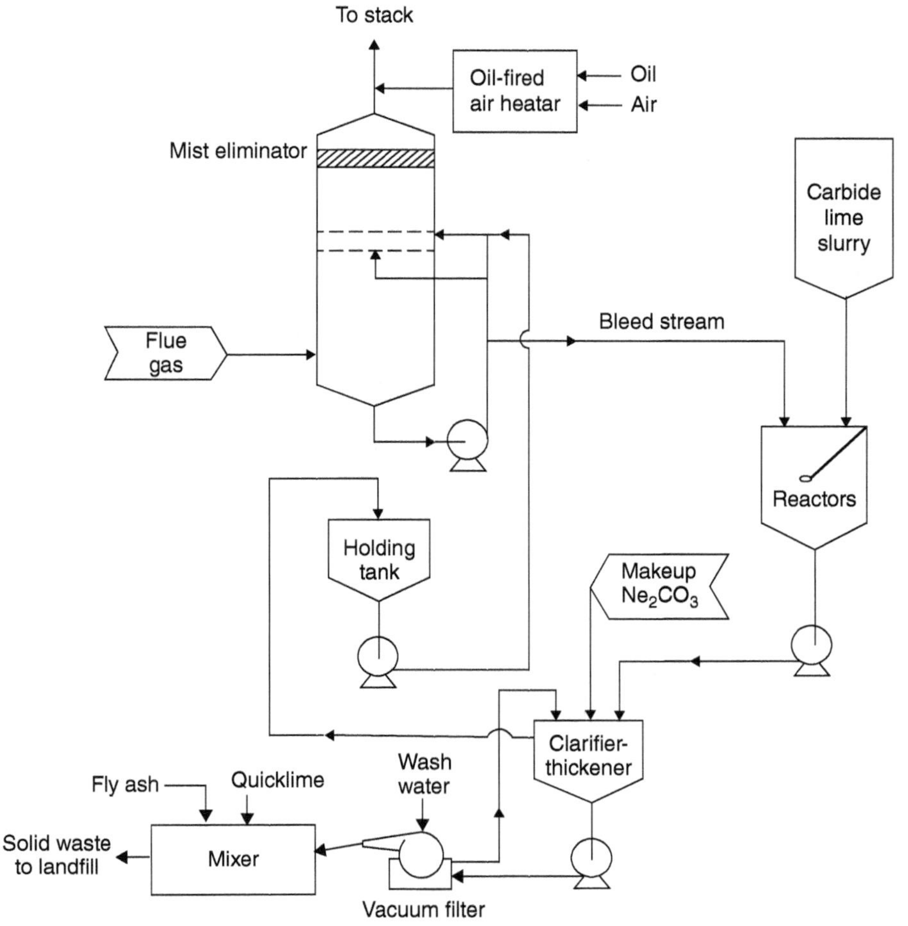

Figure 15.26 Generalized process flow of the dual-alkali system at LG&E Cane Run No. 6 (U.S. EPA, 1983).

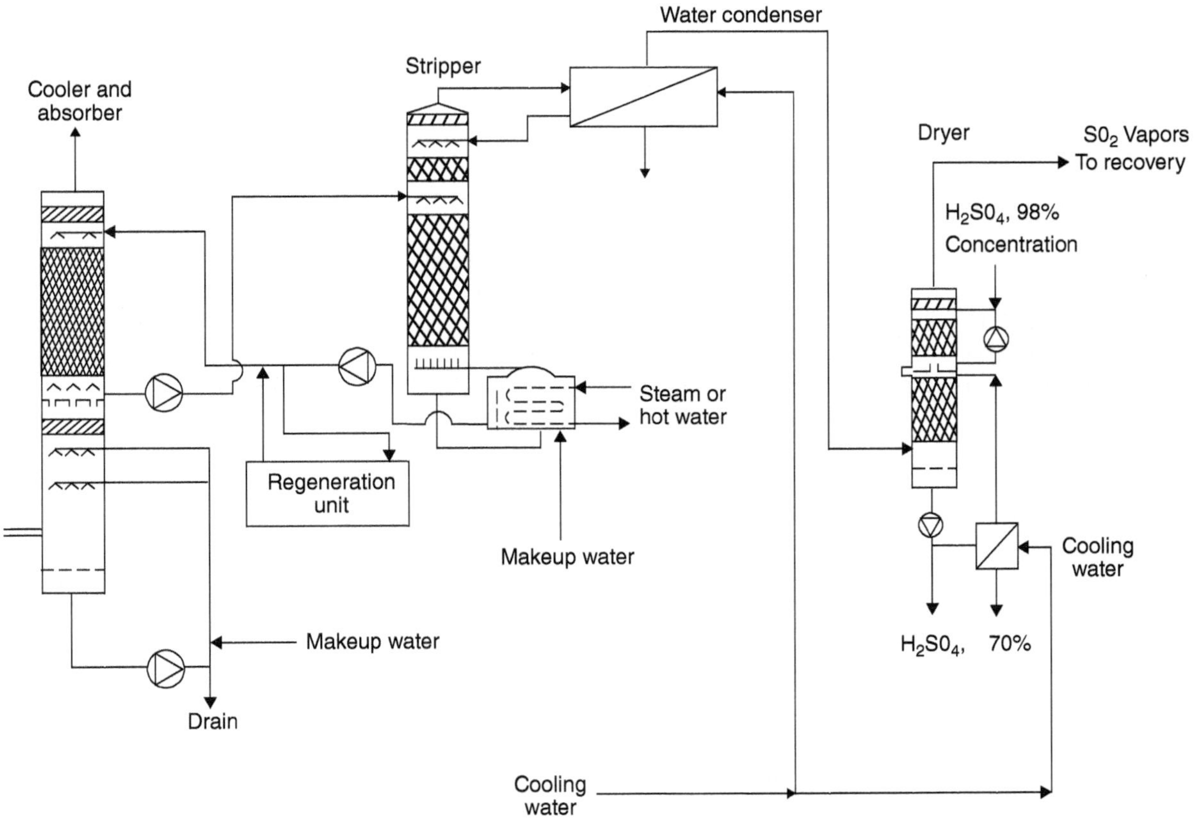

Figure 15.27 Typical citrate SO_2 control system (producing concentrated SO_2 vapor) (U.S. EPA, 1981).

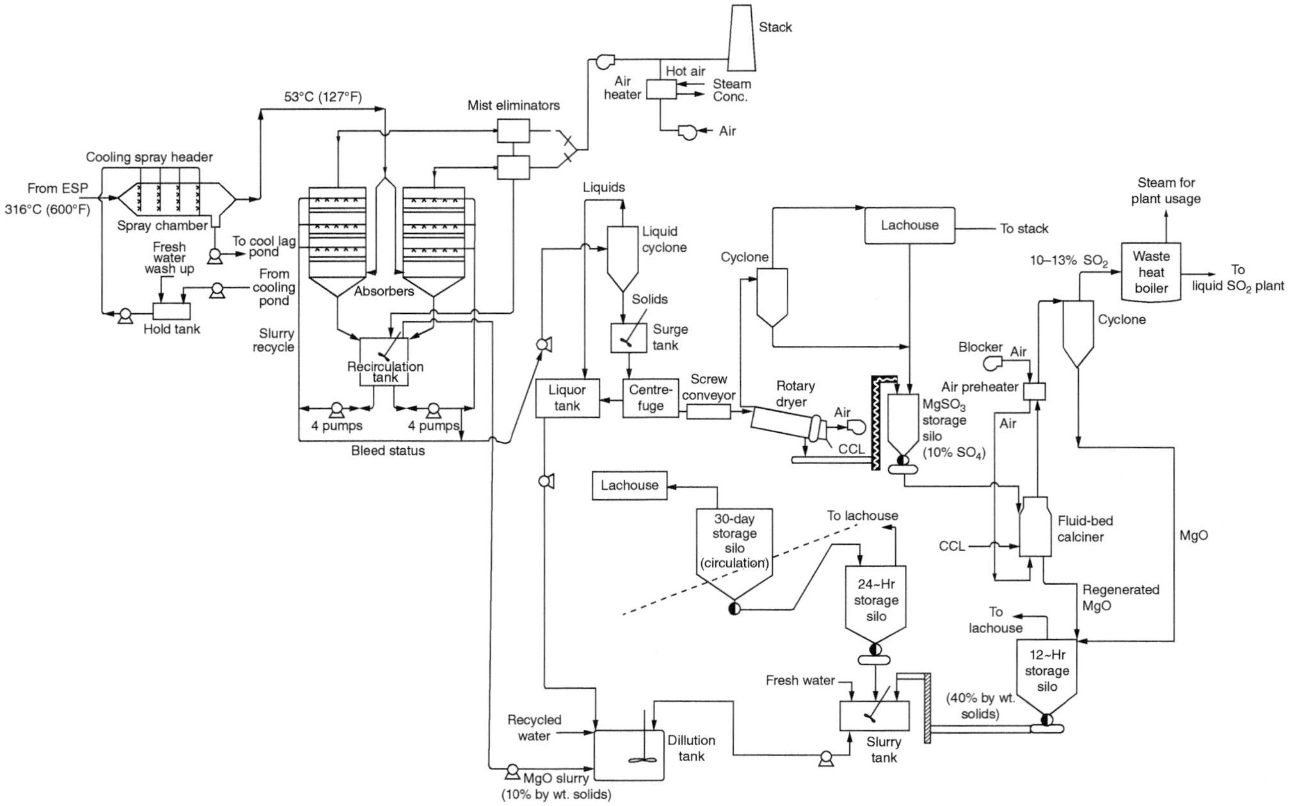

Figure 15.28 A magnesium oxide FGD system using TCA absorbers and a fluid bed calciner (U.S. EPA, 1981).

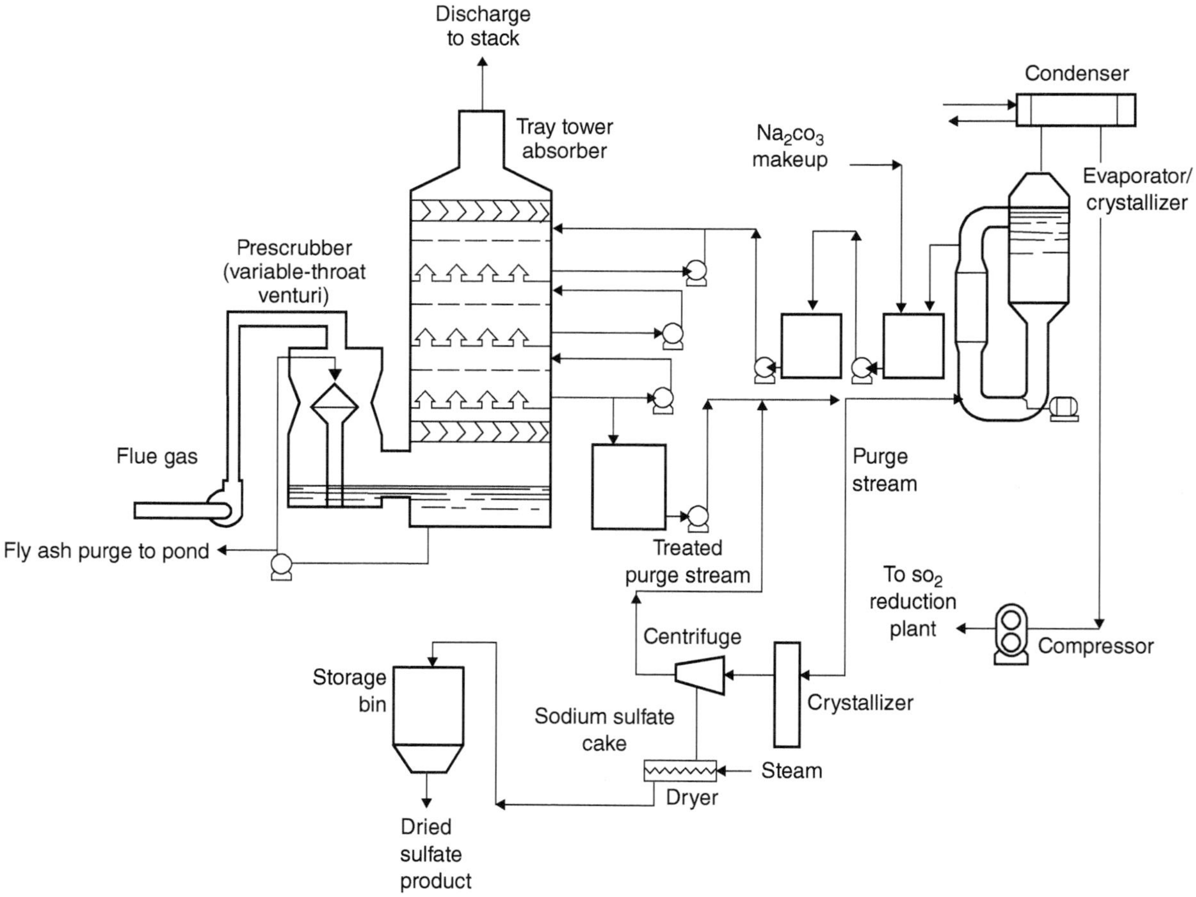

Figure 15.29 Typical Wellman–Lord SO_2 control system (U.S. EPA, 1981).

and the potential cost associated with disposal of the waste stream or product recovery. In general, for air pollution control systems, the scrubbing solutions are nontoxic in nature. The contaminants must be transferred from the gas phase to the liquid phase. This mass transfer occurs through a thin film on both sides of the gas and liquid phases. The contaminant passes through two thin films, one on the gas phase and another on the liquid phase. Mass transfer is established by concentration gradient between the bulk gas phase and the thin gas film and subsequently between the liquid-phase thin film and the bulk liquid. This mass transfer process can be theoretically evaluated to understand the performance of a wet scrubbing system. In reality, to determine the efficiency of a wet scrubber, the theoretical evaluation is combined with empirical relationships to quantify the performance of wet scrubbers. The diffusion of pollutants into the thin gas film and then into the thin aqueous film determines the effectiveness of the wet scrubbers, which depends on the characteristics of the pollutants, temperature, and the gas to solvent flow rates.

15.4.3.2 Types of Wet Scrubbers

Wet scrubbers can be classified based on the gas and liquid flow such as countercurrent, crosscurrent, and cocurrent. In countercurrent systems, gas enters from the bottom of the scrubber and flows toward the top of the scrubber, and the liquid enters the top of the scrubber and exits at the bottom of the tower. Countercurrent systems are more common than the cocurrent or crosscurrent flow systems. Since the fresh liquid with lowest pollutant concentration encounters the gas stream with the lowest concentration at the top of the tower, relatively higher removal efficiency can be achieved. At the bottom of the tower, the highest pollutant concentration in the gas stream encounters the highest pollutant concentration in the liquid stream; thus the driving force for the mass transfer is in favor throughout the height of the tower in a countercurrent system. In a crosscurrent flow system, the liquid is sprayed vertically from the top, while the gas passes horizontally, intercepting the liquid. In a cocurrent system, the gas stream and liquid flow are in the same direction.

15.4.4 Packed Towers

The packed tower shell may be metal or plastic or reinforced fiberglass (Figure 15.30). The interior shell can be treated with polymers to withstand the corrosive nature of the gas and liquid streams to be treated. The liquid distribution needs to be uniform across the cross section for the packed tower to be effective in removing contaminants. Spray nozzles are commonly used for achieving proper distribution of the water across the tower. A mist eliminator is located at the exit of the gas streams to capture any liquid droplets escaping with the gas stream. A proper structural support to the whole tower and to the packing materials is necessary for the integrity of the system. If the height of packing is greater than about 20 ft, the packing needs additional support to maintain the integrity of the packing.

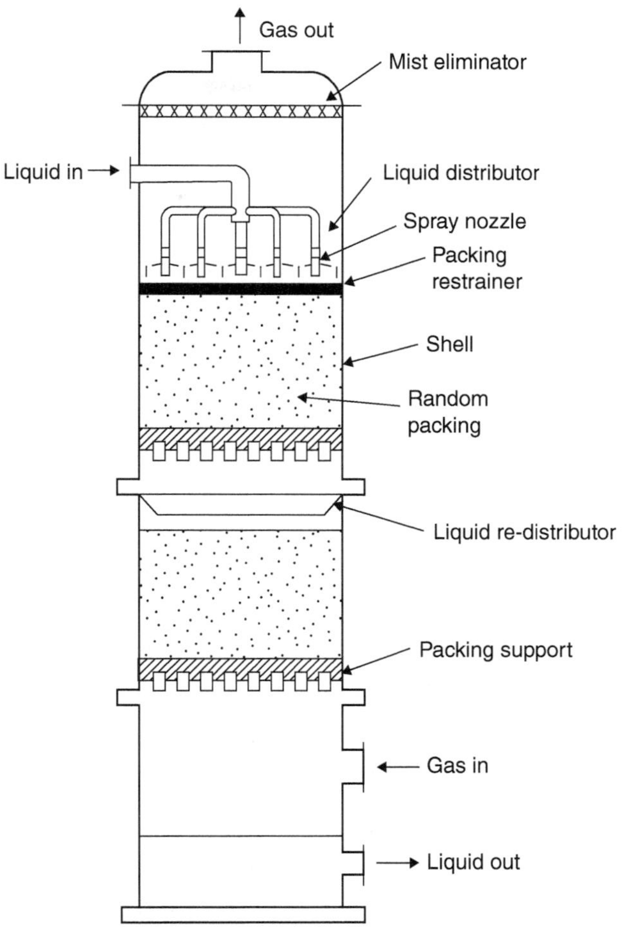

Figure 15.30 Packed tower for gas absorption (U.S. EPA, 2002).

It is also essential to distribute the incoming gas uniformly at the bottom of the tower. This can be achieved by having the gas enter at the bottom of the tower with open space between the packing and the bottom of the tower. A uniform gas flow can be achieved using a perforated plate to distribute the gas streams.

Packed towers are filled with packing materials that provide a large surface area to facilitate mass transfer of contaminant from gas stream to the liquid streams. Packed towers in general are very effective in removing especially gases soluble in water. Packing materials are made up of lightweight materials resistant to acid or alkaline, providing maximum surface area per unit volume of packing. They include Raschig rings, Berl saddles, Intalox saddles, Pall rings, and tellerettes (Figure 15.31).

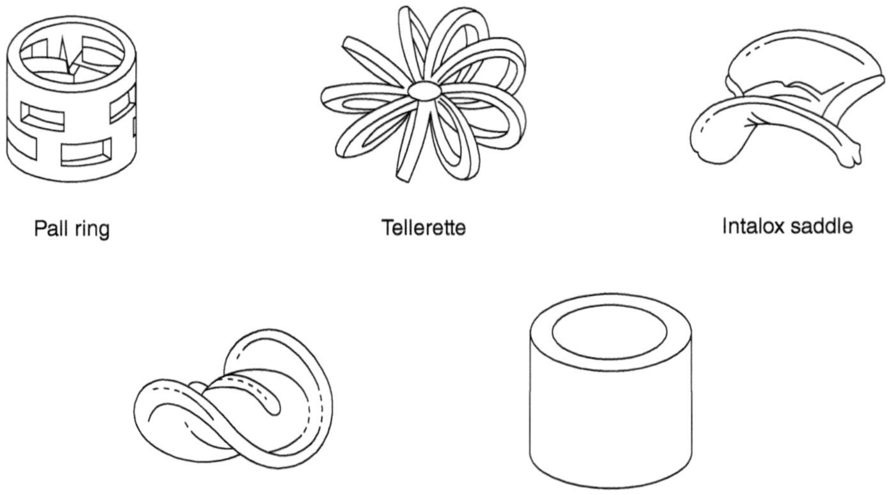

Figure 15.31 Random packing material (U.S. EPA, 2002).

Table 15.14 Typical gas to liquid ratios for wet scrubbers (U.S. EPA, 1982).

Scrubber type	Liquid to gas ratio ($l\,m^{-3}$)
Venturi	0.70–1.00
Cyclonic spray tower	0.70–1.30
Spray tower	1.30–2.70
Moving bed	1.30–2.70
Impingement plate	0.40–0.70
Packed bed	0.10–0.50

They are generally called gas absorption systems. The liquid to gas ratio is one of the critical design parameters. Excessive liquid flow or very low liquid will limit the contaminant transport. The liquid to gas ratio for a typical packed tower varies between 0.1 and 0.5 lm^{-3} of air. Table 15.14 provides liquid to gas ratio information for various types of wet scrubbing systems.

15.4.4.1 Packed Tower Operation

The pressure drop in a packing tower is due to the gas and liquid flow and the level of resistance offered by the packing materials. A packing material with high void ratio but with equal amount of surface area per volume will be preferred over low void ratio packing materials. It is reported that about 0.5–1 in. of water pressure drop per foot of packed tower can be expected. Improper design of packed towers will cause flooding of the packed tower, thus increasing the pressure drop and decreasing the mass transfer. In general, packed towers are operated at about 50–60% flood conditions. A minimum liquid flow rate is also required to wet the packing material sufficiently for effective mass transfer to occur between gas and liquid.

The effluent from the packed tower can be recycled back into the tower depending on the situation. If the scrubbing liquid used is an expensive solvent, then a solvent recovery system with a recycling pump is more economical. If water is in the scrubbing liquid, then the effluent may directly be taken to a wastewater treatment plant. The design of packed tower includes gas to liquid ratio, the velocity of gas in the tower, the height of packing, the diameter of packing, and the pressure drop of the tower. The design procedure involves several steps, and the readers of this chapter are recommended to refer to an air pollution engineering textbook for additional information on the design parameters.

15.4.5 Volatile Organic Compounds

Adsorption of contaminant on a solid material like activated carbon has been used for decades to clean the water or airstreams. In air pollution control, gas streams containing low concentration of VOC can be removed by adsorption, incineration, condensation, and other techniques. The two common techniques are adsorption and incineration. In adsorption, the vapor can be recovered as a product, and in the incineration process, the VOC is destroyed. If one prefers to recover products, then adsorption system is the choice. There are two types of adsorption: physical adsorption and chemisorption. In physical adsorption, the vapor molecules are held by the adsorption surface with weak forces and the adsorption is exothermic. The adsorption bed gets saturated by the contaminant molecules and no longer will the contaminant be removed from the gas stream. Once the saturated bed is heated, the contaminant molecules get released from the adsorbent surface. The contaminant can also be released by exerting vacuum to release the adsorbed molecules. Heating with steam or with hot inert gas is common practice in industries. In chemisorption, the contaminant molecules change its original chemical nature, and the original contaminant molecule will not be recovered. Chemisorption is commonly used in chemical industry process streams than in air pollution control systems. The efficiency of an adsorption system can be between 95 and 99% and can handle contaminants of concentrations up to 10 000 ppm. However, most of the adsorption systems handle contaminant concentration much lower due to safety issues of explosion and exposure to workers.

Adsorption systems can be fixed bed or moving beds. Fixed bed systems are a batch system where a bed will be exposed to the incoming contaminant gas stream, and once saturated, the gas stream will be switched to a clean second adsorption bed. While the second bed is in operation, the first bed is in regeneration mode. In the regeneration mode, the adsorption bed is exposed to high temperature steam or hot inert gas to release the adsorbed contaminant molecules. Once the molecules are released, the bed undergoes cooling and then ready for adsorption. Thus, more than one bed is used to handle a continuous gas flow from a plant exhaust. This type of fixed bed systems is more popular to clean gas streams from continuous operations. Canisters and 55-gal drums filled with activated carbon are used in many plants to avoid the direct emission of VOC to the ambient air. These canister-type systems will be exchanged by the vendors after its length of life. The length of life will be determined by the amount of adsorbent contained in the canister and the amount of gas flow it could handle for the concentration of the contaminant.

The adsorption systems can be classified into Langmuir isotherm and Freundlich isotherm. In Langmuir theory, the ratio of contaminant partial pressure to the adsorption capacity (P/a) is a linear function of the partial pressure. The adsorption capacity is used to determine the volume of adsorbent required to remove a certain quantity of adsorbate (contaminant). These values are available from vendors for specific type of adsorbents and for specific type of contaminants. The Langmuir isotherms consider that the contaminant molecules are captured on the surface of the adsorbent in a single molecular layer. The available adsorption surface area per volume of adsorbent is a critical parameter in selecting an adsorbent. For example, activated carbon adsorbent can have a surface area between 600 and 1600 $m^2\ g^{-1}$. This tremendous surface area per volume of adsorbent is what makes the adsorbents very suitable for air pollution applications. The adsorbents are very porous and the contaminant molecules can travel into the interior surfaces and be adsorbed.

The working capacity of an adsorbent is much smaller than the theoretical adsorption capacity. The inherent moisture in the adsorbents, the moisture in the gas stream, the loss due to adsorption zone, and the loss due to the heat wave (exothermic) all reduce the theoretical adsorption capacity of the adsorbents. The common adsorbents used are activated carbon, bone char, molecular sieves, iron oxide, and silica gels. For example, silica gels are commonly used to remove moisture from gas streams, whereas activated carbon is the most popular adsorbent and less expensive (5–6 dollars per kilogram of adsorbent).

The fixed bed systems are about 12–36 in. in depth. The higher the depth of the bed, the higher will be the pressure drop of the system. The higher the pressure drop, the higher will be the operating cost. The fixed bed design involves the determination of the volume of adsorbent necessary to capture the amount for contaminant. Once the volume of adsorbent needed is determined, with the known bed depth, the length and width of the bed can be calculated. In general length is twice the width for rectangular fixed absorbers. There are complex equations available to calculate the pressure drop from a fixed bed system. One of the equations is given below, where the mesh size of carbon is restricted to 4 × 6 if the applicable velocity range is 60–140 fpm. The velocity in fixed adsorption systems should be less than 100 fpm. The VOC level maintained in the system is about 20–25% of the low explosive level (LEL) of specific compounds to avoid any potential explosion.

Moving adsorption beds handle continuous gas flow. The moving beds have three stages: in the first stage fresh adsorbent is introduced into the system, the adsorbent is generated in the second stage by hot steam or hot inert gas, and in the third stage the adsorption material is cooled and ready for its use. The freshly regenerated adsorbent is pumped into the first stage for adsorption. The activated carbon used in these systems must be able to withstand the multiple cycles without losing its integrity.

Incineration is another technique by which VOC can be destroyed and it is very effective in oxidizing the compounds to elemental state. The destruction efficiency can be higher than 95%. There are two types of oxidation: thermal oxidation and catalytic oxidation (Figures 15.32 and 15.33). In thermal oxidation, polluted air and necessary fuel are oxidized using additional burner air to complete thermal oxidation. A simple oxidizer is nothing more than a circular insulated vessel where the polluted air and the fuel and the burner air are mixed and combusted at a desired temperature. The temperature to be

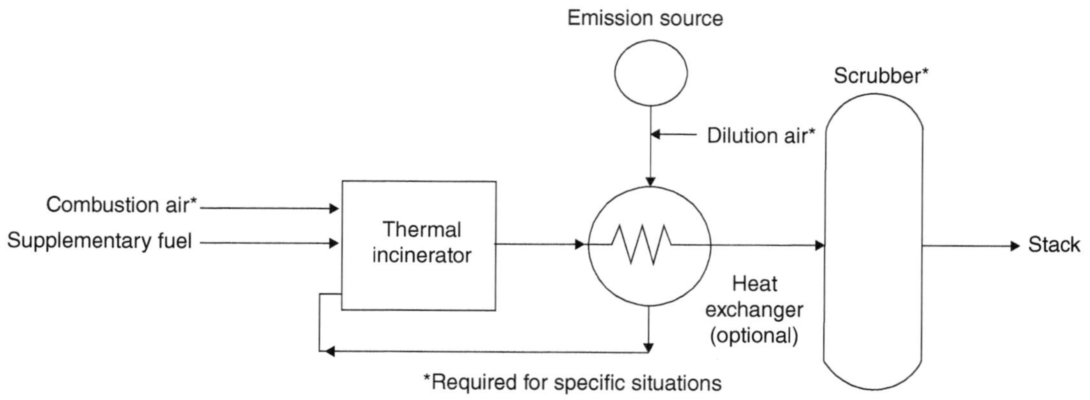

Figure 15.32 Schematic diagram of a thermal incinerator (U.S. EPA, 1991).

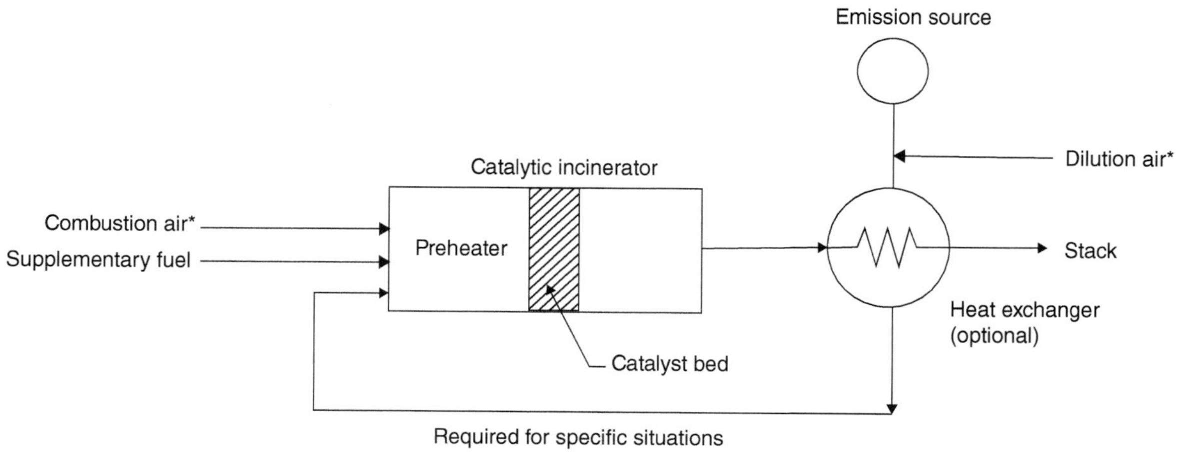

Figure 15.33 Schematic diagram of a catalytic incinerator system (U.S. EPA, 1991).

maintained can be determined by theoretical evaluations or dictated by the local or regional regulatory agencies. The combustion involves 3 T's: the correct temperature, sufficient residence time, and turbulence. For proper mixing the thermal oxidizer design involves determination of the amount of fuel required to maintain the required temperature. The sizing of the oxidizer includes diameter and length to maintain the residence time and the velocity through the system. A typical mass balancing and heat balancing of the system will yield the necessary design parameters to treat a particular flow rate. It is always good to check whether correct amount of fuel is provided to maintain the design temperature before its operation. Thermal oxidation systems are popular because they are simple to build, operate, and maintain. However, the fuel cost may be expensive to destroy a small amount of pollutant. The exhaust gas from a thermal oxidizer has high heat content, and this heat can be recovered from the system to preheat the incoming burner air or to generate low pressure steam or to use the hot gas for drying products or to generate hot water for use. A heat exchanger is necessary to recover the thermal energy. Adding a heat exchanger will increase the capital cost of the equipment.

In a catalytic oxidation system, the oxidation temperature of the compound is significantly reduced due to the presence of a catalyst. It reduces the amount of fuel required for destroying the same amount of compound. The temperature maintained in a catalytic oxidation system is about 500–700 F compared with 1200–1500 F in a thermal oxidizer. The catalytic system is much smaller in size because the residence time is about 10 times lower than that of the thermal oxidation system. Thermal oxidation system requires a residence time of 0.5–1 s, whereas catalytic systems need only 0.05–0.1 s. Thus, catalytic systems are favored in plants where space is limited. Catalytic oxidation systems are preferred and commonly used to destroy gaseous compounds from soil venting operations. Honeycombed catalysts are used to maintain a low pressure drop. For treating higher volume of gases, pellets are used, which will have higher pressure drops.

The volume of catalyst for a honeycombed system is determined based on space velocity, which typically ranges from 50 000 to 300 000 h^{-1}. Space velocity is the ratio of volume flow rate to the volume of catalyst. For a typical honeycombed system, the pressure drop is in the range of 1–3 in. of water.

Cooling the gas stream to condense the VOC is another technique used effectively to remove VOC from gas streams. It is very popular in chemical and oil refining industries where concentrated streams are condensed to recover valuable product. Condensation technique is energy intensive and may not be cost effective for gas streams with very low VOC levels, like in air pollution control systems.

15.4.6 Regenerative Thermal Oxidizer

Regenerative thermal oxidizer consists of two columns packed with ceramic or other heat transfer materials. The columns are connected by a burner at the top. The gas flows through the first column and oxidized by the burner. The hot flue gas from the burner exits through the second column. The heat from the flue gas increases the temperature of the second column as oxidation continues. Once the second column reaches a set temperature, the gas flow is changed from column 1 to 2. The gas passing through the second column gets preheated such that it needs very small amount of burner assistance for complete combustion. The hot flue gas passes through the column 1 and continue to heat the column until a set temperature is reached. At that time, flow will be switched to column 2. Thus, in the regenerative thermal oxidizing system, heat from the hot flue gas is effectively recovered to preheat the incoming air.

Acknowledgment

I would like to thank Katrina Moreira for organizing the figures, tables, and formatting. I would also like to thank Pankajam Ganesan for assisting with editing.

References

40 CFR 50 (2017). *National Primary and Secondary Ambient Air Quality Standards*. Washington, DC: Office of Federal Register.

40 CFR 60 (2017). *Standards of Performance for New Stationary Sources*. Washington, DC: Office of Federal Register.

Energy Information Administration (EIA) (2008). Greenhouse Gases, Climate Change, and Energy. https://www.eia.gov/energyexplained/index.php?page=environment_how_ghg_affect_climate (accessed 17 May 2018).

Theodore, L. (2008). *Air Pollution Control Equipment Calculations*. Hoboken, NJ: Wiley.

Theodore, L. and DePaola, V. (1980). Predicting cyclone efficiency. *JAPCA* 30: 1132–1133.

U.S. EPA (1981). *Control Techniques for Sulfur Dioxide Emissions from Stationary Sources*, 2ee. Research Triangle Park, NC: U.S. EPA.

U.S. EPA. *Control Techniques for Particulate Emissions from Stationary Sources*, Volume 1, EPA-450/3-81-005a. Research Triangle Park, NC: U.S. EPA, 1982.

U.S. EPA. *Full Scale Dual Alkali FGD Demonstration at Louisville Gas and Electric Company*, EPA-600/S7-83-039. Research Triangle Park, NC: U.S. EPA, 1983.

U.S. EPA. *Control Technologies for Hazardous Air Pollutants*, EPA-625/6-91/014. Research Triangle Park, NC: U.S. EPA, 1991.

U.S. EPA. *Stationary Sources Control Techniques Document for Fine Particulate Matter*, 1998.

U.S. EPA. *Control of Gaseous Emissions*, Student Manual, Jan 2000.

U.S. EPA. *EPA Air Pollution Control Cost Manual*, EPA-452/B-02-001. Research Triangle Park, NC: U.S. EPA, 2002

U.S. EPA. Particulate Matter Emissions, 2017. https://cfpub.epa.gov/roe/indicator_pdf.cfm?i=19 (accessed 23 January 2017).

Further Reading

Calvert, S., J. Goldschmidt, D. Leith, and D. Metha. *Wet Scrubber Handbook*, Vol I, EPA-R1-72-118a. Research Triangle Park, NC: U.S. EPA, 1972

Calvert, S., J. Goldshimidt, D. Leith, and N. Jhaveri. *Feasibility of Flux/Condensation Scrubbing for Fine Particle Collection*, EPA-650/2-73-036. 1973. Washington, DC: U.S. Environmental Protection Agency.

Cooper, C.D. and Alley, F.C. (2011). *Air Pollution Control: A Design Approach*, 4ee. Long Grove, IL: Waveland Press.

Corbitt, R.A. (1998). Standard Handbook Environmental Engineering, Second Edition, McGraw Hill.

Crawford, M. (1976). *Air Pollution Control Theory*. New York: McGraw-Hill, Inc.

Dupont, R.R., Ganesan, K., and Theodore, L. (2017). *Pollution Prevention: Sustainability, Industrial Ecology, and Green Engineering*, 2ee. Boca Raton, FL: CRC Press.

Ganesan, K., Theodore, L., and Dupont, R.R. (1996). *Air Toxics: Problems and Solutions*. Amsterdam, The Netherlands: OPA.

Henzel D.S., Laseke, B.A., Smith, E.D., and Swenson, D.O. *Limestone FGD Scrubbers: Users Handbook*, EPA-600/8-81-017. U.S. EPA, Washington DC, 1981.

Munson, J.S. (1968). Dry mechanical collectors. *Chemical Engineering* 75 (22): 147–151.

Neveril, R.B. *Capital and Operating Costs of Selected Air Pollution Control Systems*, EPA-450/5-80-002. U.S. EPA, Washington, DC, 1978

Nissen, W.I., Croker, L., Oden, L.L., et al. (1985). *Clean Power from Coal: The Bureau of Mines Citrate Process*. Bulletin 686. Washington, DC: U.S. Department of Interior, US DOI.

Noll, K. (1998). *Fundamentals of Air Quality Systems*. American Academy of Environmental Engineers Publication.

Pilot, M.J. and D.F. Meyer. *University of Washington Electrostatic Spry Scubber Evaluation*, EPA-600/2-76-100. Research Triangle Park, NC: U.S. EPA, April 1976. PB 252653.

Srivastava, R.K. *Controlling SO_2 Emissions: A Review of Technologies*, EPA-600/R-00-093. National Risk Management Research Lab, Research Triangle Park, NC, 2000.

Theodore, L. and Buonicore, A.J. (1976). *Industrial Air Pollution Control Equipment*. Cleveland, OH: CRC Pres.

Theodore L., and Reynolds, J. ESP bus sections failures: Design considerations, *JAPCA*, 33 (12), 1983.

U.S. EPA. *Operation and Maintenance Manual for Electrostatic Precipitators*, EPA-625/1-85-017. Research Triangle Park, NC: U.S. EPA, 1985.

U.S. EPA. *Operation and Maintenance Manual for Fabric Filters*, EPA-625/1-86-020. Research Triangle Park, NC: U.S. EPA, 1986.

U.S. EPA. *Sulfur Dioxide Control Technology Series: Flue Gas Desulfurization – Wellman-Lord Process*. Technology Transfer Summary Report, EPA-625/8-79-001. Research Triangle Park, NC: U.S. EPA, 1979.

Wark, K., Warner, C.F., and Davis, W.T. (1998). *Air Pollution*, 3ee. Upper Saddle River, NJ: Prentice-Hall.

16

Atmospheric Aerosols and Their Measurement

Christian M. Carrico

Department of Civil and Environmental Engineering, New Mexico Institute of Mining and Technology, Socorro, NM, USA

16.1 Overview of Particulate Matter in the Atmosphere

An aerosol is a colloidal suspension of particulate matter (PM) – solid and/or liquid – that is suspended in a gaseous carrier, most typically air. A cubic meter of air with a mass of approximately 1.2 kg or 1.2E9 µg at sea level may contain only a few µg of PM mass in a pristine environment and 100 µg or higher in a highly polluted urban area. Despite their small mass fraction in the air, the atmospheric impacts of aerosols upon air quality are profound and include cloud formation, meteorological interactions, visibility impairment, atmospheric chemistry effects, and human health impacts (Seinfeld and Pandis, 2016). Not all aerosol effects are necessarily detrimental. For example, transport of aeolian dust is a major pathway for delivery of mineral species and nutrients to oceans and other continents. Aerosols due to volcanic eruptions, another natural source, have long been recognized as having a climate influence. Following the eruption of Mt. Pinatubo in 1992, the Earth cooled for a couple years by about 0.5 °C due to the stratospheric injection of the precursor SO_2 and dust. Aerosols from human sources have recently been identified as an important component of the climate system (IPCC, 2013).

Aerosols have numerous and diverse anthropogenic and natural sources, and their physicochemical properties are intimately tied to the source type. Often, aerosols (as well as a number of toxic gas-phase species such as carbon monoxide) result as products of incomplete combustion, whether from nature or humans. The combustion of solid, liquid, or gaseous hydrocarbon fuels – and the impurities contained within – results in emissions of PM. Considerable effort has gone into researching the properties, effects, and control methods for combustion aerosols, and these efforts have greatly reduced emissions from many of these sources.

Examples of important global sources of aerosols that can be derived from either natural or human sources are smoke from biomass burning (Figure 16.1) and mineral dust aerosols. Aerosols, and particularly nanoparticles as part of the evolving world of nanotechnology, have a wide variety of industrial and medical applications as well. The focus of this chapter is aerosol effects on ambient air quality, and the myriad industrial and medical uses of aerosols are not discussed. The following examines selected aerosol properties and their measurements starting with a brief historical and regulatory review of US federal regulations.

16.2 History and Regulation

Concerns over industrial or combustion-derived "smoke," which can be equated to PM, date back centuries, but only became particularly acute and widespread during the beginning of the industrial era in the nineteenth century. Large population density, heavy industrialization, and in particular uncontrolled combustion of coal led to particular problems in London, England, as well as in emerging urban areas in other industrializing countries. The noted London Smog episode (among others of that era) of 1952 resulted in 12,000 excess deaths due to cardiopulmonary and other complications with effects persisting for 2 months following the approximately 1 week episode (Bell and Davis, 2001; Bell et al., 2004). Infamous episodes included those in Donora, Pennsylvania, and St Louis, Missouri, in the United States and numerous others in London, England, in the middle of the twentieth century.

The contributions of unfavorable meteorological conditions such as stagnant high pressure conditions with little vertical mixing were important to the London smog episode and many others. Los Angeles, California,

Handbook of Environmental Engineering, First Edition. Edited by Myer Kutz.

Figure 16.1 Plume from the High Park Fire in northern Colorado in June 2012. Smoke from biomass burning is an important global source of aerosol particles as well as gas species that impact air quality. The optical properties of the smoke particles on the microphysical level give its color and optical properties. Visible air pollution is almost always due to the interaction of aerosol particles with light.

emerged after World War II as a city with degrading air quality. The combination of rapid population and industrial growth, expanding suburban sprawl, highway construction, corresponding rapid increases in vehicle miles traveled, and influence of topography (surrounding mountains) and high solar input driving photochemistry, all contributed to the photochemical smog and particulate pollution problems of Los Angeles. The Southern California region served as a crucible for early air pollution research as well as pollution control efforts that were later modeled nationwide (Jacobs and Kelly, 2008).

The past 50 years has been a period of increasing regulatory efforts to limit impacts of PM in the United States and other industrialized countries. Comprehensive US federal regulatory efforts to mitigate air quality issues began in the middle of 1950s in the industrialized nations. In the United States, the first efforts included the Clean Air Act of 1963 with amendments in 1970, forming the backbone of the US regulatory structure. Amendments to the US Clean Air Act continue approximately every 7 years, and other specific rules have been developed to mitigate emissions from numerous source categories. These efforts have resulted in remarkable improvements in air quality in general and PM haze properties in particular (Malm et al., 2002).

Emissions from many point sources have been reduced in over the last 30 years according to U.S. Environmental Protection Agency (US EPA) data. Current remaining large contributors to $PM_{2.5}$ mass on a global scale include vehicular sources at 25% of $PM_{2.5}$ (Karagulian et al., 2015). A study examining the source apportionment of the number concentrations of ultrafine particles in a US urban area attributed 40% to gasoline vehicles and another 26% to on-road and nonroad diesel vehicles (Posner and Pandis, 2015). The contribution of mobile sources is joined by highly episodic natural sources such as wildland fires and windblown dust. Concurrently, increasing concerns due to human sources surround the deteriorating status of air quality in developing economies, most notably those of South and East Asia (Tiwari et al., 2015). The newly industrializing regions of the world have entered a period not unlike that of North America and Europe a century ago during their period of rapid industrialization, population growth, and urbanization, along with myriad environmental impacts.

16.3 Particle Concentration Measurements

The concentration of PM in the atmosphere is one of the most important aerosol parameters. Concentration is the best single indicator of the severity of particulate air pollution problems. Concentrations are most commonly expressed as a mass of PM per unit volume of ambient air (e.g. μg of PM/m^3 of air). Mass concentrations of aerosols, though in the same units of mass per volume, are distinct from the density of the bulk aerosol material as they characterize the mass of the PM in a volume of air it is suspended – rather than the volume of PM. In some instances, these measurements use actual volume concentrations, while in others they are normalized to a standard volume typically taken as 1 atm and a standard temperature taken as 0, 20, or 25 °C. Conversion of PM concentration from one temperature and pressure state to a second state is given in Eq. (16.1) where T and P must be in absolute scales. Since the air volume is in the denominator with mass concentrations, this is the inverse of the relationship for adjusting volumes or volumetric flow rates to a standard flow rate or another new state.

$$[\mathrm{PM}]_2 = [\mathrm{PM}]_1 \times \left(\frac{P_2}{P_1}\right)\left(\frac{T_1}{T_2}\right) \tag{16.1}$$

Other common particulate concentration metrics include total number concentration (N_{tot}, # m^{-3} or # cm^{-3}), surface area concentration (A_{tot}, $\mu m^2 m^{-3}$), or volume concentration of PM (V_{tot}, $\mu m^3 m^{-3}$). Each has different implications, for example, mass concentration is most important for regulatory purposes, cross sectional area concentration is most important for visible haze, and number concentration is most important for cloud droplet formation. For a uniform density material, the $PM_{2.5}$ mass concentration can be found from the product of the density ρ of the PM and integrated particulate volume concentration V_{tot} for all particles with $D_p < 2.5\,\mu m$.

Historical regulations in the United States addressed the total suspended particulate material (TSP) with 24-h and annual standards up until 1987. Later, it was identified that fine-mode particles were more hazardous to health. The current National Ambient Air Quality Standards (NAAQS) in the United States set regulations for PM mass concentration limits for two size classes of particles: $PM_{2.5}$ and PM_{10}. These categorizations include all particles with aerodynamic diameters (discussed below) less than 2.5 and 10 µm, respectively. Thus $PM_{2.5}$, sometimes referred to as the fine mode, is a subset of PM_{10}, and the difference of PM_{10} and $PM_{2.5}$ is termed PM_{coarse}. A time series showing the time history changes in PM concentrations in the United States as well as the corresponding primary US NAAQS is given in Figure 16.2. In most but not all US locations, PM concentrations have decreased in concert with the NAAQS that have become increasingly stringent over time. A summary of current PM standards showing their concentration limit and averaging time is in Table 16.1. The primary air quality standards address human health, whereas the secondary standards address human welfare concerns such as atmospheric visibility or atmospheric deposition.

Traditional PM mass concentration measurements rely upon laboratory gravimetric analysis of filter-collected samples. With a chosen size cut upstream, PM is collected upon a filter for a selected sampling period (e.g. 24 or 48 h) and typically with a design flow rate of 16.7 l min^{-1} or LPM (which equates to 1 m^3 h^{-1}) or greater to maximize deposited mass and increase the measurement accuracy. A high volume PM sampler may

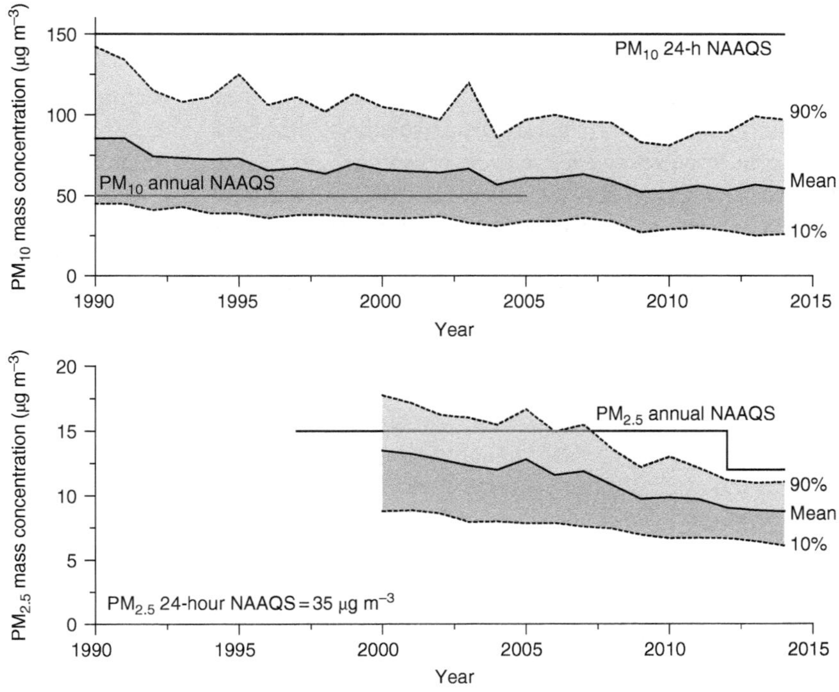

Figure 16.2 US historic average PM_{10} ($n = 193$ sites) and $PM_{2.5}$ ($n = 505$ sites) concentrations. NAAQS are shown over time as well. The PM_{10} annual standard was removed in 2005. *Source:* Data from US EPA.

Table 16.1 2016 US National Ambient Air Quality Standards for PM air quality.

Parameter	Standard type	Averaging time	Standard ($\mu g\, m^{-3}$)	How it is assessed
$PM_{2.5}$	Primary	1 yr	12.0	Annual mean, averaged over 3 yr
	Secondary	1 yr	15.0	Annual mean, averaged over 3 yr
	Primary and secondary	24 h	35	98th percentile, averaged over 3 yr
PM_{10}	Primary and secondary	24 h	150	Not to be exceeded more than once per year on an average of over 3 yr

Source: US EPA website, https://www.epa.gov/criteria-air-pollutants/naaqs-table.

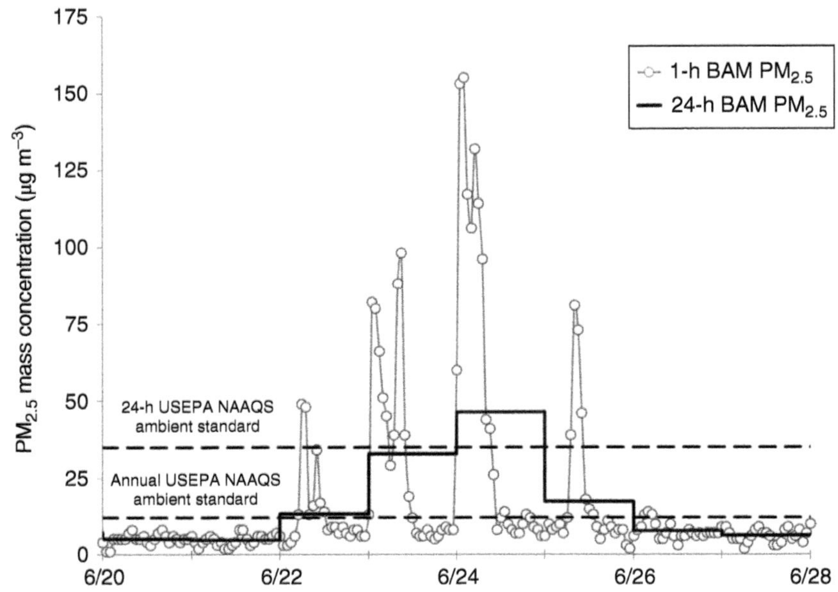

Figure 16.3 Time Series of $PM_{2.5}$ measured with a beta-attenuation monitor mass concentrations during the High Park Fire biomass burning smoke episode in Fort Collins, CO, in 2012. *Source:* Adapted from figure 10 from Carrico et al. (2016) and used with permission. Copyright 2016. Reproduced with permission of John Wiley & Sons, Inc.

sample at a flow rate of 1000 lpm. The filter is conditioned at a low RH and weighed before and after sampling at this RH to give a PM mass. If filters are shipped, they should be stored refrigerated or frozen to improve retention of volatile species, such as ammonium nitrate and some organic species. Filters ideally are weighed shortly before and after sampling to minimize the effects of adsorbed or desorbed semivolatile species. From collected filter masses, mass concentration is calculated using the actual volume of air sampled typically. Volumes may be measured with a dry gas meter or calculated from the volumetric flow rate multiplied by sampling time.

A newer continuous in situ $PM_{2.5}$ mass measurement is a filter-based instrument that employs the attenuation of beta radiation through a filter deposit to report 1-h time resolution $PM_{2.5}$ mass concentration (e.g. Met One, Inc. BAM 1020). Based on the change in beta attenuation and the volume of air sampled, a $PM_{2.5}$ mass concentration is reported. An example time series of 1- and 24-h $PM_{2.5}$ concentrations as measured by the BAM instrument during a biomass burning event, the High Park Fire in June 2012, is shown in Figure 16.3. The highly variable $PM_{2.5}$ concentrations typically peaked in the overnight hours during the times of decreased convective mixing and subsiding air masses that brought smoke generated in the mountains to the urban corridor along the Colorado Front Range (Carrico et al., 2016). Another similarly useful online technique for measuring $PM_{2.5}$ is the tapered element oscillating microbalance (e.g. Thermo Scientific Inc., model TEOM) (Val Martin et al., 2013). The TEOM using a filter element is mounted on the end of a resonator, where the resonation frequency changes as mass is deposited on the element. The TEOM technique can simultaneously make PM_{10} and $PM_{2.5}$ measurements and apply correction for volatility losses.

Particle number concentrations are most often measured with a condensation particle counter (CPC) as shown in Figure 16.4. The instrument takes the aerosol sample flowing through it and grows the particles into cloud droplets that are more easily detected due to their large size. The instrument has a saturator section where a fluid reservoir is kept at an elevated temperature saturating

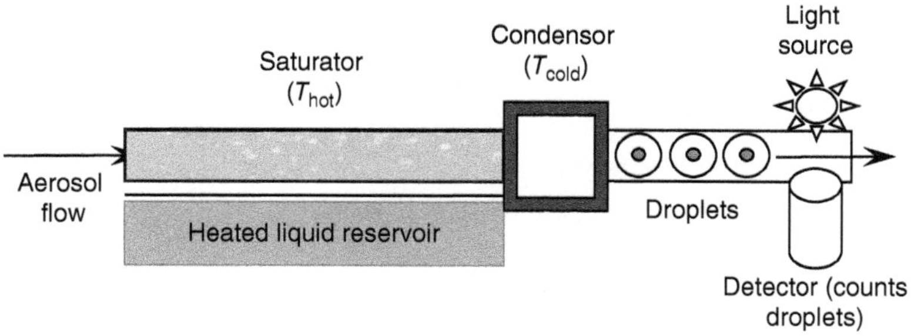

Figure 16.4 Flow diagram of a typical condensation particle counter.

the flow with the working fluid – typically an alcohol such as butanol or water is used. In the condenser section, the sample is cooled and becomes supersaturated, resulting in the particles forming cloud droplets. A diode laser and detector can then individually count the particles passing through the detection system and with flow rate calculate a number concentration. The CPC is capable of measuring total particle concentration of the entire population with D_p from a few nanometers (for ultrafine counters) up to the maximum size that can be transmitted into the instrument. It can also measure a single particle up to a concentration of 10,000 per cm^3 or more depending on instrument design.

16.4 Measuring Particle Sizing Characteristics

PM is characterized by many parameters and characteristics; one of the most important properties is particle size. Particle size affects lifetime in the atmosphere and similarly determines respiratory deposition and thus health outcomes. Particle size also drives the interaction with atmospheric radiation and thus the climate and visibility impacts. Size, usually expressed as a particle diameter (D_p) or radius of a sphere or spherical equivalent, is vital to a range of aerosol impacts. Particles are often spheres but can also take on diverse shapes from geometric (e.g. cubic for certain salts) to long fibers, to fractal agglomerations of spheres. Bio aerosols, sourced from living organisms, often take on unique shapes. Ambient aerosol particle D_p ranges over 5 orders of magnitude from a few nanometers to approximately 100 μm in diameter. Since human hairs are 50–100 μm in diameter, ambient aerosol particles are extremely small. Aside from the largest aerosol particles, individual particles are invisible to the human eye. The entire population of aerosol particles can have important effects on atmospheric visual range. Aside from the largest, it is invisible to the human eye (though a population can be quite visible as haze). Beyond this range, the lifetime of particles in the atmosphere is very short due to Brownian diffusion (very small particles) or gravitational settling (very large particles).

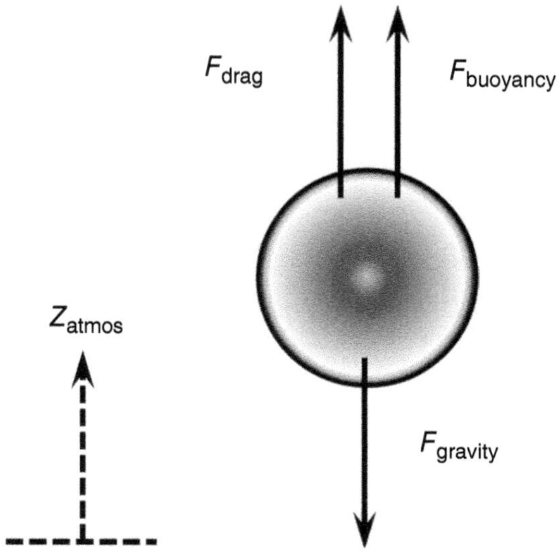

Figure 16.5 Force balance of the primary forces acting on a particle suspended in a gaseous medium in the atmosphere.

16.4.1 Terminal Settling Velocity

The terminal settling velocity is the velocity at which the drag, buoyancy, and gravitation forces on the particle are balanced (Figure 16.5), and it achieves a steady state settling velocity that is calculated using Eqs. (16.2)–(16.6) (Cooper and Alley, 2011). Here it is considered for a sphere; a shape factor would need to be introduced to account for other shapes.

$$v_t = \frac{D_p{}^2 \rho_p K_C g}{18\mu_g} \tag{16.2}$$

$$K_C = 1 + \text{Kn}\left[\alpha + \beta \exp\left(-\frac{\gamma}{\text{Kn}}\right)\right] \tag{16.3}$$

$$\text{Kn} = \frac{2\lambda_g}{D_p} \tag{16.4}$$

$$\lambda_g = \frac{\mu_g}{0.499\rho_g \overline{u_g}} \tag{16.5}$$

$$\overline{u_g} = \sqrt{\frac{8RT}{\pi M_{w,g}}} \tag{16.6}$$

where

v_t = particle terminal settling velocity ($m\,s^{-1}$)

D_p = particle geometric diameter (m)

ρ_p = density of PM ($kg\,m^{-3}$)

K_C = Cunningham slip correction factor (–)

g = gravitational acceleration ($9.8\,m\,s^{-2}$)

μ_g = gas viscosity ($kg\,m\,s^{-1}$)

Kn = Knudsen number (–)

Constants $\alpha = 1.257, \beta = 0.40, \gamma = 1.10$

λ_g = mean free path of gas molecule (m)

$\overline{u_g}$ = velocity of gas molecule ($m\,s^{-1}$)

R = universal gas constant ($8.314\,J\,K\,mol^{-1}$)

T = temperature (K)

$\overline{M_{w,g}}$ = molecular weight of gas ($28.97\,g\,mol^{-1}$ for air)

Particles with $D_p < 10\,\mu m$ generally do not settle out effectively by gravity as shown in the small terminal settling velocity for these particles (Figure 16.6).

16.4.2 Aerodynamic Diameter

One of the most important particle characteristic diameters is the aerodynamic diameter (D_a) as it determines the behavior in a flowing fluid. As discussed previously, air pollution regulations involving PM are based on aerodynamic diameters. The aerodynamic particle diameter determines the diameter of particle with unit density (i.e. $1\,g\,cm^{-3}$, equivalent to water) that has the same settling velocity as the particle of interest that may have a different density. Aerosol density, nominally taken as $1\,g\,cm^{-3}$ for water-containing aerosols, is also quite variable, even to a factor of 3 or more. Likewise, aerosol particles also differ greatly in shape including linear fibers, geometric crystalline shaped structures, spheres, and fractal soot structures of pseudospheres, among many others (Hinds, 1999). Thus aerodynamic diameter describes the equivalent unit density spherical size to which a given aerosol particle behaves in fluid flow, thus internalizing the aerodynamic differences due to diverse shape and density of particles (Eq. (16.7)) (Cooper and Alley, 2011).

$$D_a = \sqrt{\frac{18\mu v_t}{K_C \rho_w g}} \tag{16.7}$$

where

D_a = aerodynamic diameter (m)

μ = gas viscosity ($kg\,m\,s^{-1}$)

v_t = terminal settling velocity ($m\,s^{-1}$)

K_C = Cunningham slip correction factor (–)

ρ_w = density of water ($kg\,m^{-3}$)

g = gravitational acceleration ($9.8\,m\,s^{-2}$)

16.5 Ambient Aerosol Particle Size Distribution Measurements

Ambient aerosols often exhibit "modal" qualities where the distribution is composed of several modal peaks when examined as number or mass size distributions. Differently sized modes of the aerosol size distribution have distinctly different sources, sinks, lifetimes, properties, and effects on the atmosphere and humans. For example, larger "coarse" mode particles result from mechanical processes such as grinding whereas smaller, "fine" mode particles originate from chemical processes

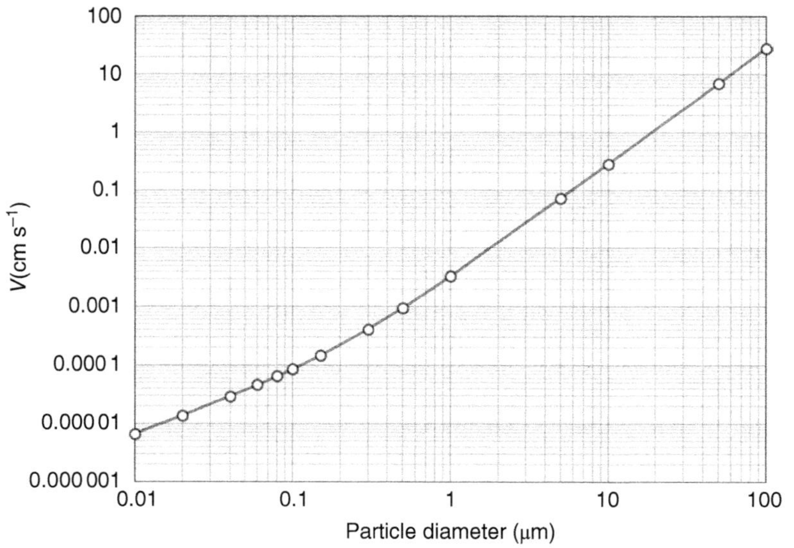

Figure 16.6 Terminal settling velocity (V, $cm\,s^{-1}$) due to gravity vs. particle diameter (μm) for unit density particles settling in air at standard temperature and pressure ($T = 298\,K$ and $P = 1\,atm$).

such as combustion. A monodisperse aerosol is a population of a single narrow size. Polydispersions are much more common in nature as aerosols typically have a broad range of sizes. Since the distribution is typically skewed with a tail in the larger sizes, log-normal distributions and statistics (geometric mean diameter D_g and geometric standard deviation σ_g) typically provide a best fit to the aerosol particle size distribution. The log-normal distribution for the number concentration of particles N is given by Eq. (16.8) (Seinfeld and Pandis, 2016), where the logarithm (either natural or base 10) of D_p is normally distributed. The major modes of the ambient aerosol distribution are discussed next.

$$\frac{dN}{d\log D_p} = \frac{N}{(2\pi)^{1/2}\log\sigma_g}\exp\left[\frac{-\left(\log D_p - \log \bar{D}_g\right)^2}{2\log^2\sigma_g}\right] \tag{16.8}$$

16.5.1 Nucleation and Aitken Modes

The Aitken mode is generally considered the population of particles with approximately $30\,\text{nm} < D_p < 100\,\text{nm}$. These are particles that are directly emitted (e.g. from combustion processes) or are formed from coagulation of the smaller nucleation mode aerosols that extend down to D_p of a few nanometers. The nucleation mode aerosols are formed from gas to particle conversion of gaseous precursor species. Together, these particles are called ultrafine particles. Interest has increased in the ultrafine portion ($D_p < 100\,\text{nm}$) of the particle population, particularly with respect to human health impacts. Though their mass is small, the ultrafine particles may dominate the number distribution. The ultrafine population has the ability to penetrate to the alveoli in the lungs and is most closely associated with the gas phase during nucleation and vapor deposition, and the reverse process of volatilization leads to the exchange of semivolatile species between the vapor and condensed phases. The chemical composition includes primary soot particles and nucleation products of gas-phase species such as volatile organic compounds and sulfuric acid. These particles are removed by coagulation into larger particles or loss due to their rapid diffusion and deposition to surfaces.

16.5.2 Accumulation Mode Particles

Particles with $0.1 < D_p < 1\,\mu\text{m}$ are often categorized as accumulation mode particles due to their persistence in the atmosphere (days to a couple of weeks). The particles are too small to settle by gravity effectively and too large to be removed by diffusion and hence accumulate in the atmosphere. These particles are the most important for effects both on health (they penetrate deep into the lungs) and on atmospheric light extinction (they are close to the wavelength of visible light and hence optically active). Removal of these particles is often by wet scavenging and precipitation. As accumulation mode particles have the longest average lifetime (days to weeks) in the atmosphere, they tend to accumulate in the atmosphere and can be transported over the continental scale.

16.5.3 Coarse Mode Particles

Coarse particles are those with $D_p > 1\,\mu\text{m}$ or sometimes considered $D_p > 2.5\,\mu\text{m}$ for regulatory purposes. Coarse particles are generated by mechanical process such as sea-salt aerosol and windblown mineral dust aerosol, which are formed from the effects of the wind on the Earth's surface. Industrial sources result from grinding and other "dust-" producing processes. Coarse particles' lifetimes are limited by their deposition velocity, which is large due to their mass (Figure 16.6). On a mass basis, coarse particles may dominate the atmospheric size distribution, particularly in background locations or near significant sources. As their lifetime is short, only the particles on the small D_p tail of the coarse distribution have significant transit distances. Nonetheless their deposition to the Earth's surface can be an important source of mass transport of material in the atmosphere to water bodies or land surfaces.

16.5.4 Measurements of Aerosol Size Distributions

Measurements of particle size typically rely upon several techniques including separation and measurement by electrical, inertial, and optical means. Electrical mobility sizing techniques (for approximate range of $3\,\text{nm} < D_p < 1\,\mu\text{m}$) rely upon a charging mechanism such as a radiation source or corona discharge, a high voltage separation means, and a counting technique such as an aerosol electrometer or a condensation particle counter as described earlier (Figure 16.7). A typical electrical technique is a differential or scanning mobility particle sizer (SMPS) that uses a concentric arrangement of a charged column and an exterior grounded cylindrical electrode (Figure 16.7). The electrodes are separated by an annular space through which the aerosol and a clean sheath flow traverse vertically from the top to the bottom of the column. The charged aerosol is size selected as a monodispersion of particles of equal mobility particles based on the flow rate and the voltage on the column. An important particle characteristic is the electrical mobility of the particles, which is a function of D_p, carrier gas properties including volumetric flow rate, column voltage, and differential mobility analyzer (DMA) column dimensions. The mobility equivalent diameter of the monodispersion results from the balance between the drag force and the electrical force on the charged particle.

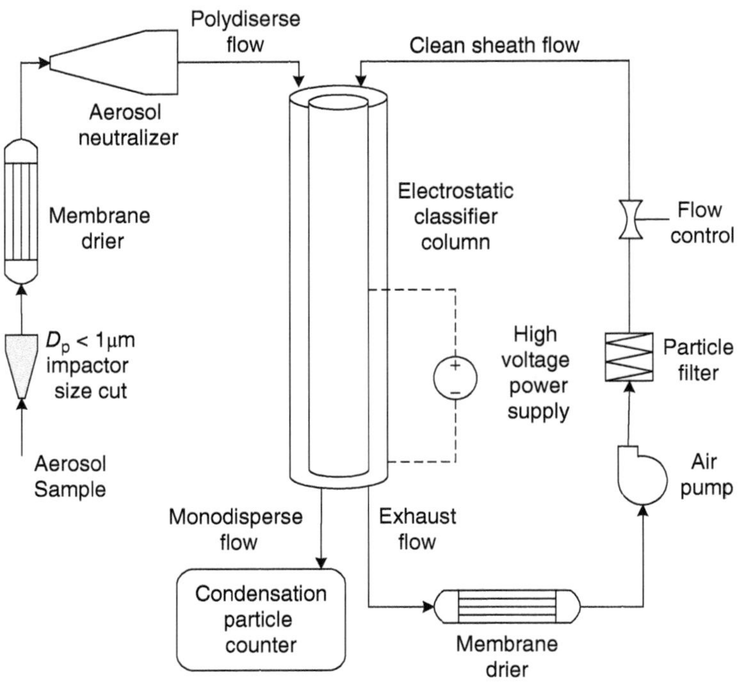

Figure 16.7 Typical differential mobility sizing spectrometer configuration with an electrostatic classifier and condensation particle counter. The filtered sheath air is recirculated, and the polydisperse and monodisperse flows are balanced.

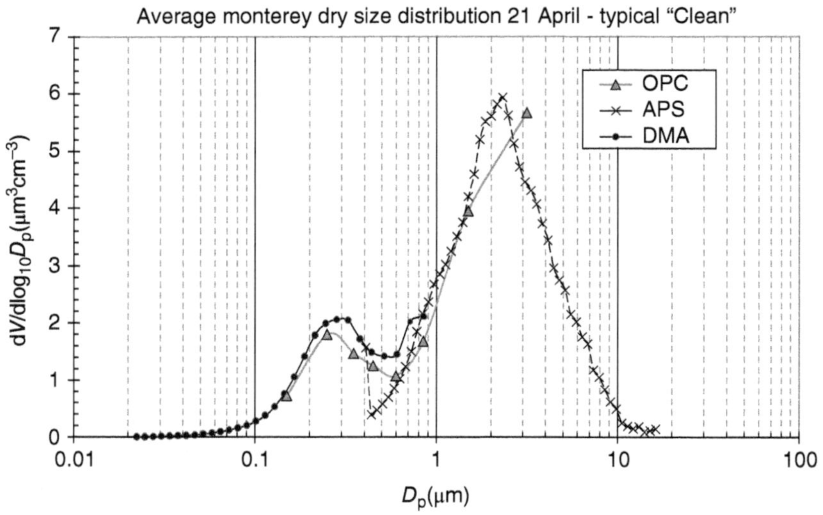

Figure 16.8 Combined size distribution of a marine aerosol measured near Monterey, CA. The measurements were made with a differential mobility particle sizer (TSI Model 3071 and Model 3010), optical particle counter (Particle Measuring Systems LASAIR 1002), and an aerodynamic particle sizer (TSI, Inc., Model 3321). The accumulation and coarse modes are clearly evident in this aerosol volume distribution, while the small bump for $D_p < 0.1\ \mu m$ is the Aitken mode. *Source:* Carrico (unpublished data, 2003). Reproduced with permission of John Wiley & Sons.

Optical techniques such as an optical spectrometer are able to measure accumulation mode particles, since they interact most strongly with visible light. Particles are counted and sized according to light scattering at a fixed angle (McMeeking et al., 2005). An aerodynamic separation technique is a time-of-flight instrument such as an aerodynamic particle sizer ($0.8\ \mu m < D_p < 20\ \mu m$), which relies upon the inertial separation of particles and timing of their flight through a fixed path length to measure their size (McMeeking et al., 2005). As the three methods rely on different techniques (electrical mobility, optical light scattering, and aerodynamic time of flight), alignment and examination of the overlap region yields additional aerosol properties such as retrieved density and refractive index (Hand and Kreidenweis, 2002). An example of the combined size distribution measured with these three methods that overlap in their sizing ranges is shown in Figure 16.8. The plot shows the volume distribution of the marine aerosol near coastal Monterey, CA. During background conditions, it is dominated by coarse mode sea-salt particles that constitute most of the volume (and thus mass) ranging approximately $0.7\ \mu m < D_p < 10\ \mu m$.

The particle size distribution can be plotted as a function of time and shown on an intensity plot (Figure 16.9) (Carrico et al., 2016). The plot shows a laboratory combustion experiment measuring the emissions from combustion of South Carolina sawgrass during the Fire Lab at Missoula Experiment IV. A fast mobility particle sizer (TSI Inc.

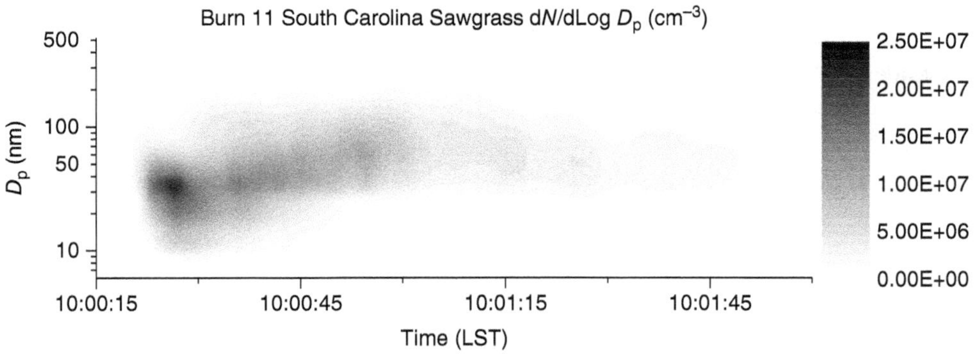

Figure 16.9 Particle size distribution evolution during the combustion of South Carolina sawgrass during a laboratory experiment. *Source:* Carrico (unpublished data 2012), following protocols of Carrico et al. (2016). Reproduced with permission of John Wiley & Sons.

Model 3091) is an electrical mobility-based sizing instrument similar to the SMPS described earlier and is capable of measuring particles with $D_p < 560$ nm at 1-s time resolution. The plot shows the number concentration $dN/d\,\mathrm{Log}D_p$ in cm^{-3} as contours as a function of particle diameter D_p on the *y*-axis and time on the *x*-axis. The combustion process starts with a strongly flaming phase producing numerous particles with $10 < D_p < 50$ nm and then decays into a smoldering phase producing lower concentrations of larger particles. Combustion of solid biomass fuels produces large concentrations of particles in tens to a few hundred nanometers (Levin et al., 2010).

16.6 Aerosol Measurements: Sampling Concerns

Aerosol sampling details are covered here briefly in general terms as it applies to many different measurement techniques discussed herein. Many sampling strategies will use an upper limit size cut at the sampling point to limit the particles measured (e.g. a device such as a cyclone or impactor that limits the upper size of particles that pass through). The size-selective inlet prevents very coarse particles such as precipitation droplets and insects as well as very large particles to enter the sampling system. This helps standardize sampling intercomparisons as the influence of a few giant particles, which have very short atmospheric lifetimes, can dominate the PM mass. The goal is an aerosol size cut as sharp as possible; size cuts are typically defined at the aerodynamic diameter with 50% passing efficiency. These are usually designed to pass through PM with diameters $D_p < 10$, 2.5, or 1 μm, denoted PM_{10}, $PM_{2.5}$, and PM_1, respectively. Currently, the size cuts typical for regulatory sampling are PM_{10} and $PM_{2.5}$. Some measurement techniques also employ sample drying to eliminate the influence of RH changes on measured aerosol properties that can also confound measurement intercomparability. The effects of ambient RH on aerosol water uptake will be discussed in more detail later.

Beyond this, the goal of the sampling strategy is to minimize the loss of sampled particles in the plumbing of the sampling train. Minimizing sample residence time is particularly worthwhile for retention of the highly diffusive ultrafine particles ($D_p < 100$ nm). However, this has to be balanced with reasonable velocities within the sampling plumbing to avoid turbulence and prevent inertial removal of coarse particles. The Reynolds number (Re) characterizes the flow regime of the gaseous sample passing through a (typically cylindrical) pipe or tube of diameter D_t (Eq. (16.11) with other parameters defined previously).

$$\mathrm{Re} = \frac{\rho_g D_t \bar{u}_g}{\mu_g} \tag{16.9}$$

Appropriate plumbing materials such as electrically conductive tubing (e.g. stainless steel, copper, carbon-impregnated silicone rubber) to minimize static charge buildup and the collection of particles on the charged surfaces are used to minimize losses. Minimizing constrictions and bends in the sampling pathway prevents inertial losses. Isokinetic sampling is desirable as it relies upon maintaining a uniform velocity profile from the atmosphere to the measurement instrument, also minimizing inertial losses of particles. Deviations from isokinetic sampling leads to over- or undersampling of particles depending on flow rate changes and particle size. Readers are referred to more comprehensive treatments of these issue in several excellent references (Hinds, 1999; Vincent, 2007).

16.7 Aerosol Formation and Aging Processes

Aerosols can be primary (directly emitted) or secondary (formed in the atmosphere from precursor gases) (Seinfeld and Pandis, 2016). Aerosol particle growth

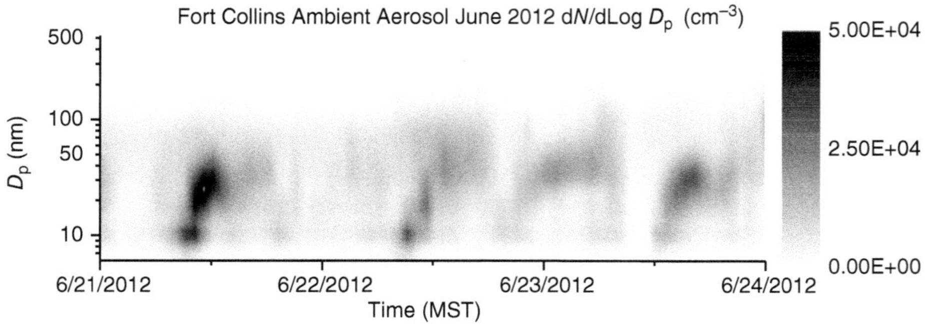

Figure 16.10 Ambient sampled aerosol size distribution in Fort Collins, CO, during June 2012. The unpolluted conditions permitted new particle growth events that occurred on successive days around midday. *Source:* Carrico (unpublished data 2012), following protocols of Carrico et al. (2016). Reproduced with permission of John Wiley & Sons.

events occur in a clean atmosphere where gas-phase species can condense to form nanometer size particles, which then can grow with time. Nucleation is the process by which new particles are formed in the atmosphere. Examples of aerosol particle growth events are shown in Figure 16.10 for relatively clean summertime conditions in Fort Collins, CO, on 3 successive days. Around midday particles with $D_p \sim 10$ nm grow to several tens of nanometers over the ensuing afternoon hours (Carrico et al., 2016).

The properties, both physical and chemical, of aerosols evolve from emission to aging in the atmosphere. Particles will coagulate to form larger particles and grow by condensation over time. Biomass smoke particle size distributions are shown in Figure 16.11, showing the relative size distributions (normalized to the integrated total number concentration) for three smoke samples from the combustion of biomass dominated by Ponderosa Pine (Carrico et al., 2016). The smoke is aged from freshly emitted (seconds old) to long range transported (days old) as shown in Figure 16.11. The transformation of the size distribution from number to mass "shifts" the distribution to the right (to larger sizes) as the mass distribution is weighted by D_p^3. The far fewer but more massive large D_p particles are much more important to the size distribution on a mass basis, whereas the typically numerous ultrafine particles are vital to the number distribution.

The plot also demonstrates the importance of aging with particle size distributions. Freshly emitted smoke aerosols were dominated by an ultrafine mode with $D_p \sim 60$ nm with a shoulder $D_p \sim 100$ nm (Carrico et al., 2016). On the mass size distribution plot in Figure 16.11b, the shoulder becomes much more important and suggests a second mode at sizes beyond the measurement range of the instrument. The smoke sampled near the High Park Fire, several hours to a fraction of a day old, shows a distribution shifted to a larger size, $D_p \sim 90$ nm. Aged smoke transported over several days from the Pacific Northwest fires to Colorado is dominated by yet larger particles with $D_p = 109$ nm. Due to the effects of dilution, coagulation, and removal mechanisms, the number concentrations are much different with total number concentrations of $N = 1\text{E}6\ cm^{-3}$ in the laboratory source tests, $N = 9\text{E}4\ cm^{-3}$ in the ambient-sampled young smoke, and $N = 4\text{E}3\ cm^{-3}$ in the aged smoke transported over long distance. Gaussian dispersion modeling is often used to model the dispersion of individual plumes as they are transported downwind of a source and diluted (Sternberg, 2015). The aging of particle populations involves coagulation, oxidation, and condensation, all of which impact the physical and chemical properties of the aerosol (generally moving the size distribution to the right, to larger sizes, and decreasing number concentrations). The relative sizing, concentration, and vertical profiles of these aerosol properties dictate how these aerosols impact human health, visibility, and climate.

16.8 Aerosol Optical Properties: Impacts on Visibility and Climate

Atmospheric "haze" or particulate "smog" (smoke + fog) is shown in the example picture near El Paso, Texas, USA (Figure 16.12). It is caused by high concentrations of particulate material or less commonly by select gas-phase species such as NO_2. The haziness of the atmosphere in optical terms is quantified by the light extinction coefficient (σ_{ext}, measured in units of inverse length such as m^{-1}). The extinction coefficient is the fraction of radiation extinguished per unit distance traveled; σ_{ext} is composed of the contributions of light scattering and light absorption by gases and particles (Eq. (16.10)).

$$\sigma_{ext} = \sigma_{sp} + \sigma_{sg} + \sigma_{ap} + \sigma_{ag} \tag{16.10}$$

where

σ_{sp} = light scattering coefficient for particles
σ_{sg} = light scattering coefficient for gases

Figure 16.11 Normalized particle number size distribution showing the transformation of a number size distribution ($dN/d\,\mathrm{Log}\,D_p$) in panel (a) to a mass size distribution ($dM/d\,\mathrm{Log}\,D_p$) in panel (b). Aging of aerosols typically shifts the size distribution to larger sizes. *Source:* Carrico (unpublished data 2012), following protocols of Carrico et al. (2016). Reproduced with permission of John Wiley & Sons.

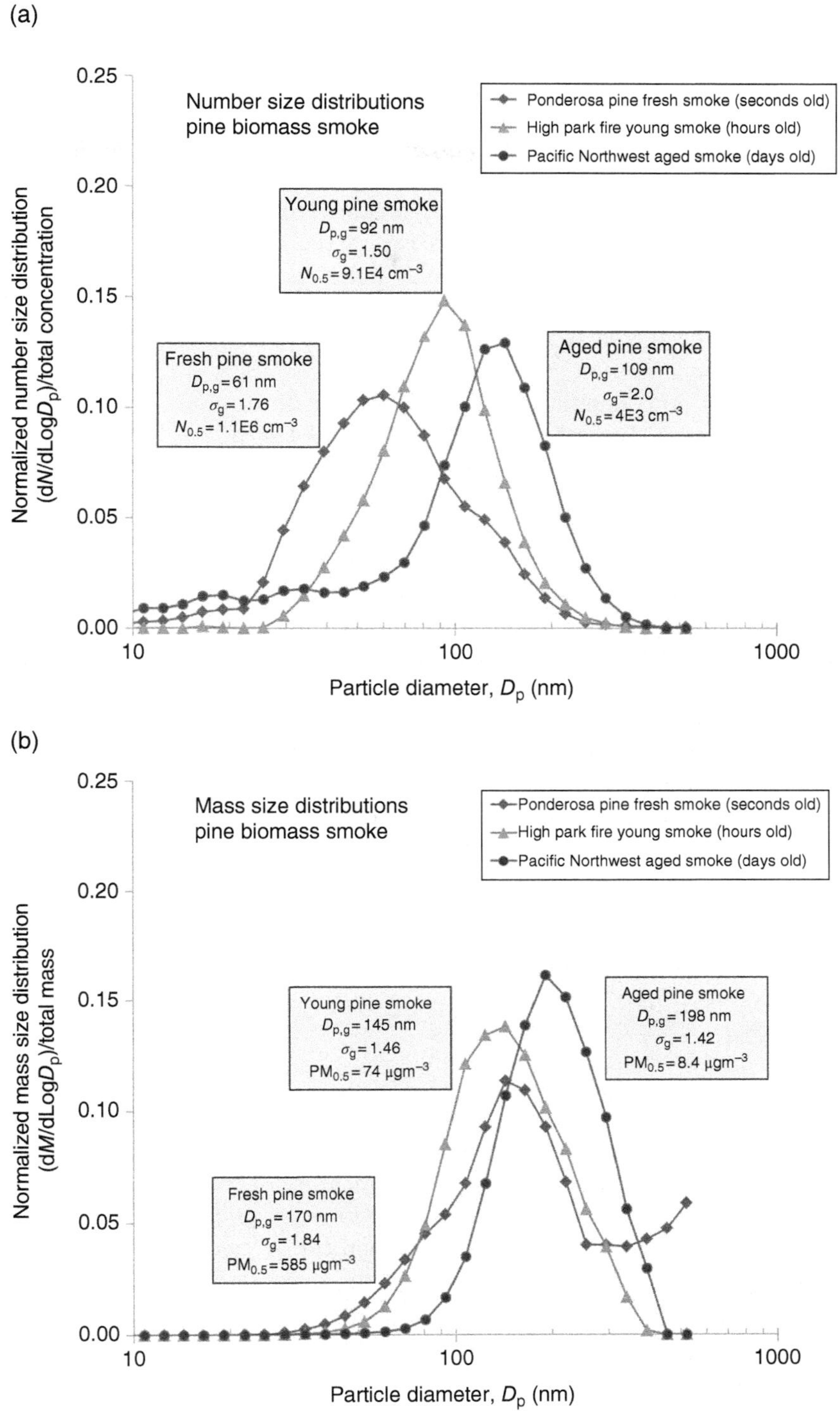

Figure 16.12 Photos of degraded visibility conditions due to anthropogenic emissions looking south near El Paso, TX, in November 2015. Despite strong progress in air quality improvements in the United States, local air quality degradation due to both anthropogenic and natural sources continues to emerge regionally in the United States.

σ_{ap} = light absorption coefficient for particles
σ_{ag} = light absorption coefficient for gases

Light scattering by particles is often dominant in the atmosphere, though a black aerosol like soot causes elevated σ_{ap} (Bond et al., 2013). Soot aerosol, black carbon, light absorbing carbon, and elemental carbon (EC) are all operationally based terms for strongly light absorbing aerosol characteristic of unburnt fuel in combustion emissions. Approximate σ_{sp} and σ_{ap} can be found from the mass concentration of $PM_{2.5}$ and black carbon (BC), respectively, and assuming mass scattering and absorption efficiencies (E_{sp} and E_{ap}), for example, ~3–4 $m^2\ g^{-1}$ and ~5–10 $m^2\ g^{-1}$ at 550 nm, respectively (Hand and Malm, 2007). Mass scattering (absorbing) efficiency is found from the ratio of σ_{sp} (σ_{ap}) to the PM (BC) mass concentrations. The efficiencies are dependent upon the particle size distribution and composition and thus have a range of about a factor of 2. For the gas phase, the most important atmospheric light absorbing species is typically NO_2, while light scattering by air molecules is termed Rayleigh scattering.

The aerosol contribution may be characterized by the aerosol single scattering albedo, the ratio of light scattering to light extinction by particles, which takes on values between 0 and 1 (Eq. (16.11)). A strongly absorbing aerosol (causes warming aloft and cooling of the surface of the Earth) will have $\omega = 0.4$ or less, while a strongly scattering aerosol (net cooling effect) will have ω approaching 1.

$$\omega = \frac{\sigma_{sp}}{\sigma_{sp} + \sigma_{ap}} \tag{16.11}$$

The wavelength dependence of light scattering is quantified by the Ångström exponent (Å), which can be approximated for discrete wavelength measurements (Eq. (16.12)). A value Å approaching zero for σ_{sp} would result from an extremely coarse mode aerosol. Å can likewise be calculated for the other light extinction parameters. For Rayleigh scattering σ_{sg}, scattering by air molecules, Å takes on a value of 4, explaining the blue color of the sky.

$$\text{Å} = -\frac{\log\left(\frac{\sigma_{sp1}}{\sigma_{sp2}}\right)}{\log\left(\frac{\lambda_1}{\lambda_2}\right)} \tag{16.12}$$

The magnitude of the light extinction coefficient controls the attenuation of light passing through the atmosphere as given by the Beer–Lambert Law (Eq. (16.13)). The light intensity I of a direct beam is a function of incident intensity I_o and decreases with a first-order dependence on the σ_{ext} and the path length L. This results in an exponential decrease in the intensity of incident light as a function of distance from the source (Figure 16.13).

$$I = I_o \times \exp(-\sigma_{ext} \times L) \tag{16.13}$$

Integrating σ_{ext} over a vertical path length in the atmosphere gives the unit less aerosol optical depth (δ), which is a measure of the column aerosol burden in optical terms (Eq. (16.14)). A number of techniques are capable of measuring δ including handheld sun photometers, satellite retrievals, and instruments deployed in the AERONET (Holben et al., 2001). Such instruments measure the wavelength-dependent intensity of atmospheric radiation for a given atmospheric and angular geometry scenario.

$$\delta = \int_{z=0}^{z=\text{top}} \sigma_{ext}\, dz \tag{16.14}$$

The reciprocal of the extinction coefficient gives an approximate measure of the visual range in the atmosphere, L_v, as characterized by the Koschmeider relationship (Eq. (16.6)). The constant of 3.9 in the numerator is approximate and is a function of the threshold of an individual's perception of contrast between a black object and white background. Viewing a scene, reaching the visual limit results in an object indistinguishable from its background. The successively further features in the view in Yosemite National

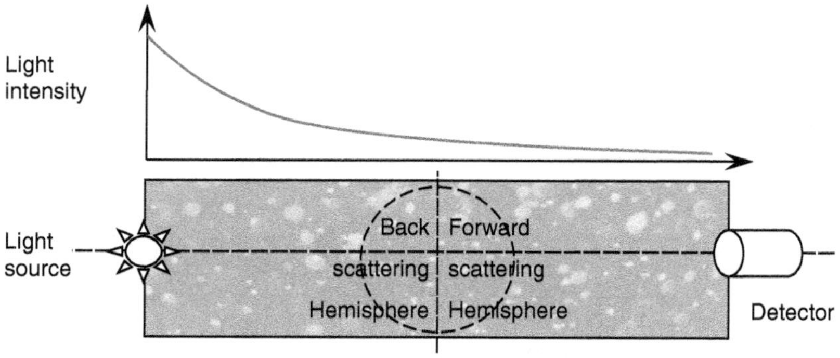

Figure 16.13 Geometry for light extinction according to the Beer–Lambert Law. The intensity of a direct beam of radiation declines exponentially as a function of the extinction coefficient and the distance traveled through the aerosol layer. Light can be scattered in the forward or backward direction, but both reduce the intensity of the incident direct beam.

Figure 16.14 Photos of visibility conditions viewing Half Dome from Turtleback Dome at Yosemite National Park in summer 2002. Visibility degradation on the left side of the photo is due to biomass burning smoke aerosols transported over long range (McMeeking et al., 2006). The distance between the two locations is 15 km and thus the visual range is approximately 15 km due to Half Dome being almost completely obscured on the left side. *Source:* Reproduced with permission of Elsevier.

Park in Figure 16.14 become harder to distinguish from the surroundings with Half Dome nearly at the visual range on the left side of the picture. The image shows the effects of biomass smoke transported over hundreds of kilometers from wildfires in Oregon in summer 2002 (McMeeking et al., 2006).

$$L_v = \frac{3.9}{\sigma_{ext}} \tag{16.15}$$

Light scattering by an aerosol may be calculated based on the contributions from the particle population. The cross-sectional area of each size of particles multiplied by the Mie light scattering efficiency and then summed over all sizes gives a calculated light scattering coefficient (Eq. (16.16)). A similar integral can be used for a continuous size distribution function.

$$\sigma_{sp} = \sum_{i=1}^{\infty} N_i K_{sp,i} \left(\frac{\pi}{4} D_i^2 \right) \tag{16.16}$$

where

N_i = number concentration of size i (m^{-3})

$K_{sp,i}$ = Mie scattering efficiency of size i (–)

D_i = midpoint diameter of bin i (m)

16.9 Measurements of Aerosol Optical Properties

Light scattering-based devices have often been used as a proxy measurement of $PM_{2.5}$ concentration and are calibrated with an aerosol of specific mass scattering efficiency. The direct measurement of aerosol light scattering has traditionally been performed by nephelometry (Figure 16.15). The instrument uses a light source (broadband, laser, or LED) and detector (photomultiplier tube) that directly measures the intensity of light scattered by aerosol passing through the instrument chamber. Due to the small fraction of light scattered over the small path length in the instrument, a sensitive detector is needed, and stray light into the instrument must be eliminated. The instrument integrates over the majority of the scattering phase function (typically ~7 to ~170°), missing light scattered in the direct forward and direct backward directions due to constraints on the instrument scattering chamber length. Variants of the instrument include a total backscatter instrument and wavelength-dependent nephelometers that use multiple wavelength-specific detectors or multiple discrete wavelength light sources.

In terms of effects on atmospheric radiation balance, the fraction of light scattered upward or the upscatter fraction β is important. Since the upscatter fraction

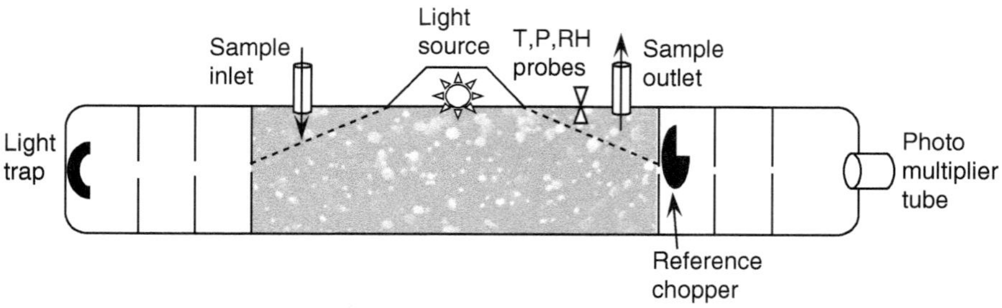

Figure 16.15 Flow diagram of a typical integrating nephelometer designed to measure in situ light scattering by particles in the aerosol sample flowing through the instrument.

depends on the particular scene geometry, a measurable quantity related to this is the backscatter fraction, b (see the scattering geometry in Figure 16.13). The backscatter fraction is the quantity of radiation scattered in the hemisphere from 90 to 170° σ_{bsp} divided by the full phase function integration of light σ_{sp} from 0 to 170° (Eq. (16.17)). The backscatter nephelometer periodically uses a backscatter shutter to block the 7–90° scattering directions, thus measuring only backward light scattering directly (Anderson et al., 1996; Heintzenberg et al., 2006).

$$b = \frac{\sigma_{bsp}}{\sigma_{sp}} \tag{16.17}$$

Other optical instruments include the integrating sphere instrument which mitigates the issues with truncation errors associated with nephelometers. A variety of light absorption measurement instruments have been developed for ambient air quality studies (Moosmüller et al., 2009). These include aethalometers, particle soot absorption photometers, photoacoustic extinctiometers, and other related instruments (Arnott et al., 1999). Newer instruments combine light extinction, scattering, and absorption measurements into a single instrument such as the photoacoustic spectrometer (Nakayama et al., 2015) or cavity ringdown and cavity-assisted phase shift methods (Cross et al., 2010; Welton et al., 2000). Open, long path-length techniques such as a transmissometer use a transmitter and detector in the open atmosphere to measure attenuation over the measured path length. Long-path transmissometers measure light extinction between a source and detector separated by a kilometer or more in the atmosphere. Light detection and ranging (LIDAR) methods emit a beam and examine the backscattered signal to determine the spatial profile of light scattering along the beam path length (Schmid et al., 2000).

16.10 Aerosol Chemical Composition

The composition of aerosol particles is vital to the optical, cloud-forming, and health impacts of these aerosols. Aerosol chemical composition is analyzed using a variety of analytical techniques, both on filter collected samples and in situ. The IMPROVE monitoring network includes filter-based samplers for collecting $PM_{2.5}$ and PM_{10} samples (Malm and Hand, 2007). Subsequently, filters are analyzed using ion chromatography for major ions, X-ray fluorescence for mineral and metal species, and thermal optical reflectance (TOR) for elemental and organic carbon species (Malm and Hand, 2007). Many chemical analysis techniques rely upon short wavelength radiation that interacts on a molecular level (e.g. X-ray diffraction, X-ray fluorescence, and others).

A "background" site at Bosque del Apache, New Mexico, when not impacted by biomass smoke or dust events, generally shows quite low PM concentrations (Figure 16.16) with the mass concentration of $PM_{2.5} = 4\,\mu g\,m^{-3}$ and an additional coarse contribution ($2.5 < D_p < 10\,\mu m$) of $6\,\mu g\,m^{-3}$. $PM_{2.5}$ composition is dominated by particulate organic material (assumes a multiplier of OM × 1.8 to account for oxygen, hydrogen, and other elements in the organic carbon compounds), followed by ammonium sulfate and soil dust species. Soil dust is reconstructed with the IMPROVE method, summing the metal and mineral species in their common oxide forms (Malm et al., 2004).

An example of the relationship between optical, physical, and chemical properties is shown in Figure 16.17, illustrating the daily profile of aerosol-extensive (dependent on the quantity in the atmosphere) and -intensive properties (dependent on the quality of the aerosol particles in the atmosphere) for the urban core of Atlanta, Georgia, USA. Light scattering and absorption are from nephelometry- and filter-deposited absorption (Radiance Research Inc., Models M903 and PSAP, respectively), $PM_{2.5}$ from 1-h in situ measurements using the oscillation frequency of a filter-deposited mass (Rupprecht and Patashnick Co. Inc., Model TEOM) and organic carbon and EC from an NDIR technique (Rupprecht and Patashnick Co. Inc., Model 5400) (Carrico et al., 2003). All the extensive aerosol properties show a morning rush hour peak, and the peak in EC concentration is very pronounced and also quite evident from the peak in EC/OC ratio. This coincides with elevated light absorption efficiency, E_{ap}, as well as a minimum in ω near 8 a.m. local

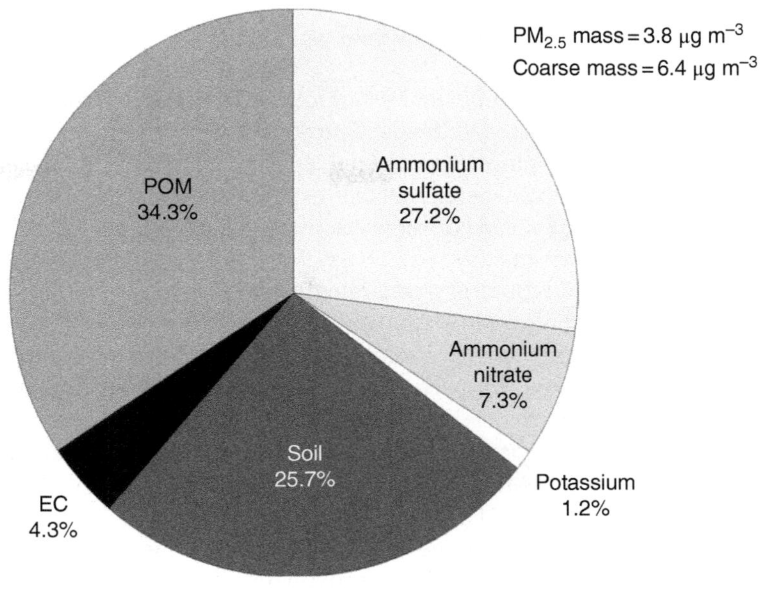

Figure 16.16 IMPROVE monitoring network $PM_{2.5}$ chemical composition from 2000 to 2014 at the Bosque del Apache site in the Rio Grande Valley in central New Mexico. IMPROVE is a collaborative association of state, tribal, and federal agencies, and international partners. US Environmental Protection Agency is the primary funding source, with contracting and research support from the National Park Service. The Air Quality Group at the University of California, Davis is the central analytical laboratory, with ion analysis provided by Research Triangle Institute, and carbon analysis provided by Desert Research Institute.

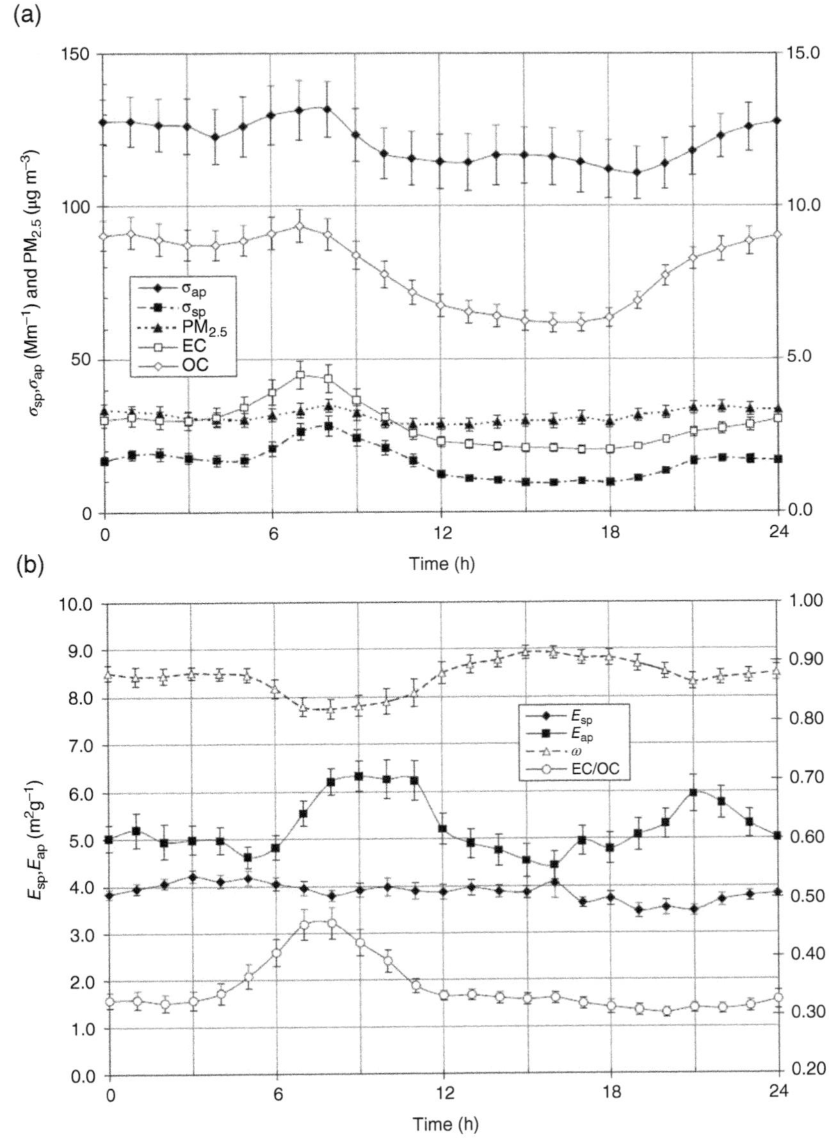

Figure 16.17 Diel variation of $PM_{2.5}$ properties measured in the urban core of Atlanta, Georgia, USA. Upper panel (a) shows extensive aerosol properties including $PM_{2.5}$ mass concentration, σ_{sp}, σ_{ap}, and organic carbon (OC) and elemental carbon (EC) concentrations. Lower panel (b) shows intensive aerosol properties including scattering and absorption mass scattering and absorption efficiencies (E_{sp}, E_{ap}), EC/OC ratio, and single scattering albedo (ω). *Source:* Reprinted from figure 3 from Carrico et al. (2003) and used with permission. Copyright 2003. Reproduced with permission of John Wiley & Sons, Inc.

time. Meteorology is important as atmospheric vertical stability also contributed to the daily profiles. The morning peak is much more pronounced due to morning atmospheric stability, whereas smaller evening peak values result from enhanced convective activity due to solar heating throughout the day.

Freshly emitted laboratory-generated biomass smoke shows a rapid evolution in sizing and optical properties over the course of a combustion experiment (Figure 16.18). Sizing properties are measured with an electrical mobility particle sizer (TSI, Inc. FMPS Model 3091) and optical properties with a photoacoustic extinctiometer (Droplet Measurement Technologies, Inc., PAX). The beginning of this burn features a strongly flaming phase producing large numbers of small diameter light absorbing particles. Approximately halfway through the experiment, the fires shift to smoldering-dominated combustion which produces larger particles that scatter light more strongly. Hence the single scattering albedo shifts from ~0.3 during the flaming phase to ~0.9 during the ending smoldering phase. A more complete description of this behavior and its drivers is given in Carrico et al. (2016).

Recently developed techniques such as aerosol mass spectrometry (AMS) have evolved to give in situ real-time chemical composition with a time resolution of an hour or less and including both carbonaceous and inorganic species (Jimenez et al., 2009). An example of such a technique is shown in Figure 16.19, giving organic carbon, total aerosol mass, and major ionic species using the aerosol chemical speciation monitor (Aerodyne Research, Inc. or ACSM). These data are from the U.S. Department of Energy aerosol and radiation monitoring network site in Brazil showing the transition from the wet to dry season. The data during the wet season show less than 5 μg m^{-3} fairly constant small background of sulfate, nitrate, chloride, and ammonium species. Nonetheless organic carbon is still the major contributor to the total nonrefractory PM mass during the clean, wet season. As the dry season commences, aerosol concentrations can reach into the hundreds of μg m^{-3} with episodic hits from biomass plumes. Though inorganic

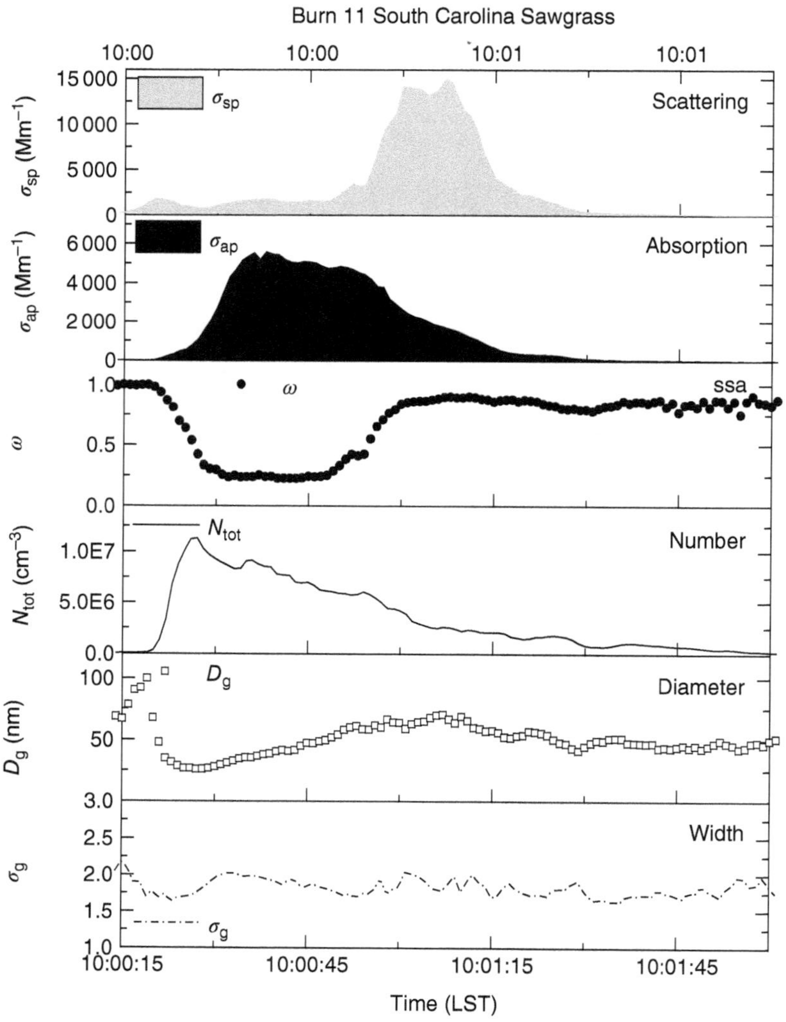

Figure 16.18 Time series of aerosol physical and optical properties for laboratory-generated smoke from combustion of South Carolina sawgrass. Time series shows (from top) particle light scattering coefficient, light absorption coefficient, aerosol single scattering albedo, total number concentration, geometric mean, and standard deviations for measurements of particles from $5 < D_p < 560$ nm. *Source:* Carrico (unpublished data 2012), following protocols of Carrico et al. (2016). Reproduced with permission of John Wiley & Sons.

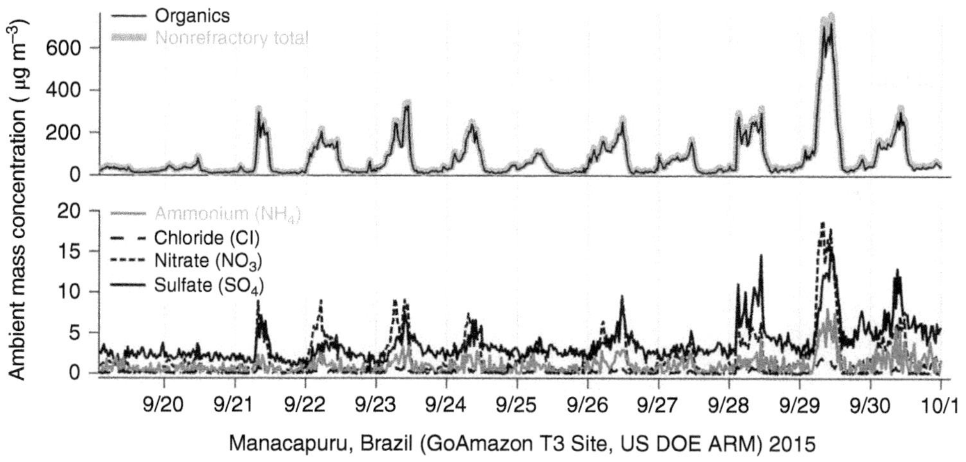

Figure 16.19 High-time resolution $PM_{2.5}$ chemical composition measured with an Aerosol Chemical Speciation Monitor (ACSM, Aerodyne, Inc.) at the T3 Monitoring site in Manacapuru, Brazil. Data were obtained from the Atmospheric Radiation Measurement (ARM) program sponsored by the US Department of Energy, Office of Science, Office of Biological and Environmental Research, and Climate and Environmental Sciences Division.

species also increase, composition becomes dominated by organic carbon from the extensive biomass burning in the region. Several similar AMS online techniques can also examine individual particle single particle composition and can determine the aerosol mixing state. Internal versus external mixing denotes how the chemical species are mixed in the aerosol. An internal mixture is a uniform mixture of all components in an individual particle, while an external mixture consists of individual single component particles mixed together proportionally to give the bulk aerosol composition.

16.11 Aerosol Hygroscopicity

Chemical composition drives numerous aerosol properties including the affinity for water. This in turn affects light attenuation as well as chemical interactions as dry particles versus aqueous droplets have much different chemistry. Aerosol particles interact with water vapor in the atmosphere and hygroscopic species uptake water with increasing RH (Carrico et al., 2005, 2008, 2010]. Hygroscopic response ranges from nearly hydrophobic (e.g. diesel soot) to strongly hygroscopic (e.g. sulfuric acid). An example of water uptake from freshly generated diesel soot is shown in Figure 16.20. The hygroscopic response is nearly flat with diameter growth factors $g(\mathrm{RH} = 90\%) < 1.02$ (the measurement uncertainty) for all cases – for low and high engine loads and denuded and undenuded samples (where the denuder removes volatile organic carbon species).

At the other end of the hygroscopicity scale, deliquescent salts exhibit a step change in particle size as the crystalline salt particle forms a saturated solution droplet at the deliquescence RH (Carrico et al., 2010). After deliquescence, with decreasing RH, the droplet follows a hysteresis loop where a supersaturated droplet only returns to a dry particle at the crystallization RH as shown in Figure 16.21 for Na_2SO_4. Such measurements are made with a hygroscopic tandem differential mobility analyzer (HTDMA), which uses two of the DMA systems shown in Figure 16.7. The first DMA selects a dry monodispersion, which is then humidified, while the second DMA measures the humidified particle size distribution.

Biomass smoke aerosols are composed of a substantial fraction of carbonaceous material. Carbonaceous aerosols range from nearly hydrophobic (e.g. fresh diesel soot emissions, Figure 16.20) to strong water uptake for specific organic species such as carboxylic acids (Prenni et al., 2003). Aerosol mixture water uptake, including biomass smoke, is complex and variable. For biomass smoke, which is a mixture of organic carbon species, EC, inorganics, and mineral species, aerosol hygroscopic growth was found ranging from very weakly hygroscopic to as strongly hygroscopic as pure salts such as ammonium sulfate (Figure 16.22) (Carrico et al., 2010). The latter cases such as rice straw and palmetto had significant salt fractions, e.g. potassium. Biomass burning source testing showed that smoke hygroscopic response can be large, dependent most strongly on the fuels combusted and hence the inorganic fraction of the aerosol (Carrico et al., 2010). In ambient aerosol with strong biomass smoke contributions, an increasing carbonaceous fraction generally led to lower hygroscopic growth (Carrico et al., 2005). Aerosol hygroscopic response during the heaviest smoke impacts was still hygroscopic, though much weaker than during unimpacted periods. Smoke measurements hygroscopicity demonstrated a dependence

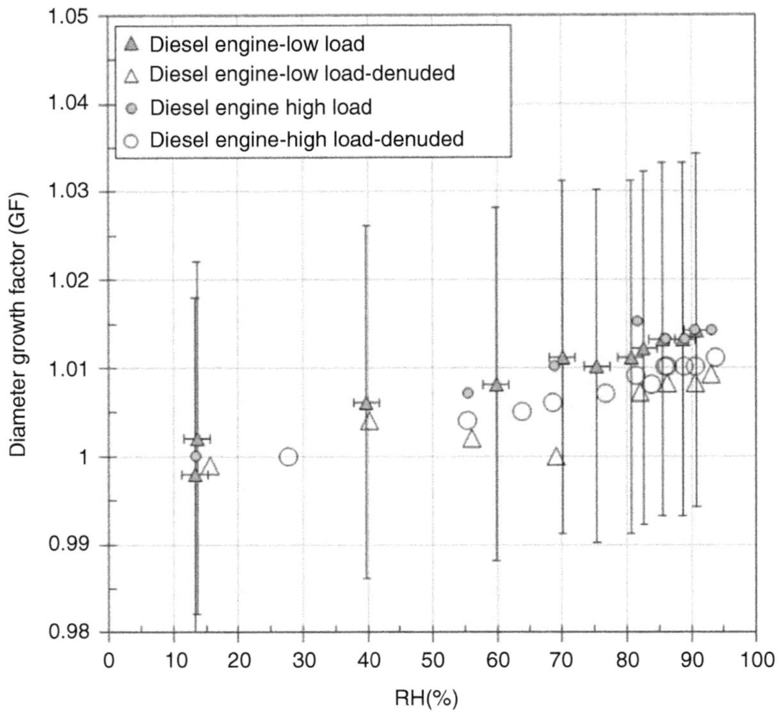

Figure 16.20 Aerosol diameter growth factor for $D_p = 100$ nm dry particles from fresh emissions from a Caterpillar, Inc., diesel engine operated at low and high loads and sampled with and without an activated carbon denuder upstream. Uncertainties in RH and growth factor measurements are shown as error bars. The slight water uptake measured for diesel engine emissions is less than the uncertainty of the measurement. *Source:* Carrico (unpublished data), following protocols in Carrico et al. (2005). Reproduced with permission of Elsevier.

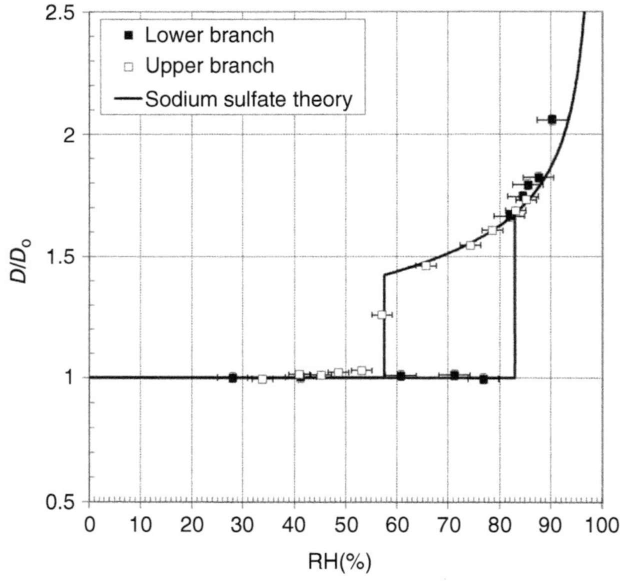

Figure 16.21 Particle water uptake (diameter D divided by dry diameter D_o) for pure Na_2SO_4 salt particles starting from a dry diameter of $D_o = 100$ nm compared to theory Tang and Munkelwitz (1994). Deliquescent salts show a step change in diameter at the deliquescence RH and exhibit hysteresis as they dehydrate and only form a solid particle at the crystallization RH. *Source:* Carrico (unpublished data 2006), following protocols of Carrico et al. (2005). Reproduced with permission of Elsevier.

on the inorganic-to-organic ratio with stronger hygroscopicity with higher inorganic ionic content (Carrico et al., 2005, 2010). It is also known that the hygroscopic response changes as the aerosol ages due to internal mixing with inorganic compounds and atmospheric oxidation (Huang et al., 2013; Martin et al., 2013).

Particle affinity for water is a key determinant of aerosol light extinction and is critically dependent on source as well as aerosol age (Carrico et al., 2003). As particles uptake water, their cross-sectional area changes as can their scattering efficiency (Eq. (16.16)). An example comparison shows the fractional increase in light scattering as a function of relative humidity, *f*(RH), for dust-dominated and polluted aerosols off of the East Asian coast during the ACE-Asia experiment (Figure 16.23). Two instruments are plotted showing reasonable agreement, a single mid-visible wavelength nephelometer (Radiance Research, Inc., Model M903) and the mid-visible channel of a three wavelength total backscatter instrument (TSI, Inc., Model 3563). Both demonstrate that at high RH > 80%, a factor of 2+ variation in light scattering response is possible, here comparing dust versus urban–industrial pollution aerosols off the coast of East Asia. From the latter instrument, also plotted on the right scale is the *f*(RH) for backscatter, which is smaller than for total scattering as the growth to larger sizes favors forward scattering. The other notable feature for the pollution-dominated aerosol is the hysteresis loop where the ambient aerosol follows a different pathway with dehumidifying conditions indicated by the upper branch of the hysteresis loop shown in Figure 16.23. As discussed previously, once hydrated, deliquescent salts form supersaturated

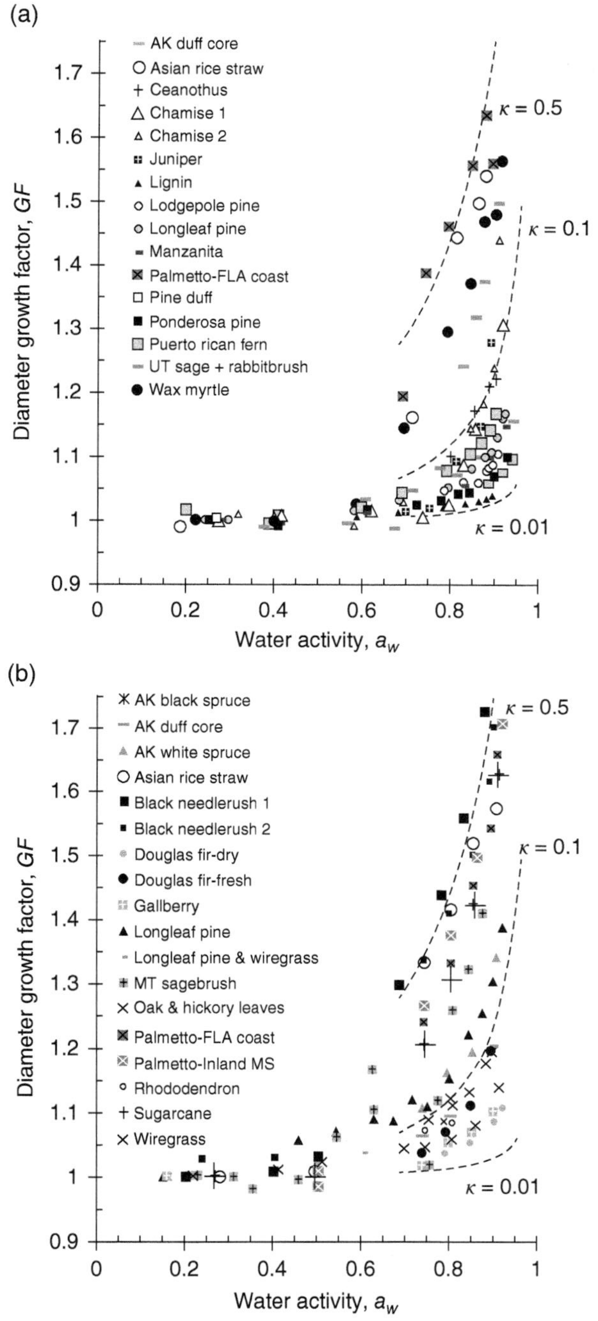

Figure 16.22 Aerosol hygroscopic diameter growth factor for $D_p = 100$ nm dry smoke particles measured for laboratory-generated biomass smoke from selected fuels. *Source:* Figure 4 from Carrico et al. (2010). https://www.atmos-chem-phys.net/10/5165/2010/. Licenced under CC BY 3.0. Reproduced with permission under the Creative Commons License 3.0 and originally published by Copernicus Publications on behalf of the European Geosciences Union.

metastable states upon decreasing RH conditions. As RH conditions approach saturation (RH = 100%), particles may activate and grow into cloud droplets as represented by the Kelvin equation.

16.12 Aerosols, Meteorology, and Climate

Aerosols are well known to interact with meteorological processes as some discussions have already focused. Pollution episodes often are contributed to by stable meteorological conditions such as the London Smog episodes discussed earlier. Cloud formation is critically dependent on the presence of aerosol particles to serve as seeds for cloud droplet growth (Petters et al., 2007). One of the most important links to weather is through the light extinction properties discussed earlier. Aerosols interact with atmospheric radiation and thus are important to energy balance of the planet and thus climate (Penner et al., 1994). Whereas the impacts on visibility are concerned more with the horizontal propagation of radiation (Figure 16.13), climate impacts are concerned with the same physics and chemistry as applied to the vertical direction in the atmosphere.

Perturbations to the radiation balance of the planet including those by aerosols, termed radiative forcing, are quantified by the radiative flux change at the top of the atmosphere (Figure 16.24) (IPCC, 2013). Attenuation of atmospheric radiation by aerosols and their cloud influences are two key means by which aerosols impact climate (Figure 16.24). Despite progress, aerosols remain the most poorly constrained and heterogeneous climate forcings as noted from the uncertainly characterization and size of the error bars.

Light extinction properties of aerosols are a major factor in aerosols' influence on the radiation balance. Light scattering aerosols backscatter some fractions of incoming solar radiation back to space depending upon the particle size, shape, fractal properties, refractive index, and scattering geometry. Likewise, absorbing aerosols prevent incoming solar radiation from reaching the surface. Absorbing aerosols have the additional effect of causing a warming aloft due to light absorption, which results in warming aloft that may also influence atmospheric vertical stability (Bond et al., 2013) (Figure 16.25). Light absorbing aerosols can enhance vertical stability leading to less dispersion of pollutants.

Aerosols are at once a climate driver and feedback, including links to hydrology and extreme ecological events such as wildfires and haboobs. As previously discussed, industrial emissions of PM in the United States have dropped in recent decades, causing a decline in most criteria pollutants. However, countervailing trends are at play, for example, related to increases in carbonaceous and dust aerosols in the southwestern United States (Hand et al., 2013; Sorooshian et al., 2013). Wildland fire and biomass smoke production is a key crossover issue encompassing climate and air quality (Rocca et al., 2014).

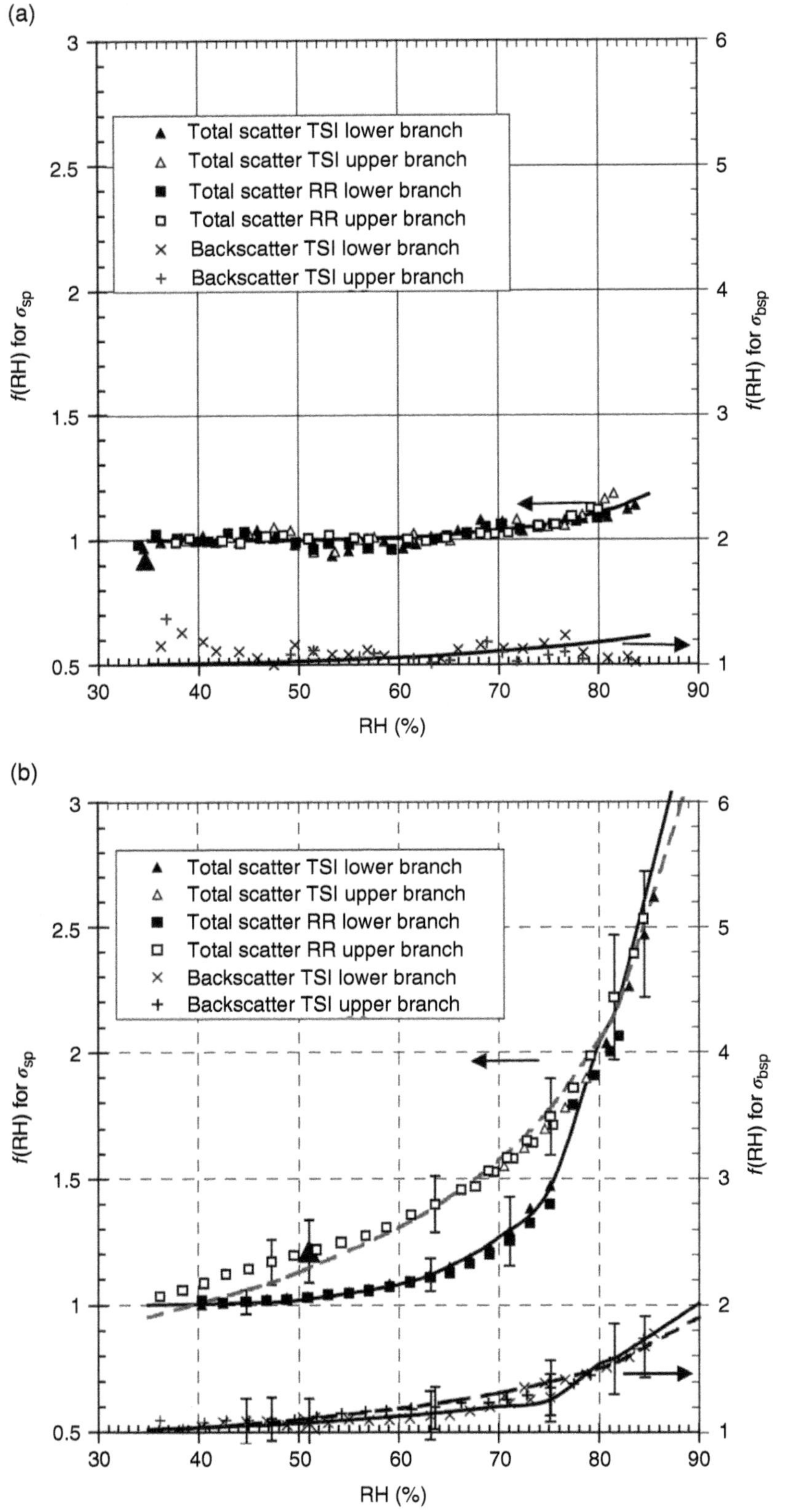

Figure 16.23 Aerosol hygroscopic response (*f(RH)*) in total scattering (σ_{sp}, left scale) and back scattering coefficients (σ_{bsp}, right scale) for a dust-dominated aerosol and a pollution-dominated aerosol during the ACE-Asia experiment off the East Asian coast (Carrico et al., 2003). Increasing and decreasing RH conditions are shown demonstrating a hysteresis loop due to metastable droplet formation. For σ_{sp} at 550 nm, two instruments are shown: TSI, Inc., and Radiance Research, Inc., integrating nephelometers. *Source:* Figure 4 from Carrico et al. (2003) reproduced with permission. Copyright 2003. Reproduced with permission of John Wiley & Sons, Inc.

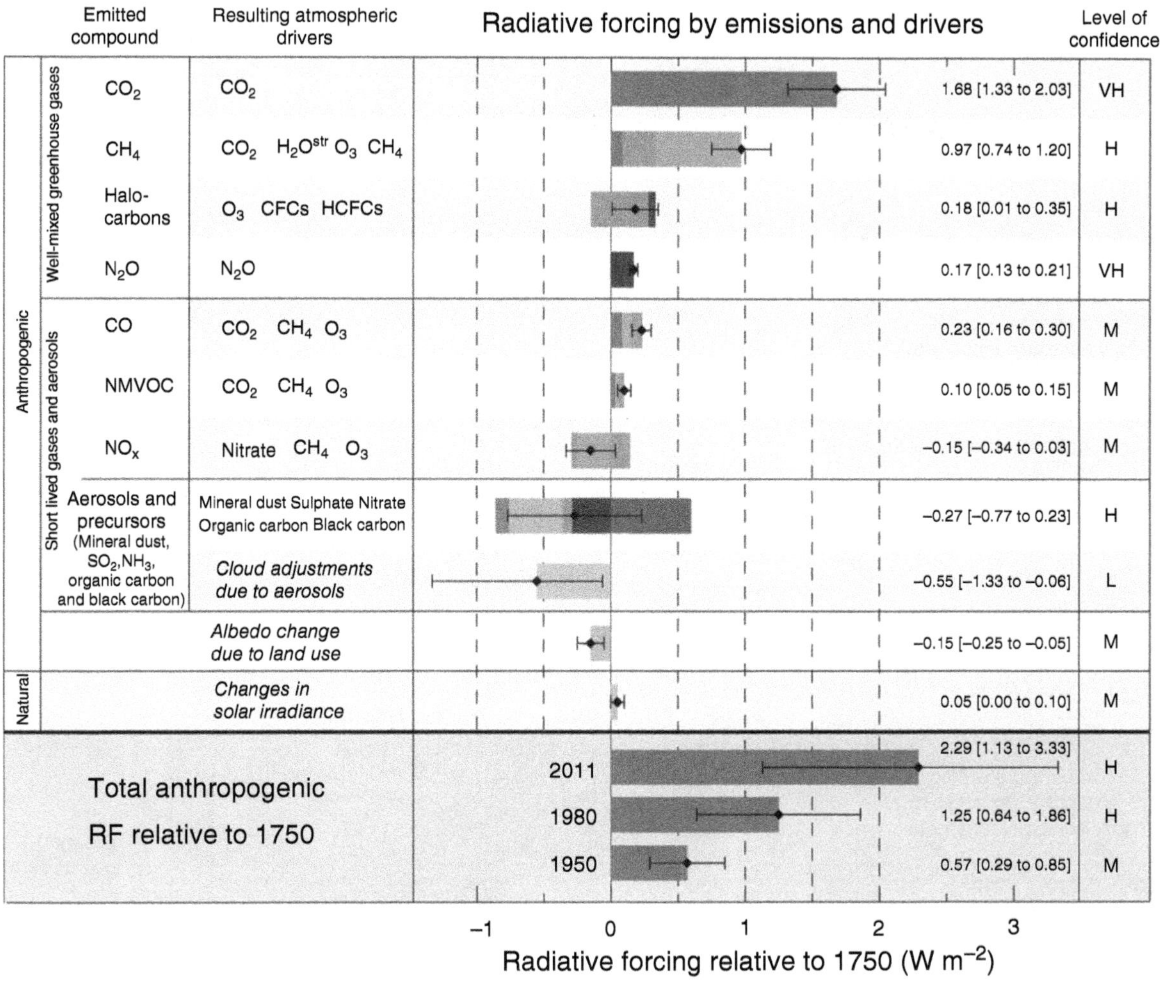

Figure 16.24 Comparison of radiative forcings due to important anthropogenic forcing agents from the IPCC Summary for Policymakers (Figure SPM.5, IPCC, 2013). Contributors for multisection bars are listed in the second column and are displayed from left to right corresponding to the bar sections from left to right. Aerosol direct (light extinction) and indirect effects both result in a net cooling influence on climate. The 2011 total anthropogenic radiative forcing was estimated approximately +2.3 W m^{-2} (a net warming).
Source: From IPCC (2013). Reproduced with permission of Intergovernmental Panel on Climate Change.

16.13 Aerosol Emission Control Technology

A number of common particle removal techniques are shared by ambient aerosol sampling and air pollution control. For example, inertial cyclones of different dimensional scales are used to remove large particles for both purposes. Particle removal mechanisms for particles can be separated into dry and wet processes and are a distinct function of particle size (Cooper and Alley, 2011). The size range of the ambient aerosol may be considered from $D_p \sim 10\,\text{nm}$ to approximately $100\,\mu\text{m}$. Particles smaller than this are removed rapidly due to diffusion to surfaces and particles larger are removed due to gravitational settling. The removal processes found in nature are similar to ones that have been employed in systems engineered to remove PM from a gas stream (Table 16.2). Removal processes rely upon settling, inertial impaction or interception, wet deposition, or diffusional losses. The processes may be enhanced by using electrical collection means such as in an electrostatic precipitator or by collecting particles in humid environments by collisions with droplets. An example of a section of an electrostatic precipitator is given in Figure 16.26. The aerosol sample is charged as it passes through oppositely charged collection plates so that particles are attracted and captured on the plates (Figure 16.26).

The efficiency of a particle collection device (η) is often defined as the mass fraction of the particles removed from the gas stream (sometimes as a function of size), whereas the penetration is the fraction that passes through the device ($Pt = 1 - \eta$). Less expensive, lower maintenance devices such as settling chambers

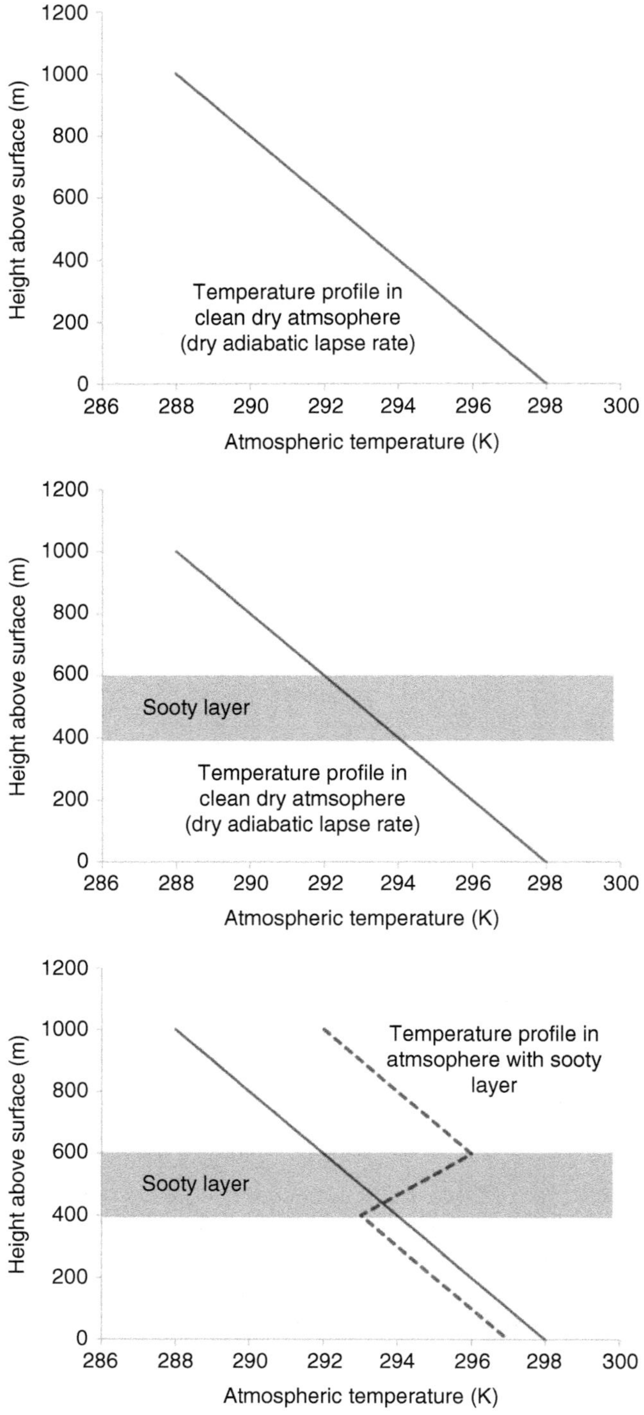

Figure 16.25 Atmospheric vertical temperature profile following the adiabatic lapse rate. With the additional of a strongly absorbing soot layer, temperature increases in the affected layer, while the absorption of radiation aloft causes surface cooling. Atmospheric vertical stability is enhanced by the warm layer.

and cyclones remove coarse particles with high efficiency but generally have low removal efficiency for fine mode particles. More expensive pollution control devices such as electrostatic precipitators and baghouse fabric filters have efficiencies that can exceed 99% down to the submicrometer range. As a result, many times a coarse control device such as a cyclone will precede a more efficient device such as an ESP or filter baghouse. A series of particle control devices have a combined *Pt* defined by the product given in Eq. (16.18). Optimizing particulate control involves assuring desired removal efficiency while minimizing pressure drops (pumping costs), capital costs, and operational logistical problems.

$$Pt_{\text{overall}} = \prod_{I=1}^{N} Pt_i \tag{16.18}$$

Table 16.2 Removal mechanisms of particle: air pollution control devices and analogs in nature.

Physical process	Natural phenomena	Removal technique
Gravitational settling	Dust deposition on surfaces	Settling chambers
Diffusional removal	Coagulation of particles	Charcoal canister adsorption
Precipitation	Wet scavenging by rain on snow events	Wet scrubber removal
Filtration	Upper airways of pulmonary system	Bag house filter
Condensational	Cloud droplet formation	Mist chamber
Inertial impaction	Wet scavenging of interstitial aerosol by precipitation droplets	Inertial separators such as cyclones and impactors
Electrostatic removal	Static electricity buildup	Electrostatic precipitator

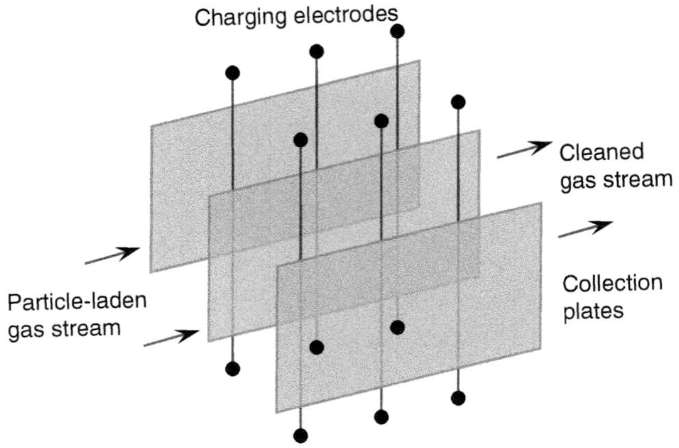

Figure 16.26 Schematic of a section of an electrostatic precipitator. Particles are given an electrical charge via charging electrodes and then captured on oppositely charged plates.

16.14 Summary and Conclusion

Aerosols have emerged in the last half a century as a fertile field of research and an active area of concern primarily related to environmental significance. Early concerns were driven by human health impacts; more recently concerns arose related to visibility in scenic areas, atmospheric chemistry effects, deposition to pristine ecosystems, and climate perturbations. The significance to climate is twofold, via direct scattering and absorption of electromagnetic radiation and indirectly by the paramount role in the formation, lifetime, and properties of clouds. Ambient aerosol particles represent a diverse population in terms of size, shape, structure, chemical composition, and optical properties. These in turn determine such properties as hygroscopicity and thermal volatility. A wide range of measurement techniques has been developed to probe and diagnose the effects of aerosols. The reduction in emissions of harmful aerosols in the United States due to regulatory efforts is one of the most important environmental successes in the last half a century. The continuing maturing of the field of aerosol research is anticipated over the next decades, and a greater understanding of aerosols' global significance will result.

References

Anderson, T.L., Covert, D.S., Marshall, S.F. et al. (1996). Performance characteristics of a high-sensitivity, three-wavelength, total scatter/backscatter nephelometer. *Journal of Atmospheric and Oceanic Technology* 13 (5): 967–986. doi: 10.1175/1520-0426(1996)013<0967:pcoahs>2.0.co;2.

Arnott, W.P., Moosmuller, H., Rogers, C.F. et al. (1999). Photoacoustic spectrometer for measuring light absorption by aerosol: instrument description. *Atmospheric Environment* 33 (17): 2845–2852.

Bell, M.L. and Davis, D.L. (2001). Reassessment of the lethal London fog of 1952: novel indicators of acute and chronic consequences of acute exposure to air pollution. *Environmental Health Perspectives* 109: 389–394. doi: 10.2307/3434786.

Bell, M.L., Davis, D.L., and Fletcher, T. (2004). A retrospective assessment of mortality from the London smog episode of 1952: the role of influenza and pollution. *Environmental Health Perspectives* 112 (1): 6–8. doi: 10.1289/ehp.6539.

Bond, T.C., Doherty, S.J., Fahey, D.W. et al. (2013). Bounding the role of black carbon in the climate system: a scientific assessment. *Journal of Geophysical Research: Atmospheres* 118 (11): 5380–5552. doi: 10.1002/jgrd.50171.

Carrico, C., Bergin, M., Xu, J. et al. (2003). Urban aerosol radiative properties: measurements during the 1999 Atlanta supersite experiment. *Journal of Geophysical Research: Atmospheres* 108 (D7). doi: 10.1029/2001JD001222.

Carrico, C., Kreidenweis, S., Malm, W. et al. (2005). Hygroscopic growth behavior of a carbon-dominated aerosol in Yosemite National Park. *Atmospheric Environment* 39 (8): 1393–1404. doi: 10.1016/j.atmosenv.2004.11.029.

Carrico, C., Kus, P., Rood, M. et al. (2003). Mixtures of pollution, dust, sea salt, and volcanic aerosol during ACE-Asia: radiative properties as a function of relative humidity. *Journal of Geophysical Research: Atmospheres* 108 (D23). doi: 10.1029/2003JD003405.

Carrico, C., Petters, M., Kreidenweis, S. et al. (2008). Aerosol hygroscopicity and cloud droplet activation of extracts of filters from biomass burning experiments. *Journal of Geophysical Research: Atmospheres* 113 (D8). doi: 10.1029/2007JD009274.

Carrico, C., Petters, M., Kreidenweis, S. et al. (2010). Water uptake and chemical composition of fresh aerosols generated in open burning of biomass. *Atmospheric Chemistry and Physics* 10 (11): 5165–5178. doi: 10.5194/acp-10-5165-2010.

Carrico, C.M., Prenni, A.J., Kreidenweis, S.M. et al. (2016). Rapidly evolving Ultrafine and fine mode biomass smoke physical properties: comparing laboratory and field results. *Journal of Geophysical Research: Atmospheres* 121 (10): 5750–5768. doi: 10.1002/2015JD024389.

Cooper, C.D. and Alley, F.C. (2011). *Air Pollution Control: A Design Approach*, 4e, 839. Long Grove, IL: Waveland Press.

Cross, E.S., Onasch, T.B., Ahern, A. et al. (2010). Soot particle studies instrument inter-comparison project overview. *Aerosol Science and Technology* 44 (8): 592–611. doi: 10.1080/02786826.2010.482113.

Hand, J. and Kreidenweis, S. (2002). A new method for retrieving particle refractive index and effective density from aerosol size distribution data. *Aerosol Science and Technology* 36 (10): 1012–1026. doi: 10.1080/02786820290092276.

Hand, J. and Malm, W. (2007). Review of aerosol mass scattering efficiencies from ground-based measurements since 1990. *Journal of Geophysical Research: Atmospheres* 112 (D18). doi: 10.1029/2007JD008484.

Hand, J., Schichtel, B., Malm, W., and Frank, A.N. (2013). Spatial and temporal trends in $PM_{2.5}$ organic and elemental carbon across the United States. *Advances in Meteorology* 2013. doi: 10.1155/2013/367674.

Heintzenberg, J., Wiedensohler, A., Tuch, T.M. et al. (2006). Intercomparisons and aerosol calibrations of 12 commercial integrating nephelometers of three manufacturers. *Journal of Atmospheric and Oceanic Technology* 23 (7): 902–914. doi: 10.1175/jtech1892.1.

Hinds, W. (1999). *Aerosol Technology Properties, Behavior, and Measurement of Airborne Particles*, 2e. New York, NY: Wiley.

Holben, B.N., Tanre, D., Smirnov, A. et al. (2001). An emerging ground-based aerosol climatology: aerosol optical depth from AERONET. *Journal of Geophysical Research: Atmospheres* 106 (D11): 12067–12097. doi: 10.1029/2001jd900014.

Huang, Y., Wu, S., Dubey, M., and French, N. (2013). Impact of aging mechanism on model simulated carbonaceous aerosols. *Atmospheric Chemistry and Physics* 13 (13): 6329–6343. doi: 10.5194/acp-13-6329-2013.

IPCC (2013). *Climate Change 2013: The Physical Science Basis. Contribution of Working Group I to the Fifth Assessment Report of the Intergovernmental Panel on Climate Change*, 1535. Cambridge, UK and New York, NY: Intergovernmental Panel on Climate Change.

Jacobs, C. and Kelly, W. (2008). *Smogtown: The Lung-Burning History of Pollution in Los Angeles, Overlook Press*. Woodstock and New York, NY: Peter Mayer Publishers, Inc.

Jimenez, J., Canagaratna, M.R., Donahue, N.M. et al. (2009). Evolution of organic aerosols in the atmosphere. *Science* 326 (5959): 1525–1529. doi: 10.1126/science.1180353.

Karagulian, F., Belis, C.A., Dora, C.F.C. et al. (2015). Contributions to cities' ambient particulate matter (PM): a systematic review of local source contributions at global level. *Atmospheric Environment* 120: 475–483. doi: 10.1016/j.atmosenv.2015.08.087.

Levin, E., McMeeking, G.R., Carrico, C.M. et al. (2010). Biomass burning smoke aerosol properties measured during Fire Laboratory at Missoula Experiments (FLAME). *Journal of Geophysical Research: Atmospheres* 115. doi: 10.1029/2009JD013601.

Malm, W. and Hand, J. (2007). An examination of the physical and optical properties of aerosols collected in the IMPROVE program. *Atmospheric Environment* 41 (16): 3407–3427. doi: 10.1016/j.atmosenv.2006.12.012.

Malm, W., Schichtel, B., Ames, R., and Gebhart, K. (2002). A 10-year spatial and temporal trend of sulfate across the United States. *Journal of Geophysical Research: Atmospheres* 107 (D22). doi: 10.1029/2002JD002107.

Malm, W., Schichtel, B., Pitchford, M. et al. (2004). Spatial and monthly trends in speciated fine particle concentration in the United States. *Journal of Geophysical Research: Atmospheres* 109 (D3). doi: 10.1029/2003JD003739.

Martin, M., Tritscher, T., Juranyi, Z. et al. (2013). Hygroscopic properties of fresh and aged wood burning particles. *Journal of Aerosol Science* 56: 15–29. doi: 10.1016/j.jaerosci.2012.08.006.

McMeeking, G., Kreidenweis, S., Carrico, C. et al. (2005). Observations of smoke-influenced aerosol during the Yosemite Aerosol Characterization Study: size distributions and chemical composition. *Journal of Geophysical Research: Atmospheres* 110 (D9). doi: 10.1029/2004JD005389.

McMeeking, G., Kreidenweis, S.M., Lunden, M. et al. (2006). Smoke-impacted regional haze in California during the summer of 2002. *Agricultural and Forest Meteorology* 137 (1–2): 25–42. doi: 10.1016/j.agrformet.2006.01.011.

Moosmüller, H., Chakrabarty, R.K., and Arnott, W.P. (2009). Aerosol light absorption and its measurement: a review. *Journal of Quantitative Spectroscopy & Radiative Transfer* 110 (11): 844–878. doi: 10.1016/j.jqsrt.2009.02.035.

Nakayama, T., Suzuki, H., Kagamitani, S. et al. (2015). Characterization of a three wavelength photoacoustic soot spectrometer (PASS-3) and a photoacoustic extinctiometer (PAX). *Journal of the Meteorological Society of Japan* 93 (2): 285–308. doi: 10.2151/jmsj.2015-016.

Penner, J.E., Charlson, R.J., Hales, J.M. et al. (1994). Quantifying and minimizing uncertainty of climate forcing by anthropogenic aerosols. *Bulletin of the American Meteorological Society* 75 (3): 375–400. doi: 10.1175/1520-0477(1994)075<0375:qamuoc>2.0.co;2.

Petters, M., Prenni, A., Kreidenweis, S., and DeMott, P. (2007). On measuring the critical diameter of cloud condensation nuclei using mobility selected aerosol. *Aerosol Science and Technology* 41 (10): 907–913. doi: 10.1080/02786820701557214.

Posner, L.N. and Pandis, S.N. (2015). Sources of ultrafine particles in the Eastern United States. *Atmospheric Environment* 111: 103–112. doi: 10.1016/j.atmosenv.2015.03.033.

Prenni, A., De Mott, P., and Kreidenweis, S. (2003). Water uptake of internally mixed particles containing ammonium sulfate and dicarboxylic acids. *Atmospheric Environment* 37 (30): 4243–4251. doi: 10.1016/S1352-2310(03)00559-4.

Rocca, M.E., Brown, P.M., MacDonald, L.H., and Carrico, C.M. (2014). Climate change impacts on fire regimes and key ecosystem services in Rocky Mountain forests. *Forest Ecology and Management* 327: 290–305. doi: 10.1016/j.foreco.2014.04.005.

Schmid, B., Livingston, J.M., Russell, P.B. et al. (2000). Clear-sky closure studies of lower tropospheric aerosol and water vapor during ACE-2 using airborne sunphotometer, airborne in-situ, space-borne, and ground-based measurements. *Tellus Series B: Chemical and Physical Meteorology* 52 (2): 568–593. doi: 10.1034/j.1600-0889.2000.00009.x.

Seinfeld, J.H. and Pandis, S.N. (2016). Atmospheric Chemistry and Physics, 3rd Edition, 1152. New York, NY: Wiley.

Sorooshian, A., Shingler, T., Harpold, A. et al. (2013). Aerosol and precipitation chemistry in the southwestern United States: spatiotemporal trends and interrelationships. *Atmospheric Chemistry and Physics* 13 (15): 7361–7379. doi: 10.5194/acp-13-7361-2013.

Sternberg, S.P.K. (2015). *Air Pollution: Engineering, Science, and Policy*. Glen Allen, VA: College Publishing, Inc.

Tang, I.P. and Munkelwitz, H.R. (1994). Water activities, densities, and refractive indices of aqueous sulfates and sodium nitrate droplets of atmospheric importance. *Journal of Geophysical Research: Atmospheres* 99 (D9): 18,801–18,808.

Tiwari, S., Hopke, P.K., Pipal, A.S. et al. (2015). Intra-urban variability of particulate matter (PM2.5 and PM10) and its relationship with optical properties of aerosols over Delhi, India. *Atmospheric Research* 166: 223–232. doi: 10.1016/j.atmosres.2015.07.007.

Vincent, J.H. (2007). *Aerosol Sampling: Science, Standards, Instrumentation and Applications*, 616. West Sussex, England: Wiley.

Welton, E., Voss, K.J., Gordon, H.R. et al. (2000). Ground-based lidar measurements of aerosols during ACE-2: instrument description, results, and comparisons with other ground-based and airborne measurements. *Tellus Series B: Chemical and Physical Meteorology* 52 (2): 636–651. doi: 10.1034/j.1600-0889.2000.00025.x.

17

Indoor Air Pollution

Shelly L. Miller

Department of Mechanical Engineering, University of Colorado, Boulder, CO, USA

17.1 Introduction

Because over 80% of our time is spent indoors, we are significantly influenced by the air we breathe in the indoor environments in which we spend our time. The indoor air environment is not an exact reflection of outdoor conditions. There are many unique characteristics and activities that happen indoors including cigarette smoking, cooking, building material emissions, and consumer product use, which are examples of determinants of indoor but not outdoor air quality. Table 17.1 summarized the typical air pollutant concentration in indoor environments. The main factors that influence the quality of the air we breathe indoors are:

- Hazardous pollutants emitted both indoors and outdoors.
- Meteorology and climate, which modifies both outdoor air pollution and the indoor environment.
- Building ventilation conditions including infiltration, natural ventilation, and mechanical ventilation.
- Pollutant decay and removal processes in indoor air and on surfaces.

Degraded indoor air quality (IAQ) adversely impacts human health. Exposure and health risk is dominated by indoor environments and is second in importance to sanitation (clean water) worldwide. Indoor air is the dominant source of radiation and toxic organics exposure in the United States through exposure to radon and consumer products, cleaning products, and material emissions. Indoor air is also a major mode of infectious disease transmission, for example, tuberculosis (TB) is spread through the coughing and sneezing of infected persons.

Pollutant emissions indoors also damage materials within buildings, including damage to cultural and historical artifacts, and damage to electronics. For example, the paintings in the Sistine Chapel were degraded over centuries due to candle use and were cleaned to reveal striking colors and detail not seen in a long time. Equipment such as electronics, semiconductors, and disk drives can be damaged by particulate and acid gas deposition.

The risk associated with poor IAQ is large compared with other risks. Table 17.2 summarizes the risk of premature death (increasing the chance of death by 1 in a million) and compares typical risks with those associated with IAQ. There are risks associated with breathing outdoor air and drinking water, but people may decide to accept different levels of risk depending on the route of exposure. Determining the level of risk that is acceptable is very complicated, and the level that is acceptable for IAQ has not been determined. We typically accept the indoor air risks because we are uninformed and because these risks are balanced by the benefits of living indoors (shelter from the weather, for example).

An important question that still has not been addressed adequately is how should society address IAQ problems, especially those that occur in homes. Many environmental regulations are applied to public goods such as air or water quality, but homes are considered a private good and so by extension is the IAQ in the home. Commercial buildings, such as large office buildings or retail stores and schools, are considered public buildings, so there are more standards and regulations for these types of indoor environments. It might be more effective to spend money to mitigate IAQ risks when they are much larger than those from outdoor air pollution, but we do not have yet a clear mechanism to do so.

In the era of climate change, energy conservation in buildings is very important, and yet the impact of energy conservation on indoor environments may be substantial if not handled correctly. Forty-one percent of the primary energy in the United States is used in the

Handbook of Environmental Engineering, First Edition. Edited by Myer Kutz.

Table 17.1 Typical indoor pollutant concentrations.

Pollutant	Concentration	Location
Carbon monoxide	2.5–28 ppm	Offices, restaurants, bars
Nitrogen dioxide	0.005–0.317 ppm	Homes with gas stoves
Respirable particles	100–700 $\mu g\,m^{-3}$	Smoking restaurants, residences
	20–60 $\mu g\,m^{-3}$	Nonsmoking restaurants, residences
Total suspended particles	39–66 $\mu g\,m^{-3}$	Homes, public buildings
Asbestos	0–2 × 10^4 fibers m^{-3}	Normal activities
	20 × 10^6 fibers m^{-3}	During maintenance
Formaldehyde	0.06–1.67 ppm	Homes with chipboard walls
Ozone	0.002–0.068 ppm	Photocopying machine room
Radon	0.005–0.94 $pCi\,l^{-1}$	House in Boston
	25–34 $pCi\,l^{-1}$	House in Florida
Radon daughters	0.003–0.013 WL[a]	House in NY, NJ
	0.005–0.05 WL	House in Florida
Benzo(*a*)pyrene	7.1–21 $ng\,m^{-3}$	Sports arena
Dimethylnitrosamine	0.11–0.24 $\mu g\,m^{-3}$	Bar
Carbon dioxide	0.086%	Lecture hall
	0.06–0.25%	School room
	0.9%	Nuclear submarines
Bioaerosols (culturable particles)	20–700 CFU m^{-3}	Schools, hospitals, residences

Source: Adapted from Wadden and Scheff (1983).
[a] WL, working level.

Table 17.2 Activities that increase your chance of dying early.

Activity	Cause of death
Smoking 1.4 cigarettes	Heart disease, cancer
Drinking 0.5 l wine	Cirrhosis of the liver
Living 2 months with a smoker	Cancer, heart disease
Traveling:	
10 mi by bicycle, 150 mi by car, 1000 mi by jet	Accident
Living 2 days in New York or Boston	Air pollution
Living 4 days in typical US residence	Lung cancer from radon exposure
Eating 100 charcoal-broiled steaks	Cancer from benzopyrene
Drinking Miami tap water for 1 year	Cancer from chloroform
Living within 5 mi of nuclear reactor for 50 years	Accident => radiation exposure => cancer

Source: From Wilson (1979).

building sector (DOE, 2011a). Small changes in building operation could have big environmental/economic impact. Traditional energy efficiency actions in buildings include tightening the building shell so that energy has a much harder time leaking out. Tight buildings minimize infiltration of outdoor air and outdoor air pollutants. This can cause IAQ problems if infiltration is the only means of ventilation. However, it can be beneficial since it reduces the amount of outdoor air pollutants infiltrating indoors. There is a conflict between air quality and energy economics existing in building codes, especially the codes that are not requiring added ventilation in tight buildings. Many codes and standards are starting to address this issue and require ventilation to be added if a home is built or retrofitted to be energy efficient and tight. There is a need for proper building design, construction, and ventilation guidelines to avoid the exposure of inhabitants to unhealthy environments.

To improve IAQ a system of standards or guidelines would be useful. In the United States, we often use the US Environmental Protection Agency (EPA) National Ambient Air Quality Standards as a guideline, since these apply to the general population and are health-based standards. IAQ has also been specified by the American Conference of Governmental Industrial Hygienists for industrial environments based on 8–10-h work days and a 40-h week. Recommended occupational standards are in the form of threshold limit values (TLVs). These limits, however, apply to healthy, younger working individuals. Currently, there are no legally enforceable health-related national standards for living and recreational spaces or transportation modes. The World Health Organization (WHO) has addressed IAQ as has Health Canada in the form of guidelines. Finally, there are some useful guidelines provided by the American Society of Heating, Refrigerating and Air-Conditioning Engineers.

17.1.1 Pollutants of Main Concern

Nitrogen dioxide (NO_2) is a respiratory irritant, for example, it causes increased susceptibility to pulmonary infection and may contribute to emphysema. It is formed from nitrogen in combustible fuels or from the N_2 in air used in the combustion process. In combustion processes that release NOx ($NO + NO_2$) outdoors, most NOx is NO. But in indoor air combustion, both NO and NO_2 are equally likely.

Carbon monoxide (*CO*) causes neurotoxic effects, culminating in coma and death in the worst cases of CO poisoning. It competes for oxygen sites on blood hemoglobin and is preferentially accepted, so that CO exposure leads to impaired uptake and transport of oxygen through the body. It is formed by incomplete combustion of carbonaceous fuels.

Sulfur dioxide (SO_2) is an upper respiratory tract irritant. It has the ability to trigger asthma, for example. It results from the oxidation of sulfur (S) during the combustion process, and S is a trace element in many fuels. It is moderately soluble so it is easily taken up by the lining of the human lung and turns into an acid.

Particles have a wide variety of health effects including respiratory tract irritation, lung cancer, and death. The effects vary with composition. Soot is a human carcinogen because of adsorbed polycyclic aromatic hydrocarbons (PAH). Some heavy metals found in particles are human carcinogens, such as arsenic and cadmium. The health effects also vary with particle size. Ash particles are larger and tend toward the coarse end of the size distribution. Soot and toxic metals are present in submicron particles and can be deposited deep in the respiratory tract. Particles come from combustion of noncombustible impurities in fuel (ash) and incomplete combustion by-products (soot). They are also formed by chemical reactions between gas-phase pollutants and these reactions can happen both indoors and outdoors, usually involving O_3 and VOCs. Particles also become airborne indoors by resuspension of dust.

Organics are commonly found in indoor environments emitted from consumer products and building materials. The effects vary with chemical, but in general they can be respiratory irritants and cause central nervous system effects, and some cause cancer such as formaldehyde and benzene.

17.1.2 Sources

The main sources of air pollutants indoors are outdoor air; combustion processes such as cooking, furnishings, and building materials; consumer products; soil vapor intrusion (radon, organic compounds); and chemical reactions. Outdoor air and combustion sources are discussed in more detail below.

17.1.2.1 Outdoor Air

Air pollutants that are dominant outdoors and not indoors are ozone and sulfur dioxide. Ozone is a strong lung oxidant and very reactive so that the levels indoors are reduced due to reactions on indoor surfaces and are typically a third of what the levels would be outdoors. In many urban, highly populated areas, ozone levels are very high so that even if the indoor concentrations are reduced, they are still of human health concern. There are very few indoor ozone sources, but two of concern are ozone generators and copier machines. Ozone generators are used to clean and disinfect indoor spaces, such as hotel rooms, but should only be used when the spaces are unoccupied. Sulfur dioxide is mainly from combustion of coal containing sulfur, which is commonly mined and burned in the Midwest and eastern parts of the United States. It is a strong acidic gas and causes ecological damage through acid rain and direct deposition. It is measured indoors as a tracer for outdoor air. Other air pollutants, including volatile organic compounds (VOCs), particulate matter (PM), and nitrogen oxides, can be elevated both indoors and outdoors.

17.1.2.1.1 Combustion

Combustion is a chemical conversion process in which molecules having high internal chemical energy – the fuel – are converted to low internal energy reaction products by an oxidant, usually air or oxygen. Energy in equals energy out, so the loss in internal energy when the fuel is converted to a reaction product must be balanced by energy released, which is thermal energy or heat (Equation 17.1). Combustion is almost always faster at higher temperatures:

$$\text{Fuel} + \text{air} \rightarrow \text{products} + \text{heat} \tag{17.1}$$

We use the heat generated from this chemical conversion process to warm our homes and cook our food. Consequently, we also release the products from this chemical conversion process to our indoor environments. If a combustion process is complete, then the only products generated from oxidizing a hydrocarbon fuel (fuel made up of hydrogen and carbon) are CO_2 and H_2O. However, many combustion reactors are not optimized to produce complete combustion, and the fuel can also contain impurities, and finally constituents in the oxidant, for example, the N_2 in air, are also oxidized to form air pollutants. Combustion conditions in indoor environments are often not very good, e.g. wood burning in a fireplace, quenching of a gas cooking flame on a pot containing cold water, or an idling cigarette.

Heating processes used indoors that are likely to emit air pollutants are faulty vented furnaces, unvented combustion appliances such as fireplaces, and wood-burning stoves and fireplaces. Cooking processes also emit air pollutants, such as gas-fired cooking ranges, cooking of food (frying, broiling, etc.), biomass cook stoves, and use of charcoal briquettes indoors. A variety of fuels are used for indoor combustion. Natural gas (CH_4) is commonly used in many parts of the United States for heating. Wood is used in fireplaces or in developing countries for heating and cooking. Meat is cooked and oils are used during cooking. Pollutants that are generated include NO*x*, CO, SO_2, PAH, VOCs (e.g. formaldehyde, acetaldehyde, benzene), and PM (both organic and inorganic soot).

Cigarettes are also combusted indoors. Smoking indoors emits thousands of toxic chemicals, and second-hand smoke (SHS) is a known human carcinogen and causes tens of thousands of respiratory diseases including cardiopulmonary diseases every year.

17.2 Completely Mixed Flow Reactor Model

17.2.1 Material Balance

The concept of material balance is used extensively in IAQ to understand the important processes that influence pollutant concentrations. Most of the problems environmental engineers work on involve characterizing and controlling some pollutant. That pollutant may be particles suspended in air or a gas-phase chemical compound. *Everything has to go somewhere* is a simple way to express a fundamental environmental engineering principle. This phrase describes the *law of conservation of mass*, which says that when a chemical reaction takes place, matter is neither created nor destroyed. This concept allows us to track pollutants from where they are emitted to where they are deposited, inhaled, or reacted using mass balance equations.

The first step in a material, or mass, balance is to define the region to be analyzed. This region can be an entire city, or a coal-fired power plant, or a room in a house. Then we identify the flow of material across the region's boundary as well as the emission, accumulation, and removal of mass within the region. To analyze these types of problems, we first account for the pollutant within the system we are analyzing according to the principles of material conservation (Figure 17.1).

Frequently, this expression can be simplified. The most common simplification is when steady-state (or equilibrium) conditions can be assumed. Equilibrium means that nothing is changing with time. For example, the system has had its input rate held constant for a long enough time that any transient responses have had a chance to die out. Pollutant concentrations are constant. Thus, nothing is accumulating in the boundary and the accumulation rate term is set equal to zero.

17.2.2 Completely Mixed Flow Reactor (CMFR) Model

Much analyses and modeling of IAQ are based on the completely mixed flow reactor (CMFR) model shown in Figure 17.2. This model consists of a reactor with air flowing into and out of it. The contents of the container are assumed *well mixed*. This mixing within the container is assumed to occur *instantaneously*, which means that the concentration of each species is uniform

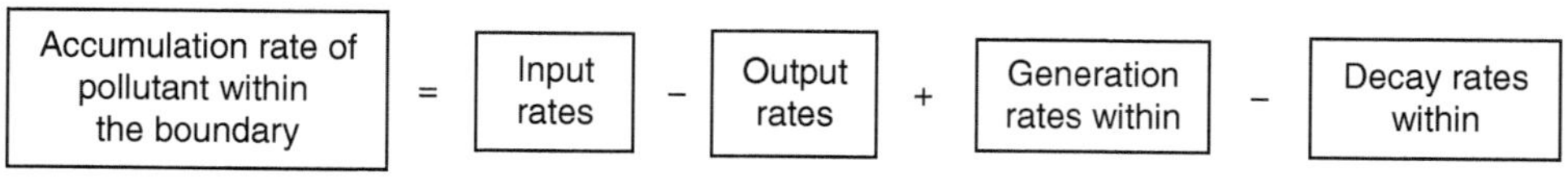

Figure 17.1 Concept of a material balance.

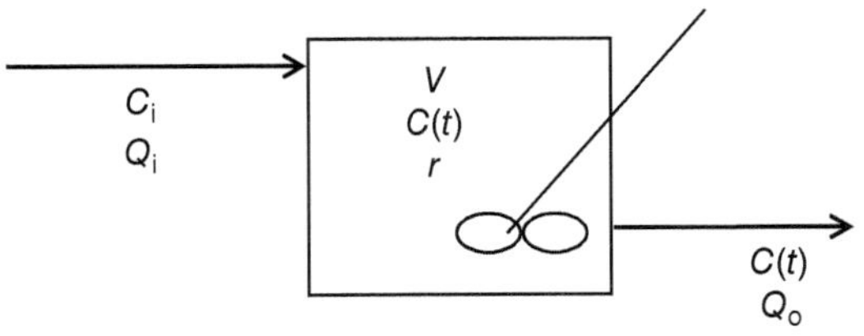

Figure 17.2 Schematic of a completely mixed flow reactor model.

throughout the container and is equal to the concentration flowing out of the container.

How to apply a material balance to a CMFR:

1) Draw a picture of the reactor with the material balance boundary.
2) Select the pollutant of interest and appropriate units for overall material balance.
3) List all known values of flows, concentrations, volumes, etc.
4) List all equations for reactions that occur inside the material balance area limits.

A material balance on the species of interest in the CMFR model yields the following differential equation:

Accumulation *Flow in* *Net rate of production*

$$\frac{d(VC(t))}{dt} = Q_iC_i - Q_oC(t) + rV \tag{17.2}$$

Flow out

where V is the container volume, Q_i is the airflow rate into the container, Q_o is the flow rate out of the container, C_i is the inlet concentration of the species, $C(t)$ is the species concentration within the container, and r is the *net rate* of production of the species within the container per unit volume. This term includes any generation of species and also any removal of species by processes such as deposition, filtration, or chemical reaction. The unit for each term in the above differential equation is **quantity of material per time**.

When applying this equation to the indoor environment, we may need to account for processes such as indoor emissions or reactions with surfaces. Usually we can assume that the total amount of fluid in the reactor remains constant, so $Q_i = Q_o = Q$.

Assuming the volume is constant,

$$\frac{dC}{dt} = Q\left[\frac{C_i - C}{V}\right] + r \tag{17.3}$$

In reactor theory, the *characteristic time* or *residence time* is given by V/Q. In the indoor environment, a primary parameter of interest is the air exchange rate, which is given by the inverse of the characteristic time:

$$\lambda_v = \frac{Q}{V} \qquad \tau = \frac{1}{\lambda_v} \tag{17.4}$$

At steady state, the accumulation rate of pollutant within the system is zero:

$$\frac{dC(t)}{dt} = 0$$

17.2.3 Solution to First-order ODE with Constant Coefficients

Throughout environmental engineering, one commonly encounters differential equations that can be cast into the following form:

$$\frac{dC(t)}{dt} = S - LC(t) \text{ and } C(t=0) = C_o \tag{17.5}$$

where $C(t)$ is the unknown pollutant concentration, S is the sum of the sources of pollutant into the system, L is the sum of the rates of losses from the system by first-order processes, S/L is the steady-state concentration, and C_o is the initial concentration.

The general solution to this equation is

$$C(t) = \underbrace{C_o e^{-Lt}}_{E_1} + \underbrace{\frac{S}{L}\left(1 - e^{-Lt}\right)}_{E_2} \tag{17.6}$$

E_1 and E_2 are associated with example below.

The characteristic time for this system to relax from its initial condition to steady state is given by

$$\tau = \frac{1}{L} \tag{17.7}$$

Example: Simulate Equation 17.6, where E1 is 1st term on Right Hand Side and E2 is 2nd term. E1 and E2 represent growth and decay terms, respectively.

$C_o = 0.5$
$S/L = 1$
$L = 2$
$S = 2$

Time (h)	*E1*	*E2*	*C*(*t*)
0.00	0.50	0.00	0.50
0.08	0.42	0.15	0.58
0.17	0.36	0.28	0.64
0.25	0.30	0.39	0.70
0.33	0.26	0.49	0.74
0.42	0.22	0.57	0.78
0.50	0.18	0.63	0.82
0.58	0.16	0.69	0.84
0.67	0.13	0.74	0.87
0.75	0.11	0.78	0.89
0.83	0.09	0.81	0.91
0.92	0.08	0.84	0.92
1.00	0.07	0.86	0.93
1.08	0.06	0.89	0.94
1.17	0.05	0.90	0.95
1.25	0.04	0.92	0.96
1.33	0.03	0.93	0.97
1.42	0.03	0.94	0.97

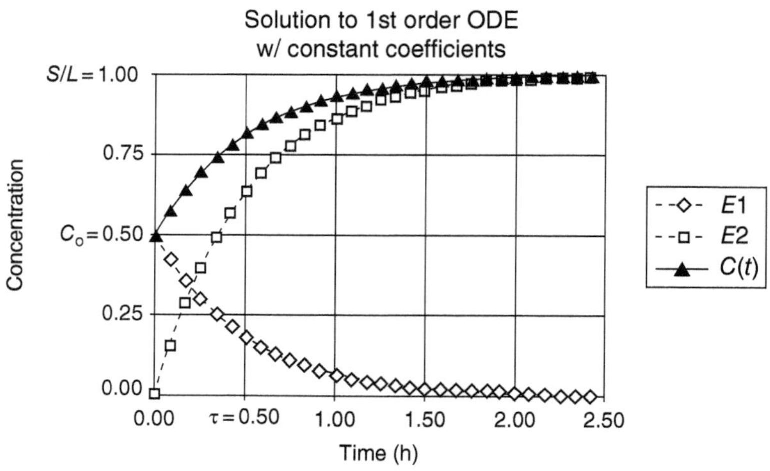

17.3 Deposition Velocity

Many pollutants physically and chemically interact with indoor surfaces, including walls, carpeting, furniture, and even people. This interaction will influence the indoor concentrations of many pollutants. Such interactions can actually be beneficial since pollutants that deposit or react on indoor surfaces can no longer be inhaled and cause potential adverse health effects. On the other hand, some pollutants can come back off surfaces and be inhaled. Particles deposited on surfaces can become resuspended, or gases sorbed onto surfaces can desorb. Surface interaction is the main mechanism for material damage to objects in buildings such as artifacts or electronics. Below is an example of a surface reaction that generates a more harmful pollutant product compared with the reactants:

$$2NO_2 + H_2O + \text{surface} \rightarrow HNO_2 + HNO_3 \quad (17.8)$$

HNO_2 gas causes more problems than NO_2 since it is an acid gas.

While deposition is also an important removal mechanism for outdoor air pollutants and is the cause of significant environmental problems such as acid rain, it is different indoors because of large surface-to-volume ratios.

Accurate modeling of the concentrations and fates of indoor pollutants requires knowing the rates and outcomes of pollutant interactions with indoor surfaces. Currently, most IAQ modeling parameterizes pollutant–surface interactions as a first-order loss process. The loss rate is commonly evaluated as the product of two terms: the surface-to-volume ratio of the indoor space and the deposition velocity. All loss processes are termed generally as *deposition*. Deposition velocity varies with airflow conditions, shape and orientation of the surface, the surface material, and the pollutant species.

Particles will attach firmly to any surface they contact. The primary adhesive force is van der Waals forces, so that once a particle strikes a surface, a lot of energy needs to be added to make a particle bounce back off of a surface. Walking on floors or carpets provides enough energy to resuspend many deposited particles, but particles adhering to walls and ceilings are assumed to be removed from the air. Gases on the other hand typically do one of two things: they react upon deposition and form a new chemical (such as ozone oxidation) or just stick and can then be desorbed depending on the environmental conditions (semi-VOCs such as flame retardants).

17.3.1 Factors Governing Deposition Process

Deposition is governed by the rate of pollutant *transport* to a surface and the probability of *transformation* following collisions with surface. These two processes operate in series and the overall loss rate is governed by the slower phenomenon.

Factors governing transport are indoor airflow conditions (bulk flow), very-near surface flows, surface temperatures, and details of surface (shape and orientation). Transport to the vicinity of the surface may occur by a variety of mechanisms including advection, diffusion, and, for particles, gravitational settling. Transport is the most important factor for particles and highly reactive gases. There are two components of transport near the surface:

i) *Aerodynamic component*: Transport of pollutant through the surface layer to the immediate vicinity of the surface.
ii) *Surface layer component*: Transport of the material through a tiny sublayer just adjacent to the surface to the ultimate adhesive substrate/surface.

Factors governing transformation include surface composition and what has previously deposited on surface. Upon interaction at the surface, we assume that the pollutant is removed and/or transformed. Desorbing may happen under some environmental conditions. For most gases, the reactions that take place on the surface are not instantaneous, so transformation is often times the most important factor at play and not transport.

17.3.2 Definition of Deposition Velocity

Deposition can be thought of conceptually analogous to electrical or heat flow through a series of resistances. The transfer of material to the surface takes place through three resistances – the aerodynamic resistance (r_a), the surface layer resistance (r_s), and the transformation resistance (r_t):

$$v_d = \left(r_a + r_s + r_t\right)^{-1} \tag{17.9}$$

Deposition velocity, v_d, is defined as net flux of a species to a surface divided by the concentration C of that species away from the surface in the air:

$$v_d = \frac{\text{species flux to surface}}{\text{concentration away from surface } (C)} = \frac{\text{mass / area-time}}{\text{mass / volume}} [=] \frac{L}{T} \tag{17.10}$$

The *flux* of anything refers to the flow of species per unit area per unit time. C should be determined at a position that is sufficiently far from the surface so that the concentration does not vary with position and is representative of the bulk concentration. Deposition is like an air filter operating indoors that removes 100% of the depositing species at a flow rate of $Q = v_d S$. For a pollutant that has an average deposition velocity, v_d, onto a surface of area, S, the rate of pollutant mass loss by deposition would be $= S \times v_d \times C$.

17.3.3 Rate of Losses by Surface Interaction

If the deposition velocity of a species indoors is known, then the rate of species loss from indoor air by surface interaction may be calculated. Consider a first-order irreversible process that removes species by surface interaction. The relationship between deposition velocity and the rate of change of the indoor concentration due to surface loss is

$$\left(\frac{d(CV)}{dt}\right)_{\text{surface loss}} = -C\int_S v_d(s)\,ds = -CS\overline{v_d} \tag{17.11}$$

where V is the volume of room, S is the total surface area, $v_d(s)$ is the deposition velocity of species onto surface at position s, ds is the area of a differential surface element, and $\overline{v_d}$ is the area-weighted mean deposition velocity.

17.3.4 Measuring Deposition Velocity

Table 17.3 summarizes some of the more commonly used methods for estimating deposition velocity. The first two methods listed in Table 17.3 are indirect methods and are based on a representation of a room or building as a CMFR.

To determine deposition velocity from analysis of transient data, the indoor concentration of the pollutant must first be elevated to a level significantly above background. Then the concentration decay is monitored as a function of time. The ventilation rate must be known,

Table 17.3 Methods for determining deposition velocity (Nazaroff et al. 1993).

Method	Application
Collect and analyze transient concentration data	Ozone, combustion appliances
Collect and analyze equilibrium concentration data	Ozone, radon decay products
Measure deposited flux and resultant airborne concentration	Fine particles, radon decay products
Conduct a theoretical mass transport analysis	Particles and reactive gases, radon decay products

Source: Reproduced with permission of ASTM International.

and so it is typically measured at the same time using tracer gas. A linear least-squares regression is fitted to the natural logarithm of concentration versus time. The slope of this line is used to infer the deposition velocity.

To infer deposition velocity from equilibrium data, a steady-state concentration of the pollutant is established, and the outdoor concentration must also be approximately steady state. The method works when entry from outdoor air is the dominant indoor source of a species like ozone and there are no transient indoor sources.

The most direct method to determine deposition velocity is to measure the flux density and the airborne concentration and use Equation (17.10). This approach usually only works with pollutants such as particles since it is difficult to measure the surface accumulation of gas-phase pollutants, which is what is needed. Also long sampling periods are usually required.

Deposition is an important process, but not particularly well understood, and its description is partly qualitative. Because it is such a complicated phenomenon, there is no complete theory that accounts for all factors governing deposition. Table 17.4 lists some typical deposition velocity values that have been measured.

17.3.5 Limitations

There are a few limitations that should be considered when measuring deposition velocity. The indoor environment must be well mixed and the deposition velocity can vary spatially. Deposition can also be enhanced by promoting mass transfer to surfaces such as with active mixing using fans. Some reactions at surfaces are a function of time, leading to time dependency that must be accounted for.

Table 17.4 Measurement and model predictions for deposition velocities.

Species	v_d (cm s^{-1})	Comments
Fine sulfate particles	0.003–0.005	Measured in telephone buildings
0.5-μm-diameter particles	0.0001–0.002	5 museums
Radon decay product Po-218	0.1–0.5	Basement
Ozone	0.015, 0.062	Stainless steel room, bedroom
NO_2	0.000231 s^{-1}	Home with gas range and oven

Source: Adapted from Wadden and Scheff (1983) and Nazaroff et al. (1993).

17.4 Ultraviolet Germicidal Irradiation

Ultraviolet germicidal irradiation (UVGI) is mainly used for disinfecting air in two configurations: ventilation duct irradiation and upper-room air irradiation. Animal studies compellingly demonstrated that ventilation duct irradiation prevented the transmission of TB from hospitalized TB patients to guinea pig colonies (Riley et al., 1957, 1962). Initial observations of the efficacy of upper-room air irradiation to control airborne infection were reported in the early 1900s by Wells (Wells and Fair, 1935; Wells, 1955). For upper-room air irradiation, germicidal lamps are suspended from the ceiling or attached to the walls; the bottom of the lamp is usually shielded or louvered to direct radiation upward above a predetermined height. The objective of this configuration is to inactivate airborne infectious agents in the upper part of the room while minimizing radiation exposure to persons in the lower part of the room. *Inactivation* in this context means the loss of the ability to replicate and form colonies. Commercially available germicidal lamps contain mercury vapors under low pressure that emit energy in the shortwave UV-C wavelength range, 100–290 nm, with about 90% of the total spectral power emitted at 254 nm.

Shortwave UV-C radiation damages a microorganism's DNA. The main mode of inactivation occurs when a photon is absorbed and pyrimidine dimers between adjacent thymine bases are formed. UVGI is used to disinfect air, water, and surfaces. Figure 17.3 from Wikipedia shows how a thymine dimer is formed. Figure 17.4 shows the action spectrum for bactericidal action of ultraviolet radiation against *Staphylococcus aureus* (modified from Gates, 1930), plotted with the absorbance spectrum for

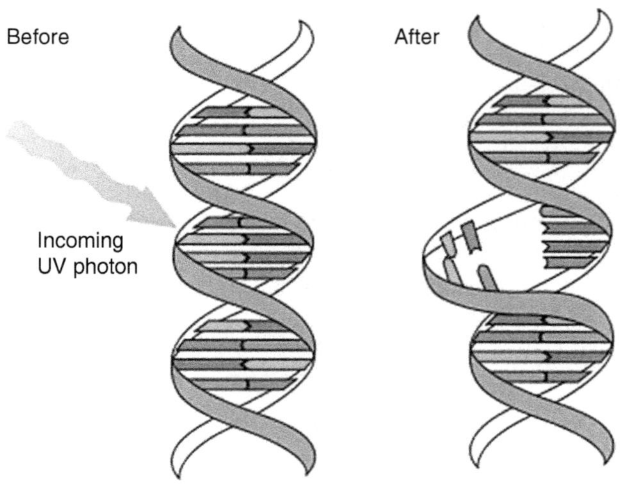

Figure 17.3 Formation of thymine dimer in DNA upon UV radiation exposure.

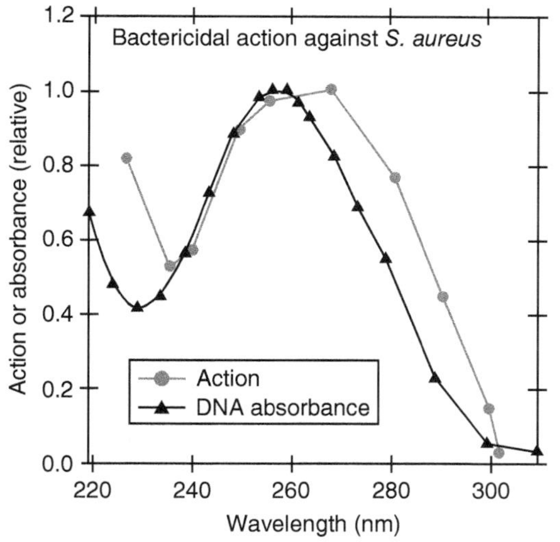

Figure 17.4 Bactericidal action spectrum and DNA absorbance spectrum.

DNA (modified from Tsuboi, 1950), from http://www.photobiology.info/Gorton.html.

17.4.1 Upper-room UVGI

Upper-room UVGI systems are typically installed where unsuspected cases of airborne disease transmission may occur. They are to be used without protective clothing, so the occupied part of the room must have low irradiation levels, achieved by the use of louvers on the fixtures. Effective air disinfection depends on good vertical mixing between the lower and upper parts of the room, since the infectious aerosol is generated in the occupied zone but must be transported to the upper zone to be inactivated. Usually mixing fans are recommended for use such as paddle fans on the ceiling. Figure 17.5 is a typical fixture used in an upper-room UVGI system (Xu et al., 2003, 2005).

17.4.2 In-duct UVGI

Lamps are installed in ventilation ducts and irradiate the entire cross section of a duct at very high levels. Effective room disinfection depends on circulating the maximum amount of room air through the duct. Inactivation within the duct depends on the velocity at which the air moves past the UV lamps.

17.4.3 Cooling Coil UVGI

Although not used to disinfect the air, UVGI is also used to disinfect surfaces inside HVAC systems such as cooling coils and drip pans. UV-C lamps are installed just downstream of the cooling coils on the supply side, irradiating the surfaces of the coil. Particle deposition and biofilm growth on cooling and heating coils can degrade performance and result in increased energy and maintenance costs (Figure 17.6). Heat exchanger surfaces are an ideal site for biofilms due to the presence of adequate nutrients (debris inherent on coil surfaces) and moisture.

Figure 17.6 Coil fouling blocks airflow. *Source*: Photo from epb.lbl.gov/coilfouling.

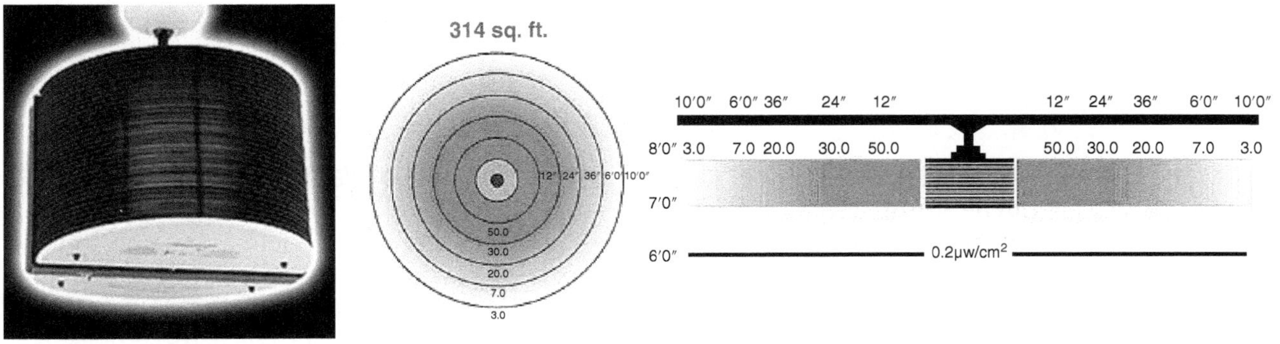

Figure 17.5 Ceiling mounted pendant fixture from Lumalier and coverage area for the fixture http://www.lumalier.com/component/content/article/106.html).

17.4.4 UVGI Effectiveness

Effectiveness quantifies the impact of UVGI on airborne microorganism concentrations. Effectiveness for inactivating bacteria can be estimated by comparing the measured culturable airborne bacteria concentration with the UVGI system on, $C_{\text{UV on}}$ (CFU m^{-3}), with the airborne culturable bacteria concentration without UVGI, $C_{\text{UV off}}$ (CFU m^{-3}):

$$E = 1 - \frac{C_{\text{UV on}}}{C_{\text{UV off}}} \quad (17.12)$$

17.4.5 UVGI Inactivation Rate

A CMFR model is needed to evaluate UVGI inactivation rates. The model is based on the assumption of complete mixing, which results in uniform airborne microorganism concentrations throughout the volume of the room. Inactivation rates for bacteria can be estimated by applying a material balance to airborne culturable bacteria within a room with volume V (m^3). Culturable bacteria are emitted from a source at a generation rate G (colony-forming units (CFU) h^{-1}). Ventilation air flows through the room removing airborne bacteria, reducing the indoor culturable bacteria concentration. The ventilation rate is expressed as the volumetric airflow rate through the room divided by the volume of the room (ACH, h^{-1}). In addition, the UVGI system inactivates airborne bacteria when operated. This inactivation rate can be approximated with a first-order decay model, denoted ACH_{UV} (h^{-1}). Gravitational settling, natural die-off, and other natural decay mechanisms for bacteria are denoted ACH_{N} (h^{-1}).

The rate of change of the indoor airborne culturable bacteria concentration with the UVGI system on $C_{\text{UV on}}(t)$ (CFU m^{-3}) is given by

$$\frac{dC_{\text{UV on}}(t)}{dt} = \frac{G}{V} - (\text{ACH} + \text{ACH}_{\text{UV}} + \text{ACH}_{\text{N}})C_{\text{UV on}}(t) \quad (17.13)$$

Solving Equation 17.13, the culturable airborne bacteria concentration as a function of time is given by

$$C_{\text{UV on}}(t) = C_o e^{-(\text{ACH}+\text{ACH}_{\text{UV}}+\text{ACH}_{\text{N}})t} \quad (17.14)$$

A similar expression can be written for the culturable airborne bacteria concentration without UVGI, $C_{\text{UV off}}(t)$.

To derive the overall removal rate, $\text{ACH} + \text{ACH}_{\text{UV}} + \text{ACH}_{\text{N}}$, the natural log form of Equation 17.14 is linearly fit to data (natural log transformed) collected during experiments with the UVGI system operating using a least-squares method. The slope of the line is equal to the overall removal rate. Similarly, data from experiments without the UVGI system operating is linearly fit to derive the removal rate due to ventilation and natural decay only, $\text{ACH} + \text{ACH}_{\text{N}}$. The UVGI inactivation rate, ACH_{UV}, is ultimately determined by calculating the difference between these two rates.

17.4.6 Environmental Factors

Relative humidity (RH) has been shown to impact UVGI air disinfection performance. Xu et al. (2005) showed that the effectiveness decreased by almost a factor of 2 when the RH was increased above 75%.

Photoreactivation (PR) can also impact performance by facilitating the recovery of UVGI-damaged genetic material in airborne bacteria. Light-activated DNA photolyase enzymes, which can repair certain types of UV-induced lesions, facilitate this recovery. It has been suggested that short wavelength light, in the near-UV range, is the main factor inducing PR. PR may affect UVGI system performance adversely in environments with incandescent light, fluorescent light, or sunlight.

17.4.7 Design Considerations

It is important to consider how much UVGI should be used in an upper-room design and where the lamps are placed in the room. Xu et al. (2005) showed that the radiation should be spread evenly throughout the room to get the best effectiveness. A general rule is that one 30-W fixture for every 18 m^2 (200 ft^2) of floor area or for every seven people in a room, whichever is greater. Note that the UV-C output of a 30-W fixture is around 10 W UV-C (Macher, 1993). Given that ceiling heights and UV output from fixtures vary, an alternative guideline is to design UV installations based on the *volume* of the room that is irradiated and the total UV-C wattage applied. Xu et al. (2005) recommends that the UV-C wattage volume power distribution be at least 6.3 W UV-C m^{-3} irradiated upper-room space. Room airflow is another important consideration when designing an appropriate system. Vertical mixing must be ensured, whether it be by ventilation design or mixing fans.

17.5 Filtration of Particles and Gases

Specific gases and particles can be removed from an airstream by passing the air through filters. A filter is typically a porous structure made up of granular or fibrous material that tends to retain particles or gases as the carrier gas passes through the voids of the filter. Filters separate particles and/or gases from the carrier gas. They are commonly employed as stand-alone air cleaners, in heating, ventilation, and air-conditioning (HVAC) systems, or in industry to remove contaminants from waste or material streams. Filtration can be used to remove pollutants at three points: (i) at the source of the contaminants (range hood), (ii) at the receptor of the contaminants (personal respirators), (iii) and from the room air (air cleaners, ventilation system).

The EPA has published a report summarizing the available information for residential air cleaners (http://www.epa.gov/iaq/pubs/residair.html, 2009).

17.5.1 Evaluating Performance

In evaluating the performance of a room air filter, it is important to distinguish between *single-pass efficiency* and *effectiveness*. The former parameter characterizes the likelihood that particles or gases will be removed from air that passes through a filter. The latter parameter describes the impact of device operation on room air concentrations; it is the more relevant indicator for the control of indoor air contaminants. The effectiveness of air filtration depends on single-pass filter efficiency. However, it also depends on the airflow rate through the filter and on other features of the indoor environment, such as the relative positions of the source, control device, and receptor, and the indoor airflow patterns. Of main concern is whether the pollutant actually passes through the filter. And for the filter to have an impact, its removal rate must be large enough relative to other removal mechanisms (deposition onto surfaces, ventilation).

Single-Pass Filter Efficiency

Single-pass efficiency, denoted η_F ($0 \leq \eta_F \leq 1$), is defined by the following equation:

$$\eta_F = \frac{C_{in} - C_{out}}{C_{in}} \tag{17.15}$$

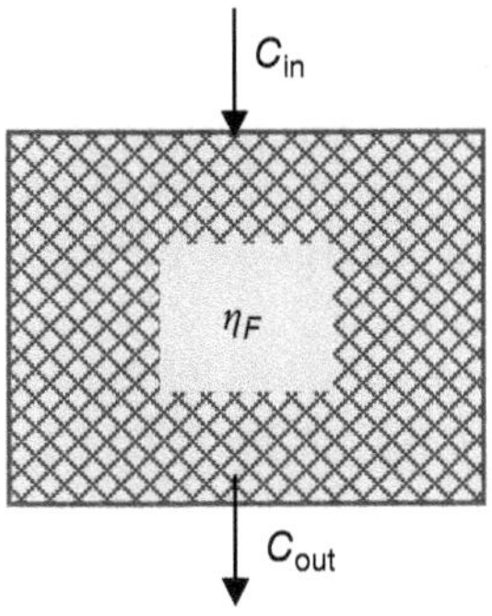

where C_{in} is the concentration of particles/gases in air as it enters the device and C_{out} is the concentration of particles/gases in air as it leaves the device.

The single-pass filter efficiency for particulate air filters is a function of particle size and of the media used in the device. Minimum single-pass efficiency occurs in the accumulation mode of the particle size distribution, at a particle diameter of about 0.3 μm. High efficiency particle air (HEPA) filters have a single-pass efficiency of ≥99.97% for 0.3-μm particles. However, the use of HEPA media does not ensure a high effectiveness for indoor air pollution control (Miller-Leiden et al., 1996).

Velocity of the air at the face of a filter, just before the entrance, is called the face velocity, $U_o = Q/A$, where Q is the volumetric flow through the filter and A is the cross-sectional area of the filter exposed to the airstream. The air velocity inside the filter, U, is greater than U_o because the volume for air is reduced by the volume of the fibers: $U = Q/A(1 - \alpha)$. The volume fraction of the fibers or grains is α. For example, for fibrous filters, α = fiber volume/total volume = 1-porosity and is typically between 0.01 and 0.3 (Hinds, 1999).

Single-pass efficiency is not constant over time. For gases, the efficiency can decrease over time as the sorption capacity of the media is used up; some types of media can be regenerated. Once the media is used up, breakthrough will occur, and the contaminant will no longer be removed from the inlet stream and will be present in the outlet stream. For particle fiber filters, both efficiency and pressure drop increase with the accumulation of collected particles in the filter. Initially this is beneficial; but eventually the pressure drop becomes excessive, and the filter is clogged. Filters with a small value of α can accommodate the greatest loading without clogging. There is no problem with breakthrough of dust particles.

When characterizing or testing filter performance, three parameters are important:

- Pressure drop.
- Capacity to hold accumulated particles or gases (breakthrough time).
- Efficiency vs. particle size (for particle filters).

In a portable particulate air cleaner, a built-in fan draws particle-laden air through filter media or an electrostatic field, blowing treated air back into the room. In evaluating the performance of an air cleaner, a useful parameter is the *clean air delivery rate* (CADR), which is equal to the single-pass efficiency times the airflow rate through the device (CADR = $\eta_F \times Q_f$). The CADR is the airflow rate that represents the effective amount of particle clean air produced by the air cleaner. The CADR does not directly indicate the benefit of using an air-cleaning device.

A parameter that does directly describe the impact of a control device on IAQ is the *effectiveness*, defined as the fractional reduction in pollutant concentration that results from application of a control device. Consider PM present in a home at a concentration C_{ref}. The use of an air cleaner reduces the PM concentration to C_{filt}.

We define the effectiveness, E, which ranges in value from 0 to 1, where 1 represents ideal performance:

$$E = \frac{C_{\text{ref}} - C_{\text{filt}}}{C_{\text{filt}}} \tag{17.16}$$

Using a CMFR model, and assuming all of the parameters are constant with time, the effectiveness can be related to the CADR by

$$E = \frac{\text{CADR}}{\text{CADR} + Q_O + kV} \tag{17.17}$$

The Association of Home Appliance Manufacturers (AHAM) has developed a certification program for labeling particle air cleaners (http://aham.org). The AHAM test method ANSI/AHAM AC-1-2002 determines the CADR for three types of PM – dust, tobacco smoke, and pollen (AHAM, 2002). The program provides labeling for certified appliances (Figure 17.7). The label indicates the CADR for removing dust, tobacco smoke, and pollen particles in cubic feet per minute. AHAM also provides recommendations about CADR levels that are appropriate for a given room or building volume. These recommendations are based on a criterion of 80% smoke reduction or an effectiveness of 0.8. For example, for a 140-m^3 room, the CADR is approximately 600 $m^3\ h^{-1}$, and for a 340-m^3 building (e.g. a 1500-ft^2 house), the CADR is 1300 $m^3\ h^{-1}$.[1]

Figure 17.7 Example of a label from the Association of Home Appliance Manufacturers (AHAM) certification program for labeling particle air cleaners. *Source*: Adapted from AHAM (2012, p. 58).

Miller-Leiden and colleagues found that commercially available portable air cleaners operated at 72 $m^3\ h^{-1}$ in a 36-m^3 ventilated (2 air changes per hour) room reduced airborne PM concentrations by 30–40% (Miller-Leiden et al., 1996); the CADR for this scenario is estimated to be 30–50 $m^3\ h^{-1}$. Offermann et al. studied the removal rates of a variety of air cleaners for environmental tobacco smoke (Offermann et al., 1985). Results showed that commercially available electrostatic precipitators (ESPs) and high single-pass efficiency pleated-fiber filters were more than 90% effective in a 35-m^3 room, removing particles from 0.1 to 1.3 μm at substantial CADRs, 200–300 $m^3\ h^{-1}$. The results from these studies illustrate the range of effectiveness that can be achieved and how important it is to size an air cleaner appropriately.

17.5.2 Particle Filtration

Three main types of particulate filtration technology are fiber filters, ESPs, and cyclones. Fiber filters and ESPs are commonly used in buildings. Fiber filters should be used when high efficiencies are needed and the volumes to treat are relatively low. ESPs are used when high efficiencies are needed and large gas volumes are treated. Fiber filters have higher operating cost but lower capital cost than ESPs. Cyclones are used for removing coarse particles, at high concentrations. These are not commonly used in indoor environments. Cyclones are mechanical filters so they are relatively inexpensive.

There are many different kinds of fiber filters, from crude mat prefilters used in HVAC systems to HEPA filters used in clean rooms to fabric filters used in industrial applications. Many materials are used, with glass fibers being the most common. The filters are either cleanable or throwaway. They can be configured into flat panels or extended surface area shapes such as pleats. Cleanable fabric filters called baghouse filters are used in industrial applications; these fabric filters are in cylindrical tubes (or bags) and hung in multiple rows to provide a large surface area. Sometimes fibers are created with permanent electrical charge to enhance particle collection.

There are four mechanisms by which particles can be deposited on a fiber in a filter:

- **Interception** – Occurs when a particle follows a gas streamline that happens to come within a two-particle radius of the surface of a fiber. The particle touches the fiber and is captured because of its finite size.
- **Inertial Impaction** – A particle, because of its inertia, is unable to adjust quickly enough to a changing streamline near a fiber; thus the particle crosses streamlines and hits the fiber.

- **Diffusion** – Brownian motion of small particles greatly enhances the chance that the particle will hit a fiber while traveling past it.
- **Electrostatic attraction** – A charged particle will be attracted to oppositely charged fibers by coulombic attraction. Increasing the charge on either particles or fibers increases collection efficiency.

Particles larger than 1 μm are removed by impaction and direct interception; particles below 0.1 are removed by diffusion and electrostatic precipitation (Hinds, 1999).

ESPs are based on the mutual attraction between particles of one charge and a collecting electrode of opposite charge. They can handle large gas volumes and have high collection efficiencies even for submicron-size particles and low energy consumption. A typical design is for wires located between rows of parallel plates, which are the collecting electrode. Flow is usually horizontal and passageways narrow (20–25 cm). Wires are charged at 20–60 kV below ground potential (note that to avoid excessive ozone formation, center electrode is usually positively charged). The very high voltage differential between electrodes causes electrons to pass from center wire into the passing gas stream. Electrons attach themselves to the gas molecules to form negative ions. Negative ions move toward grounded outer plates; positive ions return to the center wire. Negatively charged gas ions collide with particles flowing through the device. Charged particles move to the plate electrodes and are collected. The speed at which this movement takes place is called the *drift velocity* (which depends on electrical force and drag force). Small particles are removed because of low resistance to airflow; large particles are removed because they can get supercharged and are then easily collected. ESPs have similar efficiency as a function of particle size relationships as for fiber filters. ESPs must be cleaned frequently to avoid sparking and ozone formation. Portable air cleaners using ESP technology can usually be cleaned by removing the filter and rinsing with water or washing in the dishwasher.

Cyclones employ centrifugal force by spinning the gas stream to separate the particles from carrier gas. These cleaners are used mostly for industrial applications, and in indoor environments they are used to limit dust emissions from hardwood sanding – some hardwood floor sanders are fit with cyclones to help collect the generated dust. Circular motion is typically attained by a tangential gas inlet, which gradually blends in the inlet stream over 180° into a cylinder. The centrifugal force due to high rate of spin flings the particles to the outer walls of the cylinder; the particles slide down walls and into collector. The gradually cleaned gas reverses its downward spiral and ascends out the top of the device. Cyclones are used to remove coarse particles (10 μm or larger) and do not remove very small particles.

An air filter's minimum efficiency reporting value (MERV) is the parameter that is reported when a filter has been tested using the ASHRAE Standard 52.2 test procedure (ASHRAE, 2017). In summary, it is a measure of the single-pass efficiency of a filter and is typically used to test HVAC filters. The MERV is for particles in the range of 0.3–10 μm. For example, a standard residential HVAC filter made of aluminum mesh has a MERV rating of 2–3 and removes particles such as dust mites, sanding dust, or spray paint dust. Superior residential filtration has a MERV rating of 12, is a nonsupported microfine fiberglass or synthetic media, and typically controls 1–3 μm particles and legionella. The highest MERV rating is 20 for clean rooms.

17.5.3 Gas Adsorption

Adsorption is a separation process based on the ability of certain solids to remove gaseous components or water vapor preferentially from an airflow stream. An *adsorbent* is the solid adsorbing medium; an *adsorbate* refers to the gas adsorbed.

Adsorption processes are either:

- *Physical adsorption* – Gas molecules adhere to a surface as the result of intermolecular attractive forces between them (this is an exothermic process). This process is reversible. Efficiency is directly proportional to the amount of solid surface available; buildup is not restricted to a monomolecular layer.
- *Chemisorption* – Chemical interaction occurs between the adsorbate and the adsorbing medium. The bonding force is much stronger than for physical adsorption (and more exothermic). This process is frequently irreversible.

Adsorbent medium is manufactured to have enormous surface area, both externally and internally, for pollutants to stick to. Chemical specificity is also required of the adsorbent; *like absorbs like* is a common phrase used as a rule of thumb for selecting filtration media. One media commonly used for gas-phase filtration is charcoal, or activated carbon (AC). AC has a surface area of 100–1000 $m^2\ g^{-1}$, compared with 1-mm glass beads that have 0.004 $m^2\ g^{-1}$. AC has an affinity for organic compounds with high molecular weight that are nonpolar or neutral. For example, using AC for benzene removal can result in 70–90% efficiency, whereas it only provides 10–30% efficiency for formaldehyde. Humidity affects AC performance, since water will condense in the pores at high RH (70–80%). AC also works for very reactive gases such as ozone. Two common inorganic sorbents used to adsorb polar compounds are silica gels and alumina. Silica gels come in many different pore sizes and surface areas. For example, Silica Gel 100 has a surface area and pore size of 300 $m^2 g^{-1}$ and 10 nm, while Silica Gel 40 has a surface area and pore size of 750 $m^2 g^{-1}$ and 4 nm. Silica gel is also used as a desiccant because it actually adsorbs water in preference to hydrocarbons; thus wet silica gels do not adsorb hydrocarbons very well and they are not a good sorbent for humid environments. Alumina has surface areas and pore sizes of around 155 $m^2 g^{-1}$ and 5.8 nm. One characteristic of alumina is that it can be modified by changing the surface pH from acidic to basic, and as a result it will sorb a wider polarity range compared with silica gel (NIOSH, 2003).

Bench-scale filter tests can be used to predict how much contaminant can be removed with a sorbent filtration system. To do these tests, an experimental setup must be used that flows contaminated air through a media-filled filter bed, and contaminant levels are measured up and downstream of the bed. The data from these tests are used to estimate the amount of gas adsorbed per unit of adsorbent at equilibrium and at a specific temperature (called adsorption isotherm). An equation can be fit analytically to the experimental data. Breakthrough tests must also be conducted to determine how long the media can be used before it needs replacing.

17.6 Ventilation and Infiltration

Outdoor air dilutes indoor air pollutant concentrations by flowing *intentionally* through buildings as ventilation air or *unintentionally* as infiltration (and exfiltration). In general, using ventilation to dilute the indoor air is not as satisfactory for air pollution control as is local exhaust ventilation (however, in many situations local exhaust is not an option) or other source control approaches. Ventilation is also used for heat and humidity control.

Ventilation air is air that is intentionally controlled. It provides the best possibility for control of air exchange rate and air distribution in a building. There are two categories of ventilation:

- *Forced ventilation (or mechanical ventilation),* which is intentional air exchange powered by fans and blowers and distributed through intake/exhaust vents. Usually it is mandatory in larger buildings. Ventilation systems must be properly designed, installed, and operated/maintained.
- *Natural ventilation,* which is unpowered airflow through design openings such as windows. It is induced by pressures from wind and/or temperature differences between the indoor environment and outdoors. Airflow through open windows and doors can in many situations provide adequate ventilation for contaminant dilution and temperature control.

Infiltration air is unintentional and uncontrolled. Infiltration is the least reliable method to provide adequate ventilation and air distribution in a building. Infiltration into (and exfiltration out of) a building is driven by wind, indoor–outdoor temperature differences, and/or appliance-induced (for example, fireplaces, stoves, dryers) pressures across a building envelope (building envelope refers to the external shell of a building that is in contact with the outdoors).

How air is exchanged in US buildings depends on the type of building. Small buildings such as single-family residences, restaurants, shops, and other small commercial buildings typically use natural ventilation and infiltration. Large buildings such as apartment buildings, hospitals, industrial plants, and office buildings use mechanical ventilation and infiltration.

17.6.1 Driving Mechanisms

The flow of air between two points is due to the occurrence of a pressure difference between the two points. This pressure difference results in a force on the air, causing airflow from the high pressure zone to the low pressure zone. Natural ventilation and infiltration are driven by pressure differences across the building envelope. A material balance on air is used to the derive equations that describe the amount of air flowing into and out of a building. If density differences are small, volumetric airflow can also be balanced. Air density does vary with temperature and elevation (for an elevation rise from 0 to 5000 ft, or temperature change from −20 to 70 °F, the air density will drop by ~20%).

When wind impinges on a building, it creates a distribution of static pressures on the building's exterior surface that depends on the wind direction and location on the building exterior but does not depend on the pressure inside the building. The pressure is positive on the windward building surface with respect to ambient pressure, and it is negative on the leeward building surfaces. Wind usually depressurizes a house, leading to infiltration into a building on the windward side and exfiltration on the other sides (Figure 17.8).

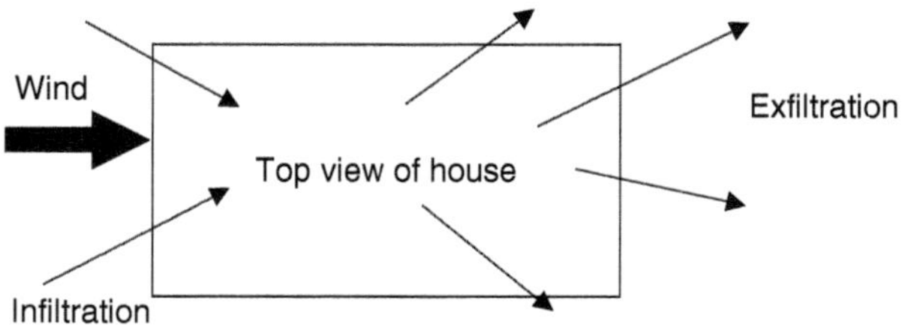

Figure 17.8 Wind depressurizes a house leading to infiltration on windward side.

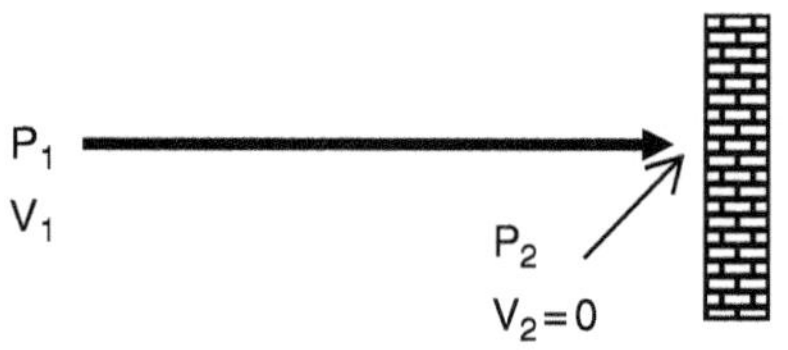

Figure 17.9 Bernoulli's equation applied to the windward side of a house.

To derive the pressure difference caused by wind, we apply Bernoulli's equation as the wind blows against building shell (Figure 17.9).

$$P+\frac{1}{2}\rho v^2+\rho gh=\text{constant} \tag{17.18}$$

$$P_1+\frac{1}{2}\rho v_1^2=P_2$$

The dynamic pressure term $(1/2)\rho v_1^2$ for this application of Bernoulli's is actually multiplied by a coefficient called the surface pressure coefficient C_s, which varies with location around the building envelope and wind direction. This modifying factor is needed because the pressure developed on the windward wall surface is not actually exactly the dynamic pressure $(1/2)\rho v_1^2$, because the air slips around the sides and over the top of the building, resulting in a surface pressure that is actually somewhat lower than the dynamic pressure:

$$\Delta P_{\text{wind}}=C_s\frac{1}{2}\rho v_1^2 \tag{17.19}$$

where ΔP_{wind} is the pressure on wall relative to the wind freestream pressure (Pa); C_s is the surface pressure coefficient (~$O(1)$), which varies from positive to negative as the angle of impingement varies from 0 to 180°; ρ is the air density (usually $1.2\,\text{kg}\,\text{m}^{-3}$); and v_1 is the freestream air velocity ($\text{m}\,\text{s}^{-1}$). A typical value is ΔP_{wind} ~ 7 Pa for $C_p=0.5$, $v_1=5\,\text{m}\,\text{s}^{-1}$.

Wind speed incident on a building is generally lower than average meteorological wind speed because of the effects of building height, terrain, and shielding by trees and bushes. Wind speed is 0 at ground level and increases with height up to an altitude of ~600 m above ground level. Meteorological measurements are typically made at a height of 10 m in open areas. The wind speed profile can be approximated by the power-law formulation:

$$v(z)=v_{10}\left(\frac{z}{10}\right)^p \quad \text{for } z<200\text{ m} \tag{17.20}$$

where z is height above the ground in m, v_{10} is the observed wind speed at 10 m, and the value of p ranges between 0.07 and 0.55, depending on the meteorological conditions of the atmosphere.

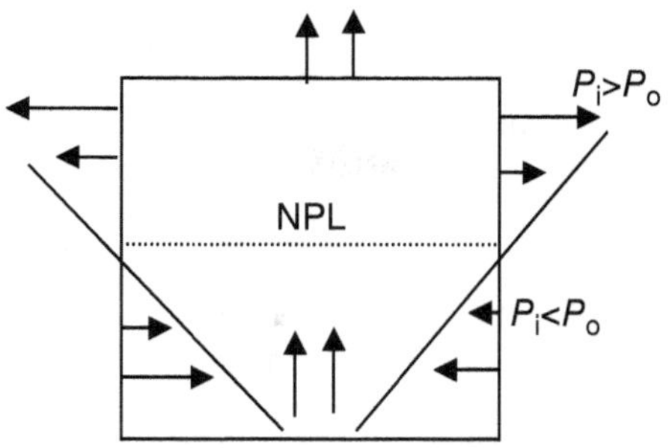

Figure 17.10 Stack effect for no wind winter conditions. NPL is the neutral pressure level.

17.6.1.1 Temperature

Temperature differences between indoors and outdoors cause density differences and therefore pressure differences. These pressure differences induce what is called the *stack effect*. During the heating season, the warmer inside air rises and flows out of the building near its top. It is replaced by colder outdoor air that enters the building near its base. During the cooling season, this directional flow is reversed and somewhat smaller since temp differences are usually smaller. Thus during the heating season, $T_i > T_o$ and pressure is positive outward at the upper part of building and positive inward across the lower part of building. The height above the ground at which the indoor and outdoor pressures are equal is called the neutral pressure level (NPL) (Figure 17.10). (Note: the stack effect can also induce soil gas entry.)

To derive the pressure difference caused by temperature, we use the principle of hydrostatics, $(dp/dz)=-\rho g$, and the ideal gas law,

$$\Delta P_{\text{temp}}=\sim 0.04 z_n \Delta T \tag{17.21}$$

where $\Delta P_{\text{temp}}=P_o-P_i$ (Pa), z_n is the height above the NPL (m), and $\Delta T=T_o-T_i$ (K).

Typical values are ΔP_{temp} ~ 0.2 Pa for a temperature difference of 10 °C at a height of 0.5 m and ΔP_{temp} ~ 1.6 Pa for a temperature difference of 20 °C at a height of 2 m.

17.6.2 Residential Ventilation Practices

Homes are ventilated mainly by infiltration plus some from natural ventilation (open windows). Some homes also have a forced-air central H(V)AC. (The V is in parentheses because some systems deliberately bring in outside air called ventilation air and some do not.) The practices are different in different regions of the country; for example, in the South homes have central air conditioning, whereas homes in the North have central heating. If there is a central H(V)AC system, it is sized for the most severe weather in the area. It is operated only intermittently (a 10–20% duty cycle is typical). These systems also have very crude filtration capabilities, with typical household filters having a MERV of 6–8. The airflow rate when the central H(V)AC system is operating is between 5 and 10 air changes per hour.

What about mixing characteristics of residences? Mixing happens pretty rapidly between rooms connected by open doors on the same floor; when doors are closed, the room itself mixes, but the air does not mix with the air in the other rooms of the home. Typically the upper stories of a residence are poorly connected to the lower floors.

Infiltration is driven by pressure differences from both indoor–outdoor temperature differences and the wind. Also, unbalanced mechanical/thermal flows, such as a bathroom fan or fireplace operation, can induce infiltration. Infiltration/air leakage occurs where two dissimilar materials meet such as at windows and door frames, where the walls meet the floor, and at electrical and plumbing penetrations. Air leakage and H(V)AC operation is coupled because leakage occurs through ducts too. Typical residential infiltration rates are ~0.6 ACH, which is the median for US homes.

Using mainly infiltration to ventilate is backwards for a number of reasons. First, the air exchange rate will be low when driving forces are low and higher when driving forces are large. Typically the driving forces are lower in the summer when the indoor–outdoor temperature difference is not as large, so less fresh air will enter the residence in the summer compared with the winter. In the winter, more fresh air will enter the residence, but it will often be colder air than the air inside, so more energy will be used to heat and the energy costs will be higher. Residences can be especially drafty in cold climates due to extreme temperature differences. Condensation occurs with exfiltration in cold climates, resulting in material damage risks to the building.

Also, house owners have very limited control over the quality and quantity of fresh air that enters their house (can only open windows, for example). There are limited opportunities for controlling indoor concentrations of outdoor pollutants by infiltration.

It would be desirable to give homeowners a higher degree of air quality control. This could be achieved with a tighter building envelope and the use of a mechanical ventilation system. Also, control devices could be used such as efficient filters on air intakes in polluted areas. Heat recovery devices could be used in colder climates. In fact, this is happening in many regions in the country to save on energy while still providing adequate ventilation air (California Energy Commission, 2015).

17.6.3 Large Building Ventilation Practices

Large buildings are ventilated using mechanical ventilation systems. Typically in these systems, 80% of the air is recirculated and 20% of the air is makeup air from outside. Makeup air is a ventilation term used to indicate the supply of outdoor air to a building to replace air removed by exhaust ventilation and combustion processes. Recirculated air refers to the air that has been exhausted from a room or building and is appropriately cleaned (sometimes) and supplied back into the room or building. These systems couple thermal control with ventilation. This can be a disadvantage because it does not allow for utilizing energy efficiently and having good air quality at the same time. In many systems, an economizer is used that increases the makeup air when T_o is favorable (or close to the indoor temperature). This is important for cooling air with large internal heat loads (i.e. building that houses computers).

The fresh air intake should be located remote from any contaminating sources such as exhaust stacks, furnace exhausts, or parking areas. Usually the air is filtered to protect the equipment and provide maximum heat exchange efficiency. It is not a common practice to filter the air to remove pollutants. Filtration is only used to keep the mechanical system clean from larger particles that might damage the fans, for example. Makeup air registers should be located in a room to provide cross ventilation, promote air distribution, and supply clean air in the breathing zone of the room. Many supply and exhaust registers are located such that short circuiting can happen, such as in the ceiling. This may not be the best configuration for promoting good IAQ. The exhaust location can also be far away from the pollutant source. In many buildings, air is exhausted from bathrooms, labs, or kitchen facilities. Typically a small positive pressure is maintained to reduce leakage into the space.

17.6.4 Airflow Through Openings

The relationship between airflow through an opening in the building envelope, such as a door or window, and the pressure difference between the indoor and outdoor environment across this opening is called the leakage function of the opening. Air leakage is a physical property of the building envelope that depends on the envelope design, construction, and deterioration over time. Leakage is measured by imposing a uniform pressure difference over the entire building envelope and measuring the airflow rate required to maintain this difference. Such a distribution would never really occur naturally, but it does provide a useful measure of the airtightness of a building. The form of the leakage function depends on the geometry of the opening.

A relationship between flow and pressure difference can be derived based on a solution to the Navier–Stokes equation for the idealized case of infinite parallel plates with several simplifying assumptions such as steady, fully developed laminar flow:

$$Q = C_D A \sqrt{\frac{2\Delta P}{\rho}} \quad (17.22)$$

where Q is the airflow rate ($m^3 s^{-1}$), C_D is the discharge coefficient for the opening (no units) that depends on the opening geometry and the Reynolds (Re) number of the flow, A is the cross-sectional area of the opening (m^2), ρ is the air density ($kg\,m^{-3}$), and ΔP is the pressure difference across opening (Pa). The Reynolds number is $Re = (Ud/\nu)$, where U is the velocity through the opening, such as a crack ($cm\,s^{-1}$); d is the width of opening (cm); and ν is the dynamic viscosity of air ($0.15\,cm^2\,s^{-1}$). For Re << 1 in the case of low velocity and a small opening, $C_D \propto \Delta P^{0.5}$. Therefore $Q \propto \Delta P$ and airflow resistance is dominated by fluid drag. For $Re \gg 1$ in the case of high velocity and large openings, C_D is constant $(O(1)) \rightarrow Q \propto \Delta P^{0.5}$, and resistance is dominated by inertia.

Based on experimental observations, for buildings in general, $Q \propto \Delta P^n$, where $0.6 < n < 0.7$ (limits are $0.5 < n < 1$). Cracks with larger flow resistances, greater flow lengths, or narrower widths tend to have an exponent closer to 1 compared to gaps with less resistance (the expression is not great over a broad range of ΔP).

Pressure differences are added together and the airflow rate through each opening is determined due to this total pressure difference. Airflow rate through openings cannot be summed since flow is not linearly related to pressure difference. Equation 17.23 details a model to superimpose flows from wind and stack effects:

$$Q_{ws} = \sqrt{\left(Q_w^2 + Q_s^2\right)} \quad (17.23)$$

where Q_{ws} is infiltration from both wind and stack effects, Q_w is infiltration from wind, and Q_s is infiltration from stack effects.

17.6.4.1 Natural Ventilation Flow Rates

Natural ventilation is the intentional flow of outdoor air because of wind and thermal pressures through controllable openings (windows, doors, skylights, roof ventilators). There are many factors that affect natural ventilation rates due to wind, including average wind speed, prevailing direction, seasonal and daily variations, and local obstructions (buildings, trees, shrubbery). Wind speed is typically lower in summer than winter, and directional frequency is also a function of season. Inlets should face directly into prevailing wind to get the best flow, and they are usually designed for wind speeds of one-half the seasonal average:

$$Q = C_v A U_w \quad (17.24)$$

where Q is the airflow rate ($m^3 s^{-1}$), C_v is the effectiveness of openings (no units), A is the cross-sectional area of inlet openings (m^2), U_w is the wind speed ($m\,s^{-1}$), and ΔP is the pressure difference across opening (Pa). C_v is assumed to be 0.5–0.6 for perpendicular winds and 0.25–0.35 for diagonal winds.

17.6.4.2 Flow Caused by Temperature Differences

If building internal resistance is not significant, the flow caused by the stack effect can be expressed by

$$Q = KA\left[2g\Delta h_{NPL}\left(\frac{T_i - T_o}{T_i}\right)\right]^{0.5} \quad (17.25)$$

where Q = airflow rate ($m^3 s^{-1}$); K is the discharge coefficient for openings (no units); A is the cross-sectional area of inlet openings (m^2); T_i, T_o are temperature indoors and outdoors, respectively (K); and Δh_{NPL} is the height from midpoint of lower opening to the NPL (m).

Equation 17.25 applies when $T_i > T_o$. K accounts for all viscous effects such as surface drag and interfacial mixing.

Estimation of Δh_{NPL} is hard. If one door represents a large fraction (e.g. 90%) of the total opening area in the envelope, the NPL is at the midheight of that aperture and Δh_{NPL} is one-half its height. Flow is then bidirectional and K is calculated according to $K = 0.40 + 0.0025|T_i - T_o|$.

Otherwise, $K = 0.65$ should be used when flow is unidirectional through the opening, and there are enough other openings to balance the flow.

17.6.5 Measuring Air Leakage

Air leakage characterizes the relationship between the pressure difference across the building envelope and the airflow rate through envelope. It is a physical property of a building and is determined by its design, construction, and seasonal effects. It is not the same as air exchange rate! Airtightness does influence air exchange rate; it is useful for comparing buildings to one another or to airtightness standards, for evaluating design and construction quality, and for studying effectiveness of retrofits. No simple relationship exists between air leakage and air exchange rate. We usually measure the leakage area of a house and then predict infiltration rate using an empirical model. This model can be used estimate infiltration, but not total ventilation.

Air leakage varies little with time and weather conditions.

Fan pressurization is used to measure air leakage. A large fan or blower is mounted in a door or window using a blower door. The blower door consists of a big fan, mounted on plywood panel with nylon sheeting to close the opening between the door frame and the fan, and is installed in place of exterior door or window. All other doors and windows are closed. Fan speed is adjustable, and a display gives a speed in RPM. The fan is calibrated at the manufacturer. A large and roughly uniform pressure difference is induced across the building shell. A micromanometer is used to measure the indoor–outdoor pressure difference. The airflow required to maintain this pressure difference is then measured. Measurement of pressure difference and speed yields a known flow rate, Q. It is important to measure Q vs. ΔP by the varying speed of the fan; and it is common to include both + and $-\Delta P$. The leakier the building, the more airflow that is necessary to induce a specific indoor–outdoor pressure difference.

Results of a blower door test consist of several combinations of pressure difference and airflow rate data. Usually the data is fit to a curve in the form of $Q = c\Delta P^n$, where c is the empirical parameter related to leakage area, n is the size of individual leaks, and Q is the flow rate. Usually two separate curve fits are done for the depressurization and pressurization, yielding two sets of c and n; the average is then calculated and used. In some cases, the predicted airflow rate is converted to an equivalent effective leakage area, L:

$$L = \frac{c\left(4\,\text{Pa}\right)^n}{\sqrt{(2/\rho)\left(4\,\text{Pa}\right)}} \tag{17.26}$$

A typical leakage of a home is 200 cm^2.

An empirical infiltration model can be used to predict Q from data on L, ΔT, and U_w and information on shielding plus building height to predict Q (ASHRAE, 2009a).

17.6.6 Ventilation Efficiency

The ventilation efficiency, ε_V, expresses the efficiency in extracting contaminants generated in indoor environments by ventilation and is defined as the ratio of two concentrations:

$$\varepsilon_V = \frac{C_e - C_o}{C_r - C_o} \tag{17.27}$$

where C_e is the contaminant concentration in the exhaust air, C_o is the contaminant concentration in the outdoor air, and C_r is the contaminant concentration at the location of interest, r.

17.6.7 Air Exchange Effectiveness

The air exchange effectiveness, E_A, expresses how the air is mixed within a room:

$$E_A = \frac{\tau_n}{\tau_a} \tag{17.28}$$

where τ_n is the nominal time constant, which equals the shortest time required to replace the air within the space at a given flow rate and space volume ($=Q/V$) and has the units of 1/time, and τ_a is the average residence time of the room air, which equals the *average* time it takes to replace the air present in the space.

The air exchange effectiveness is equal to one for piston-type (or plug-flow) ventilation, whereas for complete mixing, it is equal to 0.5. Short-circuiting of air will give rise to an effectiveness that is lower than 0.5.

17.6.8 Air Age

Consider the air in a room as made up of a large number of very small volume elements. Each element enters the room at a specific time and then circulates through the room for a specific length of time before leaving the room by ventilation. Define a fixed point in time, t_o. Each air element is assigned an age, defined as the difference between t_o and the time when the element entered the building from outdoors.

Using the decay method (17.7.1.1) to measure air age,

$$C(t) = C_o e^{-\frac{t}{\tau}} \tag{17.29}$$

where $C(t)$ is the concentration at a given time t, C_o is the initial concentration for decay test, and τ is the air age of that point.

17.7 Ventilation Measurements

There are three main methods for measuring the ventilation rate in a building:

1) *Tracer gas techniques* – These do not measure ventilation rate directly; rather they measure the concentration evolution of a tracer gas compound. A model framework is needed to interpret the data.
2) *Air velocity measurements* – These measure flow velocities directly from ventilation supply grills. These data are then converted to volumetric flows.
3) *Fan pressurization* – This method measures the air leakage of a building, usually with a blower door, which can be related to ventilation rate through empirically derived models.

17.7.1 Tracer Gas Techniques

The basic method for conducting a tracer gas study is the following:

- Release a chemical tracer or marker into the indoor air.
- Measure the concentration of the tracer as it evolves over time.
- To determine the ventilation rate, interpret the results using a model.

There are three different implementations: tracer decay, constant injection, and constant concentration. Data from a tracer gas study is interpreted using a mass balance on the tracer gas within the building. Assuming the outdoor concentration is zero, the mass balance becomes

$$V\frac{dC(t)}{dt} = E(t) - Q(t)C(t) \qquad (17.30)$$

with initial condition $C(t=0) = C_0$, where V is the volume of space being tested (m^3), $C(t)$ is the tracer gas concentration at t ($\mu g\,m^{-3}$), E is the tracer gas injection rate at time t ($\mu g\,h^{-1}$), and $Q(t)$ is the airflow rate out of building at time t ($m^3\,h^{-1}$).

In the above equation, density differences between indoor and outdoor air are generally ignored. Assumptions are only source of tracer is by injection, and only removal is by ventilation (no deposition onto indoor surfaces). Another major assumption is that the tracer is instantaneously mixed within the room volume. Note that if CO_2 is used as the tracer, then the outdoor concentration is not zero and must be accounted for in the mass balance.

Tracer gases should be chemically inert and nontoxic (e.g. this makes CO difficult to use), not normally present at significant levels (e.g. this makes it hard to use CO_2), and relatively straightforward to measure accurately over one order of magnitude of concentration or more. Some commonly used tracers for building studies are:

- Sulfur hexafluoride, SF_6 (can measure with GC/ECD down to ppb/ppt levels and w/IR absorption down to ppm levels).
- Perfluorocarbons (measure with GC/ECD down to ppt levels or less), for example, bromotrifluoromethane (R13B1).
- Other tracers include radon, CO_2, CO, etc.

17.7.1.1 Tracer Gas Decay

A small amount of tracer gas is injected into room air. It is allowed to mix thoroughly with the interior air. A CMFR model is applied to interpret the data and assumes complete mixing. After injection, the mass balance becomes

$$V\frac{dC(t)}{dt} = -Q(t)C(t) \qquad (17.31)$$

The concentration of tracer gas as a function of time is measured. (The rule of thumb is that 5–10 samples spaced over at least 1 characteristic time.) The model equation describing the concentration is

$$C(t) = C_{\text{initial}} e^{-\frac{Q}{V}t} = C_{\text{initial}} e^{-\lambda_V t} \qquad (17.32)$$

where C_{initial} is the concentration in the room after tracer gas injection and can be estimated by the mass of tracer gas/volume of room.

Figure 17.11 shows the concentration of CO as it decays due to infiltration/exfiltration in a home. To use these concentrations as a tracer gas to determine the infiltration rate, make a plot of the natural logarithm of $C(t)$ vs. t, as shown in Figure 17.12. You should see a straight line with a negative slope. Extract the best estimate of the slope. The negative of the slope is the air exchange rate. (To find the slope, usually do a least-squares fit of the data to a straight line.)

The equation for the line is $\ln C(t) = \ln C_{\text{initial}} - \lambda_V t$.

There are advantages to using this method. Because logarithms of concentration are used to interpret data, only relative concentrations are needed; this can simplify calibration of the gas analyzer because the absolute value

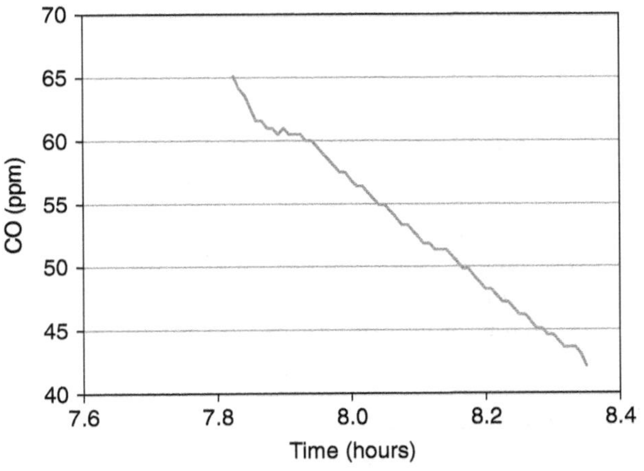

Figure 17.11 Concentration of CO versus time measured in a study of an unvented gas fireplace.

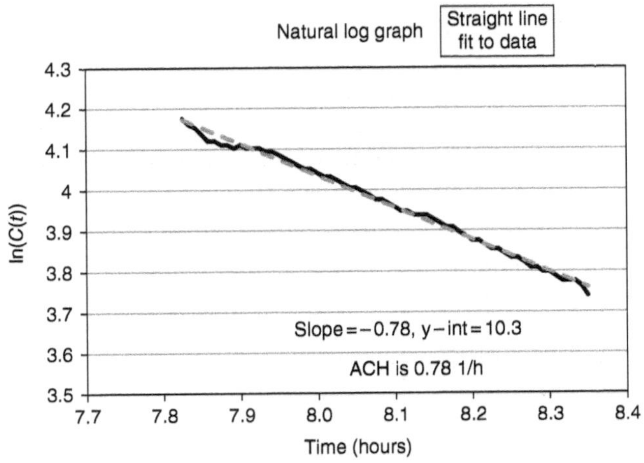

Figure 17.12 Natural log of the concentration of CO versus time, from Figure 17.11.

of the concentrations is not needed. Also the injection rate (or injection mass) does not need to be measured. You do need to make sure you inject enough so that you are within the measurement range of your instrument.

There are also disadvantages. A serious problem is imperfect mixing of tracer gas within interior air, especially at initial injection. A single room is usually well mixed, but it will take some time for the gas to mix into the space depending on the air exchange rate, temperature, air currents, etc. Little analysis has been done on the magnitude of errors due to poor mixing, unfortunately. Also, of concern is what happens if the air exchange rate changes with time during the measurements.

17.7.1.2 Constant Injection

In the constant injection method, tracer is injected into the indoor space at a constant rate. The concentration is measured over time. Data is interpreted using a CMFR model. The solution to the mass balance is

$$C(t)=\frac{E}{Q}\left(1-e^{-\lambda_V t}\right) \tag{17.33}$$

After sufficient time, the transient term of the mass balance goes to zero and the concentration reaches steady state (generation = removal):

$$Q=\frac{E}{C_{ss}} \tag{17.34}$$

The emission rate, E, of the tracer, must be known. Using Equation 17.34, the ventilation rate Q is found by knowing the steady-state concentration C_{ss}. The above equation is only valid when the ventilation rate, Q, is constant! Therefore, it is only appropriate for systems at or near equilibrium. It is useful in spaces with mechanical ventilation or with high air exchange rates.

In theory, the time dependence of the air exchange rate could be estimated using the transient equation for $C(t)$ either Equation 17.33 or the original CMFR differential equation. This is hard to do in practice, however, because it requires measurement of the absolute tracer concentrations and injection rates. An advantage to this method is that because injection is continuous, no initial mixing period is required. The volume of space must be known and sometimes this is hard to estimate; note that this is not needed in the tracer gas decay method. Again, there could be poor mixing of air, which can be a problem as with the decay method. This method can be used for measuring the "average" air exchange rate over a long period of time.

Dietz et al. (1986) developed a special case of the constant injection technique using permeation tubes as a tracer gas source. The tubes release tracer at an ideally constant rate into the building. An air sampling tube is packed with an adsorbent that collects the tracer at a constant rate by diffusion. After sampling for 1 week or more, the sampler is removed and analyzed by gas chromatography (GC)/ECD to determine average tracer gas concentration. These data result in a long-term average ventilation rate, depending on the time period over which sampling was conducted.

17.7.1.3 Constant Concentration

In the constant concentration method, the tracer gas injection rate is continuously adjusted to maintain constant concentration within the building. If the concentration is constant, then $dC/dt=0$ and the CMFR mass balance is (Equation 17.35)

$$Q(t)=\frac{E(t)}{C} \tag{17.35}$$

Similar to the constant injection method, because the injection is continuous, no initial mixing period is required for the tracer gas. Tracer concentration in each zone of the building can be separately controlled by injection into each zone; thus the amount of outdoor air flowing into each zone can then be determined. This method requires measurement of the absolute tracer concentrations and injection rates. It also requires knowing the volume of the space. And again, imperfect mixing of tracer into the indoor air can cause issues. Specifically in this case, it can cause delay in the response of the concentration to changes in the injection rate. This can make it impossible to keep the concentration constant, so Equation 17.35 is only approximation.

17.7.2 Air Velocity Measurements

One method often used in mechanically ventilated buildings is to measure airflow velocity directly at the ventilation supply grill. The velocity is then multiplied by the area of the outlet and then divided by the volume of the space to get the air exchange rate. This method works best for mechanical ventilation since velocities in naturally ventilated spaces are low and difficult to measure. Also, what is the area that you would use in a naturally ventilated space (e.g. window opening)? For mechanical ventilation systems, the velocity is most often measured coming out of supply grill, and then this velocity is multiplied by the area of grill. This method does not correct for ventilation inefficiencies, that is, the areas of the room that are not being ventilated – again, a mixing issue. Also care needs to be taken to ensure that the velocity is measured at multiple points at the supply grill so that any fluctuations can be accounted for. Hot-wire probes are commonly used for this method, providing one-dimensional measurements. The detection limits range from 0.2 to 1 $cm\,s^{-1}$ for these devices. A sonic vector anemometer will provide three-dimensional data with a range of 0.5–2000 $cm\,s^{-1}$ and typical random noise of 0.1 cm s^{-1}. Additional methods are laser Doppler velocimetry and particle image velocimetry, but these are mostly used in research applications.

17.7.3 Fan Pressurization

Fan pressurization is quick and inexpensive and measures air leakage of a building (as long as a blower door is available). It characterizes the airtightness of the building envelope and is mostly independent of weather conditions. Air leakage is a physical property of the building. There is no simple relationship between a building's air leakage and its air exchange rate. This method only accounts for infiltration and not ventilation (both mechanical and natural). Calculation equations are derived empirically.

To do a fan pressurization test, mount a large fan or blower in a door or window; this is the blower door device and consists of a frame, nylon sheeting, and a large fan. Doors and windows are closed, any gas sources must be extinguished, fire places must be cleaned of ash, etc. The fan induces a large, uniform pressure difference across building shell. The airflow of the fan is varied and pressure differences across the blower door are measured. The leakier the building, the more airflow is needed to account for the pressure difference. The relationship between the pressure difference across the building envelope and the airflow rate through the envelope is modeled by an equation of the form (17.36)

$$Q = c\Delta P^{n} \tag{17.36}$$

where

Q is the fan airflow rate (m^3 s^{-1}), ΔP is the pressure difference (Pa), c is an empirical parameter (units of c are determined by value of n), and n is the empirical parameter representing the size of individual leaks (unitless).

The airflow rate Q is usually reported at the reference pressure difference of 50 Pa. Effective leakage area is also reported and serves as a single measure of the building's airtightness. One equation for L is (ASHRAE, 2009a)

$$L = \frac{c\left(4\,\text{Pa}\right)^{n}}{\sqrt{\left(\frac{2}{\rho}\right)\left(4\,\text{Pa}\right)}}\left(\text{density } \rho \text{ in kg m}^{3}\text{, units of } L \text{ are m}^{2}\right) \tag{17.37}$$

17.8 Thermal Comfort and Psychrometrics

A main purpose of HVAC systems is to create conditions in a confined space that address human thermal comfort. Modern HVAC systems are used to heat, cool, humidify, or dehumidify depending on the climate. These systems are designed to satisfy the needs of human beings, which can be thought of as heat engines with food as the energy input.

Heat transfer is proportional to the temperature difference between a person and the surrounding space. In cold environments, a body loses more heat than it normally generates. In hot environments, a person does not dissipate enough heat from their body. Both of these conditions can lead to discomfort. A resting adult produces about 100 W of heat. Normalizing this to skin area results in ~58 W m^{-2}, and this is equal to 1 met or 50 kcal (h·m^2)$^{-1}$ (this is based on average male European skin surface area). This is also equal to 50 kcal (h·m^2)$^{-1}$. A normally healthy 20-year-old male has a maximum capacity of 12 met, which drops to 7 by age 70. Maximum rates for females are about 30% lower.

Thermal comfort is quite subjective and is influenced by physical, physiological, psychological, and other cognitive processes. Usually people experience comfort when body temperatures are held within narrow ranges, skin moisture is low, and the physiological effort of regulation is minimal. Most people feel comfortable when the temperature is between 22 and 27 °C, when the RH is 40–60%, and when the air motion has a velocity of about 0.25 m s^{-1} (Çengel and Boles, 2008; ASHRAE, 2010). For the relative performance of office worker performance to remain between 90 and 100%, the temperature should not swing more than −6 to 8 °C away from the optimal comfort temperature (Çengel and Boles, 2008). The temperature range in which the least amount of complaints was recorded in 690 commercial buildings was between 21 and 24 °C (ASHRAE, 2009b, from Federspiel, 1998).

Personal Parameters: To design a comfortable thermal environment, both personal and environmental parameters should be considered. Personal parameters include the surface area of the body, the metabolic requirements for a given activity within the environment, the heat produced for the work being done in the environment, and knowing the type of clothing being worn. Environmental parameters include air temperature, wet-bulb temperature, dew-point temperature, water vapor pressure, total atmospheric pressure, RH, and humidity ratio.

There is no lower humidity limit for thermal comfort. ASHRAE Standard 55 recommends that the dew-point temperature be >2 °C (ASHRAE, 2010). To prevent warm discomfort, it is recommended that the RH not exceed 60%. Elevated air speeds can be used to make the environment more comfortable beyond the maximum temperature.

Improving indoor environmental quality enhances productivity (Wyon, 2004). In one study, the performance of simulated office tasks improved with increasing outdoor air ventilation rates (Wargocki et al., 2000). Increasing ventilation rates also decreased the number of workers dissatisfied with the air quality, reduced odor intensity, and increased the perceived air freshness. There is good scientific evidence that the characteristics of indoor environments influences rates of respiratory illness, allergy and asthma symptoms, sick building symptoms, and worker performance (Fisk, 2000; IOM, 2000). It is estimated that the potential annual savings and productivity gains alone are $20–$160 billion from direct improvements in worker performance.

17.8.1 Psychrometrics

Psychrometrics uses thermodynamic properties to analyze air-conditioning applications. Psychrometric charts were developed to present the thermodynamic data needed in the form of easily readable charts.

Atmospheric air is a mixture of nitrogen, oxygen, and trace amounts of CO_2 and Ar, as well as water vapor and contaminants including PM and gaseous pollutants not normally present in clean atmospheric air. *Dry air* is the air that contains no water vapor. *Moist air* is the two-component mixture of dry air and water vapor. Dry air and water vapor for air-conditioning applications can be treated as an ideal gas with a constant specific heat, since the temperature ranges from −10 to 50 °C.

RH is the ratio of the amount of moisture that the air holds (P_v), or vapor pressure, relative to the maxiumum amount of moisture the air can hold at the same temperature ($P_{sat@T}$), also called saturation pressure:

$$\phi = \frac{P_v}{P_{sat@T}} \tag{17.38}$$

Saturation pressure increases with increasing temperature.

Absolute humidity (also called humidity ratio) is the ratio of the mass of water vapor present, m_v, to the mass of dry air, m_a:

$$\omega = \frac{m_v}{m_a} \tag{17.39}$$

The wet-bulb temperature T_{wb} is the temperature that a parcel of air would have if adiabatically cooled (adiabatic means at constant pressure) to saturation by evaporating water into it; and the parcel of air supplies all the latent heat needed.

The features of the psychrometric chart are shown in Figure 17.13. The chart helps to also visualize the air-conditioning process. Figure 17.14 shows the various

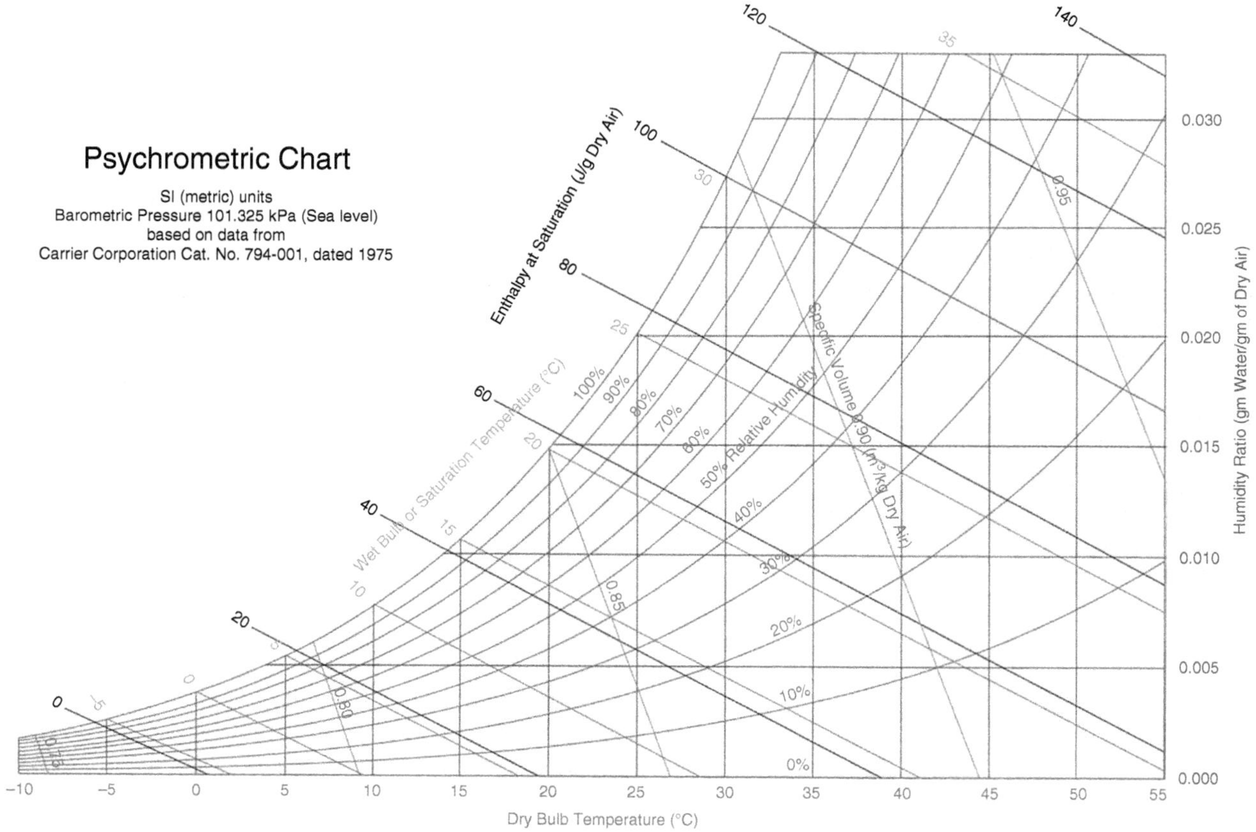

Figure 17.13 Psychrometric chart used for air-conditioning applications. *Source:* https://commons.wikimedia.org/wiki/File:PsychrometricChart.SeaLevel.SI.svg.

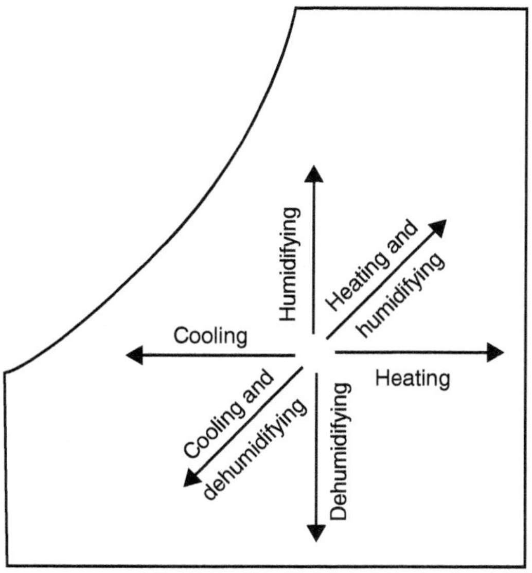

Figure 17.14 Various processes that are used to maintain a space at a desired environmental quality in the context of a psychrometric chart. *Source*: Adapted from Çengel and Boles (2008).

processes that are used to maintain a space at a desired environmental quality (from Çengel and Boles, 2008).

17.9 Energy Efficiency Retrofits

More than seven million US homes have received energy efficiency improvements since the start of the Weatherization Assistance Program in 1976. These improvements include heating and cooling upgrades, improved insulation, air sealing, caulking, weather stripping to doors and windows, and window replacement. Weatherized households are now seeing \$250–\$480 annually in energy savings. Homes that are weatherized under federal programs are required to have added ventilation (according to ASHRAE 62.2) if the home becomes too tight. This is typically done by adding a continuous exhaust ventilation system to the home, which increases ventilation by infiltration through the building shell. Many homes are not retrofitted with federal funds, but by state funds or industrial programs. Unfortunately, tightening up a home can also have negative impacts on the IAQ within a home.

17.9.1 Climate Change

Energy efficiency retrofits are applied because of climate change and the desire to save energy and reduce CO_2 emissions. The average temperature of the Earth has risen by 1.4 °F over the past 100 years. Over this same time period, human activities have released increasing amounts of carbon dioxide (CO_2) and other greenhouse gases into the atmosphere. In the intermountain west, temperatures have increased by 2 °F in the last 30 years (WWA, 2008). Climate models predict that this region will warm by 4 °F by 2050, relative to a 1950–1999 baseline.

17.9.2 Building Energy Use in the United States

Commercial (19%) and residential (22%) buildings accounted for 41% of primary energy consumption in 2010, directly translating into CO_2 emissions (DOE, 2011a). For residential buildings, space heating accounted for 45% of site energy consumption (DOE, 2011b). In Colorado, space heating accounts for more than half of household energy use (54%), compared with air conditioning, which accounts for 1% (EIA, 2009). Much of this energy is lost by escaping through unintentional openings in the homes, like cracks around doors and windows. Weatherization home improvements can save up to 30% energy.

17.9.3 Indoor Air Quality Changes with Weatherization

Outdoor pollutant levels, indoor sources, natural infiltration or mechanical ventilation, pollutant transformation, and surface deposition determine indoor pollutant levels in homes (IOM, 2011). Indoor sources produced by occupant behavior are episodic (cooking, showering) or intermittent (painting, pesticide use). Sources produced by the home construction are mainly continuous (emission from furnishing, materials, stored products). Home energy retrofits, or weatherization improvements, can improve IAQ by remediating existing hazards such as lead or radon, reducing air exchange with outdoor air lowering outdoor pollutant levels indoors, removing pollutant sources such as water leaks and unvented heaters, and adding functional ventilation and/or filtration (IOM, 2011; Nazaroff, 2013). On the other hand, weatherization can worsen IAQ by disturbing legacy pollutants such as lead or asbestos, reducing ventilation leading to an increase in indoor pollutants, introducing new formaldehyde-emitting construction materials, and failing to install mechanical venting when it is needed or installing unreliable systems.

Formaldehyde (HCHO) is a toxic air pollutant and classified as a human carcinogen, which is pervasive in homes (IARC, 2006; Hun et al., 2010). Lower ventilation rates in homes are an important determinant of HCHO levels in homes (Gilbert et al., 2005). There are numerous indoor sources of HCHO, such as pressed-wood products and consumer products (Gunschera et al., 2013). Emissions are continuous over time and levels could be significantly impacted by weatherization.

IAQ studies of weatherized residential buildings have shown that some pollutant levels increase, while others decrease. A recent study by Noris et al. (2013) conducted energy retrofits on 16 low-income multifamily apartments that were designed to save energy and improve IAQ: particulate levels, CO_2, and VOCs generally improved, whereas formaldehyde (HCHO) and nitrogen dioxide (NO_2) varied by building; larger decreases in indoor-sourced pollutants were realized with larger increases in ventilation rate, and IAQ improved more in buildings with added mechanical ventilation (excluding particles). A modeling study of weatherization impacts on IAQ (Emmerich et al. (2005) showed that airtightening by 40% (reducing leakage area from 12.5 to 7.5 ACH50) worsened occupant exposures to all pollutants and most interventions reduced some pollutants and increased others. Kitchen exhaust had the broadest positive effects.

17.9.4 Health Impacts Related to Weatherization

People spend a majority of their time indoors, and much of that time is at home. Weatherization has the potential to worsen IAQ, possibly resulting in adverse health effects. Health impacts of poor IAQ depend on the pollutant. The top four indoor air pollutants that have the highest chronic adverse health impact are PM with diameters less than 2.5 μm ($PM_{2.5}$), Second hand Tobacco Smoke, radon (smokers), and formaldehyde (Logue et al., 2012). Exposures to pollutants from indoor sources, or as the result of occupant behavior, have the potential to increase when a home becomes tighter, resulting in higher exposures. Exposure to pollutants originating outdoors would decrease. Health effects studies, which have mostly been done recently on multifamily residential buildings, have shown varying outcomes.

Wilson et al. (2013) showed that:

- Self-reported respiratory symptoms were more frequent after retrofit (−26%), residents had greater sleep disruption due to asthma (−28%), and improvements in general health were reported (reduced asthma medication usage, hypertension, sinusitis).
- There was no provision of mechanical ventilation in the retrofits.

Breysse et al. (2011) reported that:

- A green renovation of three multifamily affordable housing buildings resulted in adults reporting significant improvement in overall asthma and nonasthma respiratory health and significant improvement in nonasthma respiratory health.
- Renovations included adding continuous and spot ventilation. Energy reductions of 45% were realized.

17.9.5 What to Do Next

There is an urgent need to understand the impacts of climate change and weatherization on resident health and IAQ. It is important to understand how home energy retrofits, which are being done all over the United States, impact the complex indoor environment. Weatherization has the potential to both increase the levels of indoor sources of pollutants and decrease the levels indoors of outdoor pollutants. It is a move in the right direction for weatherization programs that use federal funds to require assessing ventilation rates, and if they do not meet the guidelines from ASHRAE 62, then mechanical ventilation must be installed. This should also probably be required for all weatherization efforts.

17.10 Health Effects of Indoor Air Pollution

During this century, severe air pollution episodes caused excess mortality (an excess number of deaths over that expected for the particular time of the year and location was recorded) and morbidity (excess illnesses) establishing that atmospheric contamination by human activities can adversely affect human health. These pollution problems can be traced back to the thirteenth century when coal began to replace wood for domestic heating and industrial uses; the impact of high-sulfur coal on air quality was dramatic. The conditions during these episodes occurred when there were low inversion levels, which concentrated the pollutants in a relatively small space and resulted in heavy fogs of toxic pollution. In all of these severe air pollution episodes, there were increased levels of SO_2 and PM. The most devastating killer smog was in London in 1952 when there were an estimated 4000 fatalities. In this incident, the inversion height was as low as 150 ft in some places, and the visibility in the center of the affected area was below 22 yd. Concentrations of SO_2 were as high as 1.3 ppm and total particles of 4.5 mg m^{-3} were measured.

Chemicals become airborne in two ways, either as little tiny particles composed of many molecules or atoms (suspended particles, PM, aerosols) or as individual molecules or atoms (gases). Particles are respirable into the lung only when they are below a certain very small size, roughly 10 μm. The largest airborne particles (>10 μm) deposit on surfaces of the nasopharynx or throat and do not enter lungs. Smaller particles are breathed into the lungs and impinge on lung surfaces. A small range of particle sizes actually remain airborne upon inhalation and are breathed in and out with respiratory movements. Gases are also breathed in and out with the respiratory movements. The amount that impinges on the walls of the lung depends on the concentration in the air. Inhaled particles that do not dissolve in the fluid coating the lung surfaces may lodge in the lungs for a long period of time, maybe even permanently, or they may be swept back up the lung passages by cilial action and eliminated from the body by coughing.

The surface of the lungs is a poor barrier against the entry of chemicals into the body. The lung surface area is very large: 750 ft^2 compared with 20 ft^2 of skin. The job of the lungs is to transfer oxygen from air to the blood and CO_2 from the blood back to the air. The surface of the lung is a very delicate thin membrane, ~1 cell thick. It separates the air in the lungs from the blood in the tissues of the lungs. This membrane allows ready passage to the bloodstream not only of oxygen but also many other chemicals that may be present as contaminants in inhaled air. Chemicals that pass through the lung surface may injure its delicate membrane and interfere with its vital function. Asbestos and materials that contain crystalline silica, cotton, dust, and coal dust cause disease by physically damaging lung surfaces. These diseases are asbestosis, silicosis, and pneumoconiosis.

17.10.1 Classes of Pollutants and Summary of Health Effects

- *Microorganisms*: Transmit infections and allergenic diseases and cause and exacerbate asthma.
- *Combustion by-products* (SHS, auto exhaust, etc.): Chemicals from combustion cause cancer, irritation, and cardiovascular disease.
- *Outdoor air* (ozone, NO, NO_2, etc.): Causes irritation, respiratory diseases, asthma, and cardiovascular disease and increases mortality.

Refer to Samet et al. (1987a, 1987b) for a substantial review on health effects and sources of indoor air pollution.

17.10.2 Radon

Radon gas is generated in soils by emanation from certain radioactive soil constituents such as radium and then transported into buildings because of pressure differences. Radon undergoes decay indoors, generating decay products. Upon inhalation and lung deposition of these decay products, lung tissue is irradiated, and the risk of lung cancer is increased. Sources of information on health effects include studies of underground miners, animal studies, and an epidemiological study in 1994 in Sweden of nonoccupational exposures. Radon exposure is the most important source of radiation exposure to public. The health risk associated with average exposures is large, and some people are exposed at 10–100 × average or more. At average concentrations, the risk developing lung cancer is 1 in 1000.

17.10.3 Volatile Organic Compounds

There are many sources of VOCs including consumer products, household cleaners, hobby materials, printed materials, industrial strength ammoniated cleaning compounds, germicidal cleaners, household pesticides, furnishings, adhesives, paints, cigarette smoking, air fresheners, furniture waxes, paints, antiperspirant, and hair spray (aerosol form). Humans also produce VOCs such as acetone and toluene. Hundreds of species are detected indoors, and commonly at 2–5 times outdoor levels, sometimes 10×. Many compounds are neurotoxic, mutagenic, carcinogenic, and odorous and cause irritation of the central nervous system. An important question that has not yet been answered is, are there synergistic effects upon exposure to multiple VOCs at the same time?

17.10.4 Nitrogen Dioxide

The dominant indoor sources of NO_2 are gas cooking ranges and unvented combustion space heaters. The current National Ambient Air Quality Standard is 53 ppb (annual average) and 100 ppb (1-h exposure). The average gas range contribution is 25 ppb and can peak at 200–400 ppb. Nitrogen dioxide is a moderate oxidant causing direct lung damage. A first response is in the sensory organs (smell, sight) at 0.08–0.26 ppm. It also reduces resistance to respiratory infection, which is indicative of respiratory damage. Many studies report higher incidence of respiratory symptoms and disease in children living with gas stoves (NOx source) versus those in homes with electric stoves.

17.10.5 Ozone

Ozone is a pulmonary irritant that affects mucous membranes and other lung tissues' respiratory function. Most ambient ozone is produced as a result of the action of sunlight on NOx and hydrocarbons, which are contained in exhaust gas from cars and combustion processes. Ozone indoors is usually derived from outdoor air. Other indoor sources are photocopying machines and electrostatic air cleaners (especially if they are dirty). For copying machines, breathing-zone concentrations up to 68 ppb have been measured under normal working conditions. Cleaning and maintenance temporally reduced emissions, but the elevated levels returned after ~2 weeks.

17.10.6 Carbon Dioxide

Carbon dioxide is produced by human metabolism and exhaled through lungs. The amount normally exhaled by an adult is 200 ml min^{-1}. Exposure of healthy individuals for prolonged periods of 1.5% CO_2 apparently causes mild metabolic stress. Exposure to 7–10% will produce unconsciousness within a few minutes. Exposure of nuclear sub crews at 0.7–1% CO_2 demonstrated a consistent increase in respiratory minute volume and cyclic changes in the acid–base balance in blood. Ventilation standards are normally set to maintain CO_2 indoor concentrations <0.5% level.

17.10.7 Formaldehyde

Formaldehyde is a widely used industrial feedstock chemical. It is emitted from particle board and urea-formaldehyde foam insulation (UFFI) (now banned). Formaldehyde emissions from particle board tend to decrease exponentially with time, and the half-life is estimated to be an average of 53 months. UFFI is produced when the two major constituents are combined with a catalyst and forced from a pressurized nozzle. The product is cured to a hardened resin. HCHO emission is related to precursor resin and breakdown reactions in the foamed product. Exposure to HCHO causes burning and tearing of eyes, and general irritation of upper respiratory passages are the first signs experienced at HCHO concentrations in 0.1–5 ppm range. Odor is sensed at 1 ppm. Concentrations of 10–20 ppm produce coughing, tightening of chest, sense of pressure in the head, and palpitation of the heart and may occur in susceptible persons at <5 ppm. Exposure above 50–100 ppm can cause serious injury such as collection of fluid in lungs (pulmonary edema), inflammation of the lungs (pneumonitis), or death. It has also caused a rare form of nasal cancer in test animals undergoing extended exposures of 2, 6, and 15 ppm. It is classified as a potential occupational carcinogen.

Measured concentrations in pre 1980 homes are detailed below:

- 10 ppb is common in polluted outdoor air. 100 ppb is serious.
- 23 Danish dwellings where chip board was used reported HCHO concentrations of 60–1670 ppb (average 460 ppb) measured in 25 rooms (Andersen et al., 1975).
- 334 mobile homes in WA state reported HCHO concentrations from 30 to 1770 ppb, and 21% of measurements were higher than 500 ppb. Particle board and plywood were the major sources (Breysse, 1981).
- Connecticut homes reported levels of 370–7400 ppb (Sardinas, et al., 1979).
- Washington state homes reported 73% of samples in 39 homes were >100 ppb and 20% >500 ppb (Breysse, 1981).

Manufacturers have voluntarily reduced HCHO emissions from pressed wood products, by changing some product formulations. Measurements show indoor concentrations are now lower. Boston homes had concentrations ranging from 5 to 130 ppb with a geometric mean of 35 (Dannemiller et al., 2013.)

An empirical model for room air concentration at equilibrium was developed by Andersen et al., 1975:

$$C = \frac{(RT - N)(aH + b)}{1 + \frac{nf}{\alpha}} \tag{17.40}$$

where

C is the air concentration in mg m^{-3}
$R = 0.064$
$a = 0.143$
$b = 0.048$
$f = 0.304$
$N = 0.764$
H is humidity in g water kg^{-1} dry air
T is the temperature in °C
n is air changes per hour
α is the area of board surface per volume of room, m^{-1}

17.10.8 Smoking

Tobacco smoke contains many potentially hazardous materials. Mainstream smoke (MS) is an aerosol drawn into the mouth of a smoker from a cigarette. Many of the pollutants are filtered out in the smoker's lungs, for example, 70% of the PM. Sidestream smoke (SS) is the material released directly into the air from the burning end of a cigarette between puffs. Second hand smoke (SHS) is the respirable matter that exists in the indoor environment as a result of smoking cigarettes and other tobacco products, generated mostly by emissions from the burning tobacco between puffs on a cigarette. Even though the major source of SHS is SS, SHS differs substantially from SS. Once the smoke from the burning cigarette tip is dispersed into the indoor environment, it undergoes dilution and physicochemical changes, such as volatilization and chemical oxidation. Of the more than 4000 chemicals identified in tobacco smoke, 300–400 have been measured in MS and SS.

Undiluted SS contains higher proportions of certain toxic chemicals than undiluted MS, for example, benzopyrene: MS 20–40 mg cig^{-1}, SS 68–136 mg cig^{-1}. Because of the lower temperature in the burning cone of the smoldering cigarette, most incomplete combustion byproducts are enriched in SS as compared with MS.

Smoke particles are likely to penetrate and deposit in the alveolar spaces of the lungs.

Smoking is the leading cause of preventable illnesses and deaths in the United States. In 1994, an estimated 48 million adults in the United States were current smokers (roughly 26% of the adult population), and 46 million adults were former smokers.

Smokers have much higher risks of lung cancer, cardiovascular diseases, respiratory diseases, and other life-threatening conditions than nonsmokers.

In 1990, an estimated 419 000 deaths in the United States were directly attributable to cigarette smoking, of which more than 150 000 were from cancer. Lung cancer (117 000 deaths), ischemic heart disease (99 000 deaths), and chronic airway obstruction (49 000 deaths) accounted for most of the deaths attributable to smoking.

17.10.8.1 Secondhand Smoke

In 1992, the US EPA categorized SHS as a group A, or known, human carcinogen. This classification is reserved for those chemicals that show the strongest evidence of a cause-and-effect relationship in contributing to cancer in humans.

The chemicals in SHS distributed between gas and particulate phases; many are human carcinogens, including benzene, acrylonitrile, and benzopyrene, and formaldehyde, or potential carcinogens, such as styrene and nickel.

Exposure of nonsmokers to SHS is pervasive. A 1996 study by Pirkle and coworkers indicates widespread exposure to ETS in the US population: 88% of American nonsmokers had detectable levels of serum cotinine, a major metabolite of nicotine, which reflects exposure to nicotine over the previous 1–2 days. They also estimated that 43% of US children aged 2 months to 11 years live in a home with at least one smoker and that 37% of adult nonsmokers live in a home with one or more smokers or report SHS exposure at work. Time–activity data gathered in California indicate that despite a lower percentage of active smokers than the US population, 20% of Californian adults smoke (Jenkins et al., 1992), 62% of the Californian population are over 12 years of age, and 46% of the nonsmokers are near SHS at some time during any given day.

There is a substantial body of scientific evidence showing that exposure to SHS increases risks of lung cancer and respiratory infections in young children. It is estimated that 3000 lung cancer deaths per year in US nonsmokers aged 35 and older are due to SHS. An estimated 150 000–300 000 cases of respiratory infections annually in infants and young children up to 18 months of age are attributed to SHS.

There is growing epidemiological evidence that passive smoking also increases the risk of heart disease in nonsmokers. Approximately 35 000–40 000 deaths from heart disease among never smokers and long-term former smokers occurred annually in the United States as a result of SHS exposure in the early 1980s. The latest estimate of the annual rate of heart disease deaths caused by SHS exposure is 37 000, about ten times greater than for lung cancer.

Other health effects investigated are nonrespiratory cancers, age at menopause, sudden infant death syndrome, and low birth weight. The National Research Council report on environmental tobacco smoke is an excellent resource (NRC, 1986).

17.10.9 Asbestos

Asbestos applies to several naturally occurring hydrated mineral silicates. The most common form in the United States is chrysotile. Asbestos particles in the ambient air appear as fibers. They occur by fallout, contact, or impact, and re-entrainment. The rate of fiber release in fallout is continuous, low level, and persistent; it may be a function of the degradation of the adhesive binder. Sixty-eight percent of the asbestos in the United States is used in the bonded form in the construction industry in products such as floor tiles, asbestos cements, and roofing shingles. Another 6% is in powder forms in insulation and acoustical products. Significantly increased risk of death from nonmalignant respiratory disease (asbestosis) has been reported for workers in insulation, textile, and other asbestos industries. All commercial forms of asbestos have been shown to be carcinogenic in man. Household contact with work-derived asbestos dust has also been associated with increased mesothelioma risk. There is a 15–20-year latent period between exposure and incidence of cancer; thus the level of exposure that causes disease is ordinarily not known.

17.10.10 Carbon Monoxide

CO is a chemical asphyxiant gas; it is colorless, odorless, and chemically nonreactive in the environment. CO is produced by incomplete combustion of C-containing fuels. Common sources of CO are using charcoal for indoor heating/cooking and maltuned unvented heaters or furnace failures (cracked heat exchangers, birds' nest). CO is toxic through interference with uptake and distribution of O_2 in the body. CO binds reversibly with hemoglobin (Hb) in the red blood cells. The affinity of Hb for CO is 250 × that for O_2 (Roughton, 1970). CO-Hb reduces the carrying capacity of the blood for O_2 and reduces the transport and delivery of O_2 to body tissues (Kleinman, 2009). The tissues with highest O_2 needs are most affected: myocardium, brain, and exercising muscles.

The following information describing the CO-Hb saturation level and corresponding health effects is adapted from the US EPA (2000), Widdop (2002), and Kleinman (2009). Typical CO-Hb levels after exposure to 35 ppm for 1 h during light activity are 1% in nonsmokers and 5% in smokers. CO poisoning is usually defined as CO-Hb ≥2% for nonsmokers and ≥9% for smokers. The following levels of CO-Hb are associated with adverse health effects as described:

- 5–10% results in changes in myocardial metabolism and possible impairment: diminution of visual perception, manual dexterity, ability to learn, headache, and fatigue.
- 20–40% results in moderate poisoning: severe headache, vomiting, fainting, and impaired neurological function.
- 40–60% results in severe poisoning: fainting, convulsions, coma, and depressed respiration.
- >70% is fatal respiratory failure.

17.10.11 Biological Agents

A myriad of biological agents may contaminate the air within a home: viruses, bacteria, actinomycetes, fungal spores, algae, amoebae, arthropod fragments and droppings, and animal and human dander. Biological agents in indoor environments cause disease infection of the respiratory tract (viruses, bacteria) and immune responses (asthma, allergies from dust mites, fungal spores). Many bacteria/viruses causing disease originate from humans such as TB, chicken pox, and measles. Others are related to the building including Legionnaires' disease (air treatment systems), hypersensitivity pneumonitis (air treatment systems), and aspergillus infections (ventilation systems).

17.11 Radon Overview

The radiation dose from inhaled decay products of radon, Rn-222, is the main source of radiation exposure for the general population. The average radon concentration in the United States is 40 Bq m^{-3}. At these levels, the risk of lung cancer caused by exposure to Rn-222 decay products is 0.3%, or roughly 10 000 cases of lung cancer every year. What is known about the health effects of radon started with studies showing that elevated cancer rates in uranium miners were associated with elevated radon exposures. General population exposures resulting from industrial processes, primarily mining and mining tailings that released radon to the outdoor and indoor air, was also found to be high (Shearer and Sill, 1969). For example, radon concentrations were found to be high inside homes and other buildings in the vicinity of Grand Junction, CO, which were built on or using the radium-rich tailings from local uranium milling. The importance of radon levels in the general building stock came to light in the 1970s. Monitoring in many countries indicated that even average indoor concentrations are significant from the point of view of environmental risk.

Radon-222 is a noble gas and is a decay product of radium-226, which in turn is part of decay chain of uranium-238. There are two other radon isotopes, the Rn-220 isotope formed in the Th-232 chain from decay of Ra-224 and the Rn-219 isotope from U-235 chain. 222 is the most important Rn isotope because it has the longest half-life of the isotopes, 3.8 days. Any material containing uranium or radium is a source of Rn-222. The half-life for Ra-226 is over 1600 years; thus Rn-222 production can be considered constant. Rn-222 formed in the soil within approximately a meter of buildings can reach the indoor environment. It is an alpha emitter as are several of its decay products. Radon decays to radionuclides that are chemically active and relatively short-lived. Typical residential concentrations are 10–100 Bq m^{-3} (0.3–3 pCi l^{-1}), while outdoor concentrations are lower, 0.04–1 pCi l^{-1}.

The 1988 book by Nazaroff and Nero is an excellent reference on radon.

17.11.1 Natural Radioactivity

Radioactivity is a nuclear process. A material is radioactive if it gives off ionizing radiation, emanations that are capable of ionizing gas molecules surrounding the material. Ionizing radiation can break atomic electron bonds when passing through matter and create ion pairs (the sun's radiation cannot do this and neither can electromagnetic waves). It involves the transmutation of elements with the result that the nucleus of atom is changed. There are three types of ionizing radiation:

- **Alpha** (positively charged, +2): An alpha emission from an atom nucleus is identical to the emission of He nuclei, which has two neutrons and two protons, but without the two electrons normally present in a neutral He atom. Its mass is 6.62×10^{-27} kg. The speed of emission is 10^7 m s^{-1}, which is five percent of the speed of light. Alphas particles can travel several centimeters in air, or 0.1–0.01 of a millimeter through solids, before they are brought to rest by collisions. The kinetic energy is very large, 5.5 MeV.
- **Beta** (negatively charged, −1): A beta emission from an atom nucleus is the emission of an electron. A neutron is transformed into a proton, an electron, and a neutrino. The electron and neutrino are emitted energetically, leaving behind a proton in the nucleus. Beta emission happens at a range of speeds depending on the nature of the nucleus, with the top speed reaching 99.95% that of the speed of light.

- **Gamma** (neutral): Electromagnetic waves are extremely short wavelengths, about 0.01 that of X-rays, originating from transformation of the energy state within nucleus. They are highly energetic photons.

When a radioactive nucleus decays by alpha or beta emission, the resulting nucleus may be unstable, and there may be a chain of successive decays until a stable configuration is reached. The most abundant radioactive material in the Earth's surface is uranium-238 (^{238}U). It undergoes 14 decays (8α and 6β) terminating at ^{206}Pb.

Alpha particles are larger and as a result do not travel very far through matter; thus they deposit energy in the close vicinity of its emission. The distance of travel is about 4 cm in air and 40 μm in cell tissue, equivalent to the thickness of a layer of dead skin. This suggests that to damage the lungs by alpha decay, inhalation must take place. If beta and gamma emissions happen outside of the body, they can still damage internal organs, since they travel millimeters and tens of centimeters through condensed matter.

17.11.2 Radioactive Transformations

The number of radioactive atoms in any radioactive material decreases continuously as some of the atoms transform. The rate at which this number decreases varies for different nuclei. This transformation process is first order with a constant rate coefficient. See Equation 17.41, where $N(t)$ is the number of radioactive atoms in a material at time t, $dN(t)$ is the number of nuclei that disintegrate in time period dt, and λ is the radioactive decay constant [=] t^{-1}:

$$\frac{dN}{dt} = -\lambda N \qquad (17.41)$$

17.11.3 Activity

The **activity**, I, is the rate of decay of a species:

$$I = \lambda N = \left(-\frac{dN}{dt}\right)_{decay} \qquad (17.42)$$

The half-life $T_{1/2}$ is the time it takes for the number of radioactive nuclei to decrease to half the number that was originally present at time $t = 0$:

$$T_{1/2} = \frac{\ln 2}{\lambda} = \frac{0.693}{\lambda} \qquad (17.43)$$

The traditional unit for activity is the curie (Ci) or picocurie (10^{-12} Ci). 1 Ci = activity of 1 g of ^{226}Ra and 1 pCi = 2.2 disintegrations per minute. The SI unit for activity is the becquerel (Bq). 1 Bq = 1 disintegration per second and 1 Bq = 27.2 pCi.

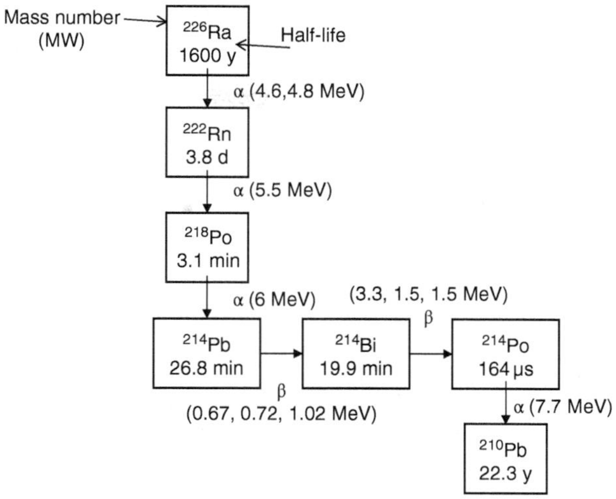

Figure 17.15 Radioactive decay chain for radon-222. The time period in each box is the half-life. The information in parentheses is the energy released with each emission.

17.11.4 Radon Decay Chain

Radioactive decays follow an almost fixed sequence (some species have 2–3 possible paths), and Figure 17.15 shows the decay chain containing Rn-222.

Each chemical element has a mass number that is the total number of protons plus neutrons, and it is roughly the atomic mass or molecular weight of the element. An alpha emission decreases the mass number by four and the atomic number by two. A beta emission increases the number of protons by one and decreases the number of neutrons by one so that the mass number does not change. Thus the chain in Figure 17.15 goes vertical with an alpha emission since the mass number changes and horizontal with a beta emission since the mass number does not change.

17.11.5 Health Concerns

There are two primary health concerns with radiation exposure. At very high doses, severe cell killing occurs, causing radiation sickness or death. At low to moderate doses, DNA is damaged, causing cancer or mutations and birth defects.

Radiation risk is quantified in terms of absorbed dose or dose equivalent. Because radioactive emissions have different energies associated with them and cause differing damage, an absorbed dose does not scale with risk when it is expressed as a rad, which equals 100 erg g^{-1}, or the unit gray (Gy), which equal 1 J K^{-1} = 100 rad. The dose equivalent that does scale with risk includes the parameter QF, or quality factor, and is expressed as radiation equivalent in people (rem is abbreviation for radiation equivalent in man), or rem = QF × rad. The QF = 20 for α particles, = 1 for β, γ. Another unit used is the sievert

Table 17.5 Relationship between dose and effects of radiation, regulations, and typical radiation exposures.

Dose equivalent	Significance
600 rem	LD_{50} (fatal to half of population if delivered rapidly)
50 rem	Max dose with no visible effects
5 rem year^{-1}	Max allowed dose for atomic workers
500 mrem year^{-1}	Max allowed dose to any individual in public (industrial)
300 mrem year^{-1}	**Effective dose equivalent from average indoor Rn levels**
170 mrem year^{-1}	Max allowed dose averaged over population (industrial)
100–150 mrem year^{-1}	Background natural radiation, excluding Rn
10 mrem year^{-1}	Chest X-ray

Source: Adapted from Nero (1988a).

(Sv) = QF × gray = 100 rem. The risk associated with 1 rem is a 1.65×10^{-4} chance of a fatal malignant disease during a lifetime. Table 17.5 summarizes the relationship between dose and some of the health effects, regulation limits, and typical exposures.

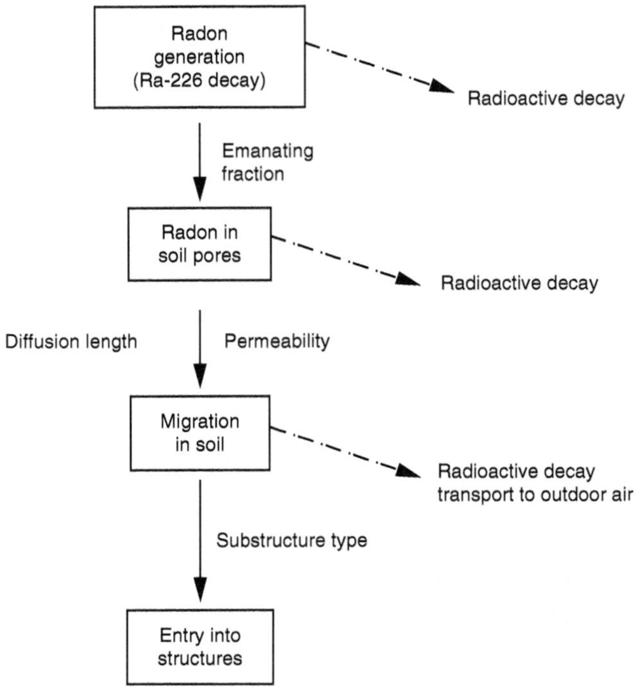

Figure 17.16 Diagram of the transport of radon indoors from the soil surrounding the substructure of a building.

17.12 Sources of Indoor Radon

Seventy percent of indoor radon comes from radon gas in the soil and rock surrounding the building substructure. The transport distance is several meters or less, and the characteristic time for transport into the indoor environment from the soil is a few weeks. The flow rate of soil gas into house is typically 0.001 1/h (compared with a typical air exchange of house 0.6 1/h). The typical thickness of soil layer adjacent to basement floor and walls that contributes Rn to the indoor environment is 70 cm. Other sources of indoor radon are water, outdoor air, and building materials. One percent is released from water used within the building. The ratio of the radon concentration in the air that comes from the water to the radon concentration in the water is 0.0001. Eighteen percent of indoor radon is from outdoor air coming inside by ventilation and infiltration and eleven percent from building materials that have been made from earthen sources containing high Ra-226 (Nazaroff and Nero, 1988). Figure 17.16 illustrates the pathway for radon to be transported indoors from the soil.

The two major factors determining indoor concentrations are the ventilation rate and the radon entry rate. How a building is ventilated affects the pressure gradient between the building and the soil that then drives bulk airflow through soil indoors. When there are high natural ventilation rates, there are typically low indoor–outdoor pressure differences. If the building has a balanced mechanical ventilation system, then there are also low pressure effects. If there is high rate of infiltration, this is typically because of high pressure driving forces, and this will drive more radon indoors. Occupant activities can also change pressure differences such as fireplace use (which would typically increase infiltration and radon transport).

Design and construction of building substructure controls the degree of movement between air in soil and air in building. Stronger coupling between indoor air and soil gas occurs when the lowest floors are typically made of either poured concrete, so in direct contact with soil, or there is a basement so that the floor lies below soil grade, or the building is slab-on-grade and the floor is built to the same level as the soil surface. Coupling is reduced for lowest floors that are wood and suspended above soil, or there is a crawl space below a wood floor. Also very important are the location of openings and penetrations through substructure.

17.12.1 Soil Characteristics

Soil characteristics play an important role in radon generation and transport. The following is a discussion of these characteristics and their typical values (Nazaroff et al., 1988). Soil grain sizes range from 2 to 2000 μm and the grain density of soil, ρ_s, is 2.65 g cm^{-3}. The soil solid fraction is made up of mineral grains that have a wide range of sizes, and the void fraction is the spaces around the grains that contain liquid and/or air. This is also known as soil porosity, and typical values are 0.4–0.6. The volume fraction of air, ε_a, is the volume of air/total volume and is ~27%. The volume fraction of water, ε_w, is called moisture content and is typically 13%. When moisture content equals the porosity, then the soil is saturated with water.

Permeability is an important parameter for advective transport. It is a measure of how readily a fluid can flow through the soil. It relates the apparent velocity of fluid flow through soil pores to the pressure gradient. It also ranges from 10^{-7} (clean gravel) to 10^{-16} (clays). At the lowest permeability, diffusion is the dominant transport process; at high permeability, convection is the dominant transport process.

Emanation coefficient, f, is the fraction of the radon that is generated from radium decay that leaves the solid soil grains and enters the pores. Typical values are between 0.05 and 0.7, and it is low for very dry or very wet soils.

17.12.2 Radon Production in the Soil

Radium-226 is the source of radon in the soil. The radium content of soil, A_{Ra}, is typically between 10 and 100 Bq kg^{-1}. Typical values in different states are:

- CO: 52 Bq kg^{-1}
- CA: 28 Bq kg^{-1}
- KY: 56 Bq kg^{-1}
- AK: 24 Bq kg^{-1}

Upon being created by the alpha decay of radium, Rn-222 atoms travel from the generation site until their energy is transferred to the host material, whether it is solid soil grains, air, or water. The distance of travel depends on what is the material made of: 0.02–0.07 μm for common minerals, 0.1 μm for water, or 63 μm in air.

17.12.3 Transport of Radon Indoors

Almost all radon indoors is transported by advection and only a small amount comes in by diffusion (5%). Many houses have a concrete foundation that blocks diffusive entry of radon, but there are plenty of cracks and design openings that allow for advective entry. The advective flows are driven by the same forces that drive general building infiltration, namely, wind and temperature differences, barometric pressure changes, and rainstorms.

Radon diffusing through soil is different than through open air since the area for diffusion is reduced because of the soil grains. In uncovered soil of infinite depth and width, assuming the radon concentration to be zero at the soil surface, the steady-state flux of radon from uncovered soil due to diffusion is ~0.03 Bq (m^2 s)$^{-1}$ (Nazaroff et al., 1988).

17.12.4 Concentration of Radon in Soil Gas

To estimate the concentration or radon in soil gas, a material balance on Rn atoms is applied to a control volume of soil (Nazaroff et al., 1988). Figure 17.17 is a drawing of the control volume of soil.

The accumulation of radon in the control volume equals the difference between the rate of radon generated and the rate of radon decay. Variable definitions and typical values are listed below for the material balance in Equation (17.44).

N_{222} = number concentration of Rn-222 atoms (# m^{-3})
V = volume of bulk soil
$\varepsilon_a V$ = volume of soil pore air
ε_a = air filled-porosity (typical value 0.40)
G = generation of Rn-222 atoms entering pore air from Ra-226 decay (# s^{-1})
D = decay of Rn-222 atoms in pore air to decay species (# s^{-1})
$\lambda_{222} = 2.1\times10^{-6}$ 1/s = Rn-222 decay constant
ρ_s = soil grain density (2650 kg m^{-3})
A_{Ra} = soil radium activity per mass soil (typical value 30 Bq kg^{-1})
f = emanation coefficient (typical value 0.2)

$$\frac{d\left(\text{number of Rn atoms}\right)}{dt} = \text{generation} - \text{decay} = G - D$$
$$N_{212} = \text{Number of Rn atoms} = N_{222}\times\varepsilon_a\times V$$
$$G = (1-\varepsilon_a)V\times\rho_s\times A_{Ra}\times f$$
$$D = \lambda_{222}\times N_{222}\times V\times\varepsilon_a \tag{17.44}$$

The steady-state solution to the differential equation from above is (there is no transport yet)

$$I_\infty = \lambda_{222}N_{222} = \left(\frac{1-\varepsilon_a}{\varepsilon_a}\right)\times\rho_s\times A_{Ra}\times f \tag{17.45}$$

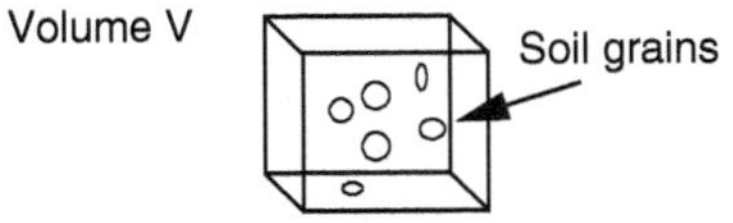

Figure 17.17 Drawing for the material balance on radon atoms in a control volume of soil.

A typical result for radon concentration in soil pore air is

$$I_\infty = 24\ \text{kBqm}^3 = 650\ \text{pCil}^{-1}\left(=24\ \text{kBq}\ \text{m}^3\right)$$

And it ranges from 100 to 10 000-pCi l^{-1}. It is much higher than the concentrations indoors and the action level for radon of 4 pCi l^{-1}.

17.13 Controlling Indoor Radon

The main objective for reducing indoor radon concentrations is to reduce the lung dose of radiation due to exposure to radon and radon decay products to an acceptable level. There are quite a few difficulties in controlling radon. For example, areas or types of buildings where high concentrations may occur must be identified. What monitoring protocol should be used, and how should the results be interpreted? What techniques might be appropriate to use to mitigate? Since homes are private environments, it makes it especially difficult. Who is responsible for what part of the radon control strategy, and how should the necessary information be communicated to homeowners/occupants? There are no regulations that directly limit indoor radon concentrations. Standards for radiation have been based on recommendations from the radiation protection community, such as The International Commission on Radiological Protection. The general standard for protection of the US public has been that no individual should suffer radiation doses of more than 0.5 rem year^{-1} and the population as a whole should not exceed an average exposure of 0.17 rem year^{-1}. A Rn decay product concentration of 20 Bq m^{-3} corresponds to an exposure rate of 0.17 rem year^{-1}.

There are generally two different strategies to reduce indoor Rn and lung cancer risk. First, aim to identify and remedy high-concentration houses (which is a small fraction of housing stock). This will substantially affect the fraction of the population at highest risk but will have only a modest effect on the average population exposure. Or efforts could be aimed at the large number of houses with concentrations within a factor of 5 or so of the average; this will substantially reduce average population exposure. Nero (1988b) recommends focusing efforts relatively rapidly on homes with unusually high concentrations; in the meantime evaluate whether or how to reduce intermediate concentrations. The EPA published "A Citizen's Guide to Radon" (EPA, 2016). They recommend testing in every residence and that homeowners should take corrective measures to reduce Rn concentrations if long-term average is above 4 pCi l^{-1} (148 Bq m^{-3}).

17.13.1 Controlling Rn Entry from Soil

The main source of radon is mass flow of soil gas containing radon into buildings through openings in the concrete basement structures due to small indoor–outdoor pressure differentials (Scott, 1988). Thus to control radon entry, soil gas flows must be reduced. There are three main approaches to controlling radon entry from soil:

1) Source removal – If the local soil is the source, source removal is not a viable mitigative option; this method is only practical in cases where the source has been introduced into the local environment by human action such as mining and mine tailings (such as in Grand Junction, CO).
2) Sealing – This increases the resistance of the building envelope to soil gas entry by sealing subsurface penetrations.
3) Soil ventilation – This reduces the soil gas flow by reducing the pressure differential between the building and the soil. This is the most important technique, usually supplemented by sealing.

A soil ventilation system consists of a perforated pipe network beneath and around the building that is maintained at a pressure lower than the building pressure, either by a small fan or by a passive vented stack. This localized reversal of the pressure difference causes air to flow out from the building through the soil to the subsurface collection system, thus stopping the entry of soil gas through the coupling of the house and soil. The average natural pressure difference between the soil and building is only a few Pa, and even a small axial fan is able to produce suctions of 25–50 Pa in the pipe network.

If the building has a concrete floor slab, subfloor ventilation can be used. An exhaust pipe is installed through the floor into a cavity in the slab that is filled with crushed stone. The opening around the pipe is sealed, and a small suction fan is placed on the end of the pipe that is located outside of the house. The suction pulls the soil gas to the cavity through the layer of soil present under the floor slab. It also draws air from the building through every crack and opening in the floor slab and through the wall–floor joint, preventing soil gas and radon from flowing from the soil into the building.

17.13.2 Indoor Air Control Techniques

The main methods that are used to control indoor air pollutants can also be applied in the case of radon. These are to add ventilation and apply air filtration. Air filtration is especially effective for Rn decay products that are attached to particles (Jonassen and McLaughlin, 1988). Since filtration removes airborne particles, it may shift partitioning of the remaining airborne radon decay products toward the unattached state, which results in a higher predicted dose per unit of alpha energy. Mechanical mixing will increase circulation of air, causing an increase in the deposition of unattached decay products.

Structures with crawl spaces often have high radon concentrations in the crawl space because of the large areas of exposed soil and low ventilation rates. Where winters are not severe, the crawl space is often unheated and basically outside the building. Concentrations can be reduced by increasing ventilation to the area, or installing a small fan, and closing the major openings in the floor to reduce air exchange between the crawl space and the living area. Structures with heated crawl spaces should probably use soil ventilation techniques.

17.14 Particles in Indoor Air

Particles suspended in a gas are called *aerosols*. Particles may be liquid or solid phase, or both. They may be pure species or complex mixtures. The diameter of a particle defines its size. Diameters of interest range from a molecular cluster (~1 nm) to 100 μm. Larger particles do not remain suspended for very long (for example, the settling velocity of a 100-μm particle is ~100 cm s^{-1}). The presence of particles does not affect the fluid mechanical properties of indoor air.

In ordinary air environments, the number of particles per volume is very large. In a typical room there are probably 10^{12} particles. A typical number concentration, N, is 10^4 # cm^{-3} = 10^{10} # m^{-3}.

A typical mass concentration $M = 140\ \mu g\,m^{-3}$ (includes all particle sizes and assumes density is 1 g cm^{-3}). The characteristic diameter $d_p \sim [(M/N)(6/\pi\rho)]^{1/3} = 0.3\ \mu m$ (about 1000× dimension of molecule).

17.14.1 Particle Size Modes in the Atmosphere

There are three main modes of particle sizes in the atmosphere: (i) nucleation, (ii) accumulation, and (iii) coarse. On a mass concentration basis, the accumulation- and coarse-mode contributions are comparable in size, but the nucleation mode is much smaller. On a surface area concentration basis, the accumulation mode dominates. On a number concentration basis, the nucleation mode is the most important, especially if there is a fresh combustion source. Otherwise, the accumulation mode may dominate. A monodisperse aerosol contains particles of a single size, while a polydisperse aerosol contains particles in a range of sizes.

The *nucleation mode*, also called ultrafine particles, includes particles less than 0.1 μm. These particles are produced by gas-to-particle conversion processes, i.e. condensation of low-volatility gases, primarily in combustion processes. They are present at significant levels only near, for example, fairly clean combustion sources. The dominant species include soot (elemental carbon), organic vapors, PAHs, and metals, and their dynamic behavior is dominated by Brownian motion.

The *accumulation mode* includes particles in the diameter range of 0.1–2.5 μm. These particles are produced by coagulation of nucleation-mode particles or condensation of low-volatility gases onto nucleation-mode particles. Coagulation is a process involving the collisions of two particles to form one larger particle. Most SHS from cigarettes is found in the accumulation mode. The particles in this mode consist of the same chemical species as the nucleation mode, plus water vapor and low vapor pressure organics. These particles have a long residence time in air: too big to diffuse rapidly but too small to be strongly affected by gravity or inertia.

The *coarse mode* includes particles in the range of 2.5–10 μm (and even higher). These are produced mechanically (grinding, abrasion) and are from both outdoor and indoor sources. The outdoor-sourced particles contain soil dust, tire, road and brake wear, and sea salt. The indoor-sourced particles contain fabric fibers, skin flakes, sneeze, cough and talking projectiles, and biogenics such as dander, pollen, fungal spores, dust mite parts, etc. Their behavior is dominated by species mass, gravitational settling onto upward-facing surfaces, and the particle size through deposition by interception.

17.14.2 Particle Size Distributions

Each particle has a distinct size, shape, and chemical composition. For example, there are soot particles that are agglomerates consisting of tiny spheres of elemental carbon, produced by combustion (looks like bunches of grapes). Particles of soil dust can be found in the air. Small human skin flakes and clothing fibers are present in occupied spaces.

The shape of particles can be quite complex. Particles are typically assumed to be spheres. This works well for liquid particles and can be a good description of some solid particles such as sea-salt articles. It is not a good approximation for fibers.

It is very difficult to describe each particle individually. Sometimes we lump all the particles together and describe by a mass or number concentration. This is insufficient because the behavior of particles is a strong function of particle size. Some chemical components of the particles may be of particular concern, and those may be associated with particles in certain size ranges. For example, deposition in the human lung following inhalation is highly dependent on particle size.

A means of providing a fairly complete description of suspended particles is to determine the total mass or total numbers of particles within certain size ranges and then further subdivide that mass into its chemical composition. For particles in size range (log dp) to (log dp + d (log dp)):

- dM = mass concentration of particles.
- dN = number concentration of particles.
- dV = volume concentration of particles.

The particle size distribution is a plot of $dM/d(\log d_p)$ versus d_p. The area under the curve between limits $d_{p,min}$ and $d_{p,max}$ is proportional to the total concentration of the parameter of interest between those limits (Figure 17.18).

To understand the meaning of data presented in this format, consider the shaded bar. The area is proportional to the mass concentration of particles in the size range between 2.5 and 4 μm.

The height of the curve for that size range is $\Delta M/\Delta \log[d_p] = 80\,\mu g\,m^{-3}$. For this size section $\Delta \log[d_p] = \log(4) - \log(2.5) = 0.6 - 0.4 = 0.2$. Thus $\Delta M = \Delta \log(d_p) \times 80\,\mu g\,m^{-3} = 0.2 \times 80 = 16\,\mu g\,m^{-3}$. This is the mass concentration associated with particles in the diameter size range of 2.5–4 μm.

17.14.3 Dynamics of Individual Particles

Air is made up of gas molecules that collide with each other randomly. The mean distance that characterizes the graininess of the gas that a molecule travels before colliding with another molecule and changing direction is the *mean free path*. The mean free path, $\lambda = 6.5 \times 10^{-6}$ cm, can be calculated from kinetic theory of gases.

The characteristic number that measures the gas graininess relative to the size of a particle is the Knudsen

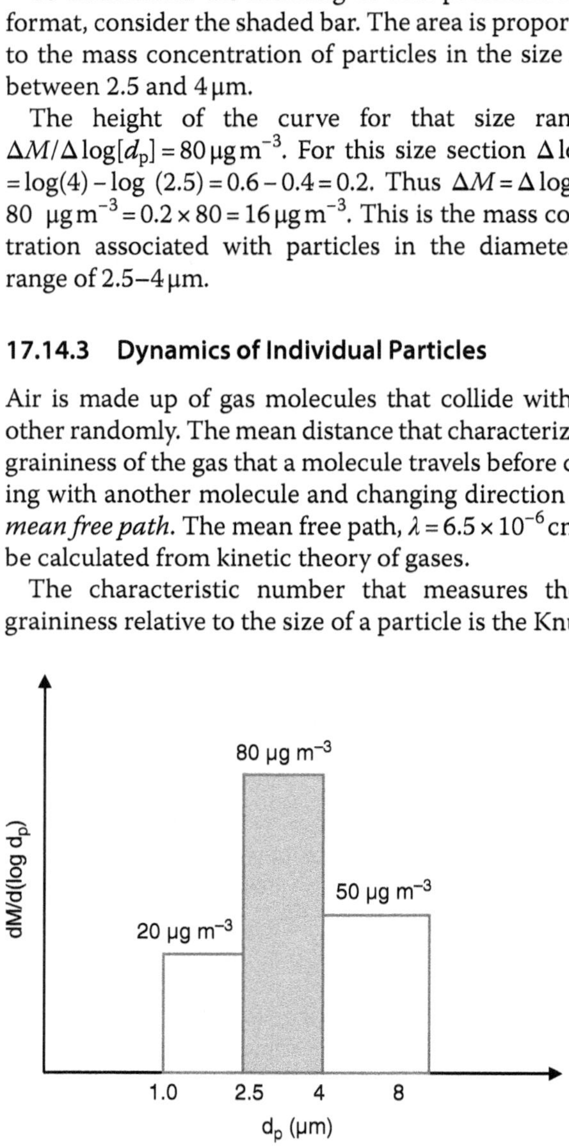

Figure 17.18 Example particle mass concentration size distribution.

number, $Kn = 2\,\lambda/d_p$. $Kn \ll 1$ indicates the continuum regime, and classical fluid mechanics can be applied. The gas looks smooth on the scale of a particle since the particle is so much bigger than the gas molecules. $Kn \gg 1$ indicates the free molecular regime. Here the particle and gas molecule are close to the same size, and the drag on the particle is much smaller than would be predicted by continuum mechanics so must correct (we denote this as slip) the drag. In the free molecular regime, we correct for slip with the Cunningham slip correction factor, $C_c = 1 + Kn(1.257 + 0.400\ \exp\ (-1.10/Kn))$. When $Kn \sim O(1)$, it indicates the transition regime, and here it is difficult to predict behavior.

17.14.3.1 Brownian Motion

Brownian motion is an erratic, zigzag motion of microscopic particles. It was first observed in 1827 by the English botanist Robert Brown, who was investigating a suspension of microscopic pollen particles in an aqueous solution. The effect was observed even in pollen samples that had been dead for more than 100 years. In 1879 a similar motion was observed for smoke particles in air. Experiments showed that the motion became more rapid and the particles moved farther in a given time interval when the temperature was raised, when the viscosity of the fluid was lowered, or when the average particle size was reduced.

The kinetic theory of matter, developed toward the end of the nineteenth century, gave a qualitative explanation for the motion of inanimate particles in solution. The atoms or molecules that make up a liquid or gas are in constant thermal motion, and the temperature of the system determines their velocity distribution. Each suspended particle collides with surrounding molecules, and each collision changes the particle's velocity by a small amount. The net effect is an erratic, random motion of the particle through the fluid. Albert Einstein and Marian Smoluchowski developed a quantitative theory of Brownian motion independently in 1905–1906; it accounts for the dependence of the effect on temperature, viscosity, and size. Einstein's theoretical explanation can be used to predict how rapidly a particle will meander from starting point due to Brownian motion:

$$x^2 = 2Dt \tag{17.46}$$

where

x^2 = mean square displacement by Brownian motion alone during time t

D = coefficient of Brownian motion (units are length^2 time^{-1}) (diffusion coefficient) $= kTC/(2\pi\mu d_p)$.

k = Boltzmann's constant = 1.38×10^{-16} erg K^{-1}.

17.14.3.2 Gravitational Settling

Particles suspended in a fluid experience a downward force due to gravity. For course particles, gravitational settling is major removal mechanism. Terminal (steady-state) settling velocity, V_{TS}, for Stokes' particles applies for particles <30 µm. Note that if particles are very large, we must correct the drag force):

$$V_{TS} = \frac{\rho_p d_p^2 g C_c}{18\mu} \tag{17.47}$$

where g is gravitational acceleration, μ is viscosity of air (1.81×10^{-5} Pa·s at 20 °C), ρ_p is density of particle, d_p is particle diameter, and C_c is Cunningham slip factor (see above).

17.14.3.3 Coagulation

Coagulation is the collision of suspended particles and the subsequent adhering to one another. Collision may occur by any mechanism that leads to relative movement between the two particles – Brownian motion, shear flows, etc. Consider an initially monodisperse aerosol, with a number concentration, N. β is the rate at which one particle "sweeps" the surrounding air of other particles. β varies only moderately with particle size and a typical value = 10^{-10} cm^3 s^{-1}.

For the case of particles undergoing Brownian motion,

$$\beta = 4\pi\left(D_1 + D_2\right)\left(r_1 + r_2\right) \tag{17.48}$$

where r_i is the particle radius of particle I and D_i is the coefficient of Brownian motion for particle i.

The total number concentration changes with time according to the following equation:

$$\frac{dN}{dt} = -\frac{1}{2}\beta N \tag{17.49}$$

Coagulation is a second-order process!

17.15 Bioaerosols

A bioaerosol refers to airborne particles suspended in a gas containing (i) intact living or dead microorganisms (as single units, in homogeneous or heterogeneous groups, or attached to other particles), (ii) microbial spores (resistant reproductive structures produced by many fungi and some bacteria), (iii) fragments of microorganisms and larger organisms (body parts of arthropods, skin scales from mammals, and pollen and plant debris), and (iv) other particles from living or dead organisms (excreta from arthropods, such as dust mites and cockroaches; allergens from microorganisms, plants, and animals; and microbial toxins).

While all bioaerosols originate from living or once-living organisms, these particles differ widely in their physical features and biological effects. Bioaerosols may cause adverse health effects because of their toxicity, infectivity, or allergenicity. Bioaerosols found indoors come mainly from human or animal occupants, indoor microbiological growth, or outdoor air that enters by infiltration, through the ventilation systems, and by other means.

17.15.1 Indoor Bioaerosols

Indoor biological contamination is defined as the presence of (i) human-sourced bioaerosol of a kind and concentration likely to cause disease or allergic reaction, (ii) inappropriate concentrations of outdoor bioaerosols, or (iii) active microbial growth on surfaces that may become aerosolized or release toxic or irritating gases and to which people may be exposed. Engineers have an important role to play in problem prevention and mitigation. Microorganisms are everywhere, so the objective is not to achieve a sterile indoor environment, but rather to control adverse effects.

Bacteria and viruses are usually brought indoors in human hosts (the major reservoir) and spread from person to person. Pets are also a source. Some bacteria (such as *Mycobacterium tuberculosis* that causes TB) and some viruses (such as the rubeola virus that causes measles or the varicella zoster virus that causes chicken pox) can be excreted and may stay as viable particles on fabrics and surfaces or enter into ventilation systems to then disperse and infect other humans. Indoor air exposures of concern are viable airborne microbes, especially viruses and bacteria, fungal spores, dead cells or cell fragments, and VOCs that are metabolic by-products of microbes, endotoxins (toxins from bacteria), and mycotoxins (toxins produced by fungi).

Viruses are replicating, noncellular particles that constitute the smallest and simplest known infectious agents. They are metabolically inert parasitic fragments of RNA and DNA. Viruses become airborne when generated by hosts via coughing, sneezing, singing, etc. They are too small to be observed with light microscopes, and the typical size is 0.02–0.3 µm in linear dimension, and they are mostly ultrafine and accumulation mode size particles.

Bacteria are single-celled organisms in the form of very small spheres (cocci), straight or curved rods (bacilli), spiral rods, or branched filaments. They are unicellular and reproduce by simple cell division, and some produce endospores (internal spores) that resist environmental stresses. Most bacteria are beneficial, but some are pathogenic. Their size is 0.1–15 μm in linear dimension, and accumulation up to coarse-mode particles.

Fungi (mold, mildew, and yeast) may be unicellular or multicellular. They do not photosynthesize (distinguishes fungi from algae) and undergo asexual reproduction via spores that may become airborne. Fungi are typically not pathogenic, but there are allergic responses to spores. Their size is 1–100 μm in linear dimension, and they are in the coarse size mode. Yeasts are unicellular fungi that reproduce primarily by budding, while molds are multicellular fungi (there are toxic molds such as *Stachybotrys chartarum*).

Dust mites are a member of the spider family and are the most widespread allergen found indoors. The allergenic reaction is due to exposure to dust mite body parts and feces (size of 10–20 μm). A dust mite has a characteristic dimension ~ 100 μm, almost visible. Their food source is human and other animal skin flakes (humans generate seven million skin flakes per minute [side note: typical viable bacterial count is 4 per skin flake!]). Dust mites are found in mattresses and bedding, upholstery, and carpets. Dust mites need microscale RH > ~75% to avoid desiccation (O'Rourke and Lebowitz, 2000). The prime dust mite season in most of the United States is late summer. They are a major asthma trigger.

Pet dander are flakes of skin and fur shed by cats and dogs and are coarse-mode particles. **Pollen** is plants' reproductive material. Pollen particles are in the coarse mode with diameters ranging from 5 to 100 μm and are usually spherical in shape.

Moisture plays a critical role in microbial growth. Most construction materials contain adequate quantities and variety of organic compounds to support microbial growth (such as wood, cellulose, organic insulation, glues, paints, and natural fiber textiles). Moisture is a major factor promoting or limiting indoor microbial growth on surfaces and in systems. Excess moisture penetrates structures, air ducts, and furnishings through diffusion, leakage, or condensation. With excess moisture the growth starts immediately, leading to building material deterioration and bioaerosol production. Environmental microbiologists use the concept of water activity, a_w, to determine whether conditions are optimal for microbial growth:

a_w = relative humidity at which moisture in material is at equilibrium

$$a_w = \frac{\text{partial pressure of water in a material}}{\text{partial pressure above pure water}} \tag{17.50}$$

The requirements of fungi for water activity:

Primary colonizers	e.g. penicillium	$0.8 \leq a_w \leq 0.85$
Secondary colonizers	e.g. aspergillus	$0.85 \leq a_w \leq 0.9$
Tertiary colonizers	e.g. stachybotrys	$0.9 \leq a_w \leq 0.95$

Temperature is not a problem for microbes in buildings; most are happy with the temperature in building climates, even if it is suboptimal for growth.

The dynamic behavior of bioaerosols is similar to those of other airborne particles. Dander, human bacterial or viral emanations, pollen, spores, and dust mite feces are considered coarse PM and are mainly affected by gravitational settling and ventilation. Viral and bacterial containing particles are fine particles affected by ventilation. Gases such as organic acids and aldehydes (endotoxins) are also mainly affected by ventilation and possibly surface deposition and chemistry.

17.15.2 Bioaerosol Sampling

There are no standard methods for sampling and analysis of microbiological agents in air, making it difficult to interpret and compare some data across studies. Sampling must take into account the fact that spore release and pollen production is highly dependent on environmental factors (T and RH, surface water, light). In outdoor air, seasonal variations may produce differences of several orders of magnitude. Air sampling can be conducted with either passive or active monitors. Passive monitors rely on gravity collection. They consist of growth media or a sticky collection surface. Active air sampling consists of a pump that draws a known quantity of air over or through a collection surface that can contain growth medium or adhesive or into a liquid in a glass impinger. Swabbing or wiping of surfaces is also used in some instances. Samples can also be collected as a function of particle size. Sampling is followed by light microscopy, cultures, antigen assays, and DNA extraction and analyses.

17.15.3 Control of Bioaerosols

Strategies for the control of bioaerosols fall into two categories, those relating to the individuals exposed and those to the physical environment. Prevention in the host can occur through immunization, avoidance of sensitization, or allergic desensitization. Physical environmental remediation includes avoiding the conditions that provide substrates for the growth of bioaerosols, controlling outdoor infiltration (avoid bringing in pollen and moisture), and removal (filtration, application of UV (UV-C) light, or substituting building materials).

Methods of source control can be categorized into four groups (WHO, 1990):

1) Proper design and construction of buildings. Use of nondeteriorating materials, materials that control moisture. Design should avoid conflicts between energy conservation and effective moisture control (sealing out rainwater by tightening building envelope).
2) Source modification. Change T or RH to control sources.
3) Maintenance, repair, and cleaning. Chemical treatment with biocides and UV irradiation. Keeping filters clean. Identifying and fixing water leaks, broken water pipes, etc.
4) Removal of pollutants from air by increasing ventilation and/or air cleaning.

17.16 Volatile Organic Compounds

VOCs are organic compounds exist in the gas phase at room temperature. They contain carbon in a reduced state (CO_2 is inorganic). Their boiling point range is from 50–100 to 240–260 °C (WHO, 1989). In outdoor air, VOCs contribute to photochemical ozone formation. In indoor air, adverse health impacts are the major concern, which include acute or those cause by short-term high doses such as sick building syndrome (SBS), odor, and irritation and chronic or those cause by a long-term cumulative dose such as cancer or neurotoxicity. Compounds of interest in indoor air include formaldehyde, benzene, chlorinated compounds, pesticides, and PAH.

The main sources of VOCs indoors are building and furniture materials, machines and appliances, copy or printing machines, the ventilation system bringing in outdoor air containing VOCs, human activities, and household products such as those used for cleaning or cooking, smoking, and other products such as glues, cosmetics, or perfume. Another major source is human metabolism and includes compounds such as acetaldehyde, ethyl alcohol, phenol, and toluene (Wang, 1975).

17.16.1 Total Volatile Organic Compounds (TVOCs)

It is very difficult to summarize in a standardized way the combined health effects of the many different VOCs to which we may be exposed indoors. Total volatile organic compound (TVOC) is an attempt to find an indicator of the perceived irritation caused by exposure to a mixture of VOCs in air at low levels. To measure TVOC, *masses* of various molecules are added. TVOC is also easily measured through chemical analysis. Compounds must be comparable (similar chemical properties) for masses to be summed through chemical analysis. Typically GC with a flame ionization or photoionization detector calibrated against toluene is used.

Mølhave (2009) reports typical TVOC concentrations in a variety of buildings. Concentrations in older houses were an order of magnitude lower than in new houses: 0.02–17 $mg\,m^{-3}$ compared with 0.5–19 $mg\,m^{-3}$. Concentrations in old schools were also lower than in new schools: 0.86 compared with 0.13–0.18 $mg\,m^{-3}$. Human complaints occurred in the homes with higher TVOC. These data suggest that the threshold at which complaints are seen is around 2 $mg\,m^{-3}$. Table 17.6 summarizes health effects caused by low level exposures to TVOCs.

17.16.2 Sick Building Syndrome

SBS occurs in buildings in which most of the inhabitants have similar complaints and symptoms when they are in building, but these complaints and symptoms disappear when they leave the building. WHO estimated that in 1982 more than 30% of new buildings were affected by indoor environmental quality problems, with no evident cause, such as excessive air pollution or defects in technical installations or construction (WHO, 1982).

Symptoms and complaints include:

- Sensory irritation of eyes, nose, and throat (pain, dryness, stinging).
- Neurological (headache, mental fatigue, reduced memory, dizziness).
- Skin irritation (pain, itching, dry).
- Nonspecific hypersensitivity (running nose and eyes, asthma-like).
- Odor and taste.

Table 17.6 Health effects caused by low level exposure to TVOCs.

Health effect	Concentration
No irritation or discomfort	<0.2 $mg\,m^{-3}$
Perception that indoor air quality is deteriorated (e.g. odor)	0.2–3 $mg\,m^{-3}$
Acute or subacute inflammatory reactions in skin or mucous membranes (e.g. headache)	3–25 $mg\,m^{-3}$
Subacute environmental stress reactions (e.g. neurotoxicity)	Occur at >25 $mg\,m^{-3}$

Source: Adapted from Mølhave (2009).

An intervention study at a call center providing a telephone directory service investigated the impact of ventilation rate and air filters on worker performance and SBS (Wargocki et al., 2004). The outdoor supply rate was adjusted to be 8 or 80% of total airflow of 3.51 h^{-1}. The supply air filters were either new or had been in place for six months. One of these independent variables was changed each week for 8 weeks. The 26 operators were blind to the conditions and each week returned questionnaires recording their environmental perceptions and SBS symptoms. Their performance was also continuously monitored by recording the average talk time every 30 min. Replacing a used filter with a clean filter reduced talk time by about 10% at the high ventilation rate, but had no effect at the low rate. Increasing the outdoor air supply rate reduced talk time by 6% with a new filter in place but increased talk time by 8% with a used filter in place. The interventions had significant effects on the perceptions and SBS symptoms. Increasing outdoor air supply rate and replacing filters can have positive effects on health, comfort, and performance.

17.17 VOC Surface Interactions

Indoor air pollutants interact with interior surfaces. Organic molecules can stick (sorb) onto indoor surfaces and then later desorb (reemit). Mass transfer is either absorption or adsorption (called sorption since sometimes we do not know exactly which process is occurring) plus desorption. These processes will change the indoor air concentration profile. How does this process impact exposure to VOCs? To determine the indoor air concentration of VOCs, we must account for sorption and desorption. Examples of VOCs that interact with surfaces are glue used in carpets, VOCs from wall paints, SHS, and solvents. Organic compounds sorbed onto building materials in a 7-year-old preschool were desorbed (reemitted) when placed in a small ventilated test chamber (Berglund et al., 1989). Para-dichlorobenzene was measured in significant concentrations in test chamber several days after source (moth repellent) was removed. Twelve months later, concentrations were still elevated indicating reemission (Tichenor and Mason, 1988). "...no rule can be deduced to predict the sorption behavior of different VOCs with different materials" (Seifert and Schmahl, 1987).

Figure 17.19 illustrates many of the factors that influence the concentrations of VOCs indoors. Most important are the following:

- Amount of material and its properties.
- Type of VOC(s).
- VOC concentration in the air.
- Airflow through the space.
- Environmental conditions including temperature, RH, and air velocity.

17.17.1 Isotherms

An isotherm expresses the equilibrium capacity of a sorbent (like flooring material, carpet) for a particular sorbate molecule (HCHO molecule) at a given gas-phase sorbate concentration. Proper isotherm and parameter values are found by fitting experimental data. The linear isotherm is usually a good model for indoor environments because indoor air concentrations are low enough.

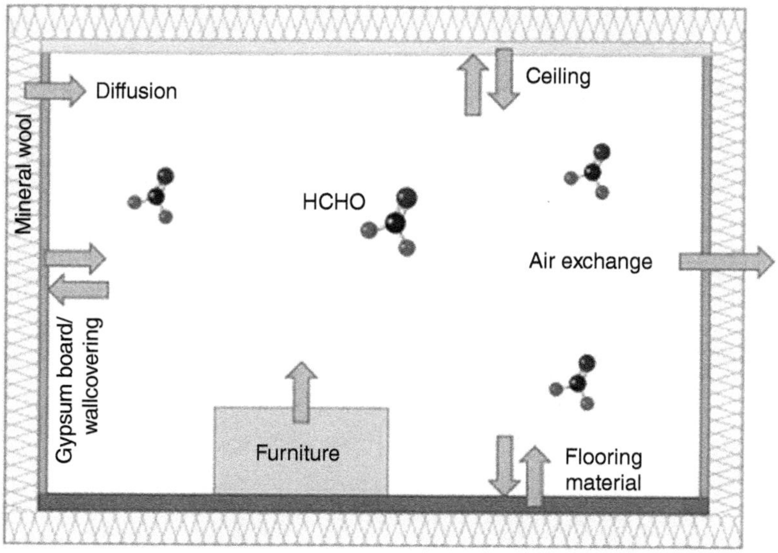

Figure 17.19 Mass balance model of VOC sorption/desorption using the Tichenor model in an indoor environment.

Note that the system is still dynamic even though it is at equilibrium: gas-phase molecules are continuously being taken up by the surface, and molecules are being released from surface. Sorption coefficients in models can vary with test conditions indicating that the sorption process is strongly influenced by environmental conditions. Tichenor et al. (1991) showed that the Langmuir isotherm may be used for sorption of smooth materials (wallboard, ceiling tile), but it does not work well for rough, complex materials (carpet).

The linear isotherm model (Tichenor et al., 1991) directly relates the concentration in the air to the mass on the surface: Applying a material balance model on VOC in indoor air (Figure 17.20) and accounting for reversible sorption results in Equation (17.12):

Direct indoor emissions at rate E (mg h^{-1}).
Surface area available for sorption is A (m^2).
Adsorption (deposition) rate coefficient k_a (m h^{-1}).
Desorption rate coefficient k_d (h^{-1}).
Ventilation at flow rate Q (m^3 h^{-1}).
Volume V (m^3).
Outdoor air concentration C_o (mg m^{-3}).
Concentration in indoor air C (mg m^{-3}).

Mass sorbed per unit area of surface M (mg m^{-2}).

$$\frac{d(CV)}{dt} = QC_o - QC + E - k_a CA + k_d MA \quad (17.51)$$

Mass balance on VOC sorbed onto surfaces is

$$\frac{d(MA)}{dt} = k_a CA - k_d MA \quad (17.52)$$

These equations are coupled first-order differential equations and can be solved analytically if coefficients are constant. The steady-state solution is shown below:

$$\begin{aligned} &QC_o - QC + E = k_a CA - k_d MA \\ &k_a CA - k_d MA = 0 \\ &\text{so, } \frac{QC_o + E}{Q} = C_{ss} \end{aligned} \quad (17.53)$$

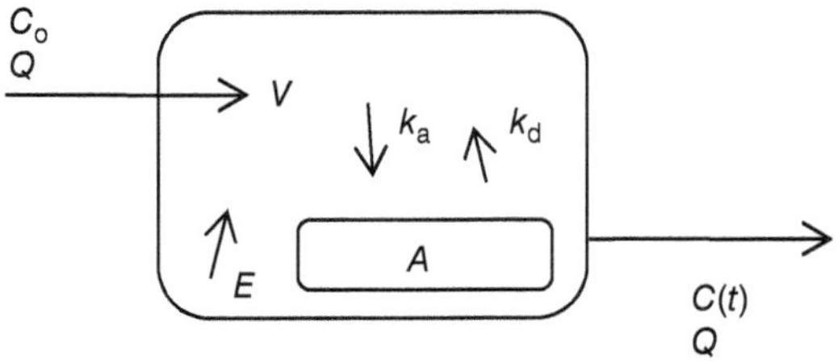

Figure 17.20 Factors that influence HCHO concentration in a real environment. *Source*: From Gunschera et al. (2013). Reproduced with permission of Elsevier.

Note that steady-state result does not depend on sorption! Sorption only affects the transient behavior.

17.18 Emissions Characterization

Exposures to air pollutants indoors are often higher than exposures that occur outdoors. Breathing indoor air is the dominant route of exposure for radon and for many VOCs. Radon comes from soil gas, whereas VOCs come from many occupant activities (smoking; Van Loy and Nazaroff, 1998), use of household products such as cleaners, and building materials such as particle board or paint (Berglund et al., 1989). Most effective control measures to reduce emissions are source or product related, such as avoiding use, substituting for a different material or product, actual product reformulation, and better manufacturing control. Consequently there is considerable interest in quantitatively characterizing emissions from indoor sources.

To characterize emissions $E(t)$ from a product or building material using experiments, the following are typically needed (Figure 17.21):

- Chamber with nonreactive/nonsorbing walls (glass, stainless steel).
- Constant temperature and RH control within the chamber and in the supply air.
- Pollutant-free air (C_{out}) supplied to the chamber at a controlled rate Q.
- Well-mixed conditions within the chamber must be maintained.

To carry out the experiments:

- Insert material or simulate product use in the chamber.
- Measure pollutant concentration in the chamber and/or in the exhaust air.
- Ideally measure $C(t)$; may also measure steady-state concentration C_{ss}.
- May only be able to measure the average concentration $\overline{C(t)}$ over some finite time period.

The material balance model for these emissions characterization experiments is shown in Equation (17.54):

$$\frac{d(CV)}{dt} = E(t) - QC(t) \quad (17.54)$$

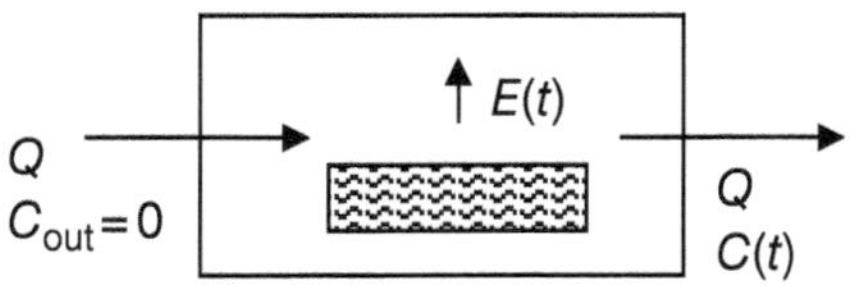

Figure 17.21 Schematic showing how to characterize VOC emissions from a product or material in a chamber.

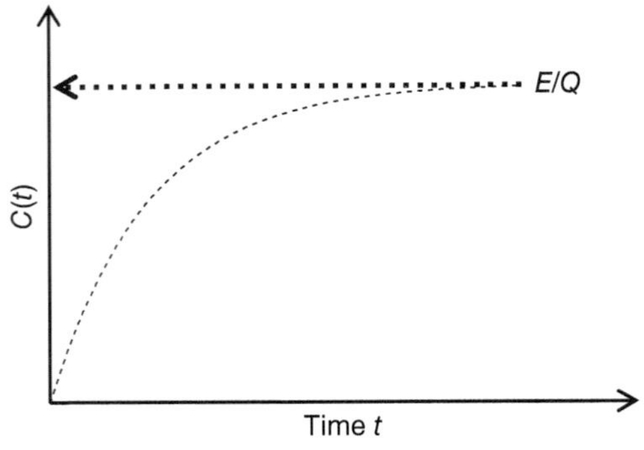

Figure 17.22 Concentration profile in experimental chamber when the emission rate is steady and not changing with time.

The goal is to find $E(t)$. If emission rate $E(t) = \text{constant}$, then it can be determined from the steady-state concentration as shown in Equation 17.55 and Figure 17.22:

$$E = QC_{ss}, \text{ where } C_{ss} = E/Q \text{ is the steady-state concentration.} \tag{17.55}$$

If, however, the emission rate $E(t)$ varies over time, the time-dependent emissions rate can be determined from the material balance equation only if $C(t)$ is known. In many product applications, or materials emissions, $E(t)$ tends to be high in the beginning and then diminishes over time as shown in Figure 17.23. Equation 17.56 is the material balance for $E(t)$ if $C(t)$ is known:

$$E(\tilde{t}) = V\frac{\mathrm{d}C(\tilde{t})}{\mathrm{d}t} + QC(\tilde{t}) \text{ at time } \tilde{t} \tag{17.56}$$

If $E(t)$ varies in time, but decays to zero over a reasonable period (less than the duration of the experiment), total mass emitted, M, can be determined according to the following procedure.

Write the equation for the total mass emitted, M:

$$M = \int_0^\infty E(t)\mathrm{d}t \tag{17.57}$$

To relate M to measured parameters, multiply the material balance equation (17.54) by dt and integrate over the measurement time period 0 to T, as shown in Equations 17.58 and 17.59 and Figure 17.24:

$$M = \int_0^T \mathrm{d}(CV) + \int_0^T Q(C(t))\mathrm{d}t \tag{17.58}$$

These two Equations (17.57) and (17.58) are valid even in the presence of sorption, provided it is fully reversible. The process would be to conduct the emissions experiment so that there is nothing left in the chamber, and then

$$\int_0^T \mathrm{d}(CV) = 0 = VC(T) - VC(t=0) \tag{17.59}$$

Thus,

$$M = Q\int_0^T C(t)\mathrm{d}t \text{ or } M = QT\overline{C(t)} \tag{17.60}$$

where $\overline{C(t)}$ is the average concentration from 0 to T.

There are many issues with emissions characterization including high cost, since the chamber is expensive and many of the instruments and/or analytical equipment can be expensive. Also the net flux of VOC from the material may depend on the environmental conditions, so it is important to achieve in chamber a concentration similar to that in application. This could be achieved by having similar loading (S/V) and air exchange rate (Q/V) in the test chamber as in application. An easier screening-type process would be useful to assess quickly the potential for pollutant emissions from materials or products. While this approach is the method of choice for research and reporting exact emission rates, a less expensive test is needed for routine use, for example, by manufacturers. For building materials, a headspace analysis in a batch reactor could be used: place material in small test vessel, made of nonreactive materials; seal up the batch reactor; wait for specified period; and then collect gas sample from the headspace in the vessel for analysis in the lab. Sometimes it is difficult to translate what this headspace analysis means for indoor air concentrations.

Another approach is the use of the small volume chamber known as Field and Laboratory Emission Cell (FLEC)

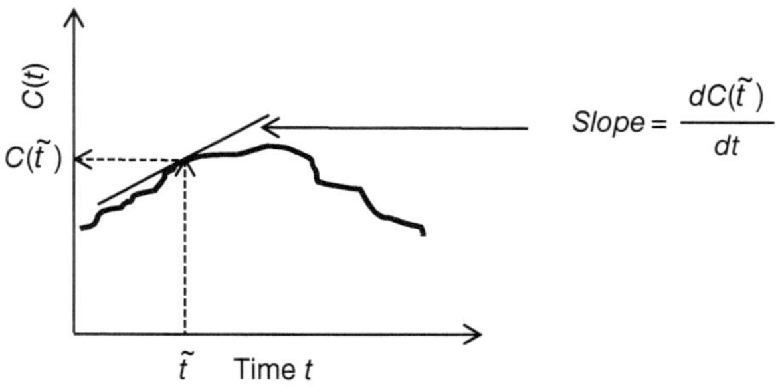

Figure 17.23 Graphic showing how emissions from a material change over time, with the higher emissions happening in the beginning of the exposure.

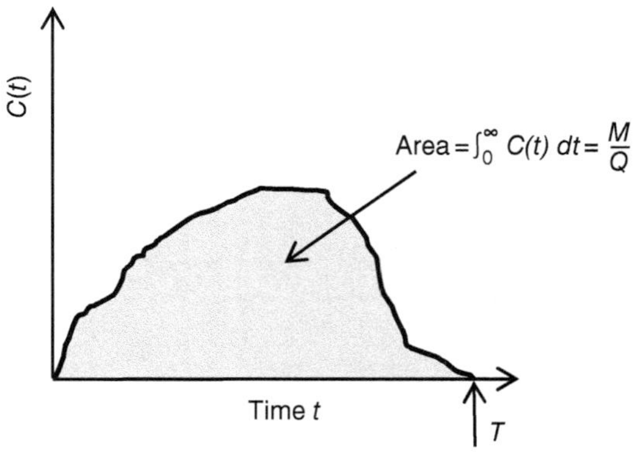

Figure 17.24 Integration of the material balance to estimate total mass emitted during the characterization experiment.

developed in Denmark (Wolkoff et al., 1993). It consists of a stainless steel cover about 30 cm in diameter. This cover is placed on the surface of a material. A clean stream of air is needed. The FLEC provides radial flow at constant velocity from perimeter toward center (not well mixed). Finally the VOC concentration in the exhaust airstream is measured.

This method can be problematic for slow, time-dependent emitters, for example, some materials emit VOCs at very low rates for several years.

17.19 Odors

Air quality is usually described by its chemical composition. In workplaces, industrial hygienists have established TLVs for single chemical compounds to which workers can be exposed and maintain safety and health. These threshold limits have been derived based on the dose–response relationship between exposure and a health endpoint. These limits are high, because they were developed for a younger, healthy population. In nonindustrial environments, there are complaints about poor IAQ, SBS, odors, etc., and it is very difficult to pinpoint that actual chemical that is causing the complaints. In some instances, only a few sensitive persons may complain, while in other situations more than half of the building occupants have symptoms. Often there is no single chemical in the indoor air that can explain the complaints, but there are many compounds that may be present at very low concentrations, levels that are not easily measured by standard instruments. Interestingly the human olfactory and chemical senses located in the nose mucous membrane are in many cases far superior to chemical analysis of the air. A natural alternative to chemical measurements is to use humans as a meter to quantify air pollution.

17.19.1 Olf and the Decipol Units

The olf is a unit used to quantify the strength of a pollution source. Danish Professor P. Ole Fanger introduced it in 1988 (Fanger, 1988). One olf is the emission rate of bioeffluents from a standard person defined as an average adult working in an office or similar nonindustrial workplace, sedentary and in thermal comfort with a hygienic standard equivalent of 0.7 baths per day and whose skin has a total area of 1.8 m^2 (Table 17.7).

The decipol is a unit used to quantify the concentration of air pollution perceived by humans, also defined by P. Ole Fanger. It is that concentration of human bioeffluents that would cause the same dissatisfaction as the actual air pollution. One decipol (dp) is the perceived air quality (PAQ) in a space with a sensory load of one olf ventilated by 10 l s^{-1}.

The following equation was derived from studies of human bioeffluents. More than 1000 sedentary men and women were studied in a thermally neutral environment

Table 17.7 Examples of typical olf levels for different pollutant sources (Fanger 1988).

Person/object	Scent emission
Sitting person	1 olf
Heavy smoker	25 olf
Athlete	30 olf
Marble	0.01 olf m^{-2}
Linoleum	0.2 olf m^{-2}
Synthetic fiber	0.4 olf m^{-2}
Rubber gasket	0.6 olf m^{-2}

Source: Reproduced with permission of Elsevier.

(Fanger, 1988). 168 men and women judged the air quality just after entering the space, and they were asked whether they would judge the air quality to be acceptable or not:

$$\begin{aligned} &\text{PD} = 395\exp\left(-1.83q^{0.25}\right) \text{ for } q > 0.32 \text{ l s}^{-1} \text{ olf} \\ &\text{PD} = 100\% \text{ for } q < 0.32 \text{ l s}^{-1} \text{ olf} \end{aligned} \tag{17.61}$$

where PD is the percentage of dissatisfied and q is the steady-state ventilation rate per olf.

The concentration of indoor air pollution depends on the source of the pollutant, its strength and emission characteristics, and the removal processes that are occurring in the indoor environment including ventilation and surface deposition. The perceived air pollution is defined as that concentration of human bioeffluents that would cause the same dissatisfaction as the actual air pollution concentration and is expressed in units of *pol*. One pol is the air pollution caused by one standard person ventilated by 1 l s^{-1} of unpolluted air. It is also expressed as a decipol, where one decipol = 0.1 olf $(\text{l s})^{-1}$. The equation below relates the percentage of dissatisfied as a function of the perceived air pollution:

$$\begin{aligned} &\text{PD} = 395\exp\left(-3.25C^{-0.25}\right) \text{ for } C < 31.3 \text{ decipol} \\ &\text{PD} = 100\% \text{ for } C > 31.3 \text{ decipol} \end{aligned} \tag{17.62}$$

where C is the perceived air pollution (decipol).

To measure the olf value of a pollutant source a panel of subjects is needed. The supply rate of outdoor air to the space is also needed. The judgment should take place immediately after the panel enters the space – this gives the first impression of the IAQ.

In the Danish town hall study, three buildings with high and one with low prevalence of SBS symptoms were studied (Zweers et al., 1990). The TVOC and air temperature correlated significantly with the panel's rating of odor intensity and odor acceptability in the rooms.

TVOC concentrations, air exchange rate, odor intensity and acceptability, and number of olfs and decipols were also significantly correlated. The threshold for reduced air quality caused by TVOCs in these four buildings was between 0.19 and 0.66 mg m^{-3}.

A study of 20 randomly selected offices and assembly halls used the olf units to quantify pollution sources (Fanger et al., 1988). The spaces were visited three times by 54 judges, who assessed the acceptability of the air. Ventilation rates, CO_2, CO, PM, and TVOC were measured, but these measurements but did not explain the large variations in PAQ. For each occupant in the 15 offices, there were ~7 olfs from other pollution sources on average, ~2 olfs from the building materials, 3 olfs from the ventilation systems, and 2 olfs by smoking. More than 30% of subjects found the air quality in the offices unacceptable. The authors suggested that the way to improve the air quality was to remove the pollution sources in the spaces and in the ventilation systems.

Acknowledgments

I want to express my gratitude to my mentor and doctoral dissertation advisor Professor William W Nazaroff, who first taught me about indoor air quality and instilled in me a passion for addressing urban air pollution problems to improve public health. This chapter was based on the course notes from Professor Nazoroff's indoor air class at the University at California, Berkeley, and the notes I have developed for my indoor air pollution course at the University of Colorado Boulder. I also sincerely appreciate all the feedback my students have provided and questions that they have asked during class, which has made me a better teacher. Finally thanks to my many colleagues who continue to research indoor air quality and move the field of inquiry further so that we may all inhabit healthy buildings.

Note

1 Assumes an air exchange rate (Q_O/V) of 1 air change per hour (h^{-1}) and a particle loss rate k of 0.1 h^{-1}.

References

AHAM (2002). Association of Home Appliance Manufacturers Method for Measuring Performance of Portable Household Electric Cord-Connected Room Air Cleaners, ANSI/AHAM AC-1-2002. https://www.google.com/url?sa=t&rct=j&q=&esrc=s&source=web&cd=1&ved=0ahUKEwjRwu79q4PZAhVD1IMKHQaYCkgQFggnMAA&url=http%3A%2F%2Fd1.amobbs.com%2Fbbs_upload782111%2Ffiles_46%2Fourdev_678322NWL2LE.doc&usg=AOvVaw0Inf5iY8JtCPkb0xQ6Kezz

AHAM (2012). *Air Cleaner Certification Program Procedural Guide Version 3.0.* Washington, DC: Association of Home Appliance Manufacturers.

Andersen, I., Lundqvist, G.R., and Molhave, L. (1975). Indoor air pollution due to chipboard used as a construction material. *Atmospheric Environment* 9: 1121–1127.

ASHRAE (2009a). Chapter 16: Ventilation and infiltration. In: *2009 ASHRAE Handbook: Fundamentals*. Atlanta, GA: ASHRAE.

ASHRAE (2009b). Chapter 9: Thermal Comfort. In: *2009 ASHRAE Handbook: Fundamentals*, 9.1–9.30. Atlanta, GA: ASHRAE.

ASHRAE (2010). Standard 55-2010: Thermal Environmental Conditions for Human Occupancy. Atlanta: American Society for Heating, Refrigerating and Air-Conditioning Engineers.

ASHRAE (2017). Standard 52.2-2017 – Method of Testing General Ventilation Air-Cleaning Devices for Removal Efficiency by Particle Size (ANSI Approved).

Berglund, B., Johansson, I., and Lindvall, T. (1989). Volatile organic compounds from building materials in a simulated chamber study. *Environment International* 15: 383–388.

Breysse, P.A. (1981). The health cost of "tight" homes. *Journal of the American Medical Association* 245: 267–268.

Breysse, J., Jacobs, D.E., Weber, W. et al. (2011). Health outcomes and green renovation of affordable housing. *Public Health Reports* 126 (Suppl 1): 64–75.

California Energy Commission. (2015). 2016 Building Energy Efficiency Standards for Residential and Nonresidential Buildings, Title 24, Part 6 CEC-400-2015-037-CMF. http://www.energy.ca.gov/2015publications/CEC-400-2015-037/CEC-400-2015-037-CMF.pdf (accessed 7 February 2018).

Çengel, Y.A. and Boles, M.A. (2008). *Thermodynamics: An Engineering Approach*, 7e. New York: McGraw Hill Chapter 14.

Dannemiller, K.C., Murphy, J.S., Dixon, S.L., Pennell, K.G., Suuberg, E.M., Jacobs, D.E., Sandel, M. (2013). Formaldehyde concentrations in household air of asthma patients determined using colorimetric detector tubes. *Indoor Air* 23: 285–294.

Dietz, R.N., Goodrich, R.W., Cote, E.A., and Wieser, R.F. (1986). Detailed description and performance of a passive perfluorocarbon tracer system for building ventilation and air exchange measurements. In: *Measured Air Leakage of Buildings*, ASTM STP 904 (ed. H.R. Trechsel and P.L. Lagus), 203–264. Philadelphia: American Society for Testing and Materials.

DOE (2011a). 2011 Buildings Energy Data Book, Chapter 1: Buildings Sector. http://buildingsdatabook.eren.doe.gov/ChapterIntro1.aspx (accessed 6 February 2018).

DOE (2011b). 2011 Buildings Energy Data Book, Chapter 2: Residential Sector. http://buildingsdatabook.eren.doe.gov/ChapterIntro2.aspx (accessed 6 February 2018).

EIA (2009). Household energy use in Colorado, http://www.eia.gov/consumption/residential/reports/2009/state_briefs/pdf/co.pdf (accessed 6 February 2018).

Emmerich, S, Howard-Reed, C., and Gupte, A. (2005). Modeling the IAQ impact of HHI interventions in inner-city housing, NIST. NISTIR 7212.

EPA (2016). A citizen's Guide to Radon: The Guide to Protecting Yourself and your Family from Radon. https://www.epa.gov/radon/citizens-guide-radon-guide-protecting-yourself-and-your-family-radon (accessed March 2017).

Fanger, P.O. (1988). Introduction of the olf and the decipol units to quantify air pollution perceived by human indoors and outdoors. *Energy and Buildings* 12: 1–6.

Fanger, P.O., Lauridsen, J., Bluyssen, P., and Clausen, G. (1988). Air pollution sources in offices and assembly halls, quantified by the olf unit. *Energy and Buildings* 12: 7–19.

Federspiel, C.C. (1998). Statistical analysis of unsolicited thermal sensation complaints in buildings. *ASHRAE Transactions* 104 (1): 912–923.

Fisk, W.J. (2000). Health and productivity gains from better indoor environments and their relationship with building energy efficiency. *Annual Review of Energy and the Environment* 25 (1): 537–566.

Gates, F.L. (1930). A study of the bactericidal action of ultra violet light: III. The absorption of ultra violet light by bacteria. *The Journal of General Physiology* 14 (1): 31–42.

Gilbert, N.L., Guay, M., David Miller, J. et al. (2005). Levels and determinants of formaldehyde, acetaldehyde, and acrolein in residential indoor air in Prince Edward Island, Canada. *Environmental Research* 99: 11–17.

Gunschera, J., Mentese, S., Salthammer, T., and Andersen, J.R. (2013). Impact of building materials on indoor formaldehyde levels: effect of ceiling tiles, mineral fiber insulation and gypsum board. *Building and Environment* 64: 138–145.

Hinds, C. (1999). *Aerosol Technology*, 2e. New York: Wiley (filters).

Hun, D.E., Corsi, R.L., Morandi, M.T., and Siegel, J.A. (2010). Formaldehyde in residences: long-term indoor concentrations and influencing factors. *Indoor Air* 20: 196–203.

IARC (2006). Formaldehyde, 2-butoxyethanol and 1-tert-butoxypropan-2-ol. *IARC Monographs on the Evaluation of Carcinogenic Risks to Humans* 88: 1–478.

Institute of Medicine (IOM) (2000). Clearing the Air, Asthma and Indoor Exposures. National Academies Press. Washington, DC.

Institute of Medicine (IOM) (2011). *Climate Change, the Indoor Environment, and Health* (ed. D. Butler and J. Spengler). Washington, DC: National Academies Press.

Jenkins, P.L., Philips, T.J., Mulberg, E.J., and Hui, S.P. (1992). Activity patters of Californians: use of and proximity to indoor pollutant sources. *Atmospheric Environment. Part A. General Topics* 26: 2141–2148.

Jonassen, N. and McLaughlin, J.P. (1988). Removal of radon and radon progeny from indoor air. In: *Radon and Its Decay Products in Indoor Air* (ed. W.W. Nazaroff and A.V. Nero), 435–458. New York: Wiley.

Kleinman, M. (2009). Carbon Monoxide. In: *Environmental Toxicants* (ed. L. Lippmann), 499. New York: Van Nostrand Reinhold.

Logue, J.M., Price, P.N., Sherman, M.H., and Singer, B.C. (2012). A method to estimate the chronic health impact of air pollutants in U.S. residences. *EHP* 120: 216–222.

Macher, J.M. (1993). The use of germicidal lamps to control tuberculosis in healthcare facilities. *Infection Control and Hospital Epidemiology* 14: 723–729.

Miller-Leiden, S., Lohascio, C., and Nazaroff, W.W. (1996). Effectiveness of in-room air filtration and dilution ventilation for tuberculosis infection control. *Journal of the Air & Waste Management Association* 46: 869–882.

Mølhave, L. (2009). Volatile organic compounds and the Sick Building Syndrome. In: *Environmental Toxicants* (ed. L. Lippmann), 242. New York: Van Nostrand Reinhold.

Nazaroff, W.W. (2013). Exploring the consequences of climate change for indoor air quality. Reprinted with permission from *Climate Change, the Indoor Environment, and Health. Environmental Research Letters* 8: 015022 http://doi.org/10.1088/1748-9326/8/1/015022.

Nazaroff, W.W. and Nero, A.V. (1988). *Radon and its Decay Products in Indoor Air*. New York: Wiley.

Nazaroff, W.W., Moed, B.A., and Sextro, R.G. (1988). Soil as a source of indoor radon: generation, migration, and entry. In: *Radon and its Decay Products in Indoor Air* (ed. W.W. Nazaroff and A.V. Nero), 57–112. New York: Wiley.

Nazaroff, W.W., Gadgil, A.J., and Weschler, C.J. (1993). Critique of the use of deposition velocity in modeling indoor air quality. In: *Modeling of Indoor Air Quality and Exposure*, ASTM, STP 1205 (ed. N.L. Nagda), 81–104. Philadelphia.

Nero, A.V. (1988a). Controlling indoor air pollution. *Scientific American* 258 (5): 42–48.

Nero, A.V. (1988b). Elements of a strategy for control of indoor radon. In: *Radon and Its Decay Products in Indoor Air* (ed. W.W. Nazaroff and A.V. Nero), 435–458. New York: Wiley.

NIOSH (2003). Guidance for filtration and air-cleaning systems to protect building environments from airborne chemical biological, or radiological attacks. Department of Health and Human Services, Centers for Disease Control and Prevention, National Institute for Occupational Safety and Health, April 2003.

Noris, F., Adamkiewicz, G., and Delp, W.W. (2013). Indoor environmental quality benefits of apartment energy retrofits. *Building and Environment* 68: 170–178.

NRC (1986). *Environmental Tobacco Smoke*. Washington D.C: National Research Council, National Academy Press.

Offermann, F.J., Sextro, R.G., Fisk, W.J. et al. (1985). Control of respirable particles in indoor air with portable air cleaners. *Atmospheric Environment (1967)* 19 (11): 1761–1771.

O'Rourke, M.K. and Lebowitz, M.D. (2000). Indoor bioaerosol Contamination. In: *Environmental Toxicants: Human exposures and TheirHealth Effects* (ed. M. Lippmann), 449–354. New York: Van Nostrand Reinhold.

Riley, R.L., Wells, W.F., Mills, C.C. et al. (1957). Air hygiene in tuberculosis: quantitative studies of infectivity and control in a pilot ward. *American Review of Tuberculosis and Pulmonary Disease* 75: 420–431.

Riley, R.L., Mills, C.C., O'Grady, F. et al. (1962). Infectiousness of air from a tuberculosis ward. Ultraviolet irradiation of infected air: comparative infectiousness of different patients. *American Review of Respiratory Disease* 85: 511–525.

Roughton, F.J.W. (1970). The equilibrium of carbon monoxide with human hemoglobin in whole blood. *Annals of the New York Academy of Sciences* 174 (1): 177–188.

Samet, J.M., Marbury, M.C., and Spengler, J.D. (1987a). Health effects and sources of indoor air pollution. Part I. *The American Review of Respiratory Disease* 136: 1486–1508.

Samet, J.M., Marbury, M.C., and Spengler, J.D. (1987b). Health effects and sources of indoor air pollution. Part II. American Review of Respiratory. *Diseases* 137: 221–242.

Sardinas, A.V., Most, R.S., Giulietti, M.A., and Honcher, P. (1979). Health effects associated with urea-formaldehyde foam insulation in Connecticut. *Journal of Environmental Health* 41: 270–272.

Scott, A.G. (1988). Preventing radon entry. In: *Radon and its Decay Products in Indoor Air* (ed. W.W. Nazaroff and A.V. Nero), 407–433. New York: Wiley.

Seifert, B. and Schmahl, H.-J. (1987). Quantification of sorption effects for selected organic substances present in indoor air. In: *Proceedings of the 4th International Conference on Indoor Air Quality and Climate*, vol. 1, 252–256. Berlin (West): Institute for Soil, Water, and Air Hygiene.

Shearer, S.D. Jr. and Sill, C.W. (1969). Evaluation of atmospheric radon in the vicinity of uranium mill tailings. *Health Physics* 17 (1): 77–88.

Tichenor, B.A. and Mason, M.A. (1988). Organic emissions from consumer products and building materials to the indoor environment. *Japca* 38 (3): 264–268.

Tichenor, B.A., Guo, Z., Dunn, J.E. et al. (1991). The interaction of vapour phase organic compounds with indoor sinks. *Indoor Air* 1: 23–35.

Tsuboi, K.K. (1950). Mouse liver nucleic acids II. Ultra-violet absorption studies. *Biochimica et Biophysica Acta* 6: 202–209.

US EPA (2000). Air Quality Criteria for Carbon Monoxide (Final Report, 2000). US Environmental Protection Agency, Office of Research and Development, National Center for Environmental Assessment, Washington Office, Washington, DC, EPA 600/P-99/001F.

Van Loy, M. and Nazaroff, W.W. (1998). Nicotine as a marker for environmental tobacco smoke: implications of sorption on indoor surface materials. *JAWMA* 48: 959–968.

Wadden, R.A. and Scheff, P.A. (1983). *Indoor Air Pollution*, 2–3. New York: Wiley.

Wang, T.C. (1975). A study of bioeffluents in a college classroom. *ASHRAE Transactions* 81 (1): 32–44.

Wargocki, P., Wyon, D.P., Sundell, J. et al. (2000). The effects of outdoor air supply rate in an office on perceived air quality, sick building syndrome (SBS) symptoms and productivity. *Indoor Air* 10 (4): 222–236.

Wargocki, P., Wyon, D.P., and Fanger, P.O. (2004). The performance and subjective response of call-center operators with new and used supply air filters at two outdoor air supply rates. *Indoor Air* 14 (suppl 8): 7–16.

Wells, W.F. (1955). *Airborne contagion and air hygiene.* Cambridge, MA: Harvard University Press.

Wells, W.F. and Fair, G.M. (1935). Viability of *B. coli* exposed to ultraviolet radiation in air. *Science* 82: 280–281.

WHO (1982). Indoor air pollutants. Exposure and health effects assessment/euro-reports and studies, No. 78, Copenhagen, WHO Regional Office.

Widdop, B. (2002). Analysis of carbon monoxide. *Annals of Clinical Biochemistry* 39 (4): 378–391.

Wilson, R. (1979). Analyzing the risks of daily life. *Technology Review* 81: 40–46.

Wilson, J., Dixon, S.L., and Jacobs, D.E. (2013). Watts-to-wellbeing: does residential energy conservation improve health? *Energy Efficiency* (5): 1–10.

Wolkoff, P., Clausen, P.A., Nielsen, P.A., and Gunnarsen, L. (1993). Documentation of field and laboratory emission cell "FLEC": identification of emission processes from carpet, linoleum, paint, and sealant by modeling. *Indoor Air* 3 (4): 291–297.

World Health Organization (1989). Indoor air quality: organic pollutants. 855–858.

World Health Organization (1990). Indoor air quality: biological contaminants, No. 31. WHO Regional Office for Europe.

WWA (2008). Climate Change in Colorado, A Synthesis to Support Water Resources Management and Adaptation. Report for the Colorado Water Conservation Board. University of Colorado Boulder, http://wwa.colorado.edu/publications/reports/WWA_ClimateChangeColoradoReport_2008.pdf (accessed 6 February 2018).

Wyon, D.P. (2004). The effects of indoor air quality on performance and productivity. *Indoor Air* 14 (s7): 92–101.

Xu, P., Peccia, J., Fabian, P. et al. (2003). Efficacy of ultraviolet germicidal irradiation of upper-room air in inactivating bacterial spores and mycobacteria in full-scale studies. *Atmospheric Environment* 37: 405–419.

Xu, P., Kujundzic, E., Peccia, J. et al. (2005). Impact of environmental factors on efficacy of upper-room air ultraviolet germicidal irradiation for inactivating airborne mycobacteria. *Environmental Science & Technology* 39 (24): 9656–9664.

Zweers, T., Skov, P., Valbjorn, O., and Molhave, L. (1990). The effect of ventilation and air pollution on perceived indoor air quality in five town halls. *Energy and Buildings* 14: 175–181.

18

Environmental Noise Pollution

Sharad Gokhale

Civil Engineering Department, Indian Institute of Technology Guwahati, Guwahati, India

18.1 Introduction

Environmental noise is the sound that is heard in the ambient.[1] Louder and intermittent noise annoys and can have adverse effects on humans, and, therefore, noise pollution has come to the forefront of the environment-related problems. Several human-made sources of noise exist in the environment. Noise has local scale impacts because it is an energy generated, transmitted, and lost in a short time within a small spatial scale (Allen et al., 2009; Weber, 2009). It is considered to be not threatening because it has not caused any catastrophic hazards as yet unlike a few well-known air pollution disasters around the world that have induced. It, being an energy, does not leave a residual effect behind. But its health hazards upon longer and regular exposure are well understood. Humans who are regularly exposed for a long time to higher noise are often at health risks. Its immediate effects are temporary and annoying but could be stress causing and upon longer exposure could be physiologically and psychologically damaging (Schwela et al., 2005).

Several countries have, from time to time, enforced regulatory limits on the noise in ambient environments to which people are exposed and also on the noise originated from various sources such as equipment and vehicles at manufacturing stages. There is an emerging trend, due to growing concern, to include noise impacts in the environmental impact assessments of new facilities or of the expansion of existing facilities. Hence, noise impacts of transportation and other infrastructural projects on the environment need consideration at the planning and developing stages (Kiely, 2007). Many developed countries have adopted the use of physical noise screens, also called as physical barriers, for a variety of noise-causing facilities to protect the environment. For example, a wall made of an absorptive material is installed along the major roads to reduce noise effectively from reaching the community or trees are planted roadside or around the industrial plant. Measures like planting trees around the noise-making facility are practically easy to adopt to reduce environmental noise.

This chapter is mainly focused on the environmental noise including its outdoor sources and effects, its science, its propagation through atmospheric medium, its relationship with the sources and the characteristics, important factors that affect it, methods of its measurement and analysis, important factors that attenuate it, and followed by noise management options.

18.2 Environmental Noise

Environmental noise refers to the noise that is generated outdoors within the community from households, public functions, loud speakers, road vehicles, trains, flying aircrafts, schools, playgrounds, sports stadiums, construction activities, and so on, which directly impact day-to-day activities of people. The accumulative noise resulting from these sources refers to community noise. The road traffic noise has become a nuisance in urban areas and almost unavoidable. It is stressful, annoying, and even threatening in the long run. Both in the developing and developed economies, rapid urbanization has made transport an essential part of our daily routine. In urban centers, noise is considered to be the most annoying where people by virtue of their residences, workplaces, schools, and commercial activities are often exposed to higher and variable noise, mainly originated from unprecedented growth in road traffic and commercial activities. Urban centers attract more people and more traffic, resulting in frequent traffic congestion. All contribute to higher noise besides other environmental issues. In developing countries, however, inadequate infrastructure, poor traffic management, pressure honking, and poor inspection and maintenance of vehicles in addition to higher traffic add more to the

Handbook of Environmental Engineering, First Edition. Edited by Myer Kutz.

environmental noise. People of residential and public buildings in urban areas, often in close proximity of trafficked roads, receive excessive noise pollution. People using urban centers, as a result, are developing urban stress. This may, due to regular exposure to high noise, lead to the diseases of higher orders such as cardiovascular and psychological effects. Public surveys have revealed worldwide that noise is the most annoying local problem (Kiely, 2007).

Environmental noise also refers to the nonoccupational noise. Noise at a receiver is the result of the medium through which it propagates. In ambient environment, the medium is the atmosphere. Therefore, there are several atmospheric characteristics that affect the noise during its propagation and play important roles in determining the noise annoyance. Besides, there are several surface elements such as trees, buildings, road surfaces, topography, etc., which obstruct the free propagation of noise-making noise analysis and assessment a complex task. In urban centers, trafficked roads are surrounded by dense buildings forming street canyons. A situation like this makes understanding noise propagation even more complicated as the emitted noise strikes multiple times to different elements. Moreover, road surfaces and building facades are hard and the noise is often reflected multiple times.

In the environment, some noises are steady while some vary constantly with time. Steady noise does not annoy as much as constantly varying noise (unsteady) does. Noise is also continuous or intermittent or impulsive. Intermittent noise is more annoying. Therefore, ill effects of noise depend upon loudness or magnitude of the noise, duration for which it lasts, and the time when it occurs. Noise varying in loudness disturbs more and is more annoying. The duration of noise whether it is short or longer, similarly, intense or normal, may cause different level of impacts upon exposure. Most often, annoyance varies with the time of occurrences as different activities are carried out throughout the day.

The continual efforts of reducing noise emissions from vehicles and various machines by developing technologies are proving counterproductive due to rapid increase in vehicular population and urban activities. As a result, high environmental noise and the exposure to it are increasing day by day. If noise continues to increase, it may lead to unsustainable environmental development as it has cumulative adverse health impacts (Joshi et al., 2003; Jarup et al., 2005). Noise has increased the annoyance level and sleep disturbance in our environment. In developing countries, this issue is even more crucial due to mixed use environment. Environmental noise management is, therefore, challenging but essential and need of the hour.

18.3 Effects on Human Health and Environment

The main short-term effect of noise is stress, referred as urban stress (Ising et al., 2004). Its long-term effects could be physiological and psychological. It is also known to have effects on plants and animals besides humans (Parris, 2015). A few studies have reported that chronic exposure to high noise can initiate subcortical stress reactions that may result in the deregulation of stress hormones and may lead to hypertension, accumulation of intra-abdominal fats, and insulin resistance (Passchier-Vermeer and Passchier, 2000; Ising et al., 2004; Luck et al., 2004; Bluhm et al., 2007; Davies, 2009). Several effects of noise and those caused by air pollution are similar (www.transportenvironment.org). For example, cardiovascular diseases such as heart rate variability, hypertension, coronary atherosclerosis, myocardial infarction, cognitive functioning, stroke, and also neurobehavioral ones like annoyance and so on (Schwartz et al., 2005; de Kluizenaar et al., 2007; Beelen et al., 2009; Selander et al., 2009; Sorenson et al., 2011; Gan et al., 2012a). A study has reported that there exists a correlation between a few air pollutants and noise both originated from road traffic (Beelen et al., 2009). It would mean that the health problems caused by air pollution could also be enhanced or may be exasperated by noise pollution. In trafficked corridors, pedestrians, cyclists, and motorbikers are often at risks. Even the in-vehicle exposure to outside noise may be significant during driving.

The noise level and the duration for which it lasts determine the potential to hearing damage, ringing in the ears, also called as tinnitus, and permanent hearing damage due to repeated exposure to higher noise (Basner et al., 2014). The hearing loss may be temporary or permanent (https://en.wikipedia.org/wiki/Noise-induced_hearing_loss). In addition, the factors such as individual's susceptibility to noise and age factor may play important roles (Henderson et al., 1993).

Work and speech interference is temporary and a common effect of noise. As mentioned earlier, an annoyance that varies with people to people is the most serious effect of noise. One of the major health outcomes of environmental noise is the decreased quality of sleep (Berglund and Lindvall, 1995), leading to various types of diseases, including nonauditory such as changes of behavior and deterioration in performance (www.noiseandhealth.org). Psychosocial health and effects on well-being due to long-term exposure to road traffic noise are also documented (Ohrstrom et al., 1998). The main psychological effects are the increased blood pressure, cardiac arrhythmia, finger pulse amplitude, heart rate, and respiratory and body movement changes (Berglund and Lindvall, 1995). According to a study carried out by Griefahn (2000) and Bluhm et al. (2004), people who live near streets with busy traffic or airports always keep windows closed and spend less time in their gardens (www.noiseandhealth.org). Such residents even get fewer visitors than to those living in relatively quieter areas (Griefahn, 2000). Noise is also said to have ecological impacts on species that are sensitive to noise (Parris, 2015). Consequent to these effects and awareness among the people in urban centers of developed countries, noise has also lowered the real estate values (Wilhelmsson, 2000).

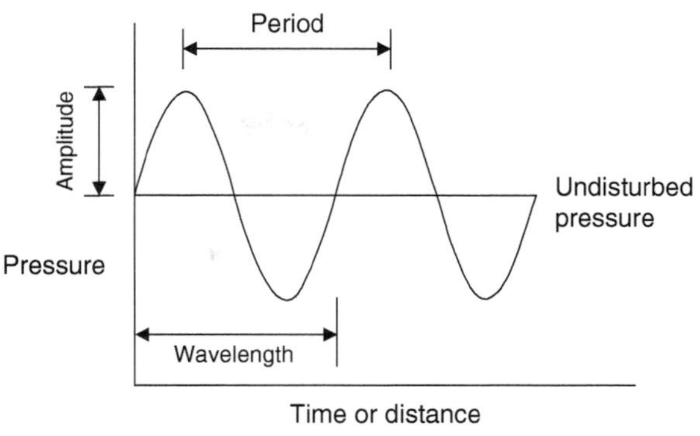

Figure 18.1 Pure sound wave.

18.4 Sound Propagation in Environment

Sounds are vibrations, which propagate through a medium causing a pressure variation in the medium. The disturbance caused to the medium due to pressure variation is heard by our ears. The sound is, therefore, any pressure variation that humans can detect or hear. For environmental noise, medium is the atmosphere. The wavelike phenomenon propagating through the atmosphere disturbs the atmospheric pressure. In a mixed urban environment, there are several sources producing sounds of varying loudness and, thus, sounds are not pure. Moreover, during propagation, sounds are obstructed by several obstacles. As a result, the sound waves in the environment do not produce any specific wave pattern. The simplest sound wave from a pure tone produces the sinusoidal pattern, which is used to determine the properties of sound such as *amplitude*, *frequency*, and *wavelength*. The sound vibrations are characterized by these properties.

The amplitude of sound that is propagating through the atmosphere means the fluctuation (due to disturbance it causes to the medium, as discussed earlier) from the mean atmospheric pressure. The amount of fluctuation determines the loudness of sound. Sound vibrations of certain amplitude are only audible to human ears. Therefore, along with the amplitude, the frequency of sound is also important. The frequency of sound is an important property. It means the periodic vibrations of the sound. It determines the pitch and is expressed in hertz (Hz). Both properties are thus important to make vibrations heard by the human ear. It varies according to person to person, age of a person, the condition of hearing loss, and physiology. The range of frequency of hearing varies from 20 to 20 000 Hz (20 kHz) (Kiely, 2007). As discussed earlier, sound in the environment is not purely sinusoidal in pattern and, therefore, with time

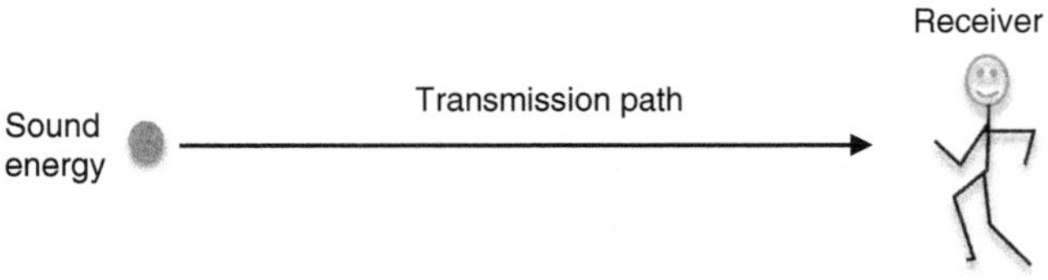

Figure 18.2 Environmental elements of the sound.

it varies in both frequency and amplitude. Its wavelength has spatial period, which means the distance over which the wave structure is repeated. It is characterized by the inverse of the spatial frequency. Figure 18.1 shows pure sound waves with the three important properties. Figure 18.2 shows the environmental elements of sound.

In the atmospheric medium, the wavelike disturbance propagates in the direction in which the particles are displaced, meaning the energy (as sound) emitted from a source is transported in the same direction as the particles oscillate back and forth, that is, parallel to the direction of energy transmission (Lamancusa, 2000). The sound waves, which travel through the atmosphere at a speed, are also affected by temperature and humidity since these parameters change the atmospheric air density. This is, however, more important in the case of far field (at far distance) impacts of noise.

Air is a nondispersive medium and, therefore, the speed at which the sound energy is transported and it propagates is same (Nicholau et al., 2009). But meteorological conditions of variable wind speed, temperature, and atmospheric turbulence can cause fluctuations of high range over a short time. If the transmission path is long, then fluctuations are large. Obstructions such as trees and buildings affect the free propagation of sound. Besides, the atmospheric absorption of sound depending upon the air density, ground absorption, and reflections from hard pavements all affect the impact of noise on the environment. Rains also play a role. Its effect may not be

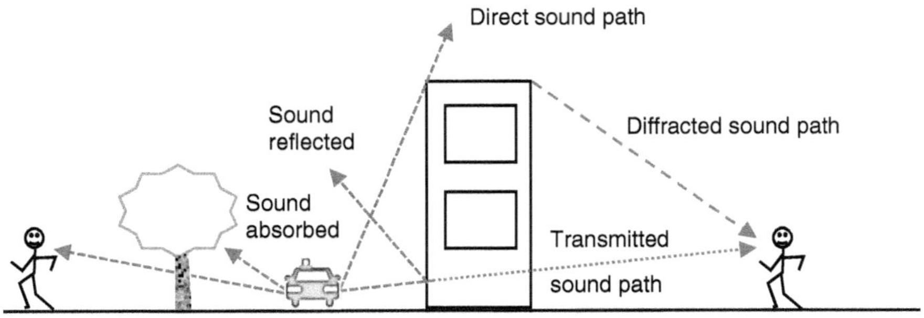

Figure 18.3 Different paths of sound when it propagates through the environment.

significant, but it changes the humidity (Lamancusa, 2009), which affects wind and temperature gradients, which may affect free propagation and eventually the noise level.

18.4.1 Factors Affecting Sound Propagation

Sound propagation in a mixed urban environment is complicated because it is obstructed during interaction with trees, buildings, road or ground surfaces, building facades, boundary walls, and so on, which cause multiple reflections of the sound. Some amount of sound is absorbed in the atmospheric air and trees, some amount is reflected, and some amount is transmitted through the buildings or such obstructions. Some amount of sound from the direct path is also diffracted, as shown in Figure 18.3. Certain meteorological conditions like refraction and turbulence create nonuniformity in the propagation medium (Lamancusa, 2009). Sound propagation is also affected by parameters like relative humidity, temperature, density of atmospheric air, and wind speed.

Sound pressure is, thus, directly affected by the presence of such elements in the mixed urban environment, and there is a loss of sound energy during propagation. Thus, atmospheric properties like pressure, temperature, relative humidity, wind speed, ground surface, topography, trees, and such other natural and artificial obstructions are important to consider in sound propagation studies. These are also called as attenuation factors since these factors weaken the sound intensity.

18.4.2 Source Effect on Sound Propagation

When the sound is generated, waves spread geometrically in the direction of the sound energy transported and the wavefront expands. The expansion of the sound wavefront is dependent on the strength and the type of the source from which the sound waves are originated. For example, some stationary sources and mobile sources would produce a different spread of wavefront. Figure 18.4 shows the wavefront expansion from the source in the direction of energy transport.

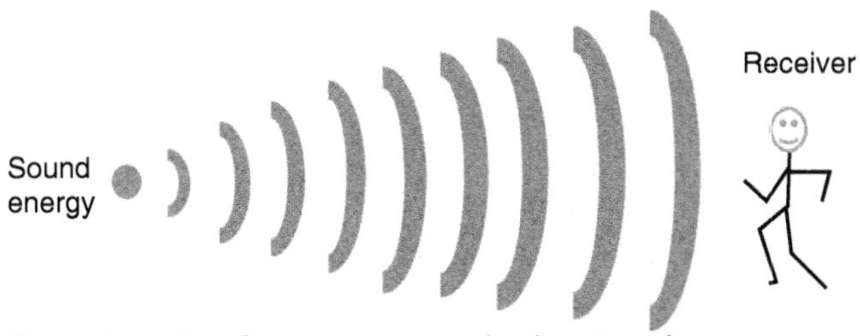

Figure 18.4 Wavefront expansion in the direction of energy transport.

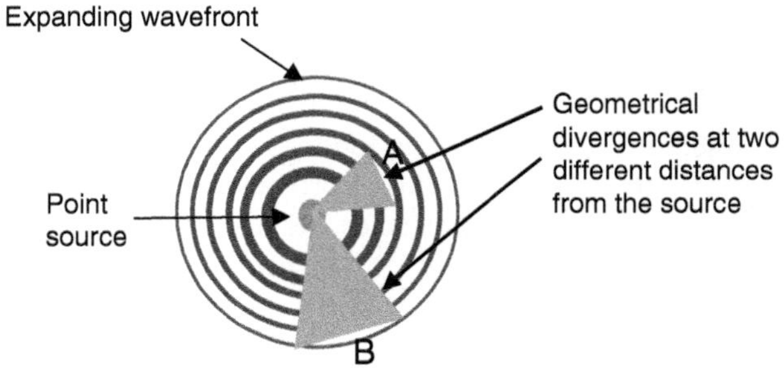

Figure 18.5 Spherical expansion of the sound wavefront.

18.4.3 Types of Wave Fronts

The geometrical shape of wavefront depends upon the type of sources from where the sound is generated. For example, a point source, fan, also referred to as the stationary source (fixed), generates the wavefront that spherically radiates sound equally in all directions. It is also called as the spherical divergence or spherical spreading. Whereas, a mobile (moving) source, for example, vehicles on the road, also referred to as the line source, generates the wavefront having cylindrical spreading. This knowledge is useful in understanding the relationship between the sound propagation and the distance from the source (Chaudhary, 2009). Figure 18.5 illustrates the expansion (spreading) of the sound wavefront.

18.4.4 Mathematics of Sound Waves

The sound-wave-created disturbance causes a small fluctuation in the atmospheric pressure. It may, therefore, be assumed to be linear (Lamancusa, 2000).

$$\Delta_p^2 = \frac{1}{c^2} \cdot \frac{\delta^2 p}{\delta t^2} \tag{18.1}$$

where c is the speed of sound, p the sound pressure, t the time, and Δ_p is the change in pressure quantity.

Mathematically, the propagation of sound is studied by treating it a simple plane wave or a spherical wave. In the plane wave, the wavefront radiates in a plane and the pressure varies in one dimension. It indicates that the plane vibrates uniformly and exhibits a constant sound pressure anywhere close to the surface. Let x be the direction in which the pressure variation is taking place. Eq. (18.1) can then be written as Eq. (18.2):

$$\frac{\delta^2 p}{\delta x^2} = \frac{1}{c^2} \cdot \frac{\delta^2 p}{\delta t^2} \tag{18.2}$$

Most sources of sound, particularly of point type, radiate spherical wavefronts. The spherical waves radiate like the light emitted from a bulb. The sound pressure decreases in the direction in which the wavefront propagates from the source. At any distance from the source, sound pressure is constant anywhere on the sphere (www.sengpielaudio.com). Let r be the radius of the sphere from the originating point source. The equation for the spherical wavefront can be written as, given in Eq. (18.3) (Lamancusa, 2000),

$$\frac{\delta^2 (r \cdot p)}{\delta t^2} = c^2 \cdot \frac{\delta^2 (r \cdot p)}{\delta t^2} \tag{18.3}$$

Equation (18.3) is valid under the assumption that sound is uniformly radiating from a point source. It indicates that sound pressure decreases as the r increases from the source.

Both Eqs. (18.2) and (18.3) produce similar solutions. This indicates that the propagation of the waves radiating outward in both plane and spherical wavefronts is similar except in the case of spherical wavefront; the amount of sound pressure depends upon the distance from the source. However, in the case of spherical propagation, if the value of r is large, the wavefront becomes similar to the plane wavefront. This can be understood by relating the acoustic particle velocity with the pressure in the propagation of both wavefronts.

Real sources in the environment do not radiate uniformly in all the directions and may, therefore, be directional. It has been perceived that the sound produced with a high frequency radiates directionally, and low frequency sound radiates uniformly. This also means that different sources producing sounds at different frequencies have nonidentical radiation patterns whether the wave radiates uniformly or not. Even if the wave propagation is directional, the pattern a particular sound exhibits is nonidentical to the pattern produced by the sound wave from any other sources. The law that holds true for the sound waves is that along any radial line, sound pressure is always proportional to the distance from the source.

18.5 Characteristics of Sound

We discussed that sound is a wavelike phenomenon or disturbance that propagates through air medium in the case of environmental noise. This disturbance fluctuates the properties of the medium like the velocity of sound, density of air, temperature of air, and mainly atmospheric pressure. Therefore, the sound that is heard is a pressure variation in the atmosphere. Knowing how much is the sound in the environment is about knowing how much disturbance due to the sound wave has occurred. How to determine the amount of disturbance then? It is easy to find out the change in the pressure relative to the undisturbed atmospheric pressure. With this the resulting noise reaching the receptor can be quantified.

Recall Figure 18.1 of pure tone sound waves. The amount by which the atmospheric pressure is fluctuated or the amount of pressure generated by the waves about the atmospheric pressure is called the sound pressure. It is the variation in pressure relative to above or below ambient air pressure. This fluctuation in the pressure is detected or measured with a microphone. The specific properties that help understand the physics of sound are discussed below. These properties are sound power, sound intensity, and sound pressure (Anderson et al., 2015).

18.5.1 Sound Power

Sound power is defined as the amount of energy, generated at a source, that is transmitted to the air medium per unit time in the direction of the propagation of the wave (Gieras et al., 2005). Sound power is transmitted through a medium. Because pressure is in phase with velocity for a plane wave, it does not depend on distance. It is expressed in watts (W) or newton meter per second (N-m s^{-1}) and denoted by W (Anderson et al., 2015).

$$\text{The sound power}\,(\mathrm{W}) = \frac{W}{t}$$

where W is the sound power and t is the time.

18.5.2 Sound Intensity

Sound intensity means the strength of sound or the concentration of sound at a fixed position in space after the wave has propagated. It is defined as the sound power transmitted per unit surface area, normal to the direction of propagation of the sound wave (Kiely, 2007). It can also be defined as the energy transmitted per unit surface area per unit time (Hassan, 2009). It is expressed in watts per square meter ($\mathrm{W\,m^{-2}}$) or newton meter per second per square meter ($\mathrm{N\text{-}m\,s^{-1}\,m^{-2}}$) (Nguemaleu and Montheu, 2014). The definition indicates that sound intensity or pressure disturbance is high in small areas.

$$\text{The sound intensity } (I) = \frac{\text{sound power } (W)}{\text{surface area } (A)}$$

where A is the surface area normal to the direction of propagation.

18.5.3 Sound Pressure

Sound pressure is defined as the amount of deviation occurred in the atmospheric pressure caused by a sound wave (Cao, 2011). It is the local pressure fluctuation from the atmospheric pressure. In physics, it is defined as the difference between the instantaneous pressure at a point in space and the static pressure of any medium or atmospheric pressure in the atmospheric environment (Hassan, 2009; Cao, 2011). It is the force of sound on a surface area perpendicular to the direction of sound. It is expressed as $\mathrm{N\,m^{-2}}$ or Pa. The sound pressure (p) and sound intensity (I) are related.

18.5.4 Sound Frequency

Sound frequency is an important property of sound and is defined as the number of sound-wave cycles per unit time. It is measured in cycle per second or hertz as 1 cycle per second is equal to 1 hertz. Frequency of sound is related to wavelength by the speed of sound (Sinha and Samuel, 2007). The wavelength is the distance that is required to complete a cycle of frequency. It also refers to the pitch of sound.

$$\text{Sound frequency } (f) = \frac{c}{\lambda}$$

where c is the speed of sound and λ is the wavelength.

18.5.5 Sound Divergence

Sound divergence means the spreading of waves, which decreases the sound pressure in the far field with the increasing distance (Kim, 2010). The sound divergence is different for a different type of sources. For defined sources, spreading of sound wave is geometrical. Geometrical divergence varies with the type of sources. For example, stationary sources or point sources such as a ceiling fan produce a sound wave having a spherical divergence; similarly, mobile sources such as a moving vehicle produce a sound wave having a cylindrical divergence. Figure 18.6a and b show the geometrical divergences for a point source and a line source, respectively.

18.6 Relationship Between Characteristics

18.6.1 Sound Pressure and Sound Intensity

To determine the relationship between sound pressure and sound intensity, we must know what impedance is. Impedance in acoustics is defined as the ratio of sound pressure (p) to particle velocity (v) (Kim, 2010). When sound propagates through air medium, it causes the particles to oscillate in the same direction and, thus, produce a longitudinal wave. The particle velocity is resulted from the oscillating of particles when sound propagates through air medium (Thompson, 2009). For a plane wave, impedance is proportional to the density of air times the speed of sound (Kathy and Werff, 2007). Even for a spherical wave, for large distances, impedance is density of air times the speed of sound (Kathy and Werff, 2007). Because in this case, the influence of the curvature of a wave is reduced to negligible. The large distance in sound studies means the distance measured in wavelengths is large. However, in the near field when the distance is small, impedance is complex and the relationship of pressure and the velocity becomes complicated.

The general definition of sound intensity (I) for any geometry is given by Eq. (18.4):

$$I = \lim_{T \to \infty} \frac{1}{T} \int_0^T (p \cdot v)\,\mathrm{d}t \qquad (18.4)$$

For a plane wave, sound power is equal to the work done ($\mathrm{d}W$) on a plane wave (or element of area "A") moving at a distance $\mathrm{d}x$ in time $\mathrm{d}t$. It is given by Eq. (18.5) (Rennie, 2014):

$$\frac{\mathrm{d}W}{\mathrm{d}t} = \frac{p \cdot A \cdot \mathrm{d}x}{\mathrm{d}t} = p \cdot A \cdot v \qquad (18.5)$$

Thus, v is the velocity at which the planer element is moving. As discussed earlier, the sound intensity is the sound power per unit area (www.engineeringtoolbox.com).

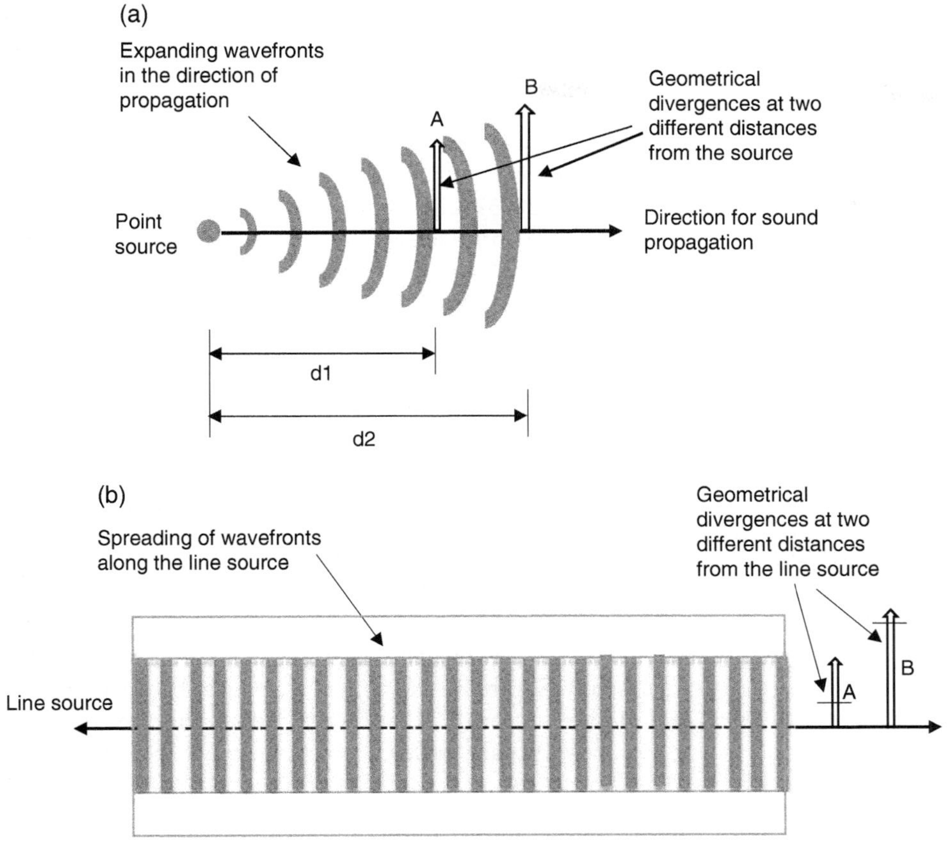

Figure 18.6 (a) Spherical expansion of the sound wavefront. (b) Cylindrical spreading of the sound wavefront.

Therefore, if we divide both sides of Eq. (18.5) by A, we get as given in Eq. (18.6):

$$\frac{dW}{dt \cdot A} = p \cdot v \tag{18.6}$$

or the above equation can be written as Eq. (18.7):

$$I = p \cdot v \tag{18.7}$$

Since for a plane wave, the values of each are instantaneous, the above equation can be written as Eq. (18.8) (Anderson et al., 2015):

$$I(t) = p(t) \cdot v(t) \tag{18.8}$$

We know that velocity at any instant of time is in phase with the instantaneous pressure. That means the instantaneous velocity is proportional to pressure ($v(t) \propto p(t)$). And from Eq. (18.8), velocity is the ratio of sound intensity $I(t)$ to sound pressure $p(t)$. Therefore, from Eq. (18.8), the sound intensity becomes, given by Eq. (18.9),

$$I(t) \propto p(t)^2 \tag{18.9}$$

From the definition, impedance offers resistance to the sound pressure while sound is propagating over the surface area. It is, therefore, defined as the ratio of pressure over a surface area in sound wave to the rate of particle flow across the surface equals the density of air times the speed of sound (de Arcas, 2007; Kathy and Werff, 2007). It is written as in Eq. (18.10):

$$\frac{p}{v} = \rho \cdot c \tag{18.10}$$

From Eq (18.8), we know $p = I/v$, substitute this in Eq. (18.10), we get Eq. (18.11):

$$I = (\rho \cdot c)v^2 \tag{18.11}$$

And, since

$$v \propto p \tag{18.11a}$$

$$v = \frac{1}{\rho \cdot c} \cdot p \tag{18.12}$$

Equation (18.12) is in same form as Eq. (18.11). Therefore, the sound intensity for a plane wave is, as given in Eq. (18.13),

$$I = \frac{p^2}{\rho \cdot c} \tag{18.13}$$

Equation (18.13) represents that sound intensity is proportional to sound pressure squared (Taherzadeh, 2014). Therefore, sound intensity is same not only for a plane wave but also for a spherical wave. For a longtime average, p^2 becomes p_{avg}^2.

Recall Eq. (18.4); we can write the average sound pressure, as given by Eq. (18.14):

$$p_{avg}^2 = \lim_{T \to \infty} \frac{1}{T} \int_0^T p^2 dt \qquad (18.14)$$

And the root mean square (rms) pressure is given by Eq. (18.15):

$$p_{rms} = \sqrt{\frac{1}{T} \int_0^T p(t)^2 dt} \qquad (18.15)$$

Assume that the source is nondirectional and the wavefront is spherically radiated. The intensity is uniform over the sphere surrounding the source. As discussed before, we know that sound pressure generates power (W) that is propagated through an area (A) and hence the sound intensity is $I = W/A$ (watt per square meter) $W m^{-2}$. For a spherical wave, the surface area would be $4.\pi.r^2$. The sound intensity is given by Eq. (18.16):

$$I = \frac{W}{4.\pi.r^2} \qquad (18.16)$$

This is also called as the "inverse square law" (Rau and Wooten, 1980). It indicates that intensity decreases with the square of the distance from the source generating a fixed amount of power (W) (Hassan, 2009), as given in Figure 18.6a and b. The sound pressure amplitude is proportional to the square root of the sound intensity (Taherzadeh, 2014).

18.6.2 Sound Pressure and Frequency

As understood from the definition, frequency of sound is the number of sound pressure variations per second. Environmental sounds are generated at different frequencies but sounds of all the frequencies are not audible. The audible range of frequencies is 0.015–15 kHz. Sound with frequency more than 8 kHz is not common in outdoor environment. It is important to consider the frequency of sounds along with the sound pressure for assessment or for working out a control strategy, because frequency affects perceived loudness (Singal, 2005). Therefore, the knowledge of frequency is useful in control measures at the sources or during transmission through an effective design of frequency-dependent properties of absorbing materials (Singal, 2005). The range of audible sound frequencies is usually expressed in octave bands in which the range of frequency is divided into a few intervals between a given frequency and twice that frequency. Therefore, octave bands refer to the intervals of frequencies. The interval of frequencies can be done arbitrarily depending upon the range we want to study or work with. But a fixed number of bands are widely used in sound studies.

To cover the whole audio range, the scale on both sides of the reference frequency is usually divided by fractions of octaves like 1/1 and 1/3 bandwidths. Table 18.1 shows the typical frequency bandwidths and octave intervals. Usually the filters having constant bandwidth are preferred at all frequencies irrespective of the filter center frequency as shown in Table 18.1 (Singal, 2005; Kiely, 2007).

It may be noted that 1/3 octave covers a wide band of frequencies and hence preferred for environmental noise analysis (IEC: 225–1966). Frequency is an inherent property of the sound that determines the pitch, which is generally affected by several environmental factors. Since sound waves can be measured in terms of frequency and wavelength, sound quality varies largely due to these two properties.

The commonly used reference sound pressure (Nguemaleu and Montheu, 2014) in air is 20 μPa (2×10^{-5} Pa) for it is the threshold of human hearing (Anderson et al., 2015). It also means the softest noise just heard by human ear. This reference sound pressure may be related to the sound frequency of 0.015 kHz or lower than that that is barely heard by the human ear. As the

Table 18.1 1/1 and 1/3 octave bands with a center frequency for an audible range.

Octave bands	Audible range of frequencies (kHz) for humans						
Octave band center frequency	0.125	0.250	0.500	1.000	2.000	4.000	8.000
Octave band limits	0.088–0.176	0.176–0.353	0.353–0.707	0.707–1.414	1.414–2.825	2.825–5.650	5.650–11.300
1/3 Octave band	0.160 0.200	0.315 0.400	0.63 0.800	1.250 1.600	2.500 3.150	5.000 6.300	

frequency increases, sound pressure increases and a high-pitch sound is heard. The relationship between the sound pressure and the frequency is used in developing various weighting scales for microphones for detecting sound pressure variations at different frequencies.

18.6.3 Sound Pressure and Distance

We discussed earlier that sound intensity is inversely proportional to the square of the distance (www.sengpielaudio.com). At a large distance, it is directly proportional to the square of the pressure (Taherzadeh, 2014). Therefore, the strength of the sound wave propagation depends upon the distance from the source where the sound is generated (Puri, 2012). Sound pressure, therefore, dampens with the increase in distance between the source and the receptor. Figure 18.5 shows the expansion of the wavefront describing that as the distance from the source increases, the surface area of the wavefront also increases, which as a result decreases the sound pressure in the direction of propagation as well as along the radiating spherical wave surface from the source. As a consequence, sound pressure reduces as the distance of the receptor increases from the source where the sound has been generated. This reduction in sound pressure with distance is regarded as the sound loss during propagation and often expressed by the inverse square law (see Eq. 18.16).

18.6.4 Source and Sound Divergence

Geometrical spreading or divergence of sound depends on the type of source (Finegold et al., 2007) and the dimension of the source relative to the distance of the receptor. To assess noise at a receptor at some distance, it is essential to determine the sound pressure at the receptor with reference to the sound power generated at the source (Mailing, 2007). If the source is smaller than the distance of the receptor from it, it is considered to be a point source. Point sources have spherical geometrical divergences. The sound pressure in the environment depends on several environmental characteristics as they affect the free propagation or transmission of the sound waves from the source to the receptor. We know that sound intensity is sound pressure squared. And sound pressure is inversely proportional to the distance squared (Gieras et al., 2005). Hence, as the distance between the source and the receptor increases, sound pressure decreases (Hassan, 2009).

Sound energy is transmitted through the environment when it is released at a source and spreads as a result of the expansion of the sound wavefronts. The geometrical spreading is not affected by the frequency of sound and so it is independent of frequency. There are two types of geometrical spreading: (i) spherical and (ii) cylindrical. The spherical spreading of sound occurs from sources that are stationary (point) sources. For example, fan, concrete mixer, diesel generator (DG) set, drilling, and so on, in all such sources sound radiates equally in all directions much like a light bulb radiates light equally in all directions. Figure 18.6a describes the spherical spreading of the sound wavefront from a point source P_s. Figure 18.6b describes the cylindrical spreading of the sound wavefront from a line source where the sound is considered to be generated per unit length. It is, therefore, constant along the length at some distance across the line source L_s, for example, at distance A, from the line source (Figure 18.6b), sound pressure is constant along the line source. The sound pressure will reduce at B distance from the line source.

It is observed from the both sound propagation patterns that sound pressure in the case of a point source (spherical spreading) reduces according to the square of the distance and in the case of line source (geometrical spreading) directly with the distance (www.sengpielaudio.com).

18.7 Environmental Noise Levels

We discussed that sound pressure disturbances that we hear are too small relative to the atmospheric pressure. Further, the range of sound pressure we hear, from lowest to loudest, is too large. Typical sound pressure magnitudes ranging from just barely audible or threshold of hearing of 20 μPa to the threshold of pain of approximately 60 Pa are quite large, the corresponding sound intensities of which is of the order of 10^{14} factor. At such a large range of pressure and high magnitude of corresponding sound intensity, it could be a difficult task to deal with the sounds generated of different powers and intensities in a mixed environment. As a result, environmental noise cannot easily be analyzed and handled in pressure or intensity unit. This large range is instead compressed into a more meaningful scale so that it can be easily read and presented. This is done by expressing the sound pressure and sound intensity in terms of decibel. Decibel is one-tenth of a bel (Taherzadeh, 2014). The decibel relates any quantity to its corresponding reference quantity, which is generally the threshold of hearing because that is the smallest detectable sensation. For example, S is any quantity related to a corresponding reference quantity S_0. The S quantity in terms of decibel is expressed by Eq. (18.17):

$$S(\text{dB}) = 10\log\left(\frac{S}{S_0}\right) \tag{18.17}$$

From the above definition, the decibel is a logarithmic unit used to describe the ratio of the signal level (www.engineeringtoolbox.com) such as power, pressure, or intensity to the respective reference level.

Sound power, pressure, and intensity when expressed in decibel become environmental noise levels – sound power level (L_w), sound pressure level (L_p), and sound intensity level (L_I), respectively. For example, sound power level is expressed as given in Eq. (18.18) (Mailing, 2007):

$$L_W\,(\text{dB}) = 10\log\left(\frac{W}{W_0}\right) \tag{18.18}$$

The W_0 is the reference sound power. Similarly, sound intensity and sound pressure are expressed as Eqs. (18.19) and (18.20), respectively:

$$L_I\,(\text{dB}) = 10\log\left(\frac{I}{I_0}\right) \tag{18.19}$$

$$L_p\,(\text{dB}) = 10\log\left(\frac{p}{p_0}\right)^2 \tag{18.20}$$

We know that sound intensity is sound pressure squared. If we take antilog of Eq. (18.20), we get

$$10^{L_p/10} = \left(\frac{p}{p_0}\right)^2 \tag{18.21}$$

The unit of all these levels is decibel because of the transformation of power, intensity, and pressure into decibel, i.e. (L_w), (L_p), and (L_I), are in decibel. The sound pressure level (L_p) is considered to be the environmental noise level, which is recorded by microphone as discussed earlier.

Since the noise levels are logarithmic function, two noise levels recorded over a time at the same location or at two different locations cannot be arithmetically added. But it can be done by adding sound intensities or sound pressure squared, which is called as the equivalent noise level (L_{eq}). It represents the constant noise level or sound pressure level that would have produced the same total energy as the actual sound level over the given time (Kiely, 2007). The L_{eq} can be applied to the noise levels continuously fluctuating over time and also to the noise levels that are discrete. It is expressed by Eq. (18.22) from Eq. (18.20):

$$L_{eq} = 10\log_{10}\left(\frac{1}{T}\int_0^T 10^{L_i/10}\,dt\right) \tag{18.22}$$

where T is the time period over which the equivalent noise level is to be determined and L_i is the noise level in the ith observation (sample).

For discrete n observations, the equivalent noise becomes, as given by Eq. (18.23),

$$L_{eq} = 10\log_{10}\left(\frac{1}{T}\sum_{i=1}^{n} f_i \cdot 10^{L_i/10}\right) \tag{18.23}$$

where f_i is the fraction of time that the sound pressure level is in the ith time interval and n is the number of observations (Kuehn, 2010).

18.8 Measurement and Analysis of Ambient Noise

Noise quality assessment is important for investigating various management options. Several brands of sophisticated sound level meters are available in the market that are used for the measurement of noise levels in decibels in the ambient as well as indoor environment for assessment. There are a few principle standards used worldwide to evaluate the noise quality from various sources. Several machineries, vehicles, and so on are restricted to some noise limits at manufacturing stages. There are restrictions (limits) set on noise-producing things, for example, DG, different vehicles, fire crackers, and construction vehicles and machineries. But over the years with the use and poor inspection and maintenance, these machineries and vehicles produce higher noise. As a result, ambient noise levels are often higher. Several countries, therefore, have regulatory limits for daytime and nighttime of the ambient noise levels for different zones in the city.

To develop proper understanding of the noise originated from a variety of sources with the essential causative factors and develop noise mitigation strategies, measurements with the instruments alone are not enough. It is important to understand the science of its origin, the medium through which it propagates, the relationship between the source and the receiver (Hassan, 2009), and the factors that attenuate noise in the environment so that proper mitigation measures can be developed and designed. The ISO 9613-1 and 9613-2 international standards outline the methods of noise calculations and various possible attenuations of sound during propagation outdoors (ISO, 9613-1, 1993, 9613-2, 1996). In urban centers, it is complicated to assess noise as noise propagation is affected by road surface, surrounded buildings, trees, topography, reflecting surfaces, and so on. There are several principle standards used worldwide in evaluating and forecasting noise levels in the mixed environment.

Sound level meters and its functions are specified in International Electrotechnical Committee (IEC), IEC651 for conventional sound level meters and IEC 804 for integrating sound level meters (Giovanni, 1999). These functions are also specified in several national standards such as BS5969 and BS6698. These standards ensure the accuracies at specified reference conditions, for which the sound level meters are usually manufactured to meet the compliance at different specific precision levels. The sound level meters are classified as type 0, type 1, and type 2 (Kiely, 2007). Type 1 refers to the reference condition of 0.7 dB, which is a precision grade standards and hence applicable for accurate field measurements. Type 2 refers to the reference condition of 1.5 dB, which is an industrial grade standards and hence applicable for non-critical survey work. Given the accuracy of type 1 sound level meters, it is recommended to use it for industrial as well as environmental measurements for compliance purposes.

Sound level meters are usually developed with *A*, *B*, and *C* frequency weightings to describe the responses in the environment. Of which, the *A* weighting scale is commonly used to describe the human's audible response. We know that, as discussed earlier, loudness is important in noise assessment, and it depends upon the sound frequency. Low frequency sounds are normally not heard, while high frequency sounds are heard at more or less same pitch or loudness. The *A* weighted sound levels are known to have correlation with the hearing damage or speech interference, and therefore, *A* is the most common weighting scale used widely for environmental noise assessment (Singal, 2005). According to the response of human ear, this scale has been designed to weight the components of noise. A range of sound level meters are available in the market with or without frequency octave band 1/1 or 1/3 bandwidths (Barron, 2003). A sound level meter (Amprobe SM-20A) is shown in Figure 18.7. The reason

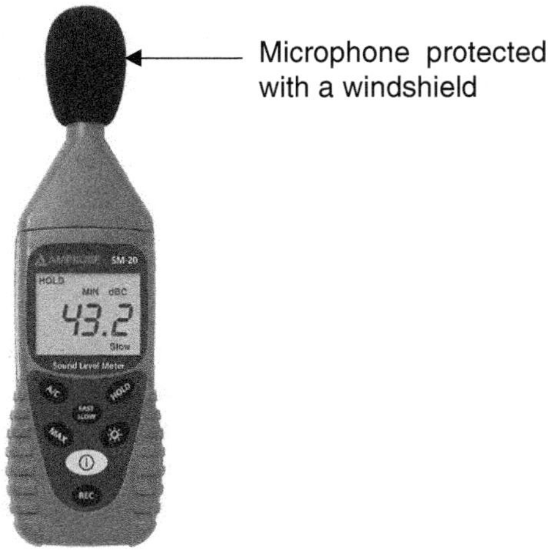

Figure 18.7 A sound level meter (Amprobe SM-20A). *Source:* Adapted from www.amprobe.com (accessed May 2016).

for including this sound level meter is that it has been used to measure the ambient noise levels for illustrating the data analysis, which is discussed in the next section. The microphone is an important element and needs special care during fieldwork. It is protected with a windshield. It is often installed on tripod at about 1.2–1.5 m height above the ground level with its microphone usually directed toward the source under study.

18.8.1 Roadside Noise

With this sound level meter, a noise measurement study was carried out at IIT Guwahati near a normal trafficked roadway to study the effect of pressure honking. Every 1 min noise level in dB was recorded for a period of 1 h on Monday, 24 August 2016 from 10 to 11 a.m. This data has been analyzed for equivalent noise levels and statistical noise levels with both honking and without honking. During the measurement campaign, it was observed that the pressure horn was used only once at 57th min, which increased the normal level of noise to over 90 dB. The variation of the observed noise levels with time (recorded every 1 min) has been presented in Figure 18.8a and b, respectively, for the data without honking and with honking. It has been observed that the peak noise level without honking was up to about 72 dB, which increased to over 90 dB during honking. For assessment purposes, the figure also shows the lines of L_{eq} calculated with Eq. (18.22).

Besides L_{eq}, data are often analyzed for statistical noise levels to interpret the data distribution recorded over a period of time. The data recorded every second or a minute for a period of 1 h or daytime hours and nighttime hours or 24 h are statistically summarized. The summary brings out the more meaningful measures in terms of percentiles regarded as L_{10}, L_{50}, and L_{90}, which are widely used statistical measures for noise assessments. L_{10} refers to the noise level that is exceeded 10% of the time measurements were taken, L_{50} refers to the noise level that is exceeded 50% of time, and similarly L_{90} refers to the noise level exceeded 90% of time (Jones, 2008). Thus, L_{10} measures the peak noise level, L_{50} measures the median noise level, and L_{90} measures the residual noise level. L_{90} is, therefore, also used as a background noise level especially when the effect of sudden change in sound energy is to be studied, for example, pressure horns. Figure 18.8a shows these statistical noise levels along with the L_{eq} during the normal traffic when no pressure horn was observed. Similarly, Figure 18.8b describes these levels when pressure horn was observed.

It has been observed that during pressure horns, noise levels jumped by about 20 dB(A) as observed from the respective peaks. The L_{eq} value was also observed to be increased drastically by about 10 dB(A) during

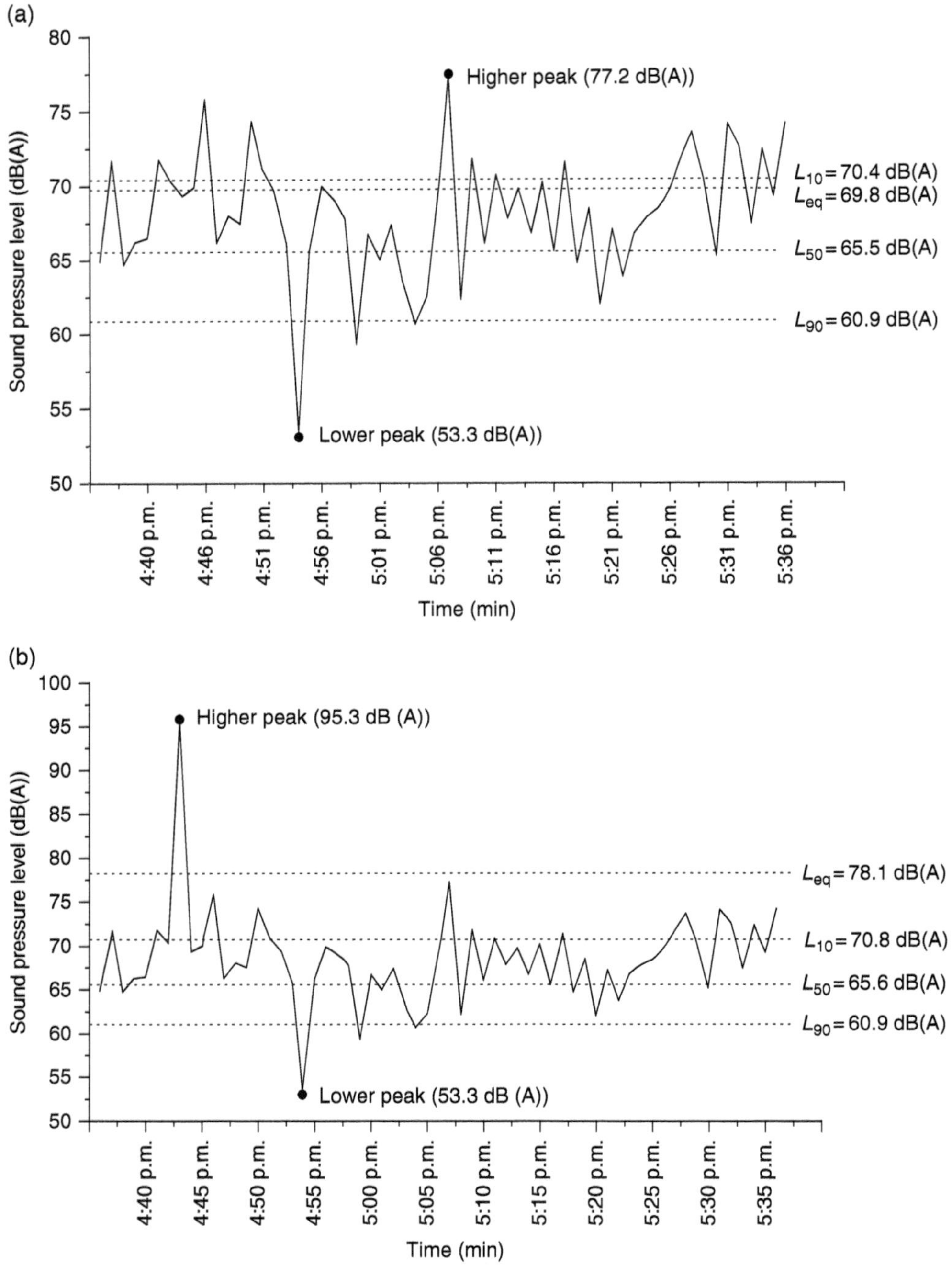

Figure 18.8 Variation of noise level recorded every minute for 1 h near a roadway when (a) no pressure honking and (b) with pressure honking.

pressure horn as compared with the traffic without pressure horns. Figure 18.9a and b showing the distribution of data reveals that in both, the range of L_{90}–L_{10} remains almost same. This indicates that during normal traffic when no pressure honking is used, the range noise levels may remain same. However, Figure 18.9b shows the heavy tail distribution, which indicates the effect of pressure horns. The L_{eq} may even exceed the L_{10}, particularly, when the honking events are minimal. In this data and analysis, one honking event has been considered to distinguish the effect on the normal variation of noise level. However, in the case of many honking events within a small period of time, L_{10} and L_{eq} may remain more or less same.

The analysis indicates the significance of the equivalent noise level as well as the statistical noise levels. Both

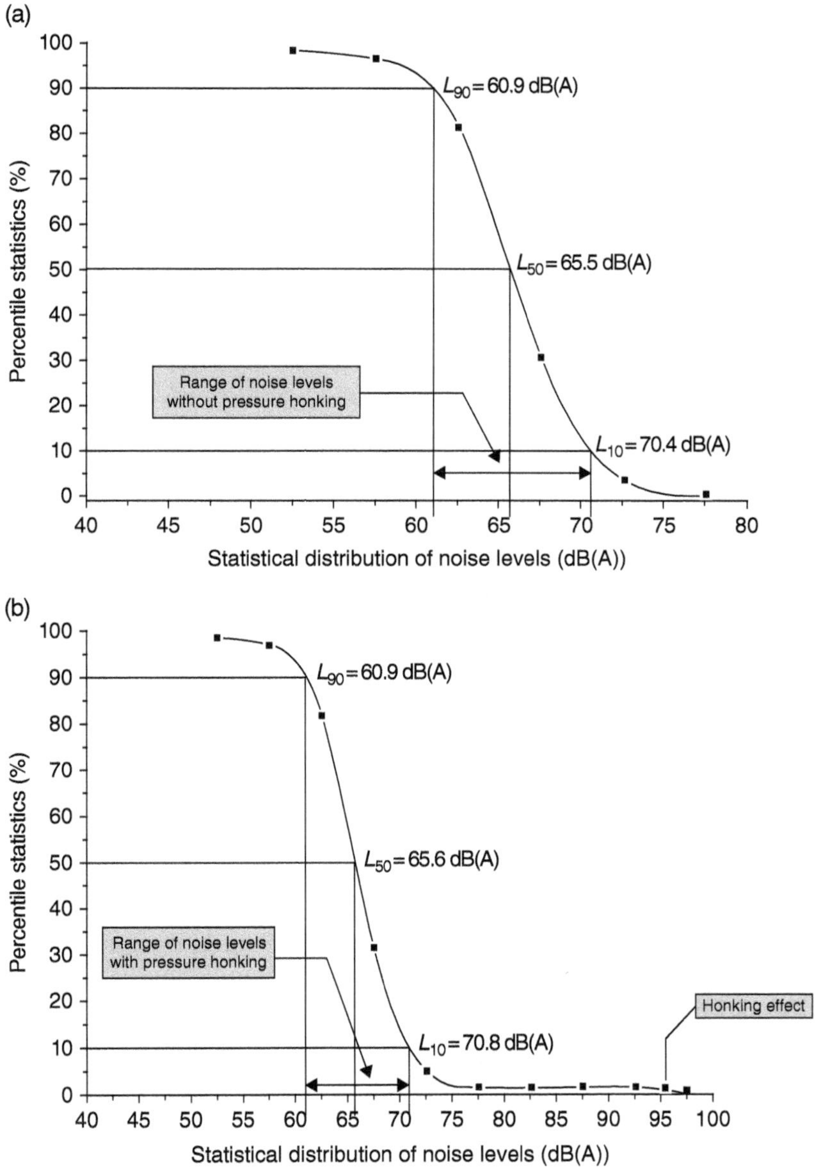

Figure 18.9 Statistical distribution of noise levels for (a) no pressure honking and (b) with pressure honking.

together can be studied to determine the effect of a specific noise-producing event. The statistical noise level in particular that indicates how frequently a particular sound level is exceeded and, therefore, can be used to quantify the time-varying noise in terms of the levels exceeded for different percentages of the measurement duration (Kiely, 2007). Such analysis reveals maximum (L_{10}) and minimum (L_{90}) noise levels including the median noise level. These levels, respectively, indicate the level exceeded for 10% of the time and 90% of the time. L_{10} thus may also be referred as the maximum noise level and L_{90} as the background noise level. These levels complement the equivalent noise level as they provide information on the noise variation range (Figure 18.9).

18.8.2 Community Noise

Community noise is referred to the noise originated from a variety of sources in the community. It is recognized as a serious public health source due to noise (WHO, 1999). Community noise is represented as day–night noise level. It is similar to 24-h equivalent noise level except that during nighttime, a penalty noise is considered to determine the nighttime noise level. This is required as during nighttime, a certain noise level may be more annoying than it is at the same level during daytime. For assessing community noise, the length of exposure to noise is also important. Because besides magnitude of the noise level, the length of exposure also plays a role in causing hearing loss. Due to several sources, community noise levels are not steady and often exhibit a large variation with time, which causes more annoyance.

The problem of noise is more annoying in the cities where there is continual growth in vehicle population, large construction activities, regular use of DG sets, unplanned townships, bus and train stations, unplanned roadside commercial developments, various social, political, and religious functions, festival celebrations, and domestic activities. Our communities, besides noise-producing sources, also have several public utilities where high noise level is not tolerated such as hospitals and schools. Urban transport is the most crucial source in the communities for the environmental noise exposure and related health effects to humans.

The social costs of noise produced from roads and rail are estimated to be about 40 billion euros a year (EC, 2011) of which over 90% is due to passenger cars and goods vehicles. According to the European Commission, the cost incurred due to noise in healthcare amounts to about 0.4% of the European Union GDP (den Boer and Schroten, 2007). This cost is expected to increase to about 20 billion euros by 2050 (EC, 2011). Traffic is the most widespread source of environmental noise (www.env-health.org). The adverse health effects of traffic noise include sociophysiological effects such as cardiovascular diseases related to heart and also impacts on mental health (Babisch, 2000). The WHO (1999) made recommendations on guidelines to protect community noise-related public health. In large cities of several developing countries such as Brazil, India, and Argentina, noise from traffic at curbside is about 100 dB(A). The main reason for such a high noise is the excess use of pressure horns.

Several countries have set the limits (noise standards) to protect the environment from the community noise. Some noise standards are recommended in terms of L_{eq} prescribed by the WHO, different countries such as Australia, the United States, Japan, and India, while some are in terms of statistical noise, L_{10}. These standards are prescribed according to the land-use patterns. For example, the WHO standards and FHWA standards (Parbat and Nagarnaik, 2007; Federal Register, 2010) are given in Table 18.2.

The guidelines are, however, not the limits or standards. Each country on the basis of these guidelines can set its own standards, i.e. limits. Some countries, as a result, have formulated the noise standards separately for daytime and nighttime. For example, in India, the Central Pollution Control Board (CPCB) has set the ambient noise standards in the year 2000 (CPCB, 2000) for daytime as well as for nighttime separately. Table 18.3 shows the noise standards for India (CPCB, 2000).

The WHO guidelines are values for the onset of health effects from noise exposure (WHO, 1999). People in the urban mixed environment usually have sporadic as well as widespread complaints, particularly, when noise levels exceed the standards by a significant margin. The prime cause of the annoyance among people in urban communities is the road traffic. Thus, the rapid growth in road traffic is affecting the environmental qualities in

Table 18.2 Noise guidelines by the WHO and FHWA to protect environment from the nuisance of noise.

	WHO		FHWA	
Sr. no.	**Land use**	**Prescribed L_{eq} dB(A)**	**Description**	**Prescribed L_{eq} dB(A)**
1	Outdoor residential areas	55 (16-h time)	Quiet and serene areas	57
2	Schools, classrooms	55	Residential areas	67
3	Playgrounds	55 (during play)	Hotels, motels, and other developed lands	72
4	Industrial, commercial, shopping, and traffic areas	70 (24-h time)	Sports facility, amphitheaters, hospitals, libraries, medical facilities, parks, places of worship, playgrounds, public meeting rooms, recreation areas, and schools (www.healthimpactproject.org; Federal Register, 2010)	67
5	Hospital, wards	30–35 (indoor 8- to 16-h time)	—	—

Table 18.3 Noise standards for India.

Land use	Ambient noise standards, L_{eq} dB(A)	
	Daytime noise level (6 a.m.–9 p.m.)	Nighttime noise level (9 p.m.–6 a.m.)
Industrial areas	75	70
Commercial areas	65	55
Residential areas	55	45
Silence zones (areas up to 100 m around hospital, schools, educational institutes, etc.)	50	40

cities and towns (Singal, 2005). Traffic fleet is comprised of many types of vehicles. Among mainly heavy diesel-driven vehicles contribute more to higher level of noise (Bodsworth and Lawrence, 1978). Besides, honking, as discussed earlier, remains a critical issue particularly in developing countries. Pressure horns alone produce an annoying level of noise and therefore a cause of urban stress.

The calculation in the case of daytime and nighttime noise equivalent noise levels is done little differently. Here, instead it is calculated separately for the daytime and nighttime for the hourly recorded noise level for the period of 16 h from 6 a.m. to 10 p.m. and for the period of 9 h from 10 p.m. to 6 a.m., respectively. The community noise from all sources is the day–night noise level (L_{dn}). As discussed, it is similar to 24-h equivalent noise level with an exception of using additional noise in dB(A) to the nighttime noise level. Equation (18.24) illustrates the L_{dn}:

$$L_{dn} = 10\log_{10}\left[\frac{1}{24}\left(\sum_{i=1}^{16}\cdot 10^{L_i/10} + \sum_{j=17}^{24}\cdot 10^{(L_j+10)/10}\right)\right] \tag{18.24}$$

where L_i is the equivalent noise level for the ith hour during the day and similarly L_j for the jth hour during the night.

Equation (18.24) can also be simplified, as given in Eq. (18.25):

$$L_{dn} = 10\log_{10}\left\{\frac{1}{24}\left[16\left(10^{L_{day}/10}\right) + 8\left(10^{(L_{night}+10)/10}\right)\right]\right\} \tag{18.25}$$

where L_{day} is the equivalent noise level for daytime hours (6 a.m.–10 p.m.) (Rau and Wooten, 1980) and similarly L_{night} is the equivalent noise level for the nighttime hours (10 p.m.–6 a.m.) (Rau and Wooten, 1980).

18.9 Environmental Noise Management

Noise management accomplishes noise control, which is often wrongly interpreted as the reduction of noise alone. But the prime goal of the noise management is to minimize the exposure of people to noise (Singal, 2005) and identify the sources within the community, which contribute to higher noise regularly so that various possible mitigation options can be evaluated. It also means creating acceptable noise environment in the areas and the hotspots of the cities where people are exposed to higher levels of noise and for longer duration (Singal, 2005). For developmental projects, a comprehensive environmental impact assessment is carried out of which noise is an important and integral component. The impact assessment study is about evaluating the current status of noise levels and forecasting impacts of the activities of the proposed projects. Social surveys are carried out to determine annoyance due to noise within the population that is likely to be affected due to the project. Based on current and forecast noise levels, cause-and-effect relationship can be established. This helps in analyzing the range of options to reduce impacts of noise, which forms an important part of environmental noise management. Noise impacts can be reduced (refer to Figure 18.2) at the source of noise, in the path during its transmission, and at the receiver by various means.

Traffic is the main cause of noise in the urban centers of our cities. Due to unprecedented growth in traffic almost in all over the world, noise pollution besides air pollution in urban areas is emerging as a big challenge to transport and environmental planners. Combined technical as well as policy measures are required to create an acceptable noise environment. For this, several options that can control noise at the source and that can be adopted practically such as alternate routes to reduce traffic load on urban roadways, restriction of heavy vehicles during peak hours, regulation on honking and traffic speed, relocation of activities that attract more traffic, etc. need consideration. Environmental noise control has been discussed in detailed in the next section.

18.9.1 Engineered Attenuation

Controlling noise at the source is always not possible. However, in case of traffic, the major source of noise in urban centers, regular inspection and maintenance (I/M) of the vehicles can keep the noise under control. Most often even the materials used for road pavements contribute to noise when a vehicle pass over, called as rolling noise. For example, concrete roads generate more noise and bitumen pavements produce less noise. The design of road infrastructure and planning of road networks need the inclusion of noise impact assessment. Similarly, during operation, the speed of traffic, its flow pattern, and traffic composition all can have effects on noise levels. Heavy vehicles, in particular diesel driven, contribute more noise than smaller petrol-driven vehicles. Devices like silencers provided at the engine exhausts can reduce noise. However, with the continuous increase of vehicular population in cities and particularly in urban centers, noise pollution has emerged annoying and causing health risks. Because of heterogeneity in the mixed urban environment, it may not be possible to control noise to the acceptable level.

Control measures adopted in the path are to prevent transmission of sound waves. Physical noise barriers are used between the sound-producing source and the receptors to lessen the sound energy from reaching to the receptor. The material used for making barriers determine the amount of attenuation, i.e. how much noise is reflected from the barrier after the sound wave strikes, how much is transmitted through the barrier, and how much is absorbed into the barrier. In mixed urban environment, there may be multiple reflections as sound originated from the moving traffic may strike to the surfaces of several elements – roads, surrounding buildings – and get reflected several times within the roadside microenvironment. Barriers can be made of absorptive materials such as either porous or nonporous materials or may be of both in combination. Barriers can be placed either near to the source or near to the receivers, and its height can be adjusted in order to prevent the direct sound from reaching the receivers. This is also possible when the direct path of sound is blocked by placing the barrier. The idea is to reduce reflections, full transmission, and full absorption within the material and to provide a barrier in such a way that sound is reduced to a comfortable level at the receivers.

We had discussed earlier that sound is attenuated by geometrical spreading. Besides, during transmission sound is also attenuated by natural environmental factors.

18.9.2 Natural Attenuation

Though attenuation caused by natural factors cannot be controlled, proper consideration of it in noise forecasting helps design noise barriers. Recall the discussion on sound wave propagation. Two different geometrical spreading will have different levels of attenuation. For example, with reference to Eq. (18.16) of sound intensity, if the distance of receiver from the source is doubled, it reduces sound pressure level by 6 dB in the case of spherical spreading, i.e. for a point source, and the reduction of 3 dB in the case of cylindrical spreading, i.e. for a line source.

Further, atmospheric properties can attenuate noise because the sound energy will be lost or reduced due to dissipation in air medium. It means some amount of sound will be absorbed by the atmosphere. It can further change due to temperature and ambient humidity. It is believed that when temperature increases meaning humidity decreases, sound absorption may decrease. This also depends upon the frequency at which the sound has originated. For higher frequencies and lower humidity, attenuation is more.

Other than atmospheric properties, meteorological conditions can also attenuate noise. For example, during winter fog when density of air is high, noise attenuation is high. Wind speed due to its constant variation may affect the noise levels and attenuate it to some extent. On the downwind side particularly the impact of noise may be more as compared with the upwind side. Attenuation caused by meteorological condition may be in the range of ±6 dB for frequencies up to 500 Hz and ±10 dB for frequencies more than 500 Hz (Herbert et al., 1989; Kiely, 2007).

Ground surface or topography in the case of a large area can have a significant effect on noise. For example, ground with grass and vegetation can absorb the sound, reducing the effective noise reaching to the receiver. But surfaces made of materials such as concrete will mostly reflect sound waves due to less absorption of it. Again, the frequencies of sound are important in the noise attenuation due to ground surfaces.

Even trees are used, referred as greenbelt as a noise barrier (Santra et al., 1998; Singal, 2005). Trees of dense foliage may also reduce noise to a great extent by absorbing the sound and depending upon the tree crown and height reduction of noise can significantly vary.

Consideration of the attenuation due to various factors into noise forecasting and accurate noise assessment can be done, and effective noise barriers can be designed.

Note

1 It is the external atmosphere to which public has access.

References

Allen, R.W., Davies, H., Cohen, M.A. et al. (2009). The spatial relationship between traffic-generated air pollution and noise in 2 US cities. *Environmental Research* 109: 334–342.

Anderson, B.E., Blotter, J.D., Gee, K.L., and Sommerfeldt, S.D. (2015). Acoustical measurements. In: *Mechanical Engineers Handbook*, 1–44. New York: Wiley.

de Arcas, G. (2007). Practical considerations in the verification of personal sound exposure meters. *Metrologia* 44 (3): 177–181.

Babisch, W. (2000). Traffic noise and cardiovascular disease: epidemiological review and synthesis. *Noise & Health* 8: 9–32.

Barron, R.F. (2003). *Industrial Noise Control and Acoustics*, Chapter 3. New York: Marcel Dekker, Inc.

Basner, M., Babisch, W., Davis, A. et al. (2014). Auditory and non-auditory effects of noise on health. *Lancet* 383 (9925): 1325–1332.

Beelen, R., Hoek, G., Houthuijs, D. et al. (2009). The joint association of air pollution and noise from road traffic with cardiovascular morality in a cohort study. *Occupational and Environmental Medicine* 66: 243–250.

Berglund, B. and Lindvall, T. (1995). *Community Noise. Archives of the Center for Sensory Research.* Stockholm: Stockholm University and Karolinska Institute.

Bluhm, G.L., Berglind, N., Nordling, E., and Rosenlund, M. (2007). Road traffic noise and hypertension. *Journal of Occupational and Environmental Medicine* 64: 122–126.

Bluhm, G., Nording, E., and Berglind, N. (2004). Road traffic noise and annoyance – an increasing environmental health problem. *Noise Health* 6 (24): 43–49.

Bodsworth, B. and Lawrence, A. (1978). The contribution of heavy vehicles to urban traffic noise. *Applied Acoustics* 11 (1): 57–68.

den Boer L.C., Schroten A. (2007). *Traffic Noise Reduction in Europe Health Effects, Social Costs and Technical Policy Options to Reduce Road and Rail Traffic Noise*, CE Delft Report. Commissioned by the European Federation for Transport and Environment (T&E), Brussels.

Cao, G. (2011). Acoustical measurement and fan fault diagnosis system based on LabVIEW. In: *Practical Applications and Solutions Using LabVIEW™ Software.* Rijeka, Croatia: InTech.

Chaudhary, R.B. (2009). Exposure of educational institutions to traffic-induced noise at Chittagong city Bangladesh. *International Journal of Vehicle Noise and Vibration* 5 (4): 287–299.

CPCB (2000). Noise Standards. Central Pollution Control Board. www.cpcb.nic.in/air-quality-standards/ (accessed 26 January 2018).

Davies, H.W. (2009). Correlations between co-exposures to noise and air pollution from traffic sources. *Occupational and Environmental Medicine* 66: 347–350.

EC (2011). *Roadmap to a Single European Transport Area Towards a Competitive and Resource Efficient Transport System* (White paper). Brussels: European Commission.

Federal Register (2010). Procedures for abatement of highway traffic noise and construction noise. Federal Highway Administration, 75FR39820, Document No. 2010-15848, pp. 39820–39839.

Finegold, L.S., Muzet, A.G., and Bernard, F.B. (2007). Chapter 24: Sleep disturbance due to transportation noise exposure. In: *Handbook of Noise and Vibration Control* (ed. M.J. Crocker). Hoboken, NJ: Wiley.

Gan, Q.W., Davies, W.H., Koehoorn, M., and Brauer, M. (2012). Association of long-term exposure to community noise and traffic related air pollution with coronary heart disease mortality. *American Journal of Epidemiology* 175 (9): 898–906.

Gieras, J.F., Wong, C., and Lai, J.C. (2005). Chapter 8: Acoustics and vibration instrumentation. In: *Noise of Polyphase Electric Motors. Electrical and Computer Engineering.* Boca Raton, FL: CRC Press.

Giovanni, B. (1999). Level meters. In: *Wiley Encyclopedia of Electrical and Electronics Engineering.* New York: Wiley.

Griefahn, B. (2000). Noise effects not only the ears. But can damage to health be objectively evaluated. *MMW Fortschritte der Medizin* 142 (14): 26–29 [in German].

Hassan, O.A.B. (2009). An overview of building acoustics and vibration. In: *Building Acoustics and Vibration – Theory and Practice*, 1–136. Singapore: World Scientific.

Henderson, D., Subramaniam, M., and Boettcher, F.A. (1993). Individual susceptibility to noise-induced hearing loss: an old topic revisited. *Ear and Hearing* 14 (3): 152–168.

Herbert A.G. et al. (1989). Sound and vibration analysis and control. *Kempe's Engineering Yearbook.* London: Morgan-Grampian Book Publishing Co.

Ising, H., Lange-Asschenfeldt, H., Moriske, H.J. et al. (2004). Low frequency noise and stress – bronchitis and cortisol in children exposed chronically to traffic noise and exhaust fumes. *Noise & Health* 6: 21–28.

ISO 9613-1 (1993). *Acoustics – Attenuation of Sound During Propagation Outdoors – Part 1: Calculation of the Absorption of Sound by the Atmosphere.* 1 (1 June 1993). Geneva: ISO.

ISO 9613-2 (1996). *Acoustics – Attenuation of Sound During Propagation outdoors – Part 2: General Method of Calculation.* 1 (15 December 1996). Geneva: ISO.

Jarup, L., Dudley, M.L., Babisch, W. et al. (2005). Hypertension and Exposure to Noise near Airports (HYENA): study design and noise exposure assessment. *Environmental Health Perspectives* 113: 1473–1478.

Jones, D. (2008). Chapter 4: Acoustical treatment for indoor areas. In: *Handbook of Sound Engineers*, 4e, 67–94. Amsterdam: Elsevier.

Joshi, S.K., Devkota, S., Chamling, S., and Shrestha, S. (2003). Environmental noise induces hearing loss in Nepal. *Medical Journal, Kathmandu University* 1 (3): 177–183.

Kathy, R. and Werff, V. (2007). Test-retest reliability of wideband reflectance measures in infants under screening and diagnostic test conditions. *Ear and Hearing* 28 (5): 669–681.

Kiely, G. (2007). *Environmental Engineering*. New Delhi: Tata McGraw Hill Education Private Ltd.

Kim, Y.-H. (2010). Radiation, scattering and diffraction. In: *Sound Propagation an Impedance Based Approach*. Chichester, UK: Wiley.

de Kluizenaar, Y., Ganesvoort, R.T., Miedema, H.M.E., and De Jong, P.E. (2007). Hypertension and road traffic noise exposure. *Journal of Occupational and Environmental Medicine* 49: 484–492.

Kuehn, J. (2010). *Instrumentation Reference Book* (ed. W. Boyes), 4, Chapter 4, 593–614. Amsterdam: Elsevier.

Lamancusa, J.S. (2000). The physics of Sound. In *Engineering Noise Control*, Chapter 5. http://www.mne.psu.edu/lamancusa/me458/5_physics.pdf (accessed 23 January 2018).

Lamancusa, J.S. (2009). Outdoor sound propagation. In *Engineering Noise Control*, Chapter 10. http://www.mne.psu.edu/lamancusa/me458/10_osp.pdf (accessed 23 January 2018).

Luck, S.L., Hagerty, B.M., Gillespie, B., and Ziemba, R.A. (2004). Acute effects of noise on blood pressure and heart rate. *Archives of Environmental Health* 59: 392–399.

Mailing, G. Jr. (2007). Noise. In: *Springer Handbook of Acoustics*. New York: Springer.

Nguemaleu, R.A.C. and Montheu, L. (2014). Chapter 4 – Computing noise pollution. In: *Roadmap to Greener Computing*. Boca Raton, FL: CRC Press.

Nicholau, V., Miholca, C., and Andrei, M. (2009). Fuzzy rules of sound speed influence on ultrasonic sensing in outdoor environment. *3rd International Workshop on Soft Computing Applications* (7 July 2009).

Ohrstrom, E., Agge, A., and Bjorkman, M. (1998). Sleep disturbances before and after reduction in road traffic noise. In: *Noise Effects 98. Proceedings of the 7th International Congress on Noise as a Public Health Problem, Sydney 1998*, vol. 2 (ed. N. Carter and R.S.F. Job), 451–454.

Parbat, D.K. and Nagarnaik, P.B. (2007). Assessment and ANN modeling of noise levels at major road intersections in an Indian intermediate city. *Journal of Research in Science, Computing and Engineering* 4 (3): 39–49.

Parris, K.M. (2015). Chapter 16: Ecological impacts of road noise and options for mitigations. In: *Ecology of Roads: A Practitioner's Guide to Impacts and Mitigations* (ed. R. van der Ree, C. Gribl and D. Smith), 151–158. New York: Wiley-Blackwell.

Passchier-Vermeer, W. and Passchier, W.F. (2000). Noise exposure and public health. *Environmental Health Perspectives* 108 (first suppl): 123–131.

Puri, J. (2012). Sources, measurement and mitigation of sound levels in transformers. *Transformers Analysis Design and Measurement*, Chapter 11. CRC Press: Boca Raton, FL.

Rau, J.G. and Wooten, D.C. (1980). *Environmental Impact Analysis Handbook*. New York: McGraw Hills.

Rennie J. (2014). Why is sound intensity proportional to the square of sound pressure not to sound pressure alone? http://physics.stackexchange.com/users/1325/john-rennie (accessed 23 January 2018).

Santra, S.C., Chakraborty, D., and Roy, B. (1998). Urban traffic noise abatement with vegetational barriers. *Journal of Acoustic Society of India* 26 (3,4): 1–10.

Schwartz, J., Litonjua, A., Suh, H. et al. (2005). Traffic related pollution and heart rate variability in a panel of elderly subjects. *Thorax* 60: 455–461.

Schwela, D., Kephalopoulos, S., and Prasher, D. (2005). Confounding or aggravating factors in noise-induced health effects – air pollutants and other stressors. *Noise & Health* 7 (28): 41–50.

Selander, J., Nilsson, M.E., Bluhm, G. et al. (2009). Long-term exposure to road traffic noise and myocardial infarction. *Epidemiology* 21: 396–404.

Singal, S.P. (2005). *Noise Pollution and Control Strategy*. New Delhi: Narosa Publishing House.

Sinha K. C. Samuel L. (2007). *Transportation Decision Making – Principles of Project Evaluation and Programming*, Chapter 11. Wiley: New York.

Sorenson, M., Hvidberg, M., Anderson, Z.J. et al. (2011). Road traffic noise and stroke – a prospective cohort study. *European Heart Journal* 32 (6): 737–744.

Taherzadeh S. (ed). (2014). Section 1 – Noise basics. In: *Noise Control*, Chapter 1. Wiley, Chichester, UK.

Thompson D. (2009). Sound radiation from wheels and track, *Railway Noise and Vibration – Mechanisms, Modelling and Means of Control* (1), Chapter 6. Elsevier: Amsterdam.

Weber, S. (2009). Spatio-temporal covariation of urban particle number concentration and ambient noise. *Atmospheric Environment* 43: 5518–5525.

WHO (1999). Guidelines for community noise. http://www.who.int/docstore/peh/noise/guidelines2.html (accessed September 2016).

Wilhelmsson, M. (2000). The impact of traffic noise on the values of single-family houses. *Journal of Environmental Planning and Management* 43 (6): 799–815.

19

Hazardous Waste Management

Clayton J. Clark II[1] *and Stephanie Luster-Teasley*[2]

[1] *Department of Civil & Environmental Engineering, FAMU-FSU College of Engineering, Florida A&M University, Tallahassee, FL, USA*
[2] *Department of Civil, Architectural, & Environmental Engineering, College of Engineering, North Carolina A&T State University, Greensboro, NC, USA*

19.1 Fundamentals

Hazardous waste is the classification of waste that has the potential to poses a substantial denture to human, plant or animal life. The U.S. Environmental Protection Agency (US EPA) first defined the term in 1976 as part of the nation's first hazardous waste law. This classification applies to solids, sludges, liquids, or gases that by nature of their chemical reactivity can cause danger to the environment or human health if released. They include chemicals which are toxic, explosive, or corrosive but do not include waste chemicals which may be radioactive or infectious. Radioactive and infections wastes, while hazardous, have special handling and management requirements.

Numerous examples of hazardous waste exposure to humans, animals, and in the environment. Examples such as the Love Canal, DDT bioaccumulation impacting Antarctic penguins, trichloroethylene plumes contaminating groundwater, lead in paint and gasoline, and mercury contamination of fish from contaminated waterways located near industries. Two recent examples of hazardous waste spills were the result of accidental releases of contaminants. In April 2010, an explosion on the British Petroleum (BP) oil rig called Deepwater Horizon killed 11 workers and released oil at a rate of 1000–60 000 barrels of oil per day. From April through September, attempts to seal the well failed leading to contamination of over 68 000 square miles of ocean and an oil slick spanning the coasts of Mississippi, Alabama, Louisiana, and Florida. In total, it is estimated that the explosion resulted in 4.9 million barrels of crude oil released. Figure 19.1 depicts a map showing the footprint of the oil spill based on satellite images taken between 25 April and 16 July 2010.

In 2014, Duke Energy, the largest electric power holding company in the U.S., reported that it spilled between 50 000 and 82 000 tons of coal ash into the Dan River between the city of Eden, North Carolina, and Danville, Virginia, polluting the waterway and drinking water source for the regional area (Figure 19.2). A security guard who noticed unusually low water in the ash pond at the coal plant led to the discovery of the spill. Most of the water had escaped and contaminated the river before anyone at Duke Energy noticed (Catawba Riverkeeper Foundation, 2015; CBS Associated Press, 2015; US EPA, 2015a). Coal ash is the remaining material after coal is burned at a power plant. It contains heavy metals and other toxic compounds such as arsenic, boron, chromium, selenium, mercury, and lead, it is a serious threat to aquatic ecosystems and local drinking water if ingested. Heavy metals are known to impact brain function, blood, disruption to cell functions, and can lead to the development of cancer. This spill was caused by a break in a 48-in stormwater pipe located underneath Duke Energy's unlined 27-acre, 155-million-gallon ash pond, ultimately draining an estimated 24–27 million gallons of contaminated water into the Dan River. The occurrence signifies the third largest coal ash spill in U.S. history.

19.1.1 Definitions and Classifications

To start understanding hazardous waste, it is important to clearly define the terminology used when discussing this type of waste and how it is classified. Table 19.1 lists fundamental hazardous waste terminology.

Hazardous waste is regulated by the federal government and can be found in Title 40 of the Code of Federal Regulation (CFR), Parts 260–271. The EPA will evaluate a chemical to measure the corrosivitity, ignitability, reactivity, and toxicity of the waste. These terms are defined in Table 19.2.

Once classified, the chemical is itemized into a specific hazardous waste listing. They are grouped into four listing types (Table 19.3).

Handbook of Environmental Engineering, First Edition. Edited by Myer Kutz.

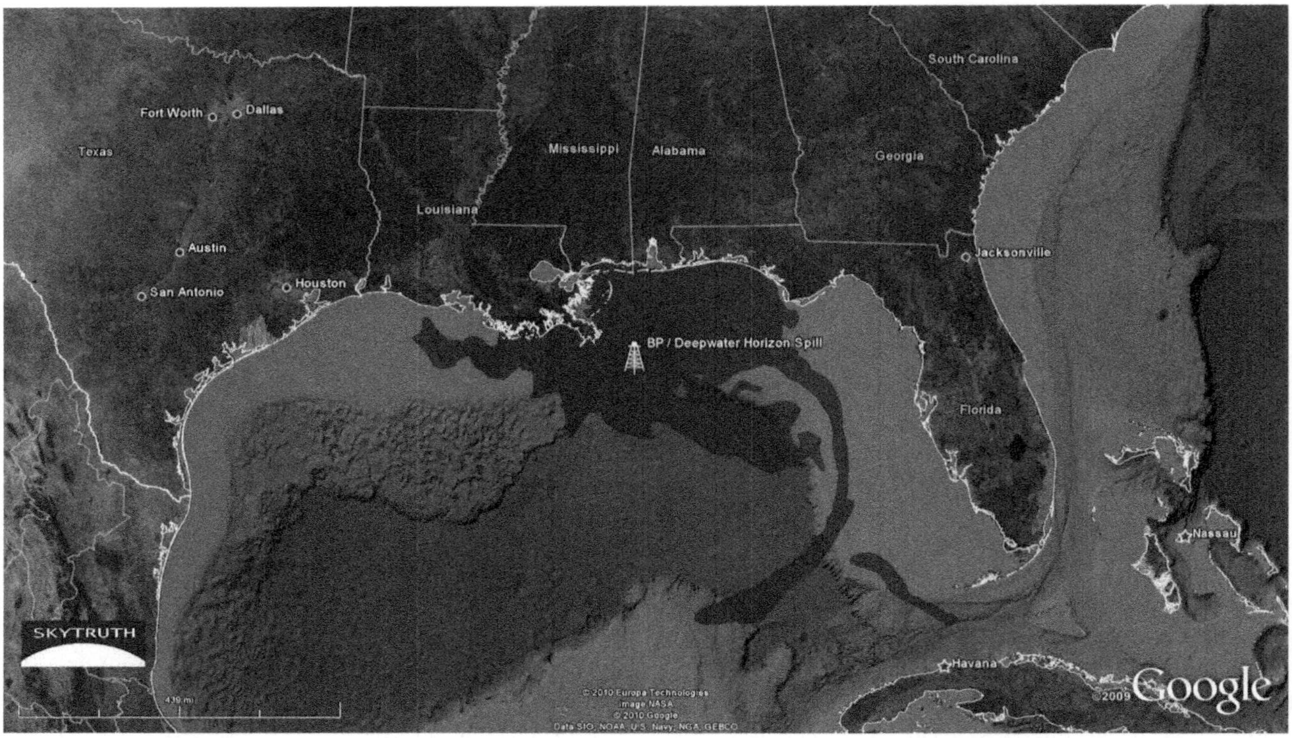

Figure 19.1 Map of BP oil spill impacting 68 000 square miles. *Source:* From Google Images (2010).

Figure 19.2 Image of the coal ash breach from Duke Power. *Source:* Reproduced with permission of Small Unmanned Aerial Systems Lab at the Wake Forest Center for Energy, Environment, and Sustainability.

Table 19.1 Fundamental hazardous waste terminology.

Term	Definition
Hazard	A chemical with characteristics or properties that results in a heightened potential threat
Environmental risk	The likelihood or probability of injury, disease or death caused by an environmental factor
Risk perception	A judgement or view of the significance of a risk to lead to injury, disease or death
Hazardous waste code	Codes that are assigned to waste based on the risk evaluation. The classifications include T – toxic, H – acutely hazardous, I – ignitability, C – corrosively, and R – reactivity
Human toxicity hazard classifications	This classification identifies the human health risk from exposure to a hazardous chemical or compound
Exposure	The amount of time or duration a hazard is present and impacts a group
Toxic release inventory (TRI)	A list or inventory that provides information to the public about hazardous waste and toxic chemicals

Table 19.2 Classifications for hazardous waste.

Corrosive/corrosivity	A chemical that demonstrates properties such as a pH ≤2 or >12.5; a liquid that can corrode steel at a rate >6.35 mm yr^{-1} at 55 °C
Ignitability	A chemical in liquid or gas form that can burn or ignite; a solid or gas where under standard pressure and temperature can ignite as a result of friction or chemical reaction; an oxidizer; a liquid that has a flash point <60 °C and contains <24% alcohol
Reactivity	A chemical that is potentially explosive when exposed to water; it is typically unstable and demonstrates a high propensity to react violently without exploding or igniting
Toxicity	The toxicity of a waste is determined by the Toxicity Characteristic Leaching Procedure (TCLP) (SW-846 Test Method 1311)

Table 19.3 Classification categories for hazardous waste.

Listing	Description
F	Waste from chemicals that have been used for it's intended purpose by the industry and post-use is not useable again and hence becomes waste for disposal
K	Waste generated by the following industries: • Wood preservation, • Organic chemicals manufacturing, • Pesticides manufacturing, • Petroleum refining, • Veterinary pharmaceuticals manufacturing, • Inorganic pigment manufacturing, • Inorganic chemicals manufacturing, • Explosives manufacturing, • Iron and steel production, • Primary aluminum production, • Secondary lead processing, • Ink formulation, and • Coking (processing of coal to produce coke).
P	Acutely toxic commercial chemicals that are unused and being discarded
U	Unused commercial chemicals that are being discarded that are acutely toxic and have one or more of the four hazardous waste characteristics of corrosivity, ignitability, reactivity, or toxicity

19.2 Legal Framework

The **Resource Conservation and Recovery Act (RCRA)** is our nation's primary law governing the disposal of solid and hazardous waste. Congress passed RCRA on 21 October 1976 to address the increasing problems the nation faced from our growing volume of municipal and industrial waste. RCRA, which amended the Solid Waste Disposal Act of 1965, set national goals for:

- Protecting human health and the environment from the potential hazards of waste disposal.
- Conserving energy and natural resources.
- Reducing the amount of waste generated.
- Ensuring that wastes are managed in an environmentally-sound manner.

RCRA gives EPA the authority to control hazardous waste from the "cradle-to-grave." This includes the generation, transportation, treatment, storage, and disposal of hazardous waste. RCRA also set forth a framework for the management of non-hazardous solid wastes. The 1986 amendments to RCRA enabled EPA to address environmental problems that could result from underground tanks storing petroleum and other hazardous substances (Davis and Cornwell, 2008).

To achieve these goals, RCRA established three distinct, yet interrelated, programs:

- The solid waste program, under RCRA Subtitle D, encourages states to develop comprehensive plans to manage nonhazardous industrial solid waste and municipal solid waste, sets criteria for municipal solid waste landfills and other solid waste disposal facilities, and prohibits the open dumping of solid waste.
- The hazardous waste program, under RCRA Subtitle C, establishes a system for controlling hazardous waste from the time it is generated until its ultimate disposal – in effect, from "cradle to grave."
- The underground storage tank (UST) program, under RCRA Subtitle I, regulates underground storage tanks containing hazardous substances and petroleum products.

RCRA encompasses several different program areas to assure compliance with statutes and regulations in the management of hazardous wastes and underground storage tanks. They include:

19.2.1 Hazardous Waste

EPA and its regulatory partners inspect facilities that generate, transport, treat, store or dispose of hazardous waste to verify compliance with applicable regulations. EPA and states' main focus areas of compliance are:

1) Identification of hazardous waste
2) Hazardous waste generators
3) Hazardous waste transporters
4) Owners and operators of hazardous waste treatment, storage and disposal

19.2.2 Used Oil

Petroleum or synthetic oil becomes used oil if contaminated with physical (e.g. metal particles) or chemical (e.g. additives) impurities during use. Used oil is regulated as hazardous waste when it is contaminated following use. EPA and its regulatory partners conduct inspections of recycled used oil facilities to assure compliance with applicable regulations. Compliance monitoring involves reviewing a facility's compliance with:

- storage requirements
- transportation requirements
- burning requirements
- processing requirements
- re-refining of the used oil requirements

For parties that merely generate used oil, the regulations establish storage standards.

For a party considered a used oil processor, re-refiner, burner, or marketer (one who generates and sells off-specification used oil directly to a used oil burner), additional tracking and paperwork requirements must be satisfied.

Relevant rules	Monitoring responsibilities
Standards for the management of used oil, 40 CFR Part 279	State authorized programs or EPA direct implementation

Used oil that contains more than 1000 parts per million (ppm) of total halogens is presumed to have been mixed with a regulated halogenated hazardous waste (i.e. spent halogenated solvents), and is therefore subject to applicable hazardous waste regulations. A person may rebut this presumption by demonstrating, through analysis or other documentation that the used oil has not been mixed with halogenated hazardous waste. One way of doing this is to show that the used oil does not contain significant concentrations of halogenated hazardous constituents (50 FR 49176; 29 November 1985). If the presumption is successfully rebutted, meaning the oil is not considered to have been mixed with a regulated hazardous waste, then it will be subject to the used oil management standards instead of the hazardous waste regulations (US EPA, 2001).

19.2.3 Universal Wastes

Universal wastes cover a variety of wastes including batteries, pesticides, lamp bulbs, and mercury-containing equipment such as light switches. Compliance monitoring involves reviewing a facility's compliance with:

- storage requirements
- transportation requirements
- marking requirements
- processing requirements

Relevant rules	Monitoring responsibilities
Standards for universal waste management, 40 CFR Part 273	State authorized programs or EPA direct implementation

19.2.4 Mixed Wastes

Mixed waste contains both radioactive and hazardous waste components. EPA's regulations provide flexibility to generators and facilities that manage low-level mixed waste (LLMW) and technologically-enhanced, naturally-occurring, and/or accelerator-produced radioactive material (NARM) containing hazardous waste. LLMW is exempt from some RCRA storage and treatment regulations, and LLMW or eligible NARM from RCRA hazardous waste transportation and disposal regulations.

These wastes are exempt from RCRA Subtitle C requirements, including permitting, provided they meet specific conditions. The exempt wastes must then be managed as radioactive waste in accordance with Nuclear Regulatory Commission (NRC) regulations or Nuclear Regulatory Commission Agreement State Regulations.

Compliance monitoring involves reviewing a facility's compliance with the regulations that apply to generators, transporters, and owners and operators of hazardous waste treatment, storage and disposal facilities to the extent the mixed waste is not exempt from those regulations.

Relevant rules	Monitoring responsibilities
Standards for the management of specific hazardous wastes and specific types of hazardous waste management facilities, 40 CFR Part 266	EPA and Nuclear Regulatory Commission

19.2.5 Land Disposal

Disposal is the placement of waste into or on the land. Disposal facilities are usually designed to permanently contain the waste and prevent the release of harmful pollutants to the environment. The most common hazardous waste disposal practice is placement in a land disposal unit such as a landfill, surface impoundment, waste pile, or land treatment unit. Compliance monitoring involves reviewing a facility's compliance with:

- waste treatment standards
- disposal prohibitions
- dilution prohibitions
- storage prohibitions

Relevant rules	Monitoring responsibilities
Land disposal restrictions, 40 CFR Part 268	State authorized programs or EPA direct implementation

19.2.6 Hazardous Waste Injection

Injection wells have a range of activities including long term storage, enhancing oil production, mining, and waste disposal which can pollute ground water. Class I wells inject hazardous and non-hazardous wastes into deep, isolated rock formations that are thousands of feet below the lowermost underground sources of drinking water (USDW). Hazardous waste disposal wells are stringently regulated by the Safe Drinking Water Act and RCRA. Most Class I hazardous wells are located at industrial facilities. Only a few Class I wells are at commercial operations that can accept hazardous waste generated offsite. Compliance monitoring under RCRA involves reviewing a facility's compliance with the regulations that apply to treatment, storage and disposal facilities.

Relevant rules	Monitoring responsibilities
Underground injection, 40 CFR Part 265, Subpart R; Miscellaneous units, 40 CFR Part 264, Subpart X	State authorized programs or EPA direct implementation

Class IV wells are shallow wells used to dispose hazardous or radioactive wastes into or above a geologic formation that contains an underground source of drinking water (USDW). In 1984, EPA banned the use of Class IV injection wells. These wells may only operate as part of an EPA- or state-authorized ground water clean-up action. Less than 32 waste clean-up sites with Class IV wells exist in the United States. The only allowable Class IV wells are used to clean up ground water contaminated by hazardous chemicals with methods such as "pump-and-treat" subsurface remediation.

19.2.7 Hazardous Wastes Imports/Exports

The Hazardous Waste Imports/Exports program requires prior notification of shipment of wastes, collects export manifests documenting individual shipments of waste, and receives export annual reports from the regulated community. The program also assists in developing and handling international and border aspects of civil RCRA enforcement matters.

Compliance monitoring involves reviewing a facility's compliance with the requirements applicable to transporters, such as:

- an EPA identification number
- transfer facility requirements
- manifesting and record keeping
- actions to be taken in the event of hazardous waste discharges or spills

Related resources:

- *The Border Center web page on hazardous waste Exit*
- *Criteria and Process for Denial of Requests to Import Waste to a U.S. Facility*

Relevant rules	Monitoring responsibilities
Standards applicable to transporters of hazardous waste, 40 CFR Part 263	EPA

19.2.8 Permitting Program

Permits are required for generators and transportation, storage and disposal facilities. Compliance monitoring involves reviewing a facility's compliance with the requirements of its permit and with the regulations applicable to the facility type.

Relevant rules	Monitoring responsibilities
EPA administered permit programs: the hazardous waste permit program, 40 CFR Parts 270	State authorized programs or EPA direct implementation

EPA's hazardous waste permitting program established by the Resource Conservation and Recovery Act (RCRA) helps ensure the safe treatment, storage and disposal of hazardous wastes by establishing specific requirements that must be followed when managing hazardous wastes. Permits for the treatment, storage or disposal of hazardous wastes are issued by authorized states or by EPA regional offices, especially for disposal of large volumes of hazardous waste at treatment, storage and disposal facilities (TSDFs). Considering these risks, Congress felt that TSDF management activities needed to be closely regulated to prevent spills, accidents and mechanical failures (US EPA, 2016). As a result, TSDFs are required to obtain permission, in the form of an RCRA permit, which establishes the administrative and technical conditions under which waste at the facility must be managed.

A RCRA permit is a legally binding document that establishes the waste management activities a facility can conduct and the conditions under which it can conduct them. The permit includes applicable EPA regulations from *40 CFR parts 260 through 270*, and also:

- Outlines facility design and operation
- Lays out safety standards
- Describes facility performance activities (e.g. monitoring and reporting)

Permits typically require facilities to develop emergency plans, find insurance and financial backing and train employees to handle hazards. Permits can also include *specific facility requirements*, such as groundwater monitoring. The permitting agency (i.e. either the state or EPA) has the authority to issue or deny permits and is responsible for monitoring the facility. They ensure that the facility is in compliance with the conditions stated in the permit. According to RCRA and its regulations, a TSDF cannot operate without a permit, with a few exceptions.

All facilities that currently treat, store or dispose of hazardous wastes or plan to do so must obtain an RCRA permit.

- TSDFs must receive a permit before they begin construction to demonstrate they can manage hazardous waste safely and responsibly. The permitting agency (i.e. the state or EPA) reviews the permit application and decides whether to grant or deny the facility an RCRA permit. Permits are typically granted for a period up to 10 years.
- Permitted operating TSDFs must submit a new permit application six months before their existing permit expires.
- TSDFs operating under interim status must also apply for a permit. Congress granted "interim status" to facilities that already existed when RCRA was enacted. Interim status allows existing facilities to continue operating while their permit applications are being reviewed.

19.2.8.1 Who Does Not Need a RCRA Permit?

Some facilities are not required to obtain an RCRA permit when handling hazardous waste provided they meet certain conditions specified in the regulations. These exceptions include:

- Businesses that generate hazardous waste and then store for short periods of time before transporting the waste off site.
- Transporters of hazardous waste.
- Entities performing containment activities during an immediate response to an emergency.

In order for a state to assume the regulatory lead as the implementing agency, it must be authorized by EPA to do so. RCRA requires authorization to ensure state programs are at least equivalent to and consistent with the federal rules. EPA establishes minimum federal standards to prevent less stringent state regulatory programs. A state that has received final authorization, known as an authorized state, implements and enforces its hazardous waste regulations. Authorized state regulations act "in lieu of" federal regulations.

The authorization status of states affects which provisions can be addressed in an RCRA hazardous waste permit. States are obligated to address all relevant permit provisions for which they have received final authorization from EPA. In addition, EPA is obligated to address all relevant provisions promulgated under the authority of the 1984 Hazardous and Solid Waste Amendments (which provide for implementation by the EPA prior to a state's authorization) when a state is not authorized for the provisions. This situation may call for a joint permit from the state and EPA.

19.2.9 Underground Storage Tanks

EPA and its regulatory partners conduct inspections of underground storage tanks (USTs) to assure compliance with technical standards and corrective actions when a release has occurred. Compliance monitoring involves reviewing a UST's:

- compliance with general requirements
- spill and overfill protection
- release detection system
- corrosion protection system
- reporting and record keeping
- tank closure system

An underground storage tank system (UST) is a tank and any underground piping connected to the tank that has at least 10% of its combined volume underground. Approximately 563 000 underground storage tanks (USTs) nationwide store petroleum or hazardous substances. The greatest potential threat from a leaking UST is contamination of groundwater, the source of drinking water for nearly half of all Americans. EPA, states, and tribes work in partnership with industry to protect the environment and human health from potential releases.

Relevant rules	Monitoring responsibilities
Technical standards and corrective action requirements for owners and operators of underground Storage Tanks (USTs), 40 CFR Part 280; Approval of state underground storage tank program, 40 CFR Part 281	State authorized programs or EPA direct implementation

19.2.10 Solid Waste

The *solid waste program*, under RCRA Subtitle D, encourages states to develop comprehensive plans to

- manage nonhazardous industrial solid waste and municipal solid waste,
- set criteria for municipal solid waste landfills and other solid waste disposal facilities, and
- prohibit the open dumping of solid waste.

The **Federal Hazardous and Solid Waste Amendments (HSWA)** – are the 1984 amendments to RCRA that focused on waste minimization and phasing out land disposal of hazardous waste as well as corrective action for releases. Some of the other mandates of this law include increased enforcement authority for EPA, more stringent hazardous waste management standards, and a comprehensive underground storage tank program (US EPA, 2015b).

The **Comprehensive Environmental Response, Compensation, and Liability Act (CERCLA)** or also known as **Superfund** – provides a Federal "Superfund" to clean up uncontrolled or abandoned hazardous-waste sites as well as accidents, spills, and other emergency releases of pollutants and contaminants into the environment (Davis and Cornwell, 2008). Through CERCLA, EPA was given power to seek out those parties responsible for any release and assure their cooperation in the cleanup. EPA cleans up orphan sites when potentially responsible parties cannot be identified or located, or when they fail to act. Through various enforcement tools, EPA obtains private party cleanup through orders, consent decrees, and other small party settlements. EPA also recovers costs from financially viable individuals and companies once a response action has been completed. EPA is authorized to implement the Act in all 50 states and U.S. territories. Superfund site identification, monitoring, and response activities in states are coordinated through the state environmental protection or waste management agencies.

The Superfund Amendments and Reauthorization Act (SARA) of 1986 reauthorized CERCLA to continue cleanup activities around the country. Several site-specific amendments, definitions clarifications, and technical requirements were added to the legislation, including additional enforcement authorities. Also, Title III of SARA authorized the Emergency Planning and Community Right-to-Know Act (EPCRA).

19.2.11 Hazardous Waste Management Facility

Hazardous waste management facilities receive hazardous wastes for treatment, storage or disposal. These facilities are often referred to as treatment, storage and disposal facilities (TSDFs), and their activities are described below:

- **Treatment** – Using various processes, such as incineration or oxidation, to alter the character or composition of hazardous wastes. Some treatment processes enable waste to be recovered and reused in manufacturing settings, while other treatment processes dramatically reduce the amount of hazardous waste.
- **Storage** – Temporarily holding hazardous wastes until they are treated or disposed. Hazardous waste is commonly stored prior to treatment or disposal, and must be stored in containers, tanks, containment buildings, drip pads, waste piles, or surface impoundments that comply with the Resource Conservation and Recovery Act (RCRA) regulations. EPA has established a method for determining the compatibility of storing multiple chemicals as mixtures, including the potential hazards

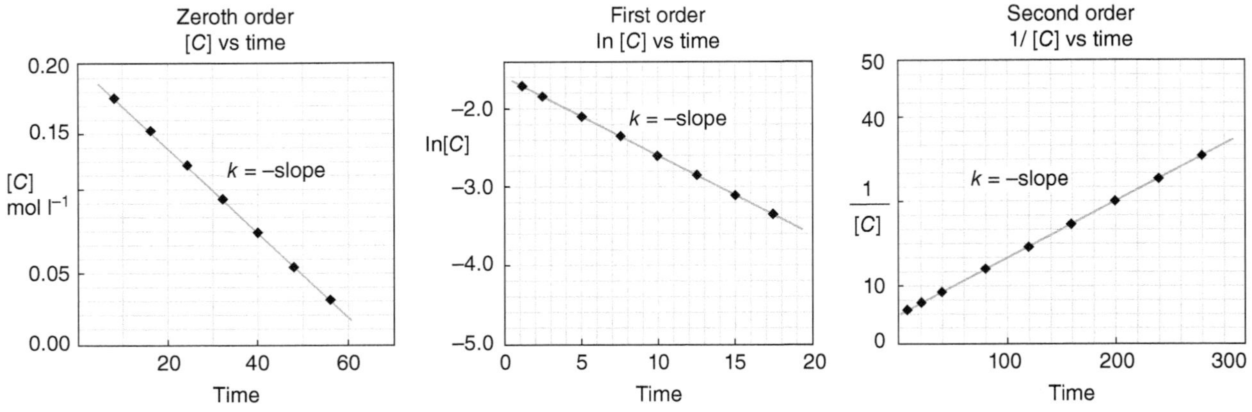

Figure 19.3 EPA's chemical compatibility chart. *Source:* From EPA-600/2-80-076 (US EPA, 1980).

from mixing these wastes (see Figure 19.3). In addition, the regulatory requirements for these types of storage units are found in title 40 of the Code of Federal Regulations (CFR) in:
 - *part 264* for permitted facilities and
 - *part 265* for interim status facilities
- **Disposal** – Permanently containing hazardous wastes. The most common type of disposal facility is a landfill, where hazardous wastes are disposed of in carefully constructed units designed to protect groundwater and surface water resources.

19.2.11.1 What Is a Hazardous Waste Management Unit?

Under RCRA, Congress authorized EPA to promulgate regulations establishing design and operating requirements for hazardous waste management units. The requirements are intended to protect human health and the environment from the risks posed by hazardous waste.

Specific regulations have been developed for various types of hazardous waste management units under Subtitle C of RCRA in *40 CFR parts 264, 265 and 266.*

These units include:

Containers

A hazardous waste container is any portable device in which a hazardous waste is stored, transported, treated, or otherwise handled. The most common hazardous waste container is the 55-gallon drum. Other examples of containers are tanker trucks, railroad cars, buckets, bags, and even test tubes.

Tanks

Tanks are stationary devices constructed of non-earthen materials used to store or treat hazardous waste. Tanks are constructed of a wide variety of materials including steel, plastic, fiberglass, and concrete.

Drip Pads

A drip pad is an engineering structure consisting of a curbed, free-draining base, constructed of non-earthen materials, and designed to convey wood preservative chemical drippage from treated wood, precipitation, and surface water run-on to an associated collection system at wood preserving plants.

Containment Buildings

Containment buildings are completely enclosed, self-supporting structures (i.e. they have four walls, a roof, and a floor) used to store or treat non-containerized hazardous waste.

Incinerators

Incinerators are enclosed devices that use controlled flame combustion for the thermal treatment of hazardous waste. When performed properly, this process destroys toxic organic constituents in hazardous waste and reduces the volume of waste that needs to be disposed.

Boilers and Industrial Furnaces

Boilers are enclosed devices that use controlled flame combustion to recover and export energy in the form of steam, heated fluids, or heated gases. Industrial furnaces are enclosed units that are integral parts of a manufacturing process and use thermal treatment to recover materials or energy from hazardous waste. Examples of an industrial furnace are cement kilns, aggregate kilns, and halogen acid furnaces (produce acid from halogenated hazardous waste).

Landfills

Landfills are excavated or engineered sites where non-liquid hazardous waste is deposited for final disposal and covered. These units are selected and designed to minimize the chance of release of hazardous waste into the environment. Design standards for hazardous waste landfills require:

- Double liner
- Double leachate collection and removal systems
- Leak detection system
- Run on, runoff, and wind dispersal controls
- Construction quality assurance program

Operators must also comply with inspection, monitoring, and release response requirements. Since landfills are permanent disposal sites and are closed with waste in place, they are subject to closure and post-closure care requirements including:

- Installing and maintaining a final cover
- Continuing operation of the leachate collection and removal system until leachate is no longer detected
- Maintaining and monitoring the leak detection system
- Maintaining ground water monitoring
- Preventing storm water run on and runoff
- Installing and protecting surveyed benchmarks

Surface Impoundments
Surface impoundments are natural topographic depressions, man-made excavations, or diked areas formed primarily of earthen materials (although lined with man-made materials) that are used to treat, store, or dispose of liquid hazardous waste. Hazardous waste surface impoundments are required to be constructed with:

- Double liner system
- Leachate collection and removal systems
- Leak detection system

To ensure proper installation and construction, regulations require the unit to have and follow a construction quality assurance program. The regulations also outline monitoring, inspection, response action, and closure requirements.

Waste Piles
Waste piles are non-containerized piles of solid, non-liquid hazardous waste that are used for temporary storage or treatment. Waste piles are required to have:

- A double liner system
- Double leachate collection and removal systems
- Leak detection system
- Run-on, run-off and wind dispersal controls
- Construction quality assurance

Land Treatment Units
Land treatment units use naturally occurring soil microbes and sunlight to treat hazardous waste. This is accomplished by applying the hazardous waste directly on the soil surface or incorporating it into the upper layers of the soil in order to degrade, transform, or immobilize the hazardous constituents. Land treatment units rely upon the physical, chemical, and biological processes occurring in the topsoil layers to contain the waste. Because of this, the units are not required to have liner systems or a leachate collection and removal systems. Before hazardous waste can be placed in a land treatment unit, operators must complete a treatment demonstration to demonstrate the unit's effectiveness and ability to treat the hazardous waste. Once operational, operators must monitor the unit (unsaturated zone monitoring) to ensure that all hazardous constituents are being treated adequately. Unit closure consists primarily of placing a vegetative cover over the unit and certifying that hazardous constituent levels in the treatment zone do not exceed background levels.

Injection Wells
Injection wells are regulated primarily under the Safe Drinking Water Act and EPA's Underground Injection Control (UIC) Program.

See also:

- *40 CFR part 264 subpart X*
- *40 CFR part 265 subpart R*

Miscellaneous and Other Units
Miscellaneous hazardous waste management units include units that do not fit within the definition of the more typical waste management units described in *40 CFR part 264* (containers, tanks, incinerators, etc.). These include geologic repositories other than injection wells, such as:

- Salt dome formations
- Salt bed formations
- Underground mines
- Underground caves
- Thermal treatment units other than incinerators, boilers or industrial furnaces

Because these units vary greatly, they are subject to environmental performance standards, not prescribed technology-based standards. The standards require that these miscellaneous units must be located, designed, constructed, operated, maintained, and closed in a manner that ensures the protection of human health and the environment.

19.3 Fate and Transport

Hazardous chemicals can be released into the environment. The mass balance for any chemical in any control volume during any time interval can be written as (Hemond and Fechner-Levy, 2000):

$$\begin{aligned}\text{Change in storage of mass} &= \text{mass transported in}\\ &\quad - \text{mass transported out} + \text{mass produced } \textit{by} \text{ sources}\\ &\quad - \text{mass eliminated by sinks}\end{aligned}$$

The mass balance expression in a control volume can also be written in terms of rates in mass per time $[\mathrm{M\,T^{-1}}]$:

$$\begin{aligned}&\text{Rate of change in storage of mass}\\ &\quad = \text{mass transport rate in} - \text{mass transport rate out}\\ &\quad + \text{mass production rate by sources}\\ &\quad - \text{mass elimination rate by sinks}\end{aligned}$$

The orders of reaction can be used to describe the fate of hazardous contaminants in the environment, including their rate of degradation. The rates are most often encountered include:

$$\text{Rate} = \frac{dC}{dt} = k \quad \text{Zeroth order}$$

$$\frac{dC}{dt} = -kC \quad \text{First order}$$

$$\frac{dC}{dt} = -kC^2 \quad \text{Second order}$$

The rate constant, k, for a first-order reaction may be determined by plotting $\ln C$ as a function of time. For second-order reactions, the rate constant is determined by plotting $1/C$ as function of time. Below are visual representations of examples of these plots (Figure 19.4).

Integration of the first-order rate equation above will yield the following relationship:

$$C = C_0 e^{-kt}$$

where C is the concentration at time t [M L^{-3}]; t is the time; C_0 is the initial concentration [M L^{-3}]; k is the first order degradation rate constant [$1\,\text{T}^{-1}$].

From this relationship, an expression can be developed for what is referred to as the half-life ($t_{1/2}$). The half-life represents the amount of time it takes for a parent compound to decay to half of its initial concentration.

$$\frac{C}{C_0} = 0.5 = e^{-kt_{1/2}}$$

$$\ln(0.5) = -kt_{\frac{1}{2}}$$

$$t_{\frac{1}{2}} = -\frac{\ln(0.5)}{k} = \frac{0.693}{k}$$

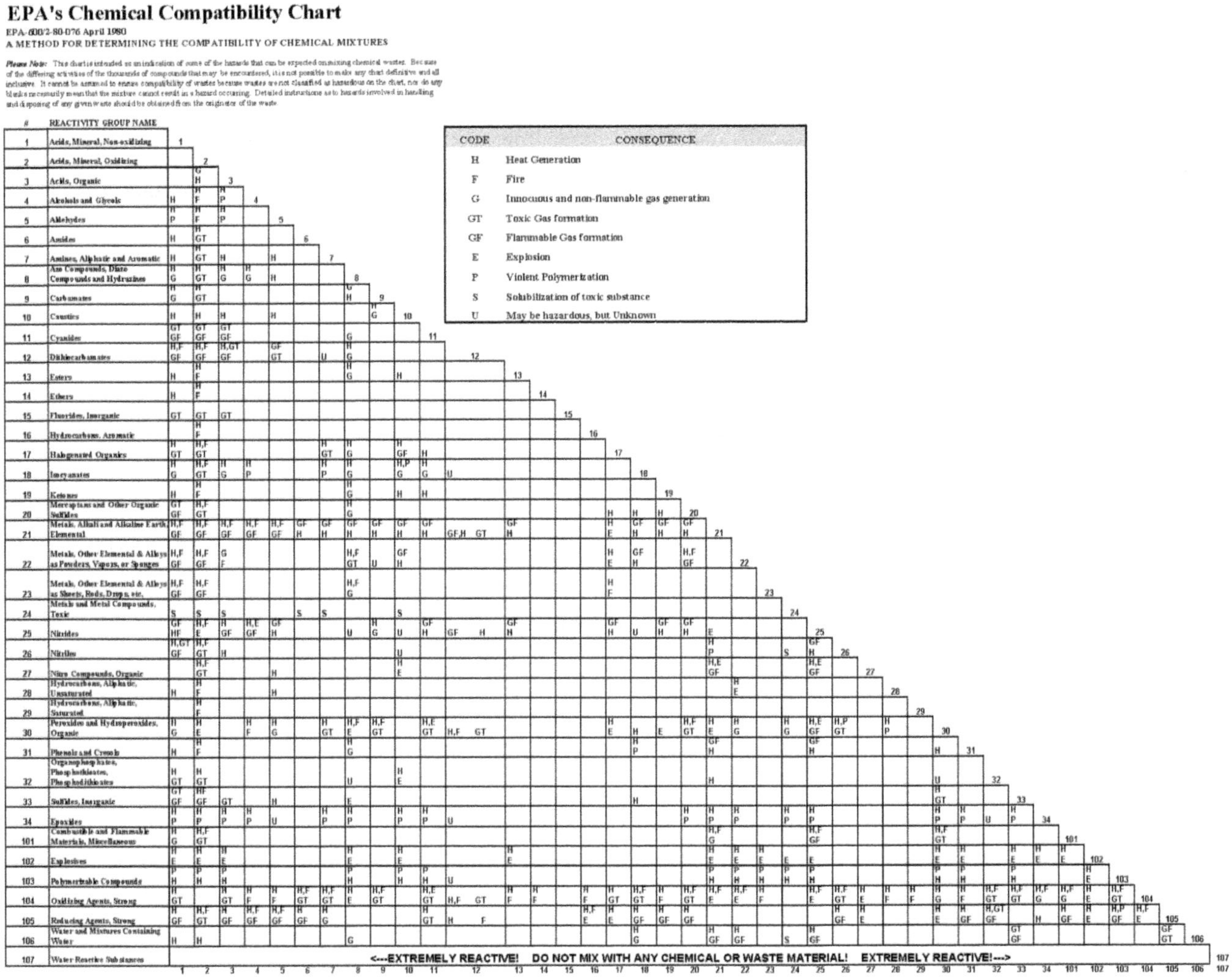

Figure 19.4 Reaction kinetics plots for zeroth order, first order and second order.

Half-lives provide exactly the same information as first-order decay rate constants. For different decay processes, half-lives may be easily compared to determine which decay or degradation mechanism is the most significant. These values may also be compared with a transport time to determine whether a high or low chemical concentration will remain at a point downgradient, downstream, or downwind of a chemical source (Hemond and Fechner-Levy, 2000).

As hazardous contaminants may be released in the environment, their fate and transport are affected by a diverse number of complex chemical and microbiological processes. The movement hazardous contaminants may be retarded, or their concentration may attenuate.

Retardation refers to processes that impede the transport of contaminants by removing or immobilizing them from a free state, such as in an aqueous solution or air. In retardation processes, such as sorption or precipitation, the immobilized contaminants are not transformed and the processes are generally reversible.

Attenuation generally refers to two types of processes: (i) transformation or (ii) irreversible removal. Attenuation processes differ from retardation processes in that it reduces the mass and/or concentration of a contaminant. This removal can occur by transference of the contaminant to another medium, like volatilization. However, more commonly, attenuation processes transform the molecular structure of the contaminant, such as oxidation-reduction.

Some processes actually increase the mobility of chemical contaminants in the environment, especially in the subsurface. These processes are categorized as mobility enhancement phenomena and may include complexation of metallic ions or dissolution of organic contaminants (Table 19.4).

19.4 Toxicology

Over 70 000 chemicals as sold in the United States meaning there is the need for understanding the potential threats to human and other biota is important. Management of chemicals found to be hazardous is paramount to protecting ecological and public health based on exposure routes, reducing the potential for risk, and understanding the toxicity of the chemicals if released into the environment.

There are four primary ways in which humans or animals can be directly exposed to hazardous waste in the environment, these include dermal, inhalation, and ingestion (Table 19.5).

Table 19.5 Methods of exposure.

Ingestion: Dust, food groundwater, soil, sediment, surface water
Inhalation: Sediment (particulate matter), air (gases or vapor), groundwater (water contact indoor or outdoor from water droplets or aerosols), surface water (water contact indoor or outdoor from water droplets or aerosols)
Dermal contact: Groundwater, surface water, sediment, soil/dust

Table 19.4 Summary of natural processes affecting the fate of hazardous constituents in the subsurface environment.

Process	Class of chemical	Effect
Sorption	Organic/inorganic	Retardation
Precipitation	Inorganic	Retardation
Ion exchange	Inorganic	Retardation
Filtration	Organic/inorganic	Retardation
Chemical oxidation-reduction	Organic/inorganic	Transformation/retardation
Biological uptake	Organic/inorganic	Retardation
Biodegradation	Organic	Transformation
Hydrolysis	Organic	Transformation
Volatilization	Organic	Elimination by intermedium transfer
Dissolution	Organic/inorganic	Mobility enhancement
Cosolvation	Organic	Mobility enhancement
Ionization	Organic	Mobility enhancement
Complexation	Inorganic	Retardation mobility enhancement
Immiscible phase	Organic	Various partitioning

Source: From LaGrega et al. (2010).

Table 19.6 Carcinogen classifications.

Human toxicity hazards	**Carcinogens**
	Toxicants that affect organs:
	Neuro – brain or nervous system
	Hepato – blood
	Nephro – kidney
	Cardio – heart
	Toxicants that affect systems:
	Pulmonary
	Hematological
	Endocrine
	Immunology
	Dermal
	Ocular
	DNA (mutagens and teratogens)
	Reproductive
Environmental toxicity hazards	Aquatic, avian, amphibian, mammalian (non-human)
Physical hazards	Explosives, corrosives, oxidizers, reducers, pH (acids and bases), violent reactions with water
Global hazards	Acid rain, global warming, ozone depletion, chemical or biological threats, extreme weather, water scarcity, loss of biodiversity, persistence, bioaccumulation

Exposure to hazardous chemicals can result in neurological impacts, infertility, birth defects, mutations, chronic disease, acute disease, and birth defects, just to name a few. Environmental toxicity encompasses the field of study investigating the effects of chemicals on living organisms. Classification of hazardous waste includes determining if the waste is carcinogenic or non-carcinogenic. Carcinogenic chemicals, or carcinogens, promote uncontrolled growth of living cells leading to the development to tumors. Whereas, non-carcinogenic, or non-carcinogens, do not appear at this time to impact living cells. Carcinogens can be further classified based on the organs they impact (Table 19.6).

The factors that affect toxicity of a hazardous waste include the form and chemical activity, dose, exposure route, the species exposed, the ability for the chemical to be absorbed, metabolism, distribution within the body, and excretion of the chemical and the presence of other chemicals. Currently, methods to protect human health and the environment is closely tied to the level of the risk and controlling exposures.

The dose of a chemical is defined as the amount of a chemical ingested, inhaled, or the subject comes into contact that can interact with the subject's biological receptors or metabolic processes. Therefore the dose, can be little or great, and the response the subject is based on the chemical toxicity. The dose response is a measure of concentration of a chemical over time to determine the effect the chemical will have on a subject. Table 19.7 provides common toxicology terms used when evaluating hazardous waste exposure.

Table 19.7 Toxicology vocabulary.

Exposure and human response	
Dose	The amount of a chemical ingested, inhaled, or the subject comes into contact that can interact with the subject's biological receptors or metabolic processes
Right-to-know act	In the US, workers have the right to know what type of chemicals they will be exposed to prior to working with a chemical; Material Safety Data Sheets must be available for anyone who wants to review chemical safety information
Carcinogens	Chemicals that promote uncontrolled growth of living cells leading to the development to tumors
Suspected carcinogens	Chemicals where the findings are inconclusive
Synergistic toxicity	Chemical exposure to multiple chemicals where in combination intensifies or increases the carcinogenic potential
Bioaccumulation	Accumulation of a chemical in the environment or living things
Acute toxicity	Death or an adverse response resulting from short-term exposure (hours to days)
Chronic toxicity	Response resulting from long-term exposure (weeks to years)
Susceptible population	Sensitive segments of a population that must receive distinct consideration in determining toxic effects of chemicals or substances (e.g. Embryonic, juvenile, elderly and or ill segment)

19.5 Environmental Audits

A systematic, documented, periodic, and objective review by a regulated entity of facility operations and practices related to meeting environmental requirements

19.5.1 Compliance Audits

These audits stresses regulatory compliance and verification, and was the initial driving motivation for development of environmental auditing.

Its major goals include determination of what specific regulatory requirements are imposed upon an operation, determination of whether the operation is in compliance, and specific pinpointing of possible violations in time to take proactive measures.

Provides a "snapshot" of plant operations and procedures, identifying instances of either compliance audit is generally superficial.

19.5.2 Liability Definition Audit

Identifies environmental problems that could reduce value of a property or expose the buyer to liability. Typically done for prospective buyers of real estate and for proposed mergers and acquisitions.

Known as part of "due diligence" and is relevant in the gathering of facts and information in a prudent manner and applying technical knowledge to identify where problems exist.

19.5.3 Management Audits

This is an audit that determines whether an adequate compliance management "system" is established, implemented, and used correctly to integrate environmental compliance into everyday operating procedures.

This audit examines cultural, management, and operational elements to include internal policies, human resources, training programs budgeting and planning systems, monitoring and reporting systems, and informational systems.

Detects potential systematic breakdowns that could manifest themselves as an environmental problem.

19.5.4 Waste Contractor Audit

Uses features of both a compliance audit and a liability definition audit to analyze commercial facilities used to store, treat, and dispose of hazardous waste.

19.5.5 Waste Minimization Audit

Examine waste generated by a facility with the objective of identifying viable actions to reuse, recycle, or otherwise reduce the quantity and toxicity of each waste stream.

19.5.6 Risk Definition Audit

Analyzes the operations of facilities that handle hazardous materials and substances. These audits serve to help in obtaining insurance coverage and are required by some governments as part of catastrophe prevention planning. These results also provide environmental guidance for hazards not specifically addressed in current environmental regulations.

19.5.6.1 Hazardous Waste Generator Categories

A generator is any person who produces a hazardous waste as listed or characterized in part 261 of title 40 of the Code of Federal Regulations (CFR). Recognizing that generators produce waste in different quantities, EPA established three categories of generators in the regulations:

- Conditionally exempt small quantity generators,
- Small quantity generators, and
- Large quantity generators.

The volume of hazardous waste each generator produces in a calendar month determines which regulations apply to that generator.

19.5.6.1.1 Conditionally Exempt Small Quantity Generators

Conditionally Exempt Small Quantity Generators (CESQG) generate 100 kg or less per month of hazardous waste or 1 kg or less per month of acutely hazardous waste. Requirements for CESQGs include:

- CESQGs must identify all the hazardous waste generated.
- CESQGs may not accumulate more than 1000 kg of hazardous waste at any time.
- CESQGs must ensure that hazardous waste is delivered to a person or facility who is
- authorized to manage it.

19.5.6.1.2 Small Quantity Generators

Small Quantity Generators (SQGs) generate more than 100 kg, but less than 1000 kg, of hazardous waste per month. Major requirements for SQGs include:

- SQGs may accumulate hazardous waste on-site for 180 days without a permit (or 270 days if shipping a distance greater than 200 miles).
- The quantity of hazardous on-site waste must never exceed 6000 kg.
- SQGs must comply with the hazardous waste manifest and pre-transport requirements at 40 CFR part 262, subpart B and 40 CFR §§262.30 through 265.33.
- SQGs must manage hazardous waste in tanks or containers, subject to the requirements found at 40 CFR §265.201 and 40 CFR part 265 subpart I except for 40 CFR §§265.176 and 265.178, respectively.
- SQGs must comply with the preparedness and prevention requirements at 40 CFR part 265 subpart C, and the land disposal restriction requirements at 40 CFR part 268.
- There must always be at least one employee available to respond to an emergency. This employee is the emergency coordinator responsible for coordinating all emergency response measures. SQGs are not required to have detailed, written contingency plans.

19.5.6.1.3 Large Quantity Generators (LQGs)

Large Quantity Generators (LQGs) generate 1000 kg per month or more of hazardous waste or more than 1 kg per month of acutely hazardous waste. Major requirements for LQGs include:

- LQGs may only accumulate waste on-site for 90 days. Certain exceptions apply.
- LQGs do not have a limit on the amount of hazardous waste accumulated on-site.
- Hazardous waste generated must be managed in tanks, containers, drip pads or containment buildings subject to the requirements found at 40 CFR part 265, subparts J, I, W, and DD, respectively.
- LQGs must comply with the hazardous waste manifest and pre-transport requirements at 40 CFR part 262 subpart B and 40 CFR §§262.30 through 265.33.
- LQGs must comply with the preparedness and prevention requirements at 40 CFR part 265 subpart C, the contingency plan and emergency procedures at 40 CFR part 265 subpart D, and the land disposal restriction requirements at 40 CFR part 268.
- LQGs must submit a biennial hazardous waste report.

19.6 General Overall Site Remediation Procedure

19.6.1 Preliminary Assessment/Site Inspection (PA/SI)

The Preliminary Assessment (PA) and Site Inspection (SI) are used by EPA to evaluate the potential for a release of hazardous substances from a site. These measures outline the investigations of site conditions. If the release of hazardous substances requires immediate or short-term response actions, these are addressed under the Emergency Response program of Superfund.

19.6.1.1 Preliminary Assessment

The PA are designed to determine whether a sites poses little or no threat to human health and the environment or if it does pose a threat, whether the threat requires further investigation. These investigations collect readily available information about a site and its surrounding area. Their purpose are also designed to distinguish, based on limited data, between sites that pose little or no threat to human health and the environment and sites that may pose a threat and require further investigation. The assessment also identifies sites requiring assessment for possible response actions. If the PA results in a recommendation for further investigation, a Site Inspection is performed.

19.6.1.2 Site Inspection

The Site Inspection (SI) portion of the remedial procedure identifies sites that enter the **NPL Site Listing Process** and provides the data needed for Hazard Ranking System (HRS) scoring and documentation. Investigators typically collect environmental and waste samples to determine what hazardous substances are present at a site. The SI is used to determine if substances are being released to the environment and assess if they have reached nearby targets. The SI is conducted in one or two stages. The first stage, or "Focused SI," tests hypotheses developed during the PA and can yield information sufficient to prepare an HRS scoring package. If further information is necessary to document an HRS score, an expanded SI (second stage) is conducted. Information collected during the PA and SI is used to calculate an HRS score. Sites with an HRS score of 28.50 or greater are eligible for listing on the NPL

19.6.2 NPL Site Listing

The National Priorities List Site Listing is a list of the most serious sites identified for possible long-term cleanup. The mechanisms for placing sites on the NPL include: (i) EPA's Hazard Ranking System (HRS); (ii) States or Territories are allowed to designate one top-priority site regardless of score; (iii) Listing is allowed if a site meets all three of the following requirements:

- Agency for Toxic Substances and Disease Registry (ATSDR) of the U.S. Public Health Service has issued a health advisory that recommends removing people from the site
- EPA determines the site poses a significant threat to public health; and
- EPA anticipates it will be more cost-effective to use its remedial authority than to use its emergency removal authority to respond to the site.

19.6.2.1 Hazard Ranking System (HRS)

Principal mechanism EPA uses to place uncontrolled waste sites on the National Priorities List. The system is a numerically based screening system that uses information from initial, limited investigations (PA/SI) to assess the relative potential of sites to pose a threat to human health or the environment. HRS scores do not determine the priority in funding EPA remedial response actions, because the information collected to develop HRS scores is not sufficient to determine either the extent of contamination or the appropriate response for a particular site. The sites with the highest scores do not necessarily come to the EPA's attention first, since this would require stopping work at sites where response actions were already underway. EPA relies on more detailed studies in the RI/FS which typically follows listing.

19.6.3 Remediation/Clean Up Process

The Superfund cleanup process begins with site discovery or notification to EPA of possible releases of hazardous substances. Sites are discovered by various parties, including citizens, State agencies, and EPA Regional offices. Once discovered, sites are entered into the Comprehensive Environmental Response, Compensation, and Liability Information System (CERCLIS). CERCLIS is EPA's computerized inventory of potential hazardous substance release sites

19.6.4 Remediation Investigation/Feasibility Study

19.6.4.1 Scoping

Scoping is the initial planning phase of the process Remediation Investigation/Feasibility Study (RI/FS). This phase usually begins with the collection of existing site data, including data from previous investigations such as the Preliminary Assessment (PA) and Site Inspection (SI). From this info, site management planning is undertaken to: preliminarily identify boundaries of the study area; identify likely remedial action objectives and whether interim actions may be necessary or appropriate; and establish whether the site may best be remedied as one or several separate operable units.

When the overall management strategy is agreed upon, the RI/FS for a specific project or the site as whole is planned. Typical scoping activities include: initiating the identification and discussion of potential Applicable or Relevant and Appropriate Requirements (ARARs) with the support agency; determining the types of decisions to be made and identifying the data and other information needed to support these decisions; assembling a technical advisory committee to assist in activities, serve as a review board for important deliverables, and monitor progress during the study; and preparing the work plan, sampling and analysis plan, health and safety plan, and community relations plan.

19.6.4.2 Site Characterization

This phase of the procedure occurs when field sampling and laboratory analyses are initiated in the RI/FS. A preliminary site characterization summary is prepared to provide the lead agency with information on the site early in the process before preparation of the full remedial investigation (RI) report. This summary is useful in determining the feasibility of potential technologies and in assisting both the lead and support agencies with the initial identification of Applicable or Relevant and Appropriate Requirements; and can also be sent to the Agency for Toxic Substances and Disease Registry to assist in performance of health assessments.

A baseline risk assessment is developed to identify the existing or potential risks that may be posed to human health and the environment by the site, and also provides valuable input to the development and evaluation of alternatives during the feasibility study (FS).

19.6.4.3 Development and Screening of Alternatives Cleanup Process

This portion of the RI/FS usually begins during scoping when likely response scenarios may first be identified, and requires: identification of remedial action objectives; identification of potential treatment, resource recovery, and containment technologies that will satisfy these objectives; screening of the technologies based on their effectiveness, implementability, and cost; and assemblage of the technologies and their associated containment or disposal requirements into alternatives for the contaminated media at the site or for the operable unit. Once potential alternatives have been developed, it may be necessary to screen out certain options to reduce the number of alternatives that will be analyzed.

19.6.4.4 Treatability Investigations

Treatability investigations are conducted primarily to: provide sufficient data to allow treatment alternatives to be fully developed and evaluated during the Detailed Analysis phase; support the remedial design and remedial action of selected alternative; and reduce cost and performance uncertainties for treatment alternatives to acceptable levels so that a remedy can be selected.

19.6.4.5 Detailed Analysis

In the last phase of the RI/FS, the Detailed Analysis, once sufficient data are available, alternatives are evaluated in detail with respect to nine evaluation criteria that EPA has developed to address the statutory requirements and preferences of CERCLA. The nine criteria include: overall protection of human health and the environment; compliance with ARARs; long-term effectiveness and permanence; reduction of toxicity, mobility, or volume; state acceptance; short-term effectiveness; implementability; cost; and community acceptance.

The alternatives are analyzed individually against each criterion and then compared against one another to determine their respective strengths and weaknesses and to identify the key trade-offs that must be balanced for the site. The results of the detailed analysis are summarized so that an appropriate remedy consistent with CERCLA can be selected.

References

Catawba Riverkeeper Foundation (2015). Duke Energy Dan River Coal Ash Spill Updates: What We Know, What We Need to Know. Retrieved from https://www.nytimes.com/2014/03/01/us/coal-ash-spill-reveals-transformation-of-north-carolina-agency.html?_r=0hp.myway.com/onlinemapfinder/ttab02/index.html?p2=^BA5^xdm003^TTAB02^us&n=782af154&st=tab&ptb=34297EF1-B97B-4699-9905-6CF5AC07D564&si=CN6C0qKao84CFcZahgodo64BEw (accessed 30 January 2018).

CBS Associated Press (Producer) (2015). Duke energy fined $102 million in coal ash spill. *CBS News* (14 May). Retrieved from http://www.cbsnews.com/news/duke-energy-fined-102-million-in-coal-ash-spill/ (accessed 21 January 2018).

Davis, M.L. and Cornwell, D.A. (2008). *Introduction to Environmental Engineering*. New York: McGraw-Hill Companies.

Google Images (2010). SkyTruth: BP/Gulf Oil Spill – 68,000 Square Miles of Direct Impact, 1600 × 908 Search by image. Retrieved from SkyTruth: BP/Gulf Oil Spill – 68,000 Square Miles of Direct Impact blog.skytruth.org.

Hemond, H.F. and Fechner-Levy, E.J. (2000). *Chemical Fate and Transport in the Environment*, 2e. San Diego, CA: Academic Press.

LaGrega, M.D., Buckingham, P.L., and Evans, J.C. (2010). *Hazardous Waste Management*. Long Grove, IL: Waveland Press.

U.S. Environmental Protection Agency (1980). EPAs Chemical Compatibility Chart, EPA-600/2-80-076 in Cincinnati, OH (April). Retrieved from https://nepis.epa.gov/ (accessed 21 January 2018).

U.S. Environmental Protection Agency (2001). RCRA, Superfund, & EPCRA Call Center Training Module: Introduction to Used Oil (40 CFR Part 266, Subpart E and Part 279), EPA530-K-02-0251. https://www.epa.gov/sites/production/files/2015-09/documents/uoil.pdf (accessed 21 January 2018).

U.S. Environmental Protection Agency (2015a). EPA's Response to the Duke Energy Coal Ash Spill in Eden, NC (8 June). Retrieved from http://www2.epa.gov/dukeenergy-coalash (accessed 21 January 2018).

U.S. Environmental Protection Agency (2015b). Summary of the Resource Conservation and Recovery Act (29 December). Retrieved from: https://www.epa.gov/laws-regulations/summary-resource-conservation-and-recovery-act (accessed 21 January 2018).

U.S. Environmental Protection Agency (2016). EPA Hazardous Waste Permitting (9 August). Retrieved from https://www.epa.gov/hwpermitting (accessed 21 January 2018).

20

Waste Minimization and Reuse Technologies

Bora Cetin[1] *and Lin Li*[2]

[1]*Department of Civil, Construction, and Environmental Engineering, Iowa State University, Ames, IA, USA*
[2]*Department of Civil and Environmental Engineering, Jackson State University, Jackson, MS, USA*

20.1 Introduction

Substantial upgrades to the current infrastructure in the United States are desperately needed. The American Society of Civil Engineers' 2013 Report Card assigned a score of D to the United States' roads for being "poor: at risk," citing 32% of major highways in the United States in poor or mediocre condition (ASCE, 2017). As population increases so does vehicle traffic; this, combined with aging highways, has caused an increase in the demand of construction aggregates (FHWA, 2004).

Aggregates are made of crushed stone, gravel, and sand and are typically used as base materials for foundations and roadways and in concrete and asphalt mixes (Mamlouk and Zaniewski, 2011). Approximately 2 billion tons of aggregates are produced each year, domestically, and production is expected to increase to more than 2.5 billion tons by 2020 (FHWA, 2004). Aggregates distribute loads to underlying soil, reducing cracking in pavement, rutting, and deformation in the base/subbase layer (Tutumluer and Pan, 2008; Haider et al., 2014).

Coarse aggregates are mainly used in the construction of highway base/subbase layers to develop a pavement layer that can not only freely drain but also provide load distribution to the underlying soil, reducing cracking in asphalt, rutting, and deformation in the base/subbase layer of highways (Tutumluer and Pan, 2008). In many cases, these materials can be replaced with reclaimed highway paving materials, secondary materials, suitable waste materials, and construction debris that are normally disposed in landfills and can generate millions of dollars savings to taxpayers. Reuse in construction has several benefits, including reduction in solid waste disposal costs incurred by industry, reduction in landfill requirements, minimization of damage to natural resources caused by excavating earthen materials for construction, obtaining added value from waste materials, conservation of production energy, and ultimately providing sustainable construction and economic growth.

Legislations have been promulgated in many states that remove barriers to large-scale beneficial reuse of recycled materials to reduce construction costs and increase sustainability. As a result, there is a policy shift, both nationally and at the state level, aimed at substantially increasing the use of such materials in geotechnical construction. Most commonly reused waste materials in geotechnical and geoenvironmental applications include coal combustion products (CCPs), recycled concrete aggregates (RCAs), construction and demolition (C&D) materials, foundry sand, slags, reclaimed asphalt materials (RPMs), recycled asphalt pavement (RAPs) materials, and used tires.

These materials are mostly used as a soil-stabilizing agent in highway sublayers (base/subbase/subgrade), a soil amendment in highway embankments and agricultural applications, and a remediation of contaminated groundwater and soil, landfill liner, and landfill cover soils. This chapter will mainly focus on the use of recycled waste materials in stabilizing highway sublayers and as a soil amendment in highway embankment constructions. Geotechnical performances and environmental impacts of these designs will be discussed. The recycled materials that will be covered in this chapter are fly ash and RCAs.

20.2 Type of Recycled Waste Materials

As mentioned previously this chapter will provide information about recycling of fly ash and RCAs in geotechnical and geoenvironmental engineering applications. The following sections summarize the background information about these materials.

Handbook of Environmental Engineering, First Edition. Edited by Myer Kutz.

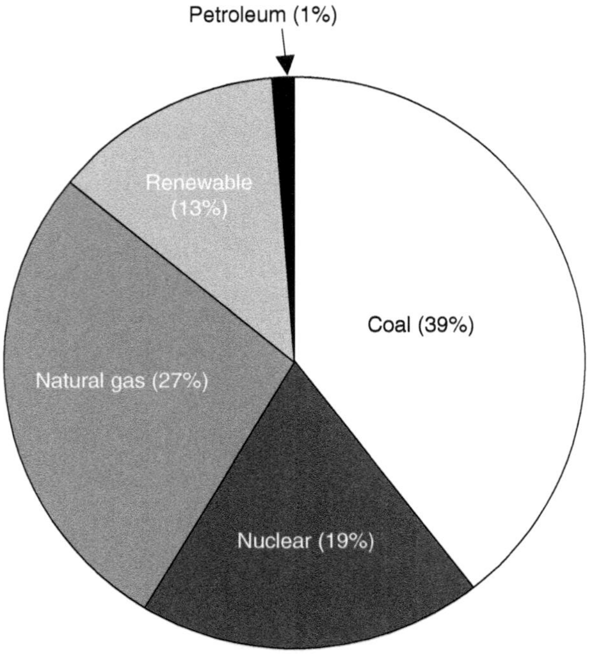

Figure 20.1 Distribution of the electricity generation from different energy sources. *Source*: From U.S. Energy Information Administration (2014).

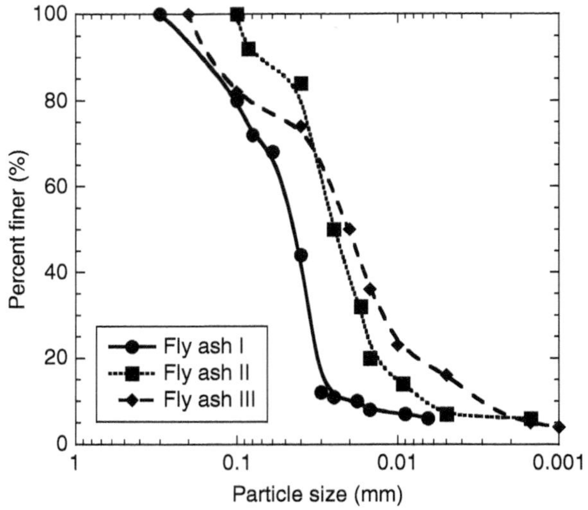

Figure 20.2 Grain size distribution of three different fly ashes used in Cetin et al. (2010). *Source*: Cetin et al. (2010). Reproduced with permission of Elsevier.

20.2.1 Fly Ash

Approximately 39% of the electricity in the United States is generated through coal power plants (Figure 20.1). Nearly 114 million tons of coal combustion by-products are produced annually. Fly ash is the most plentiful (55 million tons) coal combustion by-product (ACAA, 2013). It is estimated that 30 million tons of fly ashes are landfilled. Placement of fly ashes in landfills is a waste of valuable land space, and its design and construction is expensive. In addition, piling large amounts of fly ash at one location may create environmental problems on terrestrial and aquatic resources such as surface water and groundwater (Morar, 2008). Therefore, it is very important to find a way to recycle these fly ashes. Use of this material in different geotechnical and geoenvironmental engineering designs and applications poses a great potential to minimize this environmental challenging problem.

Fly ash is the finely divided mineral residue resulting from combustion of ground or powdered coal at electric generating plants (ASTM C618-12el, 2012). It consists of very fine silt size particles (0.074–0.005 mm) and is spherical in shape. Figure 20.2 shows an example of typical grain size distribution of fly ashes.

20.2.1.1 Application Areas

Fly ash is a pozzolan, that is, siliceous or siliceous–aluminous material that becomes cementitious when combined with an activator including lime, Portland cement, cement kiln dust, and lime kiln dust (LKD). Fly ash has been reused in not only geotechnical engineering field but also other engineering areas for many years. For instance, it has been used at large portions by the concrete and cement industry. However, this chapter will mainly focus on recycling fly ashes in geotechnical and geoenvironmental applications. Detailed information regarding these applications will be explained in the next sections.

20.2.2 Recycled Concrete Aggregates

In place of natural aggregates, crushed waste concrete, termed recycled concrete aggregate (RCA), can be used as an alternative highway subbase/base layer material. This material substitution increases project sustainability because the environmental and economic impacts associated from mining and processing of natural aggregates are eliminated. RCA use is widespread in the United States. Forty-one states recycle concrete for use as aggregate and 38 states recycle concrete as an aggregate base (FHWA, 2004). Recycling of road surfaces is also increasing in most states, with an 11% increase in 2012 over 2011 (USGS, 2013). RCA is one of the most common recycled construction materials used as base/subbase course.

RCA materials have reasonably consistent properties. Fines contents and the mortar contents of RCAs generally vary between 3–4% and 45–55%, respectively. RCAs also possess some impurities. Asphalt aggregate, aggregate with plastic fibers, brick, and wood chips are the most common impurities found in RCA materials (Edil, 2012). Tables 20.1 and 20.2 summarize the typical ranges of index and chemical properties of RCA materials.

Table 20.1 Typical ranges of index properties of RCA materials.

C_u	C_c	G_s	Abs (%)	Mortar content (%)	Impurities (%)	Gravel (%)	Sand (%)	Fines (%)	USCS	AASHTO
22–66	0.3–6	2.2–2.4	5–6.5	45–55	0.16–1.67	32–76	22–65	3–4	SW GW GP	A-1-a A-1-b

Notes: C_u, coefficient of uniformity; C_c, coefficient of curvature; G_s, specific gravity; Abs, absorption; USCS, Unified Soil Classification System (ASTM D2487-11, 2011); AASHTO, American Association of State Highway and Transportation Officials.

Table 20.2 Typical ranges of chemical compositions and total metal concentrations of RCAs.

	(%) by weight						Concentrations (mg kg^{-1})								
pH	CaO	SiO_2	Fe_2O_3	Al_2O_3	Ca	Fe	Cr	Cu	Zn	Ba	Cd	Pb	Se	Mn	As
10–12.5	11–31	35–69	2–6	5–10	8–20	~2	10–20	5–20	1–2	50–950	0.2–0.6	10–40	ND	300–700	5–50

Note: ND, nondetectable.

20.3 Recycling Applications of Fly Ash and Recycled Concrete Aggregates

20.3.1 Fly Ash

20.3.1.1 Application in Stabilizing Agent and Soil Amendment

Fly ashes are classified in two categories (ASTM C618-12el, 2012). These categories are class C and class F fly ashes. Table 20.3 summarizes the typical physical properties and chemical compositions of class C and class F fly ash and the criteria in ASTM C618 used to classify the fly ashes as class C and class F. Fly ashes, which do not comply with the requirements for class C or F fly ash, are classified as off-specification fly ash.

Class F fly ash is produced from burning anthracite and bituminous coals and contains siliceous and aluminous material. It possesses little or no cementitious value and requires addition of materials that are rich in CaO content such as lime, cement, cement kiln dust,

Table 20.3 Physical properties and chemical compositions of the fly ashes.

	Percent composition		Specifications	
Parameter	Typical class C	Typical class F	ASTM C618 class C	ASTM C618 class F
Silicon dioxide (SiO_2)	39.9	54.9		
Aluminum oxide (Al_2O_3)	16.7	25.8		
Iron oxide (Fe_2O_3)	5.8	6.9		
$SiO_2 + Al_2O_3 + Fe_2O_3$	62.4	87.6	50 Min	70 Min
Calcium oxide (CaO)	24.3	9.7		
Magnesium oxide (MgO)	4.6	1.8		
Sulfur trioxide (SO_3)	3.3	0.6	5.0 Max	4.0 Max
CaO/SiO_2	0.61	0.18		
pH	—	—		
Loss on ignition	6.0	6.0	6.0 Max	10.0 Max
Moisture content	—	—	3.0 Max	4.0 Max
Specific gravity	—	—	5 Max	5 Max
Fineness[a]	—	—	34 Max	34 Max

[a] Amount retained on #325 sieve.

LKD, etc. (Bin-Shafique, 2002). On the other hand, class C fly ash is generally produced from lignite and subbituminous coals. It consists of relatively high amount of CaO and pozzolan contents (SiO_2, Al_2O_3, Fe_2O_3) and possesses self-cementitious properties (Baker, 1987). Off-specification fly ashes may possess self-cementitious or noncementitious properties based on their total CaO content, and it is speculated that the fly ash has CaO content lower than 10% by dry weight and is not capable of initiating cementitious reactions without inclusion of activator (Arora and Aydilek, 2005; Cetin et al., 2010).

Due to high cost and potential environmental impact of fly ash landfill, recycling of fly ash gains significant attention by the regulatory agencies. As a result, many innovative ways to reuse fly ash have been developed and still being researched. One of these ways is to reuse fly ash as a stabilizing agent in the design of pavement sublayers including base, subbase, subgrade, and as a soil amendment in highway embankments (Arora and Aydilek, 2005; Li et al., 2008; Li et al., 2009; Cetin et al., 2010).

20.3.1.2 Geotechnical Performance

There have been many studies that investigate the use fly ashes in stabilization of highways base/subbase/subgrade layer and unpaved roads. In one of these studies, Arora and Aydilek (2005) investigated the engineering properties of class F fly ash amended soils as highway base materials. This study indicated that in the use of class F fly ash for stabilization purposes, one must add activators that are rich in CaO content such as lime and cement. Figure 20.3a shows the impact of cement addition on the unconfined compressive strength of cohesive soils in this study. It is shown that cement inclusion up to a certain percentage (5%) continuously increases the strength of class F soil mixture regardless of curing time. However, such trend was not observed for the same mixtures prepared with lime content ranged from 4 to 10% by dry weight (Figure 20.3b).

Based on the results of this study, it must be noted that strength gain may not be possible with all activators that are rich in CaO content within soils mixed with noncementitious class F fly ash. On the other hand, Vishwanathan et al. (1997) claimed that the strength of silty and sandy soils increased when they were mixed with lime and class F fly ashes. Cetin et al. (2010) also studied improving the engineering properties of a road surface gravel (RSG) material with LKD-activated noncementitious off-specification fly ashes. This study focused on improvement of California bearing ratio (CBR) and resilient modulus of RSG with addition of fly ash and LKD. Fly ash percentages were selected as 10 and 20% to cover the typical range used in soil stabilization (Edil et al., 2002; Bin-Shafique et al., 2004). Around 2.5 and 5% LKD by weight were added in the mixtures as an activator.

CBR of RSG before stabilization was 24, which was significantly lower than a generally accepted limit for base applications (Asphalt Institute, 2003), and thus required stabilization for use in highway construction as a base/subbase layer. In all cases, CBR of stabilized mixtures is higher than that of RSG (Table 20.4). This study proved that mixing of soil with only noncementitious fly ash is not sufficient enough to increase the strength. Inclusion of source with high CaO content, in this case it was LKD, is necessary to start cementitious reactions. Even though Arora and Aydilek (2005) did not observe any increase in

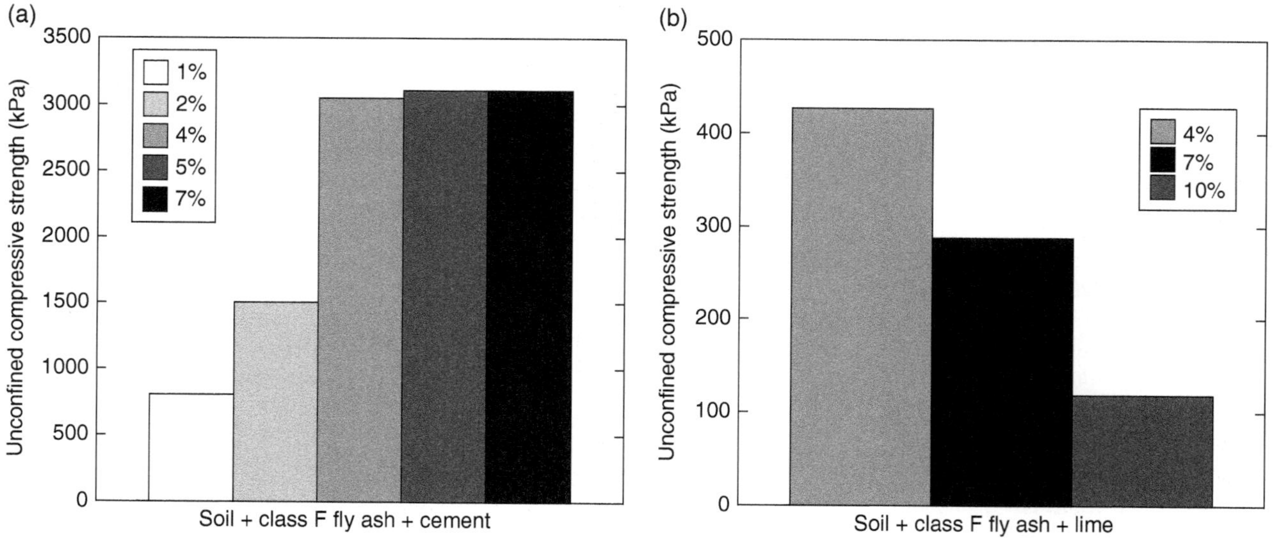

Figure 20.3 Effect of (a) cement content and (b) lime content on unconfined compressive strengths of soil mixed with 40% class F fly ash by weight. *Source*: Adopted from Arora and Aydilek (2005).

Table 20.4 CBR of RSG–fly ash–LKD mixtures (7 days cured samples).

Soil/fly ash	Fly ash content (%)	LKD content (%)	CBR
RSG	—	—	24
Fly ash I	10	2.5	108
	10	5	142
	20	2.5	112
	20	5	133
Fly ash II	10	2.5	115
	10	5	135
	20	2.5	71
	20	5	87
Fly ash III	10	2.5	69
	10	5	129
	20	2.5	84
	20	5	110

Source: Adopted from Cetin et al. (2010). Reproduced with permission of Elsevier.

(a)

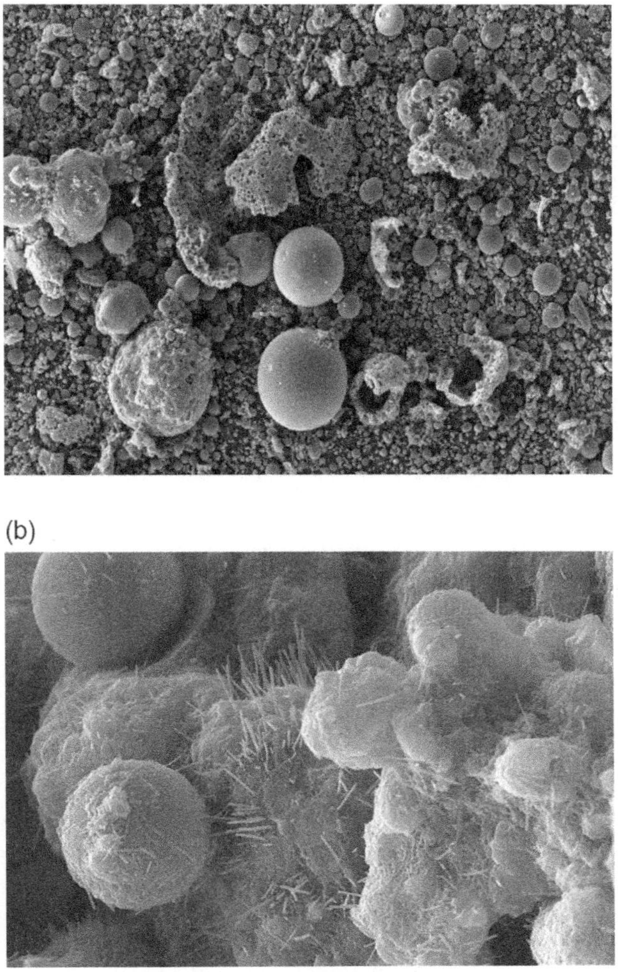

Figure 20.4 SEM photograph of (a) fly ash I and (b) RSG material amended with 10% fly ash I and 2.5% LKD by weight. *Source*: Cetin et al. (2010). Reproduced with permission of Elsevier.

the strength of soil–class F fly ash–lime mixtures indicating that cementitious reactions did not occur between class F fly ash and lime additives, this was not the case for the mixtures tested in Cetin et al. (2010). Figure 20.4 shows the coating of fly ash particles due to the hydration occurring between fly ash and LKD particles. Thus, it is also an indicator of an increase in CBR (Conner, 1990). In addition, the relatively higher amounts of calcium in RSG mixed with fly ash and LKD are evident as a result of the addition of LKD (Figure 20.5).

Curing process and curing periods are important parameters that need to be applied carefully and examined when chemical stabilization is used for improving the strength and stiffness of soils. As shown in Figure 20.6, CBR values of LKD amended soil–fly ash mixtures increase with increasing curing time. The hydration of calcium oxide (CaO) (in this case came from LKD) causes the formation of ($Ca(OH)_2$), and disassociation of ($Ca(OH)_2$) favors dissolution of silica and alumina in fly ash. This phenomenon gives rise to formation of calcium silicate hydrate (CSH) and calcium aluminate silicate hydrate (CASH) gels around soil particles. It is speculated that the release of CaO caused these increases and the temperature of the curing chamber and availability of 100% relative humidity also contributed to the cementitious reactions. The increase in CBR after 1 day of curing is relatively modest; however, CBR of the 7- and 28-day cured specimens increased up to 6 and 7 times, respectively. Similar increases in strength with increasing curing period were reported by previous researchers (Vishwanathan et al., 1997; Guney et al., 2006; Dayioglu, 2015). Cetin et al. (2010) also investigated the resilient modulus of the RSG–fly ash–LKD mixtures and found that an increase in LKD amount increased the resilient modulus of soil materials at any stress conditions (Figure 20.7). Similar trends were observed by other studies during resilient modulus and strength testing of soils stabilized with fly ash and carbide lime or cement (Consoli et al., 2001).

Based on the results from the literature, it can be speculated that class C fly ash is the most efficient coal combustion by-product to be used for improving the engineering properties of weak soils. In addition, class C fly ash can be a viable binder for stabilization of RPM for base applications (Hatipoglu et al., 2008). One study showed that CBR of RPM (70–94% approximately) increased significantly with addition of class C fly ash (Li et al., 2009). Li et al. (2009) also evaluated the impact of class C fly ash addition on resilient modulus (stiffness, M_r) of RPM materials, and it determined that the M_r of

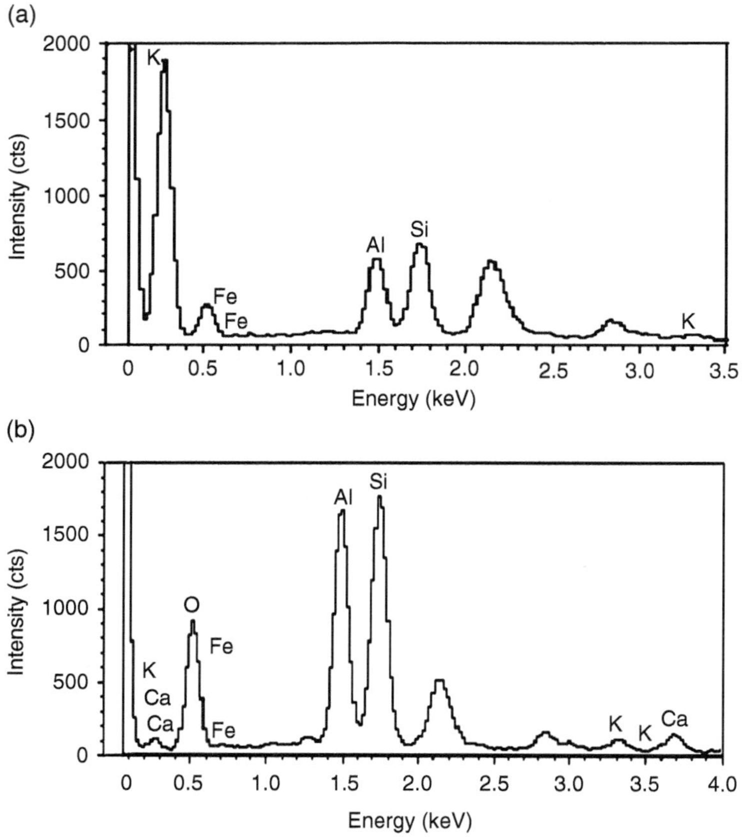

Figure 20.5 EDX plot of the SEM photograph of (a) fly ash III and (b) RSG mixed with 20% fly ash III and 5% LKD by weight. *Source*: Cetin et al. (2010). Reproduced with permission of Elsevier.

RPMs increased almost 200%. Camargo (2008) also showed that addition of 10–15% by weight of class C fly ash increases the CBR and resilient modulus of RPM and RSG by 3–6 and 9–22 times, respectively. Wen et al. (2007, 2008) investigated the impact of high carbon self-cementitious fly ash on CBR and M_r of RPM materials. Results of this study showed that CBR and M_r of RPM materials were significantly higher than that of RPM alone materials. Furthermore, CBR of these fly ash–RPM mixtures tested in these studies were mostly equal to or higher than the CBR of conventional granular aggregate base materials that were typically used such applications. In addition, these two studies also claimed that fly ash–RPM mixtures possessed lower permanent deformations under repetitive loads compared with RPMs alone.

Past literature has also proven that addition of class C fly ash not only increased the strength and the stiffness of weak soils but also prevented heaving of expansive soils (Nalbantoğlu, 2004; Chittoori and Puppala, 2011; Estabragh et al., 2013; Pal and Ghosh, 2014). Dayioglu (2015) studied the mitigation of swelling pressure of expansive shale clay with three different chemical additives, which included class C fly ash, class F fly ash, and lime. Figure 20.8 represents the impact of inclusion of these additives in clay soil. Based on the results of this study, one can conclude that lime and class C fly ash are the two efficient products to dissipate the swelling pressure of expansive soils, while the same conclusions cannot be made for the use of class F fly ash. It is believed that the swelling pressure of expansive soils is due to the release of cementitious products (Sezer et al., 2006; El-Latif, 2008; Zha et al., 2008; Estabragh et al., 2013). The hydration between fly ash, soil, and water creates cementitious products that are responsible to bind the soil particles and mitigate the swelling. Figure 20.8 also shows that the swelling pressures of the mixtures do not change significantly when they are exposed to higher curing periods at the range of 7–28 days. This could be attributed to excessive cation exchange occurring when the clay is mixed with lime and class C fly ash, which possess high CaO content and complete the hydration process very rapidly and do not require longer curing periods (Nalbantoğlu and Gucbilmez, 2001; Al-Rawas et al., 2005; Rao and Subbarao, 2012).

20.3.1.3 Durability

As explained in the previous section, the inclusion of fly ashes (class C and class F with activator) is very beneficial for improving the geotechnical performance of base/subbase/subgrade soils including sand, silt, and clay (Acosta et al., 2003; Arora and Aydilek, 2005; Edil et al., 2006; Camargo, 2008; Cetin et al., 2010; Tastan et al., 2011) and

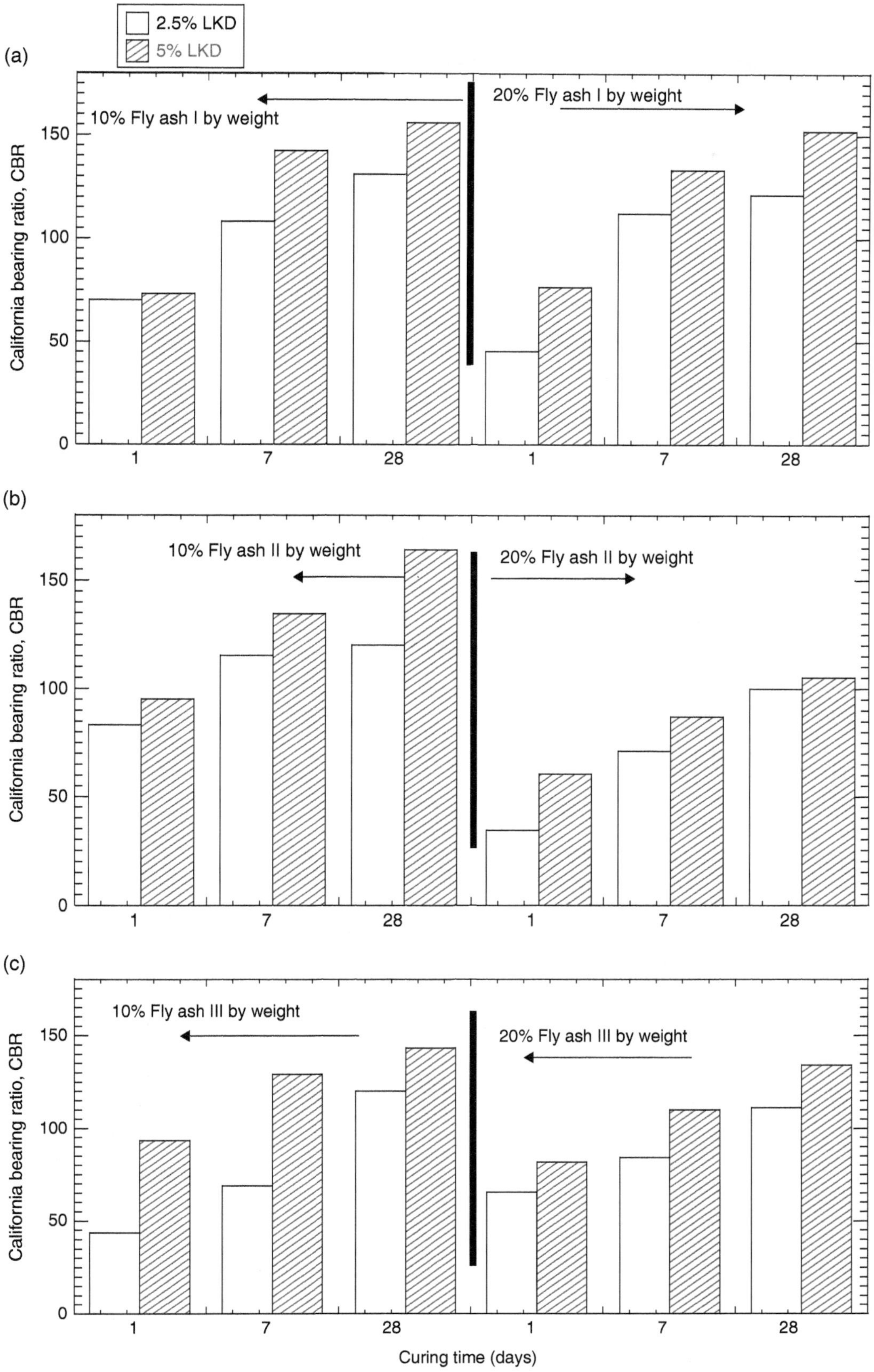

Figure 20.6 Effects of curing time on CBR of RSG mixed with (a) fly ash I, (b) fly ash II, and (c) fly ash III. *Source*: Cetin et al. (2010). Reproduced with permission of Elsevier.

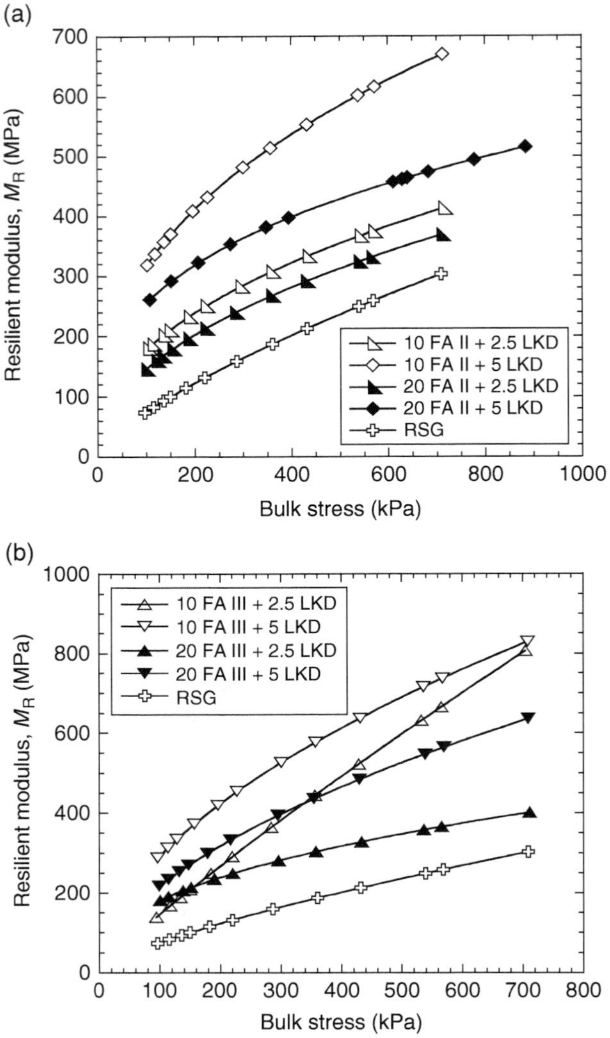

Figure 20.7 Resilient modulus of the 28-day cured specimens with varying bulk stresses: Mixtures prepared with (a) fly ash II and (b) fly ash III. Notes: FA II, fly ash II; FA III, fly ash III. *Source*: Cetin et al. (2010). Reproduced with permission of Elsevier.

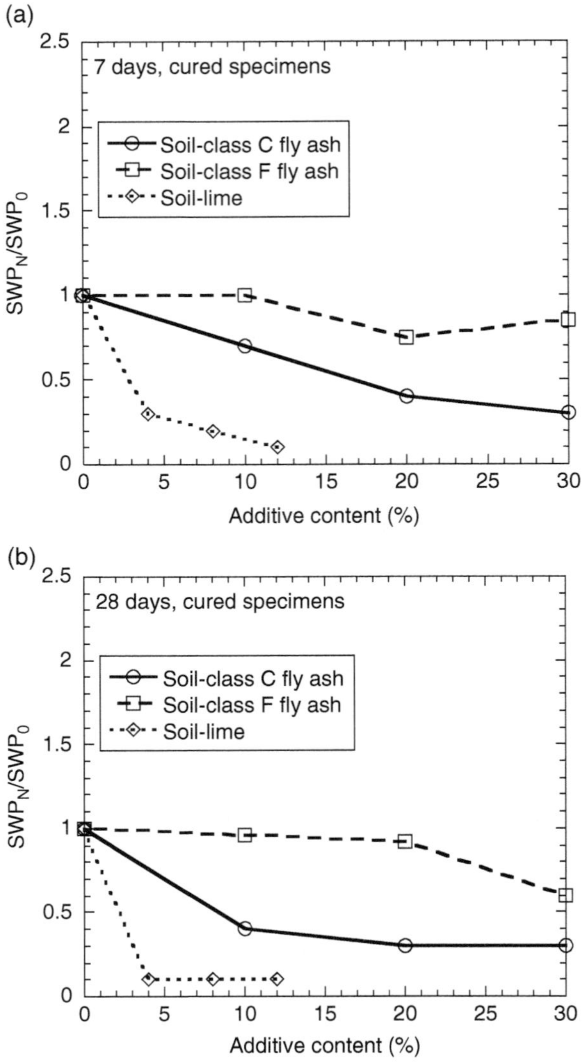

Figure 20.8 Swelling pressure ratios of clay-additive mixtures, (a) mixtures cured for 7 days, and (b) mixtures curd for 28 days. *Source*: From Dayioglu (2015).

extends the service life. However, geotechnical engineers must also measure the performance of these fly ash-stabilized materials under climatic conditions. One of the most deteriorating climatic factors is the freeze–thaw cycle process. Stabilized highway construction material should be able to resist against climatic stresses, especially freeze–thaw cycles (Cetin et al., 2010).

Pavements subjected to seasonal frost experience freezing and thawing cycles. During winter, an increase in strength and stiffness of the base and subgrade is observed. When spring comes, the base and subgrade become nearly saturated as the soils thaw and the snow and ice melt, which produce a reduction in strength and stiffness, often to values lower than prefreezing conditions. The recovery of the soil takes a long time and is partial. As a result, the weakened pavement cannot support the load for which it was originally designed and damage occurs.

Even though, there are very limited data on the engineering properties of stabilized soils with fly ash subjected to freeze–thaw cycling. Zaman and Naji (2003) focused on evaluating the effect of freeze–thaw cycles on class C fly ash-stabilized aggregate base specimens cured for 3 and 28 days. The results of this study showed that the resilient modulus of specimens cured for 3 and 28 days increased with 30 freeze–thaw cycles and 12 freeze–thaw cycles, respectively. However, a drop was observed in the resilient modulus of the 28-day cured specimens after they were exposed to freeze–thaw cycles more than 12.

Another study was performed to determine the impacts of freeze–thaw cycles on the resilient modulus of RSG material stabilized with noncementitious off-specification fly ash and LKD (Cetin et al., 2010). Soil–fly ash–LKD mixtures were prepared with 10–20% fly ashes and 2.5–5% LKD by weight as described in the previous section. All specimens were subjected to 0, 4, 8, and 12 freeze–thaw cycling and were tested for resilient modulus. It was observed that the resilient modulus of these mixtures was not affected until 4 freeze–thaw cycles (Figure 20.9). However, they lost 31–67% of their initial moduli after 12 freeze–thaw cycles. Such high changes in stiffness of these mixtures may have been observed due to the high gravel content and nonplastic nature of the coarse-grained base layer material.

A more comprehensive study to investigate the effects of freeze–thaw cycling on geotechnical engineering properties of soils was conducted by Rosa et al. (2016). In this study, various types of soils and geomaterials were stabilized with different kinds of fly ashes. These soils and geomaterials include two subgrade soils, one RSG, and two RPMs. Five different fly ashes were collected that were labeled as Columbia (class C), Dewey (off-specification, noncementitious), King (off-specification, self-cementitious), Riverside 7 (class C), and Riverside 8 (off-specification, self-cementitious) fly ash. Detailed information about the physical and chemical properties of soils, geomaterials, and fly ashes can be found in Rosa et al. (2016). Resilient modulus tests were conducted on a range of fly ash-stabilized materials after freeze–thaw cycling (0, 1, 3, 5, 10, and 12 cycles). The stabilized materials tested included fine and coarse-grained soils and recycled pavement materials. Five different fly ashes were used (Columbia and Riverside 7 [classified as class C]; Dewey, King, and Riverside 8 [classified as off-specification]) at different percentages (10, 12, 14, and 20%) and at three different water contents (7% wet of optimum to simulate wet subgrade conditions, optimum, and at field water content). Tests were also conducted on soil alone (0% fly ash) without freeze–thaw cycling to define the reference condition.

The effect of freeze–thaw cycling on resilient modulus is shown in Figure 20.10 in terms of the residual modulus, M_rN at N freeze–thaw cycles normalized by resilient modulus without freeze–thaw cycling, which is M_{ro}. For all of the mixtures, the resilient modulus decreases in response to freeze–thaw cycling and then appears to level off in approximately 1–5 cycles. The drop in modulus ranges between 7 and 50%. The average drop in resilient modulus is 28.5%. From this graph it can be concluded that for highway design, the safest way to represent the effect of freeze–thaw cycling on the resilient modulus of fly ash-stabilized pavement materials is to apply a 50% reduction.

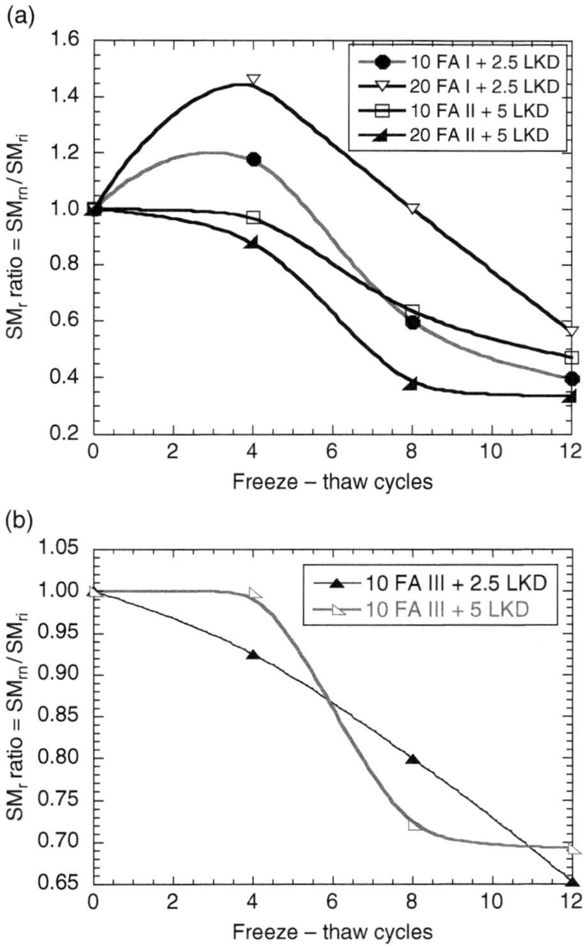

Figure 20.9 Effect of freeze and thaw cycles on SM_r values. Note: 10 FAI, 20 FAI, 10 FAII, 20 FAII, 10 FAIII, and 20 FAIII designate the specimens with 10 and 20% fly ash I, fly ash II, and fly ash III, respectively. LKD, lime kiln dust. *Source*: Cetin et al. (2010). Reproduced with permission of Elsevier.

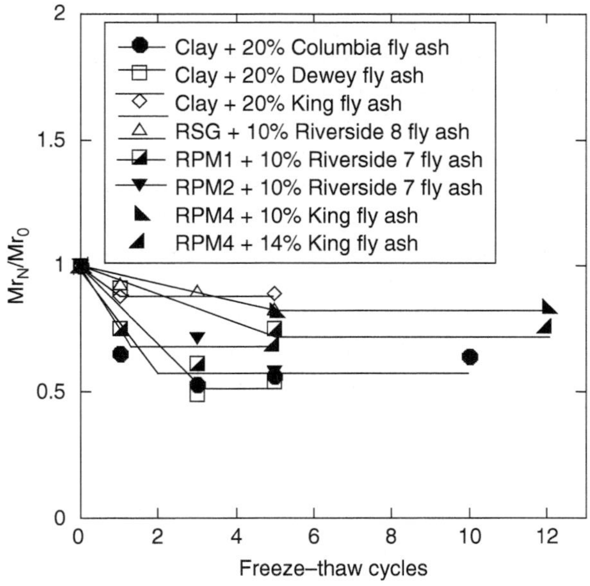

Figure 20.10 Normalized resilient modulus versus freeze–thaw cycles for all materials stabilized with fly ash. *Source*: From Rosa et al. (2016).

Similar behavior showing resilient modulus reduction after freeze–thaw cycles has been reported based on laboratory tests and field data. For example, Simonsen et al. (2002) reported resilient modulus reduction between 20 and 60% on various unstabilized coarse- and fine-grained soils after 1 freeze–thaw cycle. Culley (1971) also reported resilient modulus reduction on unstabilized clay till after 3 freeze–thaw cycles at different water contents and standard densities in a range between 41 and 66%. Lee et al. (1995) also reported resilient modulus reduction between 30 and 50% on a compacted fine-grained soil after 1 freeze–thaw cycle; the subsequent 2 freeze–thaw cycles showed insignificant effects. Also Jong et al. (1998) reported field data after winter (18 months), showing a resilient modulus reduction in the base of 35 and of 65% on the subgrade compared with the prefreezing value. In summary, based on the studies mentioned above, resilient modulus reduction of various unstabilized coarse- and fine-grained materials after 1–3 freeze–thaw cycles ranges between 20 and 66%. In this study, all fly ash stabilized showed lower resilient modulus reduction after higher number of freeze–thaw cycles (7–50%) compared with unstabilized materials, which can be attributed to the addition of self-cementing fly ash.

20.3.1.4 Environmental Suitability

20.3.1.4.1 Chemical Analysis of the Leachate

The chemical constituents of fly ashes depend mainly on composition of the coal. However, the mineralogy of the fly ashes collected from the same source can be altered based on the combustion technology used in the coal power plant (Bin-Shafique, 2002). It is known that the leaching characteristics of fly ash differ significantly between different coal combustion facilities (Torrey, 1978).

Even though fly ash is one of the best industrial by-product that can be used as a stabilizing agent or soil amendment, its potential risk to contaminate the groundwater and surface water has been questioned due to its relatively high total metal content. Fly ash contains trace metals that may have an environmental impact when it is used in geotechnical applications (McKinley et al., 2001; Bin-Shafique et al., 2006; Cetin et al., 2012, b, 2014; Komonweeraket et al., 2015).

Leachability of metals from fly ash and soil–fly ash mixtures has been studied extensively in the literature. Some of these studies go back to the 1980s. For instance, Theis et al. (1982) studied the leaching behavior of trace metals from fly ash in an aqueous system. It investigated the concentrations of certain metals (arsenic, cadmium, chromium, copper, lead, nickel, zinc) at different buffered pH of 3, 6, 9, and 12. This study summarized that leaching potential of these metals except arsenic decreased with an increase in effluent pH. Similar observations were made by Kanungo and Mohapatra (2000) and Komonweeraket et al. (2015).

Kosson et al. (1996) summarized that elements tend to leach in three main leaching patterns into the aqueous phase. These patterns are (i) cationic leaching pattern where the concentration decreases monotonically as pH increases, (ii) amphoteric leaching pattern where the concentration increases at acidic and alkaline pH, and (iii) oxyanionic pattern where concentration increases at acidic and alkaline pH. A conceptual illustration of leaching behavior of element as a function of pH is shown in Figure 20.11.

Komonweeraket et al. (2015) conducted pH-dependent small-scale laboratory water leach tests (WLTs) on soil–fly ash mixtures prepared with six different soils (sand and clayey) and seven different fly ashes (class C and off-specification). It focused on leaching of aluminum (Al), chromium (Cr), copper (Cu), iron (Fe), manganese (Mn), and zinc (Zn). This study found that leaching of all metals, except Mn, showed an amphoteric leaching pattern while Mn showed a cationic leaching pattern (Figures 20.12 and 20.13). Previous studies also found that leaching of these elements from fly ash alone tends to show amphoteric leaching patterns (Eighmy et al., 1995). It is very well known that Al is immobilized the most at neutral pH since it precipitates as $Al(OH)_3$ amorphous while its leaching is controlled by gibbsite at basic conditions, which results in dissolution of Al from this solid phase (Dijkstra et al., 2002; Dijkstra et al., 2004; Gitari et al., 2009). An increase in Al concentrations of the soil–fly ash mixtures tested in Komonweeraket et al. (2015) at pH > 9 may occur due to this chemical process.

Komonweeraket et al. (2015) determined that dissolution and precipitation of Cr(hydro)oxides have great influence on leaching potential of Cr. Figure 20.13

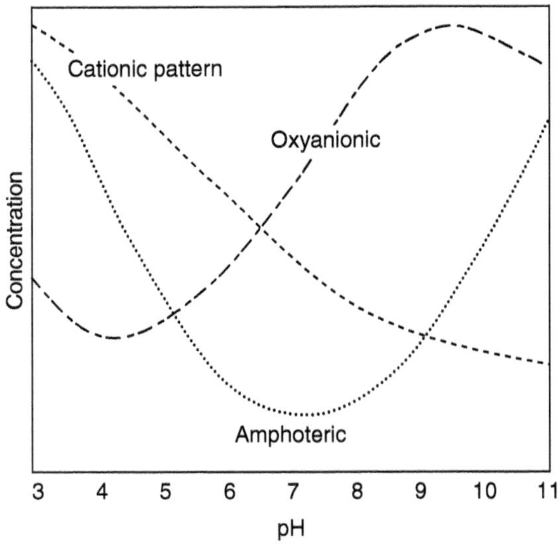

Figure 20.11 Conceptual illustration of leaching patterns as a function of pH. *Source*: Modified after Kosson et al. (1996). Reproduced with permission of Elsevier.

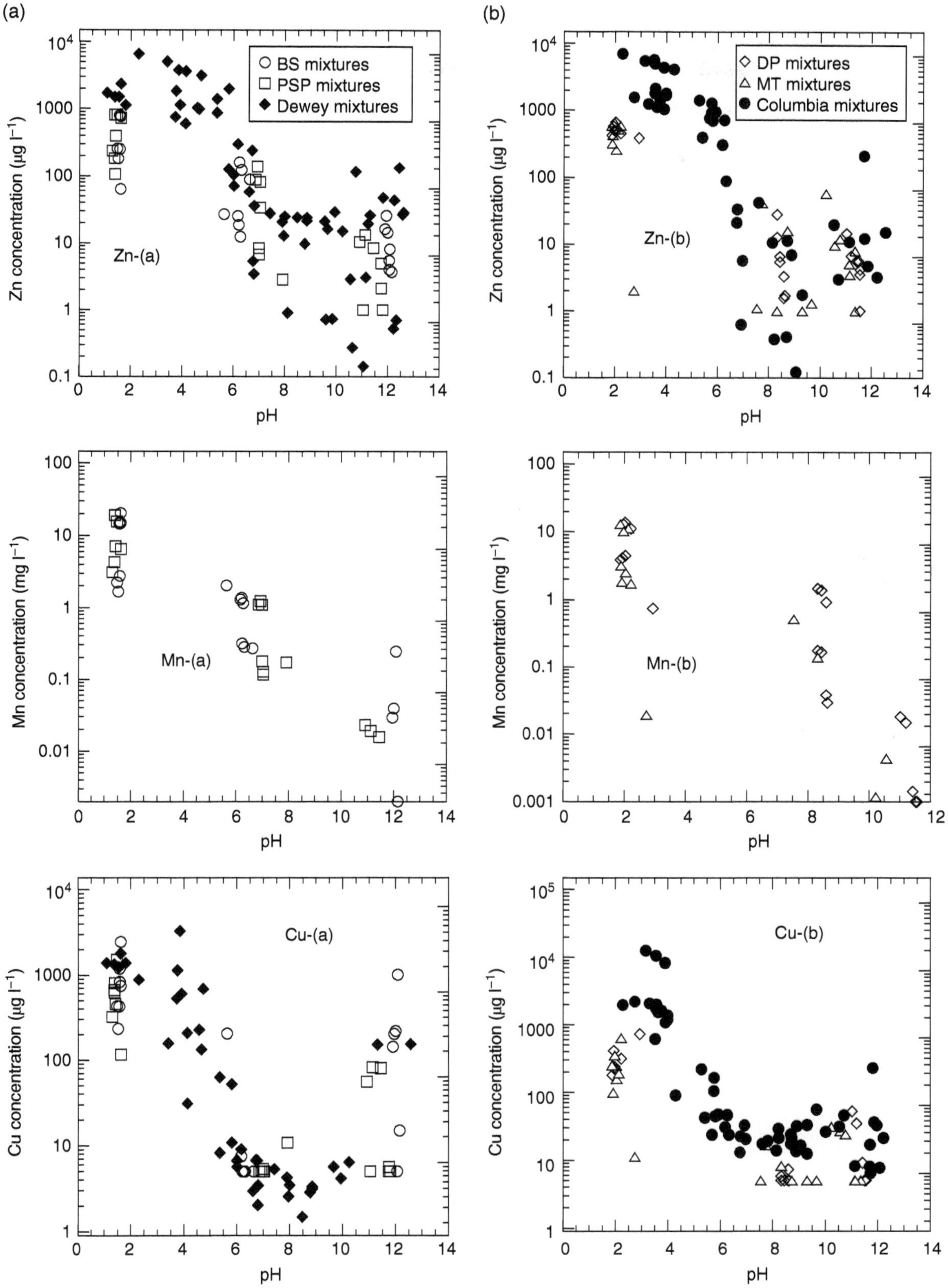

Figure 20.12 Concentrations of Zn, Mn, and Cu as a function of pH in leachates from fly ashes, soil–fly ash mixtures, and soils: (a) BS, Brandon Shores fly ash; PSP, Paul Smith precipitator fly ash; and Dewey fly ash; and (b) DP, Dickerson precipitator fly ash; MT, Morgantown fly ash; and Columbia fly ash. *Source*: Komonweeraket et al. (2015). Reproduced with permission of Elsevier.

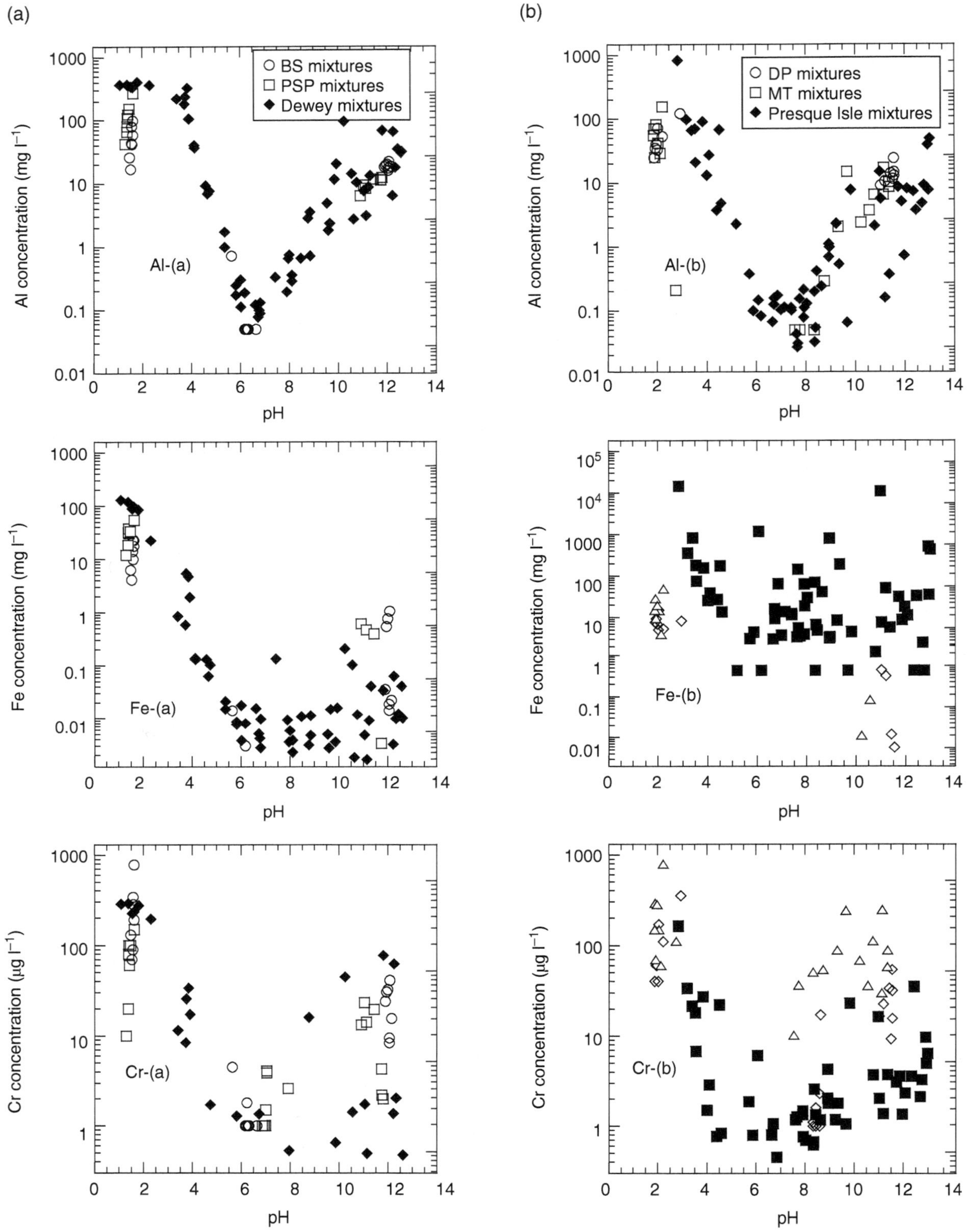

Figure 20.13 Concentrations of Al, Fe, and Cr as a function of pH in leachates from fly ashes, soil–fly ash mixtures, and soils: (a) BS, Brandon Shores fly ash; Paul Smith precipitator fly ash; and Dewey fly ash; and (b) DP, Dickerson precipitator fly ash; Morgantown fly ash; and PI, Presque Isle fly ash. *Source*: Komonweeraket et al. (2015). Reproduced with permission of Elsevier.

shows that leaching of Cr is the highest at acidic and alkaline conditions. It is speculated that leaching of Cr is controlled by the solubility of Cr carrying (hydro) oxides under these extreme pH conditions.

Fruchter et al. (1990) indicated that hematite and amorphous $Fe(OH)_3$ tend to control the leaching of Fe from fly ash samples alone. Dijkstra et al. (2002) claimed that at acidic and basic conditions, dissolution of these Fe(hydr)oxides increased and contributed on leaching of Fe. Leaching of Fe from soil–fly ash mixtures in Komonweeraket et al. (2015) was also higher at these pH conditions. Similar trend was also observed for the leaching of Cu from soil–fly ash mixtures, which experienced the highest leaching at very acidic conditions and minimum leaching at neutral pH conditions (Figure 20.12). Fruchter et al. (1990) report that leaching of Cu is solubility controlled by the dissolution/precipitation of tenorite (CuO) and $Cu(OH)_2$. Dissolution and precipitation of these solid phases are likely to influence release of Cu into aqueous solution.

The leaching of Zn is controlled by dissolution–precipitation of the oxide and hydroxide minerals such as zincite and $Zn(OH)_2$ and tends to show an amphoteric pattern (Geelhoed et al., 2002). The findings of the Komonweeraket et al. (2015) are consistent with the past studies as Zn shows an amphoteric leaching pattern for all soil–fly ash mixtures with few exceptions (Figure 20.12).

Chichester and Landsberger (1996) investigated the leaching characteristics of metals from municipal solid waste incinerator (MSWI) fly ash under dynamic conditions. This was achieved via conducting column leaching tests. Its focus was on simulating leaching of metals from a fly ash landfill. This study observed that the leached metal concentrations decreased significantly with time (at higher pore volumes of flow). Goh and Tay (1993) also conducted similar study on MSWI fly ash mixed with 5% lime by weight to investigate the impact of lime addition on the leaching of metals. Similar observations were made as reported in Chichester and Landsberger (1996) as leached metal concentrations decreased with time. Moreover, addition of lime yielded an increase in effluent pH of the MSWI fly ash and decreased concentrations of some of the certain metals (chromium and cadmium in particular) significantly.

It is very well known that in the column leaching test, there are two basic leaching patterns for fly ashes (Edil et al., 1992). These patterns are (i) first flush pattern where a high initial leaching of the metals followed by a sharp decrease to near constant concentrations after a few pore volumes of flow and (ii) the lagged flush pattern where leached metal concentrations continuously increase during the first few pore volumes of flows and start decreasing rapidly after reaching to the peak concentration. Figure 20.14 shows the illustration of these leaching patterns observed during column leach testing.

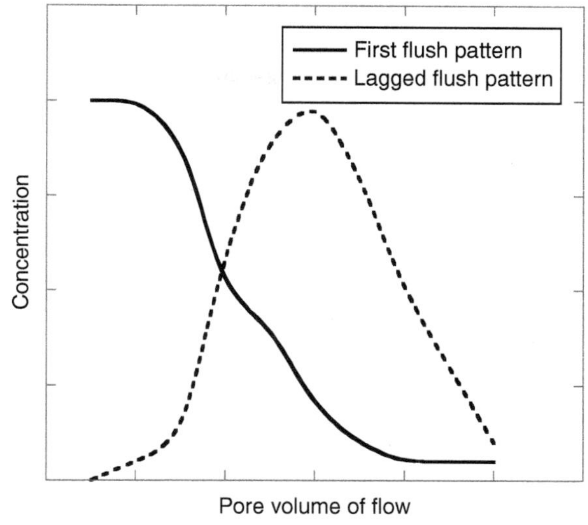

Figure 20.14 Illustrations of leaching patterns of metals from fly ashes observed during column leach testing.

Cetin et al. (2012a, b) studied the leaching behavior of aluminum, barium, boron, chromium, copper, iron, manganese, antimony, zinc, and vanadium from RSG stabilized with noncementitious off-specification fly ash and CaO-rich LKD materials. A series of column leach tests (CLTs) were conducted to investigate the leaching potential of these metals from the RSG–fly ash–LKD mixtures. Elution curves obtained from CLTs showed that metals experienced a first flush leaching pattern (Figure 20.15). Data for other metals can be found in Cetin et al. (2012a, b). It is speculated that the first flush occurs due to the release of the metals from the water-soluble fraction as well as from the sites with low adsorption energies (Bin-Shafique et al., 2006; Morar, 2008). The initially high effluent pH values of these mixtures (pH ~12) provide a possible explanation for the first flush pattern leaching of Al and Cr. At basic pHs, the availability of deprotonated (negatively charged) surface increases (Stumm and Morgan, 1996), which may have led to an increase in adsorption of cationic species and caused a decrease in the concentrations of Cr, Fe, V, Sb, and Mn into the solution.

The results of Cetin et al. (2012a, b) also indicated that an increase in LKD amount in these RSG–fly ash–LKD mixtures had great impact on effluent pH and leached metal concentrations. Increasing LKD content from 0 to 2.5% in RSG–fly ash mixtures led to a significant increase in effluent pHs. The effluent pH of the mixtures did not differ significantly between mixtures prepared with 2.5% and 55 LKD by weight (Figure 20.16). It is speculated that an increase in LKD amount increased the release of free

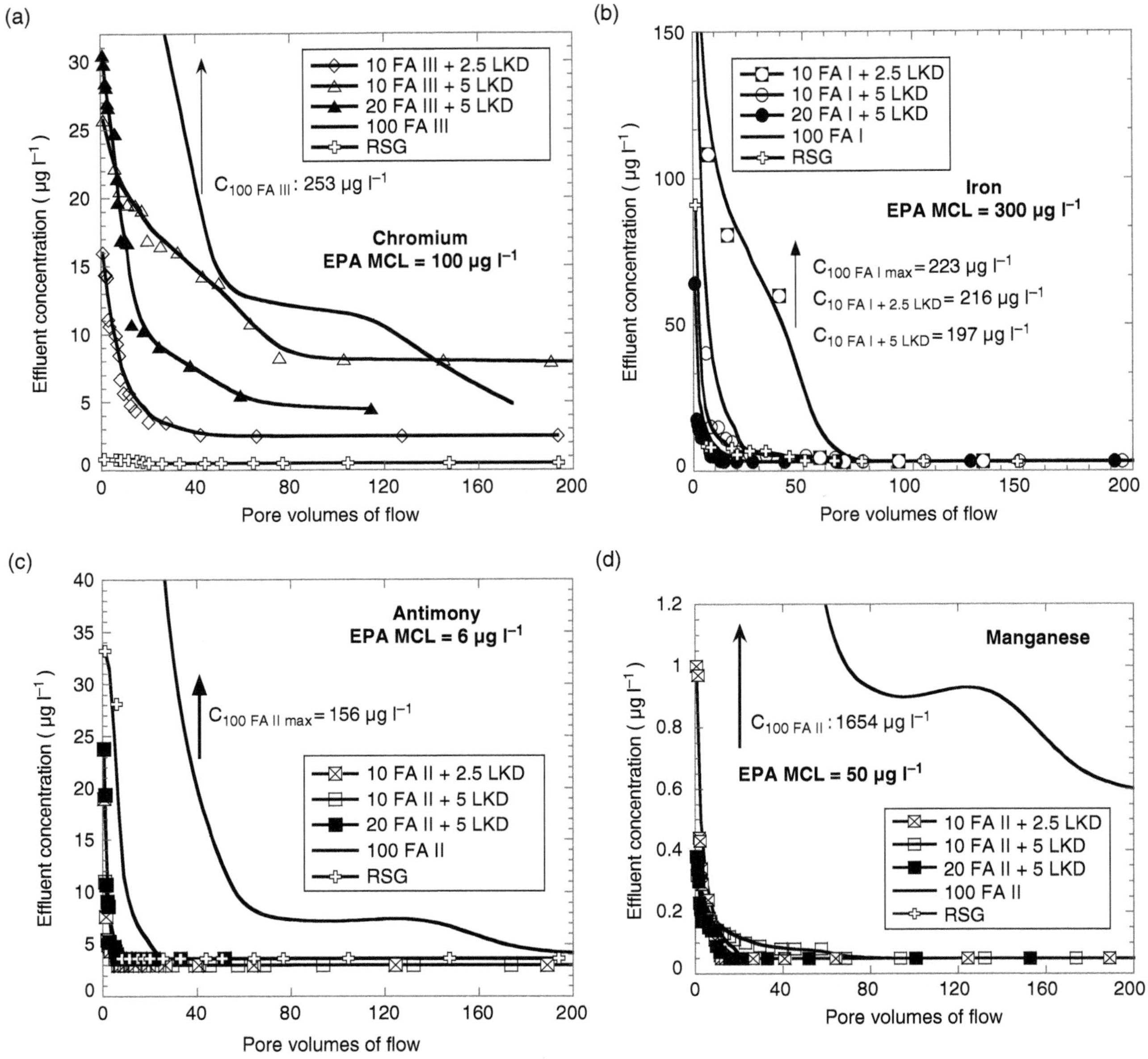

Figure 20.15 CLT elution curves for (a) chromium, (b) iron, (c) antimony, and (d) manganese. *Source*: Cetin et al. (2012a, b). Reproduced with permission of Elsevier.

lime (CaO), hydrated calcium silicate (C–S–H), and portlandite Ca $(OH)_2$, which resulted in an increase in pH values.

Table 20.5 shows the concentrations of metals (which have to comply with regulatory requirements of Environmental Protection Agency [EPA]) for RSG–fly ash–LKD mixtures compared with the US EPA maximum concentration limits for drinking waters and EPA water quality limits (WQLs) for protection of aquatic life and human health. The results of this study indicate that utilization of fly ash as a stabilizing agent with an activator (high CaO content) provides environmentally friendly sustainable geotechnical engineering applications.

Cetin and Aydilek (2013) and Cetin et al. (2014) investigated the leaching of metals from the highway embankments that were prepared with the similar soil and different noncementitious fly ashes without an activator and class C fly ash. The focus of the study was particularly on leaching of arsenic, aluminum, boron, chromium, manganese, molybdenum, selenium, and zinc from sandy soil mixed with 10, 20, and 40% different fly ashes by weight. Both studies conducted three different laboratory leaching tests, which were batch water leach tests (WLTs), toxicity characteristics leaching procedure (TCLP) tests, and column leach tests (CLTs). Laboratory results indicated that an increase in fly ash contents

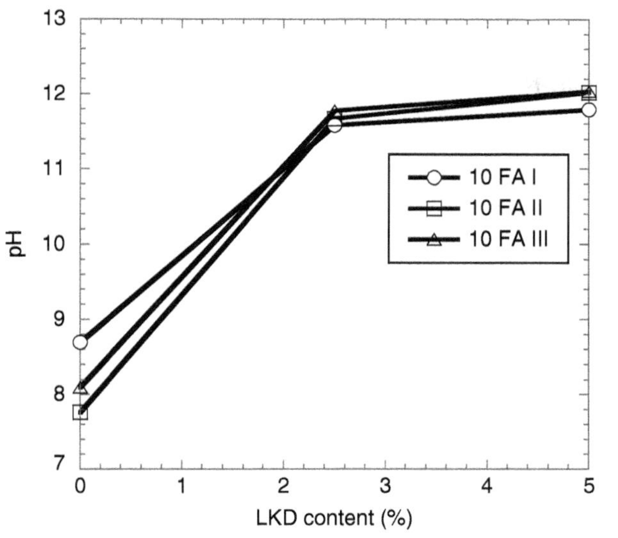

Figure 20.16 Effect of LKD content on pH of the soil mixtures. Note: 10 FA I, 10 FA II, and 10 FA III designate the specimens with 10% fly ash I, fly ash II, and fly ash III, respectively. 0% LKD content corresponds to fly ash only.

generally yield an increase in pH and leached metal concentrations regardless of the type of fly ash and metals. Some of the leached metal concentrations measured from WLTs are shown in Figure 20.17.

Figure 20.17 shows the effects of fly ash content on leaching concentrations of the Al, As, Cr, and Zn from soil–fly ash mixtures tested in Cetin et al. (2014). Results show that the leached Al concentrations increased with addition of Columbia (Co) self-cementitious fly ash and Morgantown (MT) off-specification noncementitious fly ash (Figure 20.17a). On the other hand, soil mixed with Brandon Shores (BS), Paul Smith precipitator (PSP), and Dickerson precipitator (DP) off-specification noncementitious fly ash did not experience any increase in Al concentrations with a fly ash content.

As shown in Figure 20.17b, As concentrations of soil–fly ash (PSP, MT, and DP) mixtures increase significantly when fly ash content is equal to or higher than 40% by weight. It was observed that leached As concentrations from the specimens with fly ash content higher than 40% exceeded the EPA MCL limit ($0.01\,mg\,l^{-1}$). Thus, it should be noted that the content of these types of off-specification noncementitious fly ashes in mixtures should be carefully determined to eliminate the environmental concerns caused leaching of Cr at higher concentrations with an increase in fly ash content in the mixtures (Figure 20.17d).

Komonweeraket et al. (2015) found that leached Cr concentrations for the specimens prepared with 10 and 20% off-specification noncementitious fly ashes by weight were below the detection limit indicating that these mixtures did not possess any environmental threat (Figure 20.17c). However, the soil–class C fly ash (Co) mixtures leached considerable amount Cr into the aqueous phase.

Figure 20.17d shows that leached Zn concentrations increase with BS and PSP fly ash contents, while opposite trends are observed for mixtures prepared with DP, MT,

Table 20.5 Aqueous concentrations of metals leached from RSG–fly ash–LKD mixtures.

Specimen name	Fly ash content (%)	LKD content (%)	pH	Ba ($\mu g\,l^{-1}$)	Cr ($\mu g\,l^{-1}$)	Fe ($mg\,l^{-1}$)	Cu ($\mu g\,l^{-1}$)	Zn ($\mu g\,l^{-1}$)
100 FA I	100	—	8.56	180	9	12	2.1	11
10 FA I + 2.5 LKD	10	2.5	11.5	112	6.3	4.7	1.9	21
10 FA I + 5 LKD	10	5	11.8	69	6.6	4.3	1.7	42
20 FA I + 5 LKD	20	5	11.9	162	8.7	3.5	1.8	31
100 FA II	100	—	7.53	189	4.9	14	2.8	15
10 FA II + 2.5 LKD	10	2.5	11.7	123	3.8	6.9	1.1	20
10 FA II + 5 LKD	10	5	12	61	3.9	5.5	0.13	43
20 FA II + 5 LKD	20	5	12	135	4.2	3.7	1.1	13
100 FA III	100	—	8.2	247	7.7	17	1.7	17
10 FA III + 2.5 LKD	10	2.5	11.8	151	4.4	13	1.4	18
10 FA III + 5 LKD	10	5	12	143	4.5	7.6	1.2	31
20 FA III + 5 LKD	20	5	12.1	197	6	4.4	1.4	10
RSG	—	—	4.8	56	0.6	3.2	1.1	45
US EPA MCL				2000	100	300	1300	5000
US EPA WQL				NA	570	NA	NA	NA

Source: Cetin et al. (2012a, b). Reproduced with permission of Elsevier.

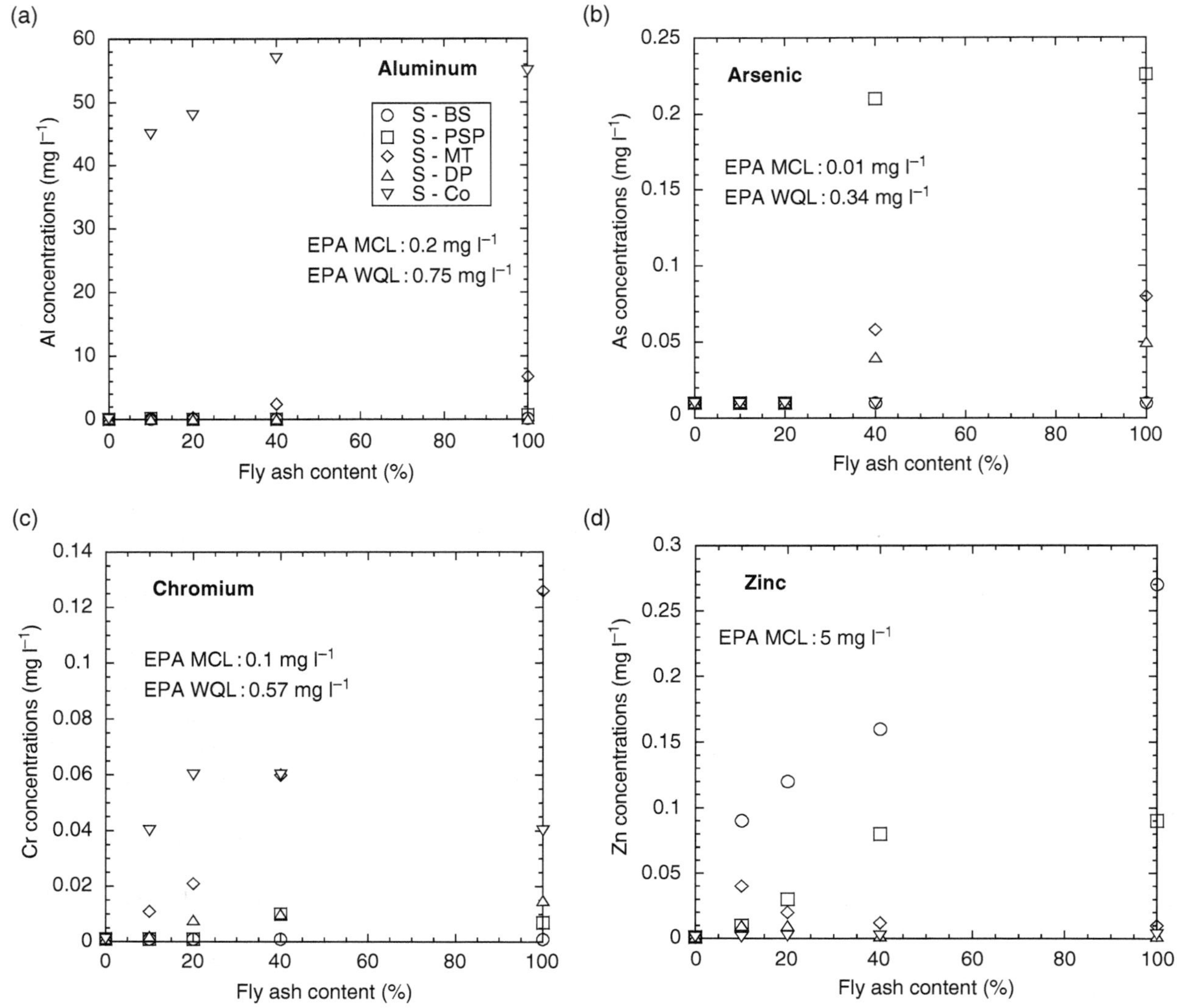

Figure 20.17 Al, As, Cr, and Zn concentrations in the WLT effluents. Notes: BS, Brandon Shores; PSP, Paul Smith precipitator; MT, Morgantown; DP, Dickerson precipitator; Co, Columbia fly ashes. *Source*: From Cetin et al. (2014).

and Co fly ashes. Komonweeraket et al. (2015) explained that effluent pH of soil–BS and soil–PSP mixtures were around 6–7.5, whereas the effluent pH of other soil–fly ash mixtures was very alkaline ranging from 9 to 12.2. It is very well known that Zn tends to show cationic leaching pattern in some cases, and its release to the aqueous solutions decreases at basic conditions. Thus, it was speculated that effluent pH had the greatest influence on the leaching of Zn from the soil–fly ash mixtures studied in Komonweeraket et al. (2015). Therefore, the mixtures prepared with DP, MT, and Co fly ashes leached lower Zn concentrations with an increase in fly ash content due to an increase in effluent pH.

20.3.1.4.2 Groundwater Numerical Modeling of Leached Metals

Based on the results of the laboratory leaching tests that have specifically investigated the leaching of metals from soil–fly ash and soil–fly ash activator mixtures, it can be indicated that the use of fly ash materials as a stabilizing agent and soil amendment provides sustainable and environmentally friendly geotechnical engineering designs. However, laboratory leaching tests are conducted under controlled environment and other processes (advection, dispersion, diffusion, biological degradation) that exist in the actual field conditions and are not incorporated during tests.

There have been many numerical models developed to simulate the leaching of metals from highway sublayers stabilized with waste/recycled materials. The following programs are the most common models that are used to simulate leaching of metals and their impacts to the groundwater for the field applications by using the data collected from the laboratory tests. These models are STUWMPP, IMPACT, and HYDRUS-2D. The US EPA Industrial Waste Management Evaluation Model

(IWEM) may also be used to evaluate leaching from pavement systems. IWEM was initially developed to simulate the leaching of metals in landfills, piles, and impoundments. However, modules specific to pavement applications are currently being developed for this model. IWEM initially developed STUWMPP model, which employs dilution–attenuation factors obtained from the Seasonal Soil Compartment (SESOIL) model to relate leaching concentrations from soils and byproducts to concentrations in underlying groundwater. SESOIL simulates the advection, diffusion, volatilization, cation exchange, and hydrolysis processes in 1D vertical transport code in the soil and groundwater. IMPACT model was specifically developed to evaluate the environmental impact of highways built with waste materials. It employs two-dimensional (2D) flow and solutes transport in the code to solve the advection–dispersion–reaction equation (ADRE). It applies the finite difference method for analyses. The following assumptions were made in IMPACT: (i) flow is assumed steady in a saturated porous media and (ii) sorption is assumed to be instantaneous and reversible. It should be noted that STUWMPP, IMPACT, and IWEM are available in the public domain.

HYDRUS-2D is a widely used software package for simulating water and solute movement in 2D variably saturated media. HYDRUS-2D uses the finite element method to solve Richards' equation for 2D variably saturated water flow and the advection–dispersion equation for solute transport (Bin-Shafique, 2002). The code has to be modified to be used to simulate the leaching concentrations of metals from highway constructions. However, it is not in the public domain.

There is a public domain numerical model; WiscLEACH was specifically developed to simulate leached metal concentrations from highway base/subbase/subgrade layers and highway embankments designed with recycled waste materials (Li et al., 2006, 2011). It combines three analytical solutions of the ADRE. WiscLEACH assumes all materials in the profile are homogeneous and isotropic. Precipitation falling on the pavement surface, the shoulders, and the surrounding ground infiltrates into the ground surface or is shed as runoff. Transport in the vadose zone beneath the fly ash layer is assumed to follow the ADRE for 1D steady-state vertical flow with 2D dispersion and linear, instantaneous, and reversible sorption. Trace elements that reach the groundwater table are transported horizontally and vertically, although the flow of groundwater is assumed to occur predominantly in the horizontal direction. Steady saturated groundwater flow is assumed, and transport in groundwater is assumed to follow the ADRE with instantaneous, reversible, and linear sorption. Chemical and biological reactions that may consume or transform trace elements are assumed to be absent. In addition, flow in the fly ash and subgrade is assumed to occur only in the vertical direction. Steady 1D unit gradient flow is assumed in the pavement layers and the vadose zone, with the net infiltration rate controlled by the least conductive layer in the profile and the annual precipitation rate. Transverse flow on top of the subgrade toward the edge of the road structure is ignored (Cetin et al., 2012). Users must put the following input data to run WiscLEACH. Input data include physical and design parameters of the pavement (thickness of asphalt layer, base/subbase layer, depth of groundwater, pavement thickness, shoulder width, point of compliance, hydraulic conductivities of each layer), annual precipitation rate, retardation factor of metals for each layer, diffusion coefficient of a metal, leached metal concentrations, effective porosity of each pavement layers, and hydraulic conductivity of the aquifer material. Figure 20.18 shows the schematic diagram of WiscLEACH developed for the highway constructions.

WiscLEACH has the following advantages over other models mentioned previously: (i) it is user friendly that can be used by engineers and regulators who do not have expertise in numerical modeling, (ii) it executes fast, (iii) it uses readily available data (pavement design parameters, climatic data bases, conventional laboratory tests data), (iv) it predicts metal concentrations in the vadose zone and groundwater, and (v) its predictions are more conservative than the ones predicted via other models.

Cetin et al. (2012b) used WiscLEACH to predict metal concentrations at different locations and years in the soil vadose zone and groundwater. Detailed information about the input used for the model can be found in Cetin et al. (2012). Figure 20.19 shows the contour plots for the predicted concentrations of Cu. The contour plots provide the predictions of the metal concentrations after 1, 2, 4, and 8 years of construction. Results show that metal concentrations decreased significantly with time and distance from the fly ash-stabilized layer surface most probably due to the dispersion of the metals in the soil vadose zone. The concentrations of Cu metals even after 1 year are much lower than peak Zn concentrations obtained from CLTs, suggesting conservative concentration measurements in CLTs. The metal concentrations mostly were adsorbed in the soil vadose zone before reaching to the groundwater. High retardation factors of subgrade would be increasing the rate of adsorption of metals before reaching to the groundwater. WiscLEACH results indicated that the metal concentrations of metals were much lower than the metal concentrations obtained from the CLTs, suggesting that the results of laboratory tests are likely to provide a conservative estimate of field metal leaching. Furthermore, it was observed that predicted leached metal concentrations even after 1 year of simulation were lower than the EPA limits. These results indicate that the use of fly ashes in such geotechnical engineering application provides environmentally friendly and sustainable design.

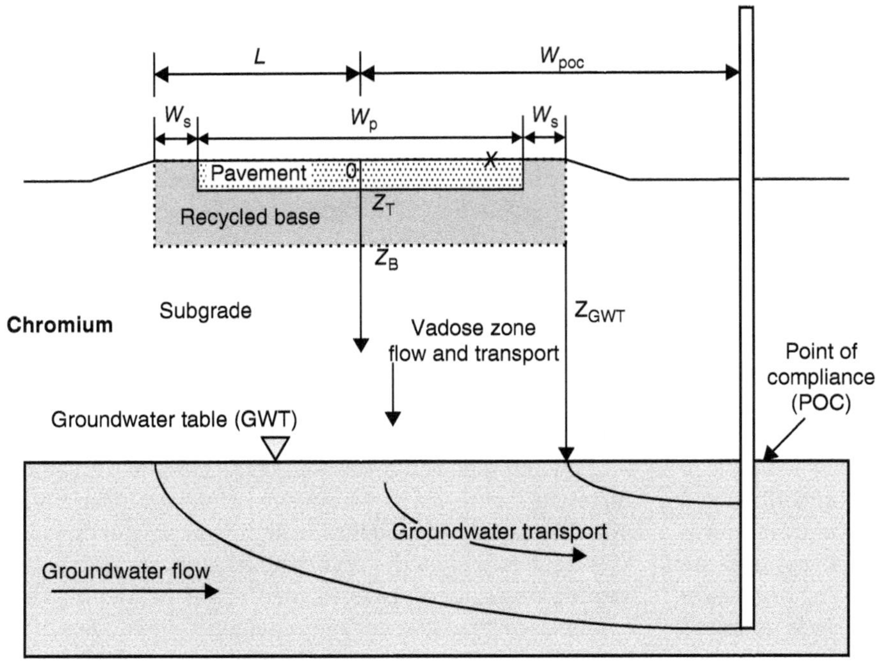

Figure 20.18 Conceptual model in WiscLEACH for predicting impacts to the vadose zone and groundwater from fly ash-stabilized highway base layer. Notes: W_s, shoulder width; W_p, pavement width; POC, point of compliance; W_{poc}, distance between the center of the pavement and point of compliance; Z_T, thickness of asphalt layer; Z_b, thickness of the entire pavement; GWT, groundwater table; and Z_{GWT}, depth of groundwater table.

20.3.2 Recycled Concrete Aggregates

20.3.2.1 Application in Pavement Sublayer Designs

RCA can be used as an alternative highway subbase/base layer material. As one of the most common recycled construction materials, RCA used as base/subbase course is widespread in the United States. Forty-one states recycle concrete for use as aggregate and 38 states recycle concrete as an aggregate base (FHWA, 2004).

20.3.2.2 Geotechnical Performance

There have been numerous studies on testing the geotechnical performance of RCA materials as highway base/subbase materials. Edil (2012) compared the resilient modulus (M_r) of RCA and natural aggregates for highway base design. RCA materials yielded 1.1–3.2 times higher summary resilient moduli (SM_r) as compared with natural aggregates at a bulk stress of 210 kPa. Bennert et al. (2000) mixed RCA materials with natural aggregates at 25, 50, and 75% by weight and reported improvement in stiffness of natural aggregates with RCA inclusion into the matrix.

Bestgen et al. (2016a) also investigated the strength, stiffness of two RCA (RCA1 and RCA2) materials, and the mixtures of these two RCA materials with four conventional graded aggregate base (GAB) materials (GAB1, GAB2, GAB3, and GAB4). It conducted a series of CBR, resilient modulus and permanent deformation tests. Each test specimen was compacted at their corresponding optimum moisture content by applying modified proctor energy. CBR values of both RCA materials used in this study were higher than that of four GAB materials (Figure 20.20). It was also observed that CBR of the RCA–GAB mixtures did not increase consistently with an increase in RCA content (Figure 20.20). Same inconsistency trends were also observed for SM_r of RCA–GAB mixtures (Figure 20.21). Average SM_r of RCA materials are higher than that of GAB materials. RCA particles possess higher angularity than would normally be found in natural unbound granular aggregates (Edil, 2012). This results in a lower particle density in the RCA matrix, and it could also be the reason for two RCAs to yield higher SM_r values than the GABs. This finding is consistent with the previous literature, which indicates that the stiffness of granular materials tends to increase with angularity of the particles (Stolle et al., 2009). Furthermore, residual mortar and cement paste on the surface of RCA particles provide rougher surface texture, which can be another reason for observing higher SM_r values for RCA (Kuo et al., 2002; FHWA, 2008). Figure 20.21 shows that SM_r of RCA–GAB mixtures did not increase consistently with RCA content as observed in CBR results. This was contradictory with the results of Bennert et al. (2000), which found that stiffness of RCA–GAB mixtures increased with higher RCA content. It is speculated that rehydration of the cement phases in RCA may have been interrupted when RCA is mixed with GAB materials in the current study. Edil (2012) also suggested that SM_r of granular materials may decrease with an increase in absorption. The absorptions of the RCA materials were significantly higher than that of the GAB materials in the study conducted by Bestgen et al. (2016). Therefore, inclusion of RCAs into GAB matrix may have

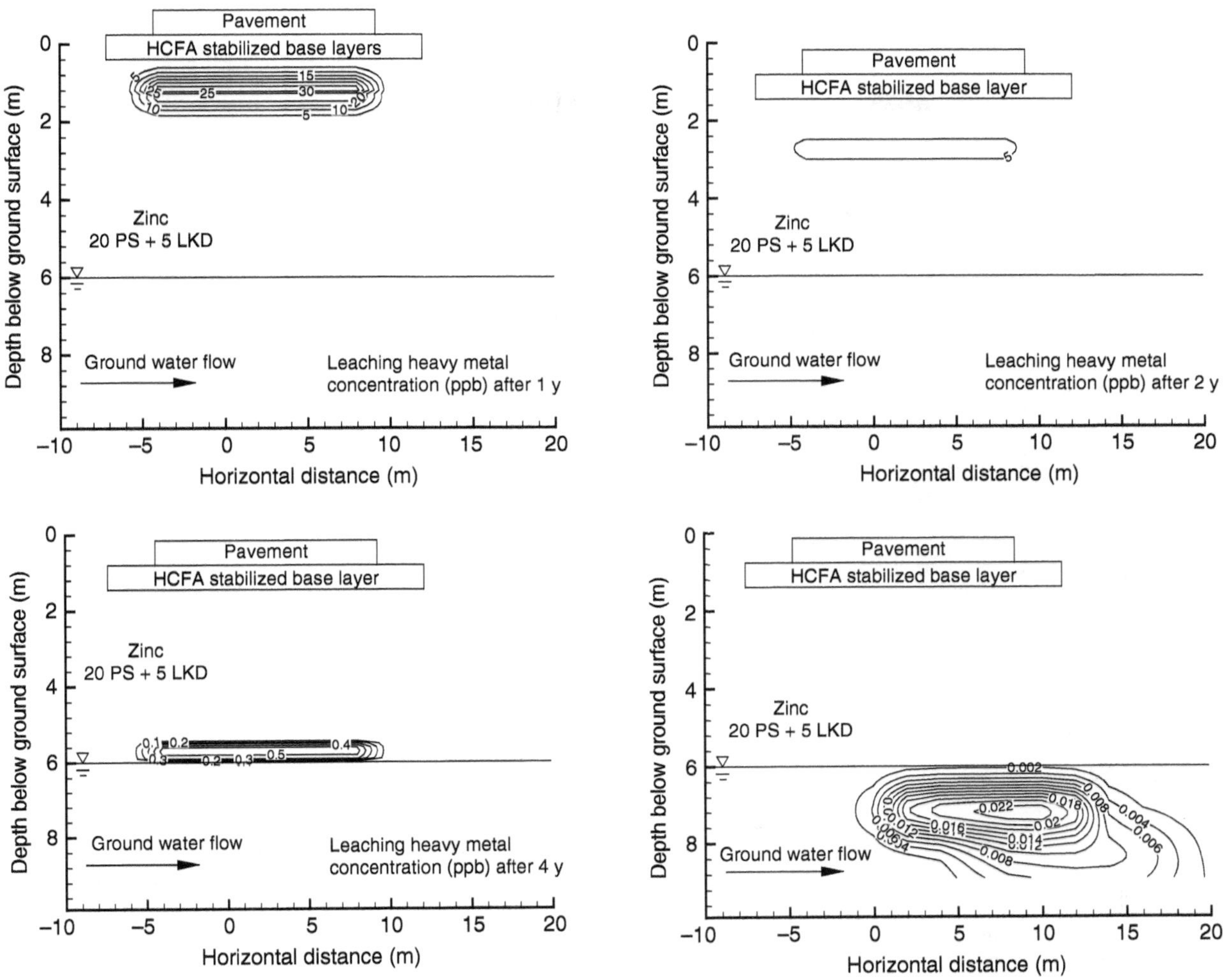

Figure 20.19 Predicted Zn concentrations in vadose zone and groundwater. Note: 20 PS + 5 LKD designate the specimens with 20% Paul Smith fly ash and 5% lime kiln dust by weight.

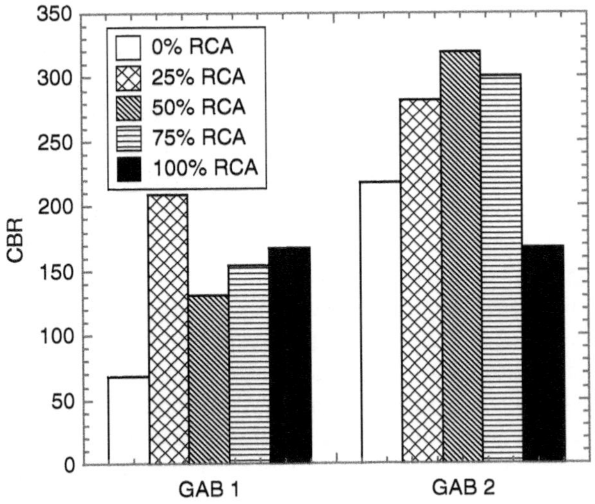

Figure 20.20 CBR of RCA and GAB materials alone and RCA–GAB mixtures. *Source*: From Bestgen et al. (2016a).

yielded a significant increase in the absorption of the RCA–GAB mixtures and resulted in a reduction in SM_r.

20.3.2.3 Durability

Geotechnical performance of RCA materials under climatic conditions has been studied for a long time. It is very important to determine the impact freeze–thaw cycles on strength and stiffness of RCA materials. Overall, it was observed that the strength and stiffness of RCA materials increased with an increase in freeze–thaw cycles (Edil, 2012; Bestgen et al., 2016a). Rosa et al. (2017) investigated the effects of freeze–thaw cycles on geomaterials (RCAs, reclaimed pavement materials, RAP materials) and soils stabilized with different fly ashes.

Figure 20.22 shows the effect of freeze–thaw cycling on the SM_r of RCA materials studied in Rosa et al. (2017). SM_r of RCA materials decreased after 5 freeze–thaw cycling. Thereafter, a consistent increase was observed in SM_r of RCA with an increase in freeze–thaw cycling.

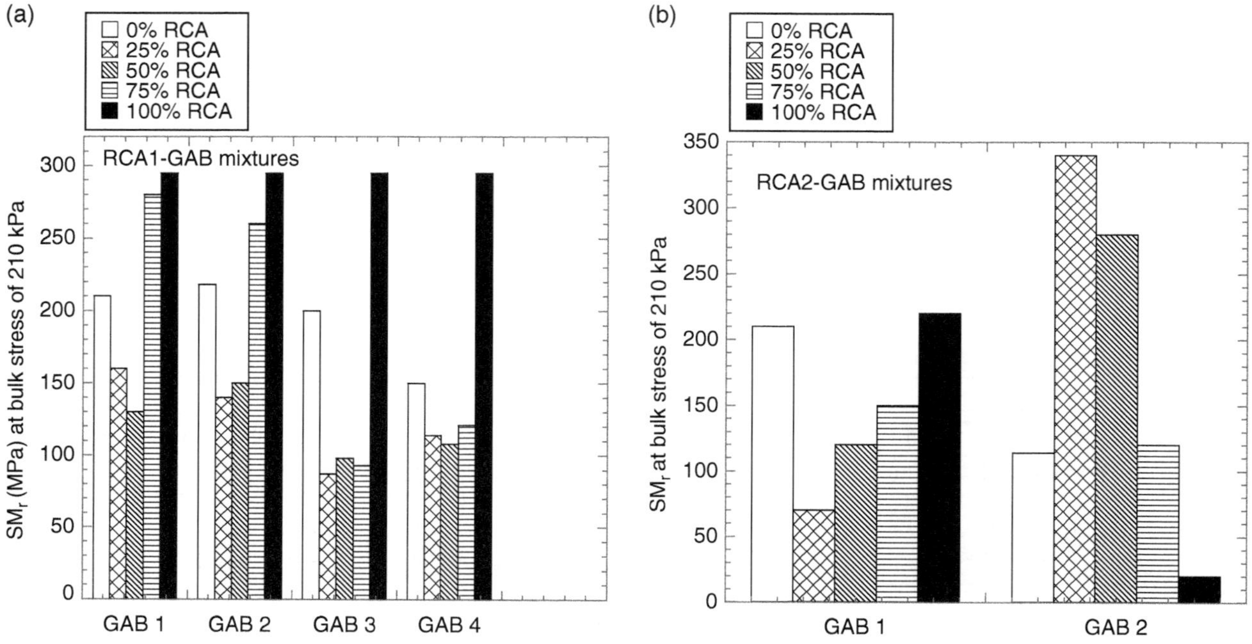

Figure 20.21 SM_r of RCA and GAB materials alone and RCA–GAB mixtures from Bestgen et al. (2016a), SM_r values calculated at a bulk stress of 208 kPa. *Source*: From Bestgen et al. (2016a).

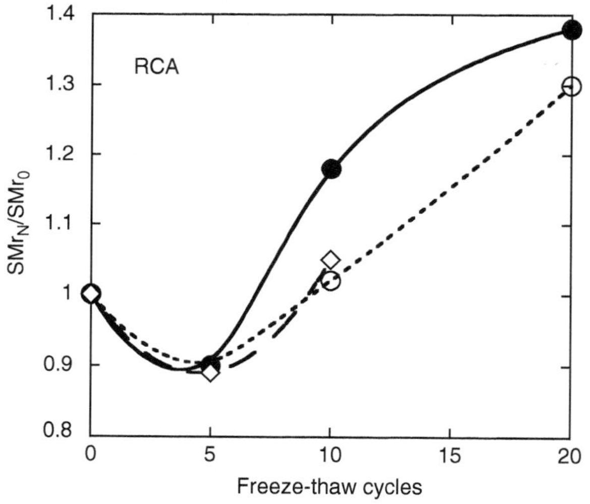

Figure 20.22 Normalized resilient modulus versus freeze–thaw cycles of RCA (recycled concrete aggregate). *Source*: From Bestgen et al. (2016).

The self-cementing properties of RCA and fines content generation over time could explain why an increase in stiffness after 5 freeze–thaw cycles occurred. These trends are consistent with other research in which the strength of subbase prepared with RCA has been found to increase with time (Arm, 2001). An increase in SM_r with time was observed and attributed to the self-cementing properties of RCA. Bestgen et al. (2016a, b) claimed that the increase in M_r of RCA materials was due to the self-cementing properties of demolished concrete.

20.3.2.4 Environmental Suitability

Previous literature showed that RCA materials possessed superior geotechnical performance (higher strength and stiffness, better resistance against climatic conditions) over conventional GAB materials. However, there are concerns about the environmental impact of RCAs used in highway base/subbase applications (Engelsen et al., 2009, 2010; Chen et al., 2013; Bestgen et al., 2016).

Past research of RCA showed high pH and concentrations of certain metals in its leachate. Engelsen et al. (2009) showed the pH of RCA to be around 12. This pH exceeds the drinking water quality standards for pH (6.5–8.5) (US EPA, 2009), indicating RCA may pose negative environmental effects. At high pH values, there is greater potential of certain metals, such as chromium and arsenic, to be released into the aqueous phase (Dijkstra et al., 2006; Engelsen et al., 2010; Chen et al., 2013; Komonweeraket et al., 2015). At the material pH of RCA, the results of the study by Engelsen et al. (2009, 2010) indicate Al, Cr, and S exceeded the US EPA drinking water maximum contaminant levels (MCL) or national secondary MCL (SMCL) for these elements that are 0.05–0.2, 0.1, and 250 mg l^{-1}, respectively. Note, MCL is for sulfate (SO_4^{2-}) not elemental S; sulfur would be present as SO_4^{2-} because the RCA leachate would most likely be in an oxidizing environment. Cr concentrations near the material pH of RCA in a study by Chen et al. (2012) also exceeded the EPA MCL. These studies show RCA leachate to possess critical concentrations of certain toxic metals.

Leaching is the process where solid phase constituents are released into the contacting aqueous phase (Van Gerven et al., 2007). Leaching is affected by physical and chemical properties of the material, leaching environment, and chemical interactions of dissolved constituents (Komonweeraket et al., 2015). Physical and chemical properties of RCA that are expected to influence leaching are particle size, carbonation, and the presence of cementitious materials. Factors related to variables in the leaching environment include liquid-to-solid ratio and extract pH (Kosson et al., 2002).

Important chemical and physical properties of RCA affecting leaching are related to particle size and "age" of the material. Fines contain more cement paste that largely controls the leaching of RCA (Engelsen et al., 2009). Previous literature determined fine particles have a tendency of possessing lower pH than coarse (Chen et al., 2012), and the pH of "aged" RCA is lower than "fresh," recently crushed concrete (Engelsen et al., 2010). Aged concrete likely contains more fines due to increased chemical and physical weathering of the material. Engelsen et al. (2010) also showed most trace metals leach more in aged RCA than in fresh RCA.

Bestgen et al. (2016b) studied the environmental suitability of RCA materials and effects of factors on leaching of metals to the groundwater and surrounding soils. It focused leaching of calcium, chromium, copper, iron, and zinc in particular. Factors investigated in this study included (i) curing time, (ii) particle size, and (iii) pH. Batch WLTs were conducted on five different RCA materials.

RCA materials were cured for 1, 7, and 28 days. Figure 20.23 shows the effluent pH of RCAs is very basic and higher than 10 regardless of the curing time that they are subjected to. Bestgen et al. (2016) indicated that

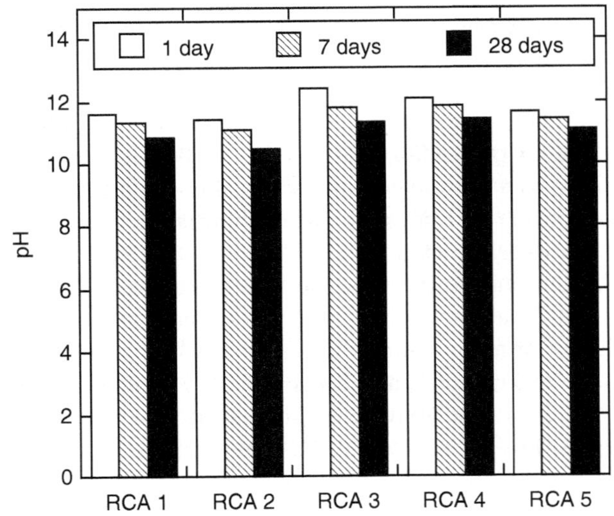

Figure 20.23 Effluent pH of RCA materials at different curing times. *Source*: From Bestgen et al. (2016b).

the leached metal concentrations were not affected by the total metal content of RCA materials. Based on the results of this study, it was observed that leachate pH and leached calcium and copper concentrations decreased when RCAs were cured for higher curing time (Figure 20.24). Such trend was not observed for the leached chromium and iron concentrations. Figure 20.24 shows that chromium and iron concentrations generally showed an initial increase then slightly stabilized. It is speculated that cement particles remaining in the RCA materials may be reactivated during longer curing periods, which may eventually cause encapsulation of particles and decrease the leaching concentrations of metals.

Important chemical and physical properties of RCA affecting leaching are related to particle size of the material. Fine particles may contain more cement paste that largely controls the leaching of RCA (Engelsen et al., 2009). The particle sizes and shape of RCAs are dependent on the crusher used in the plant or on-site, which causes different gradation of RCA when these materials are used in highway construction. Therefore, it is important to determine the impacts of particle size of RCA on the leaching of metals. The particle size fractions tested in the current study were 9.5–4.75 mm, 4.75–2.36 mm, 2.36–1.19 mm, 1.19–0.6 mm, 0.6–0.3 mm, 0.3–0.15 mm, 0.15–0.105 mm, 0.105–0.075 mm, and less than 0.075 mm.

Leaching from different particle sizes in RCA was studied for nine different fractions. pH was relatively stable in each particle size fraction, indicating the pH of the leachate is more dependent on the RCA material itself rather than particle size. Figure 20.25 shows the concentrations of Cu and Cr metals generally decrease with particle size, while no consistent relationship was found between the particle sizes and the leaching of Ca and Fe.

One of the mechanisms affecting leaching by particle size is the cement composition within each fraction. Engelsen et al. (2010) claimed that the leaching of trace metals is mostly due to the cement paste content, and it is speculated that there may be more cement paste in certain fractions than others, which may yield the results obtained from the current study.

Bestgen et al. (2016b) also conducted pH-dependent leaching tests to investigate impact of pH on leaching behavior of metals from RCA materials. pH of the RCA leachate was adjusted from 2 to 13. Figure 20.26 shows that leaching of calcium and iron follows a cationic leaching pattern where the leached Ca and Fe concentrations decrease consistently with an increase in pH. On the other hand, the leaching of chromium, copper, and zinc shows amphoteric leaching pattern where the leaching of these metals are minimum at neutral pH and maximum at extreme acidic and basic pHs. These results are consistent with the findings of previous studies (Garrabrants et al., 2004; Sanchez et al., 2009; Chen et al., 2012).

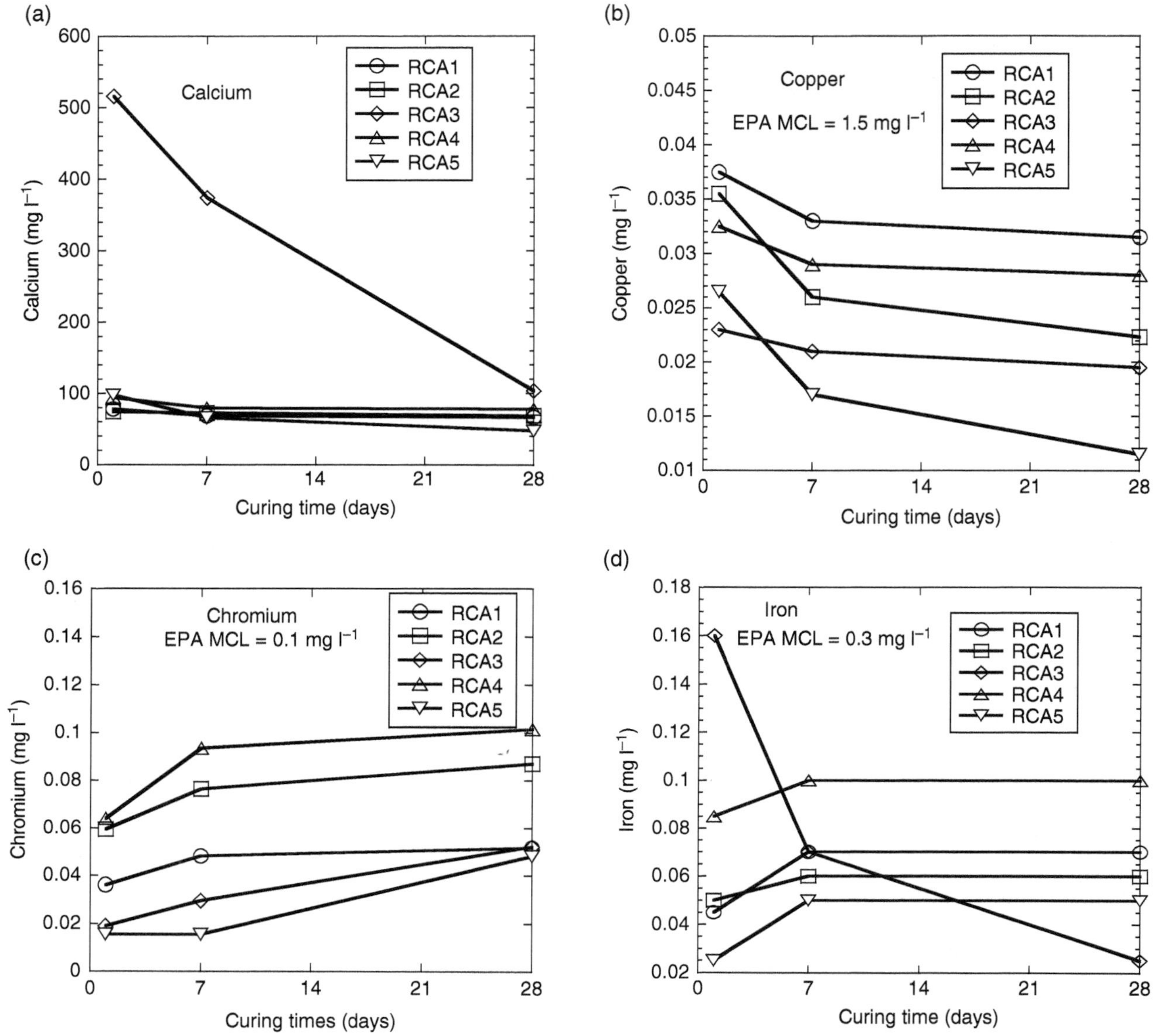

Figure 20.24 The effects of curing time on leaching of (a) calcium, (b) copper, (c) chromium, and (d) iron from RCAs. *Source*: From Bestgen et al. (2016).

EPA drinking water standards were exceeded in the acidic leaching range for Cr, Cu, and Fe. Cr concentrations exceeded the EPA MCL ($100\,\mu g\,l^{-1}$) for all RCA materials after a pH of 5. Cu exceeded the SMCL ($1\,\mu g\,l^{-1}$) at pH < 2 and Fe exceeded the EPA SMCL ($300\,\mu g\,l^{-1}$) at pH < 7. A normal field environment would have neutral pH conditions, so it is unlikely that the leaching of Cr and Cu would cause health issues. Fe concentrations may exceed the SMCL in the environment, but this optional federal standard is in place for improved taste in drinking water (US EPA, 2009), and Fe leaching from RCA would likely not be detrimental to the environment.

Leaching of major elements from wastes, under equilibrium conditions, is dominated by solubility processes (Kosson et al., 2002). Thus, the major elements studied, Ca and Fe, were likely solubility controlled. For Ca, increasing the solution pH likely caused Ca^{2+} to precipitate as a carbonate, and decreasing the pH caused dissolution of this solid (Garrabrants et al., 2004; Komonweeraket et al., 2015).

Acidic conditions caused the greatest release of Cr, Cu, Fe, and Zn as shown in Figure 20.26 due to greater dissolution of metal-bearing minerals. The pH-dependent release of Cr, Cu, Fe, and Zn is solubility controlled and likely follows the solubility of the oxide and hydroxide species for each element (Garrabrants et al., 2004; Engelsen et al., 2010), such as Cr_2O_3/$Cr(OH)_3$, CuO/$Cu(OH)_2$/Fe_2O_3/$Fe(OH)_3$, and ZnO/$Zn(OH)_2$. Cu may also be controlled by the dissolution/precipitation of carbonate species (Komonweeraket et al., 2015). The

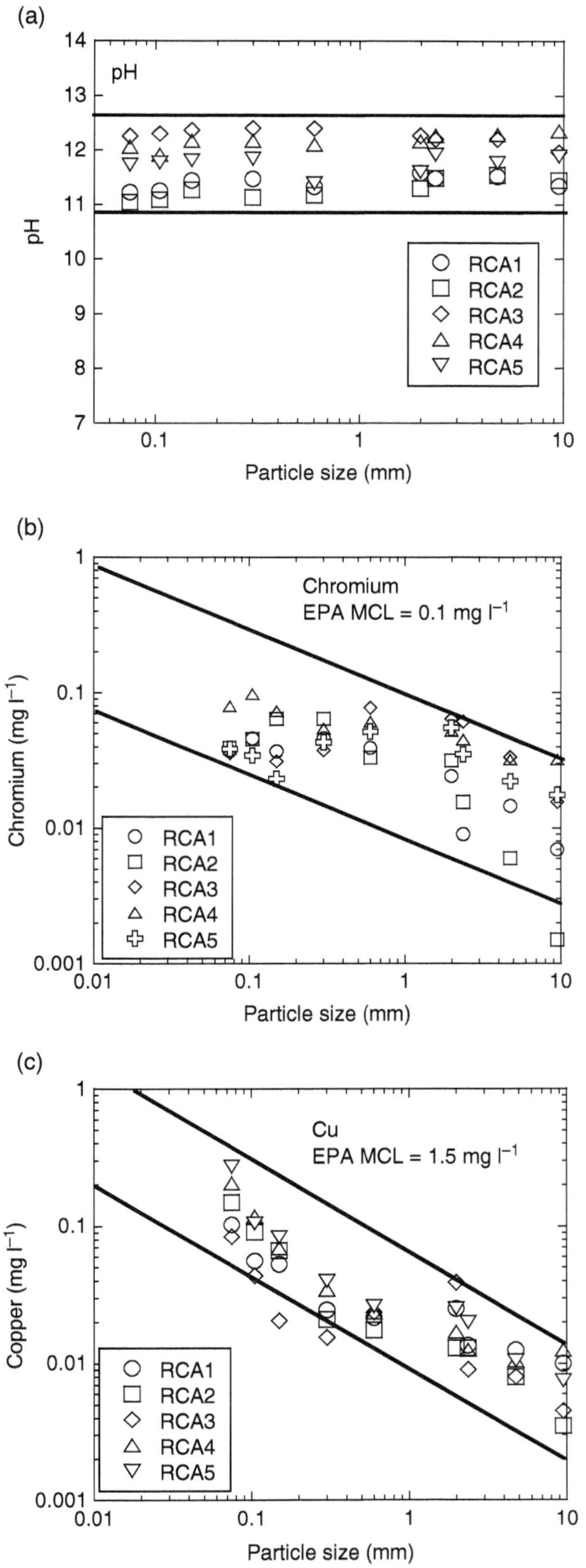

Figure 20.25 Leachate pH and leaching of (a) pH, (b) chromium, and (c) copper from different particle sizes of RCAs. *Source*: From Bestgen et al. (2016b).

solubility of the cement hydrates in RCA is also an important factor in the leaching of Fe under different pH conditions (Engelsen et al., 2009).

Overall, Bestgen et al. (2016) and other studies (Engelsen et al., 2009, 2010); Chen et al., 2012) claim that the use of RCA as roadway base/subbase layers are sustainable and environmentally friendly designs. Furthermore, it should be noted that effluent pH is the most important factor that impacts the environmental quality of the RCA materials.

20.4 Benefit of Recycling Materials Usage

20.4.1 Economic Benefit

There are several economic benefits to use industrial materials for project owners and contractors. Industrial materials are often less expensive than the virgin materials they replace, and recycling or reusing materials onsite can reduce material hauling and disposal costs. When the use of industrial material is used in the project, the total project cost can be lower that will allow more work be done with the same budget. Using fly ash in infrastructure projects can also reduce the need for new or expanded landfills, saving valuable landfill capacity.

20.4.2 Environmental Benefits

The nonrenewable virgin materials must be mined and processes with extensive energy and pollution associated with these activities. When the industrial materials are used to replace the virgin materials, it can save the energy and greenhouse gas emissions. Roads and other structures made with industrial materials can be more durable. Maintaining and replacing roads less frequently is good for the environment because it conserves natural resources and energy.

20.5 Conclusions

This chapter presents the use of industrial materials, e.g. fly ash, and RCAs in roadways and other infrastructure projects as an alternative to virgin materials and construction products. These materials are the by-products of industrial processes, which have chemical and physical properties that make them valuable resources when recycled or beneficially reused, but they are often disposed of as waste. Increasing the recycling of industrial materials in the roadway construction can save energy

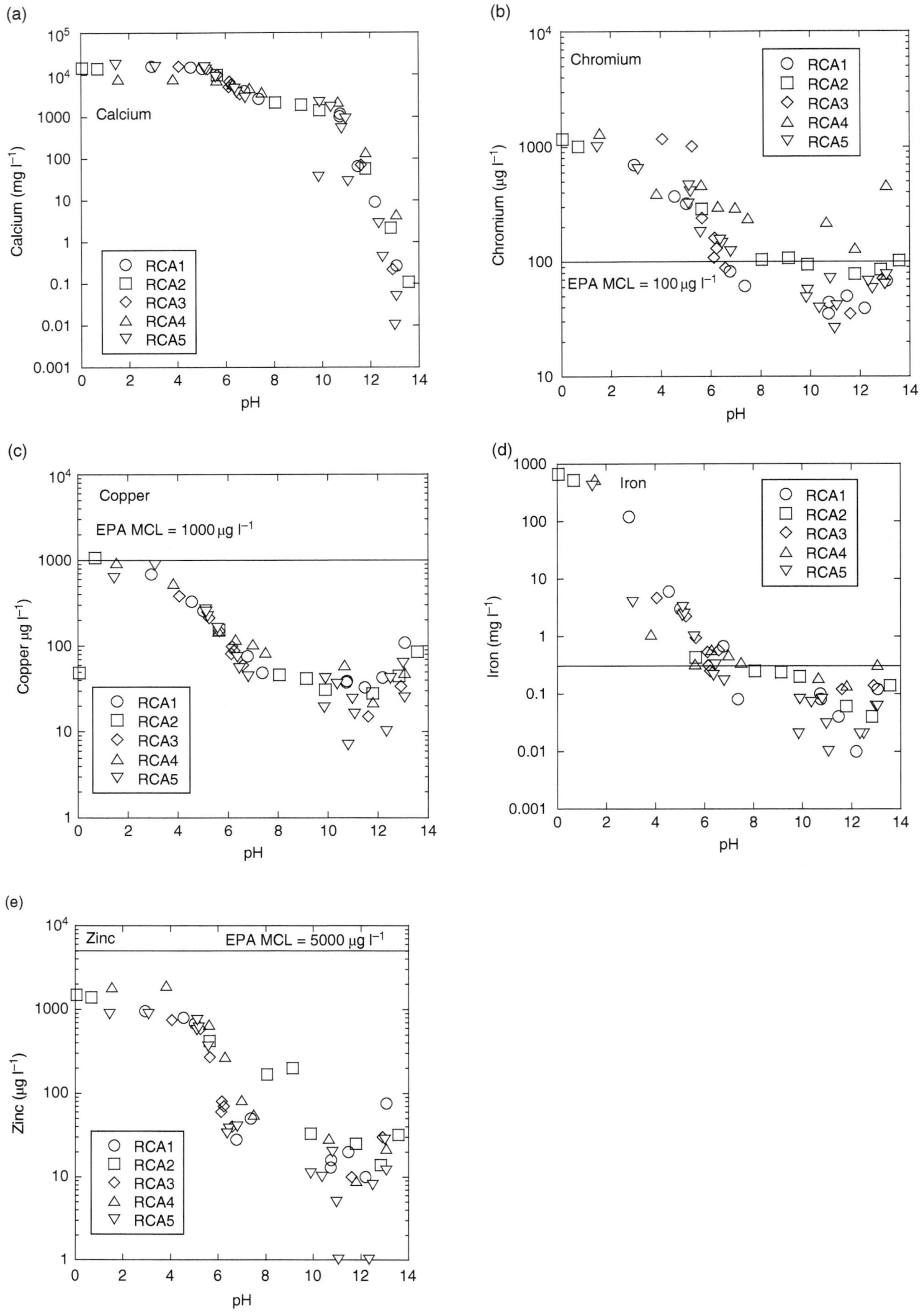

Figure 20.26 (a) Ca, (b) Cr, (c) Cu, (d) Fe, and (e) Zn from recycled concrete aggregate materials as a function of pH.

and reduce greenhouse gas emissions by managing materials more efficiently. Such industrial materials recycling is helping to green the nation's infrastructure by making roadways more durable, conserving natural resources, decreasing energy use, and reducing greenhouse gas emissions.

References

ACAA (2013). 2013 Coal Combustion Product (CCP) Production & Use Survey Report. https://www.acaa-usa.org/Portals/9/Files/PDFs/2013ReportFINAL.pdf (accessed 21 January 2018).

Acosta, H.A., Edil, T.B., and Benson, C.H. (2003). *Soil Stabilization and Drying Using Fly Ash*, Geoengineering Report No. 03-03. Department of Civil and Environmental Engineering, University of Wisconsin: Madison, WI.

Al-Rawas, A., Hago, A., and Al-Sarmi, H. (2005). Effect of lime, cement and Sarooj (artificial pozzolan) on the swelling potential of an expansive soil from Oman. *Building and Environment* 40 (5): 681–687.

Arm, M. (2001). Self-cementing properties of crushed demolished concrete in unbound layers: results from triaxial tests and field tests. *Waste Management* 21 (3): 235–239.

Arora, S. and Aydilek, A.H. (2005). Class F fly ash amended soils as highway base materials. *Journal of Materials in Civil Engineering* 17 (6): 640–649.

ASCE (2017). 2017 Report Card for America's Infrastructure. https://www.infrastructurereportcard.org/ (accessed 19 February 2018).

Asphalt Institute (2003). *Thickness Design – Highways and Streets*, Manual Series, No. 1, 110pp. Asphalt Institute, Lexington, KY

ASTM C618-12el (2012). *Standard Specification for Coal Fly Ash and Raw or Calcined Natural Pozzolan for Use in Concrete*. West Conshohocken, PA: Annual Book of ASTM Standards, ASTM International.

ASTM D2487-11 (2011). *Standard Practice for Classification of Soils for Engineering Purposes (Unified Soil Classification System)*. West Conshohocken, PA: Annual Book of ASTM Standards, ASTM International.

Baker, W.T. (1987). Production and properties of fly ash. *Utilization of Ash Workshop* (13–15 May). University of North Dakota, Grand Forks.

Bennert, T., Papp, W.J. Jr., Maher, A., and Gucunski, N. (2000). Utilization of construction and demolition debris under traffic-type loading in base and subbase applications. *Transportation Research Record* 1714: 33–39.

Bestgen, J.O., Cetin, B., and Tanyu, B.F. (2016). Effect of extraction methods and factors on leaching of metals from recycled concrete aggregates. *Environmental Science and Pollution Research* 23: 12983–13002.

Bestgen, J.O., Hatipoglu, M., Cetin, B., and Aydilek, A.H. (2016). Mechanical and environmental suitability of recycled concrete aggregate as a highway base material. *Journal of Materials in Civil Engineering – ASCE* 28: MT.1943-5533.0001564.

Bin-Shafique, M.D.S. (2002). Leaching of heavy metals from fly ash stabilized soils used in highway pavements. Ph.D. thesis, University of Wisconsin, Madison.

Bin-Shafique, S., Benson, C.H., Edil, T.B., and Hwang, K. (2006). Leachate concentrations from water leach and column leach tests on fly ash-stabilized soils. *Environmental Engineering Science* 23 (1): 53–66.

Bin-Shafique, S., Edil, T., Benson, C., and Senol, A. (2004). Incorporating a fly ash stabilized layer into pavement design-case study. *Geotechnical Engineering*, Vol. 157, No. GE4, 239–249. London: Institution of Civil Engineers.

Camargo, F.F. (2008). Strength and stiffness of recycled base materials blended with fly ash. M.S. thesis, University of Wisconsin, Madison, 123pp.

Cetin, B. and Aydilek, A.H. (2013). pH and fly ash type effect on trace metal leaching from embankment soils. *Resources, Conservation and Recycling* 80: 107–117.

Cetin, B., Aydilek, A.H., and Guney, Y. (2010). Stabilization of recycled base material using high carbon fly ash. *Resources, Conservation and Recycling* 54: 879–892.

Cetin, B., Aydilek, A.H., and Guney, Y. (2012). Leaching of trace metals from high carbon fly ash stabilized highway base layers. *Resources, Conservation and Recycling* 58: 8–17.

Cetin, B., Aydilek, A.H., and Li, L. (2012). Experimental and numerical analysis of metal leaching from fly ash-amended highway bases. *Waste Management* 32 (5): 965–978.

Cetin, B., Aydilek, A.H., and Li, L. (2014). Trace metal leaching from embankment soils amended with high-carbon fly ash. *Journal of Geotechnical and Geoenvironmental Engineering – ASCE* 140 (1): 1–13.

Chen, J., Bradshaw, S., Benson, C.H. et al. (2012). pH-dependent leaching of trace elements from recycled concrete aggregate. *Proceedings of the Geocongress 2012*, Oakland, CA, 9pp.

Chen, J., Tinjum, J.M., and Edil, T.B. (2013). Leaching of alkaline substances and heavy metals from recycled concrete aggregate used as unbound base course. *Transportation Research Record* 2349: 81–90.

Chichester, D.L. and Landsberger, S. (1996). Determination of leaching dynamics of metals from municipal solid waste incinerator fly ash using a column test. *Journal of the Air and Waste Management Association* 46: 643–649.

Chittoori, B. and Puppala, A.J. (2011). Quantitative estimation of clay mineralogy in fine-grained soils. *Journal of Geotechnical and Geoenvironmental Engineering* 137 (11): 997–1008.

Conner, J.R. (1990). *Chemical Fixation and Solidification of Hazardous Waste*. New York: Van Nostrand Reinhold.

Consoli, N.C., Prietto, P.D.M., Carraro, J.A.H. et al. (2001). Behavior of compacted soil-fly ash-carbide lime mixtures. *Journal of Geotechnical and Geoenvironmental Engineering* 127 (9): 774–782.

Culley, R. (1971). Effect of freeze-thaw cycling on stress-strain characteristics and volume change of a till subjected to repeated loading. *Canadian Geotechnical Journal* 8: 359–371.

Dayioglu, M. (2015). Improving the engineering properties of expansive Belle Fourche clay by using chemical additives. M.S. thesis, South Dakota School of Mines and Technology, Rapid City, SD.

Dijkstra, J.J., Meeussen, J.C.L., and Comans, R.N.J. (2004). Leaching of heavy metals from contaminated soils: an experimental and modeling study. *Environmental Science and Technology* 38 (16): 43–90.

Dijkstra, J.J., Van der Sloot, H.A., and Comans, R.N.J. (2002). Process identification and model development of contaminant transport in MSWI bottom ash. *Waste Management* 22 (5): 531.

Dijkstra, J.J., Van Der Sloot, H.A., and Comans, R.N.J. (2006). The leaching of major and trace elements from MSWI bottom ash as a function of pH and time. *Applied Geochemistry* 21 (2): 335.

Edil, T.B. (2012). *Recycled Unbound Materials*, Report No. 2012-35. St. Paul, MN: Minnesota Department of Transportation.

Edil, T.B., Acosta, H.A., and Benson, C.H. (2006). Stabilizing soft fine-grained soils with fly ash. *Journal of Materials – ASCE* 18: 283–294.

Edil, T.B., Benson, C., Bin-Shafique, M. et al. (2002). Field evaluation of construction alternatives for roadways over soft subgrade. *Journal of the Transportation Research Board* 1786: 36–48.

Edil, T.B., Sandstorm, L.K., and Berthouex, P.M. (1992). Interaction of inorganic leachate with compacted pozzolanic fly ash. *Journal of Geotechnical Engineering – ASCE* 118 (9): 1410–1430.

Eighmy, T.T., Eusden, J.D., Krzanowski, J.E. et al. (1995). Comprehensive approach toward understanding element speciation and leaching behavior in municipal solid-waste incineration electrostatic precipitator ash. *Environmental Science & Technology* 29 (3): 629–646.

El-Latif, M.Y.A. (2008). Swelling performance of some expansive soil treatment techniques. M.Sc thesis, Ain Shams University, Abbasseya, Cairo.

Engelsen, C.J., van der Sloot, H.A., Wibetoe, G. et al. (2009). Release of major elements from recycled concrete aggregates and geochemical modelling. *Cement and Concrete Research* 39 (5): 446–459.

Engelsen, C.J., van der Sloot, H.A., Wibetoe, G. et al. (2010). Leaching characterisation and geochemical modelling of minor and trace elements released from recycled concrete aggregates. *Cement and Concrete Research* 40 (12): 1639–1649.

Estabragh, A.R., Pereshkafti, M.R.S., Parsaei, B., and Javadi, A.A. (2013). Stabilized expansive soil behavior during wetting and drying. *International Journal of Pavement Engineering* 4 (4): 418–427.

FHWA (2004). *Transportation Applications of Recycled Concrete Aggregate*, 47pp. Washington, DC: US Department of Transportation.

FHWA (2008). *User Guideline for Byproducts and Secondary Use Materials in Pavement Construction*, Report no. FHWA-RD-97-148. McLean, VA: FHWA.

Fruchter, J.S., Ral, D., and Zachara, J.M. (1990). Identification of solubility-controlling solid phases in a large fly ash field lysimeter. *Environmental Science and Technology* 24 (8): 1173.

Garrabrants, A.C., Sanchez, F., and Kosson, D.S. (2004). Changes in constituent equilibrium leaching and pore water characteristics of a Portland cement mortar as a result of carbonation. *Waste Management* 24 (1): 19–36.

Geelhoed, J.S., Meeussen, J.C.L., Hillier, S. et al. (2002). Identification and geochemical modeling processes controlling leaching of Cr(VI) and other major elements from chromite ore processing residue. *Geochimica et Cosmochimica Acta* 66 (22): 3927–3942.

Gitari, W.M., Fatoba, O.O., Petrik, L.F., and Vadapalli, W.R.K. (2009). Leaching characteristics of selected South African fly ashes: effect of pH on the release of major and trace species. *Journal of Environmental Science and Health Part A* 44: 206–220.

Goh, A.T.C. and Tay, J.H. (1993). Municipal solid waste incinerator fly ash for geotechnical applications. *Journal of Geotechnical Engineering – ASCE* 119 (5): 811–825.

Guney, Y., Aydilek, A.H., and Demirkan, M.M. (2006). Geoenvironmental behavior of foundry sand amended mixtures for highway subbases. *Waste Management* 26: 932–945.

Haider, I., Kaya, Z., Cetin, A. et al. (2014). Drainage and mechanical behavior of highway base materials. *Journal of Irrigation and Drainage Engineering – ASCE* 140 (6): 1–14.

Hatipoglu, B., Edil, T., and Benson, C. (2008). Evaluation of base prepared from road surface gravel stabilized with fly ash. *ASCE Geotechnical Special Publication* 177: 288–295.

Jong, D.T., Bosscher, P.J., and Benson, C.H. (1998). Field assessment of changes in pavement moduli caused by freezing and thawing. *Transportation Research Record* 1615: 41–48.

Kanungo, S.B. and Mohapatra, R. (2000). Leaching behavior of various trace metals in aqueous media from two fly ash samples. *Journal of Environmental Quality* 29: 188–196.

Komonweeraket, K., Cetin, B., Aydilek, A.H. et al. (2015). Effects of pH on the leaching mechanisms of elements from fly ash mixed soils. *Fuel* 140: 788–802.

Kosson, D.S., van der Sloot, H.A., and Eighmy, T.T. (1996). An approach for estimation of contaminant release during utilization and disposal of municipal waste combustion residues. *Journal of Hazardous Materials* 47 (1–3): 43–75.

Kosson, D.S., van der Sloot, H., Sanchez, F., and Garrabrants, A.C. (2002). An integrated framework for evaluating leaching in waste management and utilization of secondary materials. *Environmental Engineering Science* 19 (3): 159–204.

Kuo, S.-S., Mahgoub, H.S., and Nazef, A. (2002). Investigation of recycled concrete made with limestone aggregate for a base course in flexible pavement. *Transportation Research Record* 1787: 99–108.

Lee, W., Bohra, N.C., Atlschaeffl, A.G., and White, T.D. (1995). Resilient modulus of cohesive soils and the effect of freeze-thaw. *Canadian Geotechnical Journal* 31: 559–568.

Li, L., Benson, C.H., and Edil, T.B. (2009). Mechanical performance of pavement geomaterials stabilized with fly ash in field applications. *Coal Combustion and Gasification Products* 1: 43–49.

Li, L., Benson, C.H., Edil, T.B., and Hatipoglu, B. (2006). Groundwater impact from coal ash in highways. *Waste and Resource Management* 159 (4): 151–162.

Li, L., Benson, C.H., Edil, T.B., and Hatipoglu, B. (2008). Sustainable construction case history: fly ash stabilization of recycled asphalt pavement material. *Geotechnical and Geological Engineering* 26: 177–187.

Li, L., Peng, B., Santos, F. et al. (2011). Groundwater impacts from leaching of coal combustion products in roadways embankment constructions. *Journal of ASTM International* 8 (8): 1–12.

Mamlouk, M.S. and Zaniewski, J.P. (2011). *Materials for Civil and Construction Engineers*. Upper Saddle River, NJ: Prentice Hall.

McKinley, J.D., Thomas, H.R., Williams, K.P., and Reid, J.M. (2001). Chemical analysis of contaminated soil strengthened by the addition of lime. *Engineering Geology* 60: 181–192.

Morar, D.L. (2008).Leaching of metals from fly ash – amended permeable reactive barriers, M.S. thesis, University of Maryland, College Park.

Nalbantoğlu, Z. (2004). Effectiveness of class C fly ash as an expansive soil stabilizer. *Construction and Building Materials* 18 (6): 377–381.

Nalbantoğlu, Z. and Gucbilmez, E. (2001). Improvement of calcareous expansive soils in semi-arid environments. *Journal of Arid Environments* 47 (4): 453–463.

Pal, S.K. and Ghosh, A. (2014). Volume change behavior of fly ash-montmorillonite clay mixtures. *International Journal of Geomechanics* 14: 59–68.

Rao, K.M. and Subbarao, G.V.R. (2012). Optimum fly ash for mechanical stabilization of expansive soils using 22 factorial experimental design. *Natural Hazards* 60: 703–713.

Rosa, M.G., Cetin, B., Edil, T.B., and Benson, C.H. (2016). Development of a test procedure for freeze-thaw durability of geomaterials stabilized with fly ash. *ASTM Geotechnical Testing Journal* 39: 938–953.

Rosa, M.G., Cetin, B., Edil, T.B., and Benson, C.H. (2017). Freeze-thaw performance of natural, recycled and stabilized materials. *Journal of Materials in Civil Engineering* 29: https://doi.org/10.1061/(ASCE)MT.1943-5533.0001844.

Sanchez, F., Langley White, M.K., and Hoang, A. (2009). Leaching from granular cement-based materials during infiltration/wetting coupled with freezing and thawing. *Journal of Environmental Management* 90 (2): 983–993.

Sezer, A., Inan, G., Yilmaz, H.R., and Ramyar, K. (2006). Utilization of a very high lime fly ash for improvement of Izmir clay. *Building and Environment* 421: 150–155.

Simonsen, E., Janoo, V., and Isacsson, U. (2002). Resilient properties of unbound road material during seasonal frost conditions. *Journal of Cold Region Engineering* 16: 28–50.

Stolle, D., Guo, P., and Liu, Y. (2009). Resilient modulus properties of granular highway materials. *Canadian Journal of Civil Engineering* 36: 639–654.

Stumm, W. and Morgan, J.J. (1996). *Aquatic Chemistry. Chemical Equilibria and Rates in Natural Waters*, 3e. New York: Wiley.

Tastan, E.O., Edil, T.B., Benson, C.H., and Aydilek, A.H. (2011). Stabilization of organic soils with fly ash. *Journal of Geotechnical and Geoenvironmental Engineering – ASCE* 137 (9): 819–833.

Theis, T.L., Halvorson, M., Levine, A. et al. (1982). *Fly Ash Attenuation Mechanisms. Report and Technical Studies on the Disposal and Utilization of Fossil-Fuel Combustion By-Products*, vol. II. Washington, DC: USEPA.

Torrey, S. (1978). *Coal Ash Utilization: Fly Ash, Bottom Ash and Slag*. Park Ridge, NJ: Noyes Data Corporation.

Tutumluer, E. and Pan, T. (2008). Aggregate morphology affecting strength and permanent deformation behavior of unbound aggregate materials. *Journal of Materials in Civil Engineering – ASCE* 20 (9): 617–627.

U.S. Energy Information Administration (2014). What is U.S. electricity generation by energy source? Washington, DC: U.S. Department of Energy. http://www.eia.gov/tools/faqs/faq.cfm?id=427&t=3 (accessed 21 January 2018).

US EPA (2009). *National Primary Drinking Water Standards*. Washington, DC: US EPA.

USGS ed. (2013). *Mineral Commodity Summaries 2013*, 198pp. Reston, VA: USGS.

Van Gerven, T., Cornelis, G., Vandoren, E., and Vandecasteele, C. (2007). Effects of carbonation and leaching on porosity in cement-bound waste. *Waste Management* 27 (7): 977–985.

Vishwanathan, R., Saylak, D., and Estakhri, C. (1997). Stabilization of subgrade soils using fly ash. *Proceedings of the Ash Utilization Symposium CAER*, Kentucky, pp. 204–211.

Wen, H., Baugh, J., and Edil, T. (2007). Use of cementitious high carbon fly ash to stabilize recycled pavement materials as pavement base material. In: *Proceedings of 86th Annual Meeting (CD ROM)*. Washington, DC: Transportation Research Board.

Wen, H., Warner, J., and Edil, T. (2008). Laboratory comparison of crushed aggregate and recycled pavement material with and without high-carbon fly ash. In: *Proceedings of 87th Annual Meeting (DVD)*. Washington, DC: Transportation Research Board.

Zaman, M. and Naji, K. (2003). Effect of freeze-thaw cycles on class C fly ash stabilized aggregate base. In: *Proceedings of 82nd Annual Meeting (CD ROM)*. Washington, DC: Transportation Research Board.

Zha, F., Liu, S., Du, Y., and Cui, K. (2008). Behavior of expansive soils stabilized with fly ash. *Natural Hazards* 47: 509–523.

21

Solid Waste Separation and Processing

Principles and Equipment

Georgios N. Anastassakis

School of Mining and Metallurgical Engineering, National Technical University of Athens, Athens, Greece

The current chapter describes the methods and equipment utilized to segregate the various components of commingled solid waste streams into single constituents for materials recycling and energy recovery from the combustible fraction. This chapter is separated into three distinct parts. The first part describes the problems from the depletion of primary material resources, the role and current level of waste materials recycling, and the prevailing concepts and processes in regard to waste processing. The scientific basis of solid waste separation techniques, as well as the similarities and differences between mineral processing and solid waste separation methods are also outlined. The second part is devoted to the detailed presentation of methods, equipment, and unit operations used in solid waste separation, such as: size reduction, classification, sorting (manual, sensor-based, robot), separation methods (gravity, magnetic, eddy current, electrostatic, ballistic, flotation), cubing-pelletizing, and baling. Finally, the third part is referred to the environmental, economic and social benefits of recycling as well as to its prospects.

21.1 Introduction

21.1.1 Resource Conservation through Reuse and Recycling

Despite the fluctuations and uneven distribution of the wealth around the world, a continuously increasing industrial and technological development has been observed since the second half of the twentieth century. This is mainly due to the improvement of life standards, especially in the developed and developing countries, and population increase. These factors result in increased demand for more products, consumption of more materials, and generation of more waste. The manufacture and utilization of products follow a certain life cycle, which is determined by the economic trends prevailing in the defined period. The industrialized countries mostly set the market rules, consumption trends, and economic cycle of products. For example, the change of commercial trends from long- to short-life products resulted in larger waste volumes and faster generation rate than in the past.

For the manufacture of most products, primary mineral and energy resources (mainly coal, oil, and natural gas) are used. These resources are confined in definite and nonrenewable deposits. Consequently, the consumption of mineral and other resources must be carried out with great concern, as their conservation is of extremely high importance for the future generations. Only then, the essence of "sustainable development" concept is fulfilled.

Sustainable development is defined as "Development that meets the needs of the present without compromising the ability of the future generations to meet their own needs" (WCED, 1987). This clearly denotes that sustainability (i) is ground on ethics, since the welfare of future generations is an ethical issue, and (ii) is a value-based relationship between people and nature, which is grounded on the three following pillars: environment, economy, and society (Renn and Goble, 1996). Historically, there has been a close correlation between economic growth and environmental degradation: as communities grow, so the environment declines. For this reason, there is a strong concern for the modern global societies to preserve the primary resources, since resource overexploitation far in past times led to collapse of local societies and civilizations (Diamond, 2005). Consequently, sustainability is closely associated with

Handbook of Environmental Engineering, First Edition. Edited by Myer Kutz.

maximization of resource employment; waste mass minimization for disposal through waste prevention, reuse, recovery, or recycling; reduction or elimination of the adverse impacts on human health, local and global environment as well; and delivery of more attractive and pleasant human settlements and social amenities.

Waste becomes an issue as people first begin to establish permanent settlements, approx. 10 000 BC. Although the problems were not intense that time, people had subconsciously applied sustainable practices, e.g. recycling and reuse of leaves and manure as fertilizers; but it is only three decades now that recycling has drawn special attention.

Despite the progress, the level of recycling and reuse of materials from wastes is not at the desired level yet because of public unawareness, public or state reluctance, absence of proper management systems, shortage of funds, cost, waste type, properties of recycled material and remanufactured product, etc. For these reasons, the use of resources largely remains unsustainable and inefficient, while waste is not properly managed in spite of the progress that has been made. Statistical data show a great divergence of the recycling rates among the European countries as well as within the United States. The recycling rate is calculated as the percentage of the recycled and composted waste to the totally generated.

According to EPA (2014), the recycling rate of municipal solid waste (MSW) in the United States has increased from 6.4% in 1960 to 34.5% in 2012. In the year 2002, the average recycling rate of MSW in the United States was 30% approximately. In this year, 11 states attained recycling rate higher than 30% (between 31.1 and 48.8%), 13 states between 20 and 30%, 9 states between 10 and 20%, while the rate for the rest was less than 10%, out of which for 3 states it was between 0.3 and 1.7% (Kaufman et al., 2004).

Corresponding data from 32 European countries for 2012 show that the average recycling rate is 32% but with significant fluctuations among the countries: 4 countries (Germany, Austria, Belgium, and Switzerland) had recycling rate higher than 50% (50–64%); 9 countries between 40 and 49.9%, 4 countries between 30 and 39.9%; 8 countries between 20 and 29.9%; 5 countries between 10 and 19.9%; and 2 countries (Romania and Turkey) less than 10% (Eurostat, 2012).

Despite the problems, recycling and energy recovery from waste is expected to be intensified in the next years as the population of even more countries worldwide becomes more aware, conscious, and sensitive to the arising environmental problems. The global efforts for continuously increasing recycling rates are related to population growth, diminishing of primary materials resources, generation of large waste volumes per year to be landfilled, restriction of the available landfill space, global demand for better quality of life, economic value hidden in waste, energy savings, etc.

21.1.2 From Landfill Mining to Circular Economy

Society has been conscious since several years ago that the increasing consumption of nonrenewable material resources cannot sustain human development evermore, resources must be utilized at their maximum, many intensive environmental problems have been generated, and some planetary boundaries, which define a safe operating space for human development, may already have been transgressed.

The consciousness of this situation led to changes in waste processing from linear approach, depicted by triangle configuration of activities, to circular, which is depicted by sequential circles (Cossu and Williams, 2015). The linear waste processing was based on the traditional take–make–consume–dispose approach. Processes such as landfill and urban mining, materials recycling, resources recovery, and waste minimization are components of the circular approach. Although there is no universal agreement on the use of these terms and their meaning may differ significantly or lexically, the following definitions seem to depict well the differences among the terms.

Landfill mining involves the activities aiming at the extraction and processing of waste materials that have been previously disposed or stocked in specific "deposits" (e.g. mine/mineral processing/metallurgical tailings, municipal landfills).

Urban mining involves the activities aiming at the systematic reclaiming of materials from anthropogenic stocks (e.g. buildings, infrastructure, industries, products in use or landfilled). These stocks, which will become available for reuse after the end of the product lifetime, represent a large resource potential, with concentrations of elements often comparable with or higher than the corresponding of the natural deposits. The strategy for urban mining has to take into account the development phase of the system "city" (Brunner, 2011). In fast-growing cities, such as some Asian, urban mining has little or no contribution, as the recovery of used materials from the city (especially construction materials) is less than the demanded from primary resources, because of the long intervening time between production and disposal of materials. In mature (steady-state) or shrinking cities, primary resources can be substituted for secondary, as the recovered urban waste materials exceed the demand. Because of the different composition and speciation, all recovered materials from anthropogenic waste are not amenable to reuse, e.g. metal aluminum from obsolete objects can be recovered, but this is not true for aluminum of kaolin used as filler in newsprints.

Resource recovery is referred to the recovery process of materials from recycling activities as well as energy recovery from combustible waste.

Materials recycling, in fact, denotes the activities that involve processing of the recovered materials flows, their subsequent transformation into new products, and their introduction into production cycles in a new way. Nowadays, urban mining is often used to describe almost any kind of materials recycling.

Waste minimization is actually a strategy aiming at the reduction of waste volume and toxicity from both production and consumption side.

Finally, *circular economy* and relative concepts (e.g. zero waste, ecodesign) are terms that adopt the circular business model and aim at keeping the products, components, and materials in use and product circles, as long as possible, by applying various recycling practices. A circular economy seeks to respect planetary boundaries through increasing the share of renewable or recyclable resources, while reducing the consumption of raw materials and energy and, at the same time, cutting emissions and material losses.

In the recent years, the term *green economy* has emerged. According to UNEP (2011), green economy results in improved human wellbeing and social equity while maintaining environmental systems. It also recognizes that an economy's potential to generate income and jobs should be largely determined by investments (public, private) moving toward the reduction of emissions and pollution, enhanced efficiency of resources and energy, and ecosystems that maintain their biodiversity and deliver a huge array of enormously valuable goods and services. As the above definition of UNEP makes clear, green economy has many links with the concept of sustainable development; also, an essential element in the concept of green economy is the necessity of reflecting the value of the environment in economic decision making.

As business and economy models are moving toward green economy, recycling is expected to beneficially contribute in several ways, as it will be further described in more details.

21.1.3 Principles of Solid Waste Separation

Basically, the existing technology highly copes, more or less, with the needs for concentrating the waste objects into separate products, as waste and minerals share the same principles for separating values from useless components. It must be noted that minerals and waste have similarities, as both are resources of useful components and use the same principles to separate useful components from gangue. Their differences are summarized to the following (Anastassakis, 2007): on a long-term basis, the composition of the feed in mineral processing plants is more constant and homogeneous than that of waste processing plants; in waste treatment, the constituents of the feed are usually liberated into single components or the liberation process is easier; if size reduction is necessary, either for liberation or for particle volume reduction, waste particle size is considerably larger than that of minerals or rocks; in contrast to raw mineral sources, the grade of the wastes in useful and recoverable materials is higher, and the flowsheet of solid waste treatment plants is simpler; and in most cases, the shape of mineral particles could be considered, more or less, spherical, especially for small particle size, while the waste particles are of variant shapes (spherical, wiry, platy, etc.). This shape variance, in some cases, hinders effective separation but in others has a beneficial effect.

Although in most cases solid wastes are efficiently separated into their constituents based on well-known scientific principles and equipment from mineral processing methods, improvement and innovation of the existing equipment are sometimes necessary because of the differences in properties and characteristics between minerals and wastes.

21.2 Size (or Volume) Reduction of Solid Waste

Solid waste is usually subjected to size reduction prior to any processing stage. Size reduction is a mechanical process that is used to obtain a uniform product of reduced size compared with that of the initial waste stream, either for direct use and/or storage or to make the materials amenable to further separation processes. Size reduction can also be applied in case of solid waste landfilling, with the major benefits being waste volume reduction and life increase of the disposal site. In practice, the terms shredding, grinding, granulation, and milling are alternatively used to describe the size (or volume) reduction process.

21.2.1 Factors Affecting the Selection of Equipment

The size reduction mechanism of solid waste, their behavior, and the selection of the proper equipment is determined by the physical properties and mechanical characteristics of solid waste streams. In most cases, these properties are significantly different than the corresponding for rocks and minerals. The latter are brittle and heterogeneous materials regarding structure discontinuities; solid waste streams may compose of species with significantly different properties (e.g. brittle and nonbrittle) but of homogeneous properties each. Waste materials, such as glass, rubble, as well as cast iron and cast nonferrous metal scraps, are considered as brittle materials. On the contrary, the following waste materials are classified as nonbrittle: paper and cardboard, wood, textiles, steel and nonferrous metal wrought alloys, plastic and rubber, domestic and bulky refuse, organic waste, etc.

Compared with minerals, size reduction of solid waste streams is relatively simpler as the waste objects are usually composed of liberated (monocomponent) materials; however, in case that waste objects have to be liberated, the process is rather easy, as it is achieved at relatively coarse particle size. Size reduction of materials is generally achieved by exerting one or combination of the following forces on the particles: compression, impact, attrition, and shear.

Compression is the most usual comminution force for brittle materials. It is realized by compressing materials between two heavy-duty metal surfaces. Crushers based on compaction force are jaw, cone, gyratory, and roll crushers. In solid waste processing, compression is a decisive force for size reduction only when waste composes of brittle materials; otherwise, it simply deforms but does not change the size of components such as metals, paper, plastic, organics, etc.

Impact is the simplest size-reduction force. It is also effective on brittle materials. As feed particles fall freely on a high-speed massif rotor, it impacts and breaks them into smaller particles. Secondarily, the material strikes onto a compact plate, where particle size is further reduced. Impactors and hammer mills are crushers based on impact force. They are characterized by high size-reduction ratio, reaching up to 20 : 1.

Attrition force causes size reduction by friction contact between particles, particle/grinding media, as well as particle/mill shell, and subsequent abrasion. It is effective for brittle materials of relatively small particle size. Rod and ball mills are mainly based on this force.

For brittle materials, shear is acting force in hammer mills (along with impact and attrition) and toothed crushing rolls. The prerequisite for successful size reduction of particles is the existence of macro- or microstructure weaknesses (e.g. layered structure, schistosity) or differences in the characteristics of adjacent grains (e.g. hardness, friability). Contrary to brittle materials, shear force is widely applied in solid waste size reduction as most constituents are soft, nonbrittle, plastic, ductile, etc., and, consequently, shear forces are effective. The various types of shredders use shear force to reduce the size of most solid waste constituents.

The most important factors that determine the selection of size reduction equipment are the physical/mechanical properties of the waste material (composition, structure, hardness, ductility, brittleness, etc.), as well as the size characteristics of feed and final product. The first factor determines the required acting force on particles and, consequently, the type of size reduction equipment. The size distribution of the feed and product determines the model and the dimensions of feed and discharge openings.

Other factors that must be taken into account upon selecting size reduction equipment are future use of product, process economics (energy requirements, maintenance, performance, etc.), safety (e.g. explosions due to shredding of closed containers with volatile substances), environmental hazards (noise, air pollution, etc.), and site considerations (circuit design and installation, product storage, etc.).

The size reduction of solid waste streams may occur in one, two, or even three stages, depending on process requirements, target, and required characteristics of the final product. A typical processing flowsheet of commingled solid waste usually includes one size-reduction stage (also called primary or coarse) to achieve a product smaller than 10–15 cm. Whenever a product of smaller size is required, secondary or even tertiary size-reduction stage is introduced. In some cases, small particle size is required either to meet the specifications (e.g. production of various forms of refuse-derived fuel (RDF)) or to recover and process various waste materials (e.g. plastic, glass, metal scrap).

21.2.2 Size Reduction Equipment

There are several types of size-reduction equipment in use in regard to solid waste unit operations. Some of them originate from size reduction of minerals, while others from solid waste unit operations. Several, and sometimes contradictory, classification schemes exist with respect to size-reduction equipment of solid waste streams. In the following, the equipment is broadly classified into impactors, hammer mills, shredders, and grinders in reference to the main working principle. The generic term "granulators" has been also used to include all the aforementioned and many other varieties of devices used in solid waste size-reduction processes.

21.2.2.1 Impactors and Hammer Mills

Impactors and hammer mills accomplish size reduction of the material primarily by impact forces, exerted on particles by fixed or free-swinging hammers that revolve mounted on a central, high-speed rotating shaft, and secondarily by hurling the material pieces at high speed against stationary surfaces. This equipment is effective mainly for size reduction of brittle materials.

The material to be crushed enters the equipment through the feed opening, located at the top or top side, and falls into the hammer circle. The first stage of size reduction is accomplished by collision of feed particles with high-speed rotating hammers, made usually of manganese steel. The second stage of breakage occurs when the particles hit the plates or the breaker bars, which line the frame of the equipment. Hammer mills are further equipped with grid bars or a grate placed at the bottom. Consequently, a third stage of size reduction occurs in hammer mills due to shear and attrition forces on oversized particles that are located between the bottom grate and hammers until their size becomes less than grid openings.

Although there are many similarities between impactors and hammer mills, there are also several differences. The main difference is that the former ones primarily rely on impact action of hammers, either fixed or free swinging, and secondarily on particles collisions, either striking each other or on stationary steel surfaces. The hammer mills rely both on centrifugal impact force of the hammers and combination of attrition and shear forces between hammers and grate.

21.2.2.1.1 Impactor

The main components of impactors are a compact rotor with fixed hammers and one or two breaking plates. Impactors have no grates or other components at the bottom to size the end product. They are constructed to break the bulky items of the refuse stream either for further size reduction or for size preparation of the material to feed a waste incinerator, since bulky materials are not suitable for incineration.

21.2.2.1.2 Hammer Mill

It is very effective for size reduction of brittle materials. Hammer mills for waste shredding are available with fixed or swing hammers, bolted on a high-speed rotating shaft (700–1200 RPM). Among the great variety of models, hammer mills are provided with major or minor differences in the design, depending on manufacturer and use.

The type with fixed hammers is suitable for size reduction of brittle materials (e.g. glass bottles, ceramics), but, under special conditions, it can be used for materials without brittle properties (wood, metal turnings, scrap metal, etc.) as shown in Figure 21.1 (Metso Minerals, 2000b). In case of mixed solid waste presorting, the passing product of hammer mill has obtained the smallest particle size compared with other devices, under the same power consumption. The grate bar or screen, at the bottom of the equipment, prevents the outflow of the particles until their size is reduced to the desired. The opening size of the grates varies broadly between 5 and 50 mm. Swing hammers type is commonly used for mixed and sorted solid waste. Hammers may also be replaced with knives or cutters. In this case, the hammers, knives, or cutters extend radially from the shaft because of the centrifugal force.

In general, hammer mills are designed with horizontal or vertical shaft. Horizontal-shaft hammer mills are either downrunning (Figure 21.2) or center-feed reversible. In center-feed mills, the material is fed down on the top of the hammer circle. The downrunning hammer mills with horizontal shaft have relatively increased cost for maintenance or replacement of the hammers, as they always rotate to the same direction. The cost depends on the throughput and the composition of solid waste. In a center-feed mill, the direction of the shaft rotation may reverse; this results in reduced replacement cost, as

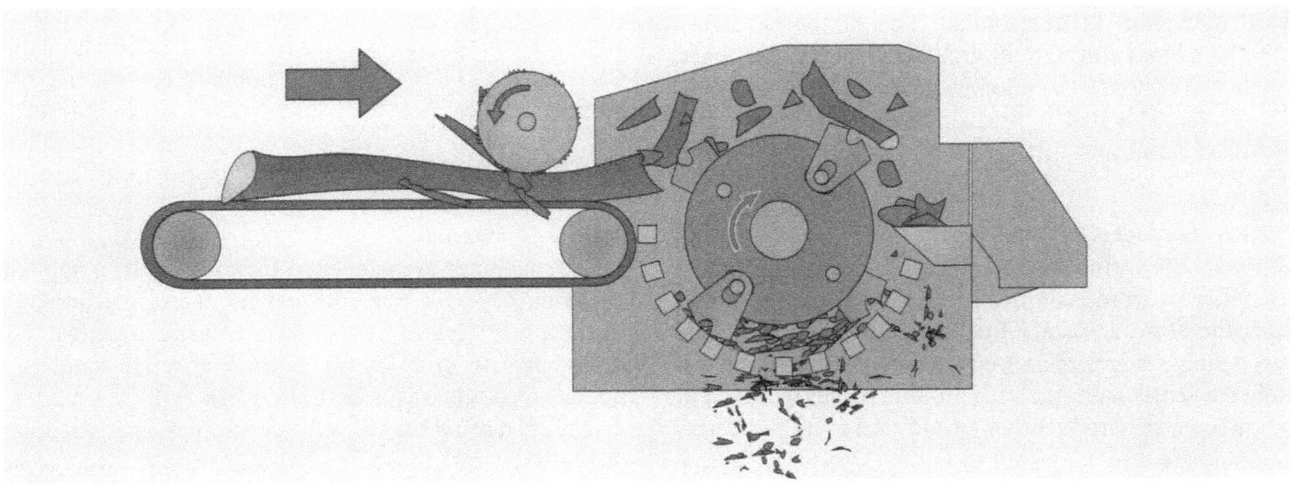

Figure 21.1 Hammer mill suitable for wood fragmentation. *Source:* Reproduced with permission of Metso Minerals (2000b).

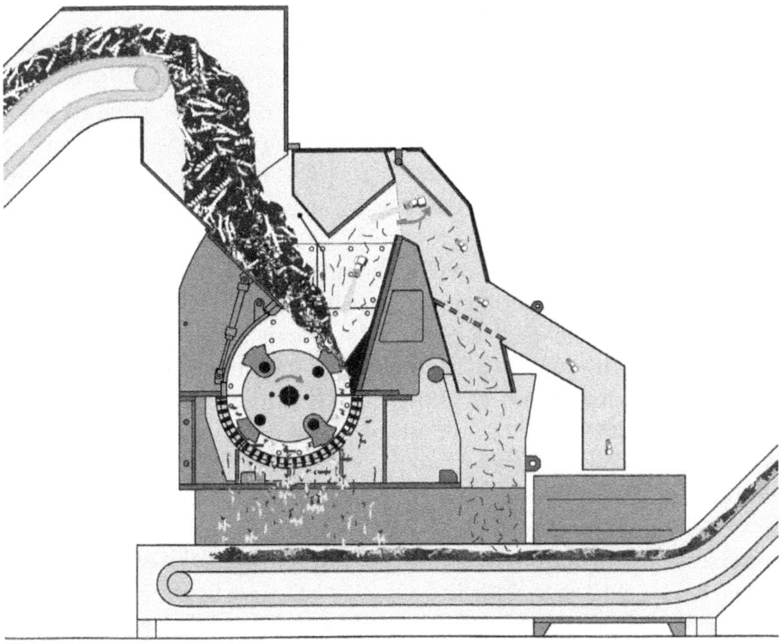

Figure 21.2 Horizontal-shaft hammer mill downrunning. *Source:* Reproduced with permission of Metso Minerals (2000c).

hammers are replaced when both cutting faces get worn. On the other side, center-feed mill along with its hopper requires more height to avoid material ejection out the feed opening of the machine.

Vertical shaft hammer mills have no grate at the bottom. The upper part of the equipment is conical and the lower cylindrical. The vertical shaft with the hammers is placed along the axis of the machine. Size reduction in this machine is achieved through repeated impacts of the material with the hammers and the breaking plate, as particles descent the conical part of the machine. The final size reduction of the infeed material is accomplished at the lower part of the machine, where the distance between hammers and breaking plate is reduced. This results in faster wear of the hammers in the cylindrical part than in the conical. The discharge of the material occurs through an annulus at the lowest part, which also controls the maximum particle size but does so less effectively than in horizontal hammer mills with a grate at the bottom.

21.2.2.2 Shredders

Although the term "shredders" is sometimes used as a general term to include the equipment that causes size reduction of solid waste, it fits best to the equipment that uses mainly shear force to reduce the size of the objects. In such a sense, hammer mills are sometimes included in the category of shredders, as, in addition to impact force, materials are also subjected to shear force for their size reduction, in case that they are equipped with stationary knives fixed at the wall. In this chapter, shredders are considered only the equipment that use shear as a principal force for solid waste size reduction. Under this consideration, flail mills, rotary shear shredders, and knife mills are the most popular equipment to shred solid waste.

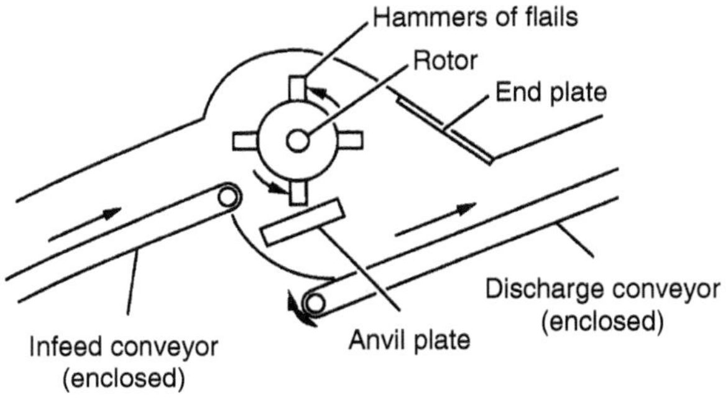

Figure 21.3 Typical configuration of flail mill. *Source:* Tchobanoglous et al. (1993). Reproduced with permission of McGraw Hill Inc.

21.2.2.2.1 Flail Mill

These machines are similar to hammer mills in appearance but they have some differences in construction and operation (Figure 21.3). Contrary to hammer mills, flail mills have no grate at the bottom and, consequently, operate as single-pass machines. Their application is primarily for tearing bags of refuse and breaking up bundles of material; in addition to tearing, they provide some mixing of the waste materials. These machines should be followed by shredder, as the product size is coarse.

21.2.2.2.2 Rotary Shear Shredder

Shear shredders (Figure 21.4) operate like scissors during materials size reduction. They are usually composed of two counterrotating parallel shafts, driven by hydraulic or electrical motors, and a series of interlocking, toothed-cutting disks mounted perpendicularly on the shafts. Shear shredders with one, three, or four shafts are also available. They are single-pass machines, while the spacing between the cutters, or the cutter and the shell in single-shaft design, determines the product size. In most cases, the product size varies between 25 and 100 mm (1–4 in). The waste material is fed between the counter-rotating shafts, and its size reduction is achieved by the shearing or tearing action of the cutting disks. In case of overload, the rotation of the shafts is automatically reversed.

(a)

(b)

Figure 21.4 (a) Rotor shear shredder and (b) details of the cutting disks. *Source:* Reproduced with permission of Metso Minerals (2000d).

Rotational speed is very low, usually ranging between 50 and 200 RPM. The low-speed hydraulic motors develop very high torque on the shafts, providing considerable shearing power to the cutting disks. This power makes shear shredders effective on difficult-to-cut materials. This equipment has been used to shred materials (metals, truck and car tires, wood, pallets, bulky objects, electrical and electronic devices, etc.) into relatively coarse particle size. On the contrary, they are not capable to shred materials into fine size (less than 10 mm). The main advantage of shear shredders is that the more they are fed with material, the more effective they are, possibly because the load presses the objects at the bottom layer to be easily grabbed between the shafts and the cutting disks.

21.2.2.2.3 Knife Shredder

Knife shredders are machines used to cut waste objects into smaller pieces for subsequent separation, or volume reduction, with consequent density increase. The design of these machines employs a series of long, sharp knives attached in special clamps on a high-speed rotor. Cutting edges or anvils are also mounted inside the casing of the machine. The material is shredded, as it is caught between rotating knives and cutting edges. Knife shredder is mostly used to reduce the size of coarse particles to a product size of the order 6–18 mm. This equipment is suitable to shred effectively soft and plastic materials, such as plastic bottles, rags, paper, etc. Hard materials (e.g. metals, glass) must have been previously removed, as they destroy the cutting edges of the knives.

For very fine shredding, knife mills have been used to reduce the size of the materials down to 1.0 mm. These machines were mainly used for RDF production, under the condition that hard and abrasive materials had been previously removed. The very fine particle size that can be achieved rendered knife mills attractive in waste shredding, but the small distance between rotating knives and cutting edges makes them amenable to increased wear. This problem is partially overcome by using cutting edges from special hard alloys but at a very high cost.

21.2.2.2.4 Bag Opener

In most treatment facilities, waste appears in bags that have to be split for liberation and efficient segregation of the content in the upcoming processing steps. In the past, this process was achieved either on the reception conveyor or in a trommel, with both ways presenting problems. The first way was not practical, as spillage of solid waste occurred; in the second way, the major problem was clogging of sieve apertures. Bag openers, which are installed at the head of the waste processing plants, are used in modern plants to tear and open the

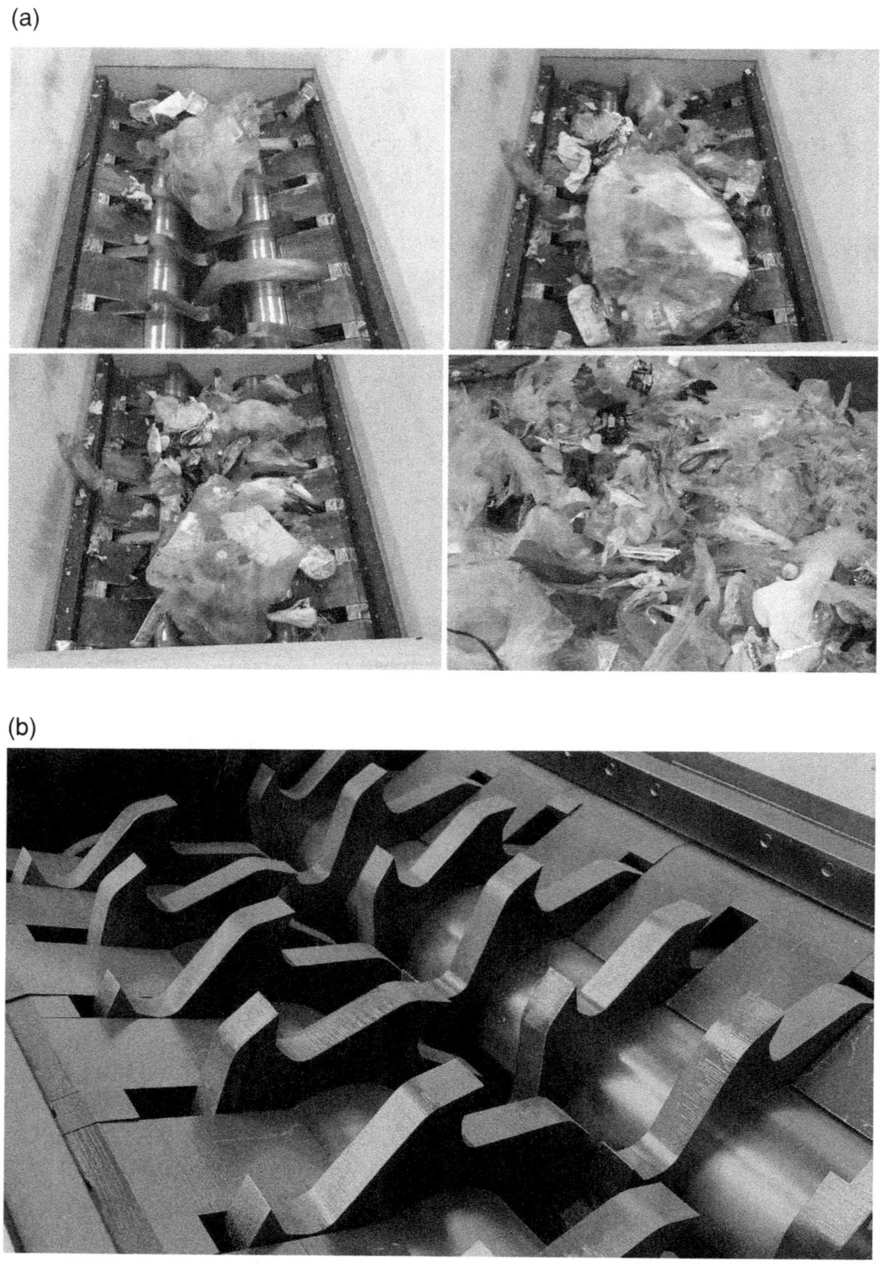

Figure 21.5 (a) Shredder bag opener with two counterrotating shafts on duty (top and bottom left images) and product (bottom right image) and (b) detail of the shafts. *Source:* Reproduced with permission of Forrec srl Recycling Systems (2016a, b).

waste-containing bags in a way that their content is neither cut nor damaged. The operation of bag openers is based either on shredding or ripping.

The shredder-bag opener is composed of two counter-rotating shafts (Figure 21.5a) and vice versa; the latter motion of the rotors is applied to reject tenacious materials for safety reasons. Cutting tools (knives or toothed blades), which are made of excellent quality alloyed steel, are fixed on the shaft (Figure 21.5b). The shape, thickness, and number of blades are defined in function of the material to be treated and of the required size at the exit of the equipment. In addition to plastic bag tearing, this machine can also work to shred relatively soft waste materials, such as small branches, wood, pallets, organics for compost, etc.

The tearing of plastic bags in the ripper is achieved with the aid of retractable or fixed tines and a comb (Figure 21.6a). The tines are made of alloyed steel and mounted on the rotating ripper head. The ripper is adjusted to provide sufficient space below the rotor to allow the heavier, flatter, and smaller materials than the opening to pass and be carried away by a conveyor, which

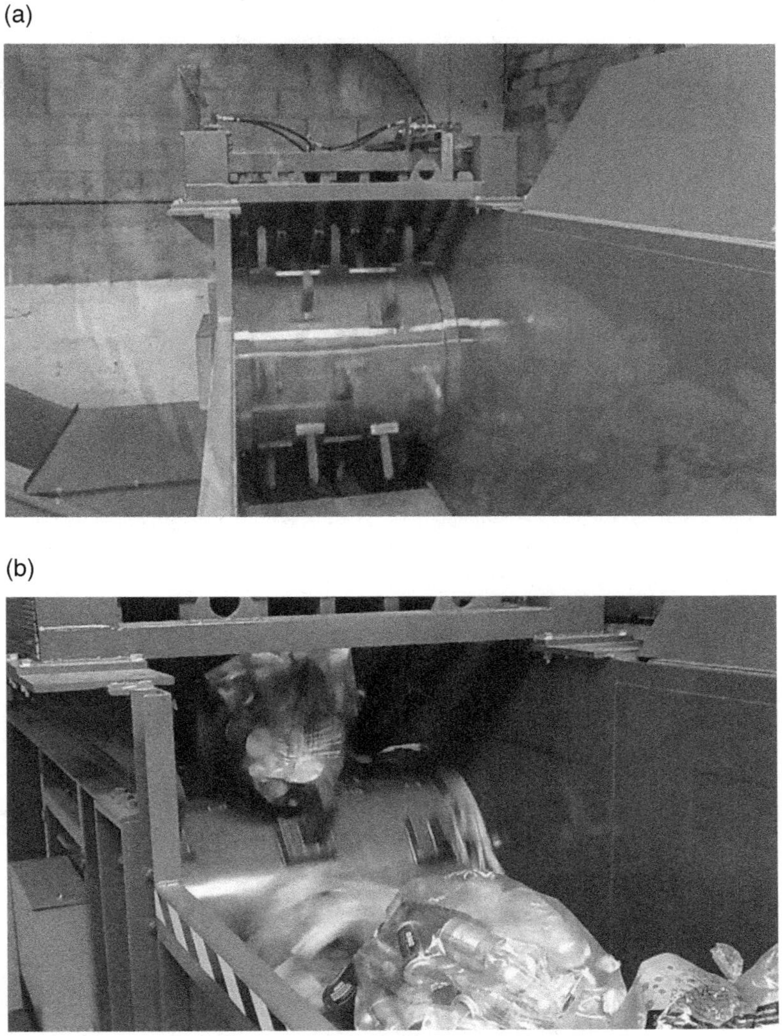

Figure 21.6 (a) Plastic bag ripper with fixed tines, and comb and (b) bag ripping. *Source:* Reproduced with permission of Matthiessen Engineering SARL (2014).

is installed below the ripper. The bags along with other light material move toward the ripper, which rotates in opposite direction to the material flow. Most of the introduced bags (about 98%) and light objects are caught by the retractable tines, lifted to the top of the rotor, and dropped into the ripper comb. As the tines pass between the comb elements, the bags are squeezed and opened (Figure 21.6b); the bagged material is liberated unaffected and, after passing through the comb elements, falls on the conveyor that carries the aforementioned unprocessed material. The tines of the rotor retract upon passing through the comb at the top of the cycle to detach any film or plastic material that might have been tangled on the tines, before the extension of the tines in a new cycle, which starts when they are just above the conveyor. Above the rotor, there is a hydraulic comb lifter to allow the passage of hard, dense materials that cannot be opened by ripping. In addition to bag shearing, rippers are also used to reduce the size of soft and light materials, such as plastic objects (Figure 21.7).

After bag opening, the material is liberated and often contains plastic films that may cause problems in the following processing steps. To avoid these problems, a film separator (grabber) is usually installed after the bag opener. Regarding its design, the grabber is composed of a rotating drum with fixed tines on its surface (Figure 21.8) to catch the plastic film, wrapping material, and other light, thin objects.

21.2.2.3 Wood Grinders

Wood grinders are used to shred large pieces of wood (e.g. yard wastes, construction debris, palettes) into chips that can be used either as fuel or for biological treatment. Tub grinder is the most commonly used equipment, but horizontal flail mills can be used as well.

Tub grinders are mobile machines, carrying diesel engines. They consist of a large-diameter tub having a slowly rotating upper section and a stationary lower section containing a flail or hammer mill (Figure 21.9). The machine is usually mounted on a trailer frame so it

Figure 21.7 Ripper for plastic and light material. *Source:* Reproduced with permission of Sherbrook-OEM (2015), ripper manufacturer: Matthiessen Engineering SARL.

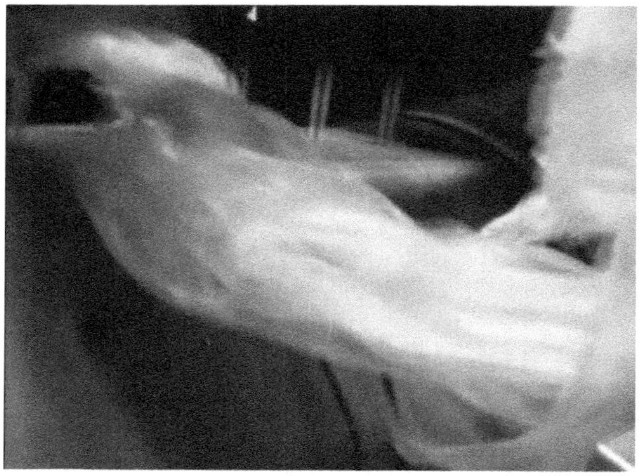

Figure 21.8 Plastic film grabber. *Source:* Reproduced with permission of Sherbrook-OEM (2015), grabber manufacturer: Matthiessen Engineering SARL.

can be easily moved. Material is fed into the top of the tub and is carried around the point where the flails or hammers intrude into the tub. The rotation of the tub ensures the continuous flow of the material to the mill. The shredded product passes a grid, which is placed below the rotor, and is directly removed by falling on a conveyor. The grid aperture determines the shredded product size.

Figure 21.9 Tub grinder. *Source:* Reproduced with permission of Vermeer (2000).

21.3 Size Separation

Size separation or classification is a process that separates the particles into fractions according to their size, in dry or wet environment. In solid waste processing, the most common process is dry screening, which is based on size differences in regard to any two dimensions of the particles. In waste treatment, screening is targeted to (i) remove the oversized materials from the waste processing stream, (ii) separate the feed into proper size fractions according to the requirements of sorting equipment, (iii) selectively sort waste components into separate products in case that selective shredding is accomplished, and (iv) remove the oversized material from combustion ash.

In terms of screening, the waste fraction with size less than that of the screen aperture is called undersize, while that of greater size is called oversize. The fraction of the undersize that passes through the apertures is called underflow (or passing), while the overall nonpassing fraction is called overflow (or remaining).

Depending on the characteristics of the waste stream, many screening devices have been used in practice, but the most common are the vibrating flat bed, rotary (trommels), and disk screens.

21.3.1 Vibrating Screen

The key element in vibrating screens is the screening surface, as the real work of particle stratification and separation is accomplished on it. Screening surface is mounted on an inclined (or horizontal) frame that is supported by bars, installed directly below the screening surface and along it. This construction permits the use of light and thin screening surfaces. The screening deck is subjected to vibrating motion with the aid of an eccentrically rotating shaft, placed directly below the supporting bars. The rotating motion of the shaft is accomplished through an electric motor and drive mechanism, which impart motion to the shaft. The configuration of the vibrating screen is shown in Figure 21.10. The screening surface may be composed of one or more decks, but in most cases no more than two. Also, it may be one deck divided mostly into two parts with different aperture size. Screens are flat surfaces with a wire mesh or metal surfaces perforated with holes.

The factors that affect the efficiency of the vibrating screens are solid waste composition, size and shape of particles, feed rate, stroke length and vibration frequency, length to width ratio, and screen slope. Vibrating screens work best with relatively dry and granular mixtures (e.g. sand, stones), separating well and offering a substantial advantage over trommels in terms of cost and space requirements. Vibrating screens are effectively used to separate fine glass, undersize material from source-separated and commingled MSW, and to process demolition and construction materials. Contrary, they are not suitable to process irregular-shaped, wet, and/or soft materials (such as paper, rags, etc.), since they cause clogging of the screen apertures, as well as when the undersize is a small fraction of the feed stream.

Figure 21.10 Vibrating screen. *Source:* Reproduced with permission of Metso Minerals (2005).

In addition to vibrating, there are also shaking (or reciprocating) screens, whose screening deck is submitted to a substantially linear motion by an eccentric mechanism. For coarse waste particles, vibrating screens prevail over reciprocating because of their higher efficiency and capacity.

21.3.2 Rotary Screen (Trommel)

Rotary screen (or trommel) is a popular screening device in solid waste processing because of its effectiveness, efficiency, and clogging resistance of the apertures due to its design and operation.

Essentially, trommel is a large-diameter rotating drum with the sidewall being the screening surface (Figure 21.11). Trommel may separate waste materials into several size fractions as the cylindrical screening surface can be composed of successive sections with different aperture size each. The screen is normally set at a sloped position between 2° and 5° that facilitates the downward motion of the materials. The waste material is fed at the elevated end of the inclined trommel and flows downward. As screen rotates, material tumbles and contacts the screening surface many times as it moves downward. Small-sized materials pass the

Figure 21.11 Trommel screen. *Source:* Reproduced with permission of Metso Minerals (2000e).

openings of the trommel, with the finer particles being removed close to the feeding point and progressive increase of the size toward the end of the length. Sometimes, metal blades (lifters) protrude into the drum of the trommel accomplishing bag breaking and better mixing of solid waste materials.

Trommel efficiency is affected both by geometrical and operational factors, such as length, diameter, angle of slope, rotational speed, and feed rate. Rotational speed is the most crucial operational factor, as it must be maintained at an optimum value. Optimum rotational speed occurs when the waste material tumbles in the drum forming a cascade, which means that waste material is partially carried up to the interior wall of the rotating drum, reaches at about the maximum height, and then falls back to the bottom. The optimum rotational speed is usually obtained at 50–80% of the critical rotating speed, which is the maximum speed that material tumbles. For rotational speed above the critical, waste material does not tumble but is centrifuged or stuck at the drum wall. Critical speed (n_c) depends on drum diameter, according to the relationship

$$n_c = \left(\frac{1}{2\pi}\right)\left(\frac{g}{r}\right)^{1/2} \tag{21.1}$$

where g = gravity acceleration
r = trommel radius

Trommel slope affects the residence time of the waste material in the drum as well as the trommel throughput; in turn, throughput depends on trommel diameter and filling factor.

The trommel screen may be installed either before or after the shredder. It is placed prior to shredding, when the feed contains high percentage of inorganic materials (e.g. stones, metals) that have to be removed, plastic bags that have to be torn off, friable materials (e.g. glass containers) that have to be broken and removed, particles with different size and shape (e.g. glass, metal, cardboard, newspaper), or increased percentage of pieces with size less than the maximum size of shredder discharge product. The trommel is installed after shredding to separate combustible sheetlike materials (organics) from predominant noncombustible granular materials (inorganics).

21.3.3 Disk Screen

Disk screens are composed of a series of parallel, horizontal, and equally spaced bars or shafts, which are installed perpendicularly to the flow of the material. Each shaft is equipped with interlocking round, oval, serrated, or star-shaped disks, installed at equal distance across the width of each bar (Figure 21.12) (West Salem Machinery, 2016). The shafts rotate, transporting wavelike the feed along the length of the machine and presenting the material to the screen openings. The undersize fraction falls in the space between the disks, while the oversize overflows the disks. In some designs, the distance between the disks may be adjusted according to the size of the material to be separated as underflow. After suitable adjustment, the same screen can be used to classify the waste material into varying size fractions.

Disk screens present several advantages over vibrating. The main advantage is that they do not favor clogging of the apertures. In case that clogging tends to occur, due to wrapping of materials around the disks, they are usually self-cleaned by rotation. The comparison of disk screen with trommel reveals that residence time, rolling around, and mixing with contaminants are reduced in disk screen because the rectangular openings keep the material moving; disk screen also prevents fines from becoming blocked by larger items or longer items and, consequently, from falling through the gaps. Materials such as metals, glass, compost and wood, tires, construction/demolition waste, and shredded plastic bottles have been separated successfully as undersize.

The screen opening is defined by the distance between the disks in one direction and the distance between the shafts in the other. The conventional disk screens have uneven openings that cause inaccurate sizing and material wrapping (Figure 21.13a). BHS has designed and patented a disk screen with even-sized openings (Figure 21.13b) that result in highly accurate material sizing and separation. The design of this screen is based on the replacement of the conventional disks with tri-shaped ones (Figure 21.14). This design is suitable to separate old corrugated cardboard (OCC) from other fiber, various plastic and metal containers, and debris. The configuration of in-line disks imparts a bouncing wavelike action on the material stream, liberating other types of fiber and contaminants from the OCC product, minimizing material wrapping and clogging, and resulting in high throughput because of the reduced downtime (Bulk Handling Systems (BHS), 2014a, b). The smaller size material passes through the void between the disks, while the OCC jumps over the disks to a bin or conveyor.

21.4 Manual-/Sensor-Based Sorting

21.4.1 Manual (or Hand) Sorting

Manual sorting was the first method used for garbage or scrap sorting, and it still remains very popular nowadays because it is the simplest, most direct, and sometimes

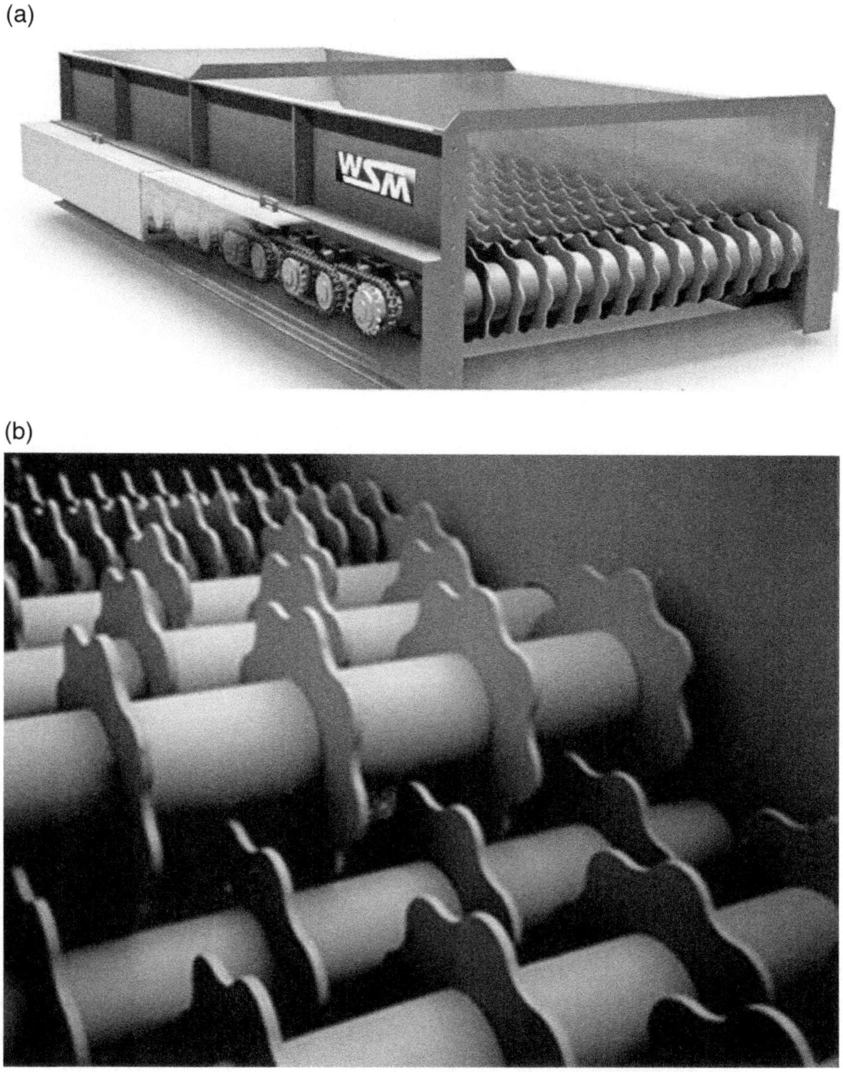

Figure 21.12 (a) Star-shaped disk screen and (b) detail of disks arrangement. *Source:* Reproduced with permission of West Salem Machinery Co. Inc (2016).

the most effective separation method. As the name denotes, the desirable components are removed from the material stream by hand. In order to distinguish the various items in a feed stream, the workers use various visual recognition tools, such as color, shape, luster, and transparency; among them, color sorting, which is based on color differences, is the most popular.

Hand sorting may involve either the removal of the recyclable items or the discards out of the waste stream (presorting). When recyclables are removed, the quality of the product is higher but the collection rate is slower. On the contrary, discards removal results in higher collection rate but in lower quality of the product; in addition, it protects from unnecessary processing lines and results in lower waste processing cost. Depending on waste composition and target, hand sorting can be applied on product stream from shredding, magnetic product to remove nonmagnetic entrapped items, and nonmagnetic product to separate items of similar density and behavior in reference to eddy-current separation, but of different composition.

Manual sorting can be practiced both for commingled and source-separated waste materials under the condition that solid waste components are of relatively large size and possess properties discernible by eye. It has been used to separate zinc, copper, brass, and stainless steel from aluminum in a nonferrous scrap stream. It has also been cited that hand sorting is capable to sort wrought from cast aluminum fractions because of their distinctive surface characteristics (Rao, 2006). Although there is no empirical evidence of the capabilities of hand sorting, it is estimated that workers in China can achieve accuracies up to 99% when sorting nonferrous automotive shred (Minter, 2008).

Hand sorting can be realized either on a static surface, e.g. ground, table, and facility floor (Figure 21.15) (Bruno, 2013), or on a belt conveyor (Figure 21.16) (Medio Multimedia, 2011). At a recovery facility with

(a)

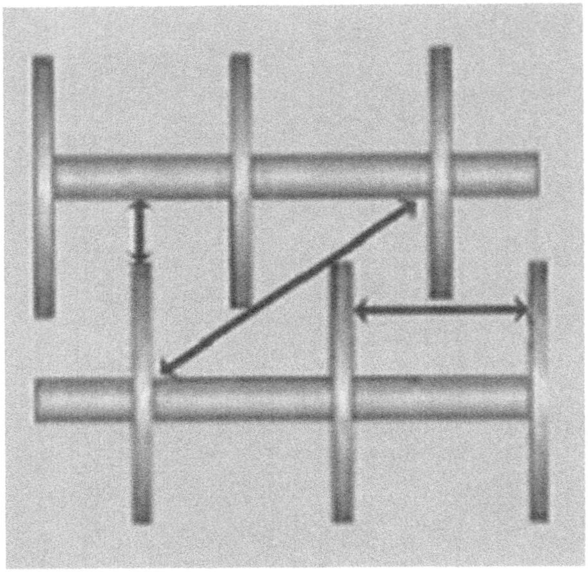

(b)

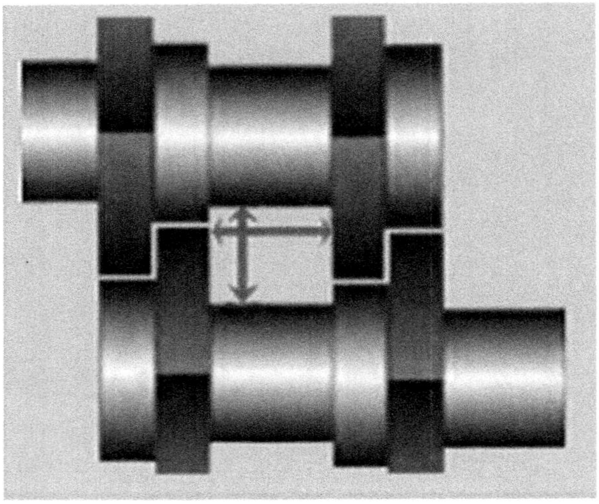

Figure 21.13 Disk screen openings (a) with uneven size and (b) with even size. *Source:* Reproduced with permission of Bulk Handling Systems (BHS) (2014a).

Figure 21.14 Disk screen with tri-shaped disks. *Source:* Reproduced with permission of Bulk Handling Systems (BHS) (2014b).

belt conveyor, the workers, placed at the side (or both sides) of the belt, separate the waste stream into individual components by handpicking and drop them into chutes that lead to storage containers. The containers are located either at the same level with the sorting line or below it, as it is presented in Figure 21.16 (Medio Multimedia, 2011; Summit Equipment, Inc., 2012a).

In addition to being unpleasant and possibly unhealthy in some cases, manual sorting is a labor-intensive process. For this reason, manual sorting is a prevalent practice for scrap separation in countries with low labor cost (e.g. some Asian countries). For example, much of the nonferrous scrap, which originates from automobiles shredding operations in Europe and the United States, arrives with container ships in the southern ports of China, where labor cost is lower compared with that of the automated processes applied in high labor cost countries. It must be pointed out that the value of scrap from the United States to Taiwan, Korea, Hong Kong, and China has grown fivefold in 5 years (Plunkert, 2010).

For efficient hand sorting, the following must be taken into consideration: (i) prior to sorting, "liberation" of the commingled waste is necessary, which means that plastic bags with stored waste must be torn and spread out on the sorting belt; (ii) the room with sorting lines must follow the rules posed by the health authorities, namely, the room must be well lighted, aerated, and air-conditioned; and (iii) sorting lines must be properly designed.

The limitations of manual sorting are referred to the high labor cost, especially for recycled commodities of relatively low selling price, and the misplacement of items. For example, in case of painted or plated items with another metal, it is difficult or impossible for the workers to identify the items correctly. As a result, automated sorting methods are applied, which are characterized by high resolution, accuracy, and efficiency.

21.4.1.1 Design Criteria of Hand-Sorting Line

In comparison with static surfaces, the use of conveyors makes hand-sorting process more efficient, under the condition that sorting facilities are properly designed. Some of the factors that determine the design of hand-sorting facilities are throughput, solid waste composition, number of components to be recovered and recycled, and characteristics of solid waste. A sorting belt conveyor is completely determined by factors such as belt width and length, speed,

(a)

(b)

Figure 21.15 Hand sorting (a) on the ground in large-scale recycling of imported shredded automobile scrap at Foshan (China) and (b) on static table in a modern establishment in Shanghai (China). *Source:* Reproduced with permission of Adam Minter, Bruno (2013).

and average waste thickness. In addition, the number of the workers must be determined. The critical factors are the length of the belt and the number of the workers.

To calculate the two aforementioned critical factors, a simple methodology is followed based on the synthesis of various practical rules, derived from experience. For the calculation, the following six data are necessary:

1) Weight of solid waste to be sorted that covers a surface of 1 m^2 in monolayer arrangement (it is determined experimentally or from diagrams).
2) Percentage of the material to be sorted in the feed.
3) Sorting capacity per worker (it is determined from field data or diagrams).
4) Belt conveyor speed.
5) Distance between workers.
6) Belt conveyor width.

Example: Let's calculate the belt conveyor length for the manual sorting of plastic bottles from a commingled waste stream. The solid waste feed rate is 10 t h^{-1} with 5% content in plastic bottles. For calculation reasons, we assume a belt conveyor speed 12 m min^{-1} with 1.0 m belt width and waste weight 15 kg m^{-1} of belt. The sorting capacity is considered to be 250 kg/8 h and worker.

(a)

(b)

Figure 21.16 Setup of hand sorting lines, (a) parallel sorting lines at Sindos (Thessaloniki, Greece) recycling center, with workers placed at both sides of the conveyors (Courtesy: Periklis Panagiotidis, from Medio Multimedia (2011)), and (b) sorting line setup with above-floor chain-belt conveyor. *Source:* Reproduced with permission of Summit Equipment, Inc. (2012a).

Calculations

1) Feed rate of conveyor belt: $15\,\text{kg}\,\text{m}^{-1} \times 12\,\text{m}\ \text{min}^{-1} \times 60\,\text{min}\,\text{h}^{-1} = 10\,800\,\text{kg}\,\text{h}^{-1} = 10.8\,\text{t}\ \text{h}^{-1}$. So, the conveyor capacity covers the hourly rate.
2) Quantity of plastic bottles: $10\,\text{t}\,\text{h}^{-1} \times 0.05 = 0.5\,\text{t}\,\text{h}^{-1}$
3) Sorting capacity is $0.250 : 8 = 0.03125\,\text{t}\,\text{h}^{-1}$ per worker
4) Number of workers = 0.5/0.03125 = 16
5) If the necessary distance between two adjacent workers to work conveniently is 1 m, then the required conveyor length is 16 m. By adding 1.5 m at the feeding point and 1.5 m at the discharge point of the belt conveyor, it derives that the total length of the conveyor is 19 m.

If the conveyor is long enough and the space is relatively limited, then two conveyors are installed.

21.4.2 Sensor-Based Sorting

With the development of sensoring technologies, sensor-based sorting systems have been widely applied in the recent years as an alternative to manual sorting of solid waste. Compared with the traditional manual sorting, the material fractions obtained are of significantly higher purity, recovery, and, consequently, value. These systems use sensors to detect specific optical, chemical, or physical properties of single particles. Basically, a typical sensor-based sorting device includes the following three units: feeding, identification (or sensor), and discharge unit.

The first part of the sorting device comprises the feeding unit, where waste particles are individualized before entering the identification unit; individualization is achieved with the use of vibrating feeders placed below the bin, conveyor belts, and chutes. The particles in the waste flow have to lay in monolayer or one-by-one arrangement, with enough free space around their boundaries to allow every single waste particle to be scanned by the sensor in the identification unit. In addition, they have to stay fixed on the surface of the conveyor belt, so that their position is traced from identification till discharge. To achieve high degree of separation, waste has to be processed by size reduction and sieving prior to sorting.

Basically, the detection of the specific attributes of the particles in the mass flow is carried out by an array of sensors that work on emission-detection principle. Radiation of known characteristics (intensity, frequency, and composition of wavelengths) is emitted to the particle surface, strikes on the surface, changes its characteristics (because of absorption, reflection, transmission, diffraction, fluorescence, etc.), and is redirected to the detector, which detects the attributes of the spectrum emitted from the surface (intensity, composition of wavelengths). In fraction of a second, the data is transmitted to the linked computer, processor analyzed, compared with preset criteria as programmed in the system software, and subsequently, a decision is made if the particle is acceptable or not upon leaving the fast-running belt. The signal processor is capable to process some decades of millions of measuring points per second. This enables the processor to identify and analyze some thousands of particles per second.

Depending on sensing technology, a large number of sensors have been developed to separate the various components in a waste flow, including color sensors, X-ray transmission sensors, laser-induced breakdown spectroscopy (LIBS) sensors, X-ray fluorescence (XRF) sensors, near infrared (NIR) sensors, magnetic induction (MI) sensors, or combination of various sensors (e.g. color, shape, induction).

After particle identification and detection of its position and size, the waste particles pass one or more arrays of air valves at the end of the conveyor. If a waste particle is defined as acceptable, the corresponding air valves are activated by the identification unit of the computer, open, and emit compressed-air pulses, ejecting the particle from the mass flow.

The process sequence for color sorting is presented in Figure 21.17.

21.4.2.1 Color Sorting

This system relies on color differences of the waste particle surface; in addition to color, other criteria being considered are brightness, transparency, and particle size and shape. The color sorting device normally consists of an illumination unit and light sensors. The material flow, which is uniformly distributed across the working width, moves toward the illumination unit uniformly lightened. Regularly arranged high-performance color sensors identify all waste components at the respective positions with high accuracy. Transmitted light systems are used to identify transparent materials (e.g. PET bottles or flakes), while reflected light systems are used for non-transparent to opaque materials (e.g. metal from metal,

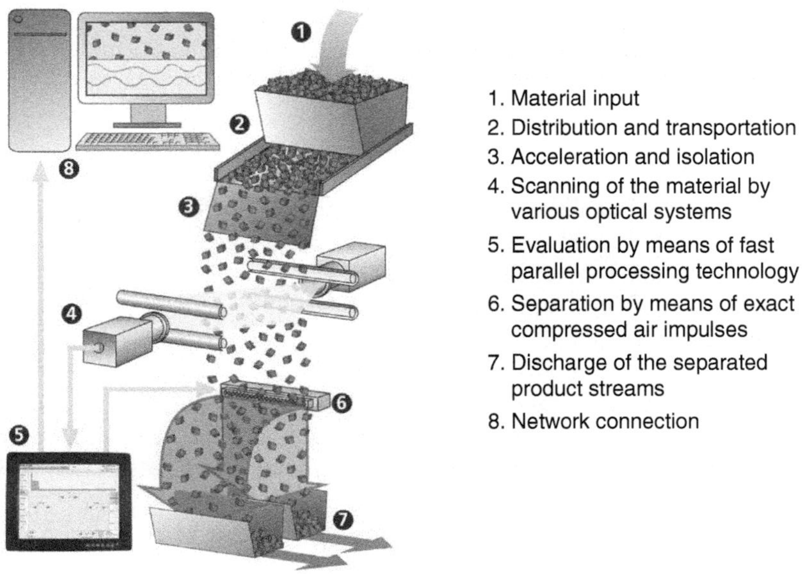

Figure 21.17 Operating principle of electronic sorting. *Source:* Reproduced with permission of Mogensen GmbH & Co. KG (2016).

e-scrap from printed circuit boards) (RTT Steinert Elektromagnetbau GmbH, 2015a). After identification, precise ejection pulses from compressed-air nozzles blow the identified items out of the material stream. In case that the material cannot be unequivocally classified by means of single-side scanning, multiside scanning is applied. The resolution limit of color sorting is approximately at least 3 mm (RTT Steinert Elektromagnetbau GmbH, 2015a). The working principle of color sorting is presented in Figure 21.17.

Color sorting systems can generally process waste with particle size ranging from 5 to 250 mm. When transparent materials are processed, the capacity ranges from 1 to 10 t h^{-1}; for metal scraps, it ranges from 2 to 200 t h^{-1}. The capacity depends on particle size and working width, which generally ranges between 0.60 and 3 m. For sized waste feed, the process efficiency is higher than 98% due to the high resolution of the sorting equipment (Mogensen GmbH & Co. KG, 2016; RTT Steinert Elektromagnetbau GmbH, 2015a). In industrial scale, color sorting has been used successfully to separate white, green, and brown glass in recycling plants, plastic bottles (PET), while it is also possible to separate nonferrous metals, e.g. copper and brass, from lead and zinc metals, as it is shown in Figure 21.18.

21.4.2.2 Near Infrared (NIR) Sorting

When materials are illuminated, they mostly reflect light in the NIR wavelength spectrum. The NIR sensor can distinguish the various materials based on the way they reflect light. It has been developed to identify and segregate plastic materials, wood, paper and cardboard in household waste, bulky refuse, industrial waste, and waste material mixtures. It can also remove chloro-substances from RDF, as well as recycle cardboard, beverage boxes, and PET bottles, and separate fine-grained plastic materials, which originate from electronic scrap or plastic bottles, by material type. NIR sensors can process particles with size range between 10 and 250 mm; depending on particle size and density, the capacity ranges from 1 to 7 t h^{-1}. The minimum resolution of NIR sensors is 3 mm (RTT Steinert Elektromagnetbau GmbH, 2015b).

21.4.2.3 X-Ray Fluorescence (XRF) Sorting

X-ray fluorescence allows the detection of existing elements in particles through photoemission. When incident X-ray radiation of sufficient energy hits an atom, the absorbed energy causes excitation of the inner shell electrons and their emission to outer shells. In case that the offered photon energy is low, the electron is unable to escape the material and, upon returning to the initial electron shell, emits photon with wavelength specific to the material. When the offered photon energy is high, the electron escapes the material and an electron vacancy is created. In this reaction, a loosely bound electron from an outer electron shell will fill the vacancy with simultaneous emission of an X-ray photon (Auger electron).

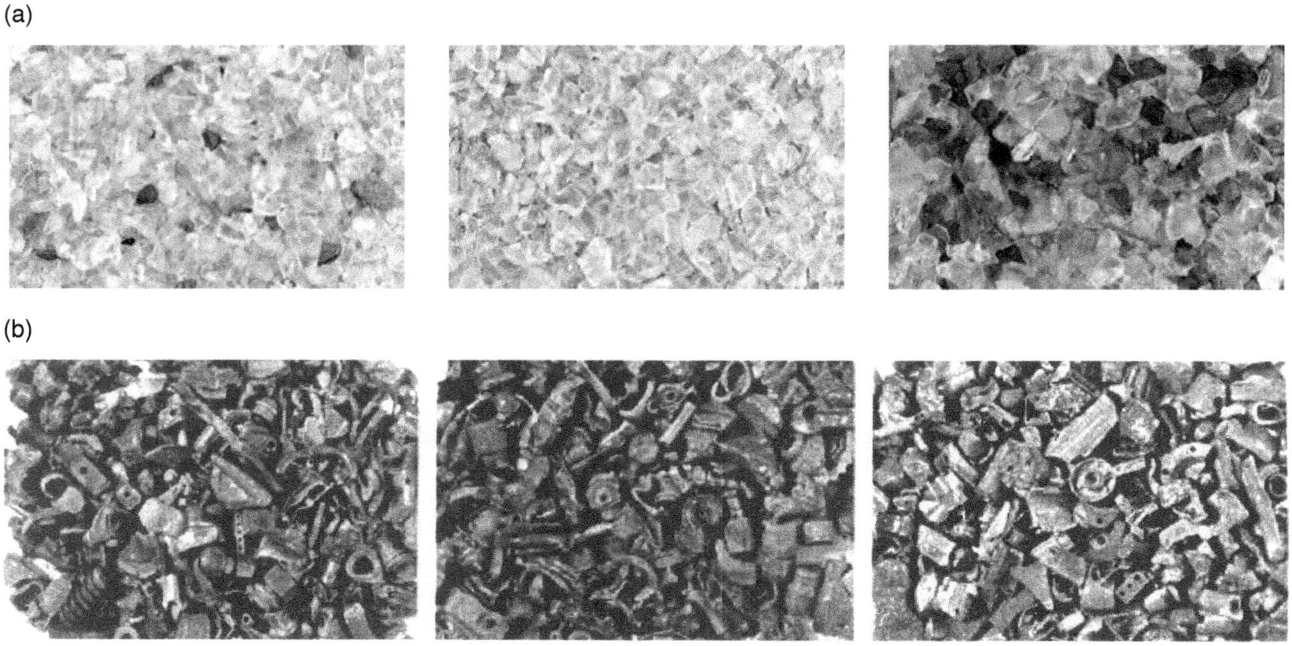

Figure 21.18 Products after optoelectronic separation (from left to right): (a) input, accept stream (white glass), reject stream (colored glass, ceramics, stone, and porcelain). *Source:* Reproduced with permission of Mogensen GmbH & Co. KG (2016) and (b) input, copper-brass stream, white metals stream. *Source:* Reproduced with permission of Mogensen GmbH & Co. KG (1999).

The emitted fluorescent X-ray photon has a characteristic energy that is equal to the binding energies difference between the shell of the transitioning electron and the shell of the vacancy. These characteristic X-rays serve as the "fingerprint" for XRF analysis given that each element has a unique spectrum of electron energy levels and, therefore, unique fluorescent X-ray photon energies.

The setup of XRF instrumentation consists of four components: excitation source, detector, data collection system, and analysis system. The excitation source is usually an X-ray tube. This tube sends an X-ray beam with various energies to the sample, which absorbs and emits the X-rays to the detector. This fluorescence can be measured with an energy dispersive X-ray sensor. The data are processed, and information is obtained about the presence of elements and their concentration. This information is combined with additional data from another sensor (e.g. electromagnetic) to identify the position and size of the object, and a set of valves is subsequently activated to precisely eject the targeted waste particle.

For accurate reading and safe sorting, however, the concentration of the element to be determined in particle mass must be at least 10–20 ppm, since XRF sorting is based on the quantitative analysis of the material.

XRF sorting has been applied in the following cases of waste material segregation: metal sorting from ferrous scrap, e.g. to recover copper from ferrous scrap and produce a clean ferrous steel fraction; aluminum cleaning to produce a more valuable clean aluminum fraction by sorting out heavy metals like copper, brass, zinc, lead, etc.; cathode ray tube (CRT) sorting to produce clean monofractions of panel glass (lead free) and funnel glass (high lead content); RDF production by sorting out PVC, glass, stones, and metals; organic sorting to produce a clean organic fraction and a clean nonorganic fraction for landfilling; PVC removal to remove flame retardants, like chlorine and bromine, out of a mixed plastic stream; and metal removal or suppression (Weiss, 2012).

Portable XRF units, which carry out sorting in real time, are currently in use; however, the relatively high cost prevents them from extensive application in scrap processing yards.

21.4.2.4 X-Ray Transmission (XRT) Sorting

When materials are exposed to X-rays, they absorb part of the incident radiation. As a result, the intensity of the transmitted radiation is reduced, and this phenomenon is referred to as transmission damping. An X-ray sensitive camera detects the intensity of the transmitted radiation; in milliseconds time, the computer software determines the difference between the incidental and transmitted radiation, and the acceptance or not of the material (Strydom, 2010). According to Lambert's law, the detected radiation is an inverse exponential function of density and thickness of the irradiated material. As material density and thickness increase, the detected radiation decreases, while the absorbed increases.

X-ray transmission is a high-speed identification technique of the materials with resolution limit about 2 mm. Contrary to other identification methods, the transmission of the beam throughout the material mass results in the detection of the entire volume of the irradiated material instead of a surface layer only and in safe conclusions about its atomic composition. Consequently, the system creates an image near to reality, which is sufficient for subtle evaluations (Steinert Elektromagnetbau GmbH, 2015a).

As waste particles show significant variance in size (thickness) and shape, it is expected that the use of a single energy wave (monochromatic imaging) to identify materials would be inaccurate. This means that waste particles of the same density but of different size and shape will transmit X-rays at different intensities (Strydom, 2010). In order to compensate the effect of different particle size and shape, dual energy X-ray transmission (DE-XRT) sorting has been used, which is a modification of a previous technique (Bokun and Osadchii, 2010).

In this process, two X-ray beams with different energy levels (low and high) hit to the material. The X-ray sensor system, which works like a line-scan camera, registers the X-rays penetrating the material and converts them into digital image data. The sensor system (Figure 21.19) consists of two channels, each capturing the image of the material at different X-ray energy levels (von Ketelhodt and Bergmann, 2010). Each particle attenuates the received X-ray radiation, thus decreasing the modulation amplitude of the sensor to varying degrees, so that these image areas appear in different shades of grey. The attenuation depends on both thickness and atomic density of the material. Images of different atomic densities are transformed into images of different spectral ranges, which make possible the classification of different color pixels according to specific atomic densities. This is accomplished almost regardless of the thickness of the material (von Ketelhodt and Bergmann, 2010).

The high resolution of the sensor (0.8–1.6 mm) is also used to evaluate the particle shape and size, material thickness, and texture of the gray-scale image that represents inclusions of material with varying densities. Thus, the X-ray transmission image processing provides a highly efficient sensor system for classifying different materials (von Ketelhodt and Bergmann, 2010).

Some applications of this technique in the recycling industry include the identification and sorting of scrap metals and plastics, estimation of the calorific value of

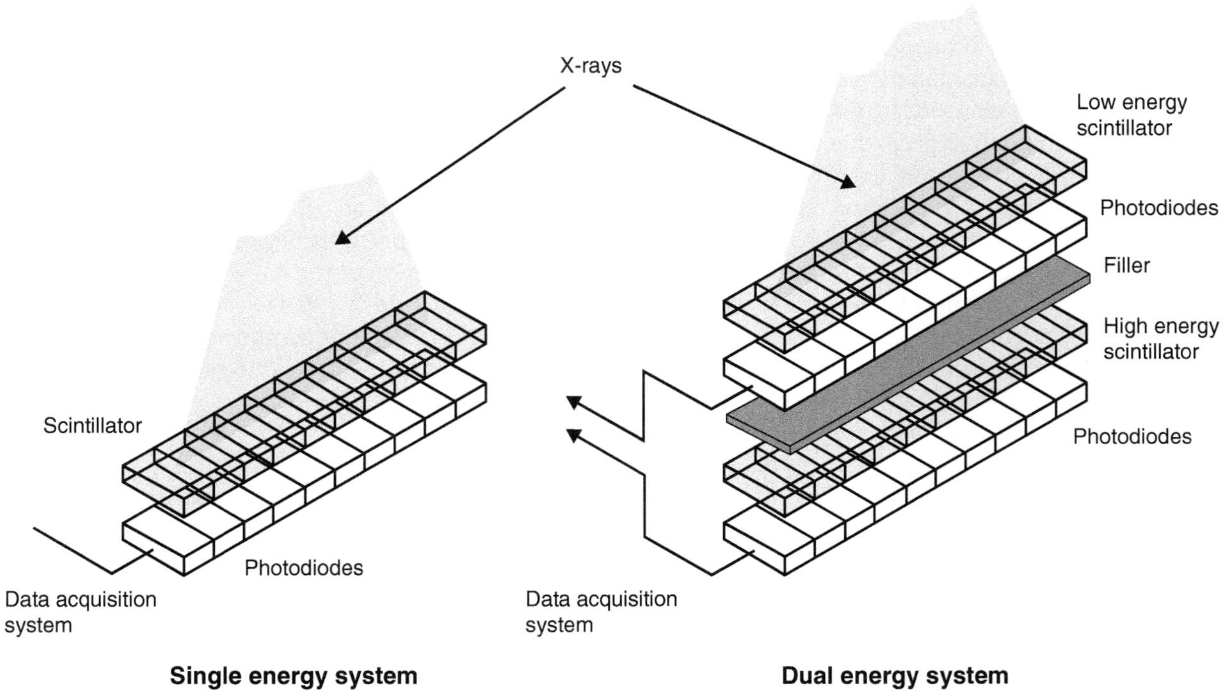

Figure 21.19 Single-energy vs. dual-energy sensing principle. *Source:* Reproduced with permission of The Southern African Institute of Mining and Metallurgy (SAIMM) (2010).

packed mixed waste, and sorting of shredder residue to recover metals, solid fuel, polymers, and inorganic filler matter (Dalmijn and de Jong, 2004; Bruno, 2015).

21.4.2.5 Laser-Induced Breakdown Spectroscopy (LIBS)

LIBS is a spectrographic technique, which utilizes a pulsed laser and optical emission spectroscopy. A typical system consists of the following major components: a neodymium-doped yttrium aluminum garnet (Nd:YAG) solid state laser, a spectrometer of wide spectrum range, a fast but highly sensitive detector, and a computer for data analysis (Dalmijn, 2003). The setup of LIBS is shown in Figure 21.20 (Applied Photonics Ltd, 2016).

In LIBS, a high energy pulsed laser is used as excitation source. The pulsed laser is focused on a small area on the surface of the material. When the laser is discharged, it ablates a small quantity of the surface material (usually in the range of nanograms to picograms), generating a plasma plume of very high temperature (approx. 10 000–20 000 K). At this temperature, the ablated material breaks down into excited ionic and atomic species, and the plasma emits continuous radiation containing no useful information about the species investigated. However, within milliseconds, the plasma expands at supersonic velocities and cools down rapidly. At this point the characteristic atomic emission lines can be collected and analyzed (Applied Photonics Ltd, 2016).

LIBS sorting technology has been applied to sort aluminum alloy (Mutz et al., 2003), separate wrought and cast alloys, sort wrought alloys by alloy family, and separate materials in automobile applications (Gesing and Harbeck, 2008). LIBS is particularly useful for high volume analyses or online industrial monitoring, as it is a very rapid technique (Gaustad et al., 2012).

However, LIBS sorting presents the following defects: the scrap must be free of coatings (e.g. paint, lubricants), since pulse laser can penetrate the surface only at small depth; potential oxide formation on the surface could cause erroneous readings; each pulsed laser is able to detect a very small surface area (much less than 1 mm^2); homogeneous distribution of the elements is required for high accuracy; and the cost is high.

21.4.2.6 Magnetic Induction (MI) Sorting

This technique is based on MI differences of particles. MI sorter can be used to recover the ferrous and nonferrous metals that have not been previously recovered by eddy current or magnetic separation, to separate one metal from the other, and to process electronic scrap. MI has been applied on primary waste streams, such as shredder residue, electronic scrap, wood chips, glass, and domestic waste (Steinert Elektromagnetbau GmbH, 2015b).

The bulk material is transported by a fast conveyor belt to the sorting zone. Below the conveyor belt, immediately in front of the main coil of the induction sorting

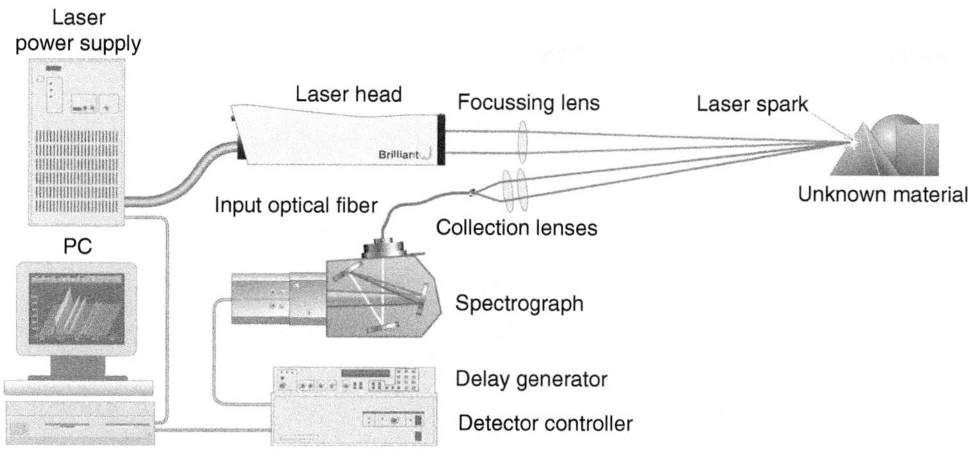

Figure 21.20 Working principle of laser-induced breakdown spectroscopy. (*Source:* Courtesy of Applied Photonics Ltd.-UK (2016), with permission.)

Figure 21.21 Magnetic induction sorting. *Source:* Reproduced with permission of Steinert Elektromagnetbau, G.H. (2015).

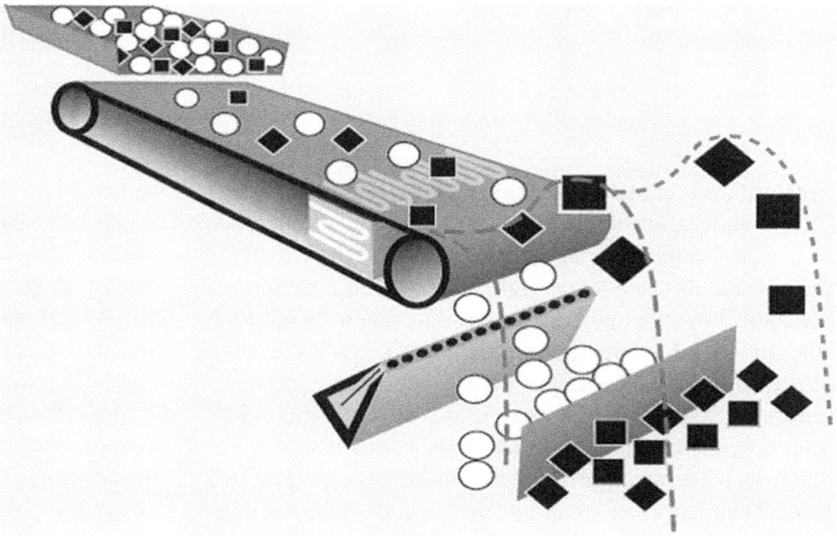

system (ISS), there is an arrangement of sensors. These sensors analyze the material over the whole width of the conveyor belt by means of MI. As soon as metallic particles are detected, electronic signals are sent to the central computerized control unit. The compressed air jets, individually controlled by programming, push the detected metals over the diverter gate, as it is shown in Figure 21.21 (Steinert Elektromagnetbau GmbH, 2015b).

21.4.3 Robot Sorting

In case of construction and demolition (C&D) waste, manual sorting is unpleasant and dangerous, because the workers are exposed to hazardous elements (e.g. dust, mold, asbestos) existing in the waste stream. In addition, hand-sorting processes often fail to minimize the participation of the value material in the reject and to maximize the purity of sorted material fractions in order to be correspondingly disposable without cost or sellable for profit. To overcome all these problems, robots have been recently designed and replaced manual sorting.

ZenRobotics Ltd. has recently developed a recycler that has been commercially used for C&D sorting. ZenRobotics Recycler (ZRR) comprises of a sensory network ("brain") on top of a conveyor, a control system, and industrial robots (Figure 21.22). The functional principle of ZRR is based on sensory fusion, data mining, machine learning, and real-time robotics in picking objects from waste streams.

Sensory fusion means combining sensory data from multiple sensors to form a coherent sense of the surroundings. Instead of following a preprogrammed routine, the robot works on the basis of inputs provided by a multisensor system (NIR spectral camera, VNIR visual range spectral camera, metal detector, 3D laser scanners,

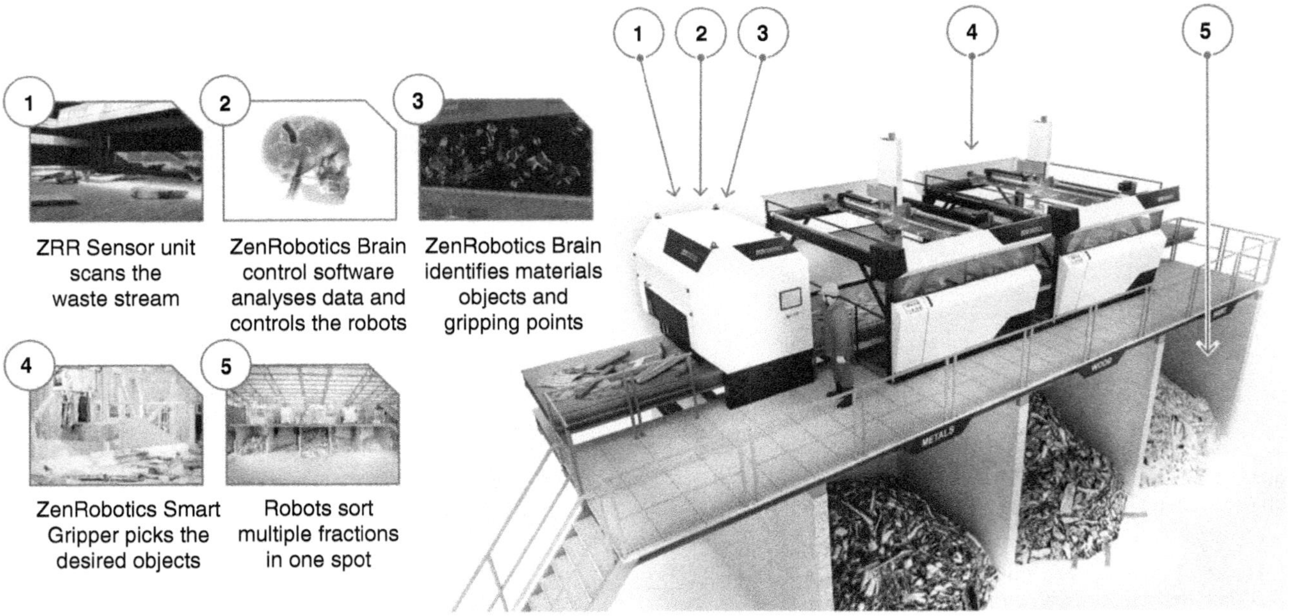

Figure 21.22 Setup of robotic recycler, (1) sensor unit, (2) brain control software, (3) identification unit of materials, objects and gripping points, (4) grippers, (5) dropoff chutes. *Source:* Reproduced with permission of ZenRobotics Ltd (2016a).

haptics, and transillumination) and analyzed by the "brain" with the use of advanced artificial intelligence (AI) algorithms. Fine sensor devices installed provide the robotic systems with high fidelity. The first step in the data flow is a high-resolution 3D sensor based on a laser line. This sensor gives an isometric 2D height map of the conveyor. The height map is used as the primary input for segmentation, that is, identifying individual "objects" on the belt. ZenRobotics Brain also applies machine learning algorithms to combine the data of two physical sensors and create a so-called virtual sensor. In addition to material recognition, high-resolution red-green-blue (RGB) cameras also offer a visual view to the waste stream for annotation. The control system manipulates the robotic hands that shift the various components of the waste stream from a conveyor into multiple chutes (Freiberg, 2011; Lukka et al., 2014).

To solve the emerging problem of robotic waste sorting that encompasses timing of movements, fine-tuning actions, and coordination, an algorithm has been designed, which is based on the functional principle of the human cerebellum. In this way, the system allows the robot to react properly to emerging situations and identify fitting solutions to the problem.

For deciding where to grasp, information is used about the point and the way that the robot could potentially grasp such as the *X*, *Y*, and *Z* coordinates, rotation, opening, and possibly other degrees of freedom. Adaptive algorithms are used to optimize the handles for estimating and maximizing the picking success for each object. For handle selection, the height map produced by the 3D sensor is the most essential input. A pneumatic gripper (Figure 21.23a) attached to an industrial robot arm picks objects of desired fractions and throws them into corresponding chutes in a ballistic trajectory (Lukka et al., 2014). The sorting line may include one or more robot arms with longitudinal and transverse movement each (Figure 21.23b).

The robot recycler is a fully automated waste sorting device; one robot can make up to 2000 picks h^{-1} (a pick every 1.8 s), lifting various-sized objects weighing up to 20 kg. A typical installation with two robot arms sorts up to 4000 picks h^{-1} and operates in two shifts, while the installation with three robot arms may work 24/7 in multiple shifts. New models can sort 3000 picks h^{-1}, while may reach 4000 picks h^{-1} at the peak. The arms of the gripper open from 50 to 500 mm, allowing sorting waste items of various sizes and shapes; the working area of each gripper is 2 × 2 m (ZenRobotics Ltd, 2016b).

According to the manufacturer, robot sorting offers several advantages over hand (manual) or optic-mechanical sorting, among which are the following: the system is far more versatile than purely mechanical sorting, as it can be adjusted to pick virtually any solid objects recognizable by the sensors, under the condition that the objects are in a suitable size range to be robot manipulated; the use of robots to pick objects renders handling of overlapping objects feasible; a robot system can also sort waste streams that omit the initial energy-consuming resizing step (nondestructive recycling), because it can manipulate objects of varying sizes and shapes; compared with hand sorting, salary costs per

(a)

(b)

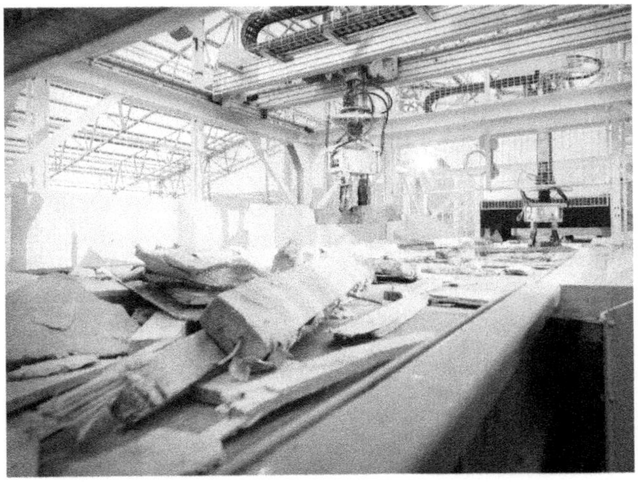

Figure 21.23 (a) Pneumatic grippers and (b) overbelt mechanism of longitudinal and transverse movement of robot arms. *Source:* Reproduced with permission of ZenRobotics Ltd (2016a).

working shift may diminish radically, so it makes sense to move from one or two shifts to three shifts; a single robot arm can pick up to 4 different fractions with high precision at the same time; sorting efficiency is very high (up to 98%); and taking into account the variable composition of waste streams, robot sorting presents unusual flexibility, as it may be adjusted very easily at the ever changing sorting conditions.

21.5 Density (or Gravity) Separation

Density (or gravity) separation is a unit operation used to separate materials, mainly based on the difference of their densities, as well as on hydro- or aerodynamic factors, such as particle size and shape, and fluid resistance on particles motion. Gravity separation is examined almost in every case of solid waste segregation, as the waste constituents have usually significant differences in densities and they are liberated at coarse particle size. In solid waste processing, dry segregation techniques prevail over those in water.

The following factors must be taken into account during equipment selection for efficient waste segregation: waste characteristics (particle size, waste stream composition, waste components density, moisture content, tendency for aggregation, etc.), product specifications, design parameters, and space requirements. The various types of air separators (air concentrators, stoners, air cyclones) are mostly used. Among the wet methods, heavy media separator is occasionally used in waste processing.

Two products are usually obtained by density separation: (i) a light fraction, which is mainly composed of organic materials (paper, plastics, food residue, garden waste, etc.), and (ii) a heavy fraction, composed mostly of inorganic materials (metals, construction debris, heavy plastics, etc.). In some cases of air separation, a middensity product is also obtained.

21.5.1 Air Separators

Air separation is a method that separates light from heavy waste components with the aid of air stream, mainly based on their density difference. In some texts, the term "air classification" is erroneously used, since, in strict terms, "classification" denotes separation according to particles size.

Depending on the direction of air that is blown into the separation chamber, air separators (or windshifters) may be horizontal, vertical, or diagonal.

Air separators with horizontal airflow are composed of a belt conveyor on which solid waste is fed. The horizontal air stream, which is blown behind the feeding point and in parallel to the conveyor, raises the light materials (paper, plastics, etc.); aided by suction, the air stream with the light fraction feeds a cyclone or expansion chamber for the separation of the solids from the conveying air stream. The heavy fraction remains on the belt and is discharged at the end point of the conveyor.

In vertical or diagonal separators, the air stream is blown upward, vertically, or with a slope (Figure 21.24) (Nihot Recycling Technology B.V., 2015, 2016). The input material is transported by conveyor into the separator. The feeding point is located in the middle, or sometimes at the top, of the duct. Solid waste feeding is realized through a rotating, air-locked device, as retaining constant aerodynamic conditions (air velocity, pressure) in the separation chamber is of high importance. Due to

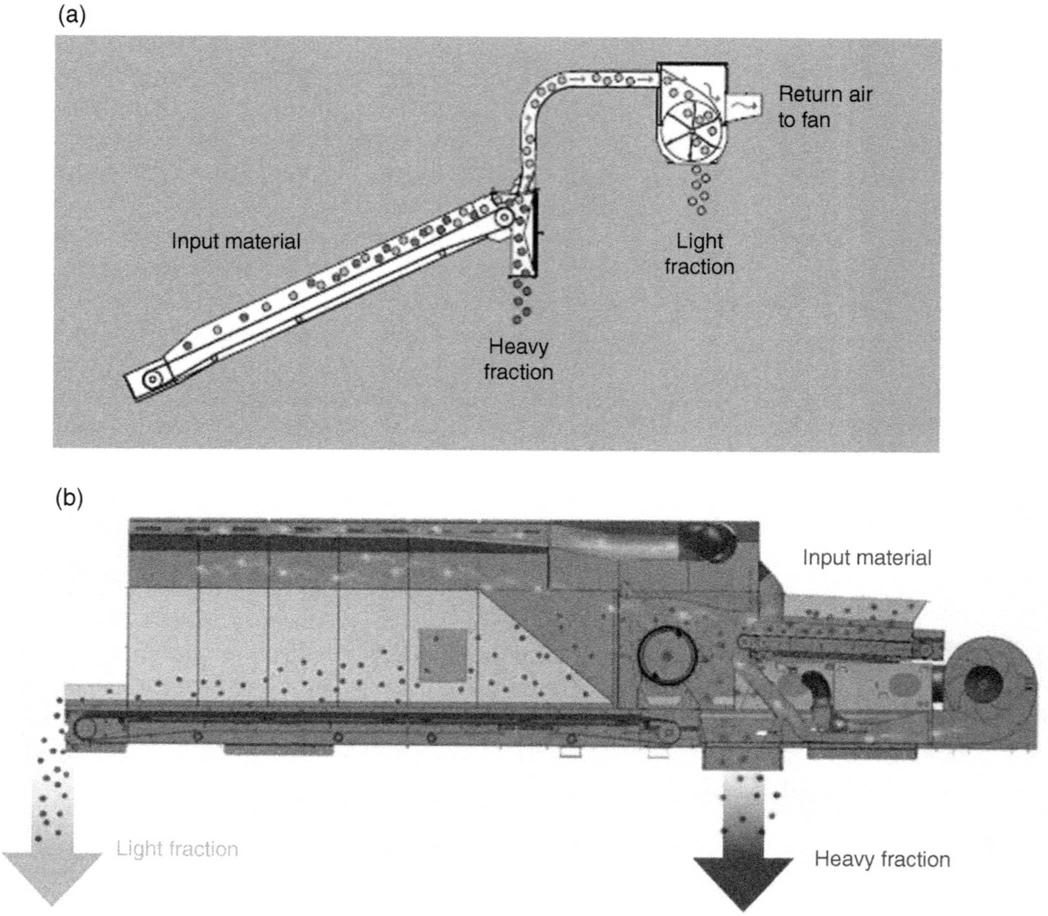

Figure 21.24 Air separators with: (a) vertical air blow, (b) diagonal air blow. *Source:* Reproduced with permission of Nihot Recycling Technology B.V. (2015) and Nihot Recycling Technology, B.V. (2016).

their weight, the heavy materials fall on the bottom of the device, from where they are removed by belt conveyor. The light materials are aspirated by controlled air through a duct into an expansion chamber (cyclone, Combi separator, or rotary air separator). In the expansion chamber, the aspirated air expands, and the light materials drop down into a rotary valve, which seals off the expansion chamber and discharges the material. The expanded air returns from the chamber to the recirculation fan. If the air contains dust, the air is directed from the fan to a dust filter. Approximately 60% of the returned air is blown back into the system through a blow nozzle.

The factors that affect separation efficiency are waste feed rate, airflow rate, cross section of the separation duct, and particles size and shape. Regarding vertical separators, there are several duct designs, among which those of straight duct, zigzag, and stacked reverse triangles (Figure 21.25). The design of the duct usually names the separator. In the first two duct designs, air velocity is constant through the entire length of the duct, as the cross section of the duct is maintained constant. In addition, zigzag separators are more effective, as their cross section favors turbulence of the air stream, tumbling of the waste, and, consequently, disintegration of aggregates. In ducts with stacked reverse triangles, the velocity of the air stream varies along the duct. This results in the differentiation of falling particles velocities and collection of particles with similar terminal velocities in the same product.

In addition to the traditional air separators (or windshifters), drum air separators have been developed to process waste streams of increased content in lightweight materials. Depending on the nature of the waste stream, drum separators may include one or two drums to obtain two or three fractions, correspondingly (Figure 21.26). Three fractions may also be obtained by processing the waste stream consecutively in an integrated separator, which is composed of single drum separator and windshifter.

In case of single-drum separator, the waste material is transported into the separation unit by a frequency controlled input conveyor. The heavy material falls

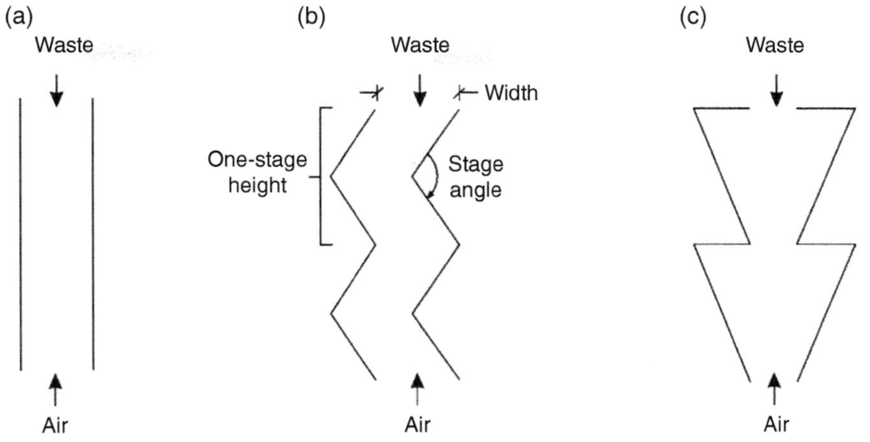

Figure 21.25 Duct designs of air concentrators with vertical airflow: (a) straight, (b) zigzag, and (c) stacked reverse triangles. *Source:* Tchobanoglous et al. (1993). Reproduced with permission of McGraw Hill Inc.

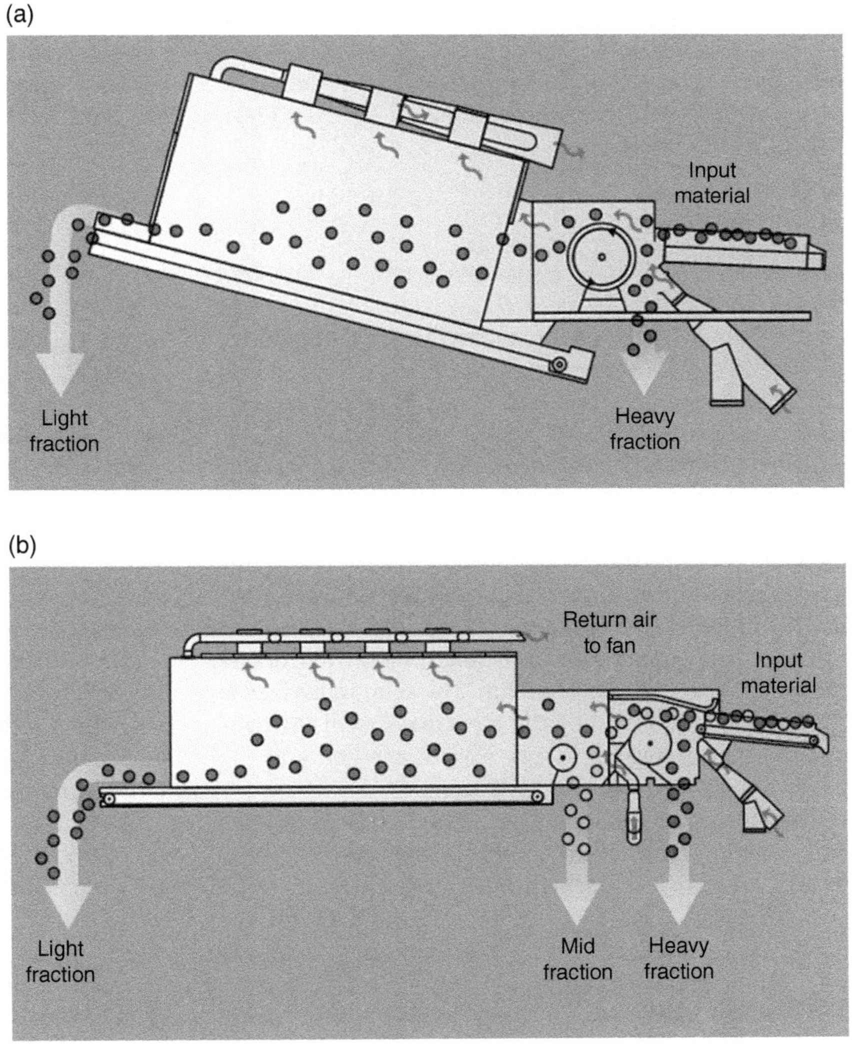

Figure 21.26 Air separators with: (a) single drum and (b) double drum. *Source:* Reproduced with permission of Nihot Recycling Technology, B.V. (2014).

down before the rotating drum, which acts as splitter. The light material is aspirated and conveyed over the drum splitter into the expansion chamber, dropped on a conveyor, and discharged outside the expansion chamber. The air returns from the expansion chamber to the recirculation fan, passes through a dust filter, and returns to the drum separation unit (Nihot Recycling Technology, B.V., 2014).

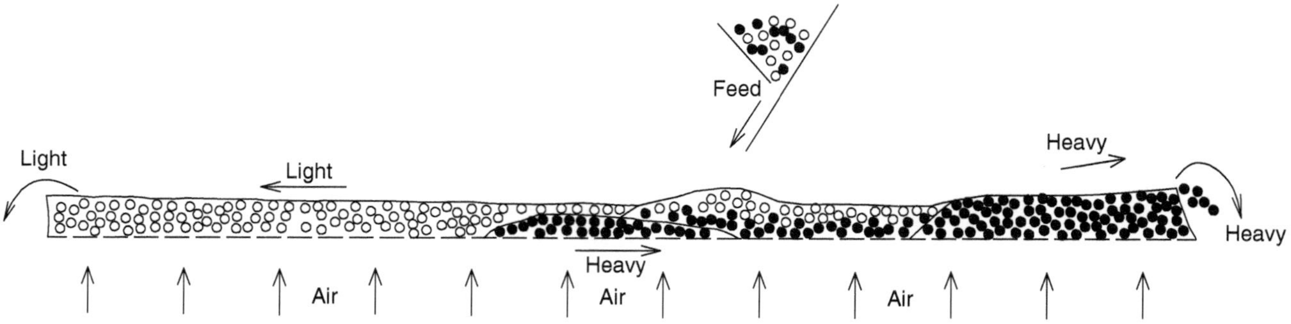

Figure 21.27 Working principle of stoner.

Depending on the nature of the waste stream, the output per drum separator may reach 35 or even 50 t h^{-1}. Air separation has been used in the recycling, cleaning, or upgrading of commercial and industrial waste, C&D waste, MSW, auto shred residue (ASR), incinerator ash, waste from electrical and electronic equipment (WEEE), compost, biomass, metals, glass, RDF/SRF, fines upgrading, etc.

21.5.2 Stoners

Essentially, stoner operates as a pneumatic jig. Solid waste is fluidized by an upward blown air-stream, and, in combination with the reciprocating motion of the sloped stoner deck, solid waste components are separated into heavy and light fraction. The operation principle is shown in Figure 21.27.

Stoner consists of a sloped deck with porous surface, through which low-pressure air is introduced. Blown air from a fan is distributed uniformly below the sorting deck; moreover, externally adjusted air valves may further regulate the air flow rate locally, according to sorting conditions. The waste material is fed between the center and the upper part of the deck, closer to the discharge point of the heavy fraction. The solid waste feed is fluidized and stratified by the air stream, with the light material occupying the upper part of the bed and the heavy remaining on the deck surface.

The stoner deck is also subjected to a straight line reciprocating motion, which affects the particles being in contact with the deck (heavies). The light material flows downhill, following the slope of the deck, as it is not affected by deck motion. On the contrary, the heavy fraction moves uphill, as it is in contact with the deck and affected by the reciprocating motion, overcomes a barrier, and is discharged on a takeaway conveyor.

Above the stoner, a dust hood is installed to collect the fluidizing air along with the dust drawn from the waste.

The separation with stoners depends on factors such as material density and particle size and shape. The operational variables of the stoner are the air flow rate, frequency and stroke of deck reciprocating motion, slope of the deck, and air valves adjustment. For effective separation, the density difference between light and heavy objects must be significant, and the feed must be classified into narrow size fractions; also, the moisture content of the waste must be low to prevent the aggregation of the light fraction and promote the fluidization of the waste.

21.5.3 Air Cyclones

In solid waste processing, air cyclones are mainly used to recover the materials from the light fraction of the conveying air streams.

Air cyclone is a devise comprising a cylindrical part (upper) and a conical (lower), as shown in Figure 21.28. The feed stream (light fraction from air separation) is introduced tangentially into the cyclone. Solid particles are subjected to centrifugal force and move to the cyclone wall. Subsequently, being in contact with the wall, they follow a spiral orbit of continuously decreasing radius until they reach the cyclone underflow, where they exit. Air is recovered from cyclone overflow. Practically, the overflow contains a small percentage of dust or very fine light material. The material from the underflow is discharged either periodically or continuously through an airtight rotating system, which secures stability of the aerodynamic conditions in the air cyclone chamber.

21.5.4 Heavy Media Separators

Heavy media separation is a wet separation method, and, as such, it has been applied in a limited extension in solid waste processing. It has been used to separate aluminum (light) from other heavier materials or metals (heavy) from plastic (light) in car recycling.

The method is based on the creation of a heavy medium suspension with density between that of heavy and light fraction. The desired density is accomplished by mixing proper quantity of finely ground heavy medium (e.g. magnetite, ferrosilicon, galena, etc.) and

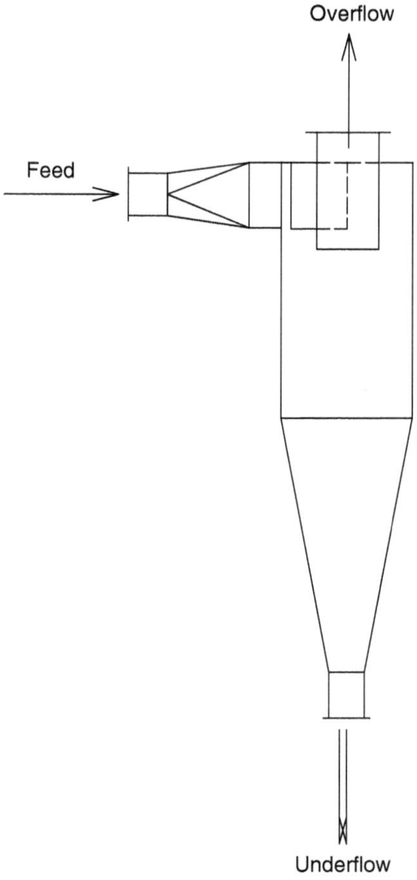

Figure 21.28 Air cyclone.

water. The shredded waste feeds the heavy medium bath. The objects of lower density than that of the suspension float (light product), while those of higher sink (heavy product).

Factors that prevent its widespread application in solid waste treatment are the complexity of the technique compared with air separation methods, the high capacity required to sustain the operation of the plant economically, and its restricted application only to materials with characteristics that are not affected by water.

21.6 Magnetic/Electrostatic Separation

Metal objects possess an appreciable amount in solid waste, with their percentage generally varying between 5 and 15%. Metal recycling is a highly ecoefficient way of reintroducing a valuable material back into the economy, as it contributes to mineral resources conservation, waste minimization, energy saving, and low production cost of metals compared with that from primary resources. Such an example is that of copper, a significant part of which originates from recycling operations, since copper can be recycled and reused countless times, without any loss of performance and difference in the quality of recycled copper in comparison with that from ore. Approximately 24 Mt of copper were produced globally in 2010, 35% of which originated from recycled copper (Glösser et al., 2013); correspondingly for EU-27, used copper from recycling amounts to 40% (Muchova et al., 2011). Taking into account the energy required for collection and sorting, the recycling of copper requires up to 85% less energy than primary production. As a consequence, there is an annual saving of 100 million MWh of electrical energy and 40 Mt of CO_2 around the world, in addition to oil, gas, or coal deposit preservation (International Copper Association Ltd, 2013). Another environmental benefit of metal recycling over landfilling is the prevention of pollution from the oxidation and dissolution of various metals in alloyed form.

Metal recovery from waste streams significantly contributes to the economic sustainability of recycling, as metal sorting is relatively easy and efficient, the unit operation cost is low, and metals price is high enough.

The methods used for metal sorting are mainly magnetic, electrostatic, and eddy-current separation, all three of which are based on the principles of electromagnetism. Magnetic separation is a well-known technology, as it has long been successfully applied in mineral processing. By contrast, the development of eddy-current separation is largely due to the applications in solid waste separation. Finally, electrostatic separation is used for the separation of a variety of materials, including nonferrous metals.

21.6.1 Magnetic Separation

21.6.1.1 General

Based on magnetic susceptibility, materials are classified into diamagnetic, paramagnetic, ferromagnetic, antiferromagnetic, and ferrimagnetic. In practical applications, materials are classified into diamagnetic, paramagnetic, and ferromagnetic. Diamagnetic materials are practically nonmagnetic, while ferromagnetic are strongly magnetic. Ferromagnetic materials can be further divided into magnetically "soft" and "hard." The magnetically "soft" materials can be magnetized but they do not remain indefinitely magnetized, contrary to the magnetically "hard" ones that retain their magnetization permanently.

Magnetic separators are used to separate and recover (i) ferrous metals from scrap, alloys, slag, mixed metals in source-separated waste (e.g. tin cans from aluminum cans), etc. and (ii) ferrous items from automobile shredding, MSW (tin cans, wires, household equipment, etc.), and combustion residues. They are also used to capture tramp iron from waste streams and prevent equipment from damage. Since magnetic items are encountered in most waste streams, magnetic separator is a basic component of the recycling plant because of its simple operation, relatively low cost, high efficiency (80–95%), high production rate, and short payback period.

In general, magnetic separators are classified into (i) standard (or low) and high intensity, according to the strength of the magnetic field, and (ii) dry and wet, according to the fluid medium (correspondingly, air or water). The magnetic properties of waste material determine the strength of the field to be applied, while the particle size determines the medium. Standard-intensity magnetic field is applied on materials with high magnetic susceptibility (ferromagnetic), while high-intensity magnetic field is applied on weakly magnetic materials (paramagnetic).

The proper strength of the magnetic field is established by either permanent magnets for standard-intensity fields, or electromagnets for high-intensity fields. Permanent magnets are made of magnetically "hard" ferromagnetic materials, which undergo special processing in a powerful magnetic field during their fabrication to "align" their internal microcrystalline structure and make them very resistant to demagnetization, even when they are not located within a magnetic field. Once, permanent magnets were made of steel and achieved low strength magnetic field, with magnetic flux density (or MI) 0.13–0.15 T (or 1300–1500 G). Nowadays, permanent magnets are made of special alloys, resulting in much stronger magnetic fields and depths. The magnetic materials used for permanent magnets are ferrites and various iron or rare-earth alloys, such as iron–aluminum–nickel–cobalt (FeAlNiCo), samarium–cobalt (SmCo), neodymium–iron–boron (NeFeB), and neodymium–iron–boron–cobalt (NdFeBCo); the aforementioned magnets resulted in fields with MI up to 0.9 T (9000 G).

In electromagnets, the magnetic field is produced by electric current. They consist of a core of magnetic or ferromagnetic material (e.g. soft iron), around which a coil has been wound. As long as electric current flows through the coil, the core remains magnetic. The magnetic field produced by electromagnet is deeper and stronger than that of permanent magnets, with MI reaching as high as 2.5 T (25 000 G). Compared with permanent magnets, the magnetic field can easily alter, or even turned off, by regulating the intensity of electric current in the coil only.

21.6.1.2 Magnet Configuration and Arrangement

Standard-intensity dry magnetic separators are commonly used in solid waste separation, because particles to be removed are relatively coarse and, usually, possess strong magnetic properties. During standard-intensity magnetic separation, all particles of the burden are subjected to the action of opposing forces, one of which is magnetic and the other mechanical or combination of forces (gravitational, centrifugal). Magnetic force affects only magnetic particles, while the mechanical forces affect all particles. Magnetic particles attach to the magnet through pinning, attraction, or elevation (Anastassakis, 2007).

The most common magnetic separators in solid waste separation and recycling are head pulley magnets, magnetic drums, suspended belt magnets, and magnetic plates (or block magnets).

21.6.1.2.1 Magnetic Head Pulley

This equipment uses the principle of magnetic particles pinning on the carrier, which carries the waste particles into the magnetic field (Figure 21.29). Substantially, it is a conveyor converted into a magnetic separator by replacing the head pulley with a strong magnetic one. Magnetic pieces will remain pinned on the belt as long as the magnetic field is effective and fall into the bin or hopper of the magnetic product as soon as the field action is attenuated or ceased.

This arrangement favors the entrapment of materials (e.g. paper, plastics, textiles, diapers, etc.) between magnetic particles and belt. Consequently, a cleaning stage is usually required to obtain a clean magnetic product. When both coarse and fine magnetic particles exist in the waste stream, the separation may be realized in two steps: first, with overhead magnets to separate the coarse magnetic fraction and, second, with head pulley magnet to efficiently separate the finer particles.

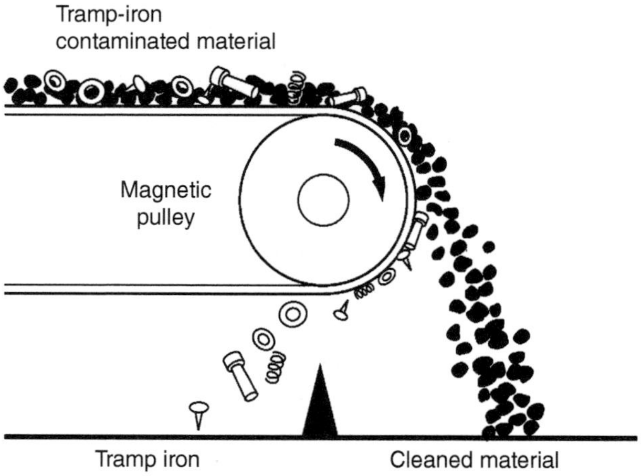

Figure 21.29 Magnetic head pulley. *Source:* Reproduced with permission from Eriez Magnetics Europe Ltd (2006a), "Copyright Eriez Manufacturing Co. All Rights Reserved."

21.6.1.2.2 Magnetic Drum

A magnetic drum is usually composed of a stationery permanent magnet (a compact piece or set of axially oriented bars of alternative polarity and equal distance) mounted inside a rotating drum. The waste material feed enters the magnetic field either on the top of the drum or below it.

In the first arrangement (Figure 21.30a), the magnetic fraction is pinned on the rotating drum and detaches the drum upon attenuation or ceasing of the field. This arrangement favors the entrapment of nonmagnetic materials (paper, plastic, etc.), and a cleaning stage of the magnetic product is possibly necessary. In the second arrangement (Figure 21.30b and c), the magnetic fraction is attracted to the drum surface or elevated, depending whether the direction of the feed is con- or countercurrent to the direction of the drum rotation. The second arrangement provides a cleaner product, especially the countercurrent rotation case.

21.6.1.2.3 Suspended (or Overbelt) Magnet

They are typically positioned before an eddy-current separator to separate ferrous metals, e.g. steel cans. The suspended magnet assembly includes a permanent magnetic core surrounded by a belt conveyor. The suspended magnet configuration is either crossbelt or in-line in reference to the waste flow (Figure 21.31).

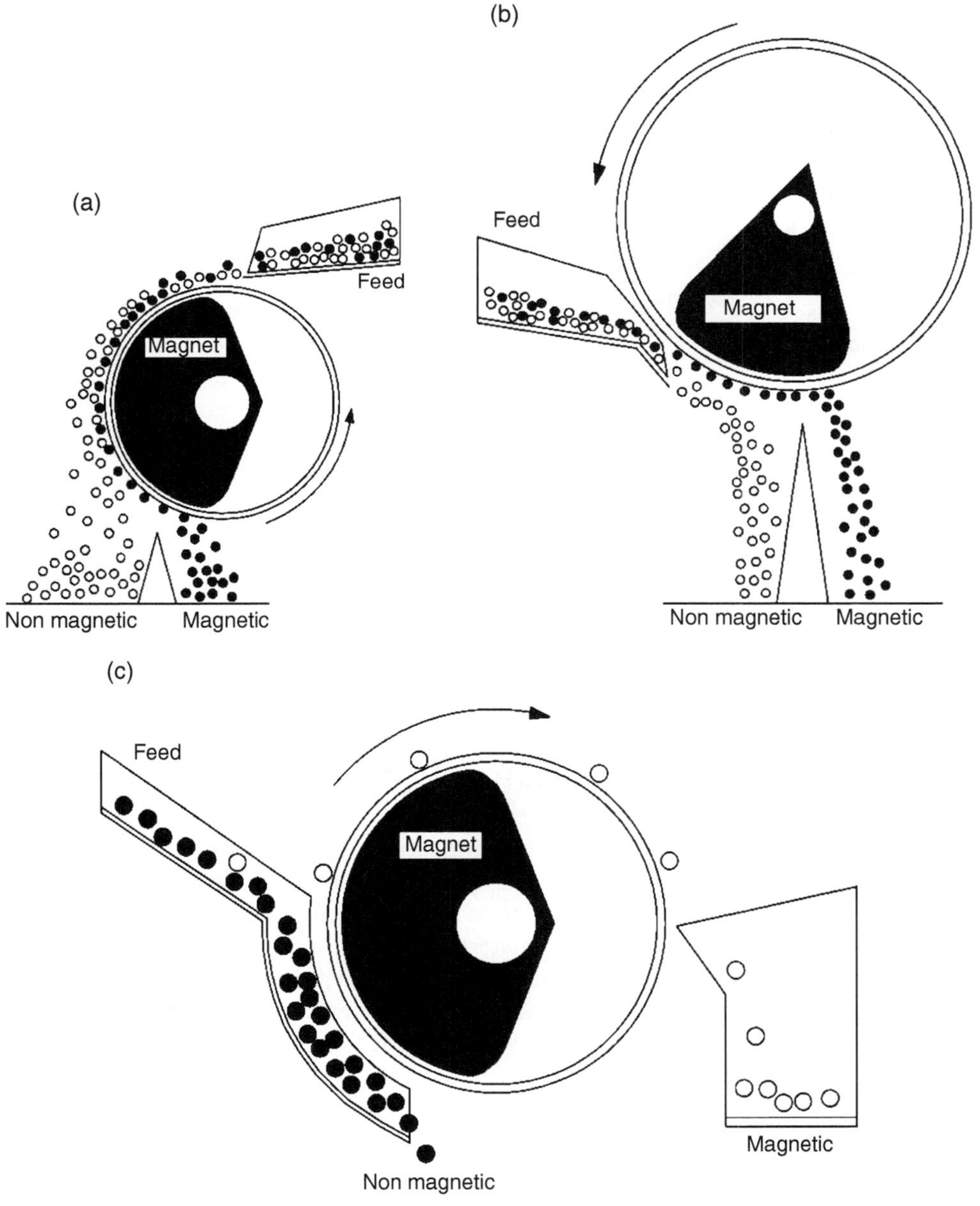

Figure 21.30 Magnetic drum separators: (a) top fed, (b) and (c) fed below the drum.

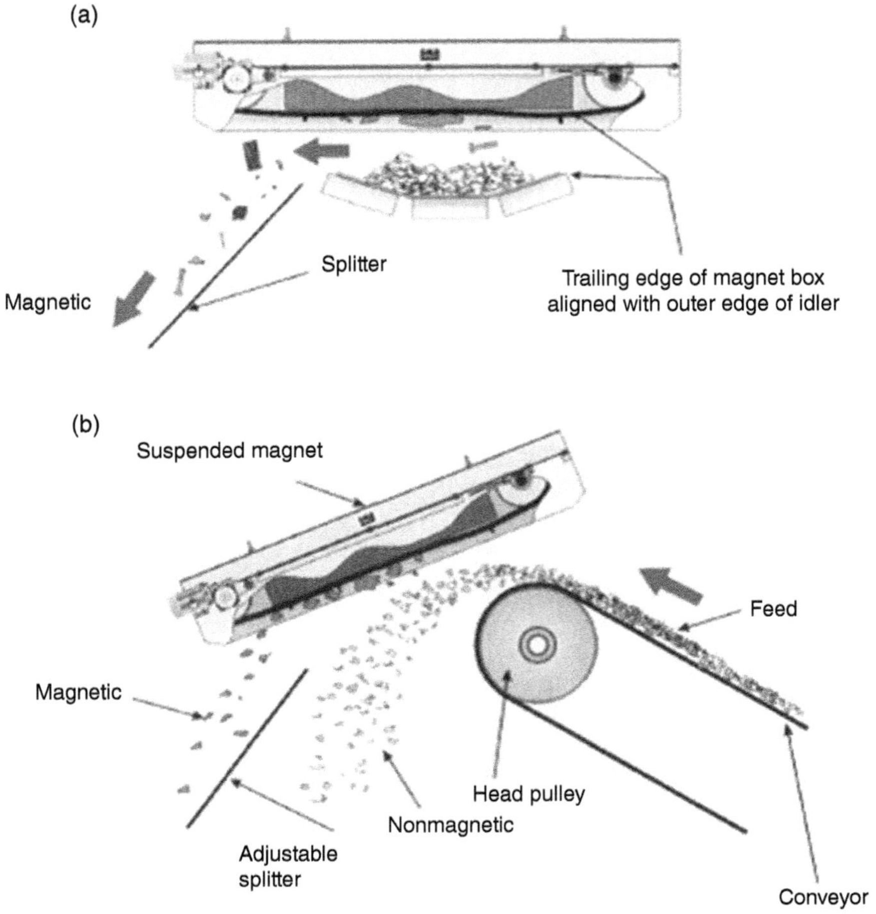

Figure 21.31 Suspended permanent magnet configurations: (a) cross-belt, (b) in-line. *Source:* Reproduced with permission from Eriez Magnetics Europe Ltd (2006b), "Copyright Eriez Manufacturing Co. All Rights Reserved."

In regard to crossbelt configuration, the magnet is placed perpendicular to the waste flow. This position requires a stronger magnet, and it is not recommended for excessive belt speed or deep material burden. The maximum size of the particles determines the distance between magnet and belt, as the distance must be two to three times the size of the largest particle to avoid interference of materials on the belt with those attached on the magnet. In turn, longer distance weakens the strength of the magnetic field, as the magnet is usually a permanent one and, consequently, waste materials with low magnetic susceptibility or small size may be lost to the nonmagnetic product. Also, the layer of the waste feed must not be thick to permit the recovery of the magnetic materials that are positioned at the lower part of the layer and diminish or eliminate the entrapment of laminated light materials (paper, plastic, etc.), although entrapment is generally not a significant problem. In crossbelt configuration, special attention must be drawn on the problem of magnetization of the rollers, below the belt, and the other metallic parts of the waste conveyor. As these components are made of ferrous metal, their magnetization exerts a counter magnetic force to the magnetic particles of the feed and prevents them from collection in the magnetic product. This problem is overcome by replacing the ferrous components with corresponding made of nonmagnetized material.

In reference to the in-line configuration, the magnet is placed in parallel to the waste flow. This arrangement succeeds a clean magnetic product, because the lifting of nonmagnetic laminated materials to the magnet is highly reduced or eliminated. This happens because the feed stream is very loose, actually suspended in the air; consequently, due to low momentum and low gravity force of the light materials, as well as the high air resistance, the light laminated material is prevented from lifting to the magnet. Similarly to the crossbelt magnets, the conveyor head pulley may be made of nonmagnetized material for efficient separation of the magnetic pieces. Also, the magnet assembly in the in-line configuration could be composed of a series of magnetic components with alternating polarity, leading to much cleaner magnetic product (Figure 21.32).

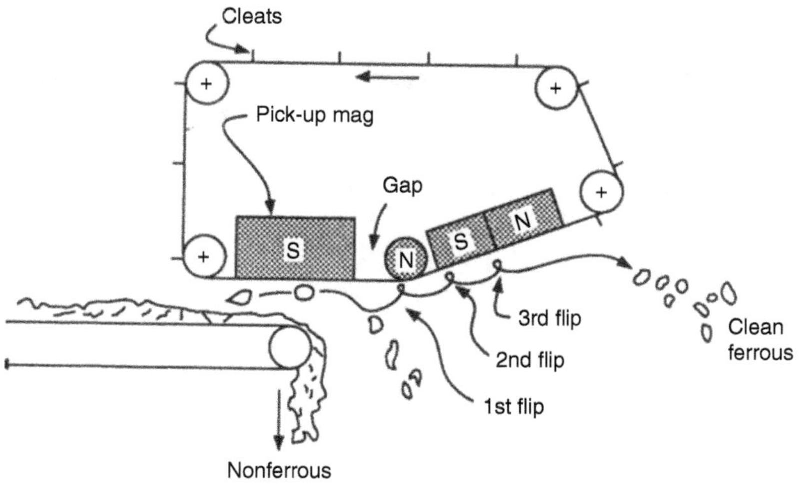

Figure 21.32 Typical magnets configuration with alternating polarity. *Source:* Glysson (1990). Reproduced with permission of McGraw Hill Inc.

Figure 21.33 Permanent block magnet for compost. *Source:* Reproduced with permission of Goudsmit Magnetic Systems (2016).

21.6.1.2.4 Magnetic Plate or Block Magnet

These magnets primarily serve as protective systems for machines or other production processes. They are used to capture fine ferrous particles or larger pieces of tramp metal from free flowing materials (e.g. plastic) or waste streams where the volume of metal contaminants is low; because of the low contaminants volume, only occasional cleaning is required. Magnetic plates are typically installed above flat or trough-shaped conveyor belts (Figure 21.33); with their simple design, these systems are easily adaptable to any type of existing installation. They usually consist of a ferrite or neodymium permanent magnet, being able to attract iron objects of various sizes, which are manually removed. Permanent magnets do not consume any energy and, as a result, they provide a cost-effective solution.

The magnetic component of the magnetic plate may also be an electromagnet. In this case, plate cleaning from magnetic objects is easier, as iron items drop off the block by turning voltage off.

21.6.2 Eddy-Current Separation

Eddy-current separators are used to separate nonferrous from ferrous metals by forces of alternating polarity; the forces are induced on particles by a high-speed magnetic rotor that spins inside a nonconductive drum, driven by conveyor belt. This separator collects the ferrous metals that stick to the magnet and rejects the nonferrous metals as the alternating magnetic fields throw them out of the product flow.

The principle of eddy-current separation relies on Faraday's law of electromagnetic induction. According to theory, when large metal particles pass through a time-varying magnetic field, eddy currents are induced into them, rendering the particles, in effect, solenoids. Following Lenz's rule, eddy currents oppose to the increase of the magnetic flux, as the particles approach the magnetic field, and tend to counterbalance the increase. Consequently, metal particles that induce eddy currents into them are repelled by the magnetic field, while the nonconducting particles are not. The time-varying magnetic field is generated by a fast-rotating electromagnet.

Eddy-current separators consist of two drums, a belt conveyor around the drums and a magnetic rotor surrounded by the head drum. The drum surrounding the magnetic rotor rotates at conventional belt conveyor speed, while the internal magnetic rotor of alternating polarity turns at much higher speed than the external (3000–4000 RPM). The rotor is usually made of ferrite or rare earth. Through the induction of eddy currents and the resulting repelling forces, the alternating magnetic field selectively repels the nonferrous metals and

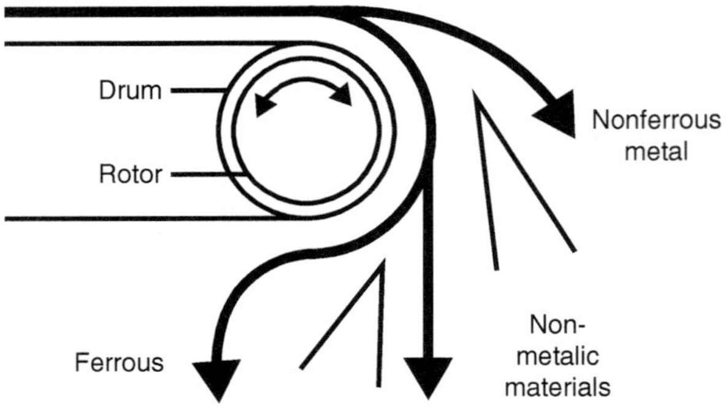

Figure 21.34 Working principle of eddy-current separator. *Source:* Reproduced with permission from Eriez Magnetics Europe Ltd (2006c), "Copyright Eriez Manufacturing Co. All Rights Reserved."

separates them from other materials. The working principle of eddy-current separator is presented in Figure 21.34. The drum and the magnetic rotor may be either concentric or eccentric (Figure 21.35). In the concentric design of the magnetic rotor, separators are provided with multiple separation points, the burden has a longer exposure to the magnetic field, and the magnetic rotor achieves better separation; as a result, even the smallest iron particles are kept by the concentric magnetic rotor and concentrated in the proper container. Separators with an eccentric design have one separation point, but the effect of the changing magnetic field is concentrated exactly on limited area, within which the burden is subjected to the forces most effectively. The eccentric pole system ensures that the magnetic field is maximized only at the moment of separation and that no other magnetic field is generated in other positions on the belt drum.

Magnetic rotor can be powered in both directions: (i) forward, which is the same direction as the head drum, and (ii) backward, which is opposite. For particle size larger than 10 mm, a forward rotation is preferable, while for smaller, a backward rotation is better. This is because smaller particles tend to rotate opposite to the direction of the head drum and the particles are caused to jump backwards out of the separation zone (Eriez Magnetics Europe Ltd, 2006d).

There are several designs of eddy-current separators, such as ramp separator (Schlömann, 1975a, b) and with vertical drum (Lungu and Schlett, 2001).

The performance of eddy-current separators depends on factors such as belt speed, rotor speed, and splitter (or splitters) position. Separation is also affected by the characteristics of the feed such as particle size, shape and conductivity, size distribution, density, and moisture. As electromechanical forces depend on particle size, for efficient separation the solid waste stream must be classified into proper size fractions. In general, eddy-current separators process solid waste with particle size between 150 and 3 mm and sometimes down to 1 mm. For small particle size (less than 10 mm), the magnetic rotor rotates in the opposite direction to the head drum for efficient separation.

The position of the eddy-current separator in the flowsheet of waste separation is after the separation of the light fraction (paper, cardboard, etc.) from the heavy fraction (metals, glass, ceramics, aggregates, etc.) by air separation and, usually, after the removal of ferrous metals by magnets. When it is required to separate nonferrous metals, there is usually a reciprocating need for ferrous metal separation. Although it is possible to recover three products (ferrous metals, nonferrous metals, nonmetallic residue) by eddy-current separation, it is preferable to segregate ferrous metals by magnetic separation prior to eddy-current separation to avoid high belt wear. This results because the magnets used in the magnetic rotor impart a strong, vibrating magnetic pull on ferrous materials, which can override eddy-current induction and result in a violent attractive force.

Eddy-current separation is suitable to separate and recover the nonferrous metals or alloys (e.g. Cu, Al, Zn, brass) from industrial and residential waste streams, as

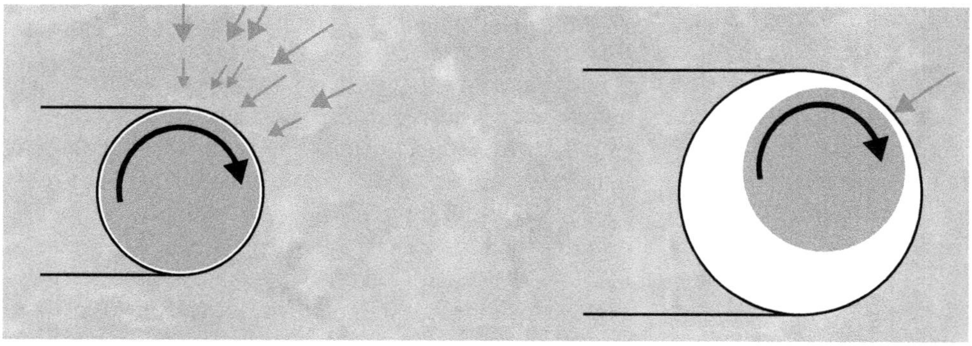

Figure 21.35 Concentric (left) and eccentric (right) design of magnetic rotor in eddy-current separators. *Source:* Reproduced with permission from Eriez Magnetics Europe Ltd (2006d), "Copyright Eriez Manufacturing Co. All Rights Reserved."

well as to separate various nonferrous metals from each other. This method has been applied to separate aluminum, copper, and zinc from nonmetallic components in car recycling; aluminum beverage cans from either presorted or mixed refuse; metals from granulated glass cullet, plastic, wood, and/or chipboard; copper from plastic; and nonferrous metals from incinerator ash. It has also been applied to separate nonferrous metals in the processing of WEEE scrap, cables, printed-circuit boards, foundry sands, crushed dross, etc.; also, when installed after a heavy media system, metal recovery is further improved.

They are able to separate particles from up to 350–400 mm down to 1–2 mm. In general, these separators have high separation efficiency (approx. 97%) and capacity that depends on material density and particle size (Table 21.1).

21.6.3 Electrostatic Separation

Electrostatic separation can be used to separate nonconducting waste material (e.g. paper, plastic, glass, etc.) from conductors (e.g. metals), or nonconductors from each other, based on differences in their electrical conductivity. In addition to particle conductivity, electrostatic separation is also influenced by particle density, size, moisture, and surface purity. Consequently, separation of waste particles is achieved under the action of electrical and mechanical forces (gravity or centrifugal).

The commercial electrostatic separators are those that render particles charged by either conductive induction or charge bombardment (corona charging).

Table 21.1 Typical capacity of eddy-current separators per meter of feed width.

Material	Size (mm)	Capacity ($t\,h^{-1}$)
PET flake	−15	2–3
Glass cullet	−40	20.0
Car fragments	−12	5.2
Car fragments	12–30	12.5
Car fragments	30–70	20.0
Shredded refuse	−50	18.0
Shredded refuse	50–150	20.0
Plastic/Al	−10	10
Copper/Pb	-8	7
Copper/PVC	-6	2.25
Ash/nonferrous	−100	21.5
Sand/nonferrous	−100	20.0

Source: Reproduced with permission from Eriez Magnetics Europe Ltd (2006d), "Copyright Eriez Manufacturing Co. All Rights Reserved."

During conductive induction, both electrical conductors and dielectric particles contact a charged roll (Figure 21.36) or inclined plate. Afterward, conductors are repelled by the roll because of their similar charge, while the dielectric components of waste stream are not influenced by the electrostatic field.

During corona charging (or high-tension separation), the separation of the waste stream is accomplished by passing solid waste through a corona discharge (Figure 21.37). The separating machine is composed of a grounded rotor, on which solid waste is fed. Corona discharge is caused by a fine wire electrode or a set of needles, in parallel to the rotor, by raising their electrical potential to a determined value, which depends on the material of the wire or needles. The type of corona (positive or negative ions) depends on the polarity of the electrode. All waste particles are charged by ion bombardment from corona discharge. The conducting particles lose their charge because of their contact with the grounded rotor, while the poor- or nonconducting particles lose their charge very slowly and, consequently, are held on the rotor surface by the image force of their surface charge.

Although electrostatic separation (conductive induction or high tension) has not been applied widely in practical applications of solid waste segregation, it is expected to be of wider use in the future in certain sectors of applications. This separation has been successfully applied to separate plastics from paper, plastics

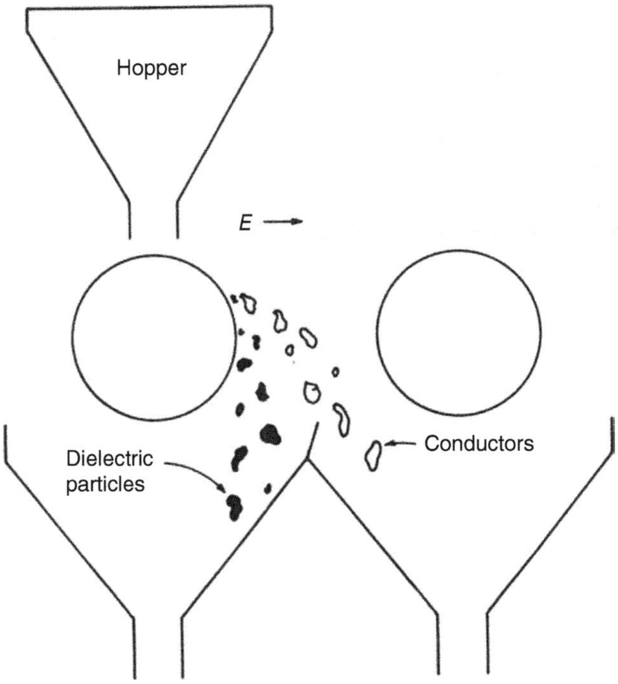

Figure 21.36 Electrostatic separation by conductive induction.

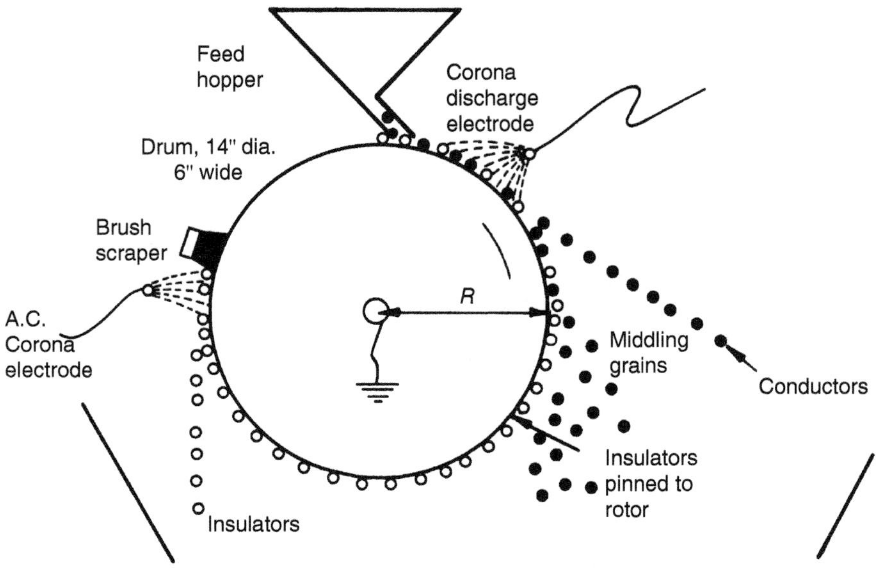

Figure 21.37 Electrostatic separation by corona discharge. *Source:* Lawer and Hopstock (1985). Reproduced with permission of SME.

from each other (Bedeković et al., 2011), shredded copper wires from their plastic insulation, glass from plastic, nonferrous metals from plastic or glass, etc.

21.7 Ballistic Separation

In addition to the classic waste separation methods, which are based on the difference of a certain physical or physicochemical property between the various components of the waste stream, there have also been developed segregation methods, which are based on a combination of physical properties, such as size, shape, density, elasticity, and friction. These methods separate organic from inorganic, as well as heavy and resilient from light and inelastic materials. Ballistic, oscillating, and "bounce and slide" separators are some of the devices based on these principles with ballistic separator being the most common. It is designed to separate solid waste at the inlet, depending on such characteristics as size, density, and shape.

The machine consists of a set (6–10) steel paddles that rotate offset against each other as a result of two crankshafts, which are located transversely on the upper and lower part of the sloped ramp (Figure 21.38). The raw material is fed into the ballistic separator and dropped onto the rotating paddles. As the paddles move, the material is rendered loose and bounces on the screen plates; as a result, the flat/large material moves up the slope, the roll back material moves down the slope, and the fines move up a small distance and fall through the screen plates in the paddles close to the feeding point (Figure 21.39).

The slope of the ramp, in most cases between 10° and 25°, can be adjusted manually or automatically (by means of a hydraulic system) to suit the material being screened. The greater the slope of the screen surface, the more the material will fall into the rolling fraction, and the cleaner the flat fraction will be, and vice versa.

The slope of the machine and the oscillatory movement of the paddles allow for the separation of the inlet flow into the following three different fractions for one separating level:

- 3D Rolling material: Bottles, cans and ferrous material, stones, heavy elements, plastic containers, wood, etc.
- Fine material: Sand, glass, plastic, cardboard, food remains, etc.
- 2D Flat material: Planar, lightweight elements, paper, cardboard, plastic film, timber, textiles, and fibrous material, etc.

Models with two or three separating levels correspondingly provide two or three 2D products of different size. The third level achieves better results (Stadler® Anlagenbau GmbH, 2016).

In order to improve the separation efficiency of rolling and planar elements, some models are equipped with fan at the bottom that helps the planar material to move to the top of the machine and/or insulation cover at the top that prevents material from being thrown.

A variety of materials can be processed with ballistic separators, e.g. residential, industrial, or crushed demolition waste. Ballistic separators are simple in construction but robust, easy in operation and adjustment, and of low operating and maintenance cost; depending on the waste material, they have high efficiency (up to 98%) and capacity (50–100 $m^3\ h^{-1}$). Because of their lower operating and maintenance cost compared with conventional and disk screens, the manufacturers claim that ballistic separators offer a screening alternative.

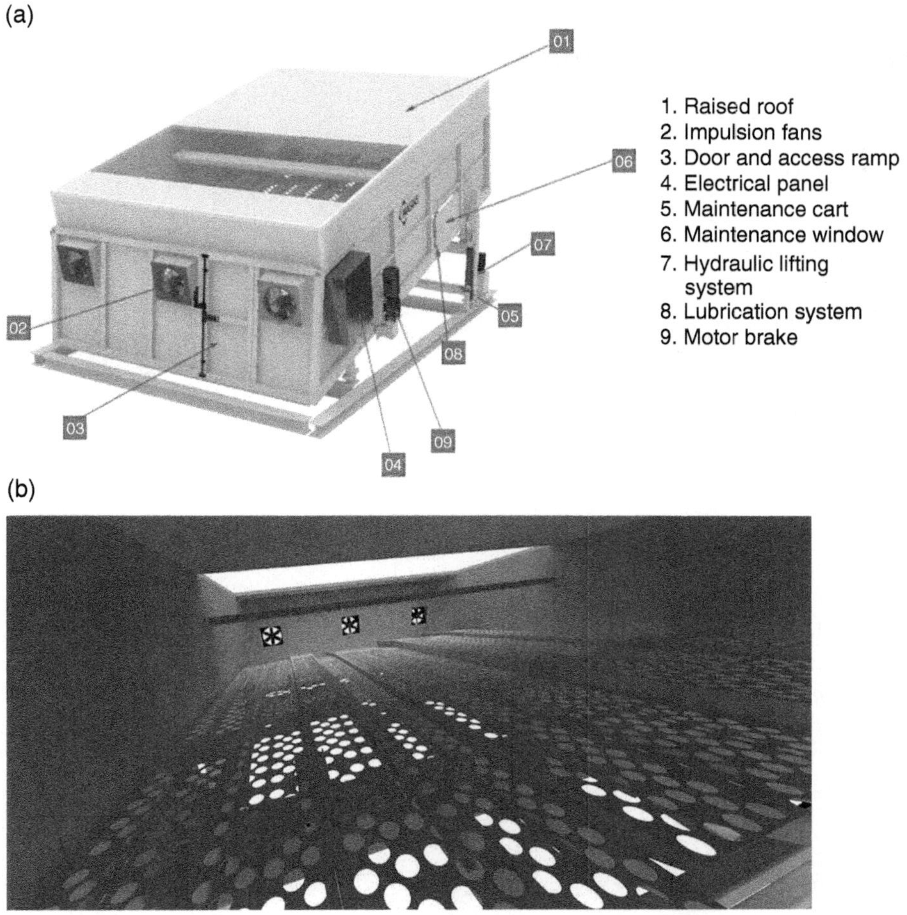

Figure 21.38 (a) Main components of a ballistic separator, (b) detail of paddles arrangement. *Source:* From MASIAS RECYCLING SL (2014a), with permission from the Copyright owner: Masias Recycling SL.

21.8 Froth Flotation

Froth flotation holds a great share among the applied methods nowadays for mineral separation. It is carried out in water and only hydrophobic minerals float after attachment on rising air bubbles in flotation cells. Its efficiency for small particle size and its selectivity are included among the various reasons for the widespread use of froth flotation in mineral processing plants. In waste separation plants, it can be applied after the separation of coarse particles with the use of dry processing methods. Froth flotation has been successfully applied at lab scale to separate fine glass particles from other inorganic components (Heginbotham, 1980) as well as plastic from each other (Drelich et al., 1998; Dodbiba et al., 2002). Its application at industrial level is very limited because waste material stream is separated usually at large particle size; wet processing of solid waste is undesirable in most cases; froth flotation has a relatively high operating cost; and materials to be recovered are of low commercial value to compensate cost and create profit. Froth flotation has been used to separate finely sized glass particles from inorganic contaminants (ceramics, stones, and metal) in a resource recovery plant in New Orleans (Glysson, 1990), using amine as collector to render the fine glass particles (−0.8 mm) hydrophobic.

21.9 Products Agglomeration (Cubing and Pelletizing)

Fluffy products (e.g. paper and plastic) have to be further treated in order to reduce their volume and, consequently, increase their density for easier transportation, storage, and processing. This is accomplished by agglomeration, which occurs as briquetting (or cubing) and pelletizing. Essentially, briquetting and pelletizing have many similarities but they are not identical. The principal difference lies in the size and shape of the agglomerates. Briquettes (or cubes) are usually of larger size and of square or rectangular cross section, while

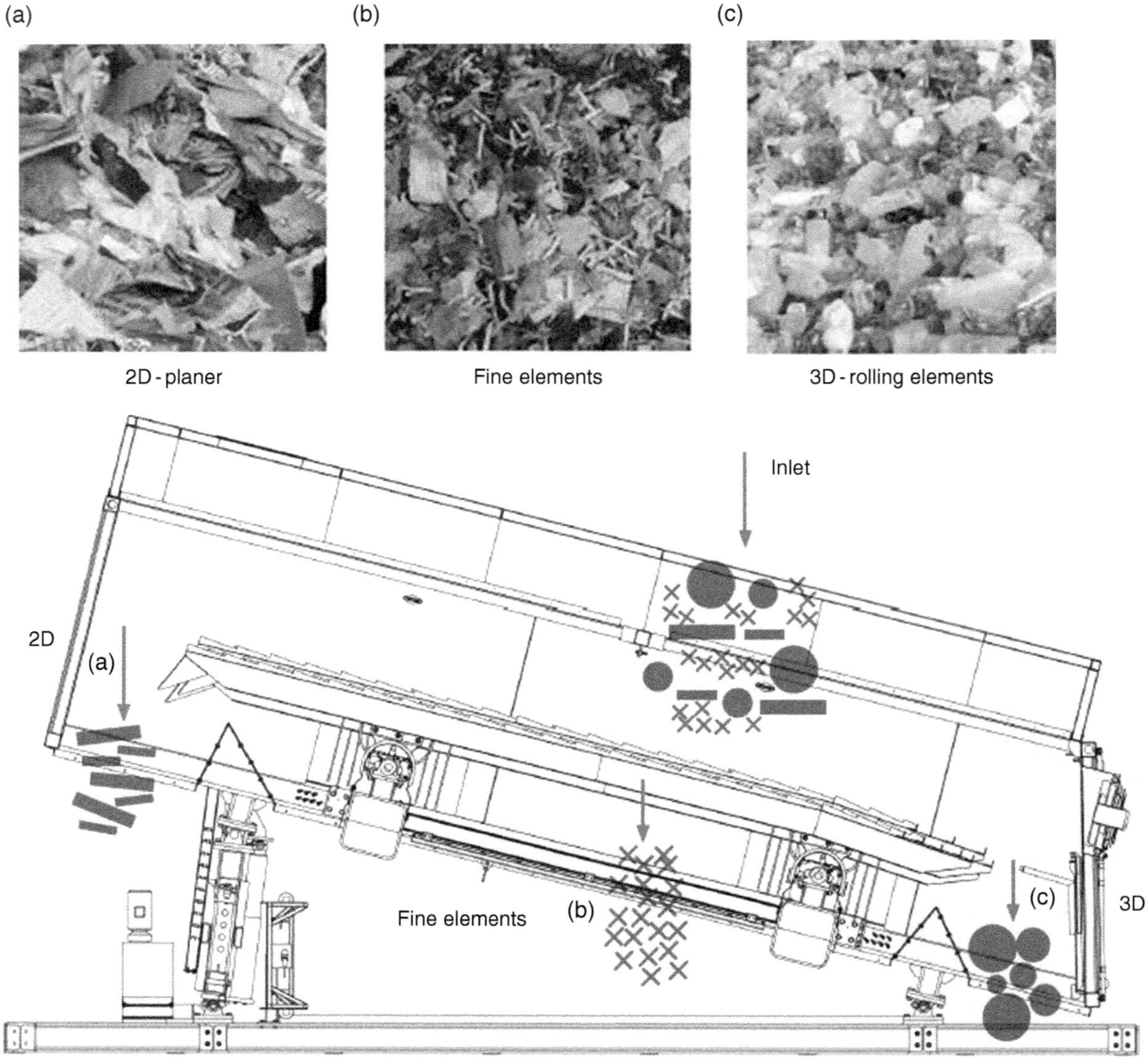

Figure 21.39 Ballistic separation. (a) 2D – planar, (b) fine, and (c) 3D – rolling elements. *Source:* From MASIAS RECYCLING SL (2014b), with permission from the Copyright owner: Masias Recycling SL.

pellets are smaller and with circular cross section (Figure 21.40). The agglomerated material is used as densified RDF for combustion, gasification, or pyrolysis.

Cubers and pelletizers consist of a cylindrical die from hard steel, perforated with holes. Inside the die there are two rollers mounted on the periphery of the die. There are two designs of the machine as regards the rotating part. In the first design, the die rotates while the rollers not, and vice versa in the second. In a third design, the rollers have been replaced by an eccentrically rotating cylinder.

The design and working principle of cubers and pelletizers is clear and well understood. Shredded RDF infeeds the machine through a short screw conveyor mounted in the casing of the main shaft. As the feed enters the agglomeration machine, it is entrapped between the die and the rollers and, consequently, is forced to pass through the die holes. As the rollers compress the material to pass, high pressure and friction are exerted against the holes wall. These effects result in temperature increase and heat creation, which, in turn, promote fusion of the materials (e.g. plastics). Besides pressure and friction, another important factor is the moisture content of the material to be agglomerated. Moisture content must be maintained at the level of 15–16%, as insufficient moisture results in weak bonds between the particles and, finally, in the disintegration of agglomerates after their extrusion. By contrast, high moisture content is also undesirable, as high temperature and pressure evaporate the moisture, and excessive stream is generated in the agglomerate. This results in its bursting and disintegration, but for a different reason

Figure 21.40 (a) Cubing of bagasse material (left) and fuel cubes (right). *Source:* Reproduced with permission of Warren&Baerg Manufacturing Inc (2013). (b) Pelletizing of municipal solid waste (left) and fuel pellet (right). *Source:* Reproduced with permission of Lundell Enterprises Inc (2015).

than previously. This is the reason that agglomeration machines are equipped with an automatic moisture control system as well as with water sprayers.

In another design, double screw presses are used to agglomerate shredded material (Figure 21.41). In addition to fluffy material, hydraulic pressing has been applied to convert cast iron borings, steel turnings, and nonferrous metal chips to high-density agglomerates.

Some advantages of product agglomeration are improved handling, reduced storage space requirements, lower transport cost, better combustion behavior, and increased heating value. The disadvantage is the additional cost, but it is counterbalanced by the benefits.

21.10 Compaction (Baling)

Compaction, also called baling, is a relatively simple process that does not cause any change in the physical or chemical properties of the product. It is essentially a densification process aiming at volume reduction of the segregated products or the waste material to be disposed. The advantages of baling are easy handling of the product (bale), lower transport cost, volume reduction, and lower space requirements in the disposal sites. Baled material does not sustain incineration. Also, it does not favor methane generation when it is landfilled.

Figure 21.41 Double screw presses for product agglomeration. *Source:* Reproduced with permission of Metso Minerals (2000a).

The densification of products and waste for landfilling is accomplished by compactors and balers.

21.10.1 Compactor

After separation and recovery of the recyclable or combustible materials, the remaining waste must be landfilled. As this material is loose, the transport cost is lower when this material is compressed in a stationary compactor. Essentially, compactors are presses, which are classified into low and high pressure, if the compaction pressure is correspondingly lower or higher than 6.8 atm. The compression can be carried out in the transport vehicle, containers, or specific chambers. After compression, solid waste may be transported for landfilling as it is or after baling.

21.10.2 Baler

Most of the recycling facilities employ a baler. Balers are used to densify urban waste, paper and corrugated cardboard, plastics (containers and bottles), aluminum, ferrous and nonferrous metals, agricultural waste, and separation residues for disposal. Baling enhances the marketability of commodities and significantly reduces transport cost and storage space. Its disadvantage is the relatively high capital cost.

Although the baler is a relatively simple device, the purchase of a baler is not an easy task, as many factors have to be considered to select the proper type. These factors are capacity, efficiency (power consumption, wire or strap usage), design and automation, size of bales, materials to be processed, safety characteristics, and cost. Depending on the case, more than one type of balers may be considered proper.

Based on the direction of the compressing ram, balers are classified into vertical and horizontal (Figure 21.42). Horizontal balers are further divided into those with single or double ram as well as with open or closed end. In any case, a baler includes the following basic parts: feed bin, compression system that includes one or more hydraulic or mechanically driven rams, compression

Figure 21.42 Twin-ram auto-tie baler. *Source:* Reproduced with permission of Summit Equipment, Inc. (2012b).

area, and discharge opening. Before being discharged, bales are fastened or tied with wire or plastic straps (Figure 21.43).

Vertical balers have a relatively low capacity and are suitable for small-scale operations. The length of the bale ranges between 0.5 and 2 m. In comparison with vertical, horizontal balers have lower height requirements and higher capacity, are used for large-scale operations, and produce larger size bales (e.g. with length ranging between 1.0 and 2.0 m). The capacity of horizontal balers depends on the density and size of the material to be processed; in most cases, the capacity ranges between 10 and 40 t h^{-1}.

When bulk products of small size have to be pressed into bales, a wrapping machine is installed along with the baler. Wrapping machine is a specific apparatus that allows the bandage of the waste bale with a plastic film and delivers in real time a packaged product with reliable properties (Figure 21.44). The machines can wrap 20–50 bales h^{-1} and even more depending on the dimensions of the bale.

The following benefits are assigned to wrapping: it prevents leakage of small-sized waste particles from the bale to the surrounding area; transport of the material is facilitated; bales are protected in case of exposure to atmospheric factors (e.g. moisture, wind); waste blocks can remain stored outdoors for a long period without degradation of their qualitative characteristics; leachate and odors are isolated; and waste transfer from the city centers to the landfill, and landfilling are facilitated.

Figure 21.43 Bale tied with straps. *Source:* Reproduced with permission of Summit Equipment, Inc. (2012c).

Figure 21.44 Bale wrapped with plastic film. *Source:* Reproduced with permission of Cross Wrap Oy Ltd (2015).

21.11 Benefits and Prospects of Recycling

The prevailing philosophy in solid waste management has fundamentally changed in the last two decades from disposal oriented to prevention and recycling oriented. This shift to the upper levels of waste hierarchy is a key factor to sustainable use of resources with concomitant environmental, economic, and social benefits.

21.11.1 Environmental Benefits

The noxious and damaging effect of uncontrollable waste disposal into the environment has been very well substantiated since a long time ago. Material reuse and energy recovery via recycling is one of the most efficient ways to reduce the deleterious effects of waste generation from human activities. Other environmental benefits from recycling are the reduction of air pollution, water usage, and mining waste.

In regard to air pollution, recycling significantly contributes to the reduction of greenhouse gas emissions (methane, carbon dioxide). As methane emission is due to the decomposition of organic matter, a key strategy to reduce methane emissions from landfills is to keep organic matter out of landfills by composting or incineration. Table 21.2 presents the greenhouse gas benefits (in million metric tons of CO_2 equivalent) for the year 2012 in association with the recovery and composting of certain materials; the equivalent benefit expressed in cars removal from the road is also shown. The calculation of the benefits from waste recycling and composting in regard to energy and greenhouse gas savings is based on WARM model, which has been developed by EPA (2014).

From Table 21.2 it derives that the greenhouse gas release in the atmosphere is reduced by 2.94 metric tons of CO_2 equivalent for each one metric ton of paper recycled. In addition, fewer trees are removed from the forests with the remaining trees capturing through absorption more carbon dioxide and releasing more oxygen to the atmosphere.

Also, the manufacture of new products from recycled materials requires less energy than from primary materials. According to Morris (1996), the energy saved (in $kJ\,kg^{-1}$) when used products are recycled into new products of the same material, in terms of low and high estimate, ranges between 14 000 (low estimation) and 35 800 (high estimation) for paper and cardboard; 42 000 and 111 000 for plastic; 900 and 5500 for glass; 201 500 and 312 100 for aluminum cans; 7000 and 37 000 for tin and bi-metal cans; 110 150 and 122 500 for other nonferrous metals; 14 500 and 21 200 for ferrous metals; 16 200 and 48 800 for tires, while 25 700 for rubber other than tires; and 58 300 for synthetic textiles, while 42 100 for cotton. By reducing the energy consumption via recycling, industry consumes less fossil fuel and generates less carbon dioxide, while global warming is inhibited.

Table 21.2 Recovered waste materials, greenhouse gas benefits, and equivalent car removal in 2012.

Material	Weight recovered (Mt)	Greenhouse gas benefits (equivalent metric Mt CO_2)	Number of cars removed from road (million)
Paper/paperboard	44.4	130.5	27
Glass	3.20	1	0.21
Metals			
Steel	5.55	9	1.9
Aluminum	0.71	6.3	1.3
Other nonferrous	1.36	5.3	1.1
Total metals	7.62	20.6	4.3
Plastics	2.80	3.2	0.67
Rubber/leather	1.35	0.7	0.145
Textiles	2.25	5.7	1.2
Wood	2.41	4.2	0.9
Other waste			
Food, other	1.74	1.4	0.29
Yard trimmings	19.6	0.8	0.17

Source: From EPA (2014).

Table 21.3 Environmental benefits from waste recycling.

	Aluminum (%)	Steel (%)	Paper (%)	Glass (%)
Energy use	90–97	47–74	23–74	4–32
Air pollution	95	85	74	20
Water pollution	97	76	35	—
Mining waste	—	97	—	80
Water use	—	40	58	50

Source: With permission from Lechter and Sheil (1986). Reproduced with permission of John Wiley & Sons.

In addition to energy savings, recycling contributes to the reduction of the ecological footprint by reducing air pollution (exhaust gases, dust), water use, water pollution, resource use, and waste for disposal (Table 21.3).

21.11.2 Economic Benefits

While the environmental benefits of recycling are widely recognized, the economic importance of recycling is less well known. By turning waste into valuable raw materials, recycling becomes a tool of economic development. This is based on the consideration that waste is a local resource and its management can contribute to local revenue, create new jobs, build new businesses and more competitive manufacturing industries, and expand the state economic base through taxation.

The basic initiative for waste resource exploitation is that discarded materials possess a value. If these materials are recovered instead of landfilled, the local community or the state has multiple economic benefits, as the landfilling cost is saved, the waste materials serve as secondary commercial commodities, and the habitants are not surcharged with the landfilling cost.

A resource management approach of material landfilling starts with the evaluation of the hidden value in waste materials, which is based on the knowledge of the amount, composition, and value of the waste commodities. A well-designed waste characterization study is the basic component to obtain accurate knowledge of waste quantity and composition. Also, efficient segregation of waste materials into various categories of recyclable and compost is necessary. Most of the reusable species can be repaired and resold, dismantled for recycling, or segregated into single streams.

The actual values of the recyclable materials are mostly determined by market rules. Factors such as energy cost, market value, availability of primary raw materials, and industrialization of developing countries influence the market values of recyclable materials greatly. For example, the energy savings from the use of certain recycled materials (metals, plastic, paper, etc.) instead of virgin increases their demand and, correspondingly, the market values. It must be pointed out that the market price of recycled materials (paper, plastic, metals) increases after further segregation. As an evidence, the following UK domestic mill price indicators for waste paper were recorded in March 2016 (all prices in UK £/metric ton): mixed paper 43–54, old cardboard 70–81, news and PAMS (newspapers/magazines) 68–73, colored best quality PAMS (papers/magazines) 117–121, and sorted office waste 126–134 (Environment Media Group Ltd, 2015). These prices hold for material ex works, usually baled or supplied according to mill specification; also, prices can vary regionally, and contractual arrangements may differ.

Although material recycling is a strategic tool for many countries to utilize efficiently their natural resources, the quantity of the landfilled waste materials still remains relatively high. Despite the increasing trend for recycling and composting in the United States since 1990, the data show that a significant fraction of MSW is still landfilled; in 2013 the total generated solid waste was 254.11 million tons, out of which only 34.3% was recycled and composted, 12.9% incinerated with energy recovery, while the landfilled waste reached 134 270 million tons (or 121.783 million metric tons), which corresponds to 52.8% of the total waste (EPA, 2015). Corresponding data of EU-28 for the year 2012 show that 42% of the total generated waste was recycled and composted (27% recycled and 15% composed), 24% incinerated, and 34% landfilled (Zerowaste Europe, 2014).

It must be pointed out that the recycling rates of the waste materials differ greatly. Data from the United States for the year 2013 (EPA, 2015) show that only paper/paperboard, yard trimmings, and nonferrous metals (other than aluminum) present recovery rates higher than 60%; the recycling rate for steel is 33%, glass 27.3%, aluminum 20%, while for the other materials below 20% (rubber and leather 16.1%, wood 15.7%, textiles 15.2%, plastics 9.2%, and food 5%).

The corresponding revenues from recycling are substantial and fast growing. In the EU, the financial turnover from the trading of seven main categories of recyclables (glass, plastic, paper/cardboard, iron/steel, copper/aluminum/nickel, precious metals, and other metals) has been almost doubled between 2004 and 2008, from 32.5 billion euros in 2004 to 60.3 billion euros in 2008 (EEA, 2011). It must be pointed out that these revenues are underestimated, as some recyclable commodities are excluded. In the United States, recycling facilities revenue is estimated to grow from \$5.6 billion in 2012 to \$6.5 billion in 2016 that means a compound annual growth rate (CAGR) of 4.0% (Harris Williams & Co Ltd, 2013). Correspondingly, the global e-waste recycling market is expected to grow from \$10.0 billion in 2012 to \$18.0 billion in 2017, representing a 12.5% CAGR (Harris Williams & Co Ltd, 2013).

21.11.3 Employment Benefits

In addition to the direct revenues, recycling and composting create more jobs of higher income than jobs related to waste landfilling or incineration. Once materials are separated, they become feedstock for processing and manufacturing plants that require more employees of unskilled, semiskilled, and skilled category. Depending on the recycled material and the country, the recycling of the same waste quantity creates 6–10 times more jobs than landfilling or incineration. These jobs are mostly generated in the private sector, at local or regional level; a few more indirect workers may be required in the public sector (city or regional level) as program administrators, managers, and advisors. The Institute for Local Self-Reliance (ILSR) worked on a scenario of manufacturing plants that could be sustained solely by waste recycling of a city with one million people (Rennie and MacLean, 1989). It was found that 6 paper mills, 1 glass-container manufacturing plant, 51 plastics processing and manufacturing plants, and 3 aluminum processing plants could be sustained, creating 1500 jobs and adding more than \$250 million dollars to the local economy.

Facts from the EU and the United States strongly support the increase of jobs as recycling rate increases. In the EU, the jobs related to recycling were increased from 230 000 in the year 2000 to 512 000 in 2008 (annual increase 10.57%), which is the second higher increase in ecoindustry sectors after that of the renewable energy sources. On a per million habitants basis, the overall employment related to the recycling of materials in European countries has increased steadily from 422 in 2000 to 611 in 2007 (annual increase 7%) (EEA, 2011).

In the United States correspondingly, it is estimated that 666 000 direct jobs were associated with MSW management in 2008, out of which the vast majority (about 86% or 574 000 jobs) are associated with recycling and composting, and only a small fraction (about 14% or 92 000 jobs) with disposal through landfilling and incineration (Goldstein and Electris, 2011). Recall that in terms of tons managed in 2008, only 84.0 million tons (Mt) was recycled and composted (about 33% of total MSW generated), while the remaining 188.5 million tons (about 67%) was disposed through landfilling or incineration (136.9 Mt landfilled, 31.6 Mt incinerated) (EPA, 2014). Projections to 2030 expect an increase to jobs associated with recycling and composting by 1.5 million in comparison with 2008, while the generated waste is expected to reach 315 Mt, out of which 236.25 Mt (or 75%) will be recycled and composted (Goldstein and Electris, 2011).

21.11.4 Prospects of Recycling

Recycling activities provide material inputs to the economy, substituting for virgin raw materials, thereby improving the resource production efficiency. There are several questions to be answered:

- How much of the global economy's material demand can be satisfied through recycling?
- What is the current share of recycled materials in resource consumption?
- How can this share be improved and to which point?

In the next section, there will be an effort to answer the previous questions and possibly some others that may further arise.

Recycling currently covers only a small proportion of the global demand for many material resources but can meet a large proportion of the economy's demand for resources, alleviating pressure on ecosystems to provide resources and assimilate waste. Already, recycling meets substantial proportions of demand for some resource groups, notably paper/cardboard and iron/steel groups.

However, even maximum recycling cannot entirely cover the global demand for resources. This is due to the accumulation of goods in a growing economy, e.g. in the construction sector, which acts as a long-term store for materials and makes them unavailable for recycling for many years. Recycled plastics and waste from electrical and electronic equipment provide the lowest coverage of demand both currently and in terms of a maximum future recycling scenario. The increase of recycling rate can help in creating a resource secure economy, but we need to accept that economic growth driven by ever-increasing material consumption is unsustainable.

At present the industrialized countries consume a huge amount of resources, with a significant proportion imported from abroad. Recycling is becoming an increasingly important source of material inputs, but there is no clear correlation between increased material recovery through recycling and reduction of resource imports. Recovered materials enter the free market like any other product and can be traded both within and outside the countries.

If the quality of a recovered material is sufficient to consider its performance equal to that of the virgin material, then it is possible to compare the amount recovered to the total demand for that resource.

Booming Asian economies and countries' directives have boosted recycling. One driver for the increasing economic importance of recycling is the globally growing demand for materials, especially in Asian economies. But in addition to rising unit prices for metal and plastic waste, another important factor is the increasing amount of recyclables sorted and placed in the market. For example, EU waste directives have significant contribution by creating obligations for recycling or recovery of ever-increasing percentages of waste and discouraging landfilling.

Recycling recorded the fastest growth rate among the largest ecoindustry subsectors (recycling, waste supply, wastewater treatment, and waste management). The growing importance of the recycling subsector is also apparent from its increasing scale in reference to the economic output of key sectors. Recycling generates more jobs of higher income than any other form of waste management, adds revenues to the economy, saves resources, and reduces environmental impacts and greenhouse gas emissions. Considering the aforementioned reasons and the global population increase, recycling is expected to further strengthen its importance.

References

Anastassakis, G.N. (2007). Solid waste disposal and recycling. In: *Environmentally Conscious Materials and Chemicals Processing* (ed. M. Kutz), 307–355. New York: Wiley.

Applied Photonics Ltd (2016). How LIBS works. http://www.appliedphotonics.co.uk (accessed 24 May 2016).

Bedeković, G., Salopek, B., and Sobota, I. (2011). Electrostatic separation of PET/PVC mixture. *Technical Gazette* 18 (2): 261–266.

Bokun, R.C. and Osadchii, S.M. (2010). Generalization of dual-energy method for material identification with X-ray Introscopy. *Journal of Surface Investigation: X-Ray, Synchrotron and Neutron Techniques* 4 (4): 591–593.

Brunner, P.H. (2011). Urban Mining – a contribution to reindustrializing the City. *Journal of Industrial Ecology* 15 (3): 339–341.

Bruno, D. (2013). How China Became the World's Junkyard Capital, The Atlantic CityLab. http://www.citylab.com/work/2013/11/how-china-became-worlds-junkyard-capital/7443/ (accessed 17 November 2015).

Bruno, E.A. (2015). Automated Sorting of Plastics for Recycling. http://infohouse.p2ric.org/ref/09/08620.pdf (accessed 2 December 2015).

Bulk Handling Systems (BHS) (2014a). Municipal solid waste systems, Brochure, p. 4.

Bulk Handling Systems (BHS) (2014b). OCC separator, Brochure, p. 2.

Cossu, R. and Williams, I.D. (2015). Urban Mining: concepts, terminology, challenges. *Waste Management* 45: 1–3.

Cross Wrap Oy Ltd (2015). CW 2200 Waste Bale Wrapper. http://www.crosswrap.com/cw_2200_bale_wrappers (accessed 11 March 2016).

Dalmijn, W. (2003). Lasers target aluminium alloy sorting. *Recycling International*: 48–51.

Dalmijn, W.L. and de Jong, T.P.R. (2004). Developments in and applications of dual energy X-ray transmission. In: *Sensorgestützte Sortierung 2004*, 1–8. Stolberg: RWTH Aachen.

Diamond, J. (2005). *Collapse – How Societies Choose to Fail or Succeed*, 571. New York: Penguin Group.

Dodbiba, G., Haruki, N., Shibayama, A. et al. (2002). Combination of sink-float separation and flotation technique for purification of shredded PET-bottle from PE or PP flakes. *International Journal of Mineral Processing* 65: 11–29.

Drelich, J., Payne, T., Kim, J.H., and Miller, J.D. (1998). Selective froth flotation of PVC from PVC/PET mixtures for the plastics recovering industry. *Polymer Engineering and Science* 38: 1378–1386.

EEA (2011). Earnings, jobs and innovation: the role of recycling in a green economy. Report no 8: p. 12.

Environment Media Group Ltd (2015). Prices, waste paper. http://www.letsrecycle.com/prices/waste-paper/ (accessed 1 April 2016).

EPA (2014). Municipal solid waste generation, recycling, and disposal in the United States: facts and figures for 2012 (EPA-530-F-14-001), pp. 6–11.

EPA (2015). Advancing Sustainable Materials Management: Facts and Figures 2013. Assessing Trends in Material Generation, Recycling and Disposal in the United States, EPA530-R-15-002, p. 3, 7.

Eriez Magnetics Europe Ltd (2006a). Permanent Magnetic Pulleys, Brochure no SB 240 UK/4, p. 2.

Eriez Magnetics Europe Ltd (2006b). Suspended Permanent Magnet, Models: CP and OP, Brochure no SB 300 UK/3, p. 2.

Eriez Magnetics Europe Ltd (2006c). Eddy Current Separators: Principle and practice, Brochure no SB 501 UK, p. 1.

Eriez Magnetics Europe Ltd (2006d). Eddy Current Separators, Models: Rev X, Rev X-E and LC, Brochure, UK, p. 3.

Eurostat (2012). How much of our municipal waste do we recycle? European Environment Agency, European Commission, p. 1.

Forrec srl Recycling Systems (2016a). Snapshots from video "The New Forrec Bags opener for municipal solid waste". https://www.youtube.com/watch?v=uxiTGqwsGaU&index=3&list=PLJFKLvSZbl_FpAIAvDFjzrq6lLwTMRncD, uploaded February 2014 (accessed 12 June 2016).

Forrec srl Recycling Systems (2016b). Multi-crusher and bag openers. http://www.forrec.eu/bag-openers-ls-1500-electric (accessed 12 June 2016).

Freiberg, T. (2011). The rise of the machines: Robot Recycling, Waste Management World, E-paper. https://waste-management-world.com/a/rise-of-the-machines-robot-recycling (accessed 2 January 2016).

Gaustad, G., Olivetti, E., and Kirchain, R. (2012). Improving Aluminum recycling: a survey of sorting and impurity removal technologies. *Resources, Conservation and Recycling* 58: 79–87.

Gesing, A. and Harbeck, H. (2008). Particle sorting of light-metal alloys and expanded use of manufacturing scrap in automotive, marine, and aerospace markets. Proceedings of Global Symposium on Recycling, Waste Treatment and Clean Technology (REWAS 2008), Cancun, Mexico.

Glösser, S., Soulier, M., and Tetcero Espinoza, L.A. (2013). Dynamic analysis of global copper flows – global stocks, postconsumer materials flows, recycling indicators, and uncertainty evaluation. *Environmental Science & Technology* 47 (12): 6564–6572.

Glysson, E.A. (1990). Solid waste. In: *Standard Handbook of Environmental Engineering* (ed. R.A. Corbitt), 8.85, 8.111–114. McGraw-Hill Inc.

Goldstein, J. and Electris, C. (2011). More jobs, less pollution: growing the recycling economy in the US, Report, Tellus Institute, pp. 10–11.

Goudsmit Magnetic Systems (2016). Magnetic separators for recycling, Brochure, p. 5.

Harris Williams & Co Ltd (2013). Waste Management and Recycling-Market Update. http://www.harriswilliams.com/sites/default/files/industry_reports/ (accessed 2 April 2016).

Heginbotham, J.H. (1980). Recovery of glass from urban refuse by froth flotation. US Bureau of Mines, IC 8826, pp. 56–60.

International Copper Association Ltd (2013). Copper Recycling, p. 1.

Kaufman, S.M., Goldstein, N., Millrath, K., and Themelis, N.J. (2004). 14th annual nationwide survey of solid waste management in the United States: the state of garbage in America. *BioCycle* 45 (1): 31–41.

von Ketelhodt, L. and Bergmann, C. (2010). Dual energy X-ray transmission sorting of coal. *The Journal of the Southern African Institute of Mining and Metallurgy* 110: 371–378.

Lawer, J.E. and Hopstock, D.M. (1985). Electrostatic and magnetic separation. *SME Mineral Processing Handbook*, N.L. Weiss, Society of Mining Engineers of the American Institute of Mining, Metallurgical, and Petroleum Engineers, Inc, New York, p. 6.8.

Lechter, R.C. and Sheil, M.T. (1986). Source separation and citizen recycling. In: *The Solid Waste Handbook*, Wiley, New York, Chapter 10 (ed. W.D. Robinson), 215–258.

Lukka, T.J., Tossavainen, T., Kujala, J.V., and Raiko, T. (2014). ZenRobotics recycler – robotic sorting using machine learning. Sensor based sorting 2014, RWTH Aachen University (11–13 March), Aachen, pp. 1–8.

Lundell Enterprises Inc (2015). LEI Densifier Model 1860. https://www.flickr.com/photos/lundellenterprises (accessed 10 March 2016).

Lungu, M. and Schlett, Z. (2001). Vertical drum Eddy-current separator with permanent magnets. *International Journal of Mineral Processing* 63: 207–216.

MASIAS RECYCLING SL (2014a). Ballistic separators. http://masiasrecycling.com/en/key-machines/ballistic/SB120 (accessed July 8, 2016).

MASIAS RECYCLING SL (2014b). Ballistic separator, Brochure, p. 4. http://masiasrecycling.com/down/MR_cat_balistico_ENG.pdf (accessed July 8, 2016).

Matthiessen Engineering SARL (2014). Mini Bag Opener SARL, Besancon, France. http://www.dailymotion.com/video/x47wga7_cop21-exposition-de-notre-mab-mini-ouvre-sacs-mini-bag-opener-mini-sackaufreisser-mini-abre-bolsas_tech (accessed 10 December 2015).

Medio Multimedia (2011). Snapshot from Enterprise Recycling Video. https://www.youtube.com/watch?v=gQpzVyTNZj4 (accessed 24 July 2016).

Metso Minerals (2000a). Lindemann DSP double screw presses, Brochure no 1238-01-02-REC, p. 1.

Metso Minerals (2000b). Lindemann KHA Hammer Mills, Brochure no. 1240-01-02-REC, p. 5.

Metso Minerals (2000c). Lindemann ZB Turning crushers, model ZB 116, Brochure no. 1232-01-02-REC, p. 5.

Metso Minerals (2000d). Lindemann RO Rotor shears, Brochure no. 1243-01-02-CS, p. 6, 8-9.

Metso Minerals (2000e). Lindemann Trommel screens, Brochure no. 1249-01-02-CS, p. 2.

Metso Minerals (2005). Nordberg TS series screens, Brochure no. 1312-04-05-CSR, p. 2.

Minter, A. (2008). Where America recycles, Washington, DC, The Atlantic, December. http://www.theatlantic.com/magazine/archive/2008/12/where-america-recyles/307187/ (accessed 17 November 2015).

Mogensen GmbH & Co. KG (1999). Mikrosort® Application Report, April 15, 1999.

Mogensen GmbH & Co. KG (2016). Sorting of Bulk Solids, MSort Brochure, pp. 4, 12.

Morris, J. (1996). Recycling versus incineration: an energy conservation analysis. *Journal of Hazardous Materials* 47: 277–293.

Muchova, L., Eder, P., and Villanueva, A. (2011). End-of-Waste Criteria for Copper and Copper Alloy Scrap: Technical Proposals. Luxembourg: Publications Office of the European Union.

Mutz, S., Julius, J., Pretz, T. et al. (2003). Sorting of secondary aluminium by automated picking. Proceedings of XXII International Mineral Processing Congress, Cape Town, p. 355.

Nihot Recycling Technology, B.V. (2014). Air separators, Brochure, p. 4.
Nihot Recycling Technology B.V. (2015). Air Technologies, Brochure, 5th print, p. 11.
Nihot Recycling Technology, B.V. (2016). SDS-i semi-mobile Windshifter, Brochure, p. 5.
Plunkert, P. (2010). Aluminium, US Geological Survey Minerals Yearbook. USGS.
Rao, S.R. (2006). *Resource Recovery and Recycling from Metallurgical Wastes*, 557. Oxford: Elsevier.
Renn, O. and Goble, R. (1996). A regional concept of qualitative growth and sustainability – support for a case study in the German state of Baden-Württemberg. *International Journal of Sustainable Development and World Ecology* 3: 1–22.
Rennie, C. and MacLean, A. (1989). *Salvaging the Future: Waste-Based Production*, 8–11. Washington, DC: ILSR.
RTT Steinert Elektromagnetbau GmbH. (2015a). Unisort® C, Brochure, pp. 2–3.
RTT Steinert Elektromagnetbau GmbH. (2015b). Unisort® P, Brochure, pp. 2–3.
Schlömann, E. (1975a). Separation of nonmagnetic metals from solid waste by permanent magnets. I. Theory. *Journal of Applied Physics* 46 (11): 5012–5021.
Schlömann, E. (1975b). Separation of nonmagnetic metals from solid waste by permanent magnets. II. Experiments on circular disks. *Journal of Applied Physics* 46 (11): 5022–5029.
Sherbrook-OEM (2015). Bag opener – Film grabber. https://www.youtube.com/watch?v=Ov0PyVG9jB8 (accessed 10 December 2015).
Stadler® Anlagenbau GmbH (2016). Ballistic separators, Brochure, p. 3.
Steinert Elektromagnetbau GmbH. (2015a). X-ray sorting system XSS, Brochure, pp. 2–3.
Steinert Elektromagnetbau GmbH. (2015b). Induction sorting system ISS®, Brochure, pp. 2–3.
Strydom, H. (2010). The Application of Dual Energy X-Ray Transmission Sorting to the Separation of Coal from Torbanite, MSc in Engineering Dissertation, Faculty of Engineering and The Built Environment, 98p. University of the Witwatersrand.
Summit Equipment, Inc. (2012a). Chain-belt conveyors. http://www.summiteq.com/Conveyors/Chainbelt Conveyors.aspx (accessed 11 June 2016).
Summit Equipment, Inc. (2012b). Twin-ram auto-tie baler, PR Series. http://www.summiteq.com/Balers/TwinRamAutoTieBalers.aspx (accessed 11 June 2016).
Summit Equipment, Inc. (2012c). Balers – Open-end baler. http://www.summiteq.com/Balers/BalerOverview.aspx (accessed 11 June 2016).
Tchobanoglous, G., Theisen, H., and Vigil, S. (1993). *Integrated Solid Waste Management: Engineering Principles and Management Issues*, 257, 559. McGraw-Hill.
UNEP (2011). Background paper for the ministerial consultations. *Green Economy, Twenty-sixth Session of the Governing Council/Global Ministerial Environment Forum*, Nairobi, Kenya, 21–24 February 2011 (UNEP/GC.26/17/Add.1).
Vermeer (2000). Tub Grinder Model TG525A, Brochure TG525A, p. 1.
Warren&Baerg Manufacturing Inc (2013). Cubing Bagasse Material for Renewable Fuel. http://www.warrenbaerg.com/news.php?ax=v&n=11&id=11&nid=31 (accessed 10 March 2016).
WCED (1987). *Our Common Future*, 45. New York: Oxford University Press.
Weiss, M. (2012). XRF – new applications in sensor-based-sorting using X-ray fluorescence. In: *Waste Management*, vol. 3 (ed. K.J. Thome-Kozmiensky and S. Thiel), 139–148. TK Verlag.
West Salem Machinery (2016). Disc Screen. http://www.westsalem.com (accessed 9 May 2016).
ZenRobotics Ltd (2016a). Photos and images. http://go.zenrobotics.com/public (accessed 21 May 2016).
ZenRobotics Ltd (2016b). Efficient waste processing 24/7. http://zenrobotics.com/zenrobotics-recycler/performance/ (accessed 2 January 2016).
Zerowaste Europe (2014). And the best waste performing country in Europe … is Estonia!. http://www.zerowasteeurope.eu (accessed 1 April 2016).

22

Waste Reduction in Metals Manufacturing

Carl C. Nesbitt

Department of Chemical Engineering, Michigan Technological University, Houghton, MI, USA

The metals industry is one of the most important sectors of the world economy. In the United States alone, the consumption of steel products approaches 120 million tons each year, while the consumption of aluminum averages 6.5 million tons per year and the copper consumption averages over 4 million tons each year. The production of metals comes from primarily two routes – production of metal from mineral resources or recycle of spent metal products. For example, on average approximately 67 million tons of steel are produced from recycled scrap (~55%), 2.9 million tons of aluminum were made from old and new scrap (~52%), and 1.3 million tons of copper were produced from scrap (~30%) in the United States each year. Almost 75% of the lead used in the United States is from secondary sources and recycling. Recycling metals directly saves our society by reducing the amount of waste generated, and energy consumed, to produce new metal products. Improving economies of individual nations typically are preceded by an increased demand for metals and metal products. An increased consumption of construction products (especially steel and copper) is one of the leading indicators that an individual country's economy is growing – as has occurred in the China and Asia markets.

The mining of ores is still the primary means of producing new metals for consumption. Every state of the United States has mining of mineral deposits (sand, gravel, salt, clays, etc.), but the majority of metal mines are in the western states. In the world, the biggest metal mining countries are in the Pacific Rim (including most South American countries, Australia, Indonesia, America, Canada, Mexico, and China), Africa (principally in South Africa, Ghana, Zimbabwe, Tanzania, Zambia, and the Democratic Republic of the Congo), and several former Soviet Republics (Kazakhstan, Uzbekistan, Tajikistan, and Turkmenistan); ores are mined, comminuted, and beneficiated to produce concentrates made of just the minerals with metal value. These concentrates are then processed by extractive processes (such as smelting or leaching) to produce a low-grade metal that can be refined or alloyed.

Recycled metals are typically separated by the consumers or the recyclers. More complex materials, as from spent automobiles and appliances, require complete shredding and an initial separation of the materials into general types (magnetic and nonmagnetic). Then the individual metal types can be melted and refined to produce metal for market. These processes, for the most part, have not changed in over 100 years. However, without exception, the amount of metal recovered from scrap has increased across the board – that is, nearly every metal consumed contains at least 10% and as much as 70% of its mass from recycled metals. That trend has risen over the years and should continue to increase as metal demand increases in Asia and North America, as energy conservation is sought, and as environmental concerns rise. Take, for instance, the recycling of automobiles. European automobile manufacturers are now required to take ownership of old or damaged automobiles for recycle into new models. The US automobile industry may face a similar fate in the next few years. Currently, there are more than 300 automobile shredding plants throughout the United States, each of which is capable of shredding up to 1000 cars per day.

The discovery and development of new reserves, the recycling of metal scrap, and the production of consumer metals are not without waste. Generally speaking, all waste can be classified by its form – solid, liquid, or vapor – and the processes (or locations) that generates it, at the mine, mill, smelter, or refinery. By far, the largest mass and volume of waste produced in metal production is in the form of solid wastes. Mining produces waste in the form of overburden that has to be removed to expose the mineral value beneath, and waste material from the separation and recovery of mineral value in concentrates has to be collected and disposed. Slags are also a necessary by-product of making pure metal ingots and sheets at smelters and refineries. Some processes, such as hydrometallurgical processing, will produce solutions rife with dissolved metals that must be treated before impounding or release. Vapors, including dusts and combustion products, are also

Handbook of Environmental Engineering, First Edition. Edited by Myer Kutz.

formed in the metal-production processes. While the amounts of metal wastes vary with the metal of interest, this chapter will discuss the types of wastes produced, steps to mitigate the release of wastes, and new developments into reuse and recycle of the wastes as valued products.

22.1 Wastes at the Mine Sites

The US Environmental Protection Agency (EPA) classifies mining and the minerals industry as the largest sector of waste production in the country. Just in the past few generations, mining techniques and processing techniques have improved to a state at which lower-grade ores can now be more economically treated. No industry demonstrates this better than the gold industry. As long ago as the California Gold Rush of the 1850s, gold ores with 10–20 troy ounces per ton (310–620 g t^{-1}) were economically processed from placer deposits. However, since the adoption of cyanidation to recover gold in the early 1900s and the advent of heap leaching in the late 1960s, gold ores with as low as 0.025 troy ounce per ton (0.8 g t^{-1}) can now be treated at a near break-even point.

This trend in treating lower-grade ores is generally seen in all metal mines. Long gone are the 1–5% copper ores or the 60% iron ores. Metal concentration steps must now be incorporated to liberate and recover the valuable metal. This is not without cost. Look at the gold mining example. At 0.03 troy ounce per ton, all of the rock will be turned into waste as the miniscule quantity of gold is extracted. As the processes to selectively recover the metal value improved, so too the techniques of waste handling had to be improved.

Environmental stewardship has been relegated to mining companies, especially in recent years. In some countries and states (most notably Wisconsin and Michigan), securities in the forms of bonds or cash payments must be made before licenses or permits will be approved for new operations. These securities are based on the expected impact of the waste piles (i.e. the probability of acid drainage generation, toxicity of contents, and total mass of earth moved). The bond is kept by the local governments and could be used to help mitigate any catastrophic failure of impoundments or any long-term environmental problems that could arise from the mining of the ores, such as acid drainage or windblown dust.

When mining companies consider developing a mineral ore deposit, as much as 10 years of planning and permitting is conducted prior to loading the first shovel of dirt. Local and federal regulatory agencies require completed environmental impact assessments, risk analyses, waste storage plans, well permits, water permits, and discharge permits for the life of the mine. As part of that planning, the mine operators must have a firm plan in place for storing the mine overburden and *poor rock*. The mining technique (open pit or underground) will also be decided based on the size and depth of the ore body, the mechanics of the surrounding rock formations, and so on. Underground mines are by definition a more selective means of extracting the ore and rarely require poor rock or waste rock piles. The mine plan is established to specifically extract ore, and little poor rock is excavated. In open pits, the poor rock piles are created by using dump trucks and bulldozers. The trucks back to the edge and end dump.

The dozers push the dumped rock over the side and *berm* the edge to give the truck drivers a sense of where the edge is. This action causes the level piles (similar to large peninsula mesas) to grow progressively outward from the mine in a planned strategic method. By the end of the mine life, these piles may be shaped to more closely resemble natural mountains and are reseeded with native flora. Although this practice is fairly modern (began in earnest in the 1980s), the impact of long-term mining projects has been lessened by the practice.

The regulatory agencies (MSHA, local and federal EPA, etc.) and regulatory laws (such as RCRA, "zero discharge" permits, etc.) have been empowered to ensure that no form of waste can be released from an individual mine site in any form (solid, liquid, or vapor) by any means (land, air, or water). During mine operations, airborne losses are mitigated through dust-control measures, which are of constant concern to the mining companies that have to transport ores and waste by haul trucks over dirt roads. As ore bodies are mined, roads are considered temporary. A road used to haul overburden or ore may be removed in the future, so paving the haul roads would not be a logical option. Today's haul roads are often tomorrow's waste rock or ore. Water trucks are a part of the operating fleet in mines, especially during the hot, dry months of the summers. Often, additives such as tree saps or polymers are added to the water to cover haulage roads to give longer-term dust suppression without harming the value or waste storage ability of the rock underneath the roads. Solid wastes from the mine (i.e. waste rock or overburden) are either stacked next to the mine or sometimes used in construction of dams and dikes for tailings impoundments.

Mine properties have to be much larger than the area of the deposit by necessity. Open-pit mines are built to include the natural angle of repose of the material to prevent slides into the mine. This practice requires that the actual mine size be a large area beyond the outlying portions of the ore deposit. Also, overburden disposal areas, space for processing plants, ore storage piles, warehousing, the maintenance garage for the large equipment, and office space must be built on the property. But by far the largest footprint belongs to the massive tailings impoundments that must sequester the mineral wastes into perpetuity. These specially designed (and permitted) areas act as the mining company's personal landfills of rock and liquid wastes.

22.1.1 Tailings Impoundments: Planning and Operating

All of the waste of a milling operation is sent to the tailings impoundment for sequestration. Mineral beneficiation requires that the gangue minerals be separated from the valued minerals. In most operations, the mass flow rate of solids in the tailings stream is not much different than the feed rate to the plant. For instance, in a gold milling operation, for every ton of ore fed to the plant, 0.1 troy ounce of gold is extracted – that is, a net reduction in the feed mass of 0.000 000 3 tons per feed ton of ore for such an operation. The waste material is then sent to the tailings impoundment for permanent storage.

As mineral processing techniques have improved, the head grade of treatable ores has diminished. In essence, the mining industry is generating more tons of tailings per ton of product but more efficiently and more economically. This means that for the same metal production, the tailings impoundment must be larger for today's operations than similar operations from several years ago.

The technology for handling the tailings has dramatically improved as well. The construction of tailings dikes has evolved to be an ongoing task of the engineers and operators of the mine site. Compaction testing of dikes and dams, density measurements, ground permeabilities, and downstream well monitoring are all part of the added duties of today's process engineers and operators. In the past, these tasks were given to consulting companies who engineered the starter dikes and built the extensions. Now, in-house engineers are trained and certified to perform these duties as part of the permitting plan of the operation.

The term *tailings* is generally defined as the solid-bearing waste product of a mineral concentrating process. More specifically, the tailings of an operation are a mixture of solid particles ranging from coarse granular (25 mm) to fine particles (300–50 μm), or even colloidal particles (<30 μm). However, it should be understood that processing ores requires very large volumes of water. Water is essential in the success of mineral separation processes, it is the best medium to transport large quantities of solids, and it eliminates dust. While tailings thickeners are used to recover a portion of the water from the waste material, the slurries that report to the tailings impoundments are typically 55–60% solids. Tailings ponds are therefore required to house the solids, as well as a large amount of process water. This conglomeration of particles and water makes tailings a complex material to handle. Add to the physical complication that the waste rock of a mine can consist of literally any mineral from quartz to clays, and it is easy to see that this material is inconsistent from mine to mine. All tailings impoundment areas must be designed specifically for the operation – that is, one size does not fit all.

In some mining processes, the tailings water may contain reagents that have to be treated before the slurries can be pounded. For instance, free cyanide or dissociable cyanide species must be treated to immobilize them in the tailings impoundment. The addition of ferrous sulfate solutions to cyanide-laden waste streams will sequester free cyanide in a stable precipitate known as *Prussian blue*, $Fe_4[Fe(CN)_6]_3$. More stringent treatment methods, such as Caro's acid (concentrated sulfuric acid and hydrogen peroxide), destroy most cyanide species such as ammonia or nitrogen and CO_2.

Waste impoundment design begins well before the operation begins construction of the plant. The size, height, and type of impoundment are predetermined by a number of factors. Designers of the tailings impoundment must keep in mind the following aspects:

- Characteristics of the tailings and liquid wastes of the processing facilities.
- Proximity to the ore body itself.
- Permeability of the soil in the area.
- Soil stability to support a large dike.
- Topography of the area.
- Volume of the holding area for the life of the mine.
- Excess volume or flood diversion trenching to handle a 100-year flood scenario.
- Applicability of installing a water retrieval system.

Properly designed tailings impoundments will not only sequester the solids but will also hold the water. As a precaution, all tailings ponds require monitoring wells downstream of and alongside the dikes. These wells constantly monitor the condition of the aquifer using conductivity probes and thermistors to measure any changes in salinity or temperature that could indicate seepage or leaks from the impoundment. Any perturbation to these measurements will send a signal to the operators that the water behind the tailings dike is leaking through the soil. Initial design of the tailings impoundment area requires testing of the permeability of the soil and ground to determine if the chosen site is proper for impoundment. In some circumstances, the ground may be made impermeable using geomembranes, clays, or other materials prior to start-up of the milling operation. However, the wells are safeguards mandated to watch for problems. Should a leak be detected, loads of clays, such as bentonite, can be dumped on dikes to stop seepage. These clays swell as they absorb water and are excellent additives to stop the seepage through a dam. Similar methods are used in ground seepage as well.

At start-up, a *starter dike* is in place to begin collecting the tailings and wastewater at the start-up of the operation. Typically, starter dikes are undersized to allow for optimal construction of the lowest portion of the diking system. Also, the total amount of tailings that will need to be impounded was predicted based on the reports of the exploration geologists who have estimated the size and grade of the ore in the deposit. Occasionally, this estimate is changed as the mine is developed, leading to an increased life beyond the life of the impoundment area. For these reasons, expansions to the tailings diking system are necessary and are most often planned in advance.

It is understood that the initial dike is not at the final height but will be the base for future expansions, or lifts, to be constructed as the operation proceeds. Often, the subsequent dikes are constructed by mixing of native soils with the coarse fraction of the tailings (separated by hydrocyclones on the dikes themselves). The new dikes are laid directly above the starter dike, allowing the level of tailings to grow subsequently. Figure 22.1 shows one operating strategy of this type.

The initial design of the dike must take into account this eventuality. The ground that will be used to support the dikes must be stable enough to not slide or degrade under the weight of the tailings and the dike. Typically, the ground is prepared by removing all plants and roots and loose material. Then layers of gravel, sand, and clays are compacted into the area to make the area more stable to support the initial and subsequent dikes to ensure tailings impoundment stability.

All lifts of the dikes must meet specifications of earthen dams. The amount of mechanical compaction, the permeability, and the size fractions of the components are carefully monitored by the engineers in charge of the lift construction.

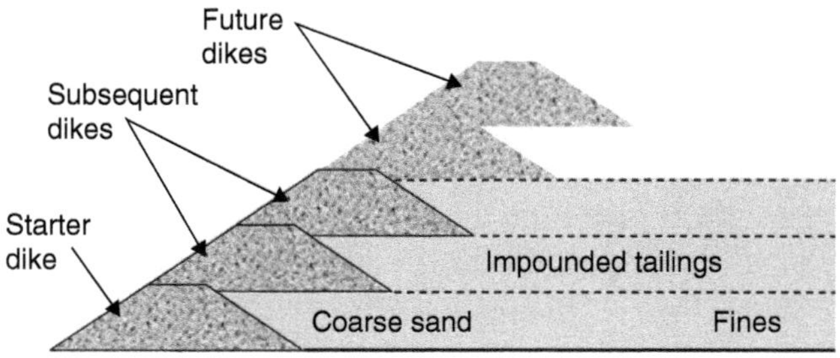

Figure 22.1 Upstream method for tailings dike construction. Tailings is naturally fractionated on the face of the dikes. Once the volume is full, another lift is placed above the starter dike.

22.1.2 Hydrometallurgical Wastes

Hydrometallurgy has grown in importance since 1940. Before that gold was principally the only metal recovered using aqueous chemical processing. The adoption of cyanidation in the 1900s caused the dramatic rise in production from hundreds of ounces per year in the late 1800s to 2005 with estimated production of 2500 metric tons per year. Likewise, copper production by leaching of ores has risen from a few thousand tons in the early 1950s (~1% of the world production) to the current estimate of 3.9 million metric tons per year (over 24% of the world copper production).

Hydrometallurgical processing has been used to treat lower-grade ores more economically than the traditional concentration/smelting operations. The processes have improved from vat leaching of copper ores to autoclaving finely ground ores to recover minute quantities of gold. The advent of heap and dump leaching in the late 1960s has dramatically lowered the grade of ores that can be processed economically.

The processing strategy used is directly dependent on the grade of ore and the ease of dissolving the metal of interest. Lixiviants range from basic cyanide solutions to strong sulfuric acid solutions. Bioleaching of ores, or ferric oxidation from bacteria, can also be used to treat ores that would have been relegated to waste heaps in the past. The lixiviant used is directly dependent on the metal to be leached. Although these reagents are typically not selective, subsequent processes have been implemented to not only increase the concentration of the metal in aqueous solutions but also eliminate the unwanted metals present in the leaching solutions. In every leaching process, there are five steps (or combinations thereof) that must be used to win the metal value from the ore:

1) Dissolution of the metal value to the aqueous phase (also known as leaching).
2) Solid–liquid separation to recover the aqueous phase apart from the spent ore.
3) Purification/concentration of the metal of value sought.
4) Recovery of the metal value by reduction or precipitation.
5) Solid–liquid separation to recover the metal apart from the solution.

Although the overall processes have evolved for specific metals, and leachants, there are principally only two leaching techniques that provide the *pregnant leach solution* for metal recovery: percolation leaching and agitated tank leaching.

Percolation leaching techniques generally include both the dissolution step and the separation of the liquid from the ore in one unit process. Percolation leaching includes in situ leaching, dump, and heap leaching. In situ leaching, most commonly used in the uranium industry, involves the direct dissolution of ore where it resides in the ore body. The lixiviant solutions are pumped into the ground to directly leach the metal-bearing minerals and then recovered by pumping the solution out of the ground. Figure 22.2 shows a typical five-well pattern used to ensure that all leach solution (injected in the center ejection well) is recovered through four adjacent withdrawal wells. Persistent rinsing of the area after the metal is recovered ensures that these wells will not remain contaminated. This process is considered only for ore bodies that meet special guidelines, such as no proximity to aquifers, readily soluble metals with weak lixiviant solutions, and impermeable host rock. Typically, there are no solid wastes from the processes. No actual solids are mined, so there are no materials to remove. Treatment methods may be required to clean the solutions for reuse or disposal.

In dump leaching, the ore is mined and stacked on an impermeable pad. It is processed by the direct application of leach solutions onto the top of the piles. Heap leaching is similar to dump leaching with the exception that the ore is mined and then crushed to produce a more uniform-sized material for the leach pad. In dump and heap leaching, the spent piles must be rinsed thoroughly to remove any toxic materials and unreacted leachants (especially cyanide in precious metal operations) and to neutralize the pH of the pile. The rinse solution is collected and treated to recover valuable metals that may remain in the piles and to remove any other metal ions species. The solution is then neutralized for reuse in rinsing the heaps or prior to discharging the water. Once completely rinsed, the piles are typically bulldozed into a more natural shape, covered with topsoil, and seeded with natural flora of the region. Often, these leached piles appear to be small hills on the landscape.

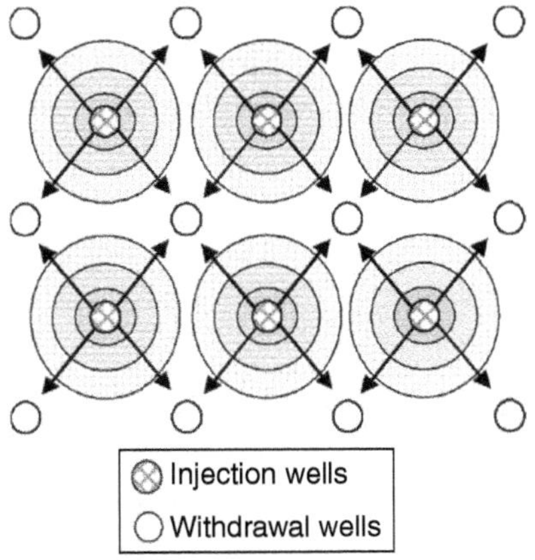

Figure 22.2 Typical arrangement of a five-hole pattern for in situ leaching of ores. The central hole is used to inject the leachant, which flows radially to the outer holes from which the metal-laden solutions are withdrawn.

In *agitated tank leaching* systems, finely ground ore must be mixed with the lixiviant until the reaction is complete. Liquid–solid separation processes are used to recover the solution and prepare the waste for disposal in a tailings impoundment. Unlike percolation leaching operations, these operations require a large impoundment area to not only handle the solids but also neutralize the chemistry and pH of the solutions used to extract the metal value.

Economically speaking, the inexpensive percolation leaching processes are reserved for lower-grade ores (that is, ores with lower net metal value.) The major costs of these processes are mining, stacking, and crushing of the ore. In agitated tank leaching operations, the ore must be mined, crushed, and finely ground. This added expense will give a better recovery of metal but will also cost considerably more than the simple percolation processes. So these processes are reserved for the higher-grade ores, which will return more revenue per ton. Tailings impoundments must handle the solids, and treatment methods must be included to neutralize the chemistry of the solutions prior to sequestering the tailings in the ponds. Also, every mine operator is responsible for preventing animals and birds from being harmed by these large man-made lakes. Great care is given to maintain fences and other methods to prevent birds from landing in the open water.

22.1.3 Acid-Rock Drainage (ARD)

Sulfide mining has been under attack for the perceived assumption that all sulfide mines (including base metal mines and high-sulfur coal mines) lead to the automatic generation of acidic mine drainage from the tailings and/or poor rock piles. In fact, acid-rock drainage (ARD) and acid-mine drainage (AMD) are the result of microorganisms that convert the soluble forms of iron and sulfur into high oxidizing solutions at low pH. *Acidithiobacillus ferrooxidans* are the primary culprit in acid-rock generation (ARG), but other microorganisms have been isolated and contribute to the problem.

The mechanism by which these bacteria produce ARD is still debated, but the fact that *Acidithiobacillus ferrooxidans* is a chemolithoautotrophic microorganism that relies on oxidizing ferrous ions (Fe^{2+}) to ferric ions (Fe^{3+}) is well documented. The bacteria are aerobic, and they catalyze the oxidation by the following reaction:

$$O_2(aq) + 4Fe^{2+} + 4H^+ \rightleftharpoons 2H_2O + 4Fe^{3+}$$

It should be noted that the bacterial presence is absolutely necessary for this reaction to proceed at low pH. As written, the reaction is thermodynamically spontaneous to the right and rapid at pH above 3.5; that is, ferric will be formed by simply aerating a ferrous solution at pH >3.5. However, at a pH above 2.4, the solubility of the ferric ion is reduced by the rapid and complete formation of ferric hydroxide ($Fe(OH)_3$) precipitate. The bacteria are required to catalyze the formation of ferric at a pH below 2 so that soluble ferric ions are produced. This is the primary reason why the presence of bacteria is required for acid-rock formation.

Understand that the bacteria will not survive, or will have very slow growth rates, in rock that does not contain the proper elements – namely, iron and sulfuric acid. The ferric that is generated by the bacteria will oxidize most sulfide minerals to solubilize the metal ions and will also oxidize the sulfide components into sulfates (hence, acid is generated from the sulfide minerals). But the bacteria first require a source of iron for their growth. That is why it is most common to find ARD generated in sulfide mineral outcrops that contain any of the common iron sulfide minerals, such as pyrite, marcasite, pyrrhotite, and chalcopyrite. The natural weathering of these minerals (albeit quite slow kinetically) will produce a small concentration of soluble ferrous ions that will allow for the natural bacteria to begin growth. As the bacteria cultures mature and expand, the net result is full-fledged production of a water that is high in oxidation potential (from the ferric ions) at a low pH (caused by the direct and indirect oxidation of the sulfides to sulfuric acid). This product water will then proceed through the rock mass and liberate all other metals and sulfides to produce the by-product of ARD, AMD, and acid drainage.

For many ores, pyrite is considered a waste mineral; that is, its value is too low to be economically recovered. The presence of pyrite in copper ores, gold ores, and coal, among others, can greatly reduce the value of the ores. For instance, in copper sulfide ores, a small amount of iron is required for optimal smelting. But too much (as would be provided by a small amount of pyrite in the concentrate) would mean a higher energy cost to smelt the concentrates. Pyrite in gold ores directly interferes with the cyanidation process. Often, the gold will actually be found in the pyrite minerals of the ore. The pyrite is readily solubilized by cyanide, so it consumes much of the cyanide, leaving none for precious metal extraction. In coals, pyrite and other sulfide minerals are the primary source of the sulfur in the coal. Most eastern coals in the United States contain high-sulfur content, almost 60–70% of which comes from the mineral pyrite. Desulfurization techniques that selectively remove inorganic sulfide minerals are the most successful at sulfur reduction of coals.

ARD is common wherever the presence of pyrite is greatest. Poor rock piles of coal mines and metal mines are notorious sites for acid generation. The Bunker Hill mine of Northern Idaho, the Berkeley Pit of Montana, and the Leviathan Mine of Eastern California are all notorious ARD sites that have pyrite in the deposits that have become exposed to the atmosphere in rock piles or the mines workings itself.

The mining companies that are harvesting the value from these mines are quite cognizant of the long-term costs that ARD poses to their business. International Network for Acid Prevention (INAP) is an organization of international mining companies dedicated to the reduction of the generation of acid from waste rock. This group is instrumental in sharing technologies for preventing and treating acid-rock runoff as it happens. Most of the successes in prevention involve preventing oxygen from contacting the waste. Storing pyrite-laden solid waste in impoundments that are covered with water, soils, plants, or impermeable geomembranes has been successfully used by some of the mining companies to keep acid generation to a minimum. Deepwater storage of wastes reduces the optimal conditions of bacterial activity. Dissolved oxygen is depleted by the aerobes themselves. As long as the flow of water is diverted so that the oxygen cannot be replenished, the area becomes anaerobic naturally. Anaerobic bacteria are known for their sulfide reduction chemistry that produces H_2S from SO_4^{2-}. These bacteria directly reduce the acid content and precipitate any soluble metal species as stable metal sulfides.

Several processes have been used to combat ARD that has already been formed. Metal precipitation and acid neutralization with lime or limestone is the most common technique used to limit the environmental impact of an acidic, metal-bearing solution from waste material. Metal hydroxide formation is quite simply achieved with moderate changes in the pH of the solution. Metal sulfide formation is typically preferred because the sulfide precipitates are more stable, more quickly formed, and much easier to remove by filtration or sedimentation. However, the sulfide sources (such as hydrogen sulfide gas, sodium sulfide, and thiourea) are very expensive on a "per pound of metal removed" basis and can produce a noxious hydrogen sulfide gas emission from the mixing tanks. The chemistry of precipitation is discussed in more detail later.

22.2 Chemical Metallurgy Wastes

Concentrates from the mines or impure metal products from metal recyclers must undergo further chemical processing steps to produce the metals for the manufacturing sector. Metallurgists classify these *chemical metallurgy* processes by the principal methods used to break the chemical bonds of the metals with the other metals or the mineral components. *Pyrometallurgy* is the field in which high temperature processing (often with the metal in a molten state) is used to produce the metal ingots and sheets that will be used to manufacture various final consumer products. *Hydrometallurgy* is the process by which purified metals are won from aqueous solutions, either by electrowinning (EW) the metal from concentrated leach solutions or by electrorefining (ER) metal anodes to produce pure cathode sheets or plates. Finally, electrometallurgy is the field in which electric current is used to either melt or plate metals. For the sake of this chapter, let us consider that electrometallurgy could actually be classified into the other two areas. That is to say, electric current in small amounts can be used to recover metal from aqueous solutions (hence it is related to hydrometallurgy) or in large amounts can be used to resistively heat and melt metal scrap (hence it is related to pyrometallurgy).

22.2.1 Pyrometallurgical Processing Wastes

Nearly every metal produced for the manufacturing sector at one point has to be processed pyrometallurgically. Iron ore concentrates are first reduced to pig iron (high carbon iron), which must be converted to steel (low carbon iron) in an open-hearth furnace, basic oxygen furnace (BOF), or steel processing plant. Gold recovered from processed ores is melted and poured into ingots at a mine. These impure bars of gold will have various other metals (most notably silver, copper, or iron) in the bars that must be separated pyrometallurgically to make 24-karat-gold bullion. Aluminum is either remelted and processed to remove alloying elements or plated as pure metal in molten salt electrolytic processes. Suffice it to say that even the melting of scrap to make ingots is a high temperature process that all metals must endure. These processes are the source of several different wastes. Solid wastes of pyrometallurgical processes, the predominant waste material, include cooled slag and dust. But gaseous wastes are also a part of the processes as well. Combustion products and gaseous reaction products must also be considered when understanding all of the processes that lead to metal production.

Steel production has changed over the past century. It is an excellent example of how the landscape has changed in the metal manufacturing sector and the volumes of waste have been reduced. After the initial discovery of blast furnaces in the late 1880s, integrated steel plants were constructed to convert iron oxide ores into pig iron in the blast furnace. The molten pig was then sent to open-hearth furnaces or poured into ladles so that the carbon in the metal (up to 6% by weight in pig iron) could be reduced to produce steel (carbon content of less than 2%). *Integrated steel plants* are self-contained, including coke ovens to produce the coke for the blast furnace from bituminous coal and auxiliary operations to collect the coke oven gas, heat from waste gases, air preheaters, and so on. The large tonnages of iron ore and coal required that they be built near the Great Lakes or other large waterways where large ships can deliver the materials cheaply. However, as the demand for steel began to push these behemoth operations to their capacity, a push to recycling more spent metal has led to today's *minimill* operations. These operations were not restricted geographically to large bodies of water. Instead of blast furnaces, they used electric arc furnaces (EAF) to melt the scrap. The molten scrap could be treated in ladles or BOFs to get the proper chemistry in the final metal. More and more of these operations are appearing as the need for recycling has increased.

Although blast furnaces and integrated operations are still important, the advent of the minimill has improved the steel manufacturing business. They are cleaner operations and produce much less solid or gaseous wastes than the previous processes. They produce steel for local consumers from local steel scrap, so transportation costs are minimal. Although this is only an iron and steel example, similar trends are happening with other metals as well. Conventional metal production from ores will continue to be the major source of most new metals, but recycling is catching on, and the benefits of reduced waste volumes are only one of the reasons. Conventional methods typically produce more wastes than the new recycling processes. For this reason we will focus primarily on the conventional processes, knowing that their overall impact has been reduced by incorporating more recycle.

22.2.1.1 Solid Wastes

22.2.1.1.1 Slags

Slags are defined as the waste products produced in the smelting or refining of molten metal products. They are produced in all metal pyrometallurgical processes. Although they are considered wastes, they are an important part of the metal-production process. Some slags are the result of additives with the feed, while some are the result of refining processes to clean the molten metal. The chemistry of the molten slags is varied to improve the metal properties during production. Slags are composed of fluxes that help to lower the melting temperature of gangue minerals that may be present in the final mineral concentrates feeding a smelter. Also, basic slags (i.e. slags that contain excess calcium, sodium, or iron oxides) help to reduce the concentration of detrimental elements (such as phosphorus, sulfur, and silicon) in pig iron and steels. The molten metal flows through the slag in droplets to improve the removal of these elements as the metal is collected at the bottom of the hearth. Slags can retain entrained amounts of the product metal of value. Because of this, much of the slag is recycled back into the furnace to reuse their chemical and thermal properties. For the most part, nonferrous metal slags are considered worthless and have little reuse; however, ferrous slags have many uses.

Given the huge steel market, the vast majority of the slags produced come from iron or steel mills. Iron and steel slag is considered a valuable commodity, worth as much as US $340 million in 2005. Approximately 60% of this commodity is from blast furnaces, and the remainder is from steel manufacturing plants that use BOFs and EAF. Pig iron is the product of blast furnace processing of iron ores. For the production of 1 ton of pig iron, a charge will consist of about 1.7 tons of iron oxide ore or other iron-bearing materials, 0.5–0.65 tons of coke, and 0.25 ton of limestone or dolomite, which is added to make the slag. The blast furnace will require from 1.8 to 2.0 tons of air to convert the coke into CO for the reduction of the iron oxides to elemental iron. Along with the 1 ton of pig iron produced, the blast furnace will produce 0.2–0.4 tons of slag, 0.05 tons of flue dust, and 2.5–3.5 tons of blast furnace gas (a combination of CO and CO_2). Given that there is from 35 to 45 million tons of pig iron produced in the United States annually, there is as much as 7–18 million tons of slag produced from this process annually.

Over the years, much has been done to make the slag more reusable. Regular cooling pits for the slag can be used to produce a slag that can be readily mixed with asphalt or cement to produce aggregates in paving asphalt and concrete. Some of the ground granular blast furnace slag (GGBFS) can be ground fine and used as feed material to make Portland cement in concrete mixtures. Air-cooled slag is used as concrete aggregate. Rapid quenching of the molten slag can produce a *popcorn slag*, which is an excellent aggregate for cement. The porous structure of the slag will securely bond with the cement matrix and produce a stronger concrete. Considering that over 1 billion tons of sand and gravel are produced each year in the United States and the average production of slag is just 40 million, the construction industry has easily swallowed up the steel and iron slag production. Only about one million tons of slag ever end up in landfills in the United States each year.

22.2.1.1.2 Casting Sands

The production of metal products requires that the molten metal be cast into near-net shapes. The steel industry uses casting sand because it is inexpensive, is readily available, efficiently absorbs heat at a predictable rate, and is easy to use. The sand is blended with water, carbon, and other admixture compounds to improve the compaction of the sand, the heat capacity, and other properties. The sand is pounded around a mold, the mold is removed, and the molten metal is poured into the sand molds. Unfortunately, the sand immediately next to the cast part is typically not reusable. Carbonization, fusion, and other factors make this material unusable and must be disposed of in landfills. Many of the companies that cast parts are implementing new programs to separate the *burned* material from undamaged sand to recycle the sand for more castings and reduce the amount shipped to landfills.

22.2.1.1.3 Dust

Particulates are a problem at many stages of the metal-production process. The amount of air or oxygen that is passed through various pyrometallurgical processes invariable will kick up the fine particles in the material. Baghouses have been designed to collect even the finest particles for disposal or recycling of vented particles (typically metal oxide particles). Various processes have been designed to treat the broad spectrum of dusts produced in the metals industry.

Fluo-solid smelters use the suspension of fine particles as a means of more efficiently processing the particles. Blast furnace feed must be agglomerated to a more stable large size to prevent the immediate expulsion of fines with the vent gases. In an EAF, the dust generated is composed primarily of zinc oxide with small amounts of iron oxide and other base metal oxides. The EAF is fed recycled steel, direct reduced iron, or pig iron. It uses electric current to resistively melt the charge. But keep in mind that zinc has a boiling point (900 °C) that is below the fusion point for steel, so any zinc galvanizing on the metal scrap will immediately be volatilized and react with the oxygen in air to produce a fine zinc oxide powder that is collected in the baghouse filters. Several specialty companies are investigating methods to salvage the zinc value from these by-products.

22.2.1.2 Gaseous Wastes

22.2.1.2.1 Sulfur Dioxide

Smelting of sulfide minerals invariably leads to the production of sulfur dioxide at some point in the process. For instance, as mentioned previously, in the world today, only 25% of the copper produced comes from the more environmentally benign hydrometallurgical process (i.e. solvent extraction and EW). An important aspect is that the aqueous processing of copper minerals obviates the need for high temperature smelting of copper. However, this implies that 75% of the world's copper comes from a sulfide operation that first concentrates the copper using flotation or some other beneficiation process (tailings is sequestered behind a tailings impoundment), matte smelting of the copper–iron–sulfur concentrates in fluo-solids smelters, or reverberatory furnaces. The matte is converted to *blister copper* by injecting air and oxygen to first produce an FeO slag and then SO_2 vapor from the melt. The Clean Air Act of 1977 has dramatically affected this practice. Long gone are the days of just sending the SO_2 up the stack. Today, the increase in copper produced hydrometallurgically has increased sulfuric acid demand and given a natural path for the SO_2 to be disposed. Nearly every smelter in the industrialized world has inserted a gas scrubber to collect the SO_2 and convert it to sulfuric acid solution (usually by the contact process). This acid availability has directly influenced the market of sulfuric acid and works in a symbiotic relationship with hydrometallurgical operations. Nearly every mine with a smelting operation has a leaching branch that uses the sulfuric acid produced to enhance metal production.

From a stoichiometric standpoint, nearly 0.35 tons of SO_2 are produced for every ton of copper generated. Nearly 0.65 tons of SO_2 are generated in the production of 1 ton of lead from galena concentrates. That means in 2005, approximately 400 000 tons of SO_2 were produced from the lead and copper production in the United States. This SO_2 made 613 000 tons of sulfuric acid or about 1.5% of the total sulfuric acid consumed in the United States. Almost all of that acid was consumed by the hydrometallurgical portion of the metal mining industry.

Carbon Dioxide. As the debate of the effect of greenhouse gases rages on, the simple fact remains that carbon dioxide production is one of the known side reactions of most metal-production operations. Carbon is an effective metal reductant. Coke is used to produce pig iron from iron oxide ores and lead from sulfide ores in blast furnaces, carbon electrodes are used to produce aluminum from bauxite leaching products, and coal is used in the reduction of zinc oxide in retorting furnaces. All told, the resulting product of metal reduction is the oxidation of carbon to carbon dioxide. It is important to keep in mind that the production of carbon dioxide has been reduced dramatically since the start of the industrial revolution of the late nineteenth century. This is best exemplified by the history of steel making in the world.

In the late 1800s, the invention of the blast furnace greatly affected the production of pig iron and consequently steel manufacture. Many historians credit this single invention with the beginning of the industrial revolution. Steel production prior to this unit process used as much as 2–3 tons of coke (fixed carbon from coal) for every ton of hot metal pig iron produced from iron oxide ores. Modern steel production has been far more efficient, with current production estimates ranging from 0.5 to 0.65 tons of coke per ton of iron produced. Add to this the recent trend away from blast furnace production from ores. Recycling of steel products (such as shredded automobiles, recycled appliances, construction materials, and other end-of-life cycle products) has led to the direct production of nearly half of all steel from recycle through minimills. The EAF (the workhorse of minimill recycling operations) does not require coke for the production of molten iron. Instead, the major waste product is small amounts of slag and EAF baghouse dust, which is primarily composed of the zinc oxide powder that is evolved from scrap that was galvanized. So in a way, the improved blast furnace operations and the advent of EAF recycling of steel have meant a dramatic drop in the carbon dioxide production of steel. Prior to 1880, steel production meant the production of as much as 11 tons of CO_2 per ton of iron, to the current rate of less than 1 ton per ton of steel consumed. The US steel industry leads this charge, but other countries (namely, Argentina, Brazil, Japan, and other major steel-producing nations) are also following suit by implementing better operational strategies in their steel production.

Lead is also produced in a blast furnace using coke as the primary reductant. Much like the iron blast furnaces, the off-gases from the smelting of lead oxides will primarily contain CO and CO_2. This gas is recycled back to the blast furnace to use the reducing power of the CO and the heating value otherwise lost. This helps to reduce the total amount of carbon that would be needed to produce lead. In many smelters, galena (the natural mineral PbS) is the feedstock. The galena is roasted to lead oxide, so some of the vapor waste from these operations will contain SO_2. As mentioned previously, this is collected and reacted to form sulfuric acid.

Primary aluminum production starts with the purified alumina (Al_2O_3) that is produced when bauxite is leached with caustic soda (NaOH). The pure Al_2O_3 is dissolved in molten cyrolite (Na_3AlF_6) held at 900 °C to about 10% by weight. Carbon electrodes are then put into the melt, and as the current is passed into the molten salt solution, the Al_2O_3 is reduced to Al liquid and O_2 gas. The oxygen from alumina reacts with the anodes to form CO_2. The molten aluminum sinks to the bottom of the melt and is tapped off and cast in pure aluminum ingots. As the Al_2O_3 becomes depleted in the molten salts, the density of the solution raises until the aluminum cannot sink. At this point, the operators notice a large spike in the applied voltage, and more alumina is charged, and the process continues.

Zinc production follows a similar path to the iron and lead processes in which carbon monoxide from carbon is the reducing agent. Carbon (coke or charcoal) is blended with zinc oxide and fed to a retort. The boiling point of pure zinc is low (~900 °C), so as the carbon is burned to form CO, the direct reduction of the zinc oxide by CO will result in the direct volatilization of the zinc. Cooling the vapor will condense the zinc, which can then be cast into ingots or alloyed with other metals.

Table 22.1 The approximate amount of CO_2 produced per ton of metal.

Metal	CO_2 per ton of metal, tons	Average metal production from ore in United States, tons per year	Annual CO_2 produced per ton of metal, tons per year
Pig iron	1.50	40 000 000	60 000 000
Lead	0.21	440 000	92 400
Aluminum	1.30	2 500 000	3 250 000
Zinc	0.67	760 000	509 200

Table 22.1 shows the relative production of CO_2 per ton of metal produced. As shown, the largest producer of CO_2 is indeed the iron and steel industry, but it should be noted that it is primarily because of the high consumption of iron and steel. Lighter metals, such as aluminum, will likely be used as a replacement for steel in hybrid automobiles and in low-density building materials. So there could be an increase in CO_2 gas production as more of the lighter density metals are consumed to replace steel.

For the most part, there are few techniques in place that can be employed to remove and collect the CO_2 from metal production. Most efforts in CO_2 reduction have been in improved smelting processes and lower carbon consumption. However, monoethanolamine (MEA) and diethanolamine (DEA) in organic solutions are used to scrub CO_2 from vent gases, but its expense precludes widespread adoption by the metals industry.

22.2.2 Hydrometallurgical Processing Wastes

As mentioned previously, most of the wastes from hydrometallurgical operations are simply treated and then sequestered in tailings impoundments. For those operations that do not have tailings impoundments, more complete process scenarios have had to be designed. For nearly every operation, pretreatment wastewaters, rinse waters, or wash waters must include neutralization and metal removal before the solutions can be discharged, impounded, or reused. The most widespread method for removing dissolved metals from solution is to precipitate them as solids for separation.

22.2.2.1 Precipitation Overview

For the majority of metal-bearing solutions, simple chemical precipitation is the most effective means of removing a substantial amount of the dissolved metals. Hydroxide precipitation is a simple, inexpensive means of reducing most of the metal ions from solution and neutralizing the pH. Sulfide precipitation using sulfide donors, such as hydrogen sulfide, sodium sulfide, thiourea, and thioacetamide, is another means of reducing soluble metals from waste streams. The precipitation process is generally the same whether the final product is metal hydroxide precipitates or metal sulfide precipitates. The final size of precipitates is a function of the following factors:

- Initial metal ion concentration.
- Rate at which precipitants are added or chemically generated.
- Precipitant concentration (i.e. the concentration of OH^- or S^{2-}).
- Aging time.

For most waste streams, the metal ion concentration is set by the process and is not variable. However, the concentration and rate of adding the precipitant can be varied. For instance, if H_2S is used as a precipitant, the actual sulfide ion concentration is defined by a complex series of equilibria:

$$H_2S(g) \rightleftharpoons H_2S(aq)$$

$$H_2S(aq) \rightleftharpoons H^+ + HS^-$$

$$HS^- \rightleftharpoons H^+ + S^=$$

As shown, hydrogen sulfide gas bubbled into water will produce an acidic pH. Consequently, the concentration of sulfide ion (S^{2-}) will be very low for even a saturate hydrogen sulfide solution. The predominant species of sulfur in solution is aqueous H_2S or HS^-. As the metals react with the sulfide species, the equilibrium between H_2S, HS^-, and S^{2-} is shifted even more. This means that the net amount of reacting species with the metal ions is quite low but will be continuously replenished as the precipitation process proceeds. Likewise, with such sulfide donors as thiourea and thioacetamide, the production of soluble sulfide species is directly dependent on the consumption of S^{2-} by precipitation.

Keep in mind that all precipitation events are due to the fact that the concentrations of the soluble species (for example, Me^{2+} and S^{2-}) have exceeded the solubility of the solid species. For the sake of discussion, let us look at a solution containing 1 g l^{-1} copper as cupric ions (Cu^{2+}). The solubility product for $Cu(OH)_2$ is defined as the equilibrium constant for the dissolution of copper hydroxide to form the individual ions of copper and hydroxide. Chemically, it is defined by the following equilibrium:

$$Cu(OH)_2(s) \rightleftharpoons Cu^{2+}(aq) + 2OH^-(aq)$$

The overall equilibrium constant of the reaction is defined as the solubility product

$$K_{SP} = \frac{a_{Cu^{2+}} \cdot (a_{OH^-})^2}{a_{Cu(OH)_2}}$$

where a_i is defined as the activity of species i.

For aqueous systems, a unit activity is expected for the solid species (i.e. we assume that the chemical reactivity of a solid in water is unchanging as long as there is solid in equilibrium with the solution). Also, for dilute concentrations, we assume that the activities are equal to the concentrations of the species. With these assumptions, we can reduce the solubility product constant equation to

$$K_{SP} = [Cu^{2+}] \cdot [OH^-]^2$$

(It is relatively simple to remember that the solubility product is the product of the soluble species.) For the solid precipitate to form, the product of the copper ion concentration and the hydroxide ion concentration squared must exceed the numerical value of K_{SP}. Thermodynamic data can provide the needed values of the free energies of the ionic species. Then the value of K_{SP} can be determined directly (the K_{SP} for $Cu(OH)_2$ is 1.58×10^{-20}). Keep in mind that the equilibrium hydroxide ion concentration is defined by the pH of the water solution. So for any metal ion concentration, we can predict the maximum pH before the precipitate will be formed.

Monhemius has gathered the K_{SP} values for several metal ion concentrations. He prepared an excellent nomograph that allows us to predict the final metal ion concentration in any water solution at a specific pH (i.e. for a specific hydroxide ion concentration). Figure 22.3 shows the diagram from Monhemius. As you can see, based on this chart, if the pH is 4 or less, no copper hydroxide precipitate will form for our $1\,g\,l^{-1}$ copper solution. However, if we increase the pH of the solution to 6, the copper ion concentration will drop to the equilibrium concentration of 10^{-4} M ($6\,mg\,l^{-1}$). Recall that these are equilibrium concentrations; that is, the final concentrations will be achieved only after a sufficiently long period of time has passed.

This brings us to consider the actual precipitation reaction and the path at which it occurs. The copper ions must come in contact with two hydroxide species to form the copper hydroxide precipitate. But to initiate this transition from aqueous species to solid compound, the ions must congregate to form nuclei for the precipitation. Most experts consider a nucleus to consist of at least three molecules. Once these nucleates are formed, the rest of the ions will simply build onto these until the equilibrium solution concentrations are achieved. So we can think of the precipitation process as consisting of two principal steps – nucleation and growth. Nucleation is the slowest step in the process and will only occur after saturation has been exceeded. It can be likened to an activation energy that must be overcome before a reaction will occur. Prior to forming nuclei, we can actually have a supersaturated solution that remains stable for some time. But once nuclei form, the solution rapidly

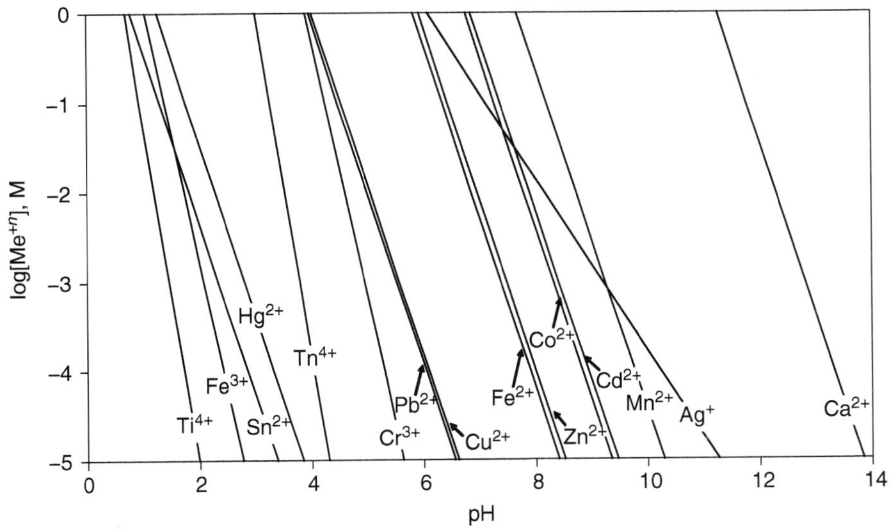

Figure 22.3 Monhemius diagram on metal ion concentration at specific pH. *Source*: Adapted from Monhemius (1977).

changes to the final equilibrium concentration as precipitates form. One way to lower the nucleation energy is to seed the process with starter nuclei. Particles recycled from a precipitation process, or particles with similar crystallinity, can be used to initiate precipitation processes without having to achieve supersaturation first.

Here is where the concentration of metal ions becomes so critical. If the metal concentration is initially low, as nucleates form, there will be little more metal ions in solution to grow the precipitates. Therefore, the precipitation results in a relatively few colloidal (i.e. ultrafine) precipitates (also known as *sols*). If the metal ion concentration is higher, then there will be plenty of extra metal ions to grow the precipitates larger from the initial colloidal nuclei. However, at too high of a concentration, the solution becomes too dense for metal ion migration. Under these conditions, a tremendous number of nuclei are formed by the metal ions in close proximity. The sheer number of nuclei consumes most of the metal ions, and again we have colloidal precipitates (gels). However, under these circumstances, an aging process can be utilized to grow larger precipitates at the expense of smaller ones. The natural tendency for precipitates will be to proceed to the lowest energy state possible. A slurry composed of a large number of colloidal particles will have a very large surface energy. The natural tendency will be for this slurry to reduce the net energy of the system by reducing the surface energy of the system. Particles coagulate and reform into larger particles, which will inherently have lower surface energies. Some smaller precipitates will actually be redissolved and reprecipitated onto larger particles (Smoluchowski's theory). The aging time is a very big factor in the size that the precipitates will ultimately achieve. The key is to allow a very long time to achieve this new equilibrium. Typically, aging can be done in less than 30 min; however, longer times will result in larger solid particles.

Why do we want larger precipitates? Keep in mind that the whole reason for this process is to reduce the metal ion concentration in waste streams. Solid particles must be extracted from the waste streams before safe discharge or reuse of the water. Larger particles are much easier to remove by decantation processes (such as thickeners) or filtration. Colloidal particles are simply too difficult to remove from water.

Now let us look at the copper solution in a sulfide precipitation scenario. The K_{SP} for CuS is as follows:

$$\text{CuS(s)} \rightleftharpoons \text{Cu}^{+2} + \text{S}_{SP}{}^{2-} = 1.26\times10^{-36}$$

Notice that the equilibrium constant is an extremely small number compared with the hydroxide K_{SP}. This shows that for the same copper ion concentration, an immeasurable amount of sulfide ions is needed to achieve saturation. In other words, it will take extremely low sulfide concentration to precipitate more of the copper than just a pH change. As we have discussed previously, the sulfide ion concentration is a function of the solution pH. That is, the sulfide ion concentration will be considerably lower at low pH (where metal ions are more stable) and become higher as the pH is raised. So like the hydroxide precipitation process, the pH can be an important variable.

Again, Monhemius has simplified the sulfide precipitation process with the diagram reprinted in Figure 22.4. Notice that the hydrogen sulfide equilibrium changes with the pH. Also notice that metals such as copper and

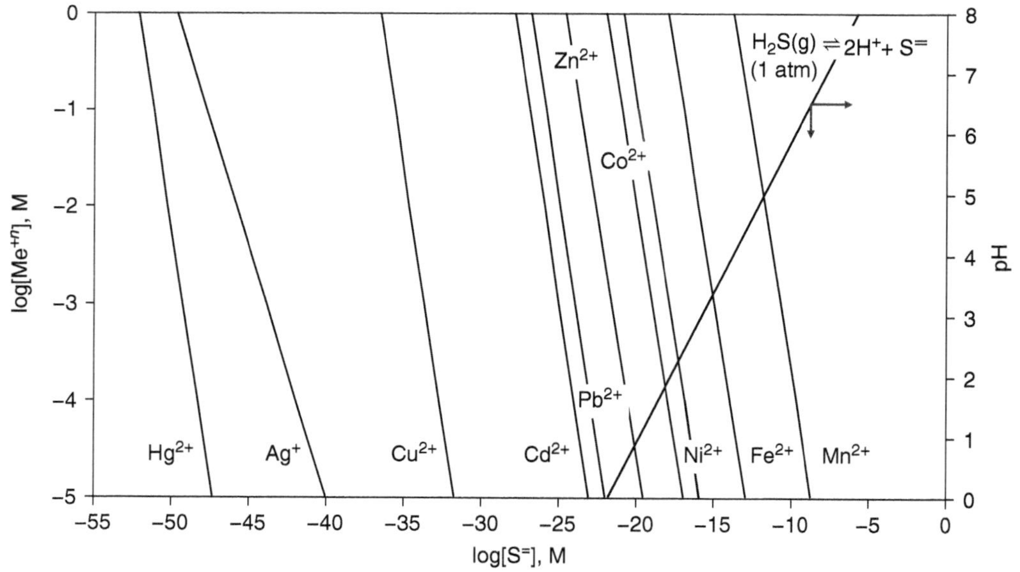

Figure 22.4 Monhemius diagram on hydrogen precipitation process. *Source*: Adapted from Monhemius (1977).

mercury form very stable sulfide precipitates at even negative pH values. Simply, this means that these sulfide species are very stable and easy to form at any pH. The iron and nickel sulfides are less stable at low pH but are very stable at high pH.

Comparing the sulfide and hydroxide precipitation processes always tends to favor the sulfide precipitation process. First, the hydroxide ions are in large concentrations immediately when a base is added. However, as we have seen, the sulfide ions that react with the metals must first pass through a series of equilibria to react. This simple pathway difference is one of the reasons sulfide precipitates are typically larger and more easily separated from the water. Also consider that hydroxide particles are simply isomers of the metal oxides that have waters of hydration incorporated in their matrix. For instance, the copper hydroxide precipitate ($Cu(OH)_2$) of our previous example is actually a hydrated copper oxide; that is, it is actually $CuO{\cdot}H_2O$. Metal sulfides are crystalline at the point of formation, so the growth is more structured and thermodynamically favored. Aging of sulfide precipitates will result in much larger particles than for hydroxide precipitates.

22.2.2.2 Spent Electrolytes for Electrorefining or Electrowinning

In both the ER and EW processes, the primary product is a pure cathode sheet. Although similar processes, ER and EW differ only in the source of the metal ions that are reduced at the cathode. In ER, impure, solid metal anodes are suspended in the electrolytic cell and oxidized to metal ions by the current. The metal ions are, in turn, reduced at the cathode. The net result is that metal ions are formed at the anode and removed at the cathode. Nickel, copper, and other metals can be cast in the impure form as anodes, which serve as the metal ion source during ER.

In EW, concentrated electrolytes from solvent extraction, ion exchange eluents, or carbon stripping are sent through an EW cell to plate the metal on cathodes. The oxidation reaction at the anode varies but typically involves the oxidation of water to form oxygen or the oxidation of ferrous ions to ferric ions. Copper and gold are the most common metals recovered by the direct plating of metal from leach solutions or concentrated electrolytes.

Despite the difference in source metals, both produce a concentrated electrolytic solution that must be recycled or treated and disposed. The refining of anodes by the electrolytic process is specifically designed to recover only one metal (copper, nickel, etc.). Consequently, all metals with a lower oxidation potential than the metal of interest will also be oxidized at the anode, but will not be reduced at the cathode. In copper refining, nickel and iron contamination in the anodes will dissolve into the electrolyte. However, because their oxidation potentials are below the copper oxidation potential, they will not plate on the cathode. Instead, they remain in the electrolyte solution and slowly build up in concentration. Arsenic is another common compound found in anodes that also dissolves and remains in the electrolyte. Periodically, these solutions have to be treated to reduce the level of contamination.

This is typically accomplished by first directly EW all of the valuable metal from the electrolytes. Keep in mind that the primary metal of interest for these processes will be the metal that has the highest oxidation (and reduction) potential. So they must be eliminated before any of the other metals can be removed. Then the solutions are distilled to remove the excess water and form the precipitate of the other metals as sulfate salts or chloride salts (depending on the acid used).

As an example, a copper ER plant will typically use electrolyte that contains $50–55\,g\,l^{-1}$ copper with a $160\,g\,l^{-1}$ sulfuric acid concentration. The copper concentration will vary only slightly as copper plates on the cathodes as an equal amount of copper is oxidized from the anodes. As the process continues, other metal values will likely increase. Arsenic levels can approach $10\,g\,l^{-1}$ (as arsenate ions), and nickel concentrates can build to $20–40\,g\,l^{-1}$ just by recycling the solutions. A portion of the solution is diverted to EW cells where the copper concentration is reduced by plating it on cathodes. Oxygen is evolved at the anode and the copper concentration is reduced as much as possible. Then the solutions are distilled to 60% sulfuric acid. At this concentration, nickel sulfate will be precipitated and can be recovered by filtration and sold. The arsenic levels are not reduced much by this process but can be reduced by adding ferric salts or other precipitating agents. The resulting solution (known as "*black acid*") can now be recycled or sold to other industries interested in soluble metals in the solution. The solids are typically sold to other metal-processing companies. Occasionally, the solids are sent to landfills or tailings impoundments for sequestration.

Similar problems occur with the constant reuse of solutions used in EW. Typically, leach solutions do not contain just the one metal of interest, but rather have any number of contaminating metals that were present in the ore.

Concentrating methods, such as solvent extraction for copper and carbon absorption for gold, will selectively strip just the metal of interest. Loaded solvents and carbon columns are stripped of their metal value in a concentrated solution that can now be electrowon to recover the metal value. The spent electrolytic solutions stripped of the copper or gold can be recycled back to recover more metal value from the solvents or carbon. Some contaminant metals may also pass through this process and continually build up in the electrolyte. For these operations, the solutions can be diluted into the original leach solutions as makeup lixiviant, or a process similar to the one just described can be used to strip contaminant metals from the electrolytes.

22.2.2.3 Pickling Liquors

Rolling mills produce two of the more important forms in which steel is sold – rods and sheets. Rods are the source materials for steel wire, construction materials (i.e. rebar and beams), steel coils, and so on. Automobile manufacturers require large shipments of steel sheets wound in large rolls. The rolling mills process hot metal to make these products, which can lead to oxidation of the metal. Ingots of the metal are sent through successively narrower rollers until the desired thickness (or gauge) of the metal is achieved. Successive rollers will produce rods of various shapes and diameters or sheets that vary from plate thicknesses to thin foils.

But one of the problems with continuously flattening or shaping metal is that it has to be hot. This means that metals such as steel and aluminum can oxidize rapidly. For aluminum this is not a problem because the oxidation layer is thin and actually protects the aluminum from further oxidation. But in steel, oxidation dulls the finish and can affect the final properties of the steel. For instance, even the slightest surface oxidation can dramatically affect the drawing processes for making steel wire. *Pickling liquors* are concentrated acid solutions (hydrochloric or sulfuric acid, primarily) that are used to remove the oxidation products, such as iron oxides or scale, that can form on hot steel products. The pickling process is an important part of the final processing of steel products for market. However, the process is tricky to operate properly. Too aggressive of an acid and the metal will be dissolved or corroded; not aggressive enough and the scale will not be removed. Several factors can influence this process:

- Acid concentration.
- Temperature of the pickling bath.
- Time that the steel remains in the bath.
- Percent of ferrous ions in the solution.
- Presence of inhibitors that slow or prevent attack on the steel surfaces.
- Agitation of the bath.

These solutions can contain as much as 60% (by volume) hydrochloric or sulfuric acid, depending on the shape of the steel and type of steel (high carbon or low carbon) to be cleaned. The solutions are used until either the acid content is too low (Baumé gravity scale), or the iron content is too high (titration technique). By the time it is ready to be discarded, pickling solutions can have very high iron concentrations (60 g l^{-1} or 0.5 lb gal^{-1} total dissolve iron). These solutions pose one of the most difficult waste disposal problems for the steel industry. There are a few other industries that can use the solutions (mainly for the acid content), but the volume of the solutions can literally flood the market.

Some of the end uses for this material are dictated by the species of iron in the solution. In the oxidized state, the ferric can be used as a flocculant in sewage treatment plants, phosphate precipitant, and pigment for paint, among other uses. Once the solutions are neutralized to a pH of about 5, the solutions can be oxidized with air (slowly) or hydrogen peroxide to create the ferric ions that are needed for these applications. In the reduced state (ferrous), the solutions can be neutralized and used as an iron supplement in livestock feed.

22.3 Conclusions

The wastes of the metals industry vary with the sector of the industry. Mining creates more wastes (primarily solids) than any of the other sectors. Tailings and waste rock are the two largest volumes of waste that metal production generates. While no attempt has been made to reduce this necessary waste, strict planning, federal and state regulations, and impoundment criteria have been adopted to lessen the impact of these wastes.

Chemical metallurgy (namely, pyrometallurgical processing and hydrometallurgical processing) also generates large amounts of solid, liquid, and gaseous wastes, but the volumes of these wastes are generally less than the mining wastes. Also, these wastes typically have a better chance at being reused (such as iron and steel slags), or better treatment processes are available to more effectively treat them. A general reduction in the production of these wastes has been ongoing, primarily due to the increased amount of recycling in recent years. The recycling trend has been market driven, to some extent, because of the increased worldwide demand for metals. The demand has exceeded mining production capacity, which is slow to develop new ore bodies with a minimal environmental impact. Energy conservation and environmental conscience are also factors in the voluntary trend of the metals industry toward more recycling.

Reference

Monhemius, A.J. (1977). Precipitation diagrams for metal hydroxides, sulphides, arsenates and phosphates. *Transactions of the Institute of Mining and Metals* 86: C202–C206.

Further Reading

Adamson, A.W. (1990). *Physical Chemistry of Surfaces*, 5e. New York: Wiley.

Ballor, N.R., Nesbitt, C.C., and Lueking, D.R. (April 20, 2006). Recovery of scrap iron metal value using biogenerated ferric iron. *Biotechnology and Bioengineering* 93 (6): 1089–1094.

Crowell, R., Nesbitt, C.C., Hendrix, J.L., and Nelson, J.H. (1986). Reactions and possible reactor design for the precipitation of metals sulfides using thioacetamide. In: *Hydrometallurgical Reactor Design and Kinetics* (ed. R.G. Bautista, R.M. Weisley and G.W. Warren), 421–439. TMS-AIME.

Freeman, H.M. (1988). *Standard Handbook of Hazardous Waste Treatment and Disposal*. New York: McGraw-Hill.

Gilchrist, J.D. (1989). *Extraction Metallurgy*, 3e. Oxford: Pergamon Press.

International Network for Acid Prevention. http://www.inap.com.au/ (accessed 15 January 2018).

Lueking D., and Nesbitt, C. C. Ferric iron biogeneration: an extreme approach, *SIM News: The Official News Magazine of the Society for Industrial Microbiology*, 54(1), 4–9 (January/February 2004).

Marsden, J. and House, I. (1992). *The Chemistry of Gold Extraction*. New York: Ellis Horwood.

McGannon, H.E. ed. (1971). *The Making, Shaping and Treating of Steel*, 9e. Pittsburgh: United States Steel Corporation.

Nesbitt, C. C., Hendrix, J. L., and Nelson, J. H. Use of Thiourea for Precipitation of Heavy Metals in Metallurgical Operation Effluents. *Extraction Metallurgy '85*, The Institute of Mining and Metallurgy, London, England, pp. 355–375, 1985.

Newton, J. (1952). *An Introduction to Metallurgy*, 2e. New York: Wiley.

Shaw, D.J. (1980). *An Introduction to Colloid and Surface Chemistry*. Oxford: Butterworth-Heinemann.

Snoeyink, V.L. and Jenkins, D. (1980). *Water Chemistry*. New York: Wiley.

Stumm, W. and Morgan, J.J. (1981). *Aquatic Chemistry: An Introduction Emphasizing Chemical Equilibria in Natural Waters*, 2e. New York: Wiley.

Uhrie, J.L., Drever, J.I., Colberg, P.J.S., and Nesbitt, C.C. (1996). In situ immobilization of heavy metals associated with uranium leach mines by bacterial sulfate reduction. *Hydrometallurgy* 43: 231–239.

United States Environmental Protection Agency (1991). Identification and Description of Mineral Processing Sectors and Waste Streams. http://www.epa.gov/epaoswer/other/mining/minedock/id.htm (accessed 15 January 2018).

United States Geological Survey (2005). Mineral Commodity Summaries. US Department of the Interior, Gale A. Norton, Secretary, US Government Printing Office, Washington, DC. http://minerals.usgs.gov/minerals/pubs/mcs/ (accessed 15 January 2018).

United States Geological Survey (2006). Mineral Commodity Summaries. US Department of the Interior, Gale A. Norton, Secretary, US Government Printing Office. Washington, DC. http://minerals.usgs.gov/minerals/pubs/mcs/ (accessed 15 January 2018).

Weiss, N.L. ed. (1985). *SME Mineral Processing Handbook*, vol. 1 and 2. New York: Society of Mining Engineers of the American Institute of Mining, Metallurgical, and Petroleum Engineers, Inc.

Wills, B.A. (1992). *Mineral Processing Technology: An Introduction to the Practical Aspects of Ore Treatment and Mineral Recovery*, 5e. Oxford: Pergamon Press.

23

Waste Reduction and Pollution Prevention for the Chemicals Industry

Methodologies, Economics, and Multiscale Modeling Approaches

Cheng Seong Khor[1,*], *Chandra Mouli R. Madhuranthakam*[2], *and Ali Elkamel*[3,4]

[1]Chemical Engineering Department, Universiti Teknologi PETRONAS, Seri Iskandar, Perak Darul Ridzuan, Malaysia
[2]Chemical Engineering Department, Abu Dhabi University, Abu Dhabi, UAE
[3]Department of Chemical Engineering, University of Waterloo, Waterloo, Ontario, Canada
[4]Department of Chemical Engineering, Khalifa University of Science and Technology, Abu Dhabi, UAE

The paradigm of green engineering that advocates environmentally conscious practices and operations has emerged as the perennial driving force of the industrial sector of the global economy. This chapter aims at providing an integrated treatment of objectives and methods for identifying and implementing waste reduction and pollution prevention (P2) initiatives in the chemical industry. We consolidate a set of tools, techniques, and best practices that would ensure the successful development of an economically sustainable P2 program with a multiscale survey and categorization of these available approaches at the levels of macro-, meso-, and microscale.

23.1 Introduction

This chapter provides a survey treatment on the subject of *waste reduction* programs and practices in chemical plant operations, extended to include the wider scope defined by the term *pollution prevention* (P2). (A discussion on the definition of the most commonly encountered terms in the literature of the subject follows.) In the context of current developments and state-of-the-art industrial practices and academic research, waste reduction and P2 can be broadly and comprehensively defined as practices that reduce or eliminate the creation of pollutants and wastes at the source. The basic paradigm of the strategies variably known as *waste reduction, waste minimization, P2*, or *cleaner production* is that avoiding the generation of wastes or pollutants can often be both more cost effective and better for the environment compared with the traditional end-of-pipe abatement solutions of controlling or disposing of pollutants only after they have been formed. This is typically accomplished via a four-pronged approach (Elkamel, 2006):

1) Increase efficiency and long-term sustainability in the utilization of raw materials, energy, water, or other resources (e.g. chemical synthesis routes with minimum or no by-products that consume sustainable or renewable sources) (Harold and Ogunnaike, 2000).
2) Substitute hazardous substances with less harmful alternatives.
3) Eliminate toxic substances from production/manufacturing processes.
4) Protect natural resources by means of conservation efforts.

23.1.1 Waste Management Hierarchy

The elements of the waste management hierarchy can be placed in the following order of preference (Allen and Rosselot, 1997; Bishop, 1999):

1) Source reduction.
2) In-process recycling (i.e. within a given production process).
3) On-site recycling (i.e. out of the process but still within the same production site).
4) Off-site recycling (i.e. within different production sites).
5) Waste treatment to reduce hazards imposed by the waste.
6) Safe environmental disposal.
7) Direct disposal into the environment.

The waste management hierarchy introduces a myriad of terms that require definitions due to subtle differences between them. To precisely define these terms is an arduous task that becomes even more complex when regulatory definitions are added to an originally operational or technical definition. Table 23.1 provides generally accepted definitions for some of the most commonly encountered terms in the literature of this field.

* Joint corresponding author with Ali Elkamel

Handbook of Environmental Engineering, First Edition. Edited by Myer Kutz.

Table 23.1 General definition for commonly encountered terms in the pollution prevention literature.

Term	Generally accepted definition(s) (and related remarks)
Source reduction Waste reduction (WAR)/waste minimization	Process or product modifications that reduce waste (Allen and Rosselot, 1997) Any activity that reduces or eliminates the generation of wastes and hazardous pollutants at the source, e.g.: i) Changes in industrial processes in the form of operating practices, technologies, inputs, and products ii) Recycling (in-process, in-line, closed-loop) that reuse and recycle wastes for the original purpose or others, such as for materials recovery or for energy production(US Environmental Protection Agency (US EPA), 1988) *Remarks*: • The term *waste reduction* and the related term *waste minimization* project broader meanings than *source reduction*. Waste reduction and waste minimization generally incorporate both source reduction and on-site recycling (i.e. the first two or three elements of the waste management hierarchy) (Allen and Rosselot, 1997). For the purpose of this article, the definition offered here will/would be adopted • While the terms *waste reduction* are more expansive than *source reduction* in certain ways, they are more restrictive in the sense that while source reduction can be employed in the context of gaseous, liquid, or solid wastes, the term *waste reduction* is generally used to refer to solid and liquid waste management activities (Allen and Rosselot, 1997)
Pollution prevention (P2)	"A multimedia concept that reduces or eliminates pollutant discharges to air, water, or land and includes the development of more environmentally acceptable products, changes in processes and practices, source reduction, beneficial use, and environmentally sound recycling" (Allen and Rosselot, 1997) *Remark*: It is generally acknowledged that P2 encompasses the first two or three elements of the waste management hierarchy for all wastes and emissions (Allen and Rosselot, 1997)
Cleaner production (CP)	• "Continuous application of an integrated preventive environmental strategy to processes and products to reduce risk to humans and the environment" (United Nations Environment Program (UNEP), 1996) • "Maximum feasible reduction of all wastes generated at production sites" (US Environmental Protection Agency (US EPA), 1992)

23.1.2 Definition of Waste Reduction and Pollution Prevention Used in This Chapter

We are inclined to adopt a broader definition of waste reduction that includes gaseous wastes as well, which will cover the definition of P2, hence, encompassing the first four elements of the waste management hierarchy (i.e. source reduction and in-process, on-site, and off-site recycling), as depicted in Figure 23.1. Thus, the terms *waste reduction* and *pollution prevention* will be used interchangeably throughout this chapter to essentially denote the same concepts. This, we hope, will enable us to capture a holistic view and coverage of the field by enabling the examination of more alternatives for reducing industrial impacts on the environment and for eliminating adverse health effects that are due to chemical plant manufacturing activities. This will also provide a true depiction of the wide-ranging extent of this tremendously exciting and meaningful area for both academic research and industrial practices. On a lighter note, we will also extensively use the shorthand term for pollution prevention, P2, in this chapter in keeping with the P2 spirit of conserving ink and paper (Freeman et al., 1992).

Additionally, the term *process* implies *product*, for instance, as in *product/process engineering*. For the sake of habitual convenience, the term *process* will be more frequently encountered throughout the chapter, but when a clear distinction between the two is warranted, an explanation will be provided.

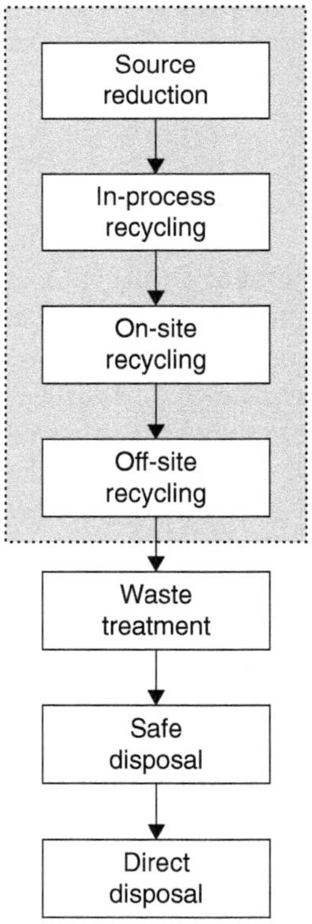

Figure 23.1 The waste management hierarchy with shaded boxes indicates elements that are included in the definition of the terms *waste reduction* and *pollution prevention* that are used interchangeably throughout this chapter. *Source*: Adapted from Allen and Rosselot (1997) and Bishop (1999).

The extensive breadth that spans the coverage of P2 has entailed a sheer amount of development and coverage of new knowledge and information in both industrial practices and academic research (Freeman, 1995; Allen and Rosselot, 1997; Allen and Shonnard, 2002). To the best of our knowledge, there has been only one comprehensive review article on the subject, that is, the 1995 survey by Freeman and coworkers of the US Environmental Protection Agency (US EPA)'s Pollution Prevention Research Branch in the Risk Reduction Engineering Laboratory (Freeman, 1995). The review focuses on P2 in the industrial sector, which is by far the sector in which major initiatives of P2 have been undertaken and will continue to be so.

23.2 Development of Pollution Prevention Programs

23.2.1 Identifying Pollution Prevention Opportunities

The life cycle of chemical manufacturing is the key toward understanding the driving forces that contribute to initiatives for P2 and the associated risk reduction. Allen and Shonnard provide a unified representation of the typical life cycle of chemical products and processes as shown in Figure 23.2 (Allen and Shonnard, 2001, 2002). In general, chemical processes evolve through life cycle phases that begin with research and development (R&D) activities, which include conceptual design (also known as product/process synthesis), followed by product/process engineering, plant operations, and, eventually, decommissioning (Pereira, 1999). In relation to this, Mulholland and Dyer of DuPont suggest a continuum as shown in Figure 23.3 that depicts the relative scale of opportunities for implementing P2 along the life cycle of a product/process, corresponding to the magnitude of benefits expected out of such initiatives (Mulholland and Dyer, 1999). It is clearly shown that the number of technology options available for reducing environmental impact is greatest early on in the life cycle, and then the options drastically decrease. In contrast, costs associated with resolving an environmental problem typically increase exponentially as the process matures and the scale of equipment becomes larger. Thus, there is considerable incentive, apart from being

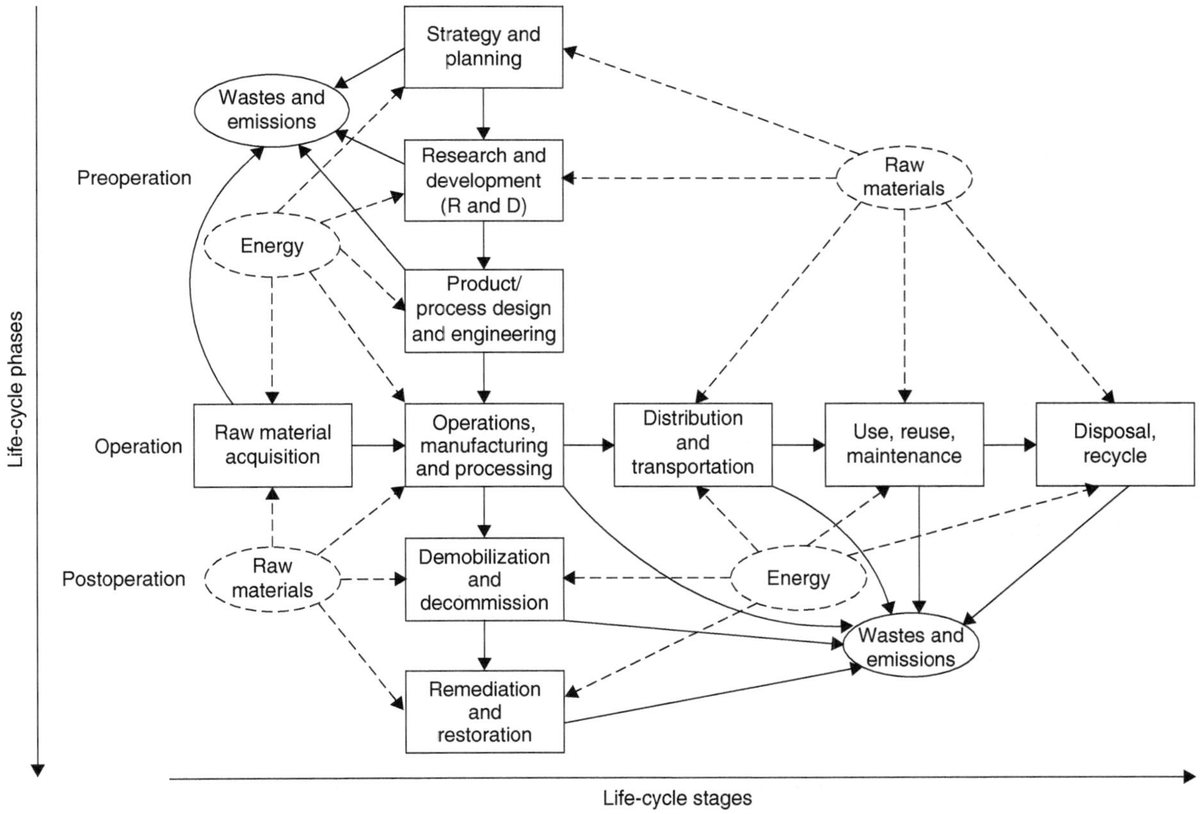

Figure 23.2 Product life cycles, including raw material extraction, material processing, distribution and transportation, use, and disposal, are depicted along the horizontal axis. Process life cycles (or process development cycle), including strategy and planning, R&D, process engineering, process operations, and decommissioning steps, are illustrated along the vertical axis. In both product and process life cycles, raw materials and energy are consumed at each stage of the life cycle, while wastes and emissions are generated. *Source:* Adapted from Allen and Shonnard (2001, 2002).

more effective, to consider a P2 implementation earlier (rather than later) in the life cycle of a product/process (Mulholland and Dyer, 1999; Pereira, 1999).

Indeed, opportunities for cost-effective solutions to P2 problems are most prevalent at the R&D and process engineering stages for new units and plants. After the basic process chemistry and basic system design philosophy are in place, significantly fewer opportunities exist, comparatively, to undertake P2-related cost-effective decisions in the later stages of the project (i.e. during detailed engineering, construction and start-up, operations and maintenance, or the dismantling phases). It is noted as well that this decision on the extent to which a P2 initiative should be implemented will also largely depend on the interplay of factors that primarily consist of corporate and business environmental goals, economics, regulatory requirements, and process maturity or product life (Moore, 1995).

Figure 23.4 provides an overview of the elements of a complete typical chemical process for consideration in targeting P2 opportunities. Although the focus for identifying P2 opportunities is necessarily on the process, other factors such as off-specification (off-spec) product, maintenance wastes, leaks, and spills can also be significant causes of pollution. Table 23.2 identifies numerous examples of potential sources of pollution and waste in chemical process plants. Once these have been identified, P2 solutions can be subsequently evaluated based on the general guidelines provided in Table 23.3 (Moore, 1995).

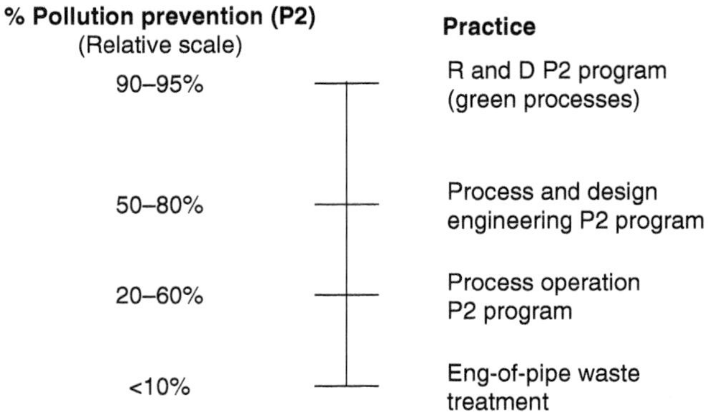

Figure 23.3 The P2 methodology continuum. *Source:* Adapted from Mulholland and Dyer (1999).

We have highlighted potential opportunities for P2 on a general basis. Next, we categorize the discussion according to P2 opportunities that potentially arise in the three most prominent phases in product/process life cycle: R&D, product/process design and engineering, and manufacturing or process operations.

Table 23.2 Potential sources of pollution and wastes.

Potential sources of air emissions
• Point-source emission: stack, vent (e.g. laboratory hood, reactor, storage tank vent), material loading and unloading operations, and others • Fugitive emissions: pumps, valves, flanges, mechanical seals, relief devices, tanks, and others • Secondary emissions: wastewater treatment unit, cooling tower, process sewer, sump, spill and leak areas, and others
Potential sources of liquid (organic or aqueous) wastes
Equipment wash solvent and water, laboratory samples, surplus chemicals, product washes and purifications, seal flushes, scrubber blowdown, cooling water, steam jets and vacuum pumps, leaks, spills, spend or used solvents, housekeeping (pad washdown), waste oils and lubricants from maintenance, and others
Potential sources of solid wastes
Spent catalysts, spent fillers, sludges, wastewater treatment biological sludge, contaminated soil, old equipment and installation, packaging material, reaction by-products, spent carbon and resins, drying aids, and others
Potential sources of groundwater contamination due to leaks or spills
Unlined ditches; process trenches; sumps; pumps, valves, and fittings; wastewater treatment ponds; product storage areas; tanks and tank farms; aboveground and underground piping; loading and unloading areas and racks and manufacturing maintenance facilities

Source: From Moore (1995).

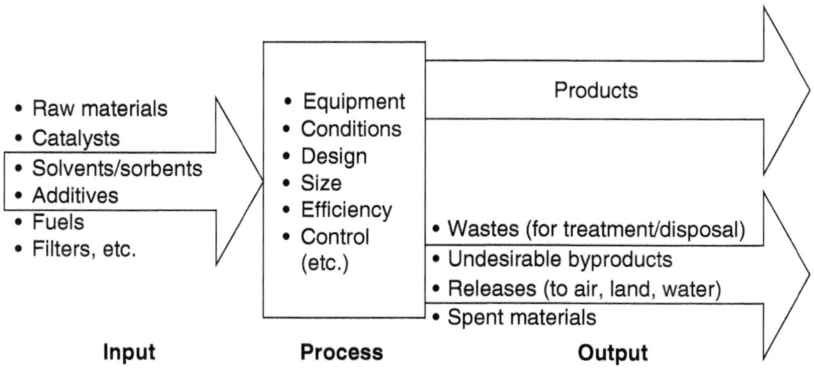

Figure 23.4 Process elements for the consideration of P2 initiatives. *Source:* Adapted from Moore (1995).

Table 23.3 Guidelines for identifying P2 solutions/opportunities.

Inventory management and operations
• Improvement in material purchasing, receiving, storage, and handling practices, which minimize materials that may exceed their shelf life, be left over or not needed, or have the potential for accidental release • Installation of the right equipment and maintaining strict preventive maintenance programs to prevent leaks, spills, or accidental releases • Effective and efficient production scheduling by reducing the number of production changes to reduce equipment cleaning and to minimize wastes generated from production transitions or turnovers/changeovers
Equipment
• Installation of equipment that produces less waste • Redesign of equipment or production lines to produce less waste and to enhance or permit recovery or recycling options • Improvement in operating efficiency of equipment • Elimination of sources of leaks and spills
Process/product
• Process changes that optimize reactions and raw material use to reduce waste generation or releases • Input material changes: material purification and material substitution that allow a change in the process to reduce waste or eliminate a hazardous constituent used in the product or during the manufacturing phase • Product substitution: developing a new product that is less hazardous in use or in ultimate disposal compared to the original • Additional automation to increase the reliability of the operations and to reduce the occurrence of process upsets and production of off-specification material
Recycling
• Participation in waste exchanges • Installation of closed-loop systems for in-process recycle • Recycling on-site at other process units or off-site for reuse • Finding new uses for previously unwanted by-products • Segregation of wastes by type to allow for recovery • Reclamation: processing of waste for resource recovery

Source: From US Environmental Protection Agency (US EPA) (1988), United Nations Environment Program (UNEP) (1996), and Moore (1995).

23.2.1.1 Pollution Prevention During the Research and Development (R&D) Stage

As emphasized earlier, the greatest opportunity for cost-effective P2 at the source is present at the R&D stage. R&D programs typically progress through three distinct phases:

1) Conceptual design to determine economic feasibility.
2) Bench-scale laboratory development studies and testing to validate the process chemistry.
3) Pilot-plant-scale development and testing to address engineering issues for process scale-up.

Conceptual Design. During the conceptual design phase, the reaction pathway synthesis is undertaken; the inherent process safety is incorporated; the general environmental impacts are investigated; and the potential waste streams are identified and studied (Mulholland and Dyer, 1999). This culminates with the formulation of relevant P2 concepts that influence the determination of the following key process parameters: (i) the operating methodology (e.g. batch vs. continuous or an operating mode between these two extremes); (ii) the process conditions (e.g. temperatures and pressures); (iii) the production procedures; (iv) the selection, integration, and design of processing equipment; and (v) the process control schemes. Naturally, the potential for P2 in subsequent stages (particularly the preliminary product/process engineering phase) is contingent upon these decisions.

Table 23.4, "Twelve principles of green chemistry," (Anastas and Warner, 1998) serves to provide a road map for chemists and chemical engineers to implement the use of sound, responsible chemistry for P2, or *green chemistry* during the conceptual design of chemical products/processes in order to reduce or eliminate the use and generation of hazardous substances (US Environmental Protection Agency (US EPA), 2006).

Additionally, in borrowing ideas from the industrial ecology concept, Allen and Shonnard advocate the relatively new approach of identifying the uses for chemical by-products in industries that extend further beyond

Table 23.4 Twelve principles of green chemistry.

1) *Prevent waste*: Design chemical syntheses to prevent waste, leaving no waste to treat or clean up
2) *Synthesize safer chemicals and products*: Design chemical products to be fully effective functionally, yet have little or no toxicity
3) *Design less hazardous chemical syntheses*: Design syntheses that use and generate substances that pose little or no toxicity to human health and the environment
4) *Use renewable feedstocks*: Use raw materials and feedstocks that are renewable rather than depleting whenever technically and economically feasible. Renewable feedstocks are often synthesized from agricultural products or are the wastes of other processes; depleting feedstocks are derived from fossil fuels (petroleum, natural gas, or coal) or are mined
5) *Use catalysts, not stoichiometric reagents*: Minimize waste by using catalysts that are highly selective for the desired reactions. Only small amounts of catalysts are required for carrying out a single reaction multiple times, and furthermore, they can be regenerated (up to a certain number of times). This is in contrast to the typically excess amount of stoichiometric reagents that are required and that only work once (i.e. consumed in every reaction and cannot be regenerated)
6) *Avoid chemical derivatives*: Avoid using unnecessary derivatization in the form of blocking or protecting–deprotecting groups or any temporary modification of physical or chemical processes. Derivatives consume additional reagents, thus, generating waste
7) *Maximize atom economy*: Design syntheses so that the final product contains the maximum proportion of the starting materials. There should be few, if any, wasted atoms
8) *Use safer solvents and reaction conditions*: Avoid using solvents, separation agents, or other auxiliary chemicals whenever possible. Use innocuous chemicals if their usage is necessary
9) *Increase energy efficiency*: Recognize the environmental and economic impacts of energy requirements and minimize them by conducting chemical reactions at ambient temperature and pressure whenever possible
10) *Design chemicals and products to degrade after use*: Design chemical products to break down to innocuous degraded substances at the end of their function or life cycle so that they do not persist and accumulate in the environment
11) *Analyze in real time to prevent pollution*: Develop the tools available to perform real-time analysis of in-process monitoring and control during syntheses to minimize or eliminate the formation of by-products, particularly hazardous substances
12) *Minimize the potential for accidents*: Design chemicals and their forms (solid, liquid, or gas) as used in a chemical process to minimize the potential for chemical accidents including explosions, fires, and releases to the environment

Source: Adapted from US Environmental Protection Agency (US EPA) (2006).

chemical manufacturing (Allen and Shonnard, 2001). Acknowledging that the principle of identifying productive uses for by-products has long been a mainstay in the chemical industry, and they promote the idea that chemical engineers should take on design tasks such as managing the heat integration between a power plant and a petroleum refinery or integrating the water usage between a semiconductor fabrication plant and a commodity chemical manufacturing facility. Indeed, such design tasks are still at the conceptual design stage of our profession. Hence, these perspectives must be broadened if we desire to make these design undertakings as common in the near future as heat integration for processes. To embark on this form of P2 initiative, we ought to begin integrating process design tools and methodologies from our traditional domain of chemical manufacturing with industries as diverse as polymer processing, semiconductor manufacturing, and pulp and paper processing.

Bench-Scale Laboratory Development Studies and Testing. During laboratory studies, the reaction chemistry is verified and validated; the waste streams are characterized; the process variables are tested; the P2 options are identified; the data are collected for the subsequent pilot-plant testing and process engineering; and the potential impact of environmental regulations are investigated (Mulholland and Dyer, 1999). Table 23.5

Table 23.5 P2 data to be collected for bench-scale laboratory development studies and testing.

Unit operation	Data required
General	• Determine corrosion rates of potential construction materials • Screen for catalytic effects of potential materials, corrosion products, and feed impurities
Reactors	• Determine reaction stoichiometry • Determine equilibrium yield • Measure catalyst selectivity, activity, and lifetime • Identify and characterize reaction by-products • Determine kinetics of major side reactions • Determine effects of recycle
Distillation and other separation processes	• Obtain vapor pressure data for products and intermediates • Obtain vapor–liquid equilibrium (VLE) data for potential entrainers, diluents, and trace compounds • Determine loading capacity and regenerative properties of absorbents

Source: Adapted from Butner (1995).

proposes relevant P2-related data to be gathered according to unit operations (Butner, 1995).

Pilot-Plant-Scale Development and Testing. During pilot-plant studies, laboratory results are verified and validated; process chemistry is established; key process variables are tested; equipment design is evaluated; and waste characteristics are defined (Mulholland and Dyer, 1999). Butner (1995) highlights the following issues to be thoroughly explored in this phase to assist the direction of P2 initiatives during the ensuing stage of process engineering:

- *Effect(s) of reactor mixing and feed distribution on the formation of by-products.* Pilot testing is often the first opportunity for conducting the process reactions in equipment that approximates subsequent scale-up to commercial reactor implementation. Therefore, it is highly imperative to meticulously inspect and analyze for unwanted by-products in the slate of products obtained, which could form as a result of residence time distributions, due to poor mixing or feed distribution, hot spots on catalysts, impurities (that are not present in laboratory grade chemicals), or catalytic effects of construction materials.
- *Fouling rates in heat exchangers.* Pilot-plant studies should be utilized to thoroughly investigate fouling due to process wastes that originate from (i) cleaning waste generated by routine heat exchanger maintenance and (ii) products of thermal decomposition formed by increased wall temperatures as a result of reductions in heat exchange coefficients.
- *Sedimentation rates and product stability.* Pilot-scale tests provide a good chance to identify the formation of sedimentation or sludge in process vessels, which indicate product stability during storage in the vessels.
- *Corrosion studies.* It is equally essential to monitor corrosion rates at this stage; although ideally, these analyses should have been initiated during the previous phase of laboratory studies.

23.2.1.2 Pollution Prevention During the Product/Process Engineering Stage

Although there are limited opportunities for major product/process changes at this stage, there still exist many opportunities for incorporating P2 into a process. In fact, as the emphasis shifts from conceptual to operational aspects of a project, it may be easier to implement P2 initiatives since we can more assuredly evaluate uncertainties that arise in the suggestions for P2. Table 23.6 offers such suggestions that are particularly suitable for the preliminary engineering phase of a project (Butner, 1995).

As a project progresses into the detailed design (also known as mechanical design or production design) stage, source reduction opportunities practically diminish chiefly because the process and preliminary plant design have become fixed, and the project on the whole becomes schedule driven. The focus at this stage departs from the process per se to the design of the associated equipment and facility, thus requiring P2 emphasis on protecting groundwater from spills (see Table 23.2) and minimizing (or even eliminating) fugitive emissions (Mulholland and Dyer, 1999).

Table 23.6 Strategies for P2 during product/process engineering.

Activity	Strategy
Materials selection	• Consider costs of waste disposal when selecting maximum allowable corrosion rates • Consider foul-resistant materials (e.g. Teflon) on heat exchange surfaces and vessels requiring frequent cleaning • Consider glass- or polymer-lined vessels when frequent cleaning is required
Cost estimation	• Incorporate "hidden" waste costs (e.g. regulatory tracking and reporting costs) in cost equations • In absence of detailed waste costs, consider *penalty function* for releases based on environmental objectives. (A detailed treatment is provided in Section 3 on P2 economics)
Piping design	• Recover waste streams separately • Minimize the length of piping runs (because this reduces material inventory) • Minimize valves and flanges (e.g. by reducing valve counts and by using welded fittings) • Route drains, vents, and relief lines to recovery or treatment • Specify bellow-seal or zero-emission valves
Instrumentation design	• Select in-line process analyzers to reduce sampling wastes • Use closed-loop (purge style) sampling ports • Install preventive maintenance monitoring equipment (e.g. vibration monitors, torque sensors) to optimize maintenance schedules • Install instrument on heat exchangers that permit real-time monitoring of fouling and leakage • Consider advanced control strategies such as model-based control (MPC)

Source: Adapted from Butner (1995).

23.2.1.3 Pollution Prevention During the Process Operations Stage

Experience documented across all industries has shown that processes that minimize waste generation at the source are the most economically optimal, especially for existing operating plants. Perhaps, the greatest success stories of P2 during process operation stage have been the application of process integration tools for heat integration via the pinch analysis or pinch technology method (Linnhoff et al., 1982) and the emerging principle of mass integration (El-Halwagi and Manousiouthakis, 1989). These have been documented by Rossiter (1994) for heat integration and Dunn and El-Halwagi (2003) for mass integration. The concepts of mass and heat integration will be revisited in this chapter under the section on quantitative methods for P2 at the mesoscale. Table 23.7 provides further examples of possible operational changes for reducing waste generation (Freeman et al., 1992).

23.2.2 An Integrated Methodology for a Successful Pollution Prevention Program Within Corporate Management Structures, Strategies, and Practices

Incorporating environmental management, particularly P2, as a part of a company's existing setup, with the hope of institutionalizing it, is a demanding task. P2 initiatives typically cut across the traditional management functional boundaries of R&D and product development, manufacturing, engineering, business management, customer and supplier relationships, and risk management. Many companies encounter difficulty in maintaining the momentum of P2 efforts, managing only to undertake portions of a P2 project, often due to a lack of serious commitment to P2 programs. Unless P2 is integrated into a company's business practices, early successes with P2 projects do not ensure long-term sustainability of the program. To overcome such a situation, Case et al. and Bishop propose general procedures for establishing and maintaining a sustainable P2 program that is applicable to the reduction of all wastes (for all medium, quantity, or toxicity), as summarized in Table 23.8 (Allen and Rosselot, 1994; Case et al., 1995).

Table 23.7 Examples of operational changes to reduce waste generation.

- Reduce raw material and product loss due to leaks, spills, drag-out, fugitive emissions, tank breathing, and off-specification (off-spec) process solution
- Minimize number of shutdowns by designing for high availability or installing more equipment (or standby equipment)
- Inspect parts before they are processed to reduce number of rejects
- Allow for enough intermediate storage to provide flexibility for reprocessing off-spec materials
- Consolidate types of equipment or chemicals to reduce the quantity and variety of waste
- Improve cleaning procedures to reduce generation of dilute mixed waste with methods such as using dry cleanup techniques, using mechanical wall wipers or squeegees, and using "pigs" or compressed gas to clean pipes and increasing drain time
- Segregate wastes to increase recoverability
- Optimize operational parameters (such as temperature, pressure, reaction time, concentration, and chemicals) to reduce by-product or waste generation
- Evaluate the need for each operational step and eliminate steps that are unnecessary
- Collect spilled or leaked material for reuse
- Design a continuous process for flexible operation such as high turndown rate rather than shutdown
- Increase level of automation or better management of process to assist in reducing wastes due to poorly operated process
- Study the potential for converting the operating mode from batch to continuous since batch processes are always at unsteady-state condition by nature and are, therefore, difficult to maintain at optimum condition

Source: Adapted from Freeman et al. (1992) and Smith and Petela (1992).

Therefore, after identifying the potential P2 initiatives, an effective work process or methodology is crucial in ensuring the successful implementation of the plans. DuPont propose a methodology, displayed in its adapted form in Figure 23.5, that provides a solid foundation for a practical P2 implementation (Mulholland and Dyer, 1999). It incorporates many of the features initially recommended by the pollution prevention assessment (PPA) framework of the US EPA, which extensively draws upon DuPont's recognized leadership and long experiences in implementing P2 projects (US Environmental Protection Agency (US EPA), 1992). Focusing on the implementation phase, the tasks that are outlined hold the key for turning the selected P2 options by a core assessment team into projects that will actually accomplish the targeted waste reductions. It is noteworthy that many P2 methodologies unnecessarily emphasize the development of an overall corporate program, failing to focus on the process operation level where waste reductions actually take place. Mulholland and Dyer highlight the following vital attributes for consideration in implementing an effective P2 program, which are often overlooked by many P2 implementation methodologies (Mulholland and Dyer, 1999):

1) The core activities vying for people's attention in the process operations stage, namely, routine production, quality and safety, maintenance, problem troubleshooting, and turnaround activities comprising start-ups and shutdowns.
2) The limited capital, time, and people resources available to execute P2 initiatives.
3) Elimination of time-consuming and unnecessarily rigorous analytical methods in performing waste assessment activities that do not commensurate with their benefits.

Table 23.8 General procedure for establishing and maintaining a successful P2 program.

1) Obtain support from top management in the form of a corporate policy statement of support
2) Initialize the program by starting to incorporate small changes throughout the company, training staff in P2, and developing a written P2 plan that describes the P2 team makeup, authority, and responsibility
3) Describe how all the groups in the company (production, engineering, maintenance, laboratory, marketing, shipping, retailing, and other functions) will cooperate to reduce waste production and energy consumption
4) Devise a plan for publicizing and gaining companywide support for the P2 program
5) Develop a plan for communicating the successes and failures of P2 programs within the company
6) Review and describe in detail the processes within the facility to determine the raw materials used and the sources of waste generation, particularly the processes that produce, use, or release hazardous or toxic materials, by clearly defining the amounts and types of substances, materials, and products under consideration; this will provide a baseline inventory to be used to set goals and evaluate progress
7) Identify potential P2 opportunities for the facility by considering the treatment, disposal, and recycling facilities and transporters that are currently used
8) Determine the cost of current waste generation and develop a (preliminary) system of proportional waste management (in terms of the ensuing pollution control and waste disposal measures) charges for the departments that generate waste
9) Describe and evaluate currently ongoing and past P2 activities at the facility
10) Present criteria for prioritizing candidate facilities, processes, and streams for P2 projects
11) Select the best P2 options for the company and implement these choices
12) Evaluate the implemented P2 program on a companywide basis as well as evaluating specific P2 projects and activities
13) Maintain and sustain the P2 program for continued growth and benefits to the company; reevaluate the program as economic situations change and/or process equipment requires upgrading

Source: Adapted from Bishop (1999) and Case et al. (1995).

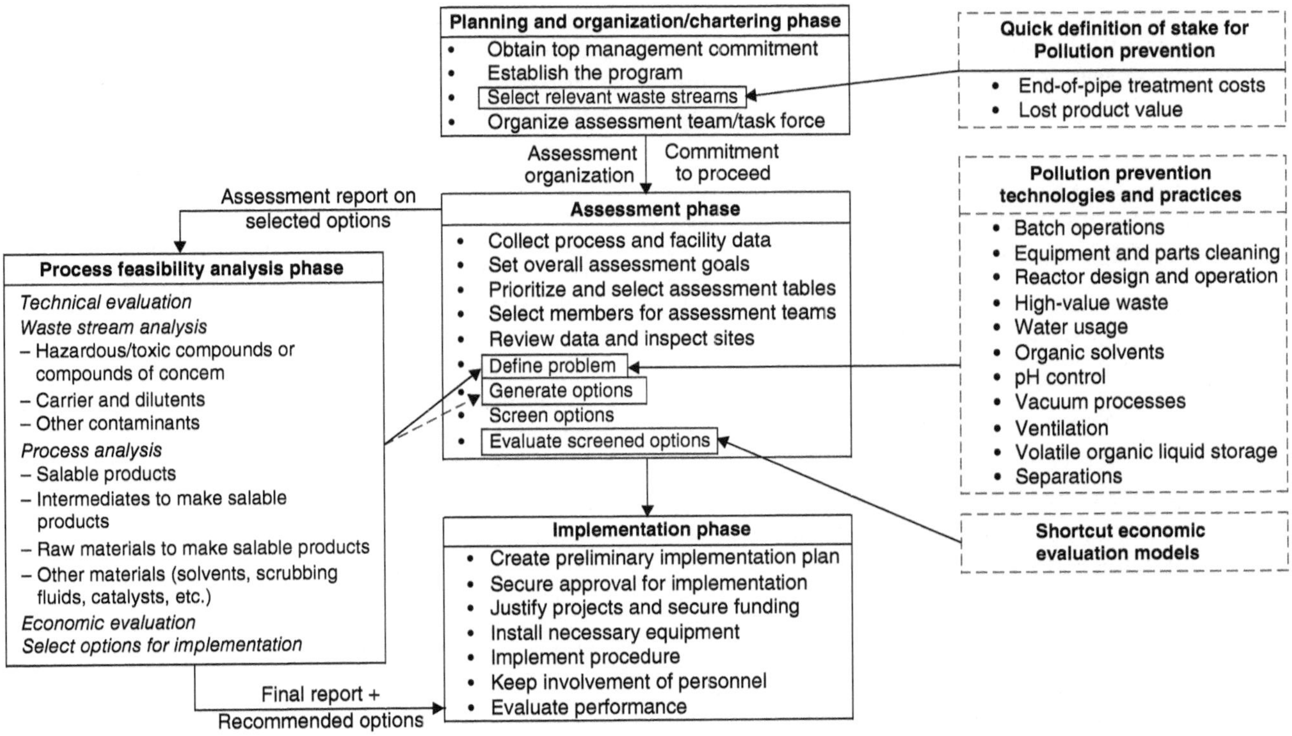

Figure 23.5 An integrated methodology leading to the successful implementation of a P2 program. *Source:* Adapted from Bishop (1999), US Environmental Protection Agency (US EPA) (1992), and Mulholland and Dyer (1999).

23.3 Economics of Pollution Prevention

23.3.1 Frameworks for Assessing and Quantifying Environmental Costs

The traditional approach to chemical process design has been to first engineer the process and only then to subsequently engineer the treatment and disposal of waste streams. However, the exponential escalation of waste management costs resulting from regulatory and societal pressures to eliminate emissions to the environment has led to capital investment and operating costs for waste management becoming a prominent fraction of the total cost of any chemical plant operation. Thus, to assess the economic viability for the implementation of a P2 program, significant efforts have been undertaken to evaluate environmental costs and benefits in assisting business decision making of determining the optimal minimum economic option, which accounts for both the process and the waste management components. Here, we present two such methods:

1) The total cost assessment (TCA) method, proposed by the American Institute of Chemical Engineers' Center for Waste Reduction Technologies (AIChE–CWRT), is perhaps the most relevant work for the design of chemical processes (Allen and Shonnard, 2001, 2002).
2) DuPont's "10-Step Method of Engineering Evaluations of Pollution Prevention" outlines the process for choosing the best approach for solving an engineering problem (Mulholland and Dyer, 1999).

Allen and Rosselot have also reported on the environmental cost estimation methodology proposed by the US EPA (Allen and Rosselot, 1997).

23.3.2 The AIChE–CWRT Total Cost Assessment (TCA) Method for Evaluating Environmental Costs

The AIChE–CWRT TCA method identifies five tiers for estimating environmental costs as listed in Table 23.9, beginning with the most tangible costs in Tier I and gradually extending to the least quantifiable costs in Tier V.

Tier I types of costs, with examples listed in Table 23.10, are effectively quantified in traditional economic analyses using conventional accounting systems. However, these systems, which focus on Tier I costs, often charge some types of environmental costs to overhead. Thus, this belies a complete estimation of project costs because it results in the possibility of these Tier II costs (see Table 23.11 for examples) being *hidden* and often difficult to distinguish from general overhead expenditures. Note also that these hidden costs are actually borne by facilities, irrespective of whether they are quantified by the facilities or are assigned to project or product lines.

The less tangible set of liability costs is designated as Tier III and includes compliance obligations, remediation obligations, obligations to compensate private parties due

Table 23.9 The AIChE–CWRT total cost assessment (TCA) method.

Tier	Type of cost
I	Costs normally captured by engineering economic evaluations
II	Administrative and regulatory environmental costs not normally assigned to individual projects
III	Liability costs
IV	Costs and benefits, *internal* to a company, associated with improved environmental performance
V	Costs and benefits, *external* to a company, associated with improved environmental performance

Source: Adapted from Allen and Shonnard (2001).

Table 23.10 Costs traditionally evaluated during financial analyses of projects (Tier I costs).

• Capital cost of equipment	• Materials
• Labor	• Supplies
• Utilities	• Structures
• Salvage value	

Source: Adapted from Allen and Shonnard (2002).

Table 23.11 Environmental costs that are often charged to overhead.

- Treatment costs
- Off-site waste management charges
- Waste treatment equipment
- Waste treatment operating expenses
- Handling costs
- Taking samples
- Sampling stormwater
- Regulatory tracking/monitoring and paperwork reporting and notification (for compliance in order to obtain permit requirements)
- Filing for permits
- Completing sample reporting forms
- Conducting waste and emission inventories
- Completing hazardous waste manifests
- Inspecting hazardous waste storage areas and keeping logs
- Preparing and updating emergency response plans
- Preparing chemical usage reports
- Reporting on P2 plans and activities
- Disposal costs
- Liability costs

Source: Adapted from Allen and Rosselot (1997), Allen and Shonnard (2002), and Butner (1995).

to personal injury, property damage, and economic loss, fines and penalties, punitive damages, and natural resource damages. The final component of Tier IV and V costs, often referred to as image or relationship costs, arises in interactions with customers, investors, insurers, suppliers, lenders, employees, regulators, and communities; they are probably the most difficult to quantify. In many cases, the Tier III–V costs are incurred by unplanned events (e.g. incidents that result in civil fines and remediation costs). Therefore, we evaluate the expected value of these costs by estimating the probability and possible timing that an event will occur with its associated cost. In the case of events that will take place in future years (e.g. costs of complying with anticipated future regulations), knowledge of the timing of event is crucial for a more precise estimation of the expected costs' present value (Allen and Shonnard, 2001, 2002).

23.3.3 DuPont's "10-Step Method of Engineering Evaluations for Pollution Prevention"

The DuPont's 10-step method (see Figure 23.6) is a generic framework that outlines the necessary steps for identifying, evaluating, and choosing the best alternative(s) for a specific engineering problem. Successful applications throughout DuPont have been reported for new process development programs, as well as environmental-related programs, testifying to its suitability for evaluating the environmental costs and benefits associated with a P2 program (Mulholland and Dyer, 1999).

As depicted in Figure 23.5 (Mulholland and Dyer, 1999), economics features prominently in two key areas: (i) the planning and organization or chartering phase and (ii) the assessment phase. In the former, we are interested in quickly defining the incentives for undertaking P2 initiatives. These incentives are largely in financial terms (e.g. prevented investment in end-of-pipe treatments, lower manufacturing costs, prevented environmental fines and penalties, improved yield, and increased plant operating time). However, they can also be less financially quantifiable when dealing with benefits derived from improved public image of the company, environmentally friendly products/processes, and fewer permitting barriers to business growth. Quantifying the financial incentives will especially help to rank the process waste streams for source reduction during the waste stream selection step, besides securing support for the program from business leadership. As described by Figure 23.5, there are two aspects to evaluate the incentives obtained from P2 initiatives: (i) the cost associated with new end-of-pipe treatment for the selected waste streams and (ii) the lost product value that goes into the waste, leaving the process either in terms of loss of raw materials alone or loss that includes the manufacturing cost.

23.4 Survey of Tools, Technologies, and Best Practices for Pollution Prevention

For the rest of the chapter, we present a more specific survey treatment of tools, technologies, and best practices for undertaking P2 by adopting a multiscale analysis of categorizing these tools into macro-, meso-, and microscale (Allen and Rosselot, 1997).

The DuPont's 10-Step method of engineering evaluations of pollution prevention: A road map to the best solution

Define problems and goals
1. Define the problem
2. Set the goals

Identify the alternatives
3. Identify all possible alternative solutions

Define the alternatives
For each alternative:
4. Develop flowsheets and facility scopes
5. Define development and operating requirements

Evaluatethe alternatives
For each alternative:
6. Estimate the new investment required
7. Estimate the change in cash costs and revenues
8. Estimate the net present value (NPV)
9. Determine all 'noneconomic' considerations

Then
10. Choose the best alternative

Each step builds on previous steps, so don't skim over the early steps

Iterative evaluations
Give insight on what's important, the detail required, potential problems and risks, and how to formulate better alternatives

Figure 23.6 The DuPont's "10-Step Method of Engineering Evaluations for Pollution Prevention." *Source*: Adapted from Mulholland and Dyer (1999).

23.4.1 Pollution Prevention at the Macroscale

Opportunities for P2 at the macroscale can be identified from three distinct perspectives (Allen and Rosselot, 1997; El-Halwagi, 1997). First, waste generation audits identify flow rates and compositions of materials in the industrial economy, which are potential P2 targets, from natural resource extraction to consumer product disposal. Second, industrial ecology studies examine the uses and the wastes associated with a particular material. Third, life cycle assessments (LCA) assess the environmental impacts due to the life cycles of individual products/processes by determining waste generation rates, energy consumption, and raw material usage.

23.4.1.1 Industrial Ecology

Audits of waste generation and management via inventory of wastes represent a first step in performing macroscale studies of P2. The next step is to integrate waste generation data with production data, recognizing that the huge amount of wastes generated annually ought not to be ignored as a potential source of reuse as industrially valuable materials. On the whole, the conversion of wastes from something undesirable into something of value eliminates the need for the release of these substances into the environment. This reduces the need for raw material extraction. In line with this, studies in the field of industrial ecology or also known as industrial metabolism examine the material efficiency of large-scale industrial systems, searching for ways that improve their efficiency. By integrating models of the uses of selected materials with data on waste generation for the same materials, it may be possible to identify targets of opportunity for recycling between industrial sectors. Studies can also be conducted to determine whether the contaminants in wastes are potentially recoverable and what factors promote and hinder the "mining" of wastes. Finally, the mix of technologies used to produce goods can be investigated to determine the most environmentally benign combination of technologies. In summary, studies of industrial ecology attempts to present macroscale views of the integration of feedstock and waste data (Allen and Rosselot, 1997).

Materials Integration Across Industrial Sectors. Substantial environmental and economic benefits are possibly derived when the waste from one facility can be used directly as feedstock in another facility: (i) the facility producing the waste is spared the burden of treatment or disposal; (ii) the facility accepting the waste can reduce the costs of feedstock materials; and (iii) the environmental burdens associated with disposal and raw material extraction are reduced. At present, arrangements for material integration between facilities are made more on an ad hoc basis than from careful planning with a thorough understanding of what types of feedstocks and wastes lend themselves to assist toward integration. However, the direct utilization of wastes is bound to be more systematically exploited as the data on waste generation and management improve and the whole field of industrial ecology matures (Allen and Rosselot, 1997).

Recovery of Valuable Materials in Waste. The factors affecting the potential for extraction and recovery of valuable contaminants in waste streams depend on (i) the concentration of the contaminants and (ii) their market value. Over and above these two factors that indicate whether a waste can be economically mined, the existence of a recycling infrastructure and the technologies currently in place influence recycling rates as well (Allen and Rosselot, 1997).

23.4.1.2 Life Cycle Assessment

Life cycle assessment is an increasingly popular macroscale approach of analytical framework tool for environmental auditing that identifies opportunities for P2 by assessing the environmental impacts associated with a product or a material. It considers an individual product or a material and traces the flows of energy, raw materials, and waste streams that were required to create, use, and dispose of the product (Allen and Rosselot, 1997).

LCA integrates environmental impacts over the entire life cycle "from cradle to grave," as it is often dubbed as. Thus, it is conducted by companies wishing to improve the environmental characteristics of their products and by companies or industrial groups intending to promote their product as environmentally superior to a competitor's product. Additionally, it is also used as a tool for product design, for corporate strategic planning and environmental planning in order to reduce possible environmental risk and thereby attain competitive advantage in the marketplace (Pereira, 1999), and for public policy making that involves government initiatives in environmental labeling, acquisition, procurement, and legislation (Allen and Rosselot, 1997; Rosselot and Allen, 2002).

The traditional main focus of chemical product and process designers on product life cycle stages from raw material extraction up to manufacturing is increasingly changing to one that considers how products will be recycled. In other words, product designers need to consider the way customers use their products, while process designers ought to avoid contamination at the sites where their processes are executed. Rosselot and Allen perceive this as design engineers having to become stewards for their products/processes throughout their life cycles (Rosselot and Allen, 2002).

We will not delve further into this subject on LCA, as it is treated elsewhere in this book as a separate dedicated chapter.

23.4.2 Pollution Prevention at the Mesoscale

Macroscale studies of P2 are valuable for the identification of P2 opportunities, typically beginning with waste audits and emission inventories. Subsequently, the building blocks of chemical processes (i.e. the unit operations) can be examined for designs and practices that contribute toward the reduction of wastes and emissions. Systematic tools are then employed for optimizing the integration of the unit operations with enhanced environmental performance in process flowsheets by exploiting suitable methods for assessing the ensuing economic performance of the optimal flowsheets (Allen and Rosselot, 1997).

23.4.2.1 Pollution Prevention for Unit Operations

As emphasized in the preceding section, it is desirable to consider the environmental ramifications of each unit operation early in the process rather than postponing until the flowsheet has been completed. An environmentally benign design is bound to be more profitable, as it will incur lower waste treatment and environmental compliance costs while converting a higher percentage of raw materials into saleable products. Allen and Shonnard stress the importance of the considerations outlined in Table 23.12 for identifying P2 opportunities for unit operations in the design of chemical processes (Allen and Shonnard, 2002) Table 23.13 (adapted from Butner (1995) and Allen and Rosselot (1997)) provides an overview of current P2-inclined operating practices and process modifications, with the indication of research needs for the necessary technology aid development (Butner, 1995; Allen and Rosselot, 1997).

23.4.2.2 Flowsheet Analysis for Pollution Prevention

Flowsheet analysis examines a process flowsheet in terms of the following: (i) the performance of the individual unit operations with improved environmental performance that make up the flowsheet and (ii) the level to which the process streams are sequenced and networked or integrated for enhanced environmental performance. The analysis of the degree of integration can be executed via the process integration of mass and energy since chemical processes and chemical plant operations essentially involve a strong interaction between mass and energy, with the overall objective of converting and processing mass via the use of energy to drive reactions, effect separations, and steer pumps and compressors (El-Halwagi, 1997).

Table 23.12 Important considerations for identifying P2 opportunities for unit operations.

1) Select raw material(s) – Nelson advocates the following suggestions to minimize human health impact and environmental damage due to any possible emissions:
   - Eliminate impurities in the feed
   - Use substitute chemicals that are less hazardous
   - Utilize waste materials from other processes
2) Carefully evaluate in-process waste generation mechanisms to direct the process designer toward environmentally sound material choices and other P2 options
3) Optimize operating conditions for each unit to achieve maximum reactor conversion and separation efficiencies
4) Minimize losses from cleaning operations:
   - Use mechanical cleaning methods to reduce solvent losses from both parts and equipment cleaning operations
   - Consider continuous processing when batch cleaning wastes are likely to be significant (e.g. with highly viscous, water-insoluble, or adherent materials)
5) Adopt the best material storage and transfer technologies to minimize releases of materials to the environment
6) Minimize energy consumption and the release of utility-related emissions
7) Enforce safe working conditions

Source: Adapted from Allen and Shonnard (2002).

This section describes a set of methods for synthesizing or restructuring (retrofitting) process flowsheets that can be generally categorized into qualitative and quantitative methods. We begin with tools that are largely qualitative in concept and implementation, focusing on the following three approaches:

1) Material flow analysis, in which mass balances are used to identify P2 opportunities.
2) Frameworks for examining flowsheets that account for modifications to existing processes in operations (i.e. design retrofitting).
3) The Douglas hierarchical design decision procedures for analyzing flowsheets, particularly in the design of new processes (Douglas, 1992, 1988).

The second set of more quantitative and rigorous flowsheet analysis tools consists of energy (heat) integration and mass integration that have been shown to contribute significantly to overall energy and mass efficiency. Two major approaches are available to implement these tools:

1) The pinch analysis or pinch technology method introduced by Linnhoff and coworkers (Linnhoff et al., 1982).
2) The mathematical programming or optimization techniques (Grossmann and Kravanja, 1995).

In this chapter, emphasis is placed on the former approach of pinch technology. It is noteworthy that the mass integration concept incorporates the use of both approaches. Also, a notable work that combines the

Table 23.13 Process modifications of unit operations for P2.

Unit operation	Changes in operating practices	Currently feasible modifications	Process modifications requiring further research and technology development
Reactors	Improve/increase selectivity through: • Better mixing of reactants • Elimination of hot and cold spots • Improved feed distribution (avoids short-circuiting) • Careful control of catalyst regeneration	• Modify catalysts to enhance selectivity or to prevent catalyst deactivation and attrition • Provide separate reactors for recycle streams to permit optimization of conversions (when warranted by conditions)	Consider: • New catalyst technologies • Changes in process chemistry • Integration of reaction and separation units
Separators	Reduce waste from boilers	• Improve separation efficiencies • Insulate the column • Change tray configuration or consider high-efficiency packing rather than conventional tray-type columns (this reduces pressure drop and decreases reboiler temperatures) • Avoid adsorptive separations when adsorbent beds cannot be readily generated • Consider low-temperature distillation columns when dealing with thermally labile process streams	Consider new separation devices that efficiently separate dilute species
Heat exchangers (HEXs)	• Use antifoulants • Use innovative cleaning devices for HEX tubes	• Use staged HEXs and adiabatic expanders to reduce HEX temperatures and reduce sludge formation • Implement heat integration • Execute online cleaning	Consider scraped-wall exchangers and evaporators with viscous materials to avoid thermal degradation of product
Vessels (e.g. storage tanks)	• Use mixers to reduce sludge formation • Add emulsifiers to dissolve tank bottoms (when the bottoms and the emulsifiers do not interfere with downstream processing), which is formed as wastes during storage • Maintain seals and paint in good condition • Reduce losses during loading and unloading • Minimize wasted container material	• Ensure easy (but safe) access to simplify cleaning • Design to ensure complete draining • Use floating roof tanks • Use high-pressure tanks • Use insulated tanks • Use variable vapor space tanks	Consider process-specific changes to eliminate need for storage of hazardous intermediates

Source: Adapted from Allen and Rosselot (1997) and Butner (1995).

three approaches of hierarchical design via process simulation using the Aspen Plus commercial simulator, pinch technology, and mathematical programming is that of Dantus and High (Dantus and High, 1996; Aspen Technology, Inc. (AspenTech), 2006).

23.4.2.2.1 Qualitative Methods

Material Flow Analysis. The simplest analytical tool for waste minimization is the mass balance. Although most process engineers are comfortable with the concept of mass balances, a number of new issues arise when mass balances are applied to P2 and waste reduction. Two key issues are (i) defining the system boundaries and (ii) determining the level of detail required in the mass balances and waste stream measurements. The former is a key aspect in determining whether an adopted strategy actually accomplishes the desired P2, thus contributing to its viability. After the system boundaries have been identified and selected, the second major decision concerns whether detailed mass balance techniques would actually yield significant information about the waste (Allen and Rosselot, 1997).

Frameworks for Examining Flowsheets of Existing Processes for Pollution Prevention Opportunities. Hazard and operability (HAZOP) studies identify potential hazards associated with each process stream via qualitative and sometimes quantitative evaluation by systematically considering possible deviations of a stream from normal operating conditions (Crowl and Louvar, 1990). A parallel analysis for P2 would entail a series of systematic questions concerning each process stream (particularly the waste streams) and cluster of unit processes, with typical questions listed in Table 23.14. In addition, Allen and Rosselot suggest the following framework for evaluating an existing process flowsheet for P2 opportunities (Allen and Rosselot, 1997):

1) Establish the system boundaries of the process.
2) Execute a waste stream audit using the material flow analysis, primarily for determining the composition of the waste stream(s) and the associated mechanism of waste formation.
3) Examine systematically the P2 options for each waste stream.
4) Examine systematically the P2 options for each unit operation.
5) Carry out an appraisal on the impact of each potential P2 program on the whole process or on the finished product.

Hierarchical Design Procedures for Analyzing Flowsheets for the Design of New Processes. Chemical processes are often synthesized by employing the Douglas hierarchical design procedure with waste minimization (as summarized in Table 23.15) (Douglas, 1985, 1988, 1990, 1992). This procedure provides opportunities to perform environmental assessments by identifying P2 options via a series of heuristic rules at each level in the design process. The hierarchy is organized in such a way that decisions that affect the generation of wastes and emissions at each level limit the decisions in the subsequent levels below it. Accordingly, computer-aided analysis tools providing the integrated capabilities indicated in Figure 23.7 will therefore be necessary to efficiently link chemical process and product design with critical environmental processes in a larger system analysis. At the earliest stage of design, choices are made among competing alternatives for reaction pathways, raw materials, solvents, and energy sources. Next, technological choices are made regarding the equipment for carrying out these reaction and separation steps. A flowsheet containing connections between these unit operations is then generated, and the material and energy requirements are computed (based on reasonable estimates) using commercial process simulation software. Opportunities for energy and mass integration are then evaluated in efforts to reduce energy consumption and waste generation. In the final process design step, the flowsheet is optimized to determine process operating parameters that maximize profit (or minimize production and operating costs, accordingly) and minimize environmental impacts (Mihelcic et al., 2003).

Table 23.14 Typical questions in P2 assessment analogous to HAZOP analyses.

- Would a change in process configuration or topology reduce waste generation?
- What changes in operating procedures might reduce wastes?
- Would changes in raw materials or process chemistry be effective in eliminating a waste stream?
- Would a change in process conditions (e.g. temperature, pressure, reactant concentrations) reduce waste generation?
- Would improvements in process control be effective?
- Would the waste be a raw material for another process?

Source: Adapted from Allen and Rosselot (1997) and Allen and Shonnard (2002).

23.4.2.2.2 Quantitative Methods

Process Energy Integration/Heat Integration. The core principle in process energy integration is to use the heat from streams that need to be cooled for heating streams that need their temperature/energy level or content raised. Heat transfer between streams in a process prevents pollution by reducing the need for fuels and for cooling tower operation. Heat integration is generally accomplished via the method of heat exchanger network synthesis (HENS) pioneered by Linnhoff and coworkers (Linnhoff and Hindmarsh, 1983; Linnhoff et al., 1982, 1983). In HENS, all of the heating and cooling requirements for a process are systematically examined to compute the extent to which streams that need to have their temperature raised can be heated by streams that need to be cooled (Linnhoff and Hindmarsh, 1983; Linnhoff et al., 1982, 1983; Douglas, 1988; Smith, 2005).

Process Mass Integration. Mass integration is a systematic methodology that provides fundamental understanding of the global flow of mass within a process and employs this understanding for identifying performance targets and for optimizing the generation and routing of species throughout the process. P2 denotes a mass allocation objective that is at the heart of mass integration. In fact, mass integration is more general and more involved than energy integration. Due to the overriding mass objectives of most processes, mass integration can potentially provide much stronger impact on a process than energy integration. Since both integration branches are compatible, mass integration coupled with energy integration essentially provides a systematic framework for understanding the big picture of a process, for identifying performance targets, and for developing solutions to improve process efficiency, which automatically translates as a P2 initiative (El-Halwagi and Spriggs, 1998).

Table 23.15 Levels of Douglas's hierarchical procedure for process synthesis with waste minimization.

Design level	Waste minimization/reduction problem and the associated strategy
1) Identify the input information of the type of problem: the material to be manufactured	
2) Specify the input–output structure of the flowsheet	
For problems caused by:	
• Reaction chemistry: change the chemistry	
• Air oxidation to NO*x*: change to O_2 in recycle CO_2 oxidations	
• Spent catalysts: regenerate the catalyst	
3) Design the recycle structure of the flowsheet	
For problems caused by:	
• Adding reactor diluents to shift the product distribution or to shift the equilibrium conversion: change the diluent	
• Adding heat carriers: change the heat carrier	
• Adding reactor solvents: change the solvent	
4) Specify/stipulate the separation system	
4a) General structure: phase splits	
4b) Vapor recovery system	
For problems caused by:	
• Absorber solvents: change the solvent	
• Regeneration of adsorption beds: change the bed stripping agent	
• Removing spent adsorbents: change to absorption or condensation	
• Use of reactive absorbers to remove toxic materials	
4c) Liquid recovery system	
For problems caused by:	
• Stripping agents: change the agent	
• Extraction solvents: change the solvent	
• Crystallizer (recycle and) purge streams (almost pure water): reuse (through mass integration techniques) the purged water elsewhere in the process	
• Crystallizer purge streams (not almost pure water): remove the contaminants and recycle the water or look for a different separation system	
• Reactive crystallization by-products: identify different separation technique	
• Spent absorbents: regenerate the adsorbent	
4d) Solid recovery system	
For problems caused by cake washing:	
• Same as for crystallizer	
• Filter mother liquor streams	
5) Process energy and mass integration	
5a) Integrate process heating and cooling requirements	
5b) Identify process waste recycling and water reuse opportunities	
6) Evaluate opportunities	
6a) Different decisions	
6b) Reactor configuration	
6c) Complex distillation columns	
7) Address flexibility and control	
8) Address process safety	

Source: Adapted from Allen and Shonnard (2002) and Douglas (1990, 1992).

Mass integration is properly defined as the use of materials that would otherwise be wasted, analogous to the notion of heat integration, which is the use of *heat* that would otherwise be wasted. Three tools are available for determining process configurations that result in mass integration (Allen and Shonnard, 2002):

1) The source–sink mapping, which is the simplest and most visual and intuitive of the methods for identifying candidate streams for mass integration.
2) Strategy for determining optimum mixing, segregation, and recycling.

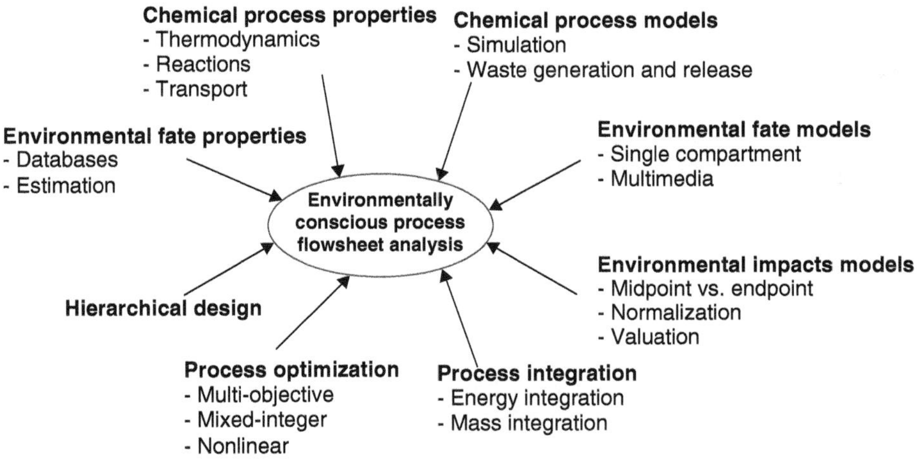

Figure 23.7 Integrated methods and capabilities for environmentally conscious design and assessment for waste reduction and P2. *Source*: Adapted from Mihelcic et al. (2003).

3) The mass exchange network (MEN) synthesis, which is the mass integration analog for heat exchange network (HEN) synthesis that is developed by Vasilios Manousiouthakis and coworkers at the University of California, Los Angeles (UCLA) (El-Halwagi and Manousiouthakis, 1989).

Source–Sink Mapping. Source–sink mapping is used to determine whether waste streams can be used as feedstocks. The first step in creating a source–sink diagram is to identify the sources and sinks of the material for which integration is desired. As an example, for the problem of water utilization (treatment and reuse) in petroleum refineries, we need to identify wastewater streams as the *sources* of water and the processes that require water as the *sinks* of water. The flow rates of the sources and sinks must be known, with the latter able to accept a range of flow rates. Contaminants and their associated concentrations that are present in the source streams and that potentially pose significant problems for the sinks need to be identified, as well as the tolerances of each sink for these contaminants. Some processes may require very pure feed, rendering infeasibility to the waste streams that contain the feed material as well as some contaminants. Nevertheless, many processes are still able to make use of material that contains impurities, while some sinks possess widely flexible tolerances (Allen and Shonnard, 2002).

Strategies for Optimization of Segregation, Mixing, and Recycling of Streams. Processes with increased number of sources and sinks that demand more complex analyses, beyond the capability of source–sink mapping, require the employment of mathematical programming techniques combined with process simulation packages for the segregation, mixing, and recycling of streams. These involve linear and nonlinear optimization approaches (Allen and Shonnard, 2002).

Mass Exchange Network (MEN) Synthesis. Unlike source–sink mapping and optimization of segregation, mixing, and recycling, MENs achieve mass integration by executing direct exchange of mass between streams through the systematic generation of a network of mass exchangers, which preferentially transfer compounds that exist as pollutants in the streams where they are originally present to streams where they have a positive value (this will be explained later in the section). MEN synthesis can be adopted for any countercurrent, direct-contact mass transfer operation such as absorption, desorption, or leaching. These concepts are illustrated by Figures 23.8 and 23.9.

Analogous to energy integration in which heat transfer is limited by energy balances and a positive driving force, the limits to mass transfer are enforced by mass balance constraints and equilibrium constraints:

1) The total mass transferred by the rich stream (i.e. the stream from which a material is to be removed) must be equal to that received by the lean stream (i.e. the stream receiving the material).
2) Mass transfer is possible only if a positive driving force exists for all rich stream/lean stream matches.

The MEN concept has also been successfully extended to novel process design concepts of wastewater

Figure 23.8 Schematic of a single mass exchanger for environmental process design. *Source*: Adapted from Dunn and El-Halwagi (2003).

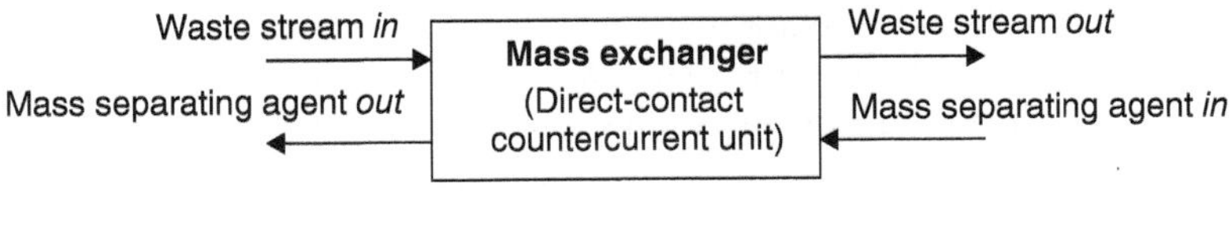

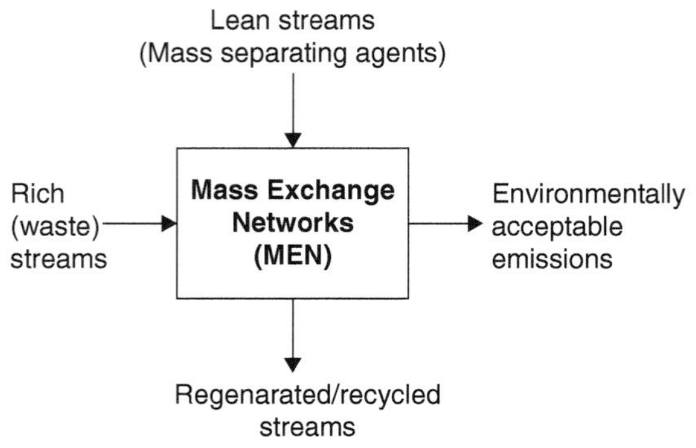

Figure 23.9 Mass exchange network (MEN) synthesis. *Source*: Adapted from Dunn and El-Halewagi (2003).

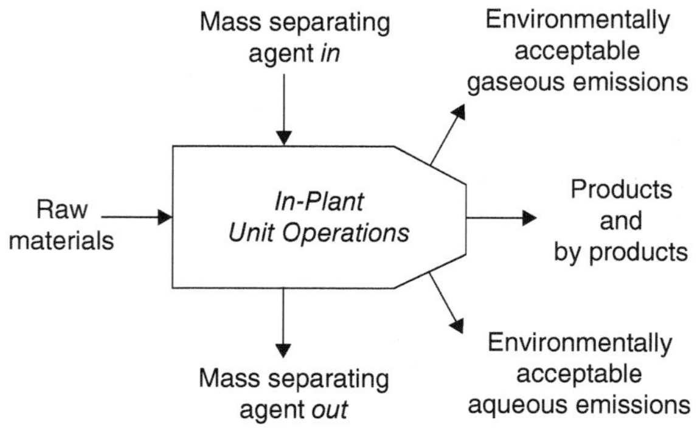

Figure 23.10 Schematic representation of waste interception and allocation networks (WINs). *Source*: Adapted from Dunn and El-Halwagi (2003).

minimization systems, waste interception and allocation networks (WINs), heat-induced waste minimization networks (HIWAMINs), and energy-induced waste minimization networks (EIWAMINs). The latter two approaches serve to improve process economics by simultaneously addressing issues of waste reduction and energy conservation in process design (Dunn and El-Halwagi, 1993; Wang and Smith, 1994; Huang and Edgar, 1995; Huang and Fan, 1995; El-Halwagi et al., 1996; Dunn and Srinivas, 1997). Figures 23.10 and 23.11 display the schematic representations of the design methodologies of WIN and HIWAMIN together with EIWAMIN, respectively.

A commendable amount of work on the synthesis of MEN for P2 is available in the literature for the dephenolization of petroleum refinery wastewaters, in which phenol is a pollutant in the water effluent streams of the catalytic cracking units, desalter wash water, and spent sweetening waters, but in other streams, phenol is present as a valuable additive (Allen et al., 1992; El-Halwagi et al., 1992). Other relevant works include that of Dunn and El-Halwagi (1993) for reducing the emission of hydrogen sulfide from pulp and paper plants, the

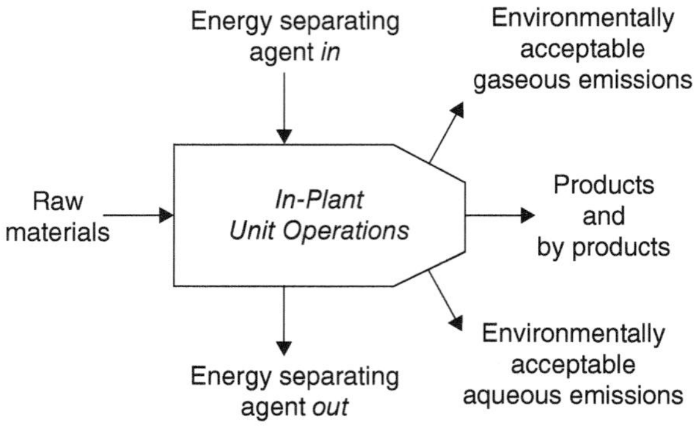

Figure 23.11 Schematic representation of heat-induced waste minimization networks (HIWAMINs) and energy-induced waste minimization networks (EIWAMINs). *Source*: Adapted from Dunn and El-Halwagi (2003).

concept-oriented papers by Gupta and Manousiouthakis (1994) and Papalexandri et al. (1994), and Hallale and Fraser (1998) for water minimization. Due to the limited space here, the interested reader is referred to the review article by Dunn and El-Halwagi (2003) and the comprehensive text by El-Halwagi (1997) for a detailed exposition on the application of MENs at industrial scale and for academic research.

23.4.3 Pollution Prevention at the Microscale

Many of the macroscale and mesoscale P2 approaches rely on a molecular-level understanding of chemical and physical process at the microscale. For example, the synthesis of new catalysts to achieve higher yields with less wastes relies on a fundamental understanding of surface chemistry. There are a large number of methods available for undertaking P2 at the microscale via the molecular-level redesign of chemical products and processes. In line with the emergence of the power of computational tools that employ mathematical programming or optimization techniques, we mention here two methods that adopt such tools:

1) Systematic design of substitute materials for environmentally harmful chemicals such as solvents.
2) Systematic analysis of reaction pathways.

The former approach utilizes *group contribution theory*, in which molecules are systematically designed to incorporate a distribution of functional groups resulting in desired properties. It relies on an understanding of the ways in which the building blocks of a molecule (i.e. its functional groups) affect its properties. The objectives of the second approach of reaction pathway synthesis are to reduce both the use of toxic precursors and the formation of unwanted by-products. In spite of current significant limitations confronting these techniques, they nonetheless hold much promise for further improvement into robust methods for the systematic identification and evaluation of desired synthesis routes (Allen and Rosselot, 1997).

23.5 Concluding Remarks

This chapter strives to compile and consolidate a broad set of tools, techniques, and best practices for identifying and subsequently implementing P2 programs, with a multiscale analysis of macro-, meso-, and microscale employed to systematically categorize these approaches. It is hoped that over time, a shift in focus could be expected and targeted, from mere compliance to a point in which environmentalism, as driven by the P2 paradigm, will be fully justified and integrated into corporate business culture, similar to the overriding emphasis, nowadays, that is placed on the health and safety elements of the ubiquitous concept of "health, safety, and environment" in chemical plant operations.

References

Allen, D.T., Bakshani, N., and Rosselot, K.S. (1992). *Pollution Prevention: Homework & Design Problems for Engineering Curricula*. New York: American Institute of Chemical Engineers' (AIChE) Center for Waste Reduction Technologies (CWRT).

Allen, D.T. and Rosselot, K.S. (1994). Pollution prevention at the macroscale: flows of wastes, industrial ecology and life-cycle analyses. *Waste Management* 14 (3–4): 317–328.

Allen, D.T. and Rosselot, K.S. (1997). *Pollution Prevention for Chemical Engineers*. New York: Wiley.

Allen, D.T. and Shonnard, D.R. (2001). Green engineering: environmentally conscious design of chemical processes and products. *AIChE Journal* 47 (9): 1906–1910.

Allen, D.T. and Shonnard, D.R. (2002). *Green Engineering: Environmentally Conscious Design of Chemical Processes*. Upper Saddle River: Prentice Hall.

Anastas, P. and Warner, J. (1998). *Green Chemistry: Theory and Practice*. New York: Oxford University Press.

Aspen Technology, Inc. (AspenTech) (2017). Aspen Plus. http://www.aspentech.com (accessed 2 February 2018).

Bishop, P.L. (1999). *Pollution Prevention: Fundamentals and Practice*. Long Grove: Waveland Press.

Butner, R.S. (1995). Pollution prevention in process development and design. In: *Industrial Pollution Prevention Handbook* (ed. H. Freeman) Chapter 44, 329–341. New York: McGraw-Hill.

Case, L., Mendicino, L., and Thomas, D. (1995). Developing and maintaining a pollution prevention program. In: *Industrial Pollution Prevention Handbook* (ed. H. Freeman) Chapter 8. New York: McGraw-Hill.

Crowl, D.A. and Louvar, J.F. (1990). *Chemical Process Safety: Fundamentals with Applications*. Englewood Cliffs: Prentice Hall.

Dantus, M.M. and High, K.A. (1996). Economic evaluation for the retrofit of chemical processes through waste minimization and process integration. *Industrial & Engineering Chemistry Research* 35: 4566–4578.

Douglas, J.M. (1985). A hierarchical decision procedure for process synthesis. *AIChE Journal* 31 (3): 353–362.

Douglas, J.M. (1988). *Conceptual Design of Chemical Processes*. New York: McGraw-Hill.

Douglas, J.M. (1990). Synthesis of multistep reaction processes. In: *Proceedings of the Third International Conference of the Foundations of Computer-Aided Process Design (FOCAPD) at Snowmass Village, Colorado, USA on July 10–14, 1989* (ed. J.J. Siirola, I.E. Grossmann and G. Stephanopoulos), 79–103. Amsterdam: CACHE–Elsevier.

Douglas, J.M. (1992). Process synthesis for waste minimization. *Industrial & Engineering Chemistry Research* 31: 238–243.

Dunn, R.F. and El-Halwagi, M.M. (1993). Optimal recycle/reuse policies for minimizing the wastes of pulp and paper plants. *Journal of Environmental Science & Health Part A* A28 (1): 217–234.

Dunn, R.F. and El-Halwagi, M.M. (2003). Process integration technology review: background and applications in the chemical process industry. *Journal of Chemical Technology and Biotechnology* 78: 1011–1021.

Dunn, R.F. and Srinivas, B.K. (1997). Synthesis of Heat-induced Waste Minimization Networks (HIWAMINs). *Advances in Environmental Research* 1: 275–301.

El-Halwagi, M.M. (1997). *Pollution Prevention through Process Integration*. San Diego: Academic Press.

El-Halwagi, M.M., El-Halwagi, A.M., and Manousiouthakis, V. (1992). Optimal design of dephenolization networks for petroleum refinery wastes. *Transactions of the Institution of Chemical Engineers Part B* 70 (B3): 131–139.

El-Halwagi, M.M., Hamad, A.A., and Garrison, G.W. (1996). Synthesis of waste interception and allocation networks. *AIChE Journal* 42: 3087–3101.

El-Halwagi, M.M. and Manousiouthakis, V. (1989). Synthesis of mass exchange networks. *AIChE Journal* 35 (8): 1233–1244.

El-Halwagi, M.M. and Spriggs, H.D. (Fall 1998). Educational tools for pollution prevention through process integration. *Chemical Engineering Education* 246–249.

Elkamel, A. (2007). Preface of special issue on pollution prevention and waste minimization via simulation and optimization. *International Journal of Environment and Pollution* 29 (1/2/3): 1–3.

Freeman, H. (1995). *Industrial Pollution Prevention Handbook*. New York: McGraw-Hill.

Freeman, H., Harten, T., Springer, J. et al. (1992). Industrial pollution prevention: a critical review. *Journal of the Air and Waste Management Association* 42: 618–656.

Grossmann, I.E. and Kravanja, Z. (1995). Mixed-integer nonlinear programming techniques for process systems engineering. *Computers & Chemical Engineering* 19 (supplement): S189–S204.

Gupta, A. and Manousiouthakis, V. (1994). Waste reduction through multicomponent mass exchange network synthesis. *Computers & Chemical Engineering* 18 (supplement): S585–S590.

Hallale, N. and Fraser, D.M. (1998). Capital cost targets for mass exchange networks. A special case: water minimization. *Chemical Engineering Science* 53 (2): 293–313.

Harold, M.P. and Ogunnaike, B.A. (2000). Process engineering in the evolving chemical industry. *AIChE Journal* 46 (11): 2123–2127.

Huang, Y.L. and Edgar, T.F. (1995). Knowledge-based design approach for the simultaneous minimization of waste generation and energy consumption in a petroleum refinery. In: *Waste Minimization through Process Design* (ed. A.P. Rossiter), 181–196. New York: McGraw-Hill.

Huang, Y.L. and Fan, L.T. (1995). Intelligent process design and control for in-plant waste minimization. In: *Waste Minimization through Process Design* (ed. A.P. Rossiter), 165–180. New York: McGraw-Hill.

Linnhoff, B., Dunford, H., and Smith, R. (1983). Heat integration of distillation columns into overall processes. *Chemical Engineering Science* 38: 1175–1188.

Linnhoff, B. and Hindmarsh, E. (1983). The pinch design method for heat exchanger networks. *Chemical Engineering Science* 38: 745–763.

Linnhoff, B., Townsend, D.W., Boland, D. et al. (1982). *User Guide on Process Integration for the Efficient Use of Energy*. Rugby: The Institution of Chemical Engineers.

Mihelcic, J.R., Crittenden, J.C., Small, M.J. et al. (2003). Sustainability science and engineering: the emergence of a new metadiscipline. *Environment Science & Technology* 37 (23): 5314–5324.

Moore, F.L. (1995). Pollution prevention in the chemical industry. In: *Industrial Pollution Prevention Handbook*, Chapter 44 (ed. H. Freeman). New York: McGraw-Hill.

Mulholland, K.L. and Dyer, J.A. (1999). *Pollution Prevention: Methodology, Technologies, and Practices*. New York: American Institute of Chemical Engineers.

Papalexandri, K.P., Pistikopoulos, E.N., and Floudas, C.A. (1994). Mass exchange networks for waste minimization: a simultaneous approach. *Chemical Engineering Research & Design, Part A: Transactions of the Institute of Chemical Engineers* 72 (A3): 279–294.

Pereira, C.J. (1999). Environmentally friendly process. *Chemical Engineering Science* 54: 1959–1973.

Rosselot, K.S. and Allen, D.T. (2002). Flowsheet analysis for pollution prevention. In: *Green Engineering: Environmentally Conscious Design of Chemical Processes* (ed. D.T. Allen and D.R. Shonnard). Upper Saddle River: Prentice Hall.

Rossiter, A.P. (1994). Process integration and pollution prevention. *AIChE Symposium Series on Pollution Prevention via Process and Product Modifications* 90 (303): 12–22 Edited by Mahmoud M. El-Halwagi and Demetri P. Petrides. New York: American Institute of Chemical Engineers (AIChE).

Smith, R. (2005). *Chemical Process Design and Integration*. New York: Wiley.

Smith, R. and Petela, E. (1992). Waste minimization in the process industries. *The Chemical Engineer* 17: 21–23.

United Nations Environment Program (UNEP) (1996). Cleaner Production Programme (a pamphlet on UNEP's cleaner production program). Geneva: UNEP.

US Environmental Protection Agency (US EPA) (1988). Waste Minimization Opportunity Assessment Manual (EPA/625/7-88/003), Washington, DC. http://www.p2pays.org/ref/10/09929.pdf (retrieved on 25 June 2006).

US Environmental Protection Agency (US EPA) (1992). Facility Pollution Prevention Guide (EPA/600/R-92/088), Washington, DC. www.p2pays.org/ref/01/00370.pdf or http://www.epa.state.oh.us/opp/tanbook/fppgbgn.html (retrieved on 25 June 2006).

US Environmental Protection Agency (US EPA) (2018). Green Chemistry. http://www.epa.gov/greenchemistry/principles.html (accessed on 2 February 2018).

Wang, Y.P. and Smith, R. (1994). Wastewater minimization. *Chemical Engineering Science* 49: 981–1006.

24

Industrial Waste Auditing

C. Visvanathan

Environmental Engineering and Management Program, Asian Institute of Technology, Khlong Luang, Thailand

24.1 Overview

In the pursuit of "sustainable production and consumption," as the true value of natural resources and nonrenewable energy sources is being globally perceived, wastes can no longer be viewed as substances that are spendable. Researches show that wastes have been traditionally discharged into natural bodies as unwanted substances that still possess some economical value. What is useless in one context can be useful in another. Importantly, pollution problems can be significantly reduced if wastes can be reused and recycled instead of discharging them to natural bodies. There is a radical shift in the perception of opportunities with industrial waste, and the current tone is to "conserve and cultivate" rather than "deploy and deplete."

Sustainable development goal (SDG) 2015 targets to sustainably manage the industrial waste by increasing its resource use efficiency. Industrialization has impacts to the environment, and a sound technological adoption will not only minimize the environmental pressure but also reduce the cost of production. SDG 2015 aims to achieve sustainable management and efficient use of natural resource by 2030. It emphasizes on reduction of waste generation through prevention, reduction, recycling, and reuse. It aims the retrofitting of industries with technologically sound and resource efficient processes/machines. Industrial waste audit is a tool to identify such areas where industries can adopt more sustainable practice that will create a win–win scenario for both the environment and the industry.

In many countries, the manufacturing industry is one of the largest polluting sectors, and every year enormous effort and financial resources are spent worldwide to deal with industrial waste. Therefore from the industry perspective, a global change in the environmental perception has a profound significance. Industrial processes, management, goals, and ethics are changing, pressured by rising environmental awareness among government and public institutions. Difficulties are ever increasing to externalize the environmental cost of industrial activities. Gradually a stage is set to internalize the environmental cost, not by marginalizing environmental concerns, but, contrarily, by increased environmental stewardship of products and processes.

In response to the situation, the global industry is embracing proactive methods that principally focus on energy conservation and waste minimization by the application of cleaner production techniques. These techniques are progressively conceived to meet the goals of environmental and economical sustainability of industries in a more dependable way. The use of traditional "end-of-pipe" treatment of waste alone is progressively becoming inadequate to satisfy the tightening requisites of modern environmental legislations. Industrial processes are also impacted by progressive phaseout leading to a total ban on several chemicals that were regularly used in industries like ozone-depleting substances (ODS), persistent organic pollutants (POPs), etc. Several national governments have pledged in various global multilateral environmental agreements about gradual but total eradication of such substances.

Open or global market regimes are bringing additional complications on top of environmental needs by putting up a stiff pricing competition. Over and above, the questions of sustainability are becoming more pressing as several global institutional buyers are including environmental criteria (like ISO 14000 series certification or others) in the procurement specification. This is driving industries to reevaluate their activities and associated costs, which also include waste treatment cost that covers roughly about 15–30% of the total operational cost. The most assertive way is to reduce energy consumption and waste generation in the first place, which improves the overall process efficiency. Several case studies proved

Handbook of Environmental Engineering, First Edition. Edited by Myer Kutz.

that such proactive approaches reduce the overall production cost and environmental liabilities. Moreover, being "greener" is also helping companies to market their products with institutional buyers and attract wider public attention.

With increasing popularity and attention to proactive methods, structured methodologies are being developed to systematically explore, analyze, and implement energy conservation and waste minimization or cleaner production programs in industries. In many industries, such techniques are now being used as one of the management tools to monitor and control process efficiency and environmental liabilities.

Traditionally, environmental impacts from industry are mainly assessed based on the type, characteristic, and volume of waste that it generates. However, recent analyses show that higher use of energy creates significant environmental impacts when environmental issues related to energy production is taken into account (typical example is GHG emission). Therefore, energy use and waste generation in industries has been recognized as interlinked systems in the way that "the more energy is used, the more pollution is created, and the more waste is generated the more energy is required." This chapter would mainly focus on principles of waste auditing with a note that principles of energy auditing will not be much different, except some specificity for which the reader may refer to other chapters of this handbook.

24.2 Waste Minimization Programs

All manufacturing processes will require raw materials and energy to manufacture a product (or an intermediary) and generate waste in some form. There are several published literatures on different types of waste generated from various industries. The reader must be aware that each manufacturing plant is unique in the type, characteristics, and quantity of waste generation. In other words, the manufacture of specific product creates particular waste quantity and quality. Thus, waste generalization is very difficult.

Compared with extraction, disposal, and transportation, manufacturing is one of the most polluting sectors. Several driving factors are compelling the contemporary manufacturing industries to review their outlook on waste management. The reader should appreciate that such factor(s) can very well be the principal basis of taking up a waste management program and form its major objective(s). Some of the major factors are:

1) *Changing perception of industrial pollution*: There has been a tremendous rise in awareness about industrial pollution in general public and institutions. Moreover, the trend of greener consumerism has made the consumers more interested about the industrial pollution. This has forced governments to take steps to control pollution from industries. Industries are constantly pressured by legislation, governmental and nongovernmental, public organizations, and mass media.
2) *Changing legislations*: There has been continuous tightening of environmental laws and regulations for industries, and implementation has been more rigorous. As a result, waste treatment technologies now require high level of efficiency to meet the stringent discharge standards. This has increased continuous upgradation of such systems.
3) *Changing waste treatment and discharge costs:* As a result of tightening legislation and discharge standards, waste treatment cost is continuously increasing. Thus being energy and resource efficient is the only way to keep producing at the same cost.
4) *Changing availability and cost of raw material and energy*: With increasing demand of raw material and energy, the availability of these resources therefore is decreasing worldwide. This has resulted in price rise of raw materials, especially fuel. Coupled with a stiff market competition, the pricing scenario is changing rapidly.
5) *Changing traditional markets and trade barriers*: The concept of protected market is giving away to global markets that are increasingly being more competitive. This is forcing the industry to increase efficiency and reduce raw material and energy consumption. Buyers in many developed countries are progressively incorporating environmental specification (ISO 14000 series certification, eco labeling, green productivity) in their procurement processes to screen companies.
6) *Changing international commitments*: More national and local governments and institutions are engaging and committing to international bilateral and multilateral agreements (like UN-mediated multilateral environmental agreements) to curb pollution by reducing and eliminating the use of known harmful substances. This pressures industries to change their manufacturing processes and redesign their products.
7) *Marketing strategy*: Raise of awareness level among the consumers has provided ample opportunities for companies to claim such practice as a part of marketing strategy.

As a result of these driving factors, many industries are taking new perspectives and strategies in handling their waste management issues and trying to resolve them in a more sustainable way. The present trend of waste management is to balance proactive methods with traditional reactive methods. The concept of reactive method is about treating the waste once it is generated, also called end-of-pipe treatment, while the proactive method comprises energy and waste minimization, waste recycling and reuse, cleaner production, etc., that reduces the requirement of end-of-pipe treatment of waste. In brief, the main advantages of energy conservation and waste minimization are:

- Raw material consumption and energy consumption can be reduced, which in turn reduces the product cost.
- Energy consumption can be reduced, thereby reducing specific energy required for a product.
- Process efficiency can be improved, increasing product yield and quality.
- Waste generation can be reduced, thereby reducing waste treatment and disposal cost.
- Waste materials can be segregated, leading to containment of hazardous and toxic waste, which in turn improves workers' health and safety.
- By-products can be recovered from waste, in addition to recycle and reuse of waste, waste heat recovery and waste exchange (with other industries) that can generate additional income.
- Increased environmental stewardship can lead to higher attention from institutional buyers and general public helping in marketing.
- Investor confidence can be increased.

In spite of the above advantages of implementing waste minimization techniques, there are some known barriers as given below:

- Some waste minimization or cleaner production techniques may involve significant capital investment with longer payback period.
- There may be obvious risk involved to implementing new systems.
- There is a lack of proper manpower and expertise in appropriate technology.
- There is a lack of information and awareness especially among small- and medium-scale industries.
- Often there is a hesitation to change traditional way of doing things.

The response of an industry will largely depend on several factors:

- Nature of the industrial process.
- Size and structure of the firm.
- Technology and information available to the company.
- Economics of prevention.
- Attitude of the government to control industrial pollution through legislations, incentives, and penalties.

24.3 Waste Minimization Cycle

Typically, a development cycle of industrial waste minimization programs comprises six phases: inception, audit, analysis, design and development, implementation, and evaluation as illustrated in Figure 24.1. The overall goal of waste reduction and cleaner production program is "to critically investigate, evaluate, design and implement such environmentally benign processes and process improvements that would minimize consumption of resource and energy and reduce waste generation in order to reduce adverse environmental impacts and effect in overall economic benefits."

1) *Inception phase*: This phase comprises setting up the goals, commitments, methodologies, task force, time frame, and budget for the project. Such goals should be quantifiable, achievable, and usable to measure the success or failure of any waste minimization or cleaner production program in real terms. Senior management plays a key role in setting up the project framework, the resources, and the project team.
2) *Audit phase*: In audit phase, relevant factory processes, which include management processes, waste treatment processes, etc., are thoroughly investigated to obtain a complete balance sheet of the raw material and energy input and outputs, including waste. Over and above of collecting and compiling all facts and figures, the audit exercise should recommend energy and waste minimization options to attain the desired goals of the program that could be carried to the detail analysis phase. If possible, preliminary technical and economical analysis may also be carried out to prioritize the options.

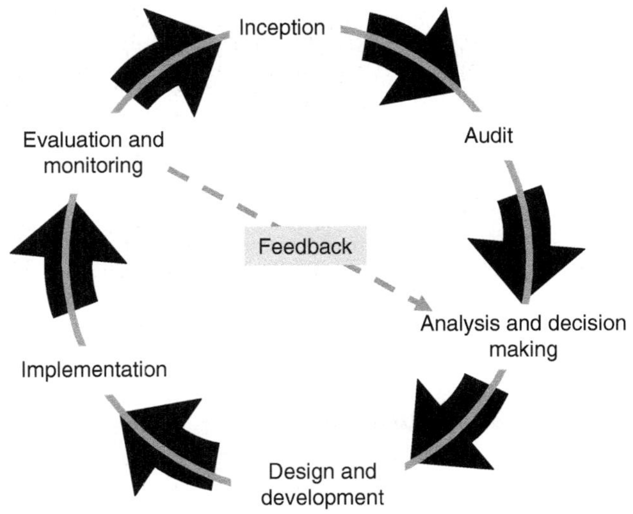

Figure 24.1 Typical waste minimization and cleaner production cycle.

3) *Analysis phase*: The analysis phase starts with detailed analysis of the findings and recommendation from the audit phase to explore the various opportunities and risks associated with the various options. It also explores further possibilities. This phase ends with making decisions on which options to be pursued and which to be dropped.
4) *Design and development phase*: Design and development phase starts with setting up the framework of design, development, and implementation of selected options. This would require planning of all actions. It is a good idea to implement different waste minimization and cleaner production improvements in stages to reduce impacts from introducing new process and process modifications. At this stage, process changes and new processes are designed and procured, and all preparatory works for implementation are undertaken.
5) *Implementation phase*: In the implementation phase, processes and process modifications are installed and integrated to the existing system, commissioned, and put to operation. All operators and workers are trained for the changes in the process and new processes.
6) *Evaluation phase*: This phase continues after the changes or new systems are fully integrated to the normal production processes. In this phase the actual results from the process modifications are monitored, evaluated, and compared with that originally conceived. This phase is often carried out as part of the routine quality assurance/quality control activities, but the outcome can lead to further actions of the waste management program.

This chapter deals with the audit phase. It discusses different aspects of waste auditing. The scope of such discussions has been limited to waste audits within a typical manufacturing industry. For the purpose of this chapter, industry would be treated in general with specific examples from different manufacturing sectors. The principles described here would generally apply to most manufacturing sectors. Some modification to the described methodologies would be necessary depending upon specific activities undertaken in the industry. Such modifications are left for the industry professionals to work out according to the requirements.

24.4 Waste Auditing

Once a waste minimization program is set to be undertaken, the physical work starts with a series of detailed surveys of ongoing activities inside the industry, starting from raw material entering the premises to finished products and by-products (including wastes) leaving it. These audits can be termed as waste audits or waste minimization audits. The principle intent of such audits is "to critically assess various inputs, processes and outputs to find methods and practices for minimizing waste and reduce the resource consumption in a more sustainable and environmentally benign way." Traditionally, industrial waste audits do not include examination of the design of the product itself, but investigate all activities of the production processes and opportunities of waste recycling/reuse including waste treatment systems. There are other terminologies that may be used like pollution prevention audit, eco audit, green audit, etc., which essentially focus on some of the common objectives of preventive approaches.

The phases of a typical waste audit process are illustrated in Figure 24.2. Note that an energy audit process can also be divided into similar phases. The rest of the chapter discusses each of these steps in detail with illustrations, examples, and workouts.

24.5 Phase I: Preparatory Works for Waste Audit

Preparatory works for a waste audit process consist of three main steps:

- Getting the management and staff involved in the program.
- Forming an audit team and appointing a team leader.
- Planning the audit exercise.

Once a waste minimization program is embarked, it is time to provide the program with personnel, technical, and financial resources. The first step is to involve stakeholders in the program to get management and staff involved directly or indirectly in the program. This should be mainly done by one of the core management groups who would ultimately be responsible for managing and monitoring the overall pollution prevention program. Normally the production management or the environmental management group has the responsibility to execute such programs. While all stakeholders can be involved in this process, more emphasis should be given on internal/direct stakeholders like the management, supervisors, and workers in taking up the initiative.

24.5.1 Getting Management and Staff Involved

Commitment from different management groups is a decisive factor in the success of a pollution prevention program. Management representatives from all relevant

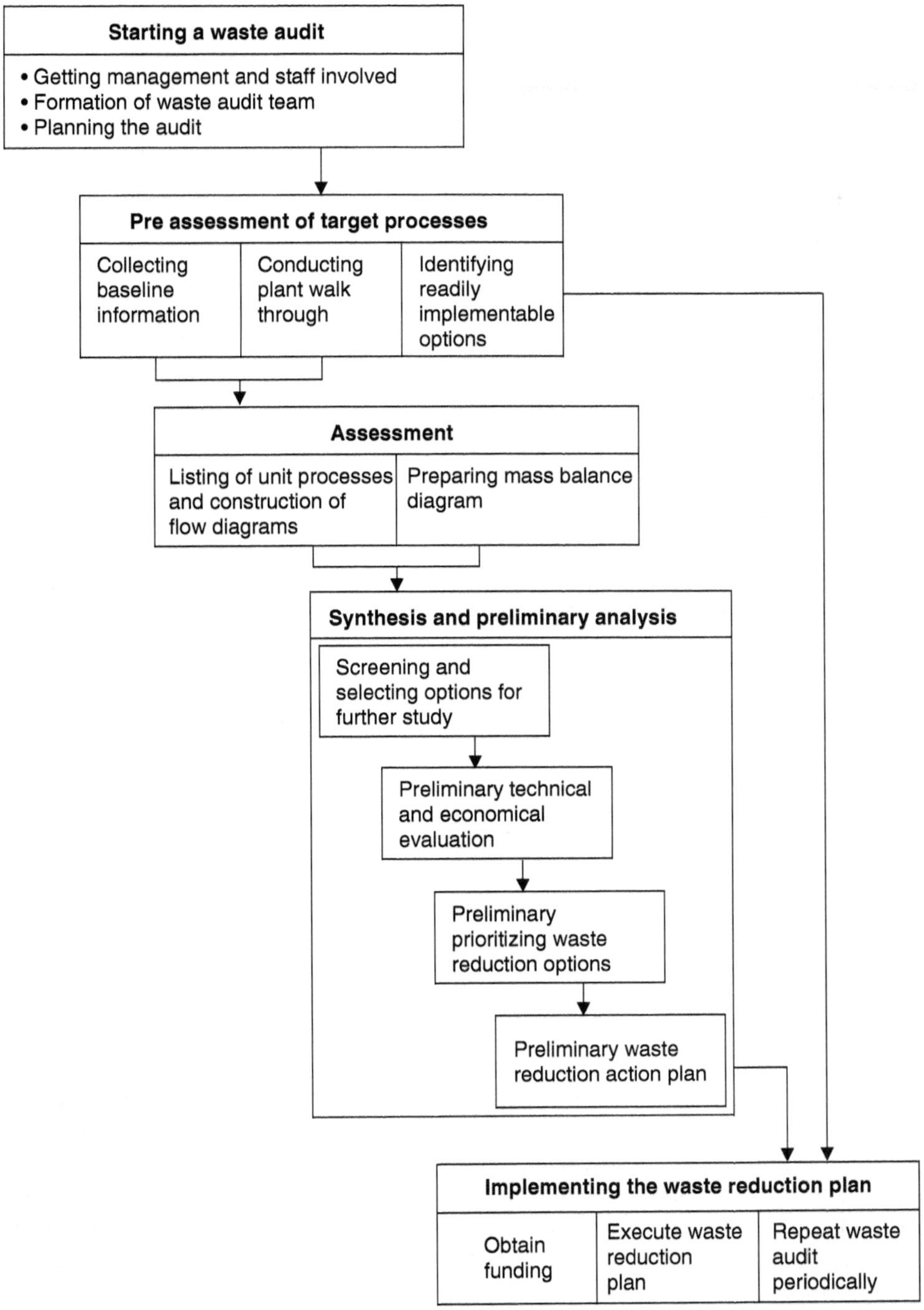

Figure 24.2 Steps of waste audit methodology.

departments should be involved in the program. In a typical manufacturing industry such departments may be:

- Executive management (typically CEO or a deputy)
- Product development and design (if present)
- Production
- Procurement and inventory control
- Operation and maintenance
- Environmental
- Marketing
- Finance
- Quality control

As the management staff would supposedly be aware of the environmental concerns and directly related to the welfare of the company, less effort would probably be required to motivate them for voluntary participation. However, it has to be confirmed that during the whole tenure of the program, there is a high level of involvement and support, even if such program may cause short-term disruption to normal activities.

Involvement of the supervisors and workers is another key factor in the success of such programs. It is the supervisors and workers of a factory who have the hands-on involvement with each of the individual activities of the factory. In order to make the best use of their day-to-day experience with the machineries, processes, and a myriad of issues with the factory processes, audit exercise should be carried out with their full involvement and support. Typically, the barrier to such an involvement is fear of certain supervisors and workers about exposure of some activities (to the management) that are not to the best interest of the company. In order to overcome such a barrier, supervisors and workers should be assured of their job security.

The exercise can also be used as an opportunity to create more awareness among the employees about the environmental issues and energy and waste minimization and cleaner production program. Rewards in the form of bonuses, prizes, or acknowledgement would motivate employees to voluntarily participate in the program. There are several common ways of raising interest and motivation that could benefit the program (Box 24.1).

Box 24.1 Ways of appraisal of interest in waste and energy minimization program.

- Use of posters or banners to inform the staff about the pollution scenario and the requirements, benefits, objectives, and goals of the upcoming waste minimization and cleaner production program.
- Publicizing success story of audit and how it leads to better occupational health and working environment.
- To publicize about the upcoming audit exercise.
- To provide some prominent identification mark (like badges) to the members of the audit team and also provide them some special privileges like free access to any place or information during the audit program.
- To organize "Environment Week" and programs like speeches, videos, skits, tree plantation, etc.
- Cash prizes/monetary incentives for the staff members coming up with some innovative idea leading to cleaner production.
- Some incentives for the machine operators/staff for quality production with minimum resources.
- Interdepartmental competition for waste minimization.
- Sharing the financial gains due to waste/energy minimization programs.

24.5.2 Formation of an Audit Team

One of the key requirements of an audit exercise is to form a proper and balanced audit team that would be responsible for all subsequent audit works. The team is normally formed by the direction of the responsible management group, in discussion with senior management and all other management groups. If required, expert advice from external consultant may be sought at this stage.

Waste audit is an interdisciplinary activity, and therefore the team should be formed with diversified expertise from representative groups or departments, which will have major contribution and interest in the program. Usually, an audit team is a subset of the project team but may include various other personnel whose contribution may be useful. For example, an audit team can be formed from:

- Environmental manager
- Plant manager
- Process or operation and maintenance engineers
- Occupational health and safety officer
- Supervisors and operators
- Laboratory technicians

The team should appoint a team leader, who will lead the audit team and coordinate the activities of all other team members. The plant manager or the environmental manager can be prospective candidates for team leader. It is recommended that a balanced audit team should be made up of three to six persons including the team leader with different expertise. This recommendation is based on typical industry structures but may be varied. In practice, the selection of the team leader and composition of the audit team will depend on the nature of the processes in the industry, scope of audit, and scale of the industry.

Each of the audit team members should be aware of the goals and objectives of the overall energy and waste minimization program and should be able to contribute to it. Duties should be delegated by the team leader to each of the members according to the strength and capabilities (of each member) and should be monitored throughout auditing phase. Members of the audit team should be relieved from the routine activities during the audit exercise.

The requirement of external consultants or experts in the team would depend on the objectives of the program and complexity of the industrial processes. If the goals are to explore simple and readily applicable waste minimization or cleaner production opportunities, e.g. reduction in general water and electricity consumption, plant engineers and supervisors may be sufficient for auditing purposes. However, if the requirement is complex like the performance of cryogenic units or distillation columns, experts in these technologies may be required. In these cases, hiring of an external consultant can be justified.

24.5.3 Planning the Audit Exercise

Before the actual audit exercise is undertaken, a substantial amount of planning work is to be done in order to carry out the audit within time and budget and least interference to normal activities of the plant. Such planning works would mainly include:

- Defining the scope of the waste audit.
- Develop an audit program.
- Prepare a specific workplan and checklists.
- Develop uniform reporting procedures.
- Inform all departments about upcoming audit programs.

However, depending on the specific need of the program, additional works may be added to the list. Each of the above tasks is discussed next.

24.5.3.1 Defining the Scope of the Waste Audit

Determining the scope of an audit is one of the fundamental steps before taking up the actual auditing process. Scope should be defined according to the goals and objectives of the waste minimization program. Goals should be quantitative, realistic, and achievable. Compliance with a set of legislative requisites within a fixed time period could be the prime goal in some instances. In other instances, there could be the need to reduce (by a certain percentage) the operation of waste treatment cost by reducing the quantity and strength of some priority pollutants (e.g. chromium, copper, or organic pollutants like phenols, cyanides, etc.). In some other cases, the goal may be to reduce consumption of chemicals (by certain tonnes per batch of product) by improving certain processes. Waste audits can also focus on very specific objectives like improving the general environmental management system in environmental management system audits and improving operational health and safety by reducing the use of toxic chemicals in operational health and safety audits, total environmental risk audit, and so forth. In every case, there should be some target sectors or processes, which have to be closely scrutinized. Identification of these target processes is crucial in defining the scope. The most common target processes of waste or energy audits are:

Production processes: All unit production processes in the plant need to be audited to check for process inputs, outputs, operating conditions and controls, and process efficiency.

Inventory processes: Inventory processes like storage and handling of chemicals, machine parts, and other accessories need to be evaluated to check loss of material while handling or storage, expired chemicals, improper or unsafe storage, leakage, spillage, etc.

Housekeeping processes: Housekeeping practices should be audited in order to check various losses during cleaning, arranging, and various types of maintenance practices. Better housekeeping practices can reduce leaks, spills, drag-out, losses in rinsing, cleaning, etc.

Waste treatment processes: All waste treatment processes like effluent treatment plants, gas treatment, dust filters, etc., should be audited to determine treatment efficiency, operating conditions and controls, etc.

Packaging processes: Packaging sometimes requires specialized systems that can generate wastes (often solid waste). Therefore, such processes must also be audited if it is identified that waste generation or energy consumption is high in such processes.

Unless the scope of the waste audit is positioned to achieve the overall goals of the program, many efforts can go waste. Table 24.1 shows how some of the common objectives are related to the target processes. Defining anything more than what is required or less may result in poor outcome. For example, if the objective of a waste minimization project is to reduce water consumption in a factory, only those processes/activities (target processes) that are using significant amount of water need to be checked. In this case, it is not particularly useful to audit inventory processes or final packaging systems that consume no water or very little water. On the other hand, when the goals are to explore opportunities of electricity

Table 24.1 Audit of target processes as per objectives of waste minimization program.

Main objective	Target processes
Legislative compliance	All processes with more emphasis on production and waste treatment processes
Reduction of toxic and hazardous wastes	Production and inventory processes where hazardous and toxic chemicals are used. Special attention to be given on transport, handling, storage, use of such chemicals, and treatment and disposal of the hazardous wastes
Operational health and safety improvement	Special attention to production processes specially with high temperature, high voltage applications, etc. Attention to be given to handling, storage, and use of hazardous and toxic chemicals. Housekeeping processes need to be checked as well
Compliance audit	Mainly the effluent treatment units only, with emphasis to check if they are meeting the required discharge standards

reduction, probably all sort of activities need to be audited with more emphasis on production and waste treatment processes that supposedly consume most of the electricity. Therefore the scope would vary depending on the objective of the program.

Defining a balanced scope is not generally a one-step procedure. Starting with preliminary scope definition, the audit team is likely to modify the scope as the audit is progressed and new information emerges, which may need extension of scope to certain target processes and eliminate certain other sectors that may not significantly contribute to the overall objective. It is suggested that the scope is reviewed at each step of the audit phase to cover all concerned areas.

24.5.3.2 Develop an Audit Program

Like all other projects, an audit exercise should stick to a program and plan to avoid any overrun. The audit team should develop a series of programs for auditing each of the target processes and time frame for each of the activities. Generally for a complete waste and energy reduction audit, the following time frames can be used for manufacturing plants:

- Small scale (less than 50 workers): 3–4 weeks
- Medium scale (50–100 workers): 4–6 weeks
- Large scale (more than 100 workers): 6–10 weeks

However, the reader should be aware that the time frame given above is much generalized, and several factors like the number of steps in the production process, the degree of complexity, the level of automation, and the quantity and characteristics of different feedstock used should be considered for determining the time frame. In the absence of any specific condition for the program, the two ways of setting up an audit program are to either follow the material flow path though the industry or audit target processes according to the significance in terms of waste reduction.

Both systems have some merits, for example, following the material flow helps keep good track of several raw material as these are transformed into products and by-products, which give a complete account of material inputs and outputs. Auditing according to priority of target processes allows a thorough investigation of major target processes that can identify more waste minimization opportunities. Development of program is a dynamic activity and can be modified as more audit exercises are undertaken and new clues evolve.

Audit timing and frequency are important factors in planning audit exercises. Timing and frequency can be fixed depending on the following:

- Type (continuous, batch, mixed) of the production process.
- Number of parameters to be audited.
- Scale of production.
- Accuracy of audited information required.

Auditing can be scheduled during peak production hours or uniformly throughout a complete production cycle depending on the nature of the production process. For example, in a typical batch production process (like textile dyeing), quantity of wastewater discharged at the end of each cycle needs to be audited, while in a continuous production process (like pulp processing in a pulp and paper industry), discharge of wastewater may be monitored at predetermined intervals. Care should be taken that while auditing the normal factory processes.

It is recommended that the major target processes are audited more than once, typically 2–3 times to obtain good representative results and assure repeatability of the information, while less important sectors can be audited once. Quality and reliability of an audit exercise would very much depend on the frequency of audit on each process, as each time a process is studied, there is an increased chance of getting new clues.

24.5.3.3 Prepare Specific Work Plan and Checklists

Each team member should prepare their own specific work plan and checklists depending upon the nature of the activities to be audited. Preparing work plan and checklists would enable auditors to focus on key attributes/factors to be audited and collect all necessary information for those processes and activities. Some basic understanding of the processes would be required to prepare good work plans and checklists.

24.5.3.4 Develop Uniform Reporting Procedures

The audit report is the final product from the audit phase. All subsequent activities would very much depend on the audit report, and its importance in the program is substantial. It is therefore recommended that the audit report should be well planned and easily comprehensible and contain all the information that may be required during the subsequent phases. The contents of the report should be determined based on the objectives of the program. A typical table of contents for a waste audit report is given in Box 24.2. Depending upon the requirement, some of the sections can be excluded. However, important sections should be retained in a good audit report.

24.5.3.5 Inform All Departments about Upcoming Audit Programs

Once the audit program is finalized, the audit team should notify all the respective departments about the upcoming audit. This would enable the departments to be prepared for the audit. It is suggested that the program be advertised at suitable locations throughout the factory to remind the supervisors and workers about the audit schedule.

Box 24.2 Table of contents for a typical waste audit report.

- Title Page
- Disclaimer, if any (disclaiming responsibility by the publisher for data or views expressed)
- Table of Contents
- Executive Summary (summarizing the complete audit exercise in the industry
- Problem Statement, Objectives, and Priorities
- Adopted Approach and Reasoning
- Process Layout, Description, and Observations
- Sources and Quantities of Pollution Load Generated
- Existing Treatment Facilities and Additional Requirements
- Observations on Proposed Treatment Systems
- Integrated Pollution Prevention Strategy
- Generation of Options, Screening
- Option Grouping and Prioritization
- Identification of Candidate Systems
- System Evaluation Criteria
- Calculations
- Method and Results of System Evaluation
- Recommendations
- Implementation Strategy
- Suggestions to the Management
- Summary and Observations
- Appendices (Tables, Graphs, Assumptions, and Calculations)

24.6 Phase II: Preassessment of Target Processes

The purpose of the preassessment is to collect all information on the target processes that may be required for detailed audits. Preassessment essentially consists of:

- Collection of baseline information
- Conducting a plant walk-through
- Identification of immediate implementable options

24.6.1 Collection of Baseline Information

Collection of information and compiling is one of the major task and purpose of waste auditing. While information on target processes would be of primary value, practical experience shows that various other information may be useful. As such, the scope of data collection can be a range over the entire cross section of the factory, which may include:

- *Organizational data:* Organizational chart, factory layout, site plan, information about the surrounding area indicating topography, water bodies, hydrology, agricultural areas, human settlements, etc.
- *Material and product data:* Specification of feedstock, process water, product and by-products with composition, instruction on usage and discharge, material safety data sheets, usable and storage life, etc. Information on quality assurance and quality control of feedstock, process water, products and by-products are required to be collected.
- *Raw material and logistic consumption data:* Feedstock, energy and water consumption records. Possible source of records of feedstock consumption can be available from stores and accounts department, water usage can be obtained from water meters or bills and energy usage from energy bills.
- *Process data:* Process flow diagrams, block diagrams, material balance diagrams, control and operational logic diagrams and instructions, manufacturer's data on each machine and process, maintenance plans and records, etc.
- *Environmental data:* Air emission, solid waste generation and effluent from different machineries including waste treatment systems, environmental directives, licenses, etc.
- *Management data:* Number of staff, their position and responsibilities, performance records, administrative instruction, occupational and safety procedures, quality assurance and quality control procedures, etc.
- *Financial data:* Product, utility and raw material cost, cost of waste treatment, operating and maintenance cost, etc.
- *Industry data:* Above sets of information for industries of similar nature. Such information, though difficult to obtain, would however allow comparison with other plants that may help in defining realistic targets and goals.

All the audit members should be well familiar with the information that would allow them to objectively carry out the audit. To an experienced auditor, even mere scanning through such information can also give him such clues like opportunities for readily implementable options, areas that would need detailed auditing, rooms to improve audit exercises, modification of scopes, etc. It is recommended that all information should be properly preserved and that sources of such information be noted to check the reliability at a later date.

All information should be checked for underlying data quality in terms of correctness and repeatability. If possible, historical data should be gathered over a period of time (suggested 2–3 years of records) and the process of data collection should be continued during the entire audit phase at predetermined intervals. It is suggested that statistical analysis be performed on the collected data to ensure their quality before these are used for analysis purposes.

24.6.2 Conducting a Plant Walk-through

A thorough walk-through of the plant is an essential part of the preassessment phase. It is recommended that all team members be involved in such audit exercise that should generally be carried out over few days with one or two sections covered in each day. Moreover it is also suggested that the audit team should be accompanied by the responsible section manager, engineer, or supervisor who are completely aware of each activity in their section. All types of management and technical information should be made available during such surveys for the audit team to compare with the actual situation. There are several benefits of such exercise:

- It is quite likely (especially in the case of large industries) that the members of the audit team may not be very familiar with the different activities carried out in each section other than their own. Therefore, this exercise would give them a chance to get conversant with many more activities and processes.
- It can reveal some obvious waste minimization or cleaner production options that may not need a detailed assessment to work out the recommendation. A "common sense approach" or simple calculations with a lesser degree of accuracy or thumb rule estimates may be sufficient for arriving at the conclusion. For instance, in situations where the steam pipes are not insulated or valves leaking, there is no need to carry out a detailed assessment of energy or steam losses to suggest insulation of the steam pipes or repair of the valves.
- By eliminating such obvious options, the auditors can narrow down the scope to those areas that require detailed assessment. This frees up some resources and the auditors can then concentrate on more intriguing issues.
- With a balanced team of auditors, personnel from other sections or departments can more critically observe activities of another department. This can give rise to lateral thinking and more avenues for improvement can be explored.

In the absence of any other plan, the best strategy for the walk-through is to follow the material flow path though the industry – from the storage of raw material, through various production processes, till it is converted to the final product and packed to be shipped. During the walk-through exercise, each member of the audit team should take a detailed note of all the activities, facts, and figures and any other information that may be useful at a later stage. It is suggested that even trivial observations are noted down as these can form some clue at a later stage. It is preferred that the auditors should prepare their own sketches, schematic arrangements, material flow diagram, block diagrams, site plans, etc., during this walk-through exercise. Even if some process diagrams are already present, often minor changes are carried out in the plant during operations. Notes of such changes should be taken. At the end, each member of the team should prepare their own report of the plant walk-through.

24.6.3 Identification of Readily Implementable Options

After carrying out the plant walk-through and making detailed notes, the audit team must discuss the various observations made and should try to identify simple and obvious measures. Such options should be simple, quickly implementable, and inexpensive. Significant waste reductions can often be achieved by such options that are based on improved operation, better handling, and tightening up of housekeeping practices. For example, simple measures such as attending to leaking hoses or installing automatic level controller may lead to significant water savings.

Segregation of waste is arguably one of the numerous measures that can effectively lead to waste reduction. It is the most central of such options and is a universal issue that needs to be addressed. Segregation of waste can offer enhanced opportunities for recycling and reuse with resultant savings in raw material costs, at the same time reducing treatment costs. Concentrated simple wastes are more likely to be of value than dilute or complex wastes. A typical audit exercise needs to address such opportunities. For example, the waste collection and storage facilities should be reviewed to determine if readily implementable waste segregation is possible.

It is suggested that readily implementable options are implemented as soon as possible. Typically, implementation of such options should be able to complete within 2–3 weeks. If implemented, the performance and impacts of the changes can be closely monitored and included in the final audit report.

It is expected that at the end of the preassessment phase, the audit team is:

- Organized and aware of detailed scope.
- Aware of all target process layouts for further audit.
- Aware of all unit operations in each of the target processes.
- Aware of sources of waste and their causes.

At the end of the preassessment, the plant personnel should be well informed of the audit purposes. Resources should be secured, and readily implemented waste reduction measures should be identified and, if possible, implemented.

24.7 Phase III: Assessment

This phase can be broadly divided into two steps:

1) Listing of unit operations and constructing process flow diagrams.
2) Preparation of material balance diagram.

24.7.1 Listing of Unit Operations and Constructing Process Flow Diagrams

This step is almost essential in any audit program as it gives a detailed insight into the production operations/process *vis-à-vis* sources of waste generation and hence enables the identification of avenues for better operating practices and waste reduction.

To develop a good representative block process diagram, the audit team should undertake a detailed check of the production units and utility areas, in order to gain understanding of all the processing operations and their interrelationships. The production or plant staff should be interviewed to know about the actual operating controls, parameters, and issues. Only after conducting a detailed walk-through and interviews with the production staff, the required baseline data should be compiled.

By connecting the individual unit operations in the form of a block diagram and highlighting the flow of materials, a process flow diagram can then be prepared. All the information related, for example, to raw materials, products/by-products, energy, water inputs, waste discharged, material and energy flows, motion and time related, etc., should be compiled during this stage and should be presented in the form of block diagram(s).

The input and output information for each unit operation should be summarized in standard units with reference to the process flow diagram. Standardized color codes and line types may be used to represent each attribute of the process diagram. Commonly, color codes are used for inputs and outputs, for example, raw material input can be represented by black lines, products by blue lines, wastes by red lines, and recycled stream by green lines. Process operations are distinguished by different line types (which may or may not be color coded), for example, the main process flow is indicated in solid lines, while intermittent operations (such as cleaning, makeup, or tank dumping) may be represented by broken lines. Similar notation may be used to distinguish batch and continuous discharges.

24.7.2 Material Balance: Process Inputs and Outputs

In the material balance exercise, a detailed account of the process inputs and outputs is made to identify the problem areas and thus the need for improvement. Material balance is important for any waste minimization project to identify and quantify previously unknown losses or emissions. Material balance is also useful for estimating the costs of additional installations and/or modifications.

By definition, the material balance includes materials entering and leaving a process. Inputs to a process or a unit operation may include raw materials, chemicals, water, air, and energy. Outputs include primary product, by-products, rejects, wastewater, and gaseous, liquid, and solid wastes that need to be stored and/or sent off-site for disposal and reusable or recyclable wastes (Figure 24.3).

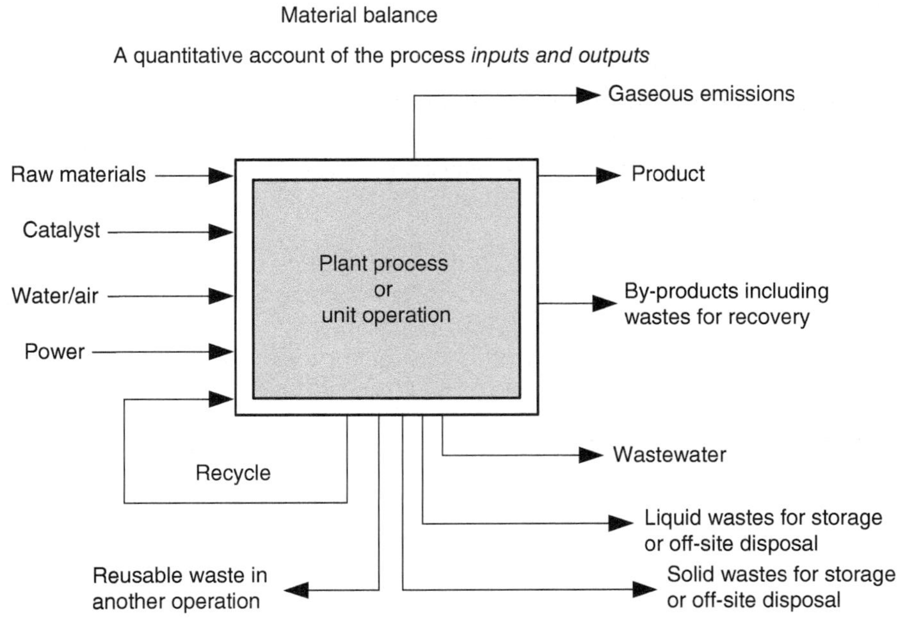

Figure 24.3 Schematic representation of a material balance sheet.

In its simplest form, a material balance is drawn up according to the mass conservation principle:

$$\text{Mass in} = \text{Mass out} + \text{Generation} - \text{Consumption} - \text{Accumulation}$$

If no chemical reactions occur and the process progresses in a steady state, the material balance gets simplified to

$$\text{Mass in} = \text{Mass out}$$

$$\text{Water in} = \text{Water out} + \text{Losses}\,(\text{evaporation, spills, etc.})$$

24.7.2.1 Sources of Information for Material Balance

There are many sources of information in establishing material balances for the various unit operations within the plant. Data may be obtained from sample analyses and measurements of raw input materials, raw material purchase records, material and emission inventories, equipment cleaning and validation procedures, batch composition records, product specifications, operating logs, standard operating procedures and manuals, etc.

Material balances are easier, more meaningful, and more accurate when they are done for individual production units, operations, or production processes. For this reason, it is important to define the material balance envelope or boundary limit accurately in addition to the tie compound. Ideally, a more accurate balance should be established primarily for the unit operation more critical from the waste generation and reduction point of view and a less accurate balance could be established for other processes.

Although it is not possible to lay precise and complete guidelines for establishing the material balance, the following guidelines might be useful:

- In the case of an extensive and complex production system, it is better to first draw up the material balance for the whole system (or even the entire production facility as such) and then concentrate on individual operations.
- When splitting up or desegregating the total system, choose the simplest individual subsystems that are critical from the waste reduction point of view.
- Choose the material balance envelope in such a way that the number of streams entering and leaving the process is the smallest possible.
- Always choose recycle streams within the envelope to start with.

For complex waste minimization audit, it might be desirable to make a preliminary or draft material balance and in the second step to evaluate and refine it. However, in the case of simple audit for small plants, the steps can be merged into one.

24.7.2.2 Selection of Priority Unit Operation

Although the material balance should be set up for all the unit operations, the unit operation that is most important from the point of view of waste generation must be identified and efforts are concentrated for that particular unit operation. This can be done by professional judgment and technical know-how of the audit team and specifically the production personnel.

24.7.2.3 Selection of Tie Compounds

A tie compound is the parameter (or substance) for which the material balance is established around a unit operation or a process. Selection of an appropriate tie compound for checking material balance is an important step. Criteria for selection of the tie compound could be:

- Expensive raw material/intermediate.
- Material common in most processing stages.
- Substance of hazardous nature.
- Substance/compound easy to measure/estimate.

A simple example of a tie compound could be water to account for most wet operations. Establishing water balance for the processes using substantial amount of water can often provide useful clues for cleaner production. In practical situations, more specific tie compounds, e.g. nickel or zinc, in electroplating shops or dyestuff in textile processing could be used. Another good example could be that of chromium in leather tanning.

Chemical oxygen demand (COD) is another very useful tie parameter that is used in the material balance exercise, especially to link the production areas with the effluent treatment plant. The audit team can estimate contribution of each process department in terms of total COD load in $kg\,day^{-1}$, knowing the volume of wastewater and the COD discharged by each department and cross-checking it with the COD load observed at the treatment facilities.

One needs not be very particular over the accuracy (of the order of 99%) of the material balance. In practice such high accuracies are rarely achievable and may not be required as well. Material balance within the tolerance range of 10% should generally be acceptable. However, if the tie compound for material balance is of hazardous nature, the higher order of accuracy should be targeted.

24.7.2.4 Steps for Preparing a Material Balance

There is a logical series of steps for preparing a constructive material balance:

1) *Determining inputs:* The inputs to the process and to each unit operation need to be quantified. As a first

step toward quantifying raw material usage, purchasing records should be examined; this readily gives an idea of the quantities involved. The raw material purchases and storage and handling should be recorded in a table format in order to derive the net input to the process.

Water is frequently used in the production process, for cooling, gas scrubbing, washouts, rinsing, and steam cleaning. The water usage needs to be accurately quantified as an input. Also some unit operations may receive recycled material from other unit operations. These also represent an input. Hence, water and recycled materials need special attention, and therefore steps 1A and 1B describe how to evaluate these two factors:

1A) *Recording water usage:* The use of water, other than for a process reaction, is a factor that should be covered in all cleaner production programs. The use of water for washing, rinsing, and cooling in process and in utility operations is often overlooked, although it represents an area where effective waste reduction can frequently be achieved simply and cheaply.

1B) *Measuring current levels of material recycling:* Some materials may be transferred from one unit to another, e.g. reuse of the final rinse in a soft drink bottle washing plant as the initial rinse, either directly or after some modifications/treatment. If recycled materials are not properly documented, double accounting may occur in the material balance particularly at the process or complete plant level, that is, a material will be quantified as an output from one process and as an input to another. Proper attention must be paid to this issue and care must be taken to avoid any such error.

2) *Quantifying outputs:* To calculate the second half of the material balance, the outputs from unit operations and the process as a whole need to be quantified. Outputs include primary product, by-products, wastewater, and gaseous (emissions to atmosphere), liquid, and solid wastes that need to be stored and/or sent off-site for disposal and reusable or recyclable wastes. It is important to identify appropriate units of measurement.

   If the product is sent off-site for sale, then the amount produced is likely to be documented in company records. However, if the product is an intermediate to be input to another process or unit operation, then the output may not be so easy to quantify. Production rates will have to be measured over a period of time. Similarly, the quantification of any by-products may require field measurement:

2A) *Accounting for wastewater:* On many sites significant quantities of both clean and contaminated water are discharged to sewer or to a watercourse. In many cases, this wastewater has environmental implications and incurs treatment costs. In addition, wastewater may wash out valuable unused raw materials from the process areas. Therefore, it is extremely important to know how much wastewater is going down the drain and what the wastewater contains. The wastewater flow, from each unit operation as well as from the entire process, needs to be quantified, sampled, and analyzed.

2B) *Gaseous emissions:* To arrive at an accurate material balance, some quantification of gaseous emissions associated with the process is necessary. For example, a tea drier exhaust may carry fine particles of tea dust. Measurement in such cases requires instruments such as thimble probe dry gas meter – vacuum pump assembly. In many instances, gaseous emissions carry some amount of hazardous materials also like volatile organic compounds and these may be accurately attended to. External assistance from experts can be sought to arrive at the material/product loss through gaseous emissions, if required.

3) *Preparing a preliminary material balance* – Preparing a material balance is designed to gain better understanding of the inputs and outputs, especially waste, of a unit operation such that areas where information is inaccurate or lacking can be identified. The initial balance should be considered as a rough assessment that has to be further refined and improved.

   The units of measurement should be standardized (liter, ton, or kilogram) on per day, per year, or per batch basis. The measured values in standard units should be summarized by reference to the process flow diagram. It may be necessary to modify the process flow diagram following the in-depth study of the plant. It is highly desirable to carry out a water balance for all water inputs and outputs to and from unit operations because water imbalances may indicate underlying problems such as leaks or spills. Similarly, a detailed material balance should be carried out for important tie compound, as agreed upon by the audit team during the planning phase.

4) *Evaluating and refining material balance:* The individual and sum totals making up the material balance should be reviewed to determine information inaccuracies. Ideally, the inputs should equal the outputs, but in practice this will rarely be the case

and some judgment will be required to determine what level of accuracy is acceptable. If there is a significant material imbalance, then further investigation is needed.

Several factors have to be taken into consideration when constructing material balances in order to avoid errors that could overstate or understate waste streams. Sometimes, all or at least a few steps of material balance may need to be repeated a few times in order to refine the material balance. These may include quantification of a few input or output streams or even hunting for some material flows that might have been totally missed in the initial stage. Additional data collection and auditing may also be required in certain cases to quantify missing links and refine the material balance.

24.8 Phase IV: Synthesis and Preliminary Analysis

Phases 1 and 2 have covered planning and undertaking waste audit, resulting in the preparation of a material balance for each unit operation. Phase 3 represents the interpretation of the material balance to identify process areas or components of concern. Figure 24.4 represents a material balance algorithm for textile industry in establishing waste reduction options.

To interpret a material balance, it is necessary to have an understanding of normal operating performance. Thus, a member of the audit team must have a good working knowledge of the process. To a trained eye, the material balance will indicate areas of concern and help to prioritize problem wastes. By using the material balance, major sources of waste may be identified, deviations from the norm in terms of waste production may be found, areas of unexplained losses may be determined, and operations that contribute to flows that exceed national or site discharge regulations may be pinpointed.

In this phase, several possible waste reduction measures are identified that can be proceeded to the analysis phase. Different waste minimization programs may require varying degrees of effort, time, and financial resources. They can be categorized into:

- Obvious waste reduction measures, including improvements in management techniques and housekeeping procedures that can be implemented cheaply and quickly.
- Long-term measures involving process modifications or process substitutions to eliminate problem wastes.

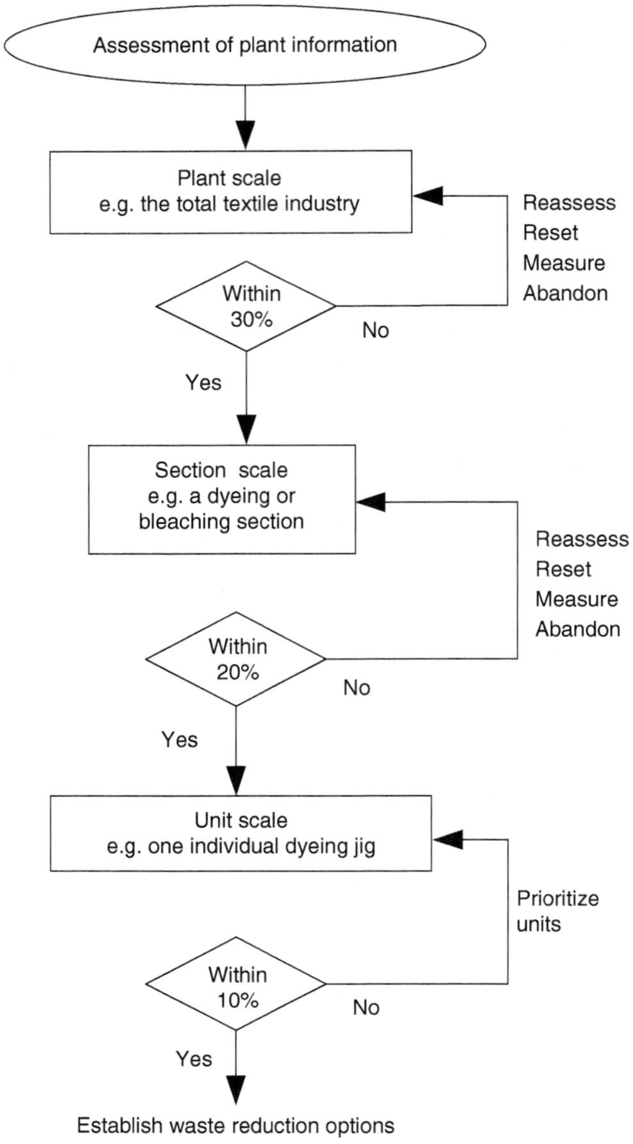

Figure 24.4 Material balance algorithm for textile industry.

24.8.1 Screening and Selecting Options for Further Study (Weighted Sum Method)

Screening and selection is recommended when a number of options have to be considered. Options can be evaluated numerically according to their merits and demerits by the *weighted sum method.* Only those options that can satisfy certain predetermined requirements should be carried forward. The weighted sum method provides a means of quantifying the important criteria that affects waste management in a particular industry. Typically, weighted sum method involves four steps:

1) *Determine what criteria would be considered in evaluation of the options.* The higher the degree of

improvements achieved, the better the result. For example, in a typical waste minimization program, such criteria can be:

- Amount in reduction in waste quantity.
- Amount of reduction in raw material consumption.
- Amount of reduction in hazardous or toxic waste.
- Improvement in health and safety condition.
- Ease of implementation.
- Cost of implementation.
- Resource and time required.

These criteria should be determined in terms of meeting the overall waste minimization objectives and goals. This should also take into consideration various types of constraints that may be present. Judgment would be required to select the criteria.

2) *Once the criteria are determined, each criterion should be given a weight.* The more important a criterion is, the higher its weight. Again, a fair amount of judgment is required to determine the weight for each criterion. Generally a relative scale of 0–10 is used to allot weight. For example, if reduction in waste treatment and disposal costs is more important, while time of implementation is relatively less important, then the reduction in waste treatment costs is given weightage of 8, while the time of implementation is given a weight of 2 or 3.
3) Each option is then rated for all the criteria. A scale of 0–10 could be used for rating. Marks are given according to the degree the criteria are satisfied, i.e. higher marks are given when the option fulfills or satisfies the criteria and low marks when the option does not suit the criteria.
4) *Each rating (for the option) is multiplied by the corresponding weight of the criterion.* An option's overall rating is the sum of all the products of ratings times the weight of the criteria.

The options that carry higher points would be carried for further analysis. Table 24.2 presents an option evaluation by weighted sum method for five options.

24.8.2 Preliminary Technical and Economical Evaluation

Once the options are screened and selected for further analysis, preliminary technical and economical evaluation may be required to set the priority of options and develop a preliminary action plan.

24.8.2.1 Technical Evaluation

The technical evaluation determines whether a proposed waste minimization option will technically be feasible and achievable within the framework of existing constraints. A technical evaluation often begins with examining the impacts of the proposed option on production processes, production schedule, product quality, extra resource requirement, real estate requirement,

Table 24.2 Workout example of weighted sum method for waste audit.

		Option rating (*R*)									
		Option 1		Option 2		Option 3		Option 4		Option 5	
Criteria	Weight (*W*)	*R*	*R*W*	*R*	*R*W*	*R*	*R*W*	*R*	*R*W*	*R*	*R*W*
Reduction in treatment/disposal costs	8	7	56	7	56	5	40	2	16	2	16
Reduction of input material costs	4	8	32	6	24	8	32	4	16	4	16
Extent of current use in industry	5	8	40	8	40	7	35	7	35	7	35
Extent on product quality (no effect = 10)	10	9	90	9	90	2	20	8	80	8	80
Low capital cost	5	2	10	5	25	4	20	7	35	8	40
Low operation and maintenance cost	5	5	25	6	30	5	25	8	40	8	40
Short implementation period	8	3	24	5	40	3	24	7	56	8	64
Ease of implementation	7	3	21	6	42	5	35	7	49	8	56
Reduction in energy bills	9	5	45	9	81	5	45	10	90	10	90
Improvement in OHS	7	10	70	3	21	10	70	2	14	2	14
Final evaluation	Sum of weighted ratings $\Sigma(W^*R)$		413		449		346		431		451
	Option ranking		4		2		5		3		1
Feasibility analysis scheduled for (date)											

operational feasibility, and safety. Several constraints like disruption to normal production schedule, shutdowns, match of specifications (between the existing equipment and the new one), lack of technical knowledge and trained manpower, etc., may be present. Moreover, psychological resistance to changes may also be an issue. It is therefore recommended that all serious changes are first tested at a laboratory scale and pilot scale. And trial runs with the prototypes and test products are undertaken before the change is implemented and integrated to the actual production process. It is also suggested that engineers, supervisors, and operators are suitably trained for the change.

Waste minimization options can also have some environmental impacts. During technical evaluation, the environmental effect of implementing the option needs to be checked. For example, if an option calls for recycling of rinse water, the effect of disposal of such water needs to be evaluated (as the concentration of solid in the recycled water would increase many times due to recycling), which may impact the receiving body.

24.8.2.2 Economic Evaluation

Economic feasibility is one of the most important criteria in determining the selection of an option. Unless forced by legislative requirement, there would be no cases where the economic merit of an option would be disregarded. Normally each organization has its own economic criteria for selection of projects. However, the three main criteria that would require to be evaluated, irrespective of other criteria, are:

1) The capital, operation, and maintenance cost of the option.
2) The benefit that it would return in comparison to the existing system.
3) The resulting payback period.

The relationship between capital cost, payback period, and likely acceptance is given in Table 24.3. Payback period can be considered as long, if it exceeds 2 years; medium, if it is between 1 and 2 years and short; and short, if less than 1 year for common waste minimization options.

Table 24.3 Relationship between capital cost, payback period, and likely acceptance.

Capital cost	Payback period	Likely acceptance
High	Long	Low
High	Medium to short	Medium
Medium	Long	Medium
Medium	Medium	Medium
Medium	Short	Medium to high
Low	Long	Medium to high
Low	Medium	High
Low	Short	High

24.8.3 Preliminary Prioritizing Waste Reduction Options

Once waste minimization options are evaluated, they would need to be prioritized for further analysis, design, and implementation. A preliminary prioritization at the audit phase will expedite the process of analysis in the next phase. Prioritization can be done by weighted sum method, in the absence of any special criteria (e.g. where some waste minimization measures become mandatory due to legislative requirements, these can then be classed as high priority options, without further analysis). Some of the criteria that can be used to prioritize the options are:

- Technical ease in implementation
- Resource and time requirement
- Impacts on production schedule
- Short-term capital requirement
- Payback period

24.8.4 Developing Preliminary Waste Reduction Action Plan

Upon suggesting the priority, the audit report should also delineate a preliminary action plan for implementation. This can be represented as a regular bar chart or in any other form that is acceptable. In normal cases, the implementation sequence should follow the order of priority. It is suggested that implementation of waste minimization options is taken up in stages so that the cumulative impact on production processes, resources, and finance can be kept low. It is probably best to initially implement low-cost options with relatively simple technical requirements followed by progressive implementation of more complex changes that may require higher capital costs. It may also be a good idea to implement an option and test the success of it before the next one is undertaken to reduce the total risk involved in undertaking changes.

24.9 Conclusion

Waste auditing is as important as any other step in a waste minimization or cleaner production project. A proper waste audit should essentially provide a platform on which the rest of the project could be built. It should be a repository of all data and information that would be required to carry out the rest of the phases. As information is the strength, so is a successful waste audit.

The principle intent of a waste audit is to critically assess various inputs, processes, and outputs to find methods and practices for minimizing waste and reduce the resource consumption in a sustainable and environmentally benign way without compromising the commercial interest of the company. The waste audit phase would typically be a data collection and information synthesis phase that explores the current situation. Traditionally, an industrial waste audit also develops a list of waste minimization options and undertakes some preliminary technical and economic feasibility studies on the identified options to recommend on these options.

In general, the first step of a systematic waste audit is to prepare for the audit by forming a team, defining the scope, and programming the audit timings and budget. Scope should be defined in line with the overall objective of the waste minimization program. The next step is to collect all baseline information, conduct plant walk-through surveys, and identify readily implementable options. This would be followed by a detailed assessment of target sectors and unit processes and performing the mass balance analysis. At the end, the data should be analyzed to build up an array of information and recommendation that would set the direction of the next phase. All data collected, information derived, and recommendation made should be presented in a form of a comprehensible and well laid out audit report.

It should be appreciated that waste audit constitutes the vital activity of data collection and investigation of the current condition of the factory and its activities. Therefore, it forms the very basis of the next series of activities. It is recommended that waste audit should be undertaken with utmost care and should be as thorough as possible. The more the effort is spend at this phase, the better will be the chance of success of the project.

Case Study of a Tapioca Starch Industry

A tapioca-manufacturing factory produces fructose syrup from tapioca. The factory is separated into two plants that produce starch milk and starch syrup. Starch milk is the raw material for the syrup production.

Phase 1: Preparatory Work for a Waste Audit

Appropriate management staffs were consulted and the audit was planned in their consensus. Factory worker and staffs were provided information about the audit objectives to gain acceptance and positive feedbacks from them. The issues and problem raised by them provided ample information to frame the audit activities.

The scope of the audit was set as follows:

1) To investigate the source of waste qualitatively and quantitatively.
2) To reduce the quantity of wastewater generation.
3) To review the existing wastewater treatment plant and propose an alternative arrangement.

In accordance with the scope and target of the audit, audit program was planned. Requisites of audits like sampling, questionnaire, checklists, work plan, reporting formats, etc., were planned before the audit exercise.

Phase 2: Preassessment of Target Process

Baseline information of the plant were collected from the respective departments of the factory, and desk study was done before the walk-through survey. Walk-through survey provided insights into the production process, raw materials consumption, water consumption, and waste and wastewater sources. Based on the baseline and walk-through survey, some of the identified information were as follows:

1) $3860\,m^3\,day^{-1}$ of water is pumped every day and coagulants and chlorine is used for its treatment.
2) Wash water from washing process and fruit water from the decantation process produced highly polluted wastewater.
3) Fruit water was recycled back for the washing process.
4) Wastewater generated from the washing process, which is composed of sand, clay and some peel of tapioca, was sent to sedimentation tank and was not managed properly creating bad odor.
5) The surrounding area of the factory faced a severe problem due to the bad odor coming from the wastewater treatment plant. Accumulation of suspended solids in the biogas pond created the bad odor.
6) Tap water in the toilet leaked continuously.
7) The sieved peel and sand were wasted from the sieving process.
8) The washed peel generated from the washing process caused odor problem.
9) The hopper machine, sieving machine, and transportation of roots were the major source of dust (Figure 24.5).

However, the baseline and walk-through survey suggested that high amount of water usage and wastewater generated was the major issue to be addressed to increase the performance of the plant significantly. Hence for the later

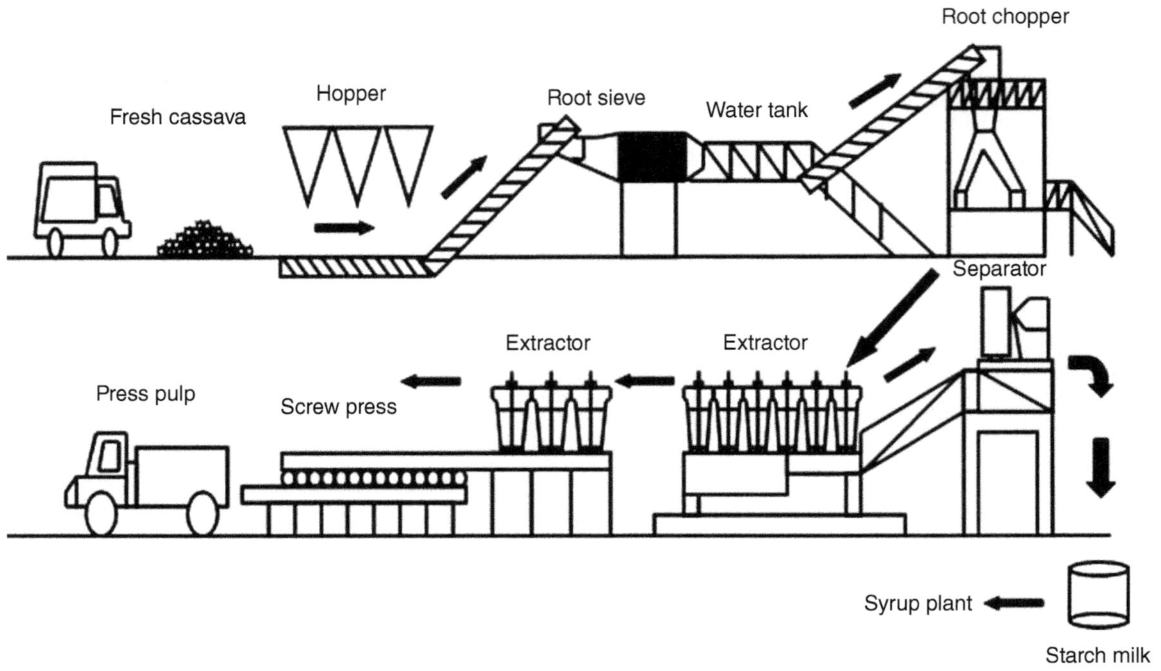

Figure 24.5 Process flow diagram of starch milk plant.

stages of the audit, water consumption and wastewater generation were prioritized.

Some of the readily implementable options identified from the walk-through survey are as follows:

1) Taking occupational health measures like wearing mask, ear plug, etc.
2) Avoiding stockpiling of roots to reduce starch loss.
3) Allocating storage area for the storage of sieve peels.
4) Installation of flow meter in the extraction unit to control the water.
5) Allocation of storage area for the pulp storage.
6) Maintenance of toilets.

Phase 3: Assessment

Unit processes involved in the production of starch milk are sieving/peeling, washing, chopping/rasping, four-stage extraction (starch and cellulose separation), and two-stage decantation.

Some of the methods adopted for quantifying data for mass balance are as follows:

1) Water parameter (pH, temp, BOD, COD, TKN, SS, etc.) analysis at individual unit processes.
2) Measurement of discharge through flow rate measurement.
3) Field observations and key informant interview (KII) to allocate the source of the solid waste.
4) Field observations of wastewater treatment plant and laboratory testing of wastewater characteristic to identify the operational problem.
5) Mapping of the problematic areas in the factory layout.

A proper block diagram was then prepared. The input and output of the process were calculated, and a mass flow diagram was also prepared (Figure 24.6).

Phase 4: Synthesis and Preliminary Analysis

After obtaining the data, it was concluded that solid waste generation meets the national standard, but wastewater production rate and characteristic were not in accordance with the standard. As seen in Table 24.1, water consumption in this plant was also found to be higher than other plants. Moreover, as fruit water was recycled back for the purpose of washing, wash water was highly polluted (Tables 24.4 and 24.5).

The audit team then analyzed the possible solution to all the issues. One of the analyses of the problem related to the high water consumption is presented below:

Wastewater from the plant originated from two process: decanter process (fruit water) and washing process (wash water). Fruit water was recycled back for the washing process that resulted in the creation of high pollutant load in the effluent of wash water. This practice failed to meet the legal effluent standard; thus modification to the process was needed. Audit team anticipated that utilizing fruit water for the washing purpose was indeed not the best option and decided to do technical and economical evaluation of other alternatives.

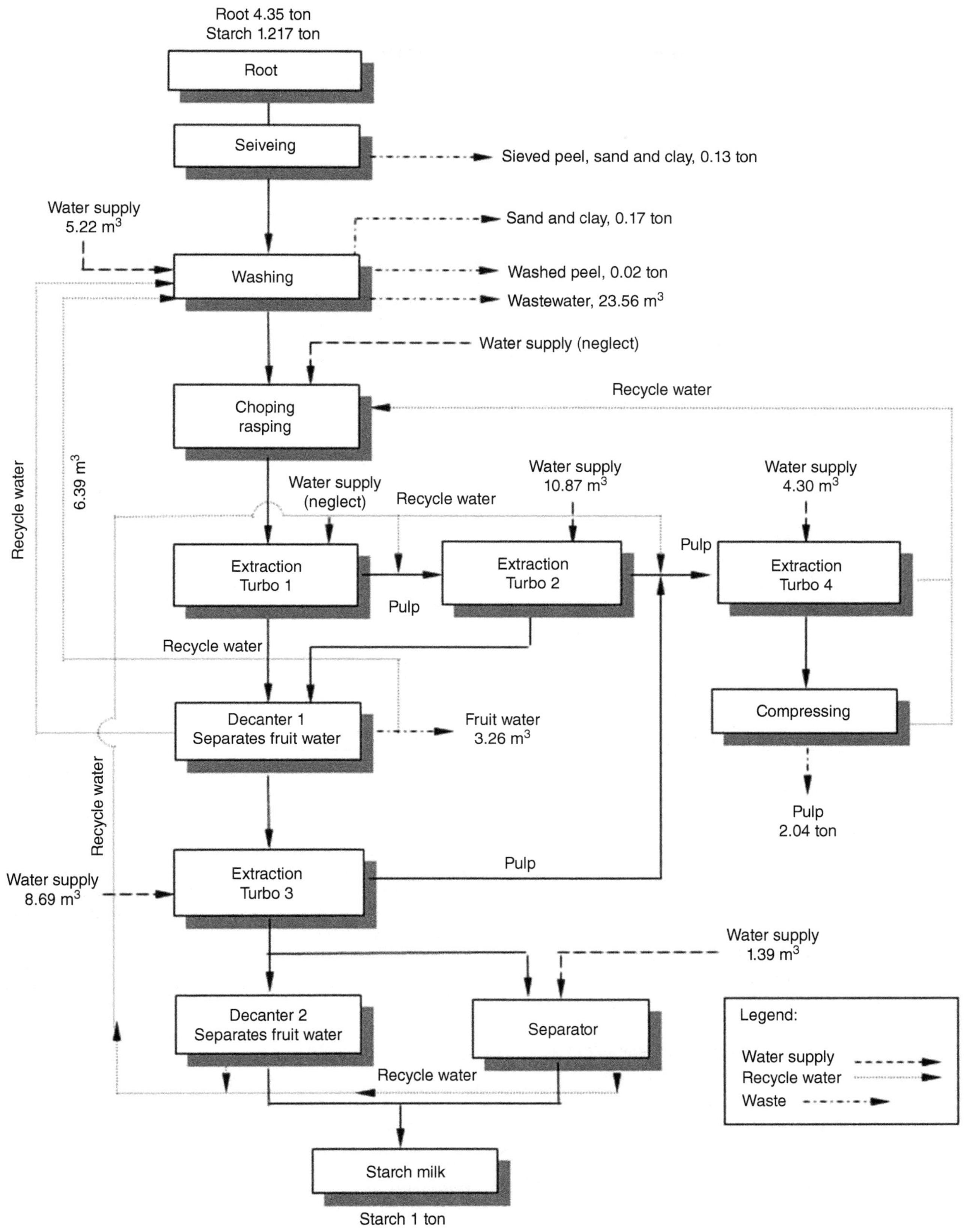

Figure 24.6 Process of starch milk production and water mass balance per ton of starch milk produced.

Table 24.4 Mass balance comparisons between tapioca-manufacturing factory and the national standard.

		Ton of unit weight/ton of root	
Status	**Type of material**	**National standard**	**Factory**
Input			
Raw material	Root of tapioca	1.0	1.0
Water consumption	Water supply	1.2	7.0
Output			
Liquid waste	Wash water	1.6	5.42
	Fruit water	1.0	0.75

Table 24.5 Comparison of wastewater characteristic loading between the national standard and tapioca-manufacturing factory.

		National standard		Factory	
Wastewater source	**Parameter**	$g\,l^{-1}$	$kg\,ton^{-1}$ root	$g\,l^{-1}$	$kg\,ton^{-1}$ root
Fruit water	BOD	20	20	15	11
	COD	40	40	23	17
	SS	6	6	7	5
	TKN	0.6	0.6	0.3	0.2
Washing water	BOD	5	8	13	70
	COD	11	17.6	19	104
	SS	4	6.4	8	43
	TKN	0.12	0.92	0.29	1.58

Table 24.6 Technical and economical evaluation.

Process	Coagulation flocculation	Sedimentation
Wash water	The COD removal efficiency by using alum (31%) was better than ferric chloride (28%). On the other hand, SS removal efficiency by ferric chloride (93%) was better than alum (83%)	The optimum time for sedimentation process was 1 h 30 min that can remove 33 and 91% of COD and SS, respectively
Fruit water	Ferric chloride removal efficiency for the COD (48%) and SS (85%) was better than alum	The optimum time of fruit water for settling was 2 h that can remove 35 and 91% of COD and SS, respectively
Economical evaluation	High initial investment and high operational cost as chemical needs to be used. Moreover skilled labor was required	Low investment and operational cost
Technical evaluation	The wash water needs to be free of any chemical. Thus using chemical might create toxic effect in the process	Sedimentation is free from chemical and can be a viable option for the pretreatment

The jar test of wash water and fruit water was done to determine the chemical type and optimum dose to reduce the pollutant concentrations through coagulation and flocculation method. Similarly, imhoff cone testing was done to determine the optimum detention time of sedimentation tank for the removal of pollution load of wash water and fruit water. Audit came up with the following evaluation (Table 24.6).

Audit team suggested the following modification to be implemented, which could reduce the existing water consumption by 50% and moreover could comply the legal standards of pollution load:

1) Recycling of washed water for the purpose of washing was proposed. A separate pretreatment unit for the treatment of wash water was proposed, while for the treatment of fruit water existing treatment unit could be used. *Previously high sediment content was the source of bad odor and low efficiency of biogas pond. The proposed solution will also mitigate this problem.*
2) Use of counter current method for the washing purpose. In this process, water was added in the final stage of the washing, which was then recycled to the former stage. The effluent from the first stage was then transferred to the treatment unit for pretreatment.
3) Treatment plant for the wash water required a fine screen (to screen the washed peel), a grit chamber (to settle sand and clay particles), and a storage tank. The water in the storage tank could be then recycled back for the washing.

Washed peel was suggested to be utilized as fertilizer.

Suggested Reading

Aldokhin, N.A., Goncharov, A.I., Grachev, M.A., and Suturn, A.N. (1990). Recycling of wastewater and solid waste at the Selenginsk Pulp and Paper Plant. *Industry and Environment* 13 (3–4): 21–23.

Cushnie, G.C. (1994). *Pollution Prevention and Control Technology for Plating Operations*. Ann Arbor: National Center for Manufacturing Sciences.

Modak, P. (1995). *Waste Minimization: A Practical Guide to Cleaner Production and Enhanced Profitability*. Ahmedabad, India: Center for Environmental Education.

Modak, P.M., Visvanathan, C., and Parasnis, M. (1995). Cleaner production audit. *ENSIC Review, AIT, Bangkok, Thailand* 32.

Nemerow, N.L. (1995). *Zero Pollution for Industry: Waste Minimization through Industrial Complexes*. New York: Wiley.

Sorrell, S. and Skea, J. (1999). *Pollution for Sale: Emission Trading and Joint Implementation*. Cornwall, UK: MPG Books Ltd.

Index

b

c

j

k

l

t